Long-Range Dependence and Self-Similarity

This modern and comprehensive guide to long-range dependence and self-similarity starts with rigorous coverage of the basics, then moves on to cover more specialized, up-to-date topics central to current research. These topics concern, but are not limited to, physical models that give rise to long-range dependence and self-similarity; central and non-central limit theorems for long-range dependent series, and the limiting Hermite processes; fractional Brownian motion and its stochastic calculus; several celebrated decompositions of fractional Brownian motion; multidimensional models for long-range dependence and self-similarity; and maximum likelihood estimation methods for long-range dependent time series. Designed for graduate students and researchers, each chapter of the book is supplemented by numerous exercises, some designed to test the reader's understanding, while others invite the reader to consider some of the open research problems in the field today.

VLADAS PIPIRAS is Professor of Statistics and Operations Research at the University of North Carolina, Chapel Hill.

MURAD S. TAQQU is Professor of Mathematics and Statistics at Boston University.

CAMBRIDGE SERIES IN STATISTICAL AND PROBABILISTIC MATHEMATICS

Editorial Board

Z. Ghahramani (Department of Engineering, University of Cambridge)
R. Gill (Mathematical Institute, Leiden University)
F. P. Kelly (Department of Pure Mathematics and Mathematical Statistics, University of Cambridge)
B. D. Ripley (Department of Statistics, University of Oxford)
S. Ross (Department of Industrial and Systems Engineering, University of Southern California)
M. Stein (Department of Statistics, University of Chicago)

This series of high-quality upper-division textbooks and expository monographs covers all aspects of stochastic applicable mathematics. The topics range from pure and applied statistics to probability theory, operations research, optimization, and mathematical programming. The books contain clear presentations of new developments in the field and also of the state of the art in classical methods. While emphasizing rigorous treatment of theoretical methods, the books also contain applications and discussions of new techniques made possible by advances in computational practice.

A complete list of books in the series can be found at www.cambridge.org/statistics.
Recent titles include the following:

19. *The Coordinate-Free Approach to Linear Models*, by Michael J. Wichura
20. *Random Graph Dynamics*, by Rick Durrett
21. *Networks*, by Peter Whittle
22. *Saddlepoint Approximations with Applications*, by Ronald W. Butler
23. *Applied Asymptotics*, by A. R. Brazzale, A. C. Davison and N. Reid
24. *Random Networks for Communication*, by Massimo Franceschetti and Ronald Meester
25. *Design of Comparative Experiments*, by R. A. Bailey
26. *Symmetry Studies*, by Marlos A. G. Viana
27. *Model Selection and Model Averaging*, by Gerda Claeskens and Nils Lid Hjort
28. *Bayesian Nonparametrics*, edited by Nils Lid Hjort et al.
29. *From Finite Sample to Asymptotic Methods in Statistics*, by Pranab K. Sen, Julio M. Singer and Antonio C. Pedrosa de Lima
30. *Brownian Motion*, by Peter Mörters and Yuval Peres
31. *Probability (Fourth Edition)*, by Rick Durrett
33. *Stochastic Processes*, by Richard F. Bass
34. *Regression for Categorical Data*, by Gerhard Tutz
35. *Exercises in Probability (Second Edition)*, by Loïc Chaumont and Marc Yor
36. *Statistical Principles for the Design of Experiments*, by R. Mead, S. G. Gilmour and A. Mead
37. *Quantum Stochastics*, by Mou-Hsiung Chang
38. *Nonparametric Estimation under Shape Constraints*, by Piet Groeneboom and Geurt Jongbloed
39. *Large Sample Covariance Matrices and High-Dimensional Data Analysis*, by Jianfeng Yao, Shurong Zheng and Zhidong Bai
40. *Mathematical Foundations of Infinite-Dimensional Statistical Models*, by Evarist Giné and Richard Nickl
41. *Confidence, Likelihood, Probability*, by Tore Schweder and Nils Lid Hjort
42. *Probability on Trees and Networks*, by Russell Lyons and Yuval Peres
43. *Random Graphs and Complex Networks (Volume 1)*, by Remco van der Hofstad
44. *Fundamentals of Nonparametric Bayesian Inferences*, by Subhashis Ghosal and Aad van der Vaart
45. *Long-Range Dependence and Self-Similarity*, by Vladas Pipiras and Murad S. Taqqu

Long-Range Dependence and Self-Similarity

Vladas Pipiras
University of North Carolina, Chapel Hill

Murad S. Taqqu
Boston University

CAMBRIDGE
UNIVERSITY PRESS

University Printing House, Cambridge CB2 8BS, United Kingdom

One Liberty Plaza, 20th Floor, New York, NY 10006, USA

477 Williamstown Road, Port Melbourne, VIC 3207, Australia

4843/24, 2nd Floor, Ansari Road, Daryaganj, Delhi – 110002, India

79 Anson Road, #06–04/06, Singapore 079906

Cambridge University Press is part of the University of Cambridge.

It furthers the University's mission by disseminating knowledge in the pursuit of education, learning, and research at the highest international levels of excellence.

www.cambridge.org
Information on this title: www.cambridge.org/9781107039469
DOI: 10.1017/9781139600347

© Vladas Pipiras and Murad S. Taqqu 2017

This publication is in copyright. Subject to statutory exception and to the provisions of relevant collective licensing agreements, no reproduction of any part may take place without the written permission of Cambridge University Press.

First published 2017

Printed in the United States of America by Sheridan Books, Inc.

A catalog record for this publication is available from the British Library.

ISBN 978-1-107-03946-9 Hardback

Cambridge University Press has no responsibility for the persistence or accuracy of URLs for external or third-party internet websites referred to in this publication and does not guarantee that any content on such websites is, or will remain, accurate or appropriate.

To Natércia and Filipa
and
to Rachelle, Yael, Jonathan, Noah, Kai and Olivia

Contents

List of Abbreviations		*page* xv
Notation		xvii
Preface		xxi
1	A Brief Overview of Time Series and Stochastic Processes	1
2	Basics of Long-Range Dependence and Self-Similarity	15
3	Physical Models for Long-Range Dependence and Self-Similarity	113
4	Hermite Processes	229
5	Non-Central and Central Limit Theorems	282
6	Fractional Calculus and Integration of Deterministic Functions with Respect to FBM	345
7	Stochastic Integration with Respect to Fractional Brownian Motion	397
8	Series Representations of Fractional Brownian Motion	437
9	Multidimensional Models	466
10	Maximum Likelihood Estimation Methods	539
A	Auxiliary Notions and Results	575
B	Integrals with Respect to Random Measures	588
C	Basics of Malliavin Calculus	602
D	Other Notes and Topics	610
Bibliography		613
Index		660

Expanded Contents

List of Abbreviations		xv
Notation		xvii
Preface		xxi

1	**A Brief Overview of Time Series and Stochastic Processes**		1
1.1	Stochastic Processes and Time Series		1
	1.1.1	Gaussian Stochastic Processes	2
	1.1.2	Stationarity (of Increments)	3
	1.1.3	Weak or Second-Order Stationarity (of Increments)	4
1.2	Time Domain Perspective		4
	1.2.1	Representations in the Time Domain	4
1.3	Spectral Domain Perspective		7
	1.3.1	Spectral Density	7
	1.3.2	Linear Filtering	8
	1.3.3	Periodogram	9
	1.3.4	Spectral Representation	9
1.4	Integral Representations Heuristics		11
	1.4.1	Representations of a Gaussian Continuous-Time Process	12
1.5	A Heuristic Overview of the Next Chapter		14
1.6	Bibliographical Notes		14
2	**Basics of Long-Range Dependence and Self-Similarity**		15
2.1	Definitions of Long-Range Dependent Series		16
2.2	Relations Between the Various Definitions of Long-Range Dependence		19
	2.2.1	Some Useful Properties of Slowly and Regularly Varying Functions	19
	2.2.2	Comparing Conditions II and III	21
	2.2.3	Comparing Conditions II and V	21
	2.2.4	Comparing Conditions I and II	23
	2.2.5	Comparing Conditions II and IV	25
	2.2.6	Comparing Conditions I and IV	28
	2.2.7	Comparing Conditions IV and III	29
	2.2.8	Comparing Conditions IV and V	29
2.3	Short-Range Dependent Series and their Several Examples		30
2.4	Examples of Long-Range Dependent Series: FARIMA Models		35
	2.4.1	FARIMA$(0, d, 0)$ Series	35

		2.4.2 FARIMA(p,d,q) Series	42
2.5	Definition and Basic Properties of Self-Similar Processes		43
2.6	Examples of Self-Similar Processes		47
	2.6.1	Fractional Brownian Motion	47
	2.6.2	Bifractional Brownian Motion	53
	2.6.3	The Rosenblatt Process	56
	2.6.4	$S\alpha S$ Lévy Motion	59
	2.6.5	Linear Fractional Stable Motion	59
	2.6.6	Log-Fractional Stable Motion	61
	2.6.7	The Telecom Process	62
	2.6.8	Linear Fractional Lévy Motion	63
2.7	The Lamperti Transformation		65
2.8	Connections Between Long-Range Dependent Series and Self-Similar Processes		67
2.9	Long- and Short-Range Dependent Series with Infinite Variance		76
	2.9.1	First Definition of LRD Under Heavy Tails: Condition A	76
	2.9.2	Second Definition of LRD Under Heavy Tails: Condition B	82
	2.9.3	Third Definition of LRD Under Heavy Tails: Codifference	82
2.10	Heuristic Methods of Estimation		84
	2.10.1	The R/S Method	84
	2.10.2	Aggregated Variance Method	88
	2.10.3	Regression in the Spectral Domain	88
	2.10.4	Wavelet-Based Estimation	93
2.11	Generation of Gaussian Long- and Short-Range Dependent Series		99
	2.11.1	Using Cholesky Decomposition	100
	2.11.2	Using Circulant Matrix Embedding	100
2.12	Exercises		106
2.13	Bibliographical Notes		108
3	**Physical Models for Long-Range Dependence and Self-Similarity**		**113**
3.1	Aggregation of Short-Range Dependent Series		113
3.2	Mixture of Correlated Random Walks		117
3.3	Infinite Source Poisson Model with Heavy Tails		120
	3.3.1	Model Formulation	120
	3.3.2	Workload Process and its Basic Properties	123
	3.3.3	Input Rate Regimes	128
	3.3.4	Limiting Behavior of the Scaled Workload Process	131
3.4	Power-Law Shot Noise Model		149
3.5	Hierarchical Model		154
3.6	Regime Switching		156
3.7	Elastic Collision of Particles		162
3.8	Motion of a Tagged Particle in a Simple Symmetric Exclusion Model		167
3.9	Power-Law Pólya's Urn		172
3.10	Random Walk in Random Scenery		177
3.11	Two-Dimensional Ising Model		180
	3.11.1	Model Formulation and Result	181

		3.11.2 Correlations, Dimers and Pfaffians	184
		3.11.3 Computation of the Inverse	198
		3.11.4 The Strong Szegö Limit Theorem	209
		3.11.5 Long-Range Dependence at Critical Temperature	212
3.12	Stochastic Heat Equation		215
3.13	The Weierstrass Function Connection		216
3.14	Exercises		221
3.15	Bibliographical Notes		223

4 Hermite Processes — 229

4.1	Hermite Polynomials and Multiple Stochastic Integrals		229
4.2	Integral Representations of Hermite Processes		232
	4.2.1 Integral Representation in the Time Domain		232
	4.2.2 Integral Representation in the Spectral Domain		233
	4.2.3 Integral Representation on an Interval		234
	4.2.4 Summary		239
4.3	Moments, Cumulants and Diagram Formulae for Multiple Integrals		241
	4.3.1 Diagram Formulae		241
	4.3.2 Multigraphs		245
	4.3.3 Relation Between Diagrams and Multigraphs		246
	4.3.4 Diagram and Multigraph Formulae for Hermite Polynomials		251
4.4	Moments and Cumulants of Hermite Processes		254
4.5	Multiple Integrals of Order Two		259
4.6	The Rosenblatt Process		262
4.7	The Rosenblatt Distribution		264
4.8	CDF of the Rosenblatt Distribution		272
4.9	Generalized Hermite and Related Processes		272
4.10	Exercises		279
4.11	Bibliographical Notes		280

5 Non-Central and Central Limit Theorems — 282

5.1	Nonlinear Functions of Gaussian Random Variables	282
5.2	Hermite Rank	285
5.3	Non-Central Limit Theorem	288
5.4	Central Limit Theorem	298
5.5	The Fourth Moment Condition	305
5.6	Limit Theorems in the Linear Case	306
	5.6.1 Direct Approach for Entire Functions	306
	5.6.2 Approach Based on Martingale Differences	311
5.7	Multivariate Limit Theorems	316
	5.7.1 The SRD Case	316
	5.7.2 The LRD Case	319
	5.7.3 The Mixed Case	321
	5.7.4 Multivariate Limits of Multilinear Processes	324
5.8	Generation of Non-Gaussian Long- and Short-Range Dependent Series	328
	5.8.1 Matching a Marginal Distribution	329

		5.8.2	Relationship Between Autocorrelations	331
		5.8.3	Price Theorem	333
		5.8.4	Matching a Targeted Autocovariance for Series with Prescribed Marginal	336
	5.9		Exercises	341
	5.10		Bibliographical Notes	342

6 Fractional Calculus and Integration of Deterministic Functions with Respect to FBM — 345

	6.1		Fractional Integrals and Derivatives	345
		6.1.1	Fractional Integrals on an Interval	345
		6.1.2	Riemann–Liouville Fractional Derivatives $\mathcal{D}$ on an Interval	348
		6.1.3	Fractional Integrals and Derivatives on the Real Line	352
		6.1.4	Marchaud Fractional Derivatives $\mathbf{D}$ on the Real Line	354
		6.1.5	The Fourier Transform Perspective	357
	6.2		Representations of Fractional Brownian Motion	359
		6.2.1	Representation of FBM on an Interval	359
		6.2.2	Representations of FBM on the Real Line	368
	6.3		Fractional Wiener Integrals and their Deterministic Integrands	369
		6.3.1	The Gaussian Space Generated by Fractional Wiener Integrals	369
		6.3.2	Classes of Integrands on an Interval	372
		6.3.3	Subspaces of Classes of Integrands	377
		6.3.4	The Fundamental Martingale	381
		6.3.5	The Deconvolution Formula	382
		6.3.6	Classes of Integrands on the Real Line	383
		6.3.7	Connection to the Reproducing Kernel Hilbert Space	384
	6.4		Applications	386
		6.4.1	Girsanov's Formula for FBM	386
		6.4.2	The Prediction Formula for FBM	388
		6.4.3	Elementary Linear Filtering Involving FBM	392
	6.5		Exercises	394
	6.6		Bibliographical Notes	395

7 Stochastic Integration with Respect to Fractional Brownian Motion — 397

	7.1		Stochastic Integration with Random Integrands	397
		7.1.1	FBM and the Semimartingale Property	397
		7.1.2	Divergence Integral for FBM	399
		7.1.3	Self-Integration of FBM	402
		7.1.4	Itô's Formulas	407
	7.2		Applications of Stochastic Integration	413
		7.2.1	Stochastic Differential Equations Driven by FBM	413
		7.2.2	Regularity of Laws Related to FBM	414
		7.2.3	Numerical Solutions of SDEs Driven by FBM	418
		7.2.4	Convergence to Normal Law Using Stein's Method	426
		7.2.5	Local Time of FBM	430
	7.3		Exercises	434
	7.4		Bibliographical Notes	435

8		**Series Representations of Fractional Brownian Motion**	437
8.1		Karhunen–Loève Decomposition and FBM	437
	8.1.1	The Case of General Stochastic Processes	437
	8.1.2	The Case of BM	438
	8.1.3	The Case of FBM	439
8.2		Wavelet Expansion of FBM	440
	8.2.1	Orthogonal Wavelet Bases	440
	8.2.2	Fractional Wavelets	445
	8.2.3	Fractional Conjugate Mirror Filters	450
	8.2.4	Wavelet-Based Expansion and Simulation of FBM	452
8.3		Paley–Wiener Representation of FBM	455
	8.3.1	Complex-Valued FBM and its Representations	455
	8.3.2	Space $\mathcal{L}_a$ and its Orthonormal Basis	456
	8.3.3	Expansion of FBM	462
8.4		Exercises	463
8.5		Bibliographical Notes	464
9		**Multidimensional Models**	466
9.1		Fundamentals of Multidimensional Models	467
	9.1.1	Basics of Matrix Analysis	467
	9.1.2	Vector Setting	469
	9.1.3	Spatial Setting	471
9.2		Operator Self-Similarity	472
9.3		Vector Operator Fractional Brownian Motions	475
	9.3.1	Integral Representations	476
	9.3.2	Time Reversible Vector OFBMs	484
	9.3.3	Vector Fractional Brownian Motions	486
	9.3.4	Identifiability Questions	492
9.4		Vector Long-Range Dependence	495
	9.4.1	Definitions and Basic Properties	495
	9.4.2	Vector FARIMA$(0, D, 0)$ Series	499
	9.4.3	Vector FGN Series	503
	9.4.4	Fractional Cointegration	504
9.5		Operator Fractional Brownian Fields	508
	9.5.1	M–Homogeneous Functions	509
	9.5.2	Integral Representations	513
	9.5.3	Special Subclasses and Examples of OFBFs	523
9.6		Spatial Long-Range Dependence	526
	9.6.1	Definitions and Basic Properties	526
	9.6.2	Examples	529
9.7		Exercises	532
9.8		Bibliographical Notes	535
10		**Maximum Likelihood Estimation Methods**	539
10.1		Exact Gaussian MLE in the Time Domain	539
10.2		Approximate MLE	542
	10.2.1	Whittle Estimation in the Spectral Domain	542
	10.2.2	Autoregressive Approximation	550

10.3	Model Selection and Diagnostics		551
10.4	Forecasting		554
10.5	R Packages and Case Studies		555
	10.5.1	The ARFIMA Package	555
	10.5.2	The FRACDIFF Package	556
10.6	Local Whittle Estimation		558
	10.6.1	Local Whittle Estimator	558
	10.6.2	Bandwidth Selection	562
	10.6.3	Bias Reduction and Rate Optimality	565
10.7	Broadband Whittle Approach		567
10.8	Exercises		569
10.9	Bibliographical Notes		570

A	**Auxiliary Notions and Results**		575
A.1	Fourier Series and Fourier Transforms		575
	A.1.1	Fourier Series and Fourier Transform for Sequences	575
	A.1.2	Fourier Transform for Functions	577
A.2	Fourier Series of Regularly Varying Sequences		579
A.3	Weak and Vague Convergence of Measures		583
	A.3.1	The Case of Probability Measures	583
	A.3.2	The Case of Locally Finite Measures	584
A.4	Stable and Heavy-Tailed Random Variables and Series		585

B	**Integrals with Respect to Random Measures**		588
B.1	Single Integrals with Respect to Random Measures		588
	B.1.1	Integrals with Respect to Random Measures with Orthogonal Increments	589
	B.1.2	Integrals with Respect to Gaussian Measures	590
	B.1.3	Integrals with Respect to Stable Measures	592
	B.1.4	Integrals with Respect to Poisson Measures	593
	B.1.5	Integrals with Respect to Lévy Measures	596
B.2	Multiple Integrals with Respect to Gaussian Measures		597

C	**Basics of Malliavin Calculus**	602
C.1	Isonormal Gaussian Processes	602
C.2	Derivative Operator	603
C.3	Divergence Integral	607
C.4	Generator of the Ornstein–Uhlenbeck Semigroup	608

D	**Other Notes and Topics**	610

Bibliography	613
Index	660

Abbreviations

ACVF	autocovariance function
ACF	autocorrelation function
AIC	Akaike information criterion
AR	autoregressive
ARMA	autoregressive moving average
BIC	Bayesian information criterion
biFBM	bifractional Brownian motion
BLUE	best linear unbiased estimator
BM	Brownian motion
CDF	cumulative distribution function
CMF	conjugate mirror filters
CRW	correlated random walk
FARIMA	fractionally integrated autoregressive moving average
FBM	fractional Brownian motion
FBF	fractional Brownian field
FBS	fractional Brownian sheet
FGN	fractional Gaussian noise
FFT	fast Fourier transform
FGN	fractional Gaussian noise
LFSM	linear fractional stable motion
LFSN	linear fractional stable noise
LRD	long-range dependence, long-range dependent
MA	moving average
MLE	maximum likelihood estimation
MRA	multi-resolution analysis
MSE	mean squared error
ODE	ordinary differential equation
OFBF	operator fractional Brownian field
OFBM	operator fractional Brownian motion
OU	Ornstein–Uhlenbeck

PDF	probability density function
RKHS	reproducing kernel Hilbert space
SDE	stochastic differential equation
SPDE	stochastic partial differential equation
SRD	short-range dependence, short-range dependent
SS	self-similar
SSSI	self-similar with stationary increments
VFBM	vector fractional Brownian motion
WN	white noise

Notation

Numbers and sequences

$n!$	n factorial
$\mathbb{Z}$	set of integers $\{\ldots, -1, 0, 1, \ldots\}$
$\mathbb{R}_+$	half-axis $(0, \infty)$
$\|x\|_2$	Euclidean norm $\|x\|_2 = (\sum_{j=1}^q |x_j|^2)^{1/2}$ for $x = (x_1, \ldots, x_q)' \in \mathbb{C}^q$
x_+, x_-	$\max\{x, 0\}, \max\{-x, 0\}$, respectively
$\Re, \Im$	real and imaginary parts, respectively
$\overline{z}$	complex conjugate of $z \in \mathbb{C}$
$x \wedge y, x \vee y$	$\min\{x, y\}, \max\{x, y\}$, respectively
$[x], \lceil x \rceil$	(floor) integer part and (ceiling) integer part of x, respectively
$\widehat{a}$	Fourier transform of sequence a
$a^\vee$	time reversion of sequence a
$a * b$	convolution of sequences a and b
$\downarrow 2, \uparrow 2$	downsampling and upsampling by 2 operations, respectively
$\ell^p(\mathbb{Z})$	space of sequences $\{a_n\}_{n \in \mathbb{Z}}$ such that $\sum_{n=-\infty}^{\infty} |a_n|^p < \infty$

Functions

$\widehat{f}$	Fourier transform of function f
$f * g$	convolution of functions f and g
$f \otimes g$	tensor product of functions f and g
$f^{\otimes k}$	kth tensor product of a function f
1_A	indicator function of a set A
$\log, \log_2$	natural logarithm (base e) and logarithm base 2, respectively
e^{ix}	complex exponential, $e^{ix} = \cos x + i \sin x$
$\Gamma(\cdot)$	gamma function
$B(\cdot, \cdot)$	beta function
$H_k(\cdot)$	Hermite polynomial of order k
$C^k(I)$	space of k–times continuously differentiable functions on an interval I
$C_b^k(I)$	space of k–times continuously differentiable functions on an interval I with the first k derivatives bounded
$L^p(\mathbb{R}^q)$	space of functions $f : \mathbb{R}^q \to \mathbb{R}$ (or $\mathbb{C}$) such that $\int_{\mathbb{R}^q} |f(x)|^p dx < \infty$
$\|f\|_{L^p(\mathbb{R}^q)}$	(semi-)norm $\left(\int_{\mathbb{R}^q} |f(x)|^p dx\right)^{1/p}$
$L^p(E, m)$	space of functions $f : E \to \mathbb{R}$ (or $\mathbb{C}$) such that $\int_E |f(x)|^p m(dx) < \infty$
$f^{\leftarrow}, f^{-1}$	(generalized) inverse of non-decreasing function f

Matrices

$\mathbb{R}^{p\times p}$, $\mathbb{C}^{p\times p}$	collections of $p \times p$ matrices with entries in $\mathbb{R}$ and $\mathbb{C}$, respectively
A', A^T	transpose of a matrix A
A^*	Hermitian transpose of a matrix A
$\det(A)$	determinant of a matrix A
Pf A	Pfaffian of a matrix A
$\|A\|$	matrix (operator) norm
$A \otimes B$	Kronecker product of matrices A and B
$A^{1/2}$	square root of positive semidefinite matrix A
$\mathrm{tr}(A)$	trace of matrix A

Probability

Var, Cov	variance, covariance, respectively
$\xrightarrow{fdd}$	convergence in the sense of finite-dimensional distributions
$\xrightarrow{d}$, $\xrightarrow{p}$	convergence in distribution and in probability, respectively
$\stackrel{d}{=}$, $\sim$	equality in distribution, as in $Z \sim \mathcal{N}(0, 1)$
$S\alpha S$	symmetric α-stable
$\|\xi\|_p$	$(\mathbb{E}\|\xi\|^p)^{1/p}$ for a random variable ξ
$\mathcal{N}(\mu, \sigma^2)$	normal (Gaussian) distribution with mean μ and variance σ^2
$\chi(X_1, \ldots, X_p)$	joint cumulant of random variables $X_1, \ldots, X_p$
$\chi_p(X)$	pth cumulant of a variable X

Time series and stochastic processes

$\gamma_X(\cdot)$	ACVF of series X
$f_X(\cdot)$	spectral density of series X
$I_X(\cdot)$	periodogram of series X
B	backshift operator
B_H or B^κ	FBM, $H = \kappa + 1/2$; vector OFBM
$Z_H^{(k)}$	Hermite process of order k
$B_{E,H}$	OFBF

Fractional calculus, Malliavin calculus and FBM

I_{a+}^α, I_{b-}^α	fractional integrals on interval
$\mathcal{D}_{a+}^\alpha$, $\mathcal{D}_{b-}^\alpha$	fractional derivatives on interval
$I_\pm^\alpha$, $\mathcal{D}_\pm^\alpha$, $\mathbf{D}_\pm^\alpha$	fractional integrals and derivatives on real line
span, $\overline{\mathrm{span}}$	linar span and closure of linear span, respectively
$\mathcal{I}_a^\kappa(\cdot)$	fractional Wiener integral
Λ_a^κ, $\|\Lambda\|_a^\kappa$	spaces of integrands
$(\cdot, \cdot)_{\mathcal{H}}$	inner product on Hilbert space $\mathcal{H}$
$\delta(\cdot)$	divergence integral ($\delta_a^\kappa(\cdot)$, in the case of FBM B^κ on an interval $[0, a]$)
D	Malliavin derivative
$\mathbb{D}^{1,p}$, $\mathbb{D}^{k,p}$	domains of the Malliavin derivative
$I_n(\cdot)$, $\widehat{I}_n(\cdot)$	multiple integrals of order n

Miscellaneous

a.e., a.s.	almost everywhere, almost surely, respectively
$\sim$	asymptotic equivalence; that is, $a(x) \sim b(x)$ if $a(x)/b(x) \to 1$
$o(\cdot), O(\cdot)$	small and big O, respectively
$\mathcal{B}(A)$	σ–field of Borel sets of a set A
$\delta_a(v)$	point mass at $v = a$
mod	modulo

Preface

We focus in this book on long-range dependence and self-similarity. The notion of long-range dependence is associated with time series whose autocovariance function decays slowly like a power function as the lag between two observations increases. Such time series emerged more than half a century ago. They have been studied extensively and have been applied in numerous fields, including hydrology, economics and finance, computer science and elsewhere. What makes them unique is that they stand in sharp contrast to Markovian-like or short-range dependent time series, in that, for example, they often call for special techniques of analysis, they involve different normalizations and they yield new limiting objects.

Long-range dependent time series are closely related to self-similar processes, which by definition are statistically alike at different time scales. Self-similar processes arise as large scale limits of long-range dependent time series, and vice versa; they can give rise to long-range dependent time series through their increments. The celebrated Brownian motion is an example of a self-similar process, but it is commonly associated with independence and, more generally, with short-range dependence. The most studied and well-known self-similar process associated with long-range dependence is fractional Brownian motion, though many other self-similar processes will also be presented in this book. Self-similar processes have become one of the central objects of study in probability theory, and are often of interest in their own right.

This volume is a modern and rigorous introduction to the subjects of long-range dependence and self-similarity, together with a number of more specialized up-to-date topics at the center of this research area. Our goal has been to write a very readable text which will be useful to graduate students as well as to researchers in Probability, Statistics, Physics and other fields. Proofs are presented in detail. A precise reference to the literature is given in cases where a proof is omitted. Chapter 2 is fundamental. It develops the basics of long-range dependence and self-similarity and should be read by everyone, as it allows the reader to gain quickly a basic familiarity with the main themes of the research area. We assume that the reader has a background in basic time series analysis (e.g., at the level of Brockwell and Davis [186]) and stochastic processes. The reader without this background may want to start with Chapter 1, which provides a brief and elementary introduction to time series analysis and stochastic processes.

The rest of the volume, namely Chapters 3–10, introduces the more specialized and advanced topics on long-range dependence and self-similarity. Chapter 3 concerns physical models that give rise to long-range dependence and/or self-similarity. Chapters 4 and 5 focus on central and non-central limit theorems for long-range dependent series, and introduce the limiting Hermite processes. Chapters 6 and 7 are on fractional Brownian motion and its stochastic calculus, and their connection to the so-called fractional calculus, the area of real analysis which extends the usual derivatives and integrals to fractional orders. Chapter 8 pursues the discussion on fractional Brownian motion by introducing several of its celebrated decompositions. Chapter 9 concerns multidimensional models, and Chapter 10 reviews the maximum likelihood estimation methods for long-range dependent time series. Chapters 3 through 10 may be read somewhat separately. They are meant to serve both as a learning tool and as a reference. Finally, Appendices A, B and C are used for reference throughout the book.

Each chapter starts with a brief overview and ends with exercises and bibliographical notes. In Appendix D, "Other notes and topics" contains further bibliographical information. The reader will find the notes extremely useful as they provide further perspectives on the subject as well as suggestions for future research.

A number of new books on long-range dependence and self-similarity, listed in Appendix D, have been published in the last ten years or so. Many features set this book apart. First, a number of topics are not treated elsewhere, including most of Chapter 8 (on decompositions) and Chapter 9 (on multidimensional models). Second, other topics provide a more systematic treatment of the area than is otherwise presently available; for example, in Chapter 3 (on physical models). Third, some specialized topics, such as in Chapters 6 and 7 (on stochastic analysis for fractional Brownian motion), reflect our perspective. Fourth, most specialized topics are up to date; for example, the classical results of Chapters 4 and 5 (on Hermite processes and non-central limit theorems) have been supplemented by recent work in the area. Finally, the book contains a substantial number of early and up-to-date references. Though even with this large number of references (more than a thousand), we had to be quite selective and could not include every single work that was relevant.

We would like to conclude this preface with acknowledgments and a tribute. A number of our former and present students have read carefully and commented on excerpts of this book, including Changryong Baek, Stefanos Kechagias and Sandeep Sarangi at the University of North Carolina, Shuyang Bai, Long Tao and Mark Veillette at Boston University, Yong Bum Cho, Heng Liu, Xuan Yang, Pengfei Zhang, Yu Gu, Junyi Zhang, Emilio Seijo and Oliver Pfaffel at Columbia University. Various parts of this volume have already been used in teaching by the authors at these institutions. As the book was taking its final shape, a number of researchers read carefully individual chapters and provided invaluable feedback. We thus express our gratitude to Solesne Bourguin, Gustavo Didier, Liudas Giraitis, Kostas Spiliopoulos, Stilian Stoev, Yizao Wang. We are grateful to Diana Gillooly at Cambridge University Press for her support and encouragement throughout the process of writing the book. We also thank the National Science Foundation and the National Security Agency for their support. We are responsible for the remaining errors and typos, and would be grateful to the readers for a quick email to either author if such are found.

Finally, we would like to pay tribute to Benoit B. Mandelbrot,[1] a pioneer in the study of long-range dependence and self-similarity, among his many other interests. He was the

---

[1] A fascinating account of Benoit B. Mandelbrot's life can be found in [681].

first to greatly popularize and draw attention to these subjects in the 1970s. It was Benoit B. Mandelbrot who introduced one of the authors (M.S.T.) to this area, who in turn passed on his interests and passion to the other author (V.P.). This volume, including the presented specialized topics, are direct fruits of the work started by Benoit B. Mandelbrot. With his passing in 2010, the scientific community lost one of its truly bright stars.

1
A Brief Overview of Time Series and Stochastic Processes

This chapter serves as a brief introduction to time series to readers unfamiliar with this topic. Knowledgeable readers may want to jump directly to Chapter 2, where the basics of long-range dependence and self-similarity are introduced. A number of references for the material of this chapter can be found in Section 1.6, below.

1.1 Stochastic Processes and Time Series

A *stochastic processes* $\{X(t)\}_{t \in T}$ is a collection of random variables $X(t)$ on some probability space $(\Omega, \mathcal{F}, \mathbb{P})$, indexed by the time parameter $t \in T$. In "*discrete time*," we typically choose for T,

$$\mathbb{Z} = \{\ldots, -1, 0, 1, \ldots\}, \ \mathbb{Z}_+ = \{0, 1, \ldots\}, \ \{1, 2, \ldots, N\}, \ldots,$$

and denote t by n. In "*continuous time*," we will often choose for T,

$$\mathbb{R}, \ \mathbb{R}_+ = [0, \infty), \ [0, N], \ldots.$$

In some instances in this volume, the parameter space T will be a subset of $\mathbb{R}^q$, $q \geq 1$, and/or $X(t)$ will be a vector with values in $\mathbb{R}^p$, $p \geq 1$. But for the sake of simplicity, we suppose in this chapter that p and q equal 1. We also suppose that $X(t)$ is real-valued.

One way to think of a stochastic process is through its law. *The law of a stochastic process* $\{X(t)\}_{t \in T}$ is characterized by its *finite-dimensional distributions* (fdd, in short); that is, the probability distributions

$$\mathbb{P}(X(t_1) \leq x_1, \ldots, X(t_n) \leq x_n), \quad t_i \in T, x_i \in \mathbb{R}, n \geq 1,$$

of the random vectors

$$\big(X(t_1), \ldots, X(t_n)\big)', \quad t_i \in T, n \geq 1.$$

Here, the prime indicates transpose, and all vectors are column vectors throughout. Thus, the finite-dimensional distributions of a stochastic process fully characterize its law and, in particular, the dependence structure of the stochastic process. In order to check that two stochastic processes have the same law, it is therefore sufficient to verify that their finite-dimensional distributions are identical. Equality and convergence in distribution is denoted by $\stackrel{d}{=}$ and $\stackrel{d}{\to}$ respectively. Thus $\{X(t)\}_{t \in T} \stackrel{d}{=} \{Y(t)\}_{t \in T}$ means

$$\mathbb{P}(X(t_1) \leq x_1, \ldots, X(t_n) \leq x_n) = \mathbb{P}(Y(t_1) \leq x_1, \ldots, Y(t_n) \leq x_n), \quad t_i \in T, x_i \in \mathbb{R}, n \geq 1.$$

A tilde $\sim$ indicates equality in distribution: for example, we write $X \sim \mathcal{N}(\mu, \sigma^2)$ if X is Gaussian with mean μ and variance σ^2.

A stochastic process is often called a *time series*, particularly when it is in discrete time and the focus is on its mean and covariance functions.

1.1.1 Gaussian Stochastic Processes

A stochastic process $\{X(t)\}_{t \in T}$ is *Gaussian* if one of the following equivalent conditions holds:

(i) The finite-dimensional distributions $Z = (X(t_1), \ldots, X(t_n))'$ are multivariate Gaussian $\mathcal{N}(b, A)$ with mean $b = \mathbb{E}Z$ and covariance matrix $A = \mathbb{E}(Z - \mathbb{E}Z)(Z - \mathbb{E}Z)'$;

(ii) $a_1 X(t_1) + \cdots + a_n X(t_n)$ is a Gaussian random variable for any $a_i \in \mathbb{R}$, $t_i \in T$;

(iii) In the case when $\mathbb{E}X(t) = 0$, for any $a_i \in \mathbb{R}$, $t_i \in T$,

$$\mathbb{E} \exp\{i(a_1 X(t_1) + \cdots + a_n X(t_n))\} = \exp\left\{-\frac{1}{2} \mathbb{E}(a_1 X(t_1) + \cdots + a_n X(t_n))^2\right\}$$

$$= \exp\left\{-\frac{1}{2} \sum_{i,j=1}^{n} a_i a_j \mathbb{E}X(t_i)X(t_j)\right\}. \quad (1.1.1)$$

The law of a Gaussian stochastic process with zero mean is determined by a *covariance function* $\mathrm{Cov}(X(t), X(s)) = \mathbb{E}X(t)X(s)$, $s, t \in T$. When the mean is not zero, the covariance function is defined as $\mathrm{Cov}(X(t), X(s)) = \mathbb{E}(X(t) - \mathbb{E}X(t))(X(s) - \mathbb{E}X(s))$, $s, t \in T$. Together with the mean function $\mathbb{E}X(t)$, $t \in T$, the covariance determines the law of the Gaussian stochastic process.

Example 1.1.1 (*Brownian motion*) Brownian motion (or *Wiener process*) is a Gaussian stochastic process $\{X(t)\}_{t \geq 0}$ with[1]

$$\mathbb{E}X(t) = 0, \quad \mathbb{E}X(t)X(s) = \sigma^2 \min\{t, s\}, \quad t, s \geq 0, \ \sigma > 0, \quad (1.1.2)$$

or, equivalently, it is a Gaussian stochastic process with independent increments $X(t_k) - X(t_{k-1})$, $k = 1, \ldots, n$, with $t_0 \leq t_1 \leq \cdots \leq t_n$ such that $X(t) - X(s) \sim \sigma \mathcal{N}(0, t - s)$, $t \geq s > 0$. Brownian motion is often denoted $B(t)$ or $W(t)$. Brownian motion $\{B(t)\}_{t \in \mathbb{R}}$ on the real line is defined as $B(t) = B_1(t)$, $t \geq 0$, and $B(t) = B_2(-t)$, $t < 0$, where B_1 and B_2 are two independent Brownian motions on the half line.

Example 1.1.2 (*Ornstein–Uhlenbeck process*) The *Ornstein–Uhlenbeck (OU) process* is a Gaussian stochastic process $\{X(t)\}_{t \in \mathbb{R}}$ with

$$\mathbb{E}X(t) = 0, \quad \mathbb{E}X(t)X(s) = \frac{\sigma^2}{2\lambda} e^{-\lambda(t-s)}, \quad t > s, \quad (1.1.3)$$

---

[1] One imposes, sometimes, the additional condition that the process has continuous paths; but we consider here only the finite-dimensional distributions.

Figure 1.1 If the process has stationary increments, then, in particular, the increments taken over the bold intervals have the same distributions.

where $\lambda > 0, \sigma > 0$ are two parameters. The OU process is the only Gaussian stationary Markov process. It satisfies the Langevin stochastic differential equation

$$dX(t) = -\lambda X(t)dt + \sigma dW(t),$$

where $\{W(t)\}$ is a Wiener process. The term $-\lambda X(t)dt$ in the equation above adds a drift towards the origin.

1.1.2 Stationarity (of Increments)

A stochastic process $\{X(t)\}_{t\in T}$ is *(strictly) stationary* if $T = \mathbb{R}$ or $\mathbb{Z}$ or $\mathbb{R}_+$ or $\mathbb{Z}_+$, and for any $h \in T$,

$$\{X(t)\}_{t\in T} \stackrel{d}{=} \{X(t+h)\}_{t\in T}. \tag{1.1.4}$$

A stochastic process $\{X(t)\}_{t\in T}$ is said to have *(strictly) stationary increments* if $T = \mathbb{R}$ or $\mathbb{Z}$ or $\mathbb{R}_+$ or $\mathbb{Z}_+$, and for any $h \in T$,

$$\{X(t+h) - X(h)\}_{t\in T} \stackrel{d}{=} \{X(t) - X(0)\}_{t\in T}. \tag{1.1.5}$$

See Figure 1.1.

Example 1.1.3 (*The OU process*) The OU process in Example 1.1.2 is strictly stationary. Its finite-dimensional distributions are determined, for $t > s$, by

$$\mathbb{E}X(t)X(s) = \frac{\sigma^2}{2\lambda}e^{-\lambda(t-s)} = \frac{\sigma^2}{2\lambda}e^{-\lambda((t+h)-(s+h))} = \mathbb{E}X(t+h)X(s+h).$$

Thus, the law of the OU process is the same when shifted by $h \in \mathbb{R}$.

An example of a stochastic process with (strictly) stationary increments is Brownian motion in Example 1.1.1.

There is an obvious connection between stationarity and stationarity of increments. If $T = \mathbb{Z}$ and $\{X_t\}_{t\in\mathbb{Z}}$ has (strictly) stationary increments, then $\Delta X_t = X_t - X_{t-1}, t \in \mathbb{Z}$, is (strictly) stationary. Indeed, for any $h \in \mathbb{Z}$,

$$\{\Delta X_{t+h}\}_{t\in\mathbb{Z}} = \{X_{t+h} - X_{t+h-1}\}_{t\in\mathbb{Z}} = \{X_{t+h} - X_h - (X_{t+h-1} - X_h)\}_{t\in\mathbb{Z}}$$

$$\stackrel{d}{=} \{X_t - X_0 - (X_{t-1} - X_0)\}_{t\in\mathbb{Z}} = \{X_t - X_{t-1}\}_{t\in\mathbb{Z}} = \{\Delta X_t\}_{t\in\mathbb{Z}}.$$

Conversely, if $\{Y_t\}_{t\in\mathbb{Z}}$ is (strictly) stationary, then $X_t = \sum_{k=1}^{t} Y_k$ can be seen easily to have (strictly) stationary increments.

If $T = \mathbb{R}$, then the *difference operator* Δ is replaced by the derivative when it exists and the sum is replaced by an integral.

1.1.3 Weak or Second-Order Stationarity (of Increments)

The probabilistic properties of (strictly) stationary processes do not change with time. In some circumstances, such as modeling, this is sometimes too strong a requirement. Instead of focusing on all probabilistic properties, one often requires instead that only second-order properties do not change with time. This leads to the following definition of (weak) stationarity.

A stochastic process $\{X(t)\}_{t\in T}$ is *(weakly or second-order) stationary* if $T = \mathbb{R}$ or $\mathbb{Z}$ and for any $t, s \in T$,

$$\mathbb{E}X(t) = \mathbb{E}X(0), \quad \text{Cov}(X(t), X(s)) = \text{Cov}(X(t-s), X(0)). \tag{1.1.6}$$

The time difference $t - s$ above is called the *time lag*. Weakly stationary processes are often called *time series*. Note that for Gaussian processes, weak stationarity is the same as strong stationarity. This, however, is not the case in general.

1.2 Time Domain Perspective

Consider a (weakly) stationary time series $X = \{X_n\}_{n\in\mathbb{Z}}$. In the time domain, one focuses on the functions

$$\gamma_X(h) = \text{Cov}(X_h, X_0) = \text{Cov}(X_{n+h}, X_n), \quad \rho_X(h) = \frac{\gamma_X(h)}{\gamma_X(0)}, \quad h, n \in \mathbb{Z}, \tag{1.2.1}$$

called the *autocovariance function* (ACVF, in short) and *autocorrelation function* (ACF, in short) of the series X, respectively. ACVF and ACF are measures of dependence in time series. Sample counterparts of ACVF and ACF are the functions

$$\widehat{\gamma}_X(h) = \frac{1}{N} \sum_{n=1}^{N-|h|} (X_{n+|h|} - \overline{X})(X_n - \overline{X}), \quad \widehat{\rho}_X(h) = \frac{\widehat{\gamma}_X(h)}{\widehat{\gamma}_X(0)}, \quad |h| \leq N - 1,$$

where $\overline{X} = \frac{1}{N}\sum_{n=1}^{N} X_n$ is the sample mean. The following are basic properties of the ACF:

- *Symmetry:* $\rho_X(h) = \rho_X(-h), h \in \mathbb{Z}$.
- *Range:* $|\rho_X(h)| \leq 1, h \in \mathbb{Z}$.
- *Interpretation:* $\rho_X(h)$ close to $-1, 0$ and 1 correspond to strong negative, weak and strong positive correlations, respectively, of the time series X at lag h.

Statistical properties of the sample AVCF and ACF are delicate, for example, $\widehat{\rho}(h)$ and $\widehat{\rho}(h-1)$ have a nontrivial (asymptotic) dependence structure.

1.2.1 Representations in the Time Domain

"Representing" a stochastic process is expressing it in terms of simpler processes. The following examples involve *representations in the time domain*.

Example 1.2.1 (*White Noise*) A time series $X_n = Z_n, n \in \mathbb{Z}$, is called a *White Noise*, denoted $\{Z_n\} \sim \text{WN}(0, \sigma_Z^2)$, if $\mathbb{E}Z_n = 0$ and

1.2 Time Domain Perspective

$$\gamma_Z(h) = \begin{cases} \sigma_Z^2, & h = 0, \\ 0, & h \neq 0, \end{cases} \quad \rho_Z(h) = \begin{cases} 1, & h = 0, \\ 0, & h \neq 0. \end{cases}$$

Example 1.2.2 (*MA(1) series*) A time series $\{X_n\}_{n \in \mathbb{Z}}$ is called a *Moving Average of order one* (MA(1) for short) if it is given by

$$X_n = Z_n + \theta Z_{n-1}, \; n \in \mathbb{Z},$$

where $\{Z_n\} \sim \text{WN}(0, \sigma_Z^2)$. Observe that

$$\gamma_X(h) = \mathbb{E} X_h X_0 = \mathbb{E}(Z_h + \theta Z_{h-1})(Z_0 + \theta Z_{-1}) = \begin{cases} \sigma_Z^2(1+\theta^2), & h = 0, \\ \sigma_Z^2 \theta, & h = 1, \\ 0, & h \geq 2, \end{cases}$$

and hence

$$\rho_X(h) = \begin{cases} 1, & h = 0, \\ \frac{\theta}{1+\theta^2}, & h = 1, \\ 0, & h \geq 2. \end{cases}$$

Example 1.2.3 (*AR(1) series*) A (weakly) stationary time series $\{X_n\}_{n \in \mathbb{Z}}$ is called *Autoregressive of order one* (AR(1) for short) if it satisfies the AR(1) equation

$$X_n = \varphi X_{n-1} + Z_n, \quad n \in \mathbb{Z},$$

where $\{Z_n\} \sim \text{WN}(0, \sigma_Z^2)$. To see that AR(1) time series exists at least for some values of φ, suppose that $|\varphi| < 1$. Then, we expect that

$$X_n = \varphi^2 X_{n-2} + \varphi Z_{n-1} + Z_n$$
$$= \varphi^m X_{n-m} + \varphi^{m-1} Z_{n-(m-1)} + \cdots + Z_n = \sum_{m=0}^{\infty} \varphi^m Z_{n-m}.$$

The time series $\{X_n\}_{n \in \mathbb{Z}}$ above is well-defined in the $L^2(\Omega)$–sense[2] because

$$\mathbb{E}\Big(\sum_{m=n_1}^{n_2} \varphi^m Z_{n-m}\Big)^2 = \sum_{m=n_1}^{n_2} \varphi^{2m} \sigma_Z^2 \to 0, \quad \text{as } n_1, n_2 \to \infty,$$

for $|\varphi| < 1$. One can easily see that it satisfies the AR(1) equation and is (weakly) stationary. Hence, the time series $\{X_n\}_{n \in \mathbb{Z}}$ is AR(1).

When $|\varphi| > 1$, AR(1) time series is obtained by reversing the AR(1) equation as

$$X_n = \varphi^{-1} X_{n+1} - \varphi^{-1} Z_{n+1}, \quad n \in \mathbb{Z},$$

and performing similar substitutions as above to obtain

---

[2] A random variable is well-defined in the $L^2(\Omega)$–sense if $\mathbb{E}|X|^2 < \infty$. A series $\sum_{n=1}^{\infty} X_n$ is well-defined in the $L^2(\Omega)$–sense if $\sum_{n=1}^{N} X_n$ converges in $L^2(\Omega)$–sense as $N \to \infty$, that is, $\mathbb{E}|\sum_{n=N_1}^{N_2} X_n|^2 \to 0$ as $N_1, N_2 \to \infty$.

$$X_n = \varphi^{-1}X_{n+1} - \varphi^{-1}Z_{n+1}$$
$$= \varphi^{-2}X_{n+2} - \varphi^{-2}Z_{n+2} - \varphi^{-1}Z_{n+1} = -\sum_{m=0}^{\infty} \varphi^{-(m+1)}Z_{n+m+1}.$$

When $|\varphi| = 1$, there is no (weakly) stationary solution to the AR(1) equation. When $\varphi = 1$, the AR(1) equation becomes $X_n - X_{n-1} = Z_n$ and the non-stationary (in fact, stationary increment) time series satisfying this equation is called *Integrated of order one* (I(1) for short). When Z_n are i.i.d., this time series is known as a *random walk*.

For $|\varphi| < 1$, for example, observe that for $h \geq 0$,

$$\gamma_X(h) = \mathbb{E}X_h X_0 = \mathbb{E}(Z_h + \cdots + \varphi^h Z_0 + \varphi^{h+1}Z_{-1} + \cdots)(Z_0 + \varphi Z_{-1} + \cdots)$$

$$= \sigma_Z^2(\varphi^h + \varphi^{h+2} + \varphi^{h+4} + \cdots) = \sigma_Z^2 \frac{\varphi^h}{1-\varphi^2}$$

and hence, since $\rho_X(-h) = \rho_X(h)$, we get for $h \in \mathbb{Z}$,

$$\rho_X(h) = \varphi^{|h|}.$$

Example 1.2.4 (*ARMA (p,q) series*)   A (weakly) stationary time series $\{X_n\}_{n \in \mathbb{Z}}$ is called *Autoregressive moving average of orders p and q* (ARMA(p, q), for short) if it satisfies the equation

$$X_n - \varphi_1 X_{n-1} - \cdots - \varphi_p X_{n-p} = Z_n + \theta_1 Z_{n-1} + \cdots + \theta_q Z_{n-q},$$

where $\{Z_n\} \sim \text{WN}(0, \sigma_Z^2)$.

ARMA(p, q) time series exists if the so-called characteristic polynomial $1 - \varphi_1 z - \cdots - \varphi_p z^p = 0$ does not have root on the unit circle $\{z : |z| = 1\}$. This is consistent with the AR(1) equation discussed above where the root $z = 1/\varphi_1$ of the polynomial $1 - \varphi_1 z = 0$ is on the unit circle when $|\varphi_1| = 1$.

Example 1.2.5 (*Linear time series*)   A time series is called *linear* if it can be written as

$$X_n = \sum_{k=-\infty}^{\infty} a_k Z_{n-k}, \qquad (1.2.2)$$

where $\{Z_n\} \sim \text{WN}(0, \sigma_Z^2)$ and $\sum_{k=-\infty}^{\infty} |a_k|^2 < \infty$. Time series in Examples 1.2.1–1.2.3 are, in fact, linear. Observe that

$$\mathbb{E}X_{n+h}X_n = \mathbb{E}\Big(\sum_{k=-\infty}^{\infty} a_k Z_{n+h-k}\Big)\Big(\sum_{k=-\infty}^{\infty} a_k Z_{n-k}\Big)$$

$$= \mathbb{E}\Big(\sum_{k'=-\infty}^{\infty} a_{k'+h} Z_{n-k'}\Big)\Big(\sum_{k=-\infty}^{\infty} a_k Z_{n-k}\Big)$$

$$= \sigma_Z^2 \sum_{k=-\infty}^{\infty} a_{k+h}a_k = \sigma_Z^2 \sum_{k=-\infty}^{\infty} a_{h-k}a_{-k} = \sigma_Z^2 (a * a^{\vee})_h,$$

where $*$ stands for the usual convolution, and $a^\vee$ denotes the time reversal of a. Since $\mathbb{E}X_{n+h}X_h$ depends only on h and $\mathbb{E}X_n = 0$, linear time series are (weakly) stationary. The variables Z_n entering the linear series (1.2.2) are known as *innovations*, especially when they are i.i.d.

Remark 1.2.6 Some of the notions above extend to continuous-time stationary processes $\{X(t)\}_{t \in \mathbb{R}}$. For example, such process is called linear when it can be represented as

$$X(t) = \int_{\mathbb{R}} a(t-u) Z(du), \quad t \in \mathbb{R}, \tag{1.2.3}$$

where $Z(du)$ is a real-valued random measure on $\mathbb{R}$ with orthogonal increments and control measure $\mathbb{E}(Z(du))^2 = du$ (see Appendix B.1, as well as Section 1.4 below), and $a \in L^2(\mathbb{R})$ is a deterministic function.

1.3 Spectral Domain Perspective

We continue considering (weakly) stationary time series $X = \{X_n\}_{n \in \mathbb{Z}}$. The material of this section is also related to the Fourier series and transform discussed in Appendix A.1.1.

1.3.1 Spectral Density

In the spectral domain, the focus is on the function

$$f_X(\lambda) = \frac{1}{2\pi} \sum_{h=-\infty}^{\infty} e^{-ih\lambda} \gamma_X(h), \quad \lambda \in (-\pi, \pi], \tag{1.3.1}$$

called the *spectral density* of the time series X. The variable λ is restricted to the domain $(-\pi, \pi]$ since the spectral density $f_X(\lambda)$ is 2π–periodic: that is, $f_X(\lambda + 2\pi k) = f_X(\lambda)$, $k \in \mathbb{Z}$. Observe also that the spectral density f_X is well-defined pointwise when $\gamma_X \in \ell^1(\mathbb{Z})$.

The variable λ enters $f_X(\lambda)$ through $e^{-ih\lambda}$, or sines and cosines. When λ is close to 0, we will talk about *low frequencies (long waves)*, and when λ is close to π, we will have *high frequencies (short waves)* in mind. Graphically, the association is illustrated in Figure 1.2.

Example 1.3.1 (*Spectral density of white noise*) If $\{Z_n\} \sim \text{WN}(0, \sigma_Z^2)$, then

$$f_Z(\lambda) = \frac{\sigma_Z^2}{2\pi},$$

that is, the spectral density $f_Z(\lambda)$, $\lambda \in (-\pi, \pi]$, is constant.

Figure 1.2 Low (left) and high (right) frequencies.

Example 1.3.2 (*Spectral density of AR(1) series*) If $\{X_n\}_{n\in\mathbb{Z}}$ is AR(1) time series with $|\varphi| < 1$ and $\gamma_X(h) = \sigma_Z^2 \varphi^{|h|}/(1-\varphi^2)$, then

$$f_X(\lambda) = \frac{\sigma_Z^2}{2\pi(1-\varphi^2)}\left(1 + \sum_{h=1}^{\infty}(e^{-ih\lambda} + e^{ih\lambda})\varphi^h\right)$$

$$= \frac{\sigma_Z^2}{2\pi(1-\varphi^2)}\left(1 + \frac{\varphi e^{-i\lambda}}{1-\varphi e^{-i\lambda}} + \frac{\varphi e^{i\lambda}}{1-\varphi e^{i\lambda}}\right) = \frac{\sigma_Z^2}{2\pi}\frac{1}{|1-\varphi e^{-i\lambda}|^2}.$$

The spectral density has the following properties:

- *Symmetry:* $f_X(\lambda) = f_X(-\lambda)$. This follows from $\gamma_X(h) = \gamma_X(-h)$. In particular, we can focus only on $\lambda \in [0, \pi]$.
- *Nonnegativeness:* $f_X(\lambda) \geq 0$. For $\gamma_X \in \ell^1(\mathbb{Z})$, this follows from

$$0 \leq \frac{1}{N}\mathbb{E}\left|\sum_{r=1}^{N} X_r e^{-ir\lambda}\right|^2 = \frac{1}{N}\mathbb{E}\left(\sum_{r,s=1}^{N} X_r X_s e^{-i(r-s)\lambda}\right)$$

$$= \frac{1}{N}\sum_{|h|<N}(N-|h|)\gamma_X(h)e^{-ih\lambda} \to \sum_{h=-\infty}^{\infty} \gamma_X(h)e^{-ih\lambda} = 2\pi f_X(\lambda),$$

as $N \to \infty$, by using dominated convergence.
- *Inverse relation:*

$$\gamma_X(h) = \int_{-\pi}^{\pi} e^{ih\lambda} f_X(\lambda) d\lambda, \quad h \in \mathbb{Z}. \tag{1.3.2}$$

1.3.2 Linear Filtering

The following is a useful result. If

$$Y_n = \sum_{k=-\infty}^{\infty} a_k X_{n-k} \tag{1.3.3}$$

with a (weakly) stationary time series $\{X_n\}_{n\in\mathbb{Z}}$ and $a = \{a_k\}_{k\in\mathbb{Z}} \in \ell^1(\mathbb{Z})$, then

$$f_Y(\lambda) = |\widehat{a}(\lambda)|^2 f_X(\lambda), \tag{1.3.4}$$

where $\widehat{a}(\lambda) = \sum_{j=-\infty}^{\infty} a_j e^{-ij\lambda}$ is the Fourier transform of a (see Appendix A.1.1). This follows from observing that

$$\gamma_Y(h) = \mathbb{E}Y_h Y_0 = \sum_{j,k=-\infty}^{\infty} a_j a_k \gamma_X(h+k-j)$$

$$= \sum_{j,k} a_j a_k \int_{-\pi}^{\pi} e^{i(h+k-j)\lambda} f_X(\lambda) d\lambda = \int_{-\pi}^{\pi} e^{ih\lambda}\left|\sum_{j=-\infty}^{\infty} a_j e^{-ij\lambda}\right|^2 f_X(\lambda) d\lambda.$$

Relation (1.3.3) transforms the series $\{X_n\}$ into the series $\{Y_n\}$ by means of a linear filter.

1.3 Spectral Domain Perspective

Example 1.3.3 (*Spectral density of AR(1) series, cont'd*)   Applying (1.3.3)–(1.3.4) to the AR(1) equation $X_n - \varphi X_{n-1} = Z_n$ with $\{Z_n\} \sim \text{WN}(0, 1)$ yields

$$|1 - \varphi e^{-i\lambda}|^2 f_X(\lambda) = f_Z(\lambda) = \frac{\sigma_Z^2}{2\pi}$$

or

$$f_X(\lambda) = \frac{\sigma_Z^2}{2\pi} \frac{1}{|1 - \varphi e^{-i\lambda}|^2},$$

the result obtained directly in Example 1.3.2.

1.3.3 Periodogram

A sample counterpart to the spectral density is defined by

$$\frac{1}{2\pi} \sum_{|h|<N} \widehat{\gamma}_X(h) e^{-ih\lambda} = \frac{1}{2\pi N} \left| \sum_{n=1}^N X_n e^{-in\lambda} \right|^2 =: \frac{I_X(\lambda)}{2\pi}, \qquad (1.3.5)$$

with the first relation holding only at the so-called *Fourier frequencies*

$$\lambda = \lambda_k = \frac{2\pi k}{N} \text{ with } k = -\left[\frac{N-1}{2}\right], \ldots, \left[\frac{N}{2}\right]. \qquad (1.3.6)$$

$I_X(\lambda)$ is known as the *periodogram*, and has the following properties:

- *Computational speed:* $I_X(\lambda_k)$ can be computed efficiently by Fast Fourier Transform (FFT) in $O(N \log N)$ steps, supposing N can be factored out in many factors.
- *Statistical properties:* $I_X(\lambda)$ is not a consistent estimator for $2\pi f_X(\lambda)$, but is asymptotically unbiased. The periodogram needs to be smoothed to become consistent.

Warning: Two definitions of the periodogram I_X are commonly found in the literature. One definition appears in (1.3.5). The other popular definition is to set the whole left-hand side of (1.3.5) for the periodogram; that is, to incorporate the denominator 2π into the periodogram. Since the two definitions are different, it is important to check which convention is used in a given source. With the definition (1.3.5), we follow the convention used in Brockwell and Davis [186].

1.3.4 Spectral Representation

A (weakly) stationary, zero mean time series $X = \{X_n\}_{n\in\mathbb{Z}}$ with a spectral density f_X can be represented as

$$X_n = \int_{-\pi}^{\pi} e^{in\lambda} Z_X(d\lambda), \qquad (1.3.7)$$

where $Z_X(d\lambda)$ is a complex-valued *random measure* such that $Z_X(-d\lambda) = \overline{Z_X(d\lambda)}$; that is, Z_X is Hermitian. Moreover,

$$\mathbb{E} Z_X(d\lambda) \overline{Z_X(d\lambda')} = 0 \qquad (1.3.8)$$

when $d\lambda \neq d\lambda'$ (i.e., having orthogonal increments), and

$$\mathbb{E}|Z_X(d\lambda)|^2 = f_X(\lambda)d\lambda. \tag{1.3.9}$$

As with most integrals, this is interpreted through discrete sums as

$$X_n \approx \sum_k e^{in\lambda_k} Z_X(d\lambda_k),$$

where $Z_X(d\lambda_k)$ are uncorrelated with variances $f_X(\lambda_k)d\lambda_k$. See Appendix B.1 for a more rigorous treatment. Thus, at frequency λ with larger value of $f_X(\lambda)$, the variance of the random coefficient at $e^{in\lambda}$ is larger. These terms dominate in the representation.

Remark 1.3.4 In writing the spectral representation (1.3.7) we assumed implicitly that the series X has a spectral density. Spectral representations, however, exist for all (weakly) stationary time series. They are written more generally as

$$X_n = \int_{(-\pi,\pi]} e^{in\lambda} Z_X(d\lambda),$$

where $Z_X(d\lambda)$ is a complex-valued random measure as above with the only difference that the property (1.3.9) is replaced by

$$\mathbb{E}|Z_X(d\lambda)|^2 = F_X(d\lambda)$$

for the so-called spectral measure F_X on $(-\pi, \pi]$. When the spectral measure F_X has a density f_X (with respect to the Lebesgue measure), f_X is the spectral density of the series X and the relation (1.3.9) holds.

Example 1.3.5 (*Spectral density of AR(1) series, cont'd*) The spectral density of AR(1) series was derived in Examples 1.3.2 and 1.3.3. Typical plots of AR(1) time series and their

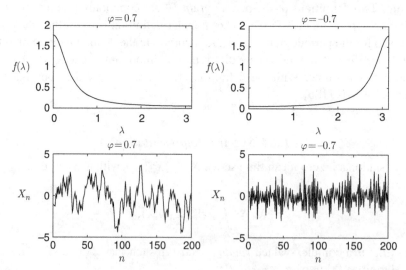

Figure 1.3 Typical plots of AR(1) spectral densities and of their sample paths in the cases $\varphi > 0$ and $\varphi < 0$.

spectral densities in the cases $\varphi > 0$ and $\varphi < 0$ are given in Figure 1.3. These plots are consistent with the idea behind spectral representation described above.

Remark 1.3.6 If $\{X_n\}_{n\in\mathbb{Z}}$ is given by its spectral representation, then by using (1.3.8) and (1.3.9),

$$\gamma_X(h) = \mathbb{E} X_h X_0 = \mathbb{E} \int_{-\pi}^{\pi} e^{ih\lambda} Z_X(d\lambda) \overline{\int_{-\pi}^{\pi} e^{i0\lambda'} Z_X(d\lambda')} = \int_{-\pi}^{\pi} e^{ih\lambda} f_X(\lambda) d\lambda,$$

which is the relation (1.3.2) connecting the spectral density to the ACVF.

Remark 1.3.7 Suppose that $f_X(\lambda) = |g_X(\lambda)|^2$ with $\overline{g_X(\lambda)} = g_X(-\lambda)$, which happens in many examples. Then,

$$X_n = \int_{-\pi}^{\pi} e^{in\lambda} Z_X(d\lambda) = \int_{-\pi}^{\pi} e^{in\lambda} g_X(\lambda) \widetilde{Z}(d\lambda),$$

where the random measure $\widetilde{Z}(d\lambda)$ satisfies

$$\mathbb{E} \widetilde{Z}(d\lambda) \overline{\widetilde{Z}(d\lambda')} = \begin{cases} d\lambda, & d\lambda = d\lambda', \\ 0, & \text{otherwise.} \end{cases}$$

Remark 1.3.8 Many notions above extend to continuous-time stationary processes $\{X(t)\}_{t\in\mathbb{R}}$. The spectral density of such a process is defined as

$$f_X(\lambda) = \frac{1}{2\pi} \int_{\mathbb{R}} e^{-i\lambda h} \gamma_X(h) dh, \quad \lambda \in \mathbb{R},$$

with the difference from (1.3.1) that it is a function for $\lambda \in \mathbb{R}$, and the sum is replaced by an integral. The inverse relation is

$$\gamma_X(h) = \int_{\mathbb{R}} e^{ih\lambda} f_X(\lambda) d\lambda, \quad h \in \mathbb{R},$$

(cf. (1.3.2)). The spectral representation reads

$$X(t) = \int_{\mathbb{R}} e^{it\lambda} Z_X(d\lambda), \quad t \in \mathbb{R},$$

where $Z_X(d\lambda)$ is a complex-valued random measure on $\mathbb{R}$ with orthogonal increments and control measure $\mathbb{E}|Z_X(d\lambda)|^2 = f_X(\lambda) d\lambda$.

1.4 Integral Representations Heuristics

The spectral representation (1.3.7) has components which are summarized through the first two columns of Table 1.1; that is, the dependence structure of X_n is transferred into the deterministic functions $e^{in\lambda}$. But one can also think of more general (integral) representations

$$X(t) = \int_E h_t(u) M(du), \quad (1.4.1)$$

where the components appearing in the last column of Table 1.1 are interpreted similarly.

Table 1.1 *Components in representations of time series and stochastic processes.*

Components in representation	Time series	Stochastic process
Dependent time series (process)	X_n	$X(t)$
Underlying space	$(-\pi, \pi]$	E
Deterministic functions	$e^{in\lambda}$	$h_t(u)$
Uncorrelated (or independent) random measure	$Z_X(d\lambda)$	$M(du)$

Example 1.4.1 (*Linear time series*) The linear time series in (1.2.2) is in fact defined through an integral representation since

$$X_n = \sum_{k=-\infty}^{\infty} a_{n-k} Z_k = \int_{\mathbb{Z}} h_n(k) M(dk) = \int_{\mathbb{Z}} h(n-k) M(dk),$$

where $h_n(k) = a_{n-k}$, $h(k) = a_k$ and $M(\{k\}) = Z_k$ are uncorrelated.

Various random measures and integral representations are defined and discussed in Appendix B.2. The following example provides a heuristic explanation of Gaussian random measures and their integrals.

Example 1.4.2 (*Gaussian random measure*) Suppose $E = \mathbb{R}$ and set $M(du) = B(du)$, viewing $B(du)$ as the increment on an infinitesimal interval du of a standard Brownian motion $\{B(u)\}_{u \in \mathbb{R}}$. Since Brownian motion has stationary and independent increments, and $\mathbb{E} B^2(u) = |u|$, one can think of the random measure $B(du)$ as satisfying

$$\mathbb{E} B(du_1) B(du_2) = \begin{cases} 0, & \text{if } du_1 \neq du_2, \\ du, & \text{if } du_1 = du_2 = du. \end{cases}$$

Thus, $B(du) \sim \mathcal{N}(0, du)$. The nonrandom measure $m(du) = \mathbb{E} B^2(du)$ is called the *control measure*. Here m is the Lebesgue control measure since $m(du) = du$. In the integral, $\int_{\mathbb{R}} h(u) B(du)$, each $B(du)$ is weighted by the nonrandom factor $h(u)$, and since the $B(du)$s are independent on disjoint intervals, one expects that

$$\int_{\mathbb{R}} h(u) B(du) \sim \mathcal{N}\left(0, \int_{\mathbb{R}} h^2(u) du\right).$$

Formally, the integral $I(h) = \int_E h(u) M(du)$ is defined first for simple functions h and then, by approximation, for all functions satisfying $\int_E h^2(u) du < \infty$.

1.4.1 Representations of a Gaussian Continuous-Time Process

Let $h \in L^2(\mathbb{R}, du)$; that is, $\int_{\mathbb{R}} h^2(u) du < \infty$. The Fourier transform of h is $\widehat{h}(x) = \int_{\mathbb{R}} e^{iux} h(x) du$ with the inverse formula $h(u) = \frac{1}{2\pi} \int_{\mathbb{R}} e^{-iux} \widehat{h}(x) dx$ (see Appendix A.1.2). It is complex-valued and Hermitian; that is, $\overline{\widehat{h}(dx)} = \widehat{h}(-dx)$. Introduce a similar transformation on $B(du)$, namely, let $\overline{\widehat{B}(dx)} = \widehat{B}(-dx)$ be complex-valued, with $\widehat{B}(dx) =$

$B_1(dx) + iB_2(dx)$, where $B_1(dx)$ and $B_2(dx)$ are real-valued, independent $\mathcal{N}(0, dx/2)$, and require $\widehat{B}$ to be Hermitian; that is, $\overline{\widehat{B}(dx)} = \widehat{B}(-dx)$. Then, $\mathbb{E}|\widehat{B}(dx)|^2 = dx$ and

$$I(h) := \int_{\mathbb{R}} h(u) B(du) \stackrel{d}{=} \frac{1}{\sqrt{2\pi}} \int_{\mathbb{R}} \widehat{h}(x) \widehat{B}(dx) =: \widehat{I}(\widehat{h}).$$

See Appendix B.1 for more details.

Example 1.4.3 (*The OU process*) Consider a stochastic process

$$X(t) = \sigma \int_{-\infty}^{t} e^{-(t-u)\lambda} B(du), \quad t \in \mathbb{R},$$

where $\sigma > 0$, $\lambda > 0$ are parameters, and $B(du)$ is a Gaussian random measure on $\mathbb{R}$ with the Lebesgue control measure du. The process $X(t)$ is Gaussian with zero mean and covariance function (for $t > s > 0$)

$$\mathbb{E} X(t) X(s) = \sigma^2 \int_{-\infty}^{s} e^{-(t-u)\lambda} e^{-(s-u)\lambda} du$$

$$= \sigma^2 e^{-(t+s)\lambda} \int_{-\infty}^{s} e^{2u\lambda} du = \frac{\sigma^2}{2\lambda} e^{-(t-s)\lambda}.$$

The process $X(t)$ is thus the OU process (see Example 1.1.2).

The integral representation above is in the time domain; that is, $I(h) = \int_{\mathbb{R}} h_t(u) B(du)$ with $h_t(u) = \sigma e^{-(t-u)\lambda} 1_{\{u<t\}}$. Observe that

$$\widehat{h_t}(x) = \int_{\mathbb{R}} e^{ixu} h_t(u) du = \sigma \int_{-\infty}^{t} e^{ixu} e^{-(t-u)\lambda} du$$

$$= \sigma e^{-t\lambda} \int_{-\infty}^{t} e^{(ix+\lambda)u} du = \frac{\sigma e^{-t\lambda} e^{(ix+\lambda)u}}{ix+\lambda} \bigg|_{u=-\infty}^{t} = \frac{e^{ixt} \sigma}{ix+\lambda}.$$

Hence, by switching to the spectral domain, the OU process can be represented as

$$X(t) = \frac{1}{\sqrt{2\pi}} \int_{\mathbb{R}} e^{ixt} \frac{\sigma}{ix+\lambda} \widehat{B}(dx).$$

Example 1.4.4 (*Brownian motion*) Brownian motion can be represented as

$$B(t) = \int_{\mathbb{R}} 1_{[0,t)}(u) B(du) = \int_{\mathbb{R}} \Big(1_{(0,\infty)}(t-u) - 1_{(0,\infty)}(-u) \Big) B(du),$$

where $B(du)$ is a Gaussian random measure with the control measure du. With $h_t(u) = 1_{[0,t)}(u)$,

$$\widehat{h_t}(x) = \int_{\mathbb{R}} e^{ixu} h_t(u) du = \int_{0}^{t} e^{ixu} du = \frac{e^{ixt} - 1}{ix}.$$

Then, switching to the spectral domain, Brownian motion can also be represented as

$$B(t) = \frac{1}{\sqrt{2\pi}} \int_{\mathbb{R}} \frac{e^{ixt} - 1}{ix} \widehat{B}(dx).$$

A reader wishing to learn more about time series analysis and stochastic processes could consult the references given in Section 1.6 below. These references are helpful in understanding better the material presented in subsequent chapters.

1.5 A Heuristic Overview of the Next Chapter

In the next chapter, we introduce basic concepts and results involving long-range dependence and self-similarity. Because the precise definitions can be rather technical, we provide here a brief and heuristic overview.

There are several definitions of long-range dependence which are, in general, not equivalent. Basically, a stationary series $\{X_n\}_{n \in \mathbb{Z}}$ is long-range dependent if its autocovariance function $\gamma_X(k) = \mathbb{E}X_k X_0 - \mathbb{E}X_k \mathbb{E}X_0$ behaves like k^{2d-1} as $k \to \infty$, where $0 < d < 1/2$. This range of d ensures that $\sum_{k=-\infty}^{\infty} \gamma_X(k) = \infty$. From a spectral domain perspective, the spectral density $f_X(\lambda)$ of $\{X_n\}_{n \in \mathbb{Z}}$ behaves as λ^{-2d} as the frequency $\lambda \to 0$. Since $d > 0$, note that the spectral density diverges as $\lambda \to 0$. A typical example is FARIMA$(0, d, 0)$ series introduced in Section 2.4.1.

We also define the related notion of self-similarity. A process $\{Y(t)\}_{t \in \mathbb{R}}$ is H–self-similar if, for any constant c, the finite-dimensional distributions of $\{Y(ct)\}_{t \in \mathbb{R}}$ are the same as those of $\{c^H Y(t)\}_{t \in \mathbb{R}}$, where H is a parameter often related to d. In fact, if the process $\{Y(t)\}$ has stationary increments, then $X_n = Y(n) - Y(n-1), n \in \mathbb{Z}$, has long-range dependence with $H = d + 1/2$. Conversely, we can obtain $Y(t)$ from X_n by using a limit theorem.

Fractional Brownian motion is a typical example of $Y(t)$. It is Gaussian, H–self-similar, and has stationary increments. We provide both time-domain and spectral-domain representations for fractional Brownian motion. We also give additional examples of non-Gaussian self-similar processes, such as the Rosenblatt process and also processes with infinite variance defined through their integral representations, for instance, linear fractional stable motion and the Telecom process.

1.6 Bibliographical Notes

There are a number of excellent textbooks on time series and their analysis. The monograph by Brockwell and Davis [186] provides a solid theoretical foundation. The classic by Priestley [833] has served generations of scientists interested in the spectral analysis of time series. For more applied and computational aspects of the time series analysis, see Cryer and Chan [271], Shumway and Stoffer [909]. Nonlinear time series are treated in Douc, Moulines, and Stoffer [327].

On the side of stochastic processes, Lindgren [635] provides an up-to-date treatment of stationary stochastic processes. The basics of Brownian motion and related stochastic calculus are treated in Karatzas and Shreve [549], Mörters and Peres [733]. A number of other facts used in this monograph are discussed in Appendices B and C.

2
Basics of Long-Range Dependence and Self-Similarity

This chapter serves as a general introduction. It is necessary for understanding the more advanced topics covered in subsequent chapters. We introduce here basic concepts and results involving long-range dependence and self-similarity.

There are various definitions of long-range dependence, not all equivalent, because of the presence of slowly varying functions. These functions vary slower than a power; for example, they can be a constant or a logarithm. We give five definitions of long-range dependence. As discussed in Section 2.2, these definitions can be equivalent depending on the choice of the slowly varying functions. The reader should read the statement of the various propositions of that section, but their proofs which are somewhat technical may be skipped in a first reading.

Short-range dependent time series are introduced in Section 2.3. A typical example is the ARMA(p, q). Examples of time series with long-range dependence, such as FARIMA$(0, d, 0)$ and FARIMA(p, d, q) are given in Section 2.4. Self-similarity is defined in Section 2.5. Examples of self-similar processes are given in Section 2.6. These include:

- Fractional Brownian motion
- Bifractional Brownian motion
- The Rosenblatt process
- $S\alpha S$ Lévy motion
- Linear fractional stable motion
- Log-fractional stable motion
- The Telecom process
- Linear fractional Lévy motion

The Lamperti transformation, relating self-similarity to stationarity is stated in Section 2.7. Section 2.8 deals with the connection between long-range dependence and self-similarity. The infinite variance case is considered in Section 2.9. Section 2.10 focuses on heuristic methods of estimation of the self-similarity parameter. These include:

- The R/S method
- The aggregated variance method
- Regression in the spectral domain
- Wavelet-based estimation

Finally, we indicate in Section 2.11 how to generate long- and short-range dependent series, using the

- Cholesky decomposition

whose complexity is $O(N^3)$ and hence relatively slow, and the

- Circulant matrix embedding

which is an exact method with the complexity $O(N \log N)$.

2.1 Definitions of Long-Range Dependent Series

We start with the notion of *long-range dependence*, also called *long memory* or *strong dependence*. It is commonly defined for second-order stationary time series. Recall from Sections 1.1.3 and 1.2 that a second-order stationary time series $X = \{X_n\}_{n \in \mathbb{Z}}$ has a constant mean $\mu_X = \mathbb{E}X_n$ and an autocovariance function

$$\gamma_X(n - m) = \mathbb{E}X_n X_m - \mathbb{E}X_n \mathbb{E}X_m = \mathbb{E}X_n X_m - \mu_X^2,$$

where "n", "m" are viewed as time and γ_X depends only on the time lag $|m - n|$. One can characterize the time series X by specifying its mean μ_X and autocovariance $\gamma_X(n), n \in \mathbb{Z}$, or one can take a "spectral domain" perspective by focusing on the spectral density $f_X(\lambda)$, $\lambda \in (-\pi, \pi]$, of the time series, if it exists (see Section 1.3). It is defined as

$$\int_{-\pi}^{\pi} e^{in\lambda} f_X(\lambda) d\lambda = \gamma_X(n), \quad n \in \mathbb{Z},$$

that is, the spectral density $f_X(\lambda)$ is a function whose Fourier coefficients are the autocovariances $\gamma_X(n)$ (up to a multiplicative constant following the convention of Appendix A.1.1).

A *white noise* time series $Z = \{Z_n\}_{n \in \mathbb{Z}}$, defined in Example 1.2.1 and denoted $\{Z_n\} \sim \mathrm{WN}(0, \sigma_Z^2)$, is a trivial example of a second-order stationary time series. It consists of uncorrelated random variables Z_n with zero mean and common variance σ_Z^2, and for which

$$\gamma_Z(n) = \begin{cases} \sigma_Z^2, & \text{if } n = 0, \\ 0, & \text{if } n \neq 0, \end{cases} \quad f_Z(\lambda) = \frac{\sigma_Z^2}{2\pi}, \text{ for } \lambda \in (-\pi, \pi].$$

White noise time series are used in the construction of correlated time series. We are interested, in particular, in time series with long-range dependence. These involve a parameter

$$d \in (0, 1/2) \tag{2.1.1}$$

and a slowly varying function L.

Definition 2.1.1 A function L is *slowly varying at infinity* if it is *positive* on $[c, \infty)$ with $c \geq 0$ and, for any $a > 0$,

$$\lim_{u \to \infty} \frac{L(au)}{L(u)} = 1. \tag{2.1.2}$$

A function L is *slowly varying at zero* if the function $L(1/u)$ is slowly varying at infinity. A function R is *regularly varying at infinity* with exponent $\rho \in \mathbb{R}$ if

$$R(u) = u^\rho L(u), \tag{2.1.3}$$

where L is slowly varying at infinity. It is *regularly varying at zero* if $R(1/u)$ is regularly varying at infinity. A slowly varying function is a regularly varying function with $\rho = 0$.

(See Bingham, Goldie, and Teugels [157] for a comprehensive treatment of slowly varying functions.) Since $R(au)/R(u) \to a^\rho$, as $u \to \infty$, the slowly varying function can typically be ignored when considering the limit. For example, the functions $L(u) = \text{const} > 0$ and $L(u) = \log u$, $u > 0$, are slowly varying at infinity.

The presence of a slowly varying function is important in practice as it yields significantly greater flexibility. Thus, in condition IV below, we consider the spectral density $f_X(\lambda) = L_4(\lambda)\lambda^{-2d}$, where L_4 is slowly varying at zero. Without L_4, we would have a very specific spectral density, namely, $f_X(\lambda) = \lambda^{-2d}$, $0 < \lambda \leq \pi$, but by choosing L_4 accordingly, we can have $f_X(\lambda) = e^{-\lambda}\lambda^{-2d}$ or even $f_X(\lambda) = |\log \lambda|\lambda^{-2d}$. But, for example, we cannot have $f_X(\lambda) = (2 + \cos(1/\lambda))\lambda^{-2d}$ since the function $2 + \cos(1/\lambda)$ is not slowly varying. The requirement that a slowly varying function is ultimately positive is part of the theory of such functions. Here we will allow a slowly varying function to be negative on $(0, c)$, $c > 0$, in order to gain flexibility in some of the conditions below.

Having defined the notion of slowly and regularly varying functions, we now turn to the various definitions of long-range dependence that can be found in the literature. Each definition involves a different condition which is stated below. Conditions I, II, III and V are expressed in the time domain and condition IV is expressed in the spectral domain. These conditions are, in general, not equivalent, as will be discussed below. They are, however, equivalent in particular cases, for example, for the FARIMA$(0, d, 0)$ series defined in Section 2.4.

Condition I. The time series $X = \{X_n\}_{n \in \mathbb{Z}}$ has a linear representation

$$X_n = \mu_X + \sum_{k=0}^{\infty} \psi_k Z_{n-k} = \mu_X + \sum_{j=-\infty}^{n} \psi_{n-j} Z_j$$

with $Z = \{Z_n\}_{n \in \mathbb{Z}} \sim \text{WN}(0, \sigma_Z^2)$ and a sequence $\{\psi_k\}_{k \geq 0}$ satisfying

$$\psi_k = L_1(k)k^{d-1}, \quad k = 0, 1, \ldots, \tag{2.1.4}$$

where L_1 is a slowly varying function at *infinity* and d is as in (2.1.1).

Condition II. The autocovariance function of the time series $X = \{X_n\}_{n \in \mathbb{Z}}$ satisfies

$$\gamma_X(k) = L_2(k)k^{2d-1}, \quad k = 0, 1, \ldots, \tag{2.1.5}$$

where L_2 is a slowly varying function at *infinity*.

Condition III. The autocovariances of the time series $X = \{X_n\}_{n \in \mathbb{Z}}$ are not absolutely summable; that is,

$$\sum_{k=-\infty}^{\infty} |\gamma_X(k)| = \infty. \tag{2.1.6}$$

Condition IV. The time series $X = \{X_n\}_{n \in \mathbb{Z}}$ has a spectral density satisfying

$$f_X(\lambda) = L_4(\lambda)\lambda^{-2d}, \quad 0 < \lambda \leq \pi, \tag{2.1.7}$$

where L_4 is a slowly varying function at *zero*.

Condition V. The time series $X = \{X_n\}_{n\in\mathbb{Z}}$ satisfies

$$\text{Var}(X_1 + \cdots + X_N) = L_5(N) N^{2d+1}, \quad N = 1, 2, \ldots, \quad (2.1.8)$$

where L_5 is a slowly varying function at infinity.

Remark 2.1.2 Since a spectral density f satisfies $f_X(\lambda) = f_X(-\lambda)$, it is enough to focus only on positive $\lambda > 0$. The convergence $\lambda \to 0$ in (2.1.7) as well as in other places below is understood to be for $\lambda > 0$. Alternatively, one could express (2.1.7) as $f_X(\lambda) = L_4(|\lambda|)|\lambda|^{-2d}$ and consider both positive and negative λ.

Remark 2.1.3 The slowly varying functions $L_1(k)$ and $L_2(k)$ in conditions I and II can be negative or zero for a finite number of ks. The slowly varying functions L_4 and L_5 in conditions IV and V are nonnegative on their domain and (strictly) positive for small and large enough arguments, respectively, by definition.

Remark 2.1.4 In many instances, for example in statistical estimation, the slowly varying functions in (2.1.4)–(2.1.7) are such that[1]

$$L(u) \sim \text{const} > 0, \quad (2.1.9)$$

at either infinity or zero. Much of the mathematical theory, however, can be developed for general, or fairly general, slowly varying functions. When assuming (2.1.9) in connection to the conditions above, we shall write $L_j(u) \sim c_j$, $j = 1, 2, 4, 5$, especially to emphasize that estimating c_js might be of interest as well, in addition to the estimation of d.

Definition 2.1.5 A second-order stationary time series $X = \{X_n\}_{n\in\mathbb{Z}}$ is called *long-range dependent* (LRD, in short) if one of the non-equivalent conditions I–V above holds. The parameter $d \in (0, 1/2)$ is called a long-range dependence (LRD) parameter. One also says that the series X has *long memory*.

As stated above, the conditions I–V are not equivalent in general and hence there is not just one accepted definition of LRD series. Which definition is to be used depends on the context. We will specify in the sequel, when necessary, the definition which is used.

We show in Section 2.2 that the conditions I–V are related as in Figure 2.1, where "$\Longrightarrow$" means implication and "$\overset{*}{\Longrightarrow}$" supposes that the slowly varying function is quasi-monotone (see Definition 2.2.11 below). The proofs and discussions in Section 2.2 shed further light on long-range dependence, and should be particularly useful to those unfamiliar with the area. The more technical parts of the proofs of the implications above are moved to Appendix A.2 in order for Section 2.2 to be more accessible.

Remark 2.1.6 The conditions above extend naturally to continuous-time stationary processes $\{X(t)\}_{t\in\mathbb{R}}$ (see Remarks 1.2.6 and 1.3.8). Condition I is stated for linear processes

---

[1] We write $a(u) \sim b(u)$ to signify $a(u)/b(u) \to 1$ as $u \to \infty$.

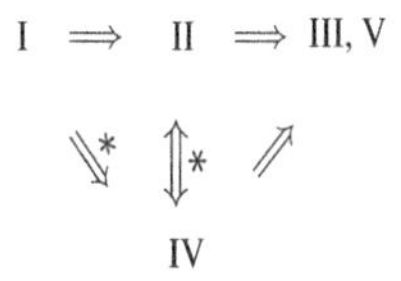

$$\text{I} \implies \text{II} \implies \text{III, V}$$
$$\Downarrow^* \quad \Updownarrow^* \quad \Uparrow$$
$$\text{IV}$$

Figure 2.1 The relations between the various definitions of LRD.

$X(t) = \int_{-\infty}^{t} a(t-u) Z(du)$ in Remark 1.2.6 requiring that $a(u) = L_1(u) u^{d-1}$. Condition II remains the same, with the only difference that $h \in \mathbb{R}$. Condition III requires $\int_{\mathbb{R}} |\gamma_X(h)| dh = \infty$. Condition IV remains exactly the same. Finally, condition V is stated for $\text{Var}(\int_0^N X(u) du)$.

2.2 Relations Between the Various Definitions of Long-Range Dependence

We first list some important properties of slowly and regularly varying functions. We then use these properties to show how the various definitions of long-range dependence are related.

2.2.1 Some Useful Properties of Slowly and Regularly Varying Functions

We consider slowly and regularly varying functions at infinity. A useful property of slowly varying functions is *Potter's bound* (Theorem 1.5.6, (i), in Bingham et al. [157]): for every $\delta > 0$, there is $u_0 = u_0(\delta)$ such that

$$\frac{L(u)}{L(v)} \leq 2 \max\left\{ \left(\frac{u}{v}\right)^\delta, \left(\frac{u}{v}\right)^{-\delta} \right\} \tag{2.2.1}$$

for $u, v \geq u_0$. Note that, without loss of generality, since it is ultimately positive, L may be replaced by $|L|$ in (2.2.1). Moreover, if $|L|$ is bounded away from 0 and infinity on every compact subset of $[0, \infty)$, then (2.2.1) implies that there is $C > 1$ such that

$$\left| \frac{L(u)}{L(v)} \right| \leq C \max\left\{ \left(\frac{u}{v}\right)^\delta, \left(\frac{u}{v}\right)^{-\delta} \right\} \tag{2.2.2}$$

for $u, v > 0$ (Theorem 1.5.6, (ii), in Bingham et al. [157] and Exercise 2.1 below). A practical consequence of (2.2.1) is that, for any arbitrarily small $\delta > 0$, there are constants C_1, C_2 and $u_1 > 0$ such that

$$C_1 u^{-\delta} \leq L(u) \leq C_2 u^\delta, \tag{2.2.3}$$

for $u \geq u_1$.

Another useful property is the so-called *uniform convergence theorem*, according to which, if R is regularly varying with exponent $\rho \in \mathbb{R}$, then

$$\sup_{a \in A} \left| \frac{R(au)}{R(u)} - a^\rho \right| \to 0, \quad \text{as } u \to \infty, \tag{2.2.4}$$

for

$$A = \begin{cases} [u_0, v_0], & 0 < u_0 \leq v_0 < \infty, & \text{if } \rho = 0, \\ (0, v_0], & 0 < v_0 < \infty, & \text{if } \rho > 0, \\ [u_0, \infty), & 0 < u_0 < \infty, & \text{if } \rho < 0, \end{cases}$$

with the assumption, when $\rho > 0$, that R is bounded on each interval of $(0, v_0]$ (Theorem 1.5.2 in Bingham et al. [157]).

The following result, known as Karamata's theorem, is very handy.

Proposition 2.2.1 *Let L be a slowly varying function at infinity and $p > -1$. Then, for*

$$c_k = L(k)k^p, \quad k \geq 1,$$

we have that

$$\sum_{k=1}^{n} c_k \sim \frac{L(n)n^{p+1}}{p+1}, \quad \text{as } n \to \infty.$$

Proof This is a standard result in the theory of regular variation. Observe that, for large enough n so that $L(n) \neq 0$,

$$S_n = \frac{1}{L(n)n^{p+1}} \sum_{k=1}^{n} c_k = \sum_{k=1}^{n} \frac{L(k)}{L(n)} \left(\frac{k}{n}\right)^p \frac{1}{n} = \int_0^1 f_n(x)dx,$$

where

$$f_n(x) = \sum_{k=1}^{n} \frac{L(k)}{L(n)} \left(\frac{k}{n}\right)^p 1_{[\frac{k-1}{n}, \frac{k}{n})}(x) = \frac{L(\lceil nx \rceil)}{L(n)} \left(\frac{\lceil nx \rceil}{n}\right)^p, \quad x \in (0, 1),$$

$1_A(x) = 1$ if $x \in A$, and $= 0$ if $x \notin A$, is the indicator function of a set A, and $\lceil a \rceil$ is the (ceiling) integer part of $a \in \mathbb{R}$.[2] We now want to apply the dominated convergence theorem. For fixed $x \in (0, 1)$, by using (2.2.4) with $\rho = 0$, $a = \lceil nx \rceil/n$ and $u = n$, we have $L(\lceil nx \rceil)/L(n) \to 1$ and hence also $f_n(x) \to x^p =: f(x)$. Since $L(k)$ defines c_k on positive integers only, we may suppose without loss of generality that L is defined on $(0, \infty)$ and satisfies the assumptions for Potter's bound (2.2.2). (If $L(k) = 0$, the Potter's bound for $L(k)/L(n) = 0$ obviously holds.) Then, by choosing a small enough δ so that $p - \delta + 1 > 0$ in Potter's bound (2.2.2), we obtain

$$|f_n(x)| \leq C\left(\frac{\lceil nx \rceil}{n}\right)^{p-\delta} \leq \begin{cases} Cx^{p-\delta}, & p - \delta < 0, \\ C, & p - \delta \geq 0. \end{cases}$$

The function in the bound is integrable on $(0, 1)$. Hence, the dominated convergence theorem yields

$$S_n \to \int_0^1 f(x)dx = \int_0^1 x^p dx = \frac{1}{p+1}. \qquad \square$$

A continuous version of the above result (Karamata's theorem) is given in the next proposition. A proof can be found in Bingham et al. [157], Proposition 1.5.8.

---

[2] That is, $\lceil a \rceil$ is the smallest integer which is greater than or equal to $a \in \mathbb{R}$.

Proposition 2.2.2 *Let L be a slowly varying function at infinity which is locally bounded in $[x_0, \infty)$, and $p > -1$. Then*

$$\int_{x_0}^{x} u^p L(u) du \sim \frac{L(x)x^{p+1}}{p+1}, \quad \text{as } x \to \infty.$$

Remark 2.2.3 We considered above slowly and regularly varying functions at infinity. If the functions are slowly and regularly varying at zero, the results above can still be used by transforming the functions to slowly and regularly varying functions at infinity, applying the results above and finally translating the results back to the zero range. For example, if $L(u)$ is a slowly varying function at zero, then $L(1/u)$ is a slowly varying function at infinity. Potter's bound (2.2.1) for $L(1/u)$ becomes: for every $\delta > 0$, there is u_0 such that

$$\frac{L(1/u)}{L(1/v)} \leq 2 \max\left\{ \left(\frac{u}{v}\right)^{\delta}, \left(\frac{u}{v}\right)^{-\delta} \right\} = 2 \max\left\{ \left(\frac{1/u}{1/v}\right)^{\delta}, \left(\frac{1/u}{1/v}\right)^{-\delta} \right\}$$

for $u, v \geq u_0$ or $1/u, 1/v \leq 1/u_0$. This is the same as: for every $\delta > 0$, there is x_0 such that

$$\frac{L(x)}{L(y)} \leq 2 \max\left\{ \left(\frac{x}{y}\right)^{\delta}, \left(\frac{x}{y}\right)^{-\delta} \right\}$$

for $x, y \leq x_0$.

In some instances, several of the conditions above characterizing long-range dependence may hold at the same time. To understand the various conditions I–V, we will compare them through a sequence of propositions. Their proofs are somewhat technical and may be skipped in a first reading.

2.2.2 Comparing Conditions II and III

Proposition 2.2.4 *Condition II implies III.*

Proof By applying Proposition 2.2.1 with $c_k = \gamma_X(k)$, $L = L_2$, $p = 2d - 1$, and using $\gamma_X(k) = \gamma_X(-k)$, $k \geq 1$, we have

$$\sum_{k=-n}^{n} |\gamma_X(k)| = |\gamma_X(0)| + 2 \sum_{k=1}^{n} |\gamma_X(k)| \sim \frac{L_2(n)n^{2d}}{d}, \quad \text{as } n \to \infty.$$

By (2.2.3), $L_2(n) \geq Cn^{-\delta}$ for large enough n, where $\delta > 0$ is arbitrarily small and C is a constant. Then, for another constant C and large enough n, $\sum_{k=-n}^{n} |\gamma_X(k)| \geq Cn^{2d-\delta} \to \infty$ by choosing $\delta < 2d$. The result now follows. $\square$

2.2.3 Comparing Conditions II and V

Another interesting consequence of Proposition 2.2.1 is related to the sample mean of the series,

$$\overline{X}_N = \frac{X_1 + \cdots + X_N}{N}. \tag{2.2.5}$$

Proposition 2.2.5 *If condition II holds, then*

$$\text{Var}(\overline{X}_N) \sim \frac{L_2(N)N^{2d-1}}{d(2d+1)}, \quad \text{as } N \to \infty. \qquad (2.2.6)$$

Proof Observe that, by using Proposition 2.2.1,

$$\text{Var}(\overline{X}_N) = \frac{1}{N^2} \sum_{j,k=1}^{N} \gamma_X(j-k) = \frac{1}{N^2} \sum_{h=-(N-1)}^{N-1} (N-|h|)\gamma_X(h)$$

$$= \frac{1}{N} \sum_{h=-(N-1)}^{N-1} \gamma_X(h) - \frac{1}{N^2} \sum_{h=-(N-1)}^{N-1} |h|\gamma_X(h)$$

$$\sim \frac{2L_2(N)N^{2d-1}}{2d} - \frac{2L_2(N)N^{2d-1}}{2d+1} = \frac{L_2(N)N^{2d-1}}{d(2d+1)}.$$

Another way to prove the result is to write

$$\text{Var}(\overline{X}_N) = \frac{1}{N}\gamma_X(0) + \frac{2}{N^2} \sum_{k=1}^{N-1} \sum_{h=1}^{k} \gamma_X(h)$$

and apply Proposition 2.2.1 twice to the last term. □

Remark 2.2.6 A wrong way would be to write

$$\text{Var}(\overline{X}_N) = \frac{1}{N} \sum_{h=-(N-1)}^{N-1} \left(1 - \frac{|h|}{N}\right)\gamma_X(h) \sim \frac{1}{N} \sum_{h=-(N-1)}^{N-1} \gamma_X(h), \qquad (2.2.7)$$

which would lead to the wrong asymptotics $L_2(N)N^{2d-1}/d$. The asymptotic equivalence in (2.2.7) cannot be justified by the dominated convergence theorem because $\sum_{h=-\infty}^{\infty} |\gamma_X(h)| = \infty$.

An immediate consequence of Proposition 2.2.5 is that condition II implies V.

Corollary 2.2.7 *Condition II implies V with the slowly varying functions related as*

$$L_5(N) \sim \frac{L_2(N)}{d(2d+1)}, \quad \text{as } N \to \infty. \qquad (2.2.8)$$

Remark 2.2.8 Proposition 2.2.1 is related to another interesting perspective on LRD. For each $N \geq 1$, form a series $X^{(N)} = \{X_n^{(N)}\}_{n \in \mathbb{Z}}$ obtained by taking averages over blocks of size N,

$$X_n^{(N)} = \frac{1}{N} \sum_{k=N(n-1)+1}^{Nn} X_k, \quad n \in \mathbb{Z}.$$

Each series $X^{(N)}$ is second order stationary. Denote its autocovariance function by $\gamma_X^{(N)}(k)$. One can show that the condition (2.1.8) with $d \in (0, 1/2)$ is equivalent to

$$\gamma_X^{(N)}(k) \sim L_5(N) N^{2d-1} \frac{1}{2}\Big(|k+1|^{2d+1} - 2|k|^{2d+1} + |k-1|^{2d+1}\Big), \quad \text{as } N \to \infty, \quad (2.2.9)$$

for each fixed $k \in \mathbb{Z}$. This is stated as Exercise 2.3 below. The expression $(|k+1|^{2d+1} - 2|k|^{2d+1} + |k-1|^{2d+1})/2$ on the right-hand side of (2.2.9) is the autocovariance function of the so-called fractional Gaussian noise (see Section 2.8), which is the series of the increments of fractional Brownian motion.

Note that the series $\{X_n\}_{n \in \mathbb{Z}}$ consisting of uncorrelated variables satisfies

$$\text{Var}(\overline{X}_N) \sim \text{Var}(X_0) N^{-1}.$$

This stands in stark contrast with Proposition 2.2.5, where the variance of the sample mean of LRD series behaves like $N^{-(1-2d)}$ as $N \to \infty$. Since $1 - 2d < 1$, this sample mean can generally be estimated less reliably than in the case of uncorrelated variables. Proposition 2.2.5 also suggests that $\overline{X}_N$ should be normalized by $N^{d-1/2}$, instead of the common normalization $N^{-1/2}$, in order to converge to a limit. If $L_2(u) \sim c_2$ as in statistical applications, then (2.2.6) becomes

$$\text{Var}(\overline{X}_N) \sim \frac{c_2 N^{2d-1}}{d(2d+1)}, \quad \text{as } N \to \infty. \quad (2.2.10)$$

Note that, for example, if the right-hand side of (2.2.10) is to be used in the construction of confidence intervals for μ_X, then both d and c_2 need to be estimated.

Note also that (2.2.10) implies

$$\text{Var}(N^{1/2-d}\overline{X}_N) \to \frac{c_2}{d(2d+1)}, \quad \text{as } N \to \infty. \quad (2.2.11)$$

The constant $c_2/d(2d+1)$ is known as the *long-run variance* of the LRD series X. (See also Remark 2.2.20 below.)

2.2.4 Comparing Conditions I and II

We now turn to showing that condition I implies II. In this regard, several observations should be made about condition I. First, the series satisfying condition I is well-defined in the $L^2(\Omega)$–sense (see the footnote to Example 1.2.3). Indeed, note that

$$\mathbb{E}\Big(\sum_{k=m_1}^{m_2} \psi_k Z_{n-k}\Big)^2 = \sigma_Z^2 \sum_{k=m_1}^{m_2} \psi_k^2,$$

where $\psi_k^2 = (L_1(k))^2 k^{2d-2}$. Using (2.2.3), we have $0 < L_1(k) \leq C k^\delta$ for large enough k where $\delta > 0$ is arbitrarily small and C is a constant. Then, for large enough m_1, m_2, the latter sum is bounded by

$$C^2 \sigma_Z^2 \sum_{k=m_1}^{m_2} k^{2d-2+2\delta}$$

and converges to 0 as $m_1, m_2 \to \infty$ for small enough δ such that $2d - 1 + 2\delta < 0$. Second, observe that, for $k \geq 0$,

$$\gamma_X(k) = \sigma_Z^2 \sum_{j=0}^{\infty} \psi_j \psi_{j+k}$$

(see Example 1.2.5). We will use:

Proposition 2.2.9 *Let L be a slowly varying function and $-1 < p < -1/2$. Then,*

$$c_j = L(j) j^p, \quad j \geq 0,$$

implies that

$$\sum_{j=0}^{\infty} c_j c_{j+k} \sim (L(k))^2 k^{2p+1} B(p+1, -2p-1), \quad \text{as } k \to \infty,$$

where

$$B(a, b) = \int_0^1 x^{a-1}(1-x)^{b-1} dx = \int_0^{\infty} y^{a-1}(1+y)^{-a-b} dy = \frac{\Gamma(a)\Gamma(b)}{\Gamma(a+b)}, \quad a, b > 0,$$
(2.2.12)

is the beta function.[3]

The function $\Gamma(z)$ in (2.2.12) is the gamma function defined by

$$\Gamma(z) = \int_0^{\infty} t^{z-1} e^{-t} dt, \quad z > 0.$$

We shall often use its following two properties:

$$\Gamma(z+1) = z\Gamma(z), \quad z > 0, \quad (2.2.13)$$

$$\Gamma(z)\Gamma(1-z) = \frac{\pi}{\sin(\pi z)}, \quad 0 < z < 1. \quad (2.2.14)$$

The relation (2.2.13), in particular, is used to extend the definition of the gamma function to $z \leq 0$ in a recursive way through $\Gamma(z) = \Gamma(z+1)/z$, $z \leq 0$. Note that with this definition, $\Gamma(k) = \infty$ when $k = 0, -1, -2, \ldots$

Proof of Proposition 2.8 The proof is similar to that of Proposition 2.2.1. Observe that, for large enough k so that $L(k) \neq 0$,

$$\frac{1}{(L(k))^2 k^{2p+1}} \sum_{j=0}^{\infty} c_j c_{j+k} = \sum_{j=0}^{\infty} \frac{L(j)}{L(k)} \frac{L(j+k)}{L(k)} \left(\frac{j}{k}\right)^p \left(1 + \frac{j}{k}\right)^p \frac{1}{k}$$

$$\sim \int_0^{\infty} u^p (1+u)^p du = \int_0^1 (1-v)^p v^{-2-2p} dv = B(p+1, -2p-1),$$

---
[3] To go from the first to the second integral in (2.2.12), set $x = y/(1+y)$.

where the last asymptotic relation follows by using Potter's bound and the dominated convergence theorem as in the proof of Proposition 2.2.1, and where the last two integrals are equal in view of the second equality in (2.2.12) with $a = p + 1$ and $b = -1 - 2p$. □

Corollary 2.2.10 *Condition I implies II. In particular,*

$$L_2(k) \sim \sigma_Z^2 (L_1(k))^2 B(d, 1 - 2d). \tag{2.2.15}$$

2.2.5 Comparing Conditions II and IV

We next turn to the comparison of conditions II and IV, which is probably the most delicate one. To be able to go from condition II to IV, we would like to be able to write

$$f_X(\lambda) = \frac{1}{2\pi} \sum_{k=-\infty}^{\infty} e^{-ik\lambda} \gamma_X(k) \tag{2.2.16}$$

for an LRD series satisfying condition II (cf. (1.3.1)). Note that, since condition II implies III, one may expect at most that (2.2.16) converges conditionally; that is, it converges but does not converge absolutely. The most general condition for this and for relating conditions II and IV involves the following definition.

Definition 2.2.11 A slowly varying function L on $[0, \infty)$ is called *quasi-monotone* if it is of bounded (finite) variation on any compact interval of $[0, \infty)$ and if, for some $\delta > 0$,

$$\int_0^x u^\delta |dL(u)| = O(x^\delta L(x)), \quad \text{as } x \to \infty. \tag{2.2.17}$$

It is known (but not obvious) that the condition (2.2.17) can be replaced by

$$\int_x^\infty u^{-\delta} |dL(u)| = O(x^{-\delta} L(x)), \quad \text{as } x \to \infty \tag{2.2.18}$$

(Corollary 2.8.2 in Bingham et al. [157]), and that "for some $\delta > 0$" can be replaced by "for all $\delta > 0$" (see the bottom of p. 108 in Bingham et al. [157]). Another popular class is the *Zygmund class* (or the class of normalized slowly varying functions) of slowly varying functions L such that, for any (equivalently, one) $\delta > 0$, $u^{-\delta} L(u)$ is ultimately decreasing[4] and $u^\delta L(u)$ is ultimately increasing. It is known (but not obvious) that the Zygmund class is a proper subset of the class of quasi-monotone slowly varying functions (Theorem 1.5.5 and Corollary 2.7.4 in Bingham et al. [157]).

Example 2.2.12 (*Quasi-monotonicity*) The slowly varying functions $\log u$, $1/\log u$ and $\log |\log u|$, $u > 0$, can be verified to be quasi-monotone. For example, for $L(u) = \log u$, $u > 0$, the integral in (2.2.18) becomes

$$\int_x^\infty u^{-\delta} d\log u = \int_x^\infty u^{-\delta-1} du = \frac{x^{-\delta}}{\delta} \leq \frac{x^{-\delta} \log x}{\delta} = \frac{x^{-\delta} L(x)}{\delta},$$

for $x > e$. Hence, the condition (2.2.18) holds.

---
[4] This means that there is a u_0 such that $u^{-\delta} L(u)$ is decreasing for $u > u_0$.

Example 2.2.13 (*Quasi-monotonicity*) Let $a > 0$ and $p \in \mathbb{R}$. The function $L(u) = (1 + a/u)^p$ is slowly varying since $L(u) \to 1$ as $u \to \infty$. The function L is also quasi-monotone since it verifies the condition (2.2.18). Indeed, observe that

$$\int_x^\infty u^{-\delta} \left| d\left(1 + \frac{a}{u}\right)^p \right| = |p|a \int_x^\infty u^{-\delta-2}\left(1 + \frac{a}{u}\right)^{p-1} du \leq C \int_x^\infty u^{-\delta-2} du$$

$$= \frac{C}{\delta+1} x^{-\delta-1} \leq C' x^{-\delta} = O\left(x^{-\delta}\left(1 + \frac{a}{x}\right)^p\right), \quad \text{as } x \to \infty,$$

where C, C' are two constants which depend on a, p, and where we used the fact that $(1 + a/x)^q \to 1$ as $x \to \infty$ for any $q \in \mathbb{R}$. This example is used in the proof of Proposition 2.2.14, which is given in Appendix A.2.

The proof of the following proposition is technical in nature and can be found in Appendix A.2. An informal derivation can be found in Remark 2.2.18 below.

Proposition 2.2.14 *Suppose L_2 in (2.1.5) is a quasi-monotone slowly varying function. Then, the series X, with autocovariance satisfying (2.1.5), has a spectral density*

$$f_X(\lambda) = \frac{1}{2\pi} \sum_{k=-\infty}^{\infty} e^{-ik\lambda} \gamma_X(k) = \frac{1}{2\pi}\left\{\gamma_X(0) + 2\sum_{k=1}^{\infty} \cos(k\lambda)\gamma_X(k)\right\}, \quad (2.2.19)$$

where the series converges conditionally for every $\lambda \in (-\pi, \pi] \setminus \{0\}$. Moreover, we have that

$$f_X(\lambda) \sim \lambda^{-2d} L_2\left(\frac{1}{\lambda}\right) \frac{1}{\pi} \Gamma(2d) \cos(d\pi), \quad \text{as } \lambda \to 0. \quad (2.2.20)$$

As a consequence, one has the following result.

Corollary 2.2.15 *If L_2 is a quasi-monotone slowly varying function, then condition* II *implies* IV. *In particular,*

$$L_4(\lambda) \sim L_2\left(\frac{1}{\lambda}\right) \frac{1}{\pi} \Gamma(2d) \cos(d\pi), \quad \text{as } \lambda \to 0. \quad (2.2.21)$$

The converses of Proposition 2.2.14 and Corollary 2.2.15 can also be proved. The statements are as follows. An informal derivation can be found in Remark 2.2.18 below.

Proposition 2.2.16 *Suppose $L_4(1/u)$ is a quasi-monotone slowly varying function on $[1/\pi, \infty)$, where L_4 in appears in (2.1.7). Then, the autocovariance function γ_X of the series X with the spectral density (2.1.7) satisfies*

$$\gamma_X(n) = 2\int_0^\pi \cos(n\lambda) f_X(\lambda) d\lambda \sim n^{2d-1} L_4\left(\frac{1}{n}\right) 2\Gamma(1-2d)\sin(d\pi), \quad \text{as } n \to \infty. \quad (2.2.22)$$

Proof This follows from $\gamma_X(n) = \int_{-\pi}^\pi e^{in\lambda} f_X(\lambda) d\lambda$ and from Proposition A.2.2 in Appendix A.2 with $p = 1 - 2d$, noting that $\cos((1-2d)\pi/2) = \sin(d\pi)$. □

2.2 Relations Between the Various Definitions of Long-Range Dependence

Corollary 2.2.17 *If the equality in (2.1.7) takes place on $(0, \pi]$ with a quasi-monotone slowly varying function $L_4(1/u)$ on $[1/\pi, \infty)$, then condition IV implies II. In particular,*

$$L_2(n) \sim L_4\left(\frac{1}{n}\right) 2\Gamma(1-2d)\sin(d\pi), \quad \text{as } n \to \infty. \tag{2.2.23}$$

Note that Corollaries 2.2.15 and 2.2.17 are consistent because of (2.2.14) and $2\sin(x)\cos(x) = \sin(2x)$.

Remark 2.2.18 The relation (2.2.20) can informally be argued as follows: by using (2.2.19) and discretization of the integral below as $\lambda \to 0$,

$$f_X(\lambda) = \frac{1}{2\pi}\sum_{k=-\infty}^{\infty} e^{-ik\lambda}\gamma_X(k) = \frac{1}{2\pi}\left\{\gamma_X(0) + 2\sum_{k=1}^{\infty}\cos(k\lambda)\gamma_X(k)\right\}$$

$$\approx \frac{1}{\pi}\sum_{k=1}^{\infty}\cos(k\lambda)L_2(k)k^{2d-1} = \lambda^{-2d}L_2\left(\frac{1}{\lambda}\right)\frac{1}{\pi}\sum_{k=1}^{\infty}\cos(k\lambda)\frac{L_2(k\lambda/\lambda)}{L_2(1/\lambda)}(k\lambda)^{2d-1}\lambda$$

$$\approx \lambda^{-2d}L_2\left(\frac{1}{\lambda}\right)\frac{1}{\pi}\int_0^{\infty}\cos(u)u^{2d-1}du = \lambda^{-2d}L_2\left(\frac{1}{\lambda}\right)\frac{1}{\pi}\Gamma(2d)\cos(d\pi),$$

where the last equality follows from Gradshteyn and Ryzhik [421], p. 437, Formula 3.761.9 (see also the relation (2.6.6) below). Similarly, the asymptotic relation in (2.2.22) can informally be derived as:

$$\gamma_X(n) = 2\int_0^{\pi}\cos(n\lambda)L_4(\lambda)\lambda^{-2d}d\lambda = 2\int_0^{n\pi}\cos(u)L_4\left(\frac{u}{n}\right)\left(\frac{u}{n}\right)^{-2d}\frac{1}{n}du$$

$$= n^{2d-1}L_4\left(\frac{1}{n}\right)2\int_0^{n\pi}\cos(u)\frac{L_4(u/n)}{L_4(1/n)}u^{-2d}du \approx n^{2d-1}L_4\left(\frac{1}{n}\right)2\int_0^{\infty}\cos(u)u^{-2d}du$$

$$= n^{2d-1}L_4\left(\frac{1}{n}\right)2\Gamma(1-2d)\cos\left(\frac{(1-2d)\pi}{2}\right) = n^{2d-1}L_4\left(\frac{1}{n}\right)2\Gamma(1-2d)\sin(d\pi),$$

by using the same relation in Gradshteyn and Ryzhik [421] as above.

Remark 2.2.19 It is known that conditions II and IV are not equivalent in general. See Examples 5.2 and 5.3 in Samorodnitsky [879], or Exercise 2.6 below. Another important related note is the following. It has sometimes been claimed in the literature that conditions II and IV are equivalent when

$$L_2(u) \sim c_2, \quad L_4(\lambda) \sim c_4, \tag{2.2.24}$$

where $c_2, c_4 > 0$ are two constants. In fact, even in this case, the two conditions are not equivalent and counterexamples can be provided for either direction. See Gubner [436]. In other words, as shown in the results above, the exact behavior of the slowly varying functions $\gamma_X(k)k^{1-2d} = L_2(k)$ and $f_X(\lambda)\lambda^{2d} = L_4(\lambda)$ is important in going from one condition to the other. This is also why, for example, condition II is written as $\gamma_X(k) = L_2(k)k^{2d-1}$ and not as $\gamma_X(k) \sim L_2(k)k^{2d-1}$.

Remark 2.2.20 If (2.2.24) holds and the slowly varying functions can be related as in (2.2.23), then

$$c_2 = c_4 2\Gamma(1 - 2d)\sin(d\pi). \qquad (2.2.25)$$

In estimation (see Section 2.10 and Chapter 10), for example, it is more advantageous to estimate c_4 in the spectral domain, for which c_2 can then be deduced from (2.2.25) and used e.g., in (2.2.10). In particular, the long-run variance of X appearing in (2.2.11) can be expressed as

$$\frac{c_2}{d(2d+1)} = \frac{c_4 2\Gamma(1-2d)\sin(d\pi)}{d(2d+1)}. \qquad (2.2.26)$$

Estimation of the long-run variance expressed through c_4 is discussed in Abadir, Distaso, and Giraitis [3]. See also Section 2.3 for the definition of *long-run variance* in the case of short-range dependence.

2.2.6 Comparing Conditions I and IV

We briefly discuss here conditions I and IV. Under condition I, the sequence $\{\psi_k\}_{k\geq 0}$ is in $\ell^2(\mathbb{Z})$ and hence has a Fourier transform

$$\psi(e^{-i\lambda}) := \widehat{\phi}(\lambda) := \sum_{k=0}^{\infty} \psi_k e^{-ik\lambda} \qquad (2.2.27)$$

convergent in $L^2(-\pi, \pi]$ (see Appendix A.1.1). Then, by Theorem 4.10.1 in Brockwell and Davis [186], the series X has a spectral density given by

$$f_X(\lambda) = \frac{\sigma_Z^2}{2\pi}|\psi(e^{-i\lambda})|^2.$$

Moreover, if the slowly varying function L_1 in (2.1.4) is quasi-monotone, Proposition A.2.1 in Appendix A.2 implies that the Fourier series (2.2.27) above converges conditionally and has a power-law like behavior around zero. This immediately yields the following result.

Proposition 2.2.21 *If L_1 is a quasi-monotone slowly varying function, then condition* I *implies* IV. *In particular,*

$$L_4(\lambda) \sim \left(L_1\left(\frac{1}{\lambda}\right)\right)^2 \frac{\Gamma(d)^2 \sigma_Z^2}{2\pi}, \quad as \ \lambda \to 0. \qquad (2.2.28)$$

Note that Corollary 2.2.15 and Proposition 2.2.21 imply that

$$L_2\left(\frac{1}{\lambda}\right) \sim \left(L_1\left(\frac{1}{\lambda}\right)\right)^2 \frac{\Gamma(d)^2 \sigma_Z^2}{2\Gamma(2d)\cos(d\pi)}. \qquad (2.2.29)$$

This is consistent with Corollary 2.2.10 if $2B(d, 1 - 2d)\Gamma(2d)\cos(d\pi) = \Gamma(d)^2$. The latter equality follows by writing $B(d, 1 - 2d) = \Gamma(d)\Gamma(1 - 2d)/\Gamma(1 - d)$ and using $\Gamma(1-d)\Gamma(d) = \pi/\sin(d\pi)$ and $\Gamma(1-2d)\Gamma(2d) = \pi/\sin(2d\pi) = \pi/(2\sin(d\pi)\cos(d\pi))$ by (2.2.14).

2.2.7 Comparing Conditions IV and III

We show that condition IV implies III. To do so, we use the following lemma.

Lemma 2.2.22 *If $\sum_{k=-\infty}^{\infty} |\gamma_X(k)| < \infty$, then the series $X = \{X_n\}_{n\in\mathbb{Z}}$ has a continuous spectral density $f_X(\lambda) = \frac{1}{2\pi} \sum_{k=-\infty}^{\infty} e^{-ik\lambda} \gamma_X(k)$.*

Proof Since $f_X(\lambda) \leq \frac{1}{2\pi} \sum_{k=-\infty}^{\infty} |\gamma_X(k)| < \infty$, it follows from the dominated convergence theorem that $\lim_{\lambda \to \lambda_0} f_X(\lambda) = f_X(\lambda_0)$. In particular, $f_X(0) = \frac{1}{2\pi} \sum_{k=-\infty}^{\infty} \gamma_X(k) < \infty$. $\square$

Suppose now that condition III fails; that is, $\sum_{k=-\infty}^{\infty} |\gamma_X(k)| < \infty$. It follows from Lemma 2.2.22 that $f_X(\lambda)$ is continuous and $f_X(0) = \frac{1}{2\pi} \sum_{k=-\infty}^{\infty} \gamma_X(k) < \infty$. But by condition IV and (2.1.1), $\lim_{\lambda \to 0} f_X(\lambda) = \lim_{\lambda \to 0} L_4(\lambda) \lambda^{-2d} = \infty$. Hence a contradiction, showing that condition IV implies III.

Corollary 2.2.23 *Condition IV implies III.*

2.2.8 Comparing Conditions IV and V

Finally, condition IV also implies V as shown in the following proposition.

Proposition 2.2.24 *Condition IV implies V. In particular,*

$$L_5(N) \sim L_4\left(\frac{1}{N}\right) \frac{2\Gamma(1-2d)\sin(d\pi)}{d(2d+1)}, \quad \text{as } N \to \infty. \tag{2.2.30}$$

Proof Write the series $\{X_n\}$ in its spectral representation as $X_n = \int_{(-\pi,\pi]} e^{in\lambda} Z(d\lambda)$, where $Z(d\lambda)$ is a complex-valued Hermitian random measure with control measure $f_X(\lambda)d\lambda$ (see Sections 1.3.4 and Appendix B.1). Then, by using $e^{i\lambda} + \cdots + e^{iN\lambda} = e^{i\lambda}(e^{iN\lambda} - 1)/(e^{i\lambda} - 1)$,

$$X_1 + \cdots + X_N = \int_{(-\pi,\pi]} e^{i\lambda} \frac{e^{iN\lambda} - 1}{e^{i\lambda} - 1} Z(d\lambda)$$

and hence

$$\text{Var}(X_1 + \cdots + X_N) = \int_{-\pi}^{\pi} \frac{|e^{iN\lambda} - 1|^2}{|e^{i\lambda} - 1|^2} f_X(\lambda) d\lambda.$$

Then,

$$\frac{N^{-1-2d}}{L_4(1/N)} \text{Var}(X_1 + \cdots + X_N) = \frac{2N^{-1-2d}}{L_4(1/N)} \int_0^{\pi} \frac{|e^{iN\lambda} - 1|^2}{|e^{i\lambda} - 1|^2} f_X(\lambda) d\lambda$$

$$= \frac{2N^{-1-2d}}{L_4(1/N)} \left(\int_0^{\epsilon} + \int_{\epsilon}^{\pi}\right) \frac{|e^{iN\lambda} - 1|^2}{|e^{i\lambda} - 1|^2} f_X(\lambda) d\lambda =: I_{1,N} + I_{2,N},$$

for $\epsilon > 0$. Note that, since the integrand above is bounded on $[\epsilon, \pi]$,

$$I_{2,N} \leq \frac{CN^{-1-2d}}{L_4(1/N)} \int_\epsilon^\pi f_X(\lambda) d\lambda \to 0,$$

as $N \to \infty$. On the other hand, by condition IV,

$$I_{1,N} = \frac{2N^{-1-2d}}{L_4(1/N)} \int_0^\epsilon \frac{|e^{iN\lambda} - 1|^2}{|e^{i\lambda} - 1|^2} \lambda^{-2d} L_4(\lambda) d\lambda$$

$$= 2 \int_0^{N\epsilon} \frac{|e^{ix} - 1|^2}{|e^{ix/N} - 1|^2} \left(\frac{x}{N}\right)^2 x^{-2d-2} \frac{L_4(x/N)}{L_4(1/N)} dx =: \int_0^\infty g_N(x) dx.$$

We have $g_N(x) \to 2|e^{ix} - 1|^2 x^{-2d-2}$, $x > 0$, as $N \to \infty$, and by using Potter's bound (2.2.1), $|g_N(x)| \leq C|e^{ix} - 1|^2 x^{-2d-2\pm\delta} =: g(x)$, $x > 0$, for any fixed $\delta > 0$. Since the function $g(x)$ is integrable on $(0, \infty)$ for sufficiently small $\delta > 0$, and $\delta \in (0, 1/2)$, the dominated convergence theorem yields

$$I_{1,N} \to 2\int_0^\infty |e^{ix} - 1|^2 x^{-2d-2} dx = 4\int_0^\infty (1 - \cos x) x^{-2d-2} dx = \frac{4}{2d+1} \int_0^\infty \sin(x) x^{-2d-1} dx$$

$$= \frac{2}{d(2d+1)} \int_0^\infty \cos(x) x^{-2d} dx = \frac{2\Gamma(1-2d)\sin(d\pi)}{d(2d+1)},$$

where we used integration by parts twice and Formula 3.761.9 in Gradshteyn and Ryzhik [421], p. 437 (see also (2.6.6)). This shows that condition IV implies V, and that (2.2.30) holds. $\square$

2.3 Short-Range Dependent Series and their Several Examples

According to condition III in Section 2.1, the autocovariances of long-range dependent series are not (absolutely) summable. This is why time series with absolutely summable autocovariances are often called *short-range dependent*.

Definition 2.3.1 A second-order stationary time series $X = \{X_n\}_{n\in\mathbb{Z}}$ is called *short-range dependent* (SRD, in short) if its autocovariances are absolutely summable; that is, if

$$\sum_{k=-\infty}^\infty |\gamma_X(k)| < \infty. \tag{2.3.1}$$

It follows from Lemma 2.2.22 that SRD series have a continuous spectral density $f_X(\lambda) = \frac{1}{2\pi} \sum_{k=-\infty}^\infty e^{-ik\lambda} \gamma_X(k)$ and, in particular, that

$$f_X(0) = \frac{1}{2\pi} \sum_{k=-\infty}^\infty \gamma_X(k) < \infty. \tag{2.3.2}$$

This is in contrast to the spectral densities of LRD series which blow up at the origin as in condition IV. Note also that, for SRD series (2.2.7) is correct and thus arguing as in the proof of Proposition 2.2.5,

2.3 Short-Range Dependent Series and their Several Examples

$$\text{Var}(\overline{X}_N) \sim \frac{1}{N} \sum_{k=-\infty}^{\infty} \gamma_X(k), \tag{2.3.3}$$

where $\overline{X}_N$ is the sample mean (Exercise 2.5). Since (2.3.2) and (2.3.3) for SRD series are obtained from the analogous expressions for LRD series by setting

$$d = 0,$$

this value of d is often associated with SRD. We also note that the constant $\sum_{k=-\infty}^{\infty} \gamma_X(k)$ appearing on the right-hand side of (2.3.3) is known as the *long-run variance* of the SRD series X. The long-run variance of LRD series was defined in (2.2.11) and (2.2.26).

There is an interesting special type of SRD series, whose autocovariances are absolutely summable as in (2.3.1), but they sum up to zero. These series have the following special name.

Definition 2.3.2 A SRD time series $X = \{X_n\}_{n \in \mathbb{Z}}$ is called *antipersistent* if

$$\sum_{k=-\infty}^{\infty} \gamma_X(k) = 0. \tag{2.3.4}$$

Observe that, for antipersistent series, (2.3.2) implies

$$f_X(0) = 0. \tag{2.3.5}$$

Examples are series with spectral density

$$f_X(\lambda) = L_4(\lambda) \lambda^{-2d}, \quad 0 < \lambda \leq \pi, \tag{2.3.6}$$

where L_4 is a slowly varying function at zero, and

$$d < 0.$$

This is the same exact expression as in condition IV of LRD series but with d being negative. As for LRD series, one could expect to relate (2.3.6) to conditions as in I, II and V for LRD series but with negative d. This is stated as Exercise 2.7.

We conclude this section with examples of several popular SRD series. In order to show that they are SRD, one just has to check that they satisfy condition (2.3.1). Examples of antipersistent series appear in Section 2.4 below.

Example 2.3.3 (*ARMA(p, q) series*) Recall from Example 1.2.4 that ARMA(p, q) time series are defined as second-order stationary series X satisfying the equation

$$X_n - \varphi_1 X_{n-1} - \cdots - \varphi_p X_{n-p} = Z_n + \theta_1 Z_{n-1} + \cdots + \theta_q Z_{n-q}, \tag{2.3.7}$$

where the coefficients $\varphi_1, \ldots, \varphi_p$ satisfy conditions given below, and $\{Z_n\}_{n \in \mathbb{Z}} \sim$ WN$(0, \sigma_Z^2)$. It is convenient to express the conditions in terms of the so-called *backshift operator* B, defined as

$$B^k X_n = X_{n-k}, \quad B^0 = I, \quad k \in \mathbb{Z}. \tag{2.3.8}$$

With this notation, the ARMA(p, q) equation becomes

$$\varphi(B) X_n = \theta(B) Z_n,$$

where
$$\varphi(z) = 1 - \varphi_1 z - \cdots - \varphi_p z^p, \quad \theta(z) = 1 + \theta_1 z + \cdots + \theta_q z^q$$
are the so-called *characteristic polynomials*. The polynomial $\varphi(z)$ can be factored out as $\varphi(z) = (1 - r_1^{-1} z) \cdot \ldots \cdot (1 - r_p^{-1} z)$. Suppose that the polynomial roots are such that $|r_m| \neq 1, m = 1, \ldots, p$.

Writing the ARMA(p, q) equation as
$$(I - r_1^{-1} B) \cdot \ldots \cdot (I - r_p^{-1} B) X_n = \theta(B) Z_n,$$
it is tempting to interpret the solution to this equation as
$$X_n = (I - r_1^{-1} B)^{-1} \cdot \ldots \cdot (I - r_p^{-1} B)^{-1} \theta(B) Z_n$$
with
$$(I - r^{-1} B)^{-1} = \begin{cases} \sum_{k=0}^{\infty} r^{-k} B^k, & \text{if } |r|^{-1} < 1, \\ -\dfrac{r B^{-1}}{I - r B^{-1}} = -r B^{-1} \sum_{k=0}^{\infty} r^k B^{-k}, & \text{if } |r|^{-1} > 1. \end{cases}$$

This time series can be written as
$$X_n = \sum_{j=-\infty}^{\infty} \psi_j B^j Z_n = \sum_{j=-\infty}^{\infty} \psi_j Z_{n-j}, \tag{2.3.9}$$
where $\psi(z) = \sum_{j=-\infty}^{\infty} \psi_j z^j$ satisfies the formal relation $\psi(z) = \theta(z) \varphi(z)^{-1}$. In fact, one can show that the series (2.3.9) is well-defined, second-order stationary and satisfies the ARMA(p, q) equation (2.3.7) (Brockwell and Davis [186]).

To show that the series (2.3.9) is SRD, observe that the function $\psi(z) = \theta(z) \varphi(z)^{-1}$ is analytic on $r^{-1} < |z| < r$ with some $r > 1$. Therefore, its Laurent series[5] $\psi(z) = \sum_{j=-\infty}^{\infty} \psi_j z^j$ converges absolutely on $r^{-1} < |z| < r$. In particular, there is $\epsilon > 0$ such that $|\psi_j| (1 + \epsilon)^{|j|} \to 0$, as $|j| \to \infty$. Then, for a constant K,
$$|\psi_j| \leq \frac{K}{(1 + \epsilon)^{|j|}}$$
and for $h > 0$,
$$|\gamma_X(h)| = \sigma_Z^2 \left| \sum_{k=-\infty}^{\infty} \psi_k \psi_{k+h} \right| \leq \sigma_Z^2 \sum_{k=-\infty}^{\infty} |\psi_k| |\psi_{k+h}| \leq \sigma_Z^2 K^2 \sum_{k=-\infty}^{\infty} \frac{1}{(1 + \epsilon)^{|k| + |k+h|}}$$
$$= \sigma_Z^2 K^2 \left(\sum_{k=0}^{\infty} \frac{1}{(1 + \epsilon)^{2k}} \right) \frac{1}{(1 + \epsilon)^h} + \sigma_Z^2 K^2 \sum_{k=1}^{h} \frac{1}{(1 + \epsilon)^h} + \sigma_Z^2 K^2 \sum_{k=h+1}^{\infty} \frac{1}{(1 + \epsilon)^{2k - h}}$$
$$= \sigma_Z^2 K^2 \left(\sum_{k=0}^{\infty} \frac{1}{(1 + \epsilon)^{2k}} + h + \sum_{k=1}^{\infty} \frac{1}{(1 + \epsilon)^{2k}} \right) s^h = (C_1 + C_2 h) s^h,$$

---

[5] A complex-valued function $\psi(z)$ may fail to be analytic at a point $z_0 \in \mathbb{C}$, and therefore cannot have a Taylor expansion around that point. It may possess an expansion in positive and negative powers of $z - z_0$ in an annular domain $r^{-1} < |z - z_0| < r$ for some $r > 1$. Such an expansion is called a *Laurent series*.

2.3 Short-Range Dependent Series and their Several Examples

where $s = 1/(1+\epsilon) \in (0, 1)$. Thus, autocovariances of ARMA(p, q) are not only absolutely summable but also at least dominated by a function with an exponential decay.

As illustration, consider the special case of the ARMA$(1, 0)$ series $X = \{X_n\}_{n\in\mathbb{Z}}$, namely, the *autoregressive series of order one* or AR(1) in Example 1.2.3. It satisfies the equation

$$X_n = \varphi X_{n-1} + Z_n, \qquad (2.3.10)$$

where $|\varphi| < 1$ and $\{Z_n\}_{n\in\mathbb{Z}} \sim \text{WN}(0, \sigma_Z^2)$. Here $\varphi_1 = \varphi$. The stationary solution of (2.3.10) is

$$X_n = \sum_{j=0}^{\infty} \varphi^j Z_{n-j},$$

with the series converging in the $L^2(\Omega)$ sense since $\sum_{j=0}^{\infty} |\varphi^j|^2 < \infty$ (Example 1.2.3). Moreover,

$$\mathbb{E}X_n = 0, \quad \text{Var}(X_n) = \frac{\sigma_Z^2}{1 - \varphi^2},$$

and the autocovariance function and the spectral density of $\{X_n\}$ are given by

$$\gamma_X(n) = \sigma_Z^2 \frac{\varphi^n}{1 - \varphi^2}, \quad f_X(\lambda) = \frac{\sigma_Z^2}{2\pi} \frac{1}{|1 - \varphi e^{-i\lambda}|^2} = \frac{\sigma_Z^2}{2\pi} \frac{1}{(1 - 2\varphi \cos\lambda + \varphi^2)} \qquad (2.3.11)$$

(see also Examples 1.3.2 and 1.3.3).

In general, for a stationary ARMA(p, q) series (2.3.7), the spectral density is

$$f_X(\lambda) = \frac{\sigma_Z^2}{2\pi} \frac{|\theta(e^{-i\lambda})|^2}{|\phi(e^{-i\lambda})|^2} \qquad (2.3.12)$$

(see Brockwell and Davis [186], Theorem 4.4.2). This is a direct consequence of the relation (2.3.9) which describes the ARMA time series $\{X_n\}$ as the output of a linear filter with white noise input and transfer function $\psi(z) = \varphi(z)^{-1}\theta(z)$ (see also Section 1.3.2). The spectral density of $\{X_n\}$ is then $f_X(\lambda) = |\psi(e^{-i\lambda})|^2 f_Z(\lambda)$ where $f_Z(\lambda) = \sigma_Z^2/2\pi$ is the spectral density of the white noise input series $\{Z_n\}$.

Example 2.3.4 (*Markov chains*) Consider a Markov chain $X = \{X_n\}_{n\geq 0}$ with a finite number of states $i = 1, \ldots, m$, and characterized by transition probabilities

$$\begin{aligned} p_{ij} &= \mathbb{P}(X_{n+1} = j | X_n = i) \\ &= \mathbb{P}(X_{n+1} = j | X_n = i, X_{n-1} = i_{n-1}, \ldots, X_0 = i_0), \quad i, j = 1, \ldots, m.\end{aligned}$$

Denote the transition matrices by

$$P = \left(p_{ij}\right)_{1 \leq i, j \leq m}, \quad P^n = \left(p_{ij}^{(n)}\right)_{1 \leq i, j \leq m},$$

where $p_{ij}^{(n)}$ denotes the probability of moving from state i to state j in n steps. Assume, in addition, that X is an irreducible Markov chain (i.e., it is possible to reach any state from any state) and aperiodic (its period equals 1). Then, there are the so-called invariant probabilities $\pi_i > 0, i = 1, \ldots, m$, such that

$$\sum_{i=1}^{m} \pi_i = 1, \quad \lim_{n \to \infty} p_{ij}^{(n)} = \pi_j, \quad \pi_i = \sum_{j=1}^{m} \pi_j p_{ji}. \qquad (2.3.13)$$

In particular, a Markov chain X with the initial distribution $\mathbb{P}(X_0 = i) = \pi_i$ is (strictly) stationary. Consider now

$$Y_n = g(X_n)$$

for some function g, which is also a stationary Markov chain under the initial distribution π for X.

We will show next that, if the matrix P has distinct characteristic roots (eigenvalues), then

$$|\gamma_Y(k)| \leq Cs^k, \quad \text{with some } 0 < s < 1. \qquad (2.3.14)$$

That is, the Markov chain has at least exponentially decreasing autocovariances and hence is SRD. Turning to the proof of (2.3.14), if the characteristic roots $\lambda_1, \ldots, \lambda_m$ (which are complex-valued when P is not symmetric) are distinct, then P can be diagonalized as

$$P = E \Lambda F' = \begin{pmatrix} e_1' \\ \vdots \\ e_m' \end{pmatrix} \begin{pmatrix} \lambda_1 & \cdots & 0 \\ \vdots & \ddots & \vdots \\ 0 & \cdots & \lambda_m \end{pmatrix} (f_1 \ldots f_m),$$

where $E = (e_1 \ldots e_m)$ consists of right eigenvectors e_i of P (possibly complex-valued) associated with eigenvalue λ_i; that is, $Pe_i = \lambda_i e_i$ and where $F = (f_1 \ldots f_m)$ consists of left eigenvectors of P satisfying $f_i' P = \lambda_i f_i'$. In matrix notation, we have $PE = E\Lambda$ and $F'P = \Lambda F'$ as well as $F' = E^{-1}$. All vectors are column vectors and prime denotes a transpose.

Then,

$$P^n = E \Lambda^n F' = \sum_{i=1}^{m} \lambda_i^n e_i f_i'.$$

Using $\sum_{j=1}^{m} p_{ij} = 1$ and (2.3.13),

$$P \begin{pmatrix} 1 \\ \vdots \\ 1 \end{pmatrix} = \begin{pmatrix} 1 \\ \vdots \\ 1 \end{pmatrix}, \quad \bar{\pi}' = \bar{\pi}' P, \quad \text{with } \bar{\pi} = \begin{pmatrix} \pi_1 \\ \vdots \\ \pi_m \end{pmatrix},$$

implying that $\lambda_1 = 1, e_1' = (1, \ldots, 1), f_1' = \bar{\pi}'$. Hence, setting

$$\Pi = (\bar{\pi} \ldots \bar{\pi})' = \begin{pmatrix} \pi_1 & \cdots & \pi_m \\ \vdots & \ddots & \vdots \\ \pi_1 & \cdots & \pi_m \end{pmatrix},$$

one gets

$$P^n = \Pi + \sum_{i=2}^{m} \lambda_i^n e_i f_i'.$$

2.4 Examples of Long-Range Dependent Series: FARIMA Models

Since $P^n \to \Pi$ and λ_is are distinct, we have $\lambda_i^n \to 0$ and hence $|\lambda_i| < 1, i = 2, \ldots, m$. Then,

$$|p_{ij}^{(n)} - \pi_j| \leq Cs^n, \quad \text{for some } 0 < s < 1, \tag{2.3.15}$$

which can be taken as $s = \max\{|\lambda_i| : i = 2, \ldots, m\}$. (Feller [361], Chapter XVI, refers to the analysis above as an algebraic treatment of Markov chains.)

Finally, observe that

$$\gamma_Y(n) = \mathbb{E}_{\bar{\pi}} g(X_n) g(X_0) - \left(\mathbb{E}_{\bar{\pi}} g(X_0)\right)^2$$

$$= \sum_{i,j=1}^m g(i)g(j)\pi_i p_{ij}^{(n)} - \sum_{i,j=1}^m g(i)g(j)\pi_i \pi_j = \sum_{i,j=1}^m g(i)g(j)\pi_i \{p_{ij}^{(n)} - \pi_j\}$$

and the result (2.3.14) follows from (2.3.15).

Relation (2.3.14) shows that Markov chains (at least with a finite number of states) are SRD. For this reason, LRD series are often called non-Markovian.

2.4 Examples of Long-Range Dependent Series: FARIMA Models

Examples of LRD series can be readily constructed by taking series satisfying condition I or IV in Section 2.1. We consider here one such celebrated class of series, namely, the FARIMA(p, d, q) series. We shall begin by introducing FARIMA$(0, d, 0)$ series, and then discuss the more general case of FARIMA(p, d, q) series. These series are LRD for $d \in (0, 1/2)$ as in Section 2.1. For completeness sake, we shall define these series for other values of d as well. For some of these values, the series is SRD.

2.4.1 FARIMA$(0, d, 0)$ Series

FARIMA$(0, d, 0)$ series will be defined as

$$X_n = (I - B)^{-d} Z_n = \sum_{j=0}^\infty b_j B^j Z_n = \sum_{j=0}^\infty b_j Z_{n-j}, \tag{2.4.1}$$

where B is the backshift operator, $\{Z_n\}_{n \in \mathbb{Z}} \sim \text{WN}(0, \sigma_Z^2)$ and b_js are coefficients in the Taylor expansion

$$(1-z)^{-d} = 1 + dz + \frac{d(d+1)}{2!} z^2 + \frac{d(d+1)(d+2)}{3!} z^3 + \cdots$$

$$= \sum_{j=0}^\infty \Big(\prod_{k=1}^j \frac{k-1+d}{k}\Big) z^j = \sum_{j=0}^\infty \frac{\Gamma(j+d)}{\Gamma(j+1)\Gamma(d)} z^j =: \sum_{j=0}^\infty b_j z^j. \tag{2.4.2}$$

(In some instances below, b_j will be denoted $b_j^{(-d)}$ in order to indicate the dependence on d, and where $(-d)$ refers to the exponent of $(I - B)$.)

The time series (2.4.1) is well-defined when $\sum_{j=0}^{\infty} b_j^2 < \infty$, which depends on the behavior of b_j as $j \to \infty$. Observe that, by Stirling's formula $\Gamma(p) \sim \sqrt{2\pi} e^{-(p-1)}(p-1)^{p-1/2}$, as $p \to \infty$, we have: for $a, b \in \mathbb{R}$,

$$\frac{\Gamma(j+a)}{\Gamma(j+b)} \sim \frac{e^{-(j+a-1)}(j+a-1)^{j+a-\frac{1}{2}}}{e^{-(j+b-1)}(j+b-1)^{j+b-\frac{1}{2}}} \quad (2.4.3)$$

$$= e^{b-a} \frac{j^{a-\frac{1}{2}}(1+\frac{a-1}{j})^{a-\frac{1}{2}}(1+\frac{a-1}{j})^j}{j^{b-\frac{1}{2}}(1+\frac{b-1}{j})^{b-\frac{1}{2}}(1+\frac{b-1}{j})^j} \sim e^{b-a} j^{a-b} \frac{e^{a-1}}{e^{b-1}} = j^{a-b}, \quad (2.4.4)$$

as $j \to \infty$. This yields

$$b_j \sim \frac{j^{d-1}}{\Gamma(d)}, \quad (2.4.5)$$

as $j \to \infty$. The formula (2.4.5) is valid for any $d \in \mathbb{R}$, but when d is zero or a negative integer, $\Gamma(d) = \infty$ and hence $1/\Gamma(d)$ should be interpreted as 0.

The time series (2.4.1) is therefore well-defined if $2(d-1)+1 = 2d-1 < 0$ or

$$d < 1/2.$$

Definition 2.4.1 A time series given by (2.4.1) is called *FARIMA(0, d, 0) series* when $d < 1/2$.

The "F" and "I" in FARIMA stand for "Fractionally" and "Integrated", respectively. The reason for this terminology will become apparent below. A FARIMA sequence is sometimes called ARFIMA, and the differencing operator $I - B$ is denoted by Δ.

Remark 2.4.2 When d is zero or a negative integer, (2.4.1) reduces to a finite sum involving the iterated differences $\Delta^{-d} = (I - B)^{-d}$. Thus,

$$d = 0 : X_n = Z_n,$$
$$d = -1 : X_n = \Delta^1 Z_n = Z_n - Z_{n-1},$$
$$d = -2 : X_n = \Delta^2 X_n = Z_n - 2Z_{n-1} + Z_{n-2}.$$

We provide next several properties of FARIMA(0, d, 0) series. It follows from its definition that FARIMA(0, d, 0), $d < 1/2$, series is stationary. Since $b_j \sim j^{d-1}/\Gamma(d)$, it is LRD when $0 < d < 1/2$, in the sense of condition I and hence II, III and V. When $d = 0$, the FARIMA(0, d, 0) series is a white noise and hence SRD. Suppose now $d < 0$ and not equal to a negative integer. Observe that

$$\sum_h |\gamma_X(h)| \leq \sigma_Z^2 \sum_h \sum_j |b_j b_{j+h}| = \sigma_Z^2 \Big(\sum_j |b_j|\Big)^2 < \infty,$$

since $|b_j| \sim j^{d-1}/|\Gamma(d)|$ are absolutely summable for $d < 0$. If d is a negative integer, then $\Gamma(d) = \infty$ and we must consider this case separately. If $d = -1$, for example, $X_n = Z_n - Z_{n-1} = \Delta Z_n$ and thus $X_n = \Delta^{-d} Z_n$ for $d = -1, -2, \ldots$. In all these cases $\gamma_X(h) = 0$

2.4 Examples of Long-Range Dependent Series: FARIMA Models

for $|h| > |d|$ and hence the series is still SRD. Thus, FARIMA$(0, d, 0)$ is SRD when $d < 0$. A more precise asymptotics for $\gamma_X(h)$ will be obtained below. The next result provides the spectral density of the series.

Proposition 2.4.3 *FARIMA$(0, d, 0)$, $d < 1/2$, series has spectral density*

$$f_X(\lambda) = \frac{\sigma_Z^2}{2\pi}\left(2\sin\frac{\lambda}{2}\right)^{-2d} = \frac{\sigma_Z^2}{2\pi}|1 - e^{-i\lambda}|^{-2d} \sim \frac{\sigma_Z^2}{2\pi}|\lambda|^{-2d}, \quad \text{as } \lambda \to 0.$$

Proof Note that, since $\sum_{j=0}^{\infty} b_j^2 < \infty$, the series $\sum_{j=0}^{\infty} b_j e^{-ij\lambda}$ converges in $L^2(-\pi, \pi]$. In fact,

$$\sum_{j=0}^{\infty} b_j e^{-ij\lambda} = (1 - e^{-i\lambda})^{-d}, \quad \text{a.e. } d\lambda. \tag{2.4.6}$$

For $d < 0$, this follows directly from the facts that the Taylor series $(1-z)^{-d} = \sum_{j=0}^{\infty} b_j z^j$ converges for $|z| = 1$ (and hence for $z = e^{-i\lambda}$) and that the $L^2(-\pi, \pi]$ and pointwise limits of a series must be the same a.e. When $0 < d < 1/2$, this argument does not apply because $\sum_{j=0}^{\infty} |b_j| = \infty$. In this case, to show (2.4.6), take $0 < r < 1$ and consider the sequence of functions $(1 - re^{-i\lambda})^{-d}$. Observe that

$$(1 - re^{-i\lambda})^{-d} = \sum_{j=0}^{\infty} b_j r^j e^{-ij\lambda} \to \sum_{j=0}^{\infty} b_j e^{-ij\lambda}$$

in $L^2(-\pi, \pi]$ as $r \to 1$, because

$$\frac{1}{2\pi} \int_{-\pi}^{\pi} \Big| \sum_{j=0}^{\infty} b_j r^j e^{-ij\lambda} - \sum_{j=0}^{\infty} b_j e^{-ij\lambda} \Big|^2 d\lambda = \sum_{j=0}^{\infty} b_j^2 (1 - r^j)^2 \to 0,$$

as $r \to 1$, by using $\sum_{j=0}^{\infty} b_j^2 < \infty$. On the other hand, $(1 - re^{-i\lambda})^{-d} \to (1 - e^{-i\lambda})^{-d}$ pointwise for all $\lambda \in (-\pi, \pi] \setminus \{0\}$. Since the $L^2(-\pi, \pi]$ and pointwise limits of a sequence must be the same a.e., we deduce (2.4.6). Finally, by using (2.4.6), Theorem 4.10.1 in Brockwell and Davis [186] yields

$$f_X(\lambda) = \frac{\sigma_Z^2}{2\pi} \Big| \sum_{j=0}^{\infty} b_j e^{-ij\lambda} \Big|^2 = \frac{\sigma_Z^2}{2\pi} \Big| (1 - e^{-i\lambda})^{-d} \Big|^2. \qquad \square$$

Corollary 2.4.4 *The FARIMA$(0, d, 0)$ series, with $d < 1/2$ but not a negative integer or zero, has autocovariances*

$$\gamma_X(k) = \sigma_Z^2 \frac{(-1)^k \Gamma(1 - 2d)}{\Gamma(k - d + 1)\Gamma(1 - k - d)} = \sigma_Z^2 \frac{\Gamma(1 - 2d)}{\Gamma(1 - d)\Gamma(d)} \frac{\Gamma(k + d)}{\Gamma(k - d + 1)} \tag{2.4.7}$$

$$\sim k^{2d-1} \sigma_Z^2 \frac{\Gamma(1 - 2d)}{\Gamma(1 - d)\Gamma(d)} = k^{2d-1} \sigma_Z^2 \frac{\Gamma(1 - 2d)\sin(d\pi)}{\pi}, \quad \text{as } k \to \infty. \tag{2.4.8}$$

When d is a negative integer or zero, the first equality in (2.4.7) continues to hold with the convention $\Gamma(0) = \Gamma(-1) = \cdots = \infty$. In particular, in this case, $\gamma_X(k) = 0$ for $k \geq 1 - d$.

Proof The first equality in (2.4.7), for any $d < 1/2$, follows from

$$\gamma_X(k) = \int_{-\pi}^{\pi} e^{ik\lambda} f_X(\lambda) d\lambda = \frac{\sigma_Z^2}{2\pi} \int_0^{2\pi} \cos(k\lambda)(2\sin(\lambda/2))^{-2d} d\lambda = \frac{\sigma_Z^2}{\pi} \int_0^{\pi} \cos(2kx)(2\sin x)^{-2d} dx$$

by using the following identity valid for $\Re \nu > 0$:

$$\int_0^{\pi} \cos(ax)(\sin x)^{\nu-1} dx = \frac{\pi \cos(a\pi/2)}{2^{\nu-1} \nu B(\frac{\nu+a+1}{2}, \frac{\nu-a+1}{2})} = \frac{\pi \cos(a\pi/2)\Gamma(\nu)2^{1-\nu}}{\Gamma((\nu+a+1)/2)\Gamma((\nu-a+1)/2)}$$

(Gradshteyn and Ryzhik [421], p. 397, Formula 3.631.8). When d is not a negative integer or zero, the second equality in (2.4.7) follows by using (2.2.13) and writing

$$\Gamma(1-k-d) = \frac{\Gamma(2-k-d)}{1-k-d} = \frac{\Gamma(3-k-d)}{(1-k-d)(2-k-d)} = \frac{\Gamma(1-d)}{(1-k-d)(2-k-d)\ldots(-d)}$$
$$= \frac{\Gamma(1-d)}{(-1)^k(d+k-1)(d+k-2)\ldots d} = \frac{\Gamma(1-d)\Gamma(d)}{(-1)^k(d+k-1)(d+k-2)\ldots d\Gamma(d)} = \frac{\Gamma(1-d)\Gamma(d)}{(-1)^k \Gamma(k+d)}.$$

The asymptotic relation in (2.4.8) follows from (2.4.4) and the equality in (2.4.8) follows from (2.2.14). □

Having an explicit form for autocovariances is useful in many instances, for example, in simulations (see Section 2.11 below). The autocorrelations $\rho_X(k) = \gamma_X(k)/\gamma_X(0)$ can be expressed as

$$\rho_X(k) = \prod_{j=1}^{k} \frac{j-1+d}{j-d}, \quad k \geq 1. \tag{2.4.9}$$

An explicit form of the partial autocorrelation coefficients is given in Exercise 2.8. These coefficients enter in the Durbin–Levinson algorithm and are useful for prediction. See also the related Exercise 2.9.

By (2.4.5), Proposition 2.4.3 and Corollary 2.4.4, the FARIMA$(0, d, 0)$, $0 < d < 1/2$, series is LRD in the sense of conditions I, IV and II. These series are thus LRD in the sense of all five conditions found in Section 2.1. Note also that, when $d < 0$, Proposition 2.4.3 shows that the series is antipersistent because in this case $\sum_{k=-\infty}^{\infty} \gamma_X(k) = 2\pi f_X(0) = 0$.

The FARIMA$(0, d, 0)$ series was defined as $X_n = (I - B)^{-d} Z_n$ in (2.4.1). Can we invert the operator $(I - B)^{-d}$ and write $(I - B)^d X_n = Z_n$? The following result provides an answer.

Proposition 2.4.5 *Let $X = \{X_n\}_{n \in \mathbb{Z}}$ be a FARIMA$(0, d, 0)$ series. If $-1 < d < 1/2$, then*

$$(I - B)^d X_n = Z_n. \tag{2.4.10}$$

The range $-1 < d < 1/2$ is motivated by the fact that the coefficients in $(I - B)^d = \sum_{j=0}^{\infty} b_j^{(d)} B^j$ satisfy

$$b_j^{(d)} \sim \frac{j^{-d-1}}{\Gamma(-d)} \tag{2.4.11}$$

2.4 Examples of Long-Range Dependent Series: FARIMA Models

in view of (2.4.5) and hence decay to zero only when $-1 < d$. The proof of the proposition is based on the fact that $(I - B)^{-d}$ and $(I - B)^d$ are linear filters operating on second-order time series.

Proof We want to show that the FARIMA series $X_n = (I - B)^{-d} Z_n$, $n \in \mathbb{Z}$, satisfies (2.4.10). Let $b_j^{(d)}$ be coefficients in $(1 - z)^d = \sum_{j=0}^{\infty} b_j^{(d)} z^j$. Write also the white noise Z_n in its spectral representation as

$$Z_n = \int_{(-\pi,\pi]} e^{in\lambda} W(d\lambda),$$

where W is a random measure with orthogonal increments having control measure $\mathbb{E}|W(d\lambda)|^2 = (\sigma_Z^2/2\pi) d\lambda$ (see Section 1.3.4 and Appendix B.1).

By Theorem 4.10.1 in Brockwell and Davis [186], concerning linear filters,

$$X_n = (I - B)^{-d} Z_n = \int_{(-\pi,\pi]} e^{in\lambda} (1 - e^{-i\lambda})^{-d} dW(\lambda).$$

(i) Consider first the case $0 < d < 1/2$. Since the coefficients $b_j^{(d)}$ in $(I - B)^d$ are absolutely summable in view of (2.4.5), Remark 1 following Theorem 4.10.1 in Brockwell and Davis [186] implies that

$$(I - B)^d X_n = \int_{(-\pi,\pi]} e^{in\lambda} (1 - e^{-i\lambda})^d (1 - e^{-i\lambda})^{-d} W(d\lambda) = \int_{(-\pi,\pi]} e^{in\lambda} W(d\lambda) = Z_n.$$
(2.4.12)

(ii) When $d = 0$, the result is trivial.

(iii) When $-1/2 < d < 0$, we can use the convergence of the series $\sum_{j=0}^{\infty} b_j^{(d)} e^{ij\lambda}$ in $L^2(-\pi, \pi]$:

$$\int_{-\pi}^{\pi} \Big| \sum_{j=0}^{\infty} b_j^{(d)} e^{ij\lambda} \Big|^2 d\lambda = 2\pi \sum_{j=0}^{\infty} |b_j^{(d)}|^2 < \infty,$$

which holds because of (2.4.11). We also have

$$\int_{-\pi}^{\pi} \Big| \sum_{j=0}^{\infty} b_j^{(d)} e^{ij\lambda} \Big|^2 |1 - e^{-i\lambda}|^{-2d} d\lambda < \infty,$$

since $|1 - e^{-i\lambda}|^{-2d}$ is bounded for $d < 0$. Then, (2.4.12) holds as well by using Theorem 4.10.1 in Brockwell and Davis [186].

(iv) When $-1 < d \leq -1/2$, we need a more refined argument because $\sum_{j=0}^{\infty} |b_j^{(d)}|^2 = \infty$ by (2.4.11). This argument also applies to the case $-1 < d < 1/2$ with $d \neq 0$. We will show that

$$\lim_{k \to \infty} \mathbb{E} \Big(\sum_{j=0}^{k} b_j^{(d)} X_{n-j} - Z_n \Big)^2 = 0.$$

Observe that (2.4.5) is valid in the range $-1 < d < 1/2$, $d \neq 0$, with $1/\Gamma(d) \neq 0$. The relation $(1-z)^d (1-z)^{-d} = 1$ yields

$$b_0^{(d)} b_0^{(-d)} = 1, \quad \sum_{j=0}^{m} b_j^{(d)} b_{m-j}^{(-d)} = 0, \; m \geq 1, \qquad (2.4.13)$$

by identifying the coefficients of z with identical powers. Then,

$$\sum_{j=0}^{k} b_j^{(d)} X_{n-j} - Z_n = \sum_{j=0}^{k} b_j^{(d)} \sum_{l=0}^{\infty} b_l^{(-d)} Z_{n-j-l} - Z_n$$

$$= \sum_{m=0}^{\infty} \Big(\sum_{j=0}^{\min\{k,m\}} b_j^{(d)} b_{m-j}^{(-d)} \Big) Z_{n-m} - Z_n = \sum_{m=k+1}^{\infty} \Big(\sum_{j=0}^{k} b_j^{(d)} b_{m-j}^{(-d)} \Big) Z_{n-m}$$

by using (2.4.13). It is then enough to show that

$$I_k := \sum_{m=k+1}^{\infty} \Big(\sum_{j=0}^{k} b_j^{(d)} b_{m-j}^{(-d)} \Big)^2 \to 0,$$

as $k \to \infty$. By using (2.4.11), for large k and a generic constant C_j,

$$I_k \leq C_1 \sum_{m=k+1}^{\infty} \Big(\sum_{j=0}^{k} (j+1)^{-d-1} (m-j)^{d-1} \Big)^2$$

$$= \frac{C_1}{k} \sum_{m=k+1}^{\infty} \Big(\sum_{j=0}^{k} \Big(\frac{j+1}{k}\Big)^{-d-1} \Big(\frac{m}{k} - \frac{j}{k}\Big)^{d-1} \frac{1}{k} \Big)^2 \frac{1}{k} \leq \frac{C_2}{k} \int_{1+\frac{1}{k}}^{\infty} \Big(\int_0^1 u^{-d-1} (x-u)^{d-1} du \Big)^2 dx$$

$$= \frac{C_2}{k} \int_{1+\frac{1}{k}}^{\infty} x^{-2} \Big(\int_0^{\frac{1}{x}} z^{-d-1} (1-z)^{d-1} dz \Big)^2 dx \leq C_3 \frac{k^{-2d-1}}{k} = C_3 k^{-2d-2} \to 0,$$

where we used $\int_0^{1/x} z^{-d-1} (1-z)^{d-1} dz \sim C x^d$, as $x \to \infty$, $\int_0^{1/x} z^{-d-1} (1-z)^{d-1} dz \sim C(x-1)^d$, as $x \to 1$, and the fact that $-1 < d < 1/2$. $\square$

Property (2.4.10) is referred to as the *invertibility* of the series X. The proposition states that X is invertible in the range $d \in (-1, 1/2)$.

In Figure 2.2, we provide several realizations of Gaussian FARIMA$(0, d, 0)$ series generated through the circulant matrix embedding method discussed in Section 2.11.2. The series in the left, middle and right plots are, respectively, antipersistent, uncorrelated (independent) and long-range dependent. Note that the realizations are consistent with the spectral domain perspective of time series outlined in Section 1.3.4. For LRD series, in particular, the spectral density diverges at the origin and hence low frequencies (long waves) dominate in the spectral representation of the series. Indeed, the LRD series in the right plot of the figure tends to stay away from the zero mean for extended periods of time. In fact, this is one of the characteristic features of LRD series. In contrast, the antipersistent series in the left plot fluctuates even more than the uncorrelated series in the middle plot.

2.4 Examples of Long-Range Dependent Series: FARIMA Models

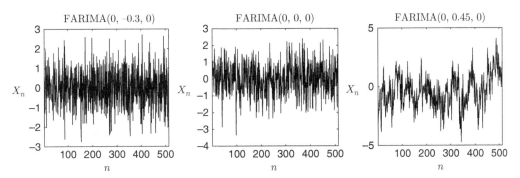

Figure 2.2 Several realizations of Gaussian FARIMA$(0, d, 0)$ series. Left: $d = -0.3$. Middle: $d = 0$. Right: $d = 0.45$.

The sequence $X_n = (I - B)^{-d} Z_n$, $n \in \mathbb{Z}$, can be viewed as the transform of the FARIMA$(0, 0, 0)$ series $\{Z_n\}$ into a FARIMA$(0, d, 0)$ series. How can we obtain the FARIMA$(0, d, 0)$ series $\{X_n\}$ from a FARIMA$(0, d', 0)$ series $\{X'_n\}$? The following proposition shows how this can be done.

Proposition 2.4.6 *Let $d < 1/2$ and $X' = \{X'_n\}_{n \in \mathbb{Z}}$ be a FARIMA$(0, d', 0)$ series with*

$$d' = d - [d + 1/2] \in \left[-\frac{1}{2}, \frac{1}{2}\right),$$

where $[\cdot]$ denotes the integer part.[6] *Then,*

$$X_n = (I - B)^{-[d+1/2]} X'_n, \quad n \in \mathbb{Z},$$

is a FARIMA$(0, d, 0)$ series.

Proof Since $d < 1/2$ and $[d + 1/2]$ is integer, $-[d + 1/2]$ is a nonnegative integer and therefore the expansion of $(1 - z)^{-[d+1/2]}$ has only a finite number of non-zero coefficients which are obviously summable. Therefore, this proposition follows as in Proposition 2.4.5 (case $0 < d < 1/2$). □

$\Delta = I - B$ is called the *differencing operator* since $\Delta X_n = (I - B) X_n = X_n - X_{n-1}$. In view of the preceding proposition, any FARIMA$(0, d, 0)$, $d < 1/2$, series can be obtained from a FARIMA$(0, d', 0)$ series with

$$-1/2 \leq d' < 1/2$$

by differencing it an integer number of times. This is why the range $[-1/2, 1/2)$ is called the *principal range* of d.

Example 2.4.7 (*FARIMA$(0, d, 0)$ series with $d < -1/2$*) Since $d = -5.7 = 0.3 - 6$, a FARIMA$(0, -5.7, 0)$ series X can be obtained by differencing $-[-5.7 + 1/2] = -[-5.2] = -(-6) = 6$ times a FARIMA$(0, 0.3, 0)$ series X'; that is, $X_n = \Delta^6 X'_n$.

---

[6] That is, $[a]$ is the largest integer which is less than or equal to $a \in \mathbb{R}$.

Similarly, since $-5.3 = -0.3 - 5$, a FARIMA$(0, -5.3, 0)$ series X can be obtained from a FARIMA$(0, -0.3, 0)$ series X' by differencing it five times; that is, $X_n = \Delta^5 X'_n$.

The preceding discussion and Proposition 2.4.6 also suggest the following natural extension of the definition of FARIMA$(0, d, 0)$ series to $d \geq 1/2$.

Definition 2.4.8 A series $X = \{X_n\}_{n \in \mathbb{Z}}$ is called FARIMA$(0, d, 0)$ for $d \geq 1/2$ if it is obtained by taking the $[d + 1/2]$th order partial sum series of FARIMA$(0, d', 0)$ series with parameter $d' = d - [d + 1/2]$ in the principal range $-1/2 \leq d' < 1/2$.

By definition, FARIMA$(0, d, 0)$ is not stationary when $d \geq 1/2$. It satisfies the relation

$$(I - B)^{[d+1/2]} X_n = (I - B)^{-(d-[d+1/2])} Z_n,$$

where $(I - B)^{[d+1/2]}$ is the differencing operator $\Delta = I - B$ iterated $[d + 1/2]$ times.

Example 2.4.9 (*FARIMA$(0, d, 0)$ series with $d \geq 1/2$*) Let Z_n, $n \in \mathbb{Z}$, be a FARIMA$(0, 0, 0)$; that is, a white noise series. The series X_n, $n \in \mathbb{Z}$, defined as $X_n = \sum_{k=1}^{n} Z_k$ for $n > 0$, $X_0 = 0$, $X_n = \sum_{k=n}^{-1} Z_k$ for $n < 0$, is a FARIMA$(0, 1, 0)$ series.

2.4.2 FARIMA(p, d, q) Series

Finally, we briefly discuss the class of stationary FARIMA(p, d, q) series. This class combines ARMA(p, q) series in Example 2.3.3 and FARIMA$(0, d, 0)$ series. As in Example 2.3.3, consider polynomials

$$\varphi(z) = 1 - \varphi_1 z - \cdots - \varphi_p z^p, \quad \theta(z) = 1 + \theta_1 z + \cdots + \theta_q z^q \tag{2.4.14}$$

and assume that they do not have common zeros. Suppose that φ does not have roots on the unit circle. Consider the series

$$X_n = \varphi^{-1}(B)\theta(B)(I - B)^{-d} Z_n =: \sum_{j=-\infty}^{\infty} b_j Z_{n-j}, \tag{2.4.15}$$

where $\{Z_n\}_{n \in \mathbb{Z}} \sim \text{WN}(0, \sigma_Z^2)$ and $\{b_k\}_{k \in \mathbb{Z}}$ are coefficients in the Laurent expansion $\varphi^{-1}(z)\theta(z)(1-z)^{-d} = \sum_{k=-\infty}^{\infty} b_k z^k$. Note that the sequence $\{b_k\}_{k \in \mathbb{Z}} \in \ell^2(\mathbb{Z})$ since b_k can be viewed as Fourier coefficients of the function $\varphi^{-1}(e^{-i\lambda})\theta(e^{-i\lambda})(1 - e^{-i\lambda})^{-d}$.

Definition 2.4.10 The series $X = \{X_n\}_{n \in \mathbb{Z}}$ in (2.4.15) is called *FARIMA(p, d, q), $d < 1/2$, series*.

The basic idea behind FARIMA(p, d, q) models is that the operator $(I - B)^{-d}$ is responsible for LRD features of the series while $\varphi^{-1}(B)\theta(B)$ accounts for SRD behavior.

A number of properties can be proved about FARIMA(p, d, q) series. As in Proposition 2.4.3, one can show that this series has a spectral density

$$f_X(\lambda) = \frac{\sigma_Z^2}{2\pi} \frac{|\theta(e^{-i\lambda})|^2}{|\varphi(e^{-i\lambda})|^2} |1 - e^{-i\lambda}|^{-2d}. \qquad (2.4.16)$$

This expression is rather obvious since the relation (2.4.15) indicates that a FARIMA(p, d, q) series can be viewed as the output of a linear filter with the transfer function $(1 - z)^{-d}$ and, as input, an ARMA(p, q) series with the spectral density given in (2.3.12).

Proposition 2.4.11 *Suppose that $\varphi(z)$ defined in (2.4.14) has roots outside the unit circle. Then, the FARIMA(p, d, q) series, $0 < d < 1/2$, is LRD in the sense of conditions I, II, III, IV and V.*

Proof Since the FARIMA(p, d, q) has spectral density (2.4.16) with $0 < d < 1/2$, it is LRD in the sense of condition IV. Since the roots of $\varphi(z)$ are outside the unit circle, the coefficients b_k in (2.4.15) are zero for $k \leq -1$ (cf. (2.3.9) and the relation preceding it). One can then show that

$$b_k \sim \frac{\theta(1)}{\varphi(1)} \frac{k^{d-1}}{\Gamma(d)} \qquad (2.4.17)$$

(Exercise 2.10; compare with (2.4.5)). Hence, FARIMA(p, d, q), $0 < d < 1/2$, series is also LRD in the sense of condition I. By Corollaries 2.2.10 and 2.2.7, and Proposition 2.2.4, the times series is LRD in the sense of conditions II, III and V as well. $\square$

Explicit autocovariance formulas are also known, for example, for FARIMA$(1, d, 0)$ series (Hosking [484]). (In the case of FARIMA$(0, d, q)$, they can be obtained directly from the case of FARIMA$(0, d, 0)$ series.) When $\theta(z)$ does not have zeroes on the unit circle and $-1 < d < 1/2$, one can show that FARIMA(p, d, q) satisfies the relation

$$\theta^{-1}(B)\varphi(B)(I - B)^d X_n = Z_n$$

(e.g., Theorem 3.4, (c), in Palma [789]); that is, $\{X_n\}$ is invertible.

2.5 Definition and Basic Properties of Self-Similar Processes

We now turn to self-similar processes. These processes are closely related to long-range dependent series though the discussion of exact relationships is postponed till Section 2.8. We shall, in the meantime, consider self-similar processes somewhat separately from long-range dependent series.

Definition 2.5.1 A stochastic process $\{X(t)\}_{t \in \mathbb{R}}$ is called *self-similar* (SS) or *H–self-similar* (H–SS) if there is $H > 0$ such that, for all $c > 0$,

$$\{X(ct)\}_{t \in \mathbb{R}} \stackrel{d}{=} \{c^H X(t)\}_{t \in \mathbb{R}}, \qquad (2.5.1)$$

where $\stackrel{d}{=}$ denotes equality of the finite-dimensional distributions.[7] Thus, for any $c > 0$ and times $t_1, \ldots, t_m$ in $\mathbb{R}$ and integer $m \geq 1$, the vector

$$(X(ct_1), \ldots, X(ct_m))'$$

has the same distribution as the vector

$$(c^H X(t_1), \ldots, c^H X(t_m))'.$$

The parameter H is called the *self-similarity parameter* (also *Hurst index* or *Hurst parameter*).

Since H needs not to be equal to 1, self-similarity should be called *self-affinity*. Since, moreover, it is not the path of $X(t)$ but its distribution which scales, one should say "statistical self-similarity" or better, "statistical self-affinity." We will continue to use "self-similarity" because this term is now widely used.

Intuitively, self-similarity means that a stochastic process scaled in time (that is, plotted with a different time scale) looks statistically the same as the original process when properly rescaled in space. Physically, expanding the time scale (large c) should require an expansion of the space scale, that is, a large factor c^H and hence the self-similarity parameter H should be positive. One also has reasons to exclude $H \leq 0$ on theoretical grounds. Note that (2.5.1) implies that, for any $c > 0$,

$$X(c) \stackrel{d}{=} c^H X(1).$$

One expects the left-hand side to converge to $X(0)$ as $c \to 0$ under mild assumptions. If $H < 0$, however, the right-hand side diverges. The case $H = 0$ could be included in Definition 2.5.1 but, under milder assumptions, it leads to the trivial stochastic process $X(t) = X(0)$ a.s. for all t (see Embrechts and Maejima [350], Theorem 1.1.2).

A self-similar process cannot be (strictly) stationary. If it were stationary, then, for any $t \in \mathbb{R}$ and $c > 0$,

$$X(t) \stackrel{d}{=} X(ct) \stackrel{d}{=} c^H X(t).$$

But the right-hand side of this relation diverges as $c \to \infty$. Even though a self-similar process is not (strictly) stationary, it can have stationary increments. In fact, our focus will often be on self-similar processes with stationary increments. One reason is that increments of self-similar processes provide models of long-range dependent series. Though the definition of stationarity of increments was given in Section 1.1.2, we recall it next to emphasize its important role in the study of SS processes.

Definition 2.5.2 A process $\{X(t)\}_{t \in \mathbb{R}}$ has *stationary increments* if, for any $h \in \mathbb{R}$,

$$\{X(t+h) - X(h)\}_{t \in \mathbb{R}} \stackrel{d}{=} \{X(t) - X(0)\}_{t \in \mathbb{R}} \tag{2.5.2}$$

---

[7] Depending on the context $\stackrel{d}{=}$ may refer either to equality of the finite-dimensional distributions or of the marginal distributions. The relation $\{X_1(t)\} \stackrel{d}{=} \{X_2(t)\}$ always means equality of the finite-dimensional distributions.

2.5 Definition and Basic Properties of Self-Similar Processes

or, equivalently, if for any $s \in \mathbb{R}$,

$$\{X(t) - X(s)\}_{t \in \mathbb{R}} \stackrel{d}{=} \{X(t-s) - X(0)\}_{t \in \mathbb{R}}.$$

If $\{X(t)\}_{t \in \mathbb{R}}$ is Gaussian with mean 0 and $X(0) = 0$, then stationarity of the increments is equivalent to

$$\mathbb{E}(X(t) - X(s))^2 = \mathbb{E}(X(t-s))^2, \quad \text{for all } s, t \in \mathbb{R} \tag{2.5.3}$$

(Exercise 2.11).

Remark 2.5.3 The relation (2.5.2) implies that, for a process $\{X(t)\}_{t \in \mathbb{R}}$ with stationary increments and any $h \in \mathbb{R}$, the process

$$Y(t) = X(t+h) - X(t), \quad t \in \mathbb{R}, \tag{2.5.4}$$

is stationary. To see that Definition 2.5.2 implies (2.5.4), note that, for any $s \in \mathbb{R}$,

$$\{Y(t+s)\}_{t \in \mathbb{R}} = \{X(t+s+h) - X(t+s)\}_{t \in \mathbb{R}}$$
$$= \{(X(t+s+h) - X(s)) - (X(t+s) - X(s))\}_{t \in \mathbb{R}}$$
$$\stackrel{d}{=} \{(X(t+h) - X(0)) - (X(t) - X(0))\}_{t \in \mathbb{R}} = \{Y(t)\}_{t \in \mathbb{R}}.$$

If $\mathbb{R}$ is replaced by $\mathbb{Z}$, then the stationarity of $\{Y(t)\}$ in (2.5.4) implies that $\{X(t)\}$ has stationary increments. Indeed for $h \in \mathbb{Z}$, say $h > 0$, observe that

$$\{X(t+h) - X(h)\}_{t \in \mathbb{Z}} = \left\{ \sum_{k=0}^{t-1} (X(k+1+h) - X(k+h)) \right\}_{t \in \mathbb{Z}} \tag{2.5.5}$$

$$\stackrel{d}{=} \left\{ \sum_{k=0}^{t-1} (X(k+1) - X(k)) \right\}_{t \in \mathbb{Z}} = \{X(t) - X(0)\}_{t \in \mathbb{Z}}. \tag{2.5.6}$$

The passage from (2.5.5) to (2.5.6) follows from viewing $k + h$ as time and 1 as the increment. Since Y is stationary, $k + h$ can be replaced by k.

Remark 2.5.4 One often chooses $h = 1$ in (2.5.4) and considers a stationary series

$$Y_n = X(n+1) - X(n), \quad n \in \mathbb{Z}.$$

Definition 2.5.5 An H–self-similar process $\{X(t)\}_{t \in \mathbb{R}}$ will be referred to as an SS or H–SS process. An H–self-similar process $\{X(t)\}_{t \in \mathbb{R}}$ with stationary increments will be referred to as SSSI or H–SSSI process.

The next result establishes several basic properties of SSSI processes. We always suppose $H > 0$.

Proposition 2.5.6 Let $X = \{X(t)\}_{t \in \mathbb{R}}$ be an H–SSSI process. Then, the following properties hold:

(a) $X(0) = 0$ a.s.

(b) If $H \neq 1$ and $\mathbb{E}|X(1)| < \infty$, then $\mathbb{E}X(t) = 0$ for all $t \in \mathbb{R}$.
(c) If $\mathbb{E}|X(1)|^\gamma < \infty$ for $0 < \gamma \leq 1$, then $H \leq 1/\gamma$.
(d) If $\mathbb{E}|X(1)|^2 < \infty$, then $\mathbb{E}(X(t))^2 = |t|^{2H}\mathbb{E}(X(1))^2$, $t \in \mathbb{R}$.
(e) If $\mathbb{E}|X(1)|^2 < \infty$, then, for $s, t \in \mathbb{R}$,

$$\mathbb{E}X(t)X(s) = \frac{\mathbb{E}X(1)^2}{2}\left\{|t|^{2H} + |s|^{2H} - |t-s|^{2H}\right\}. \tag{2.5.7}$$

(f) If $H = 1$ and $\mathbb{E}|X(1)|^2 < \infty$, then for any $t \in \mathbb{R}$, $X(t) = tX(1)$ a.s.

Proof Property (a) holds since, for $c > 0$,

$$X(0) \stackrel{d}{=} X(c \cdot 0) \stackrel{d}{=} c^H X(0)$$

and $c^H \to \infty$, as $c \to \infty$. Property (b) follows from

$$2^H \mathbb{E}X(t) = \mathbb{E}X(2t) = \mathbb{E}(X(2t) - X(t)) + \mathbb{E}X(t) = 2\mathbb{E}X(t),$$

by using self-similarity and stationarity of the increments. Property (c) follows from a similar argument leading to

$$2^{H\gamma}\mathbb{E}|X(1)|^\gamma = \mathbb{E}|X(2)|^\gamma \leq \mathbb{E}(|X(2) - X(1)| + |X(1)|)^\gamma$$
$$\leq \mathbb{E}(|X(2) - X(1)|^\gamma + |X(1)|^\gamma) = 2\mathbb{E}|X(1)|^\gamma,$$

by using the inequality $(a+b)^\gamma \leq a^\gamma + b^\gamma$, $a, b \geq 0$, $\gamma \in (0, 1]$. If $t > 0$, Property (d) follows from

$$\mathbb{E}(X(t))^2 = \mathbb{E}t^{2H}(X(1))^2 = t^{2H}\mathbb{E}(X(1))^2$$

and, if $t < 0$, from

$$\mathbb{E}(X(t))^2 = \mathbb{E}(X(t) - X(0))^2 = \mathbb{E}(X(0) - X(-t))^2 = \mathbb{E}(X(-t))^2 = (-t)^{2H}\mathbb{E}(X(1))^2.$$

Stationarity of the increments ensures that

$$\mathbb{E}(X(t) - X(s))^2 = \mathbb{E}(X(t-s) - X(0))^2 = \mathbb{E}X(t-s)^2$$

and therefore Property (e) is a consequence of

$$\mathbb{E}X(t)X(s) = \frac{1}{2}\left\{\mathbb{E}X(t)^2 + \mathbb{E}X(s)^2 - \mathbb{E}(X(t) - X(s))^2\right\}$$
$$= \frac{1}{2}\left\{\mathbb{E}X(t)^2 + \mathbb{E}X(s)^2 - \mathbb{E}(X(t-s))^2\right\}, \tag{2.5.8}$$

and Property (d). Property (f) follows from

$$\mathbb{E}(X(t) - tX(1))^2 = \mathbb{E}X(t)^2 - 2t\mathbb{E}X(t)X(1) + t^2\mathbb{E}X(1)^2$$
$$= t^2\mathbb{E}X(1)^2 - 2t \cdot t\frac{\mathbb{E}X(1)^2}{2}(t^2 + 1 - (t-1)^2) + t^2\mathbb{E}X(1)^2 = 0. \quad \square$$

With a bit more effort, one can show that for $0 < \gamma < 1$ in (c), one has $H < 1/\gamma$ and that the conclusion in (f) holds when $\mathbb{E}|X(1)| < \infty$ (Exercise 2.12).

2.6 Examples of Self-Similar Processes

2.6.1 Fractional Brownian Motion

We shall first focus on Gaussian SSSI processes. These are defined through their covariance function and hence the following result shows that the function in Proposition 2.5.6, (e), is a valid covariance function.

Proposition 2.6.1 *Let $0 < H \leq 1$. Then, the function*

$$R_H(t,s) = |t|^{2H} + |s|^{2H} - |t-s|^{2H}, \quad s, t \in \mathbb{R},$$

is positive semidefinite.[8]

Proof We have to show that $\sum_{i=1}^{n}\sum_{j=1}^{n} R_H(t_i, t_j) u_i u_j \geq 0$ for any real numbers $t_1, \ldots, t_n$ and $u_1, \ldots, u_n$. View each u_i as a mass at the points t_i, $i = 1, \ldots, n$. Now add a mass $u_0 = -\sum_{i=1}^{n} u_i$ at the origin $t_0 = 0$. Then $\sum_{i=0}^{n} u_i = 0$ and

$$\sum_{i=1}^{n}\sum_{j=1}^{n} R_H(t_i, t_j) u_i u_j = \sum_{i=1}^{n}\sum_{j=1}^{n} (|t_i|^{2H} + |t_j|^{2H} - |t_i - t_j|^{2H}) u_i u_j = -\sum_{i=0}^{n}\sum_{j=0}^{n} |t_i - t_j|^{2H} u_i u_j.$$

But, for any constant $c > 0$,

$$\sum_{i=0}^{n}\sum_{j=0}^{n} e^{-c|t_i - t_j|^{2H}} u_i u_j = \sum_{i=0}^{n}\sum_{j=0}^{n} \left(e^{-c|t_i - t_j|^{2H}} - 1\right) u_i u_j = -c\sum_{i=0}^{n}\sum_{j=0}^{n} |t_i - t_j|^{2H} u_i u_j + o(c),$$

as c tends to 0. It is therefore enough to prove that $\sum_{i=0}^{n}\sum_{j=0}^{n} e^{-c|t_i-t_j|^{2H}} u_i u_j \geq 0$. The function $e^{-c|t|^{2H}}$, $0 < H \leq 1$, is the characteristic function of a $2H$-stable random variable ξ with the scaling exponent $c^{1/2H}$; that is, $\mathbb{E}e^{it\xi} = e^{-c|t|^{2H}}$, for all $t \in \mathbb{R}$ (see Appendix A.4). It follows that

$$\sum_{j=0}^{n}\sum_{k=0}^{n} e^{-c|t_j - t_k|^{2H}} u_j u_k = \mathbb{E}\sum_{j=0}^{n}\sum_{k=0}^{n} u_j u_k e^{i(t_j - t_k)\xi} = \mathbb{E}\left|\sum_{j=0}^{n} u_j e^{it_j\xi}\right|^2 \geq 0. \quad \square$$

There is then a zero mean Gaussian process $\{Z(t)\}_{t \in \mathbb{R}}$ with the covariance function

$$\text{Cov}(Z(t), Z(s)) = \Gamma_H(t, s) = \frac{\sigma^2}{2}(|t|^{2H} + |s|^{2H} - |t-s|^{2H}),$$

where $\sigma^2 = \text{Var}(Z(1))$. Since the distribution of a zero mean Gaussian process is entirely determined by its covariance function, Proposition 2.5.6, (e), implies that, for a given H, all H–SSSI zero mean Gaussian processes differ by a multiplicative constant only. This leads to the following definition.

Definition 2.6.2 A Gaussian H–SSSI process $\{B_H(t)\}_{t \in \mathbb{R}}$ with $0 < H \leq 1$ is called *fractional Brownian motion* (FBM). It is called *standard* if $\sigma^2 = \mathbb{E}(B_H(1))^2 = 1$.

---

[8] Positive semidefiniteness is also referred to as nonnegative definiteness. See Section 9.1.1 for a definition.

The discussion above leads to the following useful corollary.

Corollary 2.6.3 *Suppose that $\{X(t)\}_{t\in\mathbb{R}}$ satisfies the following conditions:*

(i) *it is a Gaussian process with zero mean and $X(0) = 0$,*
(ii) $\mathbb{E}(X(t))^2 = \sigma^2 |t|^{2H}$ *for all $t \in \mathbb{R}$ and some $\sigma > 0$ and $0 < H \leq 1$, and*
(iii) $\mathbb{E}(X(t) - X(s))^2 = \mathbb{E}(X(t-s))^2$ *for all $s, t \in \mathbb{R}$.*

Then, $\{X(t)\}_{t\in\mathbb{R}}$ is fractional Brownian motion.

Proof A Gaussian process is determined by its mean function and covariance function. Here the mean is zero. To determine the covariance, observe that one has (2.5.8) by (iii). By (ii), this expression becomes (2.5.7), which is the covariance of fractional Brownian motion. □

Remark 2.6.4 If $H = 1$, then by the property (f) in Proposition 2.5.6, $B_H(t) = tB_H(1)$; that is, it is a straight line with a random slope. We will exclude this trivial case in the future and suppose $0 < H < 1$.

One may get a more physical understanding of FBM through its integral representations. The following result concerns integral representation in time domain. Gaussian random measures and corresponding integrals are defined in Appendix B.1. See, in particular, the integral representations in the time and spectral domains, and their relationship in (B.1.14) (see also Section 1.4.1).

Time Domain Representations of FBM

Proposition 2.6.5 *For $H \in (0, 1)$, standard FBM admits the following integral representation:*

$$\{B_H(t)\}_{t\in\mathbb{R}} \stackrel{d}{=} \left\{ \frac{1}{c_1(H)} \int_{\mathbb{R}} \left((t-u)_+^{H-1/2} - (-u)_+^{H-1/2}\right) B(du) \right\}_{t\in\mathbb{R}}, \quad (2.6.1)$$

where $x_+ = \max\{x, 0\}$, $c_1(H)^2 = \int_{\mathbb{R}} ((1-u)_+^{H-1/2} - (-u)_+^{H-1/2})^2 du$ is a normalizing constant and $B(du)$ is a Gaussian random measure on $\mathbb{R}$ with the control measure $\mathbb{E}(B(du))^2 = du$.

Proof Denote the stochastic process on the right-hand side of (2.6.1) by $X_H(t)$, without the normalizing constant. We need to show that $X_H(t)$ is FBM. It is enough to prove that X_H satisfies conditions (i)–(iii) of Corollary 2.6.3.

X_H satisfies (i) of Corollary 2.6.3 by construction if it is well-defined. For X_H to be well-defined, we have to check that the deterministic integrand of (2.6.1) is square-integrable. Potential problems may occur only around $u = 0, t$ or $-\infty$. Around $u = t$, we need to consider $((t-u)_+^{H-1/2})^2 = (t-u)_+^{2H-1}$, which is integrable since $(2H-1)+1 = 2H > 0$. The argument for the behavior around $u = 0$ is analogous. Around $u = -\infty$, $((t-u)_+^{H-1/2} - (-u)_+^{H-1/2})^2 \sim C((-u)_+^{H-3/2})^2 = C(-u)_+^{2H-3}$, which is integrable since $(2H-3)+1 = 2(H-1) < 0$.

2.6 Examples of Self-Similar Processes

To see that X_H satisfies (ii) of Corollary 2.6.3, observe that

$$\mathbb{E}(X_H(t))^2 = \int_{\mathbb{R}} \left((t-u)_+^{H-1/2} - (-u)_+^{H-1/2}\right)^2 du$$

$$= |t|^{2H} \int_{\mathbb{R}} \left((\text{sign}(t) - v)_+^{H-1/2} - (-v)_+^{H-1/2}\right)^2 dv = |t|^{2H} \int_{\mathbb{R}} \left((1-s)_+^{H-1/2} - (-s)_+^{H-1/2}\right)^2 ds,$$

where we made the changes of variables $u = |t|v$ and, if $t < 0$, $-1 - v = -s$. Similarly, X_H satisfies (iii) of Corollary 2.6.3 since

$$\mathbb{E}(X_H(t) - X_H(s))^2 = \int_{\mathbb{R}} \left((t-u)_+^{H-1/2} - (s-u)_+^{H-1/2}\right)^2 du$$

$$= \int_{\mathbb{R}} \left((t-s-v)_+^{H-1/2} - (-v)_+^{H-1/2}\right)^2 dv = \mathbb{E}(X_H(t-s))^2,$$

where we made the change of variables $s - u = -v$. $\square$

Remark 2.6.6 If $1/2 < H < 1$, then we can express (2.6.1) as

$$\{B_H(t)\}_{t \in \mathbb{R}} \stackrel{d}{=} \left\{\frac{H - 1/2}{c_1(H)} \int_{\mathbb{R}} \left(\int_0^t (s-u)_+^{H-3/2} ds\right) B(du)\right\}_{t \in \mathbb{R}}. \quad (2.6.2)$$

Remark 2.6.7 When $H = 1/2$, the integrand $((t-u)_+)^0 - ((-u)_+)^0$ is interpreted by convention as $1_{[0,t)}$ if $t \geq 0$ and $1_{(t,0]}$ if $t < 0$. Note also that with this convention, $c_1(1/2) = 1$. In this case, one gets (for $t > 0$)

$$B_{1/2}(t) = \int_0^t B(du) = B(t)$$

so that FBM reduces to Brownian motion.

Remark 2.6.8 The time domain representation (2.6.1) is not unique. For example, for any $a, b \in \mathbb{R}$,

$$\int_{\mathbb{R}} \left(a\left((t-u)_+^{H-1/2} - (-u)_+^{H-1/2}\right) + b\left((t-u)_-^{H-1/2} - (-u)_-^{H-1/2}\right)\right) B(du), \quad (2.6.3)$$

where $x_- = \max\{0, -x\}$, is another representation of FBM. If $a \neq 0, b = 0$ as in (2.6.1), the representation is called *causal* (non-anticipative) because integration over $u \leq t$ is used for $t > 0$. If $a = b$, then it is called *well-balanced*. In this case, $x_+ + x_- = |x|$, and thus

$$\{B_H(t)\}_{t \in \mathbb{R}} \stackrel{d}{=} \left\{a \int_{\mathbb{R}} \left(|t-u|^{H-1/2} - |u|^{H-1/2}\right) B(du)\right\}_{t \in \mathbb{R}}.$$

Remark 2.6.9 The two integrands in (2.6.1) cannot be separated; that is, the process $Y_H(t) = \int_{\mathbb{R}} (t-u)_+^{H-1/2} B(du)$ is not well-defined. Indeed, if it were, it would be stationary and self-similar (see the discussion preceding Definition 2.5.5). However, $Y_H(t) - Y_H(0)$ once rewritten as (2.6.1) is well-defined and yields $B_H(t)$. The subtraction of $(-u)_+^{H-1/2}$ is called an *infrared correction*.

Spectral Domain Representations of FBM

A spectral domain representation of FBM can be obtained through (B.1.14) in Appendix B.1 by computing the $L^2(\mathbb{R})$–Fourier transform of the deterministic kernel in (2.6.1) or (2.6.3). We denote the Fourier transform of a function $f \in L^2(\mathbb{R})$ as $\widehat{f}(x) = \int_{\mathbb{R}} e^{ixu} f(u) du$ (see also Appendix A.1.2).

Proposition 2.6.10 *Let $-1/2 < p < 1/2$. The $L^2(\mathbb{R})$–Fourier transform of the function*

$$f_{t,\pm}(u) = (t-u)_{\pm}^p - (-u)_{\pm}^p, \quad u \in \mathbb{R},$$

is given by

$$\widehat{f}_{t,\pm}(x) = \Gamma(p+1)\frac{e^{itx}-1}{ix}(\mp ix)^{-p} = \Gamma(p+1)\frac{e^{itx}-1}{ix}|x|^{-p}e^{\mp i\,\text{sign}(x)p\pi/2} \quad (2.6.4)$$

$$= \pm\Gamma(p+1)(e^{itx}-1)|x|^{-p-1}e^{\mp i\,\text{sign}(x)(p+1)\pi/2}. \quad (2.6.5)$$

Proof We start with two observations. Firstly, the second equality in (2.6.4) results from $\mp i = e^{\mp i\pi/2}$, so that

$$(\mp ix)^{-p} = |x|^{-p}(\mp i\,\text{sign}(x))^{-p} = |x|^{-p}e^{\mp i\,\text{sign}(x)p\pi/2}.$$

Secondly, it is known (Gradshteyn and Ryzhik [421], pp. 436–437, Formulae 3.761.4 and 3.761.9) that

$$\int_0^\infty e^{\pm iv} v^{\alpha-1} dv = \Gamma(\alpha) e^{\pm i\alpha\pi/2} \quad (2.6.6)$$

for $\alpha \in (0,1)$, where the integral is understood as $\lim_{u\to\infty} \int_0^u e^{\pm iv} v^{\alpha-1} dv$ and the convergence can be verified by applying integration by parts. Indeed, note that

$$\int_1^u e^{\pm iv} v^{\alpha-1} dv = (\mp i) \int_1^u v^{\alpha-1} de^{\pm iv} = (\mp i)\Big(u^{\alpha-1}e^{\pm iu} - e^{\pm i} - (\alpha-1)\int_1^u e^{\pm iv} v^{\alpha-2} dv\Big)$$

and the last expression has a limit as $u \to \infty$ since $\alpha < 1$ and $\int_1^\infty v^{\alpha-2} dv < \infty$.

Now consider, for example,

$$f_t(u) := f_{t,+}(u) = (t-u)_+^p - (-u)_+^p. \quad (2.6.7)$$

The cases $p < 0$ and $p > 0$ have to be considered separately. (The case $p = 0$ is trivial since it is interpreted by convention as $f_t(u) = 1_{[0,t)}(u)$.)

When $p < 0$, define a sequence of functions $f_{t,n}(u) = (t-u)_+^p 1_{\{t-u<n\}} - (-u)_+^p 1_{\{-u<n\}}$. Since $f_{t,n} \to f_t$ in $L^2(\mathbb{R})$, it is enough to show that $\widehat{f}_{t,n} \to \widehat{f}_t$ in $L^2(\mathbb{R})$, where $\widehat{f}_t = \widehat{f}_{t,+}$ is given in (2.6.4)–(2.6.5). Observe that

$$\widehat{f}_{t,n}(x) = \int_{\mathbb{R}} e^{ixu}(t-u)_+^p 1_{\{t-u<n\}} du - \int_{\mathbb{R}} e^{ixu}(-u)_+^p 1_{\{-u<n\}} du$$

$$= (e^{itx}-1)\int_{\mathbb{R}} e^{-ixs} s_+^p 1_{\{s<n\}} ds = (e^{itx}-1)\int_{\mathbb{R}} e^{-i\,\text{sign}(x)|x|s} s_+^p 1_{\{s<n\}} ds$$

$$= (e^{itx}-1)|x|^{-p-1}\int_{\mathbb{R}} e^{-i\,\text{sign}(x)v} v_+^p 1_{\{v<n|x|\}} dv = (e^{itx}-1)|x|^{-p-1}\int_0^{n|x|} e^{-i\,\text{sign}(x)v} v^p dv.$$

On the other hand, using the expression for $\widehat{f_t}(x)$ in (2.6.5) and the relation (2.6.6) with $\alpha = p+1$, we have

$$\widehat{f_t}(x) = \widehat{f_{t,+}}(x) = (e^{itx} - 1)|x|^{-p-1}\int_0^\infty e^{-i\,\mathrm{sign}(x)v}v^p dv,$$

so that

$$\widehat{f_t}(x) - \widehat{f_{t,n}}(x) = (e^{itx} - 1)|x|^{-p-1}\int_{n|x|}^\infty e^{-i\,\mathrm{sign}(x)v}v^p dv.$$

By integration by parts, $|\int_a^\infty e^{\pm iv}v^p dv| = |a^p e^{\pm ia} + p\int_a^\infty e^{\pm iv}v^{p-1}dv| \leq a^p + |p|\int_a^\infty v^{p-1}dv = 2a^p$. Hence,

$$|\widehat{f_t}(x) - \widehat{f_{t,n}}(x)| \leq C|e^{itx} - 1||x|^{-1}n^p \qquad (2.6.8)$$

and it follows that $\widehat{f_{t,n}} \to \widehat{f_t}$ in $L^2(\mathbb{R})$ as $n \to \infty$ since $p < 0$.

When $p > 0$, rewrite (2.6.7) as $f_t(u) = p\int_\mathbb{R} 1_{[0,t)}(v)(v-u)_+^{p-1}dv$ and consider

$$f_{t,n}(u) = p\int_\mathbb{R} 1_{[0,t)}(v)(v-u)_+^{p-1}1_{\{v-u<n\}}dv = p(g_t * h_n)(u),$$

where $g_t(v) = 1_{[0,t)}(v)$, $h_n(v) = (-v)_+^{p-1}1_{\{-v<n\}}$ and $*$ denotes the convolution (see Appendix A.1.2). As above, it is enough to show that $\widehat{f_{t,n}} \to \widehat{f_t}$ in $L^2(\mathbb{R})$. Observe that $\widehat{g_t}(x) = (e^{itx} - 1)/ix$ and

$$\widehat{h_n}(x) = \int_\mathbb{R} e^{ixv}(-v)_+^{p-1}1_{\{-v<n\}}dv = \int_0^n e^{-ixv}v^{p-1}dv = |x|^{-p}\int_0^{n|x|}e^{-i\,\mathrm{sign}(x)u}u^{p-1}du$$

so that

$$\widehat{f_{t,n}}(x) = p\frac{e^{itx} - 1}{ix}|x|^{-p}\int_0^{n|x|}e^{-i\,\mathrm{sign}(x)u}u^{p-1}du.$$

By using the second expression for $\widehat{f_t}(x)$ in (2.6.4) together with (2.6.6),

$$\widehat{f_t}(x) - \widehat{f_{t,n}}(x) = p\frac{e^{itx} - 1}{ix}|x|^{-p}\int_{n|x|}^\infty e^{-i\,\mathrm{sign}(x)u}u^{p-1}du.$$

It follows that $\widehat{f_{t,n}} \to \widehat{f_t}$ in $L^2(\mathbb{R})$ as $n \to \infty$, as in the case $p < 0$ above (by replacing p by $p - 1 < 0$). □

The following result provides an integral representation of FBM in the spectral domain. Hermitian Gaussian random measures are defined in Appendix B.1.

Proposition 2.6.11 *For $H \in (0,1)$, standard FBM admits the following integral representation:*

$$\{B_H(t)\}_{t\in\mathbb{R}} \stackrel{d}{=} \left\{\frac{1}{c_2(H)}\int_\mathbb{R}\frac{e^{itx} - 1}{ix}|x|^{1/2-H}\widehat{B}(dx)\right\}_{t\in\mathbb{R}}, \qquad (2.6.9)$$

where $c_2(H)^2 = \int_\mathbb{R}|e^{ix} - 1|^2|x|^{-2H-1}dx$ is a normalizing constant and $\widehat{B}(dx)$ is an Hermitian Gaussian random measure with the control measure $\mathbb{E}|\widehat{B}(dx)|^2 = dx$.

Proof We include two proofs. The first is based on Proposition 2.6.10. By Remark 2.6.8 with $a = b = 1$, the kernel $f_{t,+} + f_{t,-}$ provides a representation for FBM. By Proposition 2.6.10, its Fourier transform is

$$\widehat{f}_{t,+}(x) + \widehat{f}_{t,-}(x) = 2\Gamma(p+1)\cos(p\pi/2)\frac{e^{itx}-1}{ix}|x|^{-p},$$

since $i^p + i^{-p} = e^{ip\pi/2} + e^{-ip\pi/2} = 2\cos(p\pi/2)$. By setting $p = H - 1/2$, one obtains the representation (2.6.9) of FBM in the spectral domain (see (B.1.14) in Appendix B.1).

The second proof is based on Corollary 2.6.3. Denote the process on the right-hand side of (2.6.9) by $Y_H(t)$, without the normalizing constant. It is enough to check that Y_H satisfies conditions (i)–(iii) of Corollary 2.6.3. Condition (i) holds as long as Y_H is well-defined and real-valued. Y_H is real-valued since the deterministic integrand $g_t(x) = (e^{itx} - 1)|x|^{1/2-H}/(ix)$ satisfies $\overline{g_t(x)} = g_t(-x)$. It is well-defined since $g_t \in L^2(\mathbb{R})$.

Y_H satisfies (ii) of Corollary 2.6.3 since

$$\mathbb{E}(Y_H(t))^2 = \int_{\mathbb{R}} \left|\frac{e^{itx}-1}{ix}\right|^2 |x|^{1-2H}dx$$

$$= \int_{\mathbb{R}} |e^{i|t|x} - 1|^2 |x|^{-1-2H}dx = |t|^{2H}\int_{\mathbb{R}} |e^{iy} - 1|^2 |y|^{-1-2H}dy,$$

where we made the changes of variables $y = |t|x$. Similarly, Y_H satisfies (iii) of Corollary 2.6.3 since

$$\mathbb{E}(Y_H(t) - Y_H(s))^2 = \int_{\mathbb{R}} \left|\frac{e^{itx}-e^{isx}}{ix}\right|^2 |x|^{1-2H}dx$$

$$= \int_{\mathbb{R}} \left|\frac{e^{i(t-s)x}-1}{ix}\right|^2 |x|^{1-2H}|e^{isx}|^2 dx = \mathbb{E}(Y_H(t-s))^2. \quad \square$$

Note that the normalizing constant $c_2(H)^2$ in the proposition is there to guarantee that $\mathbb{E}B_H(1)^2 = 1$ and consequently

$$c_2(H)^2 = \int_{\mathbb{R}} \frac{|e^{ix}-1|^2}{x^2}|x|^{1-2H}dx = \int_{\mathbb{R}} |e^{ix} - 1|^2 |x|^{-2H-1}dx.$$

It can be expressed as

$$c_2(H)^2 = 4\int_0^{\infty} (1-\cos x)x^{-2H-1}dx = \frac{2}{H}\int_0^{\infty}\sin(x)x^{-2H}dx = \frac{2}{H}\Gamma(1-2H)\sin\left(\frac{(1-2H)\pi}{2}\right)$$

$$= \frac{2}{H}\Gamma(1-2H)\cos(H\pi) = \frac{\pi}{H\Gamma(2H)\sin(H\pi)} = \frac{2\pi}{\Gamma(2H+1)\sin(H\pi)} \quad (2.6.10)$$

by using (2.6.6) and, for the equality before last, (2.2.14) and $\sin(2x) = 2\sin(x)\cos(x)$.

Note also that the normalizing constants $c_1(H)$ and $c_2(H)$ can be related as follows. Let $f_t(u) = (t-u)_+^{H-1/2} - (-u)_+^{H-1/2}$ and $g_t(x) = (e^{itx} - 1)|x|^{1/2-H}/(ix)$ be the kernel functions in (2.6.1) and (2.6.9). Observe that by Proposition 2.6.10 with $p = H - 1/2$, $\widehat{f}_1(x) = \Gamma(H+1/2)g_1(x)e^{-i\text{sign}(x)(H-1/2)\pi/2}$. Then, by using Parseval's identity $\|f\|_{L^2(\mathbb{R})}^2 = \|\widehat{f}\|_{L^2(\mathbb{R})}^2/2\pi$ in (A.1.11) of Appendix A.1.2, we have

$$c_1(H)^2 = \|f_1\|_{L^2(\mathbb{R})}^2 = \frac{\|\widehat{f_1}\|_{L^2(\mathbb{R})}^2}{2\pi} = \frac{\Gamma(H+1/2)^2}{2\pi}\|g_1\|_{L^2(\mathbb{R})}^2 = \frac{\Gamma(H+1/2)^2}{2\pi}c_2(H)^2. \tag{2.6.11}$$

In particular,

$$c_1(H)^2 = \frac{\Gamma(H+1/2)^2}{\Gamma(2H+1)\sin(H\pi)} = \frac{B(H+1/2, H+1/2)}{\sin(H\pi)} = \frac{(H-1/2)B(H-1/2, 2-2H)}{2H}, \tag{2.6.12}$$

which is not obvious from the definition of the normalizing constant in the time domain. The first equality follows from (2.6.11) and (2.6.10), and the second from the definition (2.2.12) of the beta function. The last equality in (2.6.12) follows from

$$\frac{\Gamma(H+1/2)\Gamma(H+1/2)}{\Gamma(2H+1)\sin(H\pi)} = \frac{(H-1/2)^2\Gamma(H-1/2)\Gamma(H-1/2)}{2H\Gamma(2H)\sin(H\pi)}$$

$$= \frac{(H-1/2)^2\Gamma(H-1/2)\Gamma(H-1/2)\Gamma(1-2H)}{2H\Gamma(2H)\Gamma(1-2H)\sin(H\pi)}$$

$$= -\frac{(H-1/2)\Gamma(H-1/2)\Gamma(2-2H)\Gamma(H-1/2)\sin(2H\pi)}{4H\pi\sin(H\pi)}$$

$$= -\frac{(H-1/2)B(H-1/2, 2-2H)\Gamma(H-1/2)\Gamma(1-(H-1/2))\cos(H\pi)}{2H\pi}$$

$$= -\frac{(H-1/2)B(H-1/2, 2-2H)\cos(H\pi)}{2H\sin((H-1/2)\pi)} = \frac{(H-1/2)B(H-1/2, 2-2H)}{2H}.$$

Relation (2.2.14) was used for the third and fifth equalities, and the identity $\sin(2x) = 2\sin(x)\cos(x)$ was used for the fourth equality.

Figure 2.3 provides several realizations of standard FBM B_H on the interval $[0, 1]$ (in fact, on the dyadic grid $k/2^J$, $k = 0, 1, \ldots, 2^J$, with $J = 10$) which are obtained by using the circulant matrix embedding method discussed in Section 2.11.2 for generating the stationary increments of FBM. As H becomes larger, the sample paths of FBM become smoother. In fact, as noted in Section 7.1.1 of Chapter 7, the sample paths of FBM are H'–Hölder continuous for any $H' < H$ and hence indeed are smoother for larger H.

2.6.2 Bifractional Brownian Motion

Bifractional Brownian motion provides an interesting and useful extension of fractional Brownian motion.

Definition 2.6.12 *Bifractional Brownian motion* (biFBM) $B_{H,K} = \{B_{H,K}(t)\}_{t \geq 0}$, $H \in (0, 1]$, $K \in (0, 1]$, is a zero mean Gaussian process with covariance function

$$\mathbb{E}B_{H,K}(t)B_{H,K}(s) = \frac{1}{2^K}\Big((t^{2H} + s^{2H})^K - |t-s|^{2HK}\Big), \quad s, t \geq 0. \tag{2.6.13}$$

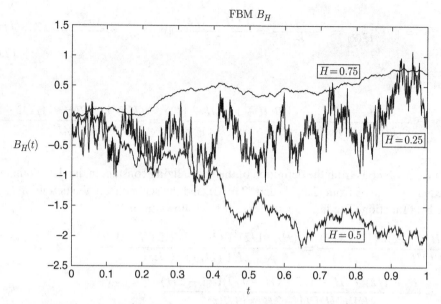

Figure 2.3 Realizations of FBM B_H with $H = 0.25$, $H = 0.5$ (Brownian motion) and $H = 0.75$.

In particular, $\mathbb{E}(B_{H,K}(t))^2 = t^{2HK}$. Note that biFBM is the usual FBM when $K = 1$. When $H = 1/2$, the covariance function (2.6.13) becomes

$$\mathbb{E}B_{1/2,K}(t)B_{1/2,K}(s) = \frac{1}{2^K}\left((t+s)^K - |t-s|^K\right), \quad s, t \geq 0, \qquad (2.6.14)$$

and can be checked to be that of $B_{K/2}(t) - B_{K/2}(-t), t \geq 0$, where $B_{K/2}$ is FBM with index $K/2$ and a suitable variance. See Section 3.12 for a physical model involving bifractional Brownian motion.

The fact that the right-hand side of (2.6.13) defines a valid covariance function is proved in Houdré and Villa [489], Proposition 2.1. It can also be shown (Bardina and Es-Sebaiy [100]) that the right-hand side of (2.6.13) is a valid covariance function when $H \in (0, 1)$, $HK \in (0, 1)$ and $K \in (1, 2)$.

The following result provides some basic properties of biFBM.

Proposition 2.6.13 *Let $B_{H,K}$, $H \in (0, 1]$, $K \in (0, 1]$, be biFBM. Then:*

(i) It is (HK)–self-similar.
(ii) It has stationary increments if and only if $K = 1$; that is, it is the usual FBM.
(iii) For all $s, t \geq 0$,

$$2^{-K}|t-s|^{2HK} \leq \mathbb{E}(B_{H,K}(t) - B_{H,K}(s))^2 \leq 2^{1-K}|t-s|^{2HK}. \qquad (2.6.15)$$

Proof Part (i) follows from the fact that, for any $c > 0$,

$$\mathbb{E}B_{H,K}(ct)B_{H,K}(cs) = c^{2HK}\mathbb{E}B_{H,K}(t)B_{H,K}(s).$$

Part (ii) follows from the relation

$$\mathbb{E}(B_{H,K}(t) - B_{H,K}(s))^2 = \frac{2}{2^K}|t-s|^{2HK} + \left(t^{2HK} + s^{2HK} - \frac{2}{2^K}(t^{2H} + s^{2H})^K\right), \quad (2.6.16)$$

which depends on $t - s$ only if and only if $K = 1$ (see also (2.5.3)). By using (2.6.16), the left-hand side of (2.6.15) in part (iii), for example, is equivalent to

$$|t-s|^{2HK} + 2^K t^{2HK} + 2^K s^{2HK} - 2(t^{2H} + s^{2H})^K \geq 0.$$

This holds for $s = 0$ and, taking $u = t/s$ when $s \neq 0$, this is equivalent to

$$f(u) := |u-1|^{2HK} + 2^K u^{2HK} + 2^K - 2(u^{2H} + 1)^K \geq 0.$$

Proving the latter inequality is a calculus exercise (see Houdré and Villa [489], Proposition 3.1, for details). □

Further insight into biFBM can be gained from the following decomposition result due to Lei and Nualart [603]. For $K \in (0, 1)$, set

$$X_K(t) = \int_0^\infty (1 - e^{-ut}) u^{-\frac{1+K}{2}} B(du), \quad t \geq 0, \quad (2.6.17)$$

where $B(du)$ is a Gaussian random measure on $(0, \infty)$ with the control measure du. It can be checked that the covariance function of X_K is given by

$$\mathbb{E} X_K(t) X_K(s) = \frac{\Gamma(1-K)}{K}\left(t^K + s^K - (t+s)^K\right), \quad s, t \geq 0. \quad (2.6.18)$$

Proposition 2.6.14 *Let $B_{H,K}$, $H \in (0,1]$, $K \in (0,1)$, be biFBM and let the process X_K be defined by (2.6.17) and independent of $B_{H,K}$. Then,*

$$\left\{2^{\frac{K-1}{2}} B_{H,K}(t) + \sqrt{\frac{K}{2\Gamma(1-K)}} X_K(t^{2H})\right\}_{t \geq 0} \stackrel{d}{=} \{B_{HK}(t)\}_{t \geq 0}, \quad (2.6.19)$$

where B_{HK} is a standard FBM with index HK.

Proof Write the covariance function (2.6.13) of biFBM as

$$\mathbb{E} B_{H,K}(t) B_{H,K}(s) = \frac{1}{2^K}\left((t^{2H} + s^{2H})^K - t^{2HK} - s^{2HK}\right) + \frac{1}{2^K}\left(t^{2HK} + s^{2HK} - |t-s|^{2HK}\right).$$

Then,

$$\mathbb{E} B_{HK}(t) B_{HK}(s) = \frac{1}{2}\left(t^{2HK} + s^{2HK} - |t-s|^{2HK}\right)$$

$$= 2^{K-1} \mathbb{E} B_{H,K}(t) B_{H,K}(s) + \frac{1}{2}\left((t^{2H})^K + (s^{2H})^K - (t^{2H} + s^{2H})^K\right)$$

$$= 2^{K-1} \mathbb{E} B_{H,K}(t) B_{H,K}(s) + \frac{K}{2\Gamma(1-K)} \mathbb{E} X_K(t^{2H}) X_K(s^{2H}).$$

This yields the decomposition (2.6.19). □

The process X_K can be shown to have a version with paths that are infinitely differentiable (Lei and Nualart [603], Theorem 2). This should not be surprising since informally differentiating under the integral sign in (2.6.17) leads to a well-defined process. Then, Proposition 2.6.14 shows that biFBM with indices H and K can be thought as FBM with index HK up to an infinitely differentiable process with a fractional time change.

2.6.3 The Rosenblatt Process

We shall give next an example of a non-Gaussian SSSI process with finite variance. We shall use the following proposition. The $L^2(\mathbb{R}^2)$–Fourier transform is discussed in Appendix A.1.2.

Proposition 2.6.15 *Let* $1/4 < p < 1/2$. *Then, the function*

$$f_t(u_1, u_2) = \int_0^t (s - u_1)_+^{p-1} (s - u_2)_+^{p-1} ds$$

is in $L^2(\mathbb{R}^2)$ *with*

$$\|f_t\|_{L^2(\mathbb{R}^2)}^2 = \frac{B(p, 1-2p)^2}{(4p-1)2p} |t|^{4p}, \qquad (2.6.20)$$

and its $L^2(\mathbb{R}^2)$–*Fourier transform is*

$$\widehat{f_t}(x_1, x_2) = \Gamma(p)^2 \frac{e^{it(x_1+x_2)} - 1}{i(x_1 + x_2)} |x_1|^{-p} |x_2|^{-p} e^{-i\,\mathrm{sign}(x_1)\frac{p\pi}{2}} e^{-i\,\mathrm{sign}(x_2)\frac{p\pi}{2}}. \qquad (2.6.21)$$

Proof It is enough to prove the result for $t = 1$. Observe that, using the assumption $1/4 < p < 1/2$,

$$\|f_1\|_{L^2(\mathbb{R}^2)}^2 = \int_{\mathbb{R}} du_1 \int_{\mathbb{R}} du_2 \Big(\int_0^1 (s - u_1)_+^{p-1} (s - u_2)_+^{p-1} ds \Big)^2$$

$$= \int_{\mathbb{R}} du_1 \int_{\mathbb{R}} du_2 \int_0^1 ds_1 \int_0^1 ds_2 (s_1 - u_1)_+^{p-1} (s_1 - u_2)_+^{p-1} (s_2 - u_1)_+^{p-1} (s_2 - u_2)_+^{p-1}$$

$$= \int_0^1 ds_1 \int_0^1 ds_2 \Big(\int_{\mathbb{R}} (s_1 - u)_+^{p-1} (s_2 - u)_+^{p-1} du \Big)^2$$

$$= 2 \int_0^1 ds_1 \int_{s_1}^1 ds_2 \Big(\int_0^\infty z^{p-1} (s_2 - s_1 + z)^{p-1} dz \Big)^2$$

$$= 2 \Big(\int_0^\infty v^{p-1} (1+v)^{p-1} dv \Big)^2 \int_0^1 ds_1 \int_{s_1}^1 ds_2 (s_2 - s_1)^{4p-2}$$

$$= 2 C_p^2 \int_0^1 ds_1 \int_{s_1}^1 ds_2 (s_2 - s_1)^{4p-2} = \frac{C_p^2}{(4p-1)2p},$$

since $4p - 1 > 0$. Here $C_p = \int_0^\infty v^{p-1}(1+v)^{p-1} dv = B(p, 1-2p)$ by (2.2.12) and the fact that $1 - 2p > 0$.

2.6 Examples of Self-Similar Processes

For the Fourier transform, let $f_{1,n}(u_1, u_2) = \int_0^1 (s - u_1)_+^{p-1} 1_{\{s - u_1 < n\}} (s - u_2)_+^{p-1} 1_{\{s - u_2 < n\}} ds$. It converges to $f_1(u_1, u_2)$ in $L^2(\mathbb{R}^2)$ by arguing as above and using the dominated convergence theorem. The Fourier transform of $f_{1,n}(u_1, u_2)$ is

$$\widehat{f_{1,n}}(x_1, x_2) = \int_\mathbb{R} \int_\mathbb{R} e^{i(x_1 u_1 + x_2 u_2)} f_{1,n}(u_1, u_2) du_1 du_2$$

$$= \int_0^1 e^{i(x_1 + x_2)s} ds \int_0^n e^{-i x_1 v_1} v_1^{p-1} dv_1 \int_0^n e^{-i x_2 v_2} v_2^{p-1} dv_2$$

$$= \frac{e^{i(x_1 + x_2)} - 1}{i(x_1 + x_2)} |x_1|^{-p} |x_2|^{-p} \int_0^{n|x_1|} e^{-i \,\text{sign}(x_1) u_1} u_1^{p-1} du_1 \int_0^{n|x_2|} e^{-i \,\text{sign}(x_2) u_2} u_2^{p-1} du_2.$$

Finally, note that this converges to $\widehat{f_1}(x_1, x_2)$ for each $x_1 \neq 0, x_2 \neq 0$, as $n \to \infty$, by using the relations (2.6.8) and (2.6.6). □

The following is a corollary of Proposition 2.6.15 (see Appendix B.2 for definitions of the multiple integrals with respect to Gaussian random measures).

Proposition 2.6.16 *Let $H_0 \in (3/4, 1)$ and $H = 2H_0 - 1 \in (1/2, 1)$. Let also $B(du)$ be a real-valued Gaussian random measure on $\mathbb{R}$ with the control measure $\mathbb{E}(B(du))^2 = du$, and $\widehat{B}(dx)$ be an Hermitian Gaussian random measure on $\mathbb{R}$ with the control measure $\mathbb{E}|\widehat{B}(dx)|^2 = dx$. Then, the process*

$$\{Z_H(t)\}_{t \in \mathbb{R}} \stackrel{d}{=} \left\{ \frac{1}{d_1^0(H_0)} \int_{\mathbb{R}^2}' \left(\int_0^t (s - u_1)_+^{H_0 - 3/2} (s - u_2)_+^{H_0 - 3/2} ds \right) B(du_1) B(du_2) \right\}_{t \in \mathbb{R}} \quad (2.6.22)$$

$$\stackrel{d}{=} \left\{ \frac{1}{d_2^0(H_0)} \int_{\mathbb{R}^2}'' \frac{e^{it(x_1 + x_2)} - 1}{i(x_1 + x_2)} |x_1|^{1/2 - H_0} |x_2|^{1/2 - H_0} \widehat{B}(dx_1) \widehat{B}(dx_2) \right\}_{t \in \mathbb{R}} \quad (2.6.23)$$

is an H–SSSI process, where the normalizing constants are related as

$$d_1^0(H_0) = \frac{\Gamma(H_0 - 1/2)^2}{2\pi} d_2^0(H_0). \quad (2.6.24)$$

Moreover, $\mathbb{E} Z_H(1)^2 = 1$ if

$$(d_1^0(H_0))^2 = \frac{2 B(H_0 - 1/2, 2 - 2H_0)^2}{(2H_0 - 1)(4H_0 - 3)} \quad (2.6.25)$$

or, equivalently, if

$$(d_2^0(H_0))^2 = \frac{2(2\Gamma(2 - 2H_0) \cos(H_0 \pi))^2}{(2H_0 - 1)(4H_0 - 3)}. \quad (2.6.26)$$

Proof By setting $p - 1 = H_0 - 3/2$, one gets $p = H_0 - 1/2 \in (1/4, 1/2)$ since $H_0 \in (3/4, 1)$. The representations (2.6.22) and (2.6.23) are well-defined and equal in distribution. The equality in distribution follows from (B.2.16) in Appendix B and Proposition 2.6.15, since the kernels in the right-hand sides of (2.6.22) and (2.6.23) are related through the

Fourier transform. Note also from Proposition 2.6.15 that the Fourier transform of the kernel in (2.6.22) produces the complex exponentials

$$e^{-i\operatorname{sign}(x_1)\frac{(H_0-1/2)\pi}{2}} e^{-i\operatorname{sign}(x_2)\frac{(H_0-1/2)\pi}{2}}$$

which do not appear in the right-hand side of (2.6.23). The exponentials can be omitted by the formula (B.2.15) for the change of variables found in Appendix B.

The process defined by the integral representation (2.6.22) can be shown to have stationary increments by using the fact that, for fixed $h \in \mathbb{R}$, the random measures $B(du)$ and $B(d(u+h))$ have the same properties. Similarly, the process is also H–SS by using the fact that, for fixed $c > 0$, the random measures $B(d(cu))$ and $c^{1/2}B(du)$ have the same properties. (Strictly speaking, it is necessary to use the change of variables formula (B.2.11) found in Appendix B.) The normalizations (2.6.25) and (2.6.26) follow from (2.6.20). □

Definition 2.6.17 The process $\{Z_H(t)\}_{t\in\mathbb{R}}$, $H \in (1/2, 1)$, in (2.6.22)–(2.6.23) is called the *Rosenblatt process*.

Remark 2.6.18 It is important to distinguish between the self-similarity parameter of fractional Brownian motion, which appears as H_0 in Proposition 2.6.16, and the self-similarity parameter H of the Rosenblatt process. Observe that the exponent $H_0 - 3/2$ appears in the representation (2.6.2) of FBM. Similarly, the exponent $1/2 - H_0$ appears in the integral representation (2.6.9) of FBM. As indicated in the proposition, we have

$$H_0 = \frac{H+1}{2} \in \left(\frac{3}{4}, 1\right) \quad \text{and} \quad H = 2H_0 - 1 \in \left(\frac{1}{2}, 1\right). \tag{2.6.27}$$

It is sometimes desirable to express the formulas in Proposition 2.6.16 by using the self-similarity parameter H instead of H_0. Thus,

$$\{Z_H(t)\}_{t\in\mathbb{R}} \stackrel{d}{=} \left\{\frac{1}{d_1(H)} \int_{\mathbb{R}^2}' \left(\int_0^t (s-u_1)_+^{-\frac{2-H}{2}} (s-u_2)_+^{-\frac{2-H}{2}} ds\right) B(du_1)B(du_2)\right\}_{t\in\mathbb{R}}$$

$$\stackrel{d}{=} \left\{\frac{1}{d_2(H)} \int_{\mathbb{R}^2}'' \frac{e^{it(x_1+x_2)} - 1}{i(x_1+x_2)} |x_1|^{-\frac{H}{2}} |x_2|^{-\frac{H}{2}} \widehat{B}(dx_1)\widehat{B}(dx_2)\right\}_{t\in\mathbb{R}} \tag{2.6.28}$$

is H–SSSI process, where the normalizing constants are related as

$$d_1(H) = \frac{\Gamma(H/2)^2}{2\pi} d_2(H). \tag{2.6.29}$$

Moreover, $\mathbb{E}Z_H(1)^2 = 1$ if

$$d_1(H)^2 = \frac{B(H/2, 1-H)^2}{(H/2)(2H-1)} \tag{2.6.30}$$

or, equivalently, if

$$d_2(H)^2 = \frac{(2\Gamma(1-H)\sin(H\pi/2))^2}{(H/2)(2H-1)}. \tag{2.6.31}$$

2.6.4 SαS Lévy Motion

This is an example of a SSSI process with infinite variance. The idea is simple. Brownian motion can be represented as $B(t) = \int_0^t B(du)$, namely as the integral of Gaussian noise. Let us replace the Gaussian noise $B(du)$ by a symmetric α-stable ($S\alpha S$) noise $M_\alpha(du)$ with $0 < \alpha < 2$ (see Appendix B.1 for a definition).

Definition 2.6.19 The $S\alpha S$ process

$$S_\alpha(t) = \int_0^t M_\alpha(du), \quad t \geq 0,$$

where $M_\alpha(du)$ is a $S\alpha S$ random measure with control measure du is called the $S\alpha S$ Lévy motion.

The $S\alpha S$ Lévy motion $S_\alpha(t)$ has independent increments because $M_\alpha(A)$ and $M_\alpha(B)$ are independent random variables if $A \cap B = \emptyset$. It is self-similar with

$$H = \frac{1}{\alpha} \in \left(\frac{1}{2}, \infty\right)$$

because for any $c > 0$, $M_\alpha(cdu) \stackrel{d}{=} c^{1/\alpha} M_\alpha(du)$. A more precise way of showing the self-similarity is by using the characteristic function. For any $t_1, \ldots, t_J \geq 0, \theta_1, \ldots, \theta_J \in \mathbb{R}$,

$$\mathbb{E} \exp\left\{i \sum_{j=1}^J \theta_j S_\alpha(ct_j)\right\} = \exp\left\{-\int_\mathbb{R} \Big|\sum_{j=1}^J \theta_j 1_{[0,ct_j)}(u)\Big|^\alpha du\right\} = \exp\left\{-\int_\mathbb{R} \Big|\sum_{j=1}^J \theta_j 1_{[0,t_j)}(u)\Big|^\alpha cdu\right\}$$

$$= \exp\left\{-\int_\mathbb{R} \Big|\sum_{j=1}^J \theta_j c^{1/\alpha} 1_{[0,t_j)}(u)\Big|^\alpha du\right\} = \mathbb{E} \exp\left\{i \sum_{j=1}^J \theta_j c^{1/\alpha} S_\alpha(t_j)\right\}.$$

Remark 2.6.20 If $\alpha = 2$, the $S\alpha S$ Lévy motion becomes Brownian motion. For $t < 0$, one defines the $S\alpha S$ Lévy motion as $S_\alpha(t) = \int_t^0 M_\alpha(du), t < 0$. One could also replace the $S\alpha S$ measure $M_\alpha(du)$ by a skewed one (see Appendix B.1).

2.6.5 Linear Fractional Stable Motion

Here is another example of SSSI process with infinite variance. The basic idea is to replace the Gaussian noise in the integral representations of FBM by a $S\alpha S$ noise (see Appendix B.1 for a definition).

Proposition 2.6.21 For $\alpha \in (0, 2), H \in (0, 1), H \neq 1/\alpha$ and $a, b \in \mathbb{R}$, the process

$$\{L_{\alpha, H}(t)\}_{t \in \mathbb{R}} \stackrel{d}{=} \left\{\int_\mathbb{R} f_t(u) M_\alpha(du)\right\}_{t \in \mathbb{R}}, \tag{2.6.32}$$

where $M_\alpha(du)$ is a $S\alpha S$ random measure on $\mathbb{R}$ with the Lebesgue control measure du, and

$$f_t(u) = a\big((t-u)_+^{H-1/\alpha} - (-u)_+^{H-1/\alpha}\big) + b\big((t-u)_-^{H-1/\alpha} - (-u)_-^{H-1/\alpha}\big), \tag{2.6.33}$$

is $S\alpha S$ and H–SSSI.

Remark 2.6.22 When $H = 1/\alpha$, the exponents in (2.6.33) become $H - 1/\alpha = 0$ and the following convention is used,

$$a_\pm^0 = (a_\pm)^0, \quad a^0 = \begin{cases} 0, & \text{if } a = 0, \\ 1, & \text{if } a \neq 0. \end{cases} \qquad (2.6.34)$$

With this convention, for $t > 0$,

$$(t-u)_+^0 - (-u)_+^0 = 1_{[0,t)}(u), \quad (t-u)_-^0 - (-u)_-^0 = -1_{(0,t]}(u). \qquad (2.6.35)$$

When $H = 1/\alpha$, the kernel function thus becomes

$$f_t(u) = (a-b)1_{[0,t)}(u), \quad u \neq 0, t.$$

Hence, when $a \neq b$ and $H = 1/\alpha$, the process $L_{\alpha, H}$ is the $S\alpha S$ Lévy motion.

Proof $L_{\alpha,H}$ is a $S\alpha S$ stochastic process as long as it is well-defined or $f_t \in L^\alpha(\mathbb{R})$. Around $u = t$, for example, $|f_t(u)|^\alpha$ behaves as (up to a constant) $(|t-u|^{H-1/\alpha})^\alpha = |t-u|^{H\alpha-1}$, which is integrable since $(H\alpha - 1) + 1 = H\alpha > 0$. Around $u = \infty$, $|f_t(u)|^\alpha$ behaves as (up to a constant) $(|u|^{H-1/\alpha-1})^\alpha = |u|^{H\alpha-1-\alpha}$, which is integrable since $(H\alpha - 1 - \alpha) + 1 = (H-1)\alpha < 0$. $L_{\alpha,H}$ has stationary increments by using the property $M_\alpha(d(u+h)) \stackrel{d}{=} M_\alpha(du)$. H–SS follows from the property $M_\alpha(d(cu)) \stackrel{d}{=} c^{1/\alpha} M_\alpha(du)$. To understand H–SS in a more rigorous way, observe that

$$\mathbb{E} \exp\left\{i \sum_{j=1}^J \theta_j L_{\alpha,H}(ct_j)\right\} = \mathbb{E} \exp\left\{i \sum_{j=1}^J \theta_j \int_\mathbb{R} f_{ct_j}(u) M_\alpha(du)\right\}$$

$$= \exp\left\{-\int_\mathbb{R} \Big|\sum_{j=1}^J \theta_j f_{ct_j}(u)\Big|^\alpha du\right\} = \exp\left\{-\int_\mathbb{R} \Big|\sum_{j=1}^J \theta_j c^H f_{t_j}(v)\Big|^\alpha dv\right\}$$

$$= \mathbb{E} \exp\left\{i \sum_{j=1}^J \theta_j c^H L_{\alpha,H}(t_j)\right\},$$

since $f_{ct}(cv) = c^{H-1/\alpha} f_t(v)$. □

Definition 2.6.23 The process $L_{\alpha,H} = \{L_{\alpha,H}(t)\}_{t \in \mathbb{R}}$, $\alpha \in (0,2)$, $H \in (0,1)$, $H \neq 1/\alpha$, in (2.6.32) is called a *linear fractional stable motion* (LFSM).

We should also note that, in contrast to the Gaussian case of FBM, there are infinitely many different $S\alpha S$ H–SSSI processes (for fixed α and H). More details can be found in Pipiras and Taqqu [820].

2.6 Examples of Self-Similar Processes

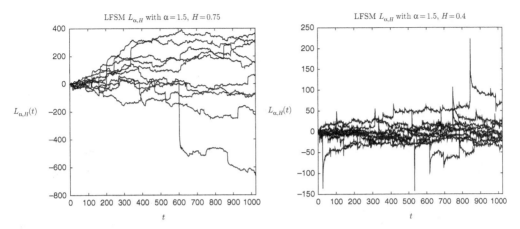

Figure 2.4 Realizations of LFSM $L_{\alpha,H}$. Left: $\alpha = 1.5$, $H = 0.75$. Right: $\alpha = 1.5$, $H = 0.4$.

Remark 2.6.24 One can replace the $S\alpha S$ measure $M_\alpha(du)$ by a skewed α-stable measure (see Appendix B.1). Note also that, since $L_{\alpha,H}(1)$ is a $S\alpha S$ random variable, its characteristic functions is $\mathbb{E}e^{i\theta L_{\alpha,H}(1)} = e^{-\sigma^\alpha |\theta|^\alpha}$ for a scale parameter $\sigma > 0$. LFSM $L_{\alpha,H}$ is called *standard* when $\sigma = 1$.

Figure 2.4 depicts sample paths of LFSM when $\alpha = 1.5$, $H = 0.75$ and $\alpha = 1.5$, $H = 0.4$. They are obtained using an approximate method based on FFT developed in Stoev and Taqqu [926]. When $H - 1/\alpha > 0$ (the left plot in the figure), the sample paths of LFSM can be shown to be continuous (e.g., Samorodnitsky and Taqqu [883], Example 12.2.3).

2.6.6 Log-Fractional Stable Motion

We used the convention (2.6.34) to conclude that LFSM is the $S\alpha S$ Lévy motion when $H = 1/\alpha$. Another way to proceed is to let $H \to 1/\alpha$. This approach leads to another $S\alpha S$ SSSI process. More specifically, let

$$a = b = 1$$

in (2.6.33), and consider

$$\frac{1}{H - 1/\alpha} L_{\alpha,H}(t) = \int_{\mathbb{R}} \left(\frac{|t-u|^{H-1/\alpha} - 1}{H - 1/\alpha} - \frac{|u|^{H-1/\alpha} - 1}{H - 1/\alpha} \right) M_\alpha(du).$$

The fact that, for any $u \neq 0$,

$$\lim_{H \to 1/\alpha} \frac{|u|^{H-1/\alpha} - 1}{H - 1/\alpha} = \lim_{H \to 1/\alpha} \frac{e^{(H-1/\alpha)\log|u|} - 1}{H - 1/\alpha} = \log|u|,$$

suggests the following definition.

Definition 2.6.25 Let $1 < \alpha < 2$ and $M_\alpha(du)$ be a $S\alpha S$ random measure on $\mathbb{R}$ with the Lebesgue control measure du. Then,

$$\int_\mathbb{R} (\log|t-u| - \log|u|) M_\alpha(du), \quad t \in \mathbb{R}, \tag{2.6.36}$$

is called a *log-fractional stable motion* (log-FSM).

It is easy to show that log-FSM is well-defined, has stationary increments and is self-similar with $H = 1/\alpha$ (see Exercise 2.15, or Samorodnitsky and Taqqu [883], Proposition 7.3.6).

Remark 2.6.26 If we set $\alpha = 2$, then log-FSM becomes Gaussian and is H–SSSI with $H = 1/2$. The resulting process must therefore be Brownian motion. The expression (2.6.36) then provides an alternative representation for Brownian motion.

2.6.7 The Telecom Process

The Telecom process appears as a limit of stochastic processes used to model computer workloads (see Section 3.3). It is a δ-stable process[9] with $1 < \delta < 2$. Its integral representation involves a δ-stable measure defined not on $\mathbb{R}$ but on $\mathbb{R} \times (0, \infty)$.

Definition 2.6.27 Let $1 < \gamma < \delta < 2$, $c > 0$ and

$$K_t(s, u) = (t-s)_+ \wedge u - (-s)_+ \wedge u, \quad t, s \in \mathbb{R}, \, u > 0,$$

where $x \wedge y = \min\{x, y\}$. The *Telecom process* is

$$Z_{\gamma,\delta}(t) = c \int_{-\infty}^\infty \int_0^\infty K_t(s, u) u^{-(1+\gamma)/\delta} M_\delta(ds, du), \quad t \in \mathbb{R}, \tag{2.6.37}$$

where $M_\delta(ds, du)$ is a δ-stable, totally skewed to the right random measure on $\mathbb{R} \times (0, \infty)$ with the control measure[10] $ds\,du$.

Properties of the kernel $K_t(s, u)$ are given in Lemma 3.3.1 of Chapter 3. Observe that, for $t > 0$,

$$Z_{\gamma,\delta}(t) = c \int_{-\infty}^0 \int_0^\infty \left(t \wedge (s+u)_+\right) u^{-(1+\gamma)/\delta} M_\delta(ds, du)$$
$$+ c \int_0^t \int_0^\infty \left((t-s) \wedge u\right) u^{-(1+\gamma)/\delta} M_\delta(ds, du). \tag{2.6.38}$$

---

[9] We use δ instead of α in this context, as is done in Section 3.3.
[10] In Section 3.3, $M_\delta(ds, du)$ is defined with control measure $ds u^{-(1+\gamma)} du$, and therefore the integral representation of the Telecom process does not contain the factor $u^{-(1+\gamma)/\delta}$.

Proposition 2.6.28 *The Telecom process is well-defined, has stationary increments and is H–self-similar with parameter*

$$H = \frac{\delta + 1 - \gamma}{\delta} \in \left(\frac{1}{\delta}, 1\right).$$

Proof The integral (2.6.37) is well-defined when $K_t(s, u)u^{-(1+\gamma)/\delta} \in L^\delta(\mathbb{R} \times \mathbb{R}_0, dsdu)$. This can be shown by considering the five regions of s and u in (3.3.11) below. Stationarity of the increments can be shown using the facts that $K_{t+h}(s, u) - K_h(s, u) = K_t(s - h, u)$ and that $M_\delta(d(s + h), du) \stackrel{d}{=} M_\delta(ds, du)$. Similarly, the H–self-similarity is a consequence of the facts that $K_{ct}(s, u)u^{-(1+\gamma)/\delta} = c^H K_t(s/c, u/c)(u/c)^{-(1+\gamma)/\delta} c^{-2/\delta}$ and that $M_\delta(d(cs), d(cu)) \stackrel{d}{=} c^{2/\delta} M_\delta(ds, du)$. □

It can be shown (Pipiras and Taqqu [820]) that LFSM and the Telecom processes, with M_δ being $S\delta S$, have different finite-dimensional distributions (for the same H and $\delta = \alpha$).

Remark 2.6.29 Suppose we set $\delta = 2$ in Definition 2.6.27 of the Telecom process. Then, the stable random measure becomes Gaussian. We can then apply Corollary 2.6.3. We conclude that the resulting process $Z_{\gamma,2}(t)$ is FBM because it is Gaussian, has stationary increments and is self-similar with

$$H = \frac{3 - \gamma}{2} \in \left(\frac{1}{2}, 1\right).$$

Thus, (2.6.37) with $\delta = 2$ provides an alternative representation of FBM.

2.6.8 Linear Fractional Lévy Motion

FBM and linear fractional stable motion have similar linear integral representations (2.6.1) and (2.6.32) with respect to Gaussian and $S\alpha S$ random measures, respectively. We introduce here a process having this type of linear integral representation but with respect to a symmetric Lévy random measure. Symmetric Lévy random measures and integrals with respect to them are defined in Appendix B.1.5. The resulting process is not exactly self-similar but it has the same covariance structure as FBM.

Definition 2.6.30 A *linear fractional Lévy motion* is defined as the process

$$\{Z_{\rho,H}(t)\}_{t\in\mathbb{R}} \stackrel{d}{=} \left\{a \int_\mathbb{R} \left((t - x)_+^{H-1/2} - (-x)_+^{H-1/2}\right) Z(dx)\right\}_{t\in\mathbb{R}}, \quad (2.6.39)$$

where $H \in (0, 1)$, $a \in \mathbb{R}$ is a constant and Z is a symmetric Lévy random measure on $\mathbb{R}$ with the Lebesgue control measure $m(dx) = dx$ and a symmetric Lévy measure ρ satisfying $\int_\mathbb{R} u^2 \rho(du) < \infty$.

By (B.1.45), a linear fractional Lévy motion has the same covariance structure as FBM; that is,

$$\mathbb{E} Z_{\rho,H}(t) Z_{\rho,H}(s) = \frac{a^2 \int_\mathbb{R} u^2 \rho(du)}{2} \left(|t|^{2H} + |s|^{2H} - |s - t|^{2H}\right), \quad s, t \in \mathbb{R}, \quad (2.6.40)$$

in addition to having zero mean $\mathbb{E} Z_{\rho,H}(t) = 0$. The process $Z_{\rho,H}$ is not expected to be exactly self-similar. Indeed, note from (B.1.43) that the characteristic function of $Z_{\rho,H}(t)$ is given by: for $\theta \in \mathbb{R}$,

$$\mathbb{E} e^{i\theta Z_{\rho,H}(t)} = \exp\left\{-\int_{\mathbb{R}}\int_{\mathbb{R}}\left(1 - \cos\left(\theta u a((t-x)_+^{H-1/2} - (-x)_+^{H-1/2})\right)\right)dx\rho(du)\right\}. \tag{2.6.41}$$

Since $\rho(du)$ does not scale – that is, it does not satisfy $\rho(d(cu)) = c^p \rho(du)$ for some $p \in \mathbb{R}$ – there is no reason a priori for $Z_{\rho,H}(t)$ and $t^H Z_{\rho,H}(1)$ to have the same characteristic function (or the same distribution).

Though $Z_{\rho,H}$ is not self-similar, it is *asymptotically self-similar at infinity* in the sense of (2.6.42) in the next proposition.

Proposition 2.6.31 *With the notation and assumptions above,*[11]

$$\frac{Z_{\rho,H}(Rt)}{R^H} \xrightarrow{fdd} \sigma_H B_H(t), \quad t \in \mathbb{R}, \tag{2.6.42}$$

as $R \to \infty$, where $\sigma_H = |a|(\int_{\mathbb{R}} u^2 \rho(du))^{1/2} c_1(H)$ with $c_1(H)$ given in (2.6.1) and (2.6.12), and B_H is a standard FBM with index H.

Proof Let $f_t(x) = a((t-x)_+^{H-1/2} - (-x)_+^{H-1/2})$ be the deterministic kernel function of a linear fractional Lévy motion appearing in (2.6.39). The characteristic function of a linear combination $\sum_{j=1}^J \theta_j Z_{\rho,H}(Rt_j)/R^H$, $\theta_j, t_j \in \mathbb{R}$, is

$$\mathbb{E}\exp\left\{i\sum_{j=1}^J \theta_j Z_{\rho,H}(Rt_j)/R^H\right\} = \mathbb{E}\exp\left\{i\int_{\mathbb{R}}\left(\sum_{j=1}^J \theta_j f_{Rt_j}(x)/R^H\right)Z(dx)\right\}$$

$$= \exp\left\{-\int_{\mathbb{R}}\int_{\mathbb{R}}\left(1-\cos\left(u\sum_{j=1}^J \theta_j f_{Rt_j}(x)/R^H\right)\right)dx\rho(du)\right\}$$

$$= \exp\left\{-\int_{\mathbb{R}}\int_{\mathbb{R}}\left(1-\cos\left(u\sum_{j=1}^J \theta_j f_{t_j}(y)/R^{1/2}\right)\right)Rdy\rho(du)\right\}$$

$$=: \exp\left\{-\int_{\mathbb{R}}\int_{\mathbb{R}} g_R(u,y)dy\rho(du)\right\},$$

after the change of variables $x = Ry$. As $R \to \infty$, $g_R(u,y) \to \frac{u^2}{2}(\sum_{j=1}^J \theta_j f_{t_j}(y))^2$ and $|g_R(u,y)| \leq \frac{u^2}{2}(\sum_{j=1}^J \theta_j f_{t_j}(y))^2$. The dominated convergence theorem implies that the characteristic function converges to

$$\exp\left\{-\frac{1}{2}\int_{\mathbb{R}} u^2 \rho(du) \int_{\mathbb{R}}\left(\sum_{j=1}^J \theta_j f_{t_j}(y)\right)^2 dy\right\},$$

which yields (2.6.42). □

---

[11] "fdd" stands for "finite-dimensional distributions".

Under additional assumptions on ρ, $Z_{\rho,H}$ is also *locally asymptotically self-similar* in the sense of (2.6.43) in the following proposition.

Proposition 2.6.32 *With the notation and assumptions above, assume in addition that the symmetric Lévy measure ρ is given by*

$$\rho(du) = 1_{\{|u|\leq 1\}} \frac{du}{|u|^{1+\alpha}},$$

where $\alpha \in (0, 2)$, and that $\widetilde{H}$ satisfying $\widetilde{H} - 1/\alpha = H - 1/2$ is such that $\widetilde{H} \in (0, 1)$. Then,

$$\frac{Z_{\rho,H}(\epsilon s)}{\epsilon^{\widetilde{H}}} \xrightarrow{fdd} L_{\alpha,\widetilde{H}}(s), \quad s \in \mathbb{R}, \tag{2.6.43}$$

where $L_{\alpha,\widetilde{H}}$ is LFSM with H, a, b in (2.6.33) replaced by $\widetilde{H}$, $aC(\alpha)^{1/\alpha}$ and 0, respectively, with $C(\alpha) = 2\int_0^\infty (1 - \cos w) w^{-1-\alpha} dw$.

Remark 2.6.33 The same convergence (2.6.43) holds for $Z_{\rho,H}(\epsilon s)$ replaced by $Z_{\rho,H}(t + \epsilon s) - Z_{\rho,H}(t)$, for fixed t (Cohen and Istas [253], Proposition 4.4.5). We also note that the constant $C(\alpha)$ above can be computed as at the end of the proof of Proposition 2.2.24.

Proof As in the proof of Proposition 2.6.31, using its notation, the characteristic function of a linear combination $\sum_{j=1}^{J} \theta_j Z_{\rho,H}(\epsilon s_j)/\epsilon^{\widetilde{H}}$, $\theta_j, s_j \in \mathbb{R}$, is

$$\mathbb{E} \exp\left\{i \sum_{j=1}^{J} \theta_j Z_{\rho,H}(\epsilon s_j)/\epsilon^{\widetilde{H}}\right\} = \exp\left\{-\int_{\mathbb{R}}\int_{\mathbb{R}} \left(1 - \cos\left(u \sum_{j=1}^{J} \theta_j f_{\epsilon s_j}(x)/\epsilon^{\widetilde{H}}\right)\right) dx \, 1_{\{|u|\leq 1\}} \frac{du}{|u|^{1+\alpha}}\right\}$$

$$= \exp\left\{-\int_{\mathbb{R}}\int_{\mathbb{R}} \left(1 - \cos\left(v \sum_{j=1}^{J} \theta_j f_{s_j}(y)\right)\right) dy \, 1_{\{|v|\leq \epsilon^{-1/\alpha}\}} \frac{dv}{|v|^{1+\alpha}}\right\},$$

after the changes of variables $x = \epsilon y$ and $u = \epsilon^{1/\alpha} v$. As $\epsilon \to 0$, the characteristic function converges to

$$\exp\left\{-\int_{\mathbb{R}}\int_{\mathbb{R}} \left(1 - \cos\left(v \sum_{j=1}^{J} \theta_j f_{s_j}(y)\right)\right) dy \frac{dv}{|v|^{1+\alpha}}\right\}$$

$$= \exp\left\{-\int_{\mathbb{R}} (1 - \cos w) \frac{dw}{|w|^{1+\alpha}} \int_{\mathbb{R}} \left|\sum_{j=1}^{J} \theta_j f_{s_j}(y)\right|^\alpha dy\right\},$$

after the change of variables $v|\sum_{j=1}^{J} \theta_j f_{s_j}(y)| = w$. This limit is the characteristic function of the respective linear combination of the limit process in (2.6.43). □

2.7 The Lamperti Transformation

We provided above some examples of SSSI processes. Suppose that we now require only that the processes are self-similar. Then many examples can be constructed using stationary processes as a starting point as the next result shows.

Theorem 2.7.1 (i) If $X = \{X(t)\}_{t \geq 0}$ is H–SS, then $Y(t) = e^{-tH}X(e^t)$, $t \in \mathbb{R}$, is stationary.
(ii) Conversely, if $Y = \{Y(t)\}_{t \in \mathbb{R}}$ is stationary, then $X(t) = t^H Y(\log t)$, $t > 0$, is H–SS.

Proof Suppose (i), namely that X is self-similar. Observe that

$$(Y(t_1+h), \ldots, Y(t_n+h))' = \left(e^{-(t_1+h)H}X(e^{t_1+h}), \ldots, e^{-(t_n+h)H}X(e^{t_n+h})\right)'$$

$$\stackrel{d}{=} (e^{-t_1 H}X(e^{t_1}), \ldots, e^{-t_n H}X(e^{t_n})')= (Y(t_1), \ldots, Y(t_n))',$$

since $\{X(e^h s)\} \stackrel{d}{=} \{e^{hH}X(s)\}$ by self-similarity.

Turning to (ii), suppose that Y is stationary. Observe that

$$(X(cs_1), \ldots, X(cs_n))' = \left(c^H s_1^H Y(\log s_1 + \log c), \ldots, c^H s_n^H Y(\log s_n + \log c)\right)'$$

$$\stackrel{d}{=} \left(c^H s_1^H Y(\log s_1), \ldots, c^H s_n^H Y(\log s_n)\right)' = \left(c^H X(s_1), \ldots, c^H X(s_n)\right)',$$

since $\{Y(t + \log c)\} \stackrel{d}{=} \{Y(t)\}$ by stationarity. □

The theorem shows in particular that there are at least as many SS processes as there are stationary processes. The transformations in the theorem are called the *Lamperti transformation* after Lamperti [589].

In the following example, we illustrate the Lamperti transformation through a well-known example.

Example 2.7.2 (*Lamperti transformation*) Let the SS process X in Theorem 2.7.1 be a standard Brownian motion, having the covariance function

$$\mathbb{E}X(t)X(s) = \min\{t, s\}, \quad s, t \geq 0.$$

The process X is self-similar with the self-similarity parameter $H = 1/2$. The Lamperti transformation yields a stationary process $Y(t) = e^{-t/2}X(e^t)$, $t \in \mathbb{R}$. The process Y is a zero mean Gaussian process with the covariance function

$$\mathbb{E}Y(t)Y(s) = e^{-t/2}e^{-s/2}\min\{e^t, e^s\} = e^{-|t-s|/2}, \quad s, t \in \mathbb{R}. \quad (2.7.1)$$

It is thus the Ornstein–Uhlenbeck (OU) process in Examples 1.1.2 and 1.4.3.

Example 2.7.3 (*Lamperti transformation*) The covariance of any H–SSSI process X with $H \in (0, 1)$ and finite variance is given in (2.5.7). Therefore, the covariance of the stationary process $Y(t)$, $t \in \mathbb{R}$, in Theorem 2.7.1, resulting from the Lamperti transformation is

$$R(t) = \mathbb{E}Y(t)Y(0) = e^{-tH}\mathbb{E}X(e^t)X(1) = \frac{\sigma^2}{2}e^{-Ht}\left(e^{2Ht} + 1 - |e^t - 1|^{2H}\right)$$

$$= \frac{\sigma^2}{2}\left(e^{Ht} + e^{-Ht} - |e^{t/2} - e^{-t/2}|^{2H}\right) = \sigma^2\left(\cosh(Ht) - 2^{2H-1}|\sinh(t/2)|^{2H}\right),$$

where $\sigma^2 = \mathbb{E} X(1)^2$. As $t \to \infty$, we have

$$R(t) = \frac{\sigma^2}{2} e^{Ht}\left(1 + e^{-2Ht} - (1 - e^{-t})^{2H}\right)$$
$$= \frac{\sigma^2}{2} e^{Ht}\left(1 + e^{-2Ht} - 1 + 2He^{-t} + O(e^{-2t})\right)$$
$$= \frac{\sigma^2}{2}\left(e^{-Ht} + 2He^{-(1-H)t} + O(e^{-(2-H)t})\right)$$

and hence $\{Y(t)\}$ is short-range dependent. If $H = 1/2$, one recovers $\mathbb{E} Y(t) Y(0) = e^{-t/2}$ when $\sigma^2 = 1$.

2.8 Connections Between Long-Range Dependent Series and Self-Similar Processes

There is often a close relationship between LRD series and SSSI processes. SSSI processes lead to LRD series by taking differences and SSSI processes are obtained as normalized limits of partial sum processes of LRD series. We discuss these connections here in some detail, focusing on series and processes with finite variance.

First, consider going from SSSI processes to LRD series. Suppose $Y = \{Y(t)\}_{t \in \mathbb{R}}$ is an H-SSSI process with $0 < H < 1$, and consider a stationary series

$$X_n = Y(n) - Y(n-1), \quad n \in \mathbb{Z}. \tag{2.8.1}$$

The resulting series has the following basic properties.

Proposition 2.8.1 *The series X defined in (2.8.1) has zero mean, $\mathbb{E} X_n^2 = \mathbb{E} Y(1)^2$, autocovariances*

$$\gamma_X(k) = \frac{\mathbb{E} Y(1)^2}{2}\left(|k+1|^{2H} - 2|k|^{2H} + |k-1|^{2H}\right) \sim \mathbb{E} Y(1)^2 H(2H-1) k^{2H-2}, \quad \text{as } k \to \infty, \tag{2.8.2}$$

and spectral density

$$f_X(\lambda) = \frac{\mathbb{E} Y(1)^2}{c_2(H)^2}|1 - e^{-i\lambda}|^2 \sum_{n=-\infty}^{\infty} |\lambda + 2\pi n|^{-1-2H} \sim \frac{\mathbb{E} Y(1)^2}{c_2(H)^2} \lambda^{1-2H}, \quad \text{as } \lambda \to 0, \tag{2.8.3}$$

where $c_2(H)$ is given in Proposition 2.6.11 and in (2.6.10).

Proof The equality in (2.8.2) follows immediately from Proposition 2.5.6, (e). For the asymptotic relation in (2.8.2), observe that, for large k and $f(x) = x^{2H}$,

$$|k+1|^{2H} - 2|k|^{2H} + |k-1|^{2H} = k^{2H-2}\frac{f(1+\frac{1}{k}) - 2f(1) + f(1-\frac{1}{k})}{\frac{1}{k^2}}$$
$$\sim k^{2H-2} f''(x)|_{x=1} = 2H(2H-1)k^{2H-2}.$$

For the spectral density, we suppose without loss of generality that $Y(t) = (\mathbb{E}Y(1)^2)^{1/2} B_H(t)$, where B_H is a standard FBM. We may do so because the spectral density depends only on the second-order properties of the process which are the same as those of standard FBM. By using Proposition 2.6.11, we can write

$$X_n = \frac{(\mathbb{E}Y(1)^2)^{1/2}}{c_2(H)} \int_{\mathbb{R}} e^{inx} \frac{(1-e^{-ix})}{ix} |x|^{\frac{1}{2}-H} \widehat{B}(dx),$$

where $\widehat{B}(dx)$ satisfies $\mathbb{E}|\widehat{B}(dx)|^2 = dx$. Then,

$$\gamma_X(k) = \mathbb{E}X_k X_0 = \frac{\mathbb{E}Y(1)^2}{c_2(H)^2} \int_{\mathbb{R}} e^{ikx} |1-e^{-ix}|^2 |x|^{-1-2H} dx$$

$$= \frac{\mathbb{E}Y(1)^2}{c_2(H)^2} \sum_{n=-\infty}^{\infty} \int_{-\pi+2\pi n}^{\pi+2\pi n} e^{ikx} |1-e^{-ix}|^2 |x|^{-1-2H} dx$$

$$= \frac{\mathbb{E}Y(1)^2}{c_2(H)^2} \int_{-\pi}^{\pi} e^{ik\lambda} |1-e^{-i\lambda}|^2 \sum_{n=-\infty}^{\infty} |\lambda+2\pi n|^{-1-2H} d\lambda = \int_{-\pi}^{\pi} e^{ik\lambda} f_X(\lambda) d\lambda,$$

from which (2.8.3) follows. □

Remark 2.8.2 Proposition 2.8.1 shows that if $1/2 < H < 1$, then the series X in (2.8.1) is LRD in the sense of conditions II and IV with

$$d = H - \frac{1}{2} \in \left(0, \frac{1}{2}\right). \tag{2.8.4}$$

Formula (2.8.4) is a standard and important relation between the LRD parameter d and the SS parameter H.

We should also note that the series (2.8.1) has a special name when Y is FBM.

Definition 2.8.3 If $Y = B_H$ is FBM, the series X in (2.8.1) is called *fractional Gaussian noise*[12] (FGN).

We now explain how SSSI processes Y arise from LRD series X. It is more instructive in this context to start with a general result concerning SS processes. This is the so-called fundamental limit theorem of Lamperti [589]. It uses the convergence to types theorem which we state next without proof. See, for example, Resnick [846], Theorem 8.7.1, p. 275, for a proof.

Theorem 2.8.4 (*Convergence to types theorem.*) Suppose that $\{X_n\}_{n\geq 0}$ is a collection of random variables, and U, V are two non-degenerate[13] random variables. Let $a_n > 0$, $\alpha_n > 0$, $b_n \in \mathbb{R}$, $\beta_n \in \mathbb{R}$.

---

[12] Typically, the word "motion" is associated with a process having stationary increments, e.g., Brownian motion, and the word "noise" is associated with a (possibly second-order) stationary process, e.g., white noise.
[13] A random variable is non-degenerate (proper) if it is not concentrated at a fixed point.

2.8 Connections Between Long-Range Dependent Series and Self-Similar Processes 69

(a) If, as $n \to \infty$,

$$\frac{X_n - b_n}{a_n} \xrightarrow{d} U, \quad \frac{X_n - \beta_n}{\alpha_n} \xrightarrow{d} V, \qquad (2.8.5)$$

then there are $A > 0$, $B \in \mathbb{R}$ such that, as $n \to \infty$,

$$\frac{\alpha_n}{a_n} \to A > 0, \quad \frac{\beta_n - b_n}{a_n} \to B, \qquad (2.8.6)$$

$$V \stackrel{d}{=} \frac{U - B}{A}. \qquad (2.8.7)$$

(b) Conversely, if (2.8.6) holds, then either relation of (2.8.5) implies the other, and (2.8.7) holds.

The following is the fundamental limit theorem of Lamperti [589].

Theorem 2.8.5 *(i) If a stochastic process Y satisfies*

$$\frac{Y(\xi t) + g(\xi)}{f(\xi)} \xrightarrow{fdd} X(t), \quad t \geq 0, \text{ as } \xi \to \infty, \qquad (2.8.8)$$

for some proper[14] process $X(t)$, deterministic function[15] $g(\xi) \in \mathbb{R}$ and some positive deterministic function $f(\xi) \to \infty$, as $\xi \to \infty$, then there is $H \geq 0$ such that

$$f(\xi) = \xi^H L(\xi), \quad g(\xi) = w(\xi)\xi^H L(\xi), \qquad (2.8.9)$$

where $L(\xi)$ is a slowly varying function at infinity and $w(\xi)$ is a function which converges to a constant w as $\xi \to \infty$. Moreover, the process $\{X(t) - w\}_{t \geq 0}$ is H–SS.

(ii) If $\{X(t) - w\}_{t \geq 0}$ is H–SS, then there are functions $f(\xi)$, $g(\xi)$ and a process Y such that the above convergence (2.8.8) holds.

Proof The proof of (ii) is easy. With $Y(t) = X(t) - w$, $g(\xi) = w\xi^H$, $f(\xi) = \xi^H$, we have, by self-similarity,

$$\left\{\frac{Y(\xi t) + g(\xi)}{f(\xi)}\right\} = \left\{\frac{X(\xi t) - w}{\xi^H} + w\right\} \stackrel{d}{=} \{(X(t) - w) + w\} = \{X(t)\}.$$

Consider now the case (i). First, we will derive (2.8.9). If $t = 1$, then

$$\frac{Y(\xi) + g(\xi)}{f(\xi)} \xrightarrow{d} X(1).$$

Similarly, with ξ replaced by $a\xi$ and $t = 1/a$, we have

$$\frac{Y(\xi) + g(a\xi)}{f(a\xi)} \xrightarrow{d} X(1/a).$$

---

[14] A process $\{X(t)\}$ is called *proper* if each random variable $X(t)$, $t > 0$, is proper (non-degenerate); that is, it is not concentrated at a fixed point.
[15] Note that unlike $Y(\xi t)$, the function $g(\xi)$ does not depend on t.

By Theorem 2.8.4, for any $a \geq 0$,

$$\frac{f(a\xi)}{f(\xi)} \to F(a) \in (0, \infty), \quad \text{as } \xi \to \infty.$$

(Note that 0 and ∞ are excluded as possible values of $F(a)$ because the limits $X(1)$ and $X(1/a)$ are proper (non-degenerate).)

Observe next that

$$\frac{f(ab\xi)}{f(\xi)} \to F(ab), \quad \frac{f(ab\xi)}{f(\xi)} = \frac{f(ab\xi)}{f(b\xi)} \frac{f(b\xi)}{f(\xi)} \to F(a)F(b).$$

This implies $F(ab) = F(a)F(b)$, or $k(x+y) = k(x) + k(y)$, $x, y \in \mathbb{R}$, with $k(x) = \log F(e^x)$. The function k is (Lebesgue) measurable as the limit of measurable functions. Then, by Theorem 1.1.8 in Bingham et al. [157], $k(x) = cx$ for a constant c. This implies that $F(a) = a^H$. Hence,

$$\frac{f(a\xi)}{f(\xi)} \to a^H \in (0, \infty), \quad \text{as } \xi \to \infty,$$

or

$$\frac{(a\xi)^{-H} f(a\xi)}{\xi^{-H} f(\xi)} \to 1, \quad \text{as } \xi \to \infty,$$

that is, $\xi^{-H} f(\xi) = L(\xi)$ is a slowly varying function. Hence, $f(\xi) = \xi^H L(\xi)$ where $H \geq 0$. (The case $H < 0$ is excluded because $f(\xi) \to \infty$.) Moreover, since

$$\frac{Y(0) + g(\xi)}{f(\xi)} \xrightarrow{fdd} X(0),$$

there is a function $w(\xi)$ such that $g(\xi) = w(\xi)\xi^H L(\xi)$ with $w(\xi) \to w \in \mathbb{R}$. Note also that (2.8.8) now yields

$$\frac{Y(\xi t)}{f(\xi)} \xrightarrow{fdd} X(t) - w =: \widetilde{X}(t).$$

Finally, to establish H–self-similarity, observe that

$$\frac{Y(a\xi t)}{\xi^H L(\xi)} \xrightarrow{fdd} \widetilde{X}(at)$$

and

$$\frac{Y(a\xi t)}{\xi^H L(\xi)} = a^H \frac{L(a\xi)}{L(\xi)} \frac{Y(a\xi t)}{(a\xi)^H L(a\xi)} \xrightarrow{fdd} a^H \widetilde{X}(t)$$

imply that $\widetilde{X}(at)$ and $a^H \widetilde{X}(t)$ have the same finite-dimensional distributions. □

Theorem 2.8.5 shows that the only processes that arise as large scale limits of processes Y are self-similar. A common example of Y and also the choice leading to SS processes with stationary increments in the limit is given by

$$Y(t) = \sum_{k=1}^{[t]} X_k, \quad (2.8.10)$$

where $\{X_k\}_{k \geq 1}$ is a stationary series, as indicated in the following corollary.

2.8 Connections Between Long-Range Dependent Series and Self-Similar Processes

Corollary 2.8.6 *(i) Let $\{X_k\}_{k\geq 1}$ be a stationary series and $\{Y(t)\}_{t\geq 0}$ be the process defined by (2.8.10). If (i) of Theorem 2.8.5 holds, then the limit $\{X(t) - w\}_{t\geq 0}$ is H–SSSI.*

(ii) If $\{X(t) - w\}_{t\geq 0}$ is H–SSSI and continuous in probability,[16] then there are functions $f(\xi)$, $g(\xi)$ and a stationary series $\{X_k\}_{k\geq 1}$ such that the process Y defined by (2.8.10) satisfies (2.8.8).

Proof (i) If (i) of Theorem 2.8.5 holds, then the limit $\{X(t) - w\}_{t\geq 0}$ is H–SS. This limit has also stationary increments because $\{X_k\}$ is stationary. (ii) For the series $\{X_k\}$, take the increments of the SSSI process, $X_k = X(k) - X(k-1)$. Moreover, take $f(\xi) = \xi^H$ and $g(\xi) = \xi^H w - w$. Then, by using Proposition 2.5.6, (a), and H–self-similarity,

$$\left\{\frac{Y(\xi t) + g(\xi)}{f(\xi)}\right\} = \left\{\frac{\sum_{k=1}^{[\xi t]} X_k + g(\xi)}{f(\xi)}\right\} = \left\{\frac{X([\xi t]) - w + \xi^H w}{\xi^H}\right\}$$

$$\stackrel{d}{=} \left\{\frac{\xi^H (X([\xi t]/\xi) - w) + \xi^H w}{\xi^H}\right\} = \left\{X\left(\frac{[\xi t]}{\xi}\right)\right\}.$$

Since X is continuous in probability,

$$X\left(\frac{[\xi t]}{\xi}\right) \stackrel{p}{\to} X(t),$$

as $\xi \to \infty$. Since the convergence in probability implies that of distributions, we obtain (2.8.8). □

To conclude this section, we shall illustrate Corollary 2.8.6 in three special cases: the first two involving FBM and the third involving the Rosenblatt process as the limiting SSSI process. We shall also replace the parameter $\xi \to \infty$ by an integer $N \to \infty$. Another special case involving LFSM is considered in Section 2.9.1.

The next result concerns FBM in the LRD regime.

Proposition 2.8.7 *Let $\{X_n\}$ be a Gaussian LRD series in the sense (2.1.8) in condition V with parameter $0 < d < 1/2$. Then, as $N \to \infty$,*

$$\frac{1}{L_5(N)^{1/2} N^{d+1/2}} \sum_{n=1}^{[Nt]} (X_n - \mathbb{E}X_n) \stackrel{fdd}{\to} B_H(t), \quad t \geq 0, \qquad (2.8.11)$$

where B_H is a standard FBM with SS parameter

$$H = d + \frac{1}{2} \in \left(\frac{1}{2}, 1\right).$$

Proof Suppose without loss of generality that $\mathbb{E}X_n = 0$ and denote the normalized partial sum process in (2.8.11) by $X^N(t)$. Because of Gaussianity, it is enough to check that $\mathbb{E}X^N(t)X^N(s) \to \mathbb{E}B_H(t)B_H(s)$, for $t > s > 0$, as $N \to \infty$. This follows from

---
[16] The process $\{Z(t)\}$ is continuous in probability if for any t, $Z(t+h) \stackrel{p}{\to} Z(t)$, as $h \to 0$

$$\mathbb{E}X^N(t)X^N(s) = \frac{1}{L_5(N)N^{2d+1}} \mathbb{E}\sum_{n=1}^{[Nt]} X_n \sum_{n=1}^{[Ns]} X_n$$

$$= \frac{1}{2L_5(N)N^{2d+1}}\left\{\mathbb{E}\Big(\sum_{n=1}^{[Nt]} X_n\Big)^2 + \mathbb{E}\Big(\sum_{n=1}^{[Ns]} X_n\Big)^2 - \mathbb{E}\Big(\sum_{n=[Ns]+1}^{[Nt]} X_n\Big)^2\right\}$$

$$= \frac{1}{2L_5(N)N^{2d+1}}\left\{\mathbb{E}\Big(\sum_{n=1}^{[Nt]} X_n\Big)^2 + \mathbb{E}\Big(\sum_{n=1}^{[Ns]} X_n\Big)^2 - \mathbb{E}\Big(\sum_{n=1}^{[Nt]-[Ns]} X_n\Big)^2\right\}$$

$$\sim \frac{1}{2}\{t^{2d+1} + s^{2d+1} - (t-s)^{2d+1}\} = \mathbb{E}B_{d+\frac{1}{2}}(t)B_{d+\frac{1}{2}}(s),$$

as $N \to \infty$, since, for example,

$$\frac{\mathbb{E}(\sum_{n=1}^{[Nt]} X_n)^2}{L_5(N)N^{2d+1}} = \frac{\mathbb{E}(\sum_{n=1}^{[Nt]} X_n)^2}{L_5([Nt])[Nt]^{2d+1}} \frac{L_5([Nt])}{L_5(N)}\left(\frac{[Nt]}{N}\right)^{2d+1} \sim t^{2d+1}$$

by using (2.1.8). □

Proposition 2.8.7 continues to hold for $d = 0$ (and BM with $H = 1/2$ in the limit), as well as for antipersistent series (with $-1/2 < d < 0$ and $0 < H < 1/2$).

The assumption of Gaussianity in Proposition 2.8.7 is not necessary to obtain FBM in the limit. The following result extends Proposition 2.8.7 to the case of linear series.

Proposition 2.8.8 *Let $\{X_n\}_{n\in\mathbb{Z}}$ be a linear series*

$$X_n = \sum_{k=-\infty}^{\infty} c_k \epsilon_{n-k}, \quad n \in \mathbb{Z}, \qquad (2.8.12)$$

with i.i.d. random variables ϵ_n, $n \in \mathbb{Z}$, with zero mean and unit variance. Suppose also that the series $\{X_n\}_{n\in\mathbb{Z}}$ is LRD in the sense (2.1.8) in condition V with parameter $0 < d < 1/2$. Then, as $N \to \infty$,

$$\frac{1}{L_5(N)^{1/2}N^{d+1/2}} \sum_{n=1}^{[Nt]} X_n \xrightarrow{fdd} B_H(t), \quad t \geq 0, \qquad (2.8.13)$$

where B_H is a standard FBM with SS parameter

$$H = d + \frac{1}{2} \in \left(\frac{1}{2}, 1\right).$$

If the sequence $\{c_k\}$ in (2.8.12) satisfies (2.1.4) in condition I, that is, $c_k = 0$, $k \leq -1$, and $c_k = L_1(k)k^{d-1}$, as $k \to \infty$, with $d \in (0, 1/2)$ and a slowly varying function $L_1(k)$, then the series $\{X_n\}_{n\in\mathbb{Z}}$ is LRD in the sense (2.1.8) (see Corollaries 2.2.10 and 2.2.7).

Proof Denote by $X^N(t)$ the normalized partial sum process on the left-hand side of (2.8.13). For $t_1, \ldots, t_J \geq 0$ and $\theta_1, \ldots, \theta_J \in \mathbb{R}$, consider

$$Z_N = \sum_{j=1}^{J} \theta_j X^N(t_j).$$

2.8 Connections Between Long-Range Dependent Series and Self-Similar Processes

It is enough to show that

$$Z_N \xrightarrow{d} \mathcal{N}(0, \sigma^2), \qquad (2.8.14)$$

with $\sigma^2 = \lim_{N \to \infty} \mathbb{E} Z_N^2$, and that

$$\mathbb{E} Z_N^2 \to \mathbb{E}\Big(\sum_{j=1}^J \theta_j B_H(t_j) \Big)^2. \qquad (2.8.15)$$

The last condition can be replaced by

$$\mathbb{E} X^N(t_j) X^N(t_k) \to \mathbb{E} B_H(t_j) B_H(t_k). \qquad (2.8.16)$$

The convergence (2.8.16) can be proved as in the proof of Proposition 2.8.7. We shall therefore prove only that the convergence (2.8.14) holds.

Let $\sigma_N = L_5(N)^{1/2} N^{d+1/2}$. Observe that

$$Z_N = \frac{1}{\sigma_N} \sum_{j=1}^J \theta_j \sum_{n=1}^{[Nt_j]} X_n = \sum_{m=-\infty}^{\infty} \frac{1}{\sigma_N} \Big(\sum_{j=1}^J \theta_j \sum_{n=1}^{[Nt_j]} c_{n-m} \Big) \epsilon_m$$

$$=: \sum_{m=-\infty}^{\infty} \Big(\sum_{j=1}^J \theta_j a_{j,m}^{(N)} \Big) \epsilon_m =: \sum_{m=-\infty}^{\infty} d_{N,m} \epsilon_m.$$

For an integer $M = M_N \to \infty$ to be chosen below, consider independent random variables $\eta_{N,1}, \ldots, \eta_{N,l_N}$ defined by

$$\eta_{N,1} = \sum_{|m|>M} d_{N,m} \epsilon_m,$$

$$\eta_{N,2} = d_{N,-M} \epsilon_{-M}, \quad \eta_{N,3} = d_{N,-M+1} \epsilon_{-M+1}, \ldots, \eta_{N,l_N} = d_{N,M} \epsilon_M.$$

(We have $l_N = 2M + 1 = 2M_N + 1$.) Note that

$$Z_N = \eta_{N,1} + \cdots + \eta_{N,l_N}.$$

To show (2.8.14), it is enough to show that the independent, non-identically distributed random variables $\eta_{N,1}, \ldots, \eta_{N,l_N}$ satisfy the Lindeberg condition, namely, for any $\delta > 0$,

$$I_N := \sum_{k=1}^{l_N} \mathbb{E}|\eta_{N,k}|^2 1_{\{|\eta_{N,k}|>\delta\}} \to 0. \qquad (2.8.17)$$

Note that

$$\sup_N \sum_{m=-\infty}^{\infty} d_{N,m}^2 = \sup_N \mathbb{E} Z_N^2 < \infty$$

by the convergence (2.8.15). Then, one can choose $M = M_N \to \infty$ such that

$$\mathbb{E}|\eta_{N,1}|^2 = \sum_{|m|>M} d_{N,m}^2 \to 0. \qquad (2.8.18)$$

To deal with the remaining variables $\eta_{N,2}, \ldots, \eta_{N,l_N}$, observe that

$$(a_{j,m}^{(N)})^2 = \frac{1}{\sigma_N^2}(c_{1-m} + \cdots + c_{[Nt_j]-m})^2 = \frac{1}{\sigma_N^2}(\sigma_N a_{j,m-1}^{(N)} + c_{[Nt_j]-m})^2$$

$$= (a_{j,m-1}^{(N)})^2 + \frac{2}{\sigma_N} a_{j,m-1}^{(N)} c_{[Nt_j]-m} + \frac{c_{[Nt_j]-m}^2}{\sigma_N^2}.$$

Summing over $m = k - l, \ldots, k$, we get after simplification, that

$$(a_{j,k}^{(N)})^2 = (a_{j,k-l-1}^{(N)})^2 + \frac{2}{\sigma_N} \sum_{m=k-l}^{k} a_{j,m-1}^{(N)} c_{[Nt_j]-m} + \frac{1}{\sigma_N^2} \sum_{m=k-l}^{k} c_{[Nt_j]-m}^2.$$

Hence,

$$(a_{j,k}^{(N)})^2 \leq (a_{j,k-l-1}^{(N)})^2 + \frac{2}{\sigma_N} \Big(\sum_{m=-\infty}^{\infty} (a_{j,m}^{(N)})^2\Big)^{1/2} \Big(\sum_{m=-\infty}^{\infty} c_m^2\Big)^{1/2} + \frac{1}{\sigma_N^2} \sum_{m=-\infty}^{\infty} c_m^2.$$

Letting $l \to -\infty$, and using the facts that $\sum_{m=-\infty}^{\infty} c_m^2 < \infty$ and $\sup_N \sum_{m=-\infty}^{\infty} (a_{j,m}^{(N)})^2 = \sup_N \mathbb{E} X^N(t_j)^2 < \infty$, we get that, for all k,

$$|a_{j,k}^{(N)}| \leq \frac{C_j}{\sigma_N^{1/2}}.$$

This also yields that, for all k,

$$|d_{N,k}| \leq \frac{C}{\sigma_N^{1/2}}. \tag{2.8.19}$$

Now, by using (2.8.19),

$$\sum_{k=2}^{l_N} \mathbb{E}|\eta_{N,k}|^2 1_{\{|\eta_{N,k}|>\delta\}} \leq \Big(\sum_{m=-\infty}^{\infty} d_{N,m}^2\Big) \mathbb{E}|\epsilon_0|^2 1_{\{|\epsilon_0|>\delta\sigma_N^{1/2}/C\}}$$

$$= (\mathbb{E} Z_N^2) \mathbb{E}|\epsilon_0|^2 1_{\{|\epsilon_0|>\delta\sigma_N^{1/2}/C\}} \to 0, \tag{2.8.20}$$

since $\sigma_N \to \infty$. Finally, the Lindeberg condition (2.8.17) follows from (2.8.18) and (2.8.20). $\square$

The third illustration involves the Rosenblatt process. The presented result is a special case of Theorem 5.3.1 in Chapter 5, whose rigorous proof is given in that chapter. We shall only explain here why the Rosenblatt process is a natural limiting process, without providing a rigorous proof.

Proposition 2.8.9 *Let $\{X_n\}_{n \in \mathbb{Z}}$ be a Gaussian series which is LRD in the sense (2.1.5) in condition II with parameter $1/4 < d < 1/2$. Then, as $N \to \infty$,*

$$\frac{1}{L_2(N)N^{2d}} \sum_{n=1}^{[Nt]} (X_n^2 - \mathbb{E} X_n^2) \xrightarrow{fdd} \beta_{2,H} Z_H(t), \quad t \geq 0, \tag{2.8.21}$$

2.8 Connections Between Long-Range Dependent Series and Self-Similar Processes

where Z_H is a standard Rosenblatt process with SS parameter

$$H = 2d \in \left(\frac{1}{2}, 1\right)$$

and the constant $\beta_{2,H}$ is such that $\beta_{2,H}^2 = 2/(H(2H-1))$.

The Rosenblatt process is a natural limiting process in (2.8.21). Indeed, this fact is easiest to see through the spectral domain. Assuming that the series $\{X_n\}$ is LRD in the sense (2.1.7) in condition IV expressed in the spectral domain. Let $L(\lambda) = L_4(\lambda)^{1/2}$ and $g(\lambda) = f_X(\lambda)^{1/2} = L(|\lambda|)|\lambda|^{-d}$, where $f_X(\lambda)$ is the spectral density function of the series $\{X_n\}$. Note that $L(\lambda)$ is a slowly varying function at zero. Now write the series $\{X_n\}$ in its spectral representation as $X_n = \int_{(-\pi,\pi]} e^{in\lambda} g(\lambda) \widehat{B}(d\lambda)$, where $\widehat{B}(d\lambda)$ is a Gaussian Hermitian random measure (see Section 1.3.4 and Appendix B.1). It is natural to expect that X_n^2 is related to a double integral with respect to the random measure $\widehat{B}(d\lambda)$. In fact, the exact relation is

$$X_n^2 - \mathbb{E}X_n^2 = \int_{(-\pi,\pi]^2}'' e^{in(\lambda_1+\lambda_2)} g(\lambda_1) g(\lambda_2) \widehat{B}(d\lambda_1) \widehat{B}(d\lambda_2).$$

This is a special case of Proposition 4.1.2, (ii), in Chapter 4, and also of the relation (B.2.23) with $p = 1$ in Appendix B.2. Denoting the normalized partial sum process in (2.8.21) by $X^N(t)$, we deduce that

$$X^N(t) \stackrel{d}{=} \frac{L_4(1/N)}{L_2(N)} \frac{1}{L(1/N)^2 N^{2d}} \int_{(-\pi,\pi]^2}'' \frac{e^{i[Nt](\lambda_1+\lambda_2)} - 1}{e^{i(\lambda_1+\lambda_2)} - 1} g(\lambda_1) g(\lambda_2) \widehat{B}(d\lambda_1) \widehat{B}(d\lambda_2)$$

$$= \frac{L_4(1/N)}{L_2(N)} \frac{1}{L(1/N)^2 N^{2d}} \int_{\mathbb{R}^2}'' \frac{e^{i\frac{[Nt]}{N}(x_1+x_2)} - 1}{e^{i\frac{x_1+x_2}{N}} - 1} g\left(\frac{x_1}{N}\right) g\left(\frac{x_2}{N}\right) 1_{(-\pi N,\pi N]^2}(x_1,x_2)$$

$$\times \widehat{B}\left(d\frac{x_1}{N}\right) \widehat{B}\left(d\frac{x_2}{N}\right)$$

$$\stackrel{d}{=} \frac{L_4(1/N)}{L_2(N)} \int_{\mathbb{R}^2}'' f_t^N(x_1,x_2) \widehat{B}(dx_1) \widehat{B}(dx_2),$$

where $\stackrel{d}{=}$ denotes equality of the finite-dimensional distributions,

$$f_t^N(x_1,x_2) = \frac{e^{i\frac{[Nt]}{N}(x_1+x_2)} - 1}{N(e^{i\frac{x_1+x_2}{N}} - 1)} \frac{g(\frac{x_1}{N})}{L(1/N)N^d} \frac{g(\frac{x_2}{N})}{L(1/N)N^d} 1_{(-\pi N,\pi N]^2}(x_1,x_2).$$

Observe that

$$f_t^N(x_1,x_2) \to \frac{e^{it(x_1+x_2)} - 1}{i(x_1+x_2)} |x_1|^{-d} |x_2|^{-d}$$

and that, by Corollary 2.2.15,

$$\frac{L_4(1/N)}{L_2(N)} \to \frac{1}{\pi} \Gamma(2d) \cos(d\pi) = \frac{1}{\pi} \Gamma(H) \cos(H\pi/2).$$

This suggests that the limit of $X_N(t)$ in the sense of finite-dimensional distributions is the process

$$\frac{\Gamma(H)\cos(H\pi/2)}{\pi} \int_{\mathbb{R}^2}'' \frac{e^{it(x_1+x_2)} - 1}{i(x_1+x_2)} |x_1|^{-d} |x_2|^{-d} \widehat{B}(dx_1) \widehat{B}(dx_2) = \frac{\Gamma(H)\cos(H\pi/2) d_2(H)}{\pi} Z_H(t),$$

where the constant $d_2(H)$ appears in (2.6.31). The multiplicative constant in the last expression reduces to $\beta_{2,H}$ by using the relation (2.2.14).

Note that the limit process in Proposition 2.8.9 is non-Gaussian and has finite variance. The result is quite striking in view of the classical central limit theorem for SRD series, and again points to fundamental differences between SRD and LRD series.

2.9 Long- and Short-Range Dependent Series with Infinite Variance

While there is a relative consensus on what SRD and LRD mean when the time series has finite variance, the same cannot be said when dealing with series having infinite variance. This is partly because there is presently no widely accepted and well developed theory for modeling with these series. More importantly, the class of infinite variance time series cannot be described through simple measures of dependence such as the covariance. In this section, we shall discuss several attempts to distinguish between SRD and LRD in the infinite variance case. We focus on α–heavy-tailed series with parameter $\alpha \in (0, 2)$ as discussed in Appendix A.4. We shall now provide a number of possible alternative (non-equivalent) definitions of LRD in the case of heavy-tailed series.

2.9.1 First Definition of LRD Under Heavy Tails: Condition A

One way to proceed is by analogy to the case of finite variance. Since second moments are infinite for heavy-tailed random variables, covariances and spectral densities cannot be used. One possible approach is to define LRD through the behavior of the partial sum process of the series. Recall from (A.4.11) and (A.4.8) in Appendix A.4 that if $\{X_n\}$ is a series of i.i.d. α–heavy-tailed variables, $\alpha \in (0, 2)$, then

$$\frac{1}{L_0(N)N^{1/\alpha}}\Big(\sum_{n=1}^{[Nt]} X_n - A([Nt])\Big) \xrightarrow{fdd} S_\alpha(t), \quad t \geq 0,$$

where $A(\cdot)$ is a deterministic function, L_0 is a slowly varying function and S_α is an α-stable Lévy motion. The case $\alpha = 2$ is associated with i.i.d. variables X_n having finite variance and Brownian motion in the limit. By analogy to the finite variance case (cf. Propositions 2.8.7 and 2.8.8), this suggests the following definition for LRD.

Condition A An α–heavy-tailed, $\alpha \in (0, 2)$, stationary time series $\{X_n\}_{n \in \mathbb{Z}}$ is LRD if it satisfies

$$\frac{1}{L(N)N^{d+1/\alpha}}\Big(\sum_{n=1}^{[Nt]} X_n - A([Nt])\Big) \xrightarrow{fdd} X_\alpha(t), \quad t \geq 0, \qquad (2.9.1)$$

with

$$d > 0,$$

where L is a slowly varying function and X_α is a (non-degenerate) stochastic process.

2.9 Long- and Short-Range Dependent Series with Infinite Variance

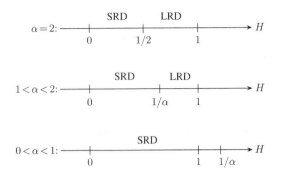

Figure 2.5 SRD and LRD regimes for various ranges of α.

When a time series satisfies (2.9.1) with $d \leq 0$, it could be called SRD. By Corollary 2.8.6 or by arguing directly, the limiting process X_α in condition A is H–SSSI with

$$H = \frac{1}{\alpha} + d,$$

so that

$$d = H - \frac{1}{\alpha}.$$

Example 2.9.1 (*Linear fractional stable noise*) Consider the LFSM process $\{L_{\alpha,H}(t)\}_{t \in \mathbb{R}}$ defined in Proposition 2.6.21, with $H \in (0, 1)$. The series $X_n = L_{\alpha,H}(n) - L_{\alpha,H}(n-1)$, $n \in \mathbb{Z}$, of its increments is called *linear fractional stable noise* (LFSN). It satisfies condition A with $d = H - 1/\alpha$. The series is LRD in the sense of this condition when $H > 1/\alpha$ which is possible only when $\alpha \in (1, 2)$ because $H \in (0, 1)$. This is illustrated in Figure 2.5 which includes the case $\alpha = 2$ where $L_{2,H}(t)$ is FBM.

Why is it not possible for LFSN to be LRD when $\alpha < 1$? A possible (heuristic) explanation is that $\alpha < 1$ corresponds to very heavy tails and thus X_n may take large values with high probability. If this were combined with LRD, there would be long strings of high values and the integral would explode.

Note that, in general, one does not expect LRD in the sense of condition A when $\alpha \in (0, 1)$. For example, if $\mathbb{E}|X_\alpha(t)|^\gamma < \infty$ with $\gamma \leq 1$, Proposition 2.5.6, (c), implies that $H \leq 1/\gamma$. Then, if X_α is α-stable, we can take $\gamma < \alpha$. By taking $\gamma < \alpha$ arbitrarily close to α, we get that $H \leq 1/\alpha$. This precludes having $H = 1/\alpha + d$ with $d > 0$. The process, therefore, cannot be LRD in the sense of condition A. A somewhat stronger statement of this fact can be found in Samorodnitsky [879], Proposition 8.1.

Here is another example of a series satisfying condition A.

Example 2.9.2 (*Heavy-tailed linear series*) Consider a linear series

$$X_n = \sum_{k=0}^{\infty} a_k \epsilon_{n-k}, \quad n \in \mathbb{Z}, \tag{2.9.2}$$

where ϵ_k, $k \in \mathbb{Z}$, are i.i.d. α–heavy-tailed random variables. Suppose for simplicity that the random variables ϵ_k are symmetric. By (A.4.5) in Appendix A.4, this is equivalent to requiring that the characteristic function of ϵ_k has the form

$$\mathbb{E}\exp\{i\theta\epsilon_k\} = \exp\left\{-(C_0 + o(1))|\theta|^\alpha h(|\theta|^{-1})\right\}, \tag{2.9.3}$$

as $\theta \to 0$, where $C_0 = \Gamma(1-\alpha)(c_1 + c_2)\cos(\alpha\pi/2)$ and h is a slowly varying function at infinity. These quantities appear in the tail probabilities

$$\mathbb{P}(\epsilon_k < -x) = \mathbb{P}(\epsilon_k > x) = cx^{-\alpha}h(x). \tag{2.9.4}$$

Because ϵ_k is symmetric, we require $c = c_1 = c_2$. See (A.4.4) and (A.4.5) in Appendix A.4. Now consider the coefficients $\{a_k\}$. As in (2.1.4), suppose that

$$a_k = L_1(k)k^{d-1}, \quad k \geq 1, \tag{2.9.5}$$

with L_1 slowly varying at infinity but d satisfying

$$0 < d < 1 - \frac{1}{\alpha}. \tag{2.9.6}$$

Observe that this implies $\alpha > 1$. One could show using (2.9.3) that under the assumptions above, the linear series (2.9.2) is well-defined and α–heavy-tailed (Exercise 2.16).

Finally, let h_α be a slowly varying function at infinity satisfying

$$h(N^{1/\alpha}h_\alpha(N)^{1/\alpha}) \sim h_\alpha(N), \tag{2.9.7}$$

as $N \to \infty$, where h is given in (2.9.4). The existence of h_α is shown following (A.4.8) in Appendix A.4. The following result shows that $\{X_n\}_{n \in \mathbb{Z}}$ satisfies condition A.

Proposition 2.9.3 *With the above notation and assumptions on the linear series $\{X_n\}_{n \in \mathbb{Z}}$, one has*

$$\frac{1}{L(N)N^{d+1/\alpha}}\sum_{n=1}^{[Nt]} X_n \xrightarrow{fdd} L_{\alpha,H}(t), \quad t \geq 0, \tag{2.9.8}$$

where $L_{\alpha,H}$ is LFSM with $H = d + 1/\alpha$, $a = 1$, $b = 0$ in (2.6.33), and where

$$L(N) = d^{-1}C_0^{1/\alpha}L_1(N)h_\alpha(N)^{1/\alpha} \tag{2.9.9}$$

is a slowly varying function at infinity.

Proof We suppose for simplicity that $a_0 = 0$. Set $a_k = 0$ for $k \leq -1$, and

$$a_N(m) = \sum_{n=1}^{N} a_{n-m}.$$

Then $X_n = \sum_{m=-\infty}^{\infty} a_{n-m}\epsilon_m$ and hence

$$\sum_{n=1}^{[Nt]} X_n = \sum_{m=-\infty}^{\infty}\left(\sum_{n=1}^{[Nt]} a_{n-m}\right)\epsilon_m = \sum_{m=-\infty}^{\infty} a_{[Nt]}(m)\epsilon_m.$$

Denote by $Y_N(t)$ the process on the left-hand side of (2.9.8) and let $A_N = L(N)N^{1/\alpha+d}$ be the normalization. Since we want to prove convergence of finite-dimensional distributions, we will consider an arbitrary linear combination of $Y_N(t_1), \ldots, Y_N(t_J)$ with $J > 0$. By using (2.9.3), for $\theta_j \in \mathbb{R}, t_j \geq 0, j = 1, \ldots, J$,

$$I_N := -\log \mathbb{E}\exp\left\{i\sum_{j=1}^{J}\theta_j Y_N(t_j)\right\}$$

$$= A_N^{-\alpha}\sum_{m=-\infty}^{\infty}(C_0 + o(1))\left|\sum_{j=1}^{J}\theta_j a_{[Nt_j]}(m)\right|^{\alpha} h\left(A_N\left|\sum_{j=1}^{J}\theta_j a_{[Nt_j]}(m)\right|^{-1}\right). \quad (2.9.10)$$

The terms $o(1)$ in (2.9.10) depend, in particular, on m. We want to show first that these terms can be taken out of the sum $\sum_{m=-\infty}^{\infty}$. To accomplish this, note that

$$A_N^{-1}a_N(m) = dC_0^{-1/\alpha}h_\alpha(N)^{-1/\alpha}L_1(N)^{-1}N^{-1/\alpha-d}\sum_{n=1}^{N}L_1(n-m)(n-m)_+^{d-1}.$$

Since h_α and L_1 are slowly varying, by using Potter's bound (2.2.2), we get that, for arbitrarily small fixed $\epsilon_1, \epsilon_2 > 0$,

$$|A_N^{-1}a_N(m)| \leq CN^{\epsilon_1}N^{-1/\alpha-d}\sum_{n=1}^{N}(n-m)_+^{d+\epsilon_2-1}$$

$$\leq CN^{-1/\alpha-d+\epsilon_1}\sum_{n=1}^{N}n^{d+\epsilon_2-1} \leq C'N^{-1/\alpha+\epsilon_1+\epsilon_2}.$$

Hence, by taking ϵ_1, ϵ_2 so that $-1/\alpha + \epsilon_1 + \epsilon_2 < 0$, we have

$$\sup_{m\in\mathbb{Z}}|A_N^{-1}a_N(m)| \to 0. \quad (2.9.11)$$

This implies, as desired, that

$$I_N = (C_0 + o(1))A_N^{-\alpha}\sum_{m=-\infty}^{\infty}\left|\sum_{j=1}^{J}\theta_j a_{[Nt_j]}(m)\right|^{\alpha} h\left(A_N\left|\sum_{j=1}^{J}\theta_j a_{[Nt_j]}(m)\right|^{-1}\right). \quad (2.9.12)$$

Using the expressions for A_N and $L(N)$, we have

$$I_N = (1 + o(1))\frac{1}{N}\sum_{m=-\infty}^{\infty}\left|d\sum_{j=1}^{J}\theta_j\frac{a_{[Nt_j]}(m)}{L_1(N)N^d}\right|^{\alpha}$$

$$\times h_\alpha(N)^{-1}h\left(h_\alpha(N)^{1/\alpha}N^{1/\alpha}\left|C_0^{-1/\alpha}d\sum_{j=1}^{J}\theta_j\frac{a_{[Nt_j]}(m)}{L_1(N)N^d}\right|^{-1}\right)$$

$$=: (1 + o(1))\int_{\mathbb{R}}f_N(u)du, \quad (2.9.13)$$

where

$$f_N(u) = \sum_{m=-\infty}^{\infty} \left| d \sum_{j=1}^{J} \theta_j \frac{a_{[Nt_j]}(m)}{L_1(N)N^d} \right|^{\alpha}$$

$$\times h_\alpha(N)^{-1} h\left(h_\alpha(N)^{1/\alpha} N^{1/\alpha} \left| C_0^{-1/\alpha} d \sum_{j=1}^{J} \theta_j \frac{a_{[Nt_j]}(m)}{L_1(N)N^d} \right|^{-1} \right) 1_{[\frac{m}{N}, \frac{m+1}{N})}(u).$$

Note that

$$\frac{a_{[Nt]}(m)}{L_1(N)N^d} = \sum_{n=1}^{[Nt]} \frac{L_1(\frac{n-m}{N}N)}{L_1(N)} \left(\frac{n}{N} - \frac{m}{N} \right)_+^{d-1} \frac{1}{N} \to \int_0^t (s-u)_+^{d-1} ds,$$

as $N \to \infty$ and $m/N \to u$. Then, by using the fact that h is a slowly varying function and the relation (2.9.7), we get that

$$f_N(u) \to \left| d \sum_{j=1}^{J} \theta_j \int_0^{t_j} (s-u)_+^{d-1} ds \right|^{\alpha} =: f(u).$$

We want to conclude that

$$\int_{\mathbb{R}} f_N(u) du \to \int_{\mathbb{R}} f(u) du. \qquad (2.9.14)$$

By the generalized dominated convergence theorem, it is enough to find functions g_N such that $|f_N(u)| \leq C g_N(u)$, and

$$g_N(u) \to g(u) \quad \text{a.e. } du \qquad (2.9.15)$$

and

$$\int_{\mathbb{R}} g_N(u) du \to \int_{\mathbb{R}} g(u) du. \qquad (2.9.16)$$

By using (2.9.11), Potter's bound (2.2.1) and the relation (2.9.7), observe that $|f_N(u)| \leq C g_N(u)$, where

$$g_N(u) = \sum_{m=-\infty}^{\infty} \left| \sum_{j=1}^{J} \theta_j \frac{a_{[Nt_j]}(m)}{L_1(N)N^d} \right|^{\alpha+\epsilon} 1_{[\frac{m}{N}, \frac{m+1}{N})}(u)$$

and $\epsilon > 0$ is arbitrarily small and fixed. As in the case of f_N above, we have

$$g_N(u) \to \left| \sum_{j=1}^{J} \theta_j \int_0^{t_j} (s-u)_+^{d-1} ds \right|^{\alpha+\epsilon} =: g(u), \qquad (2.9.17)$$

so that (2.9.15) holds. To show (2.9.16), note first that

$$\int_{\mathbb{R}} g_N(u) du = \sum_{m=-\infty}^{\infty} \left| \sum_{j=1}^{J} \theta_j \frac{a_{[Nt_j]}(m)}{L_1(N)N^d} \right|^{\alpha+\epsilon} \frac{1}{N}.$$

We will now approximate the sequence $a_k = L_1(k)k_+^{d-1}$ by the sequence $a_k^0 = k_+^{d-1}$. Set $a_N^0(m) = \sum_{n=1}^N a_{n-m}^0$. Consider

$$T_N := \sum_{m=-\infty}^\infty \left|\frac{a_N(m)}{L_1(N)N^d} - \frac{a_N^0(m)}{N^d}\right|^{\alpha+\epsilon}\frac{1}{N}.$$

Note that, by using Potter's bound (2.2.1),

$$T_N \leq \sum_{m=-\infty}^\infty \left(\sum_{n=1}^N \left|\frac{L_1(n-m)}{L_1(N)} - 1\right|\left(\frac{n}{N} - \frac{m}{N}\right)_+^{d-1}\frac{1}{N}\right)^{\alpha+\epsilon}\frac{1}{N}$$

$$\leq C \sum_{m=-\infty}^\infty \left(\sum_{n=1}^N \left(1_{\{|\frac{n}{N}-\frac{m}{N}|>A\}}\left(\frac{n}{N} - \frac{m}{N}\right)^{d+\tilde\epsilon-1} + 1_{\{|\frac{n}{N}-\frac{m}{N}|<B\}}\left(\frac{n}{N} - \frac{m}{N}\right)_+^{d-\tilde\epsilon-1}\right)\frac{1}{N}\right)^{\alpha+\epsilon}\frac{1}{N}$$

$$+ \sum_{m=-\infty}^\infty \left(\sum_{n=1}^N \left|\frac{L_1(n-m)}{L_1(N)} - 1\right|1_{\{B\leq\frac{n}{N}-\frac{m}{N}\leq A\}}\left(\frac{n}{N} - \frac{m}{N}\right)_+^{d-1}\frac{1}{N}\right)^{\alpha+\epsilon}\frac{1}{N} =: T_{N,1} + T_{N,2},$$

where $\tilde\epsilon > 0$ is arbitrarily small. The first sum $T_{N,1}$ in the bound is arbitrarily small for large enough A and small enough B since $\int_0^\infty(\int_0^1((u-v)_+^{d+\tilde\epsilon-1} + (u-v)_+^{d-\tilde\epsilon-1})du)^{\alpha+\epsilon}dv < \infty$ for small enough $\epsilon, \tilde\epsilon > 0$. The second term $T_{N,2}$ in the bound converges to zero for fixed A, B by using the uniform convergence theorem (2.2.4) for the slowly varying function L_1, and the fact that $\int_0^\infty(\int_0^1(u-v)_+^{d-1}du)^{\alpha+\epsilon}dv < \infty$ for small enough $\epsilon > 0$. This shows that $T_N \to 0$. Hence,

$$\int_\mathbb{R} g_N(u)du = (1+o(1))\sum_{m=-\infty}^\infty \left|\sum_{j=1}^J \theta_j \frac{a_{[Nt_j]}^0(m)}{N^d}\right|^{\alpha+\epsilon}\frac{1}{N}.$$

Finally, note that

$$\sum_{m=-\infty}^\infty \left|\sum_{j=1}^J \theta_j \frac{a_{[Nt_j]}^0(m)}{N^d}\right|^{\alpha+\epsilon}\frac{1}{N} = \sum_{m=-\infty}^\infty \left|\sum_{j=1}^J \theta_j \sum_{n=1}^{[Nt_j]}\left(\frac{n}{N} - \frac{m}{N}\right)_+^{d-1}\frac{1}{N}\right|^{\alpha+\epsilon}\frac{1}{N}$$

$$= (1+o(1))d^\alpha \int_\mathbb{R} \left|\sum_{j=1}^J \theta_j \int_0^{t_j}(u-v)_+^{d-1}du\right|^{\alpha+\epsilon}dv = (1+o(1))\int_\mathbb{R} g(u)du,$$

where g is defined in (2.9.17). This proves (2.9.16) and establishes (2.9.14).

In view of (2.9.10), (2.9.12), (2.9.13) and (2.9.14), it follows that

$$I_N \to \int_\mathbb{R}\left|d\sum_{j=1}^J\theta_j\int_0^{t_j}(s-v)_+^{d-1}ds\right|^\alpha dv = \int_\mathbb{R}\left|\sum_{j=1}^J\theta_j\left((t_j-v)_+^d - (-v)_+^d\right)\right|^\alpha dv$$

$$= -\log\mathbb{E}\exp\left\{i\sum_{j=1}^J\theta_j L_{\alpha,H}(t_j)\right\},$$

where $L_{\alpha,H}$ is LFSM as in the statement of the proposition. This establishes the convergence of the finite-dimensional distributions. □

2.9.2 Second Definition of LRD Under Heavy Tails: Condition B

Example 2.9.2 also suggests the following alternative definition of LRD.

Condition B Consider a linear series $X_n = \sum_{k=0}^{\infty} a_k \epsilon_{n-k}$, $n \in \mathbb{Z}$, with i.i.d. α–heavy-tailed variables ϵ_k, $k \in \mathbb{Z}$, $\alpha \in (1, 2)$, and a sequence $\{a_k\}$ satisfying

$$a_k = L_1(k) k^{d-1}, \qquad (2.9.18)$$

where L_1 is a slowly varying function at infinity. This series is LRD if

$$0 < d < 1 - \frac{1}{\alpha}. \qquad (2.9.19)$$

A drawback of condition B is that linear series form only a small sub-class of heavy-tailed series, even in the case of α-stable series. When $\{X_n\}$ is a linear series with absolutely summable coefficients $\{a_k\}$ such that $|\sum_k a_k| > 0$, one can show that Proposition 2.9.3 holds for $\alpha \in (0, 2)$ with $d = 0$ and as $N \to \infty$, the limit is stable Lévy motion (Astrauskas [53]). Such series could then be called SRD.

Example 2.9.4 (*Heavy-tailed FARIMA series*) An example is a heavy-tailed FARIMA $(0,d,0)$ series defined by $X_n = (I - B)^{-d} \epsilon_n$ with a series $\{\epsilon_n\}$ as in condition B. When the innovations ϵ_k are α-stable, the series is referred to as stable FARIMA$(0, d, 0)$. Several realizations of $S\alpha S$ FARIMA$(0, d, 0)$ series, for ds both satisfying (2.9.19) and violating (2.9.19), are given in Figure 2.6, and are interesting to compare to the Gaussian FARIMA$(0, d, 0)$ series in Figure 2.2. They are obtained using an approximate method based on FFT developed in Stoev and Taqqu [926] (cf. Figure 2.4). Extensions to heavy-tailed FARIMA(p, d, q) series are also possible. In the stable case, see Kokoszka and Taqqu [571, 572, 573], Stoev and Taqqu [926].

2.9.3 Third Definition of LRD Under Heavy Tails: Codifference

Another way to study serial dependence for heavy-tailed series is to replace covariances by other measures of dependence. This is typically done for α-stable series. One common measure of dependence is given in the following definition.

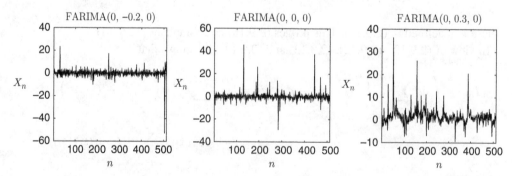

Figure 2.6 Several realizations of $S\alpha S$ FARIMA$(0, d, 0)$ series for $\alpha = 1.5$. Left: $d = -0.2$. Middle: $d = 0$. Right: $d = 0.3$.

2.9 Long- and Short-Range Dependent Series with Infinite Variance

Definition 2.9.5 Let $(X, Y)'$ be an α-stable vector[17] with $\alpha \in (0, 2)$. The *codifference* of X and Y is

$$\tau_{X,Y} = \|X\|_\alpha^\alpha + \|Y\|_\alpha^\alpha - \|X - Y\|_\alpha^\alpha,$$

where $\|\cdot\|_\alpha$ denotes a scale parameter. For α-stable series $\{X_n\}_{n \in \mathbb{Z}}$,[18] set

$$\tau(k) = \tau_{X_k, X_0}.$$

Note that when $\alpha = 2$, the codifference becomes the covariance.

The asymptotics of codifference $\tau(k)$ at large lags has been obtained for a number of candidate heavy-tailed LRD series. Let us focus on the case $\alpha \in (1, 2)$ only. For example, in the case of LFSN, it has been shown (Astrauskas, Levy, and Taqqu [54]) that

$$\tau(k) \sim C_1 k^{\alpha H - \alpha}, \qquad (2.9.20)$$

if $\overline{H}(\alpha) < H < 1$, $H \neq 1/\alpha$, and

$$\tau(k) \sim C_2 k^{H - \frac{1}{\alpha} - 1}, \qquad (2.9.21)$$

if $0 < H < \overline{H}(\alpha)$, where

$$\overline{H}(\alpha) = 1 - (\alpha(\alpha - 1))^{-1}.$$

In the case of FARIMA$(0, d, 0)$ series with α-stable innovations, the same results hold by taking $H = d + 1/\alpha$ (Kokoszka and Taqqu [571]).

Despite interesting asymptotics, it is not too clear how to define LRD using the codifference. One possibility would be as follows:

Condition C An α-stable stationary series $\{X_n\}_{n \in \mathbb{Z}}$, $\alpha \in (0, 2)$, is LRD if

$$\sum_{k=-\infty}^{\infty} |\tau(k)| = \infty. \qquad (2.9.22)$$

For LFSN, by using (2.9.20) and when $\overline{H}(\alpha) < H < 1$, $H \neq 1/\alpha$, condition C holds when $\alpha H - \alpha + 1 > 0$ or

$$H > 1 - \frac{1}{\alpha}.$$

Note that since $\overline{H}(\alpha) < 1 - 1/\alpha < 1/\alpha$, the LRD range under condition C is not the same as the LRD range $H > 1/\alpha$ under conditions A and B (in the latter case, only for FARIMA series as increments of LFSM do not have a linear representation). Another drawback of condition C is that the codifference of many other fractional stable noises (that is, increments of stable SSSI processes) does not have such nice asymptotic behavior. For example, there are fractional noises for which the codifference does not even converge to zero at large lags. See Samorodnitsky and Taqqu [883] for more details.

---

[17] One can think of an α-stable vector as having an integral representation $(\int_S f(x) M_\alpha(dx), \int_S g(x) M_\alpha(dx))'$ where $M_\alpha(dx)$ is an α-stable random measure and f, g are deterministic functions.

[18] Similarly, think of X_n as $\int_S f_n(x) M_\alpha(dx)$.

84 Basics of Long-Range Dependence and Self-Similarity

We should also mention that there are dependence measures akin to the codifference, for example, the "covariation" (Samorodnitsky and Taqqu [883]). As with the codifference, these measures of dependence are interesting but of limited practical use for defining LRD.

2.10 Heuristic Methods of Estimation

We consider here several preliminary estimation methods to detect LRD and to estimate related parameters. The methods are heuristic in the sense that a rigorous statistical theory for these estimators will not be developed in this section. The theory behind some methods can be developed as discussed in the notes in Section 2.13. For other methods, a rigorous theory presents mathematical difficulties or is not sought because of inferior performance. Many of the methods described here are widely used.

We should also note that we focus here on series with finite variance. As is common in statistical applications, all slowly varying functions appearing in definitions of LRD will be assumed constant. The focus throughout is on estimation of LRD parameter d, though estimation of the associated constants will also be mentioned whenever appropriate.

2.10.1 The R/S Method

One of the first methods which was used to detect and to estimate LRD is based on the so-called R/S statistic. The R/S statistic originates with Hurst [501] and is related to dam construction. Suppose that one would like to construct a dam which does not overflow and also has enough storage capacity for future use. Let X_n be the inflow to the dam in period n, $n = 1, \ldots, k$, and suppose that the outflow from the dam equals to a constant a which is the amount withdrawn in every period:

period	1	2	3	...	$k-1$	k
input	X_1	X_2	X_3	...	X_{k-1}	X_k
output	a	a	a	...	a	a

It is reasonable to consider a k–horizon policy where

$$a = \frac{X_1 + \cdots + X_k}{k}$$

so that the water level by period k is the same as in period 0. Then, the difference between inflow and outflow in any period $1 \leq j \leq k$ is[19]

$$S_j = \sum_{n=1}^{j} X_n - aj = \sum_{n=1}^{j} X_n - \frac{j}{k} \sum_{n=1}^{k} X_n, \quad j = 1, \ldots, k. \quad (2.10.1)$$

For convenience set $S_0 = 0$ and observe that $S_k = 0$.

The dam should also have a sufficient initial amount so that it never ends lacking water under this withdrawal policy. Hence, in order to avoid both lack of water and overflow, the dam should be at least of size

$$R(k) = \max_{0 \leq j \leq k} S_j^+ + \max_{0 \leq j \leq k} S_j^- = \max_{0 \leq j \leq k} S_j - \min_{0 \leq j \leq k} S_j, \quad (2.10.2)$$

[19] Do not confuse S_n defined in (2.10.1) with the standard deviation $S(k)$ in (2.10.5).

2.10 Heuristic Methods of Estimation

where the second equality follows from the convention $S_0 = 0$ and the "R" refers to "Range". Observe that

$$R(k) = \max_{0 \leq j \leq k}\left\{\sum_{n=1}^{j} X_n - \frac{j}{k}\sum_{n=1}^{k} X_n\right\} - \min_{0 \leq j \leq k}\left\{\sum_{n=1}^{j} X_n - \frac{j}{k}\sum_{n=1}^{k} X_n\right\}.$$

Here are two toy examples of the situation:

period	1	2	3	4
input	100	50	100	50
output	75	75	75	75
cumulative input	100	150	250	300
cumulative output	75	150	225	300
difference	25	0	25	0
$R(k)$				25

in which case $R(4) = 25$, and

period	1	2	3	4
input	100	100	50	50
output	75	75	75	75
cumulative input	100	200	250	300
cumulativeoutput	75	150	225	300
difference	25	50	25	0
$R(k)$				50

in which case $R(4) = 50$. In the latter case, the input appears more persistent leading to a higher minimum dam size. This is generally the case with LRD time series.

In the R/S method, one is interested in the behavior of

$$Q(k) = \frac{R(k)}{S(k)}, \tag{2.10.3}$$

as $k \to \infty$, where

$$S(k)^2 = \frac{1}{k}\sum_{j=1}^{k}(X_j - \overline{X}_k)^2$$

with $\overline{X}_k = \frac{1}{k}\sum_{j=1}^{k} X_j$. One divides by $S(k)$ in order to reduce the fluctuations of $R(k)$. The ratio $Q(k)$ in (2.10.3) is called the R/S (rescaled adjusted range) statistic.

In practice, given observations $X_1, \ldots, X_N$, $(N-k+1)$ replicates of $Q(k)$ are considered. These are defined as follows. Given a time series $\{X_n\}$, let $Y_0 = 0$, $Y_n = \sum_{j=1}^{n} X_j$ and

$$R(l,k) = \max_{0 \leq j \leq k}\left\{Y_{l+j} - Y_l - \frac{j}{k}(Y_{l+k} - Y_l)\right\} - \min_{0 \leq j \leq k}\left\{Y_{l+j} - Y_l - \frac{j}{k}(Y_{l+k} - Y_l)\right\},$$

$$S(l,k)^2 = \frac{1}{k}\sum_{j=l+1}^{l+k}(X_j - \overline{X}_{l,k})^2,$$

where $\overline{X}_{l,k} = \frac{1}{k}\sum_{j=l+1}^{l+k} X_j$. Then set

$$Q(l,k) = \frac{R(l,k)}{S(l,k)}. \tag{2.10.4}$$

Observe that

$$R(k) = R(0,k) = \max_{0 \le j \le k}\left\{Y_j - \frac{j}{k}Y_k\right\} - \min_{0 \le j \le k}\left\{Y_j - \frac{j}{k}Y_k\right\},$$

$S(k) = S(0,k)$ and

$$Q(k) = \frac{R(k)}{S(k)} = Q(0,k). \tag{2.10.5}$$

The asymptotics of $Q(k)$, as $k \to \infty$, can informally be derived as follows. Suppose without loss of generality that $\mathbb{E}X_n = 0$. Observe that

$$R(k) = \max_{0 \le j \le k} - \min\left\{\sum_{n=1}^{j} X_n - \frac{j}{k}\sum_{n=1}^{k} X_n\right\}$$

$$= \max_{0 \le j/k \le 1} - \min\left\{\sum_{n=1}^{k\frac{j}{k}} X_n - \frac{j}{k}\sum_{n=1}^{k} X_n\right\} = \max_{0 \le j/k \le 1} - \min\left\{Y^k\left(\frac{j}{k}\right) - \frac{j}{k}Y^k(1)\right\},$$

where

$$Y^k(t) = \sum_{n=1}^{[kt]} X_n.$$

For many time series $\{X_n\}$, one expects that there is $\rho > 0$ such that

$$k^{-\rho}Y^k(t) \xrightarrow{d} Y(t), \tag{2.10.6}$$

where the convergence in distribution happens in a function space (weak convergence). Then,

$$k^{-\rho}R(k) \xrightarrow{d} \max_{0 \le t \le 1} - \min\{Y(t) - tY(1)\}.$$

The process $\widetilde{Y}(t) = Y(t) - tY(1)$, $t \in [0,1]$, is a bridge process associated with Y, and max$-$min is its range. One similarly expects that by the weak law of large numbers,

$$S(k) \xrightarrow{p} \sqrt{\text{Var}(X_1)},$$

where $\xrightarrow{p}$ denotes convergence in probability.

Now, for SRD stationary series $\{X_n\}$, one expects that (2.10.6) holds with $\rho = 1/2$ and the limit $Y(t) = cB(t)$, where B is a standard Brownian motion and

$$c^2 = \lim_{k \to \infty}\mathbb{E}\left(\frac{1}{k^{1/2}}\sum_{n=1}^{k} X_n\right)^2 = \lim_{k \to \infty}\frac{1}{k}\sum_{n=1}^{k}\sum_{m=1}^{k}\mathbb{E}X_n X_m = \sum_{h=-\infty}^{\infty}\gamma_X(h)$$

2.10 Heuristic Methods of Estimation

with $\gamma_X(h) = \mathbb{E}X_h X_0$, is the long-run variance (see the discussion following (2.3.3)). Hence, under SRD, one expects that

$$k^{-1/2} Q(k) \overset{d}{\to} \frac{c}{\sqrt{\mathrm{Var}(X_1)}} \max_{0 \le u \le 1} - \min \{B(u) - uB(1)\}.$$

The limit involves the range of a Brownian bridge $\{B(u) - uB(1)\}_{0 \le u \le 1}$ which has been well studied (its distribution can be found in, for example, Feller [359] or Resnick [845], pp. 526–539).

On the other hand, under LRD, one expects that for many stationary time series, (2.10.6) holds with

$$\rho = H = d + 1/2$$

and the limit $Y(t) = B_H(t)$, where B_H is FBM. Hence,

$$k^{-H} Q(k) \overset{d}{\to} \frac{1}{\sqrt{\mathrm{Var}(X_1)}} \max_{0 \le u \le 1} - \min \{B_H(u) - uB_H(1)\}.$$

Thus, for large k, $Q(k) \approx k^H \xi$ or

$$\log Q(k) \approx H \log k + \log \xi,$$

for some random variable ξ, and $H = 1/2$ under SRD, and $H > 1/2$ under LRD.

In practice, one plots $(N - k + 1)$ replicates $\log Q(0, k), \ldots, \log Q(N - k, k)$ of $\log Q(k)$ versus $\log k$ and performs ordinary least squares to estimate the parameter H. The resulting plot is known as the R/S plot or the *poxplot*. See Figure 2.7.

Though R/S statistic is widely used, there is no satisfactory statistical theory for the resulting estimator of H. For example, confidence intervals are not known for the estimator. One difficulty is that $\log Q(k)$ (or $Q(k)$) is non-Gaussian, non-symmetric and dependent over k in a nontrivial way.

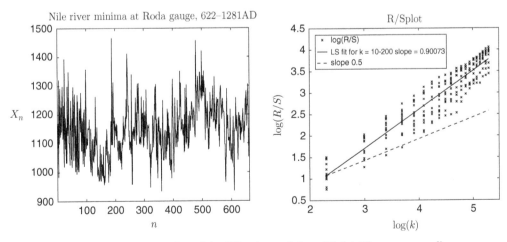

Figure 2.7 Left: Time series of the Nile river minima. Right: The corresponding R/S plot.

2.10.2 Aggregated Variance Method

This estimation method is based on the relation

$$\text{Var}(\overline{X}_N) = \text{Var}\left(\frac{1}{N}\sum_{k=1}^{N} X_k\right) \approx c_5 N^{2d-1}, \quad \text{as } N \to \infty, \tag{2.10.7}$$

where $d = 0$ for SRD series, and $0 < d < 1/2$ for LRD series. See Proposition 2.2.5. The LRD parameter d can then be estimated through the following steps:

1. For $k = 2, \ldots, N/2$, compute sample means $\overline{X}_1(k), \overline{X}_2(k), \ldots, \overline{X}_{m_k}(k)$ over non-overlapping windows of size k.
2. Calculate

$$s^2(k) = \frac{1}{m_k - 1} \sum_{j=1}^{m_k} (\overline{X}_j(k) - \overline{X}(k))^2,$$

where

$$\overline{X}(k) = \frac{1}{m_k} \sum_{j=1}^{m_k} \overline{X}_j(k) = \frac{1}{km_k} \sum_{n=1}^{km_k} X_n.$$

3. Plot $\log s^2(k)$ versus $\log k$, and estimate d and c_5 by ordinary least squares.

In view of (2.10.7), one indeed expects that, for large k, $s^2(k) \approx c_5 k^{2d-1}$ and thus

$$\log s^2(k) \approx (2d - 1) \log k + \log c_5.$$

The method has the same drawbacks as the R/S method.

2.10.3 Regression in the Spectral Domain

This estimation method is based on a relation for the spectral density f_X of a series $\{X_n\}$ around zero,

$$f_X(\lambda) \approx c_4 \lambda^{-2d}, \quad \text{as } \lambda \to 0, \tag{2.10.8}$$

where $d = 0$ corresponds to SRD series, and $0 < d < 1/2$ to LRD series. With

$$I_X(\lambda_k) = \frac{1}{N}\left|\sum_{j=1}^{N} X_j e^{-ij\lambda_k}\right|^2 \tag{2.10.9}$$

denoting the periodogram of $X_1, \ldots, X_N$ at Fourier frequencies

$$\lambda_k = \frac{2\pi k}{N} \tag{2.10.10}$$

(see Section 1.3.3), this suggests that, for $k/N \approx 0$,

$$\log \frac{I_X(\lambda_k)}{2\pi} \approx \log c_4 - 2d \log \lambda_k, \tag{2.10.11}$$

and hence that d and c_4 could be estimated by least squares regression of $\log I_X(\lambda_k)$ on $\log \lambda_k$ for $k = 1, \ldots, m$. Here, m is the number of lower frequencies used in the estimation with $m/N \approx 0$. The resulting estimator of d is called the *GPH estimator* after Geweke and Porter-Hudak [387].

We shall describe next the expected asymptotics of GPH estimator, and the resulting confidence intervals. The argument below is partly informal and, in a sense, not completely correct. It nevertheless provides good intuition and points to advantages of using methods in the spectral domain. See also Section 10.6 for another related and popular estimator in the spectral domain. The analogous estimation of the constant c_4 (or $\log c_4$), and the joint estimation of d, c_4 are left as an exercise (see Exercise 2.18).

Consider the quantity

$$w(\lambda) = \frac{1}{\sqrt{2\pi N}} \sum_{j=1}^{N} X_j e^{-ij\lambda}, \qquad (2.10.12)$$

so that $|w(\lambda)|^2 = I_X(\lambda)$. The next result establishes some of its basic properties for LRD series.

Proposition 2.10.1 *Let $\{X_n\}_{n\in\mathbb{Z}}$ be a LRD series in the sense (2.1.5) in condition II with a quasi-monotone slowly varying function L_2 and let $w(\lambda)$ be defined by (2.10.12). Then, for fixed $\lambda_k = 2\pi k/N$ $(\neq 0)$,*

$$\mathbb{E}(\Re w(\lambda_k))^2 \to \frac{f_X(\lambda_k)}{2}, \quad \mathbb{E}(\Im w(\lambda_k))^2 \to \frac{f_X(\lambda_k)}{2}, \quad \mathbb{E}\Re w(\lambda_k)\Im w(\lambda_k) \to 0, \qquad (2.10.13)$$

where f_X is the spectral density of the series, and, for fixed $\lambda_k \neq \lambda_j$ $(\neq 0)$,

$$\mathbb{E}\Re w(\lambda_k)\Re w(\lambda_j) \to 0, \quad \mathbb{E}\Im w(\lambda_k)\Im w(\lambda_j) \to 0, \quad \mathbb{E}\Re w(\lambda_k)\Im w(\lambda_j) \to 0. \quad (2.10.14)$$

Proof Using

$$\sum_{j=1}^{N} e^{-ij\lambda_k} = \frac{e^{-i\lambda_k}(e^{-iN\lambda_k} - 1)}{e^{-i\lambda_k} - 1} = \frac{e^{-i\lambda_k}(e^{-i2\pi k} - 1)}{e^{-i2\pi k/N} - 1} = 0,$$

for $\lambda_k = 2\pi k/N \neq 0$, allows writing $\sum_{j=1}^{N} X_j e^{-ij\lambda_k} = \sum_{j=1}^{N} (X_j - \mathbb{E} X_j)e^{-ij\lambda_k}$. Hence,

$$\mathbb{E}(\Re w(\lambda_k))^2 = \frac{1}{2\pi N} \sum_{m=1}^{N} \sum_{n=1}^{N} \gamma_X(m-n) \cos(m\lambda_k) \cos(n\lambda_k) =: T_1 + T_2,$$

where

$$T_1 = \frac{1}{4\pi N} \sum_{m=1}^{N} \sum_{n=1}^{N} \gamma_X(m-n) \cos((m-n)\lambda_k),$$

$$T_2 = \frac{1}{4\pi N} \sum_{m=1}^{N} \sum_{n=1}^{N} \gamma_X(m-n) \cos((m+n)\lambda_k).$$

For T_1, note that

$$T_1 = \frac{1}{4\pi}\left(\gamma_X(0) + \frac{2}{N}\sum_{h=1}^{N-1}(N-h)\gamma_X(h)\cos(h\lambda_k)\right) =: \frac{1}{2}(T_{1,1} - T_{1,2}),$$

where

$$T_{1,1} = \frac{1}{2\pi}\left(\gamma_X(0) + 2\sum_{h=1}^{N-1}\gamma_X(h)\cos(h\lambda_k)\right), \quad T_{1,2} = \frac{1}{\pi N}\sum_{h=1}^{N-1}h\gamma_X(h)\cos(h\lambda_k).$$

Now, by Proposition 2.2.14, $T_{1,1} \to f_X(\lambda_k)$. As in (A.2.6) in the proof of Proposition A.2.1 in Appendix A.2,

$$N|T_{1,2}| \leq CN^{2d}L_2(N),$$

which implies $T_{1,2} \to 0$ since $d < 1/2$. For T_2, note that

$$T_2 = \frac{1}{4\pi N}\sum_{h=-(N-1)}^{N-1}\gamma_X(h)\sum_{q=|h|+1}^{N}\cos((2q-|h|)\lambda_k).$$

Indeed, the possible values of $m - n = h$ in $\gamma_X(m-n)$ are $-(N-1), \ldots, N-1$. For example, the value $m - n = h \geq 0$ is achieved by $(m, n) = (h+1, 1), \ldots, (N, N-h)$, for which $m + n = 2m - (m - n)$ takes values $2q - h$ with $q = h+1, \ldots, N$.

By arguing as in (A.2.2) of Appendix A.2,

$$\left|\sum_{q=|h|+1}^{N}\cos((2q-|h|)\lambda_k)\right| = \left|\sum_{r=1}^{N-|h|}\cos((|h|+2r)\lambda_k)\right| = \left|\Re\left(\sum_{r=1}^{N-|h|}e^{i(|h|+2r)\lambda_k}\right)\right|$$

$$\leq \left|\sum_{r=1}^{N-|h|}e^{i(|h|+2r)\lambda_k}\right| = \left|\sum_{r=1}^{N-|h|}e^{i2r\lambda_k}\right| = \left|\frac{e^{i2\lambda_k} - e^{i2(N-|h|)\lambda_k}}{1 - e^{i2\lambda_k}}\right| \leq \frac{2}{|1 - e^{i2\lambda_k}|} = \frac{1}{\sin\lambda_k} \leq C.$$

Hence, $|T_2| \leq C' \sum_{h=0}^{N-1}|\gamma_X(h)|/N$ and $T_2 \to 0$ in view of Proposition 2.2.1. Gathering the results above, we get

$$\mathbb{E}(\Re w(\lambda_k))^2 \to \frac{f_X(\lambda_k)}{2}.$$

One can show similarly that $\mathbb{E}(\Im w(\lambda_k))^2 \to \frac{f_X(\lambda_k)}{2}$. For $\mathbb{E}\Re w(\lambda_k)\Im w(\lambda_k)$, note that

$$\mathbb{E}\Re w(\lambda_k)\Im w(\lambda_k) = \frac{1}{2\pi N}\sum_{m=1}^{N}\sum_{n=1}^{N}\gamma_X(m-n)\cos(m\lambda_k)\sin(n\lambda_k)$$

$$= \frac{1}{4\pi N}\sum_{m=1}^{N}\sum_{n=1}^{N}\gamma_X(m-n)(\sin((m+n)\lambda_k) - \sin((m-n)\lambda_k))$$

$$= \frac{1}{4\pi N}\sum_{m=1}^{N}\sum_{n=1}^{N}\gamma_X(m-n)\sin((m+n)\lambda_k),$$

since the terms with $\sin((m-n)\lambda_k)$ cancel out by symmetry. The remaining expression converges to zero by arguing as for T_2 above. The rest of the statements (2.10.14) can also be proved similarly by arguing as for T_2 above (Exercise 2.17). □

Though Proposition 2.10.1 is stated for LRD series and hence $d > 0$, the proof above works also for SRD series and hence the case $d = 0$.

Suppose now for simplicity that $\{X_n\}$ is a Gaussian series. Then, the vector

$$(\Re w(\lambda_{k_1}), \Im w(\lambda_{k_1}), \ldots, \Re w(\lambda_{k_n}), \Im w(\lambda_{k_n}))'$$

is Gaussian. By Proposition 2.10.1, for fixed distinct $\lambda_{k_1}, \ldots, \lambda_{k_n}$, it converges in distribution to a Gaussian vector

$$\left(\sqrt{\frac{f_X(\lambda_{k_1})}{2}}\eta_{k_1}, \sqrt{\frac{f_X(\lambda_{k_1})}{2}}\zeta_{k_1}, \ldots, \sqrt{\frac{f_X(\lambda_{k_n})}{2}}\eta_{k_n}, \sqrt{\frac{f_X(\lambda_{k_n})}{2}}\zeta_{k_n}\right)',$$

where $\eta_{k_1}, \zeta_{k_1}, \ldots, \eta_{k_n}, \zeta_{k_n}$ are i.i.d. $\mathcal{N}(0,1)$ random variables since their covariances vanish. On the other hand, consider the periodograms $I_X(\lambda_k)$. One has

$$\frac{I_X(\lambda_k)}{2\pi} = |w(\lambda_k)|^2 = (\Re w(\lambda_k))^2 + (\Im w(\lambda_k))^2 \xrightarrow{d} \frac{f_X(\lambda_k)}{2}(\eta_k^2 + \zeta_k^2).$$

This implies that, for fixed distinct $\lambda_{k_1}, \ldots, \lambda_{k_n}$,

$$\left(\frac{I_X(\lambda_{k_1})}{2\pi}, \ldots, \frac{I_X(\lambda_{k_n})}{2\pi}\right)' \xrightarrow{d} \left(f_X(\lambda_{k_1})\xi_{k_1}, \ldots, f_X(\lambda_{k_n})\xi_{k_n}\right)', \quad (2.10.15)$$

where $\xi_{k_1}, \ldots, \xi_{k_n}$ are independent $\chi_2^2/2$ (or standard exponential) random variables.

We shall now focus on the m Fourier frequencies closest to the origin, namely, $\lambda_j = 2\pi j/N$ for $j = 1, \ldots, m$. In view of (2.10.11), one has

$$\log \frac{I_X(\lambda_k)}{2\pi} \approx \log c_4 + d(-2\log \lambda_k).$$

Hence, we can set $y_j = \log(I_X(\lambda_j)/2\pi)$ and $z_j = -2\log \lambda_j$, let $\bar{y} = \frac{1}{m}\sum_{j=1}^m y_j$ and $\bar{z} = \frac{1}{m}\sum_{j=1}^m z_j$ and focus on the linear regression

$$y_j = \log c_4 + dz_j. \quad (2.10.16)$$

Set

$$a_j = \frac{z_j - \bar{z}}{\sum_{j=1}^m (z_j - \bar{z})^2}$$

and observe that $\sum_{j=1}^m a_j = 0$. The GPH estimator of d is the linear regression estimator of the slope in (2.10.16), namely,

$$\widehat{d}_{gph} = \frac{\sum_{j=1}^m (z_j - \bar{z})(y_j - \bar{y})}{\sum_{j=1}^m (z_j - \bar{z})^2} = \sum_{j=1}^m a_j(y_j - \bar{y}) = \sum_{j=1}^m a_j y_j = \sum_{j=1}^m a_j \log I_X(\lambda_j).$$

The next result concerns the asymptotics of the numerator and denominator of the regression weights a_j.

Proposition 2.10.2 *For $z_j = -2\log\frac{2\pi j}{N}$, $j = 1, \ldots, m$, as $m \to \infty$,*

$$z_j - \bar{z} = -2\left(\log\left(\frac{j}{m}\right) + 1\right) + o(1), \quad \sum_{j=1}^{m}(z_j - \bar{z})^2 = 4m(1 + o(1)),$$

so that

$$a_j = \frac{z_j - \bar{z}}{\sum_{j=1}^{m}(z_j - \bar{z})^2} = -\frac{1}{2m}\left(\log\left(\frac{j}{m}\right) + 1\right) + o(1), \quad (2.10.17)$$

where $o(1)$ is uniform over $j = 1, \ldots, m$.

Proof For the first relation, note that

$$-\frac{1}{2}(z_j - \bar{z}) = \log j - \frac{1}{m}\sum_{j=1}^{m}\log j = \log\left(\frac{j}{m}\right) - \frac{1}{m}\sum_{j=1}^{m}\log\left(\frac{j}{m}\right)$$

$$= \log\left(\frac{j}{m}\right) - \int_0^1 \log x\, dx + o(1) = \log\left(\frac{j}{m}\right) + 1 + o(1).$$

For the second relation, by using the first relation, note that

$$\frac{1}{m}\sum_{j=1}^{m}(z_j - \bar{z})^2 = \frac{4}{m}\sum_{j=1}^{m}\left(\log\left(\frac{j}{m}\right) + 1\right)^2 + o(1) = 4\int_0^1 (\log x + 1)^2 dx + o(1) = 4 + o(1).$$

$\square$

In view of (2.10.15) and (2.10.8), one may expect that

$$\widehat{d}_{gph} = \sum_{j=1}^{m} a_j \log \frac{I_X(\lambda_j)}{2\pi} \approx \sum_{j=1}^{m} a_j \log f_X(\lambda_j) + \sum_{j=1}^{m} a_j \log \xi_j \approx \sum_{j=1}^{m} a_j \log c\lambda_j^{-2d} + \sum_{j=1}^{m} a_j \log \xi_j$$

$$= -2d\left(\sum_{j=1}^{m} a_j \log \frac{2\pi j}{N}\right) + \sum_{j=1}^{m} a_j \log \xi_j = -2d\left(\sum_{j=1}^{m} a_j \log \frac{j}{m}\right) + \sum_{j=1}^{m} a_j \left(\log \xi_j - \mathbb{E}\log \xi_j\right),$$

since $\sum_{j=1}^{m} a_j = 0$ and where ξ_js are i.i.d. exponential random variables. By Proposition 2.10.2,

$$\sum_{j=1}^{m} a_j \log\left(\frac{j}{m}\right) \approx -\frac{1}{2}\sum_{j=1}^{m}\left(\log\left(\frac{j}{m}\right) + 1\right)\left(\log\frac{j}{m}\right)\frac{1}{m} \approx -\frac{1}{2}\int_0^1 (\log x + 1)\log x\, dx = -\frac{1}{2}.$$

Hence

$$\sqrt{m}(\widehat{d}_{gph} - d) \approx \sqrt{m}\sum_{j=1}^{m} a_j \left(\log \xi_j - \mathbb{E}\log \xi_j\right).$$

By the Lindeberg–Feller central limit theorem for independent, non-identically distributed random variables (Feller [360], Chapter VIII, Section 4), one expects that

$$\sqrt{m}\sum_{j=1}^{m} a_j \left(\log \xi_j - \mathbb{E}\log \xi_j\right) \xrightarrow{d} \mathcal{N}(0, \sigma^2),$$

where, in view of (2.10.17), the asymptotic variance is given by

$$\sigma^2 = \lim_{m\to\infty} m \sum_{j=1}^{m} a_j^2 \operatorname{Var}(\log \xi_1) = \lim_{m\to\infty} \frac{1}{4m} \sum_{j=1}^{m} \left(\log\left(\frac{j}{m}\right) + 1\right)^2 \operatorname{Var}(\log \xi_1)$$

$$= \frac{1}{4}\int_0^1 (\log x + 1)^2 dx \operatorname{Var}(\log \xi_1) = \frac{\operatorname{Var}(\log \xi_1)}{4}.$$

The exponential variable ξ_1 has the density e^{-x}, $x > 0$, and hence

$$\operatorname{Var}(\log \xi_1) = \int_0^\infty (\log x)^2 e^{-x} dx - \left(\int_0^\infty (\log x) e^{-x} dx\right)^2 = \frac{\pi^2}{6} + C^2 - C^2 = \frac{\pi^2}{6},$$

where C is the Euler constant, by using Gradshteyn and Ryzhik [421], p. 571, Formula 4.331.1 and p. 572, Formula 4.335.1. Gathering the above observations, one expects that

$$\sqrt{m}(\widehat{d}_{gph} - d) \xrightarrow{d} \mathcal{N}(0, \frac{\pi^2}{24}). \tag{2.10.18}$$

Observe that the variance does not depend on d.

The result (2.10.18) is proved rigorously in Robinson [852]. We note that the asymptotics of m in Robinson [852] is different from above. It is assumed that $m = m(N)$ is such that $m/N \to 0$. In contrast, in Proposition 2.10.1, it was assumed that λ_k and, in particular, $\lambda_m = 2\pi m/N$ are fixed as $N \to \infty$. If λ_m is fixed, then $m/N \to \text{const} > 0$. Assuming $m/N \to 0$ is more realistic since (2.10.8) is an asymptotic relation as $\lambda \to 0$. Figure 2.8 illustrates the GPH method. For further discussion, see the notes to this section in Section 2.13.

2.10.4 Wavelet-Based Estimation

We begin by describing briefly what wavelets are. More details can be found in Section 8.2 below. A wavelet $\psi : \mathbb{R} \to \mathbb{R}$ is a function with special properties. For the purpose of this

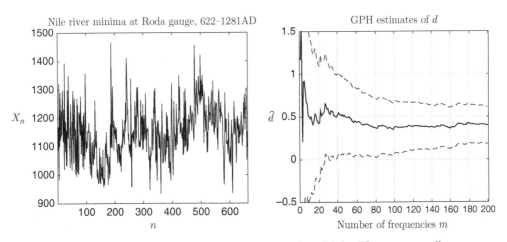

Figure 2.8 Left: Time series of the Nile river minima. Right: The corresponding GPH plot (with confidence intervals in dashed lines).

section, suppose that ψ has a compact support, is bounded, and has $Q \geq 1$ *zero moments*, that is,

$$\int_{\mathbb{R}} \psi(t) t^q dt = 0, \quad q = 0, \ldots, Q-1. \tag{2.10.19}$$

Let

$$\psi_{j,k}(t) = 2^{-j/2} \psi(2^{-j} t - k), \quad j, k \in \mathbb{Z}, \tag{2.10.20}$$

be scaled and translated copies of the wavelet ψ. Here, large j corresponds to large scales (low frequencies) and small or large negative j corresponds to small scales (high frequencies).

It is more convenient to focus here on continuous-time second-order stationary processes $X = \{X(t)\}_{t \in \mathbb{R}}$. We shall use the same notation and terminology which was used earlier. See Remark 2.1.6. For a process X, let

$$d(j, k) = \int_{\mathbb{R}} X(t) \psi_{j,k}(t) dt \tag{2.10.21}$$

be the so-called *detail (wavelet) coefficients* of X. The next result provides some basic properties of these coefficients, in the context of LRD processes X.

Proposition 2.10.3 *The following holds:*

(i) *For fixed j, the series $\{d(j, k)\}_{k \in \mathbb{Z}}$ has zero mean and is second-order stationary.*
(ii) *Detail coefficients are invariant to the addition of polynomial trends up to order $Q - 1$.*
(iii) *Let X be LRD in the sense (2.1.5) in condition II. Then, as $j \to \infty$,*

$$\mathbb{E} d(j, 0)^2 \sim C_d (2^j)^{2d} L_2(2^j), \tag{2.10.22}$$

where $C_d = \int_{\mathbb{R}^2} |s|^{2d-1} \psi(s) \psi(s+t) ds dt$.
(iv) *Let X be LRD in the sense (2.1.5) in condition II with $L_2(h) \sim c_2$, as $h \to \infty$. Suppose also that the function $\gamma_X(1/x) x^{2d-1}$ is $2Q$ times continuously differentiable at $x = 0$. Then, for fixed j,*

$$|\mathbb{E} d(j, 0) d(j, k)| \leq C |k|^{2d - 2Q - 1},$$

where the constant $C = C_{d,j}$ depends, in particular, on j.

Proof Part (i) follows from $\mathbb{E} d(j, k) = \mu_X \int_{\mathbb{R}} \psi_{j,k}(t) dt = 0$ by setting $\mu_X = \mathbb{E} X(t)$ and using (2.10.19). The second-order stationarity follows from: for $k, h \in \mathbb{Z}$,

$$\mathbb{E} d(j, k+h) d(j, k) = \int_{\mathbb{R}} \int_{\mathbb{R}} \mathbb{E} X(t) X(s) \, 2^{-j/2} \psi(2^{-j} t - (k+h)) 2^{-j/2} \psi(2^{-j} s - k) dt ds$$

$$= \int_{\mathbb{R}} \int_{\mathbb{R}} \mathbb{E} X(u + 2^j k) X(v + 2^j k) \, 2^{-j/2} \psi(2^{-j} u - h) 2^{-j/2} \psi(2^{-j} v) du dv$$

$$= \int_{\mathbb{R}} \int_{\mathbb{R}} \mathbb{E} X(u) X(v) \, 2^{-j/2} \psi(2^{-j} u - h) \, 2^{-j/2} \psi(2^{-j} v) du dv = \mathbb{E} d(j, h) d(j, 0),$$

where we used the change of variables $t = u + 2^j k$, $s = v + 2^j k$, and the second-order stationarity of the stationarity of the process $\{X(u)\}$.

For part (ii), suppose that the process is $Y(t) = \sum_{j=0}^{Q-1} a_j t^j + X(t)$ instead of $X(t)$. Then, the detail coefficients for $X(t)$ and $Y(t)$ are the same because of (2.10.19).

For part (iii), observe that

$$\mathbb{E}d(j,0)^2 = \int_{\mathbb{R}^2} \gamma_X(u-v)\psi_{j,0}(u)\psi_{j,0}(v)dudv = 2^{-j}\int_{\mathbb{R}^2}\gamma_X(u-v)\psi(2^{-j}u)\psi(2^{-j}v)dudv.$$

The change of variables $x = 2^{-j}(u-v)$ and $y = 2^{-j}v$ yields

$$\mathbb{E}d(j,0)^2 = 2^j \int_{\mathbb{R}} \gamma_X(2^j x)\Psi(x)dx,$$

where $\Psi(x) = \int_{\mathbb{R}} \psi(y)\psi(x+y)dy$. The function Ψ also has a compact support and is bounded. With $\gamma_X(h) = h^{2d-1}L_2(h)$, $h > 0$,

$$\mathbb{E}d(j,0)^2 = (2^j)^{2d}\int_{\mathbb{R}} |x|^{2d-1}L_2(2^j|x|)\Psi(x)dx$$

$$= (2^j)^{2d}L_2(2^j)\int_{\mathbb{R}} |x|^{2d-1}\frac{L_2(2^j|x|)}{L_2(2^j)}\Psi(x)dx.$$

Express the integral as $\int_\epsilon^\infty + \int_0^\epsilon$ for $\epsilon > 0$. For the first integral, by (2.2.4), $L_2(2^j|x|)/L_2(2^j) \to 1$, uniformly for $|x| \in [\epsilon, A]$, where A is arbitrary and fixed. Since Ψ has compact support, the integral is asymptotic to the right-hand side of (2.10.22), as $j \to \infty$. On the other hand, the second integral $\int_0^\epsilon$ is negligible. Indeed,

$$(2^j)^{2d}\int_0^\epsilon x^{2d-1}L_2(2^j x)|\Psi(x)|dx \leq C\int_0^{2^j\epsilon} y^{2d-1}L_2(y)dy \leq \tilde{C}\epsilon^{2d}(2^j)^{2d}L_2(2^j\epsilon)$$

by Proposition 2.2.2. For small ϵ, this is arbitrarily small with respect to $(2^j)^{2d}L_2(2^j)$. This establishes the result (iii).

For part (iv), note that $d(j,k) = \int_{\mathbb{R}} X(t)2^{-j/2}\psi(2^{-j}t - k)dt = 2^{j/2}\int_{\mathbb{R}} X(2^j t)\psi(t-k)dt$ and that the process $\{X(2^j t)\}_{t\in\mathbb{R}}$ satisfies the same assumptions as $\{X(t)\}_{t\in\mathbb{R}}$. We may thus suppose without loss of generality that $j = 0$. By the assumption on γ_X, $\gamma_X(1/x) = (1/x)^{2d-1}L_2(1/x)$ where $L_2(1/x) = \gamma_X(1/x)x^{2d-1}$ is $2Q$ times continuously differentiable. Then,

$$\gamma_X(1/x)x^{2d-1} = L_2(1/x) = b_2 + b_3 x + \cdots + b_{2Q+2}x^{2Q} + o(x^{2Q}),$$

as $x \to 0$, or

$$\gamma_X(h) = b_2 h^{2d-1} + \cdots + b_{2Q+2}h^{2d-2Q-1} + o(h^{2d-2Q-1}), \qquad (2.10.23)$$

as $h \to \infty$. With Ψ defined above, note that, for large enough k,

$$\mathbb{E}d(0,0)d(0,k) = \int_{\mathbb{R}} \gamma_X(x+k)\Psi(x)dx$$

$$= \int_{\mathbb{R}} \sum_{p=1}^{2Q+1} b_{p+1}(x+k)^{2d-p}\Psi(x)dx$$

$$+ \int_{\mathbb{R}} \left(\gamma_X(x+k) - \sum_{p=1}^{2Q+1} b_{p+1}(x+k)^{2d-p}\right)\Psi(x)dx. \qquad (2.10.24)$$

Consider now the last integral in (2.10.24). Since Ψ has a bounded support supp$\{\Psi\}$, we may suppose that $x + k$ is of the order k and hence large enough as $k \to \infty$, uniformly in $x \in$ supp$\{\Psi\}$. In particular, using (2.10.23), we have

$$\left| \gamma_X(x+k) - \sum_{p=1}^{2Q+1} b_{p+1}(x+k)^{2d-p} \right| = o(k^{2d-2Q-1}), \quad \text{as } k \to \infty,$$

uniformly in $x \in$ supp$\{\Psi\}$. Applying this relation and using the fact that Ψ is bounded, we obtain that the last integral in (2.10.24) is of the order $o(k^{2d-2Q-1})$, as $k \to \infty$.

For the first integral in (2.10.24), we first observe that ψ having Q zero moments implies that Ψ has $2Q$ zero moments. Indeed, for $q < 2Q$,

$$\int_{\mathbb{R}} x^q \Psi(x) dx = \int_{\mathbb{R}^2} (t-s)^q \psi(t) \psi(s) ds dt = \int_{\mathbb{R}^2} \sum_{m=0}^{q} \binom{q}{m} t^m (-s)^{q-m} \psi(t) \psi(s) ds dt$$

$$= \sum_{m=0}^{q} \binom{q}{m} \int_{\mathbb{R}} t^m \psi(t) dt \int_{\mathbb{R}} (-s)^{q-m} \psi(s) ds = 0,$$

since ψ has Q zero moments. Then,

$$\int_{\mathbb{R}} (x+k)^{2d-p} \Psi(x) dx = k^{2d-p} \int_{\mathbb{R}} \left(1 + \frac{x}{k}\right)^{2d-p} \Psi(x) dx = k^{2d-p} \int_{\mathbb{R}} \sum_{q=0}^{\infty} a_q \left(\frac{x}{k}\right)^q \Psi(x) dx$$

$$= k^{2d-p} \int_{\mathbb{R}} \sum_{q=2Q}^{\infty} a_q \left(\frac{x}{k}\right)^q \Psi(x) dx = O(k^{2d-p-2Q}),$$

as $k \to \infty$. Since the term with $p = 1$ dominates, one obtains $O(k^{2p-1-2Q})$ as a bound. $\square$

Proposition 2.10.3 shows that there are several advantages in using wavelets:

- invariance to polynomial trends in (ii),
- recovery of the LRD parameter d in (iii),
- strong decorrelation in (iv).

Another property, namely, that of computational efficiency related to Fast Wavelet Transform is indicated below.

Proposition 2.10.3, (iii) (assuming $L_2(h) \sim c_2$ as $h \to \infty$), suggests a straightforward way to estimate LRD parameter d. Consider

$$\overline{\mu}_j = \frac{1}{n_j} \sum_{k=1}^{n_j} |d_{j,k}|^2, \tag{2.10.25}$$

where $d_{j,k}$, $k = 1, \ldots, n_j$, are detail coefficients available at scale 2^j. They are typically approximations of the detail coefficients $d(j, k)$ defined in (2.10.21). This is because one never observes the path $X(t)$ for all $t \in \mathbb{R}$. In practice, only a sample of the discrete-time series is available. In this case, either the detail coefficients can be viewed as approximations when the discrete time series is considered a sample of a continuous-time process, or they can be viewed as the exact detail coefficients of a suitable continuous-time process constructed from discrete-time series. There is, in either case, a very fast algorithm, the so-called *Fast Wavelet Transform* (FWT), for computing these coefficients in practice (see Section 8.2).

Suppose here, in addition, that the underlying process X is Gaussian. In this case, the detail coefficients are also Gaussian. Motivated by Proposition 2.10.3, (iv), which indicates that the detail coefficients $d(j, k)$ are weakly correlated, assume also that *all* the detail coefficients $d_{j,k}$ are uncorrelated. In this case, note that, by Proposition 2.10.3, (iii), and for large scales j, the $|d_{j,k}|^2$, $1 \leq k \leq n_j$, are distributed as $C_d c_2 2^{2dj} |X_{j,k}|^2$ where $X_{j,k}$ are independent $\mathcal{N}(0, 1)$ random variables. Then

$$\log_2 \overline{\mu}_j = \log_2 \Big(\frac{1}{n_j} \sum_{k=1}^{n_j} |d_{j,k}|^2 \Big) \approx \log_2 \Big(\frac{2^{2dj} C_d c_2}{n_j} \sum_{k=1}^{n_j} |X_{j,k}|^2 \Big)$$

$$= \log_2 C_d c_2 + 2dj - \log_2 n_j + \log_2 \sum_{k=1}^{n_j} |X_{j,k}|^2. \quad (2.10.26)$$

We will compute $\log_2 \overline{\mu}_j$ from the data and use the relation (2.10.26) to estimate d and c_2 through a regression on j. To make this precise, set

$$y_j = \log_2 \overline{\mu}_j - g(j),$$

where

$$g(j) = \mathbb{E}\Big(-\log_2 n_j + \log_2 \sum_{k=1}^{n_j} |X_{j,k}|^2 \Big).$$

While $\mathbb{E} \sum_{k=1}^{n_j} |X_{j,k}|^2 = \sum_{k=1}^{n_j} \mathbb{E}|X_{j,k}|^2 = n_j$, it is not true that $\mathbb{E} \log_2 \sum_{k=1}^{n_j} |X_{j,k}|^2$ equals $\log_2 n_j$ because $\mathbb{E} \log_2 \neq \log_2 \mathbb{E}$. The term $g(j)$ will serve as a bias correction. Then

$$y_j \approx \log_2 C_d c_2 + 2dj + \epsilon_j, \quad (2.10.27)$$

with

$$\epsilon_j = -\log_2 n_j + \log_2 \sum_{k=1}^{n_j} |X_{j,k}|^2 - g(j)$$

satisfying $\mathbb{E}\epsilon_j = 0$.

The next result sheds light on $g(j)$ and $\mathrm{Var}(\epsilon_j)$.

Proposition 2.10.4 *We have*

$$g(j) = \frac{1}{\log 2}\Big(\psi_0\Big(\frac{n_j}{2}\Big) - \log \frac{n_j}{2}\Big) \sim -\frac{1}{n_j \log 2},$$

$$\sigma_j^2 := \mathrm{Var}(\epsilon_j) = \frac{\zeta\big(2, \frac{n_j}{2}\big)}{(\log 2)^2} \sim \frac{2}{n_j (\log 2)^2},$$

as $n_j \to \infty$, where $\psi_0(z) = \Gamma'(z)/\Gamma(z)$ is the digamma function and $\zeta(2, z) = \sum_{n=0}^{\infty}(z+n)^{-2}$ is a special case of the generalized Riemann zeta function.

Proof We need the relations: for $\Re\nu > 0$, $\Re\mu > 0$,

$$\int_0^\infty x^{\nu-1} e^{-\mu x} \log x \, dx = \frac{\Gamma(\nu)}{\mu^\nu}(\psi_0(\nu) - \log\mu),$$

$$\int_0^\infty x^{\nu-1} e^{-\mu x} (\log x)^2 dx = \frac{\Gamma(\nu)}{\mu^\nu}\left((\psi_0(\nu) - \log\mu)^2 + \zeta(2,\nu)\right)$$

(Gradshteyn and Ryzhik [421], p. 573, Formula 4.352.1 and p. 576, Formula 4.358.2). Since $\sum_{k=1}^{n_j} |X_{j,k}|^2$ is distributed like a chi-squared variable with n_j degrees of freedom, it has the density $x^{n_j/2-1} e^{-x/2}/(2^{n_j/2}\Gamma(n_j/2))$, $x > 0$. Then, by using $\log_2 x = \log x / \log 2$,

$$g(j) = -\log_2 n_j + \frac{1}{2^{n_j/2}\Gamma(n_j/2)} \int_0^\infty (\log_2 x) x^{n_j/2-1} e^{-x/2} dx$$

$$= -\log_2 n_j + \frac{1}{(\log 2) 2^{n_j/2}\Gamma(n_j/2)} \int_0^\infty (\log x) x^{n_j/2-1} e^{-x/2} dx$$

$$= -\log_2 n_j + \frac{1}{\log 2}(\psi_0(n_j/2) + \log 2) = \frac{1}{\log 2}\psi_0\left(\frac{n_j}{2}\right) - \frac{\log n_j - \log 2}{\log 2}$$

$$= \frac{1}{\log 2}\left(\psi_0\left(\frac{n_j}{2}\right) - \log\frac{n_j}{2}\right) \sim -\frac{1}{n_j \log 2},$$

where the last asymptotic relation follows from $\psi_0(z) = \log z - (1/2z) + O(1/z^2)$, as $z \to \infty$ (Gradshteyn and Ryzhik [421], p. 902, Formula 8.361.3).

For $\mathrm{Var}(\epsilon_j)$, observe that

$$(\log 2)^2 \mathrm{Var}(\epsilon_j) = (\log 2)^2 \mathrm{Var}\left(\log_2 \sum_{k=1}^{n_j} |X_{j,k}|^2\right)$$

$$= \frac{1}{2^{n_j/2}\Gamma(n_j/2)} \int_0^\infty (\log x)^2 x^{n_j/2-1} e^{-x/2} dx$$

$$- \left(\frac{1}{2^{n_j/2}\Gamma(n_j/2)} \int_0^\infty (\log x) x^{n_j/2-1} e^{-x/2} dx\right)^2,$$

which is $\zeta(2, n_j/2)$ by using the formulas above. The asymptotic relation follows directly from the definition of $\zeta(2, z)$ since for $z > 1$, $z^{-1} = \int_0^\infty (z+u)^{-2} du \leq \zeta(2,z) \leq \int_{-1}^\infty (z+u)^{-2} du = (z-1)^{-1}$ and hence $\zeta(2,z) \sim z^{-1}$ as $z \to \infty$. $\square$

These results suggest that d and c_2 can be estimated by a weighted least squares regression of y_j against j with weights $1/\sigma_j^2$, for some range $j = J_1, \ldots, J_2$ of large j. Since the slope in (2.10.27) is $2d$, the resulting estimator of d is

$$\widehat{d}_w = \frac{1}{2}\frac{\sum_{j=J_1}^{J_2}(1/\sigma_j^2)(z_j - \overline{z})(y_j - \overline{y})}{\sum_{j=J_1}^{J_2}(1/\sigma_j^2)(z_j - \overline{z})^2} = \frac{1}{2}\sum_{j=J_1}^{J_2} w_j y_j,$$

where

$$z_j = j, \qquad \overline{z} = \frac{\sum_{j=J_1}^{J_2}(1/\sigma_j^2) z_j}{\sum_{j=J_1}^{J_2}(1/\sigma_j^2)}$$

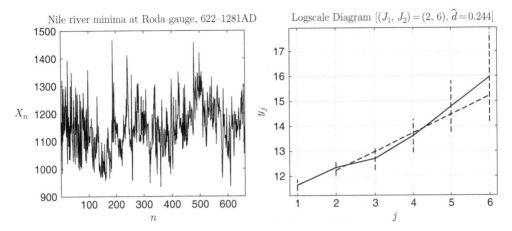

Figure 2.9 Left: Time series of the Nile river minima. Right: The corresponding log-scale diagram. The solid line connects the values of y_j, with confidence intervals indicated in vertical lines. The dashed line is the regression fit over $(J_1, J_2) = (2, 6)$.

and

$$w_j = \frac{(1/\sigma_j^2)(z_j - \bar{z})}{\sum_{j=J_1}^{J_2}(1/\sigma_j^2)(z_j - \bar{z})^2},$$

so that $\sum_{j=J_1}^{J_2} w_j = 0$. The main difference with the case of simple regression considered in Section 2.10.3 is that here, the sums are now weighted sums. One can express the regression weights in a convenient way as

$$w_j = \frac{S_0 j - S_1}{\sigma_j^2 (S_0 S_2 - S_1^2)}$$

with

$$S_k = \sum_{j=J_1}^{J_2} \frac{j^k}{\sigma_j^2}, \quad k = 0, 1, 2.$$

The plot of y_j versus j is known as the log-scale diagram. See Figure 2.9. The constant c_2 can be estimated similarly.

The variance of $\widehat{d}_w$ is approximately

$$\text{Var}(\widehat{d}_w) \approx \frac{1}{4} \frac{\sum_{j=J_1}^{J_2}((1/\sigma_j^4)(z_j - \bar{z}))^2 \sigma_j^2}{(\sum_{j=J_1}^{J_2}(1/\sigma_j^2)(z_j - \bar{z})^2)^2} = \frac{1}{4} \frac{1}{\sum_{j=J_1}^{J_2}(1/\sigma_j^2)(z_j - \bar{z})^2} = \frac{S_0}{4(S_0 S_2 - S_1^2)}$$

and $\widehat{d}_w$ is expected to be asymptotically normal since such y_j (or $\bar{\mu}_j$) is expected to satisfy a central limit theorem if the number of observations is large.

2.11 Generation of Gaussian Long- and Short-Range Dependent Series

Suppose that one would like to generate N samples $X = (X_0, \ldots, X_{N-1})'$ of a stationary, zero mean Gaussian series $\{X_n\}$ having a known covariance function $\gamma_X(n)$, $n \in \mathbb{Z}$. For notational simplicity, we write γ instead of γ_X throughout the section. The (column) vector X has the covariance structure

$$\Sigma = \begin{pmatrix} \gamma(0) & \gamma(1) & \cdots & \gamma(N-1) \\ \gamma(1) & \gamma(0) & \cdots & \gamma(N-2) \\ \vdots & \vdots & \ddots & \vdots \\ \gamma(N-1) & \gamma(N-2) & \cdots & \gamma(0) \end{pmatrix}. \quad (2.11.1)$$

We discuss below two methods for generation of X.

2.11.1 Using Cholesky Decomposition

According to the Cholesky decomposition, we can write

$$\Sigma = AA', \quad (2.11.2)$$

where A is a lower triangular matrix. Then, let Z be a $N \times 1$ vector of independent $\mathcal{N}(0, 1)$ random variables. Set

$$X = AZ. \quad (2.11.3)$$

The vector X is Gaussian, zero mean and has the desired covariance structure

$$\mathbb{E} XX' = A(\mathbb{E} ZZ')A' = AA' = \Sigma, \quad (2.11.4)$$

by using (2.11.2).

This method is of complexity $O(N^3)$ and hence relatively slow. It can be used to generate the series X of moderate sizes N only.

2.11.2 Using Circulant Matrix Embedding

This is an exact, fast and most popular synthesis method for Gaussian stationary series, with a given covariance structure. Since the method involves circulant matrices, we first recall some of their basic properties.

Recall that a matrix A is called *circulant* if it has a form

$$A = \begin{pmatrix} a_0 & a_1 & \cdots & a_{n-2} & a_{n-1} \\ a_{n-1} & a_0 & \cdots & a_{n-3} & a_{n-2} \\ \vdots & \vdots & \ddots & \vdots & \vdots \\ a_1 & a_2 & \cdots & a_{n-1} & a_0 \end{pmatrix} = \Big(a_{(k-j) \bmod n} \Big)_{1 \leq j,k \leq n},$$

that is, the first row of A is $(a_0, a_1, \ldots, a_{n-1})$ and every subsequent row is a circular shift by one element of the preceding row. The next result describes the eigenvector and eigenvalue structures of circulant matrices.

Proposition 2.11.1 *A circulant matrix A has orthonormal eigenvectors*

$$f_m = \frac{1}{\sqrt{n}} \Big(1, e^{-i\frac{2\pi m}{n}}, e^{-i\frac{2\pi 2m}{n}}, \ldots, e^{-i\frac{2\pi(n-1)m}{n}} \Big)',$$

2.11 Generation of Gaussian Long- and Short-Range Dependent Series

$m = 0, \ldots, n-1$, *and the corresponding eigenvalues*

$$\lambda_m = \sum_{j=0}^{n-1} a_j e^{-i\frac{2\pi jm}{n}}. \tag{2.11.5}$$

Proof This is easy to verify. Denoting the kth element of Af_m and f_m by $(Af_m)_k$ and $(f_m)_k$, respectively, $k = 1, \ldots, n$, observe that

$$(Af_m)_k = \frac{1}{\sqrt{n}} \sum_{l=1}^{n} a_{(l-k) \bmod n} e^{-i\frac{2\pi(l-1)m}{n}}$$

$$= \frac{1}{\sqrt{n}} \sum_{l=1}^{k-1} a_{n-k+l} e^{-i\frac{2\pi(l-1)m}{n}} + \frac{1}{\sqrt{n}} \sum_{l=k}^{n} a_{l-k} e^{-i\frac{2\pi(l-1)m}{n}}$$

$$= \frac{1}{\sqrt{n}} \sum_{j=n-k+1}^{n-1} a_j e^{-i\frac{2\pi(j-n+k-1)m}{n}} + \frac{1}{\sqrt{n}} \sum_{j=0}^{n-k} a_j e^{-i\frac{2\pi(j+k-1)m}{n}}$$

$$= \left(\sum_{j=0}^{n-1} a_j e^{-i\frac{2\pi jm}{n}}\right) \frac{1}{\sqrt{n}} e^{-i\frac{2\pi(k-1)m}{n}} = \lambda_m(f_m)_k.$$

Orthonormality is easy to prove. $\square$

Note that the eigenvectors f_m above form a discrete Fourier transform (DFT) basis. In practice, the eigenvalues λ_m can be computed using the fast Fourier transform (FFT). We have $Af_m = \lambda_m f_m$ and hence

$$AF = FG,$$

where $F = (f_0 \ldots f_{n-1})$ is the DFT matrix, $G = \text{diag}(\lambda_0, \ldots, \lambda_{n-1})$ and $*$ denotes Hermitian transpose. Recall that if F is a complex-valued matrix with entries $f_{j,k} = u_{j,k} + iv_{j,k}$, then the Hermitian transpose matrix F^* has entries $f^*_{j,k} = \overline{f}_{k,j} = u_{k,j} - iv_{k,j}$. Since $FF^* = I$, we get therefore

$$A = FGF^*.$$

We now turn to the generation of a Gaussian zero mean vector $X = (X_0, \ldots, X_{N-1})'$ having the covariance structure (2.11.1). Set $M = N - 1$ and define the following $2M \times 2M$ circulant matrix

$$\widetilde{\Sigma} = \begin{pmatrix} \gamma(0) & \gamma(1) & \ldots & \gamma(N-2) & \gamma(N-1) & \gamma(N-2) & \ldots & \gamma(1) \\ \gamma(1) & \gamma(0) & \ldots & \gamma(N-3) & \gamma(N-2) & \gamma(N-1) & \ldots & \gamma(2) \\ \vdots & \vdots & \ddots & \vdots & \vdots & \vdots & \ddots & \vdots \\ \gamma(N-1) & \gamma(N-2) & \ldots & \gamma(1) & \gamma(0) & \gamma(1) & \ldots & \gamma(N-2) \\ \vdots & \vdots & \ddots & \vdots & \vdots & \vdots & \ddots & \vdots \\ \gamma(1) & \gamma(2) & \ldots & \gamma(N-1) & \gamma(N-2) & \gamma(N-3) & \ldots & \gamma(0) \end{pmatrix}. \tag{2.11.6}$$

Note that $\widetilde{\Sigma}$ contains Σ in its upper-left corner. By the results for circulant matrices noted above,

$$\widetilde{\Sigma} = FGF^*, \tag{2.11.7}$$

where, in view of (2.11.5), the diagonal matrix $G = \operatorname{diag}(g_0, \ldots, g_{2M-1})$ has diagonal entries

$$\begin{aligned}
g_m &= \sum_{j=0}^{M} \gamma(j) e^{-i\frac{2\pi jm}{2M}} + \sum_{j=M+1}^{2M-1} \gamma(2M-j) e^{-i\frac{2\pi jm}{2M}} \\
&= \sum_{j=0}^{M} \gamma(j) e^{-i\frac{2\pi jm}{2M}} + \sum_{j=1}^{M-1} \gamma(j) e^{-i\frac{2\pi(2M-j)m}{2M}} \\
&= \gamma(0) + \gamma(M)(-1)^m + \sum_{j=1}^{M-1} \gamma(j) \left(\cos\left(\frac{-\pi jm}{M}\right) + \cos\left(\frac{\pi jm}{M}\right) \right) \\
&= \gamma(0) + \gamma(M)(-1)^m + 2 \sum_{j=1}^{M-1} \gamma(j) \cos\left(\frac{\pi jm}{M}\right).
\end{aligned} \tag{2.11.8}$$

We shall now make the following assumption which will be discussed below.

Assumption ND Suppose that the matrix $\widetilde{\Sigma}$ above is positive semidefinite. Equivalently, suppose that the eigenvalues g_m, $m = 0, \ldots, 2M-1$, are nonnegative.

Under Assumption ND there is a Gaussian vector $\widetilde{X}$ with the covariance structure $\widetilde{\Sigma}$. We will show below that such $\widetilde{X}$ can be constructed easily by using the factorization (2.11.7) of $\widetilde{\Sigma}$ above. Since $\widetilde{\Sigma}$ contains Σ in its upper-left corner, the first N elements X of $\widetilde{X}$ will then have the desired covariance structure Σ.

To construct $\widetilde{X}$ above, it is more convenient to work with complex-valued variables. Let

$$\widetilde{Y} = FG^{1/2}Z,$$

where $F = (f_0 \ldots f_{2M-1})$ is defined in Proposition 2.11.1, $G^{1/2} = \operatorname{diag}(g_0^{1/2}, \ldots, g_{2M-1}^{1/2})$ exists by using Assumption ND and $Z = Z^0 + iZ^1$ consists of two independent $\mathcal{N}(0, I_{2M})$ random vectors Z^0 and Z^1. Note that, for $m = 0, \ldots, 2M-1$,

$$\widetilde{Y}_m = \frac{1}{\sqrt{2M}} \sum_{j=0}^{2M-1} g_j^{1/2} Z_j e^{-i\frac{2\pi jm}{2M}}, \tag{2.11.9}$$

so that $\widetilde{Y}$ can be readily computed by FFT. The following result describes the covariance structure of $\widetilde{Y}$.

Proposition 2.11.2 *The Gaussian vectors $\Re(\widetilde{Y})$ and $\Im(\widetilde{Y})$ are independent, with the covariance structure*

$$\mathbb{E}\Re(\widetilde{Y})\Re(\widetilde{Y})' = \mathbb{E}\Im(\widetilde{Y})\Im(\widetilde{Y})' = \widetilde{\Sigma}.$$

Proof Note that $\mathbb{E}ZZ' = \mathbb{E}Z^0(Z^0)' - \mathbb{E}Z^1(Z^1)' = 0$, and hence

$$0 = \mathbb{E}\widetilde{Y}\widetilde{Y}' = \mathbb{E}(\Re(\widetilde{Y}) + i\Im(\widetilde{Y}))(\Re(\widetilde{Y})' + i\Im(\widetilde{Y})')$$

or
$$\mathbb{E}\Re(\widetilde{Y})\Re(\widetilde{Y})' = \mathbb{E}\Im(\widetilde{Y}))\Im(\widetilde{Y})', \quad \mathbb{E}\Re(\widetilde{Y})\Im(\widetilde{Y})' + \mathbb{E}\Im(\widetilde{Y})\Re(\widetilde{Y})' = 0.$$

On the other hand, by (2.11.7),
$$\mathbb{E}\widetilde{Y}\widetilde{Y}^* = FG^{1/2}\mathbb{E}ZZ^*G^{1/2}F^* = 2FGF^* = 2\widetilde{\Sigma},$$
since $\mathbb{E}ZZ^* = \mathbb{E}Z^0(Z^0)' + \mathbb{E}Z^1(Z^1)' = 2I_{2M}$. Hence
$$2\widetilde{\Sigma} = \mathbb{E}\widetilde{Y}\widetilde{Y}^* = \mathbb{E}(\Re(\widetilde{Y}) + i\Im(\widetilde{Y}))(\Re(\widetilde{Y})' - i\Im(\widetilde{Y})')$$
or
$$\mathbb{E}\Re(\widetilde{Y})\Re(\widetilde{Y})' + \mathbb{E}\Im(\widetilde{Y}))\Im(\widetilde{Y})' = 2\widetilde{\Sigma}, \quad \mathbb{E}\Re(\widetilde{Y})\Im(\widetilde{Y})' - \mathbb{E}\Im(\widetilde{Y})\Re(\widetilde{Y})' = 0.$$
It follows that $\mathbb{E}\Re(\widetilde{Y})\Re(\widetilde{Y})' = \mathbb{E}\Im(\widetilde{Y})\Im(\widetilde{Y})' = \widetilde{\Sigma}$ and $\mathbb{E}\Re(\widetilde{Y})\Im(\widetilde{Y})' = 0$. □

In view of Proposition 2.11.2, both $\widetilde{X} = \Re(\widetilde{Y})$ and $\widetilde{X} = \Im(\widetilde{Y})$ have the covariance structure $\widetilde{\Sigma}$. Finally, the desired vector X with the covariance structure Σ is made up of the first $N = M + 1$ elements of $\widetilde{X}$.

We have then proved the following.

Theorem 2.11.3 *Suppose that Assumption ND holds and let $\widetilde{Y} = (Y_0, \ldots, Y_{2M-1})'$ be the complex-valued vector defined in (2.11.9). Set either $\widetilde{X} = \Re(\widetilde{Y})$ or $\widetilde{X} = \Im(\widetilde{Y})$, and let the sub-vector X consist of the first $M + 1$ elements of $\widetilde{X}$. Then, X is a real-valued Gaussian vector with covariance matrix Σ.*

We shall discuss next Assumption ND. The first result shows that this assumption is expected to hold for most SRD time series of interest, at least for large enough N.

Proposition 2.11.4 *If $\{X_n\}$ is a SRD series with (strictly) positive density function $f(\lambda)$, then Assumption ND holds for large enough N.*

Proof For a SRD series, $2\pi f(\lambda) = \gamma(0) + 2\sum_{j=1}^{\infty} \gamma(j)\cos(j\lambda)$. Moreover, the truncated series
$$G_M(\lambda) := \gamma(0) + \gamma(M)\cos(M\lambda) + 2\sum_{j=1}^{M-1} \gamma(j)\cos(j\lambda)$$
converges to $2\pi f(\lambda)$ uniformly on $[0, 2\pi)$ because $\sum_{j=1}^{\infty} |\gamma(j)| < \infty$. Since f is positive, then $\inf_{\lambda \in [0,2\pi)} |G_M(\lambda)| > 0$ for large enough M. It remains to observe that $G_M(\pi m/M) = g_m, m = 0, \ldots, 2M - 1$, where g_m is given by (2.11.8). □

Note that Proposition 2.11.4 does not indicate how large N should be for Assumption ND to hold. It rather explains why one expects Assumption ND to hold. From a practical perspective, if Assumption ND does not hold for a given N (that is, some of the eigenvalues g_m are negative), the result also suggests that one should try increasing N. Regarding Proposition 2.11.4, see also Exercise 2.21.

The next result, on the other hand, provides sufficient conditions for Assumption ND to hold for a fixed M.

Proposition 2.11.5 (a) *Suppose that the sequence $\gamma(0), \ldots, \gamma(M)$ is convex, decreasing and nonnegative. Then, Assumption ND holds.*

(b) *Suppose $\gamma(k) \leq 0$, $k \in \mathbb{Z} \setminus \{0\}$. Then, Assumption ND holds for any M.*

Proof (a) Let $D_0(w) = 1$ and

$$D_j(w) = 1 + 2\sum_{k=1}^{j} \cos(\pi k w) = 1 + \frac{2}{\sin\frac{\pi w}{2}} \sum_{k=1}^{j} \cos(\pi k w) \sin\frac{\pi w}{2}$$

$$= 1 + \frac{1}{\sin\frac{\pi w}{2}} \sum_{k=1}^{j} \left(\sin\left(\pi(k+\tfrac{1}{2})w\right) - \sin\left(\pi(k-\tfrac{1}{2})w\right) \right) = \frac{\sin(\pi(j+\tfrac{1}{2})w)}{\sin\frac{\pi w}{2}}.$$

Proceeding as in summation by parts (see Exercise 2.2),

$$g_m = \gamma(0) + \gamma(M)(-1)^m + 2\sum_{j=1}^{M-1} \gamma(j) \cos\left(\frac{\pi j m}{M}\right) \tag{2.11.10}$$

$$= \gamma(0) + \gamma(M)(-1)^m + \sum_{j=1}^{M-1} \gamma(j)\left(D_j\left(\frac{m}{M}\right) - D_{j-1}\left(\frac{m}{M}\right)\right)$$

$$= \gamma(0) + \gamma(M)(-1)^m + \sum_{j=1}^{M-1} \gamma(j)D_j\left(\frac{m}{M}\right) - \sum_{j=0}^{M-2} \gamma(j+1)D_j\left(\frac{m}{M}\right)$$

$$= \gamma(0) + \gamma(M)(-1)^m + \sum_{j=1}^{M-1} \Delta\gamma(j)D_j\left(\frac{m}{M}\right) + \gamma(M)D_{M-1}\left(\frac{m}{M}\right) - \gamma(1)D_0\left(\frac{m}{M}\right)$$

$$= \gamma(0) + \gamma(M)(-1)^m + \sum_{j=0}^{M-1} \Delta\gamma(j)D_j\left(\frac{m}{M}\right) - \gamma(0)D_0\left(\frac{m}{M}\right) + \gamma(M)D_{M-1}\left(\frac{m}{M}\right)$$

$$= \gamma(M)\left((-1)^m + D_{M-1}\left(\frac{m}{M}\right)\right) + \sum_{j=0}^{M-1} \Delta\gamma(j)D_j\left(\frac{m}{M}\right),$$

where $\Delta\gamma(j) = \gamma(j) - \gamma(j+1)$ (the second term is indeed $\gamma(j+1)$, and not $\gamma(j-1)$). Proceeding once again as in summation by parts,

$$g_m = \gamma(M)\left((-1)^m + D_{M-1}\left(\frac{m}{M}\right)\right) + \sum_{j=0}^{M-1} \Delta\gamma(j)\left(K_j\left(\frac{m}{M}\right) - K_{j-1}\left(\frac{m}{M}\right)\right)$$

$$= \gamma(M)\left((-1)^m + D_{M-1}\left(\frac{m}{M}\right)\right) + \sum_{j=0}^{M-1} \Delta\gamma(j)K_j\left(\frac{m}{M}\right) - \sum_{j=0}^{M-2} \Delta\gamma(j+1)K_j\left(\frac{m}{M}\right)$$

$$= \gamma(M)\left((-1)^m + D_{M-1}\left(\frac{m}{M}\right)\right) + \Delta\gamma(M-1)K_{M-1}\left(\frac{m}{M}\right) + \sum_{j=0}^{M-2} \Delta^2\gamma(j)K_j\left(\frac{m}{M}\right),$$

where $\Delta^2\gamma(j) = \Delta\gamma(j) - \Delta\gamma(j+1)$, $K_{-1}(w) = 0$ and

$$K_j(w) = \sum_{k=0}^{j} D_k(w) = \sum_{k=0}^{j} \frac{\sin(\pi(k+\frac{1}{2})w)}{\sin\frac{\pi w}{2}} = \frac{1}{(\sin\frac{\pi w}{2})^2} \sum_{k=0}^{j} \sin\left(\pi(k+\frac{1}{2})w\right) \sin\left(\frac{\pi w}{2}\right)$$

$$= \frac{1}{2(\sin\frac{\pi w}{2})^2} \sum_{k=0}^{j} \Big(\cos(\pi k w) - \cos(\pi(k+1)w)\Big) = \frac{1-\cos(\pi(j+1)w)}{2(\sin\frac{\pi w}{2})^2}.$$

By the assumptions, $\gamma(M) \geq 0$, $\Delta\gamma(M-1) \geq 0$ and $\Delta^2\gamma(j) \geq 0$. Also $K_j(m/M) \geq 0$. Furthermore, setting $\alpha = \pi m/2M$, one gets $D_{M-1}(m/N) = \sin(\pi m - \alpha)/\sin\alpha = -(-1)^m$ if $m \neq 0$, and $D_{M-1}(0) = 1 + 2\sum_{k=1}^{M-1} 1 = 2M-1$ if $m=0$. Hence $g_m \geq 0$ for $m = 0, \ldots, 2M-1$.

(b) Note that, since $\gamma(k) \leq 0$, $k = 1, \ldots, M$, we have, in view of (2.11.10),

$$g_m \geq \gamma(0) + \gamma(M) + 2\sum_{j=1}^{M-1} \gamma(j) = g_0.$$

It is therefore enough to show that $g_0 \geq 0$. As in the proof of Proposition 2.2.5, write

$$0 \leq M \operatorname{Var}(\overline{X}_M) = \gamma(0) + 2\sum_{h=1}^{M-1} \left(1 - \frac{h}{M}\right)\gamma(h)$$

or

$$2\sum_{h=1}^{M-1} \left(1 - \frac{h}{M}\right)(-\gamma(h)) \leq \gamma(0).$$

The dominated convergence theorem yields

$$2\sum_{h=1}^{\infty} (-\gamma(h)) \leq \gamma(0).$$

In particular, since $\gamma(h) \leq 0$ for $h \geq M+1$, this yields $g_0 \geq 0$. □

Many time series models satisfy conditions in Proposition 2.11.5. In the next result, we verify that this is the case for FARIMA$(0, d, 0)$ series. Other series include fractional Gaussian noise (Exercise 2.20).

Proposition 2.11.6 (a) *When $d \in (0, 1/2)$, a FARIMA$(0, d, 0)$ series satisfies the conditions of Proposition 2.11.5, part (a), for any $M \geq 1$.*

(b) *When $d \in (-1/2, 0)$, a FARIMA$(0, d, 0)$ series satisfies the conditions of Proposition 2.11.5, part (b).*

Proof For a FARIMA$(0, d, 0)$ series, by using (2.4.9) and (2.2.13), we have

$$\gamma(k) = \gamma(k-1)\frac{k-1+d}{k-d}. \tag{2.11.11}$$

Note that $\gamma(1) > 0$ if $d \in (0, 1/2)$, and $\gamma(1) < 0$ if $d \in (-1/2, 0)$. This implies that $\gamma(k) > 0$, $k \geq 1$, when $d \in (0, 1/2)$, and $\gamma(k) < 0$, $k \geq 1$, when $d \in (-1/2, 0)$. In particular, this proves part (b).

It remains to show that, when $d \in (0, 1/2)$, $\gamma(k) - \gamma(k+1) \geq 0$ and $\gamma(k) - 2\gamma(k+1) + \gamma(k+2) \geq 0$. For this observe that, by using the recursive relation (2.11.11),

$$\gamma(k) - \gamma(k+1) = \gamma(k)\frac{1-2d}{k+1-d}$$

and similarly

$$\gamma(k) - 2\gamma(k+1) + \gamma(k+2) = \gamma(k)\frac{2(1-d)(1-2d)}{(k+1-d)(k+2-d)} \geq 0. \qquad \square$$

In Figure 2.2 of Section 2.4, we provided realizations of FARIMA$(0, d, 0)$ series for several values of d, generated by using the circulant matrix embedding method. The realizations of FBM in Figure 2.3 were obtained through the increments of FBM, that is FGN, also generated by using the circulant matrix embedding method.

2.12 Exercises

The more difficult exercises are indicated by *. Open problems are marked by **, though in some instances the line between * and ** is quite blurred.

Exercise 2.1 Prove the inequality (2.2.2).

Exercise 2.2 Show the summation by parts formulae (A.2.3) and (A.2.5) found in Appendix A.2.

Exercise 2.3 Show the equivalence of (2.1.8) and (2.2.9). *Hint:* With $Y_n = \sum_{k=1}^{n} X_k$, show and use the relation $2N^2 \gamma^{(N)}(k) = \text{Var}(Y_{(k+1)N}) - 2\text{Var}(Y_{kN}) + \text{Var}(Y_{(k-1)N})$. See Gubner [436].

Exercise 2.4 Show the relation (A.2.7) in Appendix A.2 using the results for fractional Gaussian noise. *Hint:* Approximate k^{-p} in (A.2.7) by $((k+1)^{2-p} - 2k^{2-p} + (k-1)^{2-p})/(2-p)(1-p)$.

Exercise 2.5 Show the relation (2.3.3) for SRD series.

Exercise 2.6 (i) Let $0 < \epsilon < \pi/2$ and g be a positive integrable function on $(0, \epsilon)$ satisfying (2.1.7) in condition IV. Consider the spectral density

$$f(\lambda) = g(|\lambda|)1_{(0,\epsilon)}(|x|) + g(|\pi - \lambda|)1_{(\pi-\epsilon,\pi)}(|x|), \quad -\pi < \lambda < \pi,$$

which satisfies (2.1.7) as well. Show that the corresponding autocovariance function γ satisfies $\gamma(n) = (1 + (-1)^n)\widetilde{\gamma}(n)$ with $\widetilde{\gamma}(n) = \int_{-\pi}^{\pi} \cos(n\lambda)g(\lambda)d\lambda$. Thus, condition IV does not necessarily imply condition II. (*Hint:* See Example 5.2 in Samorodnitsky [879].)

(*ii*) Let
$$g_1(\lambda) = 2^{2j}, \quad \text{if } 2^{-j} \leq \lambda \leq 2^{-j} + 2^{-2j}, \quad \text{for } j = 0, 1, 2, \ldots$$

For $\lambda = 2^{-j}$, note that $g_1(\lambda) = \lambda^{-2}$, so that $\limsup_{\lambda \to 0} \lambda^2 g_1(\lambda) > 0$. Show that, for $n \geq 1$,
$$\left| \int_0^\pi \cos(n\lambda) g_1(\lambda) d\lambda \right| \leq C n^{-1} (\log_2 n)^2.$$

Take now $f(\lambda) = g(\lambda) + g_1(\lambda)$, where $g(\lambda)$ is the spectral density of a LRD time series satisfying both conditions II and IV (e.g., a FARIMA$(0, d, 0)$ series). Conclude that the time series having f as its spectral density satisfies condition II but not condition IV. (*Hint:* See Example 5.3 in Samorodnitsky [879].)

Exercise 2.7* Compare conditions I, II, IV and V for $d < 0$.

Exercise 2.8* For a FARIMA$(0, d, 0)$, $d < 1/2$, series $X = \{X_n\}_{n \in \mathbb{Z}}$, show by induction and the Durbin–Levinson algorithm that the coefficients in the best linear predictor of X_{n+1} in terms of $X_n, \ldots, X_1$ as
$$\widehat{X}_{n+1} = \phi_{n1} X_n + \cdots + \phi_{nn} X_1$$
are given by
$$\phi_{nj} = -\binom{n}{j} \frac{\Gamma(j-d)\Gamma(n-d-j+1)}{\Gamma(-d)\Gamma(n-d+1)}, \quad j = 1, \ldots, n.$$
Deduce that the partial autocorrelation coefficients are given by $\alpha(h) = \phi_{hh} = d/(h-d)$.

Exercise 2.9* For a FARIMA$(0, d, 0)$, $d < 1/2$, series $X = \{X_n\}_{n \in \mathbb{Z}}$, show that the coefficients θ_{nj} in the innovations algorithm
$$\widehat{X}_n = \sum_{j=1}^{n} \theta_{nj}(X_{n-j} - \widehat{X}_{n-j}),$$
where $\widehat{X}_n$ is the best linear predictor of X_n in terms of $X_{n-1}, \ldots, X_0$, are given by
$$\theta_{nj} = \frac{1}{\Gamma(d)} \frac{\Gamma(j+d)}{\Gamma(j+1)} \frac{\Gamma(n+1)}{\Gamma(n+1-d)} \frac{\Gamma(n-j+1-d)}{\Gamma(n-j+1)}.$$
Hint: See Proposition 2.1 in Kokoszka and Taqqu [574].

Exercise 2.10 Prove the relation (2.4.17). *Hint:* See Kokoszka and Taqqu [571], Corollary 3.1.

Exercise 2.11 If $\{X(t)\}_{t \in \mathbb{R}}$ is Gaussian with mean 0 and $X(0) = 0$, show that stationarity of the increments is equivalent to (2.5.3).

Exercise 2.12* Show that for $0 < \gamma < 1$ in Proposition 2.5.6, (c), one has $H < 1/\gamma$ and that the conclusion in Proposition 2.5.6, (f) holds when $\mathbb{E}|X(1)| < \infty$. *Hint:* See Maejima [664], Vervaat [990].

Exercise 2.13 If the assumptions of Proposition 2.5.6, (f), hold and, in addition, the process X is right-continuous, show that almost surely, $X(t) = X(1)t$ for all $t \in \mathbb{R}$.

Exercise 2.14 Show that the finite-dimensional distributions of a standard FBM $\{B_H(t)\}_{t\in\mathbb{R}}$ converge as $H \to 0$ to those of the Gaussian process $\{(\xi_t - \xi_0)/\sqrt{2}\}_{t\in\mathbb{R}}$, where $\xi_t, t \in \mathbb{R}$, are i.i.d. $\mathcal{N}(0, 1)$ random variables. *Hint:* Show the convergence of the covariance functions.

Exercise 2.15 Show that the log-FSM (2.6.36) is well defined, has stationary increments and is self-similar with $H = 1/\alpha$.

Exercise 2.16 Show that the linear series (2.9.2) is well defined and α–heavy tailed under the assumptions (2.9.5) and (2.9.6), and the innovations ϵ_k being symmetric i.i.d. α–heavy-tailed.

Exercise 2.17 Complete the proof of Proposition 2.10.1 by showing (2.10.14).

Exercise 2.18* (i) Use the arguments developed in Section 2.10.3 to show the asymptotic normality of the estimator of the constant c_4 in (2.10.11). (ii) Similarly, derive the joint asymptotics of the estimators of d and c_4.

Exercise 2.19* Proposition 2.10.3, (iv), is stated by using time-domain conditions and proved in the time domain. State and prove this result by using conditions and arguments in the spectral domain. *Hint:* Note that, under suitable assumptions, the condition (2.10.19) can be expressed as $\widehat{\psi}(x) = Cx^{\varrho} + o(x^{\varrho})$, as $x \to 0$.

Exercise 2.20 Show that FGN satisfies the assumptions of Proposition 2.11.5, part (a), when $H > 1/2$, and part (b), when $H < 1/2$.

Exercise 2.21** Can Proposition 2.11.4 be extended to LRD series? This is essentially equivalent to showing that the convergence of the series in Propositions A.2.1 and 2.2.14 is also uniform over λ.

2.13 Bibliographical Notes

Section 2.1: We begin with a few historical remarks. The origins of long-range dependence can be traced to the works of:

- Hurst [501] who observed long-range dependence in analyzing the Nile river historical levels;

- Cox and Townsend [266] who discovered slowly decaying correlations in textile yarn measurements;
- Whittle [1007] who observed power-law correlations in the spatial context for yields in terms of plot sizes.

The books by Beran [121] and Beran, Feng, Ghosh, and Kulik [127] contain nice historical overviews of long-range dependence, with even earlier references discussing phenomena related to long-range dependence. See also Graves, Gramacy, Watkins, and Franzke [426].

Section 2.2: The results of this section relate the various definitions of long-range dependence under quite general assumptions involving slowly varying functions. As indicated in Remark 2.2.19, some of the relationships have been assumed erringly in the literature. As it can be seen from the section, most facts about slowly varying functions are drawn from the excellent and exhaustive monograph by Bingham et al. [157]. For other discussions of the various definitions of LRD, see also Terdik [955], Heyde [472], Ma [657], Guégan [437], Inoue [520]. Other conditions for relating the definition of LRD in the time and spectral domains are studied in Inoue [518].

Propositions 2.2.14 and 2.2.16 are sometimes referred to, respectively, as Abelian and Tauberian theorems (see Sections 4.3 and 4.10 in Bingham et al. [157]). Abelian theorems generally refer to results on transformations of functions and Tauberian theorems to results on functions based on their transformations, which are generally thought to require more stringent assumptions (see, for example, the discussion in the beginning of Chapter 4 in Bingham et al. [157]). Since it is a bit arbitrary what to call a function or its transformation in Propositions 2.2.14 and 2.2.16, the terminology of the Abelian and Tauberian theorems is not exactly precise.

Section 2.3: ARMA(p, q) models in Example 2.3.3 are studied using standard techniques in time series analysis (Brockwell and Davis [186]), while Markov chains in Example 2.3.4 are treated by using algebraic techniques (Feller [361], Chapter XVI). On the other hand, a Markov chain with a countable state space can exhibit long-range dependence when a return time from a state to itself has infinite variance – see Carpio and Daley [204], Oğuz and Anantharam [777].

The special role of antipersistent series seems to have been recognized early, for example, by Mandelbrot and Van Ness [682], Mandelbrot and Wallis [684]. See Bondon and Palma [174] for more recent work focusing on antipersistent series.

Section 2.4: FARIMA(p, d, q) models were introduced simultaneously by Hosking [484] and Granger and Joyeux [424]. This class of models has become centerpiece in modeling long-range dependent time series. Though see also Veitch, Gorst-Rasmussen, and Gefferth [983]. Strictly stationary solutions of ARMA equations with fractional noise are studied by Vollenbröker [992].

An interesting extension of stationary FARIMA models consists of the Gegenbauer and Gegenbauer ARMA (GARMA) models. For example, the stationary Gegenbauer time series $\{X_n\}$ satisfies the equation

$$(I - 2uB + B^2)^{d/2} X_n = Z_n, \quad n \in \mathbb{Z},$$

where $\{Z_n\} \sim WN(0, \sigma_Z^2)$ and either $|u| < 1, d < 1$ or $u = \pm 1, d < 1/2$. The spectral density of the Gegenbauer process can be expressed as

$$f_X(\lambda) = \frac{\sigma_Z^2}{2\pi} 2^{-d} |\cos(\lambda) - u|^{-d}.$$

Note that when $|u| < 1$ and $0 < d < 1$, the spectral density no longer explodes at $\lambda = 0$ as for the usual long-range dependence but at the so-called *G frequency* (Gegenbauer frequency) $\lambda_0 = \arccos u \neq 0$. In this case, the series also naturally exhibits seasonal variations with the period corresponding to the G frequency. One can show that for the same range of parameter values u and d, the autocovariance function satisfies

$$\gamma_X(n) \sim Cn^{d-1} \cos(\lambda_0 n), \quad \text{as } n \to \infty,$$

that is, it decays slowly as for the usual long-range dependence but is also characterized by the oscillating behavior. See Gray, Zhang, and Woodward [427, 428]. The GARMA series satisfy the equation of the Gegenbauer series above with the added AR and MA polynomials. For recent work on GARMA series and extensions, see McElroy and Holan [705, 706], Caporale, Cuñado, and Gil-Alana [201], Dissanayake, Peiris, and Proietti [317].

Section 2.5: The study of general self-similar processes was initiated by Lamperti [589]. See also Vervaat [990, 991], Maejima [664]. Self-similar processes are the focus of the book by Embrechts and Maejima [350]. Extensions of the notion of self-similarity for univariate processes are considered in van Harn and Steutel [971], Eglói [346].

Section 2.6: The notes below are separated according to the self-similar processes introduced in the section.

Though introduced as early as in Kolmogorov [575], Pinsker and Yaglom [811], Yaglom [1019], fractional Brownian motion (FBM) was popularized in great part by Mandelbrot and Van Ness [682] who saw the practical potential of the FBM model. What could be regarded as FBM with index $H = 0$ is introduced in Fyodorov, Khoruzhenko, and Simm [377]. Alternative forms of FBM, the so-called type I and type II FBMs, are considered in Marinucci and Robinson [690]. nth-Order FBM is introduced in Perrin, Harba, Berzin-Joseph, Iribarren, and Bonami [807]. See also the notes to Chapter 7 on FBM below.

Bifractional Brownian motion was introduced by Houdré and Villa [489], and has since been studied by a number of researchers, including Russo and Tudor [873], Tudor and Xiao [965], Lei and Nualart [603].

The Rosenblatt process was introduced by Taqqu [943] in connection to non-central limits theorems, naming it after Rosenblatt [860] who considered its marginal distribution. See Chapters 4 and 5 for more information.

Linear fractional stable motion appears first in Astrauskas [53]. Its asymptotic dependence structure is analyzed in Astrauskas et al. [54]. Various models are considered in Samorodnitsky and Taqqu [881, 882]. The linear multifractional stable motion is treated in Stoev and Taqqu [925, 928]. Many other stable self-similar processes, including the Telecom process, can be found in Pipiras and Taqqu [820].

The material on linear fractional Lévy motion is taken from Cohen and Istas [253], Section 4.4. See also Marquardt [692].

Fractional Ornstein–Uhlenbeck Lévy processes are analyzed in Wolpert and Taqqu [1009]. See also Taqqu and Wolpert [950]. Self-similar processes with stationary increments generated by point processes are studied in O'Brien and Vervaat [776].

Section 2.7: The Lamperti transformation is named after Lamperti [589]. Other interesting work related to the Lamperti transformation includes Burnecki, Maejima, and Weron [195], Flandrin, Borgnat, and Amblard [367], Genton, Perrin, and Taqqu [386], Matsui and Shieh [698].

Section 2.8: Fractional Gaussian noise was introduced in Mandelbrot and Van Ness [682]. As indicated in the section, the fundamental limit theorem is due to Lamperti [589].

The convergence of the partial sums of linear time series to fractional Brownian motion (cf. Proposition 2.8.8) was considered by Davydov [291], Gorodetskii [418].

Section 2.9: Infinite variance long memory time series are considered in Kokoszka and Taqqu [571, 572, 573], Peng [805], Burnecki and Sikora [194]. Long-range dependence and heavy tails also appear in the slightly different context of stochastic volatility models (Kulik and Soulier [581], McElroy and Jach [704]). Other definitions of long-range dependence with infinite variance are considered in Magdziarz [670], Damarackas and Paulauskas [286], Paulauskas [795, 796].

The proof of Proposition 2.9.3 is based on Astrauskas [53]. See also Barbe and McCormick [95], Peligrad and Sang [801, 802], Duffy [333] for more recent work.

Univariate stable distributions is the focus of a fascinating monograph by Uchaikin and Zolotarev [967], as well as of the classic monograph by Zolotarev [1029]. For a more recent treatment of stable distributions and processes, see Rachev and Mittnik [839], Nolan [755].

Section 2.10: As indicated in the section, the R/S method originated with Hurst [501], and has since attracted much attention, especially in the geophysics literature (e.g., Lawrence and Kottegoda [599], Montanari, Rosso, and Taqqu [731], Dmowska and Saltzman [319]). Mandelbrot and Wallis [683, 684, 686, 685] have a number of papers on the subject. A general overview about applications to hydrology can be found in Montanari [729]. Klemeš [563], however, raises doubts about the presence of long memory in hydrological time series.

The theoretical properties of the R/S method were studied by Mandelbrot [678]. See also Feller [359] who proves convergence in the case of i.i.d. data, and Hamed [448], Li, Yu, Carriquiry, and Kliemann [624] for more recent improvements. Several methods related to and improving on R/S, such as V/S, $KPSS$ and others, are considered Giraitis, Kokoszka, Leipus, and Teyssière [405].

The regression in the spectral domain was introduced by Geweke and Porter-Hudak [387]. Theoretical properties of the resulting estimator of the long-range dependence parameter were studied in depth by Robinson [852]. Other related studies include Agiakloglou, Newbold, and Wohar [15], Delgado and Robinson [299], Hurvich, Deo, and Brodsky [510], Hurvich and Deo [507], Velasco [985, 986], Kim and Phillips [555], Arteche and Orbe [49, 50]. Extensions include Andersson [31], Andrews and Guggenberger [32], Guggenberger and Sun [439], Feng and Beran [362]. See also Section 10.9 below with some related notes to Chapter 10.

The wavelet estimation methods for long-range dependent series were popularized in a series of papers by Abry and Veitch [8], Veitch and Abry [981]. But useful connections between wavelets, self-similarity and long-range dependence were recognized earlier in, for example, Flandrin [366], Tewfik and Kim [956], Dijkerman and Mazumdar [315], Wornell [1011], Abry, Veitch, and Flandrin [9]. Some earlier related work can be traced at least to Allan [19]. The work by Abry and Veitch were followed by a number of other interesting studies including Bardet [96, 97], Roughan, Veitch, and Abry [870], Veitch, Taqqu, and Abry [982], Craigmile, Guttorp, and Percival [268], Moulines, Roueff, and Taqqu [735, 736, 737], Teyssière and Abry [957], Bardet and Bibi [98], Bardet, Bibi, and Jouini [99], Faÿ, Moulines, Roueff, and Taqqu [357], Roueff and Taqqu [866, 867], Clausel, Roueff, Taqqu, and Tudor [248, 249, 250].

For a succinct introduction to the wavelet estimation methods for long-range dependent series, see Abry, Flandrin, Taqqu, and Veitch [10]. For visualization tools that can be used in the local analysis of self-similarity, see Stoev, Taqqu, Park, Michailidis, and Marron [929] and Park, Godtliebsen, Taqqu, Stoev, and Marron [793]. For an analysis of Ethernet data, see Stoev et al. [929]. Stoev and Taqqu [924] describe the wavelet estimation of the Hurst parameter in stable processes. The case of FARIMA time series with stable innovations is considered in Stoev and Taqqu [927].

Section 2.11: The circulant matrix embedding method seems to go back to Davies and Harte [290]. It was studied more extensively by Dembo, Mallows, and Shepp [301], Dietrich and Newsam [313, 314], Lowen [646], Gneiting [411], Stein [920], Craigmile [267], and Percival [806]. Extensions to multivariate (vector) time series were considered by Chan and Wood [214], Helgason, Pipiras, and Abry [463], and those to random fields by Dietrich and Newsam [313], Wood and Chan [1010], Chan and Wood [213], Stein [921, 922], Gneiting, Ševčíková, Percival, Schlather, and Jiang [413], and Helgason, Pipiras, and Abry [466].

Simulation of stationary Gaussian series from given spectral densities is considered in e.g., Chambers [211], Azimmohseni, Soltani, and Khalafi [64]. A turning-band method for the simulation of fractional Brownian fields can be found in e.g., Biermé and Moisan [149]. Midpoint displacement methods are discussed in Brouste, Istas, and Lambert-Lacroix [191].

3

Physical Models for Long-Range Dependence and Self-Similarity

In this chapter, we present a number of physical models for long-range dependence and self-similarity. What do we mean by "physical models" for long-range dependence? First, models motivated by a real-life (physical) application. Second, though these models lead to long-range dependence, they are formulated by using other principles (than long-range dependence). One of these principles is often "responsible" for the emergence of long-range dependence. For example, the infinite source Poisson model discussed in Section 3.3 below, may model an aggregate traffic of data packets on a given link in the Internet. For this particular model, long-range dependence arises from the assumed heavy tails of workload distribution for individual arrivals. We consider the following models:

- Aggregation of short-range dependent series (Section 3.1)
- Mixture of correlated random walks (Section 3.2)
- Infinite source Poisson model with heavy tails (Section 3.3)
- Power-law shot noise model (Section 3.4)
- Hierarchical model (Section 3.5)
- Regime switching (Section 3.6)
- Elastic collision of particles (Section 3.7)
- Motion of a tagged particle in a simple symmetric exclusion model (Section 3.8)
- Power-law Pólya's urn (Section 3.9)
- Random walk in random scenery (Section 3.10)
- Two-dimensional Ising model (Section 3.11)
- Stochastic heat equation (Section 3.12)
- The Weierstrass function connection (Section 3.13)

Having physical models for long-range dependence is appealing and useful for a number of reasons. For example, in such applications, the use of long-range dependent models is then justified and thus more commonly accepted. In such applications, the parameters of long-range dependence models may carry a physical meaning. For example, in the infinite source Poisson model, the long-range dependence parameter is expressed through the exponent of a heavy-tailed distribution.

3.1 Aggregation of Short-Range Dependent Series

One way long-range dependence arises is through aggregation of short-range dependent (SRD) series. We focus here on aggregation of AR(1) series only, which is one of the simplest examples of SRD series and which has attracted most attention in the literature on aggregation.

Let a be a random variable supported on $(-1, 1)$ and having a distribution function F_a. Suppose that $a^{(j)}$, $j \geq 1$, are i.i.d. copies of a. Let also $\epsilon = \{\epsilon_n\}_{n \in \mathbb{Z}}$ be a sequence of i.i.d. random variables with $\mathbb{E}\epsilon_n = 0$, $\mathbb{E}\epsilon_n^2 = \sigma_\epsilon^2$, and $\epsilon^{(j)} = \{\epsilon_n^{(j)}\}_{n \in \mathbb{Z}}$, $j \geq 1$, be i.i.d. copies of ϵ. Suppose that $a^{(j)}$, $j \geq 1$, and $\epsilon^{(j)}$, $j \geq 1$, are independent.

Consider i.i.d. AR(1) series $Y^{(j)} = \{Y_n^{(j)}\}_{n \in \mathbb{Z}}$ with random parameters $a^{(j)}$ defined by

$$Y_n^{(j)} = a^{(j)} Y_{n-1}^{(j)} + \epsilon_n^{(j)}, \quad n \in \mathbb{Z}. \tag{3.1.1}$$

Define now a series $X^{(N)} = \{X_n^{(N)}\}_{n \in \mathbb{Z}}$ obtained by aggregating $Y^{(j)}$ as follows:

$$X_n^{(N)} = \frac{1}{N^{1/2}} \sum_{j=1}^{N} Y_n^{(j)}, \quad n \in \mathbb{Z}. \tag{3.1.2}$$

For example, many macroeconomic variables (such as aggregate U.S. consumption) can be considered as an aggregation of many micro-variables.

Since the sequences $\{Y_n^{(j)}, n \in \mathbb{Z}\}$, $j = 1, 2, \ldots$ are independent, the covariances and spectral density of $\{X_n^{(N)}, n \in \mathbb{Z}\}$ do not depend on N, so that

$$\gamma_X(h) = \mathrm{Cov}(X_{n+h}^{(N)}, X_n^{(N)}) = \mathrm{Cov}(Y_{n+h}^{(1)}, Y_n^{(1)})$$

and one can denote by $f_X(\lambda)$, $\lambda \in (-\pi, \pi]$, the corresponding spectral density, if it exists.

There are two complementary ways to look at the series $X^{(N)}$.

First, one can consider $X^{(N)}$ conditionally on the values of $a^{(j)}$, $j \geq 1$. Since by (2.3.11), the AR(1) series $Y_n = aY_{n-1} + \epsilon_n$ has a spectral density (conditionally on a) given by $f_{Y|a}(\lambda) = \sigma_\epsilon^2/(2\pi|1 - ae^{-i\lambda}|^2)$ and the spectral density of the sum of independent series is the sum of the corresponding spectral densities, the series $X^{(N)}$ has a spectral density conditionally on $a^{(j)}$, $j \geq 1$, given by

$$f_{N|a}(\lambda) = \frac{\sigma_\epsilon^2}{2\pi N} \sum_{j=1}^{N} \frac{1}{|1 - a^{(j)} e^{-i\lambda}|^2}. \tag{3.1.3}$$

Note that, by the Law of Large Numbers, for each $\lambda \in (-\pi, \pi]$,

$$f_{N|a}(\lambda) \to f_X(\lambda) = \frac{\sigma_\epsilon^2}{2\pi} \mathbb{E} \frac{1}{|1 - ae^{-i\lambda}|^2} = \frac{\sigma_\epsilon^2}{2\pi} \int_{-1}^{1} \frac{F_a(dx)}{|1 - xe^{-i\lambda}|^2}, \tag{3.1.4}$$

a.s. as $N \to \infty$, under the assumption that

$$\int_{-1}^{1} \frac{F_a(dx)}{|1 - xe^{-i\lambda}|^2} < \infty. \tag{3.1.5}$$

Alternatively, one can work in the time domain. Note that, by the independence of $Y^{(j)}$, $j \geq 1$, and (2.3.11),

3.1 Aggregation of Short-Range Dependent Series

$$\gamma_X(h) = \text{Cov}(X_{n+h}^{(N)}, X_n^{(N)}) = \text{Cov}(Y_{n+h}^{(1)}, Y_n^{(1)}) = \mathbb{E}\Big(\mathbb{E}(Y_{n+h}^{(1)} Y_n^{(1)} | a^{(1)})\Big)$$
$$= \mathbb{E}\Big(\frac{\sigma_\epsilon^2 a^{|h|}}{1-a^2}\Big) = \sigma_\epsilon^2 \int_{-1}^{1} \frac{x^{|h|} F_a(dx)}{1-x^2}. \tag{3.1.6}$$

Thus, the series $X^{(N)}$ is second-order stationary, if

$$\int_{-1}^{1} \frac{F_a(dx)}{1-x^2} < \infty. \tag{3.1.7}$$

To connect the time and spectral domain perspectives, note that

$$\frac{x^{|h|}}{1-x^2} = \frac{1}{2\pi} \int_{-\pi}^{\pi} \frac{e^{ih\lambda}}{|1-xe^{-i\lambda}|^2} d\lambda$$

(which is the usual relationship between the covariance and spectral density of an AR(1) series; see (2.3.11)). We thus have from (3.1.6), the Fubini theorem and using (3.1.7) that $X^{(N)}$ has covariance

$$\gamma_X(h) = \int_{-\pi}^{\pi} e^{ih\lambda} \Big\{ \frac{\sigma_\epsilon^2}{2\pi} \int_{-1}^{1} \frac{F_a(dx)}{|1-xe^{-i\lambda}|^2} \Big\} d\lambda$$

and therefore, under (3.1.7), that $X^{(N)}$ has spectral density

$$f_X(\lambda) = \frac{\sigma_\epsilon^2}{2\pi} \int_{-1}^{1} \frac{F_a(dx)}{|1-xe^{-i\lambda}|^2}. \tag{3.1.8}$$

The series X corresponding to the spectral density $f_X(\lambda)$ in (3.1.4) and (3.1.8) is called the *aggregated series*. The next result shows that the aggregated series X is LRD in the sense (2.1.7) of condition IV if the distribution $F_a(dx)$ is suitably dense around $x = 1$.

Proposition 3.1.1 *Suppose that the distribution $F_a(dx)$ has a density*

$$f_a(x) = \begin{cases} (1-x)^{1-2d} l(1-x) \phi(x), & \text{if } x \in [0, 1), \\ 0, & \text{if } x \in (-1, 0), \end{cases} \tag{3.1.9}$$

where $d < 1/2$, $l(y)$ is a slowly varying function at $y = 0$ and $\phi(x)$ is a bounded function on $[0, 1]$ and continuous at $x = 1$ with $\phi(1) \neq 0$. Then, as $\lambda \to 0$,

$$f_X(\lambda) \sim \lambda^{-2d} l(\lambda) \frac{\sigma_\epsilon^2 \phi(1)}{2\pi} \int_0^\infty \frac{s^{1-2d}}{1+s^2} ds, \tag{3.1.10}$$

where $f_X(\lambda)$ appears in (3.1.4) and (3.1.8).

Proof Observe that, by making the change of variables $x = \cos\lambda - u\sin\lambda$ so that $dx = -(\sin\lambda) du$, $x = 0$ and $x = 1$ become $u = 1/\tan\lambda$ and

$$u = \frac{\cos\lambda - 1}{\sin\lambda} = -\frac{2\sin^2\frac{\lambda}{2}}{2\sin\frac{\lambda}{2}\cos\frac{\lambda}{2}} = -\tan\frac{\lambda}{2}.$$

Then using the identities

$$1 - x = (1 - \cos\lambda) + u\sin\lambda = 2\sin^2\frac{\lambda}{2} + u \cdot 2\sin\frac{\lambda}{2}\cos\frac{\lambda}{2} = 2\sin\frac{\lambda}{2}\Big(\sin\frac{\lambda}{2} + u\cos\frac{\lambda}{2}\Big)$$

and
$$1 - 2x\cos\lambda + x^2 = 1 - 2\cos^2\lambda + 2u\cos\lambda\sin\lambda + \cos^2\lambda - 2u\cos\lambda\sin\lambda$$
$$+ u^2\sin^2\lambda = (\sin\lambda)^2(1+u^2),$$

one gets by (3.1.8),

$$\frac{2\pi}{\sigma_\epsilon^2}f_X(\lambda) = \int_0^1 \frac{(1-x)^{1-2d}l(1-x)\phi(x)}{1-2x\cos\lambda+x^2}dx = \frac{(2\sin\frac{\lambda}{2})^{-2d}l(2\sin\frac{\lambda}{2})}{\cos\frac{\lambda}{2}}$$

$$\times \int_{-\tan\frac{\lambda}{2}}^{\frac{1}{\tan\lambda}} \frac{(\sin\frac{\lambda}{2}+u\cos\frac{\lambda}{2})^{1-2d}}{1+u^2} \frac{l(2\sin\frac{\lambda}{2}(\sin\frac{\lambda}{2}+u\cos\frac{\lambda}{2}))}{l(2\sin\frac{\lambda}{2})}$$

$$\times \phi(\cos\lambda - u\sin\lambda)du =: \frac{(2\sin\frac{\lambda}{2})^{-2d}l(2\sin\frac{\lambda}{2})}{\cos\frac{\lambda}{2}} \int_{\mathbb{R}} f_\lambda(u)du.$$

Focus now on $f_\lambda(u)$. For each $u \in \mathbb{R}$ and as $\lambda \to 0$,

$$f_\lambda(u) \to \frac{u^{1-2d}}{1+u^2}\phi(1)1_{[0,\infty)}(u)$$

and, by using Potter's bound (2.2.1) for the slowly varying function l and the boundedness of ϕ,

$$|f_\lambda(u)| \le C\frac{\max\{|\sin\frac{\lambda}{2}+u\cos\frac{\lambda}{2}|^{1-2d-\epsilon}, |\sin\frac{\lambda}{2}+u\cos\frac{\lambda}{2}|^{1-2d+\epsilon}\}}{1+u^2} \le C\frac{(1+|u|)^{1-2d+\epsilon}}{1+u^2},$$

where $\epsilon > 0$ is such that $1 - 2d \pm \epsilon > 0$ and $-2d \pm \epsilon < 0$. Since the bound above is integrable on $\mathbb{R}$, the dominated convergence theorem implies that, as $\lambda \to 0$,

$$\frac{2\pi}{\sigma_\epsilon^2}f_X(\lambda) \sim \lambda^{-2d}l(\lambda)\phi(1)\int_0^\infty \frac{u^{1-2d}du}{1+u^2}. \quad \square$$

Example 3.1.2 (*Aggregation with beta-like distribution*) One of the examples of the distributions $F_a(dx)$ which leads to LRD and which was first considered in the aggregation literature, is that with the density

$$f_a(x) = \frac{2x^{2p-1}(1-x^2)^{1-2d}}{B(p, 2-2d)}1_{(0,1)}(x), \qquad (3.1.11)$$

where $p > 0$, $0 < d < 1/2$ and B is the beta function defined in (2.2.12). One has indeed

$$\int_0^1 f_a(x)dx = \frac{1}{B(p, 2-2d)}\int_0^1 y^{p-1}(1-y)^{1-2d}dy = 1.$$

Since

$$f_a(x) = (1-x)^{1-2d}\left(\frac{2x^{2p-1}(1+x)^{1-2d}}{B(p, 2-2d)}1_{(0,1)}(x)\right),$$

we can set $l(y) \equiv 1$ and $\phi(x)$ to be equal to the contents of the bracket. Then the density (3.1.11) has the form (3.1.9) when $p \ge 1/2$. On the other hand, note also that with the density (3.1.11), it is easy to show LRD in the sense (2.1.5) in condition II directly from (3.1.6). Indeed, from (3.1.6), we get that, for $h \geq 0$,

$$\gamma_X(h) = \sigma_\epsilon^2 \int_0^1 \frac{x^h f_a(x)}{1-x^2} dx = \frac{2\sigma_\epsilon^2}{B(p, 2-2d)} \int_0^1 x^{h+2p-1}(1-x^2)^{-2d} dx$$

$$= \frac{\sigma_\epsilon^2}{B(p, 2-2d)} \int_0^1 y^{\frac{h}{2}+p-1}(1-y)^{-2d} dy = \frac{\sigma_\epsilon^2 B(p+\frac{h}{2}, 1-2d)}{B(p, 2-2d)}$$

$$= \frac{\sigma_\epsilon^2 \Gamma(1-2d)\Gamma(p+\frac{h}{2})}{B(p, 2-2d)\Gamma(p+\frac{h}{2}+1-2d)} \sim \frac{\sigma_\epsilon^2 \Gamma(1-2d)}{B(p, 2-2d)2^{2d-1}} h^{2d-1}, \quad (3.1.12)$$

as $h \to \infty$, by using (2.2.12) and (2.4.4).

We also note that the restriction to $x \in [0, 1)$ in (3.1.9) is not accidental, albeit simplifying the analysis. Exercise 3.2 considers the case where the support of $f_a(x)$ is on $x \in (-1, 0)$ along with assumptions similar to those in Proposition 3.1.1. The aggregated series X in this case is not LRD in the sense of condition IV but has instead a divergence at $\lambda = \pi$.

Another interesting direction related to aggregation is the following. Suppose X is a given LRD series with the spectral density $f_X(\lambda)$. In contrast to the preceding presentation, one could ask whether there is a distribution function $F_a(dx)$, called a *mixture distribution*, such that $f_X(\lambda)$ could be expressed as in (3.1.8). This is often referred to as a *disaggregation problem*.

Example 3.1.3 (*Disaggregation*) If X is a FARIMA$(0, d, 0)$ series with the spectral density

$$f_X(\lambda) = \frac{|1 - e^{-i\lambda}|^{-2d}}{2\pi}, \quad (3.1.13)$$

it is known that it can be disaggregated with the mixing density

$$f_a(x) = C(d) x^{d-1}(1-x)^{d-1}(1+x) 1_{[0,1)}(x), \quad (3.1.14)$$

where

$$C(d) = \frac{\Gamma(3-d)}{2\Gamma(d)\Gamma(2-2d)} = 2^{2d-2} \frac{\sin(d\pi)}{\sqrt{\pi}} \frac{\Gamma(3-d)}{\Gamma(\frac{3}{2}-d)}$$

(see Exercise 3.1 and Celov, Leipus, and Philippe [208], Proposition 5.1).

3.2 Mixture of Correlated Random Walks

A simple random walk $X_0 = 0$, $X_n = \sum_{k=1}^n \epsilon_k$, $n \in \mathbb{N}$ represents the position of a random walker at time n. Each step ϵ_k is either to the right ($\epsilon_k = 1$) with probability $1/2$ or to the left ($\epsilon_k = -1$) with probability $1/2$. X_n denotes the position at time n. In a usual random walk, the steps are independent. In correlated random walks (CRWs, in short), they are dependent.

We will assume here that if a step has been made, then the next step will be in the same direction with probability p and in the opposite direction with probability $1 - p$. (The first step is taken into either direction with probability $1/2$.) If $p = 1/2$, we fall back on the simple random walk where the steps are independent. But if $p \in (1/2, 1)$, there is persistence and the steps are correlated (though they still have zero mean). As a consequence, the resulting correlated random walk (CRW) $\{X_n\}$ has zero mean and stationary increments which are dependent. Observe also that while $\{X_n\}$ is not Markovian, $\{X_n, \epsilon_{n-1}\}$ is Markovian. Since the steps are dependent, let us compute the covariance between two steps distant by h.

Lemma 3.2.1 *For $m \geq 1$ and $h \geq 0$,*

$$\mathbb{E}(\epsilon_{m+h}\epsilon_m) = (2p-1)^h. \tag{3.2.1}$$

Proof Note first that $\mathbb{E}\epsilon_m^2 = 1$. Suppose $n \geq 1$ and let $\mathcal{F}_n = \sigma\{X_0, \ldots, X_n\} = \sigma\{X_0, \epsilon_1, \ldots, \epsilon_n\}$. Then,

$$\mathbb{E}(\epsilon_{n+1}|\mathcal{F}_n)1_{\{\epsilon_n=1\}} = \big(p(+1) + (1-p)(-1)\big)1_{\{\epsilon_n=1\}} = (2p-1)1_{\{\epsilon_n=1\}},$$
$$\mathbb{E}(\epsilon_{n+1}|\mathcal{F}_n)1_{\{\epsilon_n=-1\}} = \big(p(-1) + (1-p)(+1)\big)1_{\{\epsilon_n=-1\}} = -(2p-1)1_{\{\epsilon_n=-1\}}.$$

Hence

$$\mathbb{E}(\epsilon_{n+1}|\mathcal{F}_n) = \mathbb{E}(\epsilon_{n+1}|\mathcal{F}_n)1_{\{\epsilon_n=1\}} + \mathbb{E}(\epsilon_{n+1}|\mathcal{F}_n)1_{\{\epsilon_n=-1\}} = (2p-1)\epsilon_n$$

and by taking $n = m + h$,

$$\mathbb{E}(\epsilon_{m+h+1}\epsilon_m|\mathcal{F}_{m+h}) = \epsilon_m \mathbb{E}(\epsilon_{m+h+1}|\mathcal{F}_{m+h}) = \epsilon_{m+h}\epsilon_m(2p-1).$$

Taking expectation on both sides, one gets

$$\mathbb{E}(\epsilon_{m+h+1}\epsilon_m) = (2p-1)\mathbb{E}(\epsilon_{m+h}\epsilon_m)$$

Since $\mathbb{E}(\epsilon_{m+1}\epsilon_m) = (2p-1)\mathbb{E}\epsilon_m^2 = 2p-1$, the result follows by induction on h. □

Since $\mathbb{E}\epsilon_m = 0$ and $\mathbb{E}\epsilon_m^2 = 1$, the relation (3.2.1) equals both the correlation and the covariance between ϵ_m and ϵ_{m+h}. Observe that this covariance is indeed 0 if $p = 1/2$. Observe also that this covariance decreases exponentially fast as $n \to \infty$ and hence the sequence $\{\epsilon_m\}$ is short-range dependent. In fact, the following result holds (Exercise 3.4).

Theorem 3.2.2 *Let $\{X_n^{(p)}\}_{n\geq 0}$ be the CRW introduced above. Then*

$$\frac{1}{\sqrt{N}}X_{[Nt]}^{(p)} = \frac{1}{\sqrt{N}}\sum_{n=1}^{[Nt]} \epsilon_n \xrightarrow{fdd} \sqrt{\frac{p}{1-p}}B(t), \quad t \geq 0,$$

where $\{B(t)\}_{t\geq 0}$ is a standard Brownian motion.

To create long-range dependence and in the spirit of Section 3.1, we are going to make p random. We shall do so by choosing p according to a probability distribution μ, so that, in the corresponding enlarged probability space, for each $m \geq 1, h \geq 0$,

$$\mathbb{E}(\epsilon_{m+h}\epsilon_m) = \int_0^1 (2p-1)^h d\mu(p).$$

We shall also consider not just a single CRW but a sequence $X_{[Nt]}^{(p_j)}, j = 1, \ldots, M$, of independent copies of the CRW. Each $X^{(p_j)}$ has its own p_j chosen independently from

3.2 Mixture of Correlated Random Walks

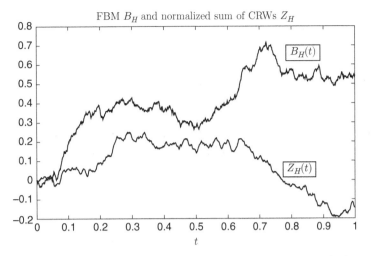

Figure 3.1 Normalized sum of CRWs $Z_H(t) = c_H(X_{[Nt]}^{(p_1)} + \cdots + X_{[Nt]}^{(p_M)})/(N^H M^{1/2})$ with $M = 100$ and $N = 1024$, and standard FBM $B_H(t)$ on the interval $[0, 1]$ when $H = 0.75$.

the distribution μ. The next theorem shows that by choosing the measure μ appropriately, $\sum_{j=1}^{M} X_{[Nt]}^{(p_j)}$ normalized, converges to FBM as $M \to \infty$ followed by $N \to \infty$. See also Figure 3.1.

Theorem 3.2.3 Let $X_{[Nt]}^{(p_1)}, \ldots, X_{[Nt]}^{(p_M)}$ be independent samples from the population of CRWs with

$$\frac{d\mu(p)}{dp} = C(H)(1-p)^{1-2H}, \quad 1/2 < p < 1, \qquad (3.2.2)$$

where $1/2 < H < 1$, $C(H) = (1-H)2^{3-2H}$ and p_j, $j \geq 1$, are i.i.d. with the distribution μ. Then,

$$\lim_{N \to \infty} c_H \frac{1}{N^H} \lim_{M \to \infty} \frac{X_{[Nt]}^{(p_1)} + \cdots + X_{[Nt]}^{(p_M)}}{\sqrt{M}} = B_H(t), \quad t \geq 0, \qquad (3.2.3)$$

where

$$c_H = \sqrt{H(2H-1)/\Gamma(3-2H)}, \qquad (3.2.4)$$

B_H is a standard FBM with the self-similarity parameter H and the convergence is in the sense of finite-dimensional distributions.

Proof The central limit theorem yields

$$\lim_{M \to \infty} (X_m^{(p_1)} + \cdots + X_m^{(p_M)})/\sqrt{M} \xrightarrow{fdd} Y_m$$

in the sense of finite-dimensional distributions in m. Here, Y_m, $m \geq 1$, is a Gaussian process with stationary increments $\Delta Y_m = Y_m - Y_{m-1}$ that have zero mean and unit variance, and in view of (3.2.1) and (3.2.2), the autocovariances

$$\gamma_{\Delta Y}(h) = \mathbb{E}(\Delta Y_{m+h}\Delta Y_m) = (1-H)2^{3-2H}\int_{1/2}^{1}(2p-1)^h(1-p)^{1-2H}dp$$

$$= (2-2H)\int_{0}^{1}x^h(1-x)^{1-2H}dx = (2-2H)\frac{\Gamma(h+1)\Gamma(2-2H)}{\Gamma(h+3-2H)} \sim \Gamma(3-2H)h^{2H-2},$$

as $h \to \infty$, by (2.2.12) and (2.4.4). Thus, the series $\{\Delta Y_m\}_{m\geq 1}$ is LRD in the sense (2.1.5) in condition II with $d = H - 1/2$ and $L_2(u) \sim \Gamma(3 - 2H)$. By Corollary 2.2.7, the series $\{\Delta Y_m\}_{m\geq 1}$ also satisfies (2.1.8) in condition V with $L_5(u) \sim \Gamma(3-2H)/(H(2H-1)) = c_H^{-2}$ with c_H given in (3.2.4). Proposition 2.8.7 then yields

$$\frac{Y_{[Nt]}}{c_H^{-1}N^H} \xrightarrow{fdd} B_H(t), \quad t \geq 0,$$

where $\{B_H(t)\}$ is a standard FBM, showing the desired convergence (3.2.3). □

Observe that since $1 - 2H < 0$, the density $C(H)(1-p)^{1-2H}$ explodes at $p = 1$, making it likely that the random walker would tend to continue walking in the same direction.

3.3 Infinite Source Poisson Model with Heavy Tails

Long-range dependence is often associated with heavy tails. (Heavy-tailed random variables are discussed in Appendix A.4.) This is apparent in many models incorporating heavy tails. One such model, called the *infinite source Poisson model*, is considered here. For completeness and illustration's sake, we consider this model for a number of different parameter regimes leading to various self-similar processes with stationary increments. One such process turns out to be FBM with the self-similarity parameter $H > 1/2$, whose increments are long-range dependent.

3.3.1 Model Formulation

In the infinite source Poisson model, workload sessions arrive to a system at Poisson arrival times and each session carries workload at some rate (reward) for a random duration time. See Figure 3.2. The basic notation and assumptions are as follows.

Arrivals:

Workload sessions start according to a Poisson process on the real line with intensity $\lambda > 0$. The arrival times are denoted $\ldots S_j, S_{j+1}, \ldots$

Durations:

The session length distribution is represented by a random variable U with distribution function $F_U(u) = \mathbb{P}(U \leq u)$ and expected value

$$\nu = \mathbb{E}U < \infty.$$

Suppose that either

$$\mathbb{E}U^2 < \infty \tag{3.3.1}$$

or U is heavy-tailed in the sense that

3.3 Infinite Source Poisson Model with Heavy Tails

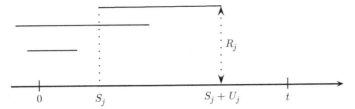

Figure 3.2 Sessions are denoted by horizontal lines. A typical session j starts at Poisson time S_j, has a workload rate R_j and lasts for a random time U_j. We are interested in the total workload from time 0 to time t.

$$\mathbb{P}(U > u) = L_U(u)\frac{u^{-\gamma}}{\gamma}, \qquad (3.3.2)$$

as $u \to \infty$, where $1 < \gamma < 2$ and L_U is a slowly varying function at infinity (which is positive for all $u > 0$). We extend the parameter range $1 < \gamma < 2$ to $1 < \gamma \leq 2$, by letting $\gamma = 2$ represent the case $\mathbb{E}U^2 < \infty$. Now let $\{U_j\}_{j \in \mathbb{Z}}$ be i.i.d. random variables with a common distribution F_U, with U_j representing session duration associated with the arrivals $S_j, j \in \mathbb{Z}$.

Rewards:
The workload rate valid during a session is given by a random variable $R > 0$ with $F_R(r) = \mathbb{P}(R \leq r)$ and

$$\mathbb{E}R < \infty.$$

We suppose either

$$\mathbb{E}R^2 < \infty \qquad (3.3.3)$$

or

$$\mathbb{P}(R > r) = L_R(r)\frac{r^{-\delta}}{\delta}, \qquad (3.3.4)$$

as $r \to 0$, where $1 < \delta < 2$ and L_R is a slowly varying function at infinity (which is positive for all $r > 0$). We extend the parameter range to $1 < \delta \leq 2$, by letting $\delta = 2$ represent the case $\mathbb{E}R^2 < \infty$. Suppose that $\{R_j\}_{j \in \mathbb{Z}}$ are i.i.d. variables with a common distribution F_R, with R_j representing the session rate associated with the arrivals $S_j, j \in \mathbb{Z}$. The workload rate is also called the *reward*.

We assume that the arrival times, durations and rewards are mutually independent.

In an application to internet traffic modeling, arrivals are associated with initiations of web sessions (file transfers), which have varying durations and workload rates. A particularly simple case is that of a fixed workload rate, say $R = 1$, $F_R(dr) = \delta_{\{1\}}(dr)$ and a duration U corresponding to the file size to be transmitted. With real data sets from Internet traffic, both file sizes and session durations are typically found to be heavy-tailed (e.g., Crovella and Bestavros [270], Guerin, Nyberg, Perrin, Resnick, Rootzén, and Stărică [438], López-Oliveros and Resnick [645]).

The cumulative workload $W_\lambda^*(t)$ from time 0 to time t is defined in Section 3.3.2 below. We are interested in the deviation of the cumulative workload from its mean $\mathbb{E}W_\lambda^*(t) = \lambda t \mathbb{E}(U)\mathbb{E}(R)$ (see (3.3.17) below), and more precisely, in the asymptotic behavior of

$$\frac{1}{b}(W_\lambda^*(at) - \mathbb{E}W_\lambda^*(at)),$$

as $\lambda, a, b \to \infty$, which describes what happens when the number of sessions increases ($\lambda \to \infty$) and when the time grows ($a \to \infty$). Different types of self-similar processes may appear in the limit depending on the respective values of the indices γ and δ, and on the respective rates at which λ, a, b tend to infinity. These are:

1) *Fractional Brownian motion* $B_H(t)$ with $1/2 < H < 1$. Recall that B_H is Gaussian with stationary dependent increments. The limit is B_H when

$$1 < \gamma < \delta = 2 \quad \text{and} \quad \frac{b}{a} \to \infty$$

with $H = \frac{3-\gamma}{2} \in (1/2, 1)$ (see Theorem 3.3.8, (a)).

2) *The Telecom process.* This is an infinite variance self-similar stable process $Z_{\gamma,\delta}(t)$ with stationary dependent increments. The limit appears when

$$1 < \gamma < \delta < 2 \quad \text{and} \quad \frac{b}{a} \to \infty$$

(Theorem 3.3.8, (b)). Because the variance is infinite, it is best to define the Telecom process through an integral representations. One can do this in two ways. First, in terms of an integral with respect to a Poisson random measure:

$$Z_{\gamma,\delta}(t) = \int_{-\infty}^{\infty} \int_0^{\infty} \int_0^{\infty} K_t(s,u) r (N(ds, du, dr) - ds u^{-(1+\gamma)} du r^{-(1+\delta)} dr), \quad (3.3.5)$$

where the kernel K_t is defined in (3.3.10) below and where the Poisson random measure $N(ds, du, dr)$ has control measure $ds u^{-(1+\gamma)} du r^{-(1+\delta)} dr$.

The factor r in the kernel together with the presence of the measure $r^{-(1+\delta)} dr$ is characteristic of a stable process with index δ (see Appendix B.1.3) and therefore one can also represent the Telecom process by an integral with respect to a δ-stable random measure $M_\delta(ds, du)$, as

$$Z_{\gamma,\delta}(t) = c_\delta \int_{-\infty}^{\infty} \int_0^{\infty} K_t(s,u) M_\delta(ds, du), \quad (3.3.6)$$

where $c_\delta > 0$ is a constant and $M_\delta(ds, du)$ has control measure $ds u^{-(1+\gamma)} du$. With the representation (3.3.6), the Telecom process is the same as that in Section 2.6.7. The Telecom process is self-similar with index

$$H = \frac{\delta + 1 - \gamma}{\delta} \in (\frac{1}{\delta}, 1).$$

As noted in Section 2.6.7, the Telecom process is different from LFSM, which is another stable self-similar process introduced in Section 2.6.5.

The second representation (3.3.6), involving $M_\delta(ds, du)$, continues to make sense if $\delta = 2$. In this case, $M_\delta(ds, du)$ is a Gaussian measure. Since $Z_{\gamma,2}(t)$ is Gaussian and H-self-similar with stationary increments, it is necessarily fractional Brownian motion $B_H(t)$ (see Corollary 2.6.3).

3) *The stable Lévy process.* This is an infinite variance process with independent stationary increments (see Section 2.6.4). It appears when

$$1 < \gamma < \delta \leq 2 \quad \text{and} \quad \frac{b}{a} \to 0$$

(see Theorem 3.3.10) and when

$$1 < \delta < \gamma \leq 2$$

(see Theorem 3.3.17). For example, in the first case, it can be represented as

$$\Lambda_\gamma(t) = \int_{-\infty}^{\infty} \int_0^{\infty} \int_0^{\infty} 1_{\{0<s<t\}} ur(N(ds, du, dr) - ds u^{-(1+\gamma)} du F_R(dr)),$$

where F_R is the distribution of R given by (3.3.3) or (3.3.4). The Poisson random measure N has control measure $ds u^{-(1+\gamma)} du F_R(dr)$. Both u and r appear in the kernel but because $\gamma < \delta$, the process is γ-stable and can be represented as

$$\Lambda_\gamma(t) = c_\gamma \int_{-\infty}^{\infty} \int_0^{\infty} 1_{\{0<s<t\}} r M_\gamma(ds, dr),$$

where $c_\gamma > 0$ and $M_\gamma(ds, dr)$ is a γ-stable random measure with control measure $ds F_R(dr)$. The presence of the indicator function $1_{\{0<s<t\}}$ yields the independent stationary increments.

4) *The Intermediate Telecom process.* This is the process $Y_{\gamma,R}(t)$ having an integral representation

$$Y_{\gamma,R}(t) = \int_{-\infty}^{\infty} \int_0^{\infty} \int_0^{\infty} K_t(s, u) r \left(N(ds, du, dr) - ds\, u^{-(1+\gamma)} du\, F_R(dr) \right),$$

where the kernel K_t is defined in (3.3.10) below and where the Poisson random measure $N(ds, du, dr)$ has control measure $ds u^{-(1+\gamma)} F_R(dr)$. It appears in a boundary case $b = a$ (see Theorem 3.3.12 below). The limit process $Y_{\gamma,R}(t)$ is not self-similar but is second-order self-similar (see Remark 3.3.13). Note the important difference between this and the Telecom process. In the Telecom process, the distribution $F_R(dr)$ is replaced by $r^{-(1+\delta)} dr$, which is not a distribution. It is the presence of $r^{-(1+\delta)} dr$ which makes the Telecom process $Z_{\gamma,\delta}(t)$ a δ-stable process with infinite variance.

5) *Brownian motion.* The limit process is Brownian motion in the case

$$\gamma = \delta = 2$$

(see Theorem 3.3.16).

3.3.2 Workload Process and its Basic Properties

The focus now is on a workload process W_λ^* defined by

$$\begin{aligned} W_\lambda^*(t) &= \text{the aggregated workload in the time interval } [0, t] \\ &= \sum_{j: S_j \leq 0} (t \wedge (S_j + U_j)_+) R_j + \sum_{j: 0 < S_j \leq t} ((t - S_j) \wedge U_j) R_j, \end{aligned} \quad (3.3.7)$$

where the first sum above represents the aggregated workload for the arrivals before time 0, and the second sum represents the aggregated workload due to the arrivals in $[0, t]$. The first sum involves the contributions of arrivals that occurred before time 0 and are still alive by time $t > 0$ and the second involves the arrivals that occurred in the time interval $(0, t]$. In the first case, the effective duration of the jth session is the smallest of $t > 0$ and $(S_j + U_j)_+$, the end-time of the jth session, whenever $S_j + U_j$ is nonnegative. In the second case, the effective duration of the jth session, up to time t, is the minimum of U_j and $t - S_j$, the amount of time lapsed since the arrival of the jth session.

For further analysis, it is convenient to express the process W_λ^* as an integral with respect to a Poisson random measure. Let $\mathbb{R}_+ = (0, \infty)$ and $N(ds, du, dr)$ denote a Poisson random measure on $\mathbb{R} \times \mathbb{R}_+ \times \mathbb{R}_+$ with intensity measure

$$n(ds, du, dr) = \lambda ds\, F_U(du) F_R(dr). \tag{3.3.8}$$

One can think of a realization of the point measure $N(ds, du, dr)$ as a collection of points $\{(S_j, U_j, R_j)\}_{j \in \mathbb{Z}}$ and the intensity measure

$$n(ds, du, dr) = \mathbb{E} N(ds, du, dr)$$

as the mean number of points in the box of size $dsdudr$ centered at s, u and r. The process W_λ^* can then be represented as

$$\begin{aligned}
W_\lambda^*(t) &= \int_{-\infty}^0 \int_0^\infty \int_0^\infty (t \wedge (s + u)_+) r N(ds, du, dr) \\
&\quad + \int_0^t \int_0^\infty \int_0^\infty ((t - s) \wedge u) r N(ds, du, dr) \\
&= \int_{-\infty}^\infty \int_0^\infty \int_0^\infty ((t - s)_+ \wedge u - (0 - s)_+ \wedge u) r N(ds, du, dr) \\
&=: \int_{-\infty}^\infty \int_0^\infty \int_0^\infty K_t(s, u) r N(ds, du, dr),
\end{aligned} \tag{3.3.9}$$

where the kernel $K_t(s, u)$ is

$$K_t(s, u) = (t - s)_+ \wedge u - (-s)_+ \wedge u. \tag{3.3.10}$$

The kernel $K_t(s, u)$ is such that

$$0 \leq K_t(s, u) = \begin{cases} 0 & \text{if } s + u \leq 0 \text{ or } s \geq t \\ s + u & \text{if } s \leq 0 \leq s + u \leq t \\ t & \text{if } s \leq 0, t \leq s + u \\ u & \text{if } 0 \leq s, s + u \leq t \\ t - s & \text{if } 0 \leq s \leq t \leq s + u. \end{cases} \tag{3.3.11}$$

It is depicted in Figure 3.3.

The following lemma lists properties of the kernel $K_t(s, u)$ which will be used in the sequel.

Lemma 3.3.1 *Let $s \in \mathbb{R}$, $t, u \in \mathbb{R}_+$ and $a > 0$. Then,*

$$K_{at}(as, au) = a K_t(s, u), \tag{3.3.12}$$

3.3 Infinite Source Poisson Model with Heavy Tails

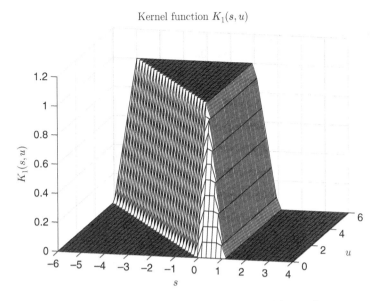

Figure 3.3 Kernel $K_t(s, u)$ in (3.3.10) with $t = 1$.

$$0 \leq K_t(s, u) = |(0, t) \cap (s, s + u)| \leq t \wedge u, \qquad (3.3.13)$$

$$K_t(s, u) = \int_0^t 1_{\{s<y<s+u\}} dy = \int_0^u 1_{\{0<y+s<t\}} dy, \qquad (3.3.14)$$

$$\int_{-\infty}^{\infty} K_t(s, u) ds = ut. \qquad (3.3.15)$$

Moreover, $K_t(s, u)$ is differentiable with respect to the variable u and

$$\frac{\partial K_t}{\partial u}(s, u) = 1_{\{0<u+s<t\}}. \qquad (3.3.16)$$

Proof Note that

$$K_t(s, u) = \int_{-s}^{t-s} 1_{\{0<y<u\}} dy = \int_0^t 1_{\{s<y<s+u\}} dy$$

so that $K_t(s, u) = |(0, t) \cap (s, s + u)| \leq t \wedge u$. Moreover, for (3.3.15),

$$\int_{-\infty}^{\infty} K_t(s, u) ds = \int_0^t \left(\int_{-\infty}^{\infty} 1_{\{s<y<s+u\}} ds \right) dy = \int_0^t u\, ds = ut.$$

One also has the second representation in (3.3.14), from which the other statements follow. □

Note that the series (3.3.7) or the integral (3.3.9) is well-defined (absolutely convergent) with probability 1. Indeed, by Campbell's theorem (e.g., Kingman [558], Section 3.2), an integral $I(f) = \int_S f(x) N(dx)$, where $N(dx)$ is a Poisson random measure on a space S with intensity measure $n(dx)$, exists with probability 1 if and only if $\int_S (|f(x)| \wedge 1) n(dx)$

$< \infty$. Moreover, if $\int_S |f(x)| n(dx) < \infty$ or f is nonnegative, then $\mathbb{E} I(f) = \int_S f(x) n(dx)$. See also Appendix B.1.4.

In the case of the integral (3.3.9), note that

$$\mathbb{E} W_\lambda^*(t) = (\mathbb{E} R)(I_1 + I_2),$$

where

$$I_1 = \int_{-\infty}^0 \int_0^\infty \Big(t \wedge (s+u)_+\Big) \lambda ds F_U(du), \quad I_2 = \int_0^t \int_0^\infty \Big((t-s) \wedge u\Big) \lambda ds F_U(du).$$

By integration by parts,

$$I_1 = \lambda \int_0^\infty dz \int_0^\infty \Big(t \wedge (u-z)_+\Big) F_U(du)$$

$$= \lambda \int_0^\infty dz \Big(\int_z^{z+t} (u-z) F_U(du) + \int_{z+t}^\infty t F_U(du) \Big)$$

$$= \lambda \int_0^\infty dz \Big(\int_z^{z+t} u F_U(du) - \int_z^{z+t} z F_U(du) + \int_{z+t}^\infty t F_U(du) \Big)$$

$$= \lambda \int_0^\infty dz \Big(- \int_z^{z+t} u d\mathbb{P}(U > u) - z \mathbb{P}(U > z) + (z+t)\mathbb{P}(U > z+t) \Big)$$

$$= \lambda \int_0^\infty dz \int_z^{z+t} \mathbb{P}(U > u) du = \lambda \int_0^\infty dz \int_z^\infty \mathbb{P}(U > u) du$$

$$\quad - \lambda \int_0^\infty dz \int_{z+t}^\infty \mathbb{P}(U > u) du$$

$$= \lambda \int_0^\infty dz \int_z^\infty \mathbb{P}(U > u) du - \lambda \int_t^\infty dz \int_z^\infty \mathbb{P}(U > u) du$$

$$= \lambda \int_0^t dz \int_z^\infty du\, \mathbb{P}(U > u)$$

and

$$I_2 = \lambda \int_0^t ds \Big(\int_0^{t-s} u F_U(du) + (t-s) \mathbb{P}(U > t-s) \Big)$$

$$= \lambda \int_0^t ds \int_0^{t-s} \mathbb{P}(U > u) du = \lambda \int_0^t dz \int_0^z du\, \mathbb{P}(U > u).$$

It follows that $I_1 + I_2 = \lambda \int_0^t dz \int_0^\infty du\, \mathbb{P}(U > u) = \lambda (\mathbb{E} U) t$, so that:

Proposition 3.3.2 *We have*

$$\mathbb{E} W_\lambda^*(t) = \lambda \mathbb{E}(U) \mathbb{E}(R) t = \lambda \nu \mathbb{E}(R) t < \infty. \tag{3.3.17}$$

We shall now involve the intensity measure $n(ds, du, dr) = \lambda ds F_U(du) F_R(dr)$ introduced in (3.3.8) and define

$$\widetilde{N}(ds, du, dr) = N(ds, du, dr) - n(ds, du, dr),$$

the so-called *compensated Poisson measure* with intensity measure $n(ds, du, dr)$. In view of (3.3.17), it is convenient to express the process $W_\lambda^*(t)$ as

$$W_\lambda^*(t) = \lambda \nu \mathbb{E}(R) t + \int_{-\infty}^{\infty} \int_0^{\infty} \int_0^{\infty} K_t(s, u) r \widetilde{N}(ds, du, dr), \qquad (3.3.18)$$

where the last integral has now zero mean.

It is now instructive to discuss briefly several other properties of the process W_λ^*.

Proposition 3.3.3 *One can write*

$$W_\lambda^*(t) = \int_0^t W_\lambda(y) dy, \qquad (3.3.19)$$

where

$$W_\lambda(y) = \int_{-\infty}^{\infty} \int_0^{\infty} \int_0^{\infty} 1_{\{s<y<s+u\}} r N(ds, du, dr). \qquad (3.3.20)$$

The process W_λ is stationary, and hence W_λ^* has stationary increments.

Proof See Exercise 3.6. □

Another important property is stated in the next proposition.

Proposition 3.3.4 *Suppose that $\mathbb{E}R^2 < \infty$. Then, $\mathbb{E}W_\lambda(y)^2 < \infty$ and*

$$\mathrm{Cov}(W_\lambda(0), W_\lambda(y)) \sim \frac{\lambda \mathbb{E}R^2 L_U(y) y^{1-\gamma}}{\gamma(\gamma-1)}, \qquad (3.3.21)$$

as $y \to \infty$. As a consequence, $\mathbb{E}W_\lambda^*(t)^2 < \infty$ and

$$\mathrm{Var}(W_\lambda^*(t)) \sim \frac{2\lambda \mathbb{E}R^2 L_U(y) t^{3-\gamma}}{\gamma(\gamma-1)(2-\gamma)(3-\gamma)}, \qquad (3.3.22)$$

as $t \to \infty$.

Proof By using the representation (3.3.20) and Campbell's theorem (e.g., Kingman [558], Section 3.2), which states that if $\int_S f(x)^2 n(dx) < \infty$ and $\int_S g(x)^2 n(dx) < \infty$, then $\mathrm{Cov}(I(f), I(g)) = \int_S f(x)g(x)n(dx) < \infty$ for $I(f) = \int_S f(x)N(dx)$ and $I(g) = \int_S g(x)N(dx)$, we have

$$\mathrm{Var}(W_\lambda(y)) = \int_{-\infty}^{\infty} \int_0^{\infty} \int_0^{\infty} 1^2_{\{s<y<s+u\}} r^2 n(ds, du, dr)$$

$$= \int_{-\infty}^{\infty} \int_0^{\infty} \int_0^{\infty} 1_{\{s<y<s+u\}} r^2 \lambda ds\, F_U(du) F_R(dr)$$

$$= \lambda \mathbb{E}R^2 \int_0^{\infty} u F_U(du) = \lambda \nu \mathbb{E}R^2 < \infty,$$

since $\int_{-\infty}^{\infty} 1_{\{s<y<s+u\}}ds = \int_{-\infty}^{\infty} 1_{\{y-u<s<y\}}ds = u$ and $\mathbb{E}R^2 < \infty$ by assumption. By Campbell's theorem again, for $y \geq 0$,

$$\text{Cov}(W_\lambda(0), W_\lambda(y)) = \int_{-\infty}^{\infty} \int_0^{\infty} \int_0^{\infty} 1_{\{s<0<s+u\}} 1_{\{s<y<s+u\}} r^2 \lambda ds\, F_U(du) F_R(dr)$$

$$= \lambda \mathbb{E}R^2 \int_0^{\infty} \left(\int_{-\infty}^0 1_{\{0<y<s+u\}} ds \right) F_U(du)$$

$$= \lambda \mathbb{E}R^2 \int_0^{\infty} \int_0^{\infty} 1_{\{0<y<u-s\}} ds\, F_U(du)$$

$$= \lambda \mathbb{E}R^2 \int_y^{\infty} (u-y) F_U(du). \tag{3.3.23}$$

By using integration by parts, (3.3.2) and Proposition 2.2.2, we further get that

$$\text{Cov}(W_\lambda(0), W_\lambda(y)) = \lambda \mathbb{E}R^2 \int_y^{\infty} \mathbb{P}(U > u) du \sim \frac{\lambda \mathbb{E}R^2 L_U(y) y^{1-\gamma}}{\gamma(\gamma-1)},$$

as $y \to \infty$. The statements concerning W_λ^* follow easily from those for W_λ, since $\text{Var}(W_\lambda^*(t)) = \int_0^t \int_0^t \text{Cov}(W_\lambda(u), W_\lambda(v)) du dv$. □

Remark 3.3.5 In view of (3.3.21), the stationary process W_λ underlying W_λ^* is LRD in the sense (2.1.5) in condition II with

$$d = \frac{2-\gamma}{2} \in (0, 1/2) \tag{3.3.24}$$

(see also Remark 2.1.6). This is an important example of how heavy tails can induce long-range dependence.

The relation (3.3.21) yields

$$\text{Var}(W_\lambda^*(t)) \sim C L_U(t) t^{2H},$$

as $t \to \infty$, where

$$H = \frac{3-\gamma}{2} \in (1/2, 1). \tag{3.3.25}$$

The parameter H in (3.3.25) will appear below as a self-similarity parameter of Gaussian processes arising as suitable limits of scaled workload process W_λ^*.

3.3.3 Input Rate Regimes

We shall now focus on the limiting behavior of the scaled workload process

$$\frac{1}{b}\left(W_\lambda^*(at) - \mathbb{E}W_\lambda^*(at) \right) = \frac{1}{b}\left(W_\lambda^*(at) - \lambda \nu \mathbb{E}(R) at \right), \tag{3.3.26}$$

as

$$\lambda, a, b \to \infty. \tag{3.3.27}$$

As will be seen below, a number of different limiting processes can be obtained depending on the speeds at which λ and a tend to infinity in (3.3.27) (the normalizing factor b will be a function of λ and a). The following three regimes will play a key role:

$$\text{fast:} \quad \frac{\lambda L_U(a)}{a^{\gamma-1}} \to \infty,$$

$$\text{slow:} \quad \frac{\lambda L_U(a)}{a^{\gamma-1}} \to 0, \qquad (3.3.28)$$

$$\text{intermediate:} \quad \frac{\lambda L_U(a)}{a^{\gamma-1}} \to c, \quad 0 < c < \infty.$$

We write
$$\overline{F}_U(a) = 1 - F_U(a).$$

Since, by (3.3.2), we have
$$\frac{\lambda L_U(a)}{a^{\gamma-1}} = \gamma \lambda a \overline{F}_U(a), \qquad (3.3.29)$$

we can reexpress the three regimes as follows:

$$\text{fast:} \quad \gamma \lambda a \overline{F}_U(a) \to \infty,$$

$$\text{slow:} \quad \gamma \lambda a \overline{F}_U(a) \to 0, \qquad (3.3.30)$$

$$\text{intermediate:} \quad \gamma \lambda a \overline{F}_U(a) \to c, \quad 0 < c < \infty.$$

The parameter λ determines the *input rate* of the Poisson arrivals. Viewing $\lambda = \lambda(a)$ as a function of a, the three regimes (3.3.28) thus impose conditions on how fast the input rate $\lambda(a)$ grows as the time scale $a \to \infty$. It will sometimes be convenient to write these three regimes as one condition,

$$\frac{\lambda L_U(a)}{a^{\gamma-1}} \to c, \qquad (3.3.31)$$

where $c \in [0, \infty]$.

We shall use the following notation: the (generalized) inverse function of a non-decreasing function $a(x)$ is the function

$$a^{\leftarrow}(t) = \inf\{x : a(x) \geq t\}.$$

For properties of $a^{\leftarrow}(t)$, see for example Resnick [847], Section 2.1.2. If $a(x)$ is continuous and strictly monotone, then one gets $a^{\leftarrow}(t)$ by solving $t = a(x)$. For example, if $a(x) = x^3$, then $a^{\leftarrow}(t) = t^{1/3}$.

There are several equivalent ways of writing the regimes (3.3.28), and several interpretations.

Proposition 3.3.6 *The condition (3.3.31) with $c \in [0, \infty]$ is equivalent to any of the following conditions (i), (ii) or (iii):*

(i) with

$$g(t) = \left(\frac{1}{\overline{F}_U}\right)^{\leftarrow}(t) = \inf\left\{x : \frac{1}{\overline{F}_U(x)} = \frac{1}{1 - F_U(x)} \geq t\right\}, \quad t > 0, \qquad (3.3.32)$$

we have
$$\frac{g(\lambda a)}{a} \to \frac{c^{1/\gamma}}{\gamma^{1/\gamma}}. \qquad (3.3.33)$$

(ii) with a stationary process $W_\lambda(y)$ defined in (3.3.20) and in the case $\mathbb{E}R^2 < \infty$, we have
$$\mathrm{Cov}(W_\lambda(0), W_\lambda(a)) \to \frac{\mathbb{E}R^2}{\gamma(\gamma-1)} c. \qquad (3.3.34)$$

(iii) with $X_{\lambda,a}$ defined in (3.3.42) and denoting the number of sessions active at both times 0 and a, we have
$$\mathbb{E}X_{\lambda,a} \to \frac{c}{\gamma(\gamma-1)}. \qquad (3.3.35)$$

Proof Consider first the equivalence of (3.3.31) and (i) when $0 < c < \infty$. Using (3.3.29), write (3.3.31) as
$$\gamma \lambda a \overline{F}_U(a) \sim c \qquad (3.3.36)$$
or
$$\frac{1}{\overline{F}_U(a)} \sim \frac{\gamma \lambda a}{c}.$$

On the other hand, $\overline{F}_U(a) \sim L_U(u)\gamma^{-1}u^{-\gamma}$ as $u \to \infty$. By (3.3.2), using Theorem 1.5.12 in Bingham et al. [157], the function g in (3.3.32), namely the inverse of $f = 1/\overline{F}_U$, can be expressed as
$$g(t) = L_g(t) t^{1/\gamma}. \qquad (3.3.37)$$

It is regularly varying with index $1/\gamma$ and slowly varying function L_g. It is, moreover, an "asymptotic inverse" of the function f; that is,
$$g(f(x)) \sim x, \quad \text{as } x \to \infty. \qquad (3.3.38)$$

Then,
$$a \sim g\left(\frac{1}{\overline{F}_U(a)}\right) \sim g\left(\frac{\gamma \lambda a}{c}\right) = L_g\left(\frac{\gamma \lambda a}{c}\right)\left(\frac{\gamma \lambda a}{c}\right)^{1/\gamma}$$
$$= L_g\left(\frac{\gamma \lambda a}{c}\right)\frac{1}{L_g(\lambda a)}\frac{\gamma^{1/\gamma}}{c^{1/\gamma}}L_g(\lambda a)(\lambda a)^{1/\gamma} \sim \frac{\gamma^{1/\gamma}}{c^{1/\gamma}} g(\lambda a) \qquad (3.3.39)$$

and thus
$$\frac{g(\lambda a)}{a} \sim \frac{c^{1/\gamma}}{\gamma^{1/\gamma}},$$
proving (3.3.33) when $0 < c < \infty$.

Consider next the case (i) with $c = 0$. We shall choose $c > 0$ and then let $c \to 0$. Then, the relation (3.3.39) becomes
$$a \sim \frac{L_g\left(\frac{\gamma \lambda a}{c}\right)}{L_g(\lambda a)} \frac{\gamma^{1/\gamma}}{c^{1/\gamma}} g(\lambda a)$$

or
$$\frac{c^{1/\gamma}}{\gamma^{1/\gamma}} \frac{L_g(\lambda a)}{L_g\left(\frac{\gamma \lambda a}{c}\right)} \sim \frac{g(\lambda a)}{a}. \tag{3.3.40}$$

By using Potter's bound (2.2.1), for $\epsilon < 1/\gamma$,
$$\frac{L_g(\lambda a)}{L_g\left(\frac{\gamma \lambda a}{c}\right)} \leq 2\left(\frac{c}{\gamma}\right)^{-\epsilon}$$

and (3.3.40) shows that $g(\lambda a)/a \to 0$.

In the case *(i)* with $c = \infty$, write as above
$$\frac{a}{g(\lambda a)} \sim \frac{\gamma^{1/\gamma}}{c^{1/\gamma}} \frac{L_g\left(\frac{\gamma \lambda a}{c}\right)}{L_g(\lambda a)}. \tag{3.3.41}$$

Applying Potter's bound (2.2.1) again, we get that $a/g(\lambda a) \to 0$.

The equivalence of (3.3.31) and *(ii)* follows directly from (3.3.29) and (3.3.21).

To see the equivalence of (3.3.31) and *(iii)*, observe that the number of sessions $X_{\lambda,a}$ active at both 0 and a can be represented as
$$X_{\lambda,a} = \int_{-\infty}^0 \int_0^\infty \int_0^\infty 1_{\{s<0<a\leq s+u\}} N(ds, du, dr). \tag{3.3.42}$$

Then, as for (3.3.23), by integration by parts,
$$\mathbb{E} X_{\lambda,a} = \lambda \int_a^\infty (u-a) F_U(du) = \lambda \int_a^\infty \mathbb{P}(U > u) du \sim \frac{\lambda L_U(a) a^{1-\gamma}}{\gamma(\gamma-1)}$$

and the equivalence follows. $\square$

The equivalence *(i)* in Proposition 3.3.6 will be useful in the results below when the limit is a stable process. The equivalences *(ii)* and *(iii)* provide physical interpretations of the regimes. For example, the fast regime, namely one where $c \to 0$, is the regime where the covariance in the underlying stationary workload process grows (case *(ii)*), and the average number of active sessions throughout $[0, a]$ increases (case *(iii)*).

3.3.4 Limiting Behavior of the Scaled Workload Process

We turn now to the behavior of scaled workload process under various regimes. We will work with characteristic functions of integrals with respect to Poisson random measures. In this regard, recall from Appendix B.1.4 that the characteristic function of $\int f(x)(N(dx) - n(dx))$ is well defined if $\int (f(x)^2 \wedge |f(x)|) n(dx) < \infty$, in which case it is given by
$$\log \mathbb{E} \exp\left\{ i \int f(x)(N(dx) - n(dx)) \right\} = \int (e^{if(x)} - 1 - if(x)) n(dx). \tag{3.3.43}$$

The following lemma about slowly varying functions and their inverse will be handy.

Lemma 3.3.7 *For a slowly varying function L and a number $\alpha > 0$, there is a slowly varying function L_α^* satisfying: for $x > 0$,*

$$L_\alpha^*(u)^{-\alpha} L\left(u^{1/\alpha} L_\alpha^*(u) x\right) \to 1, \quad \text{as } u \to \infty. \tag{3.3.44}$$

Moreover, L_α^* is unique within asymptotic equivalence.

Proof The existence of the slowly varying function L_α^* is proved following (A.4.9) in Appendix A.4 after setting $h = L$ and $h_\alpha = (L_\alpha^*)^\alpha$. The uniqueness part is also mentioned following (A.4.9). □

We shall use Lemma 3.3.7 below with $L = L_U, \alpha = \gamma$ and $L = L_R, \alpha = \delta$, in which case the corresponding slowly varying function L_α^* will be denoted $L_{U,\gamma}^*$ and $L_{R,\delta}^*$, respectively. One can use Lemma 3.3.7 also to conclude that the function g in (3.3.32) satisfies

$$g(t) \sim \gamma^{-1/\gamma} t^{1/\gamma} L_{U,\gamma}^*(t), \quad \text{as } t \to \infty. \tag{3.3.45}$$

Indeed, the function g satisfies $f(g(x)) \sim x$ in (3.3.38), where $f(x) = (1/\overline{F}_U)(x) = \gamma x^\gamma L_U(x)^{-1}$. Setting $g(t) = \gamma^{-1/\gamma} t^{1/\gamma} L_0(t)$, we deduce that $L_0(t)$ satisfies $L_0(x)^\gamma L_U(\gamma^{-1/\gamma} x^{1/\gamma} L_0(x))^{-1} \sim 1$ and hence that $L_0(x) \sim L_{U,\gamma}^*(x)$ by Lemma 3.3.7.

The following theorem concerns the behavior of a scaled workload process in the fast regime. It involves the index γ characterizing asymptotic behavior of the tail distribution of the session length (see (3.3.2)) and the index δ characterizing the asymptotic behavior of the tail distribution of the rewards (see (3.3.4)). Recall that $\delta = 2$ corresponds to finite variance rewards. The integrals with respect to Gaussian and stable measures appearing below are defined in Appendices B.1.2 and B.1.3.

Theorem 3.3.8 *Suppose that the fast regime in (3.3.28) holds, namely $\lambda L_U(a)/a^{\gamma-1} \to \infty$ as λ and $a \to \infty$.*

(a) In the case (3.3.3) of finite variance rewards,

$$1 < \gamma < \delta = 2,$$

set

$$b = \left(\gamma \lambda a^3 \overline{F}_U(a)\right)^{1/2} = \left(\lambda a^{3-\gamma} L_U(a)\right)^{1/2} \tag{3.3.46}$$

for the normalization in the scaled workload process (3.3.26). Then, as $a \to \infty$,

$$\frac{b}{a} \to \infty \tag{3.3.47}$$

and

$$\frac{1}{b}(W_\lambda^*(at) - \lambda \nu \mathbb{E}(R)at) \xrightarrow{fdd} (\mathbb{E}R^2)^{1/2} \sigma B_H(t), \tag{3.3.48}$$

where $\xrightarrow{fdd}$ denotes convergence in the sense of finite-dimensional distributions, B_H is a standard FBM with the self-similarity parameter

$$H = \frac{3-\gamma}{2} \in \left(\frac{1}{2}, 1\right),$$

3.3 Infinite Source Poisson Model with Heavy Tails

and where

$$\sigma^2 = \int_{-\infty}^{\infty} \int_0^{\infty} K_1(s,u)^2 u^{-(\gamma+1)} ds du = \frac{2}{\gamma(\gamma-1)(2-\gamma)(3-\gamma)}. \qquad (3.3.49)$$

Alternatively, the limit process can be represented as

$$(\mathbb{E}R^2)^{1/2} \int_{-\infty}^{\infty} \int_0^{\infty} K_t(s,u) M_2(ds, du), \qquad (3.3.50)$$

where $K_t(s,u)$ is the kernel defined in (3.3.10) and $M_2(ds, du)$ is a Gaussian random measure with control measure

$$ds\, u^{-(1+\gamma)} du.$$

(b) In the case (3.3.4) of infinite variance rewards, and

$$1 < \gamma < \delta < 2,$$

set

$$b = \lambda^{1/\delta} a^{(1+\delta-\gamma)/\delta} L_U(a)^{1/\delta} L_{R,\delta}^*\left(\lambda a^{1-\gamma} L_U(a)\right) \qquad (3.3.51)$$

for the normalization. Then, as $a \to \infty$,

$$\frac{b}{a} \to \infty \qquad (3.3.52)$$

and

$$\frac{1}{b}(W_\lambda^*(at) - \lambda \nu \mathbb{E}(R)at) \xrightarrow{fdd} Z_{\gamma,\delta}(t), \qquad (3.3.53)$$

where

$$Z_{\gamma,\delta}(t) = \int_{-\infty}^{\infty} \int_0^{\infty} \int_0^{\infty} K_t(s,u) r \left(N(ds, du, dr) - ds\, u^{-(1+\gamma)} du\, r^{-(1+\delta)} dr\right) \qquad (3.3.54)$$

$$\stackrel{d}{=} c_\delta \int_{-\infty}^{\infty} \int_0^{\infty} K_t(s,u) M_\delta(ds, du), \qquad (3.3.55)$$

the random measure $M_\delta(ds, du)$ is δ-stable, totally skewed to the right and has the control measure

$$ds\, u^{-(\gamma+1)} du,$$

and where

$$c_\delta = \left(\frac{2\Gamma(2-\delta)}{\delta(1-\delta)} \cos(\frac{\delta\pi}{2})\right)^{1/\delta}. \qquad (3.3.56)$$

Remark 3.3.9 The process $Z_{\gamma,\delta}$ in part (b) of the theorem is a δ-stable process, called the *Telecom process*, introduced in (3.3.5). It is H–self-similar with

$$H = \frac{\delta + 1 - \gamma}{\delta} \in \left(\frac{1}{\delta}, 1\right).$$

The stationary increments of $Z_{\gamma,\delta}$ are then LRD in the sense of condition A in Section 2.9 since

$$H - \frac{1}{\delta} = \frac{\delta - \gamma}{\gamma} > 0.$$

The stationary increments of FBM in part (a) are also LRD since $H > 1/2$.

Proof summary The idea of the proof of part (a) is as follows. Use the logarithm of the characteristic function (3.3.58) of the rescaled workload process. It is expressed as an integral. Because $a/b \to 0$, the integrand converges to the simpler form (3.3.61). Then apply the dominated convergence theorem and show that the integral can be expressed as (3.3.62) which is the logarithm of the characteristic function of a Gaussian process.

For part (b), express the logarithm of the characteristic function as (3.3.66) and compare it to (3.3.67) which turns out to be the logarithm of the characteristic function of the limit process (3.3.54). We shall now develop the proof in detail. □

Proof We first show part (a). By (3.3.2) and the fast regime condition in (3.3.28), we have

$$\left(\frac{b}{a}\right)^2 = \gamma \lambda a \overline{F}_U(a) \sim \lambda a a^{-\gamma} L_U(a) = \frac{\lambda L_U(a)}{a^{\gamma - 1}} \to \infty, \qquad (3.3.57)$$

proving (3.3.47). We now prove (3.3.48). Let $J \geq 1$ and $\theta_1, \ldots, \theta_J$ be arbitrary real numbers. By using the representation (3.3.9) of $W_\lambda^*(t)$ and (3.3.43), observe that

$$I_\lambda := \log \mathbb{E} \exp\left\{i \sum_{j=1}^J \theta_j \frac{W_\lambda^*(at_j) - \lambda \nu(\mathbb{E}R) a t_j}{b}\right\} \qquad (3.3.58)$$

$$= \log \mathbb{E} \exp\left\{i \int_{-\infty}^\infty \int_0^\infty \int_0^\infty \sum_{j=1}^J \frac{\theta_j K_{at_j}(s,u) r}{b} \widetilde{N}(ds, du, dr)\right\}$$

$$= \int_{-\infty}^\infty \int_0^\infty \int_0^\infty \left(\exp\left\{i \sum_{j=1}^J \frac{\theta_j K_{at_j}(s,u) r}{b}\right\} - 1 \right.$$

$$\left. - i \sum_{j=1}^J \frac{\theta_j K_{at_j}(s,u) r}{b}\right) \lambda ds\, F_U(du)\, F_R(dr)$$

$$= \int_{-\infty}^\infty \int_0^\infty \int_0^\infty \left(\exp\left\{i \sum_{j=1}^J \theta_j K_{t_j}(s,u) \left(\frac{ra}{b}\right)\right\} - 1 \right.$$

$$\left. - i \sum_{j=1}^J \theta_j K_{t_j}(s,u) \left(\frac{ra}{b}\right)\right) \lambda a\, ds\, F_U(d(au))\, F_R(dr),$$

where we used the scaling property (3.3.12) of K_t. By using (3.3.16), the kernel $K_t(s,u)$ is differentiable with respect to the variable u and $\frac{\partial K_t}{\partial u}(s,u) = 1_{\{0 < u+s < t\}}$. We shall now perform integration by parts on the variable u using the relation

$$\frac{d}{dx}(e^{icx} - 1 - icx) = (e^{icx} - 1)ic$$

3.3 Infinite Source Poisson Model with Heavy Tails

for constant c. Since $d\mathbb{P}(U > au) = -d(F_U(au))$ and $K_t(s, 0) = 0$, we get

$$I_\lambda = i \int_{-\infty}^{\infty} \int_0^{\infty} \int_0^{\infty} \left(\exp\left\{ i \sum_{j=1}^J \theta_j K_{t_j}(s, u)(\frac{ra}{b}) \right\} - 1 \right)$$

$$\times \sum_{j=1}^J \theta_j \frac{\partial K_{t_j}}{\partial u}(s, u)\left(\frac{ra}{b}\right) \mathbb{P}(U > au) \lambda a \, ds \, du \, F_R(dr). \quad (3.3.59)$$

We showed that, in the fast regime,

$$\frac{a}{b} = \left(\frac{a^2}{\gamma \lambda a^3 \overline{F}_U(a)} \right)^{1/2} = \left(\frac{1}{\gamma \lambda a \overline{F}_U(a)} \right)^{1/2} \to 0$$

by (3.3.30), and that

$$\left(\frac{a}{b}\right)^2 \mathbb{P}(U > au)\lambda a = \frac{a^2}{\gamma \lambda a^3 \overline{F}_U(a)} \overline{F}_U(au) \lambda a = \frac{u^{-\gamma}}{\gamma} \frac{L_U(au)}{L_U(a)} \to \frac{u^{-\gamma}}{\gamma}, \quad (3.3.60)$$

as $a \to \infty$, for any $u > 0$. In view of these relations and since $e^z - 1 \sim z$ as $z \to 0$, we obtain that the integrand $I_\lambda(s, u, r)$ of the integral I_λ converges,

$$I_\lambda(s, u, r) := i\left(\exp\left\{ i \sum_{j=1}^J \theta_j K_{t_j}(s, u)(\frac{ra}{b}) \right\} - 1 \right) \sum_{j=1}^J \theta_j \frac{\partial K_{t_j}}{\partial u}(s, u)(\frac{ra}{b}) \mathbb{P}(U > au) \lambda a$$

$$\to -\sum_{j=1}^J \theta_j K_{t_j}(s, u) \sum_{j=1}^J \theta_j \frac{\partial K_{t_j}}{\partial u}(s, u) \, r^2 \frac{u^{-\gamma}}{\gamma}, \quad (3.3.61)$$

as $\lambda, a, b \to \infty$, for any $s \in \mathbb{R}$, $u, r > 0$. We now want to apply the dominated convergence theorem to conclude the convergence of integrals.

To show that the dominated convergence theorem applies, we need to split the integration domain into $\{au \leq a_0\}$ and $\{au > a_0\}$, where a_0 is fixed. For $\{au \leq a_0\}$, by using the relation $|e^{ix} - 1| \leq |x|$, (3.3.13) and the expression for $\frac{\partial K_t}{\partial u}(s, u)$ above, we bound the integrand as

$$|I_\lambda(s, u, r)| \leq C \frac{a^3 \lambda}{b^2} 1_{\{0 < u+s < T\}} u r^2 \mathbb{P}(U > au),$$

where $T > 0$ is fixed and $C > 0$ is a constant which may change from line to line. By using $\mathbb{P}(U > au) \leq \mathbb{E}U/(au)$ from Markov's inequality and $u \leq 1$ on $\{au \leq a_0\}$ for large enough a, we get that, on $\{au \leq a_0\}$, and using (3.3.57),

$$|I_\lambda(s, u, r)| \leq C \frac{a^2}{b^2} \lambda r^2 1_{\{0 < u \leq 1\}} 1_{\{-1 < s \leq T\}} \leq C_1 \frac{r^2}{a \overline{F}_U(a)} 1_{\{0 < u \leq 1\}} 1_{\{-1 < s \leq T\}}.$$

By using (2.2.3) and for large enough a, $1/(a\overline{F}_U(a)) = \gamma a^{\gamma-1}/L_U(a) \leq C a^{\gamma-1+\epsilon}$, where ϵ is arbitrarily small. On $\{au \leq a_0\}$, this yields

$$|I_\lambda(s, u, r)| \leq C r^2 u^{1-\gamma-\epsilon} 1_{\{0 < u \leq 1\}} 1_{\{-1 < s \leq T\}}.$$

Note that the function in the bound is integrable on $(-\infty, \infty) \times (0, \infty) \times (0, \infty)$ with respect to $dsdu F_R(dr)$ since $2 - \gamma - \epsilon > 0$ for small enough ϵ. On the other hand, for $\{au \geq a_0\}$, we bound the integrand as

$$|I_\lambda(s, u, r)| \leq C \sum_{j=1}^{J} K_{t_j}(s, u) \sum_{j=1}^{J} \frac{\partial K_{t_j}}{\partial u}(s, u) \, r^2 \left(\frac{a}{b}\right)^2 \mathbb{P}(U > au)\lambda a,$$

where we used the fact that $|e^{ix} - 1| \leq |x|$ and $K_t(s, u), \frac{\partial K_t}{\partial u}(s, u) \geq 0$. By using (3.3.60),

$$\left(\frac{a}{b}\right)^2 \mathbb{P}(U > au)\lambda a = \frac{a^2}{\gamma \lambda a^3 \overline{F}_U(a)} \overline{F}_U(au)\lambda a = \frac{u^{-\gamma}}{\gamma} \frac{L_U(au)}{L_U(a)}$$

and applying Potter's bound (2.2.1) for large enough fixed a_0, we obtain for $au \geq a_0$ that

$$|I_\lambda(s, u, r)| \leq C \sum_{j=1}^{J} K_{t_j}(s, u) \sum_{j=1}^{J} \frac{\partial K_{t_j}}{\partial u}(s, u) \, r^2 u^{-\gamma \pm \epsilon},$$

where ϵ is arbitrarily small. By using the bound $K_t(s, u) \leq t \wedge u$ (see (3.3.13)), we further get that

$$|I_\lambda(s, u, r)| \leq C(T \wedge u) 1_{\{0 < s+u < T\}} r^2 u^{-\gamma \pm \epsilon},$$

where T is fixed. The function in the bound is integrable on $(-\infty, \infty) \times (0, \infty) \times (0, \infty)$ with respect to $dsdu F_R(dr)$.

Applying the dominated convergence theorem now yields

$$I_\lambda \to -\mathbb{E}R^2 \int_{-\infty}^{\infty} \int_0^{\infty} \sum_{j=1}^{J} \theta_j K_{t_j}(s, u) \sum_{j=1}^{J} \theta_j \frac{\partial K_{t_j}}{\partial u}(s, u) \frac{u^{-\gamma}}{\gamma} dsdu$$

$$= -\frac{\mathbb{E}R^2}{2} \int_{-\infty}^{\infty} \int_0^{\infty} \frac{\partial}{\partial u}\left[\left(\sum_{j=1}^{J} \theta_j K_{t_j}(s, u)\right)^2\right] \frac{u^{-\gamma}}{\gamma} dsdu$$

$$= -\frac{\mathbb{E}R^2}{2} \int_{-\infty}^{\infty} \int_0^{\infty} \left(\sum_{j=1}^{J} \theta_j K_{t_j}(s, u)\right)^2 u^{-\gamma-1} dsdu \qquad (3.3.62)$$

after integration by parts, observing that as $u \to 0$, $K_t(s, u)^2 u^{-\gamma} \leq u^2 u^{-\gamma} \to 0$. Since I_λ is the logarithm of the characteristic function, we see that the limit is a Gaussian process. This process can be represented as in (3.3.50). It is easily seen that it has stationary increments (use (3.3.14)) and is H–self-similar with parameter $H = (3 - \gamma)/2$. Hence, by Corollary 2.6.3, it is FBM.

To obtain the form of the scaling constant σ^2 in (3.3.49), observe that, by using (3.3.14) and making the change of variables $u = (y_2 - y_1)v$ in (3.3.64) below,

$$\sigma^2 = \int_{-\infty}^{\infty} \int_0^{\infty} K_1(s, u)^2 u^{-(\gamma+1)} dsdu$$

$$= \int_{-\infty}^{\infty} \int_0^{\infty} \int_0^1 \int_0^1 1_{\{s < y_1 < s+u\}} 1_{\{s < y_2 < s+u\}} u^{-(\gamma+1)} dy_1 dy_2 dsdu$$

3.3 Infinite Source Poisson Model with Heavy Tails

$$= 2 \int_{-\infty}^{\infty} \int_0^{\infty} \int_{0<y_1<y_2<1} 1_{\{s<y_1<s+u\}} 1_{\{s<y_2<s+u\}} u^{-(\gamma+1)} dy_1 dy_2 ds du$$

$$= 2 \int_{0<y_1<y_2<1} \int_{y_2-y_1}^{\infty} \left(\int_{-\infty}^{\infty} 1_{\{y_2-u<s<y_1\}} ds \right) u^{-(\gamma+1)} du\, dy_1 dy_2 \quad (3.3.63)$$

$$= 2 \int_{0<y_1<y_2<1} \int_{y_2-y_1}^{\infty} (u-(y_2-y_1)) u^{-(\gamma+1)} du\, dy_1 dy_2$$

$$= 2 \int_{0<y_1<y_2<1} (y_2-y_1)^{1-\gamma} dy_1 dy_2 \int_1^{\infty} (v-1) v^{-(\gamma+1)} dv \quad (3.3.64)$$

$$= 2 \frac{1}{(2-\gamma)(3-\gamma)} \left(\frac{1}{\gamma-1} - \frac{1}{\gamma} \right) = \frac{2}{\gamma(\gamma-1)(2-\gamma)(3-\gamma)},$$

by using, for (3.3.63),

$$1_{\{s<y_1<s+u\}} 1_{\{s<y_2<s+u\}} = 1_{\{y_1-u<s<y_1\}} 1_{\{y_2-u<s<y_2\}} = \begin{cases} 0, & \text{if } u \le y_2-y_1, \\ 1_{\{y_2-u<s<y_1\}}, & \text{if } u > y_2-y_1. \end{cases}$$

We now turn to part (b). By (3.3.51) and (3.3.44) and the fast regime condition (3.3.28),

$$\left(\frac{b}{a} \right)^{\delta} = \lambda a^{1-\gamma} L_U(a) \left(L_{R,\delta}^*(\lambda a^{1-\gamma} L_U(a)) \right)^{\delta} \to \infty, \quad (3.3.65)$$

proving (3.3.52).

We now prove the convergence of the finite-dimensional distributions (3.3.53). Using the notation I_λ introduced in (3.3.58) and arguing as above, write

$$I_\lambda = \int_{-\infty}^{\infty} \int_0^{\infty} \int_0^{\infty} h(s,u,r) \lambda ads\, F_U(d(au)) F_R(d\frac{b}{a}r), \quad (3.3.66)$$

where

$$h(s,u,r) = \exp\left\{ i \sum_{j=1}^{J} \theta_j K_{t_j}(s,u) r \right\} - 1 - i \sum_{j=1}^{J} \theta_j K_{t_j}(s,u) r.$$

On the other hand, the logarithm of the characteristic function of the limit process $Z_{\gamma,\delta}(t)$ in (3.3.54) is

$$I = \int_{-\infty}^{\infty} \int_0^{\infty} \int_0^{\infty} h(s,u,r) ds\, u^{-\gamma-1} du\, r^{-\delta-1} dr. \quad (3.3.67)$$

Write the difference between I_λ and I as

$$I_\lambda - I = I_\epsilon^1 + I_\epsilon^2 + I_\epsilon^3,$$

where, for $k = 1, 2, 3$ and $\epsilon > 0$,

$$I_\epsilon^k = \int_{-\infty}^{\infty} \int\int_{A_\epsilon^k} h(s,u,r) \left(\lambda ads\, F_U(d(au)) F_R(d\frac{b}{a}r) - ds\, u^{-\gamma-1} du\, r^{-\delta-1} dr \right)$$

with the integration domains $A_\epsilon^1 = \{u > \epsilon, r > \epsilon\}$, $A_\epsilon^2 = \{u < \epsilon < r\}$ and $A_\epsilon^3 = \{r < \epsilon\}$, involving u and r but not involving s.

We show next that, as $\lambda, a, b \to \infty$, for fixed ϵ, $I_\epsilon^1 \to 0$ and that $I_\epsilon^2, I_\epsilon^3$ are arbitrarily small for small enough ϵ. This will establish the desired convergence to the limit process given by (3.3.54).

Step 1: For I_ϵ^1, let

$$H(u,r) = \frac{1}{ur} \int_{-\infty}^{\infty} h(s,u,r) ds$$

and write

$$I_\epsilon^1 = \int\int_{A_\epsilon^1} H(u,r) \mu_\lambda(du,dr) - \int\int_{A_\epsilon^1} H(u,r) \mu(du,dr)$$

$$= \mu_\lambda(A_\epsilon^1) \int\int_{A_\epsilon^1} H(u,r) \widetilde{\mu}_\lambda(du,dr) - \mu(A_\epsilon^1) \int\int_{A_\epsilon^1} H(u,r) \widetilde{\mu}(du,dr),$$

where

$$\mu_\lambda(du,dr) = \lambda a\, u F_U(dau)\, r F_R(d(b/a)r),$$
$$\mu(du,dr) = u^{-\gamma} du\, r^{-\delta} dr$$

and $\widetilde{\mu}_\lambda$ and $\widetilde{\mu}$ are corresponding (normalized) probability measures. By using the inequality $|e^{ix} - 1| \leq |x|$, note that $|h(s,u,r)|$ is bounded by $2 \sum_{j=1}^{J} |\theta_j| K_{t_j}(s,u) r$. This implies that $H(u,r)$ is bounded by using (3.3.15). Note also that $H(u,r)$ is continuous. Then, to show that $I_\epsilon^1 \to 0$ as $\lambda, a, b \to \infty$, it is enough to prove that the normalizations converge

$$\mu_\lambda(A_\epsilon^1) \to \mu(A_\epsilon^1), \tag{3.3.68}$$

and that the normalized probability measures converge weakly

$$\widetilde{\mu}_\lambda \xrightarrow{d} \widetilde{\mu}. \tag{3.3.69}$$

For the convergence (3.3.68) of the normalizations, observe that

$$\mu_\lambda(A_\epsilon^1) = \lambda a \int_\epsilon^\infty u F_U(dau) \int_\epsilon^\infty r F_R(d\frac{b}{a}r) = \frac{\lambda a}{b} \int_{a\epsilon}^\infty u F_U(du) \int_{b\epsilon/a}^\infty r F_R(dr)$$

$$= \frac{\lambda a}{b} \Big(a\epsilon \overline{F}_U(a\epsilon) + \int_{a\epsilon}^\infty \overline{F}_U(u) du \Big) \Big(\frac{b}{a}\epsilon \overline{F}_R(\frac{b}{a}\epsilon) + \int_{b\epsilon/a}^\infty \overline{F}_R(r) dr \Big)$$

$$\sim \frac{\lambda a}{b} \Big(a\epsilon \overline{F}_U(a\epsilon) \Big(1 - \frac{1}{-\gamma + 1} \Big) \Big) \Big(\frac{b}{a}\epsilon \overline{F}_R(\frac{b}{a}\epsilon) \Big(1 - \frac{1}{-\delta + 1} \Big) \Big)$$

$$= \frac{\lambda a}{b} \Big(\frac{(a\epsilon)^{1-\gamma}}{\gamma} L_U(a\epsilon) \frac{\gamma}{\gamma - 1} \Big) \Big(\frac{1}{\delta} \Big(\frac{b}{a}\epsilon \Big)^{1-\delta} L_R(\frac{b}{a}\epsilon) \frac{\delta}{\delta - 1} \Big)$$

$$= \frac{\epsilon^{1-\delta}}{\delta - 1} \frac{\epsilon^{1-\gamma}}{\gamma - 1} \frac{L_U(a\epsilon)}{L_U(a)} \lambda a^{1+\delta-\gamma} b^{-\delta} L_U(a) L_R(\frac{b}{a}\epsilon)$$

by integration by parts and by using (3.3.2) and (3.3.4) for the regularly varying functions $\overline{F}_U, \overline{F}_R$. Note that

3.3 Infinite Source Poisson Model with Heavy Tails

$$\lambda a^{1+\delta-\gamma} b^{-\delta} L_U(a) L_R(\frac{b}{a}\epsilon) = L^*_{R,\delta}(\lambda a^{1-\gamma} L_U(a))^{-\delta}$$
$$\times L_R\left(\lambda^{1/\delta} a^{(1-\gamma)/\delta} L_U(a)^{1/\delta} L^*_{R,\delta}(\lambda a^{1-\gamma} L_U(a))\epsilon\right) \sim 1, \quad (3.3.70)$$

by using the expression for b in (3.3.51). This is of the form $L^*_{R,\delta}(y)^{-\delta} L_R(y^{1/\delta} L^*_{R,\delta}(y)\epsilon)$ where $y = \lambda a^{1-\gamma} L_U(a)$, and hence by (3.3.44), this tends to 1 as $y \to \infty$, since under the fast regime $\lambda, a, b \to \infty$ imply $y \to \infty$ by (3.3.65). It follows that

$$\mu_\lambda(A^1_\epsilon) \to \frac{\epsilon^{1-\delta}}{\delta - 1} \frac{\epsilon^{1-\gamma}}{\gamma - 1} = \int_\epsilon^\infty \int_\epsilon^\infty \mu(du, dr) = \mu(A^1_\epsilon).$$

The convergence (3.3.69) of the probability measures can be proved similarly since it is enough to show the convergence of $\widetilde{\mu}_\lambda((u_0, \infty) \times (r_0, \infty))$ to $\widetilde{\mu}((u_0, \infty) \times (r_0, \infty))$.

Step 2: Consider now I^2_ϵ. By using the inequalities $|e^{ix} - 1 - ix| \leq C|x|^{1+\kappa}$ for fixed $\kappa \in [0, 1]$ to be chosen later, $K_t(s, u) \leq u$ and the fact that $\int_{-\infty}^\infty K_t(s, u) ds = ut$ (see (3.3.15)), we have

$$\left| \int_{-\infty}^\infty \int\int_{A^2_\epsilon} h(s, u, r) \lambda a\, ds\, F_U(d(au)) F_R(d\frac{b}{a}r) \right|$$
$$\leq C\lambda a \int_\epsilon^\infty r^{1+\kappa} F_R(d\frac{b}{a}r) \int_0^\epsilon u^{1+\kappa} F_U(dau). \quad (3.3.71)$$

We need $1 + \kappa < \delta$ for the last integral in r to be well-defined. The limits of integration in the last integral in u result from $A^2_\epsilon = \{u < \epsilon < r\}$; this integral converges. The bound (3.3.71) behaves asymptotically as

$$C\lambda a\epsilon^{1+\kappa} \overline{F}_R(\frac{b}{a}\epsilon) \epsilon^{1+\kappa} \overline{F}_U(a\epsilon) = C\epsilon^{2(1+\kappa)-\gamma-\delta} \frac{L_U(a\epsilon)}{L_U(a)}$$
$$\times \lambda a^{1+\delta-\gamma} b^{-\delta} L_U(a) L_R(\frac{b}{a}\epsilon) \sim C\epsilon^{2(1+\kappa)-\gamma-\delta}, \quad (3.3.72)$$

by using (3.3.70). This is arbitrarily small for small enough ϵ if $1 + \kappa > (\gamma + \delta)/2$. One deals in a similar way with the corresponding term

$$\int_{-\infty}^\infty \int\int_{A^2_\epsilon} h(s, u, r) ds\, u^{-\gamma-1} du\, r^{-\delta-1} dr$$

for the limiting process. The computations are in fact easier and one can show that this term is arbitrarily small for small enough ϵ.

Step 3: Finally, consider I^3_ϵ. By using the inequalities $|e^{ix} - 1 - ix| \leq C|x|^2$, $K_t(s, u) \leq u \wedge t$ and the fact that $\int_{-\infty}^\infty K_t(s, u) ds = ut$ (see (3.3.15)), we have with $A^3_\epsilon = \{r < \epsilon\}$

$$\left| \int_{-\infty}^{\infty} \int\int_{A_\epsilon^3} h(s,u,r) \lambda a\, ds\, F_U(d(au)) F_R(d\frac{b}{a}r) \right|$$

$$\leq C\lambda a \int_0^\epsilon r^2 F_R(d\frac{b}{a}r) \int_0^\infty u(u \wedge T) F_U(dau),$$

where $t_1, \ldots, t_J \leq T$. Arguing as for (3.3.72), the bound behaves asymptotically as

$$C\lambda a\epsilon^2 \overline{F}_R(\frac{b}{a}\epsilon) T^2 \overline{F}_U(aT) \sim C'\epsilon^{2-\delta}.$$

This is arbitrarily small for small enough ϵ. Similarly and in an easier way, one can show that

$$\int_{-\infty}^{\infty} \int\int_{A_\epsilon^3} h(s,u,r) ds u^{-\gamma-1} du r^{-\delta-1} dr$$

is arbitrarily small for small enough ϵ.

Steps 1, 2 and 3 yield

$$I_\lambda \to I, \quad \text{as } \lambda, a, b \to \infty.$$

By using (3.3.67), we can write I in (3.3.67) as

$$I = \int_{-\infty}^{\infty} \int_0^\infty \int_0^\infty \left(e^{i\sum_{j=1}^J \theta_j K_{t_j}(s,u)r} - 1 - i\sum_{j=1}^J \theta_j K_{t_j}(s,u)r \right) ds u^{-\gamma-1} du r^{-\delta-1} dr. \tag{3.3.73}$$

This is the logarithm of the characteristic function of the limit process (3.3.54). The process (3.3.54) can also be represented as (3.3.55) by using (B.1.39) in Appendix B.1.4. □

The next theorem concerns the behavior of a scaled workload process in the slow regime.

Theorem 3.3.10 *Suppose that the slow regime in (3.3.28) holds. Suppose*

$$1 < \gamma < \delta \leq 2$$

and set

$$b = (\lambda a)^{1/\gamma} L^*_{U,\gamma}(\lambda a) \tag{3.3.74}$$

in the scaled workload process (3.3.26), for the normalization. Then, as $a \to \infty$,

$$\frac{b}{a} \to 0 \tag{3.3.75}$$

and

$$\frac{1}{b}(W_\lambda^*(at) - \lambda\nu\mathbb{E}(R)at) \xrightarrow{fdd} (\mathbb{E}R^\gamma)^{1/\gamma} \Lambda_\gamma(t), \tag{3.3.76}$$

where Λ_γ is a γ-stable Lévy process, totally skewed to the right. The limit process can be represented as

3.3 Infinite Source Poisson Model with Heavy Tails

$$(\mathbb{E}R^\gamma)^{1/\gamma} \Lambda_\gamma(t) = \int_{-\infty}^{\infty} \int_0^\infty \int_0^\infty 1_{\{0<s<t\}} ur$$
$$\times \left(N(ds, du, dr) - ds\, u^{-(1+\gamma)} du\, F_R(dr) \right) \quad (3.3.77)$$

$$\stackrel{d}{=} c_\gamma \int_{-\infty}^{\infty} \int_0^\infty 1_{\{0<s<t\}} r M_\gamma(ds, dr) \quad (3.3.78)$$

where c_γ is defined in (3.3.56) (with $\delta = \gamma$) and the random measure $M_\gamma(ds, dr)$ is γ-stable, totally skewed to the right with control measure $ds\, F_R(dr)$.

Remark 3.3.11 Let $g(t) = (1/\overline{F}_U)^{\leftarrow}(t)$ as in (3.3.32). Then, by (3.3.45),

$$g(t) \sim \gamma^{-1/\gamma} t^{1/\gamma} L_{U,\gamma}^*(t), \quad \text{as } t \to \infty,$$

where, by (3.3.44), $L_{U,\gamma}^*(t)^{-\gamma} L(t^{1/\gamma} L_{U,\gamma}^*(t)x) \to 1$ for $x > 0$. Therefore, the normalization constant b defined in (3.3.74) satisfies

$$b = (\lambda a)^{1/\gamma} L_{U,\gamma}^*(\lambda a) \sim \gamma^{1/\gamma} g(\lambda a), \quad \text{as } a \to \infty. \quad (3.3.79)$$

Proof summary The idea of the proof is as follows. Express the logarithm of the characteristic function of the rescaled workload process as an integral (3.3.81) with integrand (3.3.83). Because $a/b \to \infty$, the integrand converges as in (3.3.84). Then apply the dominated convergence theorem to get (3.3.87). It can be expressed as (3.3.88), which is the logarithm of the characteristic function of the process in (3.3.77). □

Proof By (3.3.74) and (3.3.79),

$$\frac{b}{a} \sim \frac{\gamma^{1/\gamma} g(\lambda a)}{a} \to c^{1/\gamma} = 0, \quad (3.3.80)$$

under the slow regime ($c = 0$), in view of Proposition 3.3.6, (i). Thus, (3.3.75) holds.

Let us now prove (3.3.76). This time, we will have to make a/b appear since $a/b \to \infty$. By using (3.3.43) and integration by parts, observe that

$$I_\lambda := \log \mathbb{E} \exp\left\{ i \sum_{j=1}^J \theta_j \frac{W_\lambda^*(at_j) - \lambda \nu(\mathbb{E}R) at_j}{b} \right\}$$

$$= \log \mathbb{E} \exp\left\{ i \int_{-\infty}^{\infty} \int_0^\infty \int_0^\infty \sum_{j=1}^J \frac{\theta_j K_{at_j}(s,u) r}{b} \widetilde{N}(ds, du, dr) \right\}$$

$$= \int_{-\infty}^{\infty} \int_0^\infty \int_0^\infty \left(\exp\left\{ i \sum_{j=1}^J \frac{\theta_j K_{at_j}(s,u) r}{b} \right\} - 1 \right.$$

$$\left. - i \sum_{j=1}^J \frac{\theta_j K_{at_j}(s,u) r}{b} \right) \lambda\, ds\, F_U(du)\, F_R(dr)$$

$$= i \int_{-\infty}^{\infty} \int_0^{\infty} \int_0^{\infty} \left(\exp\left\{ i \sum_{j=1}^{J} \frac{\theta_j K_{at_j}(s,u)r}{b} \right\} - 1 \right)$$

$$\times \sum_{j=1}^{J} \theta_j \frac{\partial K_{at_j}}{\partial u}(s,u) \frac{r}{b} \lambda \mathbb{P}(U > u) ds\, du\, F_R(dr),$$

as was done in (3.3.59). We can now make the change of variables $s \to bs$ and $u \to bu$. Then, $\frac{1}{b} K_{at}(s,u)$ becomes $\frac{1}{b} K_{b\frac{a}{b}t}(bs, bu) = K_{\frac{a}{b}t}(s,u)$ by (3.3.12), and $\frac{\partial K_{at}}{\partial u}(s,u) = 1_{\{0 < s+u < at\}}$ by (3.3.14) becomes $1_{\{0 < bs+bu < at\}} = 1_{\{0 < s+u < \frac{a}{b}t\}}$. With the further change of variables $s \to \frac{a}{b}s - u$, $K_{\frac{a}{b}t}(s,u)$ becomes $K_{\frac{a}{b}t}(\frac{a}{b}s - u, u)$ and $1_{\{0 < s+u < \frac{a}{b}t\}}$ becomes $1_{\{0 < s < t\}}$. Therefore,

$$I_\lambda = i \int_{-\infty}^{\infty} \int_0^{\infty} \int_0^{\infty} \left(\exp\left\{ i \sum_{j=1}^{J} \theta_j K_{\frac{a}{b}t_j}(\frac{a}{b}s - u, u)r \right\} - 1 \right)$$

$$\times \sum_{j=1}^{J} \theta_j 1_{\{0 < s < t_j\}} r\, \lambda a \mathbb{P}(U > bu) ds\, du\, F_R(dr)$$

$$=: \int_{-\infty}^{\infty} \int_0^{\infty} \int_0^{\infty} I_\lambda(s, u, r) ds\, du\, F_R(dr). \tag{3.3.81}$$

By using (3.3.74) for b, we can write[1]

$$\lambda a \mathbb{P}(U > bu) = \frac{u^{-\gamma}}{\gamma} \lambda a b^{-\gamma} L_U(bu) = \frac{u^{-\gamma}}{\gamma} \lambda a (\lambda a)^{-1} (L_{U,\gamma}^*(\lambda a))^{-\gamma} L_U(bu)$$

$$= \frac{u^{-\gamma}}{\gamma} (L_{U,\gamma}^*(\lambda a))^{-\gamma} L_U\left((\lambda a)^{1/\gamma} L_{U,\gamma}^*(\lambda a) u \right) =: \frac{u^{-\gamma}}{\gamma} f_\lambda(u) \tag{3.3.82}$$

and therefore

$$I_\lambda(s,u,r) = i \left(\exp\left\{ i \sum_{j=1}^{J} \theta_j K_{\frac{a}{b}t_j}(\frac{a}{b}s - u, u)r \right\} - 1 \right) \sum_{j=1}^{J} \theta_j 1_{\{0 < s < t_j\}} r \frac{u^{-\gamma}}{\gamma} f_\lambda(u). \tag{3.3.83}$$

Note that, by (3.3.14),

$$K_{zt}(zs - u, u) = \int_0^u 1_{\{0 < y + zs - u < zt\}} dy \to \int_0^u 1_{\{0 < s < t\}} dy = u 1_{\{0 < s < t\}},$$

as $z \to \infty$. Then, since $a/b \to \infty$ in the slow growth regime, and since $f_\lambda(u) \to 1$ as $\lambda \to \infty$ by using (3.3.44),

$$I_\lambda(s,u,r) \to i \left(\exp\left\{ i \sum_{j=1}^{J} \theta_j 1_{\{0 < s < t_j\}} ur \right\} - 1 \right) \sum_{j=1}^{J} \theta_j 1_{\{0 < s < t_j\}} r \frac{u^{-\gamma}}{\gamma} =: I(s,u,r)$$

$$\tag{3.3.84}$$

---

[1] Do not confuse the integral I_λ with its integrand $I_\lambda(s,u,r)$.

for all $s \in \mathbb{R}$, $u > 0$, $r > 0$. We want to conclude next that the convergence also takes place for the corresponding integrals; that is,

$$I_\lambda \to \int_{-\infty}^{\infty} \int_0^{\infty} \int_0^{\infty} I(s,u,r) ds du F_R(dr) =: I. \tag{3.3.85}$$

To show (3.3.85), it is enough to prove that $|I_\lambda(s,u,r)|$ is dominated by an integrable function. For this, express $f_\lambda(u)$, defined in (3.3.82), as

$$f_\lambda(u) = (L_{U,\gamma}^*(\lambda a))^{-\gamma} L_U\left((\lambda a)^{1/\gamma} L_{U,\gamma}^*(\lambda a)\right) \frac{L_U\left((\lambda a)^{1/\gamma} L_{U,\gamma}^*(\lambda a)u\right)}{L_U\left((\lambda a)^{1/\gamma} L_{U,\gamma}^*(\lambda a)\right)}$$

and consider $I_\lambda(s,u,r)$ defined in (3.3.83).

For $(\lambda a)^{1/\gamma} L_{U,\gamma}^*(\lambda a)u \geq u_0$ with large enough fixed u_0 and large enough λa, by using (3.3.44) and Potter's bound (2.2.1), we get that

$$f_\lambda(u) \leq C \max\{u^{-\epsilon}, u^{\epsilon}\},$$

where ϵ is arbitrarily small and C is a constant. By using $|e^{ix} - 1| \leq C(|x|^\kappa \wedge 1)$ with fixed $0 < \kappa < 1$ and the inequality $K_t(s,u) \leq u$, we get that

$$\left|\exp\left\{i \sum_{j=1}^{J} \theta_j K_{\frac{a}{b}t_j}(\frac{a}{b}s - u, u)r\right\} - 1\right| \leq C'(u^\kappa r^\kappa \wedge 1). \tag{3.3.86}$$

Then, for $(\lambda a)^{1/\gamma} L_{U,\gamma}^*(\lambda a)u \geq u_0$ and large enough λa,

$$|I_\lambda(s,u,r)| \leq C 1_{\{0<s<T\}} r(u^\kappa r^\kappa \wedge 1) u^{-\gamma} \max\{u^{-\epsilon}, u^{\epsilon}\} =: Cg_1(s,u,r).$$

Note that, since $(u^\kappa r^\kappa \wedge 1) \leq u^\kappa r^\kappa$ and $(u^\kappa r^\kappa \wedge 1) \leq 1$,

$$\int_{-\infty}^{\infty} \int_0^{\infty} \int_0^{\infty} g_1(s,u,r) ds du F_R(dr) \leq T\mathbb{E}(R^{1+\kappa}) \int_0^{u_1} u^{\kappa-\gamma} \max\{u^{-\epsilon}, u^\epsilon\} du$$
$$+ T\mathbb{E}(R) \int_{u_1}^{\infty} u^{-\gamma} \max\{u^{-\epsilon}, u^\epsilon\} du$$

for fixed u_1. The bound is finite if we choose $1+\kappa < \delta$, $\kappa - \gamma - \epsilon + 1 > 0$ and $-\gamma + \epsilon + 1 < 0$ or, equivalently, $\gamma + \epsilon < 1 + \kappa < \delta$ and $1 + \epsilon < \gamma$.

On $(\lambda a)^{1/\gamma} L_{U,\gamma}^*(\lambda a)u \leq u_0$, we have $u \leq u_1$ for some fixed u_1 since $(\lambda a)^{1/\gamma} L_{U,\gamma}^*(\lambda a) \to \infty$ as $a \to \infty$. Then, by using (3.3.83) and (3.3.86), we have

$$|I_\lambda(s,u,r)| \leq C(u^\kappa r^\kappa \wedge 1) 1_{\{0<s<T\}} 1_{\{u \leq u_1\}} r u^{-\gamma} f_\lambda(u) \leq C 1_{\{0<s<T\}} 1_{\{u \leq u_1\}} u^{\kappa-\gamma} r^{1+\kappa} f_\lambda(u).$$

We need to estimate $f_\lambda(u)$ which is defined in (3.3.82). In particular, we need to bound $(L_{U,\gamma}^*(\lambda a))^{-\gamma}$ by some function of u. Since $1/L_{U,\gamma}^*$ is a slowly varying function, we have $(\lambda a)^{1/\gamma} u \leq u_0/L_{U,\gamma}^*(\lambda a) \leq C(\lambda a)^{\delta_0} u_0$ where $\delta_0 > 0$ is arbitrarily small or $\lambda a \leq C' u^{-\gamma/(1-\gamma\delta_0)} = C' u^{-\gamma-\delta_1}$ where $\delta_1 > 0$ is arbitrarily small. This also implies that $(L_{U,\gamma}^*(\lambda a))^{-\gamma} \leq C(\lambda a)^{\delta_0 \gamma} \leq C' u^{-\epsilon}$ where $\delta_0 > 0$ and $\epsilon > 0$ are arbitrarily small. The other term of $f_\lambda(u)$ to bound is $L_U((\lambda a)^{1/\gamma} L_{U,\gamma}^*(\lambda a)u)$. On $(\lambda a)^{1/\gamma} L_{U,\gamma}^*(\lambda a)u \leq u_0$, this term is bounded by a constant since this is the case for the function $L_U(v) = v^\gamma \mathbb{P}(U > v)$ on $v \in [0, u_0]$. Then, combining the two bounds, on $(\lambda a)^{1/\gamma} L^*_{U,\gamma}(\lambda a) u \leq u_0$,

$$|I_\lambda(s, u, r)| \leq C 1_{\{0 < s < T\}} 1_{\{u \leq u_1\}} u^{\kappa-\gamma-\epsilon} r^{1+\kappa} =: g_2(s, u, r),$$

where u_1 is fixed and ϵ is arbitrarily small. The function $g_2(s, u, r)$ is integrable with respect to $dsdu F_R(dr)$ as long as $\kappa - \gamma - \epsilon + 1 > 0$ and $1+\kappa < \delta$ or, equivalently, $\gamma + \epsilon < 1 + \kappa < \delta$.

This establishes the convergence in (3.3.85). The limit (3.3.85) can be written as

$$I = \int_{-\infty}^{\infty} \int_{-\infty}^{\infty} \int_0^{\infty} \left(\exp\left\{ \sum_{j=1}^J \theta_j 1_{\{0<s<t_j\}} ur \right\} - 1 \right) \left(i \sum_{j=1}^J \theta_j 1_{\{0<s<t_j\}} r \right) \frac{u^{-\gamma}}{\gamma} ds du F_R(dr). \quad (3.3.87)$$

After integration by parts on the variable u, one gets

$$I = \int_{-\infty}^{\infty} \int_0^{\infty} \int_0^{\infty} \left(\exp\left\{ i \sum_{j=1}^J \theta_j 1_{\{0<s<t_j\}} ur \right\} - 1 - i \sum_{j=1}^J \theta_j 1_{\{0<s<t_j\}} ur \right) u^{-\gamma-1} ds du F_R(dr), \quad (3.3.88)$$

which is the logarithm of the characteristic function of the process in (3.3.77). The equality (3.3.78) follows by using (B.1.39) in Appendix B.1.4. □

The third theorem concerns the behavior of a scaled workload process in the intermediate regime.

Theorem 3.3.12 *Suppose the intermediate regime in (3.3.28) holds. Suppose*

$$1 < \gamma < \delta \leq 2$$

and set

$$b = a \quad (3.3.89)$$

in the scaled workload process (3.3.26). Then, as $a \to \infty$,

$$\frac{1}{b}(W^*_\lambda(at) - \lambda \nu \mathbb{E}(R)at) \xrightarrow{fdd} c^{1/(\gamma-1)} Y_{\gamma,R}\left(\frac{t}{c^{1/(\gamma-1)}}\right),$$

where c appears in (3.3.28), and

$$Y_{\gamma,R}(t) = \int_{-\infty}^{\infty} \int_0^{\infty} \int_0^{\infty} K_t(s, u) r \left(N(ds, du, dr) - ds\, u^{-(1+\gamma)} du\, F_R(dr) \right) \quad (3.3.90)$$

$$= \int_{-\infty}^{\infty} \int_0^{\infty} K_t(s, u) \left(\int_0^{\infty} r N(ds, du, dr) - \mathbb{E}(R) ds\, u^{-(1+\gamma)} du \right). \quad (3.3.91)$$

Remark 3.3.13 The limit process $Y_{\gamma,R}$ is not self-similar, because N does not have the scaling properties that a Gaussian or a stable process has. However, if we assume that the reward distribution $F_R(dr)$ has finite variance, that is $\mathbb{E}R^2 < \infty$, then

$$\text{Var}(Y_{\gamma,R}(t)) = \mathbb{E}(R^2) \int_{-\infty}^{\infty} \int_0^{\infty} K_t(s, u)^2 ds\, u^{-(1+\gamma)} du = \mathbb{E}(R^2) \sigma^2 t^{2H}, \quad H = \frac{3-\gamma}{2},$$

where σ^2 is given in (3.3.49). Thus, in this case $Y_{\gamma,R}$ is *second-order self-similar* with Hurst index H.

Remark 3.3.14 Benassi, Jaffard, and Roux [113] introduced the notion of *local asymptotic self-similarity* as another means of generalizing the class of self-similar processes. See also Section 2.6.8. It is shown in Gaigalas and Kaj [380] and with a proof more adapted to the present setting in Gaigalas [379], that the process $Y_{\gamma,R}$ is locally asymptotically self-similar with index H and with fractional Brownian motion as tangent process, in the sense that

$$\left\{ \frac{Y_{\gamma,R}(t+\lambda u) - Y_{\gamma,R}(t)}{\lambda^H}, \; u \in \mathbb{R} \right\} \xrightarrow{fdd} \{B_H(u), \; u \in \mathbb{R}\}, \quad \text{as} \quad \lambda \to 0.$$

Benassi, Cohen, and Istas [115] defined a stochastic process $X(t)$ to be *asymptotically self-similar at infinity* with index H if there exists a process $V(t)$ such that

$$\lambda^{-H} X(\lambda t) \xrightarrow{fdd} V(t), \quad \text{as } \lambda \to \infty.$$

The intermediate limit process $Y_{\gamma,R}(t)$ is asymptotically self-similar at infinity with index $H = 1/\gamma$ and the asymptotic process $V(t)$ is given by a γ-stable Lévy process, totally skewed to the right, see Gaigalas [379].

Remark 3.3.15 In the special case of fixed rewards $R \equiv 1$, $F_R(dr)$ is the delta function at $r = 1$ and the limit process (3.3.91) becomes

$$Y_\gamma(t) = \int_{-\infty}^{\infty} \int_0^{\infty} K_t(s,u) \Big(N(ds, du) - ds \, u^{-(1+\gamma)} du \Big).$$

Proof of Theorem 3.3.12 Let $c_0 = c^{1/(\gamma-1)}$. As in the proof of Theorem 3.3.10, we have[2]

$$I_\lambda := \log \mathbb{E} \exp\left\{ i \sum_{j=1}^J \theta_j \frac{W_\lambda^*(at_j) - \lambda \nu(\mathbb{E}R) at_j}{a} \right\}$$

$$= i \int_{-\infty}^{\infty} \int_0^{\infty} \int_0^{\infty} \left(\exp\left\{ i \sum_{j=1}^J \theta_j K_{at_j}(s,u) \frac{r}{a} \right\} - 1 \right)$$

$$\times \sum_{j=1}^J \theta_j \frac{\partial K_{at_j}}{\partial u}(s,u) \frac{r}{a} \lambda \mathbb{P}(U > u) ds \, du \, F_R(dr).$$

We now make the change of variables $s \to c_0 a s$ and $u \to c_0 a u$. Then, $K_{at}(s,u) = K_{c_0 a \frac{t}{c_0}}(s,u)$ becomes $K_{c_0 a \frac{t}{c_0}}(c_0 a s, c_0 a u) = c_0 a K_{\frac{t}{c_0}}(s,u)$ by the scaling property (3.3.12). Then,

$$I_\lambda = i \int_{-\infty}^{\infty} \int_0^{\infty} \int_0^{\infty} \left(\exp\left\{ i \sum_{j=1}^J \theta_j c_0 K_{t_j/c_0}(s,u) r \right\} - 1 \right) \sum_{j=1}^J \theta_j c_0 \frac{\partial K_{t_j/c_0}}{\partial u}(s,u)$$

$$\times r \lambda c_0 a \mathbb{P}(U > c_0 a u) ds \, du \, F_R(dr)$$

$$=: \int_{-\infty}^{\infty} \int_0^{\infty} \int_0^{\infty} I_\lambda(s,u,r) ds \, du \, F_R(dr).$$

---

[2] Do not confuse I_λ with the integrand $I_\lambda(s,u,r)$ defined below.

By using (3.3.28), note that $\lambda L_U(u)a^{1-\gamma} \sim c \sim c_0^{\gamma-1}$ and therefore

$$\lambda c_0 a \mathbb{P}(U > c_0 au) \sim c_0^\gamma \frac{\mathbb{P}(U > c_0 au)}{a^{-\gamma} L_U(a)} = \frac{c_0^\gamma}{\gamma} \frac{\mathbb{P}(U > c_0 au)}{\mathbb{P}(U > a)} \sim \frac{c_0^\gamma}{\gamma}(c_0 u)^{-\gamma} = \frac{u^{-\gamma}}{\gamma}.$$

Then, as $a \to \infty$,

$$I_\lambda(s, u, r) \to i\left(\exp\left\{i\sum_{j=1}^J \theta_j c_0 K_{t_j/c_0}(s, u)r\right\} - 1\right)$$

$$\times \sum_{j=1}^J \theta_j c_0 \frac{\partial K_{t_j/c_0}}{\partial u}(s, u)r \frac{u^{-\gamma}}{\gamma} =: I(s, u, r).$$

To prove that

$$I_\lambda \to \int_{-\infty}^\infty \int_0^\infty \int_0^\infty I(s, u, r) ds\, du\, F_R(dr) =: I,$$

it is enough to show that $|I_\lambda(s, u, r)|$ is dominated by an integrable function. This can be shown as in the proofs of Theorems 3.3.8, (a), and 3.3.10, and is left as Exercise 3.7. After an integration by parts, the limit I can be written as

$$I = \int_{-\infty}^\infty \int_0^\infty \int_0^\infty \left(e^{i\sum_{j=1}^J \theta_j c_0 K_{t_j/c_0}(s,u)r} - 1 - i\sum_{j=1}^J \theta_j c_0 K_{t_j/c_0}(s, u)r\right) u^{-\gamma-1} ds\, du\, F_R(dr),$$

which is the logarithm of the characteristic function of the process (3.3.90). □

The final two theorems concern the behavior of a scaled workload process in the remaining cases of parameter values, except the case $1 < \gamma = \delta < 2$ which is discussed in Remark 3.3.18 below. The distinction between fast, slow and intermediate regimes is now irrelevant. In the first theorem, we consider the case $\gamma = \delta = 2$.

Theorem 3.3.16 *Suppose*

$$\gamma = \delta = 2$$

and set

$$b = (\lambda a)^{1/2} \tag{3.3.92}$$

in the scaled workload process (3.3.26). Then, as $\lambda \to \infty$, $a \to \infty$ or $a \to \infty$, $b \to \infty$ in any arbitrary way,

$$\frac{1}{b}(W_\lambda^*(at) - \lambda \nu \mathbb{E}(R)at) \xrightarrow{fdd} (\mathbb{E}U^2 \mathbb{E}R^2)^{1/2} B(t), \tag{3.3.93}$$

where $\{B(t)\}$ is a standard Brownian motion.

Proof We only give idea of the proof. As in (3.3.58), the logarithm of the characteristic function of the scaled workload process (3.3.26) can be expressed as

3.3 Infinite Source Poisson Model with Heavy Tails

$$I_\lambda := \log \mathbb{E} \exp \left\{ i \sum_{j=1}^{J} \theta_j \frac{W^*_\lambda(at_j) - \lambda\nu(\mathbb{E}R)at_j}{b} \right\}$$

$$= \int_{-\infty}^{\infty} \int_{0}^{\infty} \int_{0}^{\infty} \left(\exp\left\{ i \sum_{j=1}^{J} \frac{\theta_j K_{at_j}(as, u)r}{b} \right\} \right.$$

$$\left. - 1 - i \sum_{j=1}^{J} \frac{\theta_j K_{at_j}(as, u)r}{b} \right) \lambda a\, ds\, F_U(du) F_R(dr)$$

$$= \int_{-\infty}^{\infty} \int_{0}^{\infty} \int_{0}^{\infty} I_\lambda(s, u, r)\, ds\, F_U(du) F_R(dr),$$

where

$$I_\lambda(s, u, r) = \left(\exp\left\{ i \sum_{j=1}^{J} \frac{\theta_j K_{at_j}(as, u)r}{b} \right\} - 1 - i \sum_{j=1}^{J} \frac{\theta_j K_{at_j}(as, u)r}{b} \right) \lambda a.$$

Since

$$K_{at}(as, u) = \int_0^u 1_{\{0 < y + as < at\}} dy \to u 1_{\{0 \le s < t\}}, \qquad (3.3.94)$$

as $a \to \infty$, and hence $K_{at}(as, u)/b \to 0$, as $a, b \to \infty$, we have, using $e^{ix} - 1 - ix \sim -x^2/2$, as $x \to 0$, and (3.3.92) that

$$I_\lambda(s, u, r) \sim -\frac{1}{2} \left(\sum_{j=1}^{J} \theta_j K_{at_j}(as, u)r \right)^2 \frac{\lambda a}{b^2} \to -\frac{1}{2} \left(\sum_{j=1}^{J} \theta_j 1_{\{0 \le s < t_j\}} \right)^2 u^2 r^2,$$

as $a, b \to \infty$. This yields

$$I_\lambda \to -\frac{1}{2} \int_{-\infty}^{\infty} \int_0^\infty \int_0^\infty \left(\sum_{j=1}^{J} \theta_j 1_{\{0 \le s < t_j\}} \right)^2 u^2 r^2 ds\, F_U(du) F_R(dr)$$

$$= -\frac{1}{2} \mathbb{E}(U^2) \mathbb{E}(R^2) \int_{-\infty}^{\infty} \left(\sum_{j=1}^{J} \theta_j 1_{\{0 \le s < t_j\}} \right)^2 ds,$$

which is the logarithm of the characteristic function of the limit process in (3.3.93). □

In the next theorem, we consider the case $1 < \delta < \gamma \le 2$.

Theorem 3.3.17 *Suppose*

$$1 < \delta < \gamma \le 2$$

and set

$$b = (\lambda a)^{1/\delta} L^*_{R,\delta}(\lambda a) \qquad (3.3.95)$$

in the scaled workload process (3.3.26). Then, as $\lambda \to \infty, a \to \infty$ or $a \to \infty, b \to \infty$ in any arbitrary way,

$$\frac{1}{b}(W^*_\lambda(at) - \lambda\nu\mathbb{E}(R)at) \xrightarrow{fdd} (\mathbb{E}U^\delta)^{1/\delta} \Lambda_\delta(t), \qquad (3.3.96)$$

where $\{\Lambda_\delta(t)\}$ is a δ-stable Lévy-stable process, totally skewed to the right. The limit process can be represented as

$$(\mathbb{E}U^\delta)^{1/\delta} \Lambda_\delta(t) = \int_{-\infty}^\infty \int_0^\infty \int_0^\infty 1_{\{0<s<t\}} ur\Big(N(ds,du,dr) - ds\, F_U(du)\, r^{-(1+\delta)}dr\Big) \tag{3.3.97}$$

$$\stackrel{d}{=} c_\delta \int_{-\infty}^\infty \int_0^\infty 1_{\{0<s<t\}} u M_\gamma(ds,du) \tag{3.3.98}$$

where c_δ is defined in (3.3.56) and the random measure $M_\delta(ds,du)$ is δ-stable, totally skewed to the right with control measure $ds\, F_U(du)$.

Proof We only give idea of the proof, which is similar to that of Theorem 3.3.10, after reversing the roles of R and U. As in (3.3.58), the logarithm of the characteristic function of the scaled workload process (3.3.26) can be expressed as

$$I_\lambda := \log \mathbb{E} \exp\left\{i \sum_{j=1}^J \theta_j \frac{W_\lambda^*(at_j) - \lambda\nu(\mathbb{E}R)at_j}{b}\right\}$$

$$= \int_{-\infty}^\infty \int_0^\infty \int_0^\infty \left(\exp\left\{i \sum_{j=1}^J \frac{\theta_j K_{at_j}(as,u)r}{b}\right\} - 1 - i \sum_{j=1}^J \frac{\theta_j K_{at_j}(as,u)r}{b}\right) \lambda a\, ds\, F_U(du) F_R(dr).$$

By integration by parts on the variable r and making the change of variables $r \to br$, we get

$$I_\lambda = \int_{-\infty}^\infty \int_0^\infty \int_0^\infty \left(\exp\left\{i \sum_{j=1}^J \frac{\theta_j K_{at_j}(as,u)r}{b}\right\} - 1\right)\left(i \sum_{j=1}^J \frac{\theta_j K_{at_j}(as,u)}{b}\right)$$
$$\times \lambda a \mathbb{P}(R>r)\, ds\, F_U(du)dr$$

$$= \int_{-\infty}^\infty \int_0^\infty \int_0^\infty \left(\exp\left\{i \sum_{j=1}^J \theta_j K_{at_j}(as,u)r\right\} - 1\right)\left(i \sum_{j=1}^J \theta_j K_{at_j}(as,u)\right)$$
$$\times \lambda a \mathbb{P}(R>br)\, ds\, F_U(du)dr$$

$$= \int_{-\infty}^\infty \int_0^\infty \int_0^\infty I_\lambda(s,u,r)\, ds\, F_U(du)dr,$$

where

$$I_\lambda(s,u,r) = \left(\exp\left\{i \sum_{j=1}^J \theta_j K_{at_j}(as,u)r\right\} - 1\right)\left(i \sum_{j=1}^J \theta_j K_{at_j}(as,u)\right)\lambda a \mathbb{P}(R>br).$$

Note that, using (3.3.95) and (3.3.44),

$$\lambda a \mathbb{P}(R>br) = \frac{r^{-\delta}}{\delta}(L_{R,\delta}^*(\lambda a))^{-\delta} L_R\Big((\lambda a)^{1/\delta} L_{R,\delta}^*(\lambda a)r\Big) \to \frac{r^{-\delta}}{\delta},$$

as $\lambda a \to \infty$. Then, using (3.3.94),

$$I_\lambda(s,u,r) \to \left(\exp\left\{i\sum_{j=1}^J \theta_j 1_{\{0\le s<t_j\}}ur\right\} - 1\right)\left(i\sum_{j=1}^J \theta_j 1_{\{0\le s<t_j\}}u\right)\frac{r^{-\delta}}{\delta}$$

and hence

$$I_\lambda \to \int_{-\infty}^\infty \int_{-\infty}^\infty \int_0^\infty \left(\exp\left\{\sum_{j=1}^J \theta_j 1_{\{0\le s<t_j\}}ur\right\} - 1\right)$$

$$\times \left(i\sum_{j=1}^J \theta_j 1_{\{0\le s<t_j\}}u\right)\frac{r^{-\delta}}{\delta}ds\, F_U(du)dr.$$

After integration by parts on the variable r, one gets

$$I = \int_{-\infty}^\infty \int_0^\infty \int_0^\infty \left(\exp\left\{i\sum_{j=1}^J \theta_j 1_{\{0<s<t_j\}}ur\right\}\right.$$

$$\left. - 1 - i\sum_{j=1}^J \theta_j 1_{\{0<s<t_j\}}ur\right)r^{-\delta-1}ds\, F_U(du)dr,$$

which is the logarithm of the characteristic function of the process in (3.3.97). The equality (3.3.98) follows by using (B.1.39) in Appendix B.1.4. □

Remark 3.3.18 As shown in Theorem 4, (ii), in Kaj and Taqqu [546], Theorem 3.3.17 holds in the case $1 < \gamma = \delta < 2$ provided $\mathbb{E}U^\delta < \infty$.

3.4 Power-Law Shot Noise Model

The workload process $W_\lambda^*(t)$ in (3.3.7) and the underlying stationary process $W_\lambda(y)$ in (3.3.19) can also be expressed as

$$W_\lambda(y) = \sum_{j=-\infty}^\infty 1_{\{S_j<y<S_j+U_j\}}R_j = \sum_{j=-\infty}^\infty g(y-S_j, U_j)R_j, \quad (3.4.1)$$

$$W_\lambda^*(t) = \int_0^t W_\lambda(y)dy = \sum_{j=-\infty}^\infty (g^*(t-S_j, U_j) - g^*(-S_j, U_j))R_j, \quad (3.4.2)$$

where S_j are Poisson arrivals, $U_j, R_j, j \in \mathbb{Z}$, are independent variables, and

$$g(y,u) = 1_{\{0<y<u\}}, \quad g^*(x,u) = \int_0^x g(y,u)dy = x_+ \wedge u. \quad (3.4.3)$$

The processes $W_\lambda(y)$ and $W_\lambda^*(t)$ are particular cases of the following more general processes, called shot noise processes.

Definition 3.4.1 Let $X_j = \{X_j(y)\}_{y\in\mathbb{R}}$, $j \in \mathbb{Z}$, and $X_j^* = \{X_j^*(y)\}_{y\in\mathbb{R}}$, $j \in \mathbb{Z}$, be two sets of i.i.d. processes, satisfying $X_j(y) = 0$, $y < 0$, and $X_j^*(t) = 0$, $t < 0$, and where, as above, $X_j^*(t) = \int_0^t X_j(y)dy$. Let S_j, $j \in \mathbb{Z}$, be Poisson arrivals with rate λ. Then, the processes

$$S_\lambda(y) = \sum_{j=-\infty}^{\infty} X_j(y - S_j) \tag{3.4.4}$$

and

$$S_\lambda^*(t) = \int_0^t S_\lambda(y) dy = \sum_{j=-\infty}^{\infty} \left(X_j^*(t - S_j) - X_j^*(-S_j) \right) \tag{3.4.5}$$

are called *shot noise processes*. The processes X_j and X_j^* are known as *shots* associated with S_λ and S_λ^*, respectively.

The idea behind a shot noise process is as follows: the impact of each Poisson arrival S_j lingers for a while according to a pattern X_j, such that at time $y > S_j$, it has an intensity value of $X_j(y - S_j)$. The total intensity at time $y > 0$ due to all the previous arrivals is then $S_\lambda(y) = \sum_{j=-\infty}^{\infty} X_j(y - S_j) = \sum_{j:S_j \le y} X_j(y - S_j)$.

Under suitable assumptions, the process S_λ is expected to be stationary and the process S_λ^* to have stationary increments. Observe, in particular, that $X_j^*(y - S_j) - X_j^*(-S_j) = X_j^*(y - S_j) - X_j^*(0 - S_j)$ so that $S_\lambda^*(0) = 0$. For example, in the case of W_λ in (3.4.1), the pattern is

$$X_j(y) = g(y, U_j) R_j = 1_{\{0 < y < U_j\}} R_j. \tag{3.4.6}$$

This is therefore a shot which takes a random value R_j and lasts for a random time U_j. Taking U_j to be heavy-tailed in (3.3.2) is what made W_λ to be long-range dependent (Proposition 3.3.4) and made the scaled workload process to converge to fractional Brownian motion with the self-similarity parameter bigger than $1/2$ (Theorem 3.3.8, (a)).

The model (3.4.6) and the corresponding heavy tails, however, is not the only mechanism leading to long-range dependence. Consider, for example, the following alternate shot noise model.

Definition 3.4.2 Suppose that the i.i.d. shots $X_j = \{X_j(y)\}_{y\in\mathbb{R}}$ have a power-law pattern in the sense that

$$X_j(y) = g(y) R_j, \tag{3.4.7}$$

where R_j, $j \in \mathbb{Z}$, are i.i.d. variables with $\mathbb{E}R_j = 0$, $0 < \mathbb{E}R_j^2 < \infty$,

$$g(y) = 0, \ y < 0, \quad \text{and} \quad g(y) = y^{-\beta} L_g(y), \tag{3.4.8}$$

where L_g is a slowly varying function at infinity which is positive, and locally bounded away from 0 and infinity in $[0, \infty)$, and

$$1/2 < \beta < 1. \tag{3.4.9}$$

The shot noise process S_λ with the shots (3.4.7)–(3.4.8) is called a *power-law shot noise process*.

For slightly different assumptions on the function g, see Exercise 3.8. Note also that the shots X_j^* for the corresponding shot noise process S_λ^* satisfy

$$X_j^*(t) = \int_0^t X_j(y)dy = \left(\int_0^t g(y)dy\right)R_j =: g^*(t)R_j, \qquad (3.4.10)$$

where g^* is given by

$$g^*(t) = \int_0^t y^{-\beta}L_g(y)dy = t^{1-\beta}L_{g^*}(t), \qquad (3.4.11)$$

and where the slowly varying function L_{g^*} satisfies

$$L_{g^*}(t) \sim \frac{L_g(t)}{1-\beta}, \quad \text{as } t \to \infty \qquad (3.4.12)$$

(Proposition 2.2.2).

Example 3.4.3 If $g(t) = t^{-\beta}(1_{\{t<2\}} + 1_{\{t\geq 2\}}\log t)$, then

$$g^*(t) = \int_0^t y^{-\beta}(1_{\{y<2\}} + 1_{\{y\geq 2\}}\log y)dy$$

$$= \frac{2^{1-\beta}}{(1-\beta)^2}(2-\beta-\log 2) + \log t\frac{t^{1-\beta}}{1-\beta} - \frac{t^{1-\beta}}{(1-\beta)^2}.$$

The following result shows that the power-law shot noise process S_λ is long-range dependent with parameter

$$d = 1 - \beta \in (0, 1/2).$$

Proposition 3.4.4 *Consider the power-law shot noise process S_λ in (3.4.4) with shots given by (3.4.7)–(3.4.9). Then, S_λ is a stationary, zero mean process with autocovariance function satisfying*

$$\text{Cov}(S_\lambda(y), S_\lambda(0)) \sim \lambda(\mathbb{E}R^2)B(1-\beta, 2\beta-1)(L_g(y))^2 y^{1-2\beta}, \qquad (3.4.13)$$

as $y \to \infty$.

Proof Represent the shot noise process as

$$S_\lambda(y) = \int_{-\infty}^\infty \int_{-\infty}^\infty g(y-s)rN(ds, dr),$$

where N is a Poisson random measure with the intensity $\lambda ds F_R(dr)$. Now observe that

$$\text{Cov}(S_\lambda(y), S_\lambda(0)) = \int_{-\infty}^\infty \int_{-\infty}^\infty g(y-s)g(-s)r^2\lambda ds F_R(dr)$$

$$= \lambda \mathbb{E}(R^2)\int_0^\infty g(z)g(y+z)dz$$

$$= \lambda \mathbb{E}(R^2)\int_0^\infty z^{-\beta}L_g(z)(y+z)^{-\beta}L_g(y+z)dz$$

$$= \lambda \mathbb{E}(R^2) y^{1-2\beta} \int_0^\infty w^{-\beta}(1+w)^{-\beta} L_g(wy) L_g((1+w)y) dw$$

$$\sim \lambda \mathbb{E}(R^2) y^{1-2\beta} (L_g(y))^2 \int_0^\infty w^{-\beta}(1+w)^{-\beta} dw$$

$$= \lambda \mathbb{E}(R^2) y^{1-2\beta} (L_g(y))^2 B(1-\beta, 2\beta-1).$$

The equivalence relation above follows from the dominated convergence theorem by using Potter's bound (2.2.2) to bound the integrand in the prelimit by $Cw^{-\beta \pm \delta}(1+w)^{-\beta+\delta}(L_g(y)^2)$, which is integrable on $(0, \infty)$ for sufficiently small $\delta > 0$. The last equality above follows from (2.2.12). □

The next result shows that a scaled power-law shot noise process S_λ^* converges to fractional Brownian motion.

Theorem 3.4.5 *Consider shot noise processes S_λ, S_λ^* in (3.4.4) and (3.4.5) with the shots given by (3.4.7)–(3.4.9). Then, as $a \to \infty$, a scaled process*

$$\frac{S_\lambda^*(at)}{a^{3/2-\beta} L_{g^*}(a)} \tag{3.4.14}$$

converges in the sense of finite-dimensional distributions to

$$S^*(t) = (\lambda \mathbb{E} R^2)^{1/2} \int_{-\infty}^\infty \left((t-s)_+^{1-\beta} - (-s)_+^{1-\beta}\right) B(du), \tag{3.4.15}$$

where $B(du)$ is a Gaussian random measure with the control measure du. The limit process S^* is FBM with the self-similarity parameter

$$H = \frac{3}{2} - \beta \in \left(\frac{1}{2}, 1\right). \tag{3.4.16}$$

Proof Let $b = a^{3/2-\beta} L_{g^*}(a)$. Using the representation

$$S_\lambda^*(at) = \int_{-\infty}^\infty \int_{-\infty}^\infty \left(g^*(at-s) - g^*(-s)\right) r N(ds, dr)$$

and (3.3.43), we have[3]

$$I_a := \log \mathbb{E} \exp\left\{i \sum_{j=1}^J \theta_j \frac{S_\lambda^*(at_j)}{b}\right\}$$

$$= \int_{-\infty}^\infty \int_{-\infty}^\infty \left(\exp\left\{i \sum_{j=1}^J \theta_j \frac{g^*(at_j - s) - g^*(-s)}{b} r\right\}\right.$$

$$\left. - 1 - i \sum_{j=1}^J \theta_j \frac{g^*(at_j - s) - g^*(-s)}{b} r\right) \lambda ds F_R(dr)$$

---
[3] Do not confuse I_a with the integrand $I_a(s, r)$ in (3.4.17) below.

$$= \int_{-\infty}^{\infty} \int_{-\infty}^{\infty} \Big(\exp \Big\{ i \sum_{j=1}^{J} \theta_j \frac{g^*(a(t_j+s)) - g^*(as)}{b} r \Big\}$$

$$- 1 - i \sum_{j=1}^{J} \theta_j \frac{g^*(a(t_j+s)) - g^*(as)}{b} r \Big) \lambda a\, ds\, F_R(dr)$$

$$=: \int_{-\infty}^{\infty} \int_{-\infty}^{\infty} I_a(s,r)\, ds\, F_R(dr). \tag{3.4.17}$$

Note that, since

$$\frac{g^*(at)}{b} = \frac{(at)^{1-\beta} L_g(at)}{a^{3/2-\beta} L_{g^*}(a)} \to 0,$$

we can use the relation $e^{ix} - 1 - ix \sim -x^2/2$ as $x \to 0$ to get

$$I_a(s,r) \sim -\frac{1}{2} \Big| \sum_{j=1}^{J} \theta_j \frac{g^*(a(t_j+s)) - g^*(as)}{b} r \Big|^2 \lambda a$$

$$= -\frac{\lambda}{2} \Big| \sum_{j=1}^{J} \theta_j \frac{g^*(a(t_j+s)) - g^*(as)}{a^{1-\beta} L_{g^*}(a)} r \Big|^2$$

$$\to -\frac{\lambda}{2} \Big| \sum_{j=1}^{J} \theta_j \big((t_j+s)_+^{1-\beta} - s_+^{1-\beta} \big) \Big|^2 r^2 =: I(s,r)$$

for any $s, r \in \mathbb{R}$. We want to conclude that

$$I_a = \int_{-\infty}^{\infty} \int_{-\infty}^{\infty} I_a(s,r)\, ds\, F_R(dr) \to \int_{-\infty}^{\infty} \int_{-\infty}^{\infty} I(s,r)\, ds\, F_R(dr) =: I$$

and it is enough to prove that $|I_a(s,r)|$ is dominated by an integrable function.

Since $|e^{ix} - 1 - ix| \leq |x|^2$ and without loss of generality, it is enough to show that, for each $t > 0$, the function $|g^*(a(t+s)) - g^*(as)|/(a^{1-\beta} L_g(a))$ is dominated by a square integrable function on $\mathbb{R}$. (We have replaced L_{g^*} by L_g in the denominator for convenience by using (3.4.12).) Observe that

$$\frac{|g^*(a(t+s)) - g^*(as)|}{a^{1-\beta} L_g(a)} = \frac{\int_{as}^{a(t+s)} g(u)\, du}{a^{1-\beta} L_g(a)} = \frac{\int_{s}^{t+s} g(au)\, du}{a^{-\beta} L_g(a)} = \int_{s_+}^{(t+s)_+} u^{-\beta} \frac{L_g(au)}{L_g(a)}\, du$$

$$\leq C \int_{s_+}^{(t+s)_+} u^{-\beta \pm \delta}\, du$$

$$= \frac{C}{1 - \beta \pm \delta} \big((t+s)_+^{1-\beta \pm \delta} - s_+^{1-\beta \pm \delta} \big),$$

where we used Potter's bound (2.2.2). The bound above is a square integrable function on $\mathbb{R}$ for small enough $\delta > 0$. □

3.5 Hierarchical Model

We present here another physical model for long-range dependence. The form of the long-range dependence will be somewhat less conventional. The model nevertheless is interesting, instructive and related to the wavelet expansions discussed in Section 8.2.

Let $U^{(n)} = \{U^{(n)}(t)\}_{t \in \mathbb{R}}$, $n \geq 0$, be a collection of independent, second-order stationary processes, having the same autocovariance function $\gamma_U(h)$, $h \in \mathbb{R}$. For $0 < a < 1$ and $k > 1$, consider a sequence of processes $X^{(n)} = \{X^{(n)}(t)\}_{t \in \mathbb{R}}$, $n \geq 0$, defined as

$$X^{(0)}(t) = U^{(0)}(t),$$
$$X^{(1)}(t) = k^{-1}X^{(0)}(k^{-1}t) + k^{-a}U^{(1)}(t)$$
$$= k^{-1}U^{(0)}(k^{-1}t) + k^{-a}U^{(1)}(t),$$
$$X^{(2)}(t) = k^{-1}X^{(1)}(k^{-1}t) + k^{-2a}U^{(2)}(t)$$
$$= k^{-2}U^{(0)}(k^{-2}t) + k^{-1-a}U^{(1)}(k^{-1}t) + k^{-2a}U^{(2)}(t)$$

and, in general,

$$X^{(n)}(t) = k^{-1}X^{(n-1)}(k^{-1}t) + k^{-na}U^{(n)}(t)$$
$$= \sum_{j=0}^{n} k^{-(n-j)-ja}U^{(j)}(k^{-(n-j)}t). \qquad (3.5.1)$$

Note that for $k \geq 1$, time flows more slowly for $X(k^{-1}t)$ than for the $X(t)$ series since, if $k = 2$ for example, then $X(10)$ occurs at $t = 10$ for $X(t)$ but it occurs at $t = 20$ for $X(k^{-1}t)$. There is a slow-down of the tempo.

The physical interpretation of the model (3.5.1) is that n represents the stage of a process and that at each stage n, the available variation $X^{(n-1)}(t)$ has its time scale extended and a new short-term variation $U^{(n)}(t)$ is introduced. An example of such process is the production of textile yarns, which are produced by repetition of thinning and pulling out. See the bibliographical notes in Section 3.15.

Observe that the process $X^{(n)}$ has the autocovariance function

$$\gamma^{(n)}(h) = \sum_{j=0}^{n}(k^{-(n-j)-ja})^2 \gamma_U(k^{-(n-j)}h) = k^{-2na}\sum_{j=0}^{n} k^{-2(1-a)j}\gamma_U(k^{-j}h), \qquad (3.5.2)$$

where the last equality follows by the change of summation parameter j to $n - j$. The relation (3.5.2) now implies that, for each h, as $n \to \infty$,

$$k^{2na}\gamma^{(n)}(h) \to \sum_{j=0}^{\infty} k^{-2(1-a)j}\gamma_U(k^{-j}h) =: \gamma_X(h), \qquad (3.5.3)$$

where the series in (3.5.3) converges absolutely since $a < 1$ and $|\gamma_U(k^{-j}h)| \leq \gamma_U(0) = \text{Var}(U^{(0)}(0))$.

What can be said about the behavior of $\gamma_X(h)$ as $h \to \infty$? The following result provides a partial answer.

Proposition 3.5.1 *Fix $0 < a < 1$ and $k > 1$ and let $\gamma_X(h)$ be the autocovariance function defined in (3.5.3). Suppose that, for some $\epsilon > 0$,*

$$|\gamma_U(h)| = O(h^{-2(1-a)-\epsilon}), \tag{3.5.4}$$

as $h \to \infty$. Then, for any $b > 0$, there is a constant $c(b)$ such that

$$\gamma_X(bk^m) \sim c(b)(bk^m)^{-2(1-a)}, \tag{3.5.5}$$

as $m \to \infty$, or equivalently as $k^m \to \infty$. Moreover,

$$c(b) = \sum_{j=-\infty}^{\infty} (k^{-j}b)^{2(1-a)} \gamma_U(k^{-j}b). \tag{3.5.6}$$

Proof Note that

$$\gamma_X(bk^m) = \sum_{j=0}^{\infty} k^{-2(1-a)j} \gamma_U(k^{-j}k^m b)$$

$$= (bk^m)^{-2(1-a)} \sum_{j=0}^{\infty} (k^{-j}k^m b)^{2(1-a)} \gamma_U(k^{-j}k^m b)$$

$$= (bk^m)^{-2(1-a)} \sum_{j=-m}^{\infty} (k^{-j}b)^{2(1-a)} \gamma_U(k^{-j}b)$$

and that

$$\sum_{j=-m}^{\infty} (k^{-j}b)^{2(1-a)} \gamma_U(k^{-j}b) \to \sum_{j=-\infty}^{\infty} (k^{-j}b)^{2(1-a)} \gamma_U(k^{-j}b) = c(b),$$

as $m \to \infty$. We have convergence by assumption (3.5.4) since for $j \to \infty$, $(k^{-j}b)^{2(1-a)} \gamma_U(k^{-j}b) = O((k^{-j}b)^{-\epsilon})$. □

The relation (3.5.5) resembles the LRD relation (2.1.5) in condition II with $h = bk^m$,

$$d = a - 1/2 \in (-1/2, 1/2)$$

and $L_2(h) \sim c(b)$. This is not exactly an LRD condition in that c depends on h through the factor b. Note also that $d \in (0, 1/2)$ only when $1/2 < a < 1$, and that there is a priori no reason for $c(b)$ to be necessarily positive.

On the other hand, $c(b)$ in (3.5.6) is almost a constant function in the following sense.

Proposition 3.5.2 *Let $1/2 < a < 1$, $k > 1$ and $c(b)$ be defined in (3.5.6). Suppose that $\gamma_U(h)$, $h \geq 0$, is a monotone decreasing function and satisfies the condition (3.5.4). Then,*

$$\frac{1}{k-1} \int_0^\infty h^{2(1-a)-1} \gamma_U(h) dh \leq c(b) \leq \frac{k}{k-1} \int_0^\infty h^{2(1-a)-1} \gamma_U(h) dh. \tag{3.5.7}$$

Proof We shall use the identities $k^j(k^{-j} - k^{-(j+1)}) = \frac{k-1}{k}$ and $k^j(k^{-(j-1)} - k^{-j}) = k-1$. The upper bound in (3.5.7) follows by expressing $c(b)$ in (3.5.6) as

$$c(b) = \frac{k}{k-1} \sum_{j=-\infty}^{\infty} (k^{-j}b)^{2(1-a)-1} \gamma_U(k^{-j}b)(k^{-j}b - k^{-(j+1)}b)$$

and noting that $h^{2(1-a)-1}\gamma_U(h)$ is monotone decreasing on $h > 0$ since $2(1-a) - 1 < 0$ when $1/2 < a$, and $\gamma_U(h)$ is monotone decreasing by assumption. The lower bound in (3.5.7) follows by writing

$$c(b) = \frac{1}{k-1} \sum_{j=-\infty}^{\infty} (k^{-j}b)^{2(1-a)-1} \gamma_U(k^{-j}b)(k^{-(j-1)}b - k^{-j}b). \qquad \Box$$

For example, when $k = 2$, the two factors $1/(k-1)$ and $k/(k-1)$ in the bound (3.5.7) are $1/(k-1) = 1$ and $k/(k-1) = 2$, respectively. The ratio of these factors is

$$\frac{k}{k-1} \Big/ \frac{1}{k-1} = k$$

and approaches 1, as $k \downarrow 1$. In particular, in the limit $k \downarrow 1$, one has

$$c(b) \sim \frac{1}{k-1} \int_0^\infty h^{2(1-a)-1} \gamma_U(h) dh =: \frac{c}{k-1}, \qquad (3.5.8)$$

where the constant c no longer depends on b. In this case, however, $c(b)$ diverges.

3.6 Regime Switching

It is part of the folklore, especially in Economics and Finance, that long-range dependence is easily confused with changes in regime (regime switching, structural change). We present below one such model, and a related convergence result. Some other models of this type are moved to exercises, and a related discussion can be found in Section 3.15.

Focus on the so-called *Markov switching model* defined as follows. Let $s^N = \{s_n^N\}_{n=1,\ldots,N}$ be a stationary Markov chain, taking the value 0 or 1, with the probability transition matrix

$$P = \begin{pmatrix} p_{00} & 1 - p_{00} \\ 1 - p_{11} & p_{11} \end{pmatrix}. \qquad (3.6.1)$$

We shall take below $p_{00} = p_{00}(N)$ and $p_{11} = p_{11}(N)$ as functions of N, which explains the use of the index N in the notation s^N. Let $\mu_0, \mu_1 \in \mathbb{R}$. Consider the series

$$X_n^N = \mu_{s_n^N} + \epsilon_n, \quad n = 1, \ldots, N, \qquad (3.6.2)$$

where $\{\epsilon_n\}$ is a sequence of i.i.d. random variables with mean 0 and variance σ_ϵ^2, and its partial sum process

$$S_N(t) = \sum_{n=1}^{[Nt]} (X_n^N - \mathbb{E}X_n^N), \quad t \in [0,1]. \qquad (3.6.3)$$

The mean $\mu_{s_n^N} = \mu_0 1_{\{s_n^N=0\}} + \mu_1 1_{\{s_n^N=1\}}$ of X_n^N is not constant. It varies randomly between two values μ_0 and μ_1. The switching mechanism is dictated by the Markov chain s_n^N, $n = 1, \ldots, N$. If one is in regime k, $k = 1, 2$, one stays there with probability p_{kk} and switches to the other regime with probability $1 - p_{kk}$.

The following auxiliary lemma will be used in analyzing the asymptotics of the partial sum process $S_N(t)$ as $N \to \infty$.

Lemma 3.6.1 *The term $\mu_{s_n^N}$ in the Markov switching model (3.6.2) satisfies: for $1 \leq m, n \leq N$,*

$$\mathrm{Cov}\left(\mu_{s_n^N}, \mu_{s_m^N}\right) = \mu' \Gamma_{|n-m|} \mu = \frac{(1-p_{00})(1-p_{11})}{(2-p_{00}-p_{11})^2} (\mu_0 - \mu_1)^2 \lambda^{|n-m|}, \quad (3.6.4)$$

where

$$\mu = \begin{pmatrix} \mu_0 \\ \mu_1 \end{pmatrix}, \quad \Gamma_j = \frac{(1-p_{00})(1-p_{11})}{(2-p_{00}-p_{11})^2} \begin{pmatrix} \lambda^j & -\lambda^j \\ -\lambda^j & \lambda^j \end{pmatrix} \quad (3.6.5)$$

and

$$\lambda = p_{00} + p_{11} - 1. \quad (3.6.6)$$

Proof Let

$$\xi_n = \begin{pmatrix} 1_{\{s_n^N=0\}} \\ 1_{\{s_n^N=1\}} \end{pmatrix},$$

so that $\mu_{s_n^N} = \mu' \xi_n$, where prime denotes the transpose. Observe further that

$$\mathrm{Cov}\left(\mu_{s_n^N}, \mu_{s_m^N}\right) = \mathrm{Cov}\left(\mu' \xi_n, \mu' \xi_m\right) = \mu' \mathrm{Cov}(\xi_n, \xi_m) \mu = \mu' \Gamma_{|n-m|} \mu, \quad (3.6.7)$$

where

$$\Gamma_j = \mathrm{Cov}(\xi_j, \xi_0) = \begin{pmatrix} \mathrm{Cov}(1_{\{s_j^N=0\}}, 1_{\{s_0^N=0\}}) & \mathrm{Cov}(1_{\{s_j^N=1\}}, 1_{\{s_0^N=0\}}) \\ \mathrm{Cov}(1_{\{s_j^N=0\}}, 1_{\{s_0^N=1\}}) & \mathrm{Cov}(1_{\{s_j^N=1\}}, 1_{\{s_0^N=1\}}) \end{pmatrix}$$

$$= \begin{pmatrix} \mathbb{P}(s_j^N=0, s_0^N=0) - \mathbb{P}(s_j^N=0)\mathbb{P}(s_0^N=0) & \mathbb{P}(s_j^N=1, s_0^N=0) - \mathbb{P}(s_j^N=1)\mathbb{P}(s_0^N=0) \\ \mathbb{P}(s_j^N=0, s_0^N=1) - \mathbb{P}(s_j^N=0)\mathbb{P}(s_0^N=1) & \mathbb{P}(s_j^N=1, s_0^N=1) - \mathbb{P}(s_j^N=1)\mathbb{P}(s_0^N=1) \end{pmatrix}.$$

Denoting by $\pi = (\pi_0, \pi_1)'$ the invariant (stationary) probabilities of s^N, and by p_{kl}^j the transition probability from state k to state l in j steps, we have

$$\Gamma_j = \begin{pmatrix} \pi_0(p_{00}^j - \pi_0) & \pi_0(p_{01}^j - \pi_1) \\ \pi_1(p_{10}^j - \pi_0) & \pi_1(p_{11}^j - \pi_1) \end{pmatrix}. \quad (3.6.8)$$

To evaluate Γ_j, we need expressions for π_0, π_1 and p_{kl}^j. One can verify that

$$\pi_0 = \frac{1 - p_{11}}{2 - p_{00} - p_{11}}, \quad \pi_1 = \frac{1 - p_{00}}{2 - p_{00} - p_{11}} \quad (3.6.9)$$

and that the probability transition matrix P in (3.6.1) is diagonalizable as

$$P = \begin{pmatrix} 1 & \pi_1 \\ 1 & -\pi_0 \end{pmatrix} \begin{pmatrix} 1 & 0 \\ 0 & \lambda \end{pmatrix} \begin{pmatrix} \pi_0 & \pi_1 \\ 1 & -1 \end{pmatrix} =: E \begin{pmatrix} 1 & 0 \\ 0 & \lambda \end{pmatrix} F', \qquad (3.6.10)$$

where λ is given by (3.6.6) (Exercise 3.10). Note that 1 and λ in (3.6.10) are the eigenvalues of the matrix P and the matrices E and F are made of the right- and left-eigenvectors of P, satisfying $F' = E^{-1}$ (cf. Example 2.3.4). The latter fact implies that

$$P^j = \begin{pmatrix} 1 & \pi_1 \\ 1 & -\pi_0 \end{pmatrix} \begin{pmatrix} 1 & 0 \\ 0 & \lambda^j \end{pmatrix} \begin{pmatrix} \pi_0 & \pi_1 \\ 1 & -1 \end{pmatrix} = \begin{pmatrix} \pi_0 + \pi_1 \lambda^j & \pi_1(1-\lambda^j) \\ \pi_0(1-\lambda^j) & \pi_1 + \pi_0 \lambda^j \end{pmatrix}$$

$$= \frac{1}{2 - p_{00} - p_{11}} \begin{pmatrix} (1-p_{11}) + (1-p_{00})\lambda^j & (1-p_{00})(1-\lambda^j) \\ (1-p_{11})(1-\lambda^j) & (1-p_{00}) + (1-p_{11})\lambda^j \end{pmatrix}$$

$$= \begin{pmatrix} p_{00}^j & p_{01}^j \\ p_{10}^j & p_{11}^j \end{pmatrix} \qquad (3.6.11)$$

(Exercise 3.10). By using (3.6.9) and (3.6.11), we obtain from (3.6.8) that Γ_j is given by (3.6.5) and (3.6.4) is finally established by using (3.6.7). $\square$

The following result provides conditions on the Markov chain s^N for the partial sum process $S_N(t)$ to converge to Brownian motion. Connections to long-range dependence are discussed following the proof.

Proposition 3.6.2 *Assume that*

$$\mu_0 \neq \mu_1 \qquad (3.6.12)$$

and that, for $k = 0, 1$,

$$p_{kk} = p_{kk}(N) = 1 - \frac{c_k}{N^{\delta_k}}, \qquad (3.6.13)$$

with $\delta_k > 0$, $0 < c_k < 1$. Suppose that the exponents

$$0 < \delta_0, \delta_1 < 1 \qquad (3.6.14)$$

are such that

$$d := \frac{1}{2}\Big(\min\{\delta_0, \delta_1\} - |\delta_0 - \delta_1|\Big) > 0. \qquad (3.6.15)$$

Then,

$$\frac{S_N(t)}{N^{d+1/2}} \xrightarrow{fdd} \sigma B(t), \quad t \in [0, 1], \qquad (3.6.16)$$

where B is a standard Brownian motion with

$$\sigma^2 = \frac{2c_0 c_1 (\mu_0 - \mu_1)^2}{(\widetilde{c})^3} \qquad (3.6.17)$$

and

$$\widetilde{c} = \begin{cases} c_0, & \text{if } \delta_0 < \delta_1, \\ c_1, & \text{if } \delta_1 < \delta_0, \\ c_0 + c_1, & \text{if } \delta_0 = \delta_1. \end{cases} \qquad (3.6.18)$$

Proof It is enough to show that, for $t_1, \ldots, t_J \in [0, 1]$ and $\theta_1, \ldots, \theta_J \in \mathbb{R}$,

$$\sum_{j=1}^{J} \theta_j \frac{S_N(t_j)}{N^{d+1/2}} \xrightarrow{d} \mathcal{N}(0, \sigma_J^2) \tag{3.6.19}$$

and that

$$\sigma_J^2 = \lim_{N \to \infty} \mathbb{E}\Big(\sum_{j=1}^{J} \theta_j \frac{S_N(t_j)}{N^{d+1/2}}\Big)^2 = \sigma^2 \mathbb{E}\Big(\sum_{j=1}^{J} \theta_j B(t_j)\Big)^2. \tag{3.6.20}$$

Since X_n^N is a stationary series, the relation (3.6.20) follows from

$$\mathbb{E}\Big(\frac{S_N(t)}{N^{d+1/2}}\Big)^2 \to \sigma^2 t, \tag{3.6.21}$$

as $N \to \infty$.

Write

$$S_N(t) = \sum_{n=1}^{[Nt]} (\mu_{s_n^N} - \mathbb{E}\mu_{s_n^N}) + \sum_{n=1}^{[Nt]} \epsilon_n =: S_{1,N}(t) + S_{2,N}(t). \tag{3.6.22}$$

Then, by the independence of s^N and $\{\epsilon_n\}$,

$$\mathbb{E}\Big(\frac{S_N(t)}{N^{d+1/2}}\Big)^2 = \mathbb{E}\Big(\frac{S_{1,N}(t)}{N^{d+1/2}}\Big)^2 + \mathbb{E}\Big(\frac{S_{2,N}(t)}{N^{d+1/2}}\Big)^2$$

and

$$\mathbb{E}\Big(\frac{S_{2,N}(t)}{N^{d+1/2}}\Big)^2 = \frac{\sigma_\epsilon^2 [Nt]}{N^{2d+1}} \to 0.$$

Hence, it is enough to show that

$$\mathbb{E}\Big(\frac{S_{1,N}(t)}{N^{d+1/2}}\Big)^2 \to \sigma^2 t. \tag{3.6.23}$$

Observe that

$$\mathbb{E}(S_{1,N}(t))^2 = \text{Var}\Big(\sum_{n=1}^{[Nt]} \mu_{s_n^N}\Big) = \text{Var}\Big(\sum_{n=1}^{[Nt]} \mu' \xi_n\Big) = \mu' \sum_{n_1,n_2=1}^{[Nt]} \text{Cov}(\xi_{n_1}, \xi_{n_2}) \mu$$

$$= \mu'\Big([Nt]\Gamma_0 + \sum_{j=1}^{[Nt]-1} ([Nt] - j)(\Gamma_j + \Gamma_j')\Big)\mu.$$

From Lemma 3.6.1 and by using

$$\sum_{j=1}^{[Nt]-1} \lambda^j = \frac{\lambda^{[Nt]} - \lambda}{\lambda - 1}, \quad \sum_{j=1}^{[Nt]-1} j\lambda^j = \frac{\lambda}{(\lambda - 1)^2}(([Nt] - 1)\lambda^{[Nt]} - [Nt]\lambda^{[Nt]-1} + 1), \tag{3.6.24}$$

we get

$$\mathbb{E}(S_{1,N}(t))^2 = V_{00}(\mu_0 - \mu_1)^2, \tag{3.6.25}$$

where
$$V_{00} = \frac{(1-p_{00})(1-p_{11})}{(2-p_{00}-p_{11})^2}\left(\frac{2[Nt]\lambda}{1-\lambda} + [Nt] + \frac{2(\lambda^{[Nt]+1}-\lambda)}{(1-\lambda)^2}\right).$$

But by (3.6.6) and (3.6.13),
$$\lambda = p_{00} + p_{11} - 1 = 1 - \frac{c_0}{N^{\delta_0}} - \frac{c_1}{N^{\delta_1}}.$$

Hence, as $N \to \infty$,
$$\frac{1}{1-\lambda} = \frac{1}{c_0 N^{-\delta_0} + c_1 N^{-\delta_1}} \sim \frac{N^{\min\{\delta_0,\delta_1\}}}{\widetilde{c}},$$

where $\widetilde{c}$ is defined in (3.6.18), and
$$\lambda^{[Nt]} = \left(1 - \frac{c_0}{N^{\delta_0}} - \frac{c_1}{N^{\delta_1}}\right)^{[Nt]} \sim \left(1 - \frac{\widetilde{c}}{N^{\min\{\delta_0,\delta_1\}}}\right)^{[Nt]} \sim \exp\left\{-\widetilde{c}[Nt]N^{-\min\{\delta_0,\delta_1\}}\right\} = o(1),$$

since $\delta_0, \delta_1 < 1$. This shows that
$$V_{00} \sim \frac{(1-p_{00})(1-p_{11})}{(2-p_{00}-p_{11})^2} \frac{2[Nt]\lambda}{1-\lambda} \sim \frac{(1-p_{00})(1-p_{11})}{(2-p_{00}-p_{11})^2} \frac{2[Nt]}{\widetilde{c}} N^{\min\{\delta_0,\delta_1\}}.$$

Now, by (3.6.13),
$$\frac{(1-p_{00})(1-p_{11})}{(2-p_{00}-p_{11})^2} \sim \frac{c_0 N^{-\delta_0} c_1 N^{-\delta_1}}{\widetilde{c}^2 N^{-2\min\{\delta_0,\delta_1\}}} \sim \frac{c_0 c_1}{\widetilde{c}^2} N^{-|\delta_0-\delta_1|}, \qquad (3.6.26)$$

since $-\delta_0 - \delta_1 + 2\min\{\delta_0,\delta_1\} = -|\delta_0 - \delta_1|$. Thus, by (3.6.15),
$$V_{00} \sim \frac{2c_0 c_1[Nt]}{\widetilde{c}^3} N^{\min\{\delta_0,\delta_1\}-|\delta_0-\delta_1|} = \frac{2c_0 c_1[Nt]}{\widetilde{c}^3} N^{2d}. \qquad (3.6.27)$$

Relations (3.6.25) and (3.6.27) now yield
$$\frac{\mathbb{E}(S_{1,N}(t))^2}{N^{2d+1}} \sim \frac{2c_0 c_1(\mu_0 - \mu_1)^2 t}{\widetilde{c}^3} = \sigma^2 t,$$

which establishes (3.6.23).

We now turn to the convergence (3.6.19). In view of the decomposition (3.6.22) and since, by the central limit theorem, as $N \to \infty$,
$$\sum_{j=1}^{J} \theta_j \frac{S_{2,N}(t_j)}{N^{1/2}} \xrightarrow{d} \mathcal{N}(0, \sigma_{2,J}^2),$$

where $0 < \sigma_{2,J}^2 < \infty$, it is enough to show that
$$\sum_{j=1}^{J} \theta_j \frac{S_{1,N}(t_j)}{N^{d+1/2}} \xrightarrow{d} \mathcal{N}(0, \sigma_J^2), \qquad (3.6.28)$$

where σ_J^2 is defined in (3.6.20). Write
$$\sum_{j=1}^{J} \theta_j \frac{S_{1,N}(t_j)}{N^{d+1/2}} = \sum_{n=1}^{N} f_n^{(N)}(s_n^N),$$

where

$$f_n^{(N)}(x) = \sum_{j=1}^{J} \theta_j \frac{1}{N^{d+1/2}} 1_{\{n \leq [Nt_j]\}} (\mu_x - \mathbb{E}\mu_{s_n^N}).$$

Recall that s_n^N may be equal to 0 or 1.

We shall apply next a central limit theorem for functionals of Markov chains. Let

$$\rho_{N,1} = \max_{2 \leq n \leq N} \sup_g \left\{ \frac{\|\mathbb{E}(g(s_n^N)|s_{n-1}^N)\|_2}{\|g(s_n^N)\|_2} : \|g(s_n^N)\|_2 < \infty \text{ and } \mathbb{E}g(s_n^N) = 0 \right\}$$

be the so-called maximal coefficient of correlation for the Markov chain $\{s_n^N\}$, where $\|X\|_2 = (\mathbb{E}X^2)^{1/2}$. Denoting $g(S_n^N) = g_0 1_{\{s_n^N=0\}} + g_1 1_{\{s_n^N=1\}}$ with $g_0, g_1 \in \mathbb{R}$, it can be seen from (3.6.9) that $\mathbb{E}g(s_n^N) = 0$ is equivalent to $g_0(1 - p_{11}) + g_1(1 - p_{00}) = 0$. A simple algebra then yields $\mathbb{E}(g(s_n^N)|s_{n-1}^N) = (p_{00} + p_{11} - 1)g(s_{n-1}^N)$ and hence

$$\rho_{N,1} = |p_{00} + p_{11} - 1| = 1 - \frac{c_0}{N^{\delta_0}} - \frac{c_1}{N^{\delta_1}}.$$

In particular, by (3.6.13),

$$\eta_N := 1 - \rho_{N,1} = \frac{c_0}{N^{\delta_0}} + \frac{c_1}{N^{\delta_1}} \sim \frac{\tilde{c}}{N^{\min\{\delta_0,\delta_1\}}}, \qquad (3.6.29)$$

as $N \to \infty$. Observe that

$$\sup_{1 \leq n \leq N} \sup_x |f_n^{(N)}(x)| \leq C_N := \frac{C}{N^{d+1/2}}$$

and also that

$$\sigma_N^2 := \text{Var}\left(\sum_{n=1}^{N} f_n^{(N)}(s_n^N)\right) \sim C',$$

by using (3.6.20). By Theorem 1 in Peligrad [800], the convergence (3.6.28) follows if

$$\frac{C_N(1 + \log \eta_N)}{\eta_N \sigma_N} \to 0. \qquad (3.6.30)$$

The condition (3.6.30) holds since C_N/η_N behaves asymptotically up to a constant as

$$N^{-\frac{1}{2}-d+\min\{\delta_0,\delta_1\}} = N^{-\frac{1}{2}+\frac{1}{2}(\min\{\delta_0,\delta_1\}+|\delta_0-\delta_1|)} = N^{-\frac{1}{2}(1-\max\{\delta_0,\delta_1\})} \to 0,$$

by using (3.6.29) and (3.6.15). □

How is the Markov switching model related to long-range dependence? By the convergence (3.6.21) established in the proof above, note that

$$\text{Var}(S_N(1)) = \text{Var}(X_1^N + \cdots + X_N^N) \sim CN^{2d+1}, \qquad (3.6.31)$$

as $N \to \infty$. When $d > 0$, this behavior is consistent with that of LRD series given in (2.1.8) in condition V. On the other hand, note that the convergence to Brownian motion in (3.6.16) suggests that the series $\{X_n^N\}$ is SRD. How can these two seemingly contradictory facts be reconciled? There is no contradiction between these two facts because the probabilistic structure of the Markov switching series $\{X_n^N\}$ depends on N. In this sense, the

Markov switching model exhibits features of both LRD and SRD series. From an estimation perspective, the Markov switching model may indeed suggest long-range dependence but common estimators of the LRD parameter d are very biased and sensitive to the choice of the tuning parameters, such as the parameter m in the GPH estimation in Section 2.10.3 (Diebold and Inoue [312], Baek, Fortuna, and Pipiras [71]).

Several aspects of the Markov switching model presented here are similar to those of other physical models for long-range dependence. The fact that the probabilities p_{00} and p_{11} approach 1, as $N \to \infty$, ensures that the underlying Markov chain s^N will tend to remain a long time in a given state. This is similar to the infinite source Poisson model in Section 3.3 where each arrival stays in the "ON" state for a long, heavy-tailed time. Thus the infinite source Poisson model is also characterized by a kind of regime switching.

The Markov switching model also appears in the spirit of models discussed in Sections 3.1 and 3.2. Its covariance structure described in Lemma 3.6.1 and the proof of Proposition 3.6.2 resembles that of AR(1) series with the autoregressive coefficient approaching 1 at a suitable rate (see also Exercise 3.11). The models of Section 3.1 consider specifically AR(1) series with a random autoregressive coefficient whose distribution has a power-law behavior around 1. Similarly, without the error terms ϵ_n in (3.6.2) and when $p_{00} = p_{11} = p$, the series X_n^N can be thought as the steps of the correlated random walk (CRW) considered in Section 3.2. While the parameter p was random and CRWs were aggregated in Section 3.2, the parameter p depends here on the sample size N and approaches 1 at a suitable rate.

3.7 Elastic Collision of Particles

Consider two particles with equal mass m moving towards each other on the one-dimensional real line $\mathbb{R}$. The first particle has constant velocity v_1, and the second has constant velocity v_2 with $v_1 \neq v_2$. What will be their velocities after an elastic collision? Elastic collisions preserve kinetic energy. Since momentum needs also to be conserved, the speeds u_1 and u_2 after collision must satisfy the following two equations:

$$\frac{1}{2}mv_1^2 + \frac{1}{2}mv_2^2 = \frac{1}{2}mu_1^2 + \frac{1}{2}mu_2^2 \quad \text{(conservation of energy)},$$

$$mv_1 + mv_2 = mu_1 + mu_2 \quad \text{(conservation of momentum)}.$$

Suppose that $v_1 \neq u_1$. Then the only solution to these equations is

$$u_1 = v_2 \quad u_2 = v_1;$$

that is, the two particles merely exchange velocities. Thus, if the second particle is immobile ($v_2 = 0$), then after collision it is the first particle which will be immobile ($u_1 = 0$).

In addition, the order of the particles is preserved since the particles cannot go through each other. Therefore, they exchange trajectories after they collide.

But two particles can collide at most once. Now imagine that there is an infinite countable number of particles with random i.i.d. velocities; that is, suppose that, if there were no collisions, their paths would be $X_i(t) = x_i + \xi_i(t)$, $i \in \mathbb{Z}$, where $X_i(0) = x_i$ and the processes ξ_is are i.i.d. and independent of x_is. The starting points x_i are random and are generated by a stationary Poisson point process on the real line with intensity $\rho > 0$. Focus on the middle

3.7 Elastic Collision of Particles

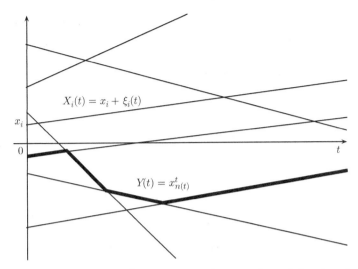

Figure 3.4 Trajectory $Y(t) = x^t_{n(t)}$ of the middle (tagged) particle when the paths $X_i(t) = x_i + \xi_i(t)$ are straight lines.

(tagged) particle, namely the one whose starting position $X_i(0)$ is below but closest to the origin (see Figure 3.4).

Let $Y(t), t \geq 0$, be the trajectory of that middle particle as it undergoes collisions. What happens to it (to its law) after adequate normalization, as $t \to \infty$?

This problem has an interesting history. Harris [455] supposed that $\{\xi_i(t)\}$ is a collection of independent Brownian motions $\{B_i(t)\}$. These are H_0–SSSI with $H_0 = 1/2$. The limit of $Y(Tt)/T^{1/4}$ as $T \to \infty$ turns out to be FBM with self-similarity parameter $H = 1/4$. Spitzer [919] supposed that the paths are straight lines with random slopes; that is, $\xi_i(t) = v_i t$ where the v_is are i.i.d. with $\mathbb{E}|v_i| < \infty$ and mean zero. These are H_0–SSSI with $H_0 = 1$. He showed that the limit of $Y(Tt)/T^{1/2}$ as $T \to \infty$ is Brownian motion; that is, FBM with $H = 1/2$. Gisselquist [407] then chose for the ξ_is another H_0–SSSI process, namely the $S\alpha S$ Lévy motion with $1 < \alpha < 2$ and hence $1/2 < H_0 = 1/\alpha < 1$. He showed that the limit of $Y(Tt)/T^{1/(2\alpha)}$ as $T \to \infty$ is again FBM with $H = 1/(2\alpha)$. In all cases, note that $H = H_0/2$.

Dürr, Goldstein, and Lebowitz [339] provide a more general perspective. They use a Poisson point process framework and show that the limit depends essentially on the function $\mathbb{E}|\xi(t)|$. They do not require that $\{\xi(t)\}$ has stationary increments.

Theorem 3.7.1 *Suppose that the starting points are generated by a Poisson point process on $\mathbb{R}$ with intensity $\rho > 0$. Let $\mathbb{P}_\xi$ denote the probability distribution of $\xi(t), t \geq 0$, and suppose that $\xi(t), t \geq 0$, satisfies the following conditions:*

$$\xi(0) = 0, \quad \mathbb{E}|\xi(t)| = \int |\xi(t)| d\mathbb{P}_\xi < \infty, \quad \mathbb{E}\xi(t) = 0.$$

Moreover, suppose that

$$\mathbb{E}|\xi(t)| = t^\alpha L(t) \tag{3.7.1}$$

for some $\alpha > 0$, where L is slowly varying at ∞, and that there is a function $c(t, s) < 0$ such that, for any $s, t \geq 0$,

$$c(t, s) = \lim_{T \to \infty} \frac{\mathbb{E}|\xi(Tt) - \xi(Ts)|}{\mathbb{E}|\xi(T)|} = \lim_{T \to \infty} \frac{\mathbb{E}|\xi(Tt) - \xi(Ts)|}{T^\alpha L(T)}. \qquad (3.7.2)$$

Then,

1. $c(t, s)$ is a homogeneous function of order α: $c(\lambda t, \lambda s) = \lambda^\alpha c(t, s)$,
2. As $T \to \infty$,

$$\frac{Y(Tt)}{\sqrt{T^\alpha L(T)}} \xrightarrow{fdd} \rho^{-1/2} Z(t), \quad t \geq 0, \qquad (3.7.3)$$

where $\{Z(t)\}$ is an $(\alpha/2)$–self-similar Gaussian process with mean 0 and covariance:

$$\mathbb{E} Z(t) Z(s) = \frac{1}{2}\big(t^\alpha + s^\alpha - c(t, s)\big), \quad s, t \geq 0. \qquad (3.7.4)$$

3. If, in addition, $\{\xi(t)\}$ has stationary increments, then $0 < \alpha \leq 1$, and $c(t, s) = |t - s|^\alpha$; that is, $\{Z(t)\}$ is FBM $\{B_{\alpha/2}(t)\}$.

The basic idea of the proof is to incorporate $\xi(t)$ in the Poisson point process framework, use the Poisson convergence to the normal and exploit the regularly varying assumption $\mathbb{E}|\xi(t)| = t^\alpha L(t)$ to obtain the limiting covariance.

Sketch of proof View $(x_i, \xi_i(\cdot)) \in \mathbb{R} \times \mathbb{R}^{[0,\infty)}$ as a realization of an abstract Poisson point process N with intensity measure

$$d\mu = \rho dx d\mathbb{P}_\xi$$

defined on $\mathbb{R} \times \mathbb{R}^{[0,\infty)}$, where $\mathbb{P}_\xi$ is the law of each process $\xi_i(\cdot)$. The non-collision particle system is $X_i(t) = x_i + \xi_i(t)$, $i \in \mathbb{Z}$, with the starting points ordered as $\ldots, X_{-1}(0) < X_0(0) \leq 0 < X_1(0) \ldots$. Now extend the order-preservation idea to define the collision process.[4] This process describes the motion of the "middle" particle, namely the one which starts at the position closest to and below 0. We use an "order-tracking" process $n(t)$ to keep track of the middle particle. Let

$$n^+(t) = |\{i : i \leq 0, X_i(t) > 0\}|, \quad n^-(t) = |\{i : i > 0, X_i(t) \leq 0\}|. \qquad (3.7.5)$$

Here, $n^+(t)$ denotes the (finite) number of particles starting below 0 that would end up above 0 at time t if there were no collision and $n^-(t)$ has the converse interpretation. Define

$$n(t) := n^+(t) - n^-(t). \qquad (3.7.6)$$

If $\ldots \leq x_{-1}^t \leq x_0^t \leq 0 < x_1^t \leq \ldots$ are the ordered positions at time t of the non-collision particle system $\{X_i(t)\}$, then the collision process is defined as

$$Y(t) = x_{n(t)}^t. \qquad (3.7.7)$$

---

[4] The particle processes $\xi_i(t)$ do not need to have continuous sample paths as in Figure 3.4. The sample paths can belong to the space of right-continuous functions having left limits for the order-preservation processes to be well-defined.

(For later reference and of independent interest, note that (3.7.7) yields

$$\{Y(t) > 0\} = \{n(t) > 0\} = \{n_+(t) > n_-(t)\}.) \tag{3.7.8}$$

Observe that $n(t) = n^+(t) - n^-(t) = N(A^+(t)) - N(A^-(t))$, where N is the Poisson random measure defined above and

$$A^+(t) = \{(x, \xi) : x \leq 0, \xi(t) + x > 0\}, \quad A^-(t) = \{(x, \xi) : x > 0, \xi(t) + x \leq 0\}$$

are measurable sets in $\mathbb{R} \times \mathbb{R}^{[0,\infty)}$. Since $A^-(t) \cap A^+(t) = \emptyset$, $n^+(t)$ and $n^-(t)$ are i.i.d. Poisson random variables for fixed t. Let us now determine the mean and variance of $n(t)$. Note that

$$\mathbb{E}n^+(t) = \int \int_{-\infty}^{0} 1_{\{\xi(t) > -x\}} \rho\, dx\, d\mathbb{P}_\xi = \rho \int \max\{\xi(t), 0\} d\mathbb{P}_\xi = \rho \mathbb{E}\xi(t)_+ = \frac{\rho}{2} \mathbb{E}|\xi(t)|,$$

where the last equality follows from the fact that $\xi(t)$ has 0 mean. Similarly, $\mathbb{E}n^-(t) = \rho \mathbb{E}|\xi(t)|/2$. Hence,

$$\mathbb{E}n(t) = \mathbb{E}n^+(t) - \mathbb{E}n^-(t) = 0$$

and

$$\mathbb{E}n(t)^2 = \text{Var}(n(t)) = \text{Var}(n^+(t)) + \text{Var}(n^-(t)) = \mathbb{E}n^+(t) + \mathbb{E}n^-(t) = 2\mathbb{E}n^+(t) = \rho \mathbb{E}|\xi(t)|,$$

since a Poisson random variable has a variance equal to its mean. Thus, the variance of $n(t)$ equals

$$\phi(t) := \text{Var}(n(t)) = \mathbb{E}n(t)^2 = \rho \mathbb{E}|\xi(t)| = \rho t^\alpha L(t). \tag{3.7.9}$$

One can show that if $\rho^{-1} n(Tt)/\sqrt{\phi(T)}$ converges to some process $Z(t)$ in finite-dimensional distributions, then so does $Y(Tt)/\sqrt{\phi(T)}$ as $T \to \infty$ (see Section 3 of Dürr et al. [339]). Hence to study the scaling limit of the collision process $\{Y(t)\}$, it is equivalent to study the limit of the simpler order-tracking process $n(t)$. But

$$\frac{n(Tt)}{\sqrt{\phi(T)}} \xrightarrow{fdd} Z(t), \tag{3.7.10}$$

where $\{Z(t)\}$ is a Gaussian process with zero mean, unit variance and covariance

$$\lim_{T \to \infty} \mathbb{E}\left(\frac{n(Tt)}{\sqrt{\phi(T)}} \frac{n(Ts)}{\sqrt{\phi(T)}}\right).$$

This is because $n(t) = n^+(t) - n^-(t)$ where $n^+(t)$ and $n^-(t)$ at fixed t are i.i.d. Poisson with mean (and variance) $\phi(t)/2$. Since $\phi(t) \to \infty$, by the standard normal approximation to the Poisson (central limit theorem), we get convergence in the sense of marginal distributions. Convergence of the finite-dimensional distributions can be obtained in a similar way. (See Exercise 3.13.)

Note that

$$\mathbb{E}n(t)n(s) = \frac{1}{2}\Big(\mathbb{E}n(t)^2 + \mathbb{E}n(s)^2 - \mathbb{E}(n(t) - n(s))^2\Big).$$

We know by (3.7.9), that $\mathbb{E}n(t)^2 = \phi(t) = \rho t^\alpha L(t)$, so

$$\lim_{T \to \infty} \frac{\mathbb{E}n(Tt)^2}{\phi(T)} = \lim_{T \to \infty} \frac{\rho(Tt)^\alpha L(Tt)}{\rho T^\alpha L(T)} = t^\alpha.$$

We are left to show that

$$\frac{\mathbb{E}(n(Tt)-n(Ts))^2}{\phi(T)} = \lim_{T\to\infty}\frac{\rho\mathbb{E}|\xi(Tt)-\xi(Ts)|}{\rho T^\alpha L(T)} = \lim_{T\to\infty}\frac{\mathbb{E}|\xi(Tt)-\xi(Ts)|}{T^\alpha L(T)} = c(t,s), \tag{3.7.11}$$

which proves (3.7.4). The first equality in (3.7.11) follows as above (see Exercise 3.14).

Since as $T \to \infty$, $Y(Tt)/\sqrt{\phi(T)}$ has the same limit $Z(t)$ as $\rho^{-1}n(Tt)/\sqrt{\phi(T)}$, we get

$$\frac{Y(Tt)}{\sqrt{T^\alpha L(T)}} \sim \frac{\rho^{-1}n(Tt)}{\sqrt{\rho^{-1}\phi(T)}} \xrightarrow{fdd} \rho^{-1/2}Z(t),$$

proving (3.7.3).

When $\{\xi(t)\}$ has stationary increments in the last part of the theorem, the covariance function in (3.7.4) becomes that of FBM with the self-simjilarity parameter $H = \alpha/2$. $\square$

Corollary 3.7.2 *Suppose $\xi(t)$, $t \geq 0$ is H–self-similar and $\mathbb{E}|\xi(1)| < \infty$. Then, as $T \to \infty$,*

$$\frac{Y(Tt)}{T^{H/2}} \xrightarrow{fdd} \rho^{-1/2}\sqrt{\mathbb{E}|\xi(1)|}\, Z(t), \quad t \geq 0,$$

where $\{Z(t)\}$ is a Gaussian process with zero mean and covariance

$$\mathbb{E}Z(t)Z(s) = \frac{1}{2}\left(t^H + s^H - \frac{\mathbb{E}|\xi(t)-\xi(s)|}{\mathbb{E}|\xi(1)|}\right), \quad s,t \geq 0. \tag{3.7.12}$$

If, in addition, $\xi(t)$, $t \geq 0$, has stationary increments, then $0 < H \leq 1$ and $\{Z(t)\}$ is the standard FBM $\{B_{H/2}(t)\}$.

Proof The result follows from Theorem 3.7.1 by observing that $\mathbb{E}|\xi(t)| = t^H \mathbb{E}|\xi(1)|$ for H–self-similar process $\{\xi(t)\}$ and hence that we have $L(t) = \mathbb{E}|\xi(1)|$. $\square$

Examples

- Newtonian particles:
  $\xi(t) = vt$: $Y(Tt)/\sqrt{T} \xrightarrow{fdd} \rho^{-1/2}\sqrt{\mathbb{E}|v|}B(t)$.
- Brownian motion:
  $\xi(t) = B(t)$: $Y(Tt)/T^{1/4} \xrightarrow{fdd} \rho^{-1/2}(2/\pi)^{1/4}B_{1/4}(t)$ since $\mathbb{E}|Z| = \sqrt{2/\pi}$ if Z is standard normal.
- Fractional Brownian motion:
  $\xi(t) = B_H(t)$ ($0 < H < 1$): $Y(Tt)/T^{H/2} \xrightarrow{fdd} \rho^{-1/2}(2/\pi)^{1/4}B_{H/2}(t)$.
- Symmetric α-stable motion:
  $\xi(t) = S_\alpha(t)$ ($1 < \alpha < 2$): $Y(Tt)/T^{1/\alpha} \xrightarrow{fdd} \rho^{-1/2}\sqrt{\mathbb{E}|S_\alpha(1)|}B_{1/(2\alpha)}(t)$.
- Integrated Brownian motion:

$$\xi(t) := \int_0^t B(s)ds = \int_0^t\int_0^s dB(r)ds :$$

Note that $\{\xi(t)\}$ is also a centered Gaussian process, but without stationary increments. It can be seen that $\{\xi(t)\}$ is 3/2–self-similar. In fact, supposing $0 \leq s \leq t$, its covariance

$\gamma(t, s)$ is

$$\gamma(t,s) = \mathbb{E}\xi(t)\xi(s) = \mathbb{E}\int_0^t B(v)dv \int_0^s B(u)du = \int_0^s \int_0^t \mathbb{E}(B(v)B(u))dvdu$$

$$= \int_0^s \int_0^t \min\{v,u\} dv du = \int_0^s \left(\int_0^u v dv + \int_u^t u dv \right) du = \frac{s^2 t}{2} - \frac{s^3}{6}$$

and, in particular, $\gamma(t, t) = t^3/3$. Hence,

$$\mathbb{E}(\xi(t) - \xi(s))^2 = \gamma(t,t) + \gamma(s,s) - 2\gamma(t,s) = \frac{t^3}{3} + \frac{2s^3}{3} - s^2 t.$$

Since $\xi(t) - \xi(s)$ is Gaussian, we have $\mathbb{E}|\xi(t) - \xi(s)| = \sqrt{(2/\pi)|t^3/3 + 2s^3/3 - s^2 t|}$, and in particular $\mathbb{E}|\xi(1)| = \sqrt{2/(3\pi)}$.

Therefore, by Corollary 3.7.2, the limit of the collision process $Y(Tt)/T^{3/4}$, properly normalized, is

$$\rho^{-1/2}(2/(3\pi))^{1/4} Z(t),$$

where $\{Z(t)\}$ is a 3/4–self-similar Gaussian process with covariance

$$\mathbb{E}Z(t)Z(s) = \frac{1}{2}\left(t^{3/2} + s^{3/2} - |t^3 + 2s^3 - 3s^2 t|^{1/2}\right), \quad 0 \leq s \leq t.$$

3.8 Motion of a Tagged Particle in a Simple Symmetric Exclusion Model

Simple exclusion is one of the canonical interacting particle systems. The corresponding exclusion system η_t, $t \geq 0$, is a Markov process with state space $\{0, 1\}^\mathbb{Z}$; that is, it consists at each t of an infinite sequence of 0s or 1s. The process describes an evolution of configurations of indistinguishable particles on $\mathbb{Z}$ with at most one particle per site. A particle at $x \in \mathbb{Z}$ waits an exponentially distributed time with mean 1, and then chooses a site $y \in \mathbb{Z}$ with probability $p(x, y)$. If y is vacant at that time, the particle at x moves to y. Otherwise, it stays at x. Whether the particle moves or not, it restarts its exponential clock. All the exponential holding times and choices related to p are independent. Since the exponential holding time is continuous, only one particle moves at a time. A basic quantity of interest is the motion of one of these particles, the so-called *tagged particle*.

In the nearest-neighbor interactions, p satisfies $p(x, y) = 0$ if $|x - y| > 1$; that is, a particle can move only to a neighboring site. The focus here is on symmetric nearest-neighbor interactions; that is, the case $p(x, x + 1) = p(x, x - 1) = 1/2$. (Asymmetric and other non-nearest-neighbor interactions are discussed in Section 3.15.) The system is assumed to start from the (equilibrium) configuration determined by a product measure ν_ρ on $\{0, 1\}^\mathbb{Z}$ with marginals ν_ρ(particle at x) $= \rho$, for all $x \in \mathbb{Z}$. The initial measure is conditioned to have a particle at 0, which is tagged and whose motion is denoted $Y(t)$. This is depicted in Figure 3.5.

The basic problem of interest is to understand the behavior of the motion $Y(T)$ of the tagged particle as $T \to \infty$, and more generally the motion viewed as a process $Y(Tt)$, $t \geq 0$, at large scales $T \to \infty$. As stated below, the limiting object is FBM with the self-similarity parameter $H = 1/4$, and the appropriate normalization of $Y(T)$ is $T^H = T^{1/4}$ and that of $Y(Tt)$ is $T^{1/4}$. The simple symmetric exclusion model can be thought as a discrete

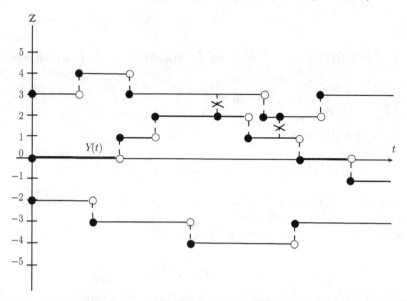

Figure 3.5 Exclusion process with the path $Y(t)$ of the tagged particle. The vertical dashed lines indicate the movement of the particles from one site to another. The vertical dashed lines with crosses indicate instances when the move was not possible because the neighboring site was occupied. A darkened circle indicates the position of the particle at that time.

analogue of the elastic collision model considered in Section 3.7 (with $\xi(t)$ being Brownian motion).

We focus on the behavior of $Y(T)$ only and state that of $Y(Tt)$ at the end of the section. We follow the approach in the original work of Arratia [46]. For other methods, see Rost and Vares [863], De Masi and Ferrari [293], and Liggett [632], Chapter VIII.

The approach of Arratia [46] is based on expressing the symmetric exclusion process in terms of another model, the so-called *stirring motions*. The system $\xi = \{\xi_t^x, x \in \mathbb{Z}, t \geq 0\}$ of random stirrings has exactly one particle per site at all times, with the particles at sites $x < y$ interchanged[5] at rate $p(x, y) = p(y, x)$ at event times determined by independent Poisson processes with intensity 1 for each pair of sites (x, y) with $p(x, y) > 0$. The probability that clocks ring at the same time is zero. Note that in the "stirring motions" model, there is no exclusion and that when the event clock rings, the interchange always occurs. The position at time t of the particle initially at site x is denoted ξ_t^x. For each x, the path ξ_t^x, $t \geq 0$, is a random walk based on p and starting at x.

For a set $A \subset \mathbb{Z}$, let

$$\eta_t^A = \{\xi_t^x : x \in A\}; \qquad (3.8.1)$$

that is, η_t^A denotes the set of the positions of random walks ξ_t^x which started at all $x \in A$. The random walks ξ_t^x started at $x \in A$ intersect. To relate these random walks to the simple exclusion process, the next step is to define their "order statistics" representing the non-intersecting motions of the exclusion process. The "order statistics" $Y_t^x (= Y_t^{x,A})$ of the

---

[5] These interchanges of particles are viewed as "stirrings", hence the term "stirring motions."

3.8 Motion of a Tagged Particle in a Simple Symmetric Exclusion Model

stirring paths ξ_t^x for $x \in A$ is defined as the paths satisfying: for all $t \geq 0$,

$$\{Y_t^x : x \in A\} = \eta_t^A,$$
$$Y_t^x < Y_t^y, \quad \text{for all } x < y, \ x, y \in A,$$
$$Y_0^x = x, \quad \text{for all } x \in A,$$

and Y_t^x, $t \geq 0$, is right-continuous with left-limits for all $x \in A$. Thus, the "order statistics" processes Y_t^x maintain their time-zero ordering. As shown in Arratia [46], these processes can also be characterized as follows. Define

$$\mu_{A,x}^+(z,t) = \left|\{y \in A : y \leq x, \xi_t^y \geq z\}\right| = \sum_{y \leq x} 1_{\{y \in A, \xi_t^y \geq z\}}, \tag{3.8.2}$$

$$\mu_{A,x}^-(z,t) = \left|\{y \in A : y > x, \xi_t^y < z\}\right| = \sum_{y > x} 1_{\{y \in A, \xi_t^y < z\}}, \tag{3.8.3}$$

that is, $\mu_{A,x}^+(z,t)$ counts the number of particles y in A initially to the left of x (x included) which moved above z by time t, and $\mu_{A,x}^-(z,t)$ is the number of particles y in A initially to the right of x (x excluded) which moved below z by time t. Then, for any $z \in \mathbb{Z}$,

$$\{Y_t^x \geq z\} = \{\mu_{A,x}^+(z,t) > \mu_{A,x}^-(z,t)\} \tag{3.8.4}$$

(Exercise 3.16). It then follows that

$$Y_t^x = \sup\{z \in \mathbb{Z} : \mu_{A,x}^+(z,t) > \mu_{A,x}^-(z,t)\}. \tag{3.8.5}$$

The definition (3.8.5) of Y_t^x is valid for symmetric random walk where $p(x,y) = p(y,x)$. In the nearest-neighbor case, at most one of the paths Y_t^x moves at any time, so that Y_t^x is the path of the tagged particle initially at $x \in A$, in the exclusion system determined by (3.8.1). Connecting this observation back to the simple symmetric exclusion process, we also have

$$Y(t) = Y_t^0, \tag{3.8.6}$$

where we take A as the initial configuration η_0 with a particle added at site 0 if necessary, thus $A = \eta_0 \cup \{0\}$. (Strictly speaking, η_0 is an element of $\{0,1\}^{\mathbb{Z}}$ but it can be identified as a subset of $\mathbb{Z}$ in the natural one-to-one manner.)

Theorem 3.8.1 *For the simple symmetric exclusion process on $\mathbb{Z}$, starting from the initial product measure with marginals $\mathbb{P}(\text{particle at } x) = \rho$, for all $x \in \mathbb{Z} \setminus \{0\}$, $\mathbb{P}(\text{particle at } 0) = 1$, the position $Y(T)$ of the tagged particle initially at the origin satisfies*

$$\frac{Y(T)}{T^{1/4}} \xrightarrow{d} \mathcal{N}\left(0, \sqrt{\frac{2}{\pi}\frac{1-\rho}{\rho}}\right), \tag{3.8.7}$$

as $T \to \infty$. Furthermore, $\mathrm{Var}(T^{-1/4}Y(T)) \to \sqrt{2/\pi}(1-\rho)/\rho$.

Proof By using the relations (3.8.6) and (3.8.4), we can relate the simple symmetric exclusion process $Y(T)$ to the stirring motions ξ_T^y as follows:

$$\{Y(T) \geq z\} = \{\mu_{A,0}^+(z,T) > \mu_{A,0}^-(z,T)\} = \{\mu_{A,0}^+(z,T) - \mu_{A,0}^-(z,T) > 0\}$$

$$= \{\sum_{y \leq 0} 1_{\{y \in A, \xi_T^y \geq z\}} - \sum_{y > 0} 1_{\{y \in A, \xi_T^y < z\}} > 0\}, \tag{3.8.8}$$

where as before $A = \eta_0 \cup \{0\}$. The idea now is to investigate all those $\xi_T^y \geq z$ for $y \leq 0$, and $\xi_T^y < z$ for $y > 0$, and then condition only on the y's belonging to A.

Let
$$N = \mu_{\mathbb{Z},0}^+(z,T) = \sum_{y \leq 0} 1_{\{\xi_T^y \geq z\}}, \tag{3.8.9}$$

where A in $\mu_{A,0}^+(z,T)$ is taken as $A = \mathbb{Z}$. It can be shown (Exercise 3.16) that, for $z \geq 1$,
$$\mu_{\mathbb{Z},0}^+(z,T) - \mu_{\mathbb{Z},0}^-(z,T) = -|\mathbb{Z} \cap (0,z)| = -(z-1), \tag{3.8.10}$$

which yields
$$\sum_{y > 0} 1_{\{\xi_T^y < z\}} = \mu_{\mathbb{Z},0}^-(z,T) = N + z - 1. \tag{3.8.11}$$

Then, label the sites contributing to (3.8.9) and (3.8.11) as
$$\{y \leq 0 : \xi_T^y \geq z\} = \{y_{-N} < y_{-N+1} < \cdots < y_{-1}\},$$
$$\{y > 0 : \xi_T^y < z\} = \{y_1 < y_2 < \cdots < y_{N+z-1}\}. \tag{3.8.12}$$

We put back $A = \eta_0 \cup \{0\}$ into the picture by considering
$$S = \sum_{i=-N}^{-1} 1_{\{y_i \in \eta_0\}} - \sum_{i=1}^{N+z-1} 1_{\{y_i \in \eta_0\}}, \tag{3.8.13}$$

where η_0 is the initial configuration (not conditioned to have particle at 0). It follows from (3.8.8)–(3.8.13) that
$$\{Y(T) \geq z\} = \{S + 1_{\{\xi_T^0 \geq z, 0 \notin \eta_0\}} > 0\}, \tag{3.8.14}$$

where the last indicator function accounts for the possibility that $0 \notin \eta_0$. In view of the result to prove, we shall take
$$z = \left\lfloor a\left(\frac{1-\rho}{\rho}\right)^{1/2}\left(\frac{2T}{\pi}\right)^{1/4} \right\rfloor, \tag{3.8.15}$$

where $a \in \mathbb{R}$ is fixed.

The right-hand side of (3.8.14) is amenable to straightforward analysis. Indeed, conditionally on N, S is just the difference of the sums of i.i.d. variables for which CLT holds. To be able to condition on N, we shall use a result established next. By Lemma 3.8.2 below, N in (3.8.9) is the sum of negatively correlated variables. Hence,
$$\text{Var}(N) = \text{Var}(\sum_{y \leq 0} 1_{\{\xi_T^y \geq z\}}) \leq \sum_{y \leq 0} \text{Var}(1_{\{\xi_T^y \geq z\}}) \leq \sum_{y \leq 0} \mathbb{E} 1_{\{\xi_T^y \geq z\}} = \mathbb{E} N.$$

Observe next that $\mathbb{E} N = \sum_{y \leq 0} \mathbb{P}(\xi_T^y \geq z) = \sum_{y \leq 0} \mathbb{P}(\xi_T^0 \geq z - y) = \sum_{v \geq 1} \mathbb{P}(\xi_T^0 \geq z - 1 + v) = \sum_{v \geq 1} \mathbb{P}(\xi_T^0 - z + 1 \geq v) = \sum_{v \geq 1} \mathbb{P}((\xi_T^0 - z + 1)_+ \geq v) = \mathbb{E}(\xi_T^0 - z + 1)_+$ by using the formula $\mathbb{E} X = \sum_{v \geq 1} \mathbb{P}(X \geq v)$ for nonnegative integer-valued random variable X. Since ξ_t^0 is the random walk $\sum_{k=1}^{P(t)} X_k$ with a Poisson process $P(t)$ having intensity 1

3.8 Motion of a Tagged Particle in a Simple Symmetric Exclusion Model

and $X_k = \pm 1$ with probability $1/2$, ξ_T^0 follows approximately the distribution $\mathcal{N}(0, T)$, as $T \to \infty$. Then, since $z = z(T) = O(T^{1/2}) = o(T^2)$ as $T \to \infty$, we have $\mathbb{E}N = \mathbb{E}(\xi_T^0 - z + 1)_+ \sim T^{1/2}\mathbb{E}\mathcal{N}(0,1)_+ = T^{1/2}\mathbb{E}|\mathcal{N}(0,1)|/2 = (T/(2\pi))^{1/2}$. By Chebyshev's inequality,

$$\mathbb{P}\left(\left|N - \left(\frac{T}{2\pi}\right)^{1/2}\right| > T^{3/8}\right) \leq \mathbb{P}\left(|N - \mathbb{E}N| > \frac{T^{3/8}}{2}\right) + \mathbb{P}\left(\left|\mathbb{E}N - \left(\frac{T}{2\pi}\right)^{1/2}\right| > T^{3/8}\right)$$
$$\leq \frac{4\text{Var}(N)}{T^{3/4}} + \mathbb{P}\left(\left|\frac{(2\pi)^{1/2}\mathbb{E}N}{T^{1/2}} - 1\right| > (2\pi)^{1/2}T^{-1/8}\right) \to 0, \quad (3.8.16)$$

by using $\text{Var}(N) \leq \mathbb{E}N \sim T^{1/2}/(2\pi)^{1/2}$ from above. The convergence of the last term to 0 above is left as Exercise 3.17.

The result (3.8.16) shows that N can be assumed to satisfy $|N - (T/(2\pi))^{1/2}| < T^{3/8}$ asymptotically. We argue next that, conditionally on any such value of N, the quantity S in (3.8.13) is asymptotically normal. Fixing $N = n$, let

$$S_{n,T} = \sum_{i=-n}^{-1} 1_{\{y_i \in \eta_0\}} - \sum_{i=1}^{n+z-1} 1_{\{y_i \in \eta_0\}},$$

where $z = z(T)$ is given in (3.8.15) and $n = n(T)$ satisfies $|n - (T/(2\pi))^{1/2}| < T^{3/8}$. Since $\mathbb{P}(y_i \in \eta_0) = \rho$, we have $\text{Var}(\sum_{i=-n}^{-1} 1_{\{y_i \in \eta_0\}}) = n\rho(1-\rho)$ and $\text{Var}(S_{n,T}) = (2n + z - 1)\rho(1-\rho)$. Since $z(T) = o(n(T))$, this yields $\text{Var}(\sum_{i=-n}^{-1} 1_{\{y_i \in \eta_0\}})/\text{Var}(S_{n,T}) = n/(2n + z - 1) \to 1/2$ and hence

$$\frac{\sum_{i=-n}^{-1} 1_{\{y_i \in \eta_0\}} - n\rho}{(\text{Var}(S_{n,T}))^{1/2}} \xrightarrow{d} \mathcal{N}(0, 1/2), \quad \frac{\sum_{i=1}^{n+z-1} 1_{\{y_i \in \eta_0\}} - (n+z-1)\rho}{(\text{Var}(S_{n,T}))^{1/2}} \xrightarrow{d} \mathcal{N}(0, 1/2).$$

For $n(T) \sim (T/(2\pi))^{1/2}$ and $z(T)$ as in (3.8.15), $\mathbb{E}S_{n,T}/(\text{Var}(S_{n,T}))^{1/2} = -(z-1)\rho/(\rho(1-\rho)(2n+z-1))^{1/2} \to -a$, so that $S_{n,T}/(\text{Var}(S_{n,T}))^{1/2} \xrightarrow{d} \mathcal{N}(-a, 1)$. We now conclude that, for any fixed $b \in \mathbb{R}$, as $T \to \infty$,

$$\sup_{n : |n - (\frac{T}{2\pi})^{1/2}| < T^{3/8}} \left|\mathbb{P}(S_{n,T} \geq b) - \int_a^\infty \frac{e^{-u^2/2}}{\sqrt{2\pi}} du\right| \to 0, \quad (3.8.17)$$

since under the supremum, $n \to \infty$ as $T \to \infty$ (b will be taken as 0 and 1 below).

The combination of (3.8.16) and (3.8.17) now yields the desired result (3.8.7). Indeed, let $E_T = \{|N - (\frac{T}{2\pi})^{1/2}| < T^{3/8}\}$ so that $\mathbb{P}(E_T) \to 1$ as $T \to \infty$ by (3.8.16). Then, by (3.8.14) with z in (3.8.15),

$$\mathbb{P}(Y(T) \geq z) \leq \mathbb{P}(S > -1) \leq \mathbb{P}(E_T^c) + \mathbb{P}(S > -1|E_T)\mathbb{P}(E_T) \to \mathbb{P}(\mathcal{N}(0,1) > a),$$

by using (3.8.17) with $b = -1$. Similarly, $\mathbb{P}(Y(T) \geq z) \geq \mathbb{P}(S > 0) \geq \mathbb{P}(S > 0|E_T)\mathbb{P}(E_T) \to \mathbb{P}(\mathcal{N}(0,1) > a)$. Therefore, $\mathbb{P}(Y(T) \geq z) \to \mathbb{P}(\mathcal{N}(0,1) > a)$, as $T \to \infty$. Since $z = z(T)$ is given by (3.8.15), we thus have shown that

$$\mathbb{P}\left(\left(\frac{\rho}{1-\rho}\right)^{1/2}\left(\frac{\pi}{2}\right)^{1/4}\frac{Y(T)}{T^{1/4}} \geq a\right) \to \mathbb{P}(\mathcal{N}(0,1) > a)$$

and hence (3.8.7).

The stated convergence of the second moments can be proved by showing that $\{T^{-1/2}Y(T)^2, t \geq 1\}$ is uniformly integrable. See Arratia [46], p. 368. □

The following auxiliary result was used in the proof above. It is proved in Arratia [46], Lemma 1, by employing the result of Harris [456] on Markov processes preserving the class of measures having positive correlations.

Lemma 3.8.2 *In a system $\{\xi_t^x, x \in \mathbb{Z}, t \geq 0\}$ of random stirrings, based on symmetric, nearest-neighbor random walk $p(x, x+1) = p(x, x-1) = 1/2$, the events $\{\xi_t^v \geq a\}$ and $\{\xi_t^w \geq b\}$ are negatively correlated for all $t \geq 0$, $v, w, a, b \in \mathbb{Z}$ with $v \neq w$.*

We conclude this section by stating the convergence in law of the corresponding scaled path of the tagged particle.

Theorem 3.8.3 *Under the assumptions and notation of Theorem 3.8.1,*

$$\left\{\frac{Y(Tt)}{T^{1/4}}\right\}_{t\geq 0} \xrightarrow{d} \left\{\left(\sqrt{\frac{2}{\pi}}\frac{1-\rho}{\rho}\right)^{1/2} B_H(t)\right\}_{t\geq 0}, \qquad (3.8.18)$$

where B_H is a standard FBM with the self-similarity parameter $H = 1/4$, and the convergence in distribution is in the space $D[0, 1]$ of right-continuous functions on $[0, 1]$ with left limits, endowed with the uniform topology.

The proof of Theorem 3.8.3 can be found in Peligrad and Sethuraman [803].

3.9 Power-Law Pólya's Urn

The correlated random walk presented in this section was introduced by Hammond and Sheffield [449]. Their informal description of the walk is as follows (Hammond and Sheffield [449], p. 697). Let μ be a probability law on the natural numbers $\mathbb{N} = \{1, 2, \ldots\}$. The walk associated with law μ will have dependent increments each of which is either -1 or 1. The series of the increments $\{X_n\}_{n \in \mathbb{Z}}$ of the random walk is such that, given the values $\{X_k\}_{k<n}$, X_n is set equal to the increment X_{n-m_n} obtained by looking back m_n steps in the series. The m_ns are random and are obtained at each vertex $n \in \mathbb{Z}$ by sampling from μ independently. This description is reminiscent of the traditional Pólya's urn process where the value of X_n represents the color of the nth ball added, and the conditional law of X_n given the past is that of a uniform sample from $\{X_1, X_2, \ldots, X_{n-1}\}$.

The look-back involves not the uniform distribution as in the traditional Pólya's urn case, but a power-law distribution μ satisfying: for $\alpha > 0$,

$$\mu\{m, m+1, \ldots\} = m^{-\alpha} L(m), \quad m \geq 1, \qquad (3.9.1)$$

where L is a (positive) slowly varying function at infinity. Thus, the tail distribution of μ is regularly varying with the index $(-\alpha)$. Denote the class of distributions μ satisfying (3.9.1) by Γ_α. For the convergence to FBM below, the parameter α will be taken as $\alpha \in (0, 1/2)$.

A more rigorous construction of the random walk is as follows. For a fixed law μ, let G_μ be the random directed graph on $\mathbb{Z}$ in which each vertex z has a unique edge pointing

3.9 Power-Law Pólya's Urn

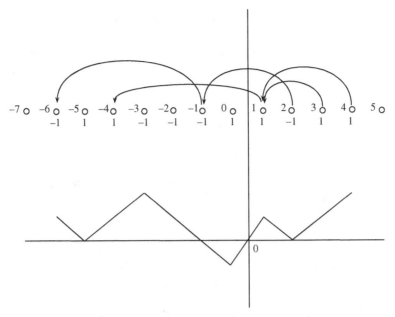

Figure 3.6 In the top plot, the ±1 values are determined by looking at the ancestral lines. In the bottom plot, they are summed and thus define the random walk.

backwards to $z - m_z$, where m_z is sampled independently according to the measure μ. The point $z - m_z$ is called the *parent* of z. The *ancestral line* of z is the decreasing sequence whose first element is z and each of whose terms is the parent of the previous one. Starting from different zs, the ancestral lines may meet at some point. See the top of Figure 3.6 where the two vertices 3 and 4 have their ancestral lines meet at 1. So each realization of G_μ partitions $\mathbb{Z}$ into connected components consisting of those zs whose ancestral lines meet.

The next result, stated here without proof (see Proposition 2 in Hammond and Sheffield [449]), describes the structure of G_μ when $\mu \in \Gamma_\alpha$.

Proposition 3.9.1 *Let $\mu \in \Gamma_\alpha$, $\alpha > 0$. If $\alpha > 1/2$, then G_μ has one component almost surely. If $\alpha < 1/2$, then G_μ has infinitely many components almost surely.*

In view of Proposition 3.9.1, we shall suppose $\alpha < 1/2$. For $p \in (0, 1)$ and $\mu \in \Gamma_\alpha$, define a law $\lambda_p = \lambda_{p,\mu}$ on functions mapping $\mathbb{Z}$ to $\{-1, 1\}$ in two steps as follows. First, determine the random components of G_μ. Then, independently give each component of G_μ a value of 1 with probability p, or -1 with probability $1 - p$, and assign the corresponding value to all vertices in that component.

To the measure λ_p, we associate a (correlated) random walk $S_p : \mathbb{Z} \to \mathbb{Z}$ by setting $S_p(0) = 0$, and

$$S_p(n) = \begin{cases} \sum_{k=1}^{n} X_p(k), & n \geq 1, \\ -\sum_{k=-n+1}^{0} X_p(k), & n \leq -1, \end{cases}$$

where $X_p : \mathbb{Z} \to \{-1, 1\}$ is a realization of the law λ_p. See Figure 3.6. Note that $S_p(n) - S_p(n-1) = X_p(n), n \in \mathbb{Z}$.

In the next result, we study the asymptotic behavior of the variance of the random walk $S_p(N)$, as $N \to \infty$. To do so, we will need to understand when ancestral lines starting from two different vertices j and k meet, in which case $X_p(j) = X_p(k)$ in the definition of the random walk $S_p(n)$. To be able to trace ancestral lines, we introduce a new random walk $\widetilde{S}_j$ as follows. Using the probability distribution μ on $\mathbb{N}$ defined in (3.9.1), let $\widetilde{S}_0 = 0$ and $\widetilde{S}_j = \sum_{l=1}^{j} Y_l$, $j \geq 1$, where Y_l are i.i.d. copies drawn from the law μ. Let also R_μ be the set of values assumed by the random walk $\widetilde{S}_j$; that is,

$$R_\mu = \left\{ m \in \mathbb{N} : m = \widetilde{S}_j \text{ for some } j \geq 0 \right\}. \tag{3.9.2}$$

Note that R_μ is a random set. Do not confuse the correlated random walk $S_p(n)$ above whose increments equal ± 1 with $\widetilde{S}_j$ whose increments are i.i.d. random variables with law μ. Set also, for $m \geq 0$,

$$q_m = \mathbb{P}(m \in R_\mu). \tag{3.9.3}$$

Note that $q_0 = 1$ since $0 \in R_\mu$. Thus, q_m denotes the probability that an ancestral line started at a vertex will visit the vertex located m steps back in the past.

Proposition 3.9.2 *Let $\alpha \in (0, 1/2)$, $\mu \in \Gamma_\alpha$ and S_p be the random walk defined above for $p \in (0, 1)$ and associated with the law μ. Then,*

$$\mathrm{Var}\left(S_p(N)\right) \sim \frac{4p(1-p)K_\alpha}{\sum_{m=0}^{\infty} q_m^2} N^{2\alpha+1} L(N)^{-2}, \tag{3.9.4}$$

where

$$K_\alpha = \frac{1}{2\alpha(2\alpha+1)} \left(\Gamma(1-2\alpha)^2 \Gamma(2\alpha) \cos(\alpha\pi) \right)^{-1}.$$

Proof Denote the Fourier transform of the sequence $\{q_m\}$ as

$$Q(t) = \sum_{m=0}^{\infty} q_m e^{-imt}. \tag{3.9.5}$$

Note that $(|Q|^2)_j = \sum_{m=0}^{\infty} q_m q_{j+m}$ is the jth Fourier coefficients of $|Q|^2 = Q\overline{Q}$. We will show first that

$$\mathrm{Var}(S_p(N)) = \frac{4p(1-p)}{(|Q|^2)_0} \left(2 \sum_{n=1}^{N-1} \sum_{j=1}^{n} (|Q|^2)_j + N(|Q|^2)_0 \right). \tag{3.9.6}$$

Observe that

$$\mathrm{Var}(S_p(N)) = \sum_{j=1}^{N} \sum_{k=1}^{N} \mathrm{Cov}(X_p(j), X_p(k)). \tag{3.9.7}$$

Two cases need to be considered for $\mathrm{Cov}(X_p(j), X_p(k))$. In one case, j and k are in different components of the random graph G_μ. Hence, conditioned on this, the covariance of $X_p(j)$ and $X_p(k)$ is 0. This case can be expressed as $A_j \cap A_k = \emptyset$, where A_j and A_k denote the ancestral lines of j and k, respectively. (Note that A_j and A_k are random sets.) In the other case, j and k are in the same component of the random graph G_μ, corresponding

3.9 Power-Law Pólya's Urn

to $A_j \cap A_k \neq \emptyset$. Note that, in this case, $X_p(j) = X_p(k) = 1$ with probability p, and $X_p(j) = X_p(k) = -1$ with probability $1 - p$. In particular, conditioned on this case, the covariance of $X_p(j)$ and $X_p(k) (= X_p(j))$ is the variance of $X_p(j)$, which is $4p(1-p)$. This shows that $\text{Cov}(X_p(j), X_p(k)) = 4p(1-p)\mathbb{P}(A_j \cap A_k \neq \emptyset)$ and hence, from (3.9.7),

$$\text{Var}(S_p(N)) = 4p(1-p) \sum_{j=1}^{N} \sum_{k=1}^{N} \mathbb{P}(A_j \cap A_k \neq \emptyset). \quad (3.9.8)$$

To compute $\mathbb{P}(A_j \cap A_k \neq \emptyset)$, that is, the probability that the ancestral lines of j and k meet, note first that $\mathbb{P}(A_j \cap A_k \neq \emptyset) = \mathbb{P}(A_j \cap A'_k \neq \emptyset)$, where A'_k is the ancestral line of k in an independent copy of G_μ. Indeed, this is easier to understand thinking of the equality $\mathbb{P}(A_j \cap A_k = \emptyset) = \mathbb{P}(A_j \cap A'_k = \emptyset)$ of the complements, since in this case, the non-intersecting ancestral lines A_j and A_k can be thought independent. To compute $\mathbb{P}(A_j \cap A'_k \neq \emptyset)$, let $|A_j \cap A'_k|$ be the number of points $z \in \mathbb{Z}$ in $A_j \cap A'_k$. Note that, for $j \leq k$, $|A_j \cap A'_k| = \sum_{h=-\infty}^{j} 1_{\{h \in A_j \cap A'_k\}}$; that is, $|A_j \cap A'_k|$ is the number of $h \leq j$ such that $h \in A_j \cap A'_k$. Hence, by using independence of A_j and A'_k,

$$\mathbb{E}|A_j \cap A'_k| = \sum_{h=-\infty}^{j} \mathbb{P}(h \in A_j)\mathbb{P}(h \in A'_k).$$

Note that, for fixed j,

$$A_j \stackrel{d}{=} j - R_\mu := \{j - m : m \in R_\mu\}, \quad (3.9.9)$$

where R_μ is defined in (3.9.2). Hence,

$$\mathbb{E}|A_j \cap A'_k| = \sum_{h=-\infty}^{j} \mathbb{P}(j - h \in R_\mu)\mathbb{P}(k - h \in R_\mu)$$

and, by using (3.9.3),

$$\mathbb{E}|A_j \cap A'_k| = \sum_{h=-\infty}^{j} q_{j-h} q_{k-h} = \sum_{m=0}^{\infty} q_m q_{k-j+m} = (|Q|^2)_{k-j}. \quad (3.9.10)$$

On the other hand, note also that

$$\mathbb{E}|A_j \cap A'_k| = \mathbb{E}\Big(|A_j \cap A'_k| \Big| A_j \cap A'_k \neq \emptyset\Big)\mathbb{P}(A_j \cap A'_k \neq \emptyset)$$

$$= \Big(1 + \sum_{m=1}^{\infty} q_m^2\Big)\mathbb{P}(A_j \cap A_k \neq \emptyset) = \Big(\sum_{m=0}^{\infty} q_m^2\Big)\mathbb{P}(A_j \cap A_k \neq \emptyset)$$

$$= (|Q|^2)_0 \, \mathbb{P}(A_j \cap A_k \neq \emptyset), \quad (3.9.11)$$

since $\mathbb{P}(A_j \cap A'_k \neq \emptyset) = \mathbb{P}(A_j \cap A_k \neq \emptyset)$ as discussed above, and given $A_j \cap A'_k \neq \emptyset$, the two ancestral lines will first meet at some point (contributing 1 in (3.9.11)) and then develop as two independent related partial sum process $\widetilde{S}$ of i.i.d. steps with distribution μ (contributing $\sum_{m=1}^{\infty} q_m^2$ for the conditional expected number).

Combining (3.9.10) and (3.9.11) yields $\mathbb{P}(A_j \cap A_k \neq \emptyset) = (|Q|^2)_{k-j}/(|Q|^2)_0$, $j \leq k$, and hence, from (3.9.8),

$$\text{Var}(S_p(N)) = \frac{4p(1-p)}{(|Q|^2)_0} \sum_{j=1}^{N} \sum_{k=1}^{N} (|Q|^2)_{|k-j|}, \qquad (3.9.12)$$

which can be written as (3.9.6) (see also the proof of Proposition 2.2.5). In view of (3.9.12) and by using Proposition 2.2.1, it is then enough to show that

$$\sum_{j=1}^{n} (|Q|^2)_j \sim n^{2\alpha} L(n)^{-2} \frac{K_\alpha(2\alpha+1)}{2} = n^{2\alpha} L(n)^{-2} \frac{1}{4\alpha} \left(\Gamma(1-\alpha)^2 \Gamma(2\alpha) \cos(\alpha\pi) \right)^{-1}, \qquad (3.9.13)$$

as $n \to \infty$. To do so, introduce $p_m = \mu\{m\}$, that is, the probability that a μ–distributed random variable takes the value m, and let $P(t) = \sum_{m=1}^{\infty} p_m e^{-imt}$ be the Fourier transform of the sequence $\{p_m\}$. By Lemma 3.9.3, $Q(t) = (1 - P(t))^{-1}$.

By using (A.2.12) in Appendix A.2 with $\alpha \in (0, 1/2)$, we have, as $t \to 0$,

$$|P(t) - 1| \sim C_\alpha t^\alpha L(t^{-1}) \left| 1 - i \tan\left(\frac{\alpha\pi}{2}\right) \right| = \Gamma(1-\alpha) t^\alpha L(t^{-1}),$$

since $C_\alpha = \Gamma(1-\alpha) \cos(\alpha\pi/2)$ (see (A.2.13)) and

$$\left| 1 - i \tan\left(\frac{\alpha\pi}{2}\right) \right| = \frac{|\cos(\alpha\pi/2) - i \sin(\alpha\pi/2)|}{\cos(\alpha\pi/2)} = \frac{|e^{-i\alpha\pi/2}|}{\cos(\alpha\pi/2)} = \frac{1}{\cos(\alpha\pi/2)}.$$

Hence, since $|Q(t)| = |P(t) - 1|^{-1}$,

$$|Q(t)| \sim \Gamma(1-\alpha)^{-1} t^{-\alpha} L(t^{-1})^{-1},$$

so that

$$|Q(t)|^2 \sim \Gamma(1-\alpha)^{-2} t^{-2\alpha} L(t^{-1})^{-2},$$

as $t \to 0$. Note by symmetry that $|Q(t)|^2 = \sum_{m=0}^{\infty} q_m^2 + 2 \sum_{j=1}^{\infty} (|Q|^2)_j \cos(jt)$ and hence, as $t \to 0$,

$$\sum_{j=1}^{\infty} (|Q|^2)_j \cos(jt) \sim \frac{\Gamma(1-\alpha)^{-2}}{2} t^{-2\alpha} L(t^{-1})^{-2}$$

$$= \frac{\ell(t^{-1})}{t^{2\alpha}} \Gamma(2\alpha) \cos(\alpha\pi) = \frac{\ell(t^{-1})}{t^{1-\alpha'}} \Gamma(1-\alpha') \sin(\alpha'\pi/2),$$

where

$$\alpha' = 1 - 2\alpha, \quad \ell(t^{-1}) = \frac{\Gamma(1-\alpha)^{-2} L(t^{-1})^{-2}}{2\Gamma(2\alpha) \cos(\alpha\pi)} = \frac{\Gamma(1-\alpha)^{-2} L(t^{-1})^{-2}}{2\Gamma(1-\alpha') \sin(\alpha'\pi/2)}.$$

By applying Theorem 4.10.1, (a), in Bingham et al. [157] with α replaced by α', we conclude that, as $n \to \infty$,

$$\sum_{j=1}^{n} (|Q|^2)_j \sim n^{1-\alpha'} \frac{\ell(n)}{1-\alpha'},$$

which is the desired relation (3.9.13) after substituting α' and the slowly varying function $\ell(n)$. □

The following auxiliary lemma was used in the proof above.

Lemma 3.9.3 *Let μ be a probability measure on $\mathbb{N}$, $p_m = \mu\{m\}$, $m \in \mathbb{N}$, and $P(t) = \sum_{m=1}^{\infty} p_m e^{-imt}$ be the Fourier transform of the sequence $\{p_m\}$. Let q_m be the probabilities (3.9.3) of a random walk $\widetilde{S}_j$ with i.i.d. steps drawn from the distribution μ visiting the sites $m \geq 0$ and $Q(t)$ be the Fourier transform (3.9.5) of the sequence $\{q_m\}$. Then,*

$$Q(t) = \sum_{j=0}^{\infty} (P(t))^j = \frac{1}{1 - P(t)}. \quad (3.9.14)$$

Proof Observe that, for $m \geq 0$,

$$q_m = \mathbb{P}(m \in R_\mu) = 1_{\{m=0\}} + \sum_{j=1}^{\infty} \mathbb{P}(\widetilde{S}_j = m) = 1_{\{m=0\}} + \sum_{j=1}^{\infty} (p^{*j})_m,$$

where p_m^{*j} denotes the mth element of the jth convolution p^{*j} of the sequence $p = \{p_m\}$. By taking the Fourier transform of the relation above, we obtain the first relation in (3.9.14). □

The scaling of the order $N^{2\alpha+1}$ derived in Proposition 3.9.2 shows that the stationary series of the increments of the correlated random walk is long-range dependent with parameter $d = \alpha$ in the sense (2.1.8) in condition V. Furthermore, as the next result shows, the scaled correlated random walk in fact converges to fractional Brownian motion. See Hammond and Sheffield [449], Proposition 3, for a proof based on a martingale central limit theorem.

Theorem 3.9.4 *Let $\alpha \in (0, 1/2)$, $\mu \in \Gamma_\alpha$ and S_p be the random walk defined above for $p \in (0, 1)$ and associated with the law μ. Let also*

$$\widetilde{c} = \Big(\frac{4p(1-p)K_\alpha}{(|Q|^2)_0}\Big)^{-1/2}$$

be the reciprocal square root of the constant appearing on the right-hand side of (3.9.4). Then,

$$\widetilde{c} N^{-\frac{1}{2}-\alpha} L(N) \Big(S_p([Nt]) - Nt(2p-1) \Big) \xrightarrow{fdd} B_H(t), \quad t \geq 0, \quad (3.9.15)$$

where B_H is a standard FBM with the self-similarity parameter

$$H = \frac{1}{2} + \alpha \in \Big(\frac{1}{2}, 1\Big). \quad (3.9.16)$$

3.10 Random Walk in Random Scenery

Kesten and Spitzer [553] considered sums of random variables with summands $Y_n = \xi(S_n)$, where

- $\{S_n, n \geq 1\}$ is a zero mean (integer-valued) random walk,
- $\xi(k)$, $k \in \mathbb{Z}$, are independent and identically distributed random variables independent of the S_ns.

The sequence $Y_n = \xi(S_n)$, $n \geq 1$, describes the motion of the random walker S_n viewed through the perspective of the random scenery $\xi(\cdot)$.[6] If one views S_n as describing the "position" of the "state" of a user at time n, and $\xi(k)$ the "reward" earned by, "cost" incurred by or "amount of work" produced by the user in state k, then $\xi(S_n)$ can also be thought as the reward earned at time n. The $\xi(S_n)$s are dependent because the random variable $\xi(S_n)$ takes the same value whenever the random walk visits the same state $\xi(k)$. By choosing the distributions as indicated in the following theorem, one obtains a self-similar process.

Theorem 3.10.1 *Suppose that, as $N \to \infty$, $N^{-1/\alpha} S_N$ converges to a SαS random variable with $1 < \alpha \leq 2$ and $N^{-1/\beta} \sum_{k=1}^{N} \xi(k)$ converges to a SβS random variable with $0 < \beta \leq 2$. Then,*

$$\frac{1}{N^{1-1/\alpha+1/(\alpha\beta)}} \sum_{n=1}^{[Nt]} \xi(S_n) \xrightarrow{fdd} \Delta_t = \int_{\mathbb{R}} \ell_t(x) X(dx), \quad t \geq 0, \quad (3.10.1)$$

where $\ell_t(x)$ is the local time of a SαS Lévy motion and $X(dx)$ is a SβS random measure on $\mathbb{R}$ with the Lebesgue control measure dx, which is independent of $\ell_t(\cdot)$. The limit process is H–SSSI with

$$H = 1 - \frac{1}{\alpha} + \frac{1}{\alpha\beta} \in \left(\frac{1}{2}, \infty\right). \quad (3.10.2)$$

The local time $\ell_t(x)$, $x \in \mathbb{R}$, of a process represents roughly the density of time in $[0, t]$ that the process spends at x. For more information about local times, see Berman [134], Section 5.12, as well as Section 7.2.5 below in the case of FBM. The local time $\ell_t(x)$ of a process $Y(\cdot)$ satisfies the following occupation time formula:

$$\int_0^t 1_A(Y(s))ds = \int_A \ell_t(x)dx, \quad (3.10.3)$$

which expresses in two different ways the total amount of time in $[0, t]$ that the process $Y(\cdot)$ spends in the set A, doing so in the left-hand side of (3.10.3) by integrating over time $s \in [0, t]$ and in the right-hand side by integrating over space $x \in \mathbb{R}$. More generally, one has

$$\int_0^t f(Y(s))ds = \int_{\mathbb{R}} f(x)\ell_t(x)dx, \quad (3.10.4)$$

for suitable functions f.

Observe now that the right-hand side of (3.10.1) has the same meaning than its left-hand side, but this meaning is expressed in terms of the continuous limit processes. Indeed, the convergence (3.10.1) can informally be derived as follows. Let $\{X(t)\}$ and $\{Y(t)\}$ be SβS

---

[6] Random walk in random scenery should not be confused with the so-called random walk in random environment, where the motion of the walk is dictated by the environment. See, for example, Kesten, Kozlov, and Spitzer [554], Sznitman [940], Bogachev [166].

and $S\alpha S$ Lévy motions approximating the partial sums S_N and $\sum_{k=1}^{N} \xi(k)$, respectively; that is,

$$\frac{1}{N^{1/\alpha}} S_{[Nt]} \approx Y(t), \quad \frac{1}{N^{1/\beta}} \sum_{k=1}^{[Nt]} \xi(k) \approx X(t). \quad (3.10.5)$$

The left-hand side of (3.10.3) is then approximated as

$$\int_0^t 1_A(Y(s))ds \approx \int_0^t 1_A\left(\frac{1}{N^{1/\alpha}} S_{[Ns]}\right) ds. \quad (3.10.6)$$

On the other hand, note that

$$\int_0^t 1_A\left(\frac{1}{N^{1/\alpha}} S_{[Ns]}\right) ds \approx \frac{1}{N} \sum_{n=1}^{[Nt]} 1_{\{\frac{1}{N^{1/\alpha}} S_n \in A\}} = \frac{1}{N} \sum_{N^{-1/\alpha} y \in A, y \in \mathbb{Z}} \sum_{n=1}^{[Nt]} 1_{\{S_n = y\}}$$

$$= \sum_{N^{-1/\alpha} y \in A, y \in \mathbb{Z}} \left(\frac{1}{N^{1-1/\alpha}} \sum_{n=1}^{[Nt]} 1_{\{S_n = y\}}\right) \frac{1}{N^{1/\alpha}}$$

$$= \sum_{N^{-1/\alpha} y \in A, y \in \mathbb{Z}} \ell_{N,t}\left(\frac{y}{N^{1/\alpha}}\right) \frac{1}{N^{1/\alpha}} \approx \int_A \ell_{N,t}(x) dx, \quad (3.10.7)$$

where

$$\ell_{N,t}(x) = \frac{1}{N^{1-1/\alpha}} \sum_{n=1}^{[Nt]} 1_{\{S_n = [N^{1/\alpha} x]\}}.$$

The relations (3.10.3), (3.10.6) and (3.10.7) suggest that $\ell_{N,t}(x)$ serves as an approximation to $\ell_t(x)$; that is,

$$\ell_{N,t}(x) \approx \ell_t(x). \quad (3.10.8)$$

Given this approximation and using the $(1/\beta)$–self-similarity of the $S\beta S$ Lévy motion $\{X(t)\}$ (Section 2.6.4), it remains to observe that

$$\frac{1}{N^{1-1/\alpha+1/(\alpha\beta)}} \sum_{n=1}^{[Nt]} \xi(S_n) = \frac{1}{N^{1-1/\alpha+1/(\alpha\beta)}} \sum_{n=1}^{[Nt]} \sum_{y \in \mathbb{Z}} \xi(y) 1_{\{S_n = y\}} = \sum_{y \in \mathbb{Z}} \ell_{N,t}\left(\frac{y}{N^{1/\alpha}}\right) \frac{\xi(y)}{N^{1/(\alpha\beta)}}$$

$$= \sum_{N^{1/\alpha} x \in \mathbb{Z}} \ell_{N,t}(x) \frac{\xi(N^{1/\alpha} x)}{N^{1/(\alpha\beta)}} \approx \sum_{N^{1/\alpha} x \in \mathbb{Z}} \ell_t(x) \frac{N^{1/\beta}}{N^{1/(\alpha\beta)}} \left(X\left(\frac{N^{1/\alpha} x}{N}\right) - X\left(\frac{N^{1/\alpha}(x-1)}{N}\right)\right)$$

$$\stackrel{d}{=} \sum_{N^{1/\alpha} x \in \mathbb{Z}} \ell_t(x) \frac{1}{N^{1/(\alpha\beta)}} \left(X(N^{1/\alpha} x) - X(N^{1/\alpha}(x-1))\right)$$

$$\stackrel{d}{=} \sum_{N^{1/\alpha} x \in \mathbb{Z}} \ell_t(x) \left(X(x) - X(x - N^{-1/\alpha})\right) \approx \int_{\mathbb{R}} \ell_t(x) dX(x).$$

The informal approach outlined above is formalized by Dombry and Guillotin-Plantard [322], and applies in a more general setting than that considered in Theorem 3.10.1.

The process $\{\Delta_t, t \geq 0\}$ in (3.10.1) is H–SSSI because the law of $\ell_{t+h}(x) - \ell_t(x)$ does not depend on $h \geq 0$ and because if $\ell_t(x)$ is the local time of an H'–SS process, then the self-similarity and the relation (3.10.3) imply

$$\{\ell_{ct}(c^{H'}x), x \in \mathbb{R}, t \geq 0\} \stackrel{d}{=} \{c^{1-H'}\ell_t(x), x \in \mathbb{R}, t \geq 0\} \qquad (3.10.9)$$

(Exercise 3.18). Consequently, for any $c > 0$,

$$\Delta_{ct} = \int_{\mathbb{R}} \ell_{ct}(x)dX(x) = \int_{\mathbb{R}} \ell_{ct}(c^{H'}x)dX(c^{H'}x)$$
$$\stackrel{d}{=} \int_{\mathbb{R}} c^{1-H'}\ell_t(x)c^{H'/\beta}dX(x) = c^{1-H'+H'/\beta}\int_{\mathbb{R}} \ell_t(x)dX(x) = c^{1-1/\alpha+1/\alpha\beta}\Delta_t,$$

since the $S\alpha S$ (resp. $S\beta S$) Lévy stable motion is self-similar with parameter $H' = 1/\alpha$ (resp. $1/\beta$).

Example 3.10.2 (*Simple random walk in Bernoulli-like scenery*) If S_N is a simple random walk and if the $\xi(k)$s with $k \in \mathbb{Z}$ take values ± 1 with probability $1/2$, then they are both in the domain of attraction of Brownian motion with $\alpha = \beta = 2$ and thus $H = 3/4$. The corresponding limit in (3.10.1) is $\int_{\mathbb{R}} \ell_t(x)B(dx)$; that is, intuitively $\int_0^t W(\widetilde{B}(s))ds$, where $W = B(dx)/dx$ is a Gaussian white noise independent of the Brownian motion $\widetilde{B}$ (see (3.10.4)).

By conditioning on ℓ_t, the finite-dimensional distributions of the process $\Delta_t = \int_{-\infty}^{\infty} \ell_t(z)X(dx), t \geq 0$, in (3.10.1) are given by

$$\mathbb{E}\exp\left\{i\sum_{j=1}^{J}\theta_j\Delta_{t_j}\right\} = \mathbb{E}\exp\left\{-C\int_{-\infty}^{\infty}\left|\sum_{j=1}^{J}\theta_j\ell_{t_j}(x)\right|^{\beta}dx\right\}, \qquad (3.10.10)$$

where C is a positive constant.

3.11 Two-Dimensional Ising Model

Another very interesting model associated with long-range dependence is the celebrated Ising[7] model, which originated in Statistical Mechanics. This is one of the few models of Statistical Mechanics which is "exactly solvable." A number of ingenious methods have been proposed to study the Ising model. We shall present here one such method, and show that the Ising model exhibits long-range dependence at the so-called critical temperature. We should add that our presentation will barely scratch the surface of what is known for the Ising model – for further notes see Section 3.15. Our goal is to provide the reader (perhaps one little familiar with Statistical Mechanics) with an idea of another interesting situation which leads to long-range dependence.

---

[7] Pronounced as [EEH sing].

3.11 Two-Dimensional Ising Model

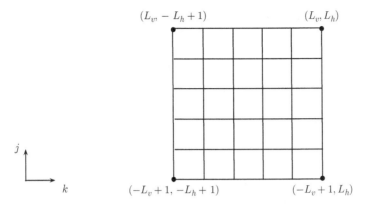

Figure 3.7 The Ising lattice.

3.11.1 Model Formulation and Result

We consider the Ising model in two dimensions. The reason is that, in one dimension, the Ising model does not exhibit long-range dependence (Exercise 3.19) and, in three dimensions, the Ising model has not been solved exactly yet (in fact, this is still one of the famous open problems in Statistical Mechanics and, more generally, Physics, though which may remain unsolved; see Cipra [245]). Thus, consider a two-dimensional square lattice (Figure 3.7), referred to as the *Ising lattice*, with

$2L_v$: the number of rows,

$2L_h$: the number of columns

("v" stands for "vertical", and "h" stands for "horizontal"). Denote a general point on the lattice as (j, k), $-L_v + 1 \leq j \leq L_v$, $-L_h + 1 \leq k \leq L_h$. Let

$$\sigma_{j,k} = \pm 1$$

be a variable associated with a lattice point (j, k).[8] We will consider below correlations between $\sigma_{0,0}$ and $\sigma_{0,N}$, and $\sigma_{0,0}$ and $\sigma_{N,N}$. In fact, we will first let $L_v, L_h \to \infty$ (the so-called thermodynamic limit), and then investigate these correlations as $N \to \infty$.

In the Ising model, the collection of variables $\{\sigma_{j,k}, -L_v + 1 \leq j \leq L_v, -L_h + 1 \leq k \leq L_h\}$ is assumed to follow the joint distribution

$$\frac{1}{Z_{L_v, L_h}} e^{-\beta \mathcal{E}}, \qquad (3.11.1)$$

where

$$\mathcal{E} = - \sum_{j=-L_v+1}^{L_v} \sum_{k=-L_h+1}^{L_h} (E^h \sigma_{j,k} \sigma_{j,k+1} + E^v \sigma_{j,k} \sigma_{j+1,k}) \qquad (3.11.2)$$

---

[8] Note that j and k in (j, k) refer to row (vertical) and column (horizontal) directions, respectively. This is in accord with the matrix notation where (j, k) is used for a matrix element $a_{j,k}$. But this is the opposite of the (x, y) convention, where for example x refers to the horizontal direction.

with some positive constants $E^h, E^v > 0$, and where

$$Z_{L_v,L_h} = \sum_{\{\sigma_{j,k}\}} e^{-\beta \mathcal{E}} \qquad (3.11.3)$$

is the normalization constant, where the sum is over all possible *configurations* (outcomes) $\{\sigma_{j,k}\}$. Since each $\sigma_{j,k}$ can take 2 values, there are $2^{(2L_v)(2L_h)}$ configurations. The parameter $\beta > 0$ is expressed as

$$\beta = \frac{1}{k_B T}, \qquad (3.11.4)$$

where $k_B = 1.38 \times 10^{-23}$ is Boltzmann's constant and $T > 0$.

The term $\mathcal{E}$ in (3.11.2) is referred to as *energy (Hamiltonian)* of a configuration $\{\sigma_{j,k}\}$. Note that the energy is determined only by interactions between neighboring sites, with the contributions E^h for horizontal interactions and E^v for vertical interactions. Note also the following effect of individual terms on the overall energy of the system. The energy is lower when neighboring sites align as $+1$ and $+1$, or as -1 and -1. Lower (negative) energy $\mathcal{E}$ translates into higher probability (3.11.1); that is, as expected in physical terms, the system favors configurations with lower energy.

The parameter $T > 0$ in (3.11.4) is referred to as *temperature* (and β as *inverse temperature*). When $T \to \infty$ (high temperature), note that $\beta \to 0$ and hence (3.11.1) approaches the uniform probability distribution on the lattice. In this case, the variables $\sigma_{j,k}$ tend to be independent and the system is in the disordered state. On the other hand, when $T \to 0$ (low temperature), note that $\beta \to \infty$ and the aligned configurations (all $+1$ or -1) are more likely. This is because the exponent in (3.11.1) will be large and positive. In this case, the system is in the ordered state.

Moving from the disordered to ordered state (large T to small T), one could expect that the system undergoes a *phase transition* at some *critical temperature* $T = T_c > 0$. We shall not define T_c in rigorous terms. Informally, the system exhibits very different characteristics for $T < T_c$ and $T > T_c$. The two-dimensional Ising model turns out to have such a phase transition at $T_c > 0$.

Here are few other notes regarding the Ising model. The distribution (3.11.1) is also known as a *Boltzmann distribution* or a *Gibbs distribution*. The normalization factor

$$Z_{L_v,L_h} = Z_{L_v,L_h}(\beta) = \sum_{\{\sigma_{j,k}\}} e^{-\beta \mathcal{E}} \qquad (3.11.5)$$

in (3.11.3) is called the *partition function*. In (3.11.5), the sum is over every value ± 1 of the variables $\sigma_{j,k}$ in the lattice. The energy $\mathcal{E}$ in (3.11.2) is often written more generally as

$$\mathcal{E} = - \sum_{j=-L_v+1}^{L_v} \sum_{k=-L_h+1}^{L_h} (E^h \sigma_{j,k}\sigma_{j,k+1} + E^v \sigma_{j,k}\sigma_{j+1,k} + H\sigma_{j,k}), \qquad (3.11.6)$$

where the terms $H\sigma_{j,k}$ in (3.11.6) account for the so-called external magnetic field[9] (see the discussion below for the reason of using "magnetic"). With (3.11.2), we thus focus only on the situation of zero magnetic field $H = 0$. (In fact, the presence of the external magnetic

---

[9] Do not confuse the magnetic field H in (3.11.6) with the Hurst parameter.

field complicates the matters considerably and the Ising model with a magnetic field $H > 0$ has not been solved explicitly yet. See, for example, McCoy [702], pp. 277–280.)

Note also that the energy $\mathcal{E}$ in (3.11.2) involves the variables σ_{j,L_h+1} and $\sigma_{L_v+1,k}$ "outside" the lattice. What exactly are these variables? This is related to the so-called *boundary conditions*. The case

$$\sigma_{j,L_h+1} = \sigma_{j,-L_h+1}, \quad \sigma_{L_v+1,k} = \sigma_{-L_v+1,k} \tag{3.11.7}$$

is referred to as that of *periodic boundary* (toroidal or doughnut-shaped) conditions. The case

$$\sigma_{j,L_h+1} = 0, \quad \sigma_{L_v+1,k} = 0 \tag{3.11.8}$$

is known as that of *free boundary* conditions. It effectively corresponds to the situation where the terms involving σ_{j,L_h+1} and $\sigma_{L_v+1,k}$ are not present in (3.11.2).

Finally, one of the original applications of the Ising model is to the phenomenon of *ferromagnetism*. The variables $\sigma_{j,k}$ model a magnetic dipole of atoms of ferromagnetic material (e.g., iron). The variables $\sigma_{j,k}$ are referred to as *spins*. The critical temperature T_c is known as the Curie temperature. Below this temperature, ferromagnetic materials exhibit spontaneous magnetization.

We are interested here in the behavior of the correlation function of the spin at the origin and the spin at position (m, n), namely

$$C(m,n) = \text{Corr}(\sigma_{0,0}, \sigma_{m,n}) = \frac{\mathbb{E}\sigma_{0,0}\sigma_{m,n} - \mathbb{E}\sigma_{0,0}\mathbb{E}\sigma_{m,n}}{\sqrt{\mathbb{E}\sigma_{0,0}^2 - (\mathbb{E}\sigma_{0,0})^2}\sqrt{\mathbb{E}\sigma_{0,0}^2 - (\mathbb{E}\sigma_{0,0})^2}} = \mathbb{E}\sigma_{0,0}\sigma_{m,n},$$

since $\mathbb{E}\sigma_{j,k} = 0$ and $\mathbb{E}\sigma_{j,k}^2 = 1 = \sigma_{j,k}^2$. Thus correlations reduce to covariances. For the sake of simplicity and illustration, we shall only consider row and diagonal covariances

$$C(0,n) = \mathbb{E}\sigma_{0,0}\sigma_{0,n}, \quad C(n,n) = \mathbb{E}\sigma_{0,0}\sigma_{n,n}. \tag{3.11.9}$$

We shall also work in the case of free boundary conditions (3.11.8). Our goal is to show the following:

Theorem 3.11.1 *At the critical temperature $T = T_c$ characterized by*

$$1 = \sinh(2\beta E_h)\sinh(2\beta E_v) \tag{3.11.10}$$

and as $n \to \infty$,

$$\lim_{L_h, L_v \to \infty} \mathbb{E}\sigma_{0,0}\sigma_{0,n} \sim \left(\frac{1+\alpha_1}{1-\alpha_1}\right)^{1/4} An^{-1/4}, \tag{3.11.11}$$

$$\lim_{L_h, L_v \to \infty} \mathbb{E}\sigma_{0,0}\sigma_{n,n} \sim An^{-1/4}, \tag{3.11.12}$$

where $A = 0.6450024\ldots$ and

$$\alpha_1 = \frac{z_h(1-z_v)}{1+z_v}, \quad z_h = \tanh(\beta E_h), \quad z_v = \tanh(\beta E_v).$$

Sketch of proof The idea is to introduce the notion of bonds in the Ising lattice, then replace this lattice by a larger one, called the *counting lattice*, to which a number of matrices are associated. The covariances of interest involve these matrices, their inverse and related determinants. Szegö's limit theorem describes the asymptotic behavior of these determinants.

The asymptotic behavior of $\mathbb{E}\sigma_{0,0}\sigma_{n,n}$ in (3.11.12) is stated in McCoy and Wu [703], (4.43), p. 265, and proved. But the asymptotic behavior of $\mathbb{E}\sigma_{0,0}\sigma_{0,n}$ in (3.11.11), stated in McCoy and Wu [703], p. 267, is only conjectured (McCoy and Wu [703], p. 266).

Whereas, as expected, the asymptotic behavior of $\mathbb{E}\sigma_{0,0}\sigma_{n,n}$ is invariant under permutation of h and v, this is not the case for $\mathbb{E}\sigma_{0,0}\sigma_{0,n}$, for which

$$\frac{1+\alpha_1}{1-\alpha_1} = \frac{1+z_v+z_h-z_h z_v}{1+z_v-z_h+z_h z_v}. \qquad (3.11.13)$$

Observe that in the special case $E_h = E_v = E$, this ratio simplifies and one has

Corollary 3.11.2 *If $E_h = E_v$, then at the critical temperature $T = T_c$, one has as $n \to \infty$,*

$$\lim_{L_h, L_v \to \infty} \mathbb{E}\sigma_{0,0}\sigma_{0,n} \sim 2^{1/8} A n^{-1/4}. \qquad (3.11.14)$$

Since, at the critical temperature T_c, the covariances behave like $n^{-1/4}$ for large n, the Ising model exhibits LRD in the sense (2.1.5) in condition II. By identifying the exponent $2d - 1$ with $-1/4$ (see (2.1.5)) and using $H = d + 1/2$ (see (2.8.4)), one gets that the Ising model exhibits LRD with

$$d = \frac{3}{8} \quad \text{and} \quad H = \frac{7}{8}. \qquad (3.11.15)$$

Theorem 3.11.1 and Corollary 3.11.2 are proved in Section 3.11.5.

3.11.2 Correlations, Dimers and Pfaffians

We shall express the correlations $C(0, n)$ and $C(n, n)$ in terms of the determinants of $n \times n$ matrices in the so-called thermodynamic limit $L_v, L_h \to \infty$ first, and then analyze them as $n \to \infty$.

Observe that

$$\mathbb{E}\sigma_{0,0}\sigma_{0,n} = \frac{1}{Z_{L_v,L_h}} \sum_{\{\sigma_{j,k}\}} \sigma_{0,0}\sigma_{0,n} e^{-\beta\mathcal{E}}. \qquad (3.11.16)$$

We shall analyze the two terms Z_{L_v,L_h} and $\sum_{\{\sigma_{j,k}\}} \sigma_{0,0}\sigma_{0,n} e^{-\beta\mathcal{E}}$ separately. The focus will be on the partition function Z_{L_v,L_h}, and similar arguments will be outlined for the second term. As for the diagonal correlation $\mathbb{E}\sigma_{0,0}\sigma_{n,n}$, we shall give its corresponding expression without proof.

3.11 Two-Dimensional Ising Model

Partition function Z_{L_v,L_h}

Observe that

$$Z_{L_v,L_h} = \sum_{\{\sigma_{j,k}\}} e^{-\beta \mathcal{E}} = \sum_{\{\sigma_{j,k}\}} e^{\beta \sum_{j=-L_v+1}^{L_v} \sum_{k=-L_h+1}^{L_h} (E_h \sigma_{j,k}\sigma_{j,k+1} + E_v \sigma_{j,k}\sigma_{j+1,k})}$$

$$= \sum_{\{\sigma_{j,k}\}} \prod_{j=-L_v+1}^{L_v} \prod_{k=-L_h+1}^{L_h} e^{\beta E_h \sigma_{j,k}\sigma_{j,k+1}} \prod_{j=-L_v+1}^{L_v} \prod_{k=-L_h+1}^{L_h} e^{\beta E_v \sigma_{j,k}\sigma_{j+1,k}}$$

$$= \sum_{\{\sigma_{j,k}\}} \prod_{j=-L_v+1}^{L_v} \prod_{k=-L_h+1}^{L_h} \left(\cosh \beta E_h + \sigma_{j,k}\sigma_{j,k+1} \sinh \beta E_h \right)$$

$$\times \prod_{j=-L_v+1}^{L_v} \prod_{k=-L_h+1}^{L_h} \left(\cosh \beta E_v + \sigma_{j,k}\sigma_{j+1,k} \sinh \beta E_v \right), \qquad (3.11.17)$$

since $\cosh a = \frac{1}{2}(e^a + e^{-a})$, $\sinh a = \frac{1}{2}(e^a - e^{-a})$ and thus

$$e^{ax} = \cosh a + x \sinh a, \quad x = \pm 1.$$

Under the free boundary conditions, we get further that

$$Z_{L_v,L_h} = (\cosh \beta E_h)^{2L_v(2L_h-1)} (\cosh \beta E_v)^{(2L_v-1)2L_h} \widetilde{Z}_{L_v,L_h}, \qquad (3.11.18)$$

where

$$\widetilde{Z}_{L_v,L_h} = \sum_{\{\sigma_{j,k}\}} \prod_{j=-L_v+1}^{L_v} \prod_{k=-L_h+1}^{L_h} \left(1 + \sigma_{j,k}\sigma_{j,k+1} z_h \right)$$

$$\times \prod_{j=-L_v+1}^{L_v} \prod_{k=-L_h+1}^{L_h} \left(1 + \sigma_{j,k}\sigma_{j+1,k} z_v \right) =: \sum_{\{\sigma_{j,k}\}} \widetilde{Z}_{L_v,L_h}(\{\sigma_{j,k}\}) \qquad (3.11.19)$$

and

$$z_h = \frac{\sinh \beta E_h}{\cosh \beta E_h} = \tanh \beta E_h, \quad z_v = \tanh \beta E_v. \qquad (3.11.20)$$

Consider now $\widetilde{Z}_{L_v,L_h}(\{\sigma_{j,k}\})$ given by (3.11.19). Expand all the products to express $\widetilde{Z}_{L_v,L_h}(\{\sigma_{j,k}\})$ as the sum over a number of terms as

$$\widetilde{Z}_{L_v,L_h}(\{\sigma_{j,k}\}) = 1 + \sigma_{-L_v+1,-L_h+1}\sigma_{-L_v+1,-L_h+2} z_h + \cdots + \sigma_{L_v-1,L_h}\sigma_{L_v,L_h} z_v$$

$$+ \cdots + \prod_{j=-L_v+1}^{L_v} \prod_{k=-L_h+1}^{L_h} \sigma_{j,k}\sigma_{j,k+1} z_h \prod_{j=-L_v+1}^{L_v} \prod_{k=-L_h+1}^{L_h} \sigma_{j,k}\sigma_{j+1,k} z_v. \qquad (3.11.21)$$

Each term in the sum (3.11.21) can be thought as containing a factor for every pair of nearest-neighbor sites (j,k) and (j',k') on the lattice. This factor is either 1, when 1 from $1 + \sigma_{j,k}\sigma_{j',k'}z$ is selected in the product, or $\sigma_{j,k}\sigma_{j',k'}z$, when $\sigma_{j,k}\sigma_{j',k'}z$ is selected from the product (z denotes z_h or z_v). For example, the first term in (3.11.21) occurs when each pair of nearest-neighbor sites contributes 1, and the last term in (3.11.21) occurs when each contributes $\sigma_{j,k}\sigma_{j',k'}z$. Each term in the sum (3.11.21) will therefore include either 1, $\sigma_{j,k}$,

$\sigma_{j,k}^2$, $\sigma_{j,k}^3$ or $\sigma_{j,k}^4$ and hence one of the five quantities can be associated with each site (j,k). Note now that the terms in the sum $\widetilde{Z}_{L_v,L_h}(\{\sigma_{j,k}\})$ with sites having $\sigma_{j,k}$ or $\sigma_{j,k}^3$ will vanish in the total sum $\sum_{\{\sigma_{j,k}\}}$ in (3.11.19) because

$$\sum_{\sigma_{j,k}=\pm 1} \sigma_{j,k} = \sum_{\sigma_{j,k}=\pm 1} \sigma_{j,k}^3 = 0.$$

Thus only

$$\text{the terms having } 1,\ \sigma_{j,k}^2 \equiv 1 \text{ or } \sigma_{j,k}^4 \equiv 1 \qquad (3.11.22)$$

will not vanish in the total sum $\sum_{\{\sigma_{j,k}\}}$.

The presence of a term $\sigma_{j,k}\sigma_{j,k+1}z_h$ in (3.11.21) can be interpreted as indicating the presence of a horizontal bond z_h connecting the sites (j,k) and $(j,k+1)$. A corresponding interpretation holds for $\sigma_{j,k}\sigma_{j+1,k}z_v$. From a different perspective, when 1, $\sigma_{j,k}^2$ or $\sigma_{j,k}^4$ is associated with the site (j,k), then zero, two or four bonds traverse the site (j,k). Therefore, the terms in the sum (3.11.21) where the conditions (3.11.22) hold can be regarded as depicting a figure on a lattice with the following properties:

(i) each bond between nearest neighbors may be used, at most, once;
(ii) an even number of bonds terminate at each site.

Figure 3.8 depicts one such figure. Observe that such figure would contribute a term

$$z_h^p z_v^q$$

to the total sum $\sum_{\{\sigma_{j,k}\}}$, where p is the number of horizontal bonds in the figure ($p = 18$ in Figure 3.8) and q is the number of vertical bonds in the figure ($q = 16$ in Figure 3.8). Statement (ii) above implies that p and q are even. Since there are

$$\sum_{\{\sigma_{j,k}\}} 1 = 2^{2L_v 2L_h}$$

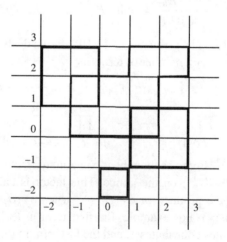

Figure 3.8 A figure (lines in bold) on the Ising lattice.

3.11 Two-Dimensional Ising Model

different configurations, one gets that

$$\tilde{Z}_{L_v,L_h} = 2^{2L_v 2L_h} \sum_{p,q} N_{p,q} z_h^p z_v^q =: 2^{2L_v 2L_h} g(z_h, z_v), \quad (3.11.23)$$

where $N_{p,q}$ is the number of figures on the lattice with the properties (i) and (ii) above, and where p, q are the numbers of horizontal and vertical bonds of the figure. By convention, $N_{0,0} = 1$. The next step is to compute the "generating function"

$$g(z_1, z_2) = \sum_{p,q} N_{p,q} z_1^p z_2^q \quad (3.11.24)$$

associated with counting of the figures as in Figure 3.8. Observe that $g(z_1, z_2) \geq 0$ since p and q are even.

Dimers

There is an ingenious way to turn the problem of computing (3.11.24) into the so-called "problem of counting closest-packed dimer coverings" on a suitable lattice, as explained in McCoy and Wu [703]. We will see later that the latter problem has a solution involving a matrix determinant which can then be studied more easily.

The first step is to replace the Ising lattice by a larger lattice, which we will call the *counting lattice*. More specifically, replace each site of the Ising lattice, as on the left side of Figure 3.9, by a six-site cluster depicted on the right side of Figure 3.9. Note that, as in the Ising lattice, the six-site cluster has four connecting lines. The difference is that one site is now replaced by six sites. The new, counting lattice has then $2K = 6(2L_v)(2L_h)$ sites and is formed by connecting six-site clusters, as depicted in Figure 3.10.

In the next step, the idea is to replace each of eight possible Ising site configuration bonds with a suitable configuration of bonds on the counting lattice. The replacement is done according to Figure 3.11. With this replacement, each figure counted in (3.11.24) is replaced by another figure in the counting lattice. In the Ising lattice, two adjacent sites may be either unconnected or connected by a bond. In the counting lattice, each site has exactly one bond with one of its neighbors indicated by a double line in Figure 3.11. Each

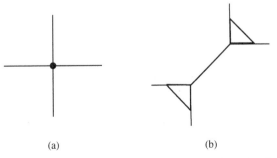

(a)   (b)

Figure 3.9 (a) A site of the Ising lattice. (b) A corresponding six-site cluster of the counting lattice.

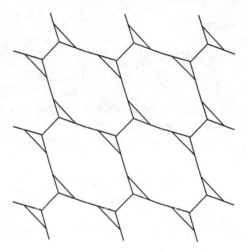

Figure 3.10 The counting lattice.

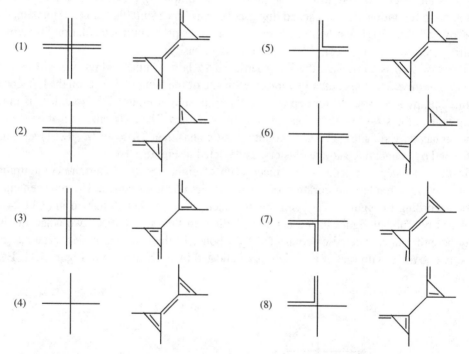

Figure 3.11 Replacement of eight possible Ising site configuration bonds by suitable configuration bonds on the counting lattice. The connecting bonds are indicated by *double* lines. Observe that a horizontal (resp. vertical) bond at an Ising site is associated with a horizontal (resp. vertical) bond connecting clusters.

such bond is called a *dimer*.[10] The figure one gets in the counting lattice is called a *closest-packed dimer configuration*.[11] Then, the problem of counting in (3.11.24) can be replaced by counting dimer configurations in the counting lattice.

To make this connection more precise, three classes of bonds need to be distinguished in the counting lattice: (1) horizontal bonds between clusters, (2) vertical bonds between clusters, and (3) bonds within a cluster. Let

$$G(z_1, z_2, z_3) = \sum_{p,q,r} N_{p,q,r} z_1^p z_2^q z_3^r \qquad (3.11.25)$$

be the corresponding generating function, where $N_{p,q,r}$ is the number of dimer configurations on the counting lattice with p bonds of type (1), q bonds of type (2) and r bonds of type (3). In particular, one can think that bonds of types (1), (2) and (3) carry weights of sizes z_1, z_2 and z_3, respectively. The generating functions (3.11.24) and (3.11.25) are then related as follows:

$$g(z_1, z_2) = G(z_1, z_2, 1), \qquad (3.11.26)$$

by assigning the weight $z_3 = 1$ to bonds within clusters. This is because a horizontal (resp. vertical) bond in the Ising lattice is associated with a horizontal (resp. vertical) bond between clusters (see Figure 3.11).

The advantage of the formulation (3.11.25) or (3.11.26) is that it can be related to the determinant of a matrix. We shall do this in two steps. In the *first step*, we shall express (3.11.26) as

$$g(z_1, z_2) = G(z_1, z_2, 1) = {\sum_p}' \overline{b}_{p_1, p_2} \overline{b}_{p_3, p_4} \cdots \overline{b}_{p_{2K-1}, p_{2K}}, \qquad (3.11.27)$$

where $\overline{B} = (\overline{b}_{pq})_{1 \leq p,q \leq 2K}$ is a suitable $2K \times 2K$ matrix (the bar does not refer here to the complex conjugate) and $\sum_p'$ is the sum over all permutations $p_1, p_2 \ldots, p_{2K}$ of $1, 2, \ldots, 2K$ satisfying

$$p_{2m-1} < p_{2m}, \ 1 \leq m < K, \quad p_{2m-1} < p_{2m+1}, \ 1 \leq m < K - 1. \qquad (3.11.28)$$

For example, if $2K = 4$, the sum is over the permutations 1234, 1324 and 1423 of 1234. Relation (3.11.28) is satisfied since these permutations are such that $p_1 < p_2$, $p_1 < p_3$ and $p_3 < p_4$, where, in the permutation 1324, for example, $p_1 = 1$, $p_2 = 3$, $p_3 = 2$, $p_4 = 4$. In the *second step*, the expression (3.11.27) will be written as the so-called *Pfaffian* of a matrix. As stated below, the Pfaffian of a matrix is the square root of its determinant. We shall now detail these two steps.

In the *first step*, the expression (3.11.27) and the permutations (3.11.28) above are naturally related to closest-packed dimer configurations on *any* lattice. Suppose a lattice (that is, *any* lattice) consists of $2K$ sites and enumerate these sites by $1, 2, \ldots, 2K$. Observe that there is a one-to-one correspondence between closest-packed dimer configurations on the lattice and permutations satisfying (3.11.28): the permutation

$$p_1 p_2 \mid p_3 p_4 \mid \cdots \mid p_{2K-1} p_{2K}$$

---

[10] In Chemistry, a dimer is a structure formed by two sub-units.
[11] One uses this term whenever each site has exactly one bond with one of its neighbors.

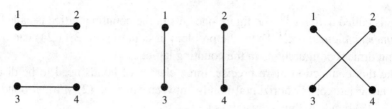

Figure 3.12 All possible closest-packed dimer configurations in a lattice of four sites.

is associated with the closest-packed dimer configuration where dimers connect the sites p_1 and p_2, p_3 and p_4, ..., p_{2K-1} and p_{2K}. Here, note that a dimer can connect any two sites. For example, with a lattice of $2K = 4$ sites, there are three closest-packed dimer configurations associated with

$$12\,|\,34, \quad 13\,|\,24 \quad \text{and} \quad 14\,|\,23, \tag{3.11.29}$$

as illustrated in Figure 3.12. Suppose, in addition, that a dimer connecting sites p and q carries a weight $\bar{b}_{p,q}$. Then, we define the weight of the closest-packed dimer configuration associated with the permutation $p_1 p_2 \,|\, p_3 p_4 \,|\, \ldots \,|\, p_{2K-1} p_{2K}$ as

$$\bar{b}_{p_1,p_2}\bar{b}_{p_3,p_4}\ldots\bar{b}_{p_{2K-1},p_{2K}}.$$

Therefore,

$$\sideset{}{'}\sum_{p} \bar{b}_{p_1,p_2}\bar{b}_{p_3,p_4}\ldots\bar{b}_{p_{2K-1},p_{2K}}$$

is exactly the sum of the weights of all closest-packed dimer configurations.

With the latter fact in mind, we now turn to the relation (3.11.27). The function $G(z_1, z_2, 1)$ can then be written as

$$G(z_1, z_2, 1) = \sideset{}{'}\sum_{p} \bar{b}_{p_1,p_2}\bar{b}_{p_3,p_4}\ldots\bar{b}_{p_{2K-1},p_{2K}},$$

where

$$2K = 6(2L_v)(2L_h)$$

is the number of sites in the counting lattice, the sum $\sum'_p$ can be viewed as over all closest-packed dimer configurations, and the weights $\bar{b}_{p,q}$ are such that

$$\bar{b}_{pq} = \begin{cases} 1, & \text{if } p \text{ and } q \text{ are neighboring sites within the same six-site cluster,} \\ z_1, & \text{if sites } p \text{ and } q \text{ connect two six-site clusters in the vertical direction,} \\ z_2, & \text{if sites } p \text{ and } q \text{ connect two six-site clusters in the horizontal direction,} \\ 0, & \text{otherwise.} \end{cases}$$

(3.11.30)

Figure 3.13 illustrates this. The weight 0 in $\bar{b}_{pq}$ ensures that only closest-packed dimer configurations with dimers connecting neighboring sites count for the function $G(z_1, z_2, 1)$. This means, in particular, that most of the elements $\bar{b}_{pq}$ are actually zero.

A matrix $\bar{B} = (\bar{b}_{pq})$ satisfying (3.11.30) can be constructed as follows. Consider first 6×6 matrices $\bar{B}(j, k; j', k')$ labeled by the Ising lattice sites (j, k) and (j', k'), where $-L_v + 1 \leq$

3.11 Two-Dimensional Ising Model   191

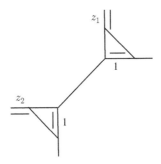

Figure 3.13 The connecting bonds between sites are indicated by double lines. The weight b_{pq} connecting two sites p and q is indiacted. It is 1, z_1 or z_2 according to whether p and q are within the same six-site cluster, connect two six-site clusters in the vertical direction or connect two six-site clusters in the horizontal direction. All other bonds have weight 0.

$j, j' \leq L_v, -L_h + 1 \leq k, k' \leq L_h$. They are zero matrices except in the following cases, where they are defined as:

$$\overline{B}(j,k; j,k) = \begin{array}{c} \\ R \\ L \\ U \\ D \\ 1 \\ 2 \end{array} \begin{array}{c} R \quad L \quad U \quad D \quad 1 \quad 2 \\ \left(\begin{array}{cccccc} 0 & 0 & 1 & 0 & 0 & 1 \\ 0 & 0 & 0 & 1 & 1 & 0 \\ 1 & 0 & 0 & 0 & 0 & 1 \\ 0 & 1 & 0 & 0 & 1 & 0 \\ 0 & 1 & 0 & 1 & 0 & 1 \\ 1 & 0 & 1 & 0 & 1 & 0 \end{array}\right) \end{array},$$

for $-L_v + 1 \leq j \leq L_v$, $-L_h + 1 \leq k \leq L_h$,

$$\overline{B}(j,k; j,k+1) = \overline{B}(j,k+1; j,k)' = \begin{array}{c} \\ R \\ L \\ U \\ D \\ 1 \\ 2 \end{array} \begin{array}{c} R \quad L \quad U \quad D \quad 1 \quad 2 \\ \left(\begin{array}{cccccc} 0 & z_1 & 0 & 0 & 0 & 0 \\ 0 & 0 & 0 & 0 & 0 & 0 \\ 0 & 0 & 0 & 0 & 0 & 0 \\ 0 & 0 & 0 & 0 & 0 & 0 \\ 0 & 0 & 0 & 0 & 0 & 0 \\ 0 & 0 & 0 & 0 & 0 & 0 \end{array}\right) \end{array},$$

for $-L_v + 1 \leq j \leq L_v$, $-L_h + 1 \leq k \leq L_h - 1$,

$$\overline{B}(j,k; j+1,k) = \overline{B}(j+1,k; j,k)' = \begin{array}{c} \\ R \\ L \\ U \\ D \\ 1 \\ 2 \end{array} \begin{array}{c} R \quad L \quad U \quad D \quad 1 \quad 2 \\ \left(\begin{array}{cccccc} 0 & 0 & 0 & 0 & 0 & 0 \\ 0 & 0 & 0 & 0 & 0 & 0 \\ 0 & 0 & 0 & z_2 & 0 & 0 \\ 0 & 0 & 0 & 0 & 0 & 0 \\ 0 & 0 & 0 & 0 & 0 & 0 \\ 0 & 0 & 0 & 0 & 0 & 0 \end{array}\right) \end{array},$$

for $-L_v + 1 \leq j \leq L_v - 1$, $-L_h + 1 \leq k \leq L_h$. Each site (j, k) of the Ising lattice corresponds to a six-site cluster in the counting lattice. Think now of $\overline{B}(j, k; j', k')$ as containing weights $\overline{b}_{pq}$ between sites in six-site clusters denoted (j, k) and (j', k') (within the same

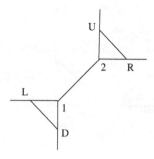

Figure 3.14 Labels for the six sites of a cluster on the counting lattice.

cluster if $(j, k) = (j', k')$). The labels $R, L, U, D, 1$ and 2 correspond to the six sites in a cluster, depicted in Figure 3.14. (R stands for "Right", L for "Left", U for "Up" and D for "Down".) Note that these matrices are exactly such that their elements $\overline{b}_{pq}$ satisfy (3.11.30). For example, within a cluster, a site R can only connect to U or 2 with the weight of 1.

Finally, the matrix $\overline{B} = (\overline{b}_{pq})_{1 \leq p,q \leq 2K}$ is made of the 6×6 blocks or submatrices $\overline{B}(j, k; j', k')$, $-L_v + 1 \leq j, j' \leq L_v$, $-L_h + 1 \leq k, k' \leq L_h$, we just defined. The exact placement of the blocks inside of the matrix $\overline{B}$ is as follows. Order the $d = (2L_v)(2L_h)$ sites (j, k) along the rows for increasing columns k; that is,

$$(-L_v + 1, -L_h + 1), \ldots, (-L_v + 1, L_h),$$
$$(-L_v + 2, -L_h + 1), \ldots, (-L_v + 2, L_h),$$
$$\ldots$$
$$(L_v, -L_h + 1), \ldots, (L_v, L_h)$$

and renumber the sites $1 \leq m \leq d$. This renumbering associates then to each pair (j, k), (j', k') a pair (m_1, m_2), $1 \leq m_1, m_2 \leq d$. In the matrix $\overline{B}$, its (m_1, m_2) block of size 6 is then defined as $\overline{B}(j, k; j', k')$.

Pfaffians

We now turn to the *second step* described in the discussion around (3.11.27), namely, to express (3.11.27) as the Pfaffian of a matrix. We first define the Pfaffian.[12]

Definition 3.11.3 Consider a $2K \times 2K$ antisymmetric real-valued matrix $A = (a_{jk})_{1 \leq j,k \leq 2K}$, that is, with elements $a_{jk} = -a_{kj}$ and $a_{jj} = 0$. Its Pfaffian is defined as

$$\text{Pf} A = {\sum_p}' \delta_p a_{p_1 p_2} a_{p_3 p_4} \cdots a_{p_{2K-1} p_{2K}}, \quad (3.11.31)$$

where, as in (3.11.28), $\sum'_p$ is the sum over all permutations $p_1, p_2 \ldots, p_{2K}$ of $1, 2, \ldots, 2K$ satisfying

$$p_{2m-1} < p_{2m}, \ 1 \leq m < K, \quad p_{2m-1} < p_{2m+1}, \ 1 \leq m < K - 1 \quad (3.11.32)$$

and where δ_p, the parity of the permutation p, is 1 if the permutation p is made up of an even number of transpositions and -1 if p is made of an odd number of transpositions.

---

[12] After a German mathematician J. F. Pfaff (1765–1825), who was a formal research supervisor of C. F. Gauss.

3.11 Two-Dimensional Ising Model

For example, if $2K = 4$, and

$$A = \begin{pmatrix} 0 & a_{12} & a_{13} & a_{14} \\ -a_{12} & 0 & a_{23} & a_{24} \\ -a_{13} & -a_{23} & 0 & a_{34} \\ -a_{14} & -a_{24} & -a_{34} & 0 \end{pmatrix},$$

then by (3.11.29),

$$\text{Pf}A = a_{12}a_{34} - a_{13}a_{24} + a_{14}a_{23}.$$

This is because 1234, 1324, 1423 involve respectively 0, 1 and 2 transpositions.

The usefulness of the Pfaffian comes from the following formula: for an antisymmetric matrix A,

$$(\text{Pf}A)^2 = \det(A) \tag{3.11.33}$$

(see, for example, pp. 47–51 in McCoy and Wu [703] for a proof). For example, if $2K = 2$ and

$$A = \begin{pmatrix} 0 & a_{12} \\ -a_{12} & 0 \end{pmatrix},$$

then one sees immediately that $(\text{Pf}A)^2 = \det(A)$ because $\text{Pf}A = a_{12}$ and $\det(A) = a_{12}^2$.

The relation (3.11.27) is not exactly the Pfaffian of the matrix $\overline{B}$ in that it does not include the parity factor δ_p. In fact, the parity factor can be introduced by suitably altering the signs of the elements of the matrix $\overline{B}$. We shall describe but not prove the assignment for doing so. See, for example, pp. 51–67 in McCoy and Wu [703] for more details.

For the sum (3.11.27) to have the parity factor δ_p, the blocks of the matrix $\overline{B}$ have to be replaced by the blocks:

$$\overline{A}(j,k; j,k) = \begin{array}{c} \\ R \\ L \\ U \\ D \\ 1 \\ 2 \end{array} \begin{pmatrix} R & L & U & D & 1 & 2 \\ 0 & 0 & -1 & 0 & 0 & 1 \\ 0 & 0 & 0 & -1 & 1 & 0 \\ 1 & 0 & 0 & 0 & 0 & -1 \\ 0 & 1 & 0 & 0 & -1 & 0 \\ 0 & -1 & 0 & 1 & 0 & 1 \\ -1 & 0 & 1 & 0 & -1 & 0 \end{pmatrix}, \tag{3.11.34}$$

for $-L_v + 1 \le j \le L_v$, $-L_h + 1 \le k \le L_h$,

$$\overline{A}(j,k; j,k+1) = -\overline{A}(j,k+1; j,k)' = \begin{array}{c} \\ R \\ L \\ U \\ D \\ 1 \\ 2 \end{array} \begin{pmatrix} R & L & U & D & 1 & 2 \\ 0 & z_1 & 0 & 0 & 0 & 0 \\ 0 & 0 & 0 & 0 & 0 & 0 \\ 0 & 0 & 0 & 0 & 0 & 0 \\ 0 & 0 & 0 & 0 & 0 & 0 \\ 0 & 0 & 0 & 0 & 0 & 0 \\ 0 & 0 & 0 & 0 & 0 & 0 \end{pmatrix}, \tag{3.11.35}$$

for $-L_v + 1 \leq j \leq L_v$, $-L_h + 1 \leq k \leq L_h - 1$,

$$\overline{A}(j,k; j+1,k) = -\overline{A}(j+1,k; j,k)' = \begin{pmatrix} & R & L & U & D & 1 & 2 \\ R & 0 & 0 & 0 & 0 & 0 & 0 \\ L & 0 & 0 & 0 & 0 & 0 & 0 \\ U & 0 & 0 & 0 & z_2 & 0 & 0 \\ D & 0 & 0 & 0 & 0 & 0 & 0 \\ 1 & 0 & 0 & 0 & 0 & 0 & 0 \\ 2 & 0 & 0 & 0 & 0 & 0 & 0 \end{pmatrix}, \quad (3.11.36)$$

for $-L_v + 1 \leq j \leq L_v - 1$, $-L_h + 1 \leq k \leq L_h$. Denote the corresponding matrix by $\overline{A}$ and its elements by $\overline{a}_{p,q}$. (The $2K \times 2K$ matrix $\overline{A}$ is defined in the same way as $\overline{B}$ but using blocks $\overline{A}(j,k; j',k')$ instead of $\overline{B}(j,k; j',k')$.) Note that the only difference between the elements of $\overline{A}$ and $\overline{B}$ is that some of the entries have different signs. It is convenient to think of the sign assignment graphically as depicted in Figure 3.15. Each bond between sites of the counting lattice now not only carries a weight (z_1, z_2, 1 or 0) but also a direction. The sign of the elements of the matrix blocks $\overline{A}(j,k; j',k')$ now corresponds to this direction. For example, the sign is positive if the connecting bond is in the forward direction, as indicated by the arrows in Figure 3.16. Thus, in the $\overline{A}(j,k; j,k)$ block, the (U, R) entry is 1 because it corresponds to the arrow $U \to R$ in Figure 3.16, but the $(U, 2)$ entry is -1 because $U \to 2$ is not in the direction of the arrow. Thus, one can show (McCoy and Wu [703], pp. 81, 51–58) that

$$g(z_1, z_2) = \sum_p{}' \delta_p \overline{a}_{p_1 p_2} \overline{a}_{p_3 p_4} \ldots \overline{a}_{p_{6d-1} p_{6d}} = \text{Pf}\,\overline{A} = (\det(\overline{A}))^{1/2}, \quad (3.11.37)$$

where the sum is as in (3.11.27), δ_p is the parity factor and $\text{Pf}\,\overline{A}$ is the Pfaffian in Definition 3.11.3. The sign at the square root in (3.11.37) is positive since $g(z_1, z_2)$ is defined in (3.11.24) with even exponents.

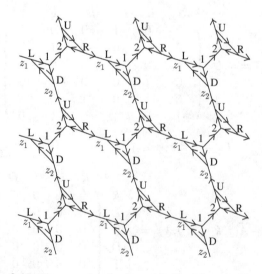

Figure 3.15 The counting lattice with directions between sites.

3.11 Two-Dimensional Ising Model

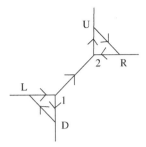

Figure 3.16 Directions in the six-site cluster of the counting lattice.

Let $c_R, c_L, c_U, c_D, c_1, c_2$ denote the columns of any 6×6 block $\overline{A}(j, k; j', k')$, and let us perform the following operations:

$$c_R - c_1, \quad c_U + c_1, \quad c_L + c_2, \quad c_D - c_2.$$

After these operations, the block $\overline{A}(j, k; j, k)$ becomes

$$\begin{array}{c} \\ R \\ L \\ U \\ D \\ 1 \\ 2 \end{array} \begin{array}{c} \begin{array}{cccccc} R & L & U & D & 1 & 2 \end{array} \\ \left(\begin{array}{cccccc} 0 & 1 & -1 & -1 & 0 & 1 \\ -1 & 0 & 1 & -1 & 1 & 0 \\ 1 & -1 & 0 & 1 & 0 & -1 \\ 1 & 1 & -1 & 0 & -1 & 0 \\ 0 & 0 & 0 & 0 & 0 & 1 \\ 0 & 0 & 0 & 0 & -1 & 0 \end{array} \right) \end{array} \qquad (3.11.38)$$

and the other blocks $\overline{A}(j, k; j', k')$ remain the same. Note that the last two rows of the block (3.11.38) are zero except the 2×2 submatrix

$$\begin{pmatrix} 0 & 1 \\ -1 & 0 \end{pmatrix} \qquad (3.11.39)$$

in the bottom-right corner. Since the determinant of the submatrix (3.11.39) equals 1, the rows and columns of the matrix $\overline{A}$ corresponding to the submatrix can be eliminated without affecting the determinant of $\overline{A}$. We thus have

$$\det(\overline{A}) = \det(A), \qquad (3.11.40)$$

where

$$A(j, k; j, k) = \begin{array}{c} \\ R \\ L \\ U \\ D \end{array} \begin{array}{c} \begin{array}{cccc} R & L & U & D \end{array} \\ \left(\begin{array}{cccc} 0 & 1 & -1 & -1 \\ -1 & 0 & 1 & -1 \\ 1 & -1 & 0 & 1 \\ 1 & 1 & -1 & 0 \end{array} \right) \end{array} \qquad (3.11.41)$$

and all other $A(j, k; j', k')$ are identical to $\overline{A}(j, k; j', k')$ with the rows and columns labeled 1 and 2 removed. Thus, we also have

$$g(z_1, z_2) = (\det(A))^{1/2}, \qquad (3.11.42)$$

where A is now a $2K \times 2K$ matrix with

$$2K = 4(2L_v)(2L_h). \qquad (3.11.43)$$

Combining (3.11.18), (3.11.23) and (3.11.42), we obtain that

$$Z_{L_v, L_h} = (\cosh \beta E_h)^{2L_v(2L_h-1)} (\cosh \beta E_v)^{(2L_v-1)2L_h} 2^{2L_v 2L_h} (\det(A))^{1/2}, \qquad (3.11.44)$$

where z_1, z_2 in A are replaced by z_h and z_v.

The right-hand side of (3.11.44) provides an expression for the denominator (partition function) of (3.11.16). One can derive similarly an expression for the numerator of (3.11.16). Proceeding as in (3.11.17) and (3.11.18), we have

$$\sum_{\{\sigma_{j,k}\}} \sigma_{0,0} \sigma_{0,n} e^{-\beta \mathcal{E}} = (\cosh \beta E_h)^{2L_v(2L_h-1)} (\cosh \beta E_v)^{(2L_v-1)2L_h}$$

$$\times \sum_{\{\sigma_{j,k}\}} \sigma_{0,0} \sigma_{0,n} \prod_{j=-L_v+1}^{L_v} \prod_{k=-L_h+1}^{L_h} \left(1 + \sigma_{j,k} \sigma_{j,k+1} z_h\right)$$

$$\times \prod_{j=-L_v+1}^{L_v} \prod_{k=-L_h+1}^{L_h} \left(1 + \sigma_{j,k} \sigma_{j+1,k} z_v\right). \qquad (3.11.45)$$

Since $\sigma_{i,j}^2 = 1$, we can write

$$\sigma_{0,0} \sigma_{0,n} = (\sigma_{0,0} \sigma_{0,1})(\sigma_{0,1} \sigma_{0,2}) \ldots (\sigma_{0,n-1} \sigma_{0,n})$$

and

$$\sigma_{0,k} \sigma_{0,k+1} (1 + \sigma_{0,k} \sigma_{0,k+1} z_h) = z_h (1 + \sigma_{0,k} \sigma_{0,k+1} z_h^{-1}),$$

so that

$$\sigma_{0,0} \sigma_{0,n} \prod_{k=0}^{n-1} (1 + \sigma_{0,k} \sigma_{0,k+1} z_h) = z_h^n \prod_{k=0}^{n-1} (1 + \sigma_{0,k} \sigma_{0,k+1} z_h^{-1}).$$

Then the relation (3.11.45) can be expressed as

$$\sum_{\{\sigma_{j,k}\}} \sigma_{0,0} \sigma_{0,n} e^{-\beta \mathcal{E}} = (\cosh \beta E_h)^{2L_v(2L_h-1)} (\cosh \beta E_v)^{(2L_v-1)2L_h} z_h^n$$

$$\times \sum_{\{\sigma_{j,k}\}} \prod_{k=0}^{n-1}(1 + \sigma_{0,k}\sigma_{0,k+1}z_h^{-1}) \prod_{j=-L_v+1}^{L_v} \prod_{k=-L_h+1}^{L_h}{}' \left(1 + \sigma_{j,k}\sigma_{j,k+1}z_h\right)$$

$$\times \prod_{j=-L_v+1}^{L_v} \prod_{k=-L_h+1}^{L_h} \left(1 + \sigma_{j,k}\sigma_{j+1,k}z_v\right), \qquad (3.11.46)$$

where $\prod_{k=-L_h+1}^{L_h}{}'$ means that the terms with $j = 0$, $k = 0, \ldots, n-1$, are omitted. The last expression is similar to (3.11.18) and (3.11.19) except for the factor z_h^n and for the fact that the weights z_h of the bonds on the horizontal line connecting $(0,0)$ and $(0,n)$ are now replaced by z_h^{-1}. This is also reflected in the fact that the term $\prod_{k=0}^{n-1}(1+\sigma_{0,k}\sigma_{0,k+1}z_h^{-1})$ now replaces the corresponding term $\prod_{k=0}^{n-1}(1 + \sigma_{0,k}\sigma_{0,k+1}z_h)$ in (3.11.19). One can then argue in the same way as for Z_{L_v,L_h}. Since Z_{L_v,L_h} is expressible as (3.11.44), by combining the expressions for the numerator and denominator of (3.11.16), we obtain that

$$\mathbb{E}\sigma_{0,0}\sigma_{0,n} = \frac{1}{Z_{L_v,L_h}} \sum_{\{\sigma_{j,k}\}} \sigma_{0,0}\sigma_{0,n} e^{-\beta \mathcal{E}} = z_h^n \left(\frac{\det(A^0)}{\det(A)}\right)^{1/2}, \qquad (3.11.47)$$

where

$$A^0 = A + \delta$$

or equivalently $\delta = A^0 - A$ with

$$\delta(0,k;0,k+1) = -\delta(0,k+1;0,k)' = \begin{pmatrix} & R & L & U & D \\ R & 0 & z_h^{-1} - z_h & 0 & 0 \\ L & 0 & 0 & 0 & 0 \\ U & 0 & 0 & 0 & 0 \\ D & 0 & 0 & 0 & 0 \end{pmatrix} \qquad (3.11.48)$$

if $0 \le k \le n-1$ and zero otherwise. Observe that the presence of δ affects only the "horizontal" edges $(0,k) \to (0,k+1)$ and $(0,k+1) \to (0,k)$ of the Ising lattice.

The matrix δ, which has dimensions $2K \times 2K$ with $2K = 4(2L_v)(2L_h)$, is zero everywhere except on its $2n$ columns and $2n$ rows. Let y be that $2n \times 2n$ submatrix of δ where it does not vanish. We have from (3.11.48) that

$$y = \begin{pmatrix} y_{RR} & y_{RL} \\ y_{LR} & y_{LL} \end{pmatrix}, \qquad (3.11.49)$$

where

$$y_{RR} = \begin{matrix} & & 00 & 01 & \cdots & 0n-1 \\ & & R & R & \cdots & R \\ 00 & R \\ 01 & R \\ \vdots & \vdots \\ 0n-1 & R \end{matrix} \begin{pmatrix} 0 & 0 & \cdots & 0 \\ 0 & 0 & \cdots & 0 \\ \vdots & \vdots & \ddots & \vdots \\ 0 & 0 & \cdots & 0 \end{pmatrix},$$

$$y_{LR} = \begin{matrix} & & 00 & 01 & \cdots & 0n-1 \\ & & R & R & \cdots & R \\ 01 & L \\ 02 & L \\ \vdots & \vdots \\ 0n & L \end{matrix} \begin{pmatrix} -(z_h^{-1} - z_h) & 0 & \cdots & 0 \\ 0 & -(z_h^{-1} - z_h) & \cdots & 0 \\ \vdots & \vdots & \ddots & \vdots \\ 0 & 0 & \cdots & -(z_h^{-1} - z_h) \end{pmatrix},$$

$$y_{RL} = \begin{matrix} & & 01 & 02 & \cdots & 0n \\ & & L & L & \cdots & L \\ 00 & R \\ 01 & R \\ \vdots & \vdots \\ 0n-1 & R \end{matrix} \begin{pmatrix} z_h^{-1} - z_h & 0 & \cdots & 0 \\ 0 & z_h^{-1} - z_h & \cdots & 0 \\ \vdots & \vdots & \ddots & \vdots \\ 0 & 0 & \cdots & z_h^{-1} - z_h \end{pmatrix},$$

$$y_{LL} = \begin{pmatrix} & & 01 & 02 & \ldots & 0n \\ & & L & L & \ldots & L \\ 01 & L & 0 & 0 & \ldots & 0 \\ 02 & L & 0 & 0 & \ldots & 0 \\ \vdots & \vdots & \vdots & \vdots & \ddots & \vdots \\ 0n & L & 0 & 0 & \ldots & 0 \end{pmatrix}.$$

In the expressions above, following δ in (3.11.48), $00, 01, \ldots, 0n-1, 01, \ldots, 0n$ refer to the Ising lattice sites with nonzero weights, and R, L refer to the site in the cluster of the counting lattice.

Let also Q be the $2n \times 2n$ submatrix of A^{-1} in this same subspace as y. Then, in view of (3.11.47),

$$(\mathbb{E}\sigma_{0,0}\sigma_{0,n})^2 = z_h^{2n} \frac{\det(A+\delta)}{\det(A)} = z_h^{2n} \det(A^{-1})\det(A+\delta)$$
$$= z_h^{2n}\det(I + A^{-1}\delta) = z_h^{2n}\det(I + Qy) = z_h^{2n}\det(y)\det(y^{-1} + Q)$$
$$= z_h^{2n}(z_h^{-1} - z_h)^{2n}\det(y^{-1} + Q) = (1 - z_h^2)^{2n}\det(y^{-1} + Q), \quad (3.11.50)$$

so that

$$(\mathbb{E}\sigma_{0,0}\sigma_{0,n})^2 = \det\Big((1 - z_h^2)(y^{-1} + Q)\Big). \quad (3.11.51)$$

3.11.3 Computation of the Inverse

To evaluate $\det(y^{-1} + Q)$ in (3.11.51), we need an expression for the inverse A^{-1} of the $2K \times 2K$ matrix A and, more specifically, for the elements of the $2n \times 2n$ submatrix Q of A^{-1} in the same subspace as y. Then, we have

$$Q = \begin{pmatrix} Q_{RR} & Q_{RL} \\ Q_{LR} & Q_{LL} \end{pmatrix}, \quad (3.11.52)$$

where

$$Q_{RR} = \begin{pmatrix} 0 & \ldots & A^{-1}(0,0;0,n-1)_{RR} \\ A^{-1}(0,1;0,0)_{RR} & \ldots & A^{-1}(0,1;0,n-1)_{RR} \\ \vdots & \ddots & \vdots \\ A^{-1}(0,n-1;0,0)_{RR} & \ldots & 0 \end{pmatrix},$$

$$Q_{LR} = \begin{pmatrix} A^{-1}(0,1;0,0)_{LR} & \ldots & A^{-1}(0,1;0,n-1)_{LR} \\ A^{-1}(0,2;0,0)_{LR} & \ldots & A^{-1}(0,2;0,n-1)_{LR} \\ \vdots & \ddots & \vdots \\ A^{-1}(0,n;0,0)_{LR} & \ldots & A^{-1}(0,n;0,n-1)_{LR} \end{pmatrix},$$

$$Q_{RL} = \begin{pmatrix} A^{-1}(0,0;0,1)_{RL} & \ldots & A^{-1}(0,0;0,n)_{RL} \\ A^{-1}(0,1;0,1)_{RL} & \ldots & A^{-1}(0,1;0,n)_{RL} \\ \vdots & \ddots & \vdots \\ A^{-1}(0,n-1;0,1)_{RL} & \ldots & A^{-1}(0,n-1;0,n)_{RL} \end{pmatrix},$$

$$Q_{LL} = \begin{pmatrix} 0 & \cdots & A^{-1}(0,1;0,n)_{LL} \\ A^{-1}(0,2;0,1)_{LL} & \cdots & A^{-1}(0,2;0,n)_{LL} \\ \vdots & \ddots & \vdots \\ A^{-1}(0,n;0,1)_{LL} & \cdots & 0 \end{pmatrix},$$

and where zeroes are on the diagonal since A^{-1} and hence Q is antisymmetric. The entries $A^{-1}(\cdot,\cdot,\cdot,\cdot)..$ are place holders for the corresponding entries of the matrix A^{-1} and will need to be evaluated.

Recall that the matrix A has dimension $2K \times 2K$ where $2K = 4(2L_h)(2L_v)$ by (3.11.43). It is made up of blocks $A(j,k; j',k')$ given by (3.11.41) if $j' = j, k' = k$, and otherwise, by (3.11.35) and (3.11.36) with the rows and columns labeled 1 and 2 removed. Note that A can thus be written as

$$\begin{aligned}
A = & I_{2L_v} \otimes I_{2L_h} \otimes \begin{pmatrix} 0 & 1 & -1 & -1 \\ -1 & 0 & 1 & -1 \\ 1 & -1 & 0 & 1 \\ 1 & 1 & -1 & 0 \end{pmatrix} \\
& + I_{2L_v} \otimes K_{2L_h} \otimes \begin{pmatrix} 0 & z_h & 0 & 0 \\ 0 & 0 & 0 & 0 \\ 0 & 0 & 0 & 0 \\ 0 & 0 & 0 & 0 \end{pmatrix} + I_{2L_v} \otimes K'_{2L_h} \otimes \begin{pmatrix} 0 & 0 & 0 & 0 \\ -z_h & 0 & 0 & 0 \\ 0 & 0 & 0 & 0 \\ 0 & 0 & 0 & 0 \end{pmatrix} \\
& + K_{2L_v} \otimes I_{2L_h} \otimes \begin{pmatrix} 0 & 0 & 0 & 0 \\ 0 & 0 & 0 & 0 \\ 0 & 0 & 0 & z_v \\ 0 & 0 & 0 & 0 \end{pmatrix} + K'_{2L_v} \otimes I_{2L_h} \otimes \begin{pmatrix} 0 & 0 & 0 & 0 \\ 0 & 0 & 0 & 0 \\ 0 & 0 & 0 & 0 \\ 0 & 0 & -z_v & 0 \end{pmatrix},
\end{aligned} \quad (3.11.53)$$

where $\otimes$ indicates the Kronecker product and

$$K_p = \begin{pmatrix} 0 & 1 & 0 & \cdots & 0 \\ 0 & 0 & 1 & \cdots & 0 \\ \vdots & \vdots & \vdots & \ddots & \vdots \\ 0 & 0 & 0 & \cdots & 1 \\ 0 & 0 & 0 & \cdots & 0 \end{pmatrix} \quad (3.11.54)$$

is an $p \times p$ matrix.[13] The nonzero blocks $A(j,k; j,k)$, $A(j,k; j,k+1)$, $A(j,k+1; j,k)$, $A(j,k; j+1,k)$ and $A(j+1,k; j,k)$ in the matrix A are captured by the respective 5 terms in the sum (3.11.53).

---

[13] Recall that the Kronecker product of an $p_1 \times q_1$ matrix U by a $p_2 \times q_2$ matrix V is the $p_1 p_2 \times q_1 q_2$ matrix

$$U \otimes V = \begin{pmatrix} u_{11} V & \cdots & u_{1q_1} V \\ \vdots & \ddots & \vdots \\ u_{p_1 1} V & \cdots & u_{p_1 q_1} V \end{pmatrix}.$$

It is, in general, not commutative.

The matrices K_p are given by (3.11.54). For computational purpose, it is convenient to replace them in (3.11.53) by the $p \times p$ matrices

$$H_p = \begin{pmatrix} 0 & 1 & 0 & \cdots & 0 \\ 0 & 0 & 1 & \cdots & 0 \\ \vdots & \vdots & \vdots & \ddots & \vdots \\ 0 & 0 & 0 & \cdots & 1 \\ 1 & 0 & 0 & \cdots & 0 \end{pmatrix}, \qquad (3.11.55)$$

which are circulant (Section 2.11). Thus, consider the following matrix

$$\widetilde{A} = I_{2L_v} \otimes I_{2L_h} \otimes \begin{pmatrix} 0 & 1 & -1 & -1 \\ -1 & 0 & 1 & -1 \\ 1 & -1 & 0 & 1 \\ 1 & 1 & -1 & 0 \end{pmatrix}$$

$$+ I_{2L_v} \otimes H_{2L_h} \otimes \begin{pmatrix} 0 & z_h & 0 & 0 \\ 0 & 0 & 0 & 0 \\ 0 & 0 & 0 & 0 \\ 0 & 0 & 0 & 0 \end{pmatrix} + I_{2L_v} \otimes H'_{2L_h} \otimes \begin{pmatrix} 0 & 0 & 0 & 0 \\ -z_h & 0 & 0 & 0 \\ 0 & 0 & 0 & 0 \\ 0 & 0 & 0 & 0 \end{pmatrix}$$

$$+ H_{2L_v} \otimes I_{2L_h} \otimes \begin{pmatrix} 0 & 0 & 0 & 0 \\ 0 & 0 & 0 & 0 \\ 0 & 0 & 0 & z_v \\ 0 & 0 & 0 & 0 \end{pmatrix} + H'_{2L_v} \otimes I_{2L_h} \otimes \begin{pmatrix} 0 & 0 & 0 & 0 \\ 0 & 0 & 0 & 0 \\ 0 & 0 & 0 & 0 \\ 0 & 0 & -z_v & 0 \end{pmatrix}.$$

(3.11.56)

The matrix $\widetilde{A}$ differs from A only at the rows and columns corresponding to the boundary sites (j, k); that is, when $j = -L_h + 1, L_h$ or $k = -L_v + 1, L_v$.

We will consider below the so-called thermodynamic limit of $L_h, L_v \to \infty$. Since the difference between A and $\widetilde{A}$ is only at the columns and rows corresponding to the (j, k) which are at boundary sites, in the thermodynamic limit, the elements of A^{-1} away from these columns and rows will be those of $\widetilde{A}^{-1}$. Since the elements in (3.11.52) are away from the boundary, we will thus suppose without loss of generality that A is actually given by (3.11.56).

The advantage of working with A given by (3.11.56) is that it now has a circulant structure which makes it easier to compute its inverse.

Lemma 3.11.4 *Let*

$$\alpha_1 = \frac{z_h(1 - z_v)}{1 + z_v}, \quad \alpha_2 = \frac{z_h^{-1}(1 - z_v)}{1 + z_v}, \qquad (3.11.57)$$

$$a_m = \frac{1}{2\pi} \int_0^{2\pi} d\theta \, e^{-im\theta} \phi(\theta), \qquad (3.11.58)$$

$$\phi(\theta) = \left(\frac{(1 - \alpha_1 e^{i\theta})(1 - \alpha_2 e^{-i\theta})}{(1 - \alpha_1 e^{-i\theta})(1 - \alpha_2 e^{i\theta})} \right)^{1/2} \qquad (3.11.59)$$

3.11 Two-Dimensional Ising Model

and δ_{ij} denote the Kronecker delta. Then,

$$\lim_{L_h, L_v \to \infty} A^{-1}(0, k; 0, k')_{RL} = \frac{1}{1 - z_h^2} (z_h \delta_{k'-k-1, 0} - a_{k'-k-1}), \qquad (3.11.60)$$

$$\lim_{L_h, L_v \to \infty} A^{-1}(0, k; 0, k')_{LR} = \frac{1}{1 - z_h^2} (-z_h \delta_{k-k'-1, 0} + a_{k-k'-1}), \qquad (3.11.61)$$

$$\lim_{L_h, L_v \to \infty} A^{-1}(0, k; 0, k')_{RR} = \lim_{L_h, L_v \to \infty} A^{-1}(0, k; 0, k')_{LL} = 0. \qquad (3.11.62)$$

Remark 3.11.5 For $a > 0$, the factor

$$g_a(e^{i\theta}) = \left(\frac{1 - ae^{\pm i\theta}}{1 - ae^{\mp i\theta}}\right)^{1/2} \qquad (3.11.63)$$

appearing twice in (3.11.59) is interpreted as

$$g_a(e^{i\theta}) = \left(\frac{1 - ae^{\pm i\theta}}{1 - ae^{\mp i\theta}}\right)^{1/2} \left(\frac{1 - ae^{\pm i\theta}}{1 - ae^{\pm i\theta}}\right)^{1/2} = \frac{1 - ae^{\pm i\theta}}{((1 - ae^{i\theta})(1 - ae^{-i\theta}))^{1/2}} = \frac{1 - ae^{\pm i\theta}}{|1 - ae^{i\theta}|}. \qquad (3.11.64)$$

If $a \neq 1$, the function $\frac{1-aw^{\pm 1}}{1-aw^{\mp 1}}$ is analytic and has nonzero real part on the unit disk $|w| \leq 1$. In this case, $g(e^{i\theta})$ can also be interpreted as

$$g_a(e^{i\theta}) = e^{\frac{1}{2}\mathrm{Log}\left(\frac{1-ae^{\pm i\theta}}{1-ae^{\mp i\theta}}\right)} = e^{\frac{1}{2}\mathrm{Log}(1-ae^{\pm i\theta})} e^{-\frac{1}{2}\mathrm{Log}(1-ae^{\mp i\theta})}, \qquad (3.11.65)$$

where Log is the principal branch of the logarithm. If $a = 1$, in view of the interpretation (3.11.64),

$$g_1(e^{i\theta}) = \frac{1 - e^{\pm i\theta}}{|1 - e^{i\theta}|} = \frac{1 - \cos\theta \pm i \sin\theta}{2 \sin\frac{\theta}{2}} = \sin\frac{\theta}{2} \pm i \cos\frac{\theta}{2}$$

$$= \pm i\left(\cos\frac{\theta}{2} \mp i \sin\frac{\theta}{2}\right) = \pm i e^{\mp i\theta/2}. \qquad (3.11.66)$$

We now turn to the proof of Lemma 3.11.4.

Proof The matrix H_p in (3.11.55) is circulant and by Proposition 2.11.1, it can be factorized using a discrete Fourier basis,

$$f_q = \frac{1}{\sqrt{p}}\left(1, e^{-i\frac{2\pi q}{p}}, e^{-i\frac{2\pi 2q}{p}}, \ldots, e^{-i\frac{2\pi(p-1)q}{p}}\right)',$$

$q = 0, \ldots, p - 1$. In this basis, by using (2.11.5) in Proposition 2.11.1, the eigenvalues of H_p are $e^{-i\frac{2\pi q}{p}}, q = 0, \ldots, p - 1$. We can thus write

$$H_p = F_p \begin{pmatrix} 1 & 0 & \cdots & 0 \\ 0 & e^{-i\frac{2\pi}{p}} & \cdots & 0 \\ \vdots & \vdots & \ddots & \vdots \\ 0 & 0 & \cdots & e^{-i\frac{2\pi(p-1)}{p}} \end{pmatrix} F_p^*, \qquad (3.11.67)$$

where $F_p = (f_0 \ldots f_{p-1})$ and F_p^* denotes the Hermitian transpose. Similarly, since H_p is real-valued,[14]

$$H_p' = F_p \begin{pmatrix} 1 & 0 & \cdots & 0 \\ 0 & e^{i\frac{2\pi}{p}} & \cdots & 0 \\ \vdots & \vdots & \ddots & \vdots \\ 0 & 0 & \cdots & e^{i\frac{2\pi(p-1)}{p}} \end{pmatrix} F_p^*. \qquad (3.11.68)$$

The matrix F_p, moreover, is unitary; that is,

$$F_p F_p^* = I_p. \qquad (3.11.69)$$

Consequently, if we denote by D_p the diagonal matrix in (3.11.67), we have $F_p^* H_p F_p = F_p^* F_p D_p F_p^* F_p = D_p$ and similarly for (3.11.68). Now, in view of (3.11.67), (3.11.68) and (3.11.69), by multiplying the left sides of the equality (3.11.56)[15] by $F_{2L_v}^* \otimes F_{2L_h}^* \otimes I_4$ and the right sides of (3.11.56) by $F_{2L_v} \otimes F_{2L_h} \otimes I_4$, we obtain[16] that

$$(F_{2L_v}^* \otimes F_{2L_h}^* \otimes I_4) A (F_{2L_v} \otimes F_{2L_h} \otimes I_4) = I_{2L_v} \otimes I_{2L_h} \otimes \begin{pmatrix} 0 & 1 & -1 & -1 \\ -1 & 0 & 1 & -1 \\ 1 & -1 & 0 & 1 \\ 1 & 1 & -1 & 0 \end{pmatrix}$$

$$+ I_{2L_v} \otimes \begin{pmatrix} 1 & 0 & \cdots & 0 \\ 0 & e^{-i\frac{2\pi}{2L_h}} & \cdots & 0 \\ \vdots & \vdots & \ddots & \vdots \\ 0 & 0 & \cdots & e^{-i\frac{2\pi(2L_h-1)}{2L_h}} \end{pmatrix} \otimes \begin{pmatrix} 0 & z_h & 0 & 0 \\ 0 & 0 & 0 & 0 \\ 0 & 0 & 0 & 0 \\ 0 & 0 & 0 & 0 \end{pmatrix}$$

$$+ I_{2L_v} \otimes \begin{pmatrix} 1 & 0 & \cdots & 0 \\ 0 & e^{i\frac{2\pi}{2L_h}} & \cdots & 0 \\ \vdots & \vdots & \ddots & \vdots \\ 0 & 0 & \cdots & e^{i\frac{2\pi(2L_h-1)}{2L_h}} \end{pmatrix} \otimes \begin{pmatrix} 0 & 0 & 0 & 0 \\ -z_h & 0 & 0 & 0 \\ 0 & 0 & 0 & 0 \\ 0 & 0 & 0 & 0 \end{pmatrix}$$

$$+ \begin{pmatrix} 1 & 0 & \cdots & 0 \\ 0 & e^{-i\frac{2\pi}{2L_v}} & \cdots & 0 \\ \vdots & \vdots & \ddots & \vdots \\ 0 & 0 & \cdots & e^{-i\frac{2\pi(2L_v-1)}{2L_v}} \end{pmatrix} \otimes I_{2L_h} \otimes \begin{pmatrix} 0 & 0 & 0 & 0 \\ 0 & 0 & 0 & 0 \\ 0 & 0 & 0 & z_v \\ 0 & 0 & 0 & 0 \end{pmatrix}$$

---

[14] Recall that $F^* = \overline{F}'$ where bar denotes "conjugate". Since H is real-valued, one has $H' = H^* = (FDF^*)^* = F^{**}D^*F^* = F\overline{D}F^*$.

[15] Recall that $\widetilde{A}$ in (3.11.56) is our A as noted before the statement of the lemma.

[16] If U_1 and U_2 are $m \times m$ and V_1 and V_2 are $p \times p$ matrices, then $(U_1 \otimes V_1)(U_2 \otimes V_2) = (U_1 U_2) \otimes (V_1 V_2)$. In particular, $(F_{2L_h}^* \otimes F_{2L_v}^*)(I_{2L_h} \otimes I_{2L_v})(F_{2L_h} \otimes F_{2L_v}) = I_{2L_h} \otimes I_{2L_v} = I_{(2L_h)(2L_v)}$ by (3.11.69).

3.11 Two-Dimensional Ising Model

$$+ \begin{pmatrix} 1 & 0 & \cdots & 0 \\ 0 & e^{i\frac{2\pi}{2L_v}} & \cdots & 0 \\ \vdots & \vdots & \ddots & \vdots \\ 0 & 0 & \cdots & e^{i\frac{2\pi(2L_v-1)}{2L_v}} \end{pmatrix} \otimes I_{2L_h} \otimes \begin{pmatrix} 0 & 0 & 0 & 0 \\ 0 & 0 & 0 & 0 \\ 0 & 0 & 0 & 0 \\ 0 & 0 & -z_v & 0 \end{pmatrix}.$$

This can be written as

$$(F_{2L_v}^* \otimes F_{2L_h}^* \otimes I_4) A (F_{2L_v} \otimes F_{2L_h} \otimes I_4)$$

$$= \mathrm{diag}\Big(A(\theta_{h,t}, \theta_{v,s}),\ \theta_{h,t} = \frac{2\pi t}{2L_h},\ t = 0, \ldots, 2L_h - 1,\ \theta_{v,s} = \frac{2\pi s}{2L_v},\ s = 0, \ldots, 2L_v - 1\Big), \tag{3.11.70}$$

where[17]

$$A(\theta_h, \theta_v) = \begin{array}{c} \\ R \\ L \\ U \\ D \end{array} \begin{pmatrix} \overset{R}{0} & \overset{L}{1 + z_h e^{-i\theta_h}} & \overset{U}{-1} & \overset{D}{-1} \\ -1 - z_h e^{i\theta_h} & 0 & 1 & -1 \\ 1 & -1 & 0 & 1 + z_v e^{-i\theta_v} \\ 1 & 1 & -1 - z_v e^{i\theta_v} & 0 \end{pmatrix} \tag{3.11.71}$$

and diag in (3.11.70) refers to placing the blocks $A(\theta_{h,t}, \theta_{v,s})$ on the diagonal. (The exact placement is $A(\theta_{h,0}, \theta_{v,0})$, $A(\theta_{h,1}, \theta_{v,0})$, ..., $A(\theta_{h,2L_h-1}, \theta_{v,0})$, $A(\theta_{h,0}, \theta_{v,1})$, $A(\theta_{h,1}, \theta_{v,1})$, ..., $A(\theta_{h,2L_h-1}, \theta_{v,1})$, and so on.)

It follows from (3.11.70) that

$$(F_{2L_v}^* \otimes F_{2L_h}^* \otimes I_4) A^{-1} (F_{2L_v} \otimes F_{2L_h} \otimes I_4) = \mathrm{diag}\Big(A^{-1}(\theta_{h,t}, \theta_{v,s})\Big)$$

and hence that

$$A^{-1} = (F_{2L_v} \otimes F_{2L_h} \otimes I_4) \mathrm{diag}\Big(A^{-1}(\theta_{h,t}, \theta_{v,s})\Big) (F_{2L_v}^* \otimes F_{2L_h}^* \otimes I_4).$$

For the blocks $A^{-1}(j, k; j', k')$ of the matrix A^{-1}, this yields[18]

$$A^{-1}(j, k; j', k') = \frac{1}{(2L_h)(2L_v)} \sum_{t=0}^{2L_h-1} \sum_{s=0}^{2L_v-1} e^{i\theta_{h,t}(k'-k) + i\theta_{v,s}(j'-j)} A^{-1}(\theta_{h,t}, \theta_{v,s}). \tag{3.11.72}$$

---

[17] As illustration, if $V = \begin{pmatrix} 0 & z \\ 0 & 0 \end{pmatrix}$, then

$$\begin{pmatrix} 1 & 0 & 0 \\ 0 & e^{i\theta_1} & 0 \\ 0 & 0 & e^{i\theta_2} \end{pmatrix} \otimes V = \begin{pmatrix} V & 0 & 0 \\ 0 & e^{i\theta_1} V & 0 \\ 0 & 0 & e^{i\theta_2} V \end{pmatrix} = \begin{pmatrix} 0 & z & & & & \\ 0 & 0 & & & & \\ & & 0 & e^{i\theta_1} z & & \\ & & 0 & 0 & & \\ & & & & 0 & e^{i\theta_2} z \\ & & & & 0 & 0 \end{pmatrix}.$$

[18] To show (3.11.72), one may suppose without loss of generality that $A^{-1}(j, k; j', k')$, $A^{-1}(\theta_{h,t}, \theta_{v,s})$ are 1×1 rather than 4×4 blocks and show that $A^{-1} = (F_{2L_v} \otimes F_{2L_h}) \mathrm{diag}\Big(A^{-1}(\theta_{h,t}, \theta_{v,s})\Big)(F_{2L_v}^* \otimes F_{2L_h}^*)$ implies (3.11.72). The latter fact is left as an exercise.

In the thermodynamic limit $L_h, L_v \to \infty$, the relation (3.11.72) becomes

$$\lim_{L_h,L_v\to\infty} A^{-1}(j,k;j',k') = \frac{1}{(2\pi)^2}\int_0^{2\pi} d\theta_h \int_0^{2\pi} d\theta_v \, e^{i(k'-k)\theta_h + i(j'-j)\theta_v} A^{-1}(\theta_h,\theta_v). \quad (3.11.73)$$

We are interested in the case $j = j' = 0$,

$$\lim_{L_h,L_v\to\infty} A^{-1}(0,k;0,k') = \frac{1}{(2\pi)^2}\int_0^{2\pi} d\theta_h \, e^{i(k'-k)\theta_h} \int_0^{2\pi} d\theta_v \, A^{-1}(\theta_h,\theta_v) \quad (3.11.74)$$

and, more specifically, in the entries at the columns and rows labeled R, L (see (3.11.60), (3.11.61) and (3.11.62)).

The inverse of $A(\theta_h, \theta_v)$ in (3.11.71) can be computed (McCoy [702], p. 358) or verified directly as

$$A^{-1}(\theta_h, \theta_v) = \frac{1}{\Delta(\theta_h, \theta_v)}$$

$$\times \begin{array}{c} \\ R \\ L \\ U \\ D \end{array} \begin{pmatrix} \begin{array}{cccc} R & L & U & D \end{array} \\ b - b^* & b + b^* - abb^* & 2 - ab^* & 2 - ab \\ -b^* - b + a^*b^*b & b^* - b & -2 + a^*b^* & 2 - a^*b \\ -2 + a^*b & 2 - ab & a^* - a & a + a^* - aa^*b \\ -2 + a^*b^* & -2 + ab^* & -a^* - a + a^*ab^* & a - a^* \end{array} \end{pmatrix},$$

(3.11.75)

where

$$a = 1 + z_h e^{-i\theta_h}, \quad b = 1 + z_v e^{-i\theta_v} \quad (3.11.76)$$

and

$$\Delta(\theta_h, \theta_v) = 4 + aa^*bb^* - (a + a^*)(b + b^*)$$
$$= (1 + z_h^2)(1 + z_v^2) - 2z_h(1 - z_v^2)\cos\theta_h - 2z_v(1 - z_h^2)\cos\theta_v. \quad (3.11.77)$$

We now want to compute $A^{-1}(0,k;0,k')_{ll'}$ at $l, l' = R, L$ in the thermodynamic limit (see (3.11.60), (3.11.61) and (3.11.62)). In view of (3.11.74), we need first to consider

$$B^{-1}(\theta_h)_{ll'} := \frac{1}{2\pi}\int_0^{2\pi} d\theta_v \, A^{-1}(\theta_h, \theta_v)_{ll'}. \quad (3.11.78)$$

The entries $A^{-1}(\theta_h, \theta_v)_{RR}$ and $A^{-1}(\theta_h, \theta_v)_{LL}$ involve $b - b^*$. Note that

$$\int_0^{2\pi} d\theta_v \frac{b - b^*}{\Delta(\theta_h, \theta_v)} = \int_0^{2\pi} d\theta_v \frac{z_h e^{-i\theta_v}}{(1 + z_h^2)(1 + z_v^2) - 2z_h(1 - z_v^2)\cos\theta_h - 2z_v(1 - z_h^2)\cos\theta_v}$$

$$- \int_0^{2\pi} d\theta_v \frac{z_h e^{i\theta_v}}{(1 + z_h^2)(1 + z_v^2) - 2z_h(1 - z_v^2)\cos\theta_h - 2z_v(1 - z_h^2)\cos\theta_v} = 0,$$

since the last two integrals are equal, by making the change of variables $\theta_v \to 2\pi - \theta_v$ in the second integral. It follows that

$$\lim_{L_h,L_v\to\infty} A^{-1}(0,k;0,k')_{RR} = \lim_{L_h,L_v\to\infty} A^{-1}(0,k;0,k')_{LL} = 0,$$

that is, the relations (3.11.62) hold.

We now turn to $A^{-1}(0,k;0,k')_{RL}$ and $A^{-1}(0,k;0,k')_{LR}$ in the thermodynamic limit. We first want to factor the quantity $\Delta(\theta_h, \theta_v)$ in (3.11.77) as a function of $e^{i\theta_v}$. Write

$$\Delta(\theta_h, \theta_v) = (1+z_h^2)(1+z_v^2) - 2z_h(1-z_v^2)\cos\theta_h - z_v(1-z_h^2)(e^{i\theta_v}+e^{-i\theta_v})$$

$$= -z_v(1-z_h^2)e^{-i\theta_v}\left(e^{2i\theta_v} - \frac{(1+z_h^2)(1+z_v^2) - 2z_h(1-z_v^2)\cos\theta_h}{z_v(1-z_h^2)}e^{i\theta_v} + 1\right) \quad (3.11.79)$$

and consider the polynomial

$$f(x) = x^2 - \frac{(1+z_h^2)(1+z_v^2) - 2z_h(1-z_v^2)\cos\theta_h}{z_v(1-z_h^2)}x + 1. \quad (3.11.80)$$

The polynomial (3.11.80) has roots x_1 and x_2 satisfying

$$x_1 x_2 = 1, \quad x_1 + x_2 = \frac{(1+z_h^2)(1+z_v^2) - 2z_h(1-z_v^2)\cos\theta_h}{z_v(1-z_h^2)}. \quad (3.11.81)$$

One can verify that

$$(1+z_h^2)(1+z_v^2) - 2z_h(1-z_v^2)\cos\theta_h - 2z_v(1-z_h^2) = (1-z_v)^2(1-\alpha_2^{-1}e^{i\theta_h})(1-\alpha_2^{-1}e^{-i\theta_h}), \quad (3.11.82)$$

where α_2 is defined in (3.11.57). This shows that $x_1 + x_2 \geq 2$ in (3.11.81), and hence that the roots x_1 and x_2 are both positive and real. Since $x_1 x_2 = 1$ in (3.11.81), we can set $x_1 = \alpha$ and $x_2 = \alpha^{-1}$ with

$$x_1 = \alpha > 1 > x_2 = \alpha^{-1} > 0. \quad (3.11.83)$$

Thus (3.11.80) becomes $f(x) = x^2 - (\alpha + \alpha^{-1})x + 1$ so that $f(e^{i\theta_v}) = e^{2i\theta_v} - (\alpha + \alpha^{-1})e^{i\theta_v} + 1$. Hence, in view of (3.11.79), we have

$$\Delta(\theta_h, \theta_v) = (e^{i\theta_v} - \alpha)(\alpha^{-1}e^{-i\theta_v} - 1)z_v(1-z_h^2). \quad (3.11.84)$$

Solving for the roots $x_1 = \alpha$ and $x_2 = \alpha^{-1}$, we can write

$$\alpha^{\pm 1} = \frac{1}{2z_v(1-z_h^2)}\left((1+z_h^2)(1+z_v^2) - z_h(1-z_v^2)(e^{i\theta_h}+e^{-i\theta_h})\right.$$
$$\left.\pm \left(\left((1+z_h^2)(1+z_v^2) - z_h(1-z_v^2)(e^{i\theta_h}+e^{-i\theta_h})\right)^2 - (2z_v(1-z_h^2))^2\right)^{1/2}\right). \quad (3.11.85)$$

The term in the square root can be written as

$$\left((1+z_h^2)(1+z_v^2) - z_h(1-z_v^2)(e^{i\theta_h}+e^{-i\theta_h})\right)^2 - \left(2z_v(1-z_h^2)\right)^2$$
$$= \left((1+z_h^2)(1+z_v^2) - z_h(1-z_v^2)(e^{i\theta_h}+e^{-i\theta_h}) - 2z_v(1-z_h^2)\right)$$
$$\times \left((1+z_h^2)(1+z_v^2) - z_h(1-z_v^2)(e^{i\theta_h}+e^{-i\theta_h}) + 2z_v(1-z_h^2)\right).$$

The relation (3.11.82) provides a factorization for the first term in the last product. One can similarly verify that the second term in the product can be written as

$$(1+z_h^2)(1+z_v^2) - 2z_h(1-z_v^2)\cos\theta_h + 2z_v(1-z_h^2) = (1+z_v)^2(1-\alpha_1 e^{i\theta_h})(1-\alpha_1 e^{-i\theta_h}), \quad (3.11.86)$$

where α_1 is defined in (3.11.57). Combining (3.11.82) and (3.11.86), we obtain from (3.11.85) that (see also McCoy and Wu [703], Eqs. (3.2)–(3.4) on pp. 86–87)

$$\alpha^{\pm 1} = \frac{1}{2z_v(1-z_h^2)}\Big((1+z_h^2)(1+z_v^2) - z_h(1-z_v^2)(e^{i\theta_h} + e^{-i\theta_h})$$
$$\pm (1-z_v^2)\big((1-\alpha_1 e^{i\theta_h})(1-\alpha_1 e^{-i\theta_h})(1-\alpha_2^{-1}e^{i\theta_h})(1-\alpha_2^{-1}e^{-i\theta_h})\big)^{1/2}\Big). \tag{3.11.87}$$

Consider now, for example, the entry RL in (3.11.75). It is

$$b + b^* - abb^* = 1 - z_v^2 - z_h e^{-i\theta_h}(1 + z_v^2 + z_v(e^{i\theta_v} + e^{-i\theta_v})). \tag{3.11.88}$$

By using (3.11.84) and (3.11.88), it follows that

$$B^{-1}(\theta_h)_{RL} = \frac{1}{2\pi}\int_0^{2\pi} d\theta_v \, A^{-1}(\theta_h, \theta_v)_{RL}$$
$$= \frac{1}{2\pi}\int_0^{2\pi} d\theta_v \, \frac{e^{i\theta_v}(1 - z_v^2 - z_h e^{-i\theta_h}(1 + z_v^2 + z_v(e^{i\theta_v} + e^{-i\theta_v})))}{(e^{i\theta_v} - \alpha)(\alpha^{-1} - e^{i\theta_v})z_v(1 - z_h^2)}$$
$$= -\frac{1}{2\pi i}\int_{|w|=1} dw \, \frac{1 - z_v^2 - z_h e^{-i\theta_h}(1 + z_v^2 + z_v(w + w^{-1}))}{(w - \alpha)(w - \alpha^{-1})z_v(1 - z_h^2)}$$
$$= -\frac{1}{2\pi i}\int_{|w|=1} dw \, \frac{1 - z_v^2 - z_h e^{-i\theta_h}(1 + z_v^2 + z_v w)}{(w - \alpha)(w - \alpha^{-1})z_v(1 - z_h^2)}$$
$$+ \frac{1}{2\pi i}\int_{|w|=1} dw \, \frac{z_h z_v e^{-i\theta_h}}{w(w - \alpha)(w - \alpha^{-1})z_v(1 - z_h^2)}.$$

Recall from (3.11.83) that $\alpha > 1 > \alpha^{-1} > 0$. Then, writing the last integral as the sum of two integrals by Cauchy's theorem,

$$B^{-1}(\theta_h)_{RL} = -\frac{1}{2\pi i}\int_{|w|=1} dw \, \frac{1 - z_v^2 - z_h e^{-i\theta_h}(1 + z_v^2 + z_v w)}{(w - \alpha)(w - \alpha^{-1})z_v(1 - z_h^2)}$$
$$+ \frac{1}{2\pi i}\int_{|w|=\epsilon} dw \, \frac{z_h z_v e^{-i\theta_h}}{w(w - \alpha)(w - \alpha^{-1})z_v(1 - z_h^2)}$$
$$+ \frac{1}{2\pi i}\int_{|w-\alpha^{-1}|=\epsilon} dw \, \frac{z_h z_v e^{-i\theta_h}}{w(w - \alpha)(w - \alpha^{-1})z_v(1 - z_h^2)},$$

for small enough $\epsilon > 0$ so that $\epsilon < \alpha^{-1} - \epsilon$. By using Cauchy's integral formula,[19] we conclude that

$$B^{-1}(\theta_h)_{RL} = -\frac{1 - z_v^2 - z_h e^{-i\theta_h}(1 + z_v^2 + z_v\alpha^{-1})}{(\alpha^{-1} - \alpha)z_v(1 - z_h^2)}.$$

---

[19] Cauchy's integral formula states that

$$f(a) = \frac{1}{2\pi i}\int_\gamma \frac{f(z)}{z - a}dz,$$

where $f: U \mapsto \mathbb{C}$ is analytic on an open subset U of $\mathbb{C}$ and γ is the boundary of a circle completely contained in U.

$$+ \frac{z_h z_v e^{-i\theta_h}}{(-\alpha)(-\alpha^{-1}) z_v (1-z_h^2)} + \frac{z_h z_v e^{-i\theta_h}}{\alpha^{-1}(\alpha^{-1}-\alpha) z_v (1-z_h^2)}$$

$$= \frac{z_h e^{-i\theta_h}}{1-z_h^2} + e^{-i\theta_h} \frac{(1-z_v^2) e^{i\theta_h} - z_h(1+z_v^2 + z_v(\alpha^{-1}+\alpha))}{z_v(1-z_h^2)(\alpha - \alpha^{-1})}$$

$$=: \frac{z_h e^{-i\theta_h}}{1-z_h^2} + e^{-i\theta_h} \widetilde{B}(\theta_h)_{RL}. \qquad (3.11.89)$$

Computing $\alpha + \alpha^{-1}$ from (3.11.87), we get that

$$1 + z_v^2 + z_v(\alpha^{-1} + \alpha) = \frac{2(1+z_v^2) - z_h(1-z_v^2)(e^{i\theta_h} + e^{-i\theta_h})}{1-z_h^2}$$

and

$$z_v(1-z_h^2)(\alpha - \alpha^{-1}) = (1-z_v^2)\Big((1-\alpha_1 e^{i\theta_h})(1-\alpha_1 e^{-i\theta_h})(1-\alpha_2^{-1} e^{i\theta_h})(1-\alpha_2^{-1} e^{-i\theta_h})\Big)^{1/2}.$$

Then, the term $\widetilde{B}(\theta_h)_{RL}$ in (3.11.89) becomes

$$\widetilde{B}(\theta_h)_{RL} = \frac{(1-z_v^2) e^{i\theta_h} + z_h^2(1-z_v^2) e^{-i\theta_h} - 2z_h(1+z_v^2)}{(1-z_h^2)(1-z_v^2)((1-\alpha_1 e^{i\theta_h})(1-\alpha_1 e^{-i\theta_h})(1-\alpha_2^{-1} e^{i\theta_h})(1-\alpha_2^{-1} e^{-i\theta_h}))^{1/2}}$$

$$= - \frac{z_h(1+z_v)^2 (1-\alpha_1 e^{-i\theta_h})(1-\alpha_2 e^{i\theta_h})}{(1-z_h^2)(1-z_v^2)((1-\alpha_1 e^{i\theta_h})(1-\alpha_1 e^{-i\theta_h})(1-\alpha_2^{-1} e^{i\theta_h})(1-\alpha_2^{-1} e^{-i\theta_h}))^{1/2}},$$

where the last equality can be verified directly using (3.11.57). Using the fact that $z_h(1+z_v)/(1-z_v) = \alpha_2^{-1}$ by (3.11.57), we can write

$$\widetilde{B}(\theta_h)_{RL} = - \frac{(1-\alpha_1 e^{-i\theta_h})(1-\alpha_2 e^{i\theta_h})}{(1-z_h^2)\alpha_2((1-\alpha_1 e^{i\theta_h})(1-\alpha_1 e^{-i\theta_h})(1-\alpha_2^{-1} e^{i\theta_h})(1-\alpha_2^{-1} e^{-i\theta_h}))^{1/2}}$$

$$= - \frac{(1-\alpha_1 e^{-i\theta_h})(1-\alpha_2 e^{i\theta_h})}{(1-z_h^2)((1-\alpha_1 e^{i\theta_h})(1-\alpha_1 e^{-i\theta_h})(1-\alpha_2 e^{i\theta_h})(1-\alpha_2 e^{-i\theta_h}))^{1/2}},$$

where we used the identity $\alpha_2^2(1-\alpha_2^{-1} e^{i\theta_h})(1-\alpha_2^{-1} e^{-i\theta_h}) = (1-\alpha_2 e^{i\theta_h})(1-\alpha_2 e^{-i\theta_h})$. With the interpretation in Remark 3.11.5, we can write the last expression as

$$\widetilde{B}(\theta_h)_{RL} = - \frac{1}{(1-z_h^2)} \Big(\frac{(1-\alpha_1 e^{-i\theta_h})(1-\alpha_2 e^{i\theta_h})}{(1-\alpha_1 e^{i\theta_h})(1-\alpha_2 e^{-i\theta_h})} \Big)^{1/2}. \qquad (3.11.90)$$

Combining (3.11.89) and (3.11.90), we obtain that

$$B^{-1}(\theta_h)_{RL} = \frac{1}{1-z_h^2} \Big(z_h e^{-i\theta_h} - e^{-i\theta_h} \Big(\frac{(1-\alpha_1 e^{-i\theta_h})(1-\alpha_2 e^{i\theta_h})}{(1-\alpha_1 e^{i\theta_h})(1-\alpha_2 e^{-i\theta_h})} \Big)^{1/2} \Big). \qquad (3.11.91)$$

In view of (3.11.74) and (3.11.78), this yields

$$\lim_{L_h, L_v \to \infty} A^{-1}(0, k; 0, k')_{RL} = \frac{z_h}{1-z_h^2} \frac{1}{2\pi} \int_0^{2\pi} e^{i(k'-k)\theta} e^{-i\theta} d\theta$$

$$-\frac{1}{(1-z_h^2)2\pi}\int_0^{2\pi} e^{i(k'-k-1)\theta}\left(\frac{(1-\alpha_1 e^{-i\theta})(1-\alpha_2 e^{i\theta})}{(1-\alpha_1 e^{i\theta})(1-\alpha_2 e^{-i\theta})}\right)^{1/2} d\theta$$

$$=\frac{z_h}{1-z_h^2}\delta_{k'-k-1,0} - \frac{1}{(1-z_h^2)2\pi}\int_0^{2\pi} e^{-i(k'-k-1)\theta}\phi(\theta)d\theta$$

$$=\frac{1}{1-z_h^2}(z_h\delta_{k'-k-1,0} - a_{k'-k-1}),$$

after defining a_n and $\phi(\theta)$ as in (3.11.58) and (3.11.59), and using $\int_0^{2\pi} e^{i\theta(k'-k-1)}d\theta = 2\pi\delta_{k'-k-1,0}$. Since the matrix $A^{-1}(\theta_h,\theta_v)$ in (3.11.75) satisfies $A^{-1}(\theta_h,\theta_v)^* = -A^{-1}(\theta_h,\theta_v)$, one has

$$B^{-1}(\theta_h)_{LR} = -B^{-1}(\theta_h)^*_{RL}$$

and one gets similarly

$$\lim_{L_h,L_v\to\infty} A^{-1}(0,k;0,k')_{LR} = \frac{1}{1-z_h^2}(-z_h\delta_{k-k'-1,0} + a_{k-k'-1}).$$

This establishes (3.11.60) and (3.11.61), and concludes the proof of the lemma. $\square$

We now want to substitute (3.11.60), (3.11.61) and (3.11.62) into the expression (3.11.51). Observe from (3.11.52) and (3.11.60)–(3.11.62) that the $2n \times 2n$ matrix Q becomes, in the thermodynamic limit,

$$\lim_{L_h,L_v\to\infty} Q = \begin{pmatrix} 0 & 0 & \cdots & 0 & \frac{z_h-a_0}{1-z_h^2} & \frac{-a_1}{1-z_h^2} & \cdots & \frac{-a_{n-1}}{1-z_h^2} \\ 0 & 0 & \cdots & 0 & \frac{-a_{-1}}{1-z_h^2} & \frac{z_h-a_0}{1-z_h^2} & \cdots & \frac{-a_{n-2}}{1-z_h^2} \\ \vdots & \vdots & \ddots & \vdots & \vdots & \vdots & \ddots & \vdots \\ 0 & 0 & \cdots & 0 & \frac{-a_{-n+1}}{1-z_h^2} & \frac{-a_{-n+2}}{1-z_h^2} & \cdots & \frac{z_h-a_0}{1-z_h^2} \\ \frac{-z_h+a_0}{1-z_h^2} & \frac{a_{-1}}{1-z_h^2} & \cdots & \frac{a_{-n+1}}{1-z_h^2} & 0 & 0 & \cdots & 0 \\ \frac{a_1}{1-z_h^2} & \frac{-z_h+a_0}{1-z_h^2} & \cdots & \frac{a_{-n+2}}{1-z_h^2} & 0 & 0 & \cdots & 0 \\ \vdots & \vdots & \ddots & \vdots & \vdots & \vdots & \ddots & \vdots \\ \frac{a_{n-1}}{1-z_h^2} & \frac{a_{n-2}}{1-z_h^2} & \cdots & \frac{-z_h+a_0}{1-z_h^2} & 0 & 0 & \cdots & 0 \end{pmatrix}.$$

Note from (3.11.49) that

$$y^{-1} = \begin{pmatrix} 0 & 0 & \cdots & 0 & \frac{-z_h}{1-z_h^2} & 0 & \cdots & 0 \\ 0 & 0 & \cdots & 0 & 0 & \frac{-z_h}{1-z_h^2} & \cdots & 0 \\ \vdots & \vdots & \ddots & \vdots & \vdots & \vdots & \ddots & \vdots \\ 0 & 0 & \cdots & 0 & 0 & 0 & \cdots & \frac{-z_h}{1-z_h^2} \\ \frac{z_h}{1-z_h^2} & 0 & \cdots & 0 & 0 & 0 & \cdots & 0 \\ 0 & \frac{z_h}{1-z_h^2} & \cdots & 0 & 0 & 0 & \cdots & 0 \\ \vdots & \vdots & \ddots & \vdots & \vdots & \vdots & \ddots & \vdots \\ 0 & 0 & \cdots & \frac{z_h}{1-z_h^2} & 0 & 0 & \cdots & 0 \end{pmatrix}.$$

Then,

$$(1-z_h^2)(y^{-1}+Q) = \begin{pmatrix} 0 & 0 & \cdots & 0 & -a_0 & -a_1 & \cdots & -a_{n-1} \\ 0 & 0 & \cdots & 0 & -a_{-1} & -a_0 & \cdots & -a_{n-2} \\ \vdots & \vdots & \ddots & \vdots & \vdots & \vdots & \ddots & \vdots \\ 0 & 0 & \cdots & 0 & -a_{-n+1} & -a_{-n+2} & \cdots & -a_0 \\ a_0 & a_{-1} & \cdots & a_{-n+1} & 0 & 0 & \cdots & 0 \\ a_1 & a_0 & \cdots & a_{-n+2} & 0 & 0 & \cdots & 0 \\ \vdots & \vdots & \ddots & \vdots & \vdots & \vdots & \ddots & \vdots \\ a_{n-1} & a_{n-2} & \cdots & a_0 & 0 & 0 & \cdots & 0 \end{pmatrix}$$

or

$$(1-z_h^2)(y^{-1}+Q) = \begin{pmatrix} 0 & -B' \\ B & 0 \end{pmatrix},$$

where $B = (a_{j-k})_{1 \le j,k \le n}$ and 0 is a $n \times n$ zero matrix. By using the formula $\det((A\ B; C\ D)) = \det(AD - CB)$ if $AB = BA$ for $n \times n$ blocks A, B, C and D, we deduce that

$$\det((1-z_h^2)(y^{-1}+Q)) = (\det(B))^2.$$

Using the expression (3.11.51) for $(\mathbb{E}\sigma_{0,0}\sigma_{0,n})^2$, this leads to the following expression which now involves the determinant of an $n \times n$ matrix:

$$S_n := \lim_{L_h, L_v \to \infty} \mathbb{E}\sigma_{0,0}\sigma_{0,n} = \det \begin{pmatrix} a_0 & a_{-1} & a_{-2} & \cdots & a_{-n+1} \\ a_1 & a_0 & a_{-1} & \cdots & a_{-n+2} \\ a_2 & a_1 & a_0 & \cdots & a_{-n+2} \\ \vdots & \vdots & \vdots & \ddots & \vdots \\ a_{n-1} & a_{n-2} & a_{n-3} & \cdots & a_0 \end{pmatrix}. \qquad (3.11.92)$$

One can similarly show (McCoy and Wu [703], p. 199, formula (3.31)) that

$$\widetilde{S}_n := \lim_{L_h, L_v \to \infty} \mathbb{E}\sigma_{0,0}\sigma_{n,n} = \det \begin{pmatrix} \widetilde{a}_0 & \widetilde{a}_{-1} & \widetilde{a}_{-2} & \cdots & \widetilde{a}_{-n+1} \\ \widetilde{a}_1 & \widetilde{a}_0 & \widetilde{a}_{-1} & \cdots & \widetilde{a}_{-n+2} \\ \widetilde{a}_2 & \widetilde{a}_1 & \widetilde{a}_0 & \cdots & \widetilde{a}_{-n+2} \\ \vdots & \vdots & \vdots & \ddots & \vdots \\ \widetilde{a}_{n-1} & \widetilde{a}_{n-2} & \widetilde{a}_{n-3} & \cdots & \widetilde{a}_0 \end{pmatrix}, \qquad (3.11.93)$$

where $\widetilde{a}_m$ is defined as a_m in (3.11.58) but using $\widetilde{\phi}$ instead of ϕ, defined by

$$\widetilde{\phi}(\theta) = \left(\frac{\sinh 2\beta E_h \sinh 2\beta E_v - e^{-i\theta}}{\sinh 2\beta E_h \sinh 2\beta E_v - e^{i\theta}} \right)^{1/2}. \qquad (3.11.94)$$

3.11.4 The Strong Szegö Limit Theorem

We are now interested in the behavior of the determinants S_n and $\widetilde{S}_n$ in (3.11.92) and (3.11.93), resp., as $n \to \infty$ at the critical temperature $T = T_c$, which is such that

$$1 = \sinh 2\beta E_h \sinh 2\beta E_v, \quad \beta = 1/T_c. \qquad (3.11.95)$$

The determinants S_n and $\widetilde{S}_n$ are defined in terms of functions $\phi(\theta)$ and $\widetilde{\phi}(\theta)$ in (3.11.59) and (3.11.94). We will need expressions of these functions at the critical temperature.

Lemma 3.11.6 *At the critical temperature T_c, $\phi(\theta)$ in (3.11.59) becomes*

$$\phi(\theta) = \left(\frac{1 - \alpha_1 e^{i\theta}}{1 - \alpha_1 e^{-i\theta}}\right)^{1/2} e^{i(-1/2)(\theta - \pi)} \tag{3.11.96}$$

and $\widetilde{\phi}(\theta)$ in (3.11.94) becomes

$$\widetilde{\phi}(\theta) = e^{i(-1/2)(\theta - \pi)}. \tag{3.11.97}$$

Proof Consider first $\widetilde{\phi}(\theta)$ in (3.11.94), which becomes by (3.11.95) and (3.11.66) in Remark 3.11.5,

$$\widetilde{\phi}(\theta) = \left(\frac{1 - e^{-i\theta}}{1 - e^{i\theta}}\right)^{1/2} = ie^{-i\theta/2} = e^{i(-1/2)(\theta - \pi)}. \tag{3.11.98}$$

Turning to the function $\phi(\theta)$ in (3.11.59), recall that the parameters α_1 and α_2 were defined in (3.11.57), and z_h and z_v in (3.11.20). We first show that if (3.11.95) holds, then $\alpha_2 = 1$. To do so, use the relations $(\cosh x)^2 - (\sinh x)^2 = 1$ and $2(\sinh x)(\cosh x) = \sinh 2x$. Then, by using $(1 - z_k^2)/(2z_k) = (\sinh 2\beta E_k)^{-1}$ for $k = h, v$, the relation (3.11.95) can be expressed as $(1 - z_h^2)(1 - z_v^2) = 4z_h z_v$. This is the same as $(1 - z_h z_v)^2 = (z_h + z_v)^2$ or $1 - z_h z_v = z_h + z_v$. The latter yields $1 = z_h^{-1}(1 - z_v)/(1 + z_v) = \alpha_2$. This now implies that

$$\alpha_1 = z_h^2 \alpha_2 = z_h^2 < 1. \tag{3.11.99}$$

Hence, by (3.11.98), $\phi(\theta)$ in (3.11.59) becomes (3.11.96). □

To obtain the asymptotic behavior of the determinants S_n and $\widetilde{S}_n$ as $n \to \infty$, we will apply the so-called *strong Szegö limit theorem*. This theorem concerns the limiting behavior, as $n \to \infty$, of the $n \times n$ Toeplitz determinant

$$D_n = \det \begin{pmatrix} c_0 & c_{-1} & \cdots & c_{-n+1} \\ c_1 & c_0 & \cdots & c_{-n+2} \\ \vdots & \vdots & \ddots & \vdots \\ c_{n-1} & c_{n-2} & \cdots & c_0 \end{pmatrix}, \tag{3.11.100}$$

where the entries c_m are the Fourier coefficients of a function C on the unit circle; that is,

$$c_m = \frac{1}{2\pi} \int_0^{2\pi} d\theta \, e^{-im\theta} C(e^{i\theta}).$$

Thus, $C(e^{i\theta}) = \sum_{m=-\infty}^{\infty} c_m e^{i\theta m}$ is the Fourier transform of the sequence $\{c_m\}$.[20] In the case of interest here, $C(e^{i\theta})$ is chosen to be either $\phi(\theta)$ or $\widetilde{\phi}(\theta)$.

The classical strong Szegö limit theorem supposes that $C(e^{i\theta})$ is "smooth" and allows one to conclude that

$$D_n \sim E(C) \mu^n, \tag{3.11.101}$$

---

[20] The notation $C(e^{i\theta})$ instead of $C(e^{-i\theta})$ found in Appendix A.1.1 is used to be consistent with the referenced literature.

3.11 Two-Dimensional Ising Model

as $N \to \infty$, where μ and $E(C)$ are two constants given by

$$\mu = \frac{1}{2\pi} \int_0^{2\pi} d\theta \, \log C(e^{i\theta}), \quad E(C) = \exp\left\{ \sum_{m=1}^{\infty} m (\log C)_m (\log C)_{-m} \right\},$$

and $(\log C)_m$ denotes the mth Fourier coefficient of the function $\log C(e^{i\theta})$. See Szegö [939]. A typical assumption for (3.11.101) to hold is that $C(\xi)$ be continuous on the unit circle $|\xi| = 1$. Note that this is not the case with the functions $\widetilde{\phi}(\theta)$ and $\phi(\theta)$ in (3.11.97) and (3.11.96). For example, note that $\widetilde{\phi}(0) = ie^{-i0} = i$ and $\widetilde{\phi}(2\pi) = ie^{-i\pi} = -i$. In the case when $C(\xi)$ is not continuous, the function $C(e^{i\theta})$ and the determinant D_N are said to have a *singularity*.

Though conjectured in the past by Fisher and Hartwig [365], the asymptotic behavior of determinants with singularities has been established only recently in Böttcher and Silbermann [176], Ehrhardt and Silbermann [347], Deift, Its, and Krasovsky [298]. We state next a corollary of Theorem 2.5 in Ehrhardt and Silbermann [347] which will be sufficient for our purposes.[21] Suppose the function $C(e^{i\theta})$ can be expressed as

$$C(e^{i\theta}) = b(e^{i\theta}) t_\beta(e^{i\theta}), \tag{3.11.102}$$

where

$$t_\beta(e^{i\theta}) = e^{i\beta(\theta - \pi)}, \quad \beta \in \mathbb{R}.$$

Note that the function $t_\beta(e^{i\theta})$ satisfies $t_\beta(e^{i0}) = e^{-i\beta\pi}$, $t_\beta(e^{i2\pi}) = e^{i\beta\pi}$, and hence is a discontinuous function on the unit circle (unless β is an integer). The function $b(e^{i\theta})$ will be assumed to be continuous on the unit circle. We will also need the Wiener-Hopf factorization of b, namely,

$$b(e^{i\theta}) = b_+(e^{i\theta}) g(b) b_-(e^{i\theta}),$$

where the factors are defined as

$$b_\pm(e^{i\theta}) = \exp\left\{ \sum_{m=1}^{\infty} e^{\pm i\theta m} (\log b)_{\pm m} \right\}, \quad g(b) = \exp((\log b)_0) \tag{3.11.103}$$

and where $(\log b)_m$ denotes the mth Fourier coefficient of the function $\log b$.

Theorem 3.11.7 (*Strong Szegö limit theorem for determinants with singularity.*) *Suppose that the function $C(e^{i\theta})$ can be expressed as (3.11.102) where $b(e^{i\theta})$ is infinitely differentiable on the unit circle. Then, the determinant D_n in (3.11.100) satisfies, as $n \to \infty$,*

$$D_n \sim E \, g(b)^n n^{-\beta^2}, \tag{3.11.104}$$

where $g(b)$ is defined in (3.11.103), and

$$E = E(b) b_+(1)^\beta b_-(1)^{-\beta} G(1+\beta) G(1-\beta). \tag{3.11.105}$$

---
[21] The result we use follows the statement of Theorem 2.5 in Ehrhardt and Silbermann [347].

In (3.11.105), $b_+(1)$, $b_-(1)$ are defined in (3.11.103),

$$E(b) = \exp\left(\sum_{m=1}^{\infty} m(\log b)_m (\log b)_{-m}\right)$$

and

$$G(1+z) = (2\pi)^{z/2} e^{-(z+1)z/2 - \gamma z^2/2} \prod_{k=1}^{\infty}\left\{\left(1+\frac{z}{k}\right)^k e^{-z+z^2/2k}\right\} \quad (3.11.106)$$

stands for the so-called Barnes G–function (γ is Euler's constant).

3.11.5 Long-Range Dependence at Critical Temperature

Applying Theorem 3.11.7 to the determinants S_n and $\widetilde{S}_n$ in (3.11.92) and (3.11.93) with the respective functions $\phi(\theta)$ and $\widetilde{\phi}(\theta)$ in (3.11.59) and (3.11.94) at the critical temperature yields the following result which concludes the proof of Theorem 3.11.1.

Theorem 3.11.8 *Let S_n and $\widetilde{S}_n$ be the determinants in (3.11.92) and (3.11.93). At the critical temperature (3.11.95), as $n \to \infty$,*

$$S_n = \lim_{L_h, L_v \to \infty} \mathbb{E}\sigma_{0,0}\sigma_{0,n} \sim \left(\frac{1+\alpha_1}{1-\alpha_1}\right)^{1/4} A n^{-1/4}, \quad (3.11.107)$$

$$\widetilde{S}_n = \lim_{L_h, L_v \to \infty} \mathbb{E}\sigma_{0,0}\sigma_{n,n} \sim A n^{-1/4}, \quad (3.11.108)$$

where α_1 is defined in (3.11.57),

$$A = G(1/2)G(3/2) = 0.6450024\ldots$$

and G is the Barnes G–function in (3.11.106).

Proof Consider first the determinant S_n with the function $\phi(\theta)$ in (3.11.96). We will apply Theorem 3.11.7. Note that $\phi(\theta)$ can be written as in (3.11.102),

$$\phi(\theta) = b(e^{i\theta}) t_\beta(e^{i\theta}),$$

where

$$b(e^{i\theta}) = \left(\frac{1-\alpha_1 e^{i\theta}}{1-\alpha_1 e^{-i\theta}}\right)^{1/2}, \quad t_\beta(e^{i\theta}) = e^{i\beta(\theta-\pi)}, \quad \beta = -1/2. \quad (3.11.109)$$

The factors (3.11.103) of b in its Wiener–Hopf factorization are

$$b_+(e^{i\theta}) = (1-\alpha_1 e^{i\theta})^{1/2}, \quad b_-(e^{i\theta}) = \frac{1}{(1-\alpha_1 e^{-i\theta})^{1/2}}, \quad g(b) = 1. \quad (3.11.110)$$

In particular,

$$b_+(1) = (1-\alpha_1)^{1/2}, \quad b_-(1) = 1/(1-\alpha_1)^{1/2}$$

and hence $b_+(1)^{-1/2}b_-(1)^{1/2} = (1-\alpha_1)^{-1/2}$. For $b(e^{i\theta})$ in (3.11.109), we have

$$\log b(e^{i\theta}) = \frac{1}{2}\log(1-\alpha_1 e^{i\theta}) - \frac{1}{2}\log(1-\alpha_1 e^{-i\theta}) = -\frac{1}{2}\sum_{k=1}^{\infty}\frac{\alpha_1^k e^{i\theta k}}{k} + \frac{1}{2}\sum_{k=1}^{\infty}\frac{\alpha_1^k e^{-i\theta k}}{k}$$

and hence the Fourier coefficients of the function $\log b$ are given by

$$(\log b)_m = \frac{1}{2\pi}\int_0^{2\pi} d\theta e^{-im\theta}\log b(e^{i\theta}) = \begin{cases} -\frac{\alpha_1^m}{2m}, & \text{if } m > 0, \\ 0, & \text{if } m = 0, \\ \frac{\alpha_1^{|m|}}{2|m|}, & \text{if } m < 0, \end{cases} \quad (3.11.111)$$

so that

$$E(b) = \exp\left(\sum_{m=1}^{\infty} m(\log b)_m(\log b)_{-m}\right) = \exp\left(-\frac{1}{4}\sum_{m=1}^{\infty}\frac{(\alpha_1^2)^m}{m}\right)$$

$$= \exp\left(\frac{1}{4}\log(1-\alpha_1^2)\right) = (1-\alpha_1^2)^{1/4}. \quad (3.11.112)$$

In view of (3.11.112), (3.11.105) and (3.11.110), we deduce (3.11.107). The result (3.11.108) for the determinant $\widetilde{S}_N$ can be obtained similarly. $\square$

Remark 3.11.9 The asymptotic behavior (3.11.108) of $\widetilde{S}_n$ was obtained in McCoy and Wu [703] directly without using the strong Szegö limit theorem as follows. In view of (3.11.97), the elements $\widetilde{a}_n$ of $\widetilde{S}_n$ can be evaluated as

$$\widetilde{a}_m = \frac{1}{2\pi}\int_0^{2\pi} d\theta\, e^{-im\theta} i e^{-i\theta/2} = \frac{2}{\pi(2m+1)}.$$

Hence, by (3.11.93),

$$\widetilde{S}_n = (2\pi^{-1})^n \det\left(\frac{1}{2m_1 - 2m_2 + 1}\right)_{0 \le m_1, m_2 \le n-1}. \quad (3.11.113)$$

The latter determinant is the special case of the so-called Cauchy determinant,

$$\widetilde{D}_n = \det\left(\frac{1}{\mu_{m_1} + \nu_{m_2}}\right)_{0 \le m_1, m_2 \le n-1}, \quad (3.11.114)$$

where $\mu_{m_1} = 2m_1 + 1$ and $\nu_{m_2} = -2m_2$. It is known (for example, McCoy and Wu [703], theorem on p. 261) that

$$\widetilde{D}_n = \frac{\prod_{0 \le m_1 < m_2 \le n-1}(\mu_{m_1} - \mu_{m_2})(\nu_{m_1} - \nu_{m_2})}{\prod_{m_1=0}^{n-1}\prod_{m_2=0}^{n-1}(\mu_{m_1} + \nu_{m_2})}. \quad (3.11.115)$$

Using this fact, (3.11.113) becomes

$$\widetilde{S}_n = (2\pi^{-1})^n \frac{\prod_{0 \le m_1 < m_2 \le n-1}(2(m_1-m_2))(2(m_2-m_1))}{\prod_{m_1=0}^{n-1}\prod_{m_2=0}^{n-1}(2(m_1-m_2)+1)}$$

$$= (2\pi^{-1})^n \prod_{0 \le m_1 < m_2 \le n-1}\frac{-4(m_1-m_2)^2}{1-4(m_1-m_2)^2}$$

$$= \left(\frac{2}{\pi}\right)^n \prod_{0 \le m_1 < m_2 \le n-1} \left(1 - \frac{1}{4(m_1 - m_2)^2}\right)^{-1}$$

$$= \begin{cases} \frac{2}{\pi}, & n = 1, \\ \left(\frac{2}{\pi}\right)^n \prod_{l=1}^{n-1} \left(1 - \frac{1}{4l^2}\right)^{l-n}, & n \ge 2. \end{cases}$$

Now use the identity (Gradshteyn and Ryzhik [421], p. 45, Formula 1.431.1)

$$\frac{\sin(\delta\pi)}{\delta\pi} = \prod_{l=1}^{\infty} \left(1 - \frac{\delta^2}{l^2}\right)$$

to write

$$\frac{2}{\pi} = \prod_{l=1}^{\infty} \left(1 - \frac{1}{4l^2}\right).$$

We get that

$$\log \widetilde{S}_n = n \sum_{l=1}^{\infty} \log\left(1 - \frac{1}{4l^2}\right) + \sum_{l=1}^{n-1} (l-n) \log\left(1 - \frac{1}{4l^2}\right)$$

$$= \sum_{l=1}^{n-1} l \log\left(1 - \frac{1}{4l^2}\right) + n \sum_{l=n}^{\infty} \log\left(1 - \frac{1}{4l^2}\right)$$

$$= -\sum_{l=1}^{n-1} \frac{1}{4l} + \sum_{l=1}^{n-1} l \left(\log\left(1 - \frac{1}{4l^2}\right) + \frac{1}{4l^2}\right) + n \sum_{l=n}^{\infty} \log\left(1 - \frac{1}{4l^2}\right)$$

$$= -\frac{1}{4} \log(n-1) - \frac{1}{4} \left(\sum_{l=1}^{n-1} \frac{1}{l} - \log(n-1)\right)$$

$$\quad + \sum_{l=1}^{n-1} l \left(\log\left(1 - \frac{1}{4l^2}\right) + \frac{1}{4l^2}\right) + n \sum_{l=n}^{\infty} \log\left(1 - \frac{1}{4l^2}\right)$$

$$= -\frac{1}{4} \log(n-1) - \frac{\gamma}{4} + \overline{A} - \frac{n}{4} \int_n^{\infty} \frac{dl}{l^2} + o(1)$$

$$= -\frac{1}{4} \log(n-1) - \frac{\gamma}{4} + \overline{A} - \frac{1}{4} + o(1), \qquad (3.11.116)$$

where

$$\gamma = \lim_{n \to \infty} \left(\sum_{l=1}^{n} l^{-1} - \log n\right) = 0.5772157\ldots$$

is Euler's constant, and

$$\overline{A} = \sum_{l=1}^{\infty} l \left(\log\left(1 - \frac{1}{4l^2}\right) + \frac{1}{4l^2}\right).$$

Thus, as $n \to \infty$,

$$\widetilde{S}_n \sim A n^{-1/4}, \qquad (3.11.117)$$

where the constant A is numerically about $A = 0.6450024\ldots$ as stated in Theorem 3.11.8.

Proof of Corollary 3.11.2 Suppose $E_h = E_v = E$ and set $c = \cosh \beta E$, $s = \sinh \beta E$, $c_2 = \cosh 2\beta E$ and $s_2 = \sinh 2\beta E$. We have

$$\frac{1+\alpha_1}{1-\alpha_1} = \frac{c^2 - s^2 + 2sc}{c^2 + s^2} = \frac{1+s_2}{\sqrt{1+s_2^2}},$$

where the last equality follows from the relations $c^2 - s^2 = 1$, $2cs = s_2$ and $c^2 + s^2 = c_2 = \sqrt{1+s_2^2}$. But (3.11.10) yields $s_2 = 1$, and therefore

$$\frac{1+\alpha_1}{1-\alpha_1} = \sqrt{2}. \qquad \square$$

3.12 Stochastic Heat Equation

The classical heat equation

$$\frac{\partial u}{\partial t} = \frac{1}{2}\frac{\partial^2 u}{\partial x^2}, \quad u(0,x) = u_0(x), \tag{3.12.1}$$

has the solution $u(t,x)$, $t \geq 0$, $x \in \mathbb{R}$, given by

$$u(t,x) = \frac{1}{\sqrt{2\pi t}} \int_{\mathbb{R}} u_0(y) e^{-\frac{(x-y)^2}{2t}} dy =: (P_t u_0)(x). \tag{3.12.2}$$

When the forcing term $f = f(t,x)$ is added to (3.12.1) as

$$\frac{\partial u}{\partial t} = \frac{1}{2}\frac{\partial^2 u}{\partial x^2} + f, \quad u(0,x) = u_0(x), \tag{3.12.3}$$

the solution can be expressed as

$$u(t,x) = (P_t u_0)(x) + \int_0^t (P_{t-s} f(s,\cdot))(x) ds$$

$$= \frac{1}{\sqrt{2\pi t}} \int_{\mathbb{R}} u_0(y) e^{-\frac{(x-y)^2}{2t}} dy + \int_0^t \frac{1}{\sqrt{2\pi(t-s)}} \int_{\mathbb{R}} e^{-\frac{(x-y)^2}{2(t-s)}} f(s,y) dy ds \tag{3.12.4}$$

for a large class of functions (distributions) f.

In a *stochastic heat equation*, the forcing f is taken as a space-time white noise. The equation is written as

$$\frac{\partial u}{\partial t} = \frac{1}{2}\frac{\partial^2 u}{\partial x^2} + \frac{\partial^2 W}{\partial t \partial x} \quad \text{or} \quad \frac{\partial u}{\partial t} = \frac{1}{2}\frac{\partial^2 u}{\partial x^2} + \dot{W}, \tag{3.12.5}$$

where $W = \{W(t,x), t \geq 0, x \in \mathbb{R}\}$ is a two-parameter Brownian motion; that is, W is a zero mean Gaussian process with covariance

$$\mathbb{E} W(t,x) W(s,y) = (t \wedge s) \begin{cases} (|x| \wedge |y|), & \text{if sign}(x) = \text{sign}(y), \\ 0, & \text{if sign}(x) \neq \text{sign}(y), \end{cases} \tag{3.12.6}$$

since $W(t,x)$ is independent of $W(s,y)$ when x and y have different signs. Stochastic heat equation is one of the basic and canonical models of stochastic partial differential equations.

Assuming $u_0(y) \equiv 0$ for simplicity and by using (3.12.4), the solution to the stochastic heat equation (3.12.5) is given by

$$u(t,x) = \int_0^t \int_{\mathbb{R}} \frac{1}{\sqrt{2\pi(t-s)}} e^{-\frac{(x-y)^2}{2(t-s)}} W(ds, dy), \qquad (3.12.7)$$

where W is a Gaussian random measure on $(0, \infty) \times \mathbb{R}$ with the control measure $ds dy$. Since the integrand is square integrable, the process $u(t,x)$ is a well-defined, mean zero Gaussian process.

Observe that, for fixed $x \in \mathbb{R}$, the covariance function of $u(t,x)$ can be computed as

$$\begin{aligned}
\mathbb{E} u(t_1,x) u(t_2,x) &= \frac{1}{2\pi} \int_0^{t_1 \wedge t_2} \int_{\mathbb{R}} \frac{1}{\sqrt{(t_1-s)(t_2-s)}} e^{-\frac{(x-y)^2}{2(t_1-s)} - \frac{(x-y)^2}{2(t_2-s)}} ds dy \\
&= \frac{1}{2\pi} \int_0^{t_1 \wedge t_2} \int_{\mathbb{R}} \frac{1}{\sqrt{(t_1-s)(t_2-s)}} e^{-\frac{(x-y)^2}{2}(\frac{1}{t_1-s} + \frac{1}{t_2-s})} ds dy \\
&= \frac{1}{\sqrt{2\pi}} \int_0^{t_1 \wedge t_2} (t_1 + t_2 - 2s)^{-1/2} ds \\
&= \frac{1}{\sqrt{2\pi}} \left((t_1 + t_2)^{1/2} - (t_1 + t_2 - 2(t_1 \wedge t_2))^{1/2} \right) \\
&= \frac{1}{\sqrt{2\pi}} \left((t_1 + t_2)^{1/2} - |t_1 - t_2|^{1/2} \right). \qquad (3.12.8)
\end{aligned}$$

Thus, for fixed $x \in \mathbb{R}$, the process $\{u(t,x)\}_{t \geq 0}$ is a bifractional Brownian motion with indices $H = 1/2$ and $K = 1/2$ introduced in Section 2.6.2.

3.13 The Weierstrass Function Connection

The complex-valued Weierstrass function

$$W^{(0)}(t) = \sum_{n=0}^{\infty} a^n e^{ib^n t}, \qquad (3.13.1)$$

where $t \in \mathbb{R}, 0 < a < 1$ and $ab \geq 1$, is a well-known example of a continuous but nowhere differentiable function. It can be modified and randomized to FBM. When b is an odd integer and $ab > 1 + (3\pi/2)$, the proof of non-differentiability of $W^{(0)}(t)$ was first given by Weierstrass [1000] (or Weierstrass [1001]). It was later generalized to the condition $ab \geq 1$ by Hardy [454]. To see its relation with FBM, set

$$a = r^{-H}, \quad b = r$$

and note that

$$0 < a < 1, \quad ab > 1 \quad \Leftrightarrow \quad r > 1, \quad 0 < H < 1.$$

We then obtain the function

$$W^{(1)}(t) = \sum_{n=0}^{\infty} r^{-Hn} e^{ir^n t},$$

3.13 The Weierstrass Function Connection

which was originally discussed by Mandelbrot [679] (pp. 388–390). He noted that its frequency spectrum (r^{-Hn} at frequency r) and its cumulative energy spectrum (see below) are similar to those of FBM with index H.

It is convenient, in this context, to represent FBM as

$$B_H(t) = c \, \Re \int_0^\infty (e^{itx} - 1) x^{-H-1/2} (dB^1(x) + i dB^2(x)), \qquad (3.13.2)$$

where B^1 and B^2 are two independent standard Brownian motions and c is some constant. The representation is equivalent to that when replacing $dB^1(x)$ and $dB^2(x)$ above by independent Gaussian random measures $B^1(dx)$ and $B^2(dx)$ on $\mathbb{R}$ having control measures dx. But working with Brownian motions will be more advantageous in some of our arguments.

It is can be checked that (3.13.2) defines an H–SSSI Gaussian process and hence is a representation of FBM (Corollary 2.6.3). The function $(x^{-H-1/2})^2 = x^{-2H-1}$ is viewed as its frequency spectrum, and its cumulative spectrum at frequencies $x \geq a$ is $\int_a^\infty x^{-2H-1} dx = c a^{-2H}$, where c is some other constant.

There are three major differences, however, between the Weierstrass function and FBM:

- the spectrum of the Weierstrass function is located at frequencies $\{r^n\}_{n \geq 0}$, whereas the spectrum of FBM is supported on $(0, \infty)$,
- the spectrum of the Weierstrass function is discrete, whereas that of FBM is continuous,
- the Weierstrass function is deterministic whereas FBM is a random process.

To deal with the first difference Mandelbrot extended the lower frequency bound $f = r^0 = 1$ to $f = 0$ (the case of FBM) by introducing the modified Weierstrass function

$$W^{(2)}(t) = \sum_{n=-\infty}^{\infty} (e^{ir^n t} - 1) r^{-Hn}, \qquad (3.13.3)$$

which converges for $0 < H < 1$. Replacing $e^{ir^n t}$ by $e^{ir^n t} - 1$ is necessary to ensure convergence as $n \to -\infty$; that is, at frequencies close to the origin, and hence to prevent the infrared (low-frequency) divergence.

Observe also the similarities between (3.13.3) and (3.13.2). $W^{(2)}$ is still nowhere differentiable, has a discrete frequency spectrum located at $\{r^n\}_{n \in \mathbb{Z}}$ and, also, is self-similar with index H, i.e., $W^{(2)}(ct) = c^H W^{(2)}(t)$, for any $t \in \mathbb{R}$, but self-similarity is restricted to factors $c = r^m$, $m \in \mathbb{Z}$. Notice that, when r tends to 1, these factors become more clustered on $\mathbb{R}_+$. Were the function $W^{(2)}(t)$ random and r close to 1, one would expect it to approximate FBM (Mandelbrot [679]).

Randomization is achieved by multiplying each term of the series $W^{(2)}(t)$ by complex random variable $\xi_n + i\eta_n$ where $\{\xi_n\}_{n \in \mathbb{Z}}$ and $\{\eta_n\}_{n \in \mathbb{Z}}$ are two independent sequences of i.i.d. real-valued random variables; that is, by considering

$$W_r^{(3)}(t) = \sum_{n=-\infty}^{\infty} (e^{ir^n t} - 1) r^{-Hn} (\xi_n + i\eta_n), \qquad (3.13.4)$$

which we call the *Weierstrass–Mandelbrot process*. We suppose that ξ_n, η_n have finite variance.

Our objective is to study the convergence of the Weierstrass–Mandelbrot process $W^{(3)}$, suitably normalized as r tends to 1. To gain some insight on the choice of normalization, consider the variance of the Weierstrass–Mandelbrot process $W^{(3)}$. Observe that if $\mathbb{E}\xi_n^2 = \mathbb{E}\eta_n^2 = 1$, then

$$\mathbb{E}|\Re W_r^{(3)}(t+s) - \Re W_r^{(3)}(s)|^2 = \frac{1}{2}\mathbb{E}|W_r^{(3)}(t+s) - W_r^{(3)}(s)|^2$$

$$= 2\sum_{n=-\infty}^{\infty}(1-\cos r^n t)r^{-2Hn} = \frac{2}{r-1} \cdot \sum_{n=-\infty}^{\infty}(1-\cos r^n t)(r^n)^{-2H-1}(r^{n+1}-r^n)$$

$$= \frac{2}{r-1} \cdot \int_0^\infty \sum_{n=-\infty}^{\infty}(1-\cos r^n t)(r^n)^{-2H-1} 1_{[r^n, r^{n+1})}(u)\, du.$$

Since, for fixed $t \geq 0$, as $r \to 1$,

$$\sum_{n=-\infty}^{\infty}(1-\cos r^n t)(r^n)^{-2H-1} 1_{[r^n, r^{n+1})}(u) \to (1-\cos ut)u^{-2H-1}, \quad \forall u > 0,$$

it follows from the dominated convergence theorem that, as $r \to 1$,

$$(r-1)\sum_{n=-\infty}^{\infty}(1-\cos r^n t)r^{-2Hn} \to \int_0^\infty (1-\cos ut)u^{-2H-1}\, du = t^{2H}\int_0^\infty \frac{(1-\cos u)}{u^{2H+1}}\, du.$$

The correct normalization is therefore $r-1$, or equivalently, $\log r$, since $\frac{\log r}{r-1} \to 1$, as $r \to 1$.

Another way of looking at the normalizer $r-1$ (or $\log r$) is as follows. Consider the modified Weierstrass function $W^{(2)}$ given in (3.13.3). The cumulative energy, i.e., the amplitude (coefficient of $(e^{ir^n t} - 1)$) squared, in the frequencies $f \geq r^m$, for some $m \in \mathbb{Z}$, is given by

$$r^{-2Hm} + r^{-2H(m+1)} + \cdots = \frac{r^{-2Hm}}{1 - r^{-2H}},$$

which explodes as $r \to 1$. The necessary correction $1 - r^{-2H}$ behaves (up to a constant) like $1 - r$ in the limit.

We are then led to consider

$$W_r(t) = (\log r)^{1/2} W_r^{(3)}(t) = (\log r)^{1/2} \sum_{n=-\infty}^{\infty}(e^{ir^n t} - 1)r^{-Hn}(\xi_n + i\eta_n). \quad (3.13.5)$$

When studying the limit behavior of (3.13.5), we want to be able to apply the central limit theorem which involves a normalization of the form $a^{1/2}$, $a \to \infty$. To make this normalization appear, express r^n as $e^{n\log r}$ and set $\log r = 1/a$. We then get

$$W_r(t) = (\log r)^{1/2} \sum_{n=-\infty}^{\infty}(e^{ie^{n\log r}t} - 1)e^{-Hn\log r}(\xi_n + i\eta_n)$$

$$= \sum_{n=-\infty}^{\infty}(e^{ie^{n/a}t} - 1)e^{-Hn/a}\frac{1}{a^{1/2}}(\xi_n + i\eta_n) = \sum_{n=-\infty}^{\infty} f_t\left(\frac{n}{a}\right)\frac{1}{a^{1/2}}(\xi_n + i\eta_n),$$

$$(3.13.6)$$

3.13 The Weierstrass Function Connection

where

$$f_t(u) = (e^{ie^u t} - 1)e^{-Hu}, \, u \in \mathbb{R}. \tag{3.13.7}$$

Notice that as r tends to 1, a tends to infinity.

As $a \to \infty$, we expect (3.13.6) to converge to $\int_{\mathbb{R}} f_t(u)(dB^1(u) + i dB^2(u))$, where B^1 and B^2 are Brownian motions. The following theorem provides sufficient conditions on the sequence $\{(\xi_n, \eta_n)\}_{n \in \mathbb{Z}}$ and on the functions f_t to imply the required convergence. Set

$$f_{t,a}(u) = \sum_{n=-\infty}^{\infty} f_t\left(\frac{n}{a}\right) 1_{[\frac{n}{a}, \frac{n+1}{a})}(u).$$

Theorem 3.13.1 *Suppose that $\{\xi_n, \eta_n, n \geq 1\}$ are independent random variables with $\mathbb{E}\xi_n^2 = \mathbb{E}\eta_n^2 = 1$, $\mathbb{E}\xi_n = \mathbb{E}\eta_n = 0$. Let $f_t : \mathbb{R} \to \mathbb{C}$ be functions such that $f_t, f_{t,a} \in L^2(\mathbb{R})$ and that*

$$\|f_t - f_{t,a}\|_{L^2(\mathbb{R})} \to 0, \quad \text{as } a \to \infty. \tag{3.13.8}$$

Then, $\sum_{n=-\infty}^{\infty} f_t(n/a)(\xi_n + i\eta_n)$ is well-defined in the $L^2(\Omega)$–sense and, as $a \to \infty$,

$$\sum_{n=-\infty}^{\infty} f_t\left(\frac{n}{a}\right) \frac{1}{a^{1/2}} (\xi_n + i\eta_n) \xrightarrow{fdd} \int_{\mathbb{R}} f_t(u)(dB^1(u) + i dB^2(u)), \tag{3.13.9}$$

where $\{B^1(u)\}_{u \in \mathbb{R}}$ and $\{B^2(u)\}_{u \in \mathbb{R}}$ are two independent standard Brownian motions and $\xrightarrow{fdd}$ means the convergence in the sense of the finite-dimensional distributions.

Proof It is enough to prove the result for fixed t and *real-valued* f. For the simplicity of notation, we drop the index t from all functions f. The assumption $f_a \in L^2(\mathbb{R})$ implies that $\|f_a\|_{L^2(\mathbb{R})}^2 = \sum_{n=-\infty}^{\infty} f(n/a)^2/a < \infty$. Hence, for example for $k_2 > k_1 \geq 1$,

$$\mathbb{E}\left|\sum_{n=k_1+1}^{k_2} f\left(\frac{n}{a}\right)(\xi_n + i\eta_n)\right|^2 = 2 \sum_{n=k_1+1}^{k_2} f\left(\frac{n}{a}\right)^2 \to 0,$$

as $k_1, k_2 \to \infty$. This shows that $\sum_{n=-\infty}^{\infty} f_t(n/a)(\xi_n + i\eta_n)$ is well-defined in the $L^2(\Omega)$–sense. Set

$$X_a = \sum_{n=-\infty}^{\infty} f\left(\frac{n}{a}\right) \frac{1}{a^{1/2}} (\xi_n + i\eta_n), \quad X = \int_{\mathbb{R}} f(u)(dB^1(u) + i dB^2(u)).$$

Since $f \in L^2(\mathbb{R})$, there is a sequence of elementary functions[22] f^j such that $\|f - f^j\|_{L^2(\mathbb{R})} \to 0$, as $j \to \infty$. Set also

$$X_a^j = \sum_{n=-\infty}^{\infty} f^j\left(\frac{n}{a}\right) \frac{1}{a^{1/2}} (\xi_n + i\eta_n), \quad X^j = \int_{\mathbb{R}} f^j(u)(dB^1(u) + i dB^2(u)).$$

By Theorem 3.1 in Billingsley [154], the series X_a converges in distribution to X if

---

[22] A function f is elementary if it has a form $f(u) = \sum_{k=1}^{n} f_k 1_{[a_k, b_k)}(u)$.

Step 1: $X^j \stackrel{d}{\to} X$, as $j \to \infty$,

Step 2: for all $j \geq 1$, $X_a^j \stackrel{d}{\to} X_j$, as $a \to \infty$,

Step 3: $\limsup_j \limsup_a \mathbb{E}|X_a^j - X_a|^2 = 0$.

Step 1 follows since X^j and X are integrals with respect to Brownian motion and $\|f - f^j\|_{L^2(\mathbb{R})} \to 0$, as $j \to \infty$. For Step 2, observe that

$$X_a^j = \int_{\mathbb{R}} f^j(u) dB_a(u),$$

where

$$B_a(u) = \begin{cases} \frac{1}{a^{1/2}} \sum_{j=1}^{[au]} (\xi_j + i\eta_j), & u \geq 0, \\ -\frac{1}{a^{1/2}} \sum_{j=[au]+1}^{0} (\xi_j + i\eta_j), & u < 0. \end{cases}$$

By the assumptions on ξ_n, η_n and the central limit theorem, B_a converges to $B^1 + iB^2$ in the sense of finite-dimensional distributions. Since f^j is an elementary function, the integral X^j depends on the process B_a through a finite number of time points only. It then converges in distribution to X^j, as $a \to \infty$. Finally, for Step 3, observe that $\mathbb{E}|X_a^j - X_a|^2 = 2\|f_a^j - f_a\|_{L^2(\mathbb{R})}^2$, where

$$f_a^j(u) = \sum_{n=-\infty}^{\infty} f^j\left(\frac{n}{a}\right) 1_{[\frac{n}{a}, \frac{n+1}{a})}(u).$$

For fixed j, by using the dominated convergence theorem and the assumption (3.13.8), we have $\|f_a^j - f_a\|_{L^2(\mathbb{R})}^2 \to \|f^j - f\|_{L^2(\mathbb{R})}^2$ as $a \to \infty$. Then, $\limsup_a \mathbb{E}|X_a^j - X_a|^2 = 2\|f^j - f\|_{L^2(\mathbb{R})}^2$, which tends to 0 as $j \to \infty$. $\square$

The function f_t in (3.13.7) characterizing the Weierstrass–Mandelbrot process satisfies the assumptions of Theorem 3.13.1. Indeed, $f_t \in L^2(\mathbb{R})$ since f_t is continuous and $|f_t(u)| \leq Ce^{-Hu}$, $|f_t(u)| \leq Ce^{(1-H)u}$. These bounds also imply that $f_{t,a} \in L^2(\mathbb{R})$. Finally, $\|f_t - f_{t,a}\|_{L^2(\mathbb{R})} \to 0$ by using the dominated convergence theorem. Consequently, applying Theorem 3.13.1 to the Weierstrass–Mandelbrot process, we get the following result.

Theorem 3.13.2 *As r tends to 1, the normalized Weierstrass–Mandelbrot process (3.13.6) converges in the sense of finite-dimensional distributions to the complex-valued fractional Brownian motion*

$$X_H(t) = \int_0^{\infty} (e^{ixt} - 1) x^{-H-\frac{1}{2}} (dB^1(x) + i dB^2(x)). \quad (3.13.10)$$

The limit of the normalized Weierstrass–Mandelbrot process (3.13.6) given in relation (3.13.9) of Theorem 3.13.1 is, in fact,

$$X'_H(t) = \int_{\mathbb{R}} (e^{ie^u t} - 1) e^{-Hu} (dB^1(u) + i dB^2(u)).$$

The processes $X'_H(t)$ and $X_H(t)$, however, have identical finite-dimensional distributions because they are both Gaussian, have mean zero and same covariance functions. One can also use the fact that, if $B(u), u \in \mathbb{R}$, is a standard real-valued Brownian motion, then so is $\mathcal{B}(x) = \int_{-\infty}^{\log x} e^{u/2} dB(u), x \geq 0$, and then make a change of variables. Observe finally that $\{\Re X_H(t)\}_{t \in \mathbb{R}}$ and $\{\Im X_H(t)\}_{t \in \mathbb{R}}$ are both fractional Brownian motions, but they are not independent.

3.14 Exercises

The symbols * and ** next to some exercises are explained in Section 2.12.

Exercise 3.1 Show that FARIMA$(0, d, 0)$ series can be viewed as an aggregated series with the mixing density (3.1.14); that is, its spectral density $f_X(\lambda)$ in (3.1.13) satisfies the relation (3.1.8) with $F_a(dx)$ having the density $f_a(x)$ given by (3.1.14). *Hint:* Use the identity $|1 - xe^{-i\lambda}|^2 = (1-x)^2(1 + 2x)(1 - \cos\lambda)/(1-x)^2)$ and a change of variables $y^{1/d} = 2x/(1-x)^2$. See also Celov et al. [208], Proposition 5.1.

Exercise 3.2 Similarly to Proposition 3.1.1 suppose that the distribution $F_a(dx)$ has a density

$$f_a(x) = (1+x)^{1-2d} l(1+x) \phi(x) 1_{(-1,0]}(x),$$

where $0 < d < 1/2$, $l(y)$ is a slowly varying function at $y = 0$ and $\phi(x)$ is a bounded function on $(-1, 0]$ and continuous at $x = -1$ with $\phi(-1) \neq 0$. Show that, as $\lambda \to \pi$,

$$f_X(\lambda) \sim |\lambda - \pi|^{-2d} l(\lambda - \pi) \frac{\sigma_\epsilon^2 \phi(-1)}{2\pi} \int_0^\infty \frac{s^{1-2d}}{1+s^2} ds,$$

where $f_X(\lambda)$ appears in (3.1.4) and (3.1.8).

Exercise 3.3** Exercise 3.1 provides an aggregation result for FARIMA$(0, d, 0)$ series. What about fractional Gaussian noise (FGN)? Can FGN be viewed as an aggregated series? A question related to this is whether fractional Brownian motion, which is not stationary but has stationary increments, can be viewed as an aggregated process in a suitable sense. The suitable sense here could mean aggregation of the integrals of continuous-time Ornstein–Uhlenbeck (OU) processes with random coefficients, since OU processes can be thought as continuous-time analogues of AR(1) series.

Exercise 3.4 Prove Theorem 3.2.2. *Hint:* Use $\gamma_\epsilon(h) = (2p-1)^h$ and follow the proof of Theorem 3.2.3.

Exercise 3.5 (*i*) Show that in the case $H = 1/2$, the limit in Theorem 3.2.3 is Brownian motion but the normalization N^H should be replaced by $\sqrt{N \log N}$. (*ii*) Suppose that μ is the uniform probability on $[1/2, 1]$. In Theorem 3.2.3, let $c_H = 1/\sqrt{2}$ and replace N^H by $\sqrt{N \log N}$. Show that the limit in Theorem 3.2.3 is then Brownian motion. (*iii*) Show that if one reverses the order of the limits in Theorem 3.2.3, then the first limit as $N \to \infty$ would yield 0, since the correct normalization would be $\sqrt{N}$.

Exercise 3.6 Prove the representation (3.3.19)–(3.3.20) for the workload process W_λ^*.

Exercise 3.7 Prove that the integrand $I_\lambda(s, u, r)$ is dominated by an integrable function in the proof of Theorem 3.3.12.

Exercise 3.8 Show that Proposition 3.4.4 and Theorem 3.4.5 hold when the function g in Definition 3.4.2 is such that $g(y) = 0$, $y < 0$, and $g(y) = (1 + y)^{-\beta} L_g(y)$, where L_g is a slowly varying function at infinity which is positive, and locally bounded away from 0 and infinity in $[0, \infty)$. That is, the shots $g(y)$ need not diverge like a power function $y^{-\beta}$ around $y = 0$.

Exercise 3.9* For the infinite source Poisson model in Section 3.3, various regimes of $\lambda \to \infty$ were considered and led to different limiting processes. Are there regimes of $\lambda \to \infty$ in the power-law shot noise model which lead to other limiting processes than fractional Brownian motion?

Exercise 3.10 Prove the results (3.6.9), (3.6.10), (3.6.11) and (3.6.27) for the Markov chain in the proof of Proposition 3.6.2.

Exercise 3.11 Consider a stationary Gaussian series $X_n^N = \phi_N X_{n-1}^N + \epsilon_n$, $n = 1, \ldots, N$, with i.i.d. $\mathcal{N}(0, \sigma_\epsilon^2)$ variables ϵ_n and $\phi_N = 1 - cN^{-d}$, where $0 < c < 1$ and $0 < d < 1/2$. As for the Markov switching model (see (3.6.31) and Proposition 3.6.2) show that

$$\text{Var}(X_1^N + \cdots + X_N^N) \sim CN^{2d+1}, \quad \frac{1}{N^{d+1/2}} \sum_{n=1}^{[Nt]} X_n^N \xrightarrow{fdd} B(t),$$

where B is Brownian motion. Other similar models can be found in Diebold and Inoue [312]. *Hint:* Use the arguments in the proof of Proposition 3.6.2 and the fact that $\mathbb{E} X_h^N X_0^N = \sigma_\epsilon^2 (\phi_N)^h / (1 - (\phi_N)^2)$.

Exercise 3.12** The aggregation result in Proposition 3.1.1, Theorem 3.2.3 on correlated random walks and Theorem 3.3.8, (a), for the infinite source Poisson model are associated with LRD series and FBM with the self-similarity parameter $H > 1/2$. How can these models be modified, especially according to some common mechanism, to be associated with antipersistent series and FBM with the self-similarity parameter $H < 1/2$? *Hint:* For the correlated random walks, a modification is suggested in Enriquez [351].

Exercise 3.13 Prove the convergence of the finite-dimensional distributions in (3.7.10).

Exercise 3.14 Prove the first equality in (3.7.11). *Hint:* Suppose $0 \leq s < t$. Then $n(t) - n(s) = N(A^+(s,t)) - N(A^-(s,t))$, where

$$A^+(s, t) = \{(x, \xi) : x + \xi(s) \leq 0, x + \xi(t) > 0\},$$
$$A^-(s, t) = \{(x, \xi) : x + \xi(s) > 0, x + \xi(t) \leq 0\}.$$

Since $A^+(s,t) \cap A^-(s,t) = \emptyset$, $N(A^+(s,t))$ and $N(A^-(s,t))$ are independent Poisson with mean

$$\int \int_\mathbb{R} 1_{\{-\xi(t) < x \leq -\xi(s)\}} \rho dx d\mathbb{P}_\xi = \rho \mathbb{E}(\xi(t) - \xi(s))_+ = \frac{\rho}{2}\mathbb{E}|\xi(t) - \xi(s)|.$$

Exercise 3.15 Prove the relation (3.8.4) for the simple symmetric exclusion model. *Hint:* Argue as in (3.7.5)–(3.7.8).

Exercise 3.16 Prove the relation (3.8.10) for the simple symmetric exclusion model.

Exercise 3.17 Prove that the second term in (3.8.16) converges to 0 as $t \to \infty$.

Exercise 3.18 Prove the relation (3.10.9) for the local time.

Exercise 3.19* Following Section 3.11.1, consider the Ising model in one dimension having the Boltzmann (Gibbs) distribution $e^{-\beta\mathcal{E}}/Z_{L_h}$, where

$$\mathcal{E} = -\sum_{k=-L_h+1}^{L_h} E^h \sigma_k \sigma_{k+1}.$$

Show that, in the thermodynamic limit,

$$\lim_{L_h \to \infty} \mathbb{E}\sigma_0 \sigma_n = (\tanh(\beta E^h))^n, \quad n \geq 0.$$

Thus, in one dimension, the Ising model exhibits short-range dependence for all inverse temperatures $\beta > 0$. *Hint:* This is a classical result for one-dimensional Ising model whose proof can be found in many sources, e.g., McCoy and Wu [703], Chapter III.

3.15 Bibliographical Notes

Section 3.1: Analysis and statistical inference for AR(1) series with random coefficients go back at least to Robinson [851]. Aggregation of short-range dependent series as one explanation of long-range dependence was put forward by Granger [422]. In particular, Granger [422] focused on the mixing density (3.1.11) and LRD was shown through the argument (3.1.12). The ideas of Granger were studied and extended by many authors. More thorough mathematical analysis of aggregation can be found in Gonçalves and Gouriéroux [417], Lippi and Zaffaroni [637], Zaffaroni [1025], Oppenheim and Viano [780], Dacunha-Castelle and Fermín [275]. This includes aggregation of series other than AR(1), and considering other densities than (3.1.11). Disaggregation problem is discussed in Dacunha-Castelle and Oppenheim [276], Dacunha-Castelle and Fermín [274], Celov et al. [208]. Estimation of the underlying mixing density from an aggregated series is considered in Leipus, Oppenheim, Philippe, and Viano [605], Beran, Schützner, and Ghosh [126]. Long-range dependence in volatility models is explained through aggregation in Zaffaroni [1026]. The superposition of Ornstein–Uhlenbeck-type processes, which can be viewed as the analogues of AR(1) series in the continuous time, was considered by Barndorff-Nielsen and Leonenko [101]. Some other works related to aggregation are Chambers [212], Leipus and Surgailis [604],

Davidson and Sibbertsen [289], Leonenko and Taufer [611], Leipus, Philippe, Puplinskaitė, and Surgailis [606], Candelpergher, Miniconi, and Pelgrin [200].

Aggregation of short-range dependent series has been geared mostly towards applications in economics. In this regard, it should be noted that a more general model than (3.1.1) is often considered, given by

$$Y_n^{(j)} = a^{(j)} Y_{n-1}^{(j)} + Z_n + \epsilon_n^{(j)}, \quad n \in \mathbb{Z}, \tag{3.15.1}$$

where the series $\{Z_n\}_{n \in \mathbb{Z}}$ is common across all $j \geq 1$ and is referred to as a macroeconomic shock. See, for example, Lewbel [621], Altissimo, Mojon, and Zaffaroni [23], Mayoral [700] to name but a few.

Section 3.2: Theorem 3.2.3 is due to Enriquez [351]. He also shows that if $M(N)$ is made to be a function of N, then the convergence to fractional Brownian motion holds as $N \to \infty$ as long as $M(N)/N^{1-2H}$ tends to ∞. Enriquez considered not only the case $H > 1/2$ but also the case $H < 1/2$. To ensure that the sum of the covariances $\sum_{n=-\infty}^{\infty} r(n)$ equals 0 when $H < 1/2$, Enriquez introduces an alternating correlated random walk, which also moves by jumps of $+1$ and -1, but where the probability of making the same jump alternates between p and 0. The account here is partly based on a presentation of this material by Shuyang Bai.

Section 3.3: A number of other models related to the infinite source Poisson model with heavy tails have been considered in the literature including the renewal-reward processes and ON/OFF models (Mandelbrot [676], Levy and Taqqu [617], Taqqu and Levy [947], Pipiras, Taqqu, and Levy [822], Gaigalas and Kaj [380], Kaj [545], Leland, Taqqu, Willinger, and Wilson [607], Taqqu, Willinger, and Sherman [952], Mikosch, Resnick, Rootzén, and Stegeman [719], Abry, Borgnat, Ricciato, Scherrer, and Veitch [12], Dombry and Kaj [323]). The asymptotic results for these related models run in parallel to those for the infinite source Poisson model. But these results in many cases preceded and inspired the work on the infinite source Poisson model, especially the analysis of the ON/OFF model with heavy tails in Mandelbrot [676] who pioneered the use of self-similar and long-range dependent models in Economics and with Leland et al. [607] which pioneered the use of self-similar and long-range dependent models in telecommunication networks.

The presentation of this section follows Kaj and Taqqu [546]. This is just one work in the otherwise vast literature on the infinite source Poisson model and its variants. The notable works include Cioczek-Georges and Mandelbrot [244], Kurtz Kurtz [585], Heath, Resnick, and Samorodnitsky [462], Konstantopoulos and Lin [576], Mikosch et al. [719], Rosenkrantz and Horowitz [862], Barakat, Thiran, Iannaccone, Diot, and Owezarski [94], Guerin et al. [438],Çağlar [196]. Variants and extensions of the model consider random transmission rates (Maulik, Resnick, and Rootzén [699], D'Auria and Resnick [288]), cluster point processes (Hohn, Veitch, and Abry [479], Faÿ, González-Arévalo, Mikosch, and Samorodnitsky [356], Fasen [354]), random fields (Kaj, Leskelä, Norros, and Schmidt [547], Biermé and Estrade [148], Kuronen and Leskelä [584]), limit theorems under control (Budhiraja, Pipiras, and Song [193]) and quite general formulations (Mikosch and Samorodnitsky [717]). Moment measures of the models are studied in Dombry and Kaj [324], Antunes, Pipiras, Abry, and Veitch [39]. This is also not to mention numerous other papers in the networking literature, where long-range dependent models are used – several books in this area are Park and Willinger [794], Sheluhin, Smolskiy, and Osin [903].

Long-range dependence aside, Araman and Glynn [41] consider related models leading to FBM with the self-similarity parameter $H < 1/2$.

Section 3.4: Power-law shot noise models go back to at least Lowen and Teich [647] (see also Chapter 9 in their monograph [648]), who also refer to earlier work by Schönfeld [891], Van der Ziel [970] for special cases related to noise in electronic devices (where, as a matter of fact, shot noise models have originated). Lowen and Teich call them fractal shot noise models and also do not consider limit laws. The presentation here is somewhat inspired by the works of Klüppelberg and Mikosch [569], Klüppelberg, Mikosch, and Schärf [570], Klüppelberg and Kühn [568], where some financial applications are considered. See also Lane [591]. A field version of the model was considered by Baccelli and Biswas [66].

Section 3.5: The presentation of this section follows Cox [265], who also refers to his earlier work in Cox and Townsend [266]. More recent related discussions can be found in Huh, Kim, Kim, and Suh [500], Militkỳ and Ibrahim [720].

Section 3.6: The material of this section is based on Diebold and Inoue [312], Baek et al. [71]. From a related practical perspective, LRD series are well known to be easily confused with series exhibiting several changes in local mean level. Indeed, apparent changes in the local mean level across a range of larger scales is a characteristic feature of LRD series. Conversely, a series composed of several changes in mean superimposed by a SRD series will exhibit LRD when using common estimators of the LRD parameter. This confusion between LRD and nonstationary-like models, such as the model involving changes in mean, has been documented well and is raised in almost all applications of LRD series. See, for example, Klemeš [563], Boes and Salas [165] in hydrology, Roughan and Veitch [869], Veres and Boda [989] in teletraffic, Diebold and Inoue [312], Granger and Hyung [423], Mikosch and Stărică [718], Smith [913] in economics and finance, Mills [721] in climatology.

A number of statistical procedures have also emerged aiming at distinguishing LRD and non-stationary models. Jach and Kokoszka [526] showed that wavelet-based, maximum likelihood tests for SRD are robust to the presence of nonstationarities. Ohanissian, Russell, and Tsay [778] used temporal aggregation to test against nonstationary models. Iacone [513], McCloskey and Perron [701] employ common estimators of the LRD parameter in the Fourier domain by selecting carefully the range of frequencies to be used in estimation. Tests for distinguishing LRD and changes in mean models are proposed in Berkes, Horváth, Kokoszka, and Shao [130], Yau and Davis [1023], Baek and Pipiras [69]. See also Qu [837].

Section 3.7: We have described the history of the elastic collision problem. Harris [455] started the investigation by looking at colliding Brownian motions, followed by Spitzer [919] who considered Newtonian particles and then Gisselquist [407] who looked at infinite variance stable particles. According to Gisselquist [407], page 235, it is Frank Spitzer, who identified in a private communication the limiting covariance in the case considered by Harris, showing that it is the covariance of FBM with $H = 1/4$. The paper of Dürr et al. [339] also deals with weak convergence. Our account is partly based on a presentation of this material by Shuyang Bai.

Section 3.8: The results presented in the section concern the simple symmetric nearest-neighbor exclusion. What happens with the motion of the tagged particle for other cases of the transition probabilities $p(x, y)$ and the dimension $d \geq 1$ of the state space $\mathbb{Z}^d$? The answer here is almost complete. Other symmetric exclusion processes on $\mathbb{Z}^d$, $d \geq 1$, were studied in Kipnis and Varadhan [560]. The asymmetric but centered exclusion processes on $\mathbb{Z}^d$, $d \geq 1$, were considered in Varadhan [973]. The asymmetric, non-centered, nearest-neighbor exclusion processes on $\mathbb{Z}$ were studied by Kipnis [559], and on $\mathbb{Z}^d$, $d \geq 3$, by Sethuraman, Varadhan, and Yau [900]. The general case of asymmetric, non-centered exclusion processes on $\mathbb{Z}$ and $\mathbb{Z}^2$ is not fully resolved, with some preliminary work reported in Sethuraman [899]. While the limit was FBM in Section 3.8, all the works referenced above involve the usual BM in the limit.

All the works discussed above concern the so-called equilibrium case (that is, when the initial configuration is in equilibrium). For related work on nonequilibrium dynamics, see Jara [530], Jara and Landim [532], Jara, Landim, and Sethuraman [533].

Renormalization and the macroscopic structure of other interacting particle systems, e.g., the voter model, were studied by Bramson and Griffeath [182], Holley and Stroock [481], among others. See also the monograph by Liggett [632].

Section 3.9: The presentation of this section follows Hammond and Sheffield [449]. The model is used in the so-called seed bank application in Blath, González Casanova, Kurt, and Spanò [162]. A fields version of the model is considered in Biermé, Durieu, and Wang [153], and an extension of the model that yields fractional Brownian motion with the self-similarity parameter $H \in (0, 1/2)$ can be found in Durieu and Wang [338].

Section 3.10: What happens in Theorem 3.10.1 if $0 < \alpha < 1$? The random walk S_N is transient in this case and Kesten and Spitzer [553] show that $N^{-1/\beta} \sum_{j=1}^{N} \xi(S_j)$ merely converges to a Lévy stable motion of index β. This makes also physical sense since the transience of S_N ensures that it stays a finite amount of time in any neighborhood instead of returning to it again and again.

Cohen and Samorodnitsky [254] expand the class of limits in (3.10.10) by taking $l_t(x)$ to be the local time of fractional Brownian motion $B_H(t)$, $0 < H < 1$. See also Jung and Markowsky [539]. As noted in the section, an even more general setting of random walks in random scenery is considered by Dombry and Guillotin-Plantard [322], allowing one to analyze a random walk and a random scenery separately. See also the references in Dombry and Guillotin-Plantard [322] for other related work. In connection to superdiffusions, similar models have been considered in Physics literature, for example, in Bouchaud, Georges, Koplik, Provata, and Redner [178].

A self-similar process arising from a random walk with random environment in random scenery is studied in Franke and Saigo [370].

Section 3.11: The Ising model was first studied in the 1920s by Ernst Ising in his Ph.D. thesis, supervised by Wilhelm Lenz.[23] Ising considered the model in one dimension and found that it does not exhibit a phase transition. Although Ising conjectured that no phase transition occurs in higher dimension, further studies pointed to the contrary, culminating in the seminal work of Onsager [779] who provided an explicit calculation for the so-called *free energy* of the two-dimensional Ising model. The Ising model has since played a special role

---

[23] The model is sometimes referred to as the Lenz–Ising model.

in Statistical Mechanics and other areas, as a source of new ideas and as one of few realistic and exactly solvable models. A nice account of the historical developments concerning the Ising model can be found in Niss [752, 753].

New research on the Ising model – as well as on other lattice and growth models – has recently explored their behavior when the edges of the lattice become infinitely small. These scaling limits are generally characterized by conformal invariance, and involve the Schramm–Loewner Evolution (SLE), the Conformal Loop Ensemble (CLE) and related probabilistic objects (see Schramm [892], Werner [1004], Lawler [598]). For example, the *interface* between the $+1$ and -1 spins in the Ising model at critical temperature has long been conjectured to converge to the SLE model with parameter $\kappa = 16/3$. This is studied rigorously by Smirnov [912].

Section 3.11 first appeared in Pipiras and Taqqu [821]; it appears here, with minor changes, with permission of the publisher. Recent studies concerning correlations, the use of dimers and large scales asymptotics in the Ising and related models include Sakai [876], Hara [453], Boutillier and de Tilière [180], Schrenk, Posé, Kranz, van Kessenich, Araújo, and Herrmann [893], Camia, Garban, and Newman [198, 199], Chen and Sakai [217].

Section 3.12: The stochastic heat equation considered in the section is one of the most basic stochastic partial differential equations (SPDEs). A number of extensions of the model have been considered. For example, the closest in spirit and its relation to self-similar processes is a stochastic heat equation with fractional white noise (in the space variable) – see a nice review paper by Balan [92] and references therein, as well as Torres, Tudor, and Viens [960], Tudor and Xiao [966] for more recent work. Other generalizations of the basic model involve a multiplicative white noise $\sigma(u)\dot{W}$ instead of $\dot{W}$ in (3.12.5). See, for example, Hairer [445], Borodin and Corwin [175], Conus, Joseph, Khoshnevisan, and Shiu [260], Hu, Huang, Nualart, and Tindel [494] for a snapshot of recent work in the linear case $\sigma(u) = u$ (also known as the parabolic Anderson model) and references therein. Yet more general SPDE models have been considered in this vast and still expanding area of research, with a large number of introductory textbooks and lecture notes available for interested readers.

Section 3.13: The Wierstrass function in the context of fractional Brownian motion was discussed originally in Mandelbrot [679], pp. 388–390. See also Berry and Lewis [135]. The material of this section is based on Pipiras and Taqqu [814], Szulga and Molz [941]. An extension to heavy-tailed and stable random processes can be found in Pipiras and Taqqu [816]. A multivariate extension of the Weierstrass–Mandelbrot function is studied in Ausloos and Berman [56].

Other notes: We have not touched upon several other topics related to physical models for long-range dependence and self-similarity. These include:

- $1/f$ noise and its origins: see, for example, Ward and Greenwood [998] and references therein.
- Anomalous diffusion, especially when related to stable laws: see, for example, Klafter and Sokolov [561]], Bouchaud and Georges [177], Uchaikin and Zolotarev [967]. Related models known as continuous time random walks were studied by Becker-Kern, Meerschaert, and Scheffler [112], Meerschaert and Scheffler [711], Meerschaert, Nane, and

Xiao [713], Jara and Komorowski [531], Magdziarz, Weron, Burnecki, and Klafter [671], Chen, Xu, and Zhu [226]. See also the notes at the end of Section 7.4.
- Particle system perspective of many self-similar processes in Bojdecki, Gorostiza, and Talarczyk [170, 171], Bojdecki and Talarczyk [169], Bojdecki, Gorostiza, and Talarczyk [172].
- Constructions involving the so-called β–regular Markov chains in Owada and Samorodnitsky [784], Jung, Owada, and Samorodnitsky [541].
- Asymptotics in multinomial occupancy schemes: see Gnedin, Hansen, and Pitman [410].
- Other miscellaneous constructions in, for example, Koutsoyiannis [578], Alvarez-Lacalle, Dorow, Eckmann, and Moses [24], Chevillon and Mavroeidis [233].

The collection of articles in Rangarajan and Ding [841] provides a number of other examples and models in physical sciences where long-range dependence is observed.

4

Hermite Processes

Limits of normalized sums of functions of Gaussian time series with long-range dependence include fractional Brownian motion, the Rosenblatt process or, more generally, the Hermite processes. We present the limit results in the next Chapter 5. We focus in the present chapter on the limit processes themselves. These limits are called *Hermite processes* because they are related to Hermite polynomials. A convenient way to represent Hermite processes is to use multiple stochastic integrals, which are defined in Appendix B.2.

We introduce Hermite polynomials in Section 4.1 and indicate how they are related to multiple stochastic integrals. We provide in Section 4.2 the integral representation of the Hermite processes in the time domain, in the spectral domain and on an interval.

Key tools are moments, cumulants and diagram formulae, which are given in Sections 4.3 and 4.4. We focus on general multiple stochastic integrals of order two in Section 4.5. Then, in Section 4.6, we treat an important example, namely the Rosenblatt process, and consider its marginal cumulative distribution function in Section 4.7 with tables in Section 4.8. Finally, in Section 4.9, we consider generalized Hermite processes.

4.1 Hermite Polynomials and Multiple Stochastic Integrals

We first define Hermite[1] polynomials.

Definition 4.1.1 The *Hermite polynomial of order* 0 is $H_0(x) = 1$. The *Hermite polynomial of order n*, $n = 1, 2, \ldots$, is defined as

$$H_n(x) = (-1)^n e^{x^2/2} \frac{d^n}{dx^n} e^{-x^2/2}, \quad x \in \mathbb{R}. \tag{4.1.1}$$

One thus has

$$H_0(x) = 1, \quad H_1(x) = x, \quad H_2(x) = x^2 - 1, \quad H_3(x) = x^3 - 3x,$$

$$H_4(x) = x^4 - 6x^2 + 3, \quad H_5(x) = x^5 - 10x^3 + 15x, \quad H_6(x) = x^6 - 15x^4 + 45x^2 - 15,$$

[1] Pronounced [er MEET].

and so on.[2] Hermite polynomials appear in the expansion

$$F(x,t) := e^{tx-\frac{t^2}{2}} = e^{x^2/2} e^{-\frac{(x-t)^2}{2}} = e^{x^2/2} \sum_{n=0}^{\infty} \frac{t^n}{n!} \left(\frac{d^n}{dt^n} e^{-\frac{(x-t)^2}{2}} \Big|_{t=0} \right) = \sum_{n=0}^{\infty} \frac{t^n H_n(x)}{n!},$$
(4.1.2)

where we used the Taylor's expansion of $e^{-\frac{(x-t)^2}{2}}$ in t, and

$$\frac{d}{dt} e^{-\frac{(x-t)^2}{2}} = \left(\frac{d}{dy} e^{-\frac{y^2}{2}} \Big|_{y=x-t} \right) \left(\frac{d}{dt}(x-t) \right) = (-1) \frac{d}{dt} e^{-\frac{y^2}{2}} \Big|_{y=x-t}$$

and, more generally,

$$\frac{d^n}{dt^n} e^{-\frac{(x-t)^2}{2}} = (-1)^n \frac{d^n}{dy^n} e^{-\frac{y^2}{2}} \Big|_{y=x-t}$$

and thus

$$\frac{d^n}{dt^n} e^{-\frac{(x-t)^2}{2}} \Big|_{t=0} = (-1)^n \frac{d^n}{dy^n} e^{-\frac{y^2}{2}} \Big|_{y=x} = (-1)^n \frac{d^n}{dx^n} e^{-\frac{x^2}{2}} = e^{-x^2/2} H_n(x),$$

yielding (4.1.2). In particular,

$$\frac{\partial^n}{\partial t^n} F(x,t) \Big|_{t=0} = H_n(x).$$
(4.1.3)

We now relate Hermite polynomials, Gaussian variables and multiple stochastic integrals. As in Appendix B.2, we denote the multiple integral with respect to a real-valued Gaussian measure in (B.2.1) as $I_k(f)$, for $f \in L^2(E^k, m^k)$, and the multiple integral with respect to an Hermitian Gaussian measure in (B.2.14) as $\widehat{I}_k(g)$, for Hermitian $g \in L^2(\mathbb{R}^k, m^k)$. The tensor product $\otimes$ is also considered in Appendix B.2.

Proposition 4.1.2 (a) *(Time domain)* Let $h \in L^2(E, m)$ with $\|h\|_{L^2(E,m)} = 1$. Then, for $k \geq 1$,

$$H_k(I_1(h)) = I_k(h^{\otimes k}).$$
(4.1.4)

(b) *(Spectral domain)* Let $g \in L^2(\mathbb{R}, m)$ be Hermitian with $\|g\|_{L^2(\mathbb{R},m)} = 1$ and m be symmetric. Then, for $k \geq 1$,

$$H_k(\widehat{I}_1(g)) = \widehat{I}_k(g^{\otimes k}).$$
(4.1.5)

Example 4.1.3 (*Hermite polynomial and multiple integral of order three*) Taking $k = 3$ in (4.1.4) and using $I_1(h) = \int_E h(u)W(du)$, for example, one has

$$H_3\left(\int_E h(u)W(du) \right) = \left(\int_E h(u)W(du) \right)^3 - 3 \int_E h(u)W(du)$$

$$= \int_{E^3}' h(u_1)h(u_2)h(u_3) W(du_1)W(du_2)W(du_3).$$

---

[2] Another, different definition of Hermite polynomials is to set $\widetilde{H}_n(x) = (-1)^n e^{x^2} \frac{d^n}{dx^n} e^{-x^2}$. The connection to the Hermite polynomials in Definition 4.1.1 is $\widetilde{H}_n(x) = 2^{n/2} H_n(\sqrt{2}x)$. Yet another way to define Hermite polynomials is to set them equal to $H_n(x)/n!$ or $H_n(x)/\sqrt{n!}$. It is important therefore to always check which definition is used.

Remark 4.1.4 If, for example, $\|h\|_{L^2(E,m)} \neq 1$, Proposition 4.1.2, (a), can be applied to $h/\|h\|_{L^2(E,m)}$ to obtain

$$\|h\|_{L^2(E,m)}^k H_k\left(\frac{I_1(h)}{\|h\|_{L^2(E,m)}}\right) = I_k(h^{\otimes k}). \tag{4.1.6}$$

Also, note that in the case of $E = \mathbb{R}$ and m being the Lebesgue measure, a Gaussian variable $I_1(h) = \int_{\mathbb{R}} h(u) B(du)$ in part (a) of Proposition 4.1.2 can be represented as $\widehat{I}_1(\widehat{g}) = \int_{\mathbb{R}} \widehat{g}(x) \widehat{B}(dx)$ in part (b) with $\widehat{g}(x) = \widehat{h}(x)/\sqrt{2\pi}$, where $\widehat{h}(x) = \int_{\mathbb{R}} e^{ixu} h(u) du$ stands for the Fourier transform of h and an Hermitian Gaussian measure $\widehat{B}(dx)$ has the Lebesgue control measure. This follows from (B.2.16).

Proof Consider part (a). The proof is by induction in k. The relation (4.1.4) is trivial for $k = 1$ since $H_1(x) = x$. Suppose the relation holds for k and consider it for $k + 1$. By using the property

$$H_{k+1}(x) = x H_k(x) - k H_{k-1}(x), \quad k \geq 1, \tag{4.1.7}$$

of Hermite polynomials (Exercise 4.1) and the induction assumption, we have that

$$H_{k+1}(I_1(h)) = I_1(h) H_k(I_1(h)) - k H_{k-1}(I_1(h)) = I_1(h) I_k(h^{\otimes k}) - k I_{k-1}(h^{\otimes (k-1)}).$$

The last expression equals $I_{k+1}(h^{\otimes(k+1)})$ by using the formula (B.2.20) with $f = h^{\otimes k}$ and $g = h$, and the fact that

$$h^{\otimes k} \otimes_1 h = h^{\otimes (k-1)} \int_E h(u) h(u) m(du) = h^{\otimes (k-1)},$$

since $\|h\|_{L^2(E,m)} = 1$. The proof of part (b) is analogous by using the formula (B.2.23). □

Example 4.1.5 (*Connection to Itô's stochastic calculus*) Take $E = \mathbb{R}$, $m(du) = du$ and $h_t(u) = 1_{[0,t)}(u)/\sqrt{t}$, $t > 0$, in Proposition 4.1.2, (a), so that

$$I_1(h_t) = \int_{\mathbb{R}} h_t(u) B(du) = \int_{\mathbb{R}} \frac{1_{[0,t)}(u)}{\sqrt{t}} B(du) = \frac{B(t)}{\sqrt{t}},$$

where $\{B(t)\}$ is a standard Brownian motion. Note, for example, that since $H_2(x) = x^2 - 1$,

$$H_2(I_1(h_t)) = H_2\left(\int_{\mathbb{R}} \frac{1_{[0,t)}(u)}{\sqrt{t}} B(du)\right) = \frac{B(t)^2}{t} - 1.$$

On the other hand, by using the relation (B.2.24) of multiple integrals to Itô's stochastic integrals and Itô's formula, one also has

$$I_2(h_t^{\otimes 2}) = \int_{\mathbb{R}^2}' \frac{1_{[0,t)}(u_1)}{\sqrt{t}} \frac{1_{[0,t)}(u_2)}{\sqrt{t}} B(du_1) B(du_2)$$

$$= \frac{2}{t} \int_0^t \left\{ \int_0^{u_2} dB(u_1) \right\} dB(u_2) = \frac{2}{t} \int_0^t B(u_2) dB(u_2)$$

$$= \frac{1}{t}(B(t)^2 - t) = \frac{B(t)^2}{t} - 1.$$

Thus, the relation (4.1.4) indeed holds for $k = 2$ and $h_t(u) = 1_{[0,t)}(u)/\sqrt{t}$.

4.2 Integral Representations of Hermite Processes

In this section, we define Hermite processes, provide their various representations and consider their basic properties.

4.2.1 Integral Representation in the Time Domain

The following is a definition of Hermite processes in time domain.

Definition 4.2.1 Let $k \geq 1$ be an integer and

$$H \in \left(\frac{1}{2}, 1\right). \tag{4.2.1}$$

Set

$$H_0 = 1 - \frac{1-H}{k} \in \left(1 - \frac{1}{2k}, 1\right), \tag{4.2.2}$$

so that $H = 1 - k(1 - H_0)$. The *Hermite process* $\{Z_H^{(k)}(t)\}_{t \in \mathbb{R}}$ *of order k* is defined as

$$Z_H^{(k)}(t) = A_{k,H} \int_{\mathbb{R}^k}' \left\{ \int_0^t \prod_{j=1}^k (s - u_j)_+^{-(\frac{1}{2} + \frac{1-H}{k})} ds \right\} B(du_1) \ldots B(du_k) \tag{4.2.3}$$

$$= a_{k,H_0} \int_{\mathbb{R}^k}' \left\{ \int_0^t \prod_{j=1}^k (s - u_j)_+^{H_0 - \frac{3}{2}} ds \right\} B(du_1) \ldots B(du_k), \quad t \in \mathbb{R}, \tag{4.2.4}$$

where $B(du)$ is a Gaussian random measure on $\mathbb{R}$ with the Lebesgue control measure du, $u_+ = \max\{u, 0\}$ and $A_{k,H}, a_{k,H_0}$ are normalizing constants. The integral $\int_0^t$ is interpreted as $-\int_t^0$ when $t < 0$. The Hermite process $\{Z_H^{(k)}(t)\}_{t \in \mathbb{R}}$ is called *standard* if $\mathbb{E}(Z_H^{(k)}(1))^2 = 1$.

The parameter H_0 is the one used in the FBM kernel and the parameter H is the self-similarity parameter of the Hermite process, as noted in Proposition 4.2.2 below. Note that the Hermite process of order one is FBM (see Section 2.6.1) and in this case $H = H_0$. The Hermite process of order two is the Rosenblatt process (see Section 2.6.3). Note also that the Hermite process of order k is defined through the kernel function

$$f_t(u_1, \ldots, u_k) = \int_0^t \prod_{j=1}^k (s - u_j)_+^{p-1} ds, \tag{4.2.5}$$

where, in view of (4.2.2),[3]

$$p = H_0 - \frac{1}{2} \in \left(\frac{1}{2} - \frac{1}{2k}, \frac{1}{2}\right). \tag{4.2.6}$$

The following result provides basic properties of Hermite processes.

---

[3] The parametrization p is well-suited for manipulating Fourier transforms.

4.2 Integral Representations of Hermite Processes

Proposition 4.2.2 *Hermite processes are well-defined, have stationary increments and are H–self-similar. The Hermite process of order k is standard when*

$$A_{k,H} = \left(\frac{H(2H-1)}{k!(B(\frac{1}{2} - \frac{1-H}{k}, \frac{2-2H}{k}))^k}\right)^{1/2} \tag{4.2.7}$$

or

$$a_{k,H_0} = \left(\frac{(1-k(1-H_0))(1-2k(1-H_0))}{k!(B(H_0-\frac{1}{2}, 2-2H_0))^k}\right)^{1/2}, \tag{4.2.8}$$

where $B(a,b)$ denotes the beta function (2.2.12).

Proof The Hermite process of order k is well-defined as long as $\|f_t\|_{L^2(\mathbb{R}^k)} < \infty$, in which case $\mathbb{E}|Z_H^{(k)}(t)|^2 = k!\|f_t\|_{L^2(\mathbb{R}^k)}^2$. Since, for example, for $t > 0$,

$$f_t(tu_1, \ldots, tu_k) = t^{(p-1)k+1} f_1(u_1, \ldots, u_k) = t^{H-\frac{k}{2}} f_1(u_1, \ldots, u_k), \tag{4.2.9}$$

it is enough to show that $\|f_1\|_{L^2(\mathbb{R}^k)} < \infty$. In fact, $\|f_1\|_{L^2(\mathbb{R}^k)}$ can be calculated easily. Arguing as in the proof of Proposition 2.6.15, observe that

$$\|f_1\|_2^2 = \int_{\mathbb{R}} du_1 \ldots \int_{\mathbb{R}} du_k \left(\int_0^1 \prod_{j=1}^k (s-u_j)_+^{p-1} ds\right)^2$$

$$= \int_0^1 ds_1 \int_0^1 ds_2 \left(\int_{\mathbb{R}} (s_1-u)_+^{p-1} (s_2-u)_+^{p-1} du\right)^k$$

$$= 2\int_0^1 ds_1 \int_{s_1}^1 ds_2 \left(\int_0^\infty z^{p-1}(s_2-s_1+z)^{p-1} dz\right)^k$$

$$= 2\left(\int_0^\infty v^{p-1}(1+v)^{p-1} dv\right)^k \int_0^1 ds_1 \int_{s_1}^1 ds_2 (s_2-s_1)^{(2p-1)k}$$

$$= 2(B(p, 1-2p))^k \int_0^1 ds_1 \int_{s_1}^1 ds_2 (s_2-s_1)^{(2p-1)k}$$

$$= \frac{2(B(p, 1-2p))^k}{((2p-1)k+1)((2p-1)k+2)} < \infty,$$

by using (4.2.6). This also yields the normalizing constants (4.2.7)–(4.2.8) by using (4.2.6) and (4.2.2). The H–self-similarity follows by using the scaling property (4.2.9) of the kernel function, and stationarity of the increments can be proved by using the fact that

$$f_{t+h}(u_1, \ldots, u_k) - f_h(u_1, \ldots, u_k) = f_t(u_1 - h, \ldots, u_k - h).$$ □

4.2.2 Integral Representation in the Spectral Domain

The following result extends the result on the Fourier transform in Proposition 2.6.15 to the kernel function (4.2.5).

Proposition 4.2.3 *Let $f_t(u_1, \ldots, u_k)$, $t, u_1, \ldots, u_k \in \mathbb{R}$, be the function defined in (4.2.5)–(4.2.6). Then,*

$$\widehat{f_t}(x_1, \ldots, x_k) = \Gamma(p)^k \frac{e^{it(x_1+\cdots+x_k)} - 1}{i(x_1 + \cdots + x_k)} \prod_{j=1}^{k} |x_j|^{-p} e^{-i\,\mathrm{sign}(x_j)\frac{p\pi}{2}}. \tag{4.2.10}$$

Proof The result can be proved in the same way as the result on the Fourier transform in Proposition 2.6.15. □

An immediate corollary of Proposition 4.2.3 is the spectral representation of Hermite processes. We write the representation using only the parametrization H_0.

Proposition 4.2.4 *The Hermite process of order k can be represented as*

$$\{Z_H^{(k)}(t)\}_{t \in \mathbb{R}} \stackrel{d}{=} \left\{ b_{k,H_0} \int_{\mathbb{R}^k}'' \frac{e^{it(x_1+\cdots+x_k)} - 1}{i(x_1 + \cdots + x_k)} \prod_{j=1}^{k} |x_j|^{\frac{1}{2}-H_0} \widehat{B}(dx_1) \ldots \widehat{B}(dx_k) \right\}_{t \in \mathbb{R}}, \tag{4.2.11}$$

where $\widehat{B}(dx)$ is an Hermitian Gaussian random measure on $\mathbb{R}$ with control measure dx. The normalizing constant b_{k,H_0} satisfies

$$b_{k,H_0} = \left(\frac{\Gamma(H_0 - 1/2)}{\sqrt{2\pi}} \right)^k a_{k,H_0}. \tag{4.2.12}$$

In particular, the Hermite process of order k is standard when

$$b_{k,H_0} = \left(\frac{(k(H_0 - 1) + 1)(2k(H_0 - 1) + 1)}{k! [2\Gamma(2 - 2H_0) \sin((H_0 - \frac{1}{2})\pi)]^k} \right)^{1/2}. \tag{4.2.13}$$

Proof The equality in distribution (4.2.11) and the relation (4.2.12) between the normalizing constants follows from (B.2.16) in Appendix B.2 and Proposition 4.2.3, since the kernels in the right-hand sides of (4.2.4) and (4.2.12) are related through the Fourier transform. Note also from Proposition 4.2.3 that the Fourier transform of the kernel in (4.2.4) produces complex exponentials

$$\prod_{j=1}^{k} e^{-i\,\mathrm{sign}(x_j)\frac{(H_0-1/2)\pi}{2}},$$

which do not appear in the right-hand side of (4.2.11). The exponentials can be omitted by the formula (B.2.15) for the change of variables found in Appendix B. Finally, the form (4.2.13) of the normalizing constant follows from (4.2.12) and (4.2.8). □

4.2.3 Integral Representation on an Interval

The representations (4.2.4) and (4.2.11) of Hermite processes are on the real line. There is also an integral representation of Hermite processes on an interval. We continue working with the parametrization H_0 only.

4.2 Integral Representations of Hermite Processes

Proposition 4.2.5 *The Hermite process of order k can be represented as*

$$\{Z_H^{(k)}(t)\}_{t\geq 0} \stackrel{d}{=}$$
$$\left\{d_{k,H_0}\int_{[0,t]^k}' \left\{\prod_{j=1}^k v_j^{\frac{1}{2}-H_0} \int_0^t s^{k(H_0-\frac{1}{2})} \prod_{j=1}^k (s-v_j)_+^{H_0-\frac{3}{2}} ds\right\} B(dv_1)\ldots B(dv_k)\right\}_{t\geq 0},$$
(4.2.14)

where $B(dv)$ is a Gaussian random measure on $\mathbb{R}$ with control measure dv. The normalizing constant d_{k,H_0} satisfies

$$d_{k,H_0} = a_{k,H_0}.\qquad(4.2.15)$$

Remark 4.2.6 When $k=1$, the expression (4.2.14) provides an integral representation on an interval of FBM with $H_0 \in (1/2, 1)$. This representation of FBM is extended to $H_0 \in (0, 1/2)$ in Chapters 6 and 7, and plays there a key role in the development of a stochastic calculus with respect to FBM.

The proof of Proposition 4.2.5 involves a number of auxiliary results, some of which have their own interest. We provide the proof next without using the usual delimiters PROOF and □ for the beginning and the end of the proof.

It is instructive to explain first the basic idea of the proof in informal terms. Interchanging the two integrals in the representation (4.2.4) and using Proposition 4.1.2, (a), we have

$$\int_{\mathbb{R}^k}' \left\{\int_0^t \prod_{j=1}^k (s-u_j)_+^{H_0-\frac{3}{2}} ds\right\} B(du_1)\ldots dB(du_k)$$
$$= \int_0^t \left\{\int_{\mathbb{R}^k}' \prod_{j=1}^k (s-u_j)_+^{H_0-\frac{3}{2}} B(du_1)\ldots dB(du_k)\right\} ds$$
$$= \int_0^t H_k\left(\int_{\mathbb{R}} (s-u)_+^{H_0-\frac{3}{2}} B(du)\right) ds.\qquad(4.2.16)$$

By finding a representation of a Gaussian process $\int_{\mathbb{R}} (s-u)_+^{H_0-\frac{3}{2}} B(du)$ on an interval and repeating the argument (4.2.16) backwards leads to the representation (4.2.14) of Hermite processes on an interval.

The Gaussian process $\int_{\mathbb{R}} (s-u)_+^{H_0-\frac{3}{2}} B(du)$ in (4.2.16) is not well-defined because $\int_{\mathbb{R}} (s-u)_+^{2H_0-3} du = \infty$. To be able to proceed as in (4.2.16), we use the idea of *regularization*. For $\epsilon > 0$, let

$$Z_{H,\epsilon}^{(k)}(t) = \int_{\mathbb{R}^k}' \left\{\int_0^t \prod_{j=1}^k (s-u_j)_+^{H_0-\frac{3}{2}} 1_{\{s-u_j>\epsilon\}} ds\right\} B(du_1)\ldots B(du_k)\qquad(4.2.17)$$

be a regularized version of $Z_H^{(k)}(t)$. By using (B.2.25) in Appendix B.2 and Proposition 4.1.2, (a), we have

$$Z_{H,\epsilon}^{(k)}(t) = \int_0^t \Big\{ \int_{\mathbb{R}^k}' \prod_{j=1}^k (s-u_j)_+^{H_0-\frac{3}{2}} 1_{\{s-u_j>\epsilon\}} B(du_1)\ldots B(du_k) \Big\} ds$$

$$= \int_0^t \|\dot{B}_{H_0,\epsilon}(s)\|_2^k H_k\Big(\frac{\dot{B}_{H_0,\epsilon}(s)}{\|\dot{B}_{H_0,\epsilon}(s)\|_2} \Big) ds, \qquad (4.2.18)$$

where $\|X\|_2 = (\mathbb{E}X^2)^{1/2}$ for a random variable X, and

$$\dot{B}_{H_0,\epsilon}(s) = \int_{\mathbb{R}} (s-u)_+^{H_0-\frac{3}{2}} 1_{\{s-u>\epsilon\}} B(du) \qquad (4.2.19)$$

is a regularized fractional Gaussian noise.

We would like to have now a representation of the regularized Gaussian process $\dot{B}_{H_0,\epsilon}$ on an interval. Such a representation, however, is not expected to have a simple analytic form. We shall therefore provide a representation of $\dot{B}_{H_0,\epsilon}$ on the positive half-axis. This will lead to a representation of Hermite processes on the positive half-axis. We shall then regularize it in a different way in order to derive the representation (4.2.14) on a finite interval.

Representation on the Positive Half-Axis

In the next auxiliary result, we establish an equivalent representation of $\dot{B}_{H_0,\epsilon}$ on the positive-half axis.

Lemma 4.2.7 *Let $H_0 < 1$. Then,*

$$\{\dot{B}_{H_0,\epsilon}(s)\}_{s\geq 0} \stackrel{d}{=} \{\dot{X}_{H_0,\epsilon}(s)\}_{s\geq 0}, \qquad (4.2.20)$$

where

$$\dot{X}_{H_0,\epsilon}(s) = \int_0^\infty z^{\frac{1}{2}-H_0}(1-sz)^{H_0-\frac{3}{2}} 1_{\{z<1/(\epsilon+s)\}} B(dz) \qquad (4.2.21)$$

and $B(dz)$ is a Gaussian random measure with control measure dz.

Proof The stationary Gaussian process $\dot{B}_{H_0,\epsilon}(s)$ has the following covariance. For $0 \leq s_1 < s_2$,

$$I := \mathbb{E}\dot{B}_{H_0,\epsilon}(s_2)\dot{B}_{H_0,\epsilon}(s_1) = \int_{\mathbb{R}} (s_2-y)_+^{H_0-\frac{3}{2}}(s_1-y)_+^{H_0-\frac{3}{2}} 1_{\{s_1-y>\epsilon\}} dy.$$

The proof follows from a series of changes of variables. Some of them are rather delicate. By making the change of variables $s_1 - y = z(s_2 - s_1)$, we get that

$$I = (s_2-s_1)^{2H_0-2} \int_0^\infty z^{H_0-\frac{3}{2}}(1+z)^{H_0-\frac{3}{2}} 1_{\{z>\epsilon/(s_2-s_1)\}} dz.$$

By making the change of variables $z = (1-u)/u$, we obtain that

$$I = (s_2-s_1)^{2H_0-2} \int_0^1 u^{1-2H_0}(1-u)^{H_0-\frac{3}{2}} 1_{\{u<(s_2-s_1)/(\epsilon+s_2-s_1)\}} du.$$

A further change of variables $u = (s_2-s_1)w/(s_1(s_2-w))$ leads to

$$I = s_1^{H_0-\frac{1}{2}} s_2^{H_0-\frac{1}{2}} \int_0^{s_1} w^{1-2H_0}(s_2-w)^{H_0-\frac{3}{2}}(s_1-w)^{H_0-\frac{3}{2}} 1_{\{w<(s_1s_2)/(\epsilon+s_2)\}} dw.$$

Finally, the change of variables $w = s_1 s_2 z$ yields

$$I = \int_0^\infty z^{1-2H_0}(1-s_2 z)^{H_0-\frac{3}{2}}(1-s_1 z)^{H_0-\frac{3}{2}} 1_{\{z<1/(\epsilon+s_1)\}} 1_{\{z<1/(\epsilon+s_2)\}} dz,$$

which is $\mathbb{E}\dot{X}_{H_0,\epsilon}(s_2)\dot{X}_{H_0,\epsilon}(s_1)$. $\square$

Lemma 4.2.7 and the relation (4.2.18) yield

$$\{Z_{H,\epsilon}^{(k)}(t)\}_{t\geq 0} \stackrel{d}{=} \left\{ \int_0^t \|\dot{X}_{H_0,\epsilon}(s)\|_2^k H_k\left(\frac{\dot{X}_{H_0,\epsilon}(s)}{\|\dot{X}_{H_0,\epsilon}(s)\|_2}\right) ds \right\}_{t\geq 0}. \tag{4.2.22}$$

Repeating the argument in (4.2.17) and (4.2.18) gives

$$\{Z_{H,\epsilon}^{(k)}(t)\}_{t\geq 0} \stackrel{d}{=} \{X_{H_0,\epsilon}^{(k)}(t)\}_{t\geq 0}, \tag{4.2.23}$$

where

$$X_{H_0,\epsilon}^{(k)}(t) = \int_{[0,\infty)^k}' \left\{ \int_0^t \prod_{j=1}^k z_j^{\frac{1}{2}-H_0}(1-sz_j)^{H_0-\frac{3}{2}} 1_{\{z_j < 1/(\epsilon+s)\}} ds \right\} B(dz_1)\ldots B(dz_k). \tag{4.2.24}$$

Letting $\epsilon \to 0$ in (4.2.23) yields the following integral representations of Hermite processes on the positive half-axis.

Proposition 4.2.8 *One has*

$$\{Z_{H_0}^{(k)}(t)\}_{t\geq 0} \stackrel{d}{=} \{X_{H_0}^{(k)}(t)\}_{t\geq 0}, \tag{4.2.25}$$

where

$$X_{H_0}^{(k)}(t) = \int_{[0,\infty)^k}' \left\{ \int_0^t \prod_{j=1}^k z_j^{\frac{1}{2}-H_0}(1-sz_j)_+^{H_0-\frac{3}{2}} ds \right\} B(dz_1)\ldots B(dz_k) \tag{4.2.26}$$

and $B(dz)$ is a Gaussian random measure on $\mathbb{R}$ with control measure dz.

Proof Relation (4.2.25) is obtained by letting $\epsilon \to 0$ in (4.2.23), and using the fact that for fixed t, $Z_{H,\epsilon}^{(k)}(t)$ and $X_{H_0,\epsilon}^{(k)}(t)$ converge in the $L^2(\Omega)$–sense to the corresponding limit random variables $Z_H^{(k)}(t)$ and $X_{H_0}^{(k)}(t)$. The latter convergence is a consequence of the fact that the corresponding deterministic integrand functions converge in $L^2(\mathbb{R}^k)$. For example, in the case of $Z_{H,\epsilon}^{(k)}(t)$ and $Z_H^{(k)}(t)$, one needs to consider

$$\int_{\mathbb{R}^k} \left(\int_0^t \prod_{j=1}^k (s-u_j)_+^{H_0-\frac{3}{2}} 1_{\{s-u_j>\epsilon\}} ds - \int_0^t \prod_{j=1}^k (s-u_j)_+^{H_0-\frac{3}{2}} ds \right)^2 du_1 \ldots du_k$$

$$= \int_{\mathbb{R}^k} \left(\int_0^t \prod_{j=1}^k (s-u_j)_+^{H_0-\frac{3}{2}} 1_{\{s-u_j\leq \epsilon\}} ds \right)^2 du_1 \ldots du_k.$$

As $\epsilon \to 0$, this last integral converges to 0 by the dominated convergence theorem and the fact that the same integral without the indicator functions is finite, by (4.2.2). The same argument can be repeated for $X_{H_0,\epsilon}^{(k)}(t)$ and $X_{H_0}^{(k)}(t)$. In this case, note in addition that

$$\int_{[0,\infty)^k} \left(\int_0^t \prod_{j=1}^k z_j^{\frac{1}{2}-H_0}(1-sz_j)_+^{H_0-\frac{3}{2}} ds \right)^2 dz_1 \ldots dz_k$$

$$= \lim_{\epsilon \to 0} \int_{[0,\infty)^k} \left(\int_0^t \prod_{j=1}^k z_j^{\frac{1}{2}-H_0}(1-sz_j)_+^{H_0-\frac{3}{2}} 1_{\{z_j<1/(\epsilon+s)\}} ds \right)^2 dz_1 \ldots dz_k$$

$$= \lim_{\epsilon \to 0} \int_{\mathbb{R}^k} \left(\int_0^t \prod_{j=1}^k (s-u_j)_+^{H_0-\frac{3}{2}} 1_{\{s-u_j>\epsilon\}} ds \right)^2 du_1 \ldots du_k$$

$$= \int_{\mathbb{R}^k} \left(\int_0^t \prod_{j=1}^k (s-u_j)_+^{H_0-\frac{3}{2}} ds \right)^2 du_1 \ldots du_k < \infty,$$

where the second equality above is a consequence of (4.2.23), so that $X_{H_0}^{(k)}(t)$ is well-defined. □

Representation on an Interval

The next step is to regularize $X_{H_0}^{(k)}$ but using a regularization different from (4.2.24). For $\epsilon > 0$, consider

$$W_{H_0,\epsilon}^{(k)}(t) = \int_{[0,\infty)^k}' \left\{ \int_0^t \prod_{j=1}^k z_j^{\frac{1}{2}-H_0}(1-sz_j)^{H_0-\frac{3}{2}} 1_{\{z_j<(1-\epsilon)/s\}} ds \right\} B(dz_1) \ldots B(dz_k). \tag{4.2.27}$$

As $\epsilon \to 0$, $W_{H_0,\epsilon}^{(k)}(t)$ converges to $X_{H_0}^{(k)}(t)$. By arguing as in (4.2.17) and (4.2.18), write

$$W_{H_0,\epsilon}^{(k)}(t) = \int_0^t \|\dot{W}_{H_0,\epsilon}(s)\|_2^k H_k\left(\frac{\dot{W}_{H_0,\epsilon}(s)}{\|\dot{W}_{H_0,\epsilon}(s)\|_2}\right) ds, \tag{4.2.28}$$

where

$$\dot{W}_{H_0,\epsilon}(s) = \int_0^\infty z^{\frac{1}{2}-H_0}(1-sz)^{H_0-\frac{3}{2}} 1_{\{z<(1-\epsilon)/s\}} B(dz). \tag{4.2.29}$$

Whereas in (4.2.24) the regularization involved $1_{\{z<1/(\epsilon+s)\}}$, here the regularization involves $1_{\{z<(1-\epsilon)/s\}}$. In the next result, we provide an integral representation of $\dot{W}_{H_0,\epsilon}$ on an interval.

Proposition 4.2.9 *Let $H_0 < 1$. Then,*

$$\{\dot{W}_{H_0,\epsilon}(s)\}_{s\geq 0} \stackrel{d}{=} \{\dot{Y}_{H_0,\epsilon}(s)\}_{s\geq 0}, \tag{4.2.30}$$

where

$$\dot{Y}_{H_0,\epsilon}(s) = s^{H_0-\frac{1}{2}} \int_0^\infty v^{\frac{1}{2}-H_0}(s-v)^{H_0-\frac{3}{2}} 1_{\{v<s/(1-\epsilon)\}} B(dv) \tag{4.2.31}$$

and $B(dv)$ is a Gaussian random measure with control measure dv.

Proof Let $0 \le s_1 < s_2$. By the change of variables $z = x/(s_1 s_2)$,

$$\mathbb{E}\dot{W}_{H_0,\epsilon}(s_2)\dot{W}_{H_0,\epsilon}(s_1) = \int_0^\infty z^{1-2H_0}(1-s_2 z)^{H_0-\frac{3}{2}}(1-s_1 z)^{H_0-\frac{3}{2}} 1_{\{z<(1-\epsilon)/s_2\}} 1_{\{z<(1-\epsilon)/s_1\}} dz$$

$$= (s_1 s_2)^{H_0-\frac{1}{2}} \int_0^\infty x^{1-2H_0}(s_2-x)^{H_0-\frac{3}{2}}(s_1-x)^{H_0-\frac{3}{2}} 1_{\{x<s_2(1-\epsilon)\}} 1_{\{x<s_1(1-\epsilon)\}} dx,$$

which is also $\mathbb{E}\dot{Y}_{H_0,\epsilon}(s_2)\dot{Y}_{H_0,\epsilon}(s_1)$. □

Proposition 4.2.9 and the relation (4.2.28) yield

$$\{W^{(k)}_{H_0,\epsilon}(t)\}_{t \ge 0} \stackrel{d}{=} \{Y^{(k)}_{H_0,\epsilon}(t)\}_{t \ge 0}, \qquad (4.2.32)$$

where

$$Y^{(k)}_{H_0,\epsilon}(t)$$

$$= \int_{[0,\infty)^k}' \left\{ \prod_{j=1}^k v_j^{\frac{1}{2}-H_0} \int_0^t s^{k(H_0-\frac{1}{2})} \prod_{j=1}^k (s-v_j)^{H_0-\frac{3}{2}} 1_{\{v_j<s(1-\epsilon)\}} ds \right\} B(dv_1)\ldots B(dv_k).$$

(4.2.33)

Letting $\epsilon \to 0$ and using Proposition 4.2.8, we obtain the integral representation (4.2.14) of Hermite processes on an interval. The proof is analogous to that of Proposition 4.2.8, and is omitted.

4.2.4 Summary

For the convenience of the reader, we gather the time, spectral, half-axis and finite interval representations of the Hermite processes in the following theorem.

Theorem 4.2.10 *Let H and H_0 satisfy (4.2.1) and (4.2.2), respectively. Let also $B(du)$ be a Gaussian random measure on $\mathbb{R}$ with control measure du, and $\widehat{B}(dx)$ be an Hermitian Gaussian random measure on $\mathbb{R}$ with control measure dx. Then, the Hermite process $Z^{(k)}_H = \{Z^{(k)}_H(t)\}_{t\in\mathbb{R}}$ has the following integral representations in the sense of equality of the finite-dimensional distributions:*

(a) Time domain representation: for $t \in \mathbb{R}$,

$$a_{k,H_0} \int_{\mathbb{R}^k}' \left\{ \int_0^t \prod_{j=1}^k (s-u_j)_+^{H_0-\frac{3}{2}} ds \right\} B(du_1)\ldots dB(du_k). \qquad (4.2.34)$$

(b) Spectral domain representation: for $t \in \mathbb{R}$,

$$b_{k,H_0} \int_{\mathbb{R}^k}'' \frac{e^{it(x_1+\cdots+x_k)}-1}{i(x_1+\cdots+x_k)} \prod_{j=1}^k |x_j|^{\frac{1}{2}-H_0} \widehat{B}(dx_1)\ldots \widehat{B}(dx_k). \qquad (4.2.35)$$

(c) *Positive half-axis representation: for* $t \geq 0$,

$$c_{k,H_0} \int'_{[0,\infty)^k} \left\{ \int_0^t \prod_{j=1}^k z_j^{\frac{1}{2}-H_0} (1-sz_j)_+^{H_0-\frac{3}{2}} ds \right\} B(dz_1) \ldots B(dz_k). \quad (4.2.36)$$

(d) *Finite interval representation: for* $t \geq 0$,

$$d_{k,H_0} \int'_{[0,t]^k} \left\{ \prod_{j=1}^k v_j^{\frac{1}{2}-H_0} \int_0^t s^{k(H_0-\frac{1}{2})} \prod_{j=1}^k (s-v_j)_+^{H_0-\frac{3}{2}} ds \right\} B(dv_1) \ldots B(dv_k). \quad (4.2.37)$$

If the Hermite process $Z_H^{(k)}$ *is standard so that* $\mathbb{E}(Z_H^{(k)}(1))^2 = 1$, *then*

$$a_{k,H_0} = c_{k,H_0} = d_{k,H_0} = \left(\frac{(1-k(1-H_0))(1-2k(1-H_0))}{k!(B(H_0-\frac{1}{2}, 2-2H_0))^k} \right)^{1/2} \quad (4.2.38)$$

$$= \left(\frac{(k(H_0-1)+1)(2k(H_0-1)+1)\Gamma(3/2-H_0)^k}{k![\Gamma(H_0-1/2)\Gamma(2-2H_0)]^k} \right)^{1/2} \quad (4.2.39)$$

and

$$b_{k,H_0} = \left(\frac{(k(H_0-1)+1)(2k(H_0-1)+1)}{k![2\Gamma(2-2H_0)\sin((H_0-\frac{1}{2})\pi)]^k} \right)^{1/2}. \quad (4.2.40)$$

Proof This follows from Definition 4.2.1, and Propositions 4.2.2, 4.2.4, 4.2.5 and 4.2.8. $\square$

It is sometimes useful to have the expressions in Theorem 4.2.10 expressed in terms of H instead of H_0. Since H and H_0 are related through (4.2.2), one has:

Corollary 4.2.11 *The Hermite process* $Z_H^{(k)} = \{Z_H^{(k)}(t)\}_{t \in \mathbb{R}}$ *has the following integral representations in the sense of equality of the finite-dimensional distributions:*

(a) *Time domain representation: for* $t \in \mathbb{R}$,

$$A_{k,H} \int'_{\mathbb{R}^k} \left\{ \int_0^t \prod_{j=1}^k (s-u_j)_+^{-(\frac{1}{2}+\frac{1-H}{k})} ds \right\} B(du_1) \ldots dB(du_k). \quad (4.2.41)$$

(b) *Spectral domain representation: for* $t \in \mathbb{R}$,

$$B_{k,H} \int''_{\mathbb{R}^k} \frac{e^{it(x_1+\cdots+x_k)}-1}{i(x_1+\cdots+x_k)} \prod_{j=1}^k |x_j|^{\frac{1-H}{k}-\frac{1}{2}} \widehat{B}(dx_1) \ldots \widehat{B}(dx_k). \quad (4.2.42)$$

(c) *Positive half-axis representation: for* $t \geq 0$,

$$C_{k,H} \int'_{[0,\infty)^k} \left\{ \int_0^t \prod_{j=1}^k z_j^{\frac{1-H}{k}-\frac{1}{2}} (1-sz_j)_+^{-(\frac{1}{2}+\frac{1-H}{k})} ds \right\} B(dz_1) \ldots B(dz_k). \quad (4.2.43)$$

(d) Finite interval representation: for $t \geq 0$,

$$D_{k,H} \int_{[0,t]^k}' \Big\{ \prod_{j=1}^{k} v_j^{\frac{1-H}{k}-\frac{1}{2}} \int_0^t s^{\frac{k}{2}-(1-H)} \prod_{j=1}^{k} (s-v_j)_+^{-(\frac{1}{2}+\frac{1-H}{k})} ds \Big\} B(dv_1)\ldots B(dv_k).$$
(4.2.44)

If $\mathbb{E}(Z_H^{(k)}(1))^2 = 1$, then

$$a_{k,H_0} = A_{k,H}, \quad b_{k,H_0} = B_{k,H}, \quad c_{k,H_0} = C_{k,H}, \quad d_{k,H_0} = D_{k,H},$$

where

$$A_{k,H} = C_{k,H} = D_{k,H} = \Big(\frac{H(2H-1)\Gamma(\frac{1}{2}+\frac{1-H}{k})^k}{k! [\Gamma(\frac{1}{2}-\frac{1-H}{k})\Gamma(\frac{2(1-H)}{k})]^k} \Big)^{1/2}$$
(4.2.45)

and[4]

$$B_{k,H} = \Big(\frac{H(2H-1)}{k! [2\Gamma(\frac{2(1-H)}{k})\sin((\frac{1}{2}-\frac{1-H}{k})\pi)]^k} \Big)^{1/2}.$$
(4.2.46)

4.3 Moments, Cumulants and Diagram Formulae for Multiple Integrals

We will provide next formulae for the moments of multiple integrals and Hermite polynomials of Gaussian variables. These results are expressed in terms of diagrams and related notions, which we recall first.

4.3.1 Diagram Formulae

For a nonempty set b, let $\mathcal{P}(b)$ denote the set of its partitions. For each pair $i, j \in b$ and for each $\pi \in \mathcal{P}(b)$, we write

$$i \sim_\pi j$$

whenever i and j belong to the same block of π. For example, if $b = \{1, 2, 3, 4\}$ and $\pi = \{\{1, 2, 3\}, \{4\}\}$, then $1 \sim_\pi 2$.

Definition 4.3.1 A *diagram* is a graphical representation of a pair of partitions $\pi, \sigma \in \mathcal{P}(b)$. It is obtained as follows. Let $\pi = \{b_1, \ldots, b_p\}$ and $\sigma = \{t_1, \ldots, t_q\}$ represent the blocks of the two partitions. Then:

1. Order the elements of each block b_i, for $i = 1, \ldots, p$.
2. Associate with each block $b_i \in \pi$ a *row* of vertices (represented as dots), in such a way that the jth vertex of the ith row corresponds to the jth element of the block b_i.
3. For every $a = 1, \ldots, q$, draw a *closed curve* around the vertices corresponding to the elements of the block $t_a \in \sigma$.

We will denote by $\Gamma(\pi, \sigma)$ the diagram of a pair of partitions π, σ. Note that the rows of the diagram indicate the blocks in π and the curves indicate the blocks in σ.

---
[4] There is a typo in the expression of (1.20) of Pipiras and Taqqu [819], namely an extra factor k.

Definition 4.3.2 A *non-flat* diagram $\Gamma(\pi, \sigma)$ is such that the closed curves defining the blocks of σ cannot join two vertices in the same row (thus having a flat or horizontal portion). We say that the diagram $\Gamma(\pi, \sigma)$ is *Gaussian*, whenever every block of σ contains exactly two elements.

When a diagram is Gaussian, one usually represents the blocks of σ not by closed curves, but by segments connecting two vertices.

Example 4.3.3 (*Non-flat (Gaussian) diagrams*) Figure 4.1, (a), provides an example of a non-flat diagram $\Gamma(\pi, \sigma)$, where $b = \{1, 2, 3, 4, 5, 6, 7\}$, $\pi = \{\{1, 2, 3\}, \{4\}, \{5, 6, 7\}\}$ and $\sigma = \{\{1, 4, 5\}, \{2, 7\}, \{3, 6\}\}$.

Figure 4.1, (b), provides an example of a non-flat Gaussian diagram $\Gamma(\pi, \sigma)$, where $b = \{1, 2, 3, 4, 5, 6\}$, $\pi = \{\{1, 2, 3\}, \{4\}, \{5, 6\}\}$ and $\sigma = \{\{1, 4\}, \{2, 5\}, \{3, 6\}\}$.

Let $n_1, \ldots, n_p \geq 1$. Denote

$$n = n_1 + \cdots + n_p, \quad [n] = \{1, \ldots, n\}, \tag{4.3.1}$$

and consider the following special partition of the set $[n]$:

$$\pi^* = \{\{1, \ldots, n_1\}, \{n_1 + 1, \ldots, n_1 + n_2\}, \ldots \{n_1 + \cdots + n_{p-1} + 1, \ldots, n\}\} \in \mathcal{P}([n]). \tag{4.3.2}$$

Now define

$$\mathcal{M}_2^0([n], \pi^*) = \{\sigma \in \mathcal{P}([n]) : \Gamma(\pi^*, \sigma) \text{ is non-flat and Gaussian}\}.$$

The moment formula for multiple integrals, called the *diagram formula*, is the following. See, for example, Peccati and Taqqu [797], Corollary 7.3.1, p. 133.

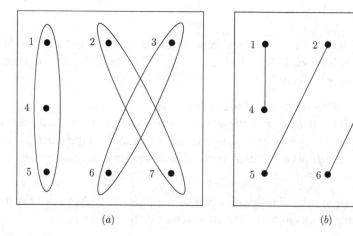

Figure 4.1 (a) A non-flat diagram. (b) A non-flat Gaussian diagram.

4.3 Moments, Cumulants and Diagram Formulae for Multiple Integrals 243

Proposition 4.3.4 (*Diagram formula.*) *For symmetric* $f_i \in L^2(E^{n_i}, m^{n_i})$, $i = 1, \ldots, p$, *and non-atomic m*,

$$\mathbb{E}(I_{n_1}(f_1) \ldots I_{n_p}(f_p)) = \sum_{\sigma \in \mathcal{M}_2^0([n], \pi^*)} \int_{E^{n/2}} f_{\sigma, p} dm^{n/2}, \qquad (4.3.3)$$

where, for every $\sigma \in \mathcal{M}_2^0([n], \pi^*)$, *the function* $f_{\sigma, p}$ *of* $n/2$ *variables, is obtained by identifying the variables* u_i *and* u_j *in the argument of*

$$f_1(u_1, \ldots, u_{n_1}) f_2(u_{n_1+1}, \ldots, u_{n_2}) \ldots f_p(u_{n_1+\cdots+n_{p-1}+1}, \ldots, u_n)$$

if and only if $i \sim_\sigma j$. *If* $\mathcal{M}_2^0([n], \pi^*) = \emptyset$ (*in particular, if n is odd*), *the right-hand side of* (4.3.3) *is interpreted as 0*.

Example 4.3.5 (*Diagram formula*) Let us evaluate $\mathbb{E} I_3(f_1) I_1(f_2) I_2(f_3)$ where f_1, f_2 and f_3 are respectively symmetric functions of 3, 1 and 2 variables. The function

$$f_1(u_1, u_2, u_3) f_2(u_4) f_3(u_5, u_6)$$

is a function of 6 variables. In this case, $b = \{1, 2, 3, 4, 5, 6\}$ and $\pi^* = \{\{1, 2, 3\}, \{4\}, \{5, 6\}\}$. One of the partitions in $\mathcal{M}_2^0([6], \pi^*)$ is $\sigma = \{\{1, 4\}, \{2, 5\}, \{3, 6\}\}$, with the corresponding diagram $\Gamma(\pi^*, \sigma)$ depicted in Figure 4.1, (*b*). This diagram is non-flat and Gaussian. It is associated to the summand

$$\int_{E^3} f_1(u_1, u_2, u_3) f_2(u_1) f_3(u_2, u_3) m(du_1) m(du_2) m(du_3) \qquad (4.3.4)$$

in the sum on the right-hand side of (4.3.3), where, according to σ, we identified u_4 with u_1, u_5 with u_2 and u_6 with u_3. To obtain the other summands, one needs to find all the partitions σ in $\mathcal{M}_2^0([6], \pi^*)$.[5] One gets

$$\mathbb{E} I_3(f_1) I_1(f_2) I_2(f_3) = \int_{E^3} \Big(f_1(u_1, u_2, u_3) f_2(u_3) f_3(u_2, u_1)$$
$$+ f_1(u_1, u_2, u_3) f_2(u_3) f_3(u_1, u_2) + f_1(u_1, u_2, u_3) f_2(u_2) f_3(u_3, u_1)$$
$$+ f_1(u_1, u_2, u_3) f_2(u_1) f_3(u_3, u_2) + f_1(u_1, u_2, u_3) f_2(u_2) f_3(u_1, u_3)$$
$$+ f_1(u_1, u_2, u_3) f_2(u_1) f_3(u_2, u_3) \Big) m(du_1) m(du_2) m(du_3) \qquad (4.3.5)$$

$$= 6 \int_{E^3} f_1(u_1, u_2, u_3) f_2(u_3) f_3(u_2, u_1) m(du_1) m(du_2) m(du_3). \qquad (4.3.6)$$

The last line is a consequence of the symmetry of the functions. We will see below that we can obtain the result immediately by using "multigraphs."

---

[5] This can be done by applying the Mathematica program described in Peccati and Taqqu [797]. One executes the command MZeroSetsEqualTwo[{3, 1, 2}], where the entry {3, 1, 2} defines the block sizes of π^*. The output indicates that there are 6 partitions σ in $\mathcal{M}_2^0([6], \pi^*)$ and lists them as {{1, 6}, {2, 5}, {3, 4}}, {{1, 5}, {2, 6}, {3, 4}}, {{1, 6}, {2, 4}, {3, 5}}, {{1, 4}, {2, 6}, {3, 5}}, {{1, 5}, {2, 4}, {3, 6}}, {{1, 4}, {2, 5}, {3, 6}}. This shows how to equalize the variables. For example, {{1, 6}, {2, 5}, {3, 4}} means "set $u_1 = u_6, u_2 = u_5, u_3 = u_4$."

Note that the diagram formula (4.3.3) concerns multiple integrals in the time domain. There is an analogous formula for multiple integrals $\widehat{I}_n(f_n)$ in the spectral domain. In this case, suppose $E = \mathbb{R}$, $m(du) = m(-du)$ and the symmetric functions f_n are Hermitian, namely, $\overline{f_n(u)} = f_n(-u)$. Then, in Proposition 4.3.4, when $i \sim_\pi j$, instead of identifying u_i and u_j, that is, setting $u_j = u_i$, one sets $u_j = -u_i$ (see also Surgailis [937], Proposition 5.1).

Example 4.3.6 (*Diagram formula in the spectral domain*) Suppose we want to evaluate $\mathbb{E}\widehat{I}_3(f_1)\widehat{I}_1(f_2)\widehat{I}_2(f_3)$ where the functions f_1, f_2 and f_3 are as in Example 4.3.5 but Hermitian and suppose $m(du) = m(-du)$. Then (4.3.5) holds with $\mathbb{E}\widehat{I}_3(f_1)\widehat{I}_1(f_2)\widehat{I}_2(f_3)$ on its left-hand side but with the arguments of f_2 and f_3 on its right-hand side having changed signs. Hence, (4.3.4) becomes

$$\int_{E^3} f_1(u_1, u_2, u_3) f_2(-u_1) f_3(-u_2, -u_3) m(du_1) m(du_2) m(du_3).$$

Similarly, the equality (4.3.6) continues to hold after the same change as for the right-hand side of (4.3.5).

The formula (4.3.3) concerns the moments of multiple integrals. There is a similar formula for the cumulants of multiple integrals.

Definition 4.3.7 For a vector of real-valued random variables $(X_1, \ldots, X_p)$ such that $\mathbb{E}|X_j|^p < \infty$, $j = 1, \ldots, p$, its *joint cumulant* is defined as

$$\chi(X_1, \ldots, X_p) = (-i)^p \frac{\partial^p}{\partial z_1 \ldots \partial z_p} \log \mathbb{E}\left(\exp\left\{ i \sum_{j=1}^p z_j X_j \right\} \right) \Big|_{z_1 = \ldots = z_p = 0}. \quad (4.3.7)$$

The usual pth cumulant of a random variable X such that $\mathbb{E}|X|^p < \infty$ is defined as

$$\chi_p(X) = \chi(\underbrace{X, \ldots, X}_{p \text{ times}}) = (-i)^p \frac{\partial^p}{\partial z^p} \log \mathbb{E}\left(\exp\{izX\} \right) \Big|_{z=0}. \quad (4.3.8)$$

One has, for example, $\chi_1(X) = \mathbb{E}X$, $\chi_2(X) = \mathbb{E}X^2 - (\mathbb{E}X)^2$, $\chi_3(X) = \mathbb{E}X^3 - 3\mathbb{E}X^2\mathbb{E}X + 2(\mathbb{E}X)^3$.

To state the result for the cumulants of multiple integrals, we need another notion related to diagrams.

Definition 4.3.8 The diagram $\Gamma(\pi, \sigma)$, $\pi, \sigma \in \mathcal{P}(b)$, is *connected* if the only partition $\rho \in \mathcal{P}(b)$ for which each block of π and each block of σ is contained in a block of ρ, is the maximal partition consisting of one block with all the elements of b.

Example 4.3.9 ((*Non-)Connected diagrams*) Figure 4.2, (a), provides an example of a non-connected diagram with $b = \{1, 2, 3, 4, 5\}$, $\pi = \{\{1, 2, 3\}, \{4\}, \{5\}\}$ and $\sigma = \{\{1, 2, 3\}, \{4, 5\}\}$, for which there exists a non-maximal partition $\rho = \{\{1, 2, 3\}, \{4, 5\}\}$. The blocks $\{1, 2, 3\}$ and $\{4, 5\}$ of this ρ contain the blocks of both π and σ, and this ρ is not the maximal partition $\{1, 2, 3, 4, 5\}$. Therefore the diagram $\Gamma(\pi, \sigma)$ is not connected.

4.3 Moments, Cumulants and Diagram Formulae for Multiple Integrals

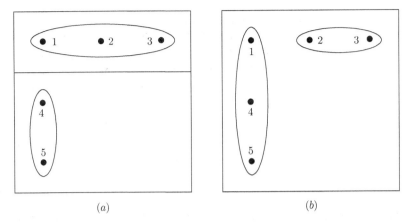

Figure 4.2 (a) Dividing a non-connected diagram. (b) A connected diagram.

Figure 4.2, (b), provides an example of a connected diagram with $b = \{1, 2, 3, 4, 5\}$, $\pi = \{\{1, 2, 3\}, \{4\}, \{5\}\}$ and $\sigma = \{\{1, 4, 5\}, \{2, 3\}\}$. The only partition ρ containing each block of π and of σ is the maximal partition $\rho = b = \{1, 2, 3, 4, 5\}$.

Here is an easy way to check geometrically if a diagram $\Gamma(\pi, \sigma)$ is connected or not. It is connected if and only if the rows of the diagrams (the blocks of π) cannot be divided into two subsets, each defining a separate diagram. Thus, in Figure 4.2, the diagram in (a) is not connected (the horizontal line divides it into two diagrams), but the diagram in (b) is connected.

Now define

$$\mathcal{M}_2([n], \pi^*) = \{\sigma \in \mathcal{P}([n]) : \Gamma(\pi^*, \sigma) \text{ is connected, non-flat and Gaussian}\}.$$

The formula for the cumulants of multiple integrals is the following (Peccati and Taqqu [797], Corollary 7.3.1, p. 133).

Proposition 4.3.10 *(Diagram formula.) For symmetric $f_i \in L^2(E^{n_i}, m^{n_i})$, $i = 1, \ldots, p$, and non-atomic m,*

$$\chi(I_{n_1}(f_1), \ldots, I_{n_p}(f_p)) = \sum_{\sigma \in \mathcal{M}_2([n], \pi^*)} \int_{E^{n/2}} f_{\sigma, p} dm^{n/2}, \tag{4.3.9}$$

where $f_{\sigma, p}$ is defined as in (4.3.3). If $\mathcal{M}_2([n], \pi^*) = \emptyset$ (in particular, if n is odd), the right-hand side of (4.3.9) is interpreted as 0.

Since $\mathcal{M}_2([n], \pi^*) \subset \mathcal{M}_2^0([n], \pi^*)$, there are fewer summands in the cumulant formula (4.3.9) than in the moment formula (4.3.3).

4.3.2 Multigraphs

There can be many diagrams. It is possible to simplify the formulae (4.3.3) and (4.3.9) by using multigraphs instead.

Definition 4.3.11 A *multigraph* $A = (V, E)$ consists of a nonempty set of points (vertices) V, and a (possibly empty) set of lines (edges) E. A multigraph may possess multiple lines connecting any two points.[6] A *labeled* (point-labeled) multigraph has its points distinguished from each other. Two labeled multigraphs will be assumed identical if for all i and j, the number of lines joining the points i and j is the same in both multigraphs.

For $p \geq 1$, let $\mathcal{A}_p$ be the set of all labeled multigraphs with p points, and whose lines do not form loops, that is, do not start and end at the same point. $A \in \mathcal{A}_p$ can be described through its p points and its *set of lines* or *pair sequence*

$$\{(i_1, j_1), \ldots, (i_q, j_q)\} \tag{4.3.10}$$

as follows: q denotes the number of lines of A and for each $s = 1, \ldots, q$, we have $i_s, j_s \in \{1, \ldots, p\}$ with $i_s \neq j_s$, a pair (i_s, j_s) symbolizing the existence of a line that joins point i_s to point j_s. Identical pairs indicate the presence of multiple lines. For example, the pair sequence of the multigraph A_1 in Figure 4.3 is $\{(1, 2), (1, 2), (3, 4), (3, 4)\}$. The requirement $i_s \neq j_s$ for all $s = 1, \ldots, q$ indicates the absence of loops.

With each multigraph $A \in \mathcal{A}_p$, we also associate a *multiplicity number* $g(A)$ defined as follows. Number each of the $\binom{p}{2}$ possible pairs of points by $u = 1, 2 \ldots, \binom{p}{2}$. Let v_u be the number of lines in A joining the pair of points numbered u. Define

$$g(A) = \prod_{u=1}^{\binom{p}{2}} \frac{1}{v_u!}, \tag{4.3.11}$$

with the convention $0! = 1$. If A has q lines, then obviously $v_1 + v_2 + \cdots + v_{\binom{p}{2}} = q$.

Definition 4.3.12 Given $p \geq 1, n_1, \ldots, n_p \geq 1$, let $\mathcal{A}(n_1, \ldots, n_p)$ denote the set of all multigraphs A of $\mathcal{A}_p$ whose p points have degrees $n_1, \ldots, n_p$ respectively. The *degree* of a point is the number of lines incident to that point; that is, the number of line with one of their endpoints at that point. For $k \geq 1$, let also $\mathcal{A}_p(k)$ denote the set of all multigraphs A of $\mathcal{A}_p$ such that all the points of A have degree k. These multigraphs are called *k–regular*.

Example 4.3.13 (*2–regular miultigraphs*) Consider the set of $p = 4$ points $V = \{1, 2, 3, 4\}$, and $k = 2$. Figure 4.3 shows all 2–regular multigraphs, that is, all multigraphs of the set $\mathcal{A}_4(2)$, where each point of the set V has degree 2. There are six such multigraphs. The respective multiplicity numbers are

$$g(A_1) = g(A_2) = g(A_3) = \frac{1}{2!2!} = \frac{1}{4}, \quad g(A_4) = g(A_5) = g(A_6) = 1.$$

4.3.3 Relation Between Diagrams and Multigraphs

With $n = n_1 + \cdots + n_p$ and the special partition π^* in (4.3.2), a partition $\sigma \in \mathcal{M}_2^0([n], \pi^*)$ or the corresponding non-flat, Gaussian diagram $\Gamma(\pi^*, \sigma)$ can naturally be associated with a multigraph $A \in \mathcal{A}(n_1, \ldots, n_p)$.

---

[6] "Multi" in "Multigraph" refers to this fact.

4.3 Moments, Cumulants and Diagram Formulae for Multiple Integrals   247

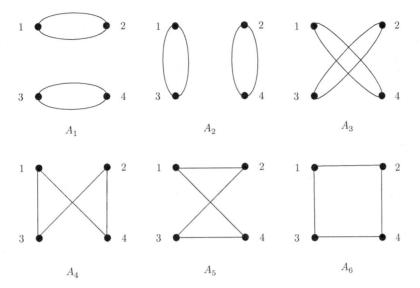

Figure 4.3 The six multigraphs of the set $\mathcal{A}_4(2)$.

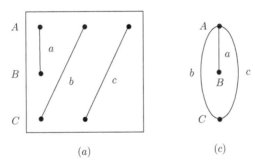

Figure 4.4 Non-flat, Gaussian diagram in (a) and the multigraph (c) associated with them.

Definition 4.3.14 We say that "a (non-flat, Gaussian) diagram $\Gamma(\pi^*, \sigma)$ is *associated* with a multigraph A", or simply, that "σ is *associated* with the multigraph A" if A is obtained from σ in the following way: Start with a diagram $\Gamma(\pi^*, \sigma)$. Suppose that π^* has p blocks and σ has q blocks corresponding to q segments. These p blocks will contribute p points to a multigraph A and these q segments will contribute q lines to A, as follows. Each segment in $\Gamma(\pi^*, \sigma)$ connecting points in the ith block $\{n_1 + \cdots + n_{i-1} + 1, \cdots, n_1 + \cdots + n_i\}$ and the jth block $\{n_1 + \cdots + n_{j-1} + 1, \cdots, n_1 + \cdots + n_j\}$ of π^* contributes a line connecting points i and j in a multigraph A. Since there are n_i segments originating from the ith block $\{n_1 + \cdots + n_{i-1} + 1, \cdots, n_1 + \cdots + n_i\}$ of π^*, the point i of the multigraph A will have degree n_i and hence A will be a multigraph from $\mathcal{A}(n_1, \ldots, n_p)$.

Example 4.3.15 (*Association of diagrams with multigraphs*) Figure 4.4 illustrates the association of diagrams with multigraphs. Consider the diagram $\Gamma(\pi^*, \sigma)$ in (a). π^* has $p = 3$ blocks denoted A, B, C and σ has $q = 3$ blocks denoted a, b, c. Hence the diagram has $q = 3$ segments. The labeling A, B, C, a, b, c of the diagram in (a) is a multigraph labeling. We included it to make the correspondence between diagram and multigraph apparent. The diagram in (a) is associated with the multigraph $A \in \mathcal{A}(3, 1, 2)$ in (c) which has 3 points A, B, C of degrees 3, 1, 2 respectively and $q = 3$ lines a, b, c. These three lines are associated with the $q = 3$ segments of the diagram $\Gamma(\pi^*, \sigma)$.

Figure 4.5 displays the diagram of Figure 4.4 in (a) and also a second diagram in (b). Both diagrams are associated with the same multigraph (c).

We defined connected diagrams in Definition 4.3.8. We need a related notion for multigraphs.

Definition 4.3.16 We will say that a multigraph $A \in \mathcal{A}_p$ is *connected* if its p points cannot be separated into two nonempty, disjoint subsets without any line connecting the points of the two subsets.

Note that connected diagrams $\Gamma(\pi^*, \sigma)$, $\sigma \in \mathcal{M}_2([n], \pi^*)$, are associated with connected multigraphs.

Example 4.3.17 *((Non-)Connected diagrams and multigraphs)* The non-flat Gaussian diagram in Figure 4.6, (a), is not connected nor is the associated multigraph in Figure 4.6, (b).

Denote by $\mathcal{A}_c(n_1, \ldots, n_p)$ ($\mathcal{A}_{p,c}(k)$, resp.) the set of all connected multigraphs $A \in \mathcal{A}(n_1, \ldots, n_p)$ ($A \in \mathcal{A}_p(k)$, resp.).

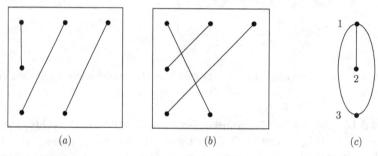

Figure 4.5 Non-flat, Gaussian diagrams (a) and (b), and the multigraph (c) associated with them.

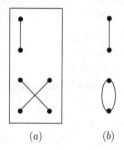

Figure 4.6 (a) A non-connected non-flat Gaussian diagram. (b) The associated non-connected multigraph.

4.3 Moments, Cumulants and Diagram Formulae for Multiple Integrals

Note that, as in Figure 4.5, different diagrams with partitions $\sigma \in \mathcal{M}_2^0([n], \pi^*)$ can be associated with the same multigraph $A \in \mathcal{A}(n_1, \ldots, n_p)$. The following auxiliary result provides the number of such σs.

Lemma 4.3.18 *The number of different[7] $\sigma \in \mathcal{M}_2^0([n], \pi^*)$ ($\mathcal{M}_2([n], \pi^*)$, resp.) associated with the same multigraph $A \in \mathcal{A}(n_1, \ldots, n_p)$ ($\mathcal{A}_c(n_1, \ldots, n_p)$, resp.) in the sense of Definition 4.3.14 is equal to*

$$\frac{n_1! \ldots n_p!}{v_1! \ldots v_{\binom{p}{2}}!}, \quad \text{or equivalently} \quad n_1! \ldots n_p! g(A), \tag{4.3.12}$$

where $g(A)$ is defined in (4.3.11).

Proof The formula (4.3.12) can be proved by induction in m, the number of pairs of points in a multigraph which have at least one connecting line. The case $m = 1$ of smallest m occurs only when the multigraph has $p = 2$ points with degrees $n_1 = n_2 =: n_0$. In this case, $v_1 = n_0$. The number of diagrams leading to the same multigraph is $n_0!$. Since

$$n_0! = \frac{n_1! n_2!}{n_0!},$$

we have (4.3.12). Suppose now that the formula (4.3.12) holds up to $m - 1$, and consider the general case of m above. Suppose without loss of generality that $v_1 \geq 1$ lines connect points i and j in a multigraph. This means that v_1 points in the block $\{n_1 + \cdots + n_{i-1} + 1, \ldots, n_1 + \cdots + n_i\}$ connect v_1 points in the block $\{n_1 + \cdots + n_{j-1} + 1, \ldots, n_1 + \cdots + n_j\}$. The number of such possibilities is

$$\binom{n_i}{v_1}\binom{n_j}{v_1} v_1! = \frac{n_i! n_j!}{(n_i - v_1)!(n_j - v_1)! v_1!}. \tag{4.3.13}$$

By the induction hypothesis, the number of possibilities to connect the rest of the points is

$$\frac{n_1! \ldots n_{i-1}!(n_i - v_1)! n_{i+1}! \ldots n_{j-1}!(n_j - v_1)! n_{j+1}! \ldots n_p!}{v_2! \ldots v_{\binom{p}{2}}!}, \tag{4.3.14}$$

since v_1 points were already connected in the blocks $\{n_1 + \cdots + n_{i-1} + 1, \ldots, n_1 + \cdots + n_i\}$ and $\{n_1 + \cdots + n_{j-1} + 1, \ldots, n_1 + \cdots + n_j\}$. Multiplying (4.3.13) and (4.3.14) leads to the formula (4.3.12). Note that the argument above does not depend on whether a multigraph is connected or not. □

The following is an immediate consequence of Lemma 4.3.18.

Corollary 4.3.19 *Let $F : \mathcal{M}_2^0([n], \pi^*) \mapsto \mathbb{R}$ be such that*

$$F(\sigma) = G(A), \tag{4.3.15}$$

---

[7] More precisely: "The number of different diagrams $\Gamma(\pi^*, \sigma)$ with $\sigma \in \mathcal{M}_2^0([n], \pi^*)$ associated with the same multigraph $A \ldots$"

where $A \in \mathcal{A}(n_1, \ldots, n_p)$ is the multigraph associated with $\sigma \in \mathcal{M}_2^0([n], \pi^*)$. Then,

$$\sum_{\sigma \in \mathcal{M}_2^0([n], \pi^*)} F(\sigma) = n_1! \ldots n_p! \sum_{A \in \mathcal{A}(n_1, \ldots, n_p)} g(A) G(A). \qquad (4.3.16)$$

This result continues to hold if throughout, $\mathcal{M}_2^0$ is replaced by $\mathcal{M}_2$ and $\mathcal{A}$ by $\mathcal{A}_c$. In particular, if $F : \mathcal{M}_2([n], \pi^*) \mapsto \mathbb{R}$, then

$$\sum_{\sigma \in \mathcal{M}_2([n], \pi^*)} F(\sigma) = n_1! \ldots n_p! \sum_{A \in \mathcal{A}_c(n_1, \ldots, n_p)} g(A) G(A). \qquad (4.3.17)$$

Applying Corollary 4.3.19 to the right-hand sides of (4.3.3) and (4.3.9), leads to the following alternative diagram formulae for the moments and cumulants of multiple integrals.

Proposition 4.3.20 *For symmetric $f_i \in L^2(E^{n_i}, m^{n_i})$, $i = 1, \ldots, p$, and non-atomic m,*

$$\mathbb{E}(I_{n_1}(f_1) \ldots I_{n_p}(f_p)) = n_1! \ldots n_p! \sum_{A \in \mathcal{A}(n_1, \ldots, n_p)} g(A) \int_{E^{n/2}} f_{A,p} dm^{n/2} \qquad (4.3.18)$$

and

$$\chi(I_{n_1}(f_1), \ldots, I_{n_p}(f_p)) = n_1! \ldots n_p! \sum_{A \in \mathcal{A}_c(n_1, \ldots, n_p)} g(A) \int_{E^{n/2}} f_{A,p} dm^{n/2}. \qquad (4.3.19)$$

Each multigraph A has p points, $q = n/2$ lines and multiplicity $g(A)$, defined in (4.3.11). Denote by $\{(i_1, j_1), \ldots, (i_{n/2}, j_{n/2})\}$ the $n/2$ lines (pair sequence) of the multigraph A. The function $f_{A,p}$ of $n/2$ variables equals

$$f_1(u_1, \ldots, u_{n_1}) f_2(u_{n_1+1}, \ldots, u_{n_2}) \ldots f_p(u_{n_1+\cdots+n_{p-1}+1}, \ldots, u_n),$$

where, for each line (i_m, j_m), $m = 1, \ldots, n/2$, one identifies any variable of f_{i_m} with any variable of f_{j_m}.

If $\mathcal{A}(n_1, \ldots, n_p) = \emptyset$ or $\mathcal{A}_c(n_1, \ldots, n_p) = \emptyset$ (in particular, if n is odd), the right-hand side of (4.3.18) or (4.3.19) is interpreted as 0.

Proof Note that, for a given (i_m, j_m), it does not matter which two variables of f_{i_m} and f_{j_m} are identified because f_is are symmetric. Note also that, up to a possible permutation of its $n/2$ variables, $f_{A,p}$ does not depend on the chosen pair sequence, that is, on the way the pair sequence is enumerated. This is because any two pair sequences describing A are the same up to a possible permutation of pairs, or indices within pairs. In particular, since $dm^{n/2}$ is symmetric in its variables, the value of $\int_{E^{n/2}} f_{A,p} dm^{n/2}$ does not depend on the chosen pair sequence.

The relation (4.3.18), for example, follows from (4.3.3) and (4.3.16) after observing the following. Any variables u_i and u_j identified for $f_{\sigma,p}$ in

$$f_1(u_1, \ldots, u_{n_1}) f_2(u_{n_1+1}, \ldots, u_{n_2}) \ldots f_p(u_{n_1+\cdots+n_{p-1}+1}, \ldots, u_n)$$

4.3 Moments, Cumulants and Diagram Formulae for Multiple Integrals

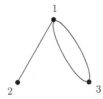

Figure 4.7 The single multigraph in $\mathcal{A}(3, 1, 2)$.

when $i \sim_\sigma j$, correspond to a pair (i, j) in the pair sequence describing A. Then, $f_{\sigma, p}$ is one possible representation for $f_{A, p}$, and

$$\int_{E^{n/2}} f_{\sigma, p} dm^{n/2} = \int_{E^{n/2}} f_{A, p} dm^{n/2},$$

so that the condition (4.3.15) holds. $\square$

Example 4.3.21 (*Diagram formula through multigraphs*) As in Example 4.3.5, let f_1, f_2 and f_3 be respective functions of 3, 1 and 2 variables. Here $n = 3 + 1 + 2 = 6$. We evaluated $\mathbb{E} I_3(f_1) I_1(f_2) I_2(f_3)$ in that example using diagrams. We will see that it can be evaluated simply by using multigraphs and applying Proposition 4.3.20. The multigraphs in $\mathcal{A}(3, 1, 2)$ have $p = 3$ points of degree $n_1 = 3$, $n_2 = 1$, $n_3 = 2$ respectively and have $q = n/2 = (3 + 1 + 2)/2 = 3$ lines. In fact, $\mathcal{A}(3, 1, 2)$ contains a single multigraph A depicted in Figure 4.7 with $g(A) = 1/2!$. This multigraph is associated with any of the six diagrams in Example 4.3.5. A pair sequence $\{(1, 3), (1, 3), (1, 2)\}$ describes A, and $f_{A, p}(v_1, v_2, v_3) = f_1(v_1, v_2, v_3) f_2(v_3) f_3(v_1, v_2)$ as defined in Proposition 4.3.20. Applying (4.3.18) now yields

$$\mathbb{E} I_3(f_1) I_1(f_2) I_2(f_3) = 6 \int_{E^3} f_1(v_1, v_2, v_3) f_2(v_3) f_3(v_1, v_2) m(dv_1) m(dv_2) m(dv_3),$$

since $n_1! n_2! n_3! / g(A) = 3! 1! 2! / 2! = 6$.

4.3.4 Diagram and Multigraph Formulae for Hermite Polynomials

By using the connection between Hermite polynomials of Gaussian variables and multiple integrals in Proposition 4.1.2, one can obtain a diagram formula for the Hermite polynomials of Gaussian variables.

Proposition 4.3.22 (*Diagram formula.*) *Let* $(X_1, \ldots, X_p)'$, $p \geq 2$, *be a Gaussian vector with* $\mathbb{E} X_j = 0$, $\mathbb{E} X_j^2 = 1$, $j = 1, \ldots, p$. *Then, for any* $n_1, \ldots, n_p \geq 1$,

$$\mathbb{E}\Big(H_{n_1}(X_1) \ldots H_{n_p}(X_p)\Big) = \sum_{\sigma \in \mathcal{M}_2^0([n], \pi^*)} \prod_{1 \leq i < j \leq p} (\mathbb{E} X_i X_j)^{\ell_{ij}(\sigma)}, \quad (4.3.20)$$

where $\ell_{ij}(\sigma)$ *is the number of connecting segments*[8] *between the blocks* i *and* j *of the partition* π^* *in the diagram* $\Gamma(\pi^*, \sigma)$.

[8] The "segments" are the "closed curves" in Definition 4.3.1.

Proof We argue first that there are $f_j \in L^2(\mathbb{R}, du)$, $j = 1, \ldots, p$, such that

$$(X_1, \ldots, X_p)' \stackrel{d}{=} (I_1(f_1), \ldots, I_1(f_p))', \qquad (4.3.21)$$

where $I_1(f) = \int_{\mathbb{R}} f(u) B(du)$ and $B(du)$ is a real-valued Gaussian random measure on $\mathbb{R}$ with control measure du. By factorizing the covariance matrix of $(X_1, \ldots, X_p)$, the following representation holds:

$$(X_1, \ldots, X_p)' \stackrel{d}{=} (\phi_{11}\epsilon_1 + \cdots + \phi_{1p}\epsilon_p, \ldots, \phi_{p1}\epsilon_1 + \cdots + \phi_{pp}\epsilon_p)', \qquad (4.3.22)$$

where ϵ_j, $j = 1, \ldots, p$, are i.i.d. $\mathcal{N}(0, 1)$ random variables. Choosing a collection of orthonormal functions $g_1, \ldots, g_p$ in $L^2(\mathbb{R}, du)$, we can represent

$$(\epsilon_1, \ldots, \epsilon_p)' \stackrel{d}{=} (I_1(g_1), \ldots, I_1(g_p))'. \qquad (4.3.23)$$

Substituting (4.3.23) into (4.3.22) leads to (4.3.21) with $f_j = \sum_{m=1}^{p} \phi_{jm} g_m$, $j = 1, \ldots, m$.

Having the representation (4.3.21), we can now apply Propositions 4.1.2 and 4.3.4 to write

$$\mathbb{E}\Big(H_{n_1}(X_1) \ldots H_{n_p}(X_p)\Big) = \mathbb{E}\Big(H_{n_1}(I_1(f_1)) \ldots H_{n_p}(I_1(f_p))\Big)$$

$$= \mathbb{E}\Big(I_{n_1}(f_1^{\otimes n_1}) \ldots I_{n_p}(f_p^{\otimes n_p})\Big) = \sum_{\sigma \in \mathcal{M}_2^0([n], \pi^*)} \int_{\mathbb{R}^{n/2}} f_{\sigma, p} dm^{n/2},$$

where $n = n_1 + \cdots + n_p$ and m is the Lebesgue measure. The function $f_{\sigma, p}$, defined in Proposition 4.3.4, is obtained by identifying u_i and u_j in the argument of

$$f_1(u_1) \ldots f_1(u_{n_1}) f_2(u_{n_1+1}) \ldots f_2(u_{n_2}) \ldots f_p(u_{n_1+\cdots+n_{p-1}+1}) \ldots f_p(u_n)$$

if and only if $i \sim_\sigma j$. The argument u_i is one of the arguments of the function f_i and the argument u_j is one of the arguments of the function f_j. It follows that

$$\int_{\mathbb{R}^{n/2}} f_{\sigma, p} dm^{n/2} = \prod_{1 \leq i < j \leq p} \left(\int_{\mathbb{R}} f_i(u) f_j(u) du \right)^{\ell_{ij}(\sigma)},$$

where $\ell_{ij}(\sigma)$ is as in (4.3.20). Since $\int_{\mathbb{R}} f_i(u) f_j(u) du = \mathbb{E} X_i X_j$, we deduce (4.3.20). □

Example 4.3.23 (*Diagram formula for Hermite polynomials*) Consider $\mathbb{E} H_2(X_1) H_2(X_2)$. Here, $\pi^* = \{\{1, 2\}, \{3, 4\}\}$ and thus $\mathcal{M}_2^0([4], \pi^*)$ contains two partitions σ, namely, $\sigma_1 = \{\{1, 3\}, \{2, 4\}\}$ and $\sigma_2 = \{\{1, 4\}, \{2, 3\}\}$. Each one of these two partitions indicates the presence of two segments connecting the blocks $\{1, 2\}$ and $\{3, 4\}$ of π^* (see Figure 4.8). Therefore,

$$\mathbb{E} H_2(X_1) H_2(X_2) = (\mathbb{E} X_1 X_2)^{\ell_{12}(\sigma_1)} + (\mathbb{E} X_1 X_2)^{\ell_{12}(\sigma_2)} = 2(\mathbb{E} X_1 X_2)^2.$$

Here is a second perspective which also illustrates the proof of Proposition 4.3.22. If $X_1 = I_1(f_1)$ and $X_2 = I_1(f_2)$, then

$$\mathbb{E} H_2(X_1) H_2(X_2) = \mathbb{E} I_2(f_1 \otimes f_1) I_2(f_2 \otimes f_2).$$

4.3 Moments, Cumulants and Diagram Formulae for Multiple Integrals

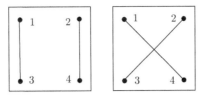

Figure 4.8 Two diagrams in Example 4.3.23.

In $f_1(u_1)f_1(u_2)f_2(u_3)f_2(u_4)$, the argument u_1 can be identified with u_3, u_2 with u_4 but also u_1 can be identified with u_4 and u_2 with u_3. One therefore obtains

$$2\left(\int_{\mathbb{R}} f_1(u)f_2(u)m(du)\right)^2 = 2(\mathbb{E}X_1X_2)^2.$$

For the cumulants $\chi(H_{n_1}(X_1), \ldots, H_{n_p}(X_p))$ of Hermite polynomials of Gaussian variables, the formula (4.3.20) holds where $\mathcal{M}_2^0([n], \pi^*)$ is replaced by $\mathcal{M}_2([n], \pi^*)$.

Proposition 4.3.24 (*Diagram formula.*) *Let* $(X_1, \ldots, X_p)'$, $p \geq 2$, *be a Gaussian vector with* $\mathbb{E}X_j = 0$, $\mathbb{E}X_j^2 = 1$, $j = 1, \ldots, p$. *Then, for any* $n_1, \ldots, n_p \geq 1$,

$$\chi\left(H_{n_1}(X_1), \ldots, H_{n_p}(X_p)\right) = \sum_{\sigma \in \mathcal{M}_2([n], \pi^*)} \prod_{1 \leq i < j \leq p} (\mathbb{E}X_iX_j)^{\ell_{ij}(\sigma)}, \qquad (4.3.24)$$

where $\ell_{ij}(\sigma)$ is the number of connecting segments between the blocks i and j of the partition π^* in the diagram $\Gamma(\pi^*, \sigma)$.

Proof The proof of this fact is analogous to that of Proposition 4.3.22 by using Proposition 4.3.10. □

The formulae (4.3.20) and (4.3.24) can be rewritten using *multigraphs*. We shall use the following additional notation. Denote by $E(A)$ the set of lines (edges) in a multigraph A. For $w \in E(A)$, set $d_1(w)$ and $d_2(w)$, with $d_1(w) < d_2(w)$, the two points connected by w. For a pair of points in a multigraph $A \in \mathcal{A}_p$ indexed by $u = 1, 2, \ldots, \binom{p}{2}$, we also denote $u(1)$ and $u(2)$, with $u(1) < u(2)$, the two points in the pair.

Proposition 4.3.25 *Let* $(X_1, \ldots, X_p)'$, $p \geq 2$, *be a Gaussian vector with* $\mathbb{E}X_j = 0$, $\mathbb{E}X_j^2 = 1$, $j = 1, \ldots, p$. *Then, for any* $n_1, \ldots, n_p \geq 1$,

$$\mathbb{E}\left(H_{n_1}(X_1) \ldots H_{n_p}(X_p)\right) = n_1! \ldots n_p! \sum_{A \in \mathcal{A}(n_1, \ldots, n_p)} g(A) \prod_{w \in E(A)} \mathbb{E}X_{d_1(w)}X_{d_2(w)}$$
$$(4.3.25)$$

and

$$\chi\left(H_{n_1}(X_1), \ldots, H_{n_p}(X_p)\right) = n_1! \ldots n_p! \sum_{A \in \mathcal{A}_c(n_1, \ldots, n_p)} g(A) \prod_{w \in E(A)} \mathbb{E}X_{d_1(w)}X_{d_2(w)},$$
$$(4.3.26)$$

where $g(A)$ is defined in (4.3.11). Moreover,

$$\prod_{w\in E(A)} \mathbb{E}X_{d_1(w)}X_{d_2(w)} = \prod_{u=1}^{\binom{p}{2}} (\mathbb{E}X_{u(1)}X_{u(2)})^{v_u},$$

where v_u appears in (4.3.11).

Proof The relations (4.3.25) and (4.3.26) follow from (4.3.20) and (4.3.24) by applying Corollary 4.3.19. $\square$

4.4 Moments and Cumulants of Hermite Processes

Let $\{Z_H^{(k)}(t)\}_{t\in\mathbb{R}}$ be the Hermite process of order k in Definition 4.2.1, and

$$\mu_p(t_1,\ldots,t_p) = \mathbb{E}Z_H^{(k)}(t_1)\ldots Z_H^{(k)}(t_p), \quad p\geq 1, t_1,\ldots,t_p \in \mathbb{R}, \tag{4.4.1}$$

be its finite-dimensional moments. In the next result, we provide a useful formula for the finite-dimensional moments $\mu_p(t_1,\ldots,t_p)$.

Proposition 4.4.1 *Let $\mu_p(t_1,\ldots,t_p)$, $p\geq 1$, $t_1,\ldots,t_p \in \mathbb{R}$, be the finite-dimensional moments of the Hermite process of order k in (4.4.1). Then,*

$$\mu_p(t_1,\ldots,t_p) = C_{p,k,H_0} \sum_{A\in\mathcal{A}_p(k)} g(A)S_D(A;t_1,\ldots,t_p), \tag{4.4.2}$$

where $g(A)$ is a multiplicity number of a multigraph A defined in (4.3.11), the constant is

$$C_{p,k,H_0} = (a_{k,H_0})^p \left(B(H_0-\frac{1}{2},2-2H_0)\right)^q (k!)^p = \left(k!(1-k(1-H_0))(1-2k(1-H_0))\right)^{p/2} \tag{4.4.3}$$

with $q = pk/2$ and

$$S_D(A;t_1,\ldots,t_p) = \int_0^{t_1} ds_1 \ldots \int_0^{t_p} ds_p \, |s_{i_1}-s_{j_1}|^{2H_0-2}\ldots|s_{i_q}-s_{j_q}|^{2H_0-2}, \tag{4.4.4}$$

where $(i_1,j_1),\ldots,(i_q,j_q)$ is a pair sequence of A.

Proof Write

$$Z_H^{(k)}(t) = a_{k,H_0}I_k(f_t),$$

where the kernel function f_t is defined in (4.2.5) with (4.2.6) and I_k denotes a multiple integral. We shall use the diagram formula (4.3.3) for the moments of multiple integrals. We want to compute

$$\mu_p(t_1,\ldots,t_p) = (a_{k,H_0})^p \mathbb{E}I_k(f_{t_1})\ldots I_k(f_{t_p}). \tag{4.4.5}$$

Recall from Section 4.3 that $[pk] = \{1,2,\ldots,pk\}$, $\pi^* = \{\{1,\ldots,k\},\{k+1,\ldots,2k\},\ldots,\{(p-1)k+1,\ldots,pk\}\}$ is the special partition of the set $[pk]$, and $\mathcal{M}_2^0([pk],\pi^*)$

denotes all partitions σ of $[pk]$ so that the diagram $\Gamma(\pi^*, \sigma)$ is non-flat and Gaussian. Then, applying the diagram formula (4.3.3),

$$\mu_p(t_1, \ldots, t_p) = (a_{k,H_0})^p \sum_{\sigma \in \mathcal{M}_2^0([pk], \pi^*)} \int_{\mathbb{R}^{pk/2}} f_{\sigma,p} d(Leb)^{pk/2},$$

where $(Leb)^{pk/2}$ denotes the Lebesgue measure on $\mathbb{R}^{pk/2}$, and $f_{\sigma,p}$ is a function of $pk/2$ variables obtained by identifying the variables u_i and u_j in

$$f_{t_1}(u_1, \ldots, u_k) f_{t_2}(u_{k+1}, \ldots, u_{2k}) \ldots f_{t_p}(u_{(p-1)k+1}, \ldots, u_{pk})$$

if and only if $i \sim_\sigma j$. In view of (4.2.5), we can also write[9]

$$\mu_p(t_1, \ldots, t_p) = (a_{k,H_0})^p \sum_{\sigma \in \mathcal{M}_2^0([pk], \pi^*)} \int_0^{t_1} ds_1 \ldots \int_0^{t_p} ds_p \left\{ \int_{\mathbb{R}^{pk/2}} g_{\sigma,p} d(Leb)^{pk/2} \right\},$$

(4.4.6)

where $g_{\sigma,p}$ is obtained by identifying the variables u_i and u_j in

$$g_{s_1}(u_1, \ldots, u_k) g_{s_2}(u_{k+1}, \ldots, u_{2k}) \ldots g_{s_p}(u_{(p-1)k+1}, \ldots, u_{pk}) \quad (4.4.7)$$

and where

$$g_s(u_1, \ldots, u_k) = \prod_{j=1}^{k} (s - u_j)_+^{H_0 - \frac{3}{2}}.$$

If $u_i = u_j = u$ are identified in (4.4.7), and u_i enters in g_{s_i} and u_j enters in g_{s_j}, then the integral $\int_{\mathbb{R}^{pk/2}} g_{\sigma,p} d(Leb)^{pk/2}$ in (4.4.6) over u yields the integral $\int_{\mathbb{R}} (s_i - u)_+^{H_0 - \frac{3}{2}} (s_j - u)_+^{H_0 - \frac{3}{2}} du$. For example, if $s_j > s_i$, by making the change of variables $s_i - u = (s_j - s_i)v$, we have

$$\int_{\mathbb{R}} (s_i - u)_+^{H_0 - \frac{3}{2}} (s_j - u)_+^{H_0 - \frac{3}{2}} du = \int_0^\infty v_+^{H_0 - \frac{3}{2}} (1 + v)_+^{H_0 - \frac{3}{2}} dv \, (s_j - s_i)^{2H_0 - 2}.$$

Since the integral is symmetric in s_i and s_j and in view of (2.2.12), we conclude that

$$\int_{\mathbb{R}} (s_i - u)_+^{H_0 - \frac{3}{2}} (s_j - u)_+^{H_0 - \frac{3}{2}} du = B(H_0 - 1/2, 2 - 2H_0) |s_i - s_j|^{2H_0 - 2}. \quad (4.4.8)$$

Observe that the function in (4.4.7) has pk arguments and hence $q = pk/2$ pairs of identified arguments. Thus, if $(i_1, j_1), \ldots, (i_q, j_q)$, $q = pk/2$, denote all the indices (i, j) of s_i and s_j associated with these pairs, we get

$$\int_{\mathbb{R}^{pk/2}} g_{\sigma,p} d(Leb)^{pk/2} = (B(H_0 - 1/2, 2 - 2H_0))^q |s_{i_1} - s_{j_1}|^{2H_0 - 2} \ldots |s_{i_q} - s_{j_q}|^{2H_0 - 2}.$$

Note that $i_m \neq j_m$, $m = 1, \ldots, q$, since $\Gamma(\pi^*, \sigma)$ is non-flat. Moreover, every index $1, \ldots, p$ appears exactly k times in the sequence $(i_1, j_1), \ldots, (i_q, j_q)$, since $\Gamma(\pi^*, \sigma)$ is

---

[9] Do not confuse $p = H_0 - 1/2$ in (4.2.5) and (4.2.6) with p in (4.4.1).

256   *Hermite Processes*

Gaussian.[10] The sequence $(i_1, j_1), \ldots, (i_q, j_q)$ can therefore be thought as a pair sequence of a multigraph $A \in \mathcal{A}_p(k)$. Then,

$$\int_0^{t_1} ds_1 \ldots \int_0^{t_p} ds_p \left\{ \int_{\mathbb{R}^{pk/2}} g_{\sigma,p} d(Leb)^{pk/2} \right\} = (B(H_0-1/2, 2-2H_0))^q S_D(A; t_1, \ldots, t_p), \tag{4.4.9}$$

where $S_D(A; t_1, \ldots, t_p)$ is defined in (4.4.4).

Every $\sigma \in \mathcal{M}_2^0([pk], \pi^*)$ is thus associated with a multigraph $A \in \mathcal{A}_p(k)$, and the relation (4.4.9) holds. However, a number of different partitions $\sigma \in \mathcal{M}_2^0([pk], \pi^*)$ lead to the same multigraph. By Lemma 4.3.18, that number is given by

$$\frac{(k!)^p}{v_1! \ldots v_{\binom{p}{2}}!} = (k!)^p g(A), \tag{4.4.10}$$

where $g(A)$ appears in (4.3.11). The numerator in (4.4.10) is $(k!)^p$ because each one of the p points of a multigraph $A \in \mathcal{A}_p(k)$ has degree k. Finally, the result (4.4.2) follows from (4.4.6), (4.4.9) and (4.4.10). □

Remark 4.4.2 One can show (Exercise 4.4) that

$$\mu_p(t_1, \ldots, t_p) = \frac{C_{p,k,H_0}}{2^q (q!)} \sum_1 \int_0^{t_1} ds_1 \ldots \int_0^{t_p} ds_p \, |s_{i_1} - s_{j_1}|^{2H_0-2} \ldots |s_{i_q} - s_{j_q}|^{2H_0-2}, \tag{4.4.11}$$

where $q = pk/2$ and the sum $\sum_1$ is over all indices $i_1, j_1, \ldots, i_q, j_q$ such that

(i) $i_1, j_1, \ldots, i_q, j_q \in \{1, 2, \ldots, p\}$,
(ii) $i_1 \neq j_1, \ldots, i_q \neq j_q$,
(iii) each number $1, 2, \ldots, p$ appears exactly k times in $(i_1, j_1, \ldots, i_q, j_q)$.

It is interesting to examine the formula (4.4.2) in the special cases of the first few values $p = 2, 3$ and 4.

When $p = 2$ and $k \geq 1$, there is only one multigraph A in $\mathcal{A}_2(k)$ with k lines connecting points 1 and 2, and with $g(A) = 1/k!$. It is depicted in Figure 4.9. Then, for $t_1, t_2 \geq 0$, the formula (4.4.2) becomes

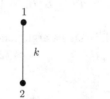

Figure 4.9 Multigraphs in $\mathcal{A}_3(k)$. The number of lines between any two points in a multigraph of the set $\mathcal{A}_2(k)$ is indicated by a number next to the line connecting the two points.

---

[10] For example: the index 2 appears exactly $k = 3$ times in the sequence
    $\{(1, 2), (2, 3), (3, 4), (4, 1), (1, 3), (2, 4)\}$.

4.4 Moments and Cumulants of Hermite Processes

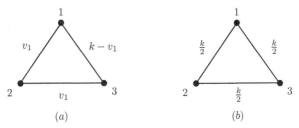

Figure 4.10 The number of lines between any two points in a multigraph of the set is indicated by a number next to the line connecting the two points.

$$\mu_2(t_1, t_2) = \frac{C_{2,k,H_0}}{k!} \int_0^{t_1} ds_1 \int_0^{t_2} ds_2 |s_1 - s_2|^{k(2H_0 - 2)}$$

$$= \frac{C_{2,k,H_0}}{k!(1 - 2k(1 - H_0))(2 - 2k(1 - H_0))} \left\{ t_1^{2 - 2k(1 - H_0)} + t_2^{2 - 2k(1 - H_0)} - |t_1 - t_2|^{2 - 2k(1 - H_0)} \right\}$$

$$= \frac{1}{2} \left\{ t_1^{2H} + t_2^{2H} - |t_1 - t_2|^{2H} \right\} \tag{4.4.12}$$

by using (4.4.3) and (4.2.1). We thus get the covariance function of an H–self-similar process with stationary increments (Proposition 2.5.6, (e)).

When $p = 3$, let v_1 be the number of lines between points 1 and 2 in a multigraph $A \in \mathcal{A}_3(k)$. Recall that every multigraph in $\mathcal{A}_3(k)$ has three points (vertices) and each point has degree k. For point 1 to have degree k, the number of lines between 1 and 3 must be $k - v_1$ and then, for point 3 to have degree k, the number of lines between the points 2 and 3 must be v_1 (Figure 4.10, (a)). In order that point 2 should have degree k as well, we must have $2v_1 = k$ and hence $v_1 = k/2$ which is an integer only when k is even. Thus, when k is even, there is only one multigraph $A \in \mathcal{A}_3(k)$ depicted in Figure 4.10, (b), and with $g(A) = 1/((k/2)!)^3$. The formula (4.4.2) becomes: when k is even,

$$\mu_3(t_1, t_2, t_3)$$
$$= \frac{C_{3,k,H_0}}{((k/2)!)^3} \int_0^{t_1} ds_1 \int_0^{t_2} ds_2 \int_0^{t_3} ds_3 |s_1 - s_2|^{\frac{k}{2}(2H_0 - 2)} |s_2 - s_3|^{\frac{k}{2}(2H_0 - 2)} |s_3 - s_1|^{\frac{k}{2}(2H_0 - 2)}$$
$$= \frac{C_{3,k,H_0}}{((k/2)!)^3} \int_0^{t_1} ds_1 \int_0^{t_2} ds_2 \int_0^{t_3} ds_3 |s_1 - s_2|^{k(H_0 - 1)} |s_2 - s_3|^{k(H_0 - 1)} |s_3 - s_1|^{k(H_0 - 1)}. \tag{4.4.13}$$

Suppose now $p = 4$ and $k \geq 1$. Let v_1, v_2 and v_3 be the number of lines in a multigraph $A \in \mathcal{A}_4(k)$ between points 1 and 2, 2 and 3, and 1 and 3, respectively. Then, since each point has degree k, the number of lines between points 1 and 4, 2 and 4, and 2 and 3 are $k - v_1 - v_3$, $k - v_1 - v_2$ and $k - v_2 - v_3$, respectively. See Figure 4.11, (a). Since point 4 has degree k, $(k - v_1 - v_3) + (k - v_1 - v_2) + (k - v_2 - v_3) = k$ or $v_1 + v_2 + v_3 = k$. This implies that the number of lines between points 1 and 4, 2 and 4, and 2 and 3 are v_2, v_3 and v_1, respectively. See Figure 4.11, (b). In the case $p = 4$, the formula (4.4.2) can now be written as

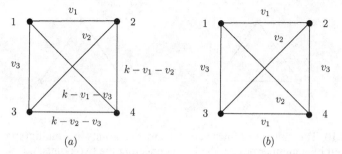

Figure 4.11 The number of lines between any two points in a multigraph of the set $\mathcal{A}_4(k)$ is indicated by a number next to the line connecting the two points.

$$\mu_4(t_1, t_2, t_3, t_4) = C_{4,k,H_0} \sum_{\substack{v_1, v_2, v_3 \geq 0 \\ v_1 + v_2 + v_3 = k}} \frac{1}{(v_1! v_2! v_3!)^2} \int_0^{t_1} ds_1 \int_0^{t_2} ds_2 \int_0^{t_3} ds_3 \int_0^{t_4} ds_4$$

$$\times |(s_1 - s_2)(s_3 - s_4)|^{v_1(2H_0 - 2)} |(s_1 - s_4)(s_2 - s_3)|^{v_2(2H_0 - 2)} |(s_1 - s_3)(s_2 - s_4)|^{v_3(2H_0 - 2)}.$$

(4.4.14)

The formulae above concern the moments of Hermite processes. There are analogous formulae for the cumulants. Let

$$\chi_p(t_1, \ldots, t_p) = \chi(Z_H^{(k)}(t_1), \ldots, Z_H^{(k)}(t_p)), \quad p \geq 1, t_1, \ldots, t_p \in \mathbb{R}, \quad (4.4.15)$$

where the joint cumulant χ is defined in (4.3.7).

Proposition 4.4.3 *Let $\chi_p(t_1, \ldots, t_p)$, $p \geq 1$, $t_1, \ldots, t_p \in \mathbb{R}$, be the joint cumulants of the Hermite process of order k in (4.4.15). Then,*

$$\chi_p(t_1, \ldots, t_p) = C_{p,k,H_0} \sum_{A \in \mathcal{A}_{p,c}(k)} g(A) S_D(A; t_1, \ldots, t_p), \quad (4.4.16)$$

where C_{p,k,H_0}, $g(A)$ and $S_D(A; t_1, \ldots, t_p)$ are as in Proposition 4.4.1.

Proof The result can be proved as Proposition 4.4.1 by using the formula (4.3.9), and by observing that the partitions $\sigma \in \mathcal{M}_2([pk], \pi^*)$ entering the formula (4.3.9) are now associated with connected multigraphs $A \in \mathcal{A}_{p,c}(k)$ (see also Proposition 4.3.25). □

When the mean is 0, the second and third cumulants are equal to the second and third moments respectively. Therefore, the formulae (4.4.12) and (4.4.13) hold as well for $\chi_2(t_1, t_2)$ and $\chi_3(t_1, t_2, t_3)$ respectively. As for χ_4, all we need is to require in addition that the multigraph be connected, thus excluding the terms $v_1 = 0$, $v_2 = 0$ or $v_3 = 0$ in (4.4.14). Therefore,

$$\chi_4(t_1, t_2, t_3, t_4) = C_{4,k,H_0} \sum_{\substack{v_1, v_2, v_3 \geq 1 \\ v_1 + v_2 + v_3 = k}} \frac{1}{(v_1! v_2! v_3!)^2} \int_0^{t_1} ds_1 \int_0^{t_2} ds_2 \int_0^{t_3} ds_3 \int_0^{t_4} ds_4$$

$$\times |(s_1 - s_2)(s_3 - s_4)|^{v_1(2H_0 - 2)} |(s_1 - s_4)(s_2 - s_3)|^{v_2(2H_0 - 2)} |(s_1 - s_3)(s_2 - s_4)|^{v_3(2H_0 - 2)}.$$

(4.4.17)

4.5 Multiple Integrals of Order Two

Consider first a general multiple integral of order two, namely,

$$I_2(f) = \int_{E^2}' f(u_1, u_2) W(du_1) W(du_2), \tag{4.5.1}$$

where W is a Gaussian random measure on $\mathbb{R}$ with control measure m and $f \in L^2(E^2, m^2)$. A number of useful results are available for such integrals. The following proposition gives its pth cumulants, where $p \geq 2$.

Proposition 4.5.1 *The pth cumulant of (4.5.1), $p \geq 2$, is*

$$\chi_p(I_2(f)) = \chi(\underbrace{I_2(f), \ldots, I_2(f)}_{p \text{ times}})$$

$$= 2^{p-1}(p-1)! \int_{E^p} f(u_1, u_2) f(u_2, u_3) \ldots f(u_{p-1}, u_p) f(u_p, u_1) m(du_1) \ldots m(du_p). \tag{4.5.2}$$

Proof Consider first the case $p = 2$. The only multigraph in $\mathcal{A}_{2,c}(2)$ with two points of degree 2 is the one depicted in Figure 4.9 with $k = 2$, for which $g(A) = 1/2!$. By (4.3.19),

$$\chi_2(I_2(f)) = 2!2! \frac{1}{2!} \int_{E^2} f(u_1, u_2)^2 m(du_1) m(du_2),$$

which proves (4.5.2) for $p = 2$.

Suppose now $p \geq 3$. We shall use again an argument based on multigraphs[11] and apply Proposition 4.3.20. Since $n_1 = n_2 = \ldots = n_p = 2$ and $n = n_1 + \cdots + n_p = 2p$, we have

$$\chi_p(I_2(f)) = (2!)^p \sum_{A \in \mathcal{A}_{p,c}(2)} g(A) \int_{E^p} f_{A,p} dm^p,$$

where $\mathcal{A}_{p,c}(2) = \mathcal{A}_c(2, \ldots, 2)$ is the set of all connected multigraphs whose $p \geq 3$ points have each degree 2. Observe that no $A \in \mathcal{A}_{p,c}(2)$ can have multiple lines and hence $g(A) = 1$. One multigraph $A \in \mathcal{A}_{p,c}(2)$ is depicted in Figure 4.12. All the other multigraphs in $\mathcal{A}_{p,c}(2)$ involve a permutation of the points. Two such permutations correspond to the same multigraph if and only if the corresponding "circles" (as in Figure 4.12) are possible rotations and flipped images of each other. There are $(p-1)!/2$ such permutations corresponding to different multigraphs. Here, $(p-1)!$ counts possible permutations after fixing a point, say 1, and $1/2$ takes into account mirror images after flipping a circle.

The multigraph in Figure 4.12 is described by a pair sequence $\{(p, 1), (1, 2), (1, 3), \ldots, (p-1, p))\}$, for which

$$f_{A,p}(u_1, \ldots, u_p) = f(u_1, u_2) f(u_2, u_3) \ldots f(u_p, u_1). \tag{4.5.3}$$

---
[11] For an argument based on diagrams instead, see Peccati and Taqqu [797], p. 139

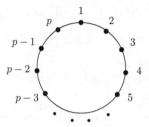

Figure 4.12 A multigraph in $\mathcal{A}_{p,c}(2)$.

For any other multigraph in $\mathcal{A}_{p,c}(2)$, $f_{A,p}$ is defined by permuting the indices in (4.5.3). Hence,

$$\int_{E^p} f_{A,p} dm^p = \int_{E^p} f(u_1, u_2) f(u_2, u_3) \ldots f(u_p, u_1) m(du_1) \ldots m(du_p)$$

for all $A \in \mathcal{A}_{p,c}(2)$. This yields

$$\chi_p(I_2(f)) = (2!)^p \frac{(p-1)!}{2} \int_{E^p} f(u_1, u_2) f(u_2, u_3) \ldots f(u_p, u_1) m(du_1) \ldots m(du_p),$$

which is what we wanted to prove. □

For symmetric $f \in L^2(E^2, m^2)$, that is, $f(u, v) = f(v, u)$ and $\int_E \int_E f(u, v)^2 m(du) m(dv) < \infty$,

$$(\mathcal{I}g)(v) = \int_E g(u) f(u, v) m(du)$$

defines an operator from $g \in L^2(E, m)$ to $\mathcal{I}g \in L^2(E, m)$, which is compact[12] and self-adjoint[13] (e.g., Conway [261], Example 2.9 and Proposition 4.7 in Chapter II). By the spectral theorem (e.g., Conway [261], Theorem 5.1 and Corollary 5.4 in Chapter II), the operator has real eigenvalues λ_k and orthonormal eigenfunctions $e_k(u)$ satisfying

$$\lambda_k e_k(v) = \int_E e_k(u) f(u, v) m(du), \quad k \geq 1, \tag{4.5.4}$$

and one has

$$(\mathcal{I}g)(v) = \int_E g(u) f(u, v) m(du) = \sum_{k=1}^{\infty} \lambda_k \left(\int_E g(u) e_k(u) m(du) \right) e_k(v). \tag{4.5.5}$$

Though we index the eigenvalues by $k \geq 1$, the number of nonzero eigenvalues can be finite.[14] Note also that the relation (4.5.5) makes sense informally since it certainly holds

---

[12] Let X and Y be normed spaces, for example, $L^2(E, m)$ as above. An operator $T : X \to Y$ is called a *compact linear operator* if T is linear and if for every bounded subset B of X, the image $T(B)$ has a compact closure $\overline{T(B)}$.

[13] Let $T : H \to H$ be a bounded linear operator on a Hilbert space H with inner product $(\cdot, \cdot)$. Its *adjoint* T^* is such that $(Tx, y) = (x, T^*y), x, y \in H$. T is *self-adjoint* if $T = T^*$. Here, $H = L^2(E, m)$ and $(g_1, g_2) = \int_E g_1(u) g_2(u) m(du)$. The operator $\mathcal{I}$ is self-adjoint because the kernel $f(u, v)$ is symmetric.

[14] As in the case $f(u, v) = h(u)h(v)$ when $(\mathcal{I}g)(v) = (\int_E g(u)h(u)du)h(v)$ has only one nonzero eigenvalue $\lambda_1 = \int_E h(u)^2 du$ with the corresponding eigenvector $e_1(v) = h(v)$.

for $g(u) = e_m(u)$, with the left-hand side equal to $\lambda_m e_m(u)$ by (4.5.4) and the right-hand side being the same by the orthonormality of the eigenfunctions $e_k(u)$, $k \geq 1$. The following basic result is useful when working with multiple integral of order two.

Lemma 4.5.2 *For symmetric $f \in L^2(E^2, m^2)$, one has*

$$I_2(f) = \sum_{k=1}^{\infty} \lambda_k (\xi_k^2 - 1), \qquad (4.5.6)$$

where the convergence is in $L^2(\Omega)$, λ_k, $k \geq 1$, are the eigenvalues appearing in (4.5.4), and $\xi_k = I_1(e_k)$, $k \geq 1$, with the eigenfunctions e_k appearing in (4.5.4). The random variables ξ_k, $k \geq 1$, are i.i.d. $\mathcal{N}(0, 1)$ random variables.

Proof Recall that $\int_E e_k(u) e_j(u) m(du) = \delta_{k,j}$ by orthogonality in $L^2(E, m)$. We first show that

$$f(u, v) = \sum_{k=1}^{\infty} \lambda_k e_k(u) e_k(v) \qquad (4.5.7)$$

a.e. $m(du)m(dv)$. Since $f \in L^2(E^2, m^2)$, by Fubini's theorem, $\int_E f(u, v)^2 m(du) < \infty$ a.e. $m(dv)$. Taking $g(u) = g_v(u) = f(u, v)$ in (4.5.5) and using (4.5.4) and (4.5.5) yields

$$\int_E f(u, v)^2 m(du) = \int_E g_v(u) f(u, v) m(du) = \sum_{k=1}^{\infty} \lambda_k \left(\int_E f(u, v) e_k(u) m(du) \right) e_k(v)$$

$$= \sum_{k=1}^{\infty} (\lambda_k e_k(v))^2 < \infty \quad \text{a.e. } m(dv). \qquad (4.5.8)$$

Since by the orthonormality of e_k in $L^2(E, m)$,

$$\sum_{k=1}^{\infty} (\lambda_k e_k(v))^2 = \int_E \left(\sum_{k=1}^{\infty} (\lambda_k e_k(v)) e_k(u) \right)^2 m(du),$$

this shows that $\sum_{k=1}^{\infty} \lambda_k e_k(u) e_k(v) = \sum_{k=1}^{\infty} (\lambda_k e_k(v)) e_k(u)$ converges in $L^2(E, m)$, a.e. $m(dv)$. Then, (4.5.5) can be written as

$$\int_E g(u) \left(f(u, v) - \sum_{k=1}^{\infty} \lambda_k e_k(u) e_k(v) \right) m(du) = 0$$

for a.e. $m(dv)$. Taking $g(u) = g_v(u) = f(u, v) - \sum_{k=1}^{\infty} \lambda_k e_k(u) e_k(v)$ gives

$$\int_E \left(f(u, v) - \sum_{k=1}^{\infty} \lambda_k e_k(u) e_k(v) \right)^2 m(du) = 0$$

a.e. $m(dv)$, which yields (4.5.7).

Moreover, by (4.5.8),

$$\sum_{k=1}^{\infty} \lambda_k^2 = \sum_{k=1}^{\infty} \lambda_k^2 \int_E e_k(v)^2 m(dv) = \int_E \sum_{k=1}^{\infty} \left(\lambda_k e_k(u)\right)^2 m(dv)$$
$$= \int_E \int_E f(u,v)^2 m(du) m(dv) < \infty,$$

since $f \in L^2(E^2, m^2)$. Since e_k are orthonormal and $\sum_{k=1}^{\infty} \lambda_k^2 < \infty$, the series (4.5.7) also converges in $L^2(E^2, m^2)$. Since $\mathbb{E}(I_2(f))^2 = 2\int_E \int_E f(u,v)^2 m(du) m(dv) < \infty$, we can then apply I_2 to both sides of (4.5.7) to get

$$I_2(f) = \sum_{k=1}^{\infty} \lambda_k I_2(e_k^{\otimes 2}) = \sum_{k=1}^{\infty} \lambda_k H_2(I_1(e_k)),$$

by using Proposition 4.1.2, (a). It remains to observe that since e_k are orthonormal, $I_1(e_k) =: \xi_k$ are i.i.d. $\mathcal{N}(0,1)$ random variables, and that $H_2(\xi_k) = \xi_k^2 - 1$. □

In particular, the characteristic function of $I_2(f)$ has the product expansion

$$\mathbb{E}e^{i\theta I_2(f)} = \prod_{k=1}^{\infty} \mathbb{E}e^{i\theta \lambda_k (\xi_k^2 - 1)} = \prod_{k=1}^{\infty} (1 - i2\theta \lambda_k)^{-1/2} e^{-i\theta \lambda_k}, \qquad (4.5.9)$$

where ξ_k^2 are i.i.d. chi-square with 1 degree of freedom. This expression can be rewritten as

$$\mathbb{E}e^{i\theta I_2(f)} = \exp\left\{-\sum_{k=1}^{\infty} \left(\frac{1}{2}\log(1 - i2\theta \lambda_k) + i\theta \lambda_k\right)\right\} = \exp\left\{\sum_{k=1}^{\infty} \left(\frac{1}{2}\sum_{p=1}^{\infty} \frac{(i2\theta \lambda_k)^p}{p} - i\theta \lambda_k\right)\right\}$$
$$= \exp\left\{\frac{1}{2}\sum_{k=1}^{\infty}\sum_{p=2}^{\infty} \frac{(i2\theta \lambda_k)^p}{p}\right\} = \exp\left\{\frac{1}{2}\sum_{p=2}^{\infty} \frac{(i2\theta)^p}{p} \sum_{k=1}^{\infty} \lambda_k^p\right\}.$$

By substituting (4.5.7) into (4.5.2), it can be shown easily that

$$\sum_{k=1}^{\infty} \lambda_k^p = \frac{\chi_p(I_2(f))}{2^{p-1}(p-1)!}, \quad p \geq 2 \qquad (4.5.10)$$

(Exercise 4.5). Hence

$$\mathbb{E}e^{i\theta I_2(f)} = \exp\left\{\frac{1}{2}\sum_{p=2}^{\infty} \frac{(i2\theta)^p}{p} \frac{\chi_p(I_2(f))}{2^{p-1}(p-1)!}\right\}, \qquad (4.5.11)$$

by using the fact that $\chi_1(I_2(f)) = \mathbb{E}I_2(f) = 0$.

4.6 The Rosenblatt Process

We shall now apply the results of Section 4.5 to the Rosenblatt process which is the Hermite process $Z_H^{(k)}$ with $k = 2$ (see Definition 4.2.1). It can be found in various forms in Proposition 2.6.16 as well as in Definition 4.2.1. We present it here again for convenience.

4.6 The Rosenblatt Process

Definition 4.6.1 The *Rosenblatt process* is the Hermite process of order two defined by

$$Z_H^{(2)}(t) = a_{2,H_0} \int_{\mathbb{R}^2}' \left\{ \int_0^t (s-u_1)_+^{H_0-\frac{3}{2}} (s-u_2)_+^{H_0-\frac{3}{2}} ds \right\} B(du_1) B(du_2), \quad t \in \mathbb{R},$$

where

$$\frac{3}{4} < H_0 < 1.$$

It is standard, that is, $\mathbb{E} Z_H^{(2)}(1)^2 = 1$, when

$$a_{2,H_0} = \frac{\sqrt{(H_0-1/2)(4H_0-3)}}{B(H_0-1/2, 2-2H_0)} = \frac{\sqrt{(H/2)(2H-1)}}{B(H/2, 1-H)}.$$

Here $H = 2H_0 - 1 \in (1/2, 1)$ is the self-similarity parameter of the Rosenblatt process.

Other representations can be found in Theorem 4.2.10.

The following proposition gives the joint cumulants and characteristic function of the Rosenblatt process.

Proposition 4.6.2 Let $Z_H^{(2)} = \{Z_H^{(2)}(t)\}_{t \in \mathbb{R}}$ be the Hermite process of order two, that is, the Rosenblatt process. For $\theta_1, \ldots, \theta_J \in \mathbb{R}$, $t_1, \ldots, t_J \in \mathbb{R}$, the pth cumulant, $p \geq 2$, and the characteristic function of a linear combination $\sum_{j=1}^J \theta_j Z_H^{(2)}(t_j)$ are given by

$$\chi_p\left(\sum_{j=1}^J \theta_j Z_H^{(2)}(t_j)\right) = 2^{-1}(p-1)! C_{p,2,H_0} \sum_{j_1,\ldots,j_p=1}^J \theta_{j_1} \ldots \theta_{j_p} S(t_{j_1}, \ldots, t_{j_p}) \quad (4.6.1)$$

and

$$\mathbb{E} \exp\left\{i \sum_{j=1}^J \theta_j Z_H^{(2)}(t_j)\right\} = \exp\left\{\frac{1}{2} \sum_{p=2}^\infty \frac{i^p C_{p,2,H_0}}{p} \sum_{j_1,\ldots,j_p=1}^J \theta_{j_1} \ldots \theta_{j_p} S(t_{j_1}, \ldots, t_{j_p})\right\}, \quad (4.6.2)$$

where $C_{p,2,H_0}$ is defined in (4.4.3) and

$$S(t_{j_1}, \ldots, t_{j_p})$$
$$= \int_0^{t_{j_1}} ds_1 \ldots \int_0^{t_{j_p}} ds_p |s_1 - s_2|^{2H_0-2} |s_2 - s_3|^{2H_0-2} \ldots |s_{p-1} - s_p|^{2H_0-2} |s_p - s_1|^{2H_0-2}. \quad (4.6.3)$$

Proof Write

$$\sum_{j=1}^J \theta_j Z_H^{(2)}(t_j) = a_{2,H_0} I_2\left(\sum_{j=1}^J \theta_j f_{t_j}\right),$$

where the kernel function

$$f_t(u_1, u_2) = \int_0^t (s-u_1)_+^{H_0-3/2} (s-u_2)_+^{H_0-3/2} ds$$

is defined in (4.2.5) with (4.2.6) and I_2 denotes a multiple integral of order two. By using (4.5.2),

$$\chi_P\Big(\sum_{j=1}^{J}\theta_j Z_H^{(2)}(t_j)\Big) = (a_{2,H_0})^p 2^{p-1}(p-1)!$$

$$\times \int_{\mathbb{R}^p} \sum_{j_1=1}^{J}\theta_{j_1} f_{t_{j_1}}(u_1, u_2) \sum_{j_2=1}^{J}\theta_{j_2} f_{t_{j_2}}(u_2, u_3) \ldots \sum_{j_p=1}^{J}\theta_{j_p} f_{t_{j_p}}(u_p, u_1) du_1 du_2 \ldots du_p$$

$$= (a_{2,H_0})^p 2^{p-1}(p-1)! \sum_{j_1,\ldots,j_p=1}^{J} \theta_{j_1}\ldots\theta_{j_p} \int_{\mathbb{R}^p} f_{t_{j_1}}(u_1, u_2) \ldots f_{t_{j_p}}(u_p, u_1) du_1 du_2 \ldots du_p$$

$$= (a_{2,H_0})^p 2^{p-1}(p-1)! \sum_{j_1,\ldots,j_p=1}^{J} \theta_{j_1}\ldots\theta_{j_p} \int_{\mathbb{R}^p} du_1\ldots du_p \int_0^{t_{j_1}} ds_1 \ldots \int_0^{t_{j_p}} ds_p$$

$$\times (s_1-u_1)_+^{H_0-\frac{3}{2}}(s_1-u_2)_+^{H_0-\frac{3}{2}}(s_2-u_2)_+^{H_0-\frac{3}{2}}(s_2-u_3)_+^{H_0-\frac{3}{2}} \ldots (s_p-u_p)_+^{H_0-\frac{3}{2}}(s_p-u_1)_+^{H_0-\frac{3}{2}}$$

$$= (a_{2,H_0})^p 2^{p-1}(p-1)! \sum_{j_1,\ldots,j_p=1}^{J} \theta_{j_1}\ldots\theta_{j_p} \int_0^{t_{j_1}} ds_1 \ldots \int_0^{t_{j_p}} ds_p$$

$$\times \Big(\int_{\mathbb{R}}(s_1-u_2)_+^{H_0-\frac{3}{2}}(s_2-u_2)_+^{H_0-\frac{3}{2}} du_2\Big)\Big(\int_{\mathbb{R}}(s_2-u_3)_+^{H_0-\frac{3}{2}}(s_3-u_3)_+^{H_0-\frac{3}{2}} du_3\Big)\times\ldots$$

$$\times \Big(\int_{\mathbb{R}}(s_{p-1}-u_p)_+^{H_0-\frac{3}{2}}(s_p-u_p)_+^{H_0-\frac{3}{2}} du_p\Big)\Big(\int_{\mathbb{R}}(s_1-u_1)_+^{H_0-\frac{3}{2}}(s_p-u_1)_+^{H_0-\frac{3}{2}} du_1\Big)$$

$$= 2^{-1}(p-1)!(a_{2,H_0})^p 2^p (B(H_0-1/2, 2-2H_0))^p \sum_{j_1,\ldots,j_p=1}^{J} \theta_{j_1}\ldots\theta_{j_p} S(t_{j_1},\ldots,t_{j_p}),$$

by using (4.4.8) and (4.6.3). Using the expression (4.4.3) for $C_{p,2,H_0}$ yields the formula (4.6.1) since $k = 2$ and $q = pk/2 = p$. The formula (4.6.2) is a consequence of (4.6.1) and

$$\mathbb{E}\exp\Big\{i\sum_{j=1}^{J}\theta_j Z_H^{(2)}(t_j)\Big\} = \exp\Big\{\sum_{p=2}^{\infty}\frac{i^p}{p!}\chi_P\Big(\sum_{j=1}^{J}\theta_j Z_H^{(2)}(t_j)\Big)\Big\}$$

since the mean (first cumulant) is zero (see also (4.5.11)). □

4.7 The Rosenblatt Distribution

We provide some results and observations on the Rosenblatt distribution, drawing from the work of Veillette and Taqqu [979]. For notational simplicity, we set

$$D = 2(1 - H_0) = 1 - H \in \Big(0, \frac{1}{2}\Big), \tag{4.7.1}$$

associated with (4.2.2) when $k = 2$.

Definition 4.7.1 The *Rosenblatt distribution* is defined as the distribution of the random variable $Z_H^{(2)}(1)$, where $Z_H^{(2)}$ is the (standard) Rosenblatt process. A random variable having the Rosenblatt distribution will be denoted as Z_D. Though for shortness sake, we will sometimes refer to Z_D as the Rosenblatt distribution itself.

By Proposition 4.6.2, the Rosenblatt distribution has cumulants $\chi_1 = 0$ and

$$\chi_p := \chi_p(Z_D) = 2^{-1}(p-1)! C_{p,2,H_0} S_p, \quad p \geq 2, \qquad (4.7.2)$$

and the characteristic function

$$\mathbb{E} e^{i\theta Z_D} = \exp\left\{\frac{1}{2} \sum_{p=2}^{\infty} \frac{i^p \theta^p C_{p,2,H_0}}{p} S_p\right\} = \exp\left\{\sum_{p=2}^{\infty} \frac{i^p \theta^p}{p!} \chi_p\right\}, \qquad (4.7.3)$$

where

$$S_p = \int_0^1 ds_1 \ldots \int_0^1 ds_p \, |s_1 - s_2|^{-D} |s_2 - s_3|^{-D} \ldots |s_p - s_1|^{-D}. \qquad (4.7.4)$$

Each moment $\mu_p = \mathbb{E} Z_D^p$, $p \geq 2$, can be expressed as a polynomial using the cumulants χ_k, $k = 1, \ldots, p$, and the *complete Bell polynomials*

$$\mu_p = B_p(0, \chi_2, \ldots, \chi_p), \qquad (4.7.5)$$

as noted, for example, in Peccati and Taqqu [797], Proposition 3.3.1. They can also be computed recursively (e.g., Smith [914]).

Thus, in order to compute any moment or cumulant, it is necessary to compute the multiple integrals S_k. The first two can be computed directly as

$$S_2 = \frac{1}{(1-D)(1-2D)}, \quad S_3 = \frac{2}{(1-D)(2-3D)} B(1-D, 1-D) \qquad (4.7.6)$$

(Exercise 4.6). For $p \geq 4$, a closed form expression for S_p is not known, which means they must be computed numerically. Computing the multiple integrals directly is intractable due to the increasing number of singularities in the integrand. This difficulty can be circumvented through the following procedure.

Denote $L^2(0,1) = L^2((0,1), dx)$ and define the integral operator $\mathcal{K}_D : L^2(0,1) \to L^2(0,1)$ as

$$(\mathcal{K}_D f)(x) = \int_0^1 |x-u|^{-D} f(u) du \qquad (4.7.7)$$

and then the sequence of functions $G_{k,D} \in L^2(0,1)$, $k \geq 1$, recursively as follows:

$$G_{1,D}(x) = \frac{(1-x)^{-D}}{\sqrt{1-D}}, \quad G_{k,D}(x) = (\mathcal{K}_D G_{k-1,D})(x), \quad k \geq 2. \qquad (4.7.8)$$

Then, we have the following alternative way to express S_p. See Proposition 2.1 in Veillette and Taqqu [979] for a proof. For $f, g \in L^2(0,1)$, $(f,g)_{L^2(0,1)} = \int_0^1 f(x)g(x)dx$ denotes their inner product.

Proposition 4.7.2 *Let η and ν be any two positive integers such that $\eta + \nu = p$. Then,*

$$S_p = (G_{\eta,D}, G_{\nu,D})_{L^2(0,1)}. \tag{4.7.9}$$

To minimize the number of integrals one needs to compute, it makes sense to choose $\eta = \nu = \frac{p}{2}$ if p is even, and $\eta = \frac{p+1}{2}$ and $\nu = \frac{p-1}{2}$ if p is odd. Proposition 4.7.2 thus reduces the problem of computing a p–dimensional integral into computing $\lceil \frac{p}{2} \rceil + 1$ one-dimensional integrals. One can also show that

$$G_{2,D}(x) = \frac{x^{1-D}}{(1-D)^{3/2}} \, {}_2F_1(D, 1, 2-D, x) + \frac{(1-x)^{1-2D} B(1-D, 1-D)}{\sqrt{1-D}},$$

where ${}_2F_1(a, b, c, x)$ is the so-called *Gauss hypergeometric function* (see (6.2.16) and (6.2.17) in Chapter 6). The function ${}_2F_1(a, b; c; x)$ is bounded for $x \in (0, 1)$ as long as $c > a + b$ (Lebedev [601], Section 9), which is true in this case. This implies that unlike $G_{1,D}$, $G_{2,D}$ is a bounded function on $(0, 1)$, since $0 < D < 1/2$.

In Section C of Veillette and Taqqu [980], the authors outline a technique for computing S_p numerically based on Proposition 4.7.2, and tabulate the first eight cumulants and moments of the Rosenblatt distribution for various values of D.

By Lemma 4.5.2, the Rosenblatt distribution can be represented as

$$Z_D = \sum_{k=1}^{\infty} \lambda_k(D)(\xi_k^2 - 1), \tag{4.7.10}$$

where $\xi_k, k \geq 1$, are i.i.d. $\mathcal{N}(0, 1)$ random variables, and $\lambda_k, k \geq 1$, are the eigenvalues (in the decreasing order) of the operator $A_D : L^2(\mathbb{R}) \to L^2(\mathbb{R})$ defined by

$$(A_D f)(v) = a_{2,H_0} \int_{\mathbb{R}} \left\{ \int_0^1 (s-u)_+^{-\frac{1+D}{2}} (s-v)_+^{-\frac{1+D}{2}} ds \right\} f(u) du, \tag{4.7.11}$$

where we used the representation (4.2.4) with $t = 1$ of the Rosenblatt distribution. The next result provides a convenient alternative characterization of the eigenvalues $\lambda_k(D), k \geq 1$.

Proposition 4.7.3 *The eigenvalues $\lambda_k(D), k \geq 1$, are the same as those of the operator $(\sigma_D)\mathcal{K}_D$ in (4.7.7), where*

$$\sigma_D = \left(\frac{1}{2}(1-D)(1-2D)\right)^{1/2}. \tag{4.7.12}$$

Proof We shall prove only one direction of the statement, namely, that the eigenvalues $\lambda_k(D), k \geq 1$, are also those of the operator $(\sigma_D)\mathcal{K}_D$. See Proposition 3.1 in Veillette and Taqqu [979] for a slightly different and complete proof. We should also note that the arguments below could be understood easier using fractional calculus of Chapters 6 and 7. But to keep the presentation self-contained, we will not use here any notions or results of fractional calculus.

If $\lambda_k(D)$ is an eigenvalue of the operator A_D, there is a function $f \in L^2(\mathbb{R})$ such that

$$a_{2,H_0} \int_{\mathbb{R}} \left\{ \int_0^1 (s-u)_+^{-\frac{1+D}{2}} (s-v)_+^{-\frac{1+D}{2}} ds \right\} f(u) du = \lambda_k(D) f(v)$$

4.7 The Rosenblatt Distribution

or

$$a_{2,H_0} \int_0^1 (s-v)_+^{-\frac{1+D}{2}} \left\{ \int_{\mathbb{R}} (s-u)_+^{-\frac{1+D}{2}} f(u)du \right\} ds = \lambda_k(D) f(v).$$

After multiplying both sides of the last relation by $(z-v)_+^{-\frac{1+D}{2}}$, $z \in (0,1)$, and integrating over $v \in \mathbb{R}$, we obtain that

$$a_{2,H_0} \int_0^1 \left\{ \int_{\mathbb{R}} (z-v)_+^{-\frac{1+D}{2}} (s-v)_+^{-\frac{1+D}{2}} dv \right\} g(s)ds = \lambda_k(D) g(z),$$

where $g(z) = \int_{\mathbb{R}} (z-v)_+^{-\frac{1+D}{2}} f(v)dv$. The inner integral over v above can be evaluated explicitly. Indeed, for example, for $z < s$, by making the changes of variables $z - v = (s-z)y$, we have

$$\int_{\mathbb{R}} (z-v)_+^{-\frac{1+D}{2}} (s-v)_+^{-\frac{1+D}{2}} dv (s-z)^{-D} \int_0^\infty y^{-\frac{1+D}{2}} (1+y)^{-\frac{1+D}{2}} dy$$

$$= (s-z)^{-D} B\left(\frac{1-D}{2}, D\right),$$

where we also used (2.2.12). Observe from (4.2.8) that

$$a_{2,H_0} = \left(\frac{1}{2}(1-D)(1-2D)\right)^{1/2} \frac{1}{B((1-D)/2, D)} = \frac{\sigma_D}{B((1-D)/2, D)}.$$

It follows that

$$\sigma_D \int_0^1 |s-z|^{-D} g(s) ds = \lambda_k(D) g(z),$$

that is, $\lambda_k(D)$ is an eigenvalue of the operator $(\sigma_D)\mathcal{K}_D$ as well. $\square$

Remark 4.7.4 The following asymptotic relationships for the eigenvalues $\lambda_k(D)$ can be established (Theorem 3.2 in Veillette and Taqqu [979]). For $0 < r < 1$,

$$\lambda_k(D) = C(D) k^{D-1}\left(1 + o\left(\frac{1}{k^r}\right)\right), \tag{4.7.13}$$

where

$$C(D) = \frac{2}{\pi^{1-D}} \sigma_D \Gamma(1-D) \sin\left(\frac{\pi D}{2}\right). \tag{4.7.14}$$

Moreover, the series

$$\sum_{k=1}^\infty (\lambda_k(D) - C(D) k^{D-1})$$

converges and equals

$$\sum_{k=1}^\infty (\lambda_k(D) - C(D) k^{D-1}) = -2^{1-D} \sigma_D \zeta(D), \tag{4.7.15}$$

where $\zeta(s) = \sum_{n=1}^\infty n^{-s}$ denotes the Riemann zeta function.

Further insight on the Rosenblatt distribution can be gained by regarding it as an infinitely divisible distribution. Recall that the characteristic function of an infinitely divisible distribution X with $\mathbb{E}X^2 < \infty$ can be written in the following form:

$$\phi(\theta) = \mathbb{E}e^{i\theta X} = \exp\left\{ia\theta - \frac{1}{2}b^2\theta^2 + \int_{-\infty}^{\infty}(e^{i\theta u} - 1 - iu\theta)\nu(du)\right\},$$

where $a \in \mathbb{R}$, $b > 0$ and ν is a positive measure on $\mathbb{R} \setminus \{0\}$ with the property that $\int \min\{u^2, 1\}\nu(du) < \infty$. This is known as the Lévy–Khintchine representation of X. Since the chi-square distribution is infinitely divisible, it is not surprising in light of (4.7.10) that the Rosenblatt distribution is also infinitely divisible.

The next result provides the Lévy–Khintchine representation of the Rosenblatt distribution.

Theorem 4.7.5 *Let Z_D have the Rosenblatt distribution with $0 < D < 1/2$ and let*

$$\lambda(D)^{-1} = (\lambda_1(D)^{-1}, \lambda_2(D)^{-1}, \ldots)$$

be the sequence of the inverses of the eigenvalues associated to the integral operator $\sigma_D \mathcal{K}_D$ defined in (4.7.7) and (4.7.12). Then, the characteristic function of Z_D can be written as

$$\phi(\theta) = \mathbb{E}e^{i\theta Z_D} = \exp\left\{\int_0^{\infty}(e^{i\theta u} - 1 - i\theta u)\nu_D(u)du\right\}, \quad (4.7.16)$$

where ν_D is supported on $(0, \infty)$ and is given by

$$\nu_D(u) = \frac{1}{2u}\sum_{n=1}^{\infty} e^{-u/(2\lambda_n(D))}, \quad u > 0. \quad (4.7.17)$$

Moreover, ν_D has the following asymptotic forms as $u \to 0^+$ and $u \to \infty$,

$$\nu_D(u) \sim \frac{2^{\frac{D}{1-D}}C(D)^{\frac{1}{1-D}}}{(1-D)}\Gamma\left(\frac{1}{1-D}\right)u^{\frac{D-2}{1-D}}, \quad u \to 0^+, \quad (4.7.18)$$

$$\nu_D(u) \sim \frac{e^{-u/(2\lambda_1)}}{2u}, \quad u \to \infty, \quad (4.7.19)$$

where $C(D)$ is defined in (4.7.14).

Proof Let

$$Z_D^{(M)} = \sum_{n=1}^{M}\lambda_n(D)(\xi_n^2 - 1).$$

We have $Z_D^{(M)} \xrightarrow{d} Z_D$, and since $Z_D^{(M)}$ is a sum of shifted i.i.d. chi-squared distributions, we can use the Lévy–Khintchine representation of a chi-square (Applebaum [40], Example 1.3.22), which is a gamma distribution with shape parameter $1/2$ and scale parameter 2:

$$\mathbb{E}e^{i\theta Z_D^{(M)}} = \prod_{n=1}^{M}e^{i\theta\lambda_n(D)(\xi_n^2-1)} = \prod_{n=1}^{M}\exp\left\{-i\theta\lambda_n(D) + \int_0^{\infty}(e^{i\theta u} - 1)\frac{e^{-u/(2\lambda_n(D))}}{2u}du\right\}.$$

$$(4.7.20)$$

Using $(1/2)\int_0^\infty e^{-u/(2\lambda)}du = \lambda$, (4.7.20) can be rewritten as

$$\prod_{n=1}^{M}\exp\left\{\int_0^\infty (e^{i\theta u} - 1 - i\theta u)\frac{e^{-u/(2\lambda_n(D))}}{2u}du\right\}$$
$$= \exp\left\{\int_0^\infty (e^{i\theta u} - 1 - i\theta u)\frac{1}{2u}G^{(M)}_{\lambda(D)^{-1}}(e^{-u/2})du\right\},$$

where

$$G^{(M)}_{\lambda(D)^{-1}}(x) = \sum_{n=1}^{M} x^{\lambda_n(D)^{-1}}.$$

Now, we let $M \to \infty$. In order to justify passing the limit through the integral, notice that

$$\left|(e^{i\theta u} - 1 - i\theta u)\frac{1}{2u}G^{(M)}_{\lambda(D)^{-1}}(e^{-u/2})\right| \le \frac{\theta^2}{4}u\, G^{(M)}_{\lambda(D)^{-1}}(e^{-u/2}) \le \frac{\theta^2}{4}u\, G_{\lambda(D)^{-1}}(e^{-u/2}),$$
(4.7.21)

where we have used the identity $|e^{iz} - 1 - z| \le z^2/2$ for $z \in \mathbb{R}$ and the definition (4.7.24) for the function $G_{\lambda(D)^{-1}}(x)$. Notice that the last bound in (4.7.21) is continuous for $0 < u < \infty$, and by (4.7.13) together with Lemma 4.7.6 below using $\alpha = 1 - D$ and $\beta = C(D)^{-1}$, we have

$$u\, G_{\lambda(D)^{-1}}(e^{-u/2}) \sim ue^{-u/(2\lambda_1)}, \quad \text{as } u \to \infty,$$
(4.7.22)

and

$$u\, G_{\lambda(D)^{-1}}(e^{-u/2}) \sim C'u(1 - e^{-u/2})^{-1/(1-D)} \sim C''u^{\frac{D}{1-D}}, \quad \text{as } u \to 0^+,$$
(4.7.23)

for some constants C' and C''. Since $0 < \frac{D}{1-D} < 1$, (4.7.22) and (4.7.23) imply that the last bound in (4.7.21) is integrable on $(0, \infty)$, and hence the dominated convergence theorem applies and

$$\mathbb{E}e^{i\theta Z_D^{(M)}} \to \mathbb{E}e^{i\theta Z_D} = \exp\left\{\int_0^\infty (e^{i\theta u} - 1 - i\theta u)\frac{1}{2u}G_{\lambda(D)^{-1}}(e^{-u/2})du\right\},$$

which verifies (4.7.16).

The final assertions (4.7.18) and (4.7.19) also follow from (4.7.13) and Lemma 4.7.6 below with $\alpha = 1 - D$ and $\beta = C(D)^{-1}$, since these imply

$$\frac{1}{2u}G_{\lambda(D)^{-1}}(e^{-u/2}) \sim \frac{1}{2u}\frac{C(D)^{-\frac{1}{1-D}}}{(1-D)}\Gamma\left(\frac{1}{1-D}\right)\left(\frac{u}{2}\right)^{-\frac{1}{1-D}}$$
$$= \frac{2^{\frac{D}{1-D}}C(D)^{\frac{1}{1-D}}}{(1-D)}\Gamma\left(\frac{1}{1-D}\right)u^{\frac{D-2}{1-D}}, \quad u \to 0^+,$$

and

$$\frac{1}{2u}G_{\lambda(D)^{-1}}(e^{-u/2}) \sim \frac{1}{2u}e^{-u/(2\lambda_1(D))}, \quad u \to \infty.$$

This concludes the proof. □

The next auxiliary lemma was used in the proof of Theorem 4.7.5 above. See Lemma 4.1 in Veillette and Taqqu [979] for a proof. Given any positive, increasing sequence $\mathbf{c} = \{c_n\}_{n=1}^{\infty}$ such that $\sum_{n=1}^{\infty} 1/c_n^2 < \infty$, define the function $G_{\mathbf{c}}(x)$ for $0 < x < 1$ as

$$G_{\mathbf{c}}(x) = \sum_{n=1}^{\infty} x^{c_n}. \qquad (4.7.24)$$

Since $\mathbf{c}^{-1} = \{c_1^{-1}, c_2^{-1}, \dots\} \in \ell^2(\mathbb{Z})$, we have $c_n \to \infty$ and thus this series converges for all $x \in (0, 1)$ since $\log x < 0$ and hence for n large enough, one has $x^{c_n} = e^{c_n \log(x)} \le c_n^{-2}$. Notice that $G_{\mathbf{c}}(0) = 0$, $G_{\mathbf{c}}(x) \to \infty$ as $x \to 1$ and $G_{\mathbf{c}}(x)$ is a continuous function for all $x \in (0, 1)$.

Lemma 4.7.6 *Suppose $\mathbf{c}$ is a positive strictly increasing sequence such that $c_n \sim \beta n^{\alpha}$, as $n \to \infty$, for some $1/2 < \alpha < 1$ and constant $\beta > 0$. Then,*

$$G_{\mathbf{c}}(x) \sim x^{c_1}, \quad \text{as } x \to 0, \qquad (4.7.25)$$

$$G_{\mathbf{c}}(x) \sim \frac{1}{\alpha \beta^{\frac{1}{\alpha}}} \Gamma\left(\frac{1}{\alpha}\right)(1-x)^{-\frac{1}{\alpha}}, \quad \text{as } x \to 1. \qquad (4.7.26)$$

Understanding the Lévy measure of a distribution has some immediate implications pertaining to its probability density funtion and distribution function. We state three such results in the following corollary. See Veillette and Taqqu [979] for their proofs, as well as Exercise 4.7.

Corollary 4.7.7 *Let Z_D denote the Rosenblatt distribution. Then:*

(i) The probability density function of Z_D is infinitely differentiable with all derivatives tending to 0 as $|x| \to \infty$.
(ii) Left tail: for $x > 0$,

$$\mathbb{P}(Z_D < -x) \le e^{-\frac{x^2}{2}}.$$

(iii) Right tail: for $\alpha > 0$,

$$\lim_{u \to \infty} \frac{\mathbb{P}(Z_D > u + \alpha)}{\mathbb{P}(Z_D > u)} = e^{-\frac{\alpha}{2\lambda_1(D)}},$$

where $\lambda_1(D)$ is the largest eigenvalue of $\sigma_D \mathcal{K}_D$ defined in (4.7.7) and (4.7.12).

The representation (4.7.10) can also be used to compute numerically the cumulative distribution function (CDF) and the probability density function (PDF) of the Rosenblatt distribution. For $M \ge 1$, define X_M and Y_M as

$$Z_D = \sum_{n=1}^{M-1} \lambda_n(\xi_n^2 - 1) + \sum_{n=M}^{\infty} \lambda_n(\xi_n^2 - 1) := X_M + Y_M. \qquad (4.7.27)$$

4.7 The Rosenblatt Distribution

Notice that Y_M has mean 0 and variance

$$\sigma_M^2 := \mathbb{E} Y_M^2 = 2 \sum_{n=M}^{\infty} \lambda_n^2. \tag{4.7.28}$$

Notice from (4.7.13) that, as $M \to \infty$,

$$\sigma_M^2 \sim 2C(D)^2 \sum_{n=M}^{\infty} n^{2D-2} \sim 2C(D)^2 (1-2D)^{-1} M^{2D-1}, \tag{4.7.29}$$

where we approximated the sum $\sum_{n=M}^{\infty} \lambda_n^2$ with an integral. This suggests that X_M alone is not a very good approximation of Z_D since this variance tends to 0 slowly with M, especially when D is close to $1/2$. As an alternative, as seen below, Y_M can be approximated by a normal distribution as $M \to \infty$. By taking advantage of this property, we can obtain accurate approximations of the distribution of Z_D.

A normal approximation of Y_M based on an Edgeworth expansion was established in Veillette and Taqqu [978]. Let $\kappa_{k,M}$ be the normalized cumulants of Y_M. These are given by

$$\kappa_{k,M} = 2^{k-1}(k-1)! \sigma_M^{-k} \sum_{n=M}^{\infty} \lambda_n(D)^k. \tag{4.7.30}$$

Notice that from the asymptotics (4.7.13) of $\lambda_n(D)$,

$$\kappa_{k,M} \sim 2^{k-1}(k-1)! \cdot 2^{-k/2}(1-2D)^{k/2} M^{k/2-kD} \cdot \int_M^{\infty} n^{k(D-1)} dn$$

$$= \frac{(k-1)!}{2} \frac{(2-4D)^{k/2}}{k-kD-1} M^{1-k/2}.$$

Let also Φ and ϕ be the CDF and PDF of a standard normal distribution, respectively. The CDF of the tail Y_M satisfies

$$\mathbb{P}(\sigma_M^{-1} Y_M \leq x) = \Phi(x) - \phi(x) \left\{ \sum_{\eta(N)} \left[\prod_{m=1}^{N} \frac{1}{k_m!} \left(\frac{\kappa_{m,M}}{m!} \right)^{k_m} \right] H_{\zeta(k_3,\ldots,k_N)}(x) \right\} + O\left(M^{-\frac{N-1}{2}} \right), \tag{4.7.31}$$

where H_n denotes the Hermite polynomial of order n, $\kappa_{k,M}$ is defined in (4.7.30), $\eta(N)$ denotes all $k_3, k_4, \ldots, k_N$ such that

$$1 \leq k_3 + 2k_4 + \cdots + (N-2)k_n \leq N - 2$$

and

$$\zeta(k_3, \ldots, k_n) = 3k_3 + 4k_4 + \cdots + Nk_N - 1.$$

The relation (4.7.31) can be used to evaluate the CDF of the tail Y_M. For X_M, methods exist already for accurately computing the CDF of a finite sum of chi-squared distributions, see for instance methods based on Laplace transform inversion (Veillette and Taqqu [977], Castaño-Martínez and López-Blázquez [207]), or Fourier transform inversion (Abate and Whitt [4]). Both tasks require computing the eigenvalues $\lambda_k(D)$. The computation of these eigenvalues, as those of the operator $(\sigma_D)\mathcal{K}_D$, is discussed in Veillette and Taqqu [979, 980].

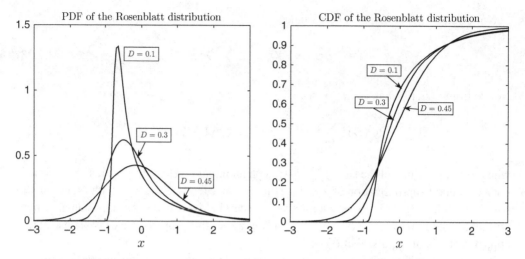

Figure 4.13 Plots of the PDF and CDF of Z_D for various D. The CDF with the steepest slope and the PDF with the highest mode correspond to $D = 0.1$.

Finally, in Figure 4.13, we present the plots of the CDF and PDF of the Rosenblatt distribution for several values of D (with $M = 50$ and $N = 5$). See Veillette and Taqqu [979, 980] for further technical details on their computation. A curious observation from Figure 4.13 is that the CDFs seem to intersect at the same point for the different considered values of D (Exercise 4.8).

4.8 CDF of the Rosenblatt Distribution

In Figure 4.13, we presented the plots of the PDF and CDF of the Rosenblatt distribution, that is, of the variable $Z_D = Z_H^{(2)}(1)$, where $D = 1 - H \in (0, 1/2)$ and $\{Z_H^{(2)}(t)\}_{t \in \mathbb{R}}$, $H \in (1/2, 1)$, is the standardized Rosenblatt process (see Definition 4.7.1). This process can appear as a limit in statistical problems. In order to obtain confidence intervals, it is necessary to compute numerically CDF of the random variable Z_D. Note that since $\{Z_H^{(2)}(t)\}_{t \in \mathbb{R}}$ is H-self-similar, one obtains the CDF of $Z_H(t)$, $t > 0$, by using the relation $\mathbb{P}(Z_H(t) < z) = \mathbb{P}(Z_H(1) < t^{-H} z)$.

Table 4.1 gives the CDF and Table 4.2 the quantiles of the random variable Z_D. They were computed numerically using the approximation described in Veillette and Taqqu [979].

4.9 Generalized Hermite and Related Processes

We describe first an interesting class of processes generalizing the Hermite processes. Throughout this section, we use the following notation: for $u = (u_1, \ldots, u_k) \in \mathbb{R}^k$ and $s \in \mathbb{R}$, we write

$$s + u = (s + u_1, \ldots, s + u_k), \quad su = (su_1, \ldots, su_k), \quad \{s > u\} = \{s > u_1, \ldots, s > u_k\}, \tag{4.9.1}$$

4.9 Generalized Hermite and Related Processes

Table 4.1 *Values of the CDF $F(x) = \mathbb{P}(Z_D \leq x)$ for various values of x and $D \in (0, 1/2)$, for the standardized Rosenblatt distribution. An entry of 0.0000 means the CDF takes a value less than 10^{-4}.*

x	D = 0.1	D = 0.2	D = 0.3	D = 0.4	D = 0.45	x	D = 0.1	D = 0.2	D = 0.3	D = 0.4	D = 0.45
−2.5	0.0000	0.0000	0.0000	0.0002	0.0020	1.0	0.8797	0.8790	0.8755	0.8645	0.8532
−2.4	0.0000	0.0000	0.0000	0.0004	0.0030	1.1	0.8901	0.8899	0.8879	0.8805	0.8725
−2.3	0.0000	0.0000	0.0000	0.0007	0.0043	1.2	0.8996	0.8997	0.8989	0.8946	0.8897
−2.2	0.0000	0.0000	0.0000	0.0013	0.0062	1.3	0.9081	0.9086	0.9087	0.9071	0.9047
−2.1	0.0000	0.0000	0.0000	0.0022	0.0088	1.4	0.9159	0.9166	0.9174	0.9180	0.9179
−2	0.0000	0.0000	0.0000	0.0037	0.0122	1.5	0.9230	0.9238	0.9253	0.9276	0.9294
−1.9	0.0000	0.0000	0.0000	0.0059	0.0167	1.6	0.9294	0.9304	0.9323	0.9360	0.9393
−1.8	0.0000	0.0000	0.0001	0.0093	0.0225	1.7	0.9353	0.9364	0.9386	0.9434	0.9479
−1.7	0.0000	0.0000	0.0004	0.0142	0.0299	1.8	0.9406	0.9418	0.9443	0.9499	0.9554
−1.6	0.0000	0.0000	0.0011	0.0210	0.0392	1.9	0.9455	0.9467	0.9494	0.9556	0.9617
−1.5	0.0000	0.0000	0.0030	0.0303	0.0506	2.0	0.9499	0.9512	0.9541	0.9606	0.9672
−1.4	0.0000	0.0000	0.0070	0.0425	0.0645	2.1	0.9540	0.9553	0.9582	0.9650	0.9719
−1.3	0.0000	0.0000	0.0148	0.0582	0.0810	2.2	0.9577	0.9590	0.9620	0.9689	0.9758
−1.2	0.0000	0.0008	0.0282	0.0777	0.1003	2.3	0.9611	0.9624	0.9654	0.9724	0.9793
−1.1	0.0000	0.0052	0.0489	0.1014	0.1227	2.4	0.9643	0.9655	0.9685	0.9754	0.9822
−1	0.0000	0.0205	0.0785	0.1293	0.1481	2.5	0.9671	0.9684	0.9713	0.9781	0.9847
−0.9	0.0011	0.0554	0.1174	0.1616	0.1766	2.6	0.9697	0.9710	0.9739	0.9805	0.9868
−0.8	0.0344	0.1134	0.1649	0.1980	0.2081	2.7	0.9721	0.9733	0.9762	0.9826	0.9886
−0.7	0.1435	0.1893	0.2195	0.2380	0.2424	2.8	0.9744	0.9755	0.9783	0.9845	0.9902
−0.6	0.2754	0.2737	0.2789	0.2810	0.2792	2.9	0.9764	0.9775	0.9802	0.9861	0.9915
−0.5	0.3864	0.3575	0.3406	0.3264	0.3182	3.0	0.9782	0.9793	0.9819	0.9876	0.9926
−0.4	0.4726	0.4349	0.4022	0.3734	0.3588	3.1	0.9799	0.9810	0.9835	0.9889	0.9936
−0.3	0.5402	0.5035	0.4619	0.4211	0.4007	3.2	0.9815	0.9825	0.9849	0.9901	0.9944
−0.2	0.5947	0.5630	0.5180	0.4687	0.4433	3.3	0.9830	0.9839	0.9862	0.9911	0.9952
−0.1	0.6400	0.6140	0.5698	0.5154	0.4861	3.4	0.9843	0.9852	0.9874	0.9921	0.9958
0	0.6784	0.6577	0.6169	0.5605	0.5285	3.5	0.9855	0.9864	0.9885	0.9929	0.9963
0.1	0.7115	0.6952	0.6591	0.6036	0.5700	3.6	0.9866	0.9875	0.9895	0.9936	0.9968
0.2	0.7403	0.7277	0.6967	0.6441	0.6102	3.7	0.9877	0.9885	0.9904	0.9943	0.9972
0.3	0.7655	0.7558	0.7300	0.6818	0.6488	3.8	0.9886	0.9894	0.9912	0.9949	0.9975
0.4	0.7878	0.7805	0.7594	0.7165	0.6853	3.9	0.9895	0.9902	0.9919	0.9954	0.9978
0.5	0.8077	0.8021	0.7853	0.7482	0.7196	4.0	0.9903	0.9910	0.9926	0.9959	0.9981
0.6	0.8253	0.8213	0.8081	0.7769	0.7514	4.1	0.9910	0.9917	0.9933	0.9963	0.9983
0.7	0.8412	0.8382	0.8282	0.8027	0.7807	4.2	0.9917	0.9924	0.9938	0.9967	0.9985
0.8	0.8554	0.8534	0.8459	0.8258	0.8074	4.3	0.9924	0.9930	0.9943	0.9970	0.9986
0.9	0.8682	0.8669	0.8616	0.8463	0.8315	4.4	0.9929	0.9935	0.9948	0.9973	0.9988
						4.5	0.9935	0.9940	0.9953	0.9976	0.9989

with analogous definitions when + is replaced by −, > is replaced by ≥, <, etc. Throughout this section, we also let I_k denote a multiple integral of order k with respect to a Gaussian random measure on $\mathbb{R}$ having the Lebesgue control measure.

We shall use the following elementary result which provides a general way to construct self-similar processes with stationary increments living in the *kth Wiener chaos*; that is, expressed through the multiple integral I_k. The proof of the result is left as Exercise 4.9.

Table 4.2 *Various quantiles of the Rosenblatt distribution for selected values of D.*

Quantile	$D = 0.1$	$D = 0.2$	$D = 0.3$	$D = 0.4$	$D = 0.45$
0.01	−0.8472	−1.0567	−1.3546	−1.7838	−2.0603
0.025	−0.8142	−0.9827	−1.7669	−1.5536	−1.7639
0.05	−0.7808	−0.9122	−1.0958	−1.3493	−1.5051
0.10	−0.7340	−0.8201	−0.9419	−1.1053	−1.6462
0.25	−0.6200	−0.6277	−0.6479	−0.6713	−0.6789
0.50	−0.3622	−0.3055	−0.2329	−0.1332	−0.0673
0.75	0.2370	0.2781	0.3666	0.5059	0.5955
0.90	1.2047	1.2031	1.2110	1.2417	1.2673
0.95	2.0015	1.9726	1.9114	1.8022	1.7262
0.975	2.8312	2.7759	2.6483	2.3858	2.1774
0.99	3.9618	3.8718	3.6579	3.1909	2.7892

Proposition 4.9.1 *Let $H \in (0, 1)$. Suppose that $\{h_t(\cdot), t \in \mathbb{R}\}$ is a family of functions defined on $\mathbb{R}^k$ satisfying:*

1. $h_t \in L^2(\mathbb{R}^k)$;
2. *For all $\lambda > 0$, there is $\beta \neq 0$ such that $h_{\lambda t}(u) = \lambda^{H+k\beta/2} h_t(\lambda^\beta u)$ for a.e. $u \in \mathbb{R}^k$ and all $t \in \mathbb{R}$;*
3. *For all $s > 0$, there exists $a \in \mathbb{R}^k$ such that $h_{t+s}(u) - h_t(u) = h_s(u + ta)$ for a.e. $u \in \mathbb{R}^k$ and all $t > 0$.*

Then,

$$Z(t) := I_k(h_t) = \int_{\mathbb{R}^k}' h_t(u_1, \ldots, u_k) B(du_1) \ldots B(du_k),$$

where $B(du)$ is a Gaussian random measure on $\mathbb{R}$ with the Lebesgue control measure du, is an H–self-similar process with stationary increments.

The following kernel functions will be used in the sequel. We let $\mathbb{R}_+ = (0, \infty)$.

Definition 4.9.2 A nonzero measurable function $g : \mathbb{R}_+^k \to \mathbb{R}$ is called a *generalized Hermite kernel* if it satisfies:

$$g(\lambda u) = \lambda^\alpha g(u), \quad \lambda > 0, u \in \mathbb{R}_+^k, \quad (4.9.2)$$

where

$$\alpha \in \left(-\frac{k+1}{2}, -\frac{k}{2}\right) \quad (4.9.3)$$

and

$$\int_{\mathbb{R}_+^k} |g(u)g(1+u)| du < \infty. \quad (4.9.4)$$

Note that (4.9.3) is equivalent to

$$0 < 2\alpha + k + 1 < 1. \quad (4.9.5)$$

4.9 Generalized Hermite and Related Processes

The next result will allow using generalized Hermite kernels in integrands for multiple integrals.

Proposition 4.9.3 *Let g be a generalized Hermite kernel on $\mathbb{R}_+^k$ defined in Definition 4.9.2. Then,*

$$h_t(u) = \int_0^t g(s-u) 1_{\{s>u\}} ds \tag{4.9.6}$$

is well-defined and belongs to $L^2(\mathbb{R}^k)$, for all $t \in \mathbb{R}$. (Note also that the convention (4.9.1) is used in (4.9.6).)

Proof To check that $h_t \in L^2(\mathbb{R}^k)$, we write

$$\int_{\mathbb{R}^k} h_t(u)^2 du = \int_{\mathbb{R}^k} du \int_0^t \int_0^t ds_1 ds_2 \, g(s_1-u) g(s_2-u) 1_{\{s_1>u\}} 1_{\{s_2>u\}}.$$

It is enough to check that the right-hand side is finite with the absolute value of the integrand. Observe that, by symmetry of s_1 and s_2, the right-hand side with the absolute value of the integrand is

$$2 \int_0^t ds_1 \int_{s_1}^t ds_2 \int_{\mathbb{R}^k} du \, |g(s_1-u) g(s_2-u)| 1_{\{s_1-u>0\}}$$

$$= 2 \int_0^t ds \int_0^{t-s} dv \int_{\mathbb{R}_+^k} dw \, |g(w) g(v+w)|$$

$$= 2 \int_0^t ds \int_0^{t-s} dv \int_{\mathbb{R}_+^k} dy \, v^k |g(vy) g(v+vy)|$$

$$= 2 \int_0^t ds \int_0^{t-s} dv \, v^{2\alpha+k} \int_{\mathbb{R}_+^k} dy \, |g(y) g(1+y)|,$$

where we used the changes of variables $s = s_1$, $v = s_2 - s_1$, $w = s_1 - u$, and then $w = vy$, and the scaling relation (4.9.2). The last integral is finite by using (4.9.4) and since $2\alpha + k + 1 > 0$ by (4.9.5). $\square$

The following result is an elementary corollary of Propositions 4.9.1 and 4.9.3.

Corollary 4.9.4 *Let g be a generalized Hermite kernel defined on $\mathbb{R}_+^k$ defined in Definition 4.9.2, and h_t, $t \in \mathbb{R}$, be defined in (4.9.6). Then, the process*

$$Z(t) = I_k(h_t), \quad t \in \mathbb{R}, \tag{4.9.7}$$

is a well-defined H–self-similar process with stationary increments, where

$$H = \alpha + \frac{k}{2} + 1 \in \left(\frac{1}{2}, 1\right). \tag{4.9.8}$$

Proof We check that the functions h_t satisfy Conditions 1–3 in Proposition 4.9.1. Condition 1 is satisfied by Proposition 4.9.3. To check self-similarity (Condition 2 with $\beta = -1$),

$$h_{\lambda t}(u) = \int_0^{\lambda t} g(s-u)1_{\{s>u\}}ds = \lambda^{\alpha+1}\int_0^t g(r-\lambda^{-1}u)1_{\{r>\lambda^{-1}u\}}\lambda dr = \lambda^{\alpha+1}h_t(\lambda^{-1}u),$$

where the second equality uses (4.9.2). The self-similarity parameter H of $I_k(h_t)$ is obtained from $\alpha + 1 = H - k/2$ (Condition 2 of Proposition 4.9.1). To check stationary increments (Condition 3 of Proposition 4.9.1), for any $t, r > 0$,

$$h_{t+r}(u) - h_t(u) = \int_t^{t+r} g(s-u)1_{\{s>u\}}ds = \int_0^r g(v+t-u)1_{\{v+t>u\}}dv = h_r(u-t). \quad \square$$

Definition 4.9.5 The process $\{Z(t)\}_{t \in \mathbb{R}}$ in (4.9.7), that is,

$$Z(t) = \int_{\mathbb{R}^k}' \left\{ \int_0^t g(s-u_1, \ldots, s-u_k)1_{\{s>u_1,\ldots,s>u_k\}}ds \right\} B(du_1)\ldots B(du_k), \quad (4.9.9)$$

where $B(du)$ is a Gaussian random measure on $\mathbb{R}$ with the Lebesgue control measure du and g is a generalized Hermite kernel defined in Definition 4.9.2, is called a *generalized Hermite process*.

Remark 4.9.6 It is known (see, e.g., Janson [529], Theorem 6.12) that if a random variable X belongs to the kth Wiener chaos, then there exist $a, b, t_0 > 0$ such that for $t \geq t_0$,

$$\exp\{-at^{2/k}\} \leq \mathbb{P}(|X| > t) \leq \exp\{-bt^{2/k}\}.$$

This shows that the generalized Hermite processes of different orders must necessarily have different laws, and the higher the order gets, the heavier the tail of the marginal distribution becomes, while they all have moments of any order. When two generalized Hermite process of the same order have the same law remains an open question.

Remark 4.9.7 The expression (4.9.9) is a time domain representation of a generalized Hermite process, analogous to (4.2.3)–(4.2.4) for the Hermite process. There is also a spectral domain representation of a generalized Hermite process, analogous to (4.2.11) for the Hermite process. Since the function $h_t \in L^2(\mathbb{R}^k)$, it has the $L^2(\mathbb{R}^k)$–Fourier transform, which can be written as

$$\widehat{h}_t(x) = \widehat{h}_t(x_1, \ldots, x_k) = \frac{e^{it(x_1+\cdots+x_k)} - 1}{i(x_1 + \cdots + x_k)}\widehat{g}(-x), \quad x = (x_1, \ldots, x_k) \in \mathbb{R}^k, \quad (4.9.10)$$

for a function $\widehat{g}(x)$ satisfying the scaling relation: for $\lambda > 0$, $\widehat{g}(\lambda x) = \lambda^{-\alpha-k}\widehat{g}(x)$ a.e. dx (see Propositions 3.12 and 3.14 in Bai and Taqqu [77]). In view of (B.2.16), the spectral representation of a generalized Hermite process is then given by

$$Z(t) = \frac{1}{(2\pi)^{k/2}}\int_{\mathbb{R}^k}'' \frac{e^{it(x_1+\cdots+x_k)} - 1}{i(x_1 + \cdots + x_k)}\widehat{g}(-x)\widehat{B}(dx_1)\ldots\widehat{B}(dx_k). \quad (4.9.11)$$

We next provide several special examples of generalized Hermite processes.

Example 4.9.8 (*Hermite kernel*) The usual Hermite processes are obtained from the generalized Hermite processes by taking the generalized Hermite kernel g as (up to a constant)

$$g(u) = g(u_1, \ldots, u_k) = \prod_{j=1}^{k} u_j^{\alpha/k}, \quad u_j > 0, \qquad (4.9.12)$$

where $\alpha/k = H_0 - 3/2$ with $H_0 \in (1 - 1/(2k), 1)$ (cf. Definition 4.2.1).

Example 4.9.9 (*Generalized Hermite kernels*) Other examples of generalized Hermite kernels and processes can be constructed by considering the following two useful classes of functions. Let $S_+^k = \{u \in \mathbb{R}_+^k : \|u\|_2 = 1\}$, where $\|u\|_2$ denotes the Euclidean norm. We say that a generalized Hermite kernel g is of *Class (B)* (B stands for "boundedness"), if on S_+^k, it is continuous a.e. and bounded. Consequently,

$$|g(u)| \le \|u\|_2^\alpha g(u/\|u\|_2) \le c\|u\|_2^\alpha,$$

for some $c > 0$. We say that a generalized Hermite kernel g on $\mathbb{R}_+^k$ having homogeneity exponent α is of *Class (L)* (L stands for "limit" as in "limit theorems"), if

1. g is continuous a.e. on $\mathbb{R}_+^k$;
2. $|g(u)| \le g^*(u)$ a.e. $u \in \mathbb{R}_+^k$, where g^* is a finite linear combination of non-symmetric Hermite kernels: $\prod_{j=1}^{k} u_j^{\gamma_j}$, where $\gamma_j \in (-1, -1/2)$, $j = 1, \ldots, k$, and $\sum_{j=1}^{k} \gamma_j = \alpha \in (-k/2 - 1/2, -k/2)$.

For example, $g^*(u)$ could be $u_1^{-3/4} u_2^{-5/8} + u_1^{-9/16} u_2^{-13/16}$ if $k = 2$. In this case, $\alpha = -11/8$.

Note that Class (B) does not include the original Hermite kernel (4.9.12) but also that Class (L) does. In fact, one can show that Class (L) contains Class (B) (Exercise 4.10 and Proposition 3.20 in Bai and Taqqu [77]). It can also be checked that if two functions g_1 and g_2 on $\mathbb{R}_+^k$ satisfy Condition 2 of Class (L) above, then $\int_{\mathbb{R}_+^k} |g_1(u) g_2(1+u)| du < \infty$ automatically holds (Exercise 4.10). In particular, the functions of Class (L) (and hence Class (B)) automatically satisfy the condition (4.9.4) of the generalized Hermite kernels.

Suppose $g(u) = \|u\|^\alpha$, where $\alpha \in (-1/2 - k/2, -k/2)$. This g belongs to Class (B) and thus also Class (L). Another example of Class (B):

$$g(u) = \frac{\prod_{j=1}^{k} u_j^{a_j}}{\sum_{j=1}^{k} u_j^b},$$

where $a_j > 0$ and $b > 0$, yielding a homogeneity exponent $\alpha = \sum_{j=1}^{k} a_j - b \in (-1/2 - k/2, -k/2)$. Another example of Class (L) but not (B):

$$g(u) = g_0(u) \vee \left(\prod_{j=1}^{k} u_j^{\alpha/k} \right).$$

where $g_0(u) > 0$ is any generalized Hermite kernel of Class (B) on $\mathbb{R}_+^k$ with homogeneity exponent α.

We conclude this section by addressing the following interesting question. According to Corollary 4.9.4, the generalized Hermite process introduced above admits a self-similarity parameter $H > 1/2$ only. Are there H–self-similar processes with stationary increments in a higher-order Wiener chaos having $H < 1/2$? To obtain such a process with $0 < H < 1/2$, we consider the following fractionally filtered kernel:

$$h_t^\beta(u) = \int_{\mathbb{R}} l_t^\beta(s) g(s-u) 1_{\{s>u\}} ds, \quad t \in \mathbb{R}, \qquad (4.9.13)$$

where g is a generalized Hermite kernel defined in Definition 4.9.2 with homogeneity exponent $\alpha \in (-k/2 - 1/2, -k/2)$, and

$$l_t^\beta(s) = \frac{1}{\beta}\left((t-s)_+^\beta - (-s)_+^\beta\right), \quad \beta \neq 0. \qquad (4.9.14)$$

One can extend it to $\beta = 0$ by writing $l_t^0(s) = 1_{[0,t)}(s)$, but this would lead us back to the generalized Hermite process case. We hence assume throughout that $\beta \neq 0$. The following proposition gives the range of β for which $I_k(h_t^\beta)$ is well-defined.

Proposition 4.9.10 *If*

$$-1 < -\alpha - \frac{k}{2} - 1 < \beta < -\alpha - \frac{k}{2} < \frac{1}{2}, \quad \beta \neq 0, \qquad (4.9.15)$$

then $h_t^\beta \in L^2(\mathbb{R}^k)$.

Proof Observe that, by making the changes of variables $s = s_1$, $v = s_2 - s_1$ and $w = s_1 - u$, and then $w = vy$ below,

$$\int_{\mathbb{R}^k} h_t^\beta(u)^2 du \leq 2 \int_{-\infty}^{\infty} ds_1 \int_{s_1}^{\infty} ds_2 \int_{\mathbb{R}^k} du \, l_t(s_1) l_t(s_2) |g(s_1 - u) g(s_2 - u)| 1_{\{s_1 > u\}}$$

$$= 2 \int_{-\infty}^{\infty} ds \int_0^{\infty} dv \int_{\mathbb{R}_+^k} dw \, l_t(s) l_t(s+v) |g(w) g(v+w)|$$

$$= 2 \int_{-\infty}^{\infty} ds \, l_t^\beta(s) \int_0^{\infty} l_t^\beta(s+v) v^{2\alpha+k} dv \int_{\mathbb{R}_+^k} dy \, |g(y) g(1+y)|.$$

We thus focus on showing $\int_{-\infty}^{\infty} ds \, l_t^\beta(s) \int_0^{\infty} l_t^\beta(s+u) u^{2\alpha+k} du < \infty$. For any $c > 0$, we have

$$\int_0^c (c-s)^{\gamma_1} s^{\gamma_2} ds = c^{\gamma_1+\gamma_2+1} \int_0^1 (1-s)^{\gamma_1} s^{\gamma_2} ds = c^{\gamma_1+\gamma_2+1} B(\gamma_1+1, \gamma_2+1),$$

for all $\gamma_1, \gamma_2 > -1$. So by noting that $\beta > -1$ and $2\alpha + k > -1$, we have

$$\int_0^{\infty} l_t^\beta(s+u) u^{2\alpha+k} du = \frac{1}{\beta} \int_0^{\infty} \left((t-s-u)_+^\beta - (-s-u)_+^\beta\right) u^{2\alpha+k} du$$

$$= \frac{1}{\beta}\left(\int_0^{t-s} (t-s-u)^\beta u^{2\alpha+k} du + \int_0^{-s} (-s-u)^\beta u^{2\alpha+k} du\right)$$

$$= \frac{B(\beta+1, 2\alpha+k+1)}{\beta}\left((t-s)_+^{\beta+\delta} - (-s)_+^{\beta+\delta}\right),$$

where
$$\delta = 2\alpha + k + 1 \in (0, 1). \quad (4.9.16)$$

We thus want to determine when the following holds:

$$\int_{\mathbb{R}} \left((t-s)_+^\beta - (-s)_+^\beta\right)\left((t-s)_+^{\beta+\delta} - (-s)_+^{\beta+\delta}\right) ds < \infty. \quad (4.9.17)$$

Suppose $t > 0$. The potential integrability problems appear near $s = -\infty, 0, t$. Near $s = -\infty$, the integrand behaves like $|s|^{2\beta+\delta-2}$, and thus we need $2\beta + \delta - 2 < -1$; near $s = 0$, the integrand behaves like $|s|^{2\beta+\delta}$, and thus $2\beta + \delta > -1$; near $s = t$, the integrand behaves like $|t-s|^{2\beta+\delta}$, and thus again $2\beta + \delta > -1$. In view of (4.9.16), these requirements are satisfied by (4.9.15). □

The next result is an immediate corollary of Propositions 4.9.1 and 4.9.10.

Corollary 4.9.11 *The process defined by $Z^\beta(t) := I_k(h_t^\beta)$, $t \in \mathbb{R}$, with h_t^β given in (4.9.13), namely,*

$Z^\beta(t)$
$$= \int_{\mathbb{R}^k}' \int_0^t \frac{1}{\beta}\left((t-s)_+^\beta - (-s)_+^\beta\right) g(s - u_1, \ldots, s - u_k) 1_{\{s > u_1, \ldots, s > u_k\}} ds\, B(du_1) \ldots B(du_k), \quad (4.9.18)$$

is an H–self-similar process having stationary increments with

$$H = \alpha + \beta + \frac{k}{2} + 1 \in (0, 1). \quad (4.9.19)$$

Remark 4.9.12 To get the antipersistent case $H < 1/2$, choose
$$\beta \in \left(-\alpha - \frac{k}{2} - 1, -\alpha - \frac{k}{2} - \frac{1}{2}\right).$$

4.10 Exercises

The symbols* and** next to some exercises are explained in Section 2.12.

Exercise 4.1 Prove the relation (4.1.7) for Hermite polynomials.

Exercise 4.2 For symmetric $f \in L^2(E^2, m^2)$, use the diagram formulae in Propositions 4.3.4 and 4.3.20 to obtain an expression for $\mathbb{E} I_2(f)^3$. Argue, in particular, that $\mathbb{E} I_2(f)^3$ is nonzero and hence $I_2(f)$ is non-symmetric in general.

Exercise 4.3 If symmetric $f_i \in L^2(E, m)$, $i = 1, \ldots, 4$, use the diagram formulae in Propositions 4.3.4 and 4.3.20 to obtain an expression for $\mathbb{E} I_1(f_1) I_1(f_2) I_1(f_3) I_1(f_4)$.

Exercise 4.4 Show that the relation (4.4.11) is equivalent to the relation (4.4.2). The former relation appears in Taqqu [944].

Exercise 4.5 By following the hint given in the text, prove the relation (4.5.10).

Exercise 4.6 Prove the two relations in (4.7.6).

Exercise 4.7* Use the Markov inequality to prove Corollary 4.7.7, (ii). *Hint:* See Corollary 4.4 in Veillette and Taqqu [979]. (The parts (i) and (iii) of Corollary 4.7.7 can be proved using auxiliary results on infinitely divisible distributions – see Corollaries 4.3 and 4.5 in Veillette and Taqqu [979].)

Exercise 4.8** Prove the conjecture stated in Section 4.7 that the CDFs of the Rosenblatt distribution Z_D cross at the same point for all Ds, as suggested by Figure 4.13.

Exercise 4.9 Prove Proposition 4.9.1.

Exercise 4.10 Prove the following properties of Classes (B) and (L) introduced in Example 4.9.9. (i) Show that Class (L) contains Class (B). (ii) If two functions g_1 and g_2 on $\mathbb{R}_+^k$ satisfy Condition 2 of Class (L), then $\int_{\mathbb{R}_+^k} |g_1(u)g_2(1+u)|du < \infty$ automatically holds.

Exercise 4.11* If X has a normal distribution, show that X^3 is indeterminate; that is, its distribution is not determined by the moments. *Hint:* Suppose without loss of generality that X has the density $e^{-x^2}/\sqrt{\pi}$ and let $d(x)$ be the density of X^3. Show then that the densities $d(x)$ and $d(x)\{1 + r(\cos(\sqrt{3}|x|^{2/3}) - \sqrt{3}\sin(\sqrt{3}|x|^{2/3}))\}$ with $|r| \leq 1/2$ have all their moments equal. See Proposition 1 in Berg [129]. (The fact that higher-order Hermite polynomials of normal variables or multiple integrals with respect to Gaussian measures may not be determinate is well known. See, e.g., Slud [910].)

4.11 Bibliographical Notes

Section 4.1: Hermite polynomials are special orthogonal polynomials that play important roles in Statistics and Probability, but also in many other areas. The monographs by Andrews, Askey, and Roy [35] and Lebedev [601] are excellent sources on the properties of Hermite and related polynomials. There are many related definitions of Hermite polynomials as noted in a footnote in Section 4.1. It is important therefore to always check which definition is used.

The connections between Hermite polynomials, multiple stochastic integrals and Gaussian variables were recognized from the very start, e.g., by Itô [524], and have been exploited and documented widely since, e.g., in Nualart [769], Major [672, 673]. The latter reference, in particular, contains interesting historical accounts.

Section 4.2: The term "Hermite processes" was coined by Murad Taqqu. The integral representations of Hermite processes in the time and spectral domains appeared first in Dobrushin and Major [321], Taqqu [945]. The integral representation on an interval was derived following Pipiras and Taqqu [819], though the representation goes back at least to Tudor [962] in the case of the Rosenblatt process, and to Nourdin, Nualart, and Tudor [766] in the general case of Hermite processes. It was subsequently used in, for example, Breton and Nourdin

[183], Chronopoulou, Tudor, and Viens [240, 241]. Representations of fractional Brownian motion are discussed further in the notes to Section 6.2 below.

Section 4.3: The presentation follows parts of Peccati and Taqqu [797] whose approach originated with Rota and Wallstrom [865]. See also the books of Janson [529], Major [673], with the latter containing interesting historical notes. In the context of Hermite processes, the diagram and related formulae have originated with Dobrushin [320], Taqqu [944], though their variants have also been known well earlier, especially in the literature of Mathematical Physics in connection to Feynman's diagrams (e.g., Polyak [828]).

Section 4.4: The moments of Hermite processes were studied originally by Taqqu [944].

Section 4.5: The representation of multiple integrals of order two in Lemma 4.5.2 goes back at least to Varberg [974]. Multiple integrals of order greater than two are known to be quite different from those of order two, for example, their cumulants do not typically determine the distribution anymore (Slud [910]).

Section 4.6: The Rosenblatt process was introduced by Taqqu [943], naming it after Rosenblatt [860] who considered its marginal distribution. The history and the developments behind the Rosenblatt process are described in Taqqu [946].

Section 4.7: As noted in the section, its material is drawn from the work of Veillette and Taqqu [979, 980]. The Rosenblatt distribution was also studied by Albin [18], Maejima and Tudor [669].

Section 4.9: The material of this section is based on Bai and Taqqu [77] who developed the notion of "generalized Hermite processes." These processes were introduced by Maejima and Tudor [668]. Similar types of self-similar processes with stationary increments of higher-order Wiener chaos were also studied by Mori and Oodaira [732], Arras [44].

5
Non-Central and Central Limit Theorems

The material of this chapter is motivated in the following question. Suppose $\{X_n\}_{n\in\mathbb{Z}}$ is a Gaussian series with long-range dependence and let $G: \mathbb{R} \mapsto \mathbb{R}$ be a deterministic function such that $\mathbb{E}G(X_n)^2 < \infty$. What is the limit of the partial sum process

$$\sum_{n=1}^{[Nt]} G(X_n), \quad t \geq 0, \tag{5.0.1}$$

as $N \to \infty$, after suitable centering and normalization? The question is important when dealing with long-range dependent series. As will be seen in this chapter, the limit is not necessarily a Gaussian (or stable) process. If the limit is not Gaussian nor stable, the corresponding limit theorem is known as a *non-central limit theorem* and the limit process is an Hermite process.

We shall also consider a number of generalizations of the setting (5.0.1), and address several related questions such as generation of non-Gaussian series.

We consider nonlinear functions of Gaussian random variables in Section 5.1 and define their Hermite rank in Section 5.2. Convergence of normalized partial sums to the Hermite processes is proved in Section 5.3. The result is stated in Theorem 5.3.1 and is called a *non-central limit theorem* because the Hermite processes are typically non-Gaussian. Section 5.4 deals with short-range dependence, where the limit of the normalized partial sums is Brownian motion. This result is stated in Theorem 5.4.1 and the proof given here is based on a method of moments following the original approach of Breuer and Major [184]. Recently, Nualart and Peccati [771] have shown that it is enough only to prove convergence of moments up to order 4, as indicated in Section 5.5.

In Section 5.6, we focus on non-central and central limit theorems for nonlinear functions of linear time series with long-range or short-range dependence. Multivariate limit theorems are considered in Section 5.7. In Section 5.8, we use nonlinear transformations of Gaussian time series to generate non-Gaussian time series and study the effect of the transformation on the dependence structure.

5.1 Nonlinear Functions of Gaussian Random Variables

In the study of partial sums (5.0.1), Hermite polynomials, defined in Definition 4.1.1, play a fundamental role. As we will see, they form a basis for the space of finite-variance, nonlinear functions of Gaussian random variables, also called Gaussian subordinated variables.

5.1 Nonlinear Functions of Gaussian Random Variables

One immediate consequence of the relation (4.1.2) for Hermite polynomials is the following classical result. It is a special case of Proposition 4.3.22 (see also Example 4.3.23). We provide a direct proof based on the properties of Hermite polynomials.

Proposition 5.1.1 *Let $(X, Y)'$ be a Gaussian vector with $\mathbb{E}X = \mathbb{E}Y = 0$ and $\mathbb{E}X^2 = \mathbb{E}Y^2 = 1$. Then, for all $n, m \geq 0$,*

$$\mathbb{E}H_n(X)H_m(Y) = \begin{cases} n!(\mathbb{E}XY)^n, & \text{if } n = m, \\ 0, & \text{if } n \neq m. \end{cases} \quad (5.1.1)$$

Proof Recall that a Gaussian random variables X with $\mathbb{E}X = 0$ has moment generating function $\mathbb{E}e^{tX} = \exp\{\frac{1}{2}\mathbb{E}(tX)^2\}$, and let $F(x, t) = e^{tx - t^2/2}$ as in (4.1.2). Then, for any $s, t \in \mathbb{R}$,

$$\mathbb{E}F(X, s)F(Y, t) = \mathbb{E}\left(\exp\left\{sX - \frac{s^2}{2}\right\}\exp\left\{tY - \frac{t^2}{2}\right\}\right) = \exp\left\{-\frac{s^2}{2} - \frac{t^2}{2}\right\}\mathbb{E}\exp\{sX + tY\}$$

$$= \exp\left\{-\frac{s^2}{2} - \frac{t^2}{2}\right\}\exp\left\{\frac{1}{2}\mathbb{E}(sX + tY)^2\right\} = \exp\{st\mathbb{E}XY\}$$

by using $\mathbb{E}X = \mathbb{E}Y = 0$ and $\mathbb{E}X^2 = \mathbb{E}Y^2 = 1$. Differentiating the left-hand side n times with respect to s and m times with respect to t, setting $s = t = 0$ and using (4.1.3) yields $\mathbb{E}H_n(X)H_m(Y)$. The same operation on the right-hand side yields $n!(\mathbb{E}XY)^n$ if $n = m$, and 0 otherwise. Indeed, this follows from

$$\exp\{st\mathbb{E}XY\} = \sum_{k=0}^{\infty} \frac{s^k t^k (\mathbb{E}XY)^k}{k!}$$

and observing that

$$\frac{\partial^{n+m}}{\partial s^n \partial t^m} s^k t^k \bigg|_{s=0, t=0} = \frac{\partial^n}{\partial s^n} s^k \bigg|_{s=0} \frac{\partial^m}{\partial t^m} t^k \bigg|_{t=0} = \begin{cases} (n!)^2, & \text{if } k = n = m, \\ 0, & \text{otherwise}. \end{cases} \quad \Box$$

Setting $X = Y$ in (5.1.1) yields the following corollary of Proposition 5.1.1. Let

$$\phi(dx) = \frac{e^{-x^2/2}}{\sqrt{2\pi}} dx \quad (5.1.2)$$

be the probability measure on $\mathbb{R}$ associated with a standard normal variable X and $L^2(\phi)$ be the space of measurable, square-integrable functions with respect to $\phi(dx)$. Then, $G \in L^2(\mathbb{R}, \phi)$ if and only if $\mathbb{E}G(X)^2 < \infty$. The space $L^2(\mathbb{R}, \phi)$ is naturally equipped with the inner product

$$(G_1, G_2)_{L^2(\mathbb{R},\phi)} = \int_\mathbb{R} G_1(x)G_2(x)\phi(dx) = \int_\mathbb{R} G_1(x)G_2(x)\frac{e^{-x^2/2}}{\sqrt{2\pi}}dx = \mathbb{E}G_1(X)G_2(X). \quad (5.1.3)$$

Corollary 5.1.2 *For any $n, m \geq 0$, and standard normal random variable X,*

$$\mathbb{E}H_n(X)H_m(X) = (H_n, H_m)_{L^2(\mathbb{R},\phi)} = \begin{cases} n!, & \text{if } n = m, \\ 0, & \text{if } n \neq m. \end{cases} \quad (5.1.4)$$

In other words, $\{H_n(x)\}_{n\geq 0}$ is a collection of orthogonal functions in the space $L^2(\mathbb{R}, \phi)$.

The collection $\{H_n(x)\}_{n\geq 0}$ not only consists of orthogonal functions but it is also a basis for the space $L^2(\mathbb{R}, \phi)$. This is stated in the next result.

Proposition 5.1.3 *The collection $\{H_n(x)\}_{n\geq 0}$ is a basis for the space $L^2(\mathbb{R}, \phi)$.*

Proof Let X be a standard normal random variable. It is enough to show that, for $G \in L^2(\mathbb{R}, \phi)$ and all $n \geq 0$,

$$(G, H_n)_{L^2(\mathbb{R},\phi)} = \mathbb{E}G(X)H_n(X) = 0, \qquad (5.1.5)$$

implies that $G(x) = 0$ a.e. $\phi(dx)$. If (5.1.5) holds, then

$$\mathbb{E}G(X)F(X, z) = 0,$$

where $F(x, z)$ is defined in (4.1.2) and we can take $z \in \mathbb{C}$. But this means that, for any $z \in \mathbb{C}$,

$$\mathbb{E}G(X)F(X,z)e^{\frac{z^2}{2}} = \mathbb{E}G(X)e^{zX} = \frac{1}{\sqrt{2\pi}} \int_{\mathbb{R}} e^{zx} G(x) e^{-x^2/2} dx = 0.$$

In particular, the $L^2(\mathbb{R})$–Fourier transform of the function $G(x)e^{-x^2/2}$ is zero. Hence, taking the inverse Fourier transform, $G(x)e^{-x^2/2} = 0$ a.e. dx. This yields $G(x) = 0$ a.e. $\phi(dx)$. □

A consequence of Proposition 5.1.3 is that every function $G \in L^2(\mathbb{R}, \phi)$ has a series expansion in Hermite polynomials:

$$G(x) = \sum_{n=0}^{\infty} g_n H_n(x), \qquad (5.1.6)$$

where the convergence takes place in $L^2(\mathbb{R}, \phi)$. Note that, since H_n are orthogonal but not orthonormal, we have

$$g_n = \frac{1}{n!}(G, H_n)_{L^2(\mathbb{R},\phi)} = \frac{1}{n!} \int_{\mathbb{R}} G(x) H_n(x) \frac{e^{-x^2/2} dx}{\sqrt{2\pi}} = \frac{1}{n!} \mathbb{E}G(X)H_n(X), \qquad (5.1.7)$$

where X is a standard normal variable. The relation (5.1.6) can be written as

$$G(X) = \sum_{n=0}^{\infty} g_n H_n(X), \qquad (5.1.8)$$

where the convergence takes place in $L^2(\Omega)$. Indeed, note that

$$\mathbb{E}\Big(G(X) - \sum_{n=0}^{N} g_n H_n(X)\Big)^2 = \Big\| G - \sum_{n=1}^{N} g_n H_n \Big\|^2_{L^2(\mathbb{R},\phi)} \to 0,$$

by using (5.1.6). The following is an immediate corollary.

Proposition 5.1.4 Let $(X, Y)'$ be a Gaussian vector with $\mathbb{E}X = \mathbb{E}Y = 0$ and $\mathbb{E}X^2 = \mathbb{E}Y^2 = 1$. Suppose $G_1, G_2 \in L^2(\mathbb{R}, \phi)$ and let $g_{1,n}$ and $g_{2,n}$, $n \geq 0$, be the coefficients in the Hermite expansions of G_1 and G_2, respectively, as in (5.1.6). Then,

$$\mathbb{E}G_1(X)G_2(Y) = \sum_{n=0}^{\infty} g_{1,n} g_{2,n} n! \, (\mathbb{E}XY)^n \qquad (5.1.9)$$

and

$$\mathrm{Cov}(G_1(X), G_2(Y)) = \sum_{n=1}^{\infty} g_{1,n} g_{2,n} n! \, (\mathbb{E}XY)^n. \qquad (5.1.10)$$

Proof As in (5.1.8), $G_1(X) = \sum_{n=0}^{\infty} g_{1,n} H_n(X)$ and $G(Y) = \sum_{m=0}^{\infty} g_{2,m} H_m(Y)$, where the convergence takes place in $L^2(\Omega)$. Then, by Proposition 5.1.1,

$$\mathbb{E}G_1(X)G_2(Y) = \sum_{n=0}^{\infty} \sum_{m=0}^{\infty} g_{1,n} g_{2,m} \mathbb{E} H_n(X) H_m(Y) = \sum_{n=0}^{\infty} g_{1,n} g_{2,n} n! \, (\mathbb{E}XY)^n,$$

which is (5.1.9). Moreover, since $\mathbb{E}G_1(X) = \mathbb{E}G_1(X)H_0(X) = g_{1,0}$, one has $\mathbb{E}G_1(X)\mathbb{E}G_2(Y) = g_{1,0} g_{2,0}$, implying (5.1.10). □

The rate of convergence of the series (5.1.6), through the asymptotic behavior of the coefficients g_n, is considered in Exercise 5.1.

5.2 Hermite Rank

Taking $G_1 = G_2$ in (5.1.10) and using the index k instead of n yield

$$\mathbb{E}G(X)G(Y) = \sum_{k=0}^{\infty} g_k^2 k! \, (\mathbb{E}XY)^k \quad \text{and} \quad \mathrm{Cov}(G(X), G(Y)) = \sum_{k=1}^{\infty} g_k^2 k! \, (\mathbb{E}XY)^k. \quad (5.2.1)$$

In particular, for a Gaussian stationary series $\{X_n\}_{n \in \mathbb{Z}}$ with $\mathbb{E}X_n = 0$, $\mathbb{E}X_n^2 = 1$ and autocovariance function γ_X, the autocovariance of a stationary series $\{G(X_n)\}_{n \in \mathbb{Z}}$ can be expressed as

$$\mathrm{Cov}(G(X_n), G(X_0)) = \sum_{k=1}^{\infty} g_k^2 k! \, (\gamma_X(n))^k. \qquad (5.2.2)$$

If the Gaussian series $\{X_n\}_{n \in \mathbb{Z}}$ is long-range dependent in the sense (2.1.5) in condition II, we have $\gamma_X(n) = L_2(n) n^{2d-1}$ for a slowly varying function L_2 and each entry $(\gamma_X(n))^k$ in (5.2.2) becomes $(L_2(n) n^{2d-1})^k$. The dominating series term in (5.2.2) is then determined by the smallest k for which the coefficient g_k is nonzero. This index k has a special name.

Definition 5.2.1 Let ϕ be the standard normal density, $G \in L^2(\mathbb{R}, \phi)$ and g_k, $k \geq 0$, be the coefficients in its Hermite expansion (5.1.6). The *Hermite rank* r of G is defined as the smallest index $k \geq 1$ for which $g_k \neq 0$; that is,

$$r = \min\{k \geq 1 : g_k \neq 0\}. \qquad (5.2.3)$$

Remark 5.2.2 One sometimes defines the Hermite rank as

$$r = \min\{k \geq 0 : g_k \neq 0\}. \tag{5.2.4}$$

This definition and (5.2.3) are equivalent, except in the case when the function G is such that $\mathbb{E}G(X) \neq 0$ where $X \sim \mathcal{N}(0, 1)$ (e.g., G is a constant $c \neq 0$). Indeed, since $g_0 = \mathbb{E}G(X)H_0(X) = \mathbb{E}G(X) \neq 0$, the Hermite rank equals 0 according to the definition (5.2.4). According to the definition (5.2.3), the Hermite rank of G would be greater than or equal to 1, including the possibility of infinity for a function G equal to a constant $c \neq 0$, since $g_k = (k!)^{-1}\mathbb{E}G(X)H_k(X) = c(k!)^{-1}\mathbb{E}H_k(X) = 0$ for all $k \geq 1$.

One can also define the Hermite rank of G as

$$r = \min\{k \geq 0 : \mathbb{E}[(G(X) - \mathbb{E}G(X))H_k(X)] \neq 0\}. \tag{5.2.5}$$

This definition is equivalent to (5.2.3) since $\mathbb{E}[(G(X) - \mathbb{E}G(X))H_k(X)]$ equals $\mathbb{E}G(X) - \mathbb{E}G(X) = 0$ if $k = 0$ and equals $\mathbb{E}G(X)H_k(X)$ if $k \geq 1$.

Example 5.2.3 Suppose $G(x) = x^2$ and $X \sim \mathcal{N}(0, 1)$. Then $\mathbb{E}G(X)H_0(X) = \mathbb{E}X^2 \neq 0$, $\mathbb{E}G(X)H_1(X) = \mathbb{E}X^3 = 0$ and $\mathbb{E}G(X)H_2(X) = \mathbb{E}X^2(X^2 - 1) = \mathbb{E}X^4 - \mathbb{E}X^2 \neq 0$. Therefore, the Hermite rank of G equals 2 according to (5.2.3) and (5.2.5), and equals to 0 according to (5.2.4). Note that all three definitions imply that $G(x) - \mathbb{E}G(X) = x^2 - 1$ has Hermite rank 2. It is therefore best to consider functions with their mean subtracted, which is what one does in the context of (non)central limit theorems.

Hence the functions $x^{2n} - \mathbb{E}X^{2n}$, $n \geq 1$, have Hermite rank 2 and the functions x^n, $n \geq 1$, n odd, have Hermite rank 1, according to the three definitions.

If $\{X_n\}$ is long-range dependent (LRD), is the sequence $\{G(X_n)\}$ LRD? Not necessarily. The following proposition provides a sufficient condition.

Proposition 5.2.4 *Let $\{X_n\}_{n \in \mathbb{Z}}$ be a stationary Gaussian series with $\mathbb{E}X_n = 0$ and $\mathbb{E}X_n^2 = 1$, which is LRD in the sense (2.1.5) in condition II. Let also $G \in L^2(\mathbb{R}, \phi)$ be a function with Hermite rank r. Then, as $n \to \infty$,*

$$\text{Cov}(G(X_n), G(X_0)) \sim g_r^2 r! (L_2(n))^r n^{(2d-1)r}. \tag{5.2.6}$$

In particular, if

$$\frac{1}{2}\left(1 - \frac{1}{r}\right) = \frac{r-1}{2r} < d < \frac{1}{2}, \tag{5.2.7}$$

then the stationary series $\{G(X_n)\}_{n \in \mathbb{Z}}$ is LRD in the sense (2.1.5) in condition II with LRD parameter d_G given by

$$d_G = rd - \frac{r-1}{2} = r\left(d - \frac{1}{2}\right) + \frac{1}{2} \in \left(0, \frac{1}{2}\right) \tag{5.2.8}$$

and slowly varying function L_G satisfying

$$L_G(u) \sim g_r^2 r! (L_2(u))^r. \tag{5.2.9}$$

Proof Since G has Hermite rank r, the expansion (5.2.2) starts at $k = r$ and hence

$$\text{Cov}(G(X_n), G(X_0)) = \sum_{k=r}^{\infty} g_k^2 k! (\gamma_X(n))^k.$$

But as $n \to \infty$, the leading term in this sum is the term with $k = r$. Therefore,

$$\text{Cov}(G(X_n), G(X_0)) \sim g_r^2 r! (\gamma_X(n))^r.$$

Since, by assumption, $\gamma_X(n) = L_2(n) n^{2d-1}$, we get (5.2.6) or equivalently

$$\text{Cov}(G(X_n), G(X_0)) = L_G(n) n^{2d_G - 1},$$

where L_G and d_G satisfy (5.2.9) and (5.2.8), respectively. The condition (5.2.7) ensures that $d_G \in (0, 1/2)$; that is, $\{G(X_n)\}_{n \in \mathbb{Z}}$ is LRD in the sense (2.1.5) in condition II. □

The next result is an immediate consequence of Propositions 5.2.4 and 2.2.5.

Corollary 5.2.5 *Set*

$$H_G = d_G + \frac{1}{2} \tag{5.2.10}$$

and

$$(L_H(u))^2 = \frac{L_G(u)}{d_G(2d_G + 1)}. \tag{5.2.11}$$

Under the assumptions and notation of Proposition 5.2.4 and when (5.2.7) holds, one has, as $N \to \infty$,

$$\text{Var}\left(\sum_{n=1}^N G(X_n)\right) \sim \frac{2 g_r^2 r! (L_2(N))^r N^{r(2d-1)+2}}{(r(2d-1)+1)(r(2d-1)+2)} \sim \frac{2 L_G(N) N^{2d_G+1}}{2 d_G(2d_G + 1)} = (L_H(N))^2 N^{2H_G}. \tag{5.2.12}$$

The relation (5.2.12) suggests, in particular, the normalization

$$L_H(N) N^{H_G} = L_H(N) N^{d_G + 1/2} = L_H(N) N^{r(d - 1/2) + 1}$$

for the convergence of partial sums (5.0.1). Table 5.1 displays the interval (5.2.7) and the value and range of d_G in (5.2.8) for the first few values of r. It also displays $H_G = d_G + 1/2$ whose range is always $(1/2, 1)$, and the range of $H_0 = d + 1/2$. The parameter $H_0 = d + 1/2$,

Table 5.1 *Range of d which ensures that $\{G(X_n)\}$ is LRD. Corresponding values of H_0, d_G and H_G.*

r	d	$H_0 = d + \frac{1}{2}$	d_G	$H_G = d_G + \frac{1}{2}$	d_G	H_G
1	$(0, \frac{1}{2})$	$(\frac{1}{2}, 1)$	d	$d + \frac{1}{2}$	$(0, \frac{1}{2})$	$(\frac{1}{2}, 1)$
2	$(\frac{1}{4}, \frac{1}{2})$	$(\frac{3}{4}, 1)$	$2d - \frac{1}{2}$	$2d$	$(0, \frac{1}{2})$	$(\frac{1}{2}, 1)$
3	$(\frac{1}{3}, \frac{1}{2})$	$(\frac{5}{6}, 1)$	$3d - 1$	$3d - \frac{1}{2}$	$(0, \frac{1}{2})$	$(\frac{1}{2}, 1)$
4	$(\frac{3}{8}, \frac{1}{2})$	$(\frac{7}{8}, 1)$	$4d - \frac{3}{2}$	$4d - 1$	$(0, \frac{1}{2})$	$(\frac{1}{2}, 1)$

related to the exponent in (2.1.8), is sometimes used instead of d. The parameter H_G will be denoted H in the next section.

5.3 Non-Central Limit Theorem

We now turn to the study of the limits of partial sums (5.0.1). This section concerns the situation when the limit of (5.0.1) is non-Gaussian in general, and Section 5.4 concerns the case when the limit is Gaussian.

In this section, the limit will involve the process

$$\beta_{k,H} Z_H^{(k)}(t) = \beta_{k,H}^0 \int_{\mathbb{R}^k}^{\prime\prime} \frac{e^{it(x_1+\cdots+x_k)} - 1}{i(x_1 + \cdots + x_k)} \prod_{j=1}^{k} |x_j|^{-\frac{1}{2} + \frac{1-H}{k}} \widehat{B}(dx_1) \ldots \widehat{B}(dx_k), \quad (5.3.1)$$

where $Z_H^{(k)} = \{Z_H^{(k)}(t)\}_{t \geq 0}$, $1/2 < H < 1$, is a standard Hermite process, defined in (4.2.42),

$$\beta_{k,H}^0 = \left(2\sin\left(\left(\frac{1}{2} - \frac{1-H}{k}\right)\pi\right) \Gamma\left(\frac{2(1-H)}{k}\right)\right)^{-k/2} \quad (5.3.2)$$

and

$$\beta_{k,H} = \frac{\beta_{k,H}^0}{B_{k,H}} = \left(\frac{k!}{H(2H-1)}\right)^{1/2}, \quad (5.3.3)$$

where $B_{k,H}$ is given in (4.2.46). Since $\mathbb{E} Z_H^{(k)}(t)^2 = t^{2H}$, the process (5.3.1) has variance $\mathbb{E}(\beta_{k,H} Z_H^{(k)}(t))^2 = (\beta_{k,H})^2 t^{2H}$. As usual, $\widehat{B}(dx)$ is an Hermitian Gaussian measure with control measure dx. Note also that we specify the Hermite process by its spectral representation as in (4.2.11).

The following is the main result of this section.

Theorem 5.3.1 *(Non-central limit theorem.) Let $\{X_n\}_{n \in \mathbb{Z}}$ be a Gaussian stationary series which is LRD in the sense (2.1.5) in condition II with $d \in (0, 1/2)$. Suppose that $\mathbb{E} X_n = 0$, $\mathbb{E} X_n^2 = 1$. Let G be a function with Hermite rank $k \geq 1$ in the sense of Definition 5.2.1. If*

$$d \in \left(\frac{1}{2}\left(1 - \frac{1}{k}\right), \frac{1}{2}\right), \quad (5.3.4)$$

then

$$\frac{1}{(L_2(N))^{k/2} N^{k(d-1/2)+1}} \sum_{n=1}^{[Nt]} (G(X_n) - \mathbb{E} G(X_n)) \xrightarrow{fdd} g_k \beta_{k,H} Z_H^{(k)}(t), \quad t \geq 0, \quad (5.3.5)$$

where $\xrightarrow{fdd}$ denotes the convergence of finite-dimensional distributions and g_k is the first nonzero coefficient in the Hermite expansion of G in Definition 5.2.1. The process $Z_H^{(k)} = \{Z_H^{(k)}(t)\}_{t \geq 0}$ is the standard Hermite process (4.2.42) of order k with the self-similarity parameter

$$H = k\left(d - \frac{1}{2}\right) + 1 \in \left(\frac{1}{2}, 1\right). \quad (5.3.6)$$

The constant $\beta^0_{k,H}$ given in (5.3.2) equals

$$\beta^0_{k,H} = \Big(2\sin(d\pi)\Gamma(1-2d)\Big)^{-k/2} \tag{5.3.7}$$

and the normalization $\beta_{k,H}$ given in (5.3.3) equals

$$\beta_{k,H} = \Big(\frac{k!}{(k(d-1/2)+1)(k(2d-1)+1)}\Big)^{1/2}. \tag{5.3.8}$$

Remark 5.3.2 By Corollary 5.2.5,

$$\text{Var}\Big(\sum_{n=1}^{N} G(X_n)\Big) \sim \frac{2g_k^2 k!(L_2(N))^k N^{k(2d-1)+2}}{(k(2d-1)+1)(k(2d-1)+2)},$$

as $N \to \infty$. The normalization $(L_2(N))^{k/2} N^{k(d-1/2)+1}$ and the constant $g_k \beta_{k,H}$ in (5.3.5) are consistent with this asymptotic result.

Proof We use a number of results found following the proof. By the reduction theorem (Theorem 5.3.3 below), it is enough to prove the result for $G(x) - \mathbb{E}G(X_n) = H_k(x)$, where $H_k(x)$ is the Hermite polynomial of order k.

Step 1 (setup): Let

$$S_{N,k}(t) = \frac{1}{A(N)} \sum_{n=1}^{[Nt]} H_k(X_n)$$

be the corresponding partial sum process, where $A(N) = (L_2(N))^{k/2} N^{k(d-1/2)+1}$ is the normalization.

Now write the series X_n in its spectral representation as

$$X_n = \int_{\mathbb{R}} e^{in\lambda} \widehat{B}_F(d\lambda),$$

where $\widehat{B}_F$ is an Hermitian Gaussian random measure on $\mathbb{R}$ with a symmetric control measure $F(d\lambda)$. The measure $F(d\lambda)$ is the spectral measure of the series $\{X_n\}$ which is concentrated on $(-\pi, \pi]$ and satisfies

$$\gamma_X(n) = \int_{\mathbb{R}} e^{in\lambda} F(d\lambda) \tag{5.3.9}$$

(see Remark 1.3.4). By using Proposition 4.1.2, (b), we have

$$H_k(X_n) = \int_{\mathbb{R}^k}'' e^{in(\lambda_1+\cdots+\lambda_k)} \widehat{B}_F(d\lambda_1) \ldots \widehat{B}_F(d\lambda_k).$$

Then,

$$S_{N,k}(t) = \frac{1}{A(N)} \int_{\mathbb{R}^k}'' \sum_{n=1}^{[Nt]} e^{in(\lambda_1+\cdots+\lambda_k)} \widehat{B}_F(d\lambda_1) \ldots \widehat{B}_F(d\lambda_k)$$

$$= \frac{1}{A(N)} \int_{\mathbb{R}^k}'' e^{i(\lambda_1+\cdots+\lambda_k)} \frac{e^{i[Nt](\lambda_1+\cdots+\lambda_k)}-1}{e^{i(\lambda_1+\cdots+\lambda_k)}-1} \widehat{B}_F(d\lambda_1) \ldots \widehat{B}_F(d\lambda_k).$$

By making the change of variables $\lambda_1 = x_1/N, \ldots, \lambda_k = x_k/N$, and using the change of variables formula (B.2.15), the process $S_{N,k}(t)$ has the same distribution as (we use the same notation)

$$S_{N,k}(t) = \int_{\mathbb{R}^k}'' K_{N,t}(x_1,\ldots,x_k)\widehat{B}_{F_N}(dx_1)\ldots\widehat{B}_{F_N}(dx_k),$$

where

$$K_{N,t}(x_1,\ldots,x_k) = \frac{e^{i\frac{[Nt]}{N}(x_1+\cdots+x_k)} - 1}{N(e^{i\frac{1}{N}(x_1+\cdots+x_k)} - 1)} \quad (5.3.10)$$

and $\widehat{B}_{F_N}$ is an Hermitian Gaussian random measure on $\mathbb{R}$ with a control measure satisfying

$$F_N(A) = \frac{N^{1-2d}}{L_2(N)} F(N^{-1}A), \quad A \in \mathcal{B}(\mathbb{R}). \quad (5.3.11)$$

For the convergence of finite-dimensional distributions, we consider a linear combination

$$\sum_{j=1}^{J} \theta_j S_{N,k}(t_j) = \int_{\mathbb{R}^k}'' k_N(x_1,\ldots,x_k)\widehat{B}_{F_N}(dx_1)\ldots\widehat{B}_{F_N}(dx_k),$$

where $\theta_j \in \mathbb{R}$, $t_j \geq 0$, $j = 1,\ldots,J$,

$$k_N(x_1,\ldots,x_k) = \sum_{j=1}^{J} \theta_j K_{N,t_j}(x_1,\ldots,x_k).$$

Step 2 (goal): We want to conclude that

$$\sum_{j=1}^{J} \theta_j S_{N,k}(t_j) \xrightarrow{d} \int_{\mathbb{R}^k}'' k_0(x_1,\ldots,x_k)\widehat{B}_{F_0}(dx_1)\ldots\widehat{B}_{F_0}(dx_k), \quad (5.3.12)$$

where

$$k_0(x_1,\ldots,x_k) = \sum_{j=1}^{J} \theta_j K_{t_j}(x_1,\ldots,x_k), \quad K_t(x_1,\ldots,x_k) = \frac{e^{it(x_1+\cdots+x_k)} - 1}{i(x_1+\cdots+x_k)} \quad (5.3.13)$$

and $\widehat{B}_{F_0}$ is an Hermitian Gaussian random measure with a control measure

$$\mathbb{E}|\widehat{B}_{F_0}(dx)|^2 = \frac{|x|^{-2d}dx}{2\sin(d\pi)\Gamma(1-2d)} =: F_0(dx).$$

Since the Hermite process is normalized, namely, $\mathbb{E}(Z_H^{(k)}(1))^2 = 1$, this would yield the desired convergence (5.3.5).

Step 3 (method): To show (5.3.12), we shall apply Proposition 5.3.6 below with F_N and F_0 as defined above, and

$$\widehat{K}_N(x_1,\ldots,x_k) = k_N(x_1,\ldots,x_k), \quad \widehat{K}_0(x_1,\ldots,x_k) = k_0(x_1,\ldots,x_k).$$

We now verify the assumptions of that proposition.

(i) The sequence F_N converges vaguely[1] to F_0 by Proposition 5.3.4.

(ii) The measure F_N is without atoms using Theorem 9.6 in Zygmund [1030], Chapter III, Section 9.

(iii) The sequence $k_N(x_1, \ldots, x_k)$ converges to $k_0(x_1, \ldots, x_k)$ uniformly in any rectangle $[-A, A]^k$ since, for fixed t, $K_{N,t}(x_1, \ldots, x_k)$ converges to $K_t(x_1, \ldots, x_k)$ in (5.3.13) uniformly in any rectangle $[-A, A]^k$. For example, if $t = 1$, the latter can be seen from

$$|K_{N,1}(x_1, \ldots, x_k) - K_1(x_1, \ldots, x_k)| = \left| \frac{e^{i(x_1+\cdots+x_k)} - 1}{N(e^{i\frac{1}{N}(x_1+\cdots+x_k)} - 1)} - \frac{e^{i(x_1+\cdots+x_k)} - 1}{i(x_1 + \cdots + x_k)} \right|$$

$$= |e^{i(x_1+\cdots+x_k)} - 1| \left| \frac{N(e^{i\frac{1}{N}(x_1+\cdots+x_k)} - 1) - i(x_1 + \cdots + x_k)}{N(e^{i\frac{1}{N}(x_1+\cdots+x_k)} - 1)(x_1 + \cdots + x_k)} \right|$$

$$\leq C_1 |e^{i(x_1+\cdots+x_k)} - 1| \frac{|\frac{1}{N}(x_1 + \cdots + x_k)^2|}{|N\frac{1}{N}(x_1 + \cdots + x_k)||x_1 + \cdots + x_k|} \leq \frac{C_2}{N},$$

for all $(x_1, \ldots, x_k) \in [-A, A]^k$, since $(x_1 + \cdots + x_k)/N$ is arbitrarily small for all $(x_1, \ldots, x_k) \in [-A, A]^k$.

(iv) It remains to check that the condition (5.3.26) below holds. Let

$$\mu_N(dx_1, \ldots, dx_k) = |k_N(x_1, \ldots, x_k)|^2 F_N(dx_1) \ldots F_N(dx_k),$$
$$\mu_0(dx_1, \ldots, dx_k) = |k_0(x_1, \ldots, x_k)|^2 F_0(dx_1) \ldots F_0(dx_k)$$

be measures on $\mathcal{B}(\mathbb{R}^k)$. The measures μ_N are finite and concentrated on $(-N\pi, N\pi]^k$. Since F_N converge vaguely to F_0 and k_N converge to k_0 uniformly on compact intervals, the measures μ_N converge vaguely to μ_0. In view of (A.3.5) in Appendix A.3, to show (5.3.26), it is enough to prove that μ_N converges not only vaguely but also weakly (with the limit being μ_0). To do so, we shall apply Lemma A.3.1 in Appendix A.3 with $\ell = k$ and $c_N = N$. We need to show that

$$\phi_N(s_1, \ldots, s_k) = \int_{\mathbb{R}^k} e^{i(\frac{[Ns_1]}{N}x_1 + \ldots + \frac{[Ns_k]}{N}x_k)} |k_N(x_1, \ldots, x_k)|^2 F_N(dx_1) \ldots F_N(dx_k)$$

tends to a continuous limit as $N \to \infty$. But by (5.3.10),

$$\phi_N(s_1, \ldots, s_k) = \sum_{j_1, j_2=1}^{J} \theta_{j_1} \theta_{j_2} \int_{\mathbb{R}^k} e^{i(\frac{[Ns_1]}{N}x_1 + \cdots + \frac{[Ns_k]}{N}x_k)}$$
$$\times K_{N,t_{j_1}}(x_1, \ldots, x_k) \overline{K_{N,t_{j_2}}(x_1, \ldots, x_k)} F_N(dx_1) \ldots F_N(dx_k)$$

$$= \sum_{j_1, j_2=1}^{J} \theta_{j_1} \theta_{j_2} \frac{1}{N^2} \sum_{n_1=0}^{[Nt_{j_1}]-1} \sum_{n_2=0}^{[Nt_{j_2}]-1} \int_{\mathbb{R}^k} e^{i(\frac{[Ns_1]}{N}x_1 + \cdots + \frac{[Ns_k]}{N}x_k)}$$
$$\times e^{i\frac{n_1}{N}(x_1+\cdots+x_k)} e^{-i\frac{n_2}{N}(x_1+\cdots+x_k)} F_N(dx_1) \ldots F_N(dx_k).$$

By (5.3.11),

---

[1] The notions of vague and weak convergence are discussed in Appendix A.3.

$$\phi_N(s_1,\ldots,s_k) = \sum_{j_1,j_2=1}^{J} \theta_{j_1}\theta_{j_2} \frac{1}{N^2} \sum_{n_1=0}^{[Nt_{j_1}]-1} \sum_{n_2=0}^{[Nt_{j_2}]-1} \int_{\mathbb{R}^k} e^{i([Ns_1]+n_1-n_2)\lambda_1} \ldots e^{i([Ns_k]+n_1-n_2)\lambda_k}$$

$$\times \frac{1}{(L_2(N))^k N^{k(2d-1)}} F(d\lambda_1)\ldots F(d\lambda_k)$$

$$= \sum_{j_1,j_2=1}^{J} \theta_{j_1}\theta_{j_2} \frac{1}{N^2} \sum_{n_1=0}^{[Nt_{j_1}]-1} \sum_{n_2=0}^{[Nt_{j_2}]-1} \frac{\gamma_X([Ns_1]+n_1-n_2)}{L_2(N)N^{2d-1}} \cdots \frac{\gamma_X([Ns_k]+n_1-n_2)}{L_2(N)N^{2d-1}},$$

where $\gamma_X(\cdot)$ is the autocovariance (5.3.9) given in (2.1.5). One can show (Exercise 5.4, (b)) that, for each $s_1,\ldots,s_k \in \mathbb{R}$,

$$\phi_N(s_1,\ldots,s_k) \to \sum_{j_1,j_2=1}^{J} \theta_{j_1}\theta_{j_2} \int_0^{t_{j_1}} dx_1 \int_0^{t_{j_2}} dx_2 |s_1+x_1-x_2|^{2d-1} \ldots |s_k+x_1-x_2|^{2d-1}.$$

(5.3.14)

The limit function is continuous (Exercise 5.4, (b)), and Lemma A.3.1 applies. □

The next results were used in the proof of Theorem 5.3.1 above. The first result allowed to replace G in the partial sum process by an Hermite polynomial.

Theorem 5.3.3 (*Reduction theorem.*) *In the setting and under the assumptions of Theorem 5.3.1, let g_k be the first nonzero coefficient in the Hermite expansion of G and $H_k(x)$ be the Hermite polynomial of order k. If the convergence (5.3.5) holds for $G(x) = g_k H_k(x)$, then it also holds for the general function G with Hermite rank k.*

Proof Since G has Hermite rank k, we have

$$G(x) = g_k H_k(x) + \sum_{j=k+1}^{\infty} g_j H_j(x) =: g_k H_k(x) + G^*(x).$$

Note that, by Corollary 5.2.5,

$$\mathrm{Var}\Big(\sum_{n=1}^{[Nt]} G^*(X_n)\Big) \sim C(L_2([Nt]))^{k+1}[Nt]^{(k+1)(2d-1)+2},$$

as $N \to \infty$. This implies that

$$\frac{1}{(L_2(N))^{k/2} N^{k(d-1/2)+1}} \sum_{n=1}^{[Nt]} G^*(X_n) \xrightarrow{p} 0,$$

for each t. The result now follows from Billingsley [154], Theorem 3.1. □

The next result shows that a properly normalized spectral measure of the series $\{X_n\}$ converges vaguely. The notion of vague convergence is discussed in Appendix A.3.

Proposition 5.3.4 *Let $\{X_n\}_{n\in\mathbb{Z}}$ by a stationary series which is LRD in the sense (2.1.5) in condition II with $d \in (0, 1/2)$. Let F be the spectral measure of $\{X_n\}_{n\in\mathbb{Z}}$, and define*

5.3 Non-Central Limit Theorem

$$F_N(A) = \frac{N^{1-2d}}{L_2(N)} F(N^{-1}A), \quad A \in \mathcal{B}(\mathbb{R}), \ N \geq 1. \tag{5.3.15}$$

Then,

$$F_N \xrightarrow{v} F_0 \tag{5.3.16}$$

in the sense of vague convergence, where

$$\frac{F_0(dx)}{dx} = C|x|^{-2d} \tag{5.3.17}$$

and

$$C^{-1} = 2\sin(d\pi)\Gamma(1-2d). \tag{5.3.18}$$

Proof a)[2] Consider a measure μ_N defined by

$$\mu_N(dx) = |K_N(x)|^2 F_N(dx),$$

where

$$K_N(x) = \frac{1}{N}\sum_{j=0}^{N-1} e^{i\frac{j}{N}x} = \begin{cases} \frac{e^{ix}-1}{N(e^{i\frac{x}{N}}-1)}, & \text{if } \frac{x}{2\pi N} \notin \mathbb{Z}, \\ 1, & \text{if } \frac{x}{2\pi N} \in \mathbb{Z}. \end{cases}$$

We now want to apply Lemma A.3.1 in Appendix A.3 to show that the measure μ_N converges weakly. We will use this fact to conclude the convergence of F_N in (5.3.16).

Take $\ell = 1$ and $C_N = N$ in Lemma A.3.1. Note that the function ϕ_N in (A.3.6) can be expressed as

$$\phi_N(t) = \int_{\mathbb{R}} e^{i\frac{[Nt]}{N}x} \left|\frac{1}{N}\sum_{j=0}^{N-1} e^{i\frac{j}{N}x}\right|^2 \frac{N^{1-2d}}{L_2(N)} F\left(d\frac{x}{N}\right) = \int_{\mathbb{R}} e^{i[Nt]\lambda} \left|\frac{1}{N}\sum_{j=0}^{N-1} e^{ij\lambda}\right|^2 \frac{N^{1-2d}}{L_2(N)} F(d\lambda)$$

$$= \frac{1}{N^2 N^{2d-1} L_2(N)} \sum_{j_1,j_2=0}^{N-1} \int_{\mathbb{R}} e^{i([Nt]+j_1-j_2)\lambda} F(d\lambda) = \frac{1}{N^2 N^{2d-1} L_2(N)} \sum_{j_1,j_2=0}^{N-1} \gamma_X([Nt]+j_1-j_2)$$

$$= \frac{1}{N^2 N^{2d-1} L_2(N)} \sum_{j=-(N-1)}^{N-1} (N-|j|) \gamma_X([Nt]+j) = \frac{1}{N} \sum_{j=-(N-1)}^{N-1} \left(1 - \frac{|j|}{N}\right) \frac{\gamma_X([Nt]+j)}{N^{2d-1} L_2(N)},$$

where $\gamma_X(n) = \int_{\mathbb{R}} e^{in\lambda} F(d\lambda)$ is the autocovariance function. One can show (Exercise 5.4, (a)) that, for every t, as $N \to \infty$,

$$\phi_N(t) \to \int_{-1}^{1} (1-|x|)|t+x|^{2d-1} dx =: \phi(t). \tag{5.3.19}$$

We have $\phi(t) = \phi(-t), t \in \mathbb{R}$. One can show (Exercise 5.4, (a)) that, for $t > 0$,

$$\phi(t) = \frac{1}{2d(2d+1)} \left\{(t+1)^{2d+1} + |t-1|^{2d+1} - 2t^{2d+1}\right\}. \tag{5.3.20}$$

Since the function $\phi(t)$ is continuous at $t = 0$, Lemma A.3.1 shows that

$$\mu_N \xrightarrow{w} \mu_0,$$

---

[2] The proof in this first part is similar to the end of the proof of Theorem 5.3.1.

where μ_0 is a finite measure having $\phi(t)$ for its Fourier transform:

$$\phi(t) = \int_{\mathbb{R}} e^{itx}\mu_0(dx). \tag{5.3.21}$$

b) We show next that F_N also converges locally weakly. For $B \in \mathcal{B}(\mathbb{R})$, note that

$$F_N(B) = \int_B |K_N(x)|^{-2}\mu_N(dx).$$

The functions $|K_N(x)|^2$ converge uniformly on compact intervals to $|K_0(x)|^2$, where

$$K_0(x) = \begin{cases} \frac{e^{ix}-1}{ix}, & \text{if } x \neq 0, \\ 1, & \text{if } x = 0. \end{cases}$$

Since $|K_0(x)|^2$ is continuous and non-vanishing in $[-\pi/2, \pi/2]$, the weak convergence of μ_N to μ_0 implies that, for every $B \in \mathcal{B}([-\pi/2, \pi/2])$ such that $\mu_0(\partial B) = 0$,

$$\lim_{N \to \infty} F_N(B) = \int_B |K_0(x)|^{-2}\mu_0(dx) =: F_0(B). \tag{5.3.22}$$

This defines a finite measure F_0 on $\mathcal{B}([-\pi/2, \pi/2])$.

c) We want to conclude next that, for $B_1, B_2 \in \mathcal{B}([-\pi/2, \pi/2])$ satisfying $B_1 = tB_2 = \{tx : x \in B_2\}$ with some $t > 0$,

$$F_0(B_1) = F_0(tB_2) = t^{1-2d}F_0(B_2). \tag{5.3.23}$$

Observe that

$$F_N(tB) = \frac{N^{1-2d}}{L_2(N)}F(N^{-1}tB) = \frac{N^{1-2d}}{\left[\frac{N}{t}\right]^{1-2d}}\frac{L_2\left(\left[\frac{N}{t}\right]\right)}{L_2(N)}\frac{\left[\frac{N}{t}\right]^{1-2d}}{L_2\left(\left[\frac{N}{t}\right]\right)}F\left(\left[\frac{N}{t}\right]^{-1}\frac{\left[\frac{N}{t}\right]}{\frac{N}{t}}B\right)$$

$$= \frac{N^{1-2d}}{\left[\frac{N}{t}\right]^{1-2d}}\frac{L_2\left(\left[\frac{N}{t}\right]\right)}{L_2(N)}F_{\left[\frac{N}{t}\right]}\left(\frac{\left[\frac{N}{t}\right]}{\frac{N}{t}}B\right).$$

For any $t > 0$ and $B \in \mathcal{B}([-\pi/2, \pi/2])$ such that $F_0(\partial B) = 0$ and $F_0(\partial(tB)) = 0$, this implies by (5.3.22) that

$$F_0(tB) = \lim_{N \to \infty} F_N(tB) = t^{1-2d}F_0(B). \tag{5.3.24}$$

This proves (5.3.23) when $F_0(\partial B_1) = F_0(\partial B_2) = 0$. We now want to remove these restrictions. In particular, (5.3.23) holds with $B_2 = (a, b)$ for $-\pi/2 \leq a, at < 0 < b, bt \leq \pi/2$ as long as $F_0(\{a\}) = F_0(\{at\}) = F_0(\{b\}) = F_0(\{bt\}) = 0$. Since the set $\{x \in [-\pi/2, \pi/2] : F_0(\{x\}) > 0\}$ is at most countable, the relation (5.3.23) holds for intervals $B_2 = (a, b)$ for $-\pi/2 \leq a, at < 0 < b, bt \leq \pi/2$, except perhaps for a countable number of a, b and t. For any fixed t, a and b satisfying $-\pi/2 \leq a, at < 0 < b, bt \leq \pi/2$, we can select $a_n \downarrow a$, $b_n \uparrow b$ and $t_n \uparrow t$ so that the relation (5.3.23) holds with $B_2 = (a_n, b_n)$ and $B_1 = (a_n t_n, b_n t_n)$. Since F_0 is a measure, and $(a_n, b_n) \uparrow (a, b)$ and $(a_n t_n, b_n t_n) \uparrow (at, bt)$, passing in the limit yields (5.3.23) with $B_2 = (a, b)$ for all

$-\pi/2 \leq a, at < 0 < b, bt \leq \pi/2$. Since intervals generate Borel sets, this shows (5.3.23) without the restrictions $F_0(\partial B_1) = F_0(\partial B_2) = 0$.

d) F_0 was defined on $\mathcal{B}([-\pi/2, \pi/2])$. We now want to define it for every bounded set $B \in \mathcal{B}(\mathbb{R})$. We do it as follows. If $B \in \mathcal{B}(\mathbb{R})$ satisfies $B \subset [-K\pi/2, K\pi/2]$, set

$$F_0(B) = K^{1-2d} F_0\left(\frac{1}{K} B\right).$$

This definition does not depend on K because of (5.3.23). It coincides with the definition when $B \in \mathcal{B}([-\pi/2, \pi/2])$. The set measure F_0 can be extended to a locally finite measure F_0 by setting, for example,

$$F_0(B) = \lim_{K \to \infty} F_0(B \cap [-K, K]).$$

The relation holds for all bounded $B \in \mathcal{B}(\mathbb{R})$ and, taking $t = 1$ in (5.3.24), this implies that F_N tends locally weakly to F_0.

e) The relation (5.3.23) holds for all $B_1, B_2 \in \mathcal{B}(\mathbb{R})$ and hence, for all $t > 0$ and $B \in \mathcal{B}(\mathbb{R})$,

$$F_0(tB) = t^{1-2d} F_0(B).$$

In particular, by taking $B = [0, 1]$,

$$F_0([0, t]) = t^{1-2d} F_0([0, 1]).$$

Since $1 - 2d > 0$, this shows that the measure F_0 is absolutely continuous with respect to the Lebesgue measure on $\mathcal{B}([0, \infty))$, and that

$$\frac{F_0(dx)}{dx} = Cx^{-2d}, \quad x \geq 0,$$

for some constant C. A similar argument yields an analogous relation for $x < 0$. The constant C is the same since F_N and hence F_0 are symmetric. This yields (5.3.17).

f) Finally, for the constant C in (5.3.18), we have from (5.3.21) and (5.3.22) that

$$\phi(t) = \int_{\mathbb{R}} e^{itx} \mu_0(dx) = \int_{\mathbb{R}} e^{itx} |K_0(x)|^2 F_0(dx) = C \int_{\mathbb{R}} e^{itx} |K_0(x)|^2 |x|^{-2d} dx$$

$$= 2C \int_{\mathbb{R}} e^{itx} \frac{1 - \cos x}{x^2} |x|^{-2d} dx = 4C \int_0^\infty \cos(tx)(1 - \cos x) x^{-2d-2} dx =: 4C\psi_d(t).$$

If $0 < -2d - 1 < 1$ or $-1 < d < -1/2$, $\psi_d(t)$ can be evaluated for $t > 0$ by using Formula 3.762.3 in Gradshteyn and Ryzhik [421]:

$$\psi_d(t) = \int_0^\infty \cos(tx) x^{-2d-2} dx - \int_0^\infty \cos(tx) \cos(x) x^{-2d-2} dx$$

$$= \cos\left(\frac{(2d+1)\pi}{2}\right) \Gamma(-2d-1) t^{2d+1}$$

$$\quad - \cos\left(\frac{(2d+1)\pi}{2}\right) \Gamma(-2d-1) \frac{1}{2}\left\{(t+1)^{2d+1} + |t-1|^{2d+1}\right\}$$

$$= \frac{1}{2} \sin(d\pi) \Gamma(-2d-1) \left\{(t+1)^{2d+1} + |t-1|^{2d+1} - 2t^{2d+1}\right\} =: \tilde{\psi}_d(t). \quad (5.3.25)$$

On the other hand, for fixed $t > 0$, the functions $\psi_d(t)$ and $\tilde{\psi}_d(t)$ are analytic for complex-valued d satisfying $-2 < \Re(-2d-1) < 1$ or $-1 < \Re(d) < 1/2$, and $\Re(-2d-1) \neq 0, -1$ or $\Re(d) \neq -1/2, 0$. Since they coincide for $-1 < d < -1/2$, they must coincide for $0 < d < 1/2$. By comparing (5.3.25) and (5.3.20), and using the relation (2.2.13), we deduce that

$$\frac{1}{2d(2d+1)}\left\{(t+1)^{2d+1} + |t-1|^{2d+1} - 2t^{2d+1}\right\}$$
$$= 2C \sin(d\pi)\Gamma(-2d-1)\left\{(t+1)^{2d+1} + |t-1|^{2d+1} - 2t^{2d+1}\right\}$$

or

$$C = \frac{1}{2\sin(d\pi)(-2d)(-2d-1)\Gamma(-2d-1)} = \frac{1}{2\sin(d\pi)\Gamma(1-2d)}. \qquad \square$$

The next proposition was used to show the convergence of multiple integrals. Its proof is based on the following classical lemma (Theorem 3.2, Billingsley [154]).

Lemma 5.3.5 *Suppose that:*
(1) For each M, $X_{N,M} \xrightarrow{d} X_M$ as $N \to \infty$,
(2) $X_M \xrightarrow{d} X$ as $M \to \infty$,
(3) $\limsup_M \limsup_N \mathbb{P}(|X_{N,M} - Y_N| > \epsilon) = 0$, for all $\epsilon > 0$.
Then, $Y_N \xrightarrow{d} X$ as $N \to \infty$.

Finally, the next proposition was used in the proof of Theorem 5.3.1.

Proposition 5.3.6 *Let F_N, $N \geq 1$, F_0 be symmetric, locally finite measures without atoms on $(\mathbb{R}, \mathcal{B}(\mathbb{R}))$ so that $F_N \xrightarrow{v} F_0$ in the sense of vague convergence. Let W_{F_N} and W_{F_0} be Hermitian Gaussian random measures on $\mathbb{R}$ with control measures F_N and F_0, respectively. Let also $\widehat{K}_N(x_1, \ldots, x_k)$ be a sequence of Hermitian, measurable functions on $\mathbb{R}^k$ tending to a continuous function $\widehat{K}_0(x_1, \ldots, x_k)$ uniformly in any rectangle $[-A, A]^k$, $A > 0$. Moreover, suppose that $\widehat{K}_0$ and $\widehat{K}_N$ satisfy the relation*

$$\lim_{A \to \infty} \sup_{N \geq 0} \int_{\mathbb{R}^k \setminus [-A,A]^k} |\widehat{K}_N(x_1, \ldots, x_k)|^2 F_N(dx_1) \ldots F_N(dx_k) = 0. \qquad (5.3.26)$$

Then,

$$\int_{\mathbb{R}^k}'' \widehat{K}_N(x_1, \ldots, x_k) W_{F_N}(dx_1) \ldots W_{F_N}(dx_k) \xrightarrow{d} \int_{\mathbb{R}^k}'' \widehat{K}_0(x_1, \ldots, x_k) W_{F_0}(dx_1) \ldots W_{F_0}(dx_k), \qquad (5.3.27)$$

where the last integral exists.

Proof Let $\widehat{H}^k$ consist of functions g which are Hermitian symmetric in the sense of (B.2.12) in Appendix B.2 and have a representation (B.2.13) with bounded sets A_j. Let $\overline{H}_{F_0}^k$ consist of those functions in $\widehat{H}^k$ for which $F_0(\partial A_j) = 0$, $\pm j = 1, \ldots, n$, where A_j are the sets in their representation (B.2.13). Since $F_N \xrightarrow{v} F_0$ by assumption and $F_0(\partial A_j) = 0$, $\pm j = 1, \ldots, n$, we have (Appendix A.3) that $F_N(A_j) \to F_0(A_j)$, $\pm j = 1, \ldots, n$. This implies that a Gaussian vector $(W_{F_N}(A_j), \pm j = 1, \ldots, n)$

5.3 Non-Central Limit Theorem

converges in distribution to a Gaussian vector $(W_{F_0}(A_j), \pm j = 1, \ldots, n)$. For $g \in \overline{H}^k_{F_0}$, $\int''_{\mathbb{R}^k} g(x_1, \ldots, x_k) W_F(dx_1) \ldots W_F(dx_k)$ is a polynomial of $W_F(A_j)$s (where $F = F_0$ or F_N). Since the $W_{F_N}(A_j)$s converge in distribution to $W_F(A_j)$s, it follows that

$$\int''_{\mathbb{R}^k} g(x_1, \ldots, x_k) W_{F_N}(dx_1) \ldots W_{F_N}(dx_k) \xrightarrow{d} \int''_{\mathbb{R}^k} g(x_1, \ldots, x_k) W_{F_0}(dx_1) \ldots W_{F_0}(dx_k). \quad (5.3.28)$$

To show (5.3.27), it is enough to prove that, for any $\epsilon > 0$, there is $g \in \overline{H}^k_{F_0}$ such that

$$\int_{\mathbb{R}^k} |\widehat{K}_N(x_1, \ldots, x_k) - g(x_1, \ldots, x_k)|^2 F_N(dx_1) \ldots F_N(dx_k) < \epsilon, \quad (5.3.29)$$

for $N = 0$ and $N > N(\epsilon)$. Indeed, denote a multiple integral with respect to W_{F_N} by $\widehat{I}_{F_N}(\cdot)$. Note that the integral $\widehat{I}_{F_0}(\widehat{K}_0)$ is well-defined by using (5.3.26) for $N = 0$ and the fact that $\widehat{K}_0$ is continuous (hence square integrable on bounded sets). By (5.3.29) for $N = 0$, there is a sequence $g_M \in \overline{H}^k_{F_0}$ such that $\widehat{I}_{F_0}(g_M) \to \widehat{I}_{F_0}(\widehat{K}_0)$ in $L^2(\Omega)$ and hence in distribution. In fact, by (5.3.29), the sequence $g_M \in \overline{H}^k_{F_0}$ can be chosen such that

$$\limsup_M \limsup_N \mathbb{E} |\widehat{I}_{F_N}(g_M) - \widehat{I}_{F_N}(\widehat{K}_N)|^2 = 0.$$

By (5.3.28), for each fixed M, $\widehat{I}_{F_N}(g_M) \to \widehat{I}_{F_0}(g_M)$ in distribution. Hence, the convergence $\widehat{I}_{F_N}(\widehat{K}_N) \to \widehat{I}_{F_0}(\widehat{K}_0)$ in (5.3.27) follows from Lemma 5.3.5 with

$$X_{N,M} = \widehat{I}_{F_N}(g_M), \quad X_M = \widehat{I}_{F_0}(g_M), \quad Y_N = \widehat{I}_{F_N}(\widehat{K}_N) \quad \text{and} \quad X = \widehat{I}_{F_0}(\widehat{K}_0).$$

By using (5.3.26), note that (5.3.29) is equivalent to the same relation where the integration over $\mathbb{R}^k$ is replaced by $[-A, A]^k$ (for large enough A). Then for $N \geq 1$, for example, write

$$\int_{[-A,A]^k} |\widehat{K}_N(x_1, \ldots, x_k) - g(x_1, \ldots, x_k)|^2 F_N(dx_1) \ldots F_N(dx_k) \leq 2 \Big(F_N([-A, A]) \Big)^k$$

$$\times \sup_{(x_1,\ldots,x_k) \in [-A,A]^k} \Big(|\widehat{K}_N(x_1, \ldots, x_k) - \widehat{K}_0(x_1, \ldots, x_k)|^2 + |\widehat{K}_0(x_1, \ldots, x_k) - g(x_1, \ldots, x_k)|^2 \Big). \quad (5.3.30)$$

Since $F_N([-A, A]) \to F_0([-A, A])$ (F_0 has no atoms and $F_0(\partial[-A, A]) = F_0(\{-A, A\}) = 0$) and F_0 is locally finite, we have $\sup_{N \geq 1} F_N([-A, A]) < \infty$. Since $\widehat{K}_N \to \widehat{K}_0$ and $\widehat{K}_0$ is continuous on $[-A, A]^k$, the supremum of the difference $|\widehat{K}_N(x_1, \ldots, x_k) - \widehat{K}_0(x_1, \ldots, x_k)|$ is arbitrarily small for large N. On the other hand, since $\widehat{K}_0$ is continuous on $[-A, A]^k$, we can choose $g \in \overline{H}^k_{F_0}$ such that the supremum of the difference $|\widehat{K}_0(x_1, \ldots, x_k) - g(x_1, \ldots, x_k)|$ is arbitrarily small. (To construct such g, one can take a grid of $[-A, A]^k$ and define g to be equal to a value of $\widehat{K}_0$ on the grid rectangle. The function g approximates $\widehat{K}_0$ uniformly as the grid gets finer since $\widehat{K}_0$ is continuous.) This shows that the bound (5.3.30) can be made arbitrarily small for large N and suitable g. The argument for $N = 0$ is analogous. $\square$

5.4 Central Limit Theorem

Non-central limit theorem (Theorem 5.3.1) applies for a Gaussian stationary series exhibiting long-range dependence when $(k-1)/2k < d < 1/2$, where k is the Hermite rank of a transformation of the series. What happens in the case $0 < d < (k-1)/2k$, possible when $k \geq 2$? The following result will show that in this case, the limit of partial sum processes is the usual Brownian motion (see Corollary 5.4.5 below). The result is stated under more general conditions.

Theorem 5.4.1 *Let $\{X_n\}_{n\in\mathbb{Z}}$ be a Gaussian stationary series with autocovariance function γ_X. Suppose that $\mathbb{E}X_n = 0$, $\mathbb{E}X_n^2 = 1$. Let G be a function with Hermite rank $k \geq 1$ in the sense of Definition 5.2.1, and γ_G be the autocovariance function of $\{G(X_n)\}_{n\in\mathbb{Z}}$. If*

$$\sum_{n=1}^{\infty} |\gamma_X(n)|^k < \infty \tag{5.4.1}$$

or, equivalently,

$$\sum_{n=1}^{\infty} |\gamma_G(n)| < \infty \quad \text{and} \quad \gamma_X(n) \to 0, \text{ as } n \to \infty, \tag{5.4.2}$$

then

$$\frac{1}{N^{1/2}} \sum_{n=1}^{[Nt]} (G(X_n) - \mathbb{E}G(X_n)) \xrightarrow{fdd} \sigma B(t), \quad t \geq 0, \tag{5.4.3}$$

where $\{B(t)\}_{t\geq 0}$ is a standard Brownian motion and

$$\sigma^2 = \sum_{m=k}^{\infty} g_m^2 m! \sum_{n=-\infty}^{\infty} (\gamma_X(n))^m, \tag{5.4.4}$$

where g_m are the coefficients in the Hermite expansion (5.1.6) of G.

Proof To see the equivalence of (5.4.1) and (5.4.2), note from (5.2.2) that

$$\gamma_G(n) = \sum_{m=k}^{\infty} g_m^2 m! (\gamma_X(n))^m = (\gamma_X(n))^k \sum_{m=k}^{\infty} g_m^2 m! (\gamma_X(n))^{m-k}.$$

Moreover, $|\gamma_X(n)| \leq 1$ since X_n has variance 1. This shows that

$$\sum_{n=1}^{\infty} |\gamma_G(n)| \leq \sum_{n=1}^{\infty} |\gamma_X(n)|^k \sum_{m=k}^{\infty} g_m^2 m!$$

and hence (5.4.1) implies (5.4.2). Conversely, since $\gamma_X(n) \to 0$ in (5.4.2), we have $|\gamma_G(n)| \geq |\gamma_X(n)|^k g_k^2 k!/2$ for large enough n and hence (5.4.2) implies (5.4.1).

Denote the partial sum process in (5.4.3) by

$$S_{N,t}(G) = \frac{1}{N^{1/2}} \sum_{n=1}^{[Nt]} (G(X_n) - \mathbb{E}G(X_n)),$$

where
$$G(X_n) - \mathbb{E}G(X_n) = \sum_{m=k}^{\infty} g_m H_m(X_n).$$

By using Proposition 5.1.1, we have
$$\mathbb{E}(S_{N,t}(G))^2 = \sum_{m=k}^{\infty} g_m^2 \mathbb{E}(S_{N,t}(H_m))^2$$

and
$$\mathbb{E}(S_{N,t}(H_m))^2 = \frac{1}{N} \sum_{n_1,n_2=1}^{[Nt]} \mathbb{E}H_m(X_{n_1})H_m(X_{n_2}) = \frac{m!}{N} \sum_{n_1,n_2=1}^{[Nt]} (\gamma_X(n_1 - n_2))^m$$
$$= \frac{m!}{N} \sum_{n=-[Nt]+1}^{[Nt]-1} ([Nt] - |n|)(\gamma_X(n))^m \leq tm! \sum_{n=-[Nt]+1}^{[Nt]-1} |\gamma_X(n)|^m \leq tm! \sum_{n=-\infty}^{\infty} |\gamma_X(n)|^k,$$
(5.4.5)

for $m \geq k$. As $N \to \infty$,
$$\mathbb{E}(S_{N,t}(H_m))^2 = m!t \sum_{n=-[Nt]+1}^{[Nt]-1} \left(\frac{[Nt]}{Nt} - \frac{|n|}{Nt}\right)(\gamma_X(n))^m \to m!t \sum_{n=-\infty}^{\infty} (\gamma_X(n))^m$$

and
$$\mathbb{E}(S_{N,t}(G))^2 \to t \sum_{m=k}^{\infty} g_m^2 m! \sum_{n=-\infty}^{\infty} (\gamma_X(n))^m.$$

This shows that the limiting variance exists, is finite and equals to $\sigma^2 t$ with σ^2 given by (5.4.4). Note also that, by using (5.4.5),
$$\mathbb{E}\Big(S_{N,t}(G) - S_{N,t}\Big(\sum_{m=k}^{M} g_m H_m\Big)\Big)^2 = \mathbb{E}\Big(S_{N,t}\Big(\sum_{m=M+1}^{\infty} g_m H_m\Big)\Big)^2$$
$$= \sum_{m=M+1}^{\infty} g_m^2 \mathbb{E}(S_{N,t}(H_m))^2 \leq t \sum_{n=-\infty}^{\infty} |\gamma_X(n)|^k \sum_{m=M+1}^{\infty} g_m^2 m!$$

is arbitrarily small for large enough M and all N. It is therefore enough to prove (5.4.3) for $G(x) - \mathbb{E}G(X_n) = \sum_{m=k}^{M} g_m H_m(x)$ for fixed $M < \infty$; that is, when G is a polynomial. The variance σ^2 in (5.4.4) is replaced by $\sigma^2 = \sum_{m=k}^{M} g_m^2 m! \sum_{n=-\infty}^{\infty} (\gamma_X(n))^m$.

For
$$G(x) - \mathbb{E}G(X_n) = \sum_{m=k}^{M} g_m H_m(x), \qquad (5.4.6)$$

set
$$S_N(G) = \sum_{j=1}^{J} \theta_j S_{N,t_j}(G),$$

where $\theta_j \in \mathbb{R}$ and $t_1, \ldots, t_J \geq 0$. By the method of moments (e.g., Gut [442], Theorem 8.6 in Chapter 5 and Section 4.10), it is enough to show that, for all integers $p \geq 1$,

$$\mathbb{E}\big(S_N(G)\big)^p \to \sigma^p \mathbb{E}\Big(\sum_{j=1}^J \theta_j B(t_j)\Big)^p. \tag{5.4.7}$$

This is equivalent to: for $p \geq 1$,

$$\chi_p\big(S_N(G)\big) \to \sigma^p \chi_p\Big(\sum_{j=1}^J \theta_j B(t_j)\Big), \tag{5.4.8}$$

where χ_p denotes the pth cumulant (see Definition 4.3.7). Since the limit is Gaussian and has zero mean,

$$\chi_p\Big(\sum_{j=1}^J \theta_j B(t_j)\Big) = \begin{cases} \mathbb{E}\big(\sum_{j=1}^J \theta_j B(t_j)\big)^2, & \text{if } p = 2, \\ 0, & \text{if } p = 1, 3, 4, \ldots \end{cases} \tag{5.4.9}$$

Note that, by using (5.4.6) and the multilinearity of joint cumulants,[3]

$$\chi_p(S_N(G)) = \chi_p\Big(\sum_{j=1}^J \sum_{m=k}^M \theta_j g_m \frac{1}{N^{1/2}} \sum_{n=1}^{[Nt_j]} H_m(X_n)\Big)$$

$$= \chi\Big(\sum_{j_1=1}^J \sum_{m_1=k}^M \theta_{j_1} g_{m_1} \frac{1}{N^{1/2}} \sum_{n_1=1}^{[Nt_{j_1}]} H_{m_1}(X_{n_1}), \ldots, \sum_{j_p=1}^J \sum_{m_p=k}^M \theta_{j_p} g_{m_p} \frac{1}{N^{1/2}} \sum_{n_p=1}^{[Nt_{j_p}]} H_{m_p}(X_{n_p})\Big)$$

$$= \sum_{j_1,\ldots,j_p=1}^J \sum_{m_1,\ldots,m_p=k}^M \Big(\prod_{\ell=1}^p \theta_{j_\ell} g_{m_\ell}\Big) \frac{1}{N^{p/2}} \sum_{n_1=1}^{[Nt_{j_1}]} \cdots \sum_{n_p=1}^{[Nt_{j_p}]} \chi\Big(H_{m_1}(X_{n_1}), \ldots, H_{m_p}(X_{n_p})\Big).$$

By using (4.3.26), we obtain that

$$\chi_p(S_N(G)) = \sum_{j_1,\ldots,j_p=1}^J \sum_{m_1,\ldots,m_p=k}^M \Big(\prod_{\ell=1}^p \theta_{j_\ell} g_{m_\ell}\Big) \frac{1}{N^{p/2}} \sum_{n_1=1}^{[Nt_{j_1}]} \cdots \sum_{n_p=1}^{[Nt_{j_p}]} m_1! \ldots m_p!$$

$$\times \sum_{A \in \mathcal{A}_c(m_1,\ldots,m_p)} g(A) \prod_{w \in E(A)} \gamma_X(n_{d_1(w)} - n_{d_2(w)})$$

$$= \sum_{j_1,\ldots,j_p=1}^J \sum_{m_1,\ldots,m_p=k}^M \Big(\prod_{\ell=1}^p \theta_{j_\ell} g_{m_\ell} m_\ell!\Big)$$

$$\times \sum_{A \in \mathcal{A}_c(m_1,\ldots,m_p)} g(A) \frac{1}{N^{p/2}} T_N(t_{j_1},\ldots,t_{j_p}, m_1,\ldots,m_p, A),$$

---

[3] $\chi(\sum_{j_1} a_{1,j_1} X_{1,j_1}, \ldots, \sum_{j_p} a_{p,j_p} X_{p,j_p}) = \sum_{j_1,\ldots,j_p} a_{1,j_1} \ldots a_{p,j_p} \chi(X_{1,j_1}, \ldots, X_{p,j_p})$. Also $\chi_p(X) = \chi(X, \ldots, X)$.

where $E(A)$ is the set of lines (edges) of a multigraph A, $d_1(w) < d_2(w)$ are the two points connected by the line w, and

$$T_N(t_{j_1}, \ldots, t_{j_p}, m_1, \ldots, m_p, A) = \sum_{n_1=1}^{[Nt_{j_1}]} \cdots \sum_{n_p=1}^{[Nt_{j_p}]} \prod_{w \in E(A)} \gamma_X(n_{d_1(w)} - n_{d_2(w)}). \quad (5.4.10)$$

Since for $p \geq 3$, any connected multigraph $A \in \mathcal{A}_c(m_1, \ldots, m_p)$ is nonstandard (see Definition 5.4.2 below), Lemma 5.4.3 implies that, for $p \geq 3$,

$$T_N(t_{j_1}, \ldots, t_{j_p}, m_1, \ldots, m_p, A) = o(N^{p/2}),$$

as $N \to \infty$. Thus, for $p \geq 3$, $\chi_p(S_N(G)) \to 0$, as $N \to \infty$. In view of (5.4.9), this establishes (5.4.8) for $p \geq 3$.

For $p = 1$, the convergence (5.4.8) is trivial since $\mathbb{E} S_N(G) = 0$ and hence both sides are zero. For $p = 2$, we need to show that

$$\mathbb{E}\left(S_N(G)\right)^2 \to \sigma^2 \mathbb{E}\left(\sum_{j=1}^{J} \theta_j B(t_j)\right)^2$$

$$= \sigma^2 \sum_{j_1, j_2=1}^{J} \theta_{j_1} \theta_{j_2} (\mathbb{E} B(t_{j_1}) B(t_{j_2})) = \sigma^2 \sum_{j_1, j_2=1}^{J} \theta_{j_1} \theta_{j_2} (t_{j_1} \wedge t_{j_2}), \quad (5.4.11)$$

where $x \wedge y = \min\{x, y\}$. Note that, by using Proposition 5.1.1,

$$\mathbb{E}\left(S_N(G)\right)^2 = \mathbb{E}\left(\sum_{j=1}^{J} \sum_{m=k}^{M} \theta_j g_m \frac{1}{N^{1/2}} \sum_{n=1}^{[Nt_j]} H_m(X_n)\right)^2$$

$$= \sum_{j_1, j_2=1}^{J} \sum_{m_1, m_2=k}^{M} \theta_{j_1} \theta_{j_2} g_{m_1} g_{m_2} \frac{1}{N} \sum_{n_1=1}^{[Nt_{j_1}]} \sum_{n_2=1}^{[Nt_{j_2}]} \left(\mathbb{E} H_{m_1}(X_{n_1}) H_{m_2}(X_{n_2})\right)$$

$$= \sum_{j_1, j_2=1}^{J} \sum_{m=k}^{M} \theta_{j_1} \theta_{j_2} g_m^2 m! \frac{1}{N} \sum_{n_1=1}^{[Nt_{j_1}]} \sum_{n_2=1}^{[Nt_{j_2}]} (\gamma_X(n_1 - n_2))^m$$

$$= \sum_{j_1, j_2=1}^{J} \sum_{m=k}^{M} \theta_{j_1} \theta_{j_2} g_m^2 m! \left(S_{N,1} + S_{N,2}\right),$$

where, with $a = t_{j_1} \wedge t_{j_2}$ and $b = t_{j_1} \vee t_{j_2} = \max\{t_{j_1}, t_{j_2}\}$,

$$S_{N,1} = \frac{1}{N} \sum_{n_1=1}^{[Na]} \sum_{n_2=1}^{[Na]} (\gamma_X(n_1 - n_2))^m, \quad S_{N,2} = \frac{1}{N} \sum_{n_1=1}^{[Na]} \sum_{n_2=[Na]+1}^{[Nb]} (\gamma_X(n_1 - n_2))^m.$$

We have

$$S_{N,1} = a \sum_{n=-[Na]}^{[Na]} \left(\frac{[Na]}{Na} - \frac{|n|}{Na}\right)(\gamma_X(n))^m \to a \sum_{n=-\infty}^{\infty} (\gamma_X(n))^m = (t_{j_1} \wedge t_{j_2}) \sum_{n=-\infty}^{\infty} (\gamma_X(n))^m.$$

By Lemma 5.4.4 below, $S_{N,2} \to 0$. It follows that

$$\mathbb{E}\big(S_N(G)\big)^2 \to \sum_{j_1,j_2=1}^{J} \sum_{m=k}^{M} \theta_{j_1}\theta_{j_2} g_m^2 m! (t_{j_1} \wedge t_{j_2}) \sum_{n=-\infty}^{\infty} (\gamma_X(n))^m = \sigma^2 \sum_{j_1,j_2=1}^{J} \theta_{j_1}\theta_{j_2}(t_{j_1} \wedge t_{j_2})$$

and hence that (5.4.11) holds. $\square$

The following result was used in the proof of Theorem 5.4.1 above. It will be stated here under more general conditions. It involves another notion related to multigraphs considered in Section 4.3.2.

Definition 5.4.2 A multigraph will be called *standard* if its points can be paired in such a way that no line joins points in different pairs. Otherwise, a multigraph is called *nonstandard*.

An example of a standard multigraph is given in Figure 5.1, (a). It consists of two pairs of points with no line joining the two pairs. All multigraphs with an odd number of points are nonstandard, as are connected multigraphs when the number of points is greater than two. A non-connected multigraph, however, is not necessarily standard. See Figure 5.1, (b), where the four points connected by lines making a square cannot be separated into two pairs with no line joining the two pairs.

Lemma 5.4.3 *Suppose the assumptions of Theorem 5.4.1. Fix a multigraph $A \in \mathcal{A}(m_1, \ldots, m_p)$, $m_1, \ldots, m_p \geq k$, and fix $j_1, \ldots, j_p \in \{1, \ldots, J\}$. Consider the sum $T_N(t_{j_1}, \ldots, t_{j_p}, m_1, \ldots, m_p, A)$ in (5.4.10). If A is a nonstandard multigraph, then as $N \to \infty$,*

$$T_N(t_{j_1}, \ldots, t_{j_p}, m_1, \ldots, m_p, A) = o(N^{p/2}). \tag{5.4.12}$$

Proof Denote $T_N(t_{j_1}, \ldots, t_{j_p}, m_1, \ldots, m_p, A)$ by $T_N(A)$. For simplicity, assume that $t_1 = \ldots = t_j = 1$ in (5.4.10). The proof is analogous in the general case. Also, without loss of generality, suppose that $m_1 \leq \cdots \leq m_p$. Note that

$$|T_N(A)| \leq \sum_{n_1,\ldots,n_p=1}^{N} \prod_{i=1}^{p} \prod_{\substack{w \in E(A):\\ d_1(w)=i}} |\gamma_X(n_i - n_{d_2(w)})|.$$

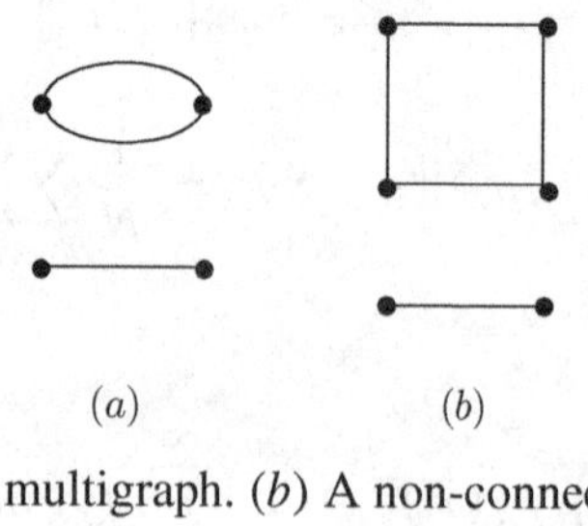

Figure 5.1 (a) A standard multigraph. (b) A non-connected and nonstandard multigraph.

For $i = 1, \ldots, p$, let $k_A(i) = \#\{w \in E(A) : d_1(w) = i\}$ be the number of lines incident to point $d_1(w) = i$ and which join points $d_2(w) > i$. By using the inequality $(a_1 \ldots a_k)^{1/k} \leq (a_1 + \cdots + a_k)/k$, $k \geq 1$, $a_i \geq 0$, between the geometric and arithmetic means, we have

$$\prod_{\substack{w \in E(A):\\ d_1(w)=i}} |\gamma_X(n_i - n_{d_2(w)})| = \left(\prod_{\substack{w \in E(A):\\ d_1(w)=i}} |\gamma_X(n_i - n_{d_2(w)})|^{k_A(i)}\right)^{1/k_A(i)}$$

$$\leq \frac{1}{k_A(i)} \sum_{\substack{w \in E(A):\\ d_1(w)=i}} |\gamma_X(n_i - n_{d_2(w)})|^{k_A(i)}.$$

Observe that

$$\sum_{n_i=1}^{N} |\gamma_X(n_i - n_{d_2(w)})|^{k_A(i)} \leq \sum_{|n| \leq N} |\gamma_X(n)|^{k_A(i)},$$

since $1 \leq n_i, n_{d_2(w)} \leq N$. Hence,

$$|T_N(A)| \leq \sum_{n_1,\ldots,n_p=1}^{N} \prod_{i=1}^{p} \frac{1}{k_A(i)} \sum_{\substack{w \in E(A):\\ d_1(w)=i}} |\gamma_X(n_i - n_{d_2(w)})|^{k_A(i)}$$

$$= \prod_{i=1}^{p} \sum_{n_i=1}^{N} \frac{1}{k_A(i)} \sum_{\substack{w \in E(A):\\ d_1(w)=i}} |\gamma_X(n_i - n_{d_2(w)})|^{k_A(i)}$$

$$\leq \prod_{i=1}^{p} \frac{1}{k_A(i)} \sum_{\substack{w \in E(A):\\ d_1(w)=i}} \sum_{|n| \leq N} |\gamma_X(n)|^{k_A(i)} = \prod_{i=1}^{p} \sum_{|n| \leq N} |\gamma_X(n)|^{k_A(i)}. \quad (5.4.13)$$

Let now $h_A(i) = k_A(i)/m_i$. If $k_A(i) = 0$, then $h_A(i) = 0$ and $\sum_{|n| \leq N} |\gamma_X(n)|^{k_A(i)} = 2N + 1 \leq CN = CN^{1-h_A(i)}$. If $k_A(i) = m_i$, then $h_A(i) = 1$ and $\sum_{|n| \leq N} |\gamma_X(n)|^{k_A(i)} = \sum_{|n| \leq N} |\gamma_X(n)|^{m_i} \leq C = CN^{1-h_A(i)}$, since $|\gamma_X(n)| \leq 1$ and $\sum_n |\gamma_X(n)|^{m_i} < \infty$ for $m_i \geq k$. Thus, if $k_A(i) = 0$ or $k_A(i) = m_i$, we have

$$\sum_{|n| \leq N} |\gamma_X(n)|^{k_A(i)} \leq CN^{1-h_A(i)}. \quad (5.4.14)$$

On the other hand, if $0 < k_A(i) < m_i$, then we want to show that

$$\sum_{|n| \leq N} |\gamma_X(n)|^{k_A(i)} = o(N^{1-h_A(i)}). \quad (5.4.15)$$

Indeed, for any fixed $\epsilon > 0$, there is $N(\epsilon)$ such that $\sum_{|n|>N(\epsilon)} |\gamma_X(n)|^{m_i} < \epsilon$. Then, by Hölder's inequality,

$$\sum_{|n|\leq N} |\gamma_X(n)|^{k_A(i)} \leq \sum_{|n|\leq N \wedge N(\epsilon)} |\gamma_X(n)|^{k_A(i)} + \sum_{N(\epsilon)<|n|\leq N} |\gamma_X(n)|^{k_A(i)}$$

$$\leq C(\epsilon) + \left(\sum_{N(\epsilon)<|n|\leq N} |\gamma_X(n)|^{m_i} \right)^{\frac{k_A(i)}{m_i}} \left(\sum_{N(\epsilon)<|n|\leq N} 1 \right)^{1-\frac{k_A(i)}{m_i}} \leq C(\epsilon) + C\epsilon^{h_A(i)} N^{1-h_A(i)}.$$

The relation (5.4.15) follows since $\epsilon > 0$ is arbitrarily small and $1 - h_A(i) > 0$.

For a nonstandard multigraph A, there is at least one i such that $0 < k_A(i) < m_i$. Combining (5.4.14) and (5.4.15), it follows from (5.4.13) that, for a nonstandard multigraph,

$$|T_N(A)| \leq o\left(N^{p-\sum_{i=1}^p h_A(i)}\right).$$

It is now enough to show that

$$\sum_{i=1}^{p} h_A(i) \geq \frac{p}{2}. \tag{5.4.16}$$

Let $p_1(w) = m_{d_1(w)}$ and $p_2(w) = m_{d_2(w)}$. They are the degrees of the points $d_1(w)$ and $d_2(w)$ joined by the line w. Since it is assumed in (5.4.10) that $d_1(w) \leq d_2(w)$ and since we assumed here that $m_1 \leq \cdots \leq m_p$, we have $p_1(w) \leq p_2(w)$. Then,

$$2\sum_{i=1}^{p} h_A(i) = 2\sum_{i=1}^{p} \frac{k_A(i)}{m_i} = 2\sum_{i=1}^{p} \sum_{\substack{w \in E(A): \\ d_1(w)=i}} \frac{1}{m_i} = 2\sum_{i=1}^{p} \sum_{\substack{w \in E(A): \\ d_1(w)=i}} \frac{1}{m_{d_1(w)}}$$

$$= 2\sum_{i=1}^{p} \sum_{\substack{w \in E(A): \\ d_1(w)=i}} \frac{1}{p_1(w)} = 2\sum_{w \in E(A)} \frac{1}{p_1(w)} \geq \sum_{w \in E(A)} \left(\frac{1}{p_1(w)} + \frac{1}{p_2(w)}\right).$$

For fixed i, the number of $w \in E(A)$ for which $d_1(w) = i$ or $d_2(w) = i$ (equivalently, $p_1(w) = m_i$ or $p_2(w) = m_i$) is exactly m_i. This implies that

$$\sum_{w \in E(A)} \left(\frac{1}{p_1(w)} + \frac{1}{p_2(w)}\right) = p$$

and hence the bound (5.4.16) follows. □

The following auxiliary lemma was also used above. Its proof is left as Exercise 5.5.

Lemma 5.4.4 Let $0 < a < b < \infty$ and $\sum_{n=1}^{\infty} |c(n)| < \infty$. Then, as $N \to \infty$,

$$\frac{1}{N} \sum_{n_1=1}^{[Na]} \sum_{n_2=[Na]+1}^{[Nb]} c(n_2 - n_1) \to 0. \tag{5.4.17}$$

The following result addresses the question raised in the beginning of this section.

Corollary 5.4.5 *Let $\{X_n\}_{n\in\mathbb{Z}}$ be a Gaussian stationary series which is LRD in the sense (2.1.5) in condition II with $d \in (0, 1/2)$. Suppose that $\mathbb{E}X_n = 0$, $\mathbb{E}X_n^2 = 1$. Let G be a function with Hermite rank $k \geq 1$ in the sense of Definition 5.2.1. If*

$$d \in \left(0, \frac{1}{2}\left(1 - \frac{1}{k}\right)\right), \tag{5.4.18}$$

then the convergence (5.4.3) to Brownian motion holds.

Proof The result follows from Theorem 5.4.1 by observing that the conditions (5.4.2) hold. Indeed, the second condition in (5.4.2) holds by condition II of LRD in (2.1.5). For the first condition in (5.4.2), Proposition 5.2.4 implies that

$$\gamma_G(n) \sim L(n) n^{k(2d-1)},$$

where L is a slowly varying function at infinity. Then, $\sum_{n=1}^{\infty} |\gamma_G(n)| < \infty$ if $k(2d-1)+1 < 0$ or $d < (1 - 1/k)/2$, which follows from the assumption (5.4.18). □

5.5 The Fourth Moment Condition

Nualart and Peccati [771] have obtained the following remarkable result: when trying to prove a central limit theorem for a multiple Wiener–Itô integral to converge to a Gaussian distribution, it is not necessary to prove convergence of all the moments to those of the Gaussian but only to prove convergence of moments (or equivalently, cumulants) up to order four. This then guarantees convergence of all the higher moments (or cumulants). There are also additional equivalent conditions involving convergence of contractions[4] $\|f_N \otimes_r f_N\|$, which are often easier to use in practice than moments.

Theorem 5.5.1 *Let $\xi_N = I_k(f_N)$ with symmetric $f_N \in L^2(E^k, m^k)$, $k \geq 2$. Suppose that*

$$\lim_{N\to\infty} k! \|f_N\|^2_{L^2(E^k, m^k)} = \lim_{N\to\infty} \mathbb{E}\xi_N^2 = 1. \tag{5.5.1}$$

Then the following conditions are equivalent:
1. $\lim_{N\to\infty} \chi_4(I_k(f_N)) = 0$.
2. *For every $r = 1, \ldots, k-1$,*

$$\lim_{N\to\infty} \|f_N \otimes_r f_N\|_{L^2(E^{2(k-r)}, m^{2(k-r)})} = 0.$$

3. *As $N \to \infty$, the sequence $\{I_k(f_N), N \geq 1\}$ converges in distribution towards $\mathcal{N}(0, 1)$.*

For a proof, see Peccati and Taqqu [797], p. 182. The condition (5.5.1) standardizes the random variables. An application to a special case of Theorem 5.4.1 can be found in Nourdin [760]. For a proof of Theorem 5.4.1 using Theorem 5.5.1, see Nourdin and Peccati [763].

---
[4] Contractions are defined in Appendix B.2. See the relation (B.2.17).

5.6 Limit Theorems in the Linear Case

In Sections 5.3 and 5.4, we presented limit theorems for partial sums of nonlinear functions of Gaussian stationary series with long-range dependence. We consider here extensions of the results to nonlinear functions of linear time series with long-range dependence. Linear time series with long-range dependence will be defined as in condition I of Section 2.1 with i.i.d. innovations $\{\mathbb{Z}_n\}$ and supposing $\mu = 0$. We thus consider:

Condition L The time series $\{X_n\}_{n \in \mathbb{Z}}$ has a linear representation

$$X_n = \sum_{k=0}^{\infty} \psi_k \epsilon_{n-k}$$

with i.i.d. innovations $\epsilon_n, n \in \mathbb{Z}$, satisfying $\mathbb{E}\epsilon_n = 0$, $\mathbb{E}\epsilon_n^2 < \infty$ and a sequence $\{\psi_k\}_{k \geq 0}$ satisfying

$$\psi_k = L_1(k) k^{d-1}, \qquad (5.6.1)$$

where $d \in (0, 1/2)$ and L_1 is a slowly varying function at infinity.

We are interested in the limits of partial sums (5.0.1) when $\{X_n\}$ satisfies condition L. Limit theorems based on two different approaches are given in Sections 5.6.1 and 5.6.2 below. In both cases, the focus is on noncentral limit theorems.

5.6.1 Direct Approach for Entire Functions

In this approach, G in (5.0.1) is supposed to be an entire function; that is, analytic on the whole complex plane, expressed as

$$G(z) = \sum_{k=0}^{\infty} c_k z^k, \quad z \in \mathbb{C}. \qquad (5.6.2)$$

We shall use the notion of the so-called *exponent rank* of G. It will play a role similar to that of the Hermite rank in the context of Gaussian series.

Definition 5.6.1 We shall say that G is of *exponent rank* $m \geq 1$ (with respect to X_1) if m is the smallest integer such that

$$\mathbb{E} G^{(m)}(X_1) \neq 0, \qquad (5.6.3)$$

where $G^{(m)}$ denotes the mth derivative of G. The corresponding *rank coefficient* is $\mathbb{E} G^{(m)}(X_1)$.

Remark 5.6.2 Note that, when X_1 is a standard Gaussian variable and G satisfies suitable smoothness and integrability assumptions, the exponent rank and the Hermite rank coincide. Indeed, by differentiating G in (5.1.6) m times, we get

$$G^{(m)}(x) = \sum_{n=m}^{\infty} g_n H_n^{(m)}(x)$$

or, by using the property $H_n^{(1)}(x) = nH_{n-1}(x)$ of Hermite polynomials,

$$G^{(m)}(x) = \sum_{n=m}^{\infty} g_n n(n-1)\ldots(n-m+1)H_{n-m}(x) = g_m m! + \sum_{k=1}^{\infty} g_{m+k}\frac{(m+k)!}{k!}H_k(x).$$

This yields $\mathbb{E}G^{(m)}(X_1) = g_m m!$. Hence both the Hermite rank and the exponent rank equal m, if m is the smallest integer such that $g_n = 0$ for all $n < m$. The preceding argument clearly holds for polynomial functions G. One can make fewer restrictions on G; see, for example, Proposition 5.8.4 below.

The next result, due to Surgailis [935], is a non-central limit theorem for partial sums of functions of linear time series. The following sketch of proof outlines, in particular, how the exponent rank arises.

Theorem 5.6.3 *Let $\{X_n\}_{n\in\mathbb{Z}}$ be a linear time series satisfying condition L above with LRD parameter d. Let G be an entire function defined by (5.6.2) such that its exponent rank (with respect to X_1) is m and*

$$\sum_{j,k=0}^{\infty} |c_j||c_k|(j!k!)^2 2^{2(j+k)}\bar{\mu}_{j+k} < \infty,$$

where $\bar{\mu}_k = \mathbb{E}|\epsilon_1|^k$, $k \geq 0$. If

$$d \in \left(\frac{1}{2}\left(1 - \frac{1}{m}\right), \frac{1}{2}\right),$$

then, as $N \to \infty$,

$$\frac{1}{(L_1(N))^m N^{m(d-1/2)+1}} \sum_{n=1}^{[Nt]}(G(X_n) - \mathbb{E}G(X_n)), \quad t \geq 0,$$

has the same limit in finite-dimensional distributions as

$$\mathbb{E}G^{(m)}(X_1)\frac{Y_{[Nt],m}}{(L_1(N))^m N^{m(d-1/2)+1}}, \quad t \geq 0,$$

where

$$Y_{[Nt],m} = \sum_{n=1}^{[Nt]} \sum_{0 \leq j_1 < j_2 < \cdots < j_m < \infty} \psi_{j_1}\epsilon_{n-j_1}\ldots\psi_{j_m}\epsilon_{n-j_m}. \quad (5.6.4)$$

Moreover, that limit is

$$\frac{(\mathbb{E}G^{(m)}(X_1))}{m!A_{m,H}}Z_H^{(m)}(t),$$

where $Z_H^{(m)} = \{Z_H^{(m)}(t)\}_{t\geq 0}$ is the standard Hermite process (4.2.42) of order m with the self-similarity parameter

$$H = m\left(d - \frac{1}{2}\right) + 1 \in \left(\frac{1}{2}, 1\right),$$

and $A_{m,H}$ is given by (4.2.7).

Sketch of Proof Note from (5.6.2) that $G(X_n) = \sum_{k=0}^{\infty} c_k (X_n)^k$. The idea of the proof consists of decomposing

$$(X_n)^k = \sum_{p_1, p_2, \ldots, p_k = 0}^{\infty} \psi_{p_1} \psi_{p_2} \cdots \psi_{p_k} \epsilon_{n-p_1} \epsilon_{n-p_2} \cdots \epsilon_{n-p_k} \quad (5.6.5)$$

in terms of the cardinality $|\{p_1, \ldots, p_k\}|$ of the set $\{p_1, \ldots, p_k\}$. When $|\{p_1, \ldots, p_k\}| = k$, the term $\psi_{p_1} \psi_{p_2} \cdots \psi_{p_k} \epsilon_{n-p_1} \epsilon_{n-p_2} \cdots \epsilon_{n-p_k}$ is not modified. When $|\{p_1, \ldots, p_k\}| < k$ and for instance equal to $k-1$ with $p_1 = p_2$, it is split in two parts:

$$\psi_{p_1} \psi_{p_2} \cdots \psi_{p_k} (\epsilon_{n-p_1} \epsilon_{n-p_2} \cdots \epsilon_{n-p_k}) = \psi_{p_1}^2 \psi_{p_3} \cdots \psi_{p_k} (\epsilon_{n-p_1}^2 \epsilon_{n-p_3} \cdots \epsilon_{n-p_k})$$
$$= \psi_{p_1}^2 \psi_{p_3} \cdots \psi_{p_k} (\mu_2 \epsilon_{n-p_3} \cdots \epsilon_{n-p_k}) + \psi_{p_1}^2 \psi_{p_3} \cdots \psi_{p_k} (\eta_{p_1}(2) \epsilon_{n-p_3} \cdots \epsilon_{n-p_k}),$$
$$(5.6.6)$$

where $\mu_\ell = \mathbb{E}\epsilon_1^\ell$ and $\eta_p(\ell) = \epsilon_p^\ell - \mu_\ell$. One then shows that the second term with $\eta_{p_1}(2)$ is negligible in the limit theorem consideration since it exhibits short-range dependence. More generally, $(X_n)^k$ can be written as

$$(X_n)^k = \sum_{\ell=0}^{k} \binom{k}{\ell} \sum_{(p)_\ell} \psi_{p_1} \epsilon_{n-p_1} \cdots \psi_{p_\ell} \epsilon_{n-p_\ell} \sum_{(V)(k-\ell)} \sum_{(q)_r : (q)_r \cap (p)_\ell = \emptyset} \psi_{q_1}^{v_1} \epsilon_{n-q_1}^{v_1} \cdots \psi_{q_r}^{v_r} \epsilon_{n-q_r}^{v_r},$$
$$(5.6.7)$$

where the summation over $(p)_\ell$ corresponds to the summation over the sets $\{p_1, \ldots, p_\ell\}$ of cardinality $|\{p_1, \ldots, p_\ell\}| = \ell$, that is, over $p_1, \ldots, p_\ell$ which take different values. The sum $\sum_{(V)(k-\ell)}$ is taken over the rest, namely, over all partitions of the set $\{1, \ldots, k-\ell\}$ of cardinality $v_1, \ldots, v_r$ such that $v_i \geq 2$, for all i. By convention, this sum equals 1 for $(V)(0)$. The sum over $(q)_r : (q)_r \cap (p)_\ell = \emptyset$ is the sum over $\{q_1, \ldots, q_r\}$ of cardinality r such that $\{q_1, \ldots, q_r\} \cap \{p_1, \ldots, p_\ell\} = \emptyset$. Thus in (5.6.8), the qs involve different indices than the ps.

As is done in (5.6.6), we now introduce centered random variables in (5.6.7) and thus write $\epsilon_{n-q_1}^{v_1} = \eta_{q_1}(v_1) + \mu_{v_1} \ldots, \epsilon_{n-q_r}^{v_r} = \eta_{q_r}(v_r) + \mu_{v_r}$. Plugging this in (5.6.7) and expanding the product, we drop the terms involving any centered random variables $\eta_{q_1}(v_1), \ldots, \eta_{q_r}(v_r)$ (e.g., the second term in (5.6.6)). These terms can be shown to be negligible since they exhibit short-range dependence. This results in

$$(X_n)_1^k = \sum_{\ell=0}^{k} \binom{k}{\ell} \sum_{(p)_\ell} \psi_{p_1} \epsilon_{n-p_1} \cdots \psi_{p_\ell} \epsilon_{n-p_\ell} \sum_{(V)(k-\ell)} \sum_{(q)_r : (q)_r \cap (p)_\ell = \emptyset} \psi_{q_1}^{v_1} \mu_{v_1} \cdots \psi_{q_r}^{v_r} \mu_{v_r}.$$
$$(5.6.8)$$

So the difference between $(X_n)^k$ in (5.6.7) and $(X_n)_1^k$ in (5.6.8) is that when there is an ϵ_{n-p}^ℓ with $\ell > 1$ in (5.6.5), it is replaced by $\mu_\ell = \mathbb{E}\epsilon_1^\ell$ in (5.6.8). Then we replace it further by

$$(Z_n)_1^k = \sum_{\ell=0}^{k} \binom{k}{\ell} \sum_{(p)_\ell} \psi_{p_1} \epsilon_{n-p_1} \cdots \psi_{p_\ell} \epsilon_{n-p_\ell} \sum_{(V)(k-\ell)} \sum_{(q)_r} \psi_{q_1}^{v_1} \mu_{v_1} \cdots \psi_{q_r}^{v_r} \mu_{v_r}. \quad (5.6.9)$$

$(Z_n)_1^k$ further modifies (5.6.8) by removing the condition $(q)_r \cap (p)_\ell = \emptyset$. In contrast to $(X_n)^k$, the notation $(Z_n)_1^k$ is a shorthand for the right-hand side of (5.6.9) and does not mean Z_n to the power k. The difference between (5.6.8) and (5.6.9) is that in (5.6.9), we also include qs that have same values as the ps.

Observe that the summands of (5.6.8) and $(Z_n)_1^k$ with $\ell > 0$ have zero mean. Then,

$$\mathbb{E}(X_n)^k = \mathbb{E}(Z_n)_1^k = \sum_{(V)(k)} \sum_{(q)_r} \psi_{q_1}^{v_1} \mu_{v_1} \cdots \psi_{q_r}^{v_r} \mu_{v_r},$$

for $k \geq 0$ and hence for $0 \leq s \leq k$,

$$\mathbb{E}(X_n)^{k-s} = \mathbb{E}(Z_n)_1^{k-s} = \sum_{(V)(k-s)} \sum_{(q)_r} \psi_{q_1}^{v_1} \mu_{v_1} \cdots \psi_{q_r}^{v_r} \mu_{v_r}. \quad (5.6.10)$$

Let us now define formally $G_1(Z_n) = \sum_{s \geq 0} c_s (Z_n)_1^s$, where $(Z_n)_1^k$ is given in (5.6.9), and prove that

$$G_1(Z_n) = \sum_{s \geq 0} c_s (Z_n)_1^s = \sum_{s \geq 0} \frac{e_s}{s!} \sum_{(p)_s} \psi_{p_1} \epsilon_{n-p_1} \cdots \psi_{p_s} \epsilon_{n-p_s}, \quad (5.6.11)$$

where

$$e_s = \mathbb{E} G^{(s)}(X_n).$$

Observe that by (5.6.10),

$$\frac{e_s}{s!} = \frac{\mathbb{E} G^{(s)}(X_n)}{s!} = \sum_{k \geq s} c_k \binom{k}{s} \mathbb{E}(X_n)^{k-s} = \sum_{k \geq s} c_k \binom{k}{s} \mathbb{E}(Z_n)_1^{k-s}.$$

Using this and again (5.6.10), note that the right-hand side of (5.6.11) can be expressed as

$$\sum_{s \geq 0} \frac{e_s}{s!} \sum_{(p)_s} \psi_{p_1} \epsilon_{n-p_1} \cdots \psi_{p_s} \epsilon_{n-p_s}$$

$$= \sum_{s \geq 0} \left(\sum_{k \geq s} c_k \binom{k}{s} \sum_{(V)(k-s)} \sum_{(q)_r} \psi_{q_1}^{v_1} \mu_{v_1} \cdots \psi_{q_r}^{v_r} \mu_{v_r} \right) \sum_{(p)_s} \psi_{p_1} \epsilon_{n-p_1} \cdots \psi_{p_s} \epsilon_{n-p_s}$$

$$= \sum_{k \geq 0} c_k \left(\sum_{s=0}^{k} \binom{k}{s} \sum_{(p)_s} \psi_{p_1} \epsilon_{n-p_1} \cdots \psi_{p_s} \epsilon_{n-p_s} \sum_{(V)(k-s)} \sum_{(q)_r} \psi_{q_1}^{v_1} \mu_{v_1} \cdots \psi_{q_r}^{v_r} \mu_{v_r} \right) = \sum_{k \geq 0} c_k (Z_n)_1^k,$$

by (5.6.9), hence proving (5.6.11).

The next step in the proof consists in showing that $\sum_{n=1}^{[Nt]} G(X_n)$ can be replaced by $\sum_{n=1}^{[Nt]} G_1(Z_n)$ and that the leading term in $\sum_{n=1}^{[Nt]} G_1(Z_n)$ is the term corresponding to $s = m$, that is by (5.6.11),

$$e_m Y_{[Nt],m} = \frac{e_m}{m!} \sum_{n=1}^{[Nt]} \sum_{(p)_m} \psi_{p_1} \epsilon_{n-p_1} \cdots \psi_{p_m} \epsilon_{n-p_m}.$$

More precisely, it is proved in Lemmas 2 and 3 of Surgailis [935] that

$$\mathbb{E} \left(\sum_{n=1}^{N} (G(X_n) - G_1(Z_n)) \right)^2 \leq CN,$$

where C is a positive constant and that

$$\mathbb{E}\Big(\sum_{n=1}^{N}\sum_{\ell\geq m+1}\frac{e_\ell}{\ell!}\sum_{(p)_\ell}\psi_{p_1}\epsilon_{p_1}\cdots\psi_{p_\ell}\epsilon_{p_\ell}\Big)^2 = \mathbb{E}\Big(\sum_{\ell\geq m+1}e_\ell Y_{N,\ell}\Big)^2 = o\Big((L_1(N))^{2m}N^{m(2d-1)+2}\Big),$$

as N tends to infinity.

To conclude, it remains to study the asymptotic behavior of the main term $e_m Y_{[Nt],m}$ and show that, as $N \to \infty$,

$$e_m \frac{Y_{[Nt],m}}{(L_1(N))^m N^{m(d-1/2)+1}} \xrightarrow{fdd} e_m Z_H^{(m)}(t).$$

The basic idea is to express $Y_{[Nt],m}$ as a discrete multiple stochastic integral as follows. Note that

$$\frac{Y_{[Nt],m}}{(L_1(N))^m N^{m(d-1/2)+1}} = \frac{1}{(L_1(N))^m N^{m(d-1/2)+1}}\sum_{n=1}^{[Nt]}\sum_{i_m<\cdots<i_2<i_1\leq n}\prod_{s=1}^{m}\psi_{n-i_s}\epsilon_{i_s}$$

$$= \sum_{i_1,i_2,\ldots,i_m}h_{t,N}(i_1,i_2,\ldots,i_m)\epsilon_{i_1}\epsilon_{i_2}\cdots\epsilon_{i_m} =: J_m(h_{t,N}),$$
(5.6.12)

where

$$h_{t,N}(i_1,\ldots,i_m) = \begin{cases} N^{-m/2}\sum_{n=1}^{[Nt]}\Big(\prod_{s=1}^{m}\frac{\psi_{n-i_s}}{L(N)N^{d-1}}1_{\{i_s\leq n\}}\Big)\frac{1}{N}, & \text{if } i_m<\cdots<i_2<i_1, \\ 0, & \text{otherwise.} \end{cases}$$
(5.6.13)

In view of (5.6.13), there is effectively no summation on diagonals in (5.6.12). The sums $J_m(h)$ as in (5.6.12) are known as discrete multiple integrals, and have similar second-order properties as multiple integrals $I_m(g)$ (with respect to Gaussian random measures). By Proposition 5.6.4, if

$$\tilde{h}_N(v_1,\ldots,v_m) = N^{m/2}h_N([v_1N],\ldots,[v_mN])$$

converges in $L^2(\mathbb{R}^m)$ to some function h, then $J_m(h_N) \xrightarrow{d} I_m(h)$. In the case (5.6.13) considered here, it can be shown[5] (and certainly seen informally) that

$$N^{m/2}h_{t,N}([Nv_1],\ldots,[Nv_m]) \to h_t(v_1,\ldots,v_m)$$
$$= \begin{cases} \int_0^t \prod_{s=1}^m (u-v_s)_+^{d-1}du, & \text{if } v_m<\cdots<v_1, \\ 0, & \text{otherwise} \end{cases}$$

in $L^2(\mathbb{R}^m)$, and hence that

$$e_m \frac{Y_{[Nt],m}}{(L_1(N))^m N^{1+m(d-1/2)}} \xrightarrow{fdd} \frac{e_m}{m!}\int_{\mathbb{R}^m}'\Big\{\int_0^t \prod_{j=1}^m (u-v_j)_+^{d-1}du\Big\}B(dv_1)\ldots B(dv_m),$$

which yields the desired result. $\square$

We used the following fact in the proof of Theorem 5.6.3. Let h be a function defined in $\mathbb{Z}^m$ such that $\sum'_{(i_1,\ldots,i_m)\in\mathbb{Z}^m}h(i_1,\ldots,i_m)^2 < \infty$, where $'$ indicates the exclusion of the

---

[5] See (4.7.21) in Giraitis, Koul, and Surgailis [406].

diagonals $i_p = i_q$, $p \neq q$, and let $\{\epsilon_i, i \in \mathbb{Z}\}$ be the standardized i.i.d. random variables. The following result relates the *discrete multiple stochastic integral*

$$J_m(h) = \sum_{(i_1,\ldots,i_m)\in\mathbb{Z}^m}{}' h(i_1,\ldots,i_m)\epsilon_{i_1}\cdots\epsilon_{i_m} \tag{5.6.14}$$

to a continuous one.

Proposition 5.6.4 *Let $h_{j,N}$ be square-summable functions defined on $\mathbb{Z}^{m_j}$, $j = 1,\ldots,k$. Let*

$$\tilde{h}_{j,N}(x_1,\ldots,x_{m_j}) = N^{m_j/2} h_{j,N}\left([Nx_1],\ldots,[Nx_{m_j}]\right), \quad j = 1,\ldots,k.$$

Suppose that there exist $h_j \in L^2(\mathbb{R}^{m_j})$, $j = 1,\ldots,k$, such that

$$\|\tilde{h}_{j,N} - h_j\|_{L^2(\mathbb{R}^{m_j})} \to 0$$

as $N \to \infty$. Then, as $N \to \infty$,

$$\left(J_{m_1}(h_{1,N}),\ldots,J_{m_k}(h_{k,N})\right) \xrightarrow{d} \left(I_{m_1}(h_1),\ldots,I_{m_k}(h_k)\right),$$

where each $I_{m_j}(h_j) = \int'_{\mathbb{R}^{m_j}} h_j(x_1,,\ldots,x_{m_j}) B(dx_1)\ldots B(dx_{m_j})$, $j = 1,\ldots,k$, denotes the m_j-tuple Wiener–Itô integral with respect to the same standard Brownian motion.

For a proof, see Surgailis [935], Lemma 7 or Giraitis et al. [406], Proposition 14.3.2. The proof is similar to that of Proposition 5.3.6 where h and h_ϵ are approximated by "simple" functions, and then the convergence of $N^{-1/2}\sum_{k=1}^{[Nu]}\epsilon_k$ to Brownian motion is used.

5.6.2 Approach Based on Martingale Differences

We shall study the asymptotic behavior of $\sum_{n=1}^{[Nt]} G(X_n)$ using a martingale difference approach. We need to introduce first some notation. Let F be the distribution of the linear time series $X_n = \sum_{k=0}^{\infty} \psi_k \epsilon_{n-k}$ and F_j the distribution of the truncated series

$$X_{n,j} = \sum_{k=0}^{j-1} \psi_k \epsilon_{n-k},$$

defined for $j \geq 0$, with the convention $X_{n,0} = 0$. Let

$$G_j(x) = \int_{\mathbb{R}} G(x+y) dF_j(y), \quad G_\infty(x) = \int_{\mathbb{R}} G(x+y) dF(y), \tag{5.6.15}$$

and

$$G_j^{(r)}(x) = \frac{d^r}{dx^r}\int_{\mathbb{R}} G(x+y) dF_j(y), \quad G_\infty^{(r)}(x) = \frac{d^r}{dx^r}\int_{\mathbb{R}} G(x+y) dF(y). \tag{5.6.16}$$

If the rth derivative $G_j^{(r)}$ of G_j exists, define

$$G_{j,\lambda}^{(r)}(x) = \sup_{|y|\leq\lambda} |G_j^{(r)}(x+y)|, \quad \lambda \geq 0.$$

We shall say that G satisfies *Condition $C(r, j, \lambda)$* if

1. $G_j^{(r)}(x)$ exists for all x and $G_j^{(r)}$ is continuous.
2. For all $x \in \mathbb{R}$,

$$\sup_{I \subset \{0,1,\ldots\}} \mathbb{E}\left\{G_{j,\lambda}^{(r)}\left(x + \sum_{i \in I} \psi_i \epsilon_i\right)\right\}^4 < \infty,$$

where the supremum is taken over all subsets I of $\{0, 1, \ldots\}$.

Let us comment on Condition $C(r, j, \lambda)$. It is satisfied if the rth derivative of G is bounded and continuous, in which case the condition does not depend on j. Moreover, if G is any polynomial, then $C(r, j, \lambda)$ holds provided that ϵ_i has finite moments of sufficiently high order.

The novelty here is that $C(r, j, \lambda)$ can hold without G being smooth. An important example is the indicator function.

Example 5.6.5 (*Condition $C(r, 1, \lambda)$ and the indicator function*) If $G(x) = 1_{\{x \leq u\}}$, for some fixed u, let us prove that G satisfies $C(r, 1, \lambda)$ for all positive λ as soon as the probability density function g of ϵ_1 has a continuous and integrable rth derivative.

Since $X_{n,1} = \psi_0 \epsilon_n$ has distribution F_1, we have

$$G_1(x) = \int_\mathbb{R} G(x+y) dF_1(y) = \int_\mathbb{R} G(x+\psi_0 y_1) g(y_1) dy_1 = \psi_0^{-1} \int_\mathbb{R} G(z) g\left(\frac{z-x}{\psi_0}\right) dz.$$

Note that

$$\frac{d^r}{dx^r}\left\{G(z) g\left(\frac{z-x}{\psi_0}\right)\right\} = \frac{(-1)^r}{\psi_0^r} G(z) g^{(r)}\left(\frac{z-x}{\psi_0}\right).$$

Since, by assumption, $\int_\mathbb{R} |g^{(r)}(y)| dy < \infty$, we get

$$G_1^{(r)}(x) = \frac{(-1)^r}{\psi_0^{r+1}} \int_\mathbb{R} G(z) g^{(r)}\left(\frac{z-x}{\psi_0}\right) dz = \frac{(-1)^r}{\psi_0^r} \int_\mathbb{R} G(x+\psi_0 y_1) g^{(r)}(y_1) dy_1.$$

Moreover, $G_1^{(r)}$ is a continuous function since $g^{(r)}$ is assumed to be a continuous function. This gives (5.6.2) of $C(r, 1, \lambda)$. Let us now check (5.6.2) of $C(r, 1, \lambda)$. For all subset I of $\{1, 2, \ldots\}$, we have

$$\mathbb{E}\left\{\sup_{|y| \leq \lambda} \left|G_1^{(r)}\left(x + \sum_{i \in I} \psi_i \epsilon_i + y\right)\right|\right\}^4$$

$$= \frac{1}{\psi_0^{4r}} \mathbb{E}\left\{\sup_{|y| \leq \lambda} \left|\int_\mathbb{R} G\left(x + \sum_{i \in I} \psi_i \epsilon_i + y + \psi_0 y_1\right) g^{(r)}(y_1) dy_1\right|\right\}^4,$$

which is bounded by $\psi_0^{-4r} \left(\int_\mathbb{R} |g^{(r)}(y_1)| dy_1\right)^4 < \infty$. Thus ensures that $G(x) = 1_{\{x \leq u\}}$ satisfies Condition $C(r, 1, \lambda)$.

Observe that the indicator function $h(x) = 1_{\{x \leq u\}}$ is allowed in the Gaussian case (Section 5.3) but not in the situation considered in Section 5.6.1. As we have just seen, it is allowed in the methodology discussed in this section.

5.6 Limit Theorems in the Linear Case

The idea here is to use a decomposition into martingale differences (also known as a mixingale decomposition), as explained in the sketch of the proof below, and to prove that the leading term is once again $Y_{[Nt],m}$, defined in (5.6.4), where here m is the power rank of G defined as follows.

Definition 5.6.6 We shall say that G is of *power rank* $m \geq 1$ (with respect to X_1) if it is the smallest integer such that

$$G_\infty^{(m)}(0) = \frac{d^m}{dx^m} \mathbb{E} G(x + X_1)\Big|_{x=0} \neq 0. \tag{5.6.17}$$

The corresponding *rank coefficient* is $G_\infty^{(m)}(0)$.

Remark 5.6.7 If G satisfies suitable smoothness and integrability assumptions, the differentiation in (5.6.17) can be taken inside the expectation leading to

$$G_\infty^{(m)}(0) = \mathbb{E} G^{(m)}(X_1),$$

that is, the exponent and power ranks of the function G are the same. This is clearly the case for polynomial functions G.

Here is the precise result due to Ho and Hsing [477].

Theorem 5.6.8 *Let $\{X_n\}_{n \in \mathbb{Z}}$ be a linear time series satisfying condition L above with LRD parameter d. Suppose that the power rank (with respect to X_1) of a function G is m. Suppose also that, for some j and λ, the condition $C(r, j, \lambda)$ holds for $r = 0, 1, \ldots, m + 2$ and $\mathbb{E} G(X_1)^2 < \infty$. If*

$$d \in \left(\frac{1}{2}\left(1 - \frac{1}{m}\right), \frac{1}{2}\right)$$

and $\mathbb{E}|\epsilon_1|^{2m \vee 8} < \infty$, then, as $N \to \infty$,

$$\frac{1}{(L_1(N))^m N^{m(d-1/2)+1}} \sum_{n=1}^{[Nt]} (G(X_n) - \mathbb{E} G(X_n)), \quad t \geq 0,$$

has the same limit in finite-dimensional distributions as

$$G_\infty^{(m)}(0) \frac{Y_{[Nt],m}}{(L_1(N))^m N^{m(d-1/2)+1}}, \quad t \geq 0,$$

where $Y_{[Nt],m}$ is defined in (5.6.4). Moreover, that limit is

$$\frac{G_\infty^{(m)}(0)}{m! A_{m,H}} Z_H^{(m)}(t),$$

where $Z_H^{(m)} = \{Z_H^{(m)}(t)\}_{t \geq 0}$ is the standard Hermite process (4.2.42) of order m with the self-similarity parameter

$$H = m\left(d - \frac{1}{2}\right) + 1 \in \left(\frac{1}{2}, 1\right),$$

and $A_{m,H}$ is given by (4.2.7).

Sketch of Proof The second parts of Theorems 5.6.3 and 5.6.8 are similar. This shows that the key is to reduce the original $\sum_{n=1}^{[Nt]} G(X_n)$ to $Y_{[Nt],m}$. It is $Y_{[Nt],m}$ which converges to the limit after suitable normalization.

The idea of the proof is to condition on the σ–fields
$$\mathcal{F}_k = \sigma\{\epsilon_i : i \leq k\},$$
using the telescoping expression
$$G(X_n) - \mathbb{E}G(X_n) = \sum_{j \geq 0} \Big(\mathbb{E}(G(X_n)|\mathcal{F}_{n-j}) - \mathbb{E}(G(X_n)|\mathcal{F}_{n-j-1})\Big), \qquad (5.6.18)$$
since the extreme summands are such that $\mathbb{E}(G(X_n)|\mathcal{F}_n) = G(X_n)$ and $\lim_{m \to -\infty} \mathbb{E}(G(X_n)|\mathcal{F}_m) = \mathbb{E}(G(X_n)|\mathcal{F}_{-\infty}) = \mathbb{E}G(X_n)$ by the martingale convergence theorem (e.g., Gut [442], Corollary 12.1 in Chapter 10). Now write
$$\mathbb{E}(G(X_n)|\mathcal{F}_{n-j}) = \mathbb{E}(G(X_{n,j} + \tilde{X}_{n,j})|\mathcal{F}_{n-j}),$$
with
$$X_{n,j} = \sum_{k=0}^{j-1} \psi_k \epsilon_{n-k} \quad \text{and} \quad \tilde{X}_{n,j} = \sum_{k \geq j} \psi_k \epsilon_{n-k}.$$
Since $\tilde{X}_{n,j}$ is $\mathcal{F}_{n-j}$–measurable and $X_{n,j}$ is independent of $\mathcal{F}_{n-j}$,
$$\mathbb{E}(G(X_n)|\mathcal{F}_{n-j}) = \int_{\mathbb{R}} G(x + \tilde{X}_{n,j}) dF_j(x) = G_j(\tilde{X}_{n,j}), \qquad (5.6.19)$$
where F_j is the distribution of $X_{n,j}$ and $G_j(y) = \mathbb{E}G(X_{n,j}+y)$. Using (5.6.18) and (5.6.19), we get that
$$\sum_{n=1}^{[Nt]}(G(X_n) - \mathbb{E}G(X_n)) = \sum_{n=1}^{[Nt]} \sum_{j_1 \geq 0} (G_{j_1}(\tilde{X}_{n,j_1}) - G_{j_1+1}(\tilde{X}_{n,j_1+1}))$$
$$= \sum_{n=1}^{[Nt]} \sum_{j_1 \geq 0} (\tilde{X}_{n,j_1} - \tilde{X}_{n,j_1+1}) G^{(1)}_{j_1+1}(\tilde{X}_{n,j_1+1})$$
$$+ \Big(G_{j_1}(\tilde{X}_{n,j_1}) - G_{j_1+1}(\tilde{X}_{n,j_1+1})$$
$$- (\tilde{X}_{n,j_1} - \tilde{X}_{n,j_1+1}) G^{(1)}_{j_1+1}(\tilde{X}_{n,j_1+1})\Big)$$
$$= \sum_{n=1}^{[Nt]} \sum_{j_1 \geq 0} \psi_{j_1} \epsilon_{n-j_1} G^{(1)}_{j_1+1}(\tilde{X}_{n,j_1+1})$$
$$+ \Big(G_{j_1}(\tilde{X}_{n,j_1}) - G_{j_1+1}(\tilde{X}_{n,j_1+1}) - \psi_{j_1} \epsilon_{n-j_1} G^{(1)}_{j_1+1}(\tilde{X}_{n,j_1+1})\Big) \qquad (5.6.20)$$
$$\approx \sum_{n=1}^{[Nt]} \sum_{j_1 \geq 0} \psi_{j_1} \epsilon_{n-j_1} G^{(1)}_{j_1+1}(\tilde{X}_{n,j_1+1}), \qquad (5.6.21)$$

after proving that the terms in the parentheses can be neglected.

We have introduced the ϵs in (5.6.21). We need now to introduce $G_\infty^{(1)}(0), \ldots, G_\infty^{(m)}(0)$. To do so, we express the summands in the remaining term (5.6.21) as

$$\psi_{j_1}\epsilon_{n-j_1} G_{j_1+1}^{(1)}(\tilde{X}_{n,j_1+1}) = \psi_{j_1}\epsilon_{n-j_1} G_\infty^{(1)}(0) + \psi_{j_1}\epsilon_{n-j_1}(G_{j_1+1}^{(1)}(\tilde{X}_{n,j_1+1}) - G_\infty^{(1)}(0))$$
$$= \psi_{j_1}\epsilon_{n-j_1} G_\infty^{(1)}(0) + \psi_{j_1}\epsilon_{n-j_1} \sum_{j_2 \geq j_1+1} (G_{j_2}^{(1)}(\tilde{X}_{n,j_2}) - G_{j_2+1}^{(1)}(\tilde{X}_{n,j_2+1})). \quad (5.6.22)$$

Focusing on the term in brackets, we write as before (see (5.6.20)),

$$G_{j_2}^{(1)}(\tilde{X}_{n,j_2}) - G_{j_2+1}^{(1)}(\tilde{X}_{n,j_2+1}) = \psi_{j_2}\epsilon_{n-j_2} G_{j_2+1}^{(2)}(\tilde{X}_{n,j_2+1})$$
$$+ \left(G_{j_2}^{(1)}(\tilde{X}_{n,j_2}) - G_{j_2+1}^{(1)}(\tilde{X}_{n,j_2+1}) - \psi_{j_2}\epsilon_{n-j_2} G_{j_2+1}^{(2)}(\tilde{X}_{n,j_2+1})\right)$$
$$\approx \psi_{j_2}\epsilon_{n-j_2} G_{j_2+1}^{(2)}(\tilde{X}_{n,j_2+1}) = \psi_{j_2}\epsilon_{n-j_2} G_\infty^{(2)}(0) + \psi_{j_2}\epsilon_{n-j_2}(G_{j_2+1}^{(2)}(\tilde{X}_{n,j_2+1}) - G_\infty^{(2)}(0)),$$
$$(5.6.23)$$

where in that last equality, we proceeded as in (5.6.22). Relations (5.6.22) and (5.6.23) yield

$$\psi_{j_1}\epsilon_{n-j_1} G_{j_1+1}^{(1)}(\tilde{X}_{n,j_1+1}) \approx \psi_{j_1}\epsilon_{n-j_1} G_\infty^{(1)}(0)$$
$$+ \psi_{j_1}\epsilon_{n-j_1} \sum_{j_2 \geq j_1+1} \left(\psi_{j_2}\epsilon_{n-j_2} G_\infty^{(2)}(0) + \psi_{j_2}\epsilon_{n-j_2}(G_{j_2+1}^{(2)}(\tilde{X}_{n,j_2+1}) - G_\infty^{(2)}(0))\right).$$

Thus, we get

$$\sum_{n=1}^{[Nt]}(G(X_n) - \mathbb{E}G(X_n)) = Y_{[Nt],1} G_\infty^{(1)}(0) + Y_{[Nt],2} G_\infty^{(2)}(0)$$
$$+ \sum_{n=1}^{[Nt]} \sum_{j_2 > j_1 \geq 0} \psi_{j_1}\epsilon_{n-j_1} \psi_{j_2}\epsilon_{n-j_2} (G_{j_2}^{(2)}(\tilde{X}_{n,j_2}) - G_\infty^{(2)}(0)), \quad (5.6.24)$$

where

$$Y_{N,r} = \sum_{n=1}^{N} \sum_{0 \leq j_1 < j_2 < \cdots < j_r} \prod_{s=1}^{r} \psi_{j_s}\epsilon_{n-j_s}.$$

Iterating, we get

$$\sum_{n=1}^{[Nt]}(G(X_n) - \mathbb{E}G(X_n)) = Y_{[Nt],1} G_\infty^{(1)}(0) + Y_{[Nt],2} G_\infty^{(2)}(0) + \cdots + Y_{[Nt],m} G_\infty^{(m)}(0) + R_N,$$
$$(5.6.25)$$

where R_N is shown to be a negligible remainder term in Ho and Hsing [477]. Hence, the first term of the expansion (5.6.25) is given by $Y_{[Nt],m} G_\infty^{(m)}(0)$, where m is the power rank. This is the same $Y_{[Nt],m}$ as in (5.6.4). One concludes by applying the last part of Theorem 5.6.3. □

5.7 Multivariate Limit Theorems

We are interested here in the limit behavior of the finite-dimensional distributions of the following vector as $N \to \infty$:

$$V_N(t) = \left(\frac{1}{A_r(N)} \sum_{n=1}^{[Nt]} (G_r(X_n) - \mathbb{E}G_r(X_n))\right)_{r=1,\ldots,R}, \quad t \geq 0, \quad (5.7.1)$$

where G_r, $r = 1, \ldots, R$, are deterministic functions and $A_r(N)$s are appropriate normalizations. We first focus on the case when $\{X_n\}_{n \in \mathbb{Z}}$ is a Gaussian stationary series. We have to distinguish between the situations where the resulting limit law for (5.7.1) is:

(a) a multivariate Gaussian process with dependent Brownian motion marginals, or
(b) a multivariate process with dependent Hermite processes as marginals, or
(c) a combination.

The cases (a), (b) and (c) will be referred to as, respectively, the SRD, LRD and mixed cases. In Section 5.7.4, we state the corresponding results for multivariate multilinear processes,

$$Y_N(t) = \left(\frac{1}{A_r(N)} \sum_{n=1}^{[Nt]} \sum_{1 \leq i_1 < \cdots < i_{k_r} < \infty} \psi_{r,i_1} \ldots \psi_{r,i_{k_r}} \epsilon_{n-i_1} \ldots \epsilon_{n-i_{k_r}}\right)_{r=1,\ldots,R}. \quad (5.7.2)$$

But first, we focus on (5.7.1).

5.7.1 The SRD Case

The first result concerns the SRD case, generalizing Theorem 5.4.1.

Theorem 5.7.1 *(SRD case.) Let $\{X_n\}_{n \in \mathbb{Z}}$ be a Gaussian stationary series with autocovariance function γ_X. Suppose that $\mathbb{E}X_n = 0$, $\mathbb{E}X_n^2 = 1$. Let G_r, $r = 1, \ldots, R$, be deterministic functions with respective Hermite ranks $k_r \geq 1$, $r = 1, \ldots, R$, in the sense of Definition 5.2.1. If*

$$\sum_{n=1}^{\infty} |\gamma_X(n)|^{k_r} < \infty, \quad r = 1, \ldots, R, [6] \quad (5.7.3)$$

then

$$V_N(t) \xrightarrow{fdd} V(t), \quad t \geq 0, \quad (5.7.4)$$

where $V_N(t)$ is given in (5.7.1) with $A_r(N) = N^{1/2}$, $r = 1, \ldots, R$. The limit process $V(t) = (\sigma_1 B_1(t), \ldots, \sigma_R B_R(t))'$ is multivariate Brownian motion, where

$$\sigma_r^2 = \sum_{m=k_r}^{\infty} g_{r,m}^2 m! \sum_{n=-\infty}^{\infty} (\gamma_X(n))^m, \quad r = 1, \ldots, R, \quad (5.7.5)$$

with $g_{r,m}$ being the coefficients in the Hermite expansion (5.1.6) of G_r, $\{B_r(t)\}_{t \in \mathbb{R}}$, $r = 1, \ldots, R$, are standard Brownian motions with the cross-covariance: for $t_1, t_2 \geq 0$,

---

[6] See (5.4.1) and (5.4.2) for an equivalent condition.

5.7 Multivariate Limit Theorems

$$\mathbb{E} B_{r_1}(t_1) B_{r_2}(t_2) = (t_1 \wedge t_2) \frac{\sigma_{r_1, r_2}}{\sigma_{r_1} \sigma_{r_2}}, \tag{5.7.6}$$

where

$$\sigma_{r_1, r_2} = \sum_{m=k_{r_1} \vee k_{r_2}}^{\infty} g_{r_1, m} g_{r_2, m} m! \sum_{n=-\infty}^{\infty} \gamma_X(n)^m \tag{5.7.7}$$

$(a \wedge b = \min\{a, b\}$ and $a \vee b = \max\{a, b\})$.

Proof The proof follows that of Theorem 5.4.1 in the univariate case. As in that proof (see the argument leading to (5.4.6)), it is enough to consider the functions

$$G_r(x) - \mathbb{E} G_r(X_n) = \sum_{m=k_r}^{M} g_{r,m} H_m(x), \tag{5.7.8}$$

where M is finite and fixed. Set

$$S_N = \sum_{r=1}^{R} \sum_{j=1}^{J_r} \theta_{r,j} V_{r,N}(t_{r,j}),$$

where $V_{r,N}(t)$ are the components of $V_N(t) = (V_{1,N}(t), \ldots, V_{R,N}(t))'$, $\theta_{r,j} \in \mathbb{R}$ and $t_{r,j} \geq 0$. By the method of moments (e.g., Gut [442], Theorem 8.6 in Chapter 5 and Section 4.10), it is enough to show that, for all integers $p \geq 1$,

$$\mathbb{E} S_N^p \to \mathbb{E} \Big(\sum_{r=1}^{R} \sum_{j=1}^{J_r} \theta_{r,j} V_r(t_{r,j}) \Big)^p, \tag{5.7.9}$$

where $V_r(t)$ are the components of the limit vector $V(t) = (V_1(t), \ldots, V_R(t))'$. This is equivalent to: for $p \geq 1$,

$$\chi_p(S_N) \to \chi_p \Big(\sum_{r=1}^{R} \sum_{j=1}^{J_r} \theta_{r,j} V_r(t_{r,j}) \Big), \tag{5.7.10}$$

where χ_p denotes the pth cumulant. Since the limit is Gaussian and has zero mean,

$$\chi_p \Big(\sum_{r=1}^{R} \sum_{j=1}^{J_r} \theta_{r,j} V_r(t_{r,j}) \Big) = \begin{cases} \mathbb{E} \Big(\sum_{r=1}^{R} \sum_{j=1}^{J_r} \theta_{r,j} V_r(t_{r,j}) \Big)^2, & \text{if } p = 2, \\ 0, & \text{if } p = 1, 3, 4, \ldots \end{cases} \tag{5.7.11}$$

As in the proof of Theorem 5.4.1 (see the argument following (5.4.9)), by using the multilinearity of joint cumulants and the relation (4.3.26),

$$\chi_p(S_N) = \chi_p \Big(\sum_{r=1}^{R} \sum_{j=1}^{J_r} \sum_{m=k_r}^{M} \theta_{r,j} g_{r,m} \frac{1}{N^{1/2}} \sum_{n=1}^{[N t_{r,j}]} H_m(X_n) \Big)$$

$$= \sum_{r_1, \ldots, r_p=1}^{R} \sum_{j_1=1}^{J_{r_1}} \cdots \sum_{j_p=1}^{J_{r_p}} \sum_{m_1=k_{r_1}}^{M} \cdots \sum_{m_p=k_{r_p}}^{M} \Big(\prod_{\ell=1}^{p} \theta_{r_\ell, j_\ell} g_{r_\ell, m_\ell} \Big)$$

$$\times \frac{1}{N^{p/2}} \sum_{n_1=1}^{[Nt_{r_1,j_1}]} \cdots \sum_{n_p=1}^{[Nt_{r_p,j_p}]} \chi\Big(H_{m_1}(X_{n_1}), \ldots, H_{m_p}(X_{n_p})\Big)$$

$$= \sum_{r_1,\ldots,r_p=1}^{R} \sum_{j_1=1}^{J_{r_1}} \cdots \sum_{j_p=1}^{J_{r_p}} \sum_{m_1=k_{r_1}}^{M} \cdots \sum_{m_p=k_{r_p}}^{M} \Big(\prod_{\ell=1}^{p} \theta_{r_\ell,j_\ell} g_{r_\ell,m_\ell}\Big)$$

$$\times \sum_{A \in \mathcal{A}_c(m_1,\ldots,m_p)} g(A) \frac{1}{N^{p/2}} T_N(t_{r_1,j_1},\ldots,t_{r_p,j_p},m_1,\ldots,m_p,A),$$

where T_Ns are defined in (5.4.10). As in the proof of Theorem 5.4.1, by using Lemma 5.4.3, we conclude that, for $p \geq 3$, $\chi_p(S_N) \to 0$, as $N \to \infty$.

For $p = 1$, the convergence (5.7.10) is trivial since $\mathbb{E}S_N = 0$ and hence both sides are zero. For $p = 2$, we need to show that

$$\mathbb{E}S_N^2 \to \mathbb{E}\Big(\sum_{r=1}^{R} \sum_{j=1}^{J_r} \theta_{r,j} V_r(t_{r,j})\Big)^2 = \sum_{r_1,r_2=1}^{R} \sum_{j_1=1}^{J_{r_1}} \sum_{j_2=1}^{J_{r_2}} \theta_{r_1,j_1} \theta_{r_2,j_2} \mathbb{E}V_{r_1}(t_{r_1,j_1}) V_{r_2}(t_{r_2,j_2})$$

$$= \sum_{r_1,r_2=1}^{R} \sum_{j_1=1}^{J_{r_1}} \sum_{j_2=1}^{J_{r_2}} \theta_{r_1,j_1} \theta_{r_2,j_2} (t_{r_1,j_1} \wedge t_{r_2,j_2}) \sum_{m=k_{r_1} \vee k_{r_2}}^{\infty} g_{r_1,m} g_{r_2,m} m! \sum_{n=-\infty}^{\infty} \gamma_X(n)^m.$$
(5.7.12)

Observe that

$$\mathbb{E}S_N^2 = \mathbb{E}\Big(\sum_{r=1}^{R} \sum_{j=1}^{J_r} \sum_{m=k_r}^{M} \theta_{r,j} g_{r,m} \frac{1}{N^{1/2}} \sum_{n=1}^{[Nt_{r,j}]} H_m(X_n)\Big)^2$$

$$= \sum_{r_1,r_2=1}^{R} \sum_{j_1=1}^{J_{r_1}} \sum_{j_2=1}^{J_{r_2}} \sum_{m_1=k_{r_1}}^{M} \sum_{m_2=k_{r_2}}^{M} \theta_{r_1,j_1} \theta_{r_2,j_2} g_{r_1,m_1} g_{r_2,m_2} \frac{1}{N} \sum_{n_1=1}^{[Nt_{r_1,j_1}]} \sum_{n_2=1}^{[Nt_{r_2,j_2}]} \mathbb{E}H_{m_1}(X_{n_1}) H_{m_2}(X_{n_2})$$

$$= \sum_{r_1,r_2=1}^{R} \sum_{j_1=1}^{J_{r_1}} \sum_{j_2=1}^{J_{r_2}} \sum_{m=k_{r_1} \vee k_{r_2}}^{M} \theta_{r_1,j_1} \theta_{r_2,j_2} g_{r_1,m} g_{r_2,m} m! \frac{1}{N} \sum_{n_1=1}^{[Nt_{r_1,j_1}]} \sum_{n_2=1}^{[Nt_{r_2,j_2}]} (\gamma_X(n_1 - n_2))^m.$$

As in the proof of Theorem 5.4.1, by using Lemma 5.4.4,

$$\mathbb{E}S_N^2 \to \sum_{r_1,r_2=1}^{R} \sum_{j_1=1}^{J_{r_1}} \sum_{j_2=1}^{J_{r_2}} \theta_{r_1,j_1} \theta_{r_2,j_2} (t_{r_1,j_1} \wedge t_{r_2,j_2}) \sum_{m=k_{r_1} \vee k_{r_2}}^{\infty} g_{r_1,m} g_{r_2,m} m! \sum_{n=-\infty}^{\infty} \gamma_X(n)^m$$

which is the desired limit (5.7.12). This limit can be expressed as

$$\sum_{r_1,r_2=1}^{R} \sum_{j_1=1}^{J_{r_1}} \sum_{j_2=1}^{J_{r_2}} \theta_{r_1,j_1} \theta_{r_2,j_2} (t_{r_1,j_1} \wedge t_{r_2,j_2}) \sigma_{r_1,r_2} = \mathbb{E}\Big(\sum_{r=1}^{R} \sum_{j=1}^{J_r} \theta_{r,j} \sigma_r B_r(t_{r,j})\Big)^2.$$

This concludes the proof. □

The next result is analogous to Corollary 5.4.5.

Corollary 5.7.2 (*SRD case.*) *Let $\{X_n\}_{n\in\mathbb{Z}}$ be a Gaussian stationary series which is LRD in the sense (2.1.5) in condition II with $d \in (0, 1/2)$. Suppose that $\mathbb{E}X_n = 0$, $\mathbb{E}X_n^2 = 1$. Let G_r, $r = 1, \ldots, R$, be deterministic functions with respective Hermite ranks $k_r \geq 1$, $r = 1, \ldots, R$, in the sense of Definition 5.2.1. If*

$$d \in \left(0, \frac{1}{2}\left(1 - \frac{1}{k_r}\right)\right), \quad r = 1, \ldots, R, \tag{5.7.13}$$

then the convergence (5.7.4) holds.

Example 5.7.3 (*Bivariate limit result in SRD case*) Under the setting of Corollary 5.7.2, suppose $d \in (0, 1/4)$, let $R = 2$ and

$$G_1(x) = aH_2(x) + bH_3(x) = bx^3 + ax^2 - 3bx - a, \quad G_2(x) = cH_3(x) = cx^3 - 3cx.$$

Here with $r = 1$, we have $k_1 = 2$ and $g_{1,2} = a$, $g_{1,3} = b$, and with $r = 2$, we have $k_2 = 3$ and $g_{2,3} = c$. In particular, for $m = k_{r_1} \vee k_{r_2} = 2 \vee 3 = 3$, we have $g_{r_1,m} g_{r_2,m} = g_{1,3} g_{2,3} = bc$. Then, in (5.7.5),

$$\sigma_1^2 = 2a^2 \sum_{n=-\infty}^{\infty} \gamma_X(n)^2 + 6b^2 \sum_{n=-\infty}^{\infty} \gamma_X(n)^3, \quad \sigma_2^2 = 6c^2 \sum_{n=-\infty}^{\infty} \gamma_X(n)^3,$$

and

$$\left(\frac{1}{N^{1/2}} \sum_{n=1}^{[Nt]} G_1(X_n), \frac{1}{N^{1/2}} \sum_{n=1}^{[Nt]} G_2(X_n)\right)' \xrightarrow{fdd} \left(\sigma_1 B_1(t), \sigma_2 B_2(t)\right)', \quad t \geq 0,$$

where the Brownian motions B_1 and B_2 have the covariance structure: for $t_1, t_2 \geq 0$,

$$\operatorname{Cov}(B_1(t_1), B_2(t_2)) = 6bc \frac{t_1 \wedge t_2}{\sigma_1 \sigma_2} \sum_{n=-\infty}^{\infty} \gamma_X(n)^3.$$

B_1 and B_2 are independent when $b = 0$.

5.7.2 The LRD Case

The next result concerns the LRD case, and extends Theorem 5.3.1. In Appendix B.2 (see (B.2.14)), $\widehat{I}_k(f)$ is used to denote a multiple integral with respect to an Hermitian Gaussian random measure on $\mathbb{R}$. To indicate its dependence on the corresponding symmetric control measure m, we shall denote the multiple integral by $\widehat{I}_{m,k}(f)$ throughout this section, and reserve the notation $\widehat{I}_k(f)$ for the case when the control measure m is the Lebesgue measure.

We also let

$$f_{H,k,t}(x_1, \ldots, x_k) = B_{k,H} \frac{e^{it(x_1+\cdots+x_k)} - 1}{i(x_1+\cdots+x_k)} \prod_{j=1}^{k} |x_j|^{\frac{1-H}{k} - \frac{1}{2}} \tag{5.7.14}$$

be the kernel function appearing in the representation (4.2.42) of a standard Hermite process of order k, where $B_{k,H}$ is the constant in (4.2.46). Note that the limit process in Theorem 5.3.1 is $g_k \beta_{k,H} \widehat{I}_k(f_{H,k,t})$. It can be seen from the proof of Theorem 5.3.1 that the limit process can also be represented as $g_k \widehat{I}_{F_0,k}(K_t)$, where K_t is given in (5.3.13) and the measure F_0 is defined by (5.3.17)–(5.3.18) (and depends on d).

Theorem 5.7.4 *(LRD case.) Let $\{X_n\}_{n \in \mathbb{Z}}$ be a Gaussian stationary series which is LRD in the sense (2.1.5) in condition II with $d \in (0, 1/2)$. Suppose that $\mathbb{E}X_n = 0$, $\mathbb{E}X_n^2 = 1$. Let G_r, $r = 1, \ldots, R$, be deterministic functions with respective Hermite ranks $k_r \geq 1$, $r = 1, \ldots, R$, in the sense of Definition 5.2.1. If*

$$d \in \left(\frac{1}{2}\left(1 - \frac{1}{k_r}\right), \frac{1}{2}\right), \quad r = 1, \ldots, R, \tag{5.7.15}$$

then

$$V_N(t) \xrightarrow{fdd} V^{(d)}(t), \quad t \geq 0, \tag{5.7.16}$$

where $V_N(t)$ is given in (5.7.1) with

$$A_r(N) = (L_2(N))^{k_r/2} N^{k_r(d-1/2)+1}, \quad r = 1, \ldots, R.$$

The limit process can be represented as

$$V^{(d)}(t) \stackrel{d}{=} \left(g_{r,k_r} \beta_{k_r,H_r} \widehat{I}_{k_r}(f_{H_r,k_r,t})\right)_{r=1,\ldots,R} \stackrel{d}{=} \left(g_{r,k_r} \widehat{I}_{F_0,k_r}(K_t)\right)_{r=1,\ldots,R} \tag{5.7.17}$$

in the sense of finite-dimensional distributions, where

$$H_r = k_r\left(d - \frac{1}{2}\right) + 1 \in \left(\frac{1}{2}, 1\right), \quad r = 1, \ldots, R, \tag{5.7.18}$$

g_{r,k_r} *is the first nonzero coefficient in the Hermite expansion of G_r in Definition 5.2.1, $\beta_{k,H}$ is the constant given in (5.3.3), $f_{H,k,t}$ is the kernel function defined in (5.7.14), K_t is given in (5.3.13) and the measure F_0 is defined by (5.3.17)–(5.3.18) (and depends on d).*

Remark 5.7.5 The individual component processes $\widehat{I}_{k_r}(f_{H_r,k_r,t})$ above are standard Hermite process of orders k_r with the self-similarity parameter H_r. They are multiple integrals with respect to the same Hermitian Gaussian random measure. It can be shown that they are dependent (Bai and Taqqu [74], Proposition 1).

Proof The proof follows closely that of Theorem 5.3.1. By the reduction theorem (Theorem 5.3.3), it is enough to prove the result for $G_r(x) - \mathbb{E}G_r(X_n) = H_{k_r}(x)$, where $H_{k_r}(x)$ is the Hermite polynomial of order k_r. As in Step 1 of that proof, one can write

$$V_{r,N}(t) = \widehat{I}_{F_N,k_r}(K_{N,t}),$$

where $K_{N,t}$ is the function given in (5.3.10) (with the number of variables k_r, suggested by the order of the multiple integral k_r) and the measure F_N satisfies (5.3.11). As in Step 2 of that proof, we would like to conclude now that

$$\sum_{r=1}^{R}\sum_{j=1}^{J_r} \theta_{r,j} V_{r,N}(t_{r,j}) = \sum_{r=1}^{R}\sum_{j=1}^{J_r} \theta_{r,j} \widehat{I}_{F_N,k_r}(K_{N,t_{r,j}})$$

5.7 Multivariate Limit Theorems

$$\to \sum_{r=1}^{R}\sum_{j=1}^{J_r} \theta_{r,j} \widehat{I}_{F_0,k_r}(K_{t_{r,j}}) = \sum_{r=1}^{R}\sum_{j=1}^{J_r} \theta_{r,j} V_r(t_{r,j}),$$

where $\theta_{r,j} \in \mathbb{R}$, $t_{r,j} \geq 0$. This can be shown as in Step 3 of that proof by using Proposition 5.7.6 below, which extends Proposition 5.3.6 to the multivariate case. $\square$

The next auxiliary result, which was used in the preceding proof, can be established in the same way as Proposition 5.3.6.

Proposition 5.7.6 *Let F_N, $N \geq 1$, F_0 be symmetric, locally finite measures without atoms on $(\mathbb{R}, \mathcal{B}(\mathbb{R}))$ so that $F_N \xrightarrow{v} F_0$ in the sense of vague convergence. Let $\widehat{K}_{N,k_r}(x_1, \ldots, x_{k_r})$, $r = 1, \ldots, R$, be a sequence of Hermitian, measurable functions on $\mathbb{R}^{k_r}$ tending to a continuous function $\widehat{K}_{0,k_r}(x_1, \ldots, x_{k_r})$ uniformly in any rectangle $[-A, A]^{k_r}$, $A > 0$. Moreover, suppose that $\widehat{K}_{0,k_r}$ and $\widehat{K}_{N,k_r}$ satisfy the relation*

$$\lim_{A \to \infty} \sup_{N \geq 0} \int_{\mathbb{R}^{k_r} \setminus [-A,A]^{k_r}} |\widehat{K}_{N,k_r}(x_1, \ldots, x_{k_r})|^2 F_N(dx_1) \ldots F_N(dx_{k_r}) = 0, \quad r = 1, \ldots, R. \tag{5.7.19}$$

Then,

$$\left(\widehat{I}_{F_N,k_r}(\widehat{K}_{N,k_r})\right)_{r=1,\ldots,R} \xrightarrow{d} \left(\widehat{I}_{F_0,k_r}(\widehat{K}_{0,k_r})\right)_{r=1,\ldots,R}, \tag{5.7.20}$$

where the last integrals exist.

Example 5.7.7 (*Bivariate limit result in LRD case*) Under the setting of Theorem 5.7.4, suppose $d \in (1/4, 1/2)$ and $L_2(u) \sim 1$ in (2.1.5) of condition II. Let also $R = 2$, $G_1(x) = H_1(x) = x$, $G_2(x) = H_2(x) = x^2 - 1$. In this case, the convergence (5.7.17) can be expressed as

$$\left(\frac{1}{N^{d+1/2}} \sum_{n=1}^{[Nt]} X_n, \frac{1}{N^{2d}} \sum_{n=1}^{[Nt]} (X_n^2 - 1)\right)' \xrightarrow{fdd} \left(\frac{1}{\sqrt{d(2d+1)}} Z_{H_1}^{(1)}(t), \frac{1}{\sqrt{d(4d-1)}} Z_{H_2}^{(2)}(t)\right)', t \geq 0,$$

where the standard fractional Brownian motion $\{Z_{H_1}^{(1)}(t)\}_{t \geq 0}$ and the standard Rosenblatt process $\{Z_{H_2}^{(2)}(t)\}_{t \geq 0}$ share the same random measure in the multiple integral representations. The components $\{Z_{H_1}^{(1)}(t)\}_{t \geq 0}$ and $\{Z_{H_2}^{(2)}(t)\}_{t \geq 0}$ are uncorrelated but dependent (as stated in Remark 5.7.5). The constant factors in the limit are consistent with (5.2.12) in Corollary 5.2.5.

5.7.3 The Mixed Case

The final result concerns the mixed case. We need to modify slightly the setting used in Theorems 5.7.1 and 5.7.4. We let

$$S_N(t) = \left(\frac{1}{A_{S,r_1}(N)} \sum_{n=1}^{[Nt]} (G_{S,r_1}(X_n) - \mathbb{E}G_{S,r_1}(X_n))\right)_{r_1=1,\ldots,R_S}, \tag{5.7.21}$$

$$L_N(t) = \left(\frac{1}{A_{L,r_2}(N)} \sum_{n=1}^{[Nt]} (G_{L,r_2}(X_n) - \mathbb{E}G_{L,r_2}(X_n))\right)_{r_2=1,\ldots,R_L}, \qquad (5.7.22)$$

where the letters S and L refer to the expected SRD and LRD behaviors of the two vector processes, respectively. We will suppose that the Hermite ranks $k_{S,r_1} \geq 1$ and $k_{L,r_2} \geq 1$ are such that $k_{S,r_1} > k_{L,r_2}$, indicating that G_{S,r_1} involves higher-order polynomials than G_{L,r_2}, which allows the $G_{S,r_1}(X_n)$s to be "less dependent" than the $G_{L,r_2}(X_n)$s.

Theorem 5.7.8 *(Mixed case.) Let $\{X_n\}_{n \in \mathbb{Z}}$ be a Gaussian stationary series which is LRD in the sense (2.1.5) in condition II with $d \in (0, 1/2)$. Suppose that $\mathbb{E}X_n = 0$, $\mathbb{E}X_n^2 = 1$. Let G_{S,r_1}, $r_1 = 1, \ldots, R_S$, be functions with respective Hermite ranks $k_{S,r_1} \geq 1$, $r_1 = 1, \ldots, R_S$, and G_{L,r_2}, $r_2 = 1, \ldots, R_L$, be functions with respective Hermite ranks $k_{L,r_2} \geq 1$, $r_2 = 1, \ldots, R_L$ in the sense of Definition 5.2.1. Suppose*

$$\frac{1}{2}\left(1 - \frac{1}{k_{L,r_2}}\right) < d < \frac{1}{2}\left(1 - \frac{1}{k_{S,r_1}}\right), \quad r_1 = 1, \ldots, R_S,\ r_2 = 1, \ldots, R_L, \qquad (5.7.23)$$

which implies $k_{S,r_1} > k_{L,r_2}$. Then,

$$(S_N(t), L_N(t))' \xrightarrow{fdd} (V(t), V^{(d)}(t))', \quad t \geq 0, \qquad (5.7.24)$$

where $V(t)$ denotes the limit of $S_N(t)$ according to Theorem 5.7.1 and $V^{(d)}(t)$ denotes the limit of $L_N(t)$ according to Theorem 5.7.4. Moreover, the processes $\{V(t)\}_{t \geq 0}$ and $\{V^{(d)}(t)\}_{t \geq 0}$ are independent.

Proof Using the reduction arguments as in the proofs of Theorems 5.7.1 and 5.7.4, we can replace $G_{S,r} - \mathbb{E}G_{S,r}$ in (5.7.21) with $\sum_{m=k_{S,r}}^{M} g_{S,r,m} H_m$, and we can replace $G_{L,r}$ in (5.7.22) with $g_{L,r,k_{L,r}} H_{k_{L,r}}$, where $k_{S,r} > k_{L,r}$ are the corresponding Hermite ranks and $g_{S,r,m}$, $g_{L,r,m}$ are the corresponding coefficients of the Hermite expansions.

For fixed finite time points t_j, $j = 1, \ldots, J$, we need to consider the joint convergence of the vector

$(S_{j,r_1,N}, L_{j,r_2,N})_{j,r_1,r_2}$

$$:= \left(\frac{1}{A_{S,r_1}(N)} \sum_{m=k_{S,r_1}}^{M} g_{S,r_1,m} S_{N,t_j}(H_m), \frac{1}{A_{L,r_2}(N)} g_{L,r_2,k_{L,r_2}} S_{N,t_j}(H_{k_{L,r_2}})\right)_{j,r_1,r_2}, \qquad (5.7.25)$$

where $j = 1, \ldots, J$, $r_1 = 1, \ldots, R_S$, $r_2 = 1, \ldots, R_L$, and $S_{N,t}(G) = \sum_{n=1}^{[Nt]} G(X_n)$.

As in the proof of Theorem 5.3.1, we can express Hermite polynomials as multiple integrals:

$$S_{j,r_1,N} = \sum_{m=k_{S,r_1}}^{M} I_m(f_{m,j,r_1,N}), \quad L_{j,r_2,N} = I_{k_{L,r_2}}(f_{j,r_2,N}),$$

where $f_{m,j,r_1,N}$, $f_{j,r_2,N}$ are some symmetric square-integrable functions.

Express the vector in (5.7.25) as $(S_N, L_N)'$, where $S_N := (S_{j,r_1,N})_{j,r_1}$, $L_N := (L_{j,r_2,N})_{j,r_2}$. By Theorem 5.7.1, S_N converges in distribution to some multivariate normal

5.7 Multivariate Limit Theorems

distribution, and by Theorem 5.7.4, L_N converges to a multivariate distribution expressed through multiple integrals. The joint convergence follows from Corollary 5.7.11 below. □

Example 5.7.9 (*Bivariate limit result in mixed case*) In the setting of Theorem 5.7.8, suppose $d \in (1/3, 1/4)$ and $L_2(u) \sim 1$ in (2.1.5) of condition II. Let also $G_1(x) = H_3(x) = x^3 - 3x$, $G_2(x) = H_2(x) = x^2 - 1$. Then, $\sigma^2 = 6 \sum_{n=-\infty}^{\infty} \gamma_X(n)^3$ and

$$\left(\frac{1}{N^{1/2}} \sum_{n=1}^{[Nt]} (X_n^3 - 3X_n), \frac{1}{N^{2d}} \sum_{n=1}^{[Nt]} (X_n^2 - 1) \right)' \stackrel{fdd}{\longrightarrow} \left(\sigma B(t), \frac{1}{\sqrt{d(4d-1)}} Z_H^{(2)}(t) \right)', \quad t \geq 0,$$

where the standard Rosenblatt process $\{Z_H^{(2)}(t)\}_{t \geq 0}$ and the standard Brownian motion $\{B(t)\}_{t \geq 0}$ are independent.

In the proof of Theorem 5.7.8, we used Corollary 5.7.11, which is based on the following theorem. It is a special case of Proposition 1.5 of Nourdin, Nualart, and Peccati [768].

Theorem 5.7.10 *Consider two vectors of multiple integrals (with respect to a Gaussian random measure):*

$$\xi_N = (I_{p_{r_1}}(f_{r_1, N}))_{r_1 = 1, \ldots, R_1}, \quad \eta_N = (I_{q_{r_2}}(g_{r_2, N}))_{r_2 = 1, \ldots, R_2}, \quad (5.7.26)$$

where $f_{r_1, N}$s and $g_{r_2, N}$s are symmetric square-integrable functions. Let ξ and η be two independent random vectors taking values in $\mathbb{R}^{R_1}$ and $\mathbb{R}^{R_2}$, respectively. Suppose that we have the following marginal convergence in distribution: as $N \to \infty$,

$$\xi_N \stackrel{d}{\to} \xi, \quad \eta_N \stackrel{d}{\to} \eta.$$

Assume that we have the pairwise contractions[7] vanishing:

$$\lim_{N \to \infty} \| f_{r_1, N} \otimes_u g_{r_2, N} \| = 0, \quad (5.7.27)$$

for all $u = 1, \ldots, p_{r_1} \wedge q_{r_2}$, $r_1 = 1, \ldots, R_1$ and $r_2 = 1, \ldots, R_2$, where $\| \cdot \|$ denotes the $L^2(\mathbb{R}^{p_{r_1} + q_{r_2} - 2u})$ norm. Then, we have the joint convergence: as $N \to \infty$,

$$(\xi_N, \eta_N)' \stackrel{d}{\to} (\xi, \eta)',$$

namely, ξ_N and η_N are asymptotically independent.

This result implies the following corollary.

Corollary 5.7.11 *Consider*

$$S_N = (I_{k_{S, r_1}}(f_{S, r_1, N}))_{r_1 = 1, \ldots, R_S}, \quad L_N = (I_{k_{L, r_2}}(f_{L, r_2, N}))_{r_2 = 1, \ldots, R_L},$$

where $k_{S, r_1} > k_{L, r_2}$, $r_1 = 1, \ldots, R_S$, $r_2 = 1, \ldots, R_L$. Suppose that, as $N \to \infty$, S_N converges in distribution to a multivariate normal law, and L_N converges in distribution to a multivariate law. Then, there are independent random vectors X and Y such that

$$(S_N, L_N)' \stackrel{d}{\to} (X, Y)'.$$

---
[7] Contractions are defined in Appendix B.2. See the relation (B.2.17).

Proof By Theorem 5.7.10, we need only to check (5.7.27). Let $(\cdot, \cdot)$ denote the inner product in a suitable dimension. By Fubini's theorem, one can write

$$\|f \otimes_u g\|^2 = \int_{\mathbb{R}^{p-u}} dx_1 \ldots dx_{p-u} \int_{\mathbb{R}^{q-u}} dx_{p-u+1} \ldots dx_{p+q-2u}$$

$$\times \left[\int_{\mathbb{R}^u} dy_1 \ldots dy_u f(x_1, \ldots, x_{p-u}, y_1, \ldots, y_u) g(x_{p-u+1}, \ldots, x_{p+q-2u}, y_1, \ldots, y_u) \right]^2$$

$$= \int_{\mathbb{R}^u} dy_1 \ldots dy_u \int_{\mathbb{R}^u} dz_1 \ldots dz_u \int_{\mathbb{R}^{p-u}} dx_1 \ldots dx_{p-u}$$

$$\times f(x_1, \ldots, x_{p-u}, y_1, \ldots, y_u) f(x_1, \ldots, x_{p-u}, z_1, \ldots, z_u)$$

$$\times \int_{\mathbb{R}^{q-u}} dx_{p-u+1} \ldots dx_{p+q-2u}$$

$$\times g(x_{p-u+1}, \ldots, x_{p+q-2u}, y_1, \ldots, y_u) g(x_{p-u+1}, \ldots, x_{p+q-2u}, z_1, \ldots, z_u)$$

$$= (f \otimes_{p-u} f, g \otimes_{q-u} g),$$

where $u = 1, \ldots, p \wedge q$, and f, g have p and q variables, respectively. We get by the Cauchy–Schwarz inequality that for $u = 1, \ldots, k_{L,r_2}, r_1 = 1, \ldots, R_S, r_2 = 1, \ldots, R_L$,

$$\|f_{S,r_1,N} \otimes_u f_{L,r_2,N}\|^2 = (f_{S,r_1,N} \otimes_{k_{S,r_1}-u} f_{S,r_1,N}, f_{L,r_2,N} \otimes_{k_{L,r_2}-u} f_{L,r_2,N})$$

$$\leq \|f_{S,r_1,N} \otimes_{k_{S,r_1}-u} f_{S,r_1,N}\| \|f_{L,r_2,N} \otimes_{k_{L,r_2}-u} f_{L,r_2,N}\| \to 0$$

because $\|f_{S,r_1,N} \otimes_{k_{S,r_1}-u} f_{S,r_1,N}\| \to 0$ for all $u = 1, \ldots, k_{S,r_1} - 1$ by the Nualart–Peccati central limit theorem [771]. Since $k_{S,r_1} > k_{L,r_2}$, this indeed covers all $u = 1, \ldots, k_{L,r_2}$. For the second term, one has by Fubini's theorem and the Cauchy–Schwarz inequality (see Peccati and Taqqu [797], Lemma 6.2.2) that

$$\|f_{L,r_2,N} \otimes_{k_{L,r_2}-u} f_{L,r_2,N}\| \leq \|f_{L,r_2,N}\|^2, \quad (5.7.28)$$

for all $u = 1, \ldots, k_{S,r_1} - 1$, and the right-hand side of (5.7.28) is bounded due to the tightness of the distribution of $I_{k_{L,r_2}}(f_{L,r_2,N})$ (Lemma 2.1 of Nourdin and Rosiński [764]). Therefore (5.7.27) holds. □

We presented above central and non-central limit theorems in the multivariate case when the underlying stationary series is Gaussian. In the linear case discussed in Section 5.6, the non-central limit behavior is typically determined by partial sums of multilinear forms (5.6.4). Multivariate results for multilinear forms are presented in the next Section 5.7.4 and are in the same spirit as those described here.

5.7.4 Multivariate Limits of Multilinear Processes

We consider here processes called *multilinear processes* (or *discrete-chaos processes*), which are defined as

$$X_n = \sum_{1 \leq i_1 < \cdots < i_k < \infty} \psi_{i_1} \ldots \psi_{i_k} \epsilon_{n-i_1} \ldots \epsilon_{n-i_k}, \quad n \in \mathbb{Z}, \quad (5.7.29)$$

where

$$\sum_{i=1}^{\infty} \psi_i^2 < \infty, \quad (5.7.30)$$

ϵ_is are i.i.d. with zero mean and unit variance, and $k \geq 1$ is the *order*. $X_n, n \in \mathbb{Z}$, is also said to belong to a *discrete chaos* of order k. Observe that the summation in (5.7.29) excludes the diagonals, in contrast to the situation considered in Theorem 5.6.3. We are thus focusing on what we may call a "discrete multiple integral."

If $\psi_i = L_1(i)i^{d-1}$ with $0 < d < 1/2$ and a slowly varying function L_1, and when $k > 1$, that is, except for linear processes, the partial sum of X_n when suitably normalized no longer converges to fractional Brownian motion. But depending on d and k, it either converges to an *Hermite process* if $\{X_n\}$ is still LRD, or it converges to Brownian motion if $\{X_n\}$ is SRD. See Giraitis et al. [406] for more details.

Consider the multilinear processes

$$X_{r,n} = \sum_{1 \leq i_1 < \cdots < i_{k_r} < \infty} \psi_{r,i_1} \cdots \psi_{r,i_{k_r}} \epsilon_{n-i_1} \cdots \epsilon_{n-i_{k_r}}, \quad r = 1, \ldots, R, \quad (5.7.31)$$

where $k_1, \ldots, k_r$ are the orders for $X_{1,n}, \ldots, X_{R,n}$ respectively, and $\{\psi_{r,i}\}$ are coefficients. Let

$$Y_{r,N}(t) = \frac{1}{A_r(N)} \sum_{n=1}^{[Nt]} X_{r,n}, \quad t \geq 0, \quad (5.7.32)$$

where $A_r(N)$ is a normalization factor such that $\lim_{N \to \infty} \text{Var}(Y_{r,N}(1)) = 1, r = 1, \ldots, R$. We want to study the limit of the following vector process as $N \to \infty$:

$$Y_N(t) = (Y_{1,N}(t), \ldots, Y_{R,N}(t))'. \quad (5.7.33)$$

Depending on $\{\psi_{r,i}\}$ and k_r, the components of $Y_N(t)$ can be either SRD, or LRD, or a mixture of SRD and LRD.

We start with some facts about multilinear processes $\{X_n\}$ in (5.7.29). Note first that the condition (5.7.30) guarantees that X_n is well-defined in $L^2(\Omega)$, since

$$\mathbb{E}X_n^2 = \sum_{1 \leq i_1 < \cdots < i_k < \infty} \psi_{i_1}^2 \cdots \psi_{i_k}^2 < \infty.$$

We use throughout a convention $\psi_i = 0$ for $i \leq 0$. One can compute the autocovariance of $\{X_n\}$ as

$$\gamma_X(n) = \sum_{1 \leq i_1 < \cdots < i_k < \infty} \psi_{n+i_1} \psi_{i_1} \cdots \psi_{n+i_k} \psi_{i_k}, \quad n \in \mathbb{Z}. \quad (5.7.34)$$

The following proposition describes the asymptotic behavior of $\gamma_X(n)$ under suitable assumptions.

Proposition 5.7.12 *Suppose $\gamma_X(n)$ is defined in (5.7.34), $\psi_i = L_1(i)i^{d-1}, i \geq 1$, with $0 < d < 1/2$ where L_1 is slowly varying at infinity. Then,*

$$\gamma_X(n) = L^*(n)n^{2d_X - 1}, \quad (5.7.35)$$

for a slowly varying function L^* and

$$d_X = k\left(d - \frac{1}{2}\right) + \frac{1}{2}. \quad (5.7.36)$$

Proof By (5.7.34), as $n \to \infty$,

$$\gamma_X(n) \sim (k!)^{-1}\left(\sum_{i=1}^{\infty} \psi_{n+i}\psi_i\right)^k$$

(the diagonal terms with $i_p = i_q$ are negligible as $n \to \infty$). Proposition 2.2.9 then yields (5.7.35) with (5.7.36) and $L^*(n) \sim (k!)^{-1} B(d, 1-2d)^k L_1(n)^{2k}$. $\square$

Remark 5.7.13 According to Proposition 5.7.12, when $d < \frac{1}{2}(1-\frac{1}{k})$, we have $\sum |\gamma_X(n)| < \infty$, and when $d > \frac{1}{2}(1-\frac{1}{k})$, we have $\sum |\gamma_X(n)| = \infty$. So for $\psi_i = L_1(i)i^{d-1}$, $0 < d < 1/2$, the quantity $\frac{1}{2}(1-\frac{1}{k})$ is the boundary between SRD and LRD.

We now define precisely what SRD and LRD mean for a multilinear process $\{X_n\}$, and from then on we use this definition whenever we talk about SRD or LRD.

Definition 5.7.14 *Let $\{X_n\}$ be a multilinear process given in (5.7.29) with coefficients ψ_i, autocovariances $\gamma_X(n)$ and order k. We say that $\{X_n\}$ is:*

(a) *SRD if for some $d \in (-\infty, \frac{1}{2}(1-\frac{1}{k}))$ and some constant $c > 0$,*

$$|\psi_i| \le c i^{d-1}, \quad i \ge 1, \quad \sum_{n=-\infty}^{\infty} \gamma_X(n) > 0; \quad (5.7.37)$$

(b) *LRD if for some $d \in (\frac{1}{2}(1-\frac{1}{k}), \frac{1}{2})$ and some L_1 slowly varying at infinity,*

$$\psi_i = L_1(i)i^{d-1}, \quad i \ge 1. \quad (5.7.38)$$

Remark 5.7.15 The ds in (5.7.37) and (5.7.38) are different. In the SRD case, $\{\psi_i\}$ is only assumed to decay faster than a power function, which implies

$$\sum_{n=0}^{\infty} |\gamma_X(n)| \le \sum_{n=0}^{\infty}\left(\sum_{i=1}^{\infty} |\psi_{n+i}\psi_i|\right)^k \le c^{2k} \sum_{n=0}^{\infty}\left(\sum_{i=1}^{\infty} (i+n)^{d-1} i^{d-1}\right)^k. \quad (5.7.39)$$

The last bound can be shown to be finite when $d < \frac{1}{2}(1-\frac{1}{k})$ (Exercise 5.8). In contrast, in the LRD case, the regular variation assumption on $\{\psi_i\}$ yields a *memory parameter* $d_X = k(d-1/2) + 1/2$ given by (5.7.36).

Next we consider the cross-covariance between two multilinear processes defined by

$$X_{1,n} = \sum_{1 \le i_1 < \cdots < i_p < \infty} \psi_{i_1} \cdots \psi_{i_p} \epsilon_{n-i_1} \cdots \epsilon_{n-i_p}, \quad (5.7.40)$$

$$X_{2,n} = \sum_{1 \le i_1 < \cdots < i_q < \infty} \phi_{i_1} \cdots \phi_{i_q} \epsilon_{n-i_1} \cdots \epsilon_{n-i_q}. \quad (5.7.41)$$

$\{X_{1,n}\}$ and $\{X_{2,n}\}$ share the same series $\{\epsilon_i\}$ but the sequences $\{\psi_i\}$ and $\{\phi_i\}$ can be different. Then, the cross-covariance is

$$\mathrm{Cov}(X_{1,n}, X_{2,0}) = \begin{cases} 0, & \text{if } p \neq q, \\ \sum_{1 \le i_1 < \cdots < i_k < \infty} \psi_{i_1} \phi_{n+i_1} \cdots \psi_{i_k} \phi_{n+i_k}, & \text{if } p = q = k, \end{cases} \quad (5.7.42)$$

for any $n \in \mathbb{Z}$.

We now state the multivariate joint convergence results for the vector process $Y_N(t)$ in (5.7.33) (the proofs can be found in Bai and Taqqu [75]). Recall that Y_N is normalized so that the asymptotic variance of every component at $t = 1$ equals 1.

Theorem 5.7.16 *(SRD case.) If all the components in Y_N defined in (5.7.33) are SRD in the sense of (5.7.37), then*

$$Y_N(t) \overset{fdd}{\longrightarrow} B(t) = (B_1(t), \ldots, B_R(t))', \quad t \ge 0,$$

where $\{B(t)\}$ is a multivariate Gaussian process with $B_1(t), \ldots, B_R(t)$ being standard Brownian motions with

$$\mathrm{Cov}(B_{r_1}(t_1), B_{r_2}(t_2)) = (t_1 \wedge t_2) \frac{\sigma_{r_1, r_2}}{\sigma_{r_1} \sigma_{r_2}} \quad (5.7.43)$$

and

$$\sigma_r^2 = \sum_{n=-\infty}^{\infty} \mathrm{Cov}(X_{r,n}, X_{r,0}), \quad \sigma_{r_1, r_2} = \sum_{n=-\infty}^{\infty} \mathrm{Cov}(X_{r_1,n}, X_{r_2,0}).$$

The normalization $A_r(N)$ in (5.7.32) satisfies $A_r(N) \sim \sigma_r \sqrt{N}$ as $N \to \infty$.

Remark 5.7.17 In view of (5.7.42) and (5.7.43), if all the components of $Y_N(t)$ have a different order, then the limit components $B_r(t)$ are uncorrelated and hence independent. Otherwise, they are in general dependent and their covariance is given by (5.7.43).

Theorem 5.7.18 *(LRD case.) If all the components in Y_N defined in (5.7.33) are LRD in the sense of (5.7.38) with $d = d_1, \ldots, d_R$, respectively, then*

$$Y_N(t) \overset{fdd}{\longrightarrow} Z^{(d,k)}(t) = (Z_{H_1}^{(k_1)}(t), \ldots, Z_{H_R}^{(k_R)}(t))', \quad t \ge 0,$$

where

$$Z_H^{(k)}(t) = A_{k,H} \int_{\mathbb{R}^k}' \Big(\int_0^t \prod_{j=1}^k (s - u_j)_+^{d-1} ds \Big) W(du_1) \ldots W(du_k), \quad (5.7.44)$$

where $W(du)$ is a Gaussian random measure on $\mathbb{R}$ with control measure du, $H = k(d - 1/2) + 1$ and $A_{k,H}$ is given by (4.2.7). The standard Hermite processes $Z_{H_r}^{(k_r)}(t)$s are sharing the same Gaussian random measure $W(du)$ in their multiple integral representations, and are dependent across r. The normalization $A_r(N)$ in (5.7.32) satisfies: for some $c_r > 0$, as $N \to \infty$,

$$A_r(N) \sim c_r N^{k_r(d_r - 1/2) + 1} L_1(N)^{k_r/2}.$$

Finally, we consider the mixed SRD and LRD case.

Theorem 5.7.19 *(Mixed case.) Break Y_N in (5.7.33) into 3 parts: $Y_N = (Y_{S_1,N}, Y_{S_2,N}, Y_{L,N})'$, where within $Y_{S_1,N}$ (R_{S_1}-dimensional) every component is SRD and has order $k_{S_1,r} = 1$, within $Y_{S_2,N}$ (R_{S_2}-dimensional) every component is SRD and has order $k_{S_2,r} \geq 2$, and within $Y_{L,N}$ (R_L-dimensional) every component is LRD. Then,*

$$Y_N(t) = (Y_{S_1,N}(t), Y_{S_2,N}(t), Y_{L,N}(t))' \xrightarrow{fdd} (W^{(1)}(t), B(t), Z^{(d,k)}(t))', \quad t \geq 0, \quad (5.7.45)$$

where $B(t) = (B_1(t), \ldots, B_{R_{S_2}}(t))'$ is the multivariate Gaussian process appearing in Theorem 5.7.16, $Z^{(d,k)}(t)$ is the multivariate Hermite process appearing in Theorem 5.7.18,

$$W^{(1)}(t) = (W(t), \ldots, W(t))', \quad (5.7.46)$$

where $W(t) = \int_0^t W(du)$ is the Brownian motion integrator for defining $Z^{(d,k)}(t)$ (see (5.7.44)). Moreover, $\{B(t)\}$ is independent of $\{W^{(1)}(t), Z^{(d,k)}(t)\}$.

Remark 5.7.20 To understand heuristically why $\{B(t)\}$ and $\{W^{(1)}(t), Z^{(d,k)}(t)\}$ are independent, note that $Y_{S_2,N}(t)$ belongs to the chaos of order greater than or equal to two, and is thus uncorrelated with $Y_{S_1,N}(t)$ which belongs to the first-order chaos, and also uncorrelated with the random noise $\{\epsilon_i\}$ which also belongs to the first-order chaos, and which after summing becomes asymptotically the Gaussian random measure W defining $Z^{(d,k)}(t)$. It is therefore simpler to establish independence in the context of Theorem 5.7.19 than in the context of Theorem 5.7.8.

5.8 Generation of Non-Gaussian Long- and Short-Range Dependent Series

Nonlinear transformations of Gaussian series provide interesting means to generate non-Gaussian stationary series. By non-Gaussian, we mean a stationary series having a particular non-Gaussian marginal distribution function. Thus, consider a stationary Gaussian series $X = \{X_n\}_{n \in \mathbb{Z}}$ with

$$\mathbb{E}X_n = 0, \quad \mathbb{E}X_n^2 = 1. \quad (5.8.1)$$

Let γ_X and ρ_X denote the autocovariance and autocorrelation functions of the series X. Because of (5.8.1), $\gamma_X(0) = 1$ and hence $\gamma_X = \rho_X$. We focus on the series $Y = \{Y_n\}_{n \in \mathbb{Z}}$ defined by

$$Y_n = G(X_n), \quad (5.8.2)$$

where $G : \mathbb{R} \mapsto \mathbb{R}$ is a deterministic function. Let γ_Y and ρ_Y denote the autocovariance and autocorrelation functions of the series Y. Let also F_Y be the marginal (cumulative) distribution function of Y.

The function G in (5.8.2) determines the marginal distribution F_Y and, as will be seen below, the autocovariance function γ_Y of the series Y. We will first discuss choices of G in terms of the marginal distribution, then study relationships between the autocovariance functions of X and Y, and finally consider the problem of generating series with a given marginal and autocovariance structure, using the construction (5.8.2).

5.8.1 Matching a Marginal Distribution

If a particular marginal distribution F_Y is sought, this may suggest a natural choice of the function G. For example, the choice

$$G(x) = x^2 \qquad (5.8.3)$$

leads to the χ_1^2 (chi-square with one degree of freedom) distribution,

$$G(x) = e^{\sigma x + \mu}, \quad \sigma > 0, \mu \in \mathbb{R}, \qquad (5.8.4)$$

leads to a log-normal distribution, and

$$G(x) = \begin{cases} 1, & \text{if } x \geq 0, \\ 0, & \text{if } x < 0, \end{cases} \qquad (5.8.5)$$

leads to the Bernoulli distribution with parameter $1/2$. This is not the only way to obtain these marginal distributions.

A marginal distribution F_Y could also be obtained through a general construction. One such standard construction is

$$G_1(x) = F_Y^{-1}(\Phi(x)), \qquad (5.8.6)$$

where $F_Y^{-1}(u) = \inf\{y : F_Y(y) \geq u\}$, $0 < u < 1$, is the generalized inverse (the quantile function) and Φ is the distribution function of a standard normal random variable. The basic idea behind (5.8.6) is that $\Phi(X)$ is a $U(0, 1)$ (uniform on $(0, 1)$) random variable for a standard normal variable X because, for $0 \leq y \leq 1$,

$$\mathbb{P}(\Phi(X) \leq y) = \mathbb{P}(X \leq \Phi^{-1}(y)) = \Phi(\Phi^{-1}(y)) = y,$$

and $F_Y^{-1}(U(0, 1))$ has distribution F_Y. Another general construction is

$$G_2(x) = F_Y^{-1}(2(\Phi(|x|) - 1/2)), \qquad (5.8.7)$$

by using the fact that $2(\Phi(|X|) - 1/2)$ is also distributed as $U(0, 1)$ for a standard normal variable X. Indeed, since $\mathbb{P}(|X| \leq a) = 2\Phi(a) - 1$, then with $0 \leq y \leq 1$ and $a = (y+1)/2$, we have

$$\mathbb{P}(2(\Phi(|X|) - 1/2) \leq y) = \mathbb{P}(\Phi(|X|) \leq a) = \mathbb{P}(|X| \leq \Phi^{-1}(a))$$
$$= 2\Phi(\Phi^{-1}(a)) - 1 = 2a - 1 = y.$$

We gather these and other basic facts about the transformations (5.8.6) and (5.8.7) in the following proposition.

Proposition 5.8.1 *Let Y be a random variable with distribution function F_Y. Let X be a standard normal variable, and G_1 and G_2 be the transformations given in (5.8.6) and (5.8.7). Then, the distribution functions of the random variables $G_1(X)$ and $G_2(X)$ are both equal to F_Y. Moreover, G_1 is a non-decreasing function and, if $G_1 \in L^2(\mathbb{R}, \phi)$ with $\phi(dx)$ defined in (5.1.2), then the first coefficient in the Hermite expansion (5.1.6) of G_1 is*

$$g_1 = \mathbb{E} G_1(X) H_1(X) = \mathbb{E} G_1(X) X = \frac{1}{\sqrt{2\pi}} \int_\mathbb{R} e^{-\frac{x^2}{2}} dG_1(x) > 0 \qquad (5.8.8)$$

and hence G_1 has Hermite rank 1. On the other hand, the function G_2 is even and, if $G_2 \in L^2(\mathbb{R}, \phi)$, then the first coefficient in the Hermite expansion of G_2 is $g_1 = 0$ and hence G_2 has Hermite rank greater than or equal to 2.

Proof The fact that $G_1(X)$ and $G_2(X)$ have distribution F_Y is argued before the proposition. G_1 is non-decreasing because Φ and F_Y^{-1} are non-decreasing. The formula (5.8.8) follows by integration by parts:

$$g_1 = \frac{1}{\sqrt{2\pi}} \int_{\mathbb{R}} G_1(x) x e^{-\frac{x^2}{2}} dx = -\frac{1}{\sqrt{2\pi}} \int_{\mathbb{R}} G_1(x) de^{-\frac{x^2}{2}} = \frac{1}{\sqrt{2\pi}} \int_{\mathbb{R}} e^{-\frac{x^2}{2}} dG_1(x) > 0,$$

since $\lim_{x \to \pm\infty} G_1(x) e^{-\frac{x^2}{2}} = 0$. The latter fact is a consequence of $\int_{\mathbb{R}} |G_1(x)| e^{-\frac{x^2}{2}} dx < \infty$, that is, $G_1 \in L^2(\mathbb{R}, \phi)$, and $G_1(x)$ being non-decreasing. The statements of the proposition about G_2 are obvious since G_2 is even. $\square$

Proposition 5.8.1 shows that the functions G_1 and G_2 in (5.8.6) and (5.8.7) are very different.

Example 5.8.2 (*Transformations for various marginals*) The function G in (5.8.3) is G_2 for the χ_1^2 distribution. Indeed, note that in this case, for $y > 0$ and for a standard normal random variable X,

$$F_Y(y) = \mathbb{P}(X^2 \leq y) = \mathbb{P}(-\sqrt{y} \leq X \leq \sqrt{y}) = \Phi(\sqrt{y}) - \Phi(-\sqrt{y}) = 2\Phi(\sqrt{y}) - 1$$

and hence

$$G_2(x) = F_Y^{-1}(2\Phi(|x|) - 1) = F_Y^{-1}(F_Y(x^2)) = x^2.$$

On the other hand, the functions G in (5.8.4) and (5.8.5) are G_1. In the case of (5.8.4), note that for $y > 0$,

$$F_Y(y) = \mathbb{P}(e^{\sigma X + \mu} \leq y) = \Phi((\log y - \mu)/\sigma)$$

and hence

$$F_Y^{-1}(u) = e^{\sigma \Phi^{-1}(u) + \mu},$$

yielding $G_1(x) = F_Y^{-1}(\Phi(x)) = e^{\sigma x + \mu} = G(x)$. In the case of (5.8.5), Y is Bernoulli with parameter $1/2$. Then, the function G is $G_1(x) = F_Y^{-1}(\Phi(x))$, where

$$F_Y(y) = \begin{cases} 0, & \text{if } y < 0, \\ 1/2, & \text{if } 0 \leq y < 1, \\ 1, & \text{if } y \geq 1, \end{cases} \qquad F_Y^{-1}(u) = \begin{cases} 0, & \text{if } 0 < u < 1/2, \\ 1, & \text{if } 1/2 \leq u < 1, \end{cases}$$

yielding

$$F_Y^{-1}(\Phi(x)) = \begin{cases} 0, & \text{if } x < 0, \\ 1, & \text{if } x \geq 0 \end{cases} = G(x).$$

5.8.2 Relationship Between Autocorrelations

We now turn to the relationships between the autocovariance functions of X and Y. By (5.2.1), we have

$$\gamma_Y(n) = \sum_{k=1}^{\infty} g_k^2 k! (\gamma_X(n))^k, \tag{5.8.9}$$

where g_k are the coefficients in the Hermite expansion (5.1.6) of G. The respective autocorrelation functions can then be related as

$$\rho_Y(n) = \sum_{k=1}^{\infty} \frac{g_k^2 k!}{\gamma_Y(0)} (\rho_X(n))^k, \tag{5.8.10}$$

since $\rho_X = \gamma_X$ by the assumption $\gamma_X(0) = 1$. We set

$$g(z) = \sum_{k=1}^{\infty} \frac{g_k^2 k!}{\gamma_Y(0)} z^k, \tag{5.8.11}$$

a function which will be used often below. We have then

$$\rho_Y(n) = g(\rho_X(n)). \tag{5.8.12}$$

Since an autocorrelation function takes values in $[-1, 1]$, we will be interested in the values of $g(z)$ for $z \in [-1, 1]$. Note that g is always well-defined for $z \in [-1, 1]$, $g(1) = \sum_{k=1}^{\infty} g_k^2 k!/\gamma_Y(0) = \text{Var}(G(X_0))/\gamma_Y(0) = 1$, $g(z)$ is non-decreasing for $z \in (0, 1)$, and $|g(z)| \leq 1$ for $z \in (-1, 1)$.

Example 5.8.3 (*Functions g for various transformations*) (a) Consider the function $G(x) = x^2$ in (5.8.3) leading to the χ_1^2 distribution. Since

$$G(x) = x^2 = 1 + (x^2 - 1) = H_0(x) + H_2(x), \tag{5.8.13}$$

the only nonzero coefficients g_k in the Hermite expansion are $g_0 = 1$, $g_1 = 1$. Then, by (5.8.9), $\gamma_Y(0) = g_2^2 2! = 2$ and hence

$$g(z) = z^2, \tag{5.8.14}$$

so that

$$\rho_Y(n) = (\rho_X(n))^2. \tag{5.8.15}$$

(b) Consider the function $G(x) = e^{\sigma x + \mu}$, $\sigma > 0$, $\mu \in \mathbb{R}$, in (5.8.4) leading to a lognormal distribution. Suppose first that $\mu = 0$. Using the identity (4.1.2), we have

$$G(x) = e^{\sigma x} = \sum_{k=0}^{\infty} \frac{e^{\sigma^2/2} \sigma^k}{k!} H_k(x), \tag{5.8.16}$$

so that $g_k = e^{\sigma^2/2} \sigma^k / k!$. Since $\mathbb{E} H_0(X) = 1$ and $\mathbb{E} H_k(X) = 0$ for $k \geq 1$, we get $\mathbb{E} e^{\sigma X} = e^{\sigma^2/2}$ and $\mathbb{E}(e^{\sigma X})^2 = \mathbb{E} e^{2\sigma X} = e^{(4\sigma^2)/2} = e^{2\sigma^2}$, so that $\gamma_Y(0) = \text{Var}(e^{\sigma X}) = \mathbb{E} e^{2\sigma X} -$

$(\mathbb{E}e^{\sigma X})^2 = e^{2\sigma^2} - e^{\sigma^2} =: c_\sigma^2$. It follows that

$$g(z) = \sum_{k=1}^\infty \frac{g_k^2 k!}{\gamma_Y(0)} z^k = \sum_{k=1}^\infty \frac{e^{\sigma^2} \sigma^{2k} k!}{c_\sigma^2 (k!)^2} z^k = \frac{e^{\sigma^2}}{c_\sigma^2} \sum_{k=1}^\infty \frac{(\sigma^2 z)^k}{k!} = \frac{e^{\sigma^2}}{c_\sigma^2}(e^{\sigma^2 z} - 1) \quad (5.8.17)$$

and hence

$$\rho_Y(n) = g(\rho_X(n)) = \frac{e^{\sigma^2}}{c_\sigma^2}(e^{\sigma^2 \rho_X(n)} - 1). \quad (5.8.18)$$

In the case of general μ, the multiplicative factor e^μ does not affect the correlation $\rho_Y(n)$ and the relation (5.8.18) continues to hold.

(c) Consider the function $G(x) = 1_{[0,\infty)}(x)$ in (5.8.5) leading to the Bernoulli distribution with parameter $1/2$. Instead of using the Hermite expansion of $G(x)$ as above (which is not that trivial for $G(x) = 1_{[0,\infty)}(x)$) to obtain the relationship between ρ_Y and ρ_X (and the corresponding function g), we shall specify the latter directly by using a well-known formula in Statistics. Observe that

$$\rho_Y(n) = \frac{\mathbb{E}Y_n Y_0 - \mathbb{E}Y_n \mathbb{E}Y_0}{\text{Var}(Y_0)} = \frac{\mathbb{E}1_{[0,\infty)}(X_n)1_{[0,\infty)}(X_0) - \mathbb{E}1_{[0,\infty)}(X_n)\mathbb{E}1_{[0,\infty)}(X_0)}{\text{Var}(1_{[0,\infty)}(X_0))}$$

$$= 4\left(\mathbb{P}(X_n \geq 0, X_0 \geq 0) - \frac{1}{4}\right), \quad (5.8.19)$$

using the fact that since X_n is centered Gaussian and hence symmetric, we have $\mathbb{E}1_{[0,\infty)}(X_n) = 1/2$. The probability in (5.8.19) is that of a bivariate, mean zero normal vector belonging to the orthant $[0,\infty)^2$. More generally, the probability P_n of a n-variate, mean zero normal variable to belong to the orthant $[0,\infty)^n$, $n \geq 1$, has been studied extensively in Statistics. It is well known (e.g., Kendall and Stuart [552], Section 15.6; Kotz, Balakrishnan, and Johnson [577], Chapter 45, Section 5.2) that, in the bivariate case $n = 2$,

$$P_2 = \frac{1}{4} + \frac{1}{2\pi} \arcsin(\rho), \quad (5.8.20)$$

where ρ is the correlation between the two components of the bivariate normal distribution. The relations (5.8.19) and (5.8.20) yield

$$\rho_Y(n) = \frac{2}{\pi} \arcsin(\rho_X(n)) =: g(\rho_X(n)). \quad (5.8.21)$$

In fact, a commonly found proof of (5.8.21) is akin to using the Price theorem as we outline in Example 5.8.6 below.

Note from (5.8.11) and the examples above that through the coefficients g_k, the transformation G determines the function g in the relationship (5.8.12) between autocorrelations ρ_Y and ρ_X. Since the same marginal distribution can be achieved through various transformations G (e.g., (5.8.6) and (5.8.7)), the corresponding functions g and relationships between autocorrelation functions will be different. Consider, for example, the χ_1^2 distribution. It can be obtained through the transformation (5.8.3) but also through the transformation

$$G(x) = F_{\chi_1^2}^{-1}(\Phi(x)), \quad (5.8.22)$$

5.8 Generation of Non-Gaussian Long- and Short-Range Dependent Series

where $F_{\chi_1^2}$ is the χ_1^2 distribution function. This is a special case of (5.8.6) which also leads to the χ_1^2 distribution. Its coefficients in the Hermite expansion can only be computed numerically. By Proposition 5.8.1, the first coefficient g_1 in the Hermite expansion of G in (5.8.22) is positive. As a consequence, the corresponding function g will take negative values (at least for negative z close to 0) and hence it will be very different from g in (5.8.14), which can only be nonnegative, associated with the transformation (5.8.13). As a further example, the log-normal distribution can be obtained through the transformation (5.8.4) but also through the transformation

$$G(x) = F_{LN}^{-1}(2(\Phi(|x|) - 1/2)), \tag{5.8.23}$$

where F_{LN} is the log-normal distribution function. The (nonnegative) function g corresponding to (5.8.23) will be different from the g in (5.8.17). For further discussion, see Section 5.8.4.

5.8.3 Price Theorem

The function G relates X_n to Y_n in the relation $Y_n = G(X_n)$ and the function g relates the autocorrelation of $\{X_n\}$ to the autocorrelation of $\{Y_n\}$. The Price theorem provides a differential equation for g, which can be solved to obtain g. In order to obtain this differential equation, we need to examine the derivatives of g. In this regard, the following auxiliary result will be useful. Recall the definition of the measure $\phi(dx)$ in (5.1.2).

Proposition 5.8.4 *Let $G \in L^2(\mathbb{R}, \phi)$ be n–times differentiable with $G^{(\ell)} = d^\ell G/dx^\ell \in L^2(\mathbb{R}, \phi)$, $\ell = 1, \ldots, n$. Suppose $\lim_{x \to \pm\infty} e^{-x^2/2} x^k G^{(l)}(x) = 0$ for $l = 1, \ldots, n-1$, and all $k \geq 1$. Then, for $\ell = 1, \ldots, n-1$, and $k \geq 0$,*

$$g_k^{(n)} = (k+1)(k+2)\ldots(k+n-\ell)g_{k+n-\ell}^{(\ell)} = (k+1)(k+2)\ldots(k+n)g_{k+n}, \tag{5.8.24}$$

where g_k and $g_k^{(\ell)}$ are the Hermite coefficients of G and $G^{(\ell)}$, respectively.

Proof Observe that by (5.1.7), (4.1.1) and integration by parts,

$$g_k^{(n)} = \frac{1}{k!}\mathbb{E}G^{(n)}(X)H_k(X) = \frac{1}{k!}\int_\mathbb{R} G^{(n)}(x)H_k(x)e^{-x^2/2}\frac{dx}{\sqrt{2\pi}}$$

$$= \frac{1}{k!}\int_\mathbb{R} G^{(n)}(x)(-1)^k\frac{d^k}{dx^k}(e^{-x^2/2})\frac{dx}{\sqrt{2\pi}} = -(-1)^k\frac{1}{k!}\int_\mathbb{R} G^{(n-1)}(x)\frac{d^{k+1}}{dx^{k+1}}(e^{-x^2/2})\frac{dx}{\sqrt{2\pi}}$$

$$= (k+1)\frac{1}{(k+1)!}\int_\mathbb{R} G^{(n-1)}(x)(-1)^{k+1}\frac{d^{k+1}}{dx^{k+1}}(e^{-x^2/2})\frac{dx}{\sqrt{2\pi}} = (k+1)g_{k+1}^{(n-1)}.$$

The relations in (5.8.24) follow by iterating the argument above. □

Proposition 5.8.4 yields the following result, known as the Price theorem, after Price [832].

Theorem 5.8.5 (*Price theorem.*) Let g be the function in (5.8.11). Under the assumptions and notation of Proposition 5.8.4, we have for $z \in (-1, 1)$,

$$\frac{d^n g}{dz^n}(z) = \frac{1}{\gamma_Y(0)} \sum_{k=0}^{\infty} (g_k^{(n)})^2 k! z^k = \frac{1}{\gamma_Y(0)} \mathbb{E} G^{(n)}(X_0) G^{(n)}(X_1)\Big|_{\gamma_X(1)=z}. \quad (5.8.25)$$

Proof The result follows from

$$\frac{d^n g}{dz^n}(z) = \sum_{k=n}^{\infty} \frac{g_k^2 k!}{\gamma_Y(0)} k(k-1)\ldots(k-n+1) z^{k-n}$$

$$= \frac{1}{\gamma_Y(0)} \sum_{\ell=0}^{\infty} g_{\ell+n}^2 (\ell+n)!(\ell+n)(\ell+n-1)\ldots(\ell+1) z^\ell$$

$$= \frac{1}{\gamma_Y(0)} \sum_{\ell=0}^{\infty} \Big(g_{\ell+n}(\ell+1)\ldots(\ell+n)\Big)^2 \ell! z^\ell$$

$$= \frac{1}{\gamma_Y(0)} \sum_{\ell=0}^{\infty} (g_\ell^{(n)})^2 \ell! z^\ell = \frac{1}{\gamma_Y(0)} \mathbb{E} G^{(n)}(X_0) G^{(n)}(X_1)\Big|_{\gamma_X(1)=z},$$

where we have used (5.8.24) and, in the last step, (5.2.1). The differentiation in the first step above is valid since $\sum_{\ell=0}^{\infty} (g_\ell^{(n)})^2 \ell! = \sum_{k=n}^{\infty} g_k^2 k! k(k-1)\ldots(k-n+1) < \infty$ by the assumption $G^{(n)} \in L^2(\mathbb{R}, \phi)$. □

We illustrate next how Theorem 5.8.5 can be used. On one hand, as the following example shows, it can be used to derive an expression for g.

Example 5.8.6 (*Functions g through Price theorem*) (a) Consider the transformation $G(x) = x^2$ in (5.8.13). If X has the $\mathcal{N}(0, 1)$ distribution, then $Y = X^2$ satisfies $\gamma_Y(0) = \text{Var}(Y) = \text{Var}(X^2) = 2$. Applying (5.8.25) with $n = 1$ and observing that $G^{(1)}(x) = 2x$ yields

$$\frac{dg}{dz}(z) = \frac{4}{2} \mathbb{E} X_0 X_1 \Big|_{\gamma_X(1)=z} = 2z.$$

Solving this differential equation for $g(z)$ yields $g(z) = z^2 + C$. The constant $C = 0$ since $g(0) = 0$. This yields $g(z) = z^2$, which agrees with (5.8.14).

(b) In the case $G(x) = e^{\sigma x}$ in (5.8.16), we get that

$$\frac{dg}{dz}(z) = \frac{\sigma^2}{c_\sigma^2} \mathbb{E} e^{\sigma X_0} e^{\sigma X_1} \Big|_{\gamma_X(1)=z}.$$

Note that, by (5.8.12) and the fact that X_n is standardized (so that $\rho_X(n) = \gamma_X(n)$), we have $\gamma_Y(n) = \mathbb{E} e^{\sigma X_n} e^{\sigma X_0} - \mathbb{E} e^{\sigma X_n} \mathbb{E} e^{\sigma X_0} = \gamma_Y(0) g(\gamma_X(n))$. Hence, after setting $n = 1$ and $\gamma_X(1) = z$,

$$\mathbb{E} e^{\sigma X_0} e^{\sigma X_1} \Big|_{\gamma_X(1)=z} = \gamma_Y(0) g(z) + (\mathbb{E} e^{\sigma X})^2.$$

5.8 Generation of Non-Gaussian Long- and Short-Range Dependent Series

As noted in Example 5.8.3, (b), $\gamma_Y(0) = e^{2\sigma^2} - e^{\sigma^2} =: c_\sigma^2$ and $\mathbb{E}e^{\sigma X} = e^{\sigma^2/2}$. Hence, the differential equation for g becomes

$$\frac{dg}{dz}(z) = \sigma^2 g(z) + \frac{\sigma^2 e^{\sigma^2}}{c_\sigma^2}.$$

The solution to this differential equation is $g(z) = Ce^{\sigma^2 z} - e^{\sigma^2}/c_\sigma^2$. The initial condition $g(0) = 0$ yields $C = e^{\sigma^2}/c_\sigma^2$ and hence

$$g(z) = \frac{\sigma^2 e^{\sigma^2}}{c_\sigma^2}(e^{\sigma^2 z} - 1),$$

which agrees with (5.8.18).

(c) Strictly speaking, the case $G(x) = 1_{[0,\infty)}(x)$ in (5.8.5) is not covered by the Price theorem since the function $G(x)$ is not differentiable at $x = 0$. But approximating G by, for example,

$$G_\epsilon(x) = \int_\mathbb{R} G(y) \frac{1}{\sqrt{2\pi}\epsilon} e^{-\frac{(x-y)^2}{2\epsilon^2}} dy = \int_{-\infty}^x \frac{1}{\sqrt{2\pi}\epsilon} e^{-\frac{z^2}{2\epsilon^2}} dz,$$

as $\epsilon \to 0$, we have for its derivative $G_\epsilon^{(1)}(x) = \frac{1}{\sqrt{2\pi}\epsilon} e^{-\frac{x^2}{2\epsilon^2}}$ and the relation (5.8.25) becomes

$$\frac{dg_\epsilon}{dz}(z) = \frac{1}{\text{Var}(G_\epsilon(X_0))} \mathbb{E} G_\epsilon^{(1)}(X_0) G_\epsilon^{(1)}(X_1) \Big|_{\gamma_X(1)=z}.$$

As $\epsilon \to 0$, $\text{Var}(G_\epsilon(X_0)) \to \text{Var}(G(X_0)) = 1/4$ and

$$\mathbb{E} G_\epsilon^{(1)}(X_0) G_\epsilon^{(1)}(X_1) \Big|_{\gamma_X(1)=z} = \int_\mathbb{R} \int_\mathbb{R} \frac{e^{-\frac{x_0^2}{2\epsilon^2}}}{\sqrt{2\pi}\epsilon} \frac{e^{-\frac{x_1^2}{2\epsilon^2}}}{\sqrt{2\pi}\epsilon} \frac{1}{2\pi(1-z^2)^{1/2}} e^{-\frac{1}{2(1-z^2)}(x_0^2 + x_1^2 - 2zx_0 x_1)} dx_0 dx_1$$

$$\to \frac{1}{2\pi(1-z^2)^{1/2}},$$

since $\frac{1}{\sqrt{2\pi}\epsilon} e^{-\frac{x_j^2}{2\epsilon^2}}$, $j = 0, 1$, act like kernel (or Dirac) functions localizing the integrand at $x_0 = x_1 = 0$. This suggests that, in the limit $\epsilon \to 0$,

$$\frac{dg}{dz}(z) = \frac{2}{\pi(1-z^2)^{1/2}}.$$

A general solution to this equation is

$$g(z) = \frac{2}{\pi} \arcsin(z) + C$$

and the constant $C = 0$ since $g(0) = 0$. The resulting function g is then the same as given in (5.8.21).

Here is another useful application of Theorem 5.8.5.

Corollary 5.8.7 *If $G \in L^2(\mathbb{R}, \phi)$ is an increasing function with $G^{(1)} \in L^2(\mathbb{R}, \phi)$, then $g(z)$ is an increasing function on $z \in (-1, 1)$.*

Proof By (5.8.25),

$$\frac{dg}{dz}(z) = \frac{1}{\gamma(0)}\mathbb{E}G^{(1)}(X_0)G^{(1)}(X_1)\Big|_{\gamma_X(1)=z}.$$

(There is no need to assume $\lim_{x\to\pm\infty} e^{-x^2/2}x^k G(x) = 0$ as this is satisfied arguing as in the proof of Proposition 5.8.1.) Since G is increasing, $G^{(1)}(x) > 0$ and hence $dg/dz(z) > 0$. □

Corollary 5.8.7 applies to the functions G in (5.8.6) when F_Y is continuous. Note also that the corollary is not obvious for $z \in (-1, 0)$ only. For $z \in (0, 1)$, the function g is always increasing by construction.

5.8.4 Matching a Targeted Autocovariance for Series with Prescribed Marginal

Consider now the problem of generating the stationary series $\{Y_n\}$ with a prescribed autocorrelation function ρ_Y and marginal distribution function F_Y. We focus on the constructions of the type $Y_n = G(X_n)$. The marginal distribution Y_n can then be obtained by using a suitable transformation G (Section 5.8.1). Such a transformation is not unique and its choice will affect the procedure for matching the autocovariance, and other order properties of the series, as well.

Suppose a transformation G was chosen to match a desired marginal F_Y. Is there an autocorrelation function ρ_X of the underlying Gaussian series $\{X_n\}$ so that $\{Y_n\}$ has the desired autocorrelation function ρ_Y? The relation (5.8.12) suggests to set

$$\rho_X(n) = g^{-1}(\rho_Y(n)), \quad (5.8.26)$$

where g^{-1} is the inverse of the function g.

The choice (5.8.26), however, may neither be possible nor yield a valid autocorrelation function. More precisely, on one hand, the choice of G and hence that of g may restrict the domain of g^{-1} for which $g^{-1}(\rho_Y(n))$ is even defined. Denote this domain by

$$\mathcal{R}_g = \{g(z) : z \in [-1, 1]\}, \quad (5.8.27)$$

which is also the range of g on $[-1, 1]$. For example, for $G(x) = x^2$ in (5.8.13), $g(z) = z^2$ so that $\mathcal{R}_g = [0, 1]$ and hence $g^{-1}(\rho_Y(n))$ is defined only for $\rho_Y(n) \geq 0$. In other words, with the transformation (5.8.13), only nonnegative autocorrelation functions ρ_Y could possibly be reached. The same conclusion holds for the general even transformation G in (5.8.7). In the case of G in (5.8.6), when F is continuous, g is increasing by Corollary 5.8.7 and hence $\mathcal{R}_g = [g(-1), g(1)]$. We have $g(1) = \sum_{k=1}^{\infty} g_k^2 k!/\gamma_Y(0) = \mathrm{Var}(G(X_0))/\gamma_Y(0) = 1$ and, similarly, $g(-1) = \mathrm{Cov}(G(X_0), G(-X_0))/\gamma_Y(0)$. Thus, in this case,

$$\mathcal{R}_g = [-g^*, 1] \quad \text{with} \quad g^* = -\frac{\mathrm{Cov}(G(X_0), G(-X_0))}{\mathrm{Var}(G(X_0))}. \quad (5.8.28)$$

There are examples where $(-g^*) > -1$ (see Example 5.8.9 below). See also Exercise 5.10.

Moreover, even if $\rho_Y(n) \in \mathcal{R}_g$ so that $\rho_X(n) = g^{-1}(\rho_Y(n))$ are well-defined, the resulting series $\rho_X(n)$ may not be a valid autocovariance function. Here is an example.

5.8 Generation of Non-Gaussian Long- and Short-Range Dependent Series

Example 5.8.8 *((Non-)Valid autocovariances)* Consider $G(x) = x^2$ in (5.8.13), the corresponding $g(z) = z^2$ in (5.8.14) and the relationship $\rho_Y(n) = (\rho_X(n))^2$ between autocorrelations in (5.8.15). Then, (5.8.26) becomes $\rho_X(n) = \pm\sqrt{\rho_Y(n)}$ or

$$\gamma_X(n) = \pm\sqrt{\frac{\gamma_Y(n)}{2}}, \qquad (5.8.29)$$

since $\mathrm{Var}(X) = \mathbb{E}X^2 - (\mathbb{E}X)^2 = 3 - 1 = 2$. Even if $\gamma_Y(n) \geq 0$ so that the right-hand side of (5.8.29) is well-defined, the left-hand side of (5.8.29) may not define a valid autocovariance structure. Indeed, consider an MA(1) autocovariance structure

$$\frac{\gamma_Y(n)}{2} = \begin{cases} 1, & \text{if } n = 0, \\ b, & \text{if } n = \pm 1, \\ 0, & \text{otherwise.} \end{cases} \qquad (5.8.30)$$

It is known (Brockwell and Davis [186], Example 1.5.1) that (5.8.30) is a valid autocovariance structure if and only if $|b| \leq 1/2$. Substituting (5.8.30) into (5.8.29) leads to

$$\gamma_X(n) = \begin{cases} 1, & \text{if } n = 0, \\ \pm\sqrt{b}, & \text{if } n = \pm 1, \\ 0, & \text{otherwise.} \end{cases} \qquad (5.8.31)$$

This is now a valid autocovariance structure if and only if $\sqrt{b} \leq 1/2$ or $0 \leq b \leq 1/4$. Thus, if $1/4 < b \leq 1/2$, $\gamma_Y(n)$ in (5.8.30) is a valid autocovariance structure but $\gamma_X(n)$ in (5.8.31) is not.

The argument above could also be repeated for moving average series of finite length $n = 0, \ldots, N-1$. It is known (Grenander and Szegö [430], Section 5.3, (a)) that $\gamma_Y(n)$, $n = 0, \ldots, N-1$, is a valid autocovariance structure if and only if $|b| \leq 1/(2\cos(\pi/(N+1)))$.

We now present a method to choose a *valid* underlying autocovariance $\rho_X = \gamma_X$, and to generate the underlying Gaussian series $\{X_n\}$ in (5.8.2). The method is a natural modification of the circulant matrix embedding method in Section 2.11.2. The series $\{X_n\}$ will be generated for $n = 0, \ldots, N-1$; that is, of length N. We suppose that a function G in (5.8.2) has been selected, and that the targeted autocorrelation $\rho_Y(n) \in \mathcal{R}_g$, where $\mathcal{R}_g$ is defined in (5.8.27).

Algorithm for generating an approximate underlying Gaussian series:

- *Step 1:* Compute
$$\widetilde{\gamma}(n) = g^{-1}(\rho_Y(n)), \quad n = 0, \ldots, N-1. \qquad (5.8.32)$$

- *Step 2:* Form a circulant matrix $\widetilde{\Sigma}$ in (2.11.6) of size $2M = 2(N-1)$ using $\widetilde{\gamma}$ instead of γ. Compute the eigenvalues $\widetilde{\lambda}_m$, $m = 0, \ldots, 2M-1$, of $\widetilde{\Sigma}$ as in (2.11.8) using $\widetilde{\gamma}$ instead of γ.[8]

- *Step 3:* Set
$$\lambda_m = \begin{cases} \widetilde{\lambda}_m, & \text{if } \widetilde{\lambda}_m \geq 0, \\ 0, & \text{if } \widetilde{\lambda}_m < 0, \end{cases} \quad m = 0, \ldots, 2M-1.$$

---

[8] The eigenvalues of $\widetilde{\Sigma}$ are denoted $\widetilde{\lambda}_m$ instead of g_m to avoid confusion with the Hermite coefficients.

Let $\Lambda = \text{diag}\{\lambda_0, \ldots, \lambda_{2M-1}\}$ and set $\widetilde{\Sigma}_0 = F\Lambda F^*$ in analogy to (2.11.7).
- *Step 4:* Generate a Gaussian series in (2.11.9) using λ_j instead of $\widetilde{\lambda}_j$. Denoting it $\widetilde{X}^0$ instead of $\widetilde{Y}$ and using Proposition 2.11.2, $\Re(\widetilde{X}^0)$ and $\Im(\widetilde{X}^0)$ have the autocovariance structure $\widetilde{\Sigma}_0$.
- *Step 5:* Take the first N values of $\Re(\widetilde{X}^0)/(\mathbb{E}\Re(\widetilde{X}^0)^2)^{1/2}$ or $\Im(\widetilde{X}^0)/(\mathbb{E}\Im(\widetilde{X}^0)^2)^{1/2}$ for the time series X_n, $n = 0, \ldots, N-1$.

Here are a few comments on the steps of the algorithm. In Step 1, $\widetilde{\gamma}(n)$ are computed either by using an explicit form of g^{-1} when it is available (as for g's in Example 5.8.3), or numerically (as discussed in Example 5.8.9 below). Since the circulant matrix $\widetilde{\Sigma}$ is symmetric, the eigenvalues $\widetilde{\lambda}_m$ in Step 2 are real-valued in general, and if some of them are negative, they are set to zero in Step 3. The rest of the argument is the same as in the circulant matrix embedding in Section 2.11.2 except that the nonnegative eigenvalues λ_m replace the eigenvalues $\widetilde{\lambda}_m$. The rescaling by $(\mathbb{E}\Re(\widetilde{X}^0)^2)^{1/2}$ or $(\mathbb{E}\Im(\widetilde{X}^0)^2)^{1/2}$ in Step 5 is there to ensure that the resulting series has variance 1.

We next shed light on the algorithm by discussing several important points. Parts of the discussion below are heuristic.

First, note that if $\widetilde{\lambda}_m \geq 0$ for all $m = 0, \ldots, 2M - 1$, then $\widetilde{\gamma}(n)$, $n = 0, \ldots, N - 1$, is a valid autocovariance structure and the algorithm generates a Gaussian series having $\widetilde{\gamma}(n)$ as autocovariance function. On the other hand, note again from Step 2 that

$$\widetilde{\lambda}_m = \widetilde{\gamma}(0) + \widetilde{\gamma}(M)(-1)^m + 2\sum_{j=1}^{M-1} \widetilde{\gamma}(j) \cos\left(\frac{\pi j m}{M}\right).$$

For large N, one expects that

$$\widetilde{\lambda}_m \approx 2\pi \widetilde{f}\left(\frac{\pi m}{M}\right) \qquad (5.8.33)$$

(see the proof of Proposition 2.11.4), where

$$\widetilde{f}(\lambda) = \frac{1}{2\pi}\left(\widetilde{\gamma}(0) + 2\sum_{j=1}^{\infty} \widetilde{\gamma}(j) \cos(j\lambda)\right).$$

One has $\widetilde{f}(\lambda) \geq 0$ if and only if $\widetilde{\gamma}(n)$, $n \in \mathbb{Z}$, is a valid autocovariance structure. Hence, for large N, the fact that $\widetilde{\lambda}_m < 0$ for some m strongly suggests that $\widetilde{\gamma}(n)$, $n = 0, \ldots, N - 1$, is not a valid autocovariance structure. The algorithm thus provides, in particular, a very fast and almost exact method to determine whether $\widetilde{\gamma}(n)$, $n = 0, \ldots, N - 1$, is a valid autocovariance structure.

Second, one can argue that the algorithm is optimal in a certain sense. Consider

$$I := \min \sum_{n=-\infty}^{\infty} |\rho_Y(n) - \rho_{app,Y}(n)|^2 w_n,$$

where the minimum is over $\rho_{app,Y}$, which is the autocorrelation function of a series $Y_{app,n} = g(X_{app,n})$ for a Gaussian series $X_{app,n}$ having autocovariance $\gamma_{app,X}$, $w_n \geq 0$ are certain weights to be chosen below and "app" stands for approximation. Supposing that $\widetilde{\gamma}(n)$ and $\gamma_{app,X}(n)$ are close and using the Taylor approximation, we get that

5.8 Generation of Non-Gaussian Long- and Short-Range Dependent Series 339

$$I = \min \sum_{n=-\infty}^{\infty} |g(\widetilde{\gamma}(n)) - g(\gamma_{app,X}(n))|^2 w_n$$

$$\approx \min \sum_{n=-\infty}^{\infty} |\widetilde{\gamma}(n) - \gamma_{app,X}(n)|^2 |g'(\widetilde{\gamma}(n))|^2 w_n = \min \sum_{n=-\infty}^{\infty} |\widetilde{\gamma}(n) - \gamma_{app,X}(n)|^2,$$

if $w_n = |g'(\widetilde{\gamma}(n))|^{-2}$. Since $\widetilde{\gamma}(n) - \gamma_{app,X}(n)$ are the Fourier coefficients of the function $\widetilde{f}(\lambda) - f_{app,X}(\lambda)$, we can further write the above as

$$I \approx \min \frac{1}{2\pi} \int_{-\pi}^{\pi} |\widetilde{f}(\lambda) - f_{app,X}(\lambda)|^2 d\lambda.$$

Since the minimum is over $f_{app,X}(\lambda) \geq 0$, the optimal choice is

$$f_{app,X}(\lambda) = \begin{cases} \widetilde{f}(\lambda), & \text{if } \widetilde{f}(\lambda) \geq 0, \\ 0, & \text{if } \widetilde{f}(\lambda) < 0. \end{cases}$$

In view of (5.8.33), this corresponds to choosing λ_m as in Step 3.

Finally, we conclude by illustrating the algorithm and some ideas above with a concrete numerical example (see also Helgason, Pipiras, and Abry [464]).

Example 5.8.9 (*Generating series with χ_1^2 marginal and AR(1) autocovariance*) Suppose a series $\{Y_n\}$ with χ_1^2 marginal and autocovariance function $\gamma_Y(n) = 2\phi^{|n|}$, $|\phi| < 1$, of AR(1) series is targeted. Suppose also that negative correlations are sought; that is, $-1 < \phi < 0$. Consider the transformation G in (5.8.22) for which negative correlations are possible (see (5.8.28)). After computing the coefficients in its Hermite expansion numerically and using (5.8.11), the plot of the corresponding function g is given in Figure 5.2, the

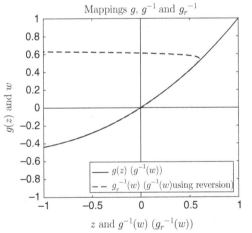
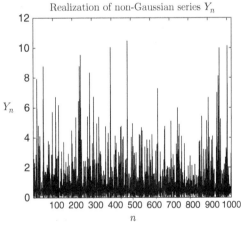

Figure 5.2 Left: The solid line is a graph $w = g(z)$ and by viewing the vertical axis as the w axis, it is a graph of $z = g^{-1}(w)$, where $w \in [-0.44, 1]$. The dashed line is the reversion approximation $g_r^{-1}(w)$ of $g^{-1}(w)$. The two coincide for $w \in [-0.44, 0.47]$ (the dashed line is obscured by the solid line), but diverge for $w > 0.47$. Right: Realization of the non-Gaussian series $\{Y_n\}$ in Example 5.8.9.

left plot. The function g is increasing with $\mathcal{R}_g = [-g^*, 1]$ in (5.8.27) and (5.8.28), where $(-g^*)$ is about -0.44. Take $\phi = -0.35$ so that $\rho_Y(n) = \phi^{|n|}$ fall in the domain $\mathcal{R}_g$. Take the sample size $N = 1024$.

Computing $\widetilde{\gamma}(n)$ in (5.8.32) requires g^{-1}. This is the function $g^{-1}(w)$, $w \in [-g^*, 1]$, seen (in solid line) in the same Figure 5.2, the left plot, but viewing the vertical (horizontal, resp.) axis as being horizontal (vertical, resp.). An explicit form of g^{-1} is not available in this example. One possibility is to solve (5.8.12) numerically for each $\rho_Y(n)$ in terms of $\rho_X(n)$, which is facilitated by the fact that g is increasing. A convenient alternative is to use the so-called *reversion* operation. The function g has a Taylor expansion in (5.8.11), with numerically computed coefficients. Denote the expansion by $g(z) = \sum_{k=1}^{\infty} b_k z^k$. It is expected that g^{-1} will also have a Taylor expansion $g^{-1}(w) = \sum_{k=1}^{\infty} d_k w^k$ (see, e.g., Henrici [467]). A substitution of the Taylor series into the equation $g(g^{-1}(w)) = w$, that is, $\sum_{j=1}^{\infty} b_j (\sum_{k=1}^{\infty} d_k w^k)^j = w$, and an identification of the coefficients yields $b_1 d_1 = 1$, $b_1 d_2 + b_2 d_1 = 0$, $b_1 d_3 + 2 b_2 d_1 d_2 + b_3 d_1^3 = 0$, ... Hence, $d_1 = b_1^{-1}$, $d_2 = -b_1^{-3} b_2$, $d_3 = b_1^{-5}(2b_2^2 - b_1 b_3)$, and so on. The latter operation on series (sequences) is known as *reversion* (see, e.g., Henrici [467]). Denote by $g_r^{-1}(w)$ the inverse function of g defined through the Taylor series and the reversion operation. A disadvantage of $g_r^{-1}(w)$ is that it does not need to coincide with $g^{-1}(w)$ over the whole range where the latter is defined (see, e.g., Henrici [467]). Indeed, Figure 5.2, the left plot, also depicts $g_r^{-1}(w)$ (dashed line): it coincides with $g^{-1}(w)$ for $w \in [-g^*, g_1]$ with $g_1 \approx 0.47$, where the dashed line is obscured by the solid line corresponding to the inverse $g^{-1}(w)$, but then diverges quickly for $w > g_1 \approx 0.47$ (the dashed line would continue below $z = -1$ if the z axis were extended to include negative values smaller than -1). We note, however, that $g_r^{-1}(w)$ can still be used to evaluate $g^{-1}(\gamma_Y(n))$ in (5.8.32) since most of the autocovariances $\gamma_Y(n)$ are expected to be small (and, in particular, belong to $[-g^*, g_1]$ where g_r^{-1} and g^{-1} coincide). In fact, in the example here, $\rho_Y(n) = 0.35^{|n|}$ belong to $[-g^*, g_1] \approx [-0.44, 0.47]$ for all $n \geq 1$.

In Figure 5.2, the right plot, depicts a realization of the non-Gaussian series $\{Y_n\}$. In Figure 5.3, for the first few lags n, we also plot the targeted autocovariance $\gamma_Y(n)$ and its approximation $\gamma_{app,Y}(n)$ obtained from the algorithm described above, as well as the difference between them.

Remark 5.8.10 The AR(1) covariance structure considered in Example 5.8.9 corresponds to SRD. Instead, we could have considered, for example, the FARIMA$(0, d, 0)$ covariance structure associated with LRD. Alternatively, if the χ_1^2 marginal is desired and just the LRD property of the resulting series $\{Y_n\}$ is sought (that is, without a specific resulting LRD covariance structure), one can consider the series $Y_n = X_n^2$ where $\{X_n\}$ is LRD. Indeed, supposing X_n is standardized, the series $\{Y_n\}$ will indeed have the χ_1^2 marginal and be LRD with the parameter $2d - 1/2 \in (0, 1/2)$ by Proposition 5.2.4, where $d \in (1/4, 1/2)$ is the LRD parameter of the series $\{X_n\}$.

Remark 5.8.11 In this section, we have focused on generation of non-Gaussian time series with prescribed marginals and targeted autocovariance structures. For estimation issues in related models, see Livsey, Lund, Kechagias, and Pipiras [640].

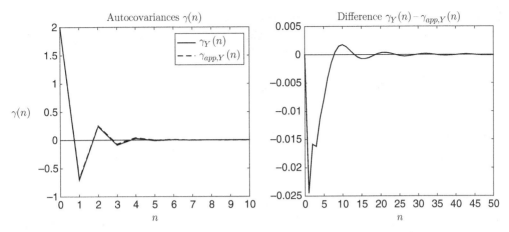

Figure 5.3 Left: The targeted autocovariance $\gamma_Y(n)$ and its approximation $\gamma_{app,Y}(n)$ in Example 5.8.9. Right: The difference $\gamma_Y(n) - \gamma_{app,Y}(n)$.

5.9 Exercises

The symbols * and ** next to some exercises are explained in Section 2.12.

Exercise 5.1 Let g_n, $n \geq 0$, be the coefficients (5.1.7) in the Hermite expansion (5.1.6) of a function $G \in L^2(\mathbb{R}, \phi)$. Show that the definition (4.1.1) of Hermite polynomials and integration by parts yield

$$g_n = -\frac{1}{n!}\mathbb{E}G'(X)H_{n-1}(X)$$

under suitable to-be-specified conditions. Conclude under these suitable conditions that $(n!)^{1/2}|g_n| \leq C/n^{1/2}$ for a constant $C > 0$. Generalize this result to obtain a bound of the kind $C/n^{r/2}$ with integer $r \geq 1$. *Hint:* See Proposition 5.8.4.

Exercise 5.2 If $G \in L^2(\mathbb{R}, \phi)$ is an even function – that is, $G(x) = G(-x)$, $x \in \mathbb{R}$ – what can its Hermite rank be? What about the Hermite rank of an odd function G satisfying $G(x) = -G(-x)$, $x \in \mathbb{R}$?

Exercise 5.3 Do functions $G(x)$, $x \in \mathbb{R}$, and $G(ax + b)$, $x \in \mathbb{R}$, for constants $a \neq 0, b$ have the same Hermite rank?

Exercise 5.4 (a) Prove the convergence (5.3.19) and that the limit can be expressed as (5.3.20). (b) Prove the more general convergence (5.3.14), and that the limit is continuous.

Exercise 5.5 Prove Lemma 5.4.4. *Hint:* If S_N denotes the left-hand side of (5.4.17), write $|S_N| \leq \int_0^a f_N(u)du$, where $f_N(u) = \sum_{n_1=1}^{[Na]} c_{N,n_1} 1_{(\frac{n_1-1}{N}, \frac{n_1}{N}]}(u)$ and $c_{N,n_1} = \sum_{n_2=1}^{[Nb]-[Na]} |c([Na] + n_2 - n_1)|$, and use the dominated convergence theorem.

Exercise 5.6** The non-central limit theorem (Theorem 5.3.1) assumes that $d \in ((1 - 1/k)/2, 1/2)$ and the central limit theorem (Theorem 5.4.1) that $d \in (0, (1-1/k)/2)$. What happens in the boundary case $d = (1 - 1/k)/2$?

Exercise 5.7 Under the assumptions of Theorem 5.6.3, let $G(z) = z^2$ and $1/4 < d < 1/2$. (*i*) Show that in this case

$$G(X_n) = \sum_{0 \leq p_1 \neq p_2 < \infty} \psi_{p_1} \psi_{p_2} \epsilon_{n-p_1} \epsilon_{n-p_2} + \sum_{0 \leq p < \infty} \psi_p^2 (\epsilon_{n-p}^2 - \mu_2) + \mathbb{E}G(X_n)$$
$$=: A_n + B_n + \mathbb{E}G(X_n),$$

where $\mu_2 = \mathbb{E}\epsilon_1^2$. (*ii*) Show that $\mathbb{E}(\sum_{n=1}^N A_n)^2 \sim L_1(N)^4 N^{4d}$ as $N \to \infty$. (*iii*) Show that the autocovariance of B_n is summable so that $\mathbb{E}(\sum_{n=1}^N B_n)^2 = O(N)$. (*iv*) Prove Theorem 5.6.3 in this case using Proposition 5.6.4.

Exercise 5.8 Prove that the bound in (5.7.39) is finite when $d < \frac{1}{2}(1 - \frac{1}{k})$. *Hint:* One way is to use a similar bound with d replaced by d^* such that $\max(0, d) < d^* < \frac{1}{2}(1 - \frac{1}{k})$, and use Proposition 2.2.9.

Exercise 5.9** The partial sum processes of the type (5.0.1) were considered throughout Sections 5.3–5.7 for $t \geq 0$. Formulate the results of these sections for partial sum processes for $t \in \mathbb{R}$.

Exercise 5.10** Prove or disprove the following conjectures. (*a*) As in (5.8.27), let $\mathcal{R}_g$ be the range of the function g, associated with a function G matching a marginal distribution F_Y. Then, $\mathcal{R}_g \subset \mathcal{R}_{g_1}$, where the function g_1 is associated with the function G_1 in (5.8.6). (*b*) Similarly and stronger, let $\mathcal{C}_g = \{\{g(\gamma_X(n))\}_{n \in \mathbb{Z}} :$ Gaussian stationary $X = \{X_n\}$ with $\mathbb{E}X_n = 0, \mathbb{E}X_n^2 = 1\}$ be the collection of possible autocovariances of the series $\{G(X_n)\}$ with a function G matching a marginal distribution F_Y and giving rise to the function g. Then, $\mathcal{C}_g \subset \mathcal{C}_{g_1}$, where the function g_1 is associated with the function G_1 in (5.8.6).

5.10 Bibliographical Notes

Section 5.1: The results in the key Proposition 5.1.1 and its consequences (e.g., Proposition 5.1.4) are sometimes referred to as *Mehler's formula*. Mehler's formula (Mehler [714]) reads, under the assumptions of Proposition 5.1.1,

$$q(x, y) = \sum_{k=0}^{\infty} \frac{(\mathbb{E}XY)^k}{k!} H_k(x) H_k(y) q(x) q(y),$$

where $q(x, y)$ is the joint density of (X, Y) and $q(x), q(y)$ are the marginal densities of X, Y. Assuming that $\mathbb{E}H_m(X)H_n(X) = \int_{\mathbb{R}} H_m(x) H_n(x) q(x) dx = n!$ if $m = n$ and $= 0$ if $m \neq n$ (which is stated as Corollary 5.1.2 but can also be proved simpler by using basic properties of Hermite polynomials), note that the Mehler's formula above implies (5.1.1) since

$$\mathbb{E}H_n(X)H_m(Y) = \int_{\mathbb{R}}\int_{\mathbb{R}} H_n(x)H_m(y)q(x,y)dxdy$$

$$= \sum_{k=0}^{\infty} \frac{(\mathbb{E}XY)^k}{k!} \int_{\mathbb{R}} H_n(x)H_k(x)q(x)dx \int_{\mathbb{R}} H_m(y)H_k(y)q(y)dy.$$

Section 5.2: The notion of Hermite rank was introduced by Taqqu [943]. A similar notion, that of "Laguerre rank" when the Gaussian distribution is replaced by a gamma distribution can be found in Leonenko, Ruiz-Medina, and Taqqu [615], together with a corresponding non-central limit theorem. Testing for Hermite rank in Gaussian subordinated processes is considered in Beran, Möhrle, and Ghosh [128].

Section 5.3: Non-central limit theorem for Gaussian stationary series was obtained by Dobrushin and Major [321], Taqqu [943, 945]. See also Rosenblatt [860, 861], Ibragimov [514], Berman [132], Maejima [661, 662]. A random field version can be found in Leonenko, Ruiz-Medina, and Taqqu [616]. The proof presented in Section 5.3 follows the approach of Dobrushin and Major [321]. The rate of convergence to the Rosenblatt-type distribution in the non-central limit theorem is studied in Anh, Leonenko, and Olenko [37].

Sly and Heyde [911] have an interesting result involving a sequence $G(X_n)$ where X_n is a linear Gaussian sequence but where $G(X_n)$ has heavy tails with exponent $0 < \alpha < 2$. Although $G(X_n)$ does not have finite variance, Hermite polynomials are still involved. They show that the limit can be an Hermite process or an α-stable Lévy motion depending on the values of the parameters. In the boundary case, the limit is a sum of independent copies of these two processes.

Section 5.4: The proof of the central limit theorem for Gaussian stationary series presented in Section 5.4 follows the approach of Breuer and Major [184]. Some earlier related work can be found in Sun [931, 932], Cuzick [272], and some further work in Maruyama [693, 694], Giraitis and Surgailis [396], Ho and Sun [478], Chambers and Slud [209, 210]. The convergence of densities in the central limit theorem for functions of Gaussian stationary series is studied in Hu, Nualart, Tindel, and Xu [496].

Section 5.5: Theorem 5.5.1 is proved in Nualart and Peccati [771] by means of stochastic calculus techniques. For books dealing with the fourth moment condition, see Peccati and Taqqu [797], Nourdin and Peccati [763], Nourdin [760] and Tudor [963]. Peccati and Zheng [799] consider functionals of a Poisson random measure.

Section 5.6: The material of Section 5.6.1 is based on Surgailis [935], Lévy-Leduc and Taqqu [618]. Other work and a related approach based on Appell polynomials, for both central and non-central limit theorems, can be found in Surgailis [934, 936, 937], Giraitis [392, 393, 394], Giraitis and Surgailis [397, 398], Avram and Taqqu [59], Surgailis and Vaičiulis [938].

Nourdin, Peccati, and Reinert [767] established the following universality result: if a sequence of off-diagonal homogeneous polynomial forms in i.i.d. standard normal random variables converges in distribution to a normal, then the convergence also holds if one replaces these i.i.d. standard normal random variables in the polynomial form by any independent standardized random variables with uniformly bounded third absolute moment. The result, which was stated for polynomial forms with a finite number of terms, was extended by Bai and Taqqu [79] to allow an infinite number of terms in the polynomial forms. Based on a contraction criterion derived from this extended universality result, they obtain a central limit theorem for long-range dependent nonlinear processes, whose memory parameter lies at the boundary between short and long memory.

The material of Section 5.6.2 follows Ho and Hsing [477], Lévy-Leduc and Taqqu [618]. Related work, for both central and non-central limit theorems, includes Ho and Hsing [476], Giraitis and Surgailis [400], Hsing [490], Wu [1012].

Section 5.7: The multivariate results involving functions of Gaussian random variable are based on Bai and Taqqu [74]. In the SRD case, the multivariate convergence to the multivariate Gaussian distribution also follows from the univariate convergence and the convergence of the covariances, as shown in Peccati and Tudor [798]. In Bai and Taqqu [74], mixed convergence was established up to Hermite order two in order to ensure that the moment-determinancy of the limit distributions. This restriction has now been eliminated due to the recent results of Nourdin et al. [768] and Nourdin and Rosiński [764], as stated in Theorem 5.7.10 and Corollary 5.7.11.

Proofs for the corresponding multivariate results for multilinear processes presented in Section 5.7.4 can be found in Bai and Taqqu [75].

Section 5.8: The material of this section is based on Helgason et al. [464]. But probabilistic modeling based on transformations of Gaussian stationary time series is quite popular in engineering mechanics (e.g., Grigoriu [431], Bocchinia and Deodatis [164], Nichols, Olson, Michalowicz, and Bucholtz [745]), signal processing (e.g., Liu and Munson [639], Scherrer, Larrieu, Owezarski, Borgnat, and Abry [887]), econometrics (e.g., Dittmann and Granger [318]) and other areas.

Other notes: Central and non-central limit theorems for Gaussian stationary vector time series are considered in Berman [133], Ho and Sun [478], Arcones [42, 43], Sánchez de Naranjo [884, 885]. Several of the references given above also cover the case of central and non-central limit theorems for Gaussian stationary random fields, including Dobrushin and Major [321], Breuer and Major [184]. For other related work, see Maejima [663, 665], Leonenko and Parkhomenko [614], Leonenko and Olenko [613], Lahiri and Robinson [588] and the textbooks by Ivanov and Leonenko [525], Leonenko [612]. Limit theorems for tempered Hermite processes can be found in Sabzikar [875].

The following papers of Bai and Taqqu contain recent results. The paper [77] introduces Generalized Hermite processes, the paper [78] deals with the convergence of long-memory discrete kth-order Volterra processes, the paper [82] studies the impact of diagonals of polynomial forms on limit theorems with long memory and the paper [76] analyzes the structure of the third moment of the generalized Rosenblatt distribution.

Recent results on limit theorems for multiple Wiener integrals have exploited the Malliavin calculus. See the monographs on the subject by Nourdin and Peccati [763], Nourdin [760]. Several notable papers are Nualart and Peccati [771], Nourdin and Peccati [762]. The basics of the Malliavin calculus are given in Appendix C, and used in Chapter 7.

6

Fractional Calculus and Integration of Deterministic Functions with Respect to FBM

There are fundamental connections between self-similarity and fractional calculus, which is an area of real analysis. These connections are explored here in the context of fractional Brownian motion (FBM).

The basics of fractional calculus can be found in Section 6.1 where we introduce the following fractional integrals and derivatives:

- Riemann–Liouville fractional integral on an interval (Definition 6.1.1)
- Riemann–Liouville fractional derivative on an interval (Definition 6.1.4)
- Fractional integral on the real line (Definition 6.1.13)
- Liouville fractional derivative on the real line (Definition 6.1.14)
- Marchaud fractional derivative on the real line (Definition 6.1.18)

Representations of FBM are obtained in terms of fractional integrals and derivatives in Section 6.2. These representations then play a central role in developing deterministic integration with respect to FBM, and in being able to address various applications involving that process.

Integrals of *deterministic* functions with respect to FBM will be called *fractional Wiener integrals*. They are introduced in Section 6.3 and defined first as integrals on an interval (Section 6.3.2) and then as integrals on the real line (Section 6.3.6). Applications involving the Girsanov formula, prediction and filtering can be found in Section 6.4. Integrals of *random* functions with respect to FBM are introduced in Chapter 7.

6.1 Fractional Integrals and Derivatives

We provide here the basics of fractional integrals and derivatives on an interval (Sections 6.1.1 and 6.1.2) and fractional integrals and derivatives on the real line (Sections 6.1.3–6.1.5).

6.1.1 Fractional Integrals on an Interval

The standard way of motivating the definition of a fractional integral is to start with an n–tuple iterated integral and show that it can be expressed as a single integral involving the parameter n. The fractional integral of order $\alpha > 0$ is then defined by replacing the integer n by the real number α in the resulting expression.

Let $a < b$ be two real numbers and ϕ be an integrable function on $[a, b]$. A multiple integral of ϕ can be expressed as

$$\int_a^{t_n} \cdots \left\{ \int_a^{t_2} \left\{ \int_a^{t_1} \phi(u) du \right\} dt_1 \right\} \cdots dt_{n-1} = \frac{1}{(n-1)!} \int_a^{t_n} \phi(u)(t_n - u)^{n-1} du, \quad (6.1.1)$$

where $t_n \in [a, b]$ and $n \geq 1$. (By convention, $(0)! = 1$ and $a^0 = 1$.) The proof is by induction. Consider $n = 2$. We have $a < u < t_1 < t_2$. Then, by changing the order of integration, we get

$$\int_a^{t_2} \int_a^{t_1} \phi(u) du dt_1 = \int_a^{t_2} \phi(u) \int_u^{t_2} dt_1 du = \int_a^{t_2} \phi(u)(t_2 - u) du,$$

so that the relation (6.1.1) holds for $n = 2$. Suppose now that it holds for $n-1$ and consider it for n. By induction and the change of the order of integration, we get in the same way that

$$\int_a^{t_n} \cdots \int_a^{t_2} \int_a^{t_1} \phi(u) du dt_1 \cdots dt_{n-1} = \frac{1}{(n-2)!} \int_a^{t_n} \int_a^{t_{n-1}} \phi(u)(t_{n-1} - u)^{n-2} du dt_{n-1}$$

$$= \frac{1}{(n-2)!} \int_a^{t_n} \phi(u) \int_u^{t_n} (t_{n-1} - u)^{n-2} dt_{n-1} du = \frac{1}{(n-1)!} \int_a^{t_n} \phi(u)(t_n - u)^{n-1} du,$$

which is the relation (6.1.1). The idea behind fractional integrals is to replace an integer n in (6.1.1) by a real number $\alpha > 0$ (hence the name "*fractional* integrals"). Observe also that $(n-1)! = \Gamma(n)$, where Γ is the gamma function. We write $L^p[a, b]$ for the space $L^p([a, b], du)$, $p > 0$.

Definition 6.1.1 Let $\phi \in L^1[a, b]$ and $\alpha > 0$. The integrals

$$(I_{a+}^\alpha \phi)(s) = \frac{1}{\Gamma(\alpha)} \int_a^s \phi(u)(s - u)^{\alpha-1} du = \frac{1}{\Gamma(\alpha)} \int_a^b \phi(u)(s - u)_+^{\alpha-1} du, \quad (6.1.2)$$

$s \in (a, b)$, and

$$(I_{b-}^\alpha \phi)(s) = \frac{1}{\Gamma(\alpha)} \int_s^b \phi(u)(u - s)^{\alpha-1} du = \frac{1}{\Gamma(\alpha)} \int_a^b \phi(u)(u - s)_+^{\alpha-1} du, \quad (6.1.3)$$

$s \in (a, b)$, are called *fractional integrals* of order α. (For any real number $x \neq 0$, $x^0 = 1$.) The fractional integral I_{a+}^α is called *left-sided* since the integration in (6.1.2) is over $[a, s]$ which is the left side of the interval $[a, b]$. The integral I_{b-}^α is called *right-sided* since the integration in (6.1.3) is over $[s, b]$. Both (6.1.2) and (6.1.3) are also called *Riemann–Liouville fractional integrals*.

Note that the integrals $I_{a+}^\alpha \phi$ and $I_{b-}^\alpha \phi$ are well-defined a.e. for functions $\phi \in L^1[a, b]$ (and hence, for functions $\phi \in L^p[a, b]$, $p > 1$, as well). Indeed, for the integral $I_{a+}^\alpha \phi$, we have

$$\Gamma(\alpha) \int_a^b (I_{a+}^\alpha |\phi|)(s) ds = \int_a^b \int_a^b |\phi(u)|(s - u)_+^{\alpha-1} du ds = \int_a^b du |\phi(u)| \int_a^b ds \, (s - u)_+^{\alpha-1}$$

$$= \alpha^{-1} \int_a^b |\phi(u)|(b - u)^\alpha du \leq \alpha^{-1}(b - a)^\alpha \int_a^b |\phi(u)| du < \infty,$$

so that $(I_{a+}^\alpha |\phi|)(s) < \infty$ and hence $(I_{a+}^\alpha \phi)(s)$ is well-defined ds-a.e. for $s \in (a, b)$. The case of the integral $I_{b-}^\alpha \phi$ is similar. Note also that, if $\alpha \geq 1$ and $\phi \in L^1[a, b]$, the integrals

$I_{a+}^\alpha \phi$ and $I_{b-}^\alpha \phi$ are well-defined at every point $s \in (a,b)$. For example, for $I_{a+}^\alpha \phi$, this follows by bounding $(s-u)^{\alpha-1}$ by $(b-a)^{\alpha-1}$.

Example 6.1.2 (*Fractional integrals of power functions*)   Let $\phi(u) = (u-a)^{-\beta}$, for $\beta < 1$. Then we have, after the change of variables $u = a + (s-a)z$,

$$(I_{a+}^\alpha \phi)(s) = \frac{1}{\Gamma(\alpha)} \int_a^s (u-a)^{-\beta}(s-u)^{\alpha-1} du = \frac{(s-a)^{\alpha-\beta}}{\Gamma(\alpha)} \int_0^1 z^{-\beta}(1-z)^{\alpha-1} dz.$$

$$= \frac{B(1-\beta,\alpha)}{\Gamma(\alpha)} (s-a)^{\alpha-\beta} = \frac{\Gamma(1-\beta)}{\Gamma(1+\alpha-\beta)} (s-a)^{\alpha-\beta},$$

where B is the beta function defined in (2.2.12). The exponent in the function ϕ has increased by α.

The following result gathers some basic properties of fractional integrals.

Proposition 6.1.3   Let $\alpha > 0$. The fractional integrals I_{a+}^α and I_{b-}^α have the following properties:

(i) *Reflection property:* if Q is the "reflection operator" defined by $(Q\phi)(u) = \phi(a+b-u)$ for $u \in [a,b]$, then

$$QI_{a+}^\alpha = I_{b-}^\alpha Q \quad \text{and} \quad QI_{b-}^\alpha = I_{a+}^\alpha Q. \tag{6.1.4}$$

(ii) *Semigroup property:* for $\phi \in L^1[a,b]$,

$$I_{a+}^\alpha I_{a+}^\beta \phi = I_{a+}^{\alpha+\beta} \phi, \quad I_{b-}^\alpha I_{b-}^\beta \phi = I_{b-}^{\alpha+\beta} \phi, \quad \alpha, \beta > 0. \tag{6.1.5}$$

The equations (6.1.5) hold at almost every point. They also hold at every point if $\alpha + \beta \geq 1$.

(iii) *Fractional integration by parts formula:* suppose $\phi \in L^p[a,b]$ and $\psi \in L^q[a,b]$ either with $\alpha \geq 1$, $p = 1$, $q = 1$, or with $0 < \alpha < 1$, $1/p + 1/q \leq 1 + \alpha$, $p > 1$, $q > 1$. Then

$$\int_a^b \phi(s)(I_{a+}^\alpha \psi)(s) ds = \int_a^b (I_{b-}^\alpha \phi)(s)\psi(s) ds. \tag{6.1.6}$$

Proof   The property (i) is elementary since by setting $v = Qu = a+b-u$,

$$\int_a^{a+b-s} \phi(u)(a+b-s-u)^{\alpha-1} du = \int_s^b \phi(a+b-v)(v-s)^{\alpha-1} dv$$

(see Figure 6.1). To show the first relation in the property (ii), note that

$$(I_{a+}^\alpha I_{a+}^\beta \phi)(s) = \frac{1}{\Gamma(\alpha)\Gamma(\beta)} \int_a^s (s-u)^{\alpha-1} \int_a^u \phi(v)(u-v)^{\beta-1} dv\, du.$$

Figure 6.1   The reflection Q moves the argument u into Qu such that $u - a = b - Qu$.

By changing the order of integration and making the change of variables $u = v + (s-v)w$,

$$(I_{a+}^\alpha I_{a+}^\beta \phi)(s) = \frac{1}{\Gamma(\alpha)\Gamma(\beta)} \int_a^s \phi(v) \int_v^s (s-u)^{\alpha-1}(u-v)^{\beta-1} du\, dv$$

$$= \frac{B(\alpha,\beta)}{\Gamma(\alpha)\Gamma(\beta)} \int_a^s \phi(v)(s-v)^{\alpha+\beta-1} dv = (I_{a+}^{\alpha+\beta}\phi)(s),$$

where $B(\alpha,\beta)$ is the beta function defined in (2.2.12). The change of variables is valid and the relation holds under the assumptions made — see the discussion following Definition 6.1.1. The second relation in (6.1.5) can be proved similarly.

The property (iii) would follow immediately if one could change the order of integration in:

$$\int_a^b \phi(s)(I_{a+}^\alpha \psi)(s) ds = \frac{1}{\Gamma(\alpha)} \int_a^b \phi(s) \int_a^s \psi(u)(s-u)^{\alpha-1} du\, ds$$

$$= \frac{1}{\Gamma(\alpha)} \int_a^b \psi(u) \int_u^b \phi(s)(s-u)^{\alpha-1} ds\, du = \int_a^b (I_{b-}^\alpha \phi)(s)\psi(s) ds.$$

Showing the validity of changing the order of integration (by Fubini's theorem) is more delicate. It is necessary to understand the spaces of functions to which functions are mapped by fractional integral operators. See Theorem 3.5 and its Corollary in Samko, Kilbas, and Marichev [878]. □

6.1.2 Riemann–Liouville Fractional Derivatives $\mathcal{D}$ on an Interval

To obtain fractional derivatives, one needs to extend the definition of fractional integrals to $\alpha < 0$. Observe, however, that we can not simply substitute $\alpha < 0$ into (6.1.2) or (6.1.3) since the resulting integrals diverge. As is the case in ordinary calculus, the definition of a derivative involves more restrictions than that of an integral. We shall suppose first that $-1 < \alpha < 0$. The operator $I_{a+}^{-\alpha}$ for $0 < \alpha < 1$ is introduced as the inverse of I_{a+}^α. In other words, consider the equation

$$(I_{a+}^\alpha \phi)(s) = \frac{1}{\Gamma(\alpha)} \int_a^s \phi(z)(s-z)^{\alpha-1} dz = f(s), \quad s \in (a,b),\ 0 < \alpha < 1, \quad (6.1.7)$$

where $\phi \in L^1[a,b]$. (The equation (6.1.7) is called *Abel's equation*.) We want to solve it for functions ϕ assuming f is well-behaved. Multiplying both sides of (6.1.7) by $(u-s)^{-\alpha}$, where $u \in (a,b)$, and integrating we get

$$\int_a^u (u-s)^{-\alpha} \int_a^s \phi(z)(s-z)^{\alpha-1} dz\, ds = \Gamma(\alpha) \int_a^u f(s)(u-s)^{-\alpha} ds. \quad (6.1.8)$$

Changing the order of integration

$$\int_a^u \phi(z) \int_z^u (u-s)^{-\alpha}(s-z)^{\alpha-1} ds\, dz = \Gamma(\alpha) \int_a^u f(s)(u-s)^{-\alpha} ds$$

and using the fact

$$\int_z^u (u-s)^{-\alpha}(s-z)^{\alpha-1} ds = B(\alpha, 1-\alpha) = \Gamma(\alpha)\Gamma(1-\alpha), \quad (6.1.9)$$

6.1 Fractional Integrals and Derivatives

which follows from (2.2.12), we obtain that

$$\int_a^u \phi(z)dz = \frac{1}{\Gamma(1-\alpha)} \int_a^u f(s)(u-s)^{-\alpha}ds. \tag{6.1.10}$$

Now, differentiating both sides with respect to u yields

$$\phi(u) = \frac{1}{\Gamma(1-\alpha)} \frac{d}{du} \int_a^u f(s)(u-s)^{-\alpha}ds.$$

The operator $I_{a+}^{-\alpha}$ for $0 < \alpha < 1$ is then defined as follows.

Definition 6.1.4 Let $0 < \alpha < 1$. The integrals

$$(\mathcal{D}_{a+}^{\alpha} f)(u) = (I_{a+}^{-\alpha} f)(u) = \frac{1}{\Gamma(1-\alpha)} \frac{d}{du} \int_a^u f(s)(u-s)^{-\alpha}ds, \tag{6.1.11}$$

$u \in (a, b)$, and

$$(\mathcal{D}_{b-}^{\alpha} f)(u) = (I_{b-}^{-\alpha} f)(u) = -\frac{1}{\Gamma(1-\alpha)} \frac{d}{du} \int_u^b f(s)(s-u)^{-\alpha}ds, \tag{6.1.12}$$

$u \in (a, b)$, are called *fractional derivatives* of order α. The fractional derivatives $\mathcal{D}_{a+}^{\alpha}$ and $\mathcal{D}_{b-}^{\alpha}$ are called left-sided and right-sided, respectively. Both (6.1.11) and (6.1.12) are also called the *Riemann–Liouville fractional derivatives*.

Remark 6.1.5 In contrast to the usual derivative, the fractional derivative is not a local operation since $(\mathcal{D}_{a+}^{\alpha} f)(u)$ does not depend only on the behavior of f in a neighborhood of u.

Remark 6.1.6 Note that, heuristically, by taking the derivative d/du inside (6.1.11), we get

$$\frac{-\alpha}{\Gamma(1-\alpha)} \int_a^u f(s)(u-s)^{-\alpha-1}ds = \frac{1}{\Gamma(-\alpha)} \int_a^u f(s)(u-s)^{-\alpha-1}ds,$$

which is exactly the fractional integral (6.1.2) with $-1 < -\alpha < 0$. Doing the same thing inside (6.1.12), we get (6.1.3), explaining the presence of the additional minus sign in (6.1.12).

The fractional derivatives $\mathcal{D}_{a+}^{\alpha} f$ and $\mathcal{D}_{b-}^{\alpha} f$ are well-defined if, for example, f is an absolutely continuous function; that is, $f(s) = f(a) + \int_a^s f'(t)dt$ for $s \in [a, b]$. Indeed, we have[1]

---

[1] It is useful to recall the formula for $\frac{d}{dx} \int_{a(x)}^{b(x)} f(x,t)dt$. Assume that the functions $f(x,t)$ and $\frac{\partial}{\partial x} f(x,t)$ are continuous in the region $x_0 \leq x \leq x_1, a(x) \leq t \leq b(x)$ of the (x,t) plane, and that $a(x)$ and $b(x)$ are both continuous with continuous derivatives for $x_0 \leq x \leq x_1$. Then, for any $x_0 < x < x_1$,

$$\frac{d}{dx} \int_{a(x)}^{b(x)} f(x,t)dt = f(x,b(x))b'(x) - f(x,a(x))a'(x) + \int_{a(x)}^{b(x)} \frac{\partial}{\partial x} f(x,t)dt.$$

This formula is known as Leibniz's rule for differentiation under the integral sign.

$$(\mathcal{D}_{a+}^\alpha f)(s) = \frac{1}{\Gamma(1-\alpha)} \frac{d}{du} \int_a^u \left(f(a) + \int_a^s f'(t)dt \right)(u-s)^{-\alpha} ds$$

$$= \frac{1}{\Gamma(1-\alpha)} \frac{d}{du} \left(f(a) \int_a^u (u-s)^{-\alpha} ds + \int_a^u dt f'(t) \int_t^u ds(u-s)^{-\alpha} \right)$$

$$= \frac{1}{(1-\alpha)\Gamma(1-\alpha)} \frac{d}{du} \left(f(a)(u-a)^{1-\alpha} + \int_a^u f'(t)(u-t)^{1-\alpha} dt \right)$$

$$= \frac{1}{\Gamma(1-\alpha)} \left(f(a)(u-a)^{-\alpha} + \int_a^u f'(t)(u-t)^{-\alpha} dt \right).$$

The case of $\mathcal{D}_{b-}^\alpha f$ is similar. These results should not be surprising. We indeed expect that if f is differentiable then it also should be fractionally differentiable with order $0 < \alpha < 1$.

Example 6.1.7 (*Fractional derivatives of power functions*) Let $0 < \alpha < 1$ and $f(s) = (s-a)^{\beta-1}$ with $s \in (a,b)$ and $\beta > 0$. Then we have that after the change of variables $s = a + (u-a)z$ and (2.2.12),

$$(\mathcal{D}_{a+}^\alpha f)(u) = \frac{1}{\Gamma(1-\alpha)} \frac{d}{du} \int_a^u (s-a)^{\beta-1}(u-s)^{-\alpha} ds = \frac{B(\beta, 1-\alpha)}{\Gamma(1-\alpha)} \frac{d}{du}(u-a)^{\beta-\alpha}$$

$$= \frac{\Gamma(\beta)}{\Gamma(1+\beta-\alpha)} \begin{cases} (\beta-\alpha)(u-a)^{\beta-\alpha-1}, & \text{if } \beta \neq \alpha, \\ 0, & \text{if } \beta = \alpha \end{cases} = \begin{cases} \frac{\Gamma(\beta)}{\Gamma(\beta-\alpha)}(u-a)^{\beta-\alpha-1}, & \text{if } \beta \neq \alpha, \\ 0, & \text{if } \beta = \alpha. \end{cases}$$

Observe that if $\beta \neq \alpha$, then the exponent $\beta - 1$ in the function f decreases by α. Now suppose that $\beta = 1$, so that $f(s) \equiv 1$ is constant. Its fractional derivative is not zero if $0 < \alpha < 1$, but is equal to

$$(\mathcal{D}_{a+}^\alpha 1)(u) = \frac{(u-a)^{-\alpha}}{\Gamma(1-\alpha)}.$$

The next result provides basic properties of fractional derivatives. We consider only $\mathcal{D}_{a+}^\alpha$. Similar relations hold for $\mathcal{D}_{b-}^\alpha$.

Proposition 6.1.8 *Let $0 < \alpha < 1$. The fractional derivative $\mathcal{D}_{a+}^\alpha$ has the following properties:*

(i) for any $\phi \in L^1[a,b]$, we have that

$$\mathcal{D}_{a+}^\alpha I_{a+}^\alpha \phi = \phi. \tag{6.1.13}$$

(ii) for any f such that $f = I_{a+}^\alpha \phi$, where $\phi \in L^1[a,b]$, we have that

$$I_{a+}^\alpha \mathcal{D}_{a+}^\alpha f = f. \tag{6.1.14}$$

(iii) if the function $I_{a+}^{1-\alpha} f$ is absolutely continuous, then

$$(I_{a+}^\alpha \mathcal{D}_{a+}^\alpha f)(s) = f(s) - \frac{(I_{a+}^{1-\alpha} f)(a)}{\Gamma(\alpha)}(s-a)^{\alpha-1}, \quad s \in (a,b), \tag{6.1.15}$$

where $(I_{a+}^{1-\alpha} f)(a) = \lim_{s \downarrow a}(I_{a+}^{1-\alpha} f)(s)$.

Remark 6.1.9 In (ii) and (iii) we have special conditions on f. This makes sense since we differentiate first before integrating.

Remark 6.1.10 The last term in (6.1.15) can be thought as the boundary term resulting from the integration I_{a+}^{α}. Its presence is naturally expected in analogy to the fundamental calculus relation $(I_{a+}^1 \mathcal{D}_{a+}^1 f)(s) = \int_a^s f'(t)dt = f(s) - f(a)$. This term is zero in (6.1.14) since $(I_{a+}^{1-\alpha} f)(a) = (I_{a+}^{1-\alpha} I_{a+}^{\alpha} \phi)(a) = (I_{a+}^1 \phi)(a) = \int_a^a \phi(u)du = 0$.

Proof For the property (i), note that, by arguing as in (6.1.8)–(6.1.10),

$$(\mathcal{D}_{a+}^{\alpha} I_{a+}^{\alpha} \phi)(u) = \frac{1}{\Gamma(1-\alpha)\Gamma(\alpha)} \frac{d}{du} \int_a^u (u-s)^{-\alpha} \int_a^s \phi(v)(s-v)^{\alpha-1} dv \, ds$$

$$= \frac{d}{du} \int_a^u \phi(v) dv = \phi(u).$$

The property (ii) follows immediately from (i). For the property (iii), note first that, by the assumption that $I_{a+}^{1-\alpha} f$ is absolutely continuous, $(I_{a+}^{1-\alpha} f)(s) = (I_{a+}^1 g)(s) + (I_{a+}^{1-\alpha} f)(a)$ for some $g \in L^1[a, b]$. By using Example 6.1.2, note that $(I_{a+}^{1-\alpha}(u-a)^{\alpha-1})(s) = \frac{\Gamma(\alpha)}{\Gamma(1)}(s-a)^{1-\alpha+\alpha-1} = \Gamma(\alpha)$. Hence, we can also write

$$\left(I_{a+}^{1-\alpha} \left(f(u) - \frac{(I_{a+}^{1-\alpha} f)(a)}{\Gamma(\alpha)}(u-a)^{\alpha-1}\right)\right)(s) = (I_{a+}^1 g)(s).$$

By applying $\mathcal{D}_{a+}^{1-\alpha}$ to both sides, and using the property (ii) above and the property (ii) of Proposition 6.1.3, we obtain

$$f(u) - \frac{(I_{a+}^{1-\alpha} f)(a)}{\Gamma(\alpha)}(u-a)^{\alpha-1} = (\mathcal{D}_{a+}^{1-\alpha} I_{a+}^1 g)(u) = (\mathcal{D}_{a+}^{1-\alpha} I_{a+}^{1-\alpha} I_{a+}^{\alpha} g)(u) = (I_{a+}^{\alpha} g)(u).$$

Then, by the property (ii),

$$\left(I_{a+}^{\alpha} \mathcal{D}_{a+}^{\alpha} \left(f(u) - \frac{(I_{a+}^{1-\alpha} f)(a)}{\Gamma(\alpha)}(u-a)^{\alpha-1}\right)\right)(s) = f(s) - \frac{(I_{a+}^{1-\alpha} f)(a)}{\Gamma(\alpha)}(s-a)^{\alpha-1}.$$

The relation (6.1.15) now follows by noting that $(\mathcal{D}_{a+}^{\alpha}(u-a)^{\alpha-1})(z) = 0$ by Example 6.1.7. □

Corollary 6.1.11 *For $\alpha > 0$, $I_{a+}^{\alpha} \phi = 0$ or $I_{b-}^{\alpha} \phi = 0$ implies that $\phi = 0$ almost everywhere.*

Proof We consider only $I_{a+}^{\alpha} \phi$. When $0 < \alpha < 1$, the result follows from Proposition 6.1.8, (i), since $\phi = \mathcal{D}_{a+}^{\alpha} I_{a+}^{\alpha} \phi = 0$ a.e. When $\alpha \geq 1$, $I_{a+}^{\alpha} \phi = 0$ implies $I_{a+}^{[\alpha]} I_{a+}^{\alpha-[\alpha]} \phi = 0$, where $[\alpha]$ is the integer part of α. Using the relation (6.1.1) and differentiating the last expression $[\alpha]$ number of times leads to $I_{a+}^{\alpha-[\alpha]} \phi = 0$. Since $0 \leq \alpha - [\alpha] < 1$, we also get $\phi = 0$ a.e. □

One may define the fractional derivative operators $\mathcal{D}_{a+}^{\alpha}$ and $\mathcal{D}_{b-}^{\alpha}$ for $\alpha \geq 1$ as well. The idea is to use the usual integer order derivative operators. For example, $(\mathcal{D}_{a+}^{1.7} f)(u)$

would be defined as $\frac{d}{du}((\mathcal{D}_{a+}^{0.7}f)(u)) = \frac{d^2}{du^2}((I_{a+}^{0.3}f)(u))$, where the equality follows from the definitions (6.1.11) and (6.1.2). Because we are interested in fractional Brownian motion we will focus on the range $\alpha \in (-1/2, 1/2)$ only.

Remark 6.1.12 Set $a = 0$. As indicated above, e.g., the Riemann–Liouville derivative $\mathcal{D}_{0+}^{\alpha}$ for $\alpha > 0$ is defined as $\frac{d^n}{du^n} I_{0+}^{n-\alpha}$ where $n = [\alpha] + 1$. If the order of the last two operations is reversed, that is, $I_{0+}^{n-\alpha} \frac{d^n}{du^n}$ is considered, one gets the so-called Caputo derivative. Note that while the Caputo derivative of a constant is always zero, this is not generally the case for the Riemann–Liouville derivative (as can be seen from Example 6.1.7 with $\beta = 1$). For more information on the Caputo derivative, see Podlubny [825], Meerschaert and Sikorskii [712]. In the context of FBM, the Riemann–Liouville derivative is sufficient for our purposes.

6.1.3 Fractional Integrals and Derivatives on the Real Line

Fractional integrals and derivatives can also be defined on the real line. Two types of fractional derivatives will be defined: the Liouville fractional derivatives and the Marchaud fractional derivatives.

Definition 6.1.13 Let $\alpha > 0$. The integrals

$$(I_+^{\alpha}\phi)(s) = \frac{1}{\Gamma(\alpha)} \int_{-\infty}^{s} \phi(u)(s-u)^{\alpha-1} du = \frac{1}{\Gamma(\alpha)} \int_{\mathbb{R}} \phi(u)(s-u)_+^{\alpha-1} du$$
$$= \frac{1}{\Gamma(\alpha)} \int_0^{\infty} \phi(s-y) y^{\alpha-1} dy, \qquad (6.1.16)$$

$s \in \mathbb{R}$, and

$$(I_-^{\alpha}\phi)(s) = \frac{1}{\Gamma(\alpha)} \int_s^{\infty} \phi(u)(u-s)^{\alpha-1} du = \frac{1}{\Gamma(\alpha)} \int_{\mathbb{R}} \phi(u)(u-s)_+^{\alpha-1} du$$
$$= \frac{1}{\Gamma(\alpha)} \int_0^{\infty} \phi(s+y) y^{\alpha-1} dy, \qquad (6.1.17)$$

$s \in \mathbb{R}$, are called *fractional integrals* of order α on the real line.

Definition 6.1.14 Let $0 < \alpha < 1$. The integrals

$$(\mathcal{D}_+^{\alpha} f)(u) = \frac{1}{\Gamma(1-\alpha)} \frac{d}{du} \int_{-\infty}^{u} f(s)(u-s)^{-\alpha} ds = \frac{1}{\Gamma(1-\alpha)} \frac{d}{du} \int_0^{\infty} f(u-y) y^{-\alpha} dy,$$
$$(6.1.18)$$

$u \in \mathbb{R}$, and

$$(\mathcal{D}_-^{\alpha} f)(u) = -\frac{1}{\Gamma(1-\alpha)} \frac{d}{du} \int_u^{\infty} f(s)(s-u)^{-\alpha} ds = \frac{1}{\Gamma(1-\alpha)} \frac{d}{du} \int_0^{\infty} f(u+y) y^{-\alpha} dy,$$
$$(6.1.19)$$

$u \in \mathbb{R}$, are called *Liouville fractional derivatives* of order α on the real line.

6.1 Fractional Integrals and Derivatives

Remark 6.1.15 Note that these definitions are similar to Definition 6.1.1 and Definition 6.1.4, respectively, but with $a = -\infty$ and $b = \infty$. The basic question, however, is: *What is the domain of these new fractional operators?*

What about the domain of the operators $I_\pm^\alpha$? Let us observe that $I_\pm^\alpha \phi$ is well-defined for $0 < \alpha < 1$ and $\phi \in L^p(\mathbb{R})$ with $1 \leq p < 1/\alpha$. Indeed, we have

$$\Gamma(\alpha)(I_+^\alpha \phi)(s) = \int_0^1 \phi(s-v) v^{\alpha-1} dv + \int_1^\infty \phi(s-v) v^{\alpha-1} dv. \tag{6.1.20}$$

The second term is bounded for all s by Hölder's inequality when $1 < p < 1/\alpha$ and simply by $\|\phi\|_{L^1(\mathbb{R})}$ when $p = 1$. The first term is bounded by the generalized Minkowski's inequality. Recall that the generalized Minkowski's inequality (e.g., Zygmund [1030], p. 19), given by

$$\left\{ \int_{\Omega_1} dx \left| \int_{\Omega_2} f(x,y) dy \right|^p \right\}^{1/p} \leq \int_{\Omega_2} dy \left\{ \int_{\Omega_1} |f(x,y)|^p dx \right\}^{1/p}, \tag{6.1.21}$$

is valid for all $p \geq 1$ and any (possibly infinite) intervals Ω_1 and Ω_2. If $\Omega_2 = (0, 2)$ and $f(x, y) = f(x)$ for $0 < y < 1$ and $f(x, y) = g(x)$ for $1 < y < 2$, then one obtains the usual Minkowski's inequality $\|f + g\|_{L^p(\Omega_1)} \leq \|f\|_{L^p(\Omega_1)} + \|g\|_{L^p(\Omega_1)}$. To show that the first term of (6.1.20) is bounded, apply (6.1.21) with $\Omega_1 = \mathbb{R}$ and $\Omega_2 = [0, 1]$ to write

$$\left\{ \int_\mathbb{R} ds \left| \int_0^1 |\phi(s-v)| v^{\alpha-1} dv \right|^p \right\}^{1/p} \leq \int_0^1 dv \left\{ \int_\mathbb{R} |\phi(s-v) v^{\alpha-1}|^p ds \right\}^{1/p}$$

$$= \int_0^1 v^{\alpha-1} dv \, \|\phi\|_{L^p(\mathbb{R})} < \infty.$$

This yields $\int_0^1 |\phi(s-v)| v^{\alpha-1} dv < \infty$ a.e. ds and hence that $\int_0^1 \phi(s-v) v^{\alpha-1} dv$ is well-defined a.e. ds.

Example 6.1.16 (*Fractional integrals and derivatives of an exponential function*) Consider the exponential function $\phi(u) = e^{\lambda u}$ with parameter $\lambda > 0$. Then, by (6.1.16) for $\alpha > 0$,

$$(I_+^\alpha \phi)(s) = \frac{1}{\Gamma(\alpha)} \int_0^\infty e^{\lambda(s-y)} y^{\alpha-1} dy = \frac{e^{\lambda s}}{\Gamma(\alpha)} \int_0^\infty e^{-\lambda y} y^{\alpha-1} dy = e^{\lambda s} \lambda^{-\alpha}, \quad s \in \mathbb{R}, \tag{6.1.22}$$

and by (6.1.18) for $0 < \alpha < 1$,

$$(\mathcal{D}_+^\alpha \phi)(u) = \frac{1}{\Gamma(1-\alpha)} \frac{d}{du} \int_0^\infty e^{\lambda(u-y)} y^{-\alpha} dy = \frac{d}{du} e^{\lambda u} \lambda^{\alpha-1} = e^{\lambda u} \lambda^\alpha, \quad u \in \mathbb{R}. \tag{6.1.23}$$

Note that $I_-^\alpha \phi$ and $\mathcal{D}_-^\alpha \phi$ are not considered since the integrals entering their definitions are infinite.

Example 6.1.17 (*Fractional integrals and derivatives of indicator functions*) We want to fractionally integrate and differentiate an indicator function. Let $a < b$ be real numbers. Then, for $\alpha > 0$,

$$\Gamma(1+\alpha)(I_{\pm}^{\alpha} 1_{[a,b)})(s) = \frac{\Gamma(1+\alpha)}{\Gamma(\alpha)} \int_{\mathbb{R}} 1_{[a,b)}(u)(u-s)_{\mp}^{\alpha-1} du$$

$$= \alpha \int_a^b (u-s)_{\mp}^{\alpha-1} du = (b-s)_{\mp}^{\alpha} - (a-s)_{\mp}^{\alpha}, \quad s \in \mathbb{R}. \quad (6.1.24)$$

Similarly, for $0 < \alpha < 1$, we have that

$$\Gamma(1-\alpha)(\mathcal{D}_{\pm}^{\alpha} 1_{[a,b)})(u) = \pm \frac{d}{du} \int_{\mathbb{R}} 1_{[a,b)}(s)(s-u)_{\mp}^{-\alpha} ds$$

$$= \pm \frac{1}{1-\alpha} \frac{d}{du}((b-u)_{\mp}^{1-\alpha} - (a-u)_{\mp}^{1-\alpha}) = (b-u)_{\mp}^{-\alpha} - (a-u)_{\mp}^{-\alpha}, \quad u \in \mathbb{R}. \quad (6.1.25)$$

The fractional integrals $I_{\pm}^{\alpha}$ have properties analogous to those of fractional integrals on an interval in Proposition 6.1.3. So, for example, we have $QI_{\pm}^{\alpha}\phi = I_{\mp}^{\alpha}Q\phi$, where the "reflection operator" Q is defined by $(Q\phi)(u) = \phi(-u)$ for $u \in \mathbb{R}$. We similarly have that, for "sufficiently good" functions ϕ and ψ, the *semigroup property* (Samko et al. [878], p. 34)

$$I_{\pm}^{\alpha} I_{\pm}^{\beta} \phi = I_{\pm}^{\alpha+\beta} \phi, \quad \alpha, \beta > 0, \quad (6.1.26)$$

and the *fractional integration by parts formula* (Samko et al. [878], p. 34)

$$\int_{\mathbb{R}} \phi(s)(I_+^{\alpha}\psi)(s) ds = \int_{\mathbb{R}} (I_-^{\alpha}\phi)(s)\psi(s) ds \quad (6.1.27)$$

hold. If $I_{\pm}^{\alpha}\phi = 0$, we also expect that $\phi = 0$ almost everywhere (cf. Corollary 6.1.11).

What about properties of fractional derivatives $\mathcal{D}_{\pm}^{\alpha}$? For example, do we have $\mathcal{D}_{\pm}^{\alpha} I_{\pm}^{\alpha}\phi = \phi$? The answer is yes, but only for functions $\phi \in L^1(\mathbb{R})$. (Indeed, as in the proof of property (i) of Proposition 6.1.8 we have that $\mathcal{D}_+^{\alpha} I_+^{\alpha}\phi = \frac{d}{du}\int_{-\infty}^{u} \phi(s) ds$ which assumes summability of the function ϕ at infinity.) It turns out that this property is not strong enough in the context of fractional Brownian motion. There is, however, another fractional derivative which allows for greater flexibility.

6.1.4 Marchaud Fractional Derivatives D on the Real Line

Observe that for the Liouville fractional derivative $\mathcal{D}_+^{\alpha} f$ with a "sufficiently good" function f, we have

$$(\mathcal{D}_+^{\alpha} f)(u) = \frac{1}{\Gamma(1-\alpha)} \frac{d}{du} \int_0^{\infty} v^{-\alpha} f(u-v) dv = \frac{1}{\Gamma(1-\alpha)} \int_0^{\infty} v^{-\alpha} f'(u-v) dv$$

$$= \frac{\alpha}{\Gamma(1-\alpha)} \int_0^{\infty} f'(u-v) dv \int_v^{\infty} \frac{d\xi}{\xi^{1+\alpha}} = \frac{\alpha}{\Gamma(1-\alpha)} \int_0^{\infty} \frac{d\xi}{\xi^{1+\alpha}} \int_0^{\xi} f'(u-v) dv$$

$$= \frac{\alpha}{\Gamma(1-\alpha)} \int_0^{\infty} \frac{f(u) - f(u-\xi)}{\xi^{1+\alpha}} d\xi. \quad (6.1.28)$$

The corresponding relation for $\mathcal{D}_-^{\alpha} f$ is given by

$$(\mathcal{D}_-^{\alpha} f)(u) = \frac{\alpha}{\Gamma(1-\alpha)} \int_0^{\infty} \frac{f(u) - f(u+\xi)}{\xi^{1+\alpha}} d\xi. \quad (6.1.29)$$

It is clear that the operators $\mathcal{D}_\pm^\alpha$ in Definition 6.1.14 and the operators defined by the right-hand side of (6.1.28) and (6.1.29) have different domains. So, for example, we may take f to behave like a constant at infinity or even grow slowly enough at infinity for the right-hand side of (6.1.28) or (6.1.29) to exist whereas $\mathcal{D}_\pm^\alpha f$ is not defined for such f since constants are not in $L^1(\mathbb{R})$.

Definition 6.1.18 Let $\alpha \in (0, 1)$. The operators

$$\mathbf{D}_\pm^\alpha f = \lim_{\epsilon \to 0} \mathbf{D}_{\pm,\epsilon}^\alpha f, \tag{6.1.30}$$

where

$$(\mathbf{D}_{\pm,\epsilon}^\alpha f)(u) = \frac{\alpha}{\Gamma(1-\alpha)} \int_\epsilon^\infty \frac{f(u) - f(u \mp \xi)}{\xi^{1+\alpha}} d\xi, \tag{6.1.31}$$

are called *Marchaud fractional derivatives* on the real line of order α. The expressions in (6.1.31) are called *truncated Marchaud fractional derivatives*.

Remark 6.1.19 There is another way to obtain Marchaud fractional derivatives. Let $-1 < \alpha < 0$. Heuristically, by substituting this negative α into the expression (6.1.17) of I_-^α we get that

$$(I_-^\alpha \phi)(s) = \frac{1}{\Gamma(-|\alpha|)} \int_0^\infty \phi(v+s) v^{-|\alpha|-1} dv. \tag{6.1.32}$$

If ϕ is not trivial, this integral diverges at $v = 0$ like

$$\frac{1}{\Gamma(-|\alpha|)} \int_0^\infty \phi(s) v^{-|\alpha|-1} dv. \tag{6.1.33}$$

But by subtracting (6.1.33) from (6.1.32), we get

$$\frac{1}{\Gamma(-|\alpha|)} \int_0^\infty (\phi(v+s) - \phi(s)) v^{-|\alpha|-1} dv = \frac{-|\alpha|}{\Gamma(1-|\alpha|)} \int_0^\infty (\phi(v+s) - \phi(s)) v^{-|\alpha|-1} dv$$

$$= \frac{|\alpha|}{\Gamma(1-|\alpha|)} \int_0^\infty (\phi(s) - \phi(v+s)) v^{-|\alpha|-1} dv$$

$$= (\mathbf{D}_-^{-\alpha} \phi)(s),$$

since $|\alpha|\Gamma(-|\alpha|) = -\Gamma(1-|\alpha|)$.

We have used a "renormalization" perspective by compensating for the divergence. Let us explain what we did in more rigorous terms. For $\epsilon > 0$, write

$$\frac{1}{\Gamma(-|\alpha|)} \int_\epsilon^\infty \phi(v+s) v^{-|\alpha|-1} dv = \frac{\phi(s+\epsilon) \epsilon^{-|\alpha|}}{|\alpha|\Gamma(-|\alpha|)} + J_\epsilon,$$

where, using $|\alpha|\Gamma(-|\alpha|) = -\Gamma(1-|\alpha|)$,

$$J_\epsilon = \frac{\phi(s+\epsilon)\epsilon^{-|\alpha|}}{\Gamma(1-|\alpha|)} - \frac{|\alpha|}{\Gamma(1-|\alpha|)} \int_\epsilon^\infty \phi(v+s) v^{-|\alpha|-1} dv.$$

(The term $\phi(s+\epsilon)\epsilon^{-|\alpha|}$ arises as the boundary terms at $v = \epsilon$ in the integration by parts of $-|\alpha| \int_\epsilon^\infty \phi(v+s) v^{-|\alpha|-1} dv = \int_\epsilon^\infty \phi(v+s) dv^{-|\alpha|}$.) Although both terms $\int_\epsilon^\infty \phi(v+$

$s)v^{-|\alpha|-1}dv$ and $\epsilon^{-|\alpha|}$ diverge as $\epsilon \to 0$, the limit of J_ϵ may exist. Moreover, since $\epsilon^{-|\alpha|} = |\alpha| \int_\epsilon^\infty v^{-|\alpha|-1}dv$, we get that

$$\lim_{\epsilon \to 0} J_\epsilon = \frac{|\alpha|}{\Gamma(1-|\alpha|)} \int_0^\infty (\phi(s) - \phi(v+s))v^{-|\alpha|-1}dv = (\mathbf{D}_-^{-\alpha}\phi)(s). \quad (6.1.34)$$

The limit of J_ϵ is called the *finite part* of the divergent integral $I_-^\alpha \phi$ in the sense of Hadamard. The notation is $\lim_{\epsilon \to 0} J_\epsilon = \text{p.f.} \, I_-^\alpha \phi$, where p.f. stands for *partie finie*. By (6.1.34), we have

$$\mathbf{D}_-^{-\alpha}\phi = \text{p.f.} \, I_-^\alpha \phi$$

(for sufficiently good function ϕ). A similar argument works for $\mathbf{D}_+^{-\alpha}$.

The following theorem is proved in Samko et al. [878], p. 125. We will use it in the sequel with $\alpha \in (0, 1/2)$ and $p = 2$. The result is the real line counterpart of Proposition 6.1.8, (*i*).

Theorem 6.1.20 *If $f = I_\pm^\alpha \phi$ with $\phi \in L^p(\mathbb{R})$, $\alpha \in (0, 1)$ and $1 \leq p < 1/\alpha$, then $\mathbf{D}_\pm^\alpha f = \phi$; that is, $\mathbf{D}_\pm^\alpha I_\pm^\alpha \phi = \phi$.*

Thus to prove that $\mathbf{D}_\pm^\alpha f = \phi$, it is sufficient to show that $f = I_\pm^\alpha \phi$, as long as $\phi \in L^p(\mathbb{R})$, $1 \leq p < 1/\alpha$.

Remark 6.1.21 Because of Theorem 6.1.20 we will also use in the sequel the notation

$$I_\pm^\alpha = \mathbf{D}_\pm^{-\alpha} \quad \text{for} \quad \alpha \in (-1, 0).$$

Thus, in the context of fractional integrals on the real line, $I_\pm^\alpha$ with $\alpha \in (-1, 0)$ will refer to the Marchaud and not the Liouville fractional derivative, unless mentioned otherwise.

Example 6.1.22 (*Marchaud fractional derivatives of indicator functions*) Let us show that, for $\alpha \in (0, 1)$ and $a < b$,

$$\Gamma(1-\alpha)(\mathbf{D}_\pm^\alpha 1_{[a,b)})(u) = (b-u)_\mp^{-\alpha} - (a-u)_\mp^{-\alpha}, \quad u \in \mathbb{R}, \quad (6.1.35)$$

which is similar to what one obtains with $\mathcal{D}_\pm^\alpha 1_{[a,b)}$ (compare (6.1.35) with (6.1.25)). By Theorem 6.1.20, it is enough to show that $I_\pm^\alpha((b-u)_\mp^{-\alpha} - (a-u)_\mp^{-\alpha}) = 1_{[a,b)}/\Gamma(1-\alpha)$. For I_-^α, we have

$$I = (I_-^\alpha((b-u)_+^{-\alpha} - (a-u)_+^{-\alpha}))(s) = \frac{1}{\Gamma(\alpha)} \int_\mathbb{R} ((b-u)_+^{-\alpha} - (a-u)_+^{-\alpha})(u-s)_+^{\alpha-1}du.$$

If $s \geq b$, then clearly $I = 0 = 1_{[a,b)}(s)/\Gamma(1-\alpha)$. If $a \leq s < b$, then

$$I = \frac{1}{\Gamma(\alpha)} \int_s^b (b-u)^{-\alpha}(u-s)^{\alpha-1}du = \frac{B(\alpha, 1-\alpha)}{\Gamma(\alpha)} = \frac{1_{[a,b)}(s)}{\Gamma(1-\alpha)}.$$

If $s < a$, we similarly get that $I = 0 = 1_{[a,b)}(s)/\Gamma(1-\alpha)$. Another way to prove (6.1.35) is by direct computations using Definition 6.1.18 (Exercise 6.1).

Remark 6.1.23 For $\alpha \geq 1$, the (left-sided) fractional derivative on the real line can be defined by extending (6.1.18), (6.1.28) or using the renormalization idea namely as

$$(\mathcal{D}_+^\alpha f)(u) = \frac{1}{\Gamma(n-\alpha)} \frac{d^n}{du^n} \int_{-\infty}^u f(s)(u-s)^{n-1-\alpha} ds, \qquad (6.1.36)$$

$$(\mathbf{D}_+^\alpha f)(u) = -\frac{1}{\Gamma(-\alpha) A_l(\alpha)} \int_0^\infty \frac{(\Delta_\xi^l f)(u)}{\xi^{1+\alpha}} d\xi, \qquad (6.1.37)$$

where $n = [\alpha] + 1$ in (6.1.36) and (6.1.37) ($[\alpha]$ denotes the integer part of $\alpha > 0$), l in (6.1.37) is some integer such that $l > \alpha$, and

$$(\Delta_\xi^l f)(u) = ((E - \tau_\xi)^l f)(u) = \sum_{k=0}^l (-1)^k \binom{l}{k} f(u - k\xi),$$

$$(Ef)(u) = f(u), \quad (\tau_\xi^k f)(u) = f(u - k\xi), \quad u, \xi \in \mathbb{R}, \ k \in \mathbb{N},$$

$$A_l(\alpha) = \sum_{k=1}^l (-1)^{k-1} \binom{l}{k} k^\alpha.$$

The notation is similar to that used in time series analysis. E stands for the identity operator, τ_ξ for a shift to the left of length ξ (backshift operator), τ_ξ^k stands for the iterated shift and Δ_ξ is the difference operator $E - \tau_\xi$. The functions $\mathcal{D}_+^\alpha f$ and $\mathbf{D}_+^{-\alpha} f$ are identical for good enough functions f (in particular, $\mathbf{D}_+^{-\alpha} f$ does not depend on the choice of l, $l > \alpha$). For more details, see Samko et al. [878], pp. 114–120.

6.1.5 The Fourier Transform Perspective

It is helpful to use Fourier transforms to express fractional integrals and derivatives. They often provide a great simplification. Let $\widehat{\phi}(x) = \int_{\mathbb{R}} e^{ixu} \phi(u) du$ denote the $L^2(\mathbb{R})$–Fourier transform of the function ϕ (see Appendix A.1.2). It is part of the folklore to write

$$\widehat{I_\pm^\alpha \phi}(x) = \widehat{\phi}(x)/(\mp ix)^\alpha, \qquad (6.1.38)$$

where

$$(\mp ix)^\alpha = e^{\alpha \log|x| \mp i \operatorname{sign}(x) \frac{\alpha\pi}{2}} = |x|^\alpha e^{\mp i \operatorname{sign}(x) \frac{\alpha\pi}{2}}.$$

Observe that the semigroup property (6.1.26) can be verified heuristically if we use (6.1.38) to take the Fourier transforms of the left-hand side of (6.1.26) to obtain the Fourier transform of the right-hand side of (6.1.26). The idea behind the relation (6.1.38) is as follows. Consider, for example, the fractional integral I_+^α with $\alpha \in (0, 1/2)$. Then

$$(I_+^\alpha \phi)(s) = \frac{1}{\Gamma(\alpha)} \int_{\mathbb{R}} \phi(u)(s-u)_+^{\alpha-1} du \qquad (6.1.39)$$

is well-defined for $\phi \in L^2(\mathbb{R})$ by the argument following Definition 6.1.14. Now observe that $I_+^\alpha \phi$ is the convolution of the functions $\phi(u)$ and $\psi_\alpha(u) = u_+^{\alpha-1}/\Gamma(\alpha)$. Since a convolution in the "time domain" becomes a product in the "spectral domain," we would expect that

$$\widehat{I_-^\alpha \phi}(x) = \widehat{\phi}(x) \widehat{\psi_a}(x).$$

(Observe, however, that the function ψ_a is neither in $L^1(\mathbb{R})$ nor in $L^2(\mathbb{R})$.) Taking the preceding relation nevertheless for granted, we obtain (6.1.38) by evaluating $\widehat{\psi}_a$ as follows:

$$\widehat{\psi}_a(x) = \frac{1}{\Gamma(\alpha)} \int_{\mathbb{R}} e^{ixu} u_+^{\alpha-1} du$$

$$= \frac{|x|^{-\alpha}}{\Gamma(\alpha)} \left(\int_{\mathbb{R}} e^{iv} v_+^{\alpha-1} dv 1_{\{x>0\}} + \int_{\mathbb{R}} e^{-iv} v_+^{\alpha-1} dv 1_{\{x<0\}} \right) = |x|^{-\alpha} e^{i\operatorname{sign}(x)\frac{\alpha\pi}{2}},$$

since $\int_{\mathbb{R}} e^{\pm iv} v_+^{\alpha-1} dv = \Gamma(\alpha) e^{\pm i\alpha\pi/2}$ for $0 < \alpha < 1$ by (2.6.6). These arguments can be made rigorous by using approximations. We have the following result, which we state without proof. See, for example, Proposition 2.3 in Pipiras and Taqqu [815].

Proposition 6.1.24 Let $\phi \in L^2(\mathbb{R})$. If $0 < \alpha < 1/2$, suppose that the function ϕ is such that

$$\int_{\mathbb{R}} |\widehat{\phi}(x)|^2 |x|^{-2\alpha} dx < \infty. \tag{6.1.40}$$

If $-1/2 < \alpha < 0$, suppose that there is $f \in L^2(\mathbb{R})$ such that $\phi = I_\pm^{-\alpha} f$. Then, for $-1/2 < \alpha < 1/2$, we have that

$$\widehat{I_\pm^\alpha \phi}(x) = \widehat{\phi}(x)/(\mp ix)^\alpha.$$

The conditions of Proposition 6.1.24 are practical. Suppose first $\alpha \in (0, 1/2)$. Since we are considering $L^2(\mathbb{R})$–Fourier transforms, we want both the function ϕ and the result of the integration $\widehat{I_\pm^\alpha \phi}(x) = \widehat{\phi}(x)/(\mp ix)^\alpha$ to be in $L^2(\mathbb{R})$, which gives the condition (6.1.40). The conditions for the case $\alpha \in (-1/2, 0)$ yield the result by allowing the conditions for $\alpha \in (0, 1/2)$ to be satisfied.

Example 6.1.25 (*Fourier transform of fractional integrals of indicator functions*) Let $\alpha \in (-1/2, 1/2)$ and $a, b \in \mathbb{R}$. We will show that the function $\phi(u) = 1_{[a,b)}(u)$, $u \in \mathbb{R}$, satisfies the conditions of Proposition 6.1.24. Indeed, if $\alpha \in (0, 1/2)$, then

$$\int_{\mathbb{R}} |\widehat{1_{[a,b)}}(x)|^2 |x|^{-2\alpha} dx = \int_{\mathbb{R}} \left| \frac{e^{ibx} - e^{iax}}{ix} \right|^2 |x|^{-2\alpha} dx < \infty.$$

If $\alpha \in (-1/2, 0)$, one can use (6.1.35) and verify by direct calculations (Exercise 6.2) that

$$1_{[a,b)} = I_\pm^{-\alpha} f_{a,b}, \tag{6.1.41}$$

where $f_{a,b} = I_\pm^\alpha 1_{[a,b)} = \mathbf{D}_\pm^{-\alpha} 1_{[a,b)}$ belongs to $L^2(\mathbb{R})$. Hence, Proposition 6.1.24 implies that

$$\widehat{I_\pm^\alpha 1_{[a,b)}}(x) = \widehat{1_{[a,b)}}(x)(\mp ix)^{-\alpha} = \frac{e^{ibx} - e^{iax}}{ix} (\mp ix)^{-\alpha}. \tag{6.1.42}$$

Remark 6.1.26 Another interesting approach to fractional integro-differentiation is via fractional differences. For $\alpha > 0$, define[2]

$$f_{\pm}^{(\alpha)}(u) = \lim_{h \to 0} \frac{(\Delta_{\pm h}^{\alpha} f)(u)}{h^{\alpha}}, \qquad (6.1.43)$$

where

$$(\Delta_h^{\alpha} f)(u) = ((E - \tau_h)^{\alpha} f)(u) = \sum_{k=0}^{\infty} (-1)^k \binom{\alpha}{k} f(u - kh),$$

$\binom{\alpha}{k}$ are the binomial coefficients, $(Ef)(u) = f(u)$ and $(\tau_h^k f)(u) = f(u - kh)$ for all $u, h \in \mathbb{R}$ and $k \geq 1$. The expressions in (6.1.43) are called *Grünwald–Letnikov fractional derivatives* of order α of a function f. For more information, see Section 20 in Samko et al. [878].

6.2 Representations of Fractional Brownian Motion

A number of integral representations of fractional Brownian motion (FBM) B_H were derived in previous chapters: the time domain representation in Proposition 2.6.5, the spectral domain representation in Proposition 2.6.11, and the finite interval representation in Proposition 4.2.5 (the latter when $H > 1/2$). We rewrite here these representations in terms of fractional integrals (derivatives), and also extend the finite interval representation to the case $H < 1/2$. We begin with and focus on the finite interval representation, which plays a central role in integration with respect to FBM.

Notation: In the context of fractional integration it is convenient to use another parametrization of FBM. Let

$$\boxed{\kappa = H - \frac{1}{2}.} \qquad (6.2.1)$$

As in (2.8.4), κ is often denoted d, but d can be confusing in the integration context. Now $H \in (0, 1)$ corresponds to $\kappa \in (-1/2, 1/2)$. We will also denote the FBM B_H in terms of the parameter κ as B^{κ}:

$$\boxed{B_H, \quad H \in (0, 1) \quad \leftrightarrow \quad B^{\kappa}, \quad \kappa \in (-1/2, 1/2).} \qquad (6.2.2)$$

The usual standard Brownian motion is denoted B^0.

While the case $\kappa = 0$ is always included below, our focus is on $\kappa \in (-1/2, 1/2) \setminus \{0\}$. In particular, in the proofs below, we will not deal with the case $\kappa = 0$, for which most of the results are trivial.

6.2.1 Representation of FBM on an Interval

We provide below the finite interval representation of FBM in terms of fractional integrals (derivatives) on an interval (these fractional integrals are defined in Section 6.1.1). The representation will involve the kernel function

---

[2] The notation $f_{\pm}^{(\alpha)}$ is not a typo!

360 *Fractional Calculus and Integration of Deterministic Functions with Respect to FBM*

$$f_t(s) = s^{-\kappa}\left(I_{a-}^{\kappa} u^{\kappa} 1_{[0,t)}(u)\right)(s), \tag{6.2.3}$$

where $a > 0$, $\kappa \in (-1/2, 1/2)$ and $s, t \in [0, a]$. By the definitions (6.1.3) and (6.1.12) of fractional integrals and derivatives, one has

$$f_t(s) = \frac{1}{\Gamma(\kappa)} s^{-\kappa} \int_0^t u^{\kappa}(u-s)_+^{\kappa-1} du, \quad \text{if } \kappa \in (0, 1/2), \tag{6.2.4}$$

$$f_t(s) = -\frac{1}{\Gamma(\kappa+1)} s^{-\kappa} \frac{d}{ds} \int_0^t u^{\kappa}(u-s)_+^{\kappa} du, \quad \text{if } \kappa \in (-1/2, 0). \tag{6.2.5}$$

We will first provide several alternative representations of the kernel function $f_t(s)$ which will be useful in the sequel.

Proposition 6.2.1 *Let $a > 0$, $\kappa \in (-1/2, 1/2)$ and $s, t \in [0, a]$. Let also $f_t(s)$ be defined in (6.2.3). Then,*

$$-\Gamma(\kappa+1) f_t(s) = s^{-\kappa} \frac{d}{ds} \int_0^t u^{\kappa}(u-s)_+^{\kappa} du \tag{6.2.6}$$

$$= (2\kappa+1) s^{-\kappa-1} \int_s^t u^{\kappa}(u-s)_+^{\kappa} du - \left(\frac{t}{s}\right)^{\kappa+1}(t-s)_+^{\kappa} \tag{6.2.7}$$

$$= \kappa s^{-\kappa} \int_s^t u^{\kappa-1}(u-s)_+^{\kappa} du - \left(\frac{t}{s}\right)^{\kappa}(t-s)_+^{\kappa} \tag{6.2.8}$$

$$= s^{\kappa} F_i\left(\frac{t}{s}\right) 1_{[0,t)}(s), \quad i = 1, 2, \tag{6.2.9}$$

where, for $z > 1$,

$$F_1(z) = (2\kappa+1) \int_1^z w^{\kappa}(w-1)^{\kappa} dw - z^{\kappa+1}(z-1)^{\kappa}, \tag{6.2.10}$$

$$F_2(z) = \kappa \int_1^z w^{\kappa-1}(w-1)^{\kappa} dw - z^{\kappa}(z-1)^{\kappa}. \tag{6.2.11}$$

Remark 6.2.2 When $s \geq t$, all sides of (6.2.6)–(6.2.8) are equal to 0, and therefore it is enough to define $F_1(z)$ and $F_2(z)$ for $z > 1$.

Remark 6.2.3 The relation (6.2.8) can informally be derived as

$$\frac{d}{ds} \int_0^t u^{\kappa}(u-s)_+^{\kappa} du = \frac{d}{ds} \int_s^t u^{\kappa}(u-s)_+^{\kappa} du = \frac{d}{ds} \int_0^t (v+s)^{\kappa} 1_{\{v<t-s\}} v^{\kappa} dv$$

$$= \kappa \int_0^t (v+s)^{\kappa-1} 1_{\{v<t-s\}} v^{\kappa} dv + \int_0^t (v+s)^{\kappa}(-\delta_{\{t-s\}}(v)) v^{\kappa} dv$$

$$= \kappa \int_s^t u^{\kappa-1}(u-s)_+^{\kappa} du - t^{\kappa}(t-s)_+^{\kappa},$$

where $\delta_{\{a\}}(v)$ denotes a point mass at $v = a$.

Proof It is enough to consider the case $0 < s < t$. When $\kappa \in (-1/2, 0)$, the relation (6.2.6) is given above. When $\kappa \in (0, 1/2)$, by differentiating inside the integral, we expect that

$$s^{-\kappa}\frac{d}{ds}\int_0^t u^\kappa(u-s)_+^\kappa du = -\kappa s^{-\kappa}\int_0^t u^\kappa(u-s)_+^{\kappa-1}du.$$

This could be proved by the definition of derivative. For example, for $h > 0$, consider

$$\int_0^t u^\kappa(u-s-h)_+^\kappa du - \int_0^t u^\kappa(u-s)_+^\kappa du = \int_{s+h}^t u^\kappa\Big((u-s-h)^\kappa - (u-s)^\kappa\Big)du$$
$$- \int_s^{s+h} u^\kappa(u-s)^\kappa du =: T_1(s,h) + T_2(s,h).$$

By the mean value theorem,

$$\frac{T_1(s,h)}{h} = \kappa \int_{s+h}^t u^\kappa(u-s^*(h))^{\kappa-1}du = \kappa \int_{s+h-s^*(h)}^{t-s^*(h)} (u+s^*(h))^\kappa u^{\kappa-1}du,$$

where $s < s^*(h) < s+h$. As $h \to 0$, $T_1(s,h)/h \to \kappa \int_0^{t-s}(u+s)^\kappa u^{\kappa-1}du = \kappa \int_0^t u^\kappa(u-s)_+^{\kappa-1}du$ by the dominated convergence theorem. On the other hand, by making the change of variables $u-s = hv$, $T_2(s,h) = -h^{\kappa+1}\int_0^1(s+hv)^\kappa v^\kappa dv \sim C(s)h^{\kappa+1}$ and hence $T_2(s,h)/h \to 0$, as $h \to 0$.

Note also that the expression (6.2.9) with $i = 1$ is the expression (6.2.7), and that (6.2.9) with $i = 2$ is the expression (6.2.8). It is therefore enough to show that $F_1(z) = F_2(z)$, for $z > 1$, and that the relation (6.2.9) holds with $i = 1$.

The relation (6.2.9) with $i = 1$ holds since, after the change of variables $u = sw$,

$$\frac{d}{ds}\int_0^t u^\kappa(u-s)_+^\kappa du = \frac{d}{ds}\int_s^t u^\kappa(u-s)^\kappa du = \frac{d}{ds}s^{2\kappa+1}\int_1^{t/s} w^\kappa(w-1)_+^\kappa dw$$

$$= (2\kappa+1)s^{2\kappa}\int_1^{t/s} w^\kappa(w-1)_+^\kappa dw - s^{2\kappa+1}\frac{t}{s^2}\left(\frac{t}{s}\right)^\kappa\left(\frac{t}{s}-1\right)^\kappa = s^{2\kappa}F_1\left(\frac{t}{s}\right).$$

Finally, for $F_1(z) = F_2(z)$, $z > 1$, note first that

$$\frac{d}{dz}F_1(z) = (2\kappa+1)z^\kappa(z-1)^\kappa - (\kappa+1)z^\kappa(z-1)^\kappa - \kappa z^{\kappa+1}(z-1)^{\kappa-1}$$
$$= \kappa z^\kappa(z-1)^\kappa - \kappa z^{\kappa+1}(z-1)^{\kappa-1} = -\kappa z^\kappa(z-1)^\kappa,$$

$$\frac{d}{dz}F_2(z) = \kappa z^{\kappa-1}(z-1)^\kappa - \kappa z^{\kappa-1}(z-1)^\kappa - \kappa z^\kappa(z-1)^{\kappa-1} = -\kappa z^\kappa(z-1)^\kappa,$$

so that $\frac{d}{dz}F_1(z) = \frac{d}{dz}F_2(z)$. To conclude that $F_1(z) = F_2(z)$, it is enough to show that $\lim_{z \downarrow 1}(F_1(z) - F_2(z)) = 0$. When $\kappa \in (0, 1/2)$, this follows from $F_1(1+) = F_2(1+) = 0$. For $\kappa \in (-1/2, 0)$, this follows from noting that

$$F_1(z) - F_2(z) = (2\kappa+1)\int_1^z w^\kappa(w-1)^\kappa dw - \kappa\int_1^z w^{\kappa-1}(w-1)^\kappa dw - z^\kappa(z-1)^{\kappa+1},$$

since $(z^{\kappa+1} - z^\kappa)(z-1)^\kappa = z^\kappa(z-1)^{\kappa+1}$, which tends to 0, as $z \downarrow 1$. □

The following three results are immediate corollaries of Proposition 6.2.1. The first corollary shows that the derivative of the kernel function $f_t(s)$ with respect to t has a particularly simple form. This result is quite useful (see, for example, the proof of Proposition 6.2.8 in the case $\kappa \in (-1/2, 0)$).

Corollary 6.2.4 *Let $a > 0$, $\kappa \in (-1/2, 1/2)$ and $s, t \in [0, a]$. Let also $f_t(s)$ be defined in (6.2.3). Then,*

$$\frac{\partial}{\partial t} f_t(s) = \frac{s^{-\kappa} t^\kappa (t-s)^{\kappa-1}}{\Gamma(\kappa)}. \tag{6.2.12}$$

Proof By using (6.2.8),

$$-\Gamma(\kappa+1) \frac{\partial}{\partial t} f_t(s) = \kappa s^{-\kappa} t^{\kappa-1}(t-s)^\kappa - \kappa \frac{t^{\kappa-1}}{s^\kappa}(t-s)^\kappa - \kappa \left(\frac{t}{s}\right)^\kappa (t-s)^{\kappa-1}$$
$$= -\kappa s^{-\kappa} t^\kappa (t-s)^{\kappa-1},$$

which yields the result. $\square$

The second corollary provides the behavior of the kernel function around the points $s = 0$ and $t = 0$.

Corollary 6.2.5 *Let $a > 0$, $\kappa \in (-1/2, 1/2)$ and $s, t \in [0, a]$. Let also $f_t(s)$ be defined in (6.2.3). Then, as $s \downarrow 0$,*

$$f_t(s) \sim \begin{cases} \frac{t^{2\kappa}}{2\Gamma(\kappa+1)} s^{-\kappa}, & \text{if } \kappa \in (0, 1/2), \\ \frac{\Gamma(1-2\kappa)}{2\Gamma(1-\kappa)} s^\kappa, & \text{if } \kappa \in (-1/2, 0), \end{cases} \tag{6.2.13}$$

and, as $s \uparrow t$,

$$f_t(s) \sim \frac{(t-s)^\kappa}{\Gamma(\kappa+1)}. \tag{6.2.14}$$

Proof The relation (6.2.13) follows from (6.2.9) after observing that as $z \to \infty$, $F_2(z) \sim \kappa \int_1^z w^{2\kappa-1} dw - z^{2\kappa} \sim -z^{2\kappa}/2$ when $\kappa \in (0, 1/2)$, and that $F_2(z) \sim \kappa \int_1^\infty w^{\kappa-1}(w-1)^\kappa dw = \kappa \int_0^\infty z^\kappa (z+1)^{\kappa-1} dz = \kappa B(\kappa+1, -2\kappa) = \kappa \frac{\Gamma(\kappa+1)\Gamma(-2\kappa)}{\Gamma(1-\kappa)} = -\frac{\Gamma(\kappa+1)\Gamma(1-2\kappa)}{2\Gamma(1-\kappa)}$ when $\kappa \in (-1/2, 0)$. The relation (6.2.14) follows from (6.2.9) after observing that as $z \downarrow 1$, $F_2(z) \sim -(z-1)^\kappa$. $\square$

The third corollary provides yet another, useful representation of the kernel function.

Corollary 6.2.6 *Let $a > 0$, $\kappa \in (-1/2, 1/2)$ and $s, t \in [0, a]$. Let also $f_t(s)$ be defined in (6.2.3). Then,*

$$-\Gamma(\kappa+1) f_t(s) = \kappa s^{-\kappa} \int_s^t (s^\kappa - u^\kappa)(u-s)_+^{\kappa-1} du - (t-s)_+^{\kappa-1}. \tag{6.2.15}$$

Proof The relation (6.2.15) can be proved by rewriting the right-hand side of (6.2.8) as follows:

$$\kappa s^{-\kappa} \int_s^t u^{\kappa-1}(u-s)_+^\kappa du - \left(\frac{t}{s}\right)^\kappa (t-s)_+^\kappa$$
$$= \kappa s^{-\kappa} \int_s^t u^{\kappa-1}\left((u-s)_+^\kappa - (t-s)_+^\kappa\right) du + \kappa s^{-\kappa} \int_s^t u^{\kappa-1} du\, (t-s)_+^\kappa - s^{-\kappa} t^\kappa (t-s)_+^\kappa$$

$$= -\kappa^2 s^{-\kappa} \int_s^t u^{\kappa-1} \Big(\int_u^t (z-s)_+^{\kappa-1} dz\Big) du + s^{-\kappa} (t^\kappa - s^\kappa)(t-s)_+^\kappa - s^{-\kappa} t^\kappa (t-s)_+^\kappa$$

$$= -\kappa^2 s^{-\kappa} \int_s^t (z-s)_+^{\kappa-1} \Big(\int_s^z u^{\kappa-1} du\Big) dz - (t-s)_+^\kappa$$

$$= -\kappa s^{-\kappa} \int_s^t (z^\kappa - s^\kappa)(z-s)_+^{\kappa-1} dz - (t-s)_+^\kappa,$$

where we changed the order of integration above. $\square$

The next alternative representation involves the celebrated Gauss hypergeometric function $_2F_1(a, b, c, z)$. Recall that this function is defined for any $a, b, c, z \in \mathbb{C}$ satisfying $c \neq 0, -1, \ldots$, and $|\arg(1-z)| < \pi$ (that is, z is not on the section of the real axis starting at $+1$ and stretching to $+\infty$; any such z would satisfy $\arg(1-z) = -\pi$). When $|z| \leq 1$, the function is given by the series

$$_2F_1(a, b, c, z) = \sum_{k=0}^\infty \frac{(a)_k (b)_k}{(c)_k} \frac{z^k}{k!}, \qquad (6.2.16)$$

where $(a)_0 = 1$ and $(a)_k = \Gamma(a+k)/\Gamma(a) = a(a+1)\ldots(a+k-1)$ is the Pochhammer symbol. For other z's (satisfying $|\arg(1-z)| < \pi$), the hypergeometric function is extended by analytic continuation. When $0 < \Re b < \Re c$, it is known that

$$_2F_1(a, b, c, z) = \frac{\Gamma(c)}{\Gamma(b)\Gamma(c-b)} \int_0^1 v^{b-1} (1-v)^{c-b-1} (1-zv)^{-a} dv, \qquad (6.2.17)$$

for all z satisfying $|\arg(1-z)| < \pi$. It is also known that, for fixed z, $_2F_1(a, b, c, z)$ is analytic in its parameters a, b and c. This and more information on the Gauss hypergeometric function can be found, for example, in Lebedev [601], Section 9.

Proposition 6.2.7 *Let $a > 0$, $\kappa \in (-1/2, 1/2)$ and $s, t \in [0, a]$. Let also $f_t(s)$ be defined in (6.2.3). Then,*

$$\Gamma(\kappa+1) f_t(s) = (t-s)_+^\kappa \, _2F_1\Big(-\kappa, \kappa, \kappa+1, 1-\frac{t}{s}\Big). \qquad (6.2.18)$$

Proof When $\kappa \in (0, 1/2)$ and $0 \leq s < t \leq a$, the representation follows from

$$\Gamma(\kappa+1) f_t(s) = \kappa s^{-\kappa} \int_s^t u^\kappa (u-s)^{\kappa-1} du = \kappa s^{-\kappa} \int_0^{t-s} (v+s)^\kappa v^{\kappa-1} dv$$

$$= \kappa s^{-\kappa} (t-s)^\kappa \int_0^1 ((t-s)w + s)^\kappa w^{\kappa-1} dw$$

$$= \kappa (t-s)^\kappa \int_0^1 w^{\kappa-1} \Big(1 - \Big(1-\frac{t}{s}\Big)w\Big)^\kappa dw$$

$$= (t-s)^\kappa \frac{\Gamma(\kappa+1)}{\Gamma(\kappa)\Gamma((\kappa+1)-\kappa)} \int_0^1 w^{\kappa-1} (1-w)^{(\kappa+1)-\kappa-1} \Big(1-\Big(1-\frac{t}{s}\Big)w\Big)^\kappa dw$$

$$= (t-s)^\kappa \, _2F_1\Big(-\kappa, \kappa, \kappa+1, 1-\frac{t}{s}\Big),$$

where we used the changes of variables $v = su$ and $v = (t - s)w$, and the integral representation (6.2.17).

When $\kappa \in (-1/2, 0)$, the relation (6.2.18) can be proved by arguing that the left- and right-hand sides are analytic functions for $|\kappa| < 1/2$, $\kappa \in \mathbb{C}$. Then, since the two sides coincide for real $\kappa \in (0, 1/2)$, they also coincide for real $\kappa \in (-1/2, 0)$. For the right-hand side, the analyticity follows from the analytic properties of the hypergeometric function mentioned above. For the left-hand side, suppose it to be defined, for example, by the representation (6.2.7), which is now considered for $|\kappa| < 1/2$, $\kappa \in \mathbb{C}$. All power functions a^κ, $a \neq 0$, entering (6.2.7) are analytic. The integral $f(\kappa) = \int_s^t u^\kappa (u-s)^\kappa du$ can be shown to be analytic for $|\kappa| < 1/2$ directly through the definition by using the arguments as those in Pipiras and Taqqu [817], Appendix A. For example, as in that appendix, one can show that

$$\frac{d}{d\kappa} f(\kappa) = \int_s^t \Big(\log u \, u^\kappa (u-s)^\kappa + \log(u-s) \, u^\kappa (u-s)^\kappa \Big) du. \qquad \Box$$

Most mathematical software, for example, MATLAB, performs calculations of Gauss hypergeometric function. Figure 6.2 plots the kernel function $f_t(s)$ for $t = 1$ and several values of κ.

We now provide the finite interval representation of FBM in terms of fractional integrals (derivatives) on an interval. It involves the function $f_t(s)$ in (6.2.3). When $\kappa \in (0, 1/2)$, the representation corresponds to that in Proposition 4.2.5, but we will provide an alternative and more direct proof.

Proposition 6.2.8 *Let $a > 0$, $\kappa \in (-1/2, 1/2)$ and B^κ be a standard FBM. Then,*

$$\{B^\kappa(t)\}_{t \in [0,a]} \stackrel{d}{=} \Big\{ \sigma_\kappa \int_0^a s^{-\kappa} \Big(I_{a-}^\kappa u^\kappa 1_{[0,t)}(u) \Big)(s) dB^0(s) \Big\}_{t \in [0,a]} \qquad (6.2.19)$$

in the sense of finite-dimensional distributions, where

$$\sigma_\kappa^2 = \frac{\pi \kappa (2\kappa + 1)}{\Gamma(1 - 2\kappa) \sin(\kappa \pi)} = \frac{\Gamma(\kappa)^2 \kappa (2\kappa + 1)}{B(\kappa, 1 - 2\kappa)}. \qquad (6.2.20)$$

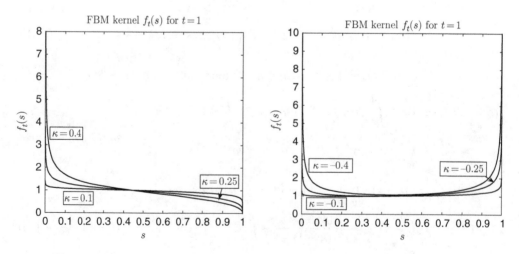

Figure 6.2 The kernel function $f_t(s)$ in (6.2.3) for $t = 1$ and several values of κ.

Remark 6.2.9 Note that the endpoint a of the interval $[0, a]$ plays little role in the representation (6.2.19). The kernel function $f_t(s)$ in (6.2.3) is supported on the interval $[0, t]$. It could be expressed alternatively as

$$f_t(s) = s^{-\kappa}\left(I_-^\kappa u^\kappa 1_{[0,t)}(u)\right)(s),$$

where I_-^κ is a fractional integral or derivative in (6.1.17) or (6.1.30). The processes on the two sides of (6.2.19) can then be indexed by $t \in [0, \infty)$, instead of $t \in [0, a]$, and the integral $\int_0^a$ replaced by $\int_0^\infty$. (Another possibility is to use I_{t-}^κ and the integral $\int_0^t$.) In any case, the integration is effectively over the interval $[0, t]$. It is for this reason that (6.2.19) is called a representation on an interval.

Proof The cases $\kappa \in (0, 1/2)$ and $\kappa \in (-1/2, 0)$ are treated separately.

The case $\kappa \in (0, 1/2)$: Since the covariance function of a standard FBM B^κ is

$$\Gamma^\kappa(u, v) = \mathbb{E}B^\kappa(u)B^\kappa(v) = \frac{1}{2}\left(|u|^{2\kappa+1} + |v|^{2\kappa+1} - |u-v|^{2\kappa+1}\right), \qquad (6.2.21)$$

with $u, v \in \mathbb{R}$, we get that $d^2\Gamma^\kappa(u, v) = \kappa(2\kappa + 1)|u-v|^{2\kappa-1}dudv$. This implies that

$$\Gamma^\kappa(t_1, t_2) = \kappa(2\kappa+1) \int_\mathbb{R} \int_\mathbb{R} 1_{[0,t_1)}(u) 1_{[0,t_2)}(v) |u-v|^{2\kappa-1}dudv. \qquad (6.2.22)$$

We will argue next[3] that, for $u, v \in [0, a]$,

$$|u-v|^{2\kappa-1} = \frac{(uv)^\kappa}{B(\kappa, 1-2\kappa)} \int_0^a s^{-2\kappa}(v-s)_+^{\kappa-1}(u-s)_+^{\kappa-1} ds, \qquad (6.2.23)$$

where B is the beta function defined in (2.2.12). Suppose without loss of generality that $v > u$. Then, the right-hand side of (6.2.23) becomes

$$\frac{(uv)^\kappa}{B(\kappa, 1-2\kappa)} \int_0^u s^{-2\kappa}(v-s)^{\kappa-1}(u-s)^{\kappa-1}ds = \frac{(uv)^\kappa}{B(\kappa, 1-2\kappa)} \int_0^u \left(\frac{v}{s}-1\right)^{\kappa-1}\left(\frac{u}{s}-1\right)^{\kappa-1}\frac{ds}{s^2}.$$

After the change of variables $1/s = w$, this becomes

$$\frac{(uv)^\kappa}{B(\kappa, 1-2\kappa)} \int_{1/u}^\infty (vw-1)^{\kappa-1}(uw-1)^{\kappa-1} dw.$$

Further changes of variables $uw - 1 = y$ and then $y = \frac{(v/u-1)}{(v/u)}z = \frac{v-u}{v}z$ lead to

$$\frac{(uv)^\kappa}{B(\kappa, 1-2\kappa)} \int_0^\infty \left(\frac{v}{u}y + \left(\frac{v}{u}-1\right)\right)^{\kappa-1} y^{\kappa-1}\frac{dy}{u}$$

$$= \frac{(uv)^\kappa}{B(\kappa, 1-2\kappa)} \int_0^\infty \left(\frac{v}{u}-1\right)^{\kappa-1}(z+1)^{\kappa-1}\left(\frac{v-u}{v}\right)^{\kappa-1} z^{\kappa-1}\frac{v-u}{v}\frac{dz}{u}$$

$$= \frac{(v-u)^{2\kappa-1}}{B(\kappa, 1-2\kappa)} \int_0^\infty (z+1)^{\kappa-1}z^{\kappa-1}dz = (v-u)^{2\kappa-1},$$

---
[3] See also the proof of Lemma 4.2.7.

since $\int_0^\infty (z+1)^{\kappa-1} z^{\kappa-1} dz = B(\kappa, 1-2\kappa)$ by using (2.2.12). This establishes (6.2.23). (One can also prove (6.2.23) by making the single change of variables $s = (uv)z/(\min(u,v)z + |u-v|)$.)

By changing the order of integration in (6.2.22) with $|u-v|^{2\kappa-1}$ substituted by (6.2.23) and using Definition 6.1.1, we obtain that, for $t_1, t_2 \in [0, a]$,

$$\Gamma^\kappa(t_1, t_2) = \frac{\Gamma(\kappa)^2 \kappa (2\kappa + 1)}{B(\kappa, 1 - 2\kappa)} \int_0^a \left(s^{-\kappa} (I_{a-}^\kappa u^\kappa 1_{[0,t_1)}(u))(s) \right) \left(s^{-\kappa} (I_{a-}^\kappa u^\kappa 1_{[0,t_2)}(u))(s) \right) ds.$$

This establishes the proposition in the case $\kappa \in (0, 1/2)$.

The case $\kappa \in (-1/2, 0)$: Let $f_t(s)$ be the kernel function defined in (6.2.3). We want to show that, for $0 \leq t_1 < t_2 \leq a$,

$$\int_0^a f_{t_1}(s) f_{t_2}(s) ds = \frac{1}{2\sigma_\kappa^2} \left(t_1^{2\kappa+1} + t_2^{2\kappa+1} - (t_2 - t_1)^{2\kappa+1} \right).$$

It is enough to prove that

$$\frac{\partial}{\partial t_2} \int_0^a f_{t_1}(s) f_{t_2}(s) ds = \frac{2\kappa + 1}{2\sigma_\kappa^2} \left(t_2^{2\kappa} - (t_2 - t_1)^{2\kappa} \right), \quad t_1 < t_2, \qquad (6.2.24)$$

and when $t_2 = t_1$,

$$\int_0^a f_{t_1}(s) f_{t_1}(s) ds = \frac{t_1^{2\kappa}}{\sigma_\kappa^2}. \qquad (6.2.25)$$

The relation (6.2.25) is left as an exercise (Exercise 6.3). By taking the differentiation inside the integral, the relation (6.2.24) would follow from

$$\int_0^a f_{t_1}(s) \frac{\partial}{\partial t_2} f_{t_2}(s) ds = \frac{2\kappa + 1}{2\sigma_\kappa^2} \left(t_2^{2\kappa} - (t_2 - t_1)^{2\kappa} \right). \qquad (6.2.26)$$

By using (6.2.9) and (6.2.12), this is equivalent to

$$\int_0^{t_1} \frac{s^\kappa F_2(\frac{t_1}{s})}{-\Gamma(\kappa + 1)} \frac{s^{-\kappa} t_2^\kappa (t_2 - s)^{\kappa-1}}{\Gamma(\kappa)} ds = \frac{2\kappa + 1}{2\sigma_\kappa^2} \left(t_2^{2\kappa} - (t_2 - t_1)^{2\kappa} \right)$$

or, after the change of variables $s = t_1 w$,

$$\int_0^1 F_2(\frac{1}{w}) t_2^\kappa (t_2 - t_1 w)^{\kappa-1} t_1 dw = -\frac{(2\kappa + 1)\Gamma(\kappa + 1)\Gamma(\kappa)}{2\sigma_\kappa^2} \left(t_2^{2\kappa} - (t_2 - t_1)^{2\kappa} \right)$$

or, after setting $b = t_2/t_1$ and simplifying the constant by using (6.2.20),

$$\int_0^1 F_2(\frac{1}{w}) (b - w)^{\kappa-1} dw = \frac{B(\kappa, 1 - 2\kappa)}{2} b^{-\kappa} \left((b-1)^{2\kappa} - b^{2\kappa} \right), \quad b > 1. \qquad (6.2.27)$$

Denote the left-hand side of (6.2.27) by $I(b)$. We want to show that it is equal to the right-hand side. By using (6.2.6) and (6.2.9) with $t = 1$,

$$I(b) = \int_0^1 \frac{d}{dw} \left(\int_0^1 u^\kappa (u - w)_+^\kappa du \right) w^{-2\kappa} (b - w)^{\kappa-1} dw.$$

6.2 Representations of Fractional Brownian Motion

By integration by parts and then changing the order of integration, we get

$$I(b) = -\int_0^1 \left(\int_0^1 u^\kappa (u-w)_+^\kappa du\right) \frac{d}{dw}\left(w^{-2\kappa}(b-w)^{\kappa-1}\right) dw$$

$$= 2\kappa \int_0^1 \left(\int_w^1 u^\kappa (u-w)^\kappa du\right) w^{-2\kappa-1}(b-w)^{\kappa-1} dw$$

$$+ (\kappa-1)\int_0^1 \left(\int_w^1 u^\kappa (u-w)^\kappa du\right) w^{-2\kappa}(b-w)^{\kappa-2} dw$$

$$= 2\kappa \int_0^1 du\, u^\kappa \int_0^u w^{-2\kappa-1}(u-w)^\kappa (b-w)^{\kappa-1} dw$$

$$+ (\kappa-1)\int_0^1 du\, u^\kappa \int_0^u w^{-2\kappa}(u-w)^\kappa (b-w)^{\kappa-2} dw.$$

By making a series of changes of variables $w = uv$, $v = 1/z$, $z = x+1$ and $x = \frac{(b/u)-1}{(b/u)} y = \frac{b-u}{b} y$ below, we further get that

$$I(b) = 2\kappa \int_0^1 du\, u^{\kappa-1} \int_0^1 v^{-2\kappa-1}(1-v)^\kappa \left(\frac{b}{u} - v\right)^{\kappa-1} dv$$

$$+ (\kappa-1)\int_0^1 du\, u^{\kappa-1} \int_0^1 v^{-2\kappa}(1-v)^\kappa \left(\frac{b}{u} - v\right)^{\kappa-2} dv$$

$$= 2\kappa \int_0^1 du\, u^{\kappa-1} \int_1^\infty (z-1)^\kappa \left(\frac{b}{u}z - 1\right)^{\kappa-1} dz$$

$$+ (\kappa-1)\int_0^1 du\, u^{\kappa-1} \int_1^\infty (z-1)^\kappa \left(\frac{b}{u}z - 1\right)^{\kappa-2} dz$$

$$= 2\kappa \int_0^1 du\, u^{\kappa-1} \int_0^\infty x^\kappa \left(\frac{b}{u}x + \frac{b}{u} - 1\right)^{\kappa-1} dx$$

$$+ (\kappa-1)\int_0^1 du\, u^{\kappa-1} \int_0^\infty x^\kappa \left(\frac{b}{u}x + \frac{b}{u} - 1\right)^{\kappa-2} dx$$

$$= 2\kappa b^{-\kappa-1} \int_0^1 (b-u)^{2\kappa} du \int_0^\infty y^\kappa (y+1)^{\kappa-1} dy$$

$$+ (\kappa-1) b^{-\kappa-1} \int_0^1 u(b-u)^{2\kappa-1} du \int_0^\infty y^\kappa (y+1)^{\kappa-2} dy.$$

In view of (2.2.12) and then by integration by parts, we obtain that

$$I(b) = 2\kappa B(\kappa+1, -2\kappa) b^{-\kappa-1} \int_0^1 (b-u)^{2\kappa} du$$

$$+ (\kappa-1) B(\kappa+1, 1-2\kappa) b^{-\kappa-1} \int_0^1 u(b-u)^{2\kappa-1} du$$

$$= B(\kappa, 1-2\kappa) \kappa b^{-\kappa-1} \left(-\int_0^1 (b-u)^{2\kappa} du - \int_0^1 u(b-u)^{2\kappa-1} du\right)$$

$$= B(\kappa, 1-2\kappa) b^{-\kappa-1} \left(\frac{1}{2}(b-1)^{2\kappa} - \frac{2\kappa+1}{2}\int_0^1 (b-u)^{2\kappa} du\right)$$

$$= B(\kappa, 1 - 2\kappa) b^{-\kappa-1} \left(\frac{1}{2}(b-1)^{2\kappa} - \frac{2\kappa+1}{2} \int_0^1 (b-u)^{2\kappa} du \right)$$

$$= B(\kappa, 1 - 2\kappa) b^{-\kappa-1} \left(\frac{1}{2}(b-1)^{2\kappa} + \frac{1}{2}(b-1)^{2\kappa+1} - \frac{1}{2} b^{2\kappa+1} \right)$$

$$= B(\kappa, 1 - 2\kappa) b^{-\kappa-1} \left(\frac{(b-1)^{2\kappa} b}{2} - \frac{b^{2\kappa} b}{2} \right) = \frac{B(\kappa, 1 - 2\kappa)}{2} b^{-\kappa} \left((b-1)^{2\kappa} - b^{2\kappa} \right),$$

which establishes (6.2.27). $\square$

6.2.2 Representations of FBM on the Real Line

In the next result, we rewrite the integral representations (2.6.1) and (2.6.9) of FBM using fractional integrals (derivatives) on the real line.

Proposition 6.2.10 *Let $\kappa \in (-1/2, 1/2)$. A standard FBM B^κ has the following representations in the sense of finite-dimensional distributions:*
(a) Time domain representation:

$$\{B^\kappa(t)\}_{t \in \mathbb{R}} \stackrel{d}{=} \left\{ \frac{1}{c_{1,\kappa}} \int_\mathbb{R} \left((t-s)_+^\kappa - (-s)_+^\kappa \right) dB^0(s) \right\}_{t \in \mathbb{R}} \qquad (6.2.28)$$

$$\stackrel{d}{=} \left\{ \frac{\Gamma(\kappa+1)}{c_{1,\kappa}} \int_\mathbb{R} (I_-^\kappa 1_{[0,t)})(s) dB^0(s) \right\}_{t \in \mathbb{R}}, \qquad (6.2.29)$$

where

$$c_{1,\kappa}^2 = \int_\mathbb{R} \left((1-u)_+^\kappa - (-u)_+^\kappa \right)^2 du = \frac{\kappa B(\kappa, 1-\kappa)}{2\kappa+1} \qquad (6.2.30)$$

and $dB^0(s) = B^0(ds)$ is a Gaussian measure with the Lebesgue control measure $\mathbb{E}(B(ds))^2 = ds$.
(b) Spectral domain representation:

$$\{B^\kappa(t)\}_{t \in \mathbb{R}} \stackrel{d}{=} \left\{ \frac{\Gamma(\kappa+1)}{(2\pi)^{1/2} c_{1,\kappa}} \int_\mathbb{R} \frac{e^{ixt}-1}{ix} |x|^{-\kappa} d\widehat{B}(x) \right\}_{t \in \mathbb{R}} \qquad (6.2.31)$$

$$\stackrel{d}{=} \left\{ \frac{\Gamma(\kappa+1)}{(2\pi)^{1/2} c_{1,\kappa}} \int_\mathbb{R} \frac{e^{ixt}-1}{ix} (ix)^{-\kappa} d\widehat{B}(x) \right\}_{t \in \mathbb{R}} \qquad (6.2.32)$$

$$\stackrel{d}{=} \left\{ \frac{\Gamma(\kappa+1)}{(2\pi)^{1/2} c_{1,\kappa}} \int_\mathbb{R} (\widehat{I_-^\kappa 1_{[0,t)}})(x) d\widehat{B}(x) \right\}_{t \in \mathbb{R}}, \qquad (6.2.33)$$

where $d\widehat{B}(x) = \widehat{B}(dx)$ is an Hermitian Gaussian random measure with the Lebesgue control measure $\mathbb{E}|\widehat{B}(dx)|^2 = dx$.

Proof The representation (6.2.28) of standard FBM is the one given in (2.6.1). The representation (6.2.29) follows from (6.1.24) when $\kappa \in (0, 1/2)$, and (6.1.35) when $\kappa \in (-1/2, 0)$. The constant $c_{1,\kappa}$ in (6.2.30) is given below (2.6.1), and the second equality in (6.2.30) follows from (2.6.12).

The spectral representation (6.2.31) is given in Proposition 2.6.11 and the normalizing factor follows from (2.6.11). Since $|(ix)^{-\kappa}|^2 = (|x|^{-\kappa})^2$, the process in (6.2.32) has the

6.3 Fractional Wiener Integrals and their Deterministic Integrands

6.3.1 The Gaussian Space Generated by Fractional Wiener Integrals

Fractional Wiener integrals on an interval $[0, a]$ are integrals of the form

$$\int_0^a f(u) dB^\kappa(u) =: \mathcal{I}_a^\kappa(f), \tag{6.3.1}$$

where f is a deterministic function and B^κ is FBM. In contrast to integration with respect to Brownian motion, FBM does not have independent increments when $\kappa \neq 0$.

We shall first discuss what kind of objects the integrals (6.3.1) are, and what purposes they serve. We focus on the case of an interval $[0, a]$ throughout, though the case of the real line will also be discussed briefly below.

As with most integrals, one can think informally of a fractional Wiener integral as

$$\mathcal{I}_a^\kappa(f) = \int_0^a f(u) dB^\kappa(u) \approx \sum_{k=0}^{K-1} f_k^K (B^\kappa(u_{k+1}^K) - B^\kappa(u_k^K)), \tag{6.3.2}$$

where $0 \leq u_0^K < u_1^K < \cdots < u_K^K \leq a$ is a partition of $[0, a]$, and f_k^K is somehow related to $f(u)$, for example, $f_k^K = f(u_k^K)$ if f is sufficiently smooth. The fractional Wiener integral is expected to be defined by (6.3.2) in the limit of $K \to \infty$. It is natural to take here the limit in the $L^2(\Omega)$–sense. Note also that the sum in (6.3.2) is a linear combination of the values of FBM on the interval $[0, a]$.

Motivated by these observations, it is natural to consider the following space of random variables.

Definition 6.3.1 The *Gaussian space* $\mathcal{H}_a^\kappa$ associated with FBM B^κ on an interval $[0, a]$ is defined as

$$\mathcal{H}_a^\kappa = \overline{\text{span}}\{B^\kappa(t), t \in [0, a]\}, \tag{6.3.3}$$

where the linear span

$$\text{span}\{B^\kappa(t), t \in [0, a]\} = \Big\{ \sum_{j=1}^J c_j B^\kappa(t_j) : c_j \in \mathbb{R}, t_j \in [0, a], j = 1, \ldots, J, J \geq 1 \Big\} \tag{6.3.4}$$

consists of all possible linear combination of FBM on the interval $[0, a]$, and $\overline{\text{span}}$ indicates its closure in $L^2(\Omega)$.

Any linear combination of FBM on $[0, a]$ belongs to the Gaussian space $\mathcal{H}_a^\kappa$. Any finite collection $X_1, \ldots, X_n$ of elements of $\mathcal{H}_a^\kappa$ forms a zero mean *Gaussian* vector $(X_1, \ldots, X_n)$. This follows from the fact that Gaussianity is preserved under the $L^2(\Omega)$–limits. The Gaussian space $\mathcal{H}_a^\kappa$ is a subset of $L^2(\Omega)$, and is naturally equipped with the $L^2(\Omega)$ inner product

$$(X, Y)_{\mathcal{H}_a^\kappa} = \mathbb{E} XY.$$

Since $\mathcal{H}_a^\kappa$ is defined as the closure in $L^2(\Omega)$, it is a *complete* inner product space; that is, a Hilbert space.[4] Recall that an inner product space is complete if every Cauchy sequence of elements in the space has a limit that is also in the space. If the space is not complete, then the limit may either not exist or be outside the space (for example, it could be a so-called "generalized process").

In view of (6.3.2), it is natural to try to associate the elements X of the Gaussian space $\mathcal{H}_a^\kappa$ with fractional Wiener integrals $\mathcal{I}_a^\kappa(f)$ and hence, more importantly, with *deterministic* functions or integrands f (and vice-versa). One natural association involves the set $\mathcal{E}$ of *elementary (step) functions* defined as

$$f(u) = \sum_{k=1}^n f_k 1_{(u_k, u_{k+1}]}(u), \qquad (6.3.5)$$

where $f_k \in \mathbb{R}$ and $0 \le u_k \le u_{k+1} \le a$, $k = 1, \ldots, n$. For the elementary function f in (6.3.5), one naturally associates the random variable

$$\mathcal{I}_a^\kappa(f) = \sum_{k=1}^n f_k (B^\kappa(u_{k+1}) - B^\kappa(u_k)), \qquad (6.3.6)$$

which is a finite combination of FBM increments. Conversely, any finite combination of FBM on the interval $[0, a]$ can be written as a fractional integral of an elementary function. Indeed, applying the summation by parts formula (A.2.5) in Appendix A.2 with $n = 1$, $a_k = 0$ for $k \ge m+1$, yields

$$\sum_{k=1}^m c_k B^\kappa(u_k) = \widehat{C}_1 B^\kappa(u_1) + \sum_{k=2}^m \widehat{C}_k (B^\kappa(u_k) - B^\kappa(u_{k-1})) = \sum_{k=1}^m \widehat{C}_k (B^\kappa(u_k) - B^\kappa(u_{k-1})),$$

after we set $u_0 = 0$ and hence $B^\kappa(u_0) = 0$, where $\widehat{C}_k = \sum_{j=k}^m c_j$. For example, $3B^\kappa(1) + 2B^\kappa(2) = 5(B^\kappa(1) - B^\kappa(0)) + 2(B^\kappa(2) - B^\kappa(1)) = \mathcal{I}_a^\kappa(f)$ with $f(u) = 5(1_{(0,1]}(u)) + 2(1_{(1,2]}(u))$. In view of (6.3.4), one then has

$$\mathrm{span}\{B^\kappa(t), t \in [0, a]\} = \{\mathcal{I}_a^\kappa(f), f \in \mathcal{E}\}$$

and by (6.3.3), we can take the closure in $L^2(\Omega)$ of both sides and get

$$\mathcal{H}_a^\kappa = \overline{\{\mathcal{I}_a^\kappa(f), f \in \mathcal{E}\}}.$$

This association between elementary functions (or their fractional Wiener integrals) and the elements of $\mathcal{H}_a^\kappa$ can naturally be extended to larger classes of functions $\mathcal{C}$ by using the following basic result. Note that a special attention is paid to whether $\mathcal{C}$ is complete. Some classes for FBM considered below will not be complete.

Proposition 6.3.2 *Let $\mathcal{E}$ be the set of elementary functions (6.3.5), $\mathcal{I}_a^\kappa$ be the map defined on elementary functions by (6.3.6) and let $\kappa \in (-1/2, 1/2)$. Suppose that $\mathcal{C}$ is a set of deterministic functions on $[0, a]$ such that*

---

[4] The notation $\mathcal{H}$ in $\mathcal{H}_a^\kappa$ is chosen so as to suggest that the space is Hilbert. An inner product space is a linear (vector) space with an inner product.

(1) $\mathcal{C}$ is an inner product space with an inner product $(f, g)_\mathcal{C}$ for $f, g \in \mathcal{C}$,
(2) $\mathcal{E} \subset \mathcal{C}$ and $(f, g)_\mathcal{C} = \mathbb{E}\mathcal{I}_a^\kappa(f)\mathcal{I}_a^\kappa(g)$, for $f, g \in \mathcal{E}$,
(3) the set $\mathcal{E}$ is dense in $\mathcal{C}$.

Then there is an isometry, denoted $\mathcal{I}_a^\kappa$, between the space $\mathcal{C}$ and a linear subspace of $\mathcal{H}_a^\kappa$ which is an extension of the map $f \to \mathcal{I}_a^\kappa(f)$, for $f \in \mathcal{E}$. Moreover, $\mathcal{C}$ is isometric to $\mathcal{H}_a^\kappa$ itself if and only if $\mathcal{C}$ is complete.

Proof To define the isometry $\mathcal{I}_a^\kappa$, let $f \in \mathcal{C}$. By the assumption (3), there is a sequence $\{f_n\} \subset \mathcal{E}$ such that $f_n \to f$ in $\mathcal{C}$. In particular, $\{f_n\}$ is Cauchy in $\mathcal{C}$ and hence, by the property (2), $\{\mathcal{I}_a^\kappa(f_n)\}$ is Cauchy in $L^2(\Omega)$. Since the space $L^2(\Omega)$ is complete, there is $X \in L^2(\Omega)$ such that $X = \lim \mathcal{I}_a^\kappa(f_n)$. Now, set $\mathcal{I}_a^\kappa(f) = X$. Since $\{\mathcal{I}_a^\kappa(f_n)\} \subset \mathcal{H}_a^\kappa$ and $\mathcal{H}_a^\kappa$ is a closed subset in $L^2(\Omega)$, we obtain that $\mathcal{I}_a^\kappa(f) \in \mathcal{H}_a^\kappa$. We can thus define the map $\mathcal{I}_a^\kappa$ from the space $\mathcal{C}$ into the space $\mathcal{H}_a^\kappa$. One can verify that this definition does not depend on the approximating sequence $\{f_n\}$ (Exercise 6.4). The construction of $\mathcal{I}_a^\kappa$ and the property (2) imply that, for $f, g \in \mathcal{C}$,

$$(f, g)_\mathcal{C} = \mathbb{E}\mathcal{I}_a^\kappa(f)\mathcal{I}_a^\kappa(g).$$

Moreover, since the map $\mathcal{I}_a^\kappa$ is linear, we conclude that $\mathcal{I}_a^\kappa$ is indeed an isometry between the space $\mathcal{C}$ and a linear subspace of $\mathcal{H}_a^\kappa$.

For the last statement of the proposition, if $\mathcal{C}$ is isometric to the full space $\mathcal{H}_a^\kappa$, then $\mathcal{C}$ is complete because the space $\mathcal{H}_a^\kappa$ is complete. Conversely, if $\mathcal{C}$ is complete, then the map is onto because $\mathcal{E}$ is dense in $\mathcal{C}$ and hence $\mathcal{C}$ is isometric to $\mathcal{H}_a^\kappa$ itself. $\square$

Definition 6.3.3 The isometry map $\mathcal{I}_a^\kappa : \mathcal{C} \mapsto \mathcal{H}_a^\kappa$ in Proposition 6.3.2 is denoted $\mathcal{I}_a^\kappa(f) = \int_0^a f(u)dB^\kappa(u)$, $f \in \mathcal{C}$. The latter integral is called a *fractional Wiener integral* of $f \in \mathcal{C}$, and $\mathcal{C}$ is called a *class of integrands*.

When $\kappa = 0$, FBM B^κ is the usual (standard) Brownian motion B^0. For $f, g \in \mathcal{E}$, one can show (Exercise 6.5) that

$$\mathbb{E}\mathcal{I}_a^0(f)\mathcal{I}_a^0(g) = \int_0^a f(u)g(u)du. \tag{6.3.7}$$

A natural space of integrands for Brownian motion is then $\mathcal{C} = L^2[0, a]$, equipped with the usual inner product $(f, g)_\mathcal{C} = \int_0^a f(u)g(u)du$. The conditions (1)–(3) of Proposition 6.3.2 are satisfied. Since $L^2[0, a]$ is complete, it is isometric to $\mathcal{H}_a^0$. The resulting integral $\mathcal{I}_a^0(f) = \int_0^a f(u)dB^0(u)$, $f \in L^2[0, a]$, is called the *Wiener integral*. It can also be viewed as $\int_0^a f(u)B(du)$ where $B(du)$ is the Gaussian random measure with the control measure $\mathbb{E}|B(du)|^2 = du$.

Several classes $\mathcal{C}$ for FBM B^κ, $\kappa \neq 0$, are constructed in Section 6.3.2 below. The Gaussian space $\mathcal{H}_a^\kappa$ and fractional Wiener integrals are natural objects to use in a number of applications. For example, since B^κ is a Gaussian process, its predicted value at time $t \geq a$ given its past $B^\kappa(s)$, $s \in [0, a]$,

$$\mathbb{E}\Big(B^\kappa(t)|B^\kappa(s), s \in [0, a]\Big), \quad t \geq a, \tag{6.3.8}$$

is an element of $\mathcal{H}_a^\kappa$ and hence is expected to be written as a fractional Wiener integral. The corresponding prediction formula for (6.3.8) is derived in Section 6.4, which also includes other applications.

6.3.2 Classes of Integrands on an Interval

In this section, we construct classes of integrands for FBM on an interval $[0, a]$ with $a > 0$. Suppose that $\{B^\kappa(u)\}_{u \in [0,a]}$ is a standard FBM on an interval $[0, a]$. Then, by the representation (6.2.19), we obtain that, for every elementary function $f \in \mathcal{E}$ and $\kappa \in (-1/2, 1/2)$,

$$\mathcal{I}_a^\kappa(f) \stackrel{d}{=} \sigma_\kappa \int_0^a s^{-\kappa}\left(I_{a-}^\kappa u^\kappa f(u)\right)(s) dB^0(s), \quad (6.3.9)$$

where σ_κ is given in (6.2.20). It follows that, for all $f, g \in \mathcal{E}$,

$$\mathbb{E}(\mathcal{I}_a^\kappa(f)\mathcal{I}_a^\kappa(g)) = \sigma_\kappa^2 \int_0^a s^{-2\kappa}\left(I_{a-}^\kappa u^\kappa f(u)\right)(s)\left(I_{a-}^\kappa u^\kappa g(u)\right)(s) ds. \quad (6.3.10)$$

- **The case** $\kappa \in (0, 1/2)$

Consider first the case $\kappa \in (0, 1/2)$. Relations (6.3.9) and (6.3.10) lead us to introduce the following class of functions on an interval $[0, a]$

Definition 6.3.4 For $\kappa \in (0, 1/2)$, let

$$\Lambda_a^\kappa = \left\{ f : u^\kappa f(u) \in L^1[0, a] \text{ and } \int_0^a s^{-2\kappa}\left((I_{a-}^\kappa u^\kappa f(u))(s)\right)^2 ds < \infty \right\}. \quad (6.3.11)$$

We assume in (6.3.11) that the function $u^\kappa f(u)$ is in $L^1[0, a]$ so that the fractional integral $(I_{a-}^\kappa u^\kappa f(u))(s)$ is well-defined a.e. for $s \in [0, a]$ (see the discussion following Definition 6.1.1). The following result describes the space Λ_a^κ, and allows one to use Proposition 6.3.2 in constructing fractional Wiener integrals.

Theorem 6.3.5 *For $\kappa \in (0, 1/2)$, the class of functions Λ_a^κ, defined by (6.3.11), is a linear space with the inner product*

$$(f, g)_{\Lambda_a^\kappa} = \sigma_\kappa^2 \int_0^a s^{-2\kappa}\left(I_{a-}^\kappa u^\kappa f(u)\right)(s)\left(I_{a-}^\kappa u^\kappa g(u)\right)(s) ds. \quad (6.3.12)$$

The set of elementary functions $\mathcal{E}$ is dense in the space Λ_a^κ. The space Λ_a^κ is not complete.

Proof Note that $(f, f)_{\Lambda_a^\kappa} = 0$ implies $f = 0$ a.e. by Corollary 6.1.11. The other properties of an inner product, that is, linearity and symmetry, are also satisfied. Since Λ_a^κ is a linear space, it is then an inner product space. The fact that $\mathcal{E}$ is dense, is proved in Theorem 4.1 of Pipiras and Taqqu [815]. Since the latter fact is expected and the proof is technical, we will not provide it here. We will instead outline the steps to prove the last statement of the theorem concerning incompleteness.

First, let $0 \leq b \leq a$. As expected, the function

$$g_b(u) = u^{-\kappa}(I_{a-}^{-\kappa} s^\kappa 1_{[0,b)}(s))(u) = u^{-\kappa}(\mathcal{D}_{a-}^\kappa s^\kappa 1_{[0,b)}(s))(u) \quad (6.3.13)$$

6.3 Fractional Wiener Integrals and their Deterministic Integrands

satisfies the inverse relation

$$s^{-\kappa}(I_{a-}^{\kappa}u^{\kappa}g_{b}(u))(s) = 1_{[0,b)}(s) \qquad (6.3.14)$$

(Exercise 6.6, (i)). Then, for $0 \leq c < b \leq a$, the function $g_{c,b}(u) = g_b(u) - g_c(u)$ satisfies

$$s^{-\kappa}(I_{a-}^{\kappa}u^{\kappa}g_{c,b}(u))(s) = 1_{[c,b)}(s). \qquad (6.3.15)$$

Second, by using (6.3.15) and since linear combinations of indicator functions are dense in $L^2[0, a]$, one can show that Λ_a^{κ} is complete if and only if, for every $\phi \in L^2[0, a]$, there is a function $f_{\phi} \in \Lambda_a^{\kappa}$ such that

$$s^{-\kappa}(I_{a-}^{\kappa}u^{\kappa}f_{\phi}(u))(s) = \phi(s) \quad \text{a.e. } ds \qquad (6.3.16)$$

(Exercise 6.6, (ii)). The basic idea now is that one cannot expect (6.3.16) to be true. Heuristically, since I_{a-}^{κ} is an operator which involves an integral, the left-hand side of (6.3.16) must satisfy some smoothness condition (for example, one can take its fractional derivative $(\mathcal{D}_{a-}^{\kappa}s^{\kappa})$ to obtain, by (6.1.13), $(\mathcal{D}_{a-}^{\kappa}s^{\kappa}s^{-\kappa}(I_{a-}^{\kappa}u^{\kappa}f_{\phi}(u))(s))(z) = (\mathcal{D}_{a-}^{\kappa}(I_{a-}^{\kappa}u^{\kappa}f_{\phi}(u)))(z) = z^{\kappa}f_{\phi}(z))$, whereas such condition need not to hold for a general $\phi \in L^2[0, a]$.

Third, one can actually provide an interesting example of a function $\phi \in L^2[0, a]$ for which (6.3.16) is not satisfied. In fact, we will argue first that there are continuous functions ψ on $[0, a]$ such that the equation

$$(I_{a-}^{\kappa}g(u))(s) = \psi(s) \quad \text{a.e. } ds \qquad (6.3.17)$$

has no solution in $g \in L^1[0, a]$. Proceeding by contradiction, if (6.3.17) had a solution $g_{\psi} \in L^1[0, a]$, then by Proposition 6.1.8, (i), $g_{\psi}(u) = (\mathcal{D}_{a-}^{\kappa}\psi)(u)$. In particular, by the definition (6.1.12),

$$g_{\psi}(u) = (\mathcal{D}_{a-}^{\kappa}\psi)(u) = -\frac{1}{\Gamma(1-\kappa)}\frac{d}{du}\int_u^a \psi(s)(s-u)^{-\kappa}ds,$$

that is, the function

$$U_{\psi}(u) = \int_0^a \psi(s)(s-u)_+^{-\kappa}ds = \int_{\mathbb{R}}\psi(s+u)1_{\{0<s+u<a\}}s_+^{-\kappa}ds \qquad (6.3.18)$$

would be differentiable a.e. du on $[0, a]$. We will show, for example, that when $a = 1$ (otherwise, use appropriate scaling), this is not the case for $U_{\psi^*}(u)$ where $\psi^*(s)$ is the Weierstrass function (see Section 3.13),

$$\psi^*(s) = \sum_{n=0}^{\infty} b^{-pn}e^{ib^n s}, \quad b > 1, \ 0 < p < \kappa < 1/2. \qquad (6.3.19)$$

The function ψ^* is complex-valued but one can focus on its real or imaginary part. As indicated in Section 3.13, these are continuous.

We want to show that $U_{\psi^*}(u)$ is not differentiable on $(0, 1)$. For $u \in (0, 1)$, we have by (6.3.18) with $a = 1$,

$$U_{\psi^*}(u) = \int_0^{1-u} s^{-\kappa}\Big(\sum_{n=0}^{\infty} b^{-pn} e^{ib^n s} e^{ib^n u}\Big) ds = \sum_{n=0}^{\infty} b^{-pn}\Big(\int_0^{1-u} s^{-\kappa} e^{ib^n s} ds\Big) e^{ib^n u}$$

$$= \sum_{n=0}^{\infty} b^{-pn}\Big(\int_0^{\infty} s^{-\kappa} e^{ib^n s} ds\Big) e^{ib^n u} - \sum_{n=0}^{\infty} b^{-pn}\Big(\int_{1-u}^{\infty} s^{-\kappa} e^{ib^n s} ds\Big) e^{ib^n u} =: y_1(u) - y_2(u).$$

One can show that the function y_2 is differentiable on $[0, 1/2]$ (Pipiras and Taqqu [815], proof of Lemma 5.7). The function y_1, on the other hand, is

$$y_1(u) = \Big(\int_0^{\infty} z^{-\kappa} e^{iz} dz\Big) \sum_{n=0}^{\infty} b^{-(p-\kappa+1)n} e^{ib^n u},$$

after the change of variables $b^n s = z$. Up to a multiplicative constant, since $b^{-(p-\kappa+1)} b = b^{\kappa-p} > 1$, y_1 is the Weierstrass function (3.13.1). It is nowhere differentiable on $[0, 1]$. We have thus constructed a continuous function ψ for which $U_\psi(u)$ is not differentiable, and hence (6.3.17) has no solution g.

Finally, and as a fourth point, if (6.3.17) does not have a solution for a function ψ, then after writing (6.3.17) as $s^{-\kappa}(I_{a-}^\kappa g(u))(s) = s^{-\kappa}\psi(s)$, (6.3.16) does not have a solution for $\phi(s) = s^{-\kappa}\psi(s)$. Moreover, since we can take a continuous function ψ as in (6.3.19), the function $\phi(s) = s^{-\kappa}\psi(s)$ is in $L^2[0, a]$ since $\kappa < 1/2$ and hence (6.3.16) does not have a solution for ϕ in $L^2[0, a]$. □

Theorem 6.3.5 and Proposition 6.3.2 imply that the space Λ_a^κ of functions f is isometric to a *strict* subspace of $\mathcal{H}_a^\kappa$. Let $\mathcal{I}_a^\kappa$ denote the isometry map. We can then define the integral with respect to FBM as follows.

Definition 6.3.6 For $\kappa \in (0, 1/2)$ and $f \in \Lambda_a^\kappa$,

$$\mathcal{I}_a^\kappa(f) \equiv \int_0^a f(u) dB^\kappa(u) \stackrel{d}{=} \sigma_\kappa \int_0^a s^{-\kappa}\Big(I_{a-}^\kappa u^\kappa f(u)\Big)(s) dB^0(s), \qquad (6.3.20)$$

where σ_κ is given in (6.2.20).

The space Λ_a^κ can be viewed as a class of deterministic integrands for FBM on an interval $[0, a]$. Note also that because of (6.3.11), we have:

Proposition 6.3.7 *The covariance relation (6.3.10) is valid for $f, g \in \Lambda_a^\kappa$.*

- **The case $\kappa \in (-1/2, 0)$**

Consider now the case $\kappa \in (-1/2, 0)$ and set $f(u) = u^{-\kappa}(I_{a-}^{-\kappa} s^\kappa \phi(s))(u)$ for some function $\phi \in L^2[0, a]$. The function f is well-defined for $\phi \in L^2[0, a]$ because the function $s^\kappa \phi(s)$ is in $L^1[0, a]$ for $\kappa > -1/2$. Indeed, $\int_0^a |s^\kappa \phi(s)| ds \leq (\int_0^a s^{2\kappa} ds)^{1/2} (\int_0^a |\phi(s)|^2 ds)^{1/2} < \infty$ for $\kappa > -1/2$ by the Cauchy-Schwarz inequality. Property (i) of Proposition 6.1.8 then implies

$$s^{-\kappa}(I_{a-}^\kappa u^\kappa f(u))(s) = s^{-\kappa}(I_{a-}^\kappa u^\kappa u^{-\kappa}(I_{a-}^{-\kappa} z^\kappa \phi(z))(u))(s) = \phi(s) \qquad (6.3.21)$$

and hence that

$$\int_0^a s^{-2\kappa}\left((I_{a-}^\kappa u^\kappa f(u))(s)\right)^2 ds = \int_0^a \phi(s)^2 ds < \infty.$$

Based on this observation we introduce the class of functions:

Definition 6.3.8 For $\kappa \in (-1/2, 0)$, let

$$\Lambda_a^\kappa = \left\{ f : f(u) = u^{-\kappa}(I_{a-}^{-\kappa} s^\kappa \phi_f(s))(u) \text{ for } \phi_f \in L^2[0, a] \right\}. \quad (6.3.22)$$

Let us compare Definition 6.3.4 for $\kappa > 0$ to Definition 6.3.8 for $\kappa < 0$. In Definition 6.3.4, we require $u^\kappa f(u) \in L^1[0, a]$, but here we require that f results from a fractional integration involving a function $\phi_f \in L^2[0, a]$.

The next result is the analogue of Theorem 6.3.5 in the case $\kappa \in (-1/2, 0)$. We will not provide a proof. The fact that $\mathcal{E}$ is dense, is proved in Pipiras and Taqqu [817], Theorem 4.2. The completeness follows easily from the fact that the space $L^2[0, a]$ is complete (Pipiras and Taqqu [817], the discussion following Theorem 4.2).

Theorem 6.3.9 *For $\kappa \in (-1/2, 0)$, the class of functions Λ_a^κ, defined by (6.3.22), is a linear space with the inner product*

$$(f, g)_{\Lambda_a^\kappa} = \sigma_\kappa^2 \int_0^a s^{-2\kappa}(I_{a-}^\kappa u^\kappa f(u))(s)(I_{a-}^\kappa u^\kappa g(u))(s) ds \quad (6.3.23)$$

$$= \sigma_\kappa^2 \int_0^a \phi_f(s)\phi_g(s) ds, \quad (6.3.24)$$

where $\phi_f, \phi_g \in L^2[0, a]$ are associated with the functions f and g, respectively, by Definition 6.3.8. The set of elementary functions $\mathcal{E}$ is dense in the space Λ_a^κ. The space Λ_a^κ is complete.

Theorem 6.3.9 and Proposition 6.3.2 imply that the space Λ_a^κ is isometric to the space $\mathcal{H}_a^\kappa$ itself. If $\mathcal{I}_a^\kappa$ is the isometry map, we have (6.3.20) for $\kappa \in (-1/2, 0)$ as well. The space Λ_a^κ can be viewed as the class of deterministic integrands for FBM on an interval $[0, a]$. Definition 6.3.6 can thus be extended to $\kappa \in (-1/2, 0)$.

Definition 6.3.10 For $\kappa \in (-1/2, 0)$ and $f \in \Lambda_a^\kappa$,

$$\mathcal{I}_a^\kappa(f) \equiv \int_0^a f(u) dB^\kappa(u) \stackrel{d}{=} \sigma_\kappa \int_0^a s^{-\kappa}\left(I_{a-}^\kappa u^\kappa f(u)\right)(s) dB^0(s) = \sigma_\kappa \int_0^a \phi_f(s) dB^0(s). \quad (6.3.25)$$

Instead of first finding a function $\phi_f \in L^2[0, a]$ so that $f(u) = u^{-\kappa}(I_{a-}^{-\kappa} s^\kappa \phi_f(s))(u)$ and hence $s^{-\kappa}(I_{a-}^\kappa u^\kappa f(u))(s) = \phi_f(s)$, the next auxiliary result allows working with the (weighted) fractional derivative $s^{-\kappa}(I_{a-}^\kappa u^\kappa f(u))(s)$ directly. The result will sometimes be used to check that a function f belongs to the space Λ_a^κ and hence can be integrated with respect to FBM.

Lemma 6.3.11 *Let $\kappa \in (-1/2, 0)$ and f be a function on $[0, a]$. Suppose that the (weighted) fractional derivative $s^{-\kappa}(I_{a-}^{\kappa} u^{\kappa} f(u))(s) =: \phi(s)$ can be evaluated directly. Then, if $\phi \in L^2[0, a]$, we have*

$$u^{-\kappa}(I_{a-}^{-\kappa} s^{\kappa} \phi(s))(u) = f(u) \tag{6.3.26}$$

and hence $f \in \Lambda_a^{\kappa}$ with $\phi_f = \phi$.

Proof If $\phi \in L^2[0, a]$, we have $u^{-\kappa}(I_{a-}^{-\kappa} s^{\kappa} \phi(s))(u) = u^{-\kappa}(I_{a-}^{-\kappa} I_{a-}^{\kappa} v^{\kappa} f(v))(u) = f(u)$ by Proposition 6.1.8, (iii), as long as $(I_{a-}^{1+\kappa} v^{\kappa} f(v))$ is absolutely continuous. But since $(I_{a-}^{\kappa} u^{\kappa} f(u))(s)$ cane be evaluated by definition, this means that the derivative of $\int_s^a u^{\kappa} f(u)(u-s)^{\kappa} du$ exists; that is, $(I_{a-}^{1+\kappa} v^{\kappa} f(v))(s)$ is differentiable (with the derivative $(I_{a-}^{\kappa} u^{\kappa} f(u))(s) = s^{\kappa} \phi(s)$, which is in $L^1[0, a]$) and hence is absolutely continuous. Thus, in this case, $f \in \Lambda_a^{\kappa}$ since ϕ_f can be taken as $\phi_f = \phi$. □

The following result provides a useful and perhaps unexpected characterization of the space Λ_a^{κ} when $\kappa \in (-1/2, 0)$.

Proposition 6.3.12 *For $\kappa \in (-1/2, 0)$,*

$$\Lambda_a^{\kappa} = \left\{ f : f(u) = (I_{a-}^{-\kappa} \psi_f)(u) \text{ for } \psi_f \in L^2[0, a] \right\}. \tag{6.3.27}$$

Proof Consider the relation

$$u^{-\kappa}(I_{a-}^{-\kappa} s^{\kappa} \phi(s))(u) = (I_{a-}^{-\kappa} \psi)(u). \tag{6.3.28}$$

We need to show that, for $\phi \in L^2[0, a]$, there is $\psi \in L^2[0, a]$ satisfying (6.3.28), and vice versa. Thus, suppose $\phi \in L^2[0, a]$. Write the left-hand side of (6.3.28) as

$$f(u) := \Gamma(-\kappa) u^{-\kappa}(I_{a-}^{-\kappa} s^{\kappa} \phi(s))(u) = u^{-\kappa} \int_0^a s^{\kappa} \phi(s)(s-u)_+^{-\kappa-1} ds$$

$$= u^{-\kappa} \int_{a^{-1}}^{\infty} (z^{-1})^{\kappa} \phi(z^{-1})(z^{-1} - u)_+^{-\kappa-1} z^{-2} dz = u^{-2\kappa-1} \int_{a^{-1}}^{\infty} z^{-1} \phi(z^{-1})(u^{-1} - z)_+^{-\kappa-1} dz$$

$$= u^{-2\kappa-1} \int_{a^{-1}}^{\infty} z^{-1} \phi(z^{-1}) 1_{[a^{-1}, \infty)}(z)(u^{-1} - z)_+^{-\kappa-1} dz$$

$$= \Gamma(-\kappa) u^{-2\kappa-1}(I_{0+}^{-\kappa} z^{-1} \phi(z^{-1}) 1_{[a^{-1}, \infty)}(z))(u^{-1}). \tag{6.3.29}$$

Setting $f(u) = 0$ for $u > a$, the relation (6.3.29) holds for all $u > 0$ (not just $u < a$). Note that, since $\phi \in L^2[0, a]$, we have $z^{-1} \phi(z^{-1}) \in L^2[a^{-1}, \infty)$ and hence $z^{-1} \phi(z^{-1}) 1_{[a^{-1}, \infty)}(z) \in L^2(0, \infty)$. By Lemma 3.2 in Samko et al. [878], p. 70, since $\kappa > -1 + 1/2$, there is $g \in L^2(0, \infty)$ such that

$$(u^{-1})^{\kappa}(I_{0+}^{-\kappa} z^{-1} \phi(z^{-1}) 1_{[a^{-1}, \infty)}(z))(u^{-1}) = (I_{0+}^{-\kappa} z^{\kappa} g(z))(u^{-1}). \tag{6.3.30}$$

Since the left-hand side of (6.3.30) equals 0 for $u > a$ (or $u^{-1} < a^{-1}$), the right-hand side equals 0 for $u > a$ (or $u^{-1} < a^{-1}$) as well, and we can write (6.3.30) as

$$(u^{-1})^{\kappa}(I_{0+}^{-\kappa} z^{-1} \phi(z^{-1}) 1_{[a^{-1}, \infty)}(z))(u^{-1}) = (I_{0+}^{-\kappa} z^{\kappa} g(z) 1_{[a^{-1}, \infty)}(z))(u^{-1}). \tag{6.3.31}$$

Substituting (6.3.31) into (6.3.29) and working backwards in (6.3.29), we get that

$$f(u) = \Gamma(-\kappa)u^{\kappa-1}(I_{0+}^{-\kappa}z^{\kappa}g(z)1_{[a^{-1},\infty)}(z))(u^{-1})$$

$$= u^{-\kappa-1}\int_{a^{-1}}^{\infty} z^{\kappa}g(z)(u^{-1}-z)_{+}^{-\kappa-1}dz = u^{-\kappa-1}\int_{0}^{a}(s^{-1})^{\kappa}g(s^{-1})(u^{-1}-s^{-1})_{+}^{-\kappa-1}s^{-2}ds$$

$$= \int_{0}^{a} s^{-1}g(s^{-1})(s-u)_{+}^{-\kappa-1}ds = \Gamma(-\kappa)(I_{a-}^{-\kappa}s^{-1}g(s^{-1}))(u) =: \Gamma(-\kappa)(I_{a-}^{-\kappa}\psi)(u),$$

where $\psi(s) = s^{-1}g(s^{-1})$. Since $g \in L^2[a^{-1}, \infty)$, we deduce that $\psi \in L^2[0, a]$, which proves (6.3.28) in one direction. The other direction can be proved analogously by working backwards in the argument above. $\square$

Remark 6.3.13 In view of (6.3.27), the space Λ_a^{κ} is often written as $\Lambda_a^{\kappa} = I_{a-}^{-\kappa}(L^2[0, a])$. The elements of the latter space can also be characterized as follows (Samko et al. [878], Theorem 13.4, p. 232): $f \in I_{a-}^{-\kappa}(L^2[0, a])$ if and only if

$$f(u)(u-a)^{\kappa} \in L^2[0, a] \quad \text{and} \quad \sup_{\epsilon > 0}\int_{\epsilon}^{a}\left|\int_{a}^{u-\epsilon}\frac{f(u)-f(s)}{(u-s)^{1-\kappa}}ds\right|^2 du < \infty. \quad (6.3.32)$$

6.3.3 Subspaces of Classes of Integrands

- **The case $\kappa \in (0, 1/2)$**

The class of integrands Λ_a^{κ}, $\kappa \in (0, 1/2)$, has a useful subspace. Suppose that the function f is such that $|f| \in \Lambda_a^{\kappa}$ so that $f \in \Lambda_a^{\kappa}$ as well. By using the relation (6.2.23) for $|u-v|^{2\kappa-1}$ and Fubini's theorem, we get that

$$\frac{B(\kappa, 1-2\kappa)}{\Gamma(\kappa)^2}\int_{0}^{a}\int_{0}^{a}f(u)f(v)|u-v|^{2\kappa-1}dudv = \int_{0}^{a}s^{-2\kappa}\left((I_{a-}^{\kappa}u^{\kappa}f(u))(s)\right)^2 ds. \quad (6.3.33)$$

Now consider the following space of functions:

Definition 6.3.14 For $\kappa \in (0, 1/2)$, set

$$|\Lambda|_a^{\kappa} = \left\{f : \int_{0}^{a}\int_{0}^{a}|f(u)||f(v)||u-v|^{2\kappa-1}dudv < \infty\right\}. \quad (6.3.34)$$

By the relation (6.3.33), $|\Lambda|_a^{\kappa}$ is a linear subspace of Λ_a^{κ}. It also becomes an inner product space with the inner product $(f, g)_{|\Lambda|_a^{\kappa}}$ defined by $(f, g)_{\Lambda_a^{\kappa}}$ for $f, g \in |\Lambda|_a^{\kappa}$ and hence can also be viewed as a class of integrands for FBM. By using the relation (6.2.23) for $|u-v|^{2\kappa-1}$ and Fubini's theorem, we have that

$$(f, g)_{|\Lambda|_a^{\kappa}} = \kappa(2\kappa+1)\int_{0}^{a}\int_{0}^{a}f(u)g(v)|u-v|^{2\kappa-1}dudv. \quad (6.3.35)$$

Since for $\kappa \in (0, 1/2)$, Λ_a^{κ} is not complete, the space $|\Lambda|_a^{\kappa}$ is not complete either (Pipiras and Taqqu [817], Theorem 5.1). We restate (6.3.35) as the following useful result.

Proposition 6.3.15 *If $\kappa \in (0, 1/2)$ and $f, g \in |\Lambda|_a^\kappa$, then the covariance structure of the integrals $\mathcal{I}_a^\kappa(f)$ and $\mathcal{I}_a^\kappa(g)$ can also be expressed as*

$$\mathbb{E}\mathcal{I}_a^\kappa(f)\mathcal{I}_a^\kappa(g) = \kappa(2\kappa + 1)\int_0^a \int_0^a f(u)g(v)|u - v|^{2\kappa-1}dudv.$$

The function space $|\Lambda|_a^\kappa$ is sometimes more convenient to work with than the space Λ_a^κ. For example, one may define on $|\Lambda|_a^\kappa$ a norm $\|\cdot\|_a$ by

$$\|f\|_a^2 = \int_0^a \int_0^a |f(u)||f(v)||u - v|^{2\kappa-1}dudv, \tag{6.3.36}$$

which is easier to manipulate than the norm induced by the inner product on Λ_a^κ; that is,

$$\|f\|_{|\Lambda|_a^\kappa}^2 = \int_0^a \int_0^a f(u)f(v)|u - v|^{2\kappa-1}dudv. \tag{6.3.37}$$

Moreover, as the next proposition states, the space $|\Lambda|_a^\kappa$ is complete under the new norm $\|\cdot\|_a$.

Proposition 6.3.16 *The space $(|\Lambda|_a^\kappa, \|\cdot\|_a)$ when $\kappa \in (0, 1/2)$ is a complete normed space.*

Proof The fact that $\|\cdot\|_a$ is a norm is elementary. To show completeness, let $\{f_n\}$ be a Cauchy sequence in $(|\Lambda|_a^\kappa, \|\cdot\|_a)$, that is, $\|f_n - f_m\|_a \to 0$, as $n, m \to \infty$. We have to find $f \in |\Lambda|_a^\kappa$ such that $\|f - f_n\|_a \to 0$, as $n \to \infty$. For this, observe that, since $|u - v|^{2\kappa-1} \geq a^{2\kappa-1}$ for $u, v \in (0, a)$, $\|g\|_a^2 \geq a^{2\kappa-1}\int_0^a \int_0^a |g(u)||g(v)|dudv = a^{2\kappa-1}(\int_0^a |g(u)|du)^2$. Thus, $\{f_n\}$ is also a Cauchy sequence in $L^1[0, a]$. Since the space $L^1[0, a]$ is complete, there is a function $f \in L^1[0, a]$ such that $f_n \to f$ in $L^1[0, a]$. There is also a subsequence n_l such that $f_{n_l} \to f$ a.e. To show that f is the desired limit in $\|\cdot\|_a$, note that, by Fatou's lemma, that $\|f\|_a = \|\liminf_l f_{n_l}\|_a \leq \liminf_l \|f_{n_l}\|_a < \infty$. Then, by Fatou's lemma again, $\|f - f_n\|_a = \|\liminf_l f_{n_l} - f_n\|_a \leq \liminf_l \|f_{n_l} - f_n\|_a \to 0$, as $n \to \infty$, since $\{f_n\}$ is Cauchy in the norm $\|\cdot\|_a$. □

The following basic result provides other useful subsets of the spaces introduced.

Proposition 6.3.17 *For $\kappa \in (0, 1/2)$,*

$$L^2[0, a] \subset L^{2/(2\kappa+1)}[0, a] \subset |\Lambda|_a^\kappa \subset \Lambda_a^\kappa. \tag{6.3.38}$$

Proof The inclusion $L^2[0, a] \subset L^{2/(2\kappa+1)}[0, a]$ is obvious, since $2/(2\kappa + 1) < 2$ for $\kappa > 0$. To show the inclusion $L^{2/(2\kappa+1)}[0, a] \subset |\Lambda|_a^\kappa$, observe that, by Hölder's inequality,

$$\|f\|_a^2 = \int_0^a \int_0^a |f(u)||f(v)||u - v|^{2\kappa-1}dudv = \int_0^a |f(u)|\Big(\int_0^a |f(v)||u - v|^{2\kappa-1}dv\Big)du$$

$$\leq \Big(\int_0^a |f(u)|^{\frac{2}{2\kappa+1}}du\Big)^{\frac{2\kappa+1}{2}}\Big(\int_0^a \Big(\int_0^a |f(v)||u - v|^{2\kappa-1}dv\Big)^{\frac{2}{1-2\kappa}}du\Big)^{\frac{1-2\kappa}{2}}$$

$$= \Gamma(2k)\Big(\int_0^a |f(u)|^{\frac{2}{2\kappa+1}}du\Big)^{\frac{2\kappa+1}{2}}\Big(\int_0^a \Big((I_{0+}^{2\kappa}|f|)(u) + (I_{a-}^{2\kappa}|f|)(u)\Big)^{\frac{2}{1-2\kappa}}du\Big)^{\frac{1-2\kappa}{2}}.$$

6.3 Fractional Wiener Integrals and their Deterministic Integrands

By using Theorem 3.5 in Samko et al. [878], p. 66, $I_{0+}^{2\kappa}$ and $I_{a-}^{2\kappa}$ are bounded operators from $L^{2/(2\kappa+1)}[0, a]$ to $L^{2/(1-2\kappa)}[0, a]$. This establishes the inclusion. $\square$

Remark 6.3.18 The end of the proof above also yields the following useful inequalities: for $\kappa \in (0, 1/2)$, there are constants C, C' such that

$$\|f\|_a \leq C \|f\|_{L^{2/(2\kappa+1)}[0,a]} \leq C' \|f\|_{L^2[0,a]}. \qquad (6.3.39)$$

- **The case $\kappa \in (-1/2, 0)$**

When $\kappa \in (-1/2, 0)$, a useful subspace of Λ_a^κ is specified in the following result. Let

$$C^\gamma[0, a] = \{f : [0, a] \to \mathbb{R} : |f(u) - f(v)| \leq C|u - v|^\gamma\} \qquad (6.3.40)$$

be the space of γ–Hölder continuous functions on $[0, a]$.

Proposition 6.3.19 Let $\kappa \in (-1/2, 0)$ and $0 < (-\kappa) < \gamma$. Then, for $f \in C^\gamma[0, a]$,

$$s^{-\kappa}(I_{a-}^\kappa u^\kappa f(u))(s) = \frac{s^{-\kappa}}{\Gamma(\kappa)} \int_0^a (f(u) - f(s)) u^\kappa (u-s)_+^{\kappa-1} du + s^{-\kappa} f(s)(I_{a-}^\kappa u^\kappa)(s) \qquad (6.3.41)$$

$$= \frac{1}{\Gamma(\kappa)} \int_0^a (f(u) - f(s))(u-s)_+^{\kappa-1} du + \frac{s^{-\kappa}}{\Gamma(\kappa)} \int_0^a f(u)(u^\kappa - s^\kappa)(u-s)_+^{\kappa-1} du$$
$$+ \frac{f(s)(a-s)^\kappa}{\Gamma(\kappa)} \qquad (6.3.42)$$

and

$$C^\gamma[0, a] \subset \Lambda_a^\kappa. \qquad (6.3.43)$$

Remark 6.3.20 Informally, the relation (6.3.41) follows from

$$-\Gamma(\kappa+1)(I_{a-}^\kappa u^\kappa f(u))(s) = \frac{d}{ds} \int_0^a f(u) u^\kappa (u-s)_+^\kappa du = -\kappa \int_0^a f(u) u^\kappa (u-s)_+^{\kappa-1} du$$

$$= -\kappa \int_0^a (f(u) - f(s)) u^\kappa (u-s)_+^{\kappa-1} du - \kappa f(s) \int_0^a u^\kappa (u-s)_+^{\kappa-1} du$$

$$= -\kappa \int_0^a (f(u) - f(s)) u^\kappa (u-s)_+^{\kappa-1} du + f(s) \frac{d}{ds} \int_0^a u^\kappa (u-s)_+^\kappa du.$$

Proof We first show the equality of the right-hand sides of (6.3.41) and (6.3.42). Observe that the right-hand side of (6.3.41) is equal to

$$\frac{s^{-\kappa}}{\Gamma(\kappa)} \int_0^a (f(u) - f(s)) s^\kappa (u-s)_+^{\kappa-1} du + \frac{s^{-\kappa}}{\Gamma(\kappa)} \int_0^a (f(u) - f(s))(u^\kappa - s^\kappa)(u-s)_+^{\kappa-1} du$$

$$+ s^{-\kappa} f(s)(I_{a-}^\kappa u^\kappa)(s) = \frac{1}{\Gamma(\kappa)} \int_0^a (f(u) - f(s))(u-s)_+^{\kappa-1} du$$

$$+ \frac{s^{-\kappa}}{\Gamma(\kappa)} \int_0^a f(u)(u^\kappa - s^\kappa)(u-s)_+^{\kappa-1} du$$

$$- \frac{f(s) s^{-\kappa}}{\Gamma(\kappa)} \int_0^a (u^\kappa - s^\kappa)(u-s)_+^{\kappa-1} du + s^{-\kappa} f(s)(I_{a-}^\kappa u^\kappa)(s)$$

380 *Fractional Calculus and Integration of Deterministic Functions with Respect to FBM*

$$= \frac{1}{\Gamma(\kappa)} \int_0^a (f(u) - f(s))(u-s)_+^{\kappa-1} du + \frac{s^{-\kappa}}{\Gamma(\kappa)} \int_0^a f(u)(u^\kappa - s^\kappa)(u-s)_+^{\kappa-1} du$$
$$+ \frac{f(s)(a-s)^\kappa}{\Gamma(\kappa)}$$

by using (6.2.15) with $t = a$ in the last step, which leads to the right-hand side of (6.3.42).

We now prove (6.3.42). Let $g(s) = \int_0^a f(u) u^\kappa (u-s)_+^\kappa du$. We need to show that g is differentiable and that

$$\frac{d}{ds} g(s) = \frac{1}{\Gamma(\kappa)} \int_0^a (f(u) - f(s))(u-s)_+^{\kappa-1} du + \frac{s^{-\kappa}}{\Gamma(\kappa)} \int_0^a f(u)(u^\kappa - s^\kappa)(u-s)_+^{\kappa-1} du$$
$$+ \frac{f(s)(a-s)^\kappa}{\Gamma(\kappa)} =: I_1 + I_2 + I_3. \qquad (6.3.44)$$

For example, for $h > 0$, observe that

$$g(s+h) - g(s) = \int_{s+h}^a f(u) u^\kappa (u-s-h)^\kappa du - \int_s^a f(u) u^\kappa (u-s)^\kappa du$$

$$= \int_{s+h}^a f(u) u^\kappa \left((u-s-h)^\kappa - (u-s)^\kappa \right) du - \int_s^{s+h} f(u) u^\kappa (u-s)^\kappa du$$

$$= -\kappa \int_{s+h}^a f(u) u^\kappa \left(\int_s^{s+h} (u-z)^{\kappa-1} dz \right) du - \int_s^{s+h} f(u) u^\kappa (u-s)^\kappa du$$

$$= -\kappa \int_s^{s+h} \left(\int_{s+h}^a f(u) u^\kappa (u-z)^{\kappa-1} du \right) dz - \int_s^{s+h} f(u) u^\kappa (u-s)^\kappa du$$

$$= -\kappa \int_s^{s+h} \left(\int_{s+h}^a (f(u) - f(z))(u-z)^{\kappa-1} du \right) z^\kappa dz$$
$$- \kappa \int_s^{s+h} \left(\int_{s+h}^a f(u)(u^\kappa - z^\kappa)(u-z)^{\kappa-1} du \right) dz$$
$$- \kappa \int_s^{s+h} f(z) z^\kappa \left(\int_{s+h}^a (u-z)^{\kappa-1} du \right) dz - \int_s^{s+h} f(u) u^\kappa (u-s)^\kappa du$$

$$= -\kappa \int_s^{s+h} \left(\int_{s+h}^a (f(u) - f(z))(u-z)^{\kappa-1} du \right) z^\kappa dz$$
$$- \kappa \int_s^{s+h} \left(\int_{s+h}^a f(u)(u^\kappa - z^\kappa)(u-z)^{\kappa-1} du \right) dz$$
$$- \int_s^{s+h} f(z) z^\kappa (a-z)^\kappa dz + \int_s^{s+h} f(z) z^\kappa (s+h-z)^\kappa dz$$
$$- \int_s^{s+h} f(u) u^\kappa (u-s)^\kappa du =: I_{1,h} + I_{2,h} + I_{3,h} + I_{4,h} + I_{5,h}.$$

$$(6.3.45)$$

To prove (6.3.44), it is enough to show that $h^{-1} I_{k,h} \to I_k$ for $k = 1, 2, 3$ as $h \to 0$, and $h^{-1}(I_{4,h} + I_{5,h}) \to 0$ as $h \to 0$. This is left as Exercise 6.7.

Consider now (6.3.43) and take $f \in C^\gamma[0, a]$. We need to show that $f \in \Lambda_a^\kappa$, that is, $s^{-\kappa}(I_{a-}^\kappa u^\kappa f(u))(s)$ or the right-hand side of (6.3.41) belongs to $L^2[0, a]$ (see Lemma

6.3.11). By using the bounds $|f(u) - f(v)| \leq C|u-v|^\gamma$ in the first term of (6.3.41) and $|f(s)| \leq C$ in the second term of (6.3.41), it is enough to prove that $g_1(s), g_2(s) \in L^2[0, a]$, where

$$g_1(s) = s^{-\kappa} \int_0^a u^\kappa (u-s)_+^{\gamma+\kappa-1} du, \quad g_2(s) = s^{-\kappa} (I_{a-}^\kappa u^\kappa)(s).$$

The function $g_2(s)$ equals the kernel function $f_t(s)$ of FBM at $t = a$ (see (6.2.3)) and hence belongs to $L^2[0, a]$. To see that $g_1 \in L^2[0, a]$, write g_1 as, after the change of variables $u = sv$,

$$g_1(s) = s^{\kappa+\gamma} \int_1^{a/s} v^\kappa (v-1)^{\gamma+\kappa-1} dv. \tag{6.3.46}$$

The function $g_1(s)$ is continuous for $s \in (0, a]$. Since a γ–Hölder function is also γ'–Hölder for $\gamma' < \gamma$, we may assuming without loss of generality that $\gamma + 2\kappa < 0$ or $\gamma < (-2\kappa)$. In addition, by assumption $\gamma + \kappa > 0$. Then, the integral in (6.3.46) converges and therefore $g_1(s) \sim Cs^{\kappa+\gamma}$ as $s \to 0$. Hence g_1 belongs to $L^2[0, a]$. □

Remark 6.3.21 Let $\kappa \in (-1/2, 0)$. By Theorem 3.5 in Samko et al. [878], p. 55, since $1 < 2 < 1/(-\kappa)$,

$$\Lambda_a^\kappa = I_{a-}^{-\kappa}(L^2[0, a]) \subset L^q[0, a] \subset L^2[0, a], \tag{6.3.47}$$

where $q = 2/(1 + 2\kappa) > 2$. By Theorem 14.2 in Samko et al. [878], p. 256, if $f \in \Lambda_a^\kappa = I_{a-}^{-\kappa}(L^2[0, a])$, then

$$w_2(f, \epsilon) = \sup_{|t| \leq \epsilon} \Big(\int_0^a |f(u) - f(u-t)|^2 du \Big)^{1/2} \leq C\epsilon^{-\kappa}, \tag{6.3.48}$$

where one sets $f(u-t) = 0$ when $u - t < 0$. This result is useful in showing that a function, which fails to satisfy (6.3.48) and hence is not regular enough, does not belong to the space of integrands Λ_a^κ.

In the following sections, we provide several examples involving fractional Wiener integrals.

6.3.4 The Fundamental Martingale

Let $\{B^\kappa(u)\}_{u \in [0, a]}$ be a standard FBM with parameter $\kappa \in (-1/2, 1/2)$. For $t \in [0, a]$, define

$$M^\kappa(t) = \frac{1}{\sigma_{2,\kappa}} \int_0^a u^{-\kappa} (I_{a-}^{-\kappa} 1_{[0,t)})(u) dB^\kappa(u), \tag{6.3.49}$$

where $\sigma_{2,\kappa} = (2\kappa + 1)\Gamma(1 + \kappa)$ is a constant chosen for convenience (see Lemma 6.4.1 below). Observe that M^κ can be written as

$$M^\kappa(t) = \frac{1}{\sigma_{2,\kappa} \Gamma(1-\kappa)} \int_0^t u^{-\kappa} (t-u)^{-\kappa} dB^\kappa(u), \tag{6.3.50}$$

since $\Gamma(1-\kappa)(I_{a-}^{-\kappa} 1_{[0,t)})(u) = (t-u)_+^{-\kappa}$. When $(-\kappa) > 0$, the last identity follows from $\Gamma(1-\kappa)(I_{a-}^{-\kappa} 1_{[0,t)})(u) = \Gamma(1-\kappa)(I_{-}^{-\kappa} 1_{[0,t)})(u) = (t-u)_+^{-\kappa} - (-u)_+^{-\kappa} = (t-u)_+^{-\kappa}$ for

$u > 0$, where we used (6.1.24); when $(-\kappa) < 0$, this follows from $\Gamma(1-\kappa)(I_{a-}^{-\kappa}1_{[0,t)})(u) = \Gamma(1-\kappa)(\mathcal{D}_{a-}^{\kappa}1_{[0,t)})(u) = \Gamma(1-\kappa)(\mathcal{D}_{-}^{\kappa}1_{[0,t)})(u) = (t-u)_+^{-\kappa} - (-u)_+^{-\kappa} = (t-u)_+^{-\kappa}$ for $u > 0$, where we used (6.1.25). Then, by the definition of the integral with respect to FBM B^κ (Definitions 6.3.6 and 6.3.10), we get that

$$M^\kappa(t) \stackrel{d}{=} \frac{\sigma_\kappa}{\sigma_{2,\kappa}} \int_0^a s^{-\kappa} \left(I_{a-}^{\kappa} u^\kappa u^{-\kappa}(I_{a-}^{-\kappa}1_{[0,t)})\right)(u))(s) dB^0(s) = \frac{\sigma_\kappa}{\sigma_{2,\kappa}} \int_0^t s^{-\kappa} dB^0(s),$$
(6.3.51)

so that the process M^κ is an integral with respect to Brownian motion and hence a Gaussian martingale with the quadratic variation[5]

$$[M^\kappa, M^\kappa]_t = \mathbb{E}(M^\kappa(t)^2) = \frac{\sigma_\kappa^2}{\sigma_{2,\kappa}^2} \int_0^t s^{-2\kappa} ds = \frac{\sigma_\kappa^2}{\sigma_{2,\kappa}^2(1-2\kappa)} t^{1-2\kappa}. \quad (6.3.52)$$

where the first equality follows from the facts that $\{M^\kappa(t)^2 - [M^\kappa, M^\kappa]_t\}_{t \in [0,a]}$ is a martingale and $[M^\kappa, M^\kappa]_0 = 0$ (see, e.g., Klebaner [562], Theorem 7.27, p. 199), and hence that $\mathbb{E}(M^\kappa(t)^2 - [M^\kappa, M^\kappa]_t) = \mathbb{E}(M^\kappa(0)^2 - [M^\kappa, M^\kappa]_0) = 0$. Similarly, if $t_1 < t_2$,

$$\mathbb{E}M^\kappa(t_1)M^\kappa(t_2) = \frac{\sigma_\kappa^2}{\sigma_{2,\kappa}^2(1-2\kappa)} t_1^{1-2\kappa}.$$

Hence $M^\kappa(t)$ is nothing else than

$$\{M^\kappa(t)\} \stackrel{d}{=} \left\{\frac{\sigma_\kappa}{\sigma_{2,\kappa}(1-2\kappa)^{1/2}} B^0(t^{1-2\kappa})\right\},$$

that is, Brownian motion with rescaled time $t \to t^{1-2\kappa}$. The process M^κ is called the *fundamental martingale* because it plays an important role, for example, in Girsanov's formula (Section 6.4.1).

6.3.5 The Deconvolution Formula

Set

$$f_\kappa(u, t) = u^{-\kappa}(I_{a-}^{-\kappa} s^\kappa 1_{[0,t)}(s))(u), \quad u, t \in [0, a],$$

for $\kappa \in (-1/2, 1/2)$. We will first show that $f_\kappa(\cdot, t) \in \Lambda_a^\kappa$, where Λ_a^κ is given in Definition 6.3.4 when $\kappa \in (0, 1/2)$ and in Definition 6.3.8 when $\kappa \in (-1/2, 0)$. We will then establish a "deconvolution formula" for FBM.

Arguing as for (6.2.8), one can verify that

$$-\Gamma(1-\kappa)f_\kappa(u,t) = \kappa u^{-\kappa} \int_u^t z^{\kappa-1}(z-u)_+^{-\kappa} dz - \left(\frac{t}{u}\right)^\kappa (t-u)_+^{-\kappa}, \quad (6.3.53)$$

---

[5] Quadratic variation of a process $X(t)$ is defined as a limit in probability

$$[X, X]_t = \lim \sum_{i=1}^n (X(t_i^n) - X(t_{i-1}^n))^2,$$

where the limit is taken over partitions $0 = t_0^n < t_1^n < \cdots < t_n^n = t$ with $\max_{1 \le i \le n}(t_i^n - t_{i-1}^n) \to 0$.

6.3 Fractional Wiener Integrals and their Deterministic Integrands

for $u, t \in [0, a]$. When $\kappa \in (-1/2, 0)$, since $1_{[0,t)} \in L^2[0, a]$, we obtain from (6.3.22) that $f_\kappa(\cdot, t) \in \Lambda_a^\kappa$ and from (6.3.21) that

$$s^{-\kappa}(I_{a-}^\kappa u^\kappa f_\kappa(u, t))(s) = 1_{[0,t)}(s), \quad s \in [0, a]. \tag{6.3.54}$$

When $\kappa \in (0, 1/2)$, the relation (6.3.54) holds as well. To see this, observe that, by using (6.1.3), (6.3.53), the relation $\int_c^d (u - c)^{\kappa-1}(d - u)^{-\kappa} du = B(1 - \kappa, \kappa) = \Gamma(\kappa)\Gamma(1 - \kappa)$ with $c < d$ (see (6.1.9)) and Fubini's theorem,

$$-\Gamma(\kappa)\Gamma(1 - \kappa)(I_{a-}^\kappa u^\kappa f_\kappa(u, t))(s)$$
$$= \kappa \int_s^t (u - s)_+^{\kappa-1} \int_u^t z^{\kappa-1}(z - u)_+^{-\kappa} dz\, du - t^\kappa \int_s^t (u - s)_+^{\kappa-1}(t - u)_+^{-\kappa} du$$
$$= \kappa \int_s^t dz\, z^{\kappa-1} \int_s^z du\, (u - s)_+^{\kappa-1}(z - u)_+^{-\kappa} - t^\kappa B(1 - \kappa, \kappa) 1_{[0,t)}(s)$$
$$= B(1 - \kappa, \kappa) 1_{[0,t)}(s) \left(z^\kappa \Big|_{z=s}^t - t^\kappa \right) = -\Gamma(\kappa)\Gamma(1 - \kappa) s^\kappa 1_{[0,t)}(s).$$

Since $1_{[0,t)} \in L^2[0, a]$, we also obtain from (6.3.54) and (6.3.11) that $f_\kappa(\cdot, t) \in \Lambda_a^\kappa$ when $\kappa \in (0, 1/2)$.

It follows from (6.3.20), (6.3.25) and (6.3.54) that, when $\kappa \in (-1/2, 1/2)$,

$$\left\{ (\sigma_\kappa)^{-1} \int_0^a f_\kappa(u, t) dB^\kappa(u) \right\}_{t \in [0,a]} \stackrel{d}{=} \left\{ \int_0^a s^{-\kappa}(I_{a-}^\kappa u^\kappa f_\kappa(u, t))(s) dB^0(s) \right\}_{t \in [0,a]}$$
$$= \left\{ \int_0^a 1_{[0,t)}(s) dB^0(s) \right\}_{t \in [0,a]} = \{B^0(t)\}_{t \in [0,a]}. \tag{6.3.55}$$

The expression involving f_κ in (6.3.55) is the *deconvolution formula* for FBM on an interval $[0, a]$. It compensates for the dependence of FBM and yields Brownian motion, a process with independent increments. It is used for example in deriving the prediction formula for FBM (see Section 6.4.2).

6.3.6 Classes of Integrands on the Real Line

Fractional Wiener integrals can similarly be defined on the real line. Denoted

$$\mathcal{I}^\kappa(f) = \int_\mathbb{R} f(u) dB^\kappa(u) \tag{6.3.56}$$

for deterministic functions f on the real line, these integrals are now elements of the Gaussian space $\overline{\text{span}}\{B^\kappa(t), t \in \mathbb{R}\}$ (cf. Definition 6.3.1). Several classes of integrands f can be defined, including

$$\Lambda^\kappa = \begin{cases} \{f : \int_\mathbb{R} ((I_-^\kappa f)(s))^2 ds < \infty\}, & \text{if } \kappa \in (0, 1/2), \\ \{f : f = I_-^{-\kappa} \phi_f\} \text{ for } \phi_f \in L^2(\mathbb{R}), & \text{if } \kappa \in (-1/2, 0), \end{cases} \tag{6.3.57}$$

$$|\Lambda|^\kappa = \left\{ f : \int_\mathbb{R} \int_\mathbb{R} |f(u)||f(v)||u - v|^{2\kappa-1} du\, dv < \infty \right\}, \quad \text{if } \kappa \in (0, 1/2), \tag{6.3.58}$$

and also
$$\widehat{\Lambda}^\kappa = \left\{ f : f \in L^2(\mathbb{R}), \int_{\mathbb{R}} |\widehat{f}(x)||x|^{-2\kappa} dx < \infty \right\}, \quad \text{if } \kappa \in (-1/2, 1/2). \quad (6.3.59)$$

The following inclusions hold for the classes of integrands (6.3.57)–(6.3.59):
$$\widehat{\Lambda}^\kappa \subset \Lambda^\kappa, \quad \kappa \in (-1/2, 1/2), \quad (6.3.60)$$
$$L^1(\mathbb{R}) \cap L^2(\mathbb{R}) \subset L^{2/(2\kappa+1)}(\mathbb{R}) \subset |\Lambda|^\kappa \subset \Lambda^\kappa, \quad \kappa \in (0, 1/2). \quad (6.3.61)$$

For example, the construction of the class Λ^κ is based on the representation (6.2.29) of a standard FBM, and the corresponding fractional Wiener integrals are defined as
$$\mathcal{I}^\kappa(f) = \int_{\mathbb{R}} f(u) dB^\kappa(u) \stackrel{d}{=} \frac{\Gamma(\kappa+1)}{c_{1,\kappa}} \int_{\mathbb{R}} (I_-^\kappa f)(s) dB^0(s), \quad (6.3.62)$$

for $f \in \Lambda^\kappa$, $\kappa \in (-1/2, 1/2)$. For $\kappa \in (-1/2, 0)$, the last integral is also $\int_{\mathbb{R}} \phi_f(s) dB^0(s)$. See Pipiras and Taqqu [815] for more information on fractional Wiener integrals on the real line.

Example 6.3.22 In analogy to the Ornstein–Uhlenbeck process in Examples 1.1.2 and 1.4.3, consider the process
$$Y^\kappa(t) = \int_{\mathbb{R}} e^{-\lambda(t-u)} 1_{\{t-u>0\}} dB^\kappa(u) = e^{-\lambda t} \int_{\mathbb{R}} e^{\lambda u} 1_{\{u<t\}} dB^\kappa(u), \quad (6.3.63)$$

where $\lambda > 0$ and $\kappa \in (-1/2, 1/2)$. The process is well-defined (for fixed t) since the integrand $f_t(u) = e^{\lambda u} 1_{\{u<t\}}$ belongs to $\widehat{\Lambda}^\kappa \subset \Lambda^\kappa$. Indeed, its Fourier transform is
$$\widehat{f_t}(x) = \int_{\mathbb{R}} e^{ixu} e^{\lambda u} 1_{\{u<t\}} du = \frac{e^{(\lambda+ix)t}}{\lambda + ix}$$

and hence $|\widehat{f_t}(x)|^2 = e^{\lambda t}/(\lambda^2 + x^2)$ satisfies $\int_{\mathbb{R}} |\widehat{f}(x)||x|^{-2\kappa} dx < \infty$. The process (6.3.63) is called the fractional Ornstein–Uhlenbeck (OU) process. It is zero mean, Gaussian and stationary.

The fractional OU process satisfies the Langevin equation
$$Y^\kappa(t) = \xi - \lambda \int_0^t Y^\kappa(s) ds + B^\kappa(t), \quad t \geq 0, \quad (6.3.64)$$

with FBM B^κ as the driving process (cf. Example 1.1.2) and $\xi = Y^\kappa(0)$. The equation (6.3.64) and the fractional OU process can be interpreted in the pathwise sense (Cheridito, Kawaguchi, and Maejima [229]). In a weaker and simpler sense, one could verify that the fractional OU process satisfies (6.3.64) for fixed t. The latter and other basic properties of the fractional OU process can be found in Exercise 6.8.

6.3.7 Connection to the Reproducing Kernel Hilbert Space

The reproducing kernel Hilbert space (RKHS) of FBM B^κ provides an alternative characterization of the Gaussian space $\mathcal{H}_a^\kappa$ defined in (6.3.3). We first recall the definition of a RKHS (Grenander [429], p. 93, Weinert [1002]). Suppose that $X = \{X(t)\}_{t \in T}$ is a zero mean stochastic process indexed by a set T with the covariance function $\Gamma(t_1, t_2)$, $t_1, t_2 \in T$, and let $\overline{\text{span}}(X)$ be the closure in $L^2(\Omega)$ of all linear combinations $\sum_{i=1}^k a_i X(t_i)$, $a_i \in \mathbb{R}$, $t_i \in T$. Then with the Hilbert space $\overline{\text{span}}(X)$ one can always associate an isometric Hilbert space $\mathbb{H}(\Gamma)$ of deterministic functions. The space $\mathbb{H}(\Gamma)$ is called the RKHS and is characterized by the following two properties:

(1) $\Gamma(\cdot, t) \in \mathbb{H}(\Gamma)$, for all $t \in T$, and
(2) $(g, \Gamma(\cdot, t))_{\mathbb{H}(\Gamma)} = g(t)$, for all $t \in T$ and $g \in \mathbb{H}(\Gamma)$.

It consists of all functions of the form $f(t) = \sum_{i=1}^k a_i \Gamma(t, t_i)$, $t \in T$, and their limits under the norm $\|f\|^2 = \sum_{i,j=1}^k a_i a_j \Gamma(t_i, t_j)$. The isometry map $\mathcal{J}$ between the Hilbert spaces $\overline{\text{span}}(X)$ and $\mathbb{H}(\Gamma)$ satisfies

$$\mathcal{J} : \sum_{i=1}^k a_i \Gamma(\cdot, t_i) \longmapsto \sum_{i=1}^k a_i X(t_i) \qquad (6.3.65)$$

and $(\mathcal{J}(g), \mathcal{J}(h))_{L^2(\Omega)} = (g, h)_{\mathbb{H}(\Gamma)}$. If the covariance function Γ can be expressed as

$$\Gamma(t_1, t_2) = \int_U f_{t_1}(u) f_{t_2}(u) d\nu(u), \qquad (6.3.66)$$

where $(U, \mathcal{U}, \nu)$ is a measure space and $\{f_t : t \in T\} \subset L^2(U, \nu)$, then the RKHS $\mathbb{H}(\Gamma)$ is characterized by

$$\mathbb{H}(\Gamma) = \left\{ g : g(t) = \int_U g^*(u) f_t(u) d\nu(u), \text{ for some } g^* \in \overline{\text{span}}\{f_t, t \in T\} \right\}, \qquad (6.3.67)$$

$$(g, h)_{\mathbb{H}(\Gamma)} = \int_U g^*(u) h^*(u) d\nu(u), \qquad (6.3.68)$$

where $\overline{\text{span}}\{f_t, t \in T\}$ is the closure in $L^2(U, \nu)$ of all linear combinations of $f_t, t \in T$ (see Grenander [429]).

Let $\mathbb{H}_a(\Gamma^\kappa)$ be the RKHS of FBM B^κ on an interval $[0, a]$, where Γ^κ stands for the covariance function of B^κ. We shall now characterize the RKHS $\mathbb{H}_a(\Gamma^\kappa)$ of FBM.

Proposition 6.3.23 *Let $\kappa \in (-1/2, 1/2)$ and B^κ be a standard FBM. Then, the RKHS $\mathbb{H}_a(\Gamma^\kappa)$ of B^κ is*

$$\mathbb{H}_a(\Gamma^\kappa) = \left\{ g : g(t) = \sigma_\kappa^2 \int_0^a g^*(s) s^{-\kappa} (I_{a-}^\kappa u^\kappa 1_{[0,t)}(u))(s) ds, \text{ for some } g^* \in L^2[0, a] \right\} \qquad (6.3.69)$$

with the inner product

$$(g, h)_{\mathbb{H}_a(\Gamma^\kappa)} = \sigma_\kappa^2 \int_0^a g^*(s) h^*(s) ds, \qquad (6.3.70)$$

where σ_κ^2 is given in (6.2.20).

Proof By Proposition 6.2.8, we have

$$\Gamma^\kappa(t_1, t_2) = \sigma_\kappa^2 \int_0^a \left(s^{-\kappa}(I_{a-}^\kappa u^\kappa 1_{[0,t_1)}(u))(s)\right)\left(s^{-\kappa}(I_{a-}^\kappa u^\kappa 1_{[0,t_2)}(u))(s)\right) ds, \quad t_1, t_2 \in [0, a]. \tag{6.3.71}$$

Then, (6.3.69)–(6.3.70) would follow from (6.3.67)–(6.3.68). We just need to show that $\overline{\text{span}}\{f_t, t \in [0, a]\} = L^2[0, a]$, where $f_t(s) = s^{-\kappa}(I_{a-}^\kappa u^\kappa 1_{[0,t)}(u))(s)$. The closure of the span is taken in $L^2[0, a]$ and hence $\overline{\text{span}}\{f_t, t \in [0, a]\} \subset L^2[0, a]$. The converse can be shown as follows. By Theorems 6.3.5 and 6.3.9, the set of elementary functions $\mathcal{E}$ is dense in Λ_a^κ. Elementary functions can be written as linear combinations of functions $1_{[0,t)}$, $t \in [0, a]$. Thus, for any function $g \in \Lambda_a^\kappa$, the function $f_g(s) = s^{-\kappa}(I_{a-}^\kappa u^\kappa g(u))(s)$ can be approximated in $L^2[0, a]$ by linear combinations of functions $f_t(s)$, $t \in [0, a]$. For any $b > 0$, there is a function $g = g_b$ such that $f_g = 1_{[0,b)}$: when $\kappa \in (0, 1/2)$, this follows from (6.3.14), and the case $\kappa \in (-1/2, 0)$ can be dealt with in a similar way. This shows that $\mathcal{E} \subset \overline{\text{span}}\{f_t, t \in [0, a]\}$. Since the linear span of indicator functions of intervals is dense in $L^2[0, a]$, it follows that $L^2[0, a] \subset \overline{\text{span}}\{f_t, t \in [0, a]\}$. □

Example 6.3.24 (*Functions associated with increments of FBM*) By (6.3.65), the random variable $B^\kappa(t_2) - B^\kappa(t_1)$, $0 \le t_1 < t_2 \le a$, in $\mathcal{H}_a^\kappa$ is represented by the function $g(t) = \Gamma^\kappa(t, t_1) - \Gamma^\kappa(t, t_2)$ in $\mathbb{H}_a(\Gamma^\kappa)$. By (6.3.71), the corresponding function g^* is given by $g^*(s) = s^{-\kappa}(I_{a-}^\kappa u^\kappa 1_{[t_1,t_2)}(u))(s)$. Since the function g depends on κ, the RKHS can be viewed as representing "the space of integrals," in contrast to the spaces introduced in Section 6.3.2, which can be regarded as "spaces of integrands" because they associate to $B^\kappa(t_2) - B^\kappa(t_1)$ the indicator function $1_{[t_1,t_2)}$ which plays the role of an integrand.

6.4 Applications

We provide here several applications illustrating the use of fractional Wiener integrals in addressing questions about FBM. These include Girsanov's formula, the prediction formula and elementary linear filtering.

6.4.1 Girsanov's Formula for FBM

Let $\{B^\kappa(u)\}_{u \in [0,a]}$ be a standard FBM and suppose that $\{M^\kappa(u)\}_{u \in [0,a]}$ is the fundamental martingale defined in Section 6.3.4. Suppose also that FBM is defined on some probability space $(\Omega, \mathcal{F}_a, P)$, where $\mathcal{F}_a$ is the σ–field generated by FBM B^κ on an interval $[0, a]$. Let

$$\mathcal{L}_a(M^\kappa) = \exp\left\{M^\kappa(a) - \frac{1}{2}\langle M^\kappa, M^\kappa \rangle_a\right\} \tag{6.4.1}$$

and, for any $r \in \mathbb{R}$, define a probability measure P_r on $(\Omega, \mathcal{F}_a)$ by the relation

$$\left.\frac{dP_r}{dP}\right|_{\mathcal{F}_a} = \mathcal{L}_a(rM^\kappa) = \exp\left\{rM^\kappa(a) - \frac{r^2}{2}\langle M^\kappa, M^\kappa \rangle_a\right\}. \tag{6.4.2}$$

Since $M^\kappa(a)$ is Gaussian with mean 0 and variance $\langle M^\kappa, M^\kappa \rangle_a$, we have $\mathbb{E}\mathcal{L}_a(rM^\kappa) = 1$ and therefore $\mathcal{L}_a(rM^\kappa)$ is well-defined. The following lemma is the key to Girsanov's formula.

6.4 Applications

Lemma 6.4.1 *Let $\{B^\kappa(u)\}_{u\in[0,a]}$ be a standard FBM with parameter $\kappa \in (-1/2, 1/2)$ and $\{M^\kappa(u)\}_{u\in[0,a]}$ be the fundamental martingale defined in Section 6.3.4. Then, for any $0 \le s \le t \le a$,*

$$\mathbb{E} M^\kappa(t) B^\kappa(s) = s. \quad (6.4.3)$$

Proof Suppose first that $\kappa \in (0, 1/2)$. Then, by (6.3.49), Proposition 6.3.7, the changes of variables $u = sy$ and $z = syx$ below, and the expression (6.2.20) for the constant σ_κ^2, we obtain

$$(2\kappa + 1)\Gamma(1+\kappa)\mathbb{E} M^\kappa(t) B^\kappa(s) = \mathbb{E}\Big(\int_0^a u^{-\kappa}(I_{a-}^{-\kappa} 1_{[0,t)})(u)dB^\kappa(u) \int_0^a 1_{[0,s)}(u)dB^\kappa(u)\Big)$$

$$= \sigma_\kappa^2 \int_0^a z^{-\kappa}\big(I_{a-}^\kappa u^\kappa u^{-\kappa}(I_{a-}^{-\kappa} 1_{[0,t)})(u)\big)(z)\, z^{-\kappa}\big(I_{a-}^\kappa u^\kappa 1_{[0,s)}(u)\big)(z)dz$$

$$= \sigma_\kappa^2 \int_0^a z^{-2\kappa} 1_{[0,t)}(z)(I_{a-}^\kappa u^\kappa 1_{[0,s)}(u))(z)dz = \frac{\sigma_\kappa^2}{\Gamma(\kappa)} \int_0^s z^{-2\kappa}\int_z^s u^\kappa(u-z)^{\kappa-1}du\, dz$$

$$= \frac{\sigma_\kappa^2}{\Gamma(\kappa)} \int_0^s du\, u^\kappa \int_0^u dz\, z^{-2\kappa}(u-z)^{\kappa-1} = \frac{\sigma_\kappa^2}{\Gamma(\kappa)} \int_0^1 dy \int_0^1 x^{-2\kappa}(1-x)^{\kappa-1}dx$$

$$= \frac{\sigma_\kappa^2}{\Gamma(\kappa)} B(\kappa, 1-2\kappa)s = (2\kappa+1)\Gamma(1+\kappa)s.$$

When $\kappa \in (-1/2, 0)$, by using the integration by parts below, we have similarly,

$$(2\kappa + 1)\Gamma(1+\kappa)\mathbb{E} M^\kappa(t) B^\kappa(s) = \sigma_\kappa^2 \int_0^a z^{-2\kappa} 1_{[0,t)}(z)(I_{a-}^\kappa u^\kappa 1_{[0,s)}(u))(z)dz$$

$$= -\frac{\sigma_\kappa^2}{\Gamma(1+\kappa)} \int_0^s z^{-2\kappa}\frac{d}{dz}\Big(\int_z^s u^\kappa(u-z)^\kappa du\Big)dz$$

$$= \frac{(-2\kappa)\sigma_\kappa^2}{\Gamma(1+\kappa)} \int_0^s z^{-2\kappa-1}\Big(\int_z^s u^\kappa(u-z)^\kappa du\Big)dz$$

$$= \frac{(-2\kappa)\sigma_\kappa^2}{\Gamma(1+\kappa)} \int_0^s du\, u^\kappa \int_0^u dz\, z^{-2\kappa-1}(u-z)^\kappa = \frac{(-2\kappa)\sigma_\kappa^2}{\Gamma(1+\kappa)} s \int_0^1 dy \int_0^1 x^{-2\kappa-1}(1-x)^\kappa dx$$

$$= \frac{(-2\kappa)\sigma_\kappa^2}{\Gamma(1+\kappa)} sB(-2\kappa, \kappa+1) = (2\kappa+1)\Gamma(1+\kappa)s. \qquad \Box$$

The following result is *Girsanov's formula* for FBM on an interval $[0, a]$.

Theorem 6.4.2 *For $\kappa \in (-1/2, 1/2)$, let $\{B^\kappa(u)\}_{u\in[0,a]}$ be a standard FBM and, for $r \in \mathbb{R}$, let P_r be the probability measure defined by (6.4.2). Then the distribution of B^κ under P_r is equal to the distribution of fractional Brownian motion with a drift r, that is $\{B^\kappa(u) + ru\}_{u\in[0,a]}$, under P.*

Proof It is enough to show that, for any $u_k \in [0, a]$ and $c_k \in \mathbb{C}$, $k = 1, \cdots, n$,

$$\mathbb{E}\exp\Big\{\sum_{k=1}^n c_k B^\kappa(u_k) + rM^\kappa(a) - \frac{r^2}{2}\langle M^\kappa, M^\kappa\rangle_a\Big\} = \mathbb{E}\exp\Big\{\sum_{k=1}^n c_k(B^\kappa(u_k) + ru_k)\Big\}.$$
(6.4.4)

Observe that $\sum_{k=1}^{n} c_k B^\kappa(u_k) + r M^\kappa(a)$ is a complex-valued Gaussian random variable with zero mean and second moment $\mathbb{E}(\sum_{k=1}^{n} c_k B^\kappa(u_k) + r M^\kappa(a))^2$ equal to

$$\sum_{k,l=1}^{n} c_k c_l \Gamma^\kappa(u_k, u_l) + 2r \sum_{k=1}^{n} c_k u_k + r^2 \langle M^\kappa, M^\kappa \rangle_a,$$

where we used Lemma 6.4.1 and where Γ^κ denotes the covariance function of FBM B^κ. Then, since for a zero mean, complex-valued Gaussian random variable X, $\mathbb{E}e^{zX} = e^{(z^2 \mathbb{E} X^2)/2}$, $z \in \mathbb{C}$ (Exercise 6.10), the left-hand side of (6.4.4) becomes

$$\exp\left\{\frac{1}{2} \sum_{k,l=1}^{n} c_k c_l \Gamma^\kappa(u_k, u_l) + r \sum_{k=1}^{n} c_k u_k\right\},$$

which is also the right-hand side of (6.4.4). $\square$

Remark 6.4.3 Theorem 6.4.2 naturally leads to the maximum likelihood estimator of the drift parameter r under the model $\{B^\kappa(u) + ru\}_{u \in [0,a]}$, namely,

$$\widehat{r}_{mle} = \arg\max_r \mathcal{L}_a(r M^\kappa), \tag{6.4.5}$$

where the likelihood $\mathcal{L}_a$ is given in (6.4.1). Solving $\frac{\partial}{\partial r} \mathcal{L}_a(r M^\kappa) = 0$ yields

$$\widehat{r}_{mle} = \frac{M^\kappa(a)}{\langle M^\kappa, M^\kappa \rangle_a}. \tag{6.4.6}$$

See Exercise 6.11 for some properties of the estimator (6.4.6).

6.4.2 The Prediction Formula for FBM

The prediction problem for FBM B^κ is to find an explicit expression, called a prediction formula, for the conditional expectation $X = \mathbb{E}(B^\kappa(t) | B^\kappa(s), s \in [0, a])$ with some fixed $0 < a < t$. As already indicated in Section 6.3.1, we have $X \in \mathcal{H}_a^\kappa = \overline{\text{span}}\{B^\kappa(u), u \in [0, a]\}$ and therefore we expect that $X = \int_0^a f(u) dB^\kappa(u)$ for some function f.

The prediction formula can heuristically be derived as follows. For $s \in [0, t]$, let

$$B^0(s) = \int_0^t u^{-\kappa} (I_{t-}^{-\kappa} z^\kappa 1_{[0,s)}(z))(u) dB^\kappa(u) \tag{6.4.7}$$

and recall from Section 6.3.5 that B^0 is the usual Brownian motion. One expects that the relation (6.4.7) is invertible, namely,

$$B^\kappa(s) = \int_0^t u^{-\kappa} (I_{t-}^{\kappa} z^\kappa 1_{[0,s)}(z))(u) dB^0(u), \tag{6.4.8}$$

and that $\sigma\{B^\kappa(s), s \in [0, a]\} = \sigma\{B^0(s), s \in [0, a]\}$ up to sets of measure zero, where $\sigma\{B^\kappa(s), s \in [0, a]\}$ is the σ-field generated by the random variables $B^\kappa(s), s \in [0, a]$. Then, since B^0 is the usual Brownian motion,

$$\mathbb{E}(B^\kappa(t) | B^\kappa(s), s \in [0, a]) = \mathbb{E}\left(\int_0^t u^{-\kappa}(I_{t-}^{\kappa} z^\kappa 1_{[0,t)}(z))(u) dB^0(u) \Big| B^0(s), s \in [0, a]\right)$$

$$= \int_0^a u^{-\kappa}(I_{t-}^{\kappa} z^\kappa 1_{[0,t)}(z))(u) dB^0(u).$$

6.4 Applications

By setting $1_{[0,t)} = 1_{[0,a)} + 1_{[a,t)}$ in this last relation and then using (6.4.8), we get

$$\mathbb{E}(B^\kappa(t)|B^\kappa(s), s \in [0,a]) = B^\kappa(a) + \int_0^a u^{-\kappa}(I^\kappa_{t-} z^\kappa 1_{[a,t)}(z))(u) dB^0(u).$$

Since

$$u^{-\kappa}(I^\kappa_{t-} z^\kappa 1_{[a,t)}(z))(u) = u^{-\kappa} I^\kappa_{a-}\left(v^\kappa v^{-\kappa} I^{-\kappa}_{a-}(I^\kappa_{t-} z^\kappa 1_{[a,t)}(z))(v)\right)(u)$$

and since from (6.3.20) and (6.3.25) for an integrable function f,

$$\int_0^a f(u) dB^\kappa(u) = \int_0^a u^{-\kappa}(I^\kappa_{a-} z^\kappa f(z))(u) dB^0(u),$$

we expect that

$$\mathbb{E}(B^\kappa(t)|B^\kappa(s), s \in [0,a]) = B^\kappa(a) + \int_0^a u^{-\kappa}\left(I^{-\kappa}_{a-}(I^\kappa_{t-} z^\kappa 1_{[a,t)}(z))\right)(u) dB^\kappa(u).$$

These computations suggest the following prediction formula for FBM.

Theorem 6.4.4 *Let $0 < a < t$, $\kappa \in (-1/2, 1/2)$ and B^κ be FBM. Then*

$$\mathbb{E}(B^\kappa(t)|B^\kappa(s), s \in [0,a]) = B^\kappa(a) + \int_0^a \Psi_t(a,u) dB^\kappa(u), \qquad (6.4.9)$$

where, for $u \in (0,a)$,

$$\Psi_t(a,u) = u^{-\kappa}\left(I^{-\kappa}_{a-}(I^\kappa_{t-} z^\kappa 1_{[a,t)}(z))\right)(u) \qquad (6.4.10)$$

$$= \frac{\sin(\kappa\pi)}{\pi} u^{-\kappa}(a-u)^{-\kappa} \int_a^t \frac{z^\kappa(z-a)^\kappa}{z-u} dz. \qquad (6.4.11)$$

Remark 6.4.5 After the change of variables $z = t - (t-a)y$, the kernel function $\Psi_t(a,u)$ can be expressed as

$$\Psi_t(a,u) = \frac{\sin(\kappa\pi)}{\pi} u^{-\kappa}(a-u)^{-\kappa} \int_0^1 \frac{(t-(t-a)y)^\kappa(t-a-(t-a)y)^\kappa}{t-u-(t-a)y}(t-a) dy$$

$$= \frac{\sin(\kappa\pi)}{\pi} u^{-\kappa}(a-u)^{-\kappa}(t-a)^{\kappa+1} t^\kappa (t-u)^{-1}$$

$$\times \int_0^1 \left(1 - \frac{(t-a)}{t}y\right)^\kappa (1-y)^\kappa \left(1 - \frac{(t-a)}{t-u}y\right)^{-1} dy.$$

By using Formula 3.211 in Gradshteyn and Ryzhik [421], p. 318, and using the fact that $B(\kappa+1, 1) = \Gamma(\kappa+1)\Gamma(1)/\Gamma(\kappa+2) = 1/(\kappa+1)$ (appearing in the formula), we get

$$\Psi_t(a,u)$$
$$= \frac{\sin(\kappa\pi)}{\pi(\kappa+1)} u^{-\kappa}(a-u)^{-\kappa}(t-a)^{\kappa+1} t^\kappa (t-u)^{-1} F_1\left(1, -\kappa, 1, \kappa+2; \frac{(t-a)}{t}, \frac{(t-a)}{t-u}\right), \qquad (6.4.12)$$

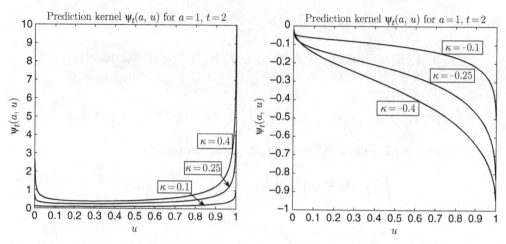

Figure 6.3 The kernel function $\Psi_t(a, u)$ in (6.4.12) for several values of κ for $a = 1$ and $t = 2$.

where F_1 is the so-called Appell's hypergeometric function. Figure 6.3 depicts the function $\Psi_t(a, u)$ computed through the formula (6.4.12).[6]

Remark 6.4.6 If $\kappa = 0$, namely if B^κ is the Brownian motion B^0, then (6.4.10) becomes $\Psi_t(a, u) = 1_{[a,t)}(u)$ and hence $\int_0^a \Psi_t(a, u) dB^0(u) = 0$. Thus, as expected,

$$\mathbb{E}(B^0(t)|B^0(s), s \in [0, a]) = B^0(a).$$

Proof Let us first verify that the right-hand sides of (6.4.10) and (6.4.11) are equal when $\kappa \in (0, 1/2)$. Observe that by (6.1.3), for $v \in (0, a)$,

$$(I_{t-}^\kappa z^\kappa 1_{[a,t)}(z))(v) = \frac{1}{\Gamma(\kappa)} \int_0^t z^\kappa 1_{[a,t)}(z)(z-v)^{\kappa-1} dz = \frac{1}{\Gamma(\kappa)} \int_a^t z^\kappa (z-v)^{\kappa-1} dz \tag{6.4.13}$$

and hence by (6.1.12),

$$\left(I_{a-}^{-\kappa}(I_{t-}^\kappa z^\kappa 1_{[a,t)}(z))\right)(u) = -\frac{1}{\Gamma(1-\kappa)\Gamma(\kappa)} \frac{d}{du} \int_u^a \int_a^t z^\kappa (z-v)^{\kappa-1} dz \, (v-u)^{-\kappa} dv \tag{6.4.14}$$

$$= -\frac{\sin(\kappa\pi)}{\pi} \frac{d}{du} \int_a^t dz \, z^\kappa \int_u^a dv \, (z-v)^{\kappa-1} (v-u)^{-\kappa},$$

since $\Gamma(1-\kappa)\Gamma(\kappa) = \pi/\sin(\kappa\pi)$. By making the change of variables $v = u + (z-u)s$ and then taking d/du inside the integral, we get

---

[6] The Appell hypergeometric function is implemented in the Matlab package "Generation of Random Variates," or in the R package "appell."

$$\left(I_{a-}^{-\kappa}(I_{t-}^{\kappa}z^{\kappa}1_{[a,t)}(z))\right)(u) = -\frac{\sin(\kappa\pi)}{\pi}\frac{d}{du}\int_{a}^{t}dz\,z^{\kappa}\int_{0}^{\frac{a-u}{z-u}}ds\,(1-s)^{\kappa-1}s^{-\kappa}$$

$$= -\frac{\sin(\kappa\pi)}{\pi}\int_{a}^{t}dz\,z^{\kappa}\left(1-\frac{a-u}{z-u}\right)^{\kappa-1}\left(\frac{a-u}{z-u}\right)^{-\kappa}\frac{d}{du}\left(\frac{a-u}{z-u}\right)$$

$$= \frac{\sin(\kappa\pi)}{\pi}(a-u)^{-\kappa}\int_{a}^{t}\frac{z^{\kappa}(z-a)^{\kappa}}{z-u}dz. \qquad (6.4.15)$$

When $\kappa \in (-1/2, 0)$, the right-hand sides of (6.4.10) and (6.4.11) are still equal. To see this, start with (6.1.12) or use Remark 6.1.6. In the latter case, observe first that (6.4.13) with $\Gamma(\kappa) = \Gamma(1+\kappa)/\kappa$ still holds for $v \in (0, a)$, since $z \neq v$ in the range of integration. Then, (6.4.14) becomes

$$\left(I_{a-}^{-\kappa}(I_{t-}^{\kappa}z^{\kappa}1_{[a,t)}(z))\right)(u) = \frac{\kappa}{\Gamma(1+\kappa)\Gamma(-\kappa)}\int_{u}^{a}\int_{a}^{t}z^{\kappa}(z-v)^{\kappa-1}dz\,(v-u)^{-\kappa-1}dv$$

$$= -\frac{\kappa\sin(\kappa\pi)}{\pi}\int_{a}^{t}dz\,z^{\kappa}\int_{u}^{a}dv\,(z-v)^{\kappa-1}(v-u)^{-\kappa-1}.$$

Observe now that, by making the change of variables $v = u + (z-u)s$ and then $s = 1/w$,

$$\kappa\int_{u}^{a}(z-v)^{\kappa-1}(v-u)^{-\kappa-1}dv = \kappa(z-u)^{-1}\int_{0}^{\frac{a-u}{z-u}}(1-s)^{\kappa-1}s^{-\kappa-1}ds$$

$$= \kappa(z-u)^{-1}\int_{\frac{z-u}{a-u}}^{\infty}(w-1)^{\kappa-1}dw = -(z-u)^{-1}\left(\frac{z-u}{a-u}-1\right)^{\kappa} = -\frac{(a-u)^{-\kappa}(z-a)^{\kappa}}{z-u}.$$

This yields the relation (6.4.15) in the case $\kappa \in (-1/2, 0)$ as well.

To verify that the integral in (6.4.9) is well-defined for $\kappa \in (-1/2, 0)$, it is enough to check that the function $s^{-\kappa}(I_{t-}^{\kappa}z^{\kappa}1_{[a,t)}(z))(s)$ belongs to $L^{2}[0, a]$. The same condition needs to be verified in the case $\kappa \in (0, 1/2)$. In both cases this can be deduced from (6.4.13). Finally, one needs to prove (6.4.9), namely that the best predictor of $B^{\kappa}(t)$, $t > a$, given $\{B^{\kappa}(s), s \in [0, a]\}$ is $\widehat{\xi}_{t} = B^{\kappa}(a) + \int_{0}^{a}\Psi_{t}(a, u)dB^{\kappa}(u)$; that is, $\mathbb{E}(B^{\kappa}(t)|B^{\kappa}(s), s \in [0, a]) = \widehat{\xi}_{t}$. By the definition of conditional expectation, this is equivalent to $\mathbb{E}B^{\kappa}(t)1_{A} = \mathbb{E}\widehat{\xi}_{t}1_{A}$ for any $A \in \sigma\{B^{\kappa}(s), s \in [0, a]\}$. Since the σ–field $\sigma\{B^{\kappa}(s), s \in [0, a]\}$ is generated by cylinder sets $C = \{B^{\kappa}(s_{j}) \in B_{j}, j = 1, \ldots, n\}$, $s_{j} \in [0, a]$, $B_{j} \in \mathcal{B}(\mathbb{R})$, it is enough to show that $\mathbb{E}B^{\kappa}(t)1_{C} = \mathbb{E}\widehat{\xi}_{t}1_{C}$. Since the vectors $(B^{\kappa}(t), B^{\kappa}(s_{j}), j = 1, \ldots, n)$ and $(\widehat{\xi}_{t}, B^{\kappa}(s_{j}), j = 1, \ldots, n)$ are zero mean Gaussian, the last relation follows from $\mathbb{E}B^{\kappa}(t)B^{\kappa}(s) = \mathbb{E}\widehat{\xi}_{t}B^{\kappa}(s)$, for all $s \in [0, a]$, or equivalently,

$$\mathbb{E}B^{\kappa}(s)(B^{\kappa}(t) - B^{\kappa}(a)) = \mathbb{E}B^{\kappa}(s)\int_{0}^{a}\Psi_{t}(a, u)dB^{\kappa}(u)$$

or

$$\mathbb{E}\int_{0}^{t}1_{[0,s)}(u)dB^{\kappa}(u)\int_{0}^{t}1_{[a,t)}(u)dB^{\kappa}(u) = \mathbb{E}\int_{0}^{a}1_{[0,s)}(u)dB^{\kappa}(u)\int_{0}^{a}\Psi_{t}(a, u)dB^{\kappa}(u). \qquad (6.4.16)$$

By Definitions 6.3.6 and 6.3.10, the left-hand side of (6.4.16) is

$$\sigma_{\kappa}^{2}\int_{0}^{t}v^{-2\kappa}(I_{t-}^{\kappa}u^{\kappa}1_{[0,s)}(u))(v)(I_{t-}^{\kappa}u^{\kappa}1_{[a,t)}(u))(v)dv$$

or

$$\sigma_\kappa^2 \int_0^a v^{-2\kappa} (I_{a-}^\kappa u^\kappa 1_{[0,s)}(u))(v)(I_{t-}^\kappa u^\kappa 1_{[a,t)}(u))(v) dv,$$

since $(I_{t-}^\kappa u^\kappa 1_{[0,s)}(u))(v) = 0$ for $a < v < t$ and $(I_{t-}^\kappa u^\kappa 1_{[0,s)}(u))(v) = (I_{a-}^\kappa u^\kappa 1_{[0,s)}(u))(v)$ for $v \in [0, a]$ by using Definitions 6.1.1 and 6.1.4. Similarly, the right-hand side of (6.4.16) is

$$\sigma_\kappa^2 \int_0^a v^{-2\kappa} (I_{a-}^\kappa u^\kappa 1_{[0,s)}(u))(v)(I_{a-}^\kappa u^\kappa \Psi_t(a, u))(v) dv,$$

which equals the left-hand side, since $(I_{a-}^\kappa u^\kappa \Psi_t(a, u))(v) = (I_{t-}^\kappa z^\kappa 1_{[a,t)}(z))(v)$ by using the expression (6.4.10). $\square$

6.4.3 Elementary Linear Filtering Involving FBM

Fractional integrals and derivatives can be also used to solve a filtering problem in a simple linear system driven by FBM. To describe the model, we let $\{B^\kappa(u)\}_{u \in [0,t]}$, $t > 0$, $\kappa \in (-1/2, 1/2)$, be a standard FBM and ξ be a fixed random variable, called a signal, which is independent of B^κ and is normally distributed with mean 0 and variance 1. Suppose that the observation process $\{Y(v)\}_{v \in [0,t]}$ is given by

$$Y(v) = \int_0^v a(u)\xi du + \int_0^v b(u) dB^\kappa(u), \tag{6.4.17}$$

where a and b are some known (deterministic) functions and where B^κ is a source of noise. The filtering problem is then to compute the best estimator $\widehat{\xi}_t$ of the signal ξ based on the observations $Y(v)$, $v \in (0, t)$. "Best" here means that

$$\mathbb{E}(\widehat{\xi}_t - \xi)^2 \leq \mathbb{E}(\widetilde{\xi}_t - \xi)^2 \tag{6.4.18}$$

for any other estimator $\widetilde{\xi}_t$ of ξ based on the same observations. It is well-known that (6.4.18) implies that $\widehat{\xi}_t = \mathbb{E}(\xi|Y(v), v \in [0, t])$ and, since the process $(\xi, \{Y(v)\}_{v \in [0,t]})$ is Gaussian, we expect that

$$\widehat{\xi}_t = \mathbb{E}(\xi|Y(v), v \in [0, t]) = \int_0^t \widehat{f}_t(u) dY(u) \tag{6.4.19}$$

for some deterministic function $\widehat{f}_t$. The problem is then to find the function $\widehat{f}_t$.

We will find the function $\widehat{f}_t$ in (6.4.19) heuristically, in the sense that we will assume that all integrals involved are well-defined. As in Section 6.4.2, (6.4.19) is equivalent to $\mathbb{E}(\int_0^t \widehat{f}_t(u) dY(u) - \xi | Y(v), v \in [0, t]) = 0$ and this holds if and only if

$$\mathbb{E}\left(\int_0^t \widehat{f}_t(u) dY(u) - \xi\right)(Y(t_2) - Y(t_1)) = 0, \tag{6.4.20}$$

for all $0 \leq t_1 < t_2 \leq t$. Since

$$\int_0^t \widehat{f}_t(u) dY(u) - \xi = \left(\int_0^t \widehat{f}_t(u) a(u) du - 1\right)\xi + \int_0^t \widehat{f}_t(u) b(u) dB^\kappa(u),$$

$$Y(t_2) - Y(t_1) = \int_0^t a(u) 1_{[t_1,t_2)}(u) du\, \xi + \int_0^t b(u) 1_{[t_1,t_2)}(u) dB^\kappa(u),$$

we obtain from the independence of ξ and B^κ that (6.4.20) is equivalent to

$$\left(\int_0^t \widehat{f}_t(u)a(u)du - 1\right)\int_0^t a(u)1_{[t_1,t_2)}(u)du$$
$$+ \mathbb{E}\int_0^t \widehat{f}_t(u)b(u)dB^\kappa(u)\int_0^t b(u)1_{[t_1,t_2)}(u)dB^\kappa(u) = 0 \qquad (6.4.21)$$

or to

$$\sigma_\kappa^2 \int_0^t s^{-2\kappa}\left(I_{t-}^\kappa u^\kappa \widehat{f}_t(u)b(u)\right)(s)\left(I_{t-}^\kappa u^\kappa b(u)1_{[t_1,t_2)}(u)\right)(s)ds$$
$$= \left(1 - \int_0^t \widehat{f}_t(u)a(u)du\right)\int_0^t a(u)1_{[t_1,t_2)}(u)du. \qquad (6.4.22)$$

By using the fractional integration by parts formula (6.1.6), we can write (6.4.22) as

$$\sigma_\kappa^2 \int_0^t \left(I_{0+}^\kappa s^{-2\kappa}(I_{t-}^\kappa u^\kappa \widehat{f}_t(u)b(u))(s)\right)(z)\, z^\kappa b(z)1_{[t_1,t_2)}(z)dz$$
$$= \int_0^t a(z)\left(1 - \int_0^t \widehat{f}_t(u)a(u)du\right)1_{[t_1,t_2)}(z)dz. \qquad (6.4.23)$$

We expect (6.4.23) to hold for all $0 \leq t_1 < t_2 \leq t$ if and only if

$$\sigma_\kappa^2 z^\kappa b(z)\left(I_{0+}^\kappa s^{-2\kappa}(I_{t-}^\kappa u^\kappa \widehat{f}_t(u)b(u))(s)\right)(z)$$
$$= a(z)\left(1 - \int_0^t \widehat{f}_t(u)a(u)du\right). \qquad (6.4.24)$$

To solve (6.4.24), one first finds $\widehat{f}_t^*$ such that

$$\sigma_\kappa^2 z^\kappa b(z)\left(I_{0+}^\kappa s^{-2\kappa}(I_{t-}^\kappa u^\kappa \widehat{f}_t^*(u)b(u))(s)\right)(z) = a(z). \qquad (6.4.25)$$

By inverting (6.4.25), it is easily seen that

$$\widehat{f}_t^*(u) = \sigma_\kappa^{-2} u^{-\kappa} b(u)^{-1}\left(I_{t-}^{-\kappa} s^{2\kappa}(I_{0+}^{-\kappa} z^{-\kappa} a(z)b(z)^{-1})(s)\right)(u). \qquad (6.4.26)$$

The function $\widehat{f}_t$ satisfying (6.4.24) is then given by

$$\widehat{f}_t(u) = P(t)\widehat{f}_t^*(u), \qquad (6.4.27)$$

where

$$P(t) = \left(1 + \int_0^t \widehat{f}_t^*(u)a(u)du\right)^{-1}. \qquad (6.4.28)$$

Indeed, after substituting (6.4.27) into (6.4.24), one gets

$$P(t) = 1 - P(t)\int_0^t \widehat{f}_t^*(u)a(u)du,$$

which leads to (6.4.28). See Section 7.4 for references on filtering involving fractional Brownian motion.

6.5 Exercises

The symbols * and ** next to some exercises are explained in Section 2.12.

Exercise 6.1 Prove the relation (6.1.35) by direct computations using Definition 6.1.18.

Exercise 6.2 Prove the relation (6.1.41) by direct computations. *Hint:* Use (6.1.35).

Exercise 6.3 Prove the relation (6.2.25).

Exercise 6.4 In the proof of Proposition 6.3.2, show that the definition of $\mathcal{I}_a^\kappa(f)$ does not depend on an approximating sequence $\{f_n\}$ of f.

Exercise 6.5 Show that the relation (6.3.7) holds for Wiener integrals $\mathcal{I}_a^0(f)$ and $\mathcal{I}_a^0(g)$ defined with respect to the usual (standard) Brownian motion and $f, g \in \mathcal{E}$.

Exercise 6.6 (*i*) Show that the function (6.3.13) satisfies the relation (6.3.14). (*ii*) Show the statement leading to (6.3.16).

Exercise 6.7* Show the convergence of $I_{i,h}$, $i = 1, \ldots, 5$, stated following (6.3.45) and under the assumption that f is γ–Hölder continuous with $0 < (-\kappa) < \gamma$.

Exercise 6.8 Consider the fractional OU process Y^κ in Example 6.3.22. (*i*) Show that it satisfies the equation (6.3.64). (*ii*) Show that it is long-range dependent with parameter $d = \kappa$ in the sense that $\mathbb{E}Y^\kappa(0)Y^\kappa(h) \sim Ch^{2\kappa-1}$, as $h \to \infty$. *Hint:* See Cheridito et al. [229].

Exercise 6.9** Various spaces of integrands were introduced on an interval $[0, a]$ in Section 6.3.2, and on the real line $\mathbb{R}$ in Section 6.3.6. How are these spaces related? For example, if f is a function on $[0, a]$ belonging to one of the spaces, does this imply that the function $f1_{[0,a]}$ defined on $\mathbb{R}$ belongs to one of the spaces for $\mathbb{R}$? And vice versa, if f is a function on $\mathbb{R}$ belonging to one of the spaces, what about its restriction $f|_{[0,a]}$ to the interval $[0, a]$?

Exercise 6.10 Suppose that $X = X_1 + iX_2$ is a zero mean, complex-valued Gaussian random variable (that is, the random bivariate vector $(X_1, X_2)'$ is Gaussian and has zero mean). Show that $\mathbb{E}e^{zX} = e^{(z^2 \mathbb{E} X^2)/2}$, $z \in \mathbb{C}$.

Exercise 6.11 Consider the maximum likelihood estimator $\widehat{r}_{mle}$ in (6.4.6). Show that, under the probability measure P_r defined by (6.4.2), the estimator $\widehat{r}_{mle}$ has the following distribution:

$$\widehat{r}_{mle} \stackrel{d}{=} \mathcal{N}\left(r, \frac{1}{ca^{1-2\kappa}}\right)$$

for a constant c (depending on κ).

6.6 Bibliographical Notes

Section 6.1: A comprehensive reference on fractional integrals and derivatives is the monograph by Samko et al. [878] (consisting of 976 pages and with parts of it in tiny font!). For a geometric interpretation of fractional calculus, see Podlubny [826]. The presentation found in Section 6.1 is motivated by applications to fractional Brownian motions, and is aimed to be concise, self-contained and intuitive.

Section 6.2: The representation of FBM on the real line and the connection to fractional integrals (derivatives) in Section 6.2.2 is known already from the celebrated work of Mandelbrot and Van Ness [682], who were one of the first to consider and certainly to popularize FBM. The term "fractional" refers precisely to the fact that FBM can be obtained from Brownian motion through fractional integration (derivation).

The representation of FBM on an interval in Section 6.2.1 appeared in Molchan [726], Molchan and Golosov [725]. See also Molchan [727] with a discussion on earlier developments. It was later rediscovered by Barton and Poor [104], Norros et al. [758], Decreusefond and Üstünel [296].

Connections between the representations of FBM on the real line and an interval are studied in Jost [537]. See also Picard [809].

Section 6.3: This section is based on the work in Pipiras and Taqqu [815, 817, 818]. Fractional Wiener integrals are also considered in many of the references to Section 7.1 on stochastic integration given below. Integrands for FBM are viewed as generalized functions in Jolis (2007), forming a Hilbert space even in the case $H > 1/2$.

RKHS associated with general self-similar processes and several applications are considered in Nuzman and Poor [775]. Similar integration theory for the so-called tempered FBM is developed in Meerschaert and Sabzikar [708]. Other work on fractional OU process considered in Example 6.3.22 can be found in Kaarakka and Salminen [543], Sun and Guo [933].

Section 6.4: The Girsanov formula for FBM and its use in statistical estimation of the drift parameter (see Remark 6.4.3) appeared in Norros et al. [758]. Extensions of the formula to more general drift processes with applications to statistical inference can be found in Le Breton [600], Kleptsyna et al. [566, 566], Brouste and Kleptsyna [190], and the monograph by Prakasa Rao [831], among others. See also Sottinen [915].

The prediction formula for FBM appeared in Molchan [726], Molchan and Golosov [725], Gripenberg and Norros [433]. For extensions, see Duncan [334], Duncan and Fink [336], Fink et al. [364].

Linear filtering problems for FBM were considered in Le Breton [600], Kleptsyna and co-authors [565, 566, 564], Ahmed and Charalambous [16], Inoue et al. [522] and others.

Other problems which involve fractional Wiener integrals, Gaussian spaces and spaces of integrands concern local independence of FBM (Norros and Saksman [757]), continuity in the Hurst parameter of the law of fractional Wiener integrals (Jolis and Viles [536]), mutual information for stochastic signals and FBM (Duncan [335]), fractional Brownian bridges (Gasbarra et al. [383]).

Other notes: In this chapter, we explored connections between FBM and fractional calculus. Fractional calculus is also related in interesting and fundamental ways to stable distributions, stable Lévy motions and the so-called anomalous diffusions. A basic connection is that the transition probability $p_t(x)$ of a stable Lévy motion satisfies a space-time fractional diffusion equation. See, for example, Meerschaert [707], Meerschaert and Sikorskii [712].

7

Stochastic Integration with Respect to Fractional Brownian Motion

Integrals of deterministic functions with respect to FBM are considered in Chapter 6. Several approaches have been proposed to define integration with respect to FBM when the integrands are *random*. We shall focus in Section 7.1 on the approach based on the Malliavin calculus (see Appendix C for basics). Due to the prevalence of the Malliavin calculus in probability theory, this approach has now become a cornerstone for dealing with integrals with respect to FBM. We focus on:

- FBM and the semimartingale property
- Divergence integral for FBM
- Self-integration of FBM
- Itô's formulas

Section 7.2 contains some applications. These involve:

- Stochastic differential equations driven by FBM
- Regularity of probability laws
- Numerical solutions of stochastic differential equations driven by FBM
- Stein's method
- The local time of FBM

As in Chapter 6 (see (6.2.1) and (6.2.2)), for notational convenience, FBM B_H with the self-similarity parameter $H \in (0, 1)$ will be denoted B^κ with $\kappa = H - 1/2 \in (-1/2, 1/2)$.

7.1 Stochastic Integration with Random Integrands

In Section 7.1.1, we include a well-known result that FBM is not a semimartingale. This shows that the classical Itô integration theory for semimartingales (e.g., Protter [834]) cannot be applied to FBM. The divergence integral for FBM based on the Malliavin calculus is considered in Section 7.1.2. In Section 7.1.3, we study the special case of self-integration. Section 7.1.4 concerns Itô's formulas.

7.1.1 FBM and the Semimartingale Property

Let $B^\kappa = \{B^\kappa(t)\}_{t \in \mathbb{R}}$ be FBM with parameter $\kappa \in (-1/2, 1/2)$. For $p > 0$, consider

$$Y_{n,p} = n^{p(\kappa+1/2)-1} \sum_{j=1}^{n} \left| B^\kappa\left(\frac{j}{n}\right) - B^\kappa\left(\frac{j-1}{n}\right) \right|^p$$

and

$$\widehat{Y}_{n,p} = n^{-1} \sum_{j=1}^{n} \left| B^\kappa(j) - B^\kappa(j-1) \right|^p.$$

By the $(\kappa + 1/2)$–self-similarity property of FBM, for each fixed n, $Y_{n,p}$ has the same distribution as $\widetilde{Y}_{n,p}$. The stationary series $\{B^\kappa(j) - B^\kappa(j-1)\}_{j \in \mathbb{Z}}$ is ergodic since it is Gaussian with autocovariance function decaying to zero or, equivalently, having a spectral density (e.g., Lindgren [635]). Then, the ergodic theorem (e.g., Lindgren [635]) implies that, as $n \to \infty$,

$$\widehat{Y}_{n,p} \to \mathbb{E}|B^\kappa(1) - B^\kappa(0)|^p = \mathbb{E}|B^\kappa(1)|^p =: c_p$$

almost surely and in $L^1(\Omega)$. Hence, $Y_{n,p} \xrightarrow{d} c_p$ and therefore $Y_{n,p} \xrightarrow{p} c_p$. This yields

$$V_{n,p} := \sum_{j=1}^{n} \left| B^\kappa\left(\frac{j}{n}\right) - B^\kappa\left(\frac{j-1}{n}\right) \right|^p \xrightarrow{p} \begin{cases} 0, & \text{if } p(\kappa + 1/2) > 1, \\ \infty, & \text{if } p(\kappa + 1/2) < 1. \end{cases} \quad (7.1.1)$$

The limit in probability of $V_{n,p}$ as $n \to \infty$ is called the pth variation of FBM. Observe that if $p_1 < p_2$, then the p_2th variation is less than the p_1th variation.

We want to argue that FBM B^κ is not a semimartingale when $\kappa \neq 0$. Arguing by contradiction, suppose that B^κ is a semimartingale. If $\kappa > 0$, we can choose $p \in (1/(\kappa + 1/2), 2)$ in (7.1.1) so that $V_{n,p} \to 0$ in probability and hence almost surely along a subsequence. This implies that the pth variation of FBM is zero, and since $p < 2$, so is its quadratic variation. Since B^κ is assumed to be a semimartingale, this implies that it must be merely a finite-variation process.[1] But since for $p \in (1, 1/(\kappa + 1/2))$, that is $p > 1$, the limit of $V_{n,p}$ is almost surely infinite, B^κ cannot have finite variation.

If $\kappa < 0$, we can choose $p > 2$ such that $p(\kappa + 1/2) < 1$. By (7.1.1), the limit of $V_{n,p}$ is almost surely infinite. This contradicts the almost-sure finiteness of the quadratic variation of B^κ, assuming that B^κ is a semimartingale. In either case, if $\kappa \neq 0$, B^κ is not a semimartingale. This fact is stated as the following result.

Proposition 7.1.1 *FBM $B^\kappa = \{B^\kappa(t)\}_{t \in \mathbb{R}}$, $\kappa \in (-1/2, 1/2)$, is not a semimartingale when $\kappa \neq 0$.*

The next result provides a useful characterization of the sample paths of FBM. The space of Hölder continuous functions is defined in (6.3.40).

Proposition 7.1.2 *FBM $B^\kappa = \{B^\kappa(t)\}_{t \in \mathbb{R}}$, $\kappa \in (-1/2, 1/2)$, has a version whose sample paths are γ–Hölder continuous for any $\gamma < \kappa + 1/2$.*

Proof The result is a consequence of the following Kolmogorov's continuity criterion (e.g., Revuz and Yor [848], Theorem 2.1 in Chapter I). Let $X = \{X(t)\}_{t \in \mathbb{R}}$ be a stochastic process. If there are constants $C > 0$, $p > 0$, $\epsilon > 0$ such that, for every s, t,

---

[1] By the Doob–Meyer decomposition (e.g., Protter [834]), a semimartingale is the sum of a local martingale (it has a finite *quadratic* variation) and an increasing finite-variation process. Finite variation refers to total or 1-variation being finite.

$$\mathbb{E}|X(t) - X(s)|^p \leq C|t - s|^{1+\epsilon},$$

then the process X has a version whose sample paths are γ–Hölder continuous for any $\gamma \in (0, \epsilon/p)$. The criterion is applied to FBM with $p > (\kappa + 1/2)^{-1}$, leading to, by the $(\kappa + 1/2)$–self-similarity of FBM,

$$\mathbb{E}|B^\kappa(t) - B^\kappa(s)|^p = C|t - s|^{p(\kappa+1/2)} = C|t - s|^{1+\epsilon},$$

where $\epsilon = p(\kappa + 1/2) - 1$. Thus, FBM has a version whose sample paths are γ–Hölder continuous for any $0 < \gamma < \epsilon/p = \kappa + 1/2 - 1/p$. Since p can be taken arbitrarily large, the result of the proposition follows. □

7.1.2 Divergence Integral for FBM

The divergence (Skorokhod) integral is defined for a general isonormal Gaussian process in Appendix C. We first recall the definition of an isonormal Gaussian process given in Definition C.1.1 of Appendix C. A stochastic process W defined by $W = \{W(h), h \in \mathcal{H}\}$ is called an *isonormal Gaussian process* if W is a centered Gaussian family of random variables such that $\mathbb{E}(W(h)W(g)) = (f, g)_{\mathcal{H}}$ for all $h, g \in \mathcal{H}$, where $\mathcal{H}$ is a real separable Hilbert space with scalar product denoted by $(\cdot, \cdot)_{\mathcal{H}}$. FBM can be seen as an isonormal process in a natural way through a suitable choice of the Hilbert space $\mathcal{H}$, as noted below. As in the case of deterministic integrands (Section 6.3), view the interval of interest as $[0, a]$, where $a > 0$. The presentation below will often consider the cases $\kappa < 0$ and $\kappa > 0$ separately.

When $\kappa < 0$, for the Hilbert space $\mathcal{H}$ and the corresponding inner product in Appendix C, one can take

$$\mathcal{H} = \Lambda_a^\kappa, \quad (f, g)_{\mathcal{H}} = (f, g)_{\Lambda_a^\kappa}, \qquad (7.1.2)$$

where Λ_a^κ is defined in (6.3.22), and $(f, g)_{\Lambda_a^\kappa}$ is given in (6.3.23)–(6.3.24). When $\kappa > 0$, the corresponding space Λ_a^κ in (6.3.11) is not complete and thus not a Hilbert space. In this case, one can take

$$\mathcal{H} = \overline{\Lambda}_a^\kappa, \quad (f, g)_{\mathcal{H}} = (f, g)_{\Lambda_a^\kappa}, \qquad (7.1.3)$$

where the bar over $\overline{\Lambda}_a^\kappa$ indicates the completion under the norm induced by the inner product $(f, g)_{\Lambda_a^\kappa}$ in (6.3.12). The elements of $\overline{\Lambda}_a^\kappa$ are not necessarily functions. They could be generalized functions such as a Dirac distribution. We nevertheless write

$$(f, g)_{\Lambda_a^\kappa} = \sigma_\kappa^2 \int_0^a s^{-2\kappa} \Big(I_{a-}^\kappa u^\kappa f(u)\Big)(s) \Big(I_{a-}^\kappa u^\kappa g(u)\Big)(s) ds \qquad (7.1.4)$$

for $\kappa < 0$ or $\kappa > 0$. Often, however, the functions f and g will be smooth. An isonormal process associated with FBM can then be defined as

$$\Big\{B^\kappa(h) = \int_0^a h(u) dB^\kappa(u), \; h \in \mathcal{H}\Big\}, \qquad (7.1.5)$$

where the integral on the right-hand side is a fractional Wiener integral.

Definition 7.1.3 The divergence integral associated with $\{B^\kappa(h), h \in \mathcal{H}\}$ is denoted $\delta_a^\kappa(X)$. In analogy to (7.1.5), we will continue writing

$$\delta_a^\kappa(X) = \int_0^a X(u) dB^\kappa(u). \tag{7.1.6}$$

In particular, if $h \in \mathcal{H}$, then $\delta_a^\kappa(h) = B^\kappa(h) = \int_0^a h(u) dB^\kappa(u)$, in the sense of the fractional Wiener integral. For example, $\delta_a^\kappa(1_{[0,t)}) = B^\kappa(t)$.

We shall now use the various definitions in Appendix C related to the Malliavin calculus. In particular, the class $\mathcal{S}$ of smooth random variables of the form $F = f(W(h_1), \ldots, W(h_n))$, where $h_1, \ldots, h_n \in \mathcal{H}$, and where $f \in C_p^\infty(\mathbb{R}^n)$; that is, f and all its derivatives have at most polynomial growth. The derivative operator D (Appendix C) is defined for random variables in $\mathcal{S}$ and, more generally, in $\mathbb{D}^{1,2}$ (see (C.2.6)). For example, for fixed $t \in [0, a]$, $(B^\kappa(t))^n = (B^\kappa(1_{[0,t)}))^n \in \mathcal{S}$ and

$$(D(B^\kappa(t))^n)(u) = D_u(B^\kappa(t))^n = n(B^\kappa(t))^{n-1} 1_{[0,t)}(u)$$

is an $\mathcal{H}$–valued random element. As mentioned in Appendix C, for example, $|B^\kappa(t)| \in \mathbb{D}^{1,2}$ but $|B^\kappa(t)| \notin \mathcal{S}$.

The domain Dom δ_a^κ of the divergence integral δ_a^κ consists of $\mathcal{H}$–valued square integrable random variables $X \in L^2(\Omega; \mathcal{H})$ satisfying conditions (i) and (ii) in Definition C.3.1. Since the elements of $\mathcal{H}$ are functions (or possibly distributions when $\kappa > 0$), we can view X as a process $X = \{X(u)\}_{u \in [0,a]}$. By one of the properties of divergence operators discussed in Appendix C, the space $\mathbb{D}^{1,2}(\mathcal{H})$ is included in Dom δ_a^κ. By (C.2.14), $\mathbb{D}^{1,2}(\mathcal{H})$ is equipped with the norm $\|\cdot\|_{1,2,\mathcal{H}}$, where

$$\|X\|_{1,2,\mathcal{H}}^2 = \mathbb{E}\|X\|_\mathcal{H}^2 + \mathbb{E}\|DX\|_{\mathcal{H}\otimes\mathcal{H}}^2. \tag{7.1.7}$$

In view of (7.1.2) and (7.1.3), the norm $\|\cdot\|_\mathcal{H}$ in (7.1.7) is induced by the inner products (6.3.23)–(6.3.24) for $\kappa \in (-1/2, 0)$ and (6.3.12) for $\kappa \in (0, 1/2)$. We shall describe next the structure of the tensor product Hilbert space $\|\cdot\|_{\mathcal{H}\otimes\mathcal{H}}$.

The Tensor Product Hilbert Space

Since $\mathcal{H}$ is defined in terms of univariate fractional integrals and derivatives, $\|\cdot\|_{\mathcal{H}\otimes\mathcal{H}}$ is naturally related to fractional integrals and derivatives of two variables. In analogy to (6.1.3), a fractional integral of a function of two variables of orders $\alpha_1 > 0$, $\alpha_2 > 0$ on the rectangle $(a_1, b_1) \times (a_2, b_2)$ is defined as

$$(I_{b_1-,b_2-}^{\alpha_1,\alpha_2} f)(s_1, s_2) = \frac{1}{\Gamma(\alpha_1)\Gamma(\alpha_2)} \int_{a_1}^{b_1} \int_{a_2}^{b_2} f(u_1, u_2)(u_1 - s_1)_+^{\alpha_1-1} (u_2 - s_2)_+^{\alpha_2-1} du_1 du_2. \tag{7.1.8}$$

In analogy to (6.1.12), a fractional derivative of a function of two variables of orders $\alpha_1 \in (0, 1)$, $\alpha_2 \in (0, 1)$ is defined as

$$(I_{b_1-,b_2-}^{-\alpha_1,-\alpha_2} f)(s_1, s_2)$$
$$= \frac{1}{\Gamma(1+\alpha_1)\Gamma(1+\alpha_2)} \frac{\partial^2}{\partial s_1 \partial s_2} \int_{a_1}^{b_1} \int_{a_2}^{b_2} f(u_1, u_2)(u_1 - s_1)_+^{-\alpha_1} (u_2 - s_2)_+^{-\alpha_2} du_1 du_2. \tag{7.1.9}$$

See Section 24.1 of Samko et al. [878] for more information on fractional integrals and derivatives of functions of several variables. Note that we are dealing here with fractional integrals over *finite* rectangles. With $b- = (b_1-, b_2-)$, $u = (u_1, u_2)$, $s = (s_1, s_2)$ and $v^\beta = v_1^\beta v_2^\beta$, $\Gamma(\beta) = \Gamma(\beta_1)\Gamma(\beta_2)$ for $v = (v_1, v_2)$, $\beta = (\beta_1, \beta_2)$, (7.1.8) and (7.1.9) can be rewritten as

$$(I_{b-}^\alpha f)(s) = \frac{1}{\Gamma(\alpha)} \int_{[a_1,b_1]\times[a_2,b_2]} f(u)(u-s)_+^{\alpha-1} du, \tag{7.1.10}$$

$$(I_{b-}^{-\alpha} f)(s) = \frac{1}{\Gamma(1+\alpha)} \frac{\partial^2}{\partial s_1 \partial s_2} \int_{[a_1,b_1]\times[a_2,b_2]} f(u)(u-s)_+^{-\alpha} du, \tag{7.1.11}$$

where $du = du_1 du_2$.

When $\kappa \in (0, 1/2)$, the bivariate analogues of Λ_a^κ in (6.3.11) and its inner product in (6.3.12) are

$$\Lambda_a^{\kappa,\otimes 2} = \left\{ f : [0,a]^2 \to \mathbb{R} : \int_{[0,a]^2} s^{-2\kappa} \left((I_{a-}^\kappa u^\kappa f(u))(s) \right)^2 ds < \infty \right\} \tag{7.1.12}$$

and

$$(f, g)_{\Lambda_a^{\kappa,\otimes 2}} = \sigma_\kappa^4 \int_{[0,a]^2} s^{-2\kappa} \left(I_{a-}^\kappa u^\kappa f(u) \right)(s) \left(I_{a-}^\kappa u^\kappa g(u) \right)(s) ds, \tag{7.1.13}$$

where $\kappa = (\kappa, \kappa)$ and $a- = (a-, a-)$. Similarly, when $\kappa \in (-1/2, 0)$, the bivariate analogues of Λ_a^κ in (6.3.22) and its inner product in (6.3.23) are

$$\Lambda_a^{\kappa,\otimes 2} = \left\{ f : f(u) = u^{-\kappa}(I_{a-}^{-\kappa} s^\kappa \phi_f(s))(u) \text{ for } \phi_f \in L^2[0,a]^2 \right\} \tag{7.1.14}$$

and

$$(f, g)_{\Lambda_a^{\kappa,\otimes 2}} = \sigma_\kappa^4 \int_{[0,a]^2} s^{-2\kappa} \left(I_{a-}^\kappa u^\kappa f(u) \right)(s) \left(I_{a-}^\kappa u^\kappa g(u) \right)(s) ds \tag{7.1.15}$$

$$= \sigma_\kappa^4 \int_{[0,a]^2} \phi_f(s) \phi_g(s) ds. \tag{7.1.16}$$

See also Lemma 6.3.11 which applies to (7.1.14) as well.

The following result describes the product space $\mathcal{H} \otimes \mathcal{H}$. The proof is left as Exercise 7.1.

Lemma 7.1.4 *Let $\mathcal{H}$ be defined in (7.1.2) or (7.1.3). If $\kappa \in (-1/2, 0)$, then*

$$\mathcal{H} \otimes \mathcal{H} = \Lambda_a^{\kappa,\otimes 2}, \quad (f, g)_{\mathcal{H}\otimes\mathcal{H}} = (f, g)_{\Lambda_a^{\kappa,\otimes 2}} \tag{7.1.17}$$

and if $\kappa \in (0, 1/2)$, then

$$\mathcal{H} \otimes \mathcal{H} = \overline{\Lambda}_a^{\kappa,\otimes 2}, \quad (f, g)_{\mathcal{H}\otimes\mathcal{H}} = (f, g)_{\Lambda_a^{\kappa,\otimes 2}}, \tag{7.1.18}$$

where the bar indicates completion under the norm induced by the inner product $(f, g)_{\mathcal{H}\otimes\mathcal{H}}$.

As in the univariate case (see (6.3.34)), when $\kappa \in (0, 1/2)$, it is convenient to work with a subspace of $\Lambda_a^{\kappa,\otimes 2}$ and hence of $\mathcal{H} \otimes \mathcal{H}$. Let

$$|\Lambda|_a^{\kappa,\otimes 2} = \left\{ f : [0,a]^2 \to \mathbb{R} : \|f\|_{|\Lambda|_a^{\kappa,\otimes 2}} < \infty \right\}, \tag{7.1.19}$$

where

$$\|f\|^2_{|\Lambda|^{\kappa,\otimes 2}_a} = \int_{[0,a]^4} |f(u)||f(v)||u-v|^{2\kappa-1}dudv \qquad (7.1.20)$$

and as above, $|u-v|^{2\kappa-1} = |u_1-v_1|^{2\kappa-1}|u_2-v_2|^{2\kappa-1}$ and $dudv = du_1du_2dv_1dv_2$. Then,

$$|\Lambda|^{\kappa,\otimes 2}_a \subset \Lambda^{\kappa,\otimes 2}_a \subset \mathcal{H} \otimes \mathcal{H}. \qquad (7.1.21)$$

If $f \in |\Lambda|^{\kappa,\otimes 2}_a$, by using the relation (6.2.23) for $|u_1-v_1|^{2\kappa-1}$ and $|u_2-v_2|^{2\kappa-1}$, it follows by the Fubini's theorem as in the univariate case (cf. (6.3.35)) that

$$(f,g)_{\Lambda^{\kappa,\otimes 2}_a} = \kappa^2(2\kappa+1)^2 \int_{[0,a]^4} f(u)g(v)|u-v|^{2\kappa-1}dudv. \qquad (7.1.22)$$

As in the univariate case, when $\kappa \in (-1/2, 0)$, a subspace of $\Lambda^{\kappa,\otimes 2}_a$ could be constructed using Hölder continuous functions of two variables.

7.1.3 Self-Integration of FBM

We now return to the divergence integral (7.1.6). We are interested in whether $X = B^\kappa = \{B^\kappa(u)\}_{u \in [0,a]}$ belongs to Dom δ^κ_a, and thus whether the divergence integral

$$\delta^\kappa_a(B^\kappa) = \int_0^a B^\kappa(u)dB^\kappa(u) \qquad (7.1.23)$$

is well-defined. The cases $\kappa \in (0, 1/2)$, $\kappa \in (-1/4, 0)$ and $\kappa \in (-1/2, -1/4)$ are considered separately in the following three examples. (The case $\kappa = -1/4$ is excluded.)

Example 7.1.5 (*The case $\kappa \in (0, 1/2)$*) When $\kappa \in (0, 1/2)$, we will show that $B^\kappa \in \mathbb{D}^{1,2}(\mathcal{H})$ and hence $B^\kappa \in$ Dom δ^κ_a. By (7.1.7), we need to verify that $\|B^\kappa\|_{1,2,\kappa} < \infty$, where

$$\|B^\kappa\|^2_{1,2,\mathcal{H}} = \mathbb{E}\|B^\kappa\|^2_\mathcal{H} + \mathbb{E}\|DB^\kappa\|^2_{\mathcal{H} \otimes \mathcal{H}}. \qquad (7.1.24)$$

We first consider the term $\mathbb{E}\|DB^\kappa\|^2_{\mathcal{H} \otimes \mathcal{H}}$. For fixed u, $D_v(B^\kappa(u)) = D_v B^\kappa(1_{[0,u)}) = 1_{[0,u)}(v)$. Viewing D as an operation on $\mathcal{H}$–valued random variables and using an approximation argument, one can show that $(DB^\kappa)(u,v) = 1_{[0,u)}(v)$ (Exercise 7.2). Then,

$$\mathbb{E}\|DB^\kappa\|^2_{\mathcal{H} \otimes \mathcal{H}} = \|1_{[0,u)}(v)\|^2_{\mathcal{H} \otimes \mathcal{H}} \le C\|1_{[0,u)}(v)\|^2_{|\Lambda|^{\kappa,\otimes 2}_a}$$

$$\le C\|1\|^2_{|\Lambda|^{\kappa,\otimes 2}_a} = C\int_{[0,a]^4} |u_1-v_1|^{2\kappa-1}|u_2-v_2|^{2\kappa-1}du_1du_2dv_1dv_2 < \infty,$$

by using (7.1.20) and (7.1.22). Similarly,

$$\mathbb{E}\|B^\kappa\|_{\mathcal{H}}^2 \leq C \int_0^a \int_0^a \mathbb{E}|B^\kappa(u)||B^\kappa(v)||u-v|^{2\kappa-1} du dv$$

$$\leq C' \int_0^a \int_0^a |u|^{\kappa+1/2}|v|^{\kappa+1/2}|u-v|^{2\kappa-1} du dv \leq C'' \int_0^a \int_0^a |u-v|^{2\kappa-1} du dv < \infty.$$

Thus, $\|B^\kappa\|_{1,2,\kappa} < \infty$ and the integral (7.1.23) is well defined. By the basic properties of divergence integrals (Appendix C), we have $\mathbb{E}\delta_a^\kappa(B^\kappa) = 0$ and $\mathbb{E}\delta_a^\kappa(B^\kappa)^2 = \mathbb{E}\|B^\kappa\|_{\mathcal{H}}^2 + \mathbb{E}(DB^\kappa, (DB^\kappa)^*)_{\mathcal{H}\otimes\mathcal{H}}$. The latter variance can be calculated more easily by writing $\delta_a^\kappa(B^\kappa) = (B^\kappa(a)^2 - a^{2\kappa+1})/2$ (see Itô's formula in Section 7.1.4).

Example 7.1.6 (*The case $\kappa \in (-1/4, 0)$*) When $\kappa \in (-1/4, 0)$, we will show similarly that $\|B^\kappa\|_{1,2,\kappa} < \infty$, where $\|B^\kappa\|_{1,2,\kappa}^2$ is given by (7.1.24) and as above $(DB^\kappa)(u, v) = 1_{[0,u)}(v)$. Hence, in this case, the integral (7.1.23) is well defined as well.

We thus want to show that $\mathbb{E}\|B^\kappa\|_{\mathcal{H}}^2 < \infty$ and $\|DB^\kappa\|_{\mathcal{H}\otimes\mathcal{H}}^2 = \|1_{[0,u)}(v)\|_{\mathcal{H}\otimes\mathcal{H}}^2 < \infty$. FBM B^κ is γ–Hölder continuous for any $\gamma < \kappa + 1/2$ (see Proposition 7.1.2). Since $(-\kappa) < \kappa + 1/2$ for $\kappa > -1/4$, we can apply Proposition 6.3.19 to write

$$s^{-\kappa}(I_{a-}^\kappa u^\kappa B^\kappa(u))(s) = \frac{s^{-\kappa}}{\Gamma(\kappa)} \int_0^a (B^\kappa(u) - B^\kappa(s))u^\kappa(u-s)_+^{\kappa-1} du$$

$$-s^{-\kappa} B^\kappa(s)(I_{a-}^\kappa u^\kappa)(s) =: \frac{X_1(s)}{\Gamma(\kappa)} - X_2(s).$$

To show $\mathbb{E}\|B^\kappa\|_{\mathcal{H}}^2 < \infty$, it is then enough to prove that $\mathbb{E}\|X_j\|_{L^2[0,a]}^2 < \infty$, $j = 1, 2$. We consider only the more difficult case $j = 1$. In this case, observe that, by the generalized Minkowski's inequality (6.1.21),

$$\mathbb{E}\|X_1\|_{L^2[0,a]}^2 = \int_0^a s^{-2\kappa} \mathbb{E}\Big(\int_0^a (B^\kappa(u) - B^\kappa(s))u^\kappa(u-s)_+^{\kappa-1} du\Big)^2 ds$$

$$\leq \int_0^a s^{-2\kappa} \Big(\int_0^a \big(\mathbb{E}(B^\kappa(u) - B^\kappa(s))^2\big)^{1/2} u^\kappa (u-s)_+^{\kappa-1} du\Big)^2 ds$$

$$= \int_0^a s^{-2\kappa} \Big(\int_0^a |u-s|^{\kappa+1/2} u^\kappa (u-s)_+^{\kappa-1} du\Big)^2 ds$$

$$= \int_0^a s^{-2\kappa} \Big(\int_0^a u^\kappa (u-s)_+^{2\kappa-1/2} du\Big)^2 ds$$

$$= \int_0^a s^{4\kappa+1} \Big(\int_1^{a/s} w^\kappa (w-1)^{2\kappa-1/2} dw\Big)^2 ds < \infty.$$

Note that the inner integral above is finite only when $\kappa > -1/4$.

Let us argue now that $\mathbb{E}\|DB^\kappa\|_{\mathcal{H}\otimes\mathcal{H}}^2 = \|1_{[0,u)}(v)\|_{\mathcal{H}\otimes\mathcal{H}}^2 < \infty$ when $\kappa \in (-1/4, 0)$. In view of (7.1.17), (7.1.15) and (7.1.11), we need to show that $s_1^{-\kappa} s_2^{-\kappa} F(s_1, s_2)$ is square integrable, where

$$F(s_1, s_2) = \frac{\partial^2}{\partial s_1 \partial s_2} \int_0^a \int_0^a 1_{[0,u_1)}(u_2) u_1^\kappa u_2^\kappa (u_1 - s_1)_+^\kappa (u_2 - s_2)_+^\kappa du_1 du_2.$$

Observe that

$$F(s_1, s_2) = \frac{\partial^2}{\partial s_1 \partial s_2} \int_0^a du_1\, u_1^\kappa (u_1 - s_1)_+^\kappa \left(\int_0^{u_1} u_2^\kappa (u_2 - s_2)_+^\kappa du_2 \right)$$

$$= \frac{d}{ds_1} \int_0^a du_1\, u_1^\kappa (u_1 - s_1)_+^\kappa \frac{d}{ds_2} \left(\int_0^{u_1} u_2^\kappa (u_2 - s_2)_+^\kappa du_2 \right)$$

$$= -\frac{s_2^{2\kappa}}{\Gamma(\kappa + 1)} \frac{d}{ds_1} \int_0^a du_1\, u_1^\kappa (u_1 - s_1)_+^\kappa F_2\left(\frac{u_1}{s_2} \right) 1_{[0, u_1)}(s_2)$$

$$=: -\frac{s_2^{2\kappa}}{\Gamma(\kappa + 1)} G(s_1, s_2),$$

where F_2 is defined in (6.2.11). (We will not prove that the derivative can be taken inside of the integral above.) We now need to show that $s_1^{-\kappa} s_2^\kappa G(s_1, s_2)$ is square integrable.

When $s_1 < s_2$, note that

$$G(s_1, s_2) = \frac{d}{ds_1} \int_{s_2}^a u_1^\kappa (u_1 - s_1)^\kappa F_2\left(\frac{u_1}{s_2} \right) du_1$$

$$= \kappa \int_{s_2}^a u_1^\kappa (u_1 - s_1)^{\kappa-1} F_2\left(\frac{u_1}{s_2} \right) du_1 = \kappa s_2^{2\kappa} \int_1^{a/s_2} w^\kappa \left(w - \frac{s_1}{s_2} \right)^{\kappa-1} F_2(w) dw.$$

Observe that, by (6.2.11), $F_2(w) \sim -(w-1)^\kappa$ as $w \downarrow 1$ and, as in the proof of Corollary 6.2.5, $F_2(w) \to C$ as $w \to \infty$. Then, $|F_2(w)| \leq C_1(w-1)^\kappa + C_2$ and hence

$$|G(s_1, s_2)| \leq C s_2^{2\kappa} \int_1^{a/s_2} \left(w - \frac{s_1}{s_2} \right)^{\kappa-1} (w-1)^\kappa dw + C' s_2^{2\kappa} \int_1^{a/s_2} \left(w - \frac{s_1}{s_2} \right)^{\kappa-1} dw$$

$$\leq C s_2^{2\kappa} \int_1^\infty \left(w - \frac{s_1}{s_2} \right)^{\kappa-1} (w-1)^\kappa dw + C' s_2^{2\kappa} \int_1^\infty \left(w - \frac{s_1}{s_2} \right)^{\kappa-1} dw$$

$$= C s_2^{2\kappa} \int_0^\infty \left(x + 1 - \frac{s_1}{s_2} \right)^{\kappa-1} x^\kappa dx + \frac{2C'}{(-\kappa)} s_2^{2\kappa} \left(1 - \frac{s_1}{s_2} \right)^\kappa$$

$$\leq C s_2^{2\kappa} \left(\frac{s_2 - s_1}{s_2} \right)^{2\kappa} \int_0^\infty (y+1)^{\kappa-1} y^\kappa dy + \frac{2C'}{(-\kappa)} s_2^{2\kappa} \left(\frac{s_2 - s_1}{s_2} \right)^\kappa$$

$$\leq c(s_2 - s_1)^{2\kappa} + c s_2^\kappa (s_2 - s_1)^\kappa =: c G_0(s_1, s_2),$$

where we made the changes of variables $w = x + 1$, and then $x = (1 - s_1/s_2) y = (s_2 - s_1) y / s_2$ above. One can now easily verify that $s_1^{-\kappa} s_2^\kappa G_0(s_1, s_2)$ is square integrable over $0 < s_1 < s_2 < a$ when $\kappa \in (-1/4, 0)$.

When $s_1 > s_2$, we only outline the proof. Arguing as in Remark 6.2.3, we can write

$$G(s_1, s_2) = \frac{d}{ds_1} \int_{s_1}^a du_1\, u_1^\kappa (u_1 - s_1)^\kappa F_2\left(\frac{u_1}{s_2} \right)$$

$$= \frac{d}{ds_1} \int_0^a dv\, (v + s_1)^\kappa F_2\left(\frac{v + s_1}{s_2} \right) v^\kappa 1_{\{v < a - s_1\}}$$

$$= \kappa \int_0^a dv\, (v + s_1)^{\kappa-1} F_2\left(\frac{v + s_1}{s_2} \right) v^\kappa 1_{\{v < a - s_1\}}$$

$$+ \int_0^a dv\, (v + s_1)^\kappa F_2^{(1)}\left(\frac{v + s_1}{s_2} \right) \frac{1}{s_2} v^\kappa 1_{\{v < a - s_1\}}$$

$$+ \int_0^a dv\, (v+s_1)^\kappa F_2\left(\frac{v+s_1}{s_2}\right) v^\kappa (-\delta_{\{a-s_1\}}(v))$$

$$= \kappa \int_{s_1}^a du\, u^{\kappa-1} F_2\left(\frac{u}{s_2}\right)(u-s_1)^\kappa + \int_{s_1}^a du\, u^\kappa F_2^{(1)}\left(\frac{u}{s_2}\right)\frac{1}{s_2}(u-s_1)^\kappa$$

$$- a^\kappa F_2\left(\frac{a}{s_2}\right)(a-s_1)^\kappa =: G_1(s_1,s_2) + G_2(s_1,s_2) + G_3(s_1,s_2),$$

where $F_2^{(1)}$ denotes the derivative of F_2 and $\delta_{\{a\}}(v)$ denotes a point mass at $v=a$. One can show that $s_1^{-\kappa} s_2^\kappa G_j(s_1,s_2)$, $j=1,2,3$, are square integrable over $0 < s_2 < s_1 < a$ when $\kappa \in (-1/4, 0)$. To do this, use the bound $|F_2(w)| \leq C_1(w-1)^\kappa + C_2$ above and also the fact that $F_2^{(1)}(w) = -\kappa w^\kappa (w-1)^{\kappa-1}$, which follows from (6.2.11).

Example 7.1.7 (*The case $\kappa \in (-1/2, -1/4)$*) When $\kappa \in (-1/2, -1/4)$, for fixed ω, the function $u \mapsto B^\kappa(u, \omega)$ is too irregular to belong to $\mathcal{H}$. We will in fact show that $\mathbb{P}(\omega : B^\kappa(\cdot, \omega) \in \mathcal{H}) = 0$ and hence the divergence integral (7.1.23) is not defined. Arguing by contradiction, suppose $B^\kappa(\cdot, \omega) \in \mathcal{H}$ on a set of ωs of positive probability. For notational simplicity, we suppress the dependence of $B^\kappa(\cdot, \omega)$ on ω. Then, by (6.3.48),

$$\int_\epsilon^a |B^\kappa(u) - B^\kappa(u-\epsilon)|^2 du \leq C\epsilon^{-2\kappa}$$

on a set of positive probability, where C is a random constant. By making the change of variables $u = \epsilon v$ and by using the $(\kappa + 1/2)$–self-similarity of FBM, this is equivalent to

$$\epsilon^{2\kappa+2} \int_1^{a/\epsilon} |B^\kappa(v) - B^\kappa(v-1)|^2 dv \leq C\epsilon^{-2\kappa}$$

or

$$\epsilon \int_1^{a/\epsilon} |B^\kappa(v) - B^\kappa(v-1)|^2 dv \leq C\epsilon^{-4\kappa-1}. \tag{7.1.25}$$

By the ergodic theorem, as $\epsilon \to 0$, the left-hand side of (7.1.25) converges almost surely to $a\mathbb{E}\int_1^2 |B^\kappa(v) - B^\kappa(v-1)|^2 dv = a\mathbb{E}(B^\kappa(1))^2 = a > 0$. But $\epsilon^{-4\kappa-1}$ is bounded from below by a positive constant as $\epsilon \to 0$ only when $-4\kappa - 1 \leq 0$ or $\kappa \geq -1/4$. This contradicts $\kappa < -1/4$.

As shown in Example 7.1.7, $B^\kappa \notin \mathcal{H}$ and hence $B^\kappa \notin \text{Dom}\,\delta_a^\kappa$. To accommodate more irregular integrands, several extensions of the divergence integral were proposed in the case $\kappa < 0$. One approach is based on the following idea.

Recall from (C.3.2) in Appendix C that the divergence integral $\delta_a^\kappa(X)$ is characterized by the property: for any $F \in \mathbb{D}^{1,2}$ (and, in particular, $F \in \mathcal{S}$),

$$\mathbb{E}(F\delta_a^\kappa(X)) = \mathbb{E}(DF, X)_\mathcal{H}. \tag{7.1.26}$$

The right-hand side of (7.1.26) requires the integrand $X \in \mathcal{H}$ a.s., which is not the case for $X = B^\kappa$ when $\kappa \in (-1/2, -1/4)$. Note, however, that the inner product in (7.1.26) can informally be written as

$$(DF, X)_{\mathcal{H}} = (DF, X)_{\Lambda_a^\kappa} = \sigma_\kappa^2 \int_0^a s^{-2\kappa} (I_{a-}^\kappa u^\kappa (DF)(u))(s)(I_{a-}^\kappa u^\kappa X(u))(s) ds$$

$$= \sigma_\kappa^2 \int_0^a u^\kappa \big(I_{0+}^\kappa s^{-2\kappa} (I_{a-}^\kappa z^\kappa (DF)(z))(s)\big)(u) X(u) du, \quad (7.1.27)$$

by using (6.3.23) and the relation (6.1.6) (when considered for fractional derivatives). The right-hand side of (7.1.27) involves a fractional derivative of DF but not of $X(u)$. This discussion suggests the following notation and definition.

In analogy to $\mathcal{H} = \Lambda_a^\kappa$ for $\kappa \in (-1/2, 0)$ in (6.3.22), let

$$\mathcal{H}_2 = \left\{ f : f(z) = z^{-\kappa} \big(I_{a-}^{-\kappa} s^{2\kappa} (I_{0+}^{-\kappa} u^{-\kappa} \psi_f(u))(s)\big)(z) \text{ for } \psi_f \in L^2[0, a] \right\}. \quad (7.1.28)$$

We have $\mathcal{H}_2 \subset \mathcal{H} = \Lambda_a^\kappa$. Indeed, by Lemma 3.2 in Samko et al. [878], p. 70,

$$\phi_f(s) := s^\kappa (I_{0+}^{-\kappa} u^{-\kappa} \psi_f(u))(s) = (I_{0+}^{-\kappa} \Psi_f(u))(s)$$

with $\Psi_f \in L^2[0, a]$. Then, by Theorem 3.5 in Samko et al. [878], p. 66, $\phi_f \in L^q[0, a]$ with $q \in 2/(1 + 2(-\kappa))$ and hence $\phi_f \in L^2[0, a]$. Thus, if $f \in \mathcal{H}_2$, we have $f(z) = z^{-\kappa} \big(I_{a-}^{-\kappa} s^\kappa \phi_f(s)\big)(z)$ with $\phi_f \in L^2[0, a]$ and hence $f \in \mathcal{H} = \Lambda_a^\kappa$. Since $\mathcal{H}_2 \subset \mathcal{H}$, $B^\kappa(\phi)$ is defined for $\phi \in \mathcal{H}_2$. Let $\mathcal{S}_{\mathcal{H}}$ be the space of random variables of the form

$$F = h(B^\kappa(\phi_1), \ldots, B^\kappa(\phi_n)), \quad (7.1.29)$$

where $n \geq 1$, $h \in C_p^\infty(\mathbb{R}^n)$ and $\phi_k \in \mathcal{H}_2$, $k = 1, \ldots, n$ (cf. (C.2.1)).

Definition 7.1.8 A process $X = \{X(u)\}_{u \in [0, a]}$ satisfying $\mathbb{E} \int_0^a X(u)^2 du < \infty$ will be said to belong to the domain $\text{Dom}^* \delta_a^\kappa$ if there exists a random variable $\delta_a^\kappa(X) \in L^2(\Omega)$ such that, for all $F \in \mathcal{S}_{\mathcal{H}}$,

$$\mathbb{E}(F \delta_a^\kappa(X)) = \sigma_\kappa^2 \mathbb{E} \int_0^a u^\kappa \big(I_{0+}^\kappa s^{-2\kappa} (I_{a-}^\kappa z^\kappa (DF)(z))(s)\big)(u) X(u) du. \quad (7.1.30)$$

Note a formal analogy between this relation and (7.1.26) and (7.1.27). One can show (Cheridito and Nualart [228]) that $\text{Dom}\, \delta_a^\kappa \subset \text{Dom}^* \delta_a^\kappa$, and that δ_a^κ restricted to $\text{Dom}\, \delta_a^\kappa$ coincides with the divergence integral.

Example 7.1.9 (*The case $\kappa \in (-1/2, 0)$ and the extended domain*) We argue here that $X(u) = B^\kappa(u)$ belongs to $\text{Dom}^* \delta_a^\kappa$ for all $\kappa \in (-1/2, 0)$, and that

$$\delta_a^\kappa(B^\kappa) = \frac{(B^\kappa(a))^2 - a^{2\kappa+1}}{2} = \frac{a^{2\kappa+1}}{2} \left(\left(\frac{B^\kappa(a)}{a^{\kappa+1/2}} \right)^2 - 1 \right) \quad (7.1.31)$$

in Definition 7.1.8. (This fact also follows from the Itô's formula stated in Proposition 7.1.14 below.) We need to show that the relation (7.1.30) holds for $F \in \mathcal{S}_{\mathcal{H}}$, $X = B^\kappa$ and $\delta_a^\kappa(B^\kappa)$ given by (7.1.31). We suppose without loss of generality that B^κ is standard; that is, $\mathbb{E} B^\kappa(1)^2 = 1$. The outline of the proof is as follows.

First, it is enough to prove (7.1.30) for $F = H_n(B^\kappa(\phi))$, where $\phi \in \mathcal{H}_2$ and H_n is the Hermite polynomial of arbitrary order $n \geq 0$. This follows from the facts that the collection $\{H_n(B^\kappa(\phi)), \phi \in \mathcal{H}\}$ is dense in $L^2(\Omega, \mathbb{P}, \sigma\{B^\kappa(u), u \in [0, a]\})$, and that $\mathcal{H}_2$ is dense in $\mathcal{H}$. We may also suppose without loss of generality that $\mathbb{E} B^\kappa(\phi)^2 = \|\phi\|_{\mathcal{H}}^2 = 1$.

Second, for $F = H_n(B^\kappa(\phi))$, since $H'_n(x) = nH_{n-1}(x)$, we have $(DF)(z) = D_z F = nH_{n-1}(B^\kappa(\phi))\phi(z)$. Hence, the relation (7.1.30) becomes

$$\mathbb{E}(H_n(B^\kappa(\phi))\delta_a^\kappa(X)) = \sigma_\kappa^2 n \int_0^a u^\kappa \left(I_{0+}^\kappa s^{-2\kappa}(I_{a-}^\kappa z^\kappa \phi(z))(s)\right)(u) \mathbb{E}(H_{n-1}(B^\kappa(\phi))B^\kappa(u))du. \tag{7.1.32}$$

We will show that this is verified with $\delta_a^\kappa(B^\kappa))$ given in (7.1.31).

Since $H_1(x) = x$ and $H_2(x) = x^2 - 1$, we can write $B^\kappa(u) = H_1(B^\kappa(u))$ and express $\delta_a^\kappa(B^\kappa))$ in (7.1.31) as $(a^{2\kappa+1}/2)H_2(B^\kappa(a)/a^{\kappa+1/2})$. Then, (7.1.32) is equivalent to

$$\frac{a^{2\kappa+1}}{2}\mathbb{E}(H_n(B^\kappa(\phi))H_2(B^\kappa(a)/a^{\kappa+1/2}))$$
$$= \sigma_\kappa^2 n \int_0^a u^\kappa \left(I_{0+}^\kappa s^{-2\kappa}(I_{a-}^\kappa z^\kappa \phi(z))(s)\right)(u) \mathbb{E}(H_{n-1}(B^\kappa(\phi))H_1(B^\kappa(u)))du. \tag{7.1.33}$$

When $n \neq 2$, both sides of (7.1.33) are equal to zero by Proposition 5.1.1, by using $\mathbb{E}B^\kappa(\phi)^2 = 1$ and $\mathbb{E}B^\kappa(a)^2 = a^{2\kappa+1}$. When $n = 2$, by using (6.1.6), the left-hand side of (7.1.33) is

$$(\mathbb{E}B^\kappa(\phi)B^\kappa(a))^2 = (\mathbb{E}B^\kappa(\phi)B^\kappa(1_{[0,a)}))^2$$
$$= \left(\sigma_\kappa^2 \int_0^a s^{-2\kappa}(I_{a-}^\kappa z^\kappa \phi(z))(s)(I_{a-}^\kappa z^\kappa 1_{[0,a)}(z))(s)ds\right)^2$$
$$= \left(\sigma_\kappa^2 \int_0^a u^\kappa \left(I_{0+}^\kappa s^{-2\kappa}(I_{a-}^\kappa z^\kappa \phi(z))(s)\right)(u)du\right)^2.$$

Similarly, the right-hand side of (7.1.33) is

$$\sigma_\kappa^2 2 \int_0^a u^\kappa \left(I_{0+}^\kappa s^{-2\kappa}(I_{a-}^\kappa z^\kappa \phi(z))(s)\right)(u) \mathbb{E}B^\kappa(\phi)B^\kappa(u)du$$
$$= 2(\sigma_\kappa^2)^2 \int_0^a u^\kappa \left(I_{0+}^\kappa s^{-2\kappa}(I_{a-}^\kappa z^\kappa \phi(z))(s)\right)(u)$$
$$\times \left(\int_0^a v^\kappa \left(I_{0+}^\kappa s^{-2\kappa}(I_{a-}^\kappa z^\kappa \phi(z))(s)\right)(v)1_{[0,u)}(v)dv\right)du$$
$$= (\sigma_\kappa^2)^2 \int_0^a \int_0^a u^\kappa \left(I_{0+}^\kappa s^{-2\kappa}(I_{a-}^\kappa z^\kappa \phi(z))(s)\right)(u)v^\kappa \left(I_{0+}^\kappa s^{-2\kappa}(I_{a-}^\kappa z^\kappa \phi(z))(s)\right)(v)dudv \tag{7.1.34}$$
$$= \left(\sigma_\kappa^2 \int_0^a u^\kappa \left(I_{0+}^\kappa s^{-2\kappa}(I_{a-}^\kappa z^\kappa \phi(z))(s)\right)(u)du\right)^2,$$

by using the symmetry of the integrand in u and v in (7.1.34) above. Thus, the relation (7.1.33) holds for $n = 2$ as well.

7.1.4 Itô's Formulas

We provide here several of Itô's formulas for FBM and involving the divergence integral. These formulas can be viewed as extensions of the deterministic calculus formula

$$F(x) = F(0) + \int_0^x F'(y)dy \tag{7.1.35}$$

to a stochastic setting involving FBM. For example, in analogy to (7.1.35), one would like to express $F(B^\kappa(t))$ through a divergence integral. The cases $\kappa > 0$ and $\kappa < 0$ need to be treated separately. We focus on the case $\kappa > 0$, and only mention a few results in the case $\kappa < 0$.

The case $\kappa > 0$ is special because of the following result. Suppose $f, g : [0, a] \to \mathbb{R}$ are α– and β–Hölder continuous functions, respectively, such that

$$\alpha + \beta > 1. \qquad (7.1.36)$$

It is known (Young [1024]) that, under (7.1.36), the Riemann–Stieltjes integral

$$(RS)\int_0^a f(s)dg(s) := \lim_{|\pi| \to 0} \sum_{k=1}^n f(s_k)(g(t_k) - g(t_{k-1})) \qquad (7.1.37)$$

exists, where π refers to a partition $0 = t_0 < t_1 < \cdots < t_n = a$, $s_k \in [t_{k-1}, t_k]$ and $|\pi| = \max_k |t_k - t_{k-1}|$ is the mesh of the partition. How is this related to FBM and Itô's formula?

FBM B^κ is β–Hölder continuous a.s. with $\beta = \kappa + 1/2 - \epsilon$ for any small $\epsilon > 0$ (Proposition 7.1.2). If $F \in C^2[0, a]$ (that is, F is twice continuously differentiable on $[0, a]$), then $F'(B^\kappa(s))$ is also α–Hölder continuous a.s. with $\alpha = \kappa + 1/2 - \epsilon$ for any small $\epsilon > 0$. Since these α and β satisfy (7.1.36) for $\kappa > 0$ and small enough $\epsilon > 0$, the Riemann–Stieltjes integral

$$(RS)\int_0^a F'(B^\kappa(s))dB^\kappa(s) \qquad (7.1.38)$$

is defined pathwise a.s. Moreover, writing

$$F(B^\kappa(t)) - F(B^\kappa(0)) = \sum_{k=1}^n (F(B^\kappa(t_k)) - F(B^\kappa(t_{k-1})))$$

$$= \sum_{k=1}^n F'(B^\kappa(s_k))(B^\kappa(t_k) - B^\kappa(t_{k-1})) \qquad (7.1.39)$$

for $\in [0, a]$, $s_k \in [t_{k-1}, t_k]$ by the mean value theorem, where $0 = t_0 < t_1 < \cdots < t_n = t$ is a partition, and passing to the limit in the partition mesh leads to the formula

$$F(B^\kappa(t)) = F(B^\kappa(0)) + (RS)\int_0^t F'(B^\kappa(s))dB^\kappa(s). \qquad (7.1.40)$$

The formula (7.1.40) is an Itô's formula for the (pathwise) Riemann–Stieltjes integral with respect to FBM B^κ when $\kappa > 0$. We should note that the assumption $\kappa > 0$ is key here. For example, in the case $\kappa = 0$ (Brownian motion), the behavior of (7.1.39) as the mesh decreases is more involved.[2] The limiting behavior of (7.1.39) is involved for $\kappa < 0$ as well (see Section 5.2.3 in Nualart [769]).

We shall rewrite next the Itô's formula (7.1.40) in terms of the divergence integral with respect to FBM. The next result relates the (pathwise) Riemann–Stieltjes integral, appearing in (7.1.40), to the divergence integral.

---

[2] In the case $\kappa = 0$, the limit of (7.1.39) with $s_k = t_{k-1}$ is the so-called Itô integral. When $B^\kappa(s_k) = B^0(s_k)$ is replaced by $(B^0(t_k) + B^0(t_{k-1}))/2$, the limit is the so-called Stratonovich integral.

Proposition 7.1.10 *Let $a > 0$, $\kappa \in (0, 1/2)$ and $G \in C^1(\mathbb{R})$ be such that*

$$|G'(x)| \leq Ce^{px^2}, \quad x \in \mathbb{R}, \tag{7.1.41}$$

for some constants $C > 0$ and $p < (4a^{2\kappa+1})^{-1}$. Then,

$$(RS)\int_0^a G(B^\kappa(u))dB^\kappa(u) = \int_0^a G(B^\kappa(u))dB^\kappa(u) + \frac{2\kappa+1}{2}\int_0^a G'(B^\kappa(u))u^{2\kappa}du, \tag{7.1.42}$$

where $\int_0^a G(B^\kappa(u))dB^\kappa(u)$ is the divergence integral.[3]

Remark 7.1.11 Since $D_s(G(B^\kappa(u))) = D_s(G(B^\kappa(1_{[0,u)}))) = G'(B^\kappa(u))1_{[0,u)}(s)$ and $u^{2\kappa} = 2\kappa \int_0^a |u-s|^{2\kappa-1} 1_{[0,u)}(s)ds$, the relation (7.1.42) can also be written as

$$(RS)\int_0^a G(B^\kappa(u))dB^\kappa(u)$$
$$= \int_0^a G(B^\kappa(u))dB^\kappa(u) + \kappa(2\kappa+1)\int_0^a\int_0^a D_s(G(B^\kappa(u)))|u-s|^{2\kappa-1}dsdu. \tag{7.1.43}$$

In fact, a more general fact holds: if a stochastic process $X \in \{X(u)\}_{u\in[0,a]}$ satisfies $X \in \mathbb{D}^{1,2}(\mathcal{H})$ and its derivative DX is such that

$$\int_0^a\int_0^a |D_s(X(u))||u-s|^{2\kappa-1}duds < \infty \text{ a.s.} \quad \text{and} \quad \mathbb{E}\|DX\|^2_{L^{1/(\kappa+1/2)}[0,a]^2} < \infty, \tag{7.1.44}$$

then

$$(RS)\int_0^a X(u)dB^\kappa(u) = \int_0^a X(u)dB^\kappa(u) + \kappa(2\kappa+1)\int_0^a\int_0^a D_s(X(u))|u-s|^{2\kappa-1}dsdu \tag{7.1.45}$$

(Proposition 5.2.3 in Nualart [769]). Relation (7.1.45) is important since it relates the Riemann–Stieltjes integral to the divergence integral. Moreover, whereas the divergence integral has always zero mean, this is not necessarily the case for the Riemann–Stieltjes integral.

Proof We first indicate a few consequences of the assumption (7.1.41), which will be used below. Note that, by using (7.1.41),

$$\mathbb{E}\sup_{u\in[0,a]}|G'(B^\kappa(u))|^2 \leq C^2\mathbb{E}e^{2p\sup_{u\in[0,a]}|B^\kappa(u)|^2} \leq C^2\mathbb{E}e^{2pa^{2\kappa+1}\sup_{u\in[0,1]}|B^\kappa(u)|^2}. \tag{7.1.46}$$

By Fernique's inequality (e.g., Bogachev [167], Theorem 2.8.5), for large enough x,

$$\mathbb{P}(\sup_{u\in[0,1]}|B^\kappa(u)| > x) \leq e^{-(\frac{1}{2\sigma^2}-\epsilon)x^2} = e^{-(\frac{1}{2}-\epsilon)x^2}, \tag{7.1.47}$$

---
[3] Sometimes, the divergence integral $\int_0^a X(u)dB^\kappa(u)$ is denoted $\int_0^a X(u)\delta B^\kappa(u)$ to avoid confusion with other integrals.

where $\sigma^2 = \sup_{u\in[0,1]} \mathbb{E}|B^\kappa(u)|^2 = \sup_{u\in[0,1]} u^{2\kappa+1} = 1$ and $\epsilon > 0$ is arbitrarily small. In view of (7.1.46) and (7.1.47),

$$\mathbb{E} \sup_{u\in[0,a]} |G'(B^\kappa(u))|^2 < \infty, \tag{7.1.48}$$

as long as $2pa^{2\kappa+1} - \frac{1}{2} < 0$ or $p < (4a^{2\kappa+1})^{-1}$. Note also that, by the mean value theorem, $|G(x) - G(0)| \le |G'(x^*)||x| \le C|x|e^{px^2}$ for $x^* \in [0, x]$ and thus, by slightly increasing p, G satisfies (7.1.41) as well. Hence, (7.1.48) also holds for G replacing G'.

By the discussion preceding the proposition, the Riemann–Stieltjes integral in (7.1.42) can be approximated by

$$S_n := \sum_{k=1}^n G(B^\kappa(t_{k-1}))(B^\kappa(t_k) - B^\kappa(t_{k-1})) = \sum_{k=1}^n G(B^\kappa(t_{k-1}))B^\kappa(1_{[t_{k-1},t_k)}),$$

where $0 = t_0 < t_1 < \cdots < t_n = a$ is a partition π_n such that $|\pi_n| \to 0$. By using the formula (C.3.4) in Appendix C with $F = G(B^\kappa(t_{k-1}))$, $W = B^\kappa$ and $h = 1_{[t_{k-1},t_k)}$, we can write $G(B^\kappa(t_{k-1}))B^\kappa(1_{[t_{k-1},t_k)})$ as $\delta_a^\kappa(G(B^\kappa(t_{k-1}))1_{[t_{k-1},t_k)}) + (DG(B^\kappa(t_{k-1})), 1_{[t_{k-1},t_k)})_\mathcal{H}$; that is, by (7.1.6),

$$\int_0^a G(B^\kappa(t_{k-1}))1_{[t_{k-1},t_k)}(u)dB^\kappa(u) + (DG(B^\kappa(t_{k-1})), 1_{[t_{k-1},t_k)})_\mathcal{H}.$$

Thus, the approximating sum S_n can be written as

$$S_n = \sum_{k=1}^n \int_0^a G(B^\kappa(t_{k-1}))1_{[t_{k-1},t_k)}(u)dB^\kappa(u) + \sum_{k=1}^n (DG(B^\kappa(t_{k-1})), 1_{[t_{k-1},t_k)})_\mathcal{H} =: S_{n,1} + S_{n,2}.$$

It is enough to show that $S_{n,1} \to \int_0^a G(B^\kappa(u))dB^\kappa(u)$ in $L^2(\Omega)$ and $S_{n,2} \to \frac{2\kappa+1}{2}\int_0^a G'(B^\kappa(u))u^{2\kappa}du$ a.s.

For $S_{n,1}$, note that $S_{n,1} = \int_0^a G_n(u)dB^\kappa(u)$ with $G_n(u) = \sum_{k=1}^n G(B^\kappa(t_{k-1}))1_{[t_{k-1},t_k)}(u)$. Then, $S_{n,1} \to \int_0^a G(B^\kappa(u))dB^\kappa(u)$ in $L^2(\Omega)$ is equivalent to $\int_0^a F_n(B^\kappa(u))dB^\kappa(u) \to 0$ in $L^2(\Omega)$, where

$$F_n(u) = G(B^\kappa(u)) - \sum_{k=1}^n G(B^\kappa(t_{k-1}))1_{[t_{k-1},t_k)}(u).$$

In view of (C.3.10), it is enough to show that $\mathbb{E}\|F_n\|_\mathcal{H}^2 \to 0$ and $\mathbb{E}\|DF_n\|_{\mathcal{H}\otimes\mathcal{H}}^2 \to 0$.

By using (6.3.39), we have

$$\mathbb{E}\|F_n\|_\mathcal{H}^2 \le C\mathbb{E}\int_0^a \left(G(B^\kappa(u)) - \sum_{k=1}^n G(B^\kappa(t_{k-1}))1_{[t_{k-1},t_k)}(u)\right)^2 du$$

$$\le Ca\mathbb{E}\sup_{|u-v|\le|\pi_n|} |G(B^\kappa(u)) - G(B^\kappa(v))|^2 =: Ca\mathbb{E}\xi_n.$$

Applying the mean value theorem, $\xi_n \le \sup_{|u|\le a} |G'(B^\kappa(u))|^2 \sup_{|u-v|\le|\pi_n|} |B^\kappa(u) - B^\kappa(v)|^2 \to 0$ a.s. and that $\xi_n \le 4\sup_{|u|\le a} |G(B^\kappa(u))|^2$, where the bound is integrable by (7.1.48) with G replacing G'. The dominated convergence theorem implies that $\mathbb{E}\|F_n\|_\mathcal{H}^2 \to 0$.

7.1 Stochastic Integration with Random Integrands

We now turn to showing $\mathbb{E}\|DF_n\|^2_{\mathcal{H}\otimes\mathcal{H}} \to 0$. By using relations of the type $D_v G(B^\kappa(u)) = G'(B^\kappa(u))1_{[0,u)}(v)$ (see (C.2.3) in Appendix C) and writing $1_{[0,u)}(v) = \sum_{k=1}^n 1_{[0,u)}(v)1_{[t_{k-1},t_k)}(u) = \sum_{k=1}^n (1_{[0,t_{k-1})}(v)1_{[t_{k-1},t_k)}(u) + 1_{[t_{k-1},u)}(v)1_{[t_{k-1},t_k)}(u)) = \sum_{k=1}^n (1_{[0,t_{k-1})}(v)1_{[t_{k-1},t_k)}(u) + 1_{\{t_{k-1}\leq v<u<t_k\}})$, we get

$$D_v(F_n(u)) = G'(B^\kappa(u))1_{[0,u)}(v) - \sum_{k=1}^n G'(B^\kappa(t_{k-1}))1_{[0,t_{k-1})}(v)1_{[t_{k-1},t_k)}(u)$$

$$= \sum_{k=1}^n (G'(B^\kappa(u)) - G'(B^\kappa(t_{k-1})))1_{[0,t_{k-1})}(v)1_{[t_{k-1},t_k)}(u)$$

$$+ \sum_{k=1}^n G'(B^\kappa(u))1_{\{t_{k-1}\leq v<u<t_k\}}$$

$$=: D_1(u,v) + D_2(u,v).$$

By using a bound analogous to the one in (6.3.39), it is enough to show that $\mathbb{E}\|D_j\|^2_{L^2[0,a]^2} \to 0$, $j=1,2$. Note that

$$\mathbb{E}\|D_1\|^2_{L^2[0,a]^2} \leq a^2 \mathbb{E}\sup_{|u-v|\leq|\pi_n|}|G'(B^\kappa(u)) - G'(B^\kappa(v))|^2 =: a^2 \mathbb{E}\xi'_n.$$

Since G' is assumed continuous, it is also uniformly continuous on any finite interval. Since the process B^κ is Hölder continuous (a.s.), it follows that $\xi'_n \to 0$ a.s. Moreover, $\xi_n \leq 4\sup_{|u|\leq a}|G'(B^\kappa(u))|^2$, where the bound is integrable by (7.1.48). Hence, by the dominated convergence theorem, $\mathbb{E}\|D_1\|^2_{L^2[0,a]^2} \to 0$. Similarly,

$$\mathbb{E}\|D_2\|^2_{L^2[0,a]^2} \leq \mathbb{E}\sup_{|u|\leq a}|G'(B^\kappa(u))|^2 \int_0^a\int_0^a \Big|\sum_{k=1}^n 1_{\{t_{k-1}\leq v<u<t_k\}}\Big|^2 dudv$$

$$= \mathbb{E}\sup_{|u|\leq a}|G'(B^\kappa(u))|^2 \sum_{k=1}^n \int_0^a\int_0^a 1_{\{t_{k-1}\leq v<u<t_k\}}dudv$$

$$= \frac{1}{2}\mathbb{E}\sup_{|u|\leq a}|G'(B^\kappa(u))|^2 \sum_{k=1}^n (t_k - t_{k-1})^2 \leq \frac{a|\pi_n|}{2}\mathbb{E}\sup_{|u|\leq a}|G'(B^\kappa(u))|^2 \to 0.$$

Finally, we show the convergence of $S_{n,2}$. By using $(DG(B^\kappa(t_{k-1})))(u) = G'(B^\kappa(t_{k-1}))1_{[0,t_{k-1})}(u)$, we have

$$S_{n,2} = \sum_{k=1}^n (G'(B^\kappa(t_{k-1}))1_{[0,t_{k-1})}(u), 1_{[t_{k-1},t_k)})_{\mathcal{H}} = \sum_{k=1}^n G'(B^\kappa(t_{k-1}))(1_{[0,t_{k-1})}(u), 1_{[t_{k-1},t_k)})_{\mathcal{H}}$$

$$= \sum_{k=1}^n G'(B^\kappa(t_{k-1}))\mathbb{E}B^\kappa(t_{k-1})(B^\kappa(t_k) - B^\kappa(t_{k-1}))$$

$$= \sum_{k=1}^n G'(B^\kappa(t_{k-1}))\frac{1}{2}\Big(t_k^{2\kappa+1} - t_{k-1}^{2\kappa+1} - (t_k - t_{k-1})^{2\kappa+1}\Big)$$

$$= \frac{1}{2} \sum_{k=1}^{n} G'(B^\kappa(t_{k-1}))\left(t_k^{2\kappa+1} - t_{k-1}^{2\kappa+1}\right) - \frac{1}{2} \sum_{k=1}^{n} G'(B^\kappa(t_{k-1}))(t_k - t_{k-1})^{2\kappa+1}$$
$$=: S_{n,2,1} + S_{n,2,2}.$$

It remains to observe that $S_{n,2,1} \to \frac{2\kappa+1}{2} \int_0^a G'(B^\kappa(u))u^{2\kappa} du$ a.s. and that

$$S_{n,2,2} \leq \frac{|\pi_n|^{2\kappa}}{2} \sum_{k=1}^{n} |G'(B^\kappa(t_{k-1}))|(t_k - t_{k-1}) \to 0 \cdot \int_0^a |G'(B^\kappa(u))| du = 0. \qquad \square$$

Combining (7.1.40) and Proposition 7.1.10 leads to the following Itô's formula for FBM and the divergence integral in the case $\kappa > 0$.

Corollary 7.1.12 (*Itô's formula for FBM with $\kappa \in (0, 1/2)$.*) Let $a > 0$, $\kappa \in (0, 1/2)$ and $F \in C^2(\mathbb{R})$ be such that

$$|F''(x)| \leq Ce^{px^2}, \quad x \in \mathbb{R}, \tag{7.1.49}$$

for some constants $C > 0$ and $p < (4a^{2\kappa+1})^{-1}$. Then, for $t \in [0, a]$,

$$F(B^\kappa(t)) = F(B^\kappa(0)) + \int_0^t F'(B^\kappa(u))dB^\kappa(u) + \frac{2\kappa+1}{2} \int_0^t F''(B^\kappa(u))u^{2\kappa} du. \tag{7.1.50}$$

The relation (7.1.50) is the most basic Itô's formula for FBM. Its extensions to the case $F(t, B^\kappa(t))$ can be derived easily from the results above (Exercise 7.3), and to the case $F(X(t))$ with $X(t) = \int_0^t Y(u)dB^\kappa(u)$ and a random process Y can be found in Alòs and Nualart [20] (see also Theorem 5.2.1 in Nualart [769]).

Example 7.1.13 (*Itô's formula for geometric FBM*) By taking $F(x) = e^{bx}$ with $b \in \mathbb{R}$ in Corollary 7.1.12 and also using (7.1.40), we obtain that

$$e^{bB^\kappa(t)} = 1 + b \int_0^t e^{bB^\kappa(u)} dB^\kappa(u) + \frac{(2\kappa+1)b^2}{2} \int_0^t e^{bB^\kappa(u)} u^{2\kappa+1} du$$
$$= 1 + b\,(RS) \int_0^t e^{bB^\kappa(u)} dB^\kappa(u).$$

By taking $F(t, x) = e^{bx+ct}$ with $b, c \in \mathbb{R}$ in Exercise 7.3, we also have

$$e^{bB^\kappa(t)+ct} = 1 + c \int_0^t e^{bB^\kappa(u)+cu} du + b \int_0^t e^{bB^\kappa(u)+cu} dB^\kappa(u)$$
$$+ \frac{(2\kappa+1)b^2}{2} \int_0^t e^{bB^\kappa(u)+cu} u^{2\kappa+1} du$$
$$= 1 + c \int_0^t e^{bB^\kappa(u)+cu} du + b\,(RS) \int_0^t e^{bB^\kappa(u)+cu} dB^\kappa(u).$$

When $\kappa < 0$, an Itô's formula for FBM is stated below. The difference from the case $\kappa > 0$ is that, for the same Itô's formula to hold, the divergence integral with respect to FBM should be considered in the larger domain $\text{Dom}^* \delta_a^\kappa$ (see Definition 7.1.8).

Proposition 7.1.14 (*Itô's formula for FBM with $\kappa \in (-1/2, 0)$.*) *Let $\kappa \in (-1/2, 0)$ and suppose that the function $F \in C^2(\mathbb{R})$ satisfies the condition (7.1.49). Then, for any $t \in [0, a]$, the process $\{F'(B^\kappa(u))1_{[0,t)}(u)\}$ belongs to Dom* δ_a^κ, and the Itô's formula (7.1.50) holds.*

Proposition 7.1.14 is stated here without proof. The proof follows the approach found in Example 7.1.9, which corresponds to the choice $F(x) = x^2$ in the Itô's formula (7.1.50). See Theorem 5.2.2 in Nualart [769].

7.2 Applications of Stochastic Integration

We present here several applications of the Malliavin calculus and stochastic integration for FBM. In Section 7.2.1, we consider stochastic differential equations driven by FBM. In Section 7.2.2, we consider the smoothness of laws related to FBM. Section 7.2.3 concerns numerical solutions of stochastic differential equations driven by FBM. In Section 7.2.4, the convergence to a normal law of quantities related to FBM is proved using Stein's method. Further comments on these and other applications can be found in Section 7.4 with the notes to this chapter. In particular, we discuss there extensions of the applications found in Section 6.4.

7.2.1 Stochastic Differential Equations Driven by FBM

We consider here stochastic differential equations (SDEs) driven by FBM B^κ: for $x_0 \in \mathbb{R}$ and $t \in [0, a]$,

$$X(t) = x_0 + \int_0^t b(X(s))ds + (RS)\int_0^t \sigma(X(s))dB^\kappa(s), \qquad (7.2.1)$$

where b and σ are real functions, and (RS) stands for the (pathwise) Riemann–Stieltjes integral (see (7.1.37)).

Suppose that $\kappa \in (0, 1/2)$.[4] If σ has a bounded partial derivative which is λ–Hölder continuous with $\lambda > (\kappa + 1/2)^{-1} - 1$, and b is Lipschitz, then there is a unique solution to the SDE (7.2.1) which is $(\kappa+1/2-\epsilon)$–Hölder continuous of order for any $\epsilon > 0$ (Nualart and Răşcanu [773]). The sufficient conditions on σ and b above are verified when $\sigma \in C_b^2(\mathbb{R})$ and $b \in C_b^1(\mathbb{R})$, where $C_b^k(\mathbb{R})$, $k \geq 1$, is the space of k–times continuously differentiable functions on $\mathbb{R}$ with the first k derivatives being bounded.

Moreover, the solution $X(t)$ to (7.2.1) can be expressed through the so-called *Doss–Sussmann transformation* as

$$X(t) = \phi(B^\kappa(t), Y(t)), \qquad (7.2.2)$$

where ϕ is characterized by

$$\frac{\partial}{\partial x}\phi(x, y) = \sigma(\phi(x, y)), \quad \phi(0, y) = y, \quad x, y \in \mathbb{R}, \qquad (7.2.3)$$

---
[4] The case $\kappa \in (-1/2, 0)$ could be considered if the integral with respect to FBM in (7.2.1) is interpreted in a suitable sense (which is different from RS). See Nourdin and Simon [765].

and Y is the solution to the random ODE

$$Y(t) = x_0 + \int_0^t a(B^\kappa(s), Y(s))ds \qquad (7.2.4)$$

with

$$a(x, y) = \frac{b(\phi(x, y))}{\frac{\partial}{\partial y}\phi(x, y)} = b(\phi(x, y)) \exp\left\{-\int_0^x \sigma'(\phi(u, y))du\right\} \qquad (7.2.5)$$

for $x, y \in \mathbb{R}$ (Klingenhöfer and Zähle [567]). The second relation in (7.2.5), which reduces to

$$\frac{\partial}{\partial y}\phi(x, y) = \exp\left\{\int_0^x \sigma'(\phi(z, y))dz\right\}, \qquad (7.2.6)$$

is stated as Exercise 7.4.

7.2.2 Regularity of Laws Related to FBM

The Malliavin calculus was developed originally to address questions about the regularity (smoothness) of laws related to Brownian motion and, more generally, Gaussian processes. A number of fundamental results are known in this direction. Consider the general framework of the Malliavin calculus recalled in Appendix C and suppose that a random variable $F \in \mathbb{D}^{1,2}$ is such that $DF/\|DF\|_{\mathcal{H}}^2 \in \text{Dom}\,\delta$. Then the law of F has a continuous and bounded density with respect to the Lebesgue measure on $\mathbb{R}$ given by

$$p(x) = \mathbb{E}\left(1_{\{F>x\}}\delta\left(\frac{DF}{\|DF\|_{\mathcal{H}}^2}\right)\right) \qquad (7.2.7)$$

(Proposition 2.1.1 in Nualart [769]; sufficient conditions are given in Exercise 2.1.1 of Nualart [769]). For example, F has a density with respect to the Lebesgue measure on $\mathbb{R}$ if

$$F \in \mathbb{D}_{\text{loc}}^{1,2} \quad \text{and} \quad \|DF\|_{\mathcal{H}} > 0 \text{ a.s.} \qquad (7.2.8)$$

(Theorem 2.1.3 in Nualart [769]). See Chapter 2 of Nualart [769] for more information.

We shall illustrate the above approach through (7.2.8) on a particular functional of FBM, following the work of Nourdin and Simon [765], and Nualart and Saussereau [774]. See Section 7.4 for further notes. The particular application of Nourdin and Simon [765] concerns the solutions to stochastic differential equations (7.2.1) driven by FBM B^κ. We are interested in whether the solution $X(t)$ to the SDE (7.2.1) given by (7.2.2)–(7.2.5) has a *density*, and shall use the criterion (7.2.8).

The following proposition involves the first condition in (7.2.8) and also provides an expression for the derivative of the solution $X(t)$.

Proposition 7.2.1 *Let $\kappa \in (0, 1/2)$ and $b, \sigma \in C_b^3(\mathbb{R})$. Then, the unique solution $X(t)$ to (7.2.1) satisfies $X(t) \in \mathbb{D}_{\text{loc}}^{1,2}$, $t \in [0, a]$, and*

$$D_s(X(t)) = \sigma(X(s)) \exp\left\{\int_s^t b'(X(u))du + (RS)\int_s^t \sigma'(X(u))dB^\kappa(u)\right\}, \qquad (7.2.9)$$

$0 \le s \le t \le a$.

Proof The proof of $X(t) \in \mathbb{D}_{\text{loc}}^{1,2}$ can be found in Nualart and Saussereau [774], Theorem 6, and will not be given here.[5] We shall now prove the relation (7.2.9).

Step 1: By (7.2.2)–(7.2.5),

$$X(t) = \phi(B^\kappa(t), Y(t)), \tag{7.2.10}$$

where ϕ is given by (7.2.3) and Y satisfies

$$Y(s) = x_0 + \int_0^s L(u)^{-1} b(\phi(B^\kappa(u), Y(u))) du = x_0 + \int_0^s L(u)^{-1} b(X(u)) du \tag{7.2.11}$$

with

$$L(u) = \frac{\partial \phi}{\partial y}(B^\kappa(u), Y(u)) = \exp\left\{ \int_0^{B^\kappa(u)} \sigma'(\phi(z, Y(u))) dz \right\}. \tag{7.2.12}$$

Fix $s \in [0, t]$. By the chain rule (C.2.7) in Appendix C, we have

$$\begin{aligned} D_s(X(t)) &= \frac{\partial \phi}{\partial x}(B^\kappa(t), Y(t)) D_s(B^\kappa(t)) + \frac{\partial \phi}{\partial y}(B^\kappa(t), Y(t)) D_s(Y(t)) \\ &= \sigma(\phi(B^\kappa(t), Y(t))) 1_{[0,t]}(s) + L(t) D_s(Y(t)) = \sigma(X(t)) + L(t) D_s(Y(t)), \end{aligned} \tag{7.2.13}$$

where we used (7.2.3) and (7.2.12). This can also be expressed as

$$N(t) = L(t)^{-1} \sigma(X(t)) + D_s(Y(t)), \tag{7.2.14}$$

where

$$N(t) = L(t)^{-1} D_s(X(t)). \tag{7.2.15}$$

We will find next expressions for the two terms in (7.2.14); that is, $L(t)^{-1} \sigma(X(t))$ and $D_s(Y(t))$ where $s < t$. We will use them to obtain a differential equation for $N(t)$ from which (7.2.9) will follow.

Step 2: We need a generalization of the formula (7.1.40). By the same argument as for (7.1.40), if $F : \mathbb{R}^2 \to \mathbb{R}$ is twice continuously differentiable and Y is a process with finite variation, we have

$$\begin{aligned} F(B^\kappa(t), Y(t)) &= F(B^\kappa(0), Y(0)) \\ &+ (RS) \int_0^t \frac{\partial F}{\partial x}(B^\kappa(s), Y(s)) dB^\kappa(s) + \int_0^t \frac{\partial F}{\partial y}(B^\kappa(s), Y(s)) dY(s). \end{aligned} \tag{7.2.16}$$

---

[5] The idea is to view FBM as defined on the canonical space $\Omega = C[0, a] = \{w : w = w(t)$ is continuous for $t \in [0, a]\}$. One then shows the Fréchet differentiability of the solution $X(\omega)$ with respect to the driving function ω, and uses this fact to deduce the so-called differentiability of the solution in the direction of the reproducing kernel Hilbert space (RKHS). Finally one applies, for example, Proposition 4.1.3 in Nualart [769] which states that such differentiable functionals belong to $\mathbb{D}_{\text{loc}}^{1,2}$.

The differentiability in the direction of the RKHS means the following. The RKHS $\mathbb{H}_a(\Gamma^\kappa)$ of FBM is defined in Section 6.3.7 and satisfies $\mathbb{H}_a(\Gamma^\kappa) \subset C[0, a]$. The differentiability of the solution $X(w)$ in the direction of the RKHS means the Fréchet differentiability of the mapping $\mathbb{H}_a(\Gamma^\kappa) \ni g \mapsto X(w + g)$, for all $w \in C[0, a]$.

By applying (7.2.16) to $L(t) = \frac{\partial \phi}{\partial y}(B^K(t), Y(t))$, we obtain

$$L(t) = \frac{\partial \phi}{\partial y}(0, x_0) + (RS)\int_0^t \frac{\partial^2 \phi}{\partial x \partial y}(B^K(u), Y(u))dB^K(u) + \int_0^t \frac{\partial^2 \phi}{\partial y^2}(B^K(u), Y(u))dY(u). \tag{7.2.17}$$

Note from (7.2.6) that $\frac{\partial \phi}{\partial y}(0, x_0) = 1$ and

$$\frac{\partial^2 \phi}{\partial x \partial y}(x, y) = \sigma'(\phi(x, y))\frac{\partial \phi}{\partial y}(x, y), \quad \frac{\partial^2 \phi}{\partial y^2}(x, y) = \int_0^x \sigma''(\phi(z, y))\frac{\partial \phi}{\partial y}(z, y)dz \frac{\partial \phi}{\partial y}(x, y). \tag{7.2.18}$$

It follows that

$$\frac{\partial^2 \phi}{\partial x \partial y}(B^K(u), Y(u)) = L(u)\sigma'(X(u)) \tag{7.2.19}$$

by (7.2.10) and

$$\frac{\partial^2 \phi}{\partial y^2}(B^K(u), Y(u)) = L(u)\int_0^{B^K(u)} \sigma''(\phi(z, Y(u)))\frac{\partial \phi}{\partial y}(z, Y(u))dz =: M(u), \tag{7.2.20}$$

where $L(u)$ is defined in (7.2.12). Hence, by (7.2.17),

$$L(t) = 1 + (RS)\int_0^t L(u)\sigma'(X(u))dB^K(u) + \int_0^t M(u)dY(u). \tag{7.2.21}$$

We also have the identity

$$L(t)^{-1}\sigma(X(t)) = L(s)^{-1}\sigma(X(s)) + \int_s^t \frac{d}{du}\Big(L(u)^{-1}\sigma(X(u))\Big)du$$

$$= L(s)^{-1}\sigma(X(s)) - \int_s^t L(u)^{-2}\sigma(X(u))dL(u) + \int_s^t L(u)^{-1}\sigma'(X(u))dX(u)$$

$$= L(s)^{-1}\sigma(X(s)) - (RS)\int_s^t L(u)^{-1}\sigma(X(u))\sigma'(X(u))dB^K(u)$$

$$\quad - \int_s^t L(u)^{-2}M(u)\sigma(X(u))dY(u)$$

$$\quad + \int_s^t L(u)^{-1}\sigma'(X(u))b(X(u))du + (RS)\int_s^t L(u)^{-1}\sigma'(X(u))\sigma(X(u))dB^K(u)$$

$$= L(s)^{-1}\sigma(X(s)) + \int_s^t \Big(\sigma'(X(u)) - L(u)^{-2}M(u)\sigma(X(u))\Big)dY(u), \tag{7.2.22}$$

where we used (7.2.21) and (7.2.1), and also (7.2.11) in the last line.

Step 3: Now consider the second term $D_s(Y(t))$ in (7.2.14). We have from (7.2.11),

$$D_s(Y(t)) = \int_s^t D_s\Big(L(u)^{-1}b(X(u))\Big)du. \tag{7.2.23}$$

Observe for the integrand that, by the chain rule,

$$D_s\Big(L(u)^{-1}b(X(u))\Big) = L(u)^{-1}b'(X(u))D_s(X(u)) - L(u)^{-2}b(X(u))D_s(L(u))$$

$$= N(u)b'(X(u)) - L(u)^{-1}D_s(L(u))\frac{dY(u)}{du}, \tag{7.2.24}$$

7.2 Applications of Stochastic Integration

since $L(u)^{-1}D_s(X(u)) = N(u)$ by the definition (7.2.15) and $dY(u) = L(u)^{-1}b(X(u))du$ by (7.2.11). To find an expression for $D_s(L(u))$, recall from (7.2.12) that $L(u) = \frac{\partial \phi}{\partial y}(B^K(u), Y(u))$. Then, by the chain rule again,

$$D_s(L(u)) = \frac{\partial^2 \phi}{\partial x \partial y}(B^K(u), Y(u))D_s(B^K(u)) + \frac{\partial^2 \phi}{\partial y^2}(B^K(u), Y(u))D_s(Y(u))$$
$$= L(u)\sigma'(X(u)) + M(u)(N(u) - L(u)^{-1}\sigma(X(u))), \qquad (7.2.25)$$

where we used (7.2.19), $D_s(B^K(u)) = 1_{[0,u)}(s) = 1$ for $u > s$, (7.2.20) and $D_s(Y(u)) = N(u) - L(u)^{-1}\sigma(X(u))$ in view of (7.2.14). Substituting (7.2.25) into (7.2.24), and then (7.2.24) into (7.2.23) yields

$$D_s(Y(t)) = \int_s^t N(u)b'(X(u))du$$
$$- \int_s^t \left(L(u)^{-1}M(u)N(u) + \sigma'(X(u)) - L(u)^{-2}M(u)\sigma(X(u)) \right) dY(u).$$
$$(7.2.26)$$

Step 4: Finally, we turn back to (7.2.14) whose right-hand side is the sum of $L(t)^{-1}\sigma(X(t))$ and $D_s(Y(t))$. Substituting the obtained expressions (7.2.22) for $L(t)^{-1}\sigma(X(t))$ and (7.2.26) for $D_s(Y(t))$ yields, after cancelations,

$$N(t) = L(s)^{-1}\sigma(X(s)) + \int_s^t N(u)b'(X(u))du - \int_s^t L(u)^{-1}M(u)N(u)dY(u)$$
$$= L(s)^{-1}\sigma(X(s)) + \int_s^t N(u)b'(X(u))du - \int_s^t L(u)^{-2}M(u)b(X(u))N(u)du$$
$$= L(s)^{-1}\sigma(X(s)) + \int_s^t \left(b'(X(u)) - L(u)^{-2}M(u)b(X(u)) \right) N(u)du, \qquad (7.2.27)$$

by using $dY(u) = L(u)^{-1}b(X(u))du$ by (7.2.11). The equation (7.2.27) has then the explicit solution given by

$$N(t) = L(s)^{-1}\sigma(X(s)) \exp \left\{ \int_s^t (b'(X(u)) - L(u)^{-2}M(u)b(X(u)))du \right\}. \qquad (7.2.28)$$

By using (7.2.15), this implies that

$$D_s(X(t)) = \sigma(X(s)) \exp \left\{ \int_s^t b'(X(u))du \right\} \frac{L(t)}{L(s)} \exp \left\{ -\int_s^t L(u)^{-1}M(u)dY(u) \right\}. \qquad (7.2.29)$$

By using the fact that the solution to (7.2.21) can be expressed as

$$L(t) = \exp \left\{ (RS) \int_0^t \sigma'(X(u))dB^K(u) + \int_0^t L(u)^{-1}M(u)dY(u) \right\} \qquad (7.2.30)$$

(Exercise 7.5), we can write

$$\frac{L(t)}{L(s)} \exp \left\{ -\int_s^t L(u)^{-1}M(u)dY(u) \right\} = \exp \left\{ (RS) \int_s^t \sigma'(X(u))dB^K(u) \right\}.$$

Together with (7.2.29), this yields the desired expression (7.2.9). □

The next result concerns the existence of a density of the solution $X(t)$ in (7.2.1) and Proposition 7.2.1. The following quantities will be used. Consider σ in (7.2.1), let $J = \sigma^{-1}\{0\}$ and let also int J be the interior of J. In the deterministic equation

$$x(t) = x_0 + \int_0^t b(x(s))ds, \qquad (7.2.31)$$

there is no term involving σ, in contrast to the equation (7.2.1) for $X(t)$. Note that since σ maps int J to 0, $\sigma(X(t))$ is 0 as long as $X(t) \in$ int J. This is the case, in particular, at the start time $t = 0$, where $X(0) = x_0$. Consider the deterministic time

$$t_x = \inf\{t \geq 0 : x(t) \notin \text{int } J\}. \qquad (7.2.32)$$

One can show that

$$t_x = \inf\{t \geq 0 : X(t) \notin \text{int } J\} \quad \text{a.s.} \qquad (7.2.33)$$

(Exercise 7.6 below, and Lemma 2 in Nourdin and Simon [765]).

Theorem 7.2.2 *Suppose $\kappa \in (0, 1/2)$ and $b, \sigma \in C_b^3(\mathbb{R})$. Then, $X(t)$ has a density with respect to the Lebesgue measure if and only if $t > t_x$.*

Proof Suppose first that $t > t_x$. In view of Proposition 7.2.1, we only need to check that the second relation in (7.2.8) holds with $F = X(t)$; that is, $\|DX(t)\|_{\mathcal{H}} > 0$ a.s. Since $t > t_x$, by using (7.2.33) and the a.s. continuity of $s \mapsto \sigma(X(s))$, the function $s \mapsto \sigma(X(s))$ does not vanish on a subset of $[0, t]$ a.s. Then, by (7.2.9), the function $s \mapsto D_s(X(t))$ does not vanish on a subset of $[0, t]$ a.s. either. By Corollary 6.1.11, the same then holds for the function $s \mapsto s^{-\kappa}(I_{a-}^{\kappa} u^{\kappa} D_u(X(t)))(s)$ and hence, by using (7.1.3) and (6.3.12),

$$\|DX(t)\|_{\mathcal{H}}^2 = \sigma_{\kappa}^2 \int_0^a s^{-2\kappa}\left((I_{a-}^{\kappa} u^{\kappa} D_u(X(t)))(s)\right)^2 ds > 0 \quad \text{a.s.}$$

Conversely and arguing by contradiction, suppose that $t \leq t_x$. Then, by uniqueness, $X(t) = x(t)$ a.s. where $x(t)$ satisfies (7.2.31). Since $x(t)$ is deterministic, $X(t)$ does not have a density. □

7.2.3 Numerical Solutions of SDEs Driven by FBM

We use here the Malliavin calculus to study the convergence of some numerical approximations of SDEs driven by FBM. Our goal is not to present a thorough analysis but rather to give an idea of the problem considered. The approach follows the work of Hu, Liu, and Nualart [495]. Several other works in this direction, usually also involving the Malliavin calculus, are discussed briefly in Section 7.4.

As in Section 7.2.2, consider the SDE on $\mathbb{R}$ given by

$$X(t) = x_0 + \int_0^t b(X(s))ds + (RS)\int_0^t \sigma(X(s))dB^{\kappa}(s), \qquad (7.2.34)$$

where B^{κ} is FBM and $b, \sigma : \mathbb{R} \to \mathbb{R}$ are continuous functions. Suppose $\kappa \in (0, 1/2)$ throughout the section. See Section 7.2.2 for sufficient conditions on b and σ for the existence and uniqueness of the solution $X(t)$ to the SDE (7.2.34).

7.2 Applications of Stochastic Integration

For $n \geq 1$, consider uniform partitions of the interval $[0, a]$ given by $t_k = ka/n$, $k = 0, \ldots, n$. Define an approximation to the SDE (7.2.34) as: for $t \in [t_k, t_{k+1})$,

$$X_n(t) = X_n(t_k) + b(X_n(t_k))(t - t_k) + \sigma(X_n(t_k))(B^\kappa(t) - B^\kappa(t_k))$$
$$+ \frac{1}{2}\sigma'(X_n(t_k))\sigma(X_n(t_k))(t - t_k)^{2\kappa+1}. \tag{7.2.35}$$

The presence of the last term in (7.2.35) ensures a faster convergence rate (see also Remark 7.2.5 below and, for motivation, consult the introduction of Hu et al. [495]). Without the last term, the approximation (7.2.35) is known as the *Euler scheme*. When $\kappa = 0$, the Euler (also known as Euler-Maruyama) scheme converges to the solution of the SDE (7.2.34) where the integral with respect to BM is in the Itô sense. Hu et al. [495] refer to (7.2.35) as a *modified Euler scheme*.

The next theorem provides the rate of convergence for the approximation scheme $X_n(t)$. See the remarks following the theorem for additional comments. Note that we suppose for simplicity that $b \equiv 0$ in (7.2.34) and (7.2.35). Recall also the notation $C_b^k(\mathbb{R})$, $k \geq 1$, from Section 7.2.2.

Theorem 7.2.3 *Suppose that $\kappa \in (0, 1/2)$, $b \equiv 0$ and $\sigma \in C_b^4(\mathbb{R})$. Let X and X_n be the solutions to the equations (7.2.34) and (7.2.35), respectively. Then, for any $p \geq 1$, there is a constant C which does not depend on n, such that*

$$\sup_{t \in [0,a]} \left(\mathbb{E}|X(t) - X_n(t)|^p \right)^{1/p} \leq C\gamma_n^{-1}, \tag{7.2.36}$$

where

$$\gamma_n^{-1} = \begin{cases} n^{-1}, & \text{if } \kappa \in (1/4, 1/2), \\ n^{-1}\sqrt{\log n}, & \text{if } \kappa = 1/4, \\ n^{-2\kappa - 1/2}, & \text{if } \kappa \in (0, 1/4). \end{cases} \tag{7.2.37}$$

Remark 7.2.4 The rate γ_n in the theorem above is exact for the error process $X(t) - X_n(t)$ in the sense that the sequence of the processes $\gamma_n(X(t) - X_n(t))$, $t \in [0, a]$, has a weak limit (Theorems 6.1 and 8.1 in Hu et al. [495]).

Remark 7.2.5 The Euler scheme is the approximation (7.2.35) without the last term. It has a slower rate of convergence $n^{2\kappa}$ for all $\kappa \in (0, 1/2)$ (Theorem 10.1 and Corollary 10.2 in Hu et al. [495]).

Proof Set $Y_n(t) = X(t) - X_n(t)$, $t \in [0, a]$. Define a function η by setting

$$\eta(s) = t_k \quad \text{if } s \in [t_k, t_{k+1}) \tag{7.2.38}$$

and recall that $b \equiv 0$. Observe that $(\kappa + 1/2) \int_0^t s^{2\kappa} ds = \frac{1}{2} t^{2\kappa+1}$ and hence

$$Y_n(t) = (RS) \int_0^t \left(\sigma(X(s)) - \sigma(X_n(\eta(s))) \right) dB^\kappa(s)$$
$$- (\kappa + 1/2) \int_0^t \sigma'(X_n(\eta(s)))\sigma(X_n(\eta(s)))(s - \eta(s))^{2\kappa} ds$$

$$= (RS) \int_0^t \Big(\sigma(X(s)) - \sigma(X_n(s)) + \sigma(X_n(s)) - \sigma(X_n(\eta(s)))\Big) dB^\kappa(s)$$

$$- (\kappa + 1/2) \int_0^t \sigma'(X_n(\eta(s)))\sigma(X_n(\eta(s)))(s - \eta(s))^{2\kappa} ds.$$

By using the relation $\sigma(y) - \sigma(x) = \int_0^1 \sigma'(\theta y + (1-\theta)x)(y-x)d\theta$, we have

$$\sigma(X(s)) - \sigma(X_n(s)) = \int_0^1 \sigma'\Big(\theta X(s) + (1-\theta)X_n(s)\Big)Y_n(s)d\theta,$$

$$\sigma(X_n(s)) - \sigma(X_n(\eta(s))) = \int_0^1 \sigma'\Big(\theta X_n(s) + (1-\theta)X_n(\eta(s))\Big)(X_n(s) - X_n(\eta(s)))d\theta.$$

Note that, by (7.2.35),

$$X_n(s) - X_n(\eta(s)) = \sigma(X_n(\eta(s)))(B^\kappa(s) - B^\kappa(\eta(s))) + \frac{1}{2}(\sigma' \cdot \sigma)(X_n(\eta(s)))(s - \eta(s))^{2\kappa+1},$$

where we write $(\sigma' \cdot \sigma)(x)$ for $\sigma'(x)\sigma(x)$. This leads to

$$Y_n(t) = (RS) \int_0^t \sigma_1(s)Y_n(s)dB^\kappa(s) + (RS) \int_0^t \sigma_2(s)(B^\kappa(s) - B^\kappa(\eta(s)))dB^\kappa(s)$$

$$+ (RS) \int_0^t \sigma_3(s)(s - \eta(s))^{2\kappa+1} dB^\kappa(s) - (\kappa + 1/2) \int_0^t \sigma_4(s)(s - \eta(s))^{2\kappa} ds,$$

(7.2.39)

where

$$\sigma_1(s) = \int_0^1 \sigma'\Big(\theta X(s) + (1-\theta)X_n(s)\Big)d\theta,$$

$$\sigma_2(s) = \int_0^1 \sigma'\Big(\theta X_n(s) + (1-\theta)X_n(\eta(s))\Big)d\theta\, \sigma(X_n(\eta(s))),$$

$$\sigma_3(s) = \frac{1}{2}\int_0^1 \sigma'\Big(\theta X_n(s) + (1-\theta)X_n(\eta(s))\Big)d\theta\, (\sigma' \cdot \sigma)(X_n(\eta(s))),$$

$$\sigma_4(s) = (\sigma' \cdot \sigma)(X_n(\eta(s))). \qquad (7.2.40)$$

Note that $Y_n(s)$ enters linearly in the first integral of (7.2.39). The first integral can be "eliminated" in the following standard way. The idea is to replace $Y_n(t)$ by $Y_n(t)\Lambda_n(t)^{-1}$, where

$$\Lambda_n(t) = 1 + (RS) \int_0^t \sigma_1(s)\Lambda_n(s)dB^\kappa(s). \qquad (7.2.41)$$

The solution to this equation is given by

$$\Lambda_n(t) = \exp\left\{(RS) \int_0^t \sigma_1(s)dB^\kappa(s)\right\} \qquad (7.2.42)$$

and also satisfies

$$\Lambda_n(t)^{-1} = 1 - (RS) \int_0^t \sigma_1(s)\Lambda_n(s)^{-1} dB^\kappa(s) \qquad (7.2.43)$$

(Exercise 7.8).[6] Writing $d(Y_n(t)\Lambda_n(t)^{-1}) = \Lambda_n(t)^{-1}dY_n(t) + Y_n(t)d(\Lambda_n(t)^{-1})$ and substituting the expressions for $dY_n(t)$ and $d(\Lambda_n(t)^{-1})$ based on (7.2.39) and (7.2.43), respectively, leads to, after cancelation,

$$d(Y_n(t)\Lambda_n(t)^{-1}) = \Lambda_n(t)^{-1}\sigma_2(t)(B^\kappa(t) - B^\kappa(\eta(t)))dB^\kappa(t)$$
$$+ \Lambda_n(t)^{-1}\sigma_3(t)(t - \eta(t))^{2\kappa+1}dB^\kappa(t)$$
$$- (\kappa + 1/2)\Lambda_n(t)^{-1}\sigma_4(t)(t - \eta(t))^{2\kappa}dt,$$

that is,

$$Y_n(t) = \Lambda_n(t) \times \Big\{ (RS)\int_0^t \Lambda_n(s)^{-1}\sigma_2(s)(B^\kappa(s) - B^\kappa(\eta(s)))dB^\kappa(s)$$
$$+ (RS)\int_0^t \Lambda_n(s)^{-1}\sigma_3(s)(s - \eta(s))^{2\kappa+1}dB^\kappa(s)$$
$$- (\kappa + 1/2)\int_0^t \Lambda_n(s)^{-1}\sigma_4(s)(s - \eta(s))^{2\kappa}ds \Big\}. \quad (7.2.44)$$

This can be expressed further as

$$Y_n(t) = E_0(t) + E_1(t) + E_2(t) + E_3(t), \quad (7.2.45)$$

where

$$E_0(t) = \Lambda_n(t)\,(RS)\int_0^t \Lambda_n(s)^{-1}\sigma_3(s)(s - \eta(s))^{2\kappa+1}dB^\kappa(s),$$

$$E_1(t) = \Lambda_n(t)\,(RS)\int_0^t \Big(\Lambda_n(s)^{-1}\sigma_2(s) - \Lambda_n(\eta(s))^{-1}\sigma_4(s)\Big)(B^\kappa(s) - B^\kappa(\eta(s)))dB^\kappa(s),$$

$$E_2(t) = \Lambda_n(t)\,(RS)\int_0^t \Lambda_n(\eta(s))^{-1}\sigma_4(s)(B^\kappa(s) - B^\kappa(\eta(s)))dB^\kappa(s)$$
$$- (\kappa + 1/2)\Lambda_n(t)\int_0^t \Lambda_n(\eta(s))^{-1}\sigma_4(s)(s - \eta(s))^{2\kappa}ds,$$

$$E_3(t) = (\kappa + 1/2)\Lambda_n(t)\int_0^t \Big(\Lambda_n(\eta(s))^{-1} - \Lambda_n(s)^{-1}\Big)\sigma_4(s)(s - \eta(s))^{2\kappa}ds. \quad (7.2.46)$$

Observe that there is no integral with σ_1 anymore in (7.2.44) and (7.2.45)–(7.2.46). It is then enough to show that

$$\Big(\mathbb{E}|E_j(t)|^p\Big)^{1/p} \le C\gamma_n^{-1}, \quad j = 0, 1, 2, 3. \quad (7.2.47)$$

To simplify the notation, we only consider below the case $p = 1$ and also write $\|Z\|_q = (\mathbb{E}|Z|^q)^{1/q}$ for a random variable Z.

Bounding $\|E_0(t)\|_1$: Observe that

$$\|E_0(t)\|_1 \le \|\Lambda_n(t)\|_2 \Big\|(RS)\int_0^t \Lambda_n(s)^{-1}\sigma_3(s)(s - \eta(s))^{2\kappa+1}dB^\kappa(s)\Big\|_2. \quad (7.2.48)$$

---

[6] Note informally that $d\Lambda_n(t)^{-1} = -\Lambda_n(t)^{-2}d\Lambda_n(t) = -\Lambda_n(t)^{-2}\sigma_1(s)\Lambda_n(t)dB^\kappa(s) = -\sigma_1(s)\Lambda_n(t)^{-1}dB^\kappa(s)$.

By Proposition 7.2.7 below, $|\Lambda_n(t)| \leq K \exp\{K\|B^\kappa\|_\beta^{1/\beta}\}$ for all $t \in [0, a]$. By Fernique's inequality (e.g., Bogachev [167], Theorem 2.8.5; the inequality is the same as (7.1.47) but with the supremum norm $\|x\|_\infty = \sup_{u\in[0,a]} |x(u)|$ replaced by the Hölder norm $\|x\|_\beta$ defined in (7.2.59)) and since $\beta > 1/2$ (or $2\beta > 1$), the random variable $\exp\{K\|B^\kappa\|_\beta^{1/\beta}\}$ has finite moments of any order. This implies that $\|\Lambda_n(t)\|_2 \leq C < \infty$ for all $t \in [0, a]$.

To bound the $L^2(\Omega)$–norm of the Riemann–Stieltjes integral in (7.2.48), we use Proposition 7.2.8 below with $p = 2$, $F(t) = \Lambda_n(t)^{-1}\sigma_3(t)$ and $\nu = 2\kappa + 1$ to obtain

$$\left\|(RS)\int_0^t \Lambda_n(s)^{-1}\sigma_3(s)(s - \eta(s))^{2\kappa+1} dB^\kappa(s)\right\|_2 \leq Cn^{-2\kappa-1} t^{\kappa+1/2} F_{1,2}, \quad (7.2.49)$$

where $F_{1,2} = \sup_{s,t\in[0,a]}(\|F(t)\|_2 \vee \|D_s F(t)\|_2)$ as given in (7.2.64). Propositions 7.2.6 and 7.2.7 can now be used to show that $F_{1,2} < \infty$. Indeed, note that, since $\sigma \in C_b^4(\mathbb{R})$, $|F(t)| \leq C|\Lambda_n(t)^{-1}|$ so that $\|F(t)\|_2 \leq C < \infty$ by arguing as for $\Lambda_n(t)$ above using Proposition 7.2.7. Similarly, note that, by the chain rule and using the definition of $\sigma_3(t)$ in (7.2.40),

$$D_s F(t) = D_s(\Lambda_n(t)^{-1})\sigma_3(t) + \Lambda_n(t)^{-1} D_s(\sigma_3(t)) = D_s(\Lambda_n(t)^{-1})\sigma_3(t)$$
$$+ \frac{\Lambda_n(t)^{-1}}{2}\left\{\int_0^1 \sigma'(\theta X_n(t) + (1-\theta)X_n(\eta(t))) d\theta\, (\sigma'\cdot\sigma)'(X_n(\eta(t))) D_s X_n(\eta(t))\right.$$
$$\left.+ \int_0^1 \sigma''(\theta X_n(t) + (1-\theta)X_n(\eta(t)))\Big(\theta D_s X_n(t) + (1-\theta)D_s X_n(\eta(t))\Big) d\theta\, (\sigma'\cdot\sigma)(X_n(\eta(t)))\right\}.$$

By using Propositions 7.2.6 and 7.2.7, $|D_s F(t)| \leq C \exp\{C\|B^\kappa\|_\beta^{1/\beta}\}$. Arguing as for $\Lambda_n(t)$ above, this yields $\|D_s F(t)\|_2 \leq C < \infty$. Hence, as stated above, $F_{1,2} < \infty$.

Gathering the bounds above, we can conclude that

$$\|E_0(t)\|_1 \leq Cn^{-2\kappa-1} \quad (7.2.50)$$

and hence that (7.2.47) holds for $j = 0$, since $n^{-2\kappa-1} \leq \gamma_n^{-1}$.

Bounding $\|E_1(t)\|_1$: Since $\sigma_2(\eta(s)) = \int_0^1 \sigma'(\theta X_n(\eta(s)) + (1-\theta)X_n(\eta(s))) d\theta\, \sigma(X_n(\eta(s))) = \int_0^1 \sigma'(X_n(\eta(s))) d\theta\, \sigma(X_n(\eta(s))) = (\sigma'\cdot\sigma)(X_n(\eta(s))) = \sigma_4(s)$, note that

$$E_1(t) = \Lambda_n(t)\,(RS)\int_0^t \Big(\Lambda_n(s)^{-1}\sigma_2(s) - \Lambda_n(\eta(s))^{-1}\sigma_2(\eta(s))\Big)(B^\kappa(s) - B^\kappa(\eta(s))) dB^\kappa(s). \quad (7.2.51)$$

By applying Proposition 7.2.7 to bound $\Lambda_n(t)$ in (7.2.51) and Proposition 7.2.10 with $F(s) \equiv 1$, $G(s) = \Lambda_n(s)^{-1}\sigma_2(s)$ to bound the Riemann–Stieltjes integral in (7.2.51), we obtain that

$$|E_1(t)| \leq C_1 e^{C_1\|B^\kappa\|_\beta^{1/\beta}} \|G\|_\beta \|B^\kappa\|_\beta^2 n^{1-3\beta} t^\beta.$$

Since $\sigma \in C_b^4(\mathbb{R})$ by assumption, we have $|G(s)| \leq C|\Lambda_n(s)^{-1}|$ and bounding $\Lambda_n(s)^{-1}$ by using Proposition 7.2.7 leads to

$$|E_1(t)| \leq C_2 e^{C_2\|B^\kappa\|_\beta^{1/\beta}} n^{1-3\beta}.$$

7.2 Applications of Stochastic Integration

By using Fernique's inequality as for $\|E_0(t)\|_1$ above, we conclude that

$$\|E_1(t)\|_1 \leq Cn^{1-3\beta} \tag{7.2.52}$$

and hence that (7.2.47) holds for $j = 1$, since $n^{1-3\beta} \leq \gamma_n^{-1}$.

Bounding $\|E_2(t)\|_1$: Since $\sigma_4(s) = (\sigma' \cdot \sigma)(X_n(\eta(s)))$ in (7.2.40), write

$$E_2(t) = \Lambda_n(t) \sum_{k=0}^{[nt/a]} F_n(t_k)$$

$$\times \left((RS) \int_{t_k}^{t_{k+1} \wedge t} (B^\kappa(s) - B^\kappa(t_k)) dB^\kappa(s) - (\kappa + 1/2) \int_{t_k}^{t_{k+1} \wedge t} (s - t_k)^{2\kappa} ds \right), \tag{7.2.53}$$

where $F_n(t) = \Lambda_n(t)^{-1}(\sigma' \cdot \sigma)(X_n(t))$. By using the relation (7.1.45) between the Riemann-Stieltjes and divergence integrals, observe that

$$(RS) \int_{t_k}^{t_{k+1} \wedge t} (B^\kappa(s) - B^\kappa(t_k)) dB^\kappa(s) = (RS) \int_{t_k}^{t_{k+1} \wedge t} B^\kappa(1_{[t_k, s)}) dB^\kappa(s)$$

$$= \int_{t_k}^{t_{k+1} \wedge t} B^\kappa(1_{[t_k, s)}) dB^\kappa(s) + \kappa(2\kappa + 1) \int_{t_k}^{t_{k+1} \wedge t} \int_{t_k}^{t_{k+1} \wedge t} D_u B^\kappa(1_{[t_k, s)}) |u - s|^{2\kappa - 1} du ds$$

$$= \int_{t_k}^{t_{k+1} \wedge t} \int_{t_k}^{s} dB^\kappa(v) dB^\kappa(s) + \kappa(2\kappa + 1) \int_{t_k}^{t_{k+1} \wedge t} \int_{t_k}^{t_{k+1} \wedge t} 1_{[t_k, s)}(u) |u - s|^{2\kappa - 1} du ds$$

$$= \int_{t_k}^{t_{k+1} \wedge t} \int_{t_k}^{s} dB^\kappa(v) dB^\kappa(s) + (\kappa + 1/2) \int_{t_k}^{t_{k+1} \wedge t} (s - t_k)^{2\kappa} ds. \tag{7.2.54}$$

Combining (7.2.53) and (7.2.54) yields

$$E_2(t) = \Lambda_n(t) \sum_{k=0}^{[nt/a]} F_n(t_k) \int_{t_k}^{t_{k+1} \wedge t} \int_{t_k}^{s} dB^\kappa(v) dB^\kappa(s)$$

and hence

$$\|E_2(t)\|_1 \leq \|\Lambda_n(t)\|_2 \left\| \sum_{k=0}^{[nt/a]} F_n(t_k) \int_{t_k}^{t_{k+1} \wedge t} \int_{t_k}^{s} dB^\kappa(v) dB^\kappa(s) \right\|_2.$$

It was shown that $\|\Lambda_n(t)\|_2 \leq C < \infty$ when bounding $\|E_0(t)\|_1$ above. On the other hand, by Proposition 7.2.9,

$$\left\| \sum_{k=0}^{[nt/a]} F_n(t_k) \int_{t_k}^{t_{k+1} \wedge t} \int_{t_k}^{s} dB^\kappa(v) dB^\kappa(s) \right\|_2 \leq C \gamma_n^{-1} t^{1/2} \|(F_n)_*\|_q,$$

where $q > 2$ and $(F_n)_*$ is defined in (7.2.66). By using Propositions 7.2.6 and 7.2.7, and arguing as in bounding $\|E_0(t)\|_1$ above, it can be shown that $|(F_n)_*| \leq C \exp\{C \|B^\kappa\|_\beta^{1/\beta}\}$, so that $\|(F_n)_*\|_q < \infty$ for any q by Fernique's inequality as above.

Gathering the bounds above, we can conclude that

$$\|E_2(t)\|_1 \leq C \gamma_n^{-1} \tag{7.2.55}$$

and hence that (7.2.47) holds for $j = 2$.

Bounding $\|E_3(t)\|_1$: By using Proposition 7.2.7, $|\Lambda_n(s)^{-1} - \Lambda_n(\eta(s))^{-1}| \le K\exp\{K\|B^\kappa\|_\beta^{1/\beta}\}|s - \eta(s)|^\beta$ for all $s \in [0, a]$ and $|\Lambda_n(t)| \le K\exp\{K\|B^\kappa\|_\beta^{1/\beta}\}$ for all $t \in [0, a]$. Hence,

$$|E_3(t)| \le Ce^{C\|B^\kappa\|_\beta^{1/\beta}} \int_0^t (s - \eta(s))^{2\kappa+\beta} ds \le C'e^{C\|B^\kappa\|_\beta^{1/\beta}} n^{-\kappa-\beta}, \tag{7.2.56}$$

since $0 \le s - \eta(s) \le n^{-1}$. As in several places above, Fernique's inequality yields

$$\|E_3(t)\|_1 \le Cn^{-\kappa-\beta} \tag{7.2.57}$$

and hence that (7.2.47) holds for $j = 3$, since $n^{-\kappa-\beta} \le \gamma_n^{-1}$. □

In the proof of Theorem 7.2.3 above we used the following auxiliary results proved in Hu et al. [495]. For a function $x : [0, a] \to \mathbb{R}$, $0 \le r < s \le a$ and $\beta > 0$, let

$$\|x\|_{r,s,\infty} = \sup_{r \le u \le s} |x(u)|, \quad \|x\|_{r,s,\beta} = \sup_{r \le u < v \le s} \frac{|x(u) - x(v)|}{|u - v|^\beta} \tag{7.2.58}$$

be the supremum and Hölder norms, respectively. When $r = 0$ and $s = a$, we denote these norms as

$$\|x\|_{0,a,\infty} = \|x\|_\infty, \quad \|x\|_{0,a,\beta} = \|x\|_\beta. \tag{7.2.59}$$

Given a process $P = \{P(t) : t \in [0, a]\}$ such that $P(t) \in \mathbb{D}^{N,2}$ for each t and for some $N \ge 1$, consider also a random variable

$$\mathscr{D}_N P = \sup_{r_0,\ldots,r_N \in [0,a]} \Big\{|P(r_0)|, |D_{r_1} P(r_0)|, \ldots, |D^N_{r_1,\ldots,r_N} P(r_0)|,$$

$$\|P\|_{0,a,\beta}, \|D_{r_1} P\|_{r_1,a,\beta}, \ldots, \|D^N_{r_1,\ldots,r_N} P\|_{r_1 \vee \ldots \vee r_N, a, \beta}\Big\}. \tag{7.2.60}$$

The next result is contained in Proposition 3.6 of Hu et al. [495].

Proposition 7.2.6 *Let $\mathscr{D}_N$ be defined as above, X be the solution of the SDE (7.2.34) and X_n be the modified Euler scheme (7.2.35). Fix $N \ge 0$, $\beta \in (1/2, 1/2 + \kappa)$ and suppose that $b \equiv 0$ and $\sigma \in C_b^{N+2}(\mathbb{R})$. Then, there is a constant $K > 0$ such that*

$$\max\{|\mathscr{D}_N X|, |\mathscr{D}_N X_n|\} \le Ke^{K\|B^\kappa\|_\beta^{1/\beta}} \tag{7.2.61}$$

for all $n \ge 1$.

The following result is stated as Eq. (4.8) in the proof of Theorem 4.1 of Hu et al. [495].

Proposition 7.2.7 *Suppose the assumptions of Theorem 7.2.3. Fix $\beta \in (1/2, 1/2 + \kappa)$. Let $\Lambda_n(t)$ be the process defined in (7.2.41)–(7.2.42). Then, there is a constant $K > 0$ such that*

$$\max\{|\mathscr{D}_2 \Lambda_n|, |\mathscr{D}_2(\Lambda_n^{-1})|\} \le Ke^{K\|B^\kappa\|_\beta^{1/\beta}} \tag{7.2.62}$$

for all $n \ge 1$.

The final three results make part of Lemmas 11.3, 11.4 and 11.5 in Hu et al. [495].

Proposition 7.2.8 *Let B^κ be FBM with index $\kappa \in (0, 1/2)$. Fix $\nu \geq 0$ and $p \geq 1/(\kappa+1/2)$. Let $F = \{F(t) : t \in [0, a]\}$ be a stochastic process whose trajectories are Hölder continuous of order $\gamma > 1/2 - \kappa$ and such that $F(t) \in \mathbb{D}^{1,q}$, $t \in [0, a]$, for some $q > p$. Then, there is a constant C (not depending on F) such that, for all $0 \leq s < t \leq a$,*

$$\left\| (RS) \int_s^t F(u)(u - \eta(u))^\nu dB^\kappa(u) \right\|_p \leq C n^{-\nu} (t-s)^{\kappa+1/2} F_{1,p}, \quad (7.2.63)$$

where

$$F_{1,p} = \sup_{s,t \in [0,a]} \left(\|F(t)\|_p \vee \|D_s F(t)\|_p \right). \quad (7.2.64)$$

Proposition 7.2.9 *Let B^κ be FBM with index $\kappa \in (0, 1/2)$. Fix $p \geq 1/(\kappa + 1/2)$. Let $F = \{F(t) : t \in [0, a]\}$ be a stochastic process such that $F(t) \in \mathbb{D}^{2,q}$, $t \in [0, a]$, for some $q > p$. Then, there is a constant C (not depending on F) such that, for all $0 \leq s < t \leq a$,*

$$\left\| \sum_{k=[ns/a]}^{[nt/a]} F(t_k) \int_{t_k \vee s}^{t_{k+1} \wedge t} \int_{t_k}^u dB^\kappa(v) dB^\kappa(u) \right\|_p \leq C \gamma_n^{-1} (t-s)^{1/2} \|F_*\|_q, \quad (7.2.65)$$

where

$$F_* = \sup_{s,t \in [0,a]} \left(|F(t)| \vee |D_s F(t)| \vee |D_r D_s F(t)| \right). \quad (7.2.66)$$

Observe that the integrals in (7.2.65) are the divergence integrals, in contrast to the Riemann–Stieltjes integral in (7.2.63).

Proposition 7.2.10 *Let B^κ be FBM with index $\kappa \in (0, 1/2)$. Let $F = \{F(t) : t \in [0, a]\}$ and $G = \{G(t) : t \in [0, a]\}$ be stochastic processes whose trajectories are Hölder continuous of order $\beta \in (1/2, 1/2 + \kappa)$. Then, there is a constant C (not depending on F and G) such that, for all $0 \leq s < t \leq a$,*

$$\left| (RS) \int_s^t F(u)(G(u) - G(\eta(u)))(B^\kappa(u) - B^\kappa(\eta(u))) dB^\kappa(u) \right|$$

$$\leq C(\|F\|_\infty + \|F\|_\beta) \|G\|_\beta \|B^\kappa\|_\beta^2 n^{1-3\beta} (t-s)^\beta. \quad (7.2.67)$$

The next example illustrates the approximations introduced above.

Example 7.2.11 (*Geometric FBM*) The process $X(t) = e^{B^\kappa(t)+t}$ satisfies the SDE

$$X(t) = 1 + \int_0^t X(s) ds + (RS) \int_0^t X(s) dB^\kappa(s) \quad (7.2.68)$$

(Example 7.1.13). Figure 7.1 depicts the trajectories of the true solution $X(t)$, the Euler approximation and the modified Euler approximation when $\kappa = 0.25$ or 0.05, $n = 1024$ and $a = 1$. Note that the modified Euler approximation follows the true solution much more closely than the Euler approximation. It should also be noted that this effect is more pronounced when κ is close to zero, as with $\kappa = 0.05$ shown in the figure, right plot. For larger κ, as for $\kappa = 0.25$, left plot, both approximations follow the true solution quite closely.

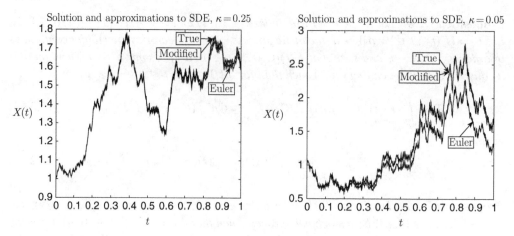

Figure 7.1 Trajectories of the solution, the Euler approximation and the modified Euler approximation to the SDE (7.2.68).

7.2.4 Convergence to Normal Law Using Stein's Method

Stein's method is an interesting and powerful way of proving convergence to a normal law, in situations when the classical central limit theorem does not apply, as well as of determining the convergence rate (see, for example, Chen, Goldstein, and Shao [218]). The Malliavin calculus has recently been combined with Stein's method to establish normal approximations when working on the Wiener space. This combination has attracted considerable attention. See, for example, the monograph by Nourdin and Peccati [763]. A short description can be found in Nualart [770]. In this section, we briefly touch upon this approach and illustrate it in the special case of FBM.

Recall that the total variation distance between the laws of two random variables X and Y is defined as

$$d_{TV}(X, Y) = \sup_{B \in \mathcal{B}(\mathbb{R})} \Big| \mathbb{P}(X \in B) - \mathbb{P}(Y \in B) \Big|, \qquad (7.2.69)$$

where $\mathcal{B}(\mathbb{R})$ denotes the σ-field of Borel sets in $\mathbb{R}$. If $d_{TV}(X_n, X) \to 0$, then X_n converges to X in distribution since we can let $B = (-\infty, x]$. The following are two fundamental results in the framework of the Malliavin calculus and Stein's method. See Nourdin and Peccati [763], Theorems 5.1.3 and 5.2.6. For definitions, see Appendix C.

Theorem 7.2.12 (*Nourdin–Peccati*) Let $F \in \mathbb{D}^{1,2}$ with $\mathbb{E}F = 0$, $\mathbb{E}F^2 = \sigma^2 > 0$ and N have the normal distribution $\mathcal{N}(0, \sigma^2)$. Then,

$$d_{TV}(F, N) \leq \frac{2}{\sigma^2} \mathbb{E}\Big| \sigma^2 - (DF, -DL^{-1}F)_{\mathcal{H}} \Big|, \qquad (7.2.70)$$

where D is the derivative and L is the generator of the Ornstein–Uhlenbeck semigroup (Appendix C).

Corollary 7.2.13 (*Nourdin–Peccati*) *Let $q \geq 2$ be an integer, and let $F = I_q(f)$ be a multiple integral of order q with $\mathbb{E}F^2 = \sigma^2 > 0$. Then,*

$$d_{TV}(F, N) \leq \frac{2}{\sigma^2}\sqrt{\mathrm{Var}\Big(\frac{1}{q}\|DF\|_{\mathcal{H}}^2\Big)} \leq \frac{2}{\sigma^2}\sqrt{\frac{q-1}{3q}(\mathbb{E}F^4 - 3\sigma^4)}. \tag{7.2.71}$$

Remark 7.2.14 In Appendix B.2, multiple integrals $F = I_q(f)$ are defined for $f \in L^2(E^q, m^q)$. As in Examples C.1.2 and C.2.7 in Appendix C, this is the white noise setting associated with the Hilbert space $\mathcal{H} = L^2(E, m)$. This is also the setting which will be used in the proofs of Corollary 7.2.13 and Proposition 7.2.15 below. But it should be noted that multiple integrals $F = I_q(f)$ can be defined for general underlying Hilbert spaces $\mathcal{H}$ and Corollary 7.2.13 holds in this general case (see Nourdin and Peccati [763]).

Proof We shall prove Corollary 7.2.13. By (C.4.2), $L^{-1}F = L^{-1}I_q(f) = -\frac{1}{q}I_q(f) = -\frac{F}{q}$, so that $(DF, -DL^{-1}F)_{\mathcal{H}} = \frac{1}{q}\|DF\|_{\mathcal{H}}^2$. By Exercise 7.9, $\sigma^2 = \frac{1}{q}\mathbb{E}\|DF\|_{\mathcal{H}}^2$. Hence, by Theorem 7.2.12,

$$d_{TV}(F, N) \leq \frac{2}{\sigma^2}\mathbb{E}\Big|\sigma^2 - (DF, -DL^{-1}F)_{\mathcal{H}}\Big| = \frac{2}{\sigma^2}\mathbb{E}\Big|\frac{1}{q}\|DF\|_{\mathcal{H}}^2 - \frac{1}{q}\mathbb{E}\|DF\|_{\mathcal{H}}^2\Big|$$

$$\leq \frac{2}{\sigma^2}\sqrt{\mathbb{E}\Big|\frac{1}{q}\|DF\|_{\mathcal{H}}^2 - \frac{1}{q}\mathbb{E}\|DF\|_{\mathcal{H}}^2\Big|^2} = \frac{2}{\sigma^2}\sqrt{\mathrm{Var}\Big(\frac{1}{q}\|DF\|_{\mathcal{H}}^2\Big)},$$

which yields the first inequality in (7.2.71). The second inequality in (7.2.71) is stated as Exercise 7.10. □

We illustrate the results above on the quadratic variation of a standard FBM B^κ. More specifically, set

$$F_n = \frac{n^{2\kappa+1}}{\sigma_n}\sum_{k=1}^{n}\Big(\Big(B^\kappa\Big(\frac{k}{n}\Big) - B^\kappa\Big(\frac{k-1}{n}\Big)\Big)^2 - n^{-2\kappa-1}\Big), \tag{7.2.72}$$

where $\mathbb{E}F_n = 0$ and where σ_n is such that $\mathbb{E}F_n^2 = 1$. The next result, which is an application of Corollary 7.2.13, shows the convergence of F_n to a normal law and establishes the rate of convergence.

Proposition 7.2.15 *Let $\kappa \in (-1/2, 1/4)$, $\kappa \neq 0$, and $\gamma(h) = \frac{1}{2}(|h+1|^{2\kappa+1} + |h-1|^{2\kappa+1} - 2|h|^{2\kappa+1})$, $h \in \mathbb{Z}$, be the autocovariance function of fractional Gaussian noise series $X_k = B^\kappa(k) - B^\kappa(k-1)$ (see Definition 2.8.3 and (2.8.2)). As above, let also N denote the standard normal random variable. Then,*

$$\lim_{n\to\infty}\frac{\sigma_n^2}{n} = 2\sum_{h=-\infty}^{\infty}(\gamma(h))^2 \tag{7.2.73}$$

and

$$d_{TV}(F_n, N) \leq C(\kappa)\begin{cases} n^{-1/2}, & \text{if } \kappa \in (-1/2, 1/8), \\ n^{-1/2}(\log n)^{3/2}, & \text{if } \kappa = 1/8, \\ n^{4\kappa-1}, & \text{if } \kappa \in (1/8, 1/4). \end{cases} \tag{7.2.74}$$

Remark 7.2.16 The result (7.2.74) shows that F_n converges in distribution to a standard normal law. The convergence can also be deduced readily by applying Theorem 5.4.1 to the partial sums $\widetilde{F}_n$ in the proof below. The assumption $\kappa < 1/4$ is essential because when $\kappa > 1/4$, $\widetilde{F}_n$ does not converge to a normal law (see Theorem 5.3.1).

Proof By the $(\kappa+1/2)$-self-similarity of FBM, the random variable F_n has the same law as

$$\widetilde{F}_n = \frac{1}{\sigma_n} \sum_{k=1}^n \left((B^\kappa(k) - B^\kappa(k-1))^2 - 1\right)$$

$$= \frac{1}{\sigma_n} \sum_{k=1}^n H_2(B^\kappa(k) - B^\kappa(k-1)) = \frac{1}{\sigma_n} \sum_{k=1}^n H_2(I_1(e_k)), \quad (7.2.75)$$

where H_2 is the Hermite polynomial of order two in Definition 4.1.1, I_1 is a single integral with respect to a Gaussian measure on $\mathbb{R}$ having the Lebesgue control measure, and $e_k(u) = c_{1,\kappa}^{-1}((k-u)_+^\kappa - (k-1-u)_+^\kappa)$, $u \in \mathbb{R}$ (Proposition 6.2.10). The representation (7.2.75) and Proposition 5.1.1 yield

$$\sigma_n^2 = \sum_{k_1,k_2=1}^n 2(\mathbb{E}I_1(e_{k_1})I_1(e_{k_2}))^2 = \sum_{k_1,k_2=1}^n 2(\gamma(k_1-k_2))^2 = 2\sum_{h=-(n-1)}^{n-1} (n-h)(\gamma(h))^2. \quad (7.2.76)$$

By (2.8.2),

$$\gamma(h) \sim Ch^{2\kappa-1}, \quad \text{as } h \to \infty,$$

and hence $(\gamma(h))^2 \sim C'h^{4\kappa-2}$, as $h \to \infty$. Since $\kappa < 1/4$ by assumption, the dominated convergence theorem now yields the relation (7.2.73).

By Proposition 4.1.2, (a),

$$\widetilde{F}_n = \frac{1}{\sigma_n} \sum_{k=1}^n I_2(e_k^{\otimes 2}),$$

where I_2 is a multiple integral of order two. We shall apply the first inequality in (7.2.71). To do so, we need to find $D\widetilde{F}_n$ and then consider $\text{Var}(\|D\widetilde{F}_n\|_\mathcal{H}^2)$, where $\mathcal{H} = L^2(\mathbb{R}, du)$. By (C.2.9),

$$(D\widetilde{F}_n)(u) = \frac{2}{\sigma_n} \sum_{k=1}^n I_1\left(e_k(\cdot)e_k(u)\right) = \frac{2}{\sigma_n} \sum_{k=1}^n I_1(e_k)e_k(u),$$

which is a function of u. Then

$$\|D\widetilde{F}_n\|_\mathcal{H}^2 = \frac{4}{\sigma_n^2} \sum_{k_1,k_2=1}^n I_1(e_{k_1})I_1(e_{k_2})(e_{k_1}, e_{k_2})_\mathcal{H},$$

where $(e_{k_1}, e_{k_2})_\mathcal{H} = \mathbb{E}I_1(e_{k_1})I_1(e_{k_2}) = \gamma(k_1 - k_2)$. Hence,

$$\mathbb{E}\|D\widetilde{F}_n\|_\mathcal{H}^2 = \frac{4}{\sigma_n^2} \sum_{k_1,k_2=1}^n (e_{k_1}, e_{k_2})_\mathcal{H}^2 = \frac{4}{\sigma_n^2} \sum_{k_1,k_2=1}^n (\gamma(k_1-k_2))^2 = \frac{4}{\sigma_n^2} \frac{\sigma_n^2}{2} = 2,$$

by the choice (7.2.76) of σ_n.

7.2 Applications of Stochastic Integration

To compute $\text{Var}(\|D\widetilde{F}_n\|_\mathcal{H}^2) = \mathbb{E}\|D\widetilde{F}_n\|_\mathcal{H}^4 - (\mathbb{E}\|D\widetilde{F}_n\|_\mathcal{H}^2)^2 = \mathbb{E}\|D\widetilde{F}_n\|_\mathcal{H}^4 - 4$, note from above that

$$\mathbb{E}\|D\widetilde{F}_n\|_\mathcal{H}^4 = \frac{16}{\sigma_n^4} \sum_{k_1,k_2,k_3,k_4=1}^n \mathbb{E}\Big(I_1(e_{k_1})I_1(e_{k_2})I_1(e_{k_3})I_1(e_{k_4})\Big)(e_{k_1},e_{k_2})_\mathcal{H}(e_{k_3},e_{k_4})_\mathcal{H}.$$

By using Proposition 4.3.20,

$$\mathbb{E}\Big(I_1(e_{k_1})I_1(e_{k_2})I_1(e_{k_3})I_1(e_{k_4})\Big)$$
$$= (e_{k_1},e_{k_2})_\mathcal{H}(e_{k_3},e_{k_4})_\mathcal{H} + (e_{k_1},e_{k_3})_\mathcal{H}(e_{k_2},e_{k_4})_\mathcal{H} + (e_{k_1},e_{k_4})_\mathcal{H}(e_{k_2},e_{k_3})_\mathcal{H}.$$

Hence,

$$\mathbb{E}\|D\widetilde{F}_n\|_\mathcal{H}^4 = \frac{16}{\sigma_n^4} \sum_{k_1,k_2,k_3,k_4=1}^n (e_{k_1},e_{k_2})_\mathcal{H}^2 (e_{k_3},e_{k_4})_\mathcal{H}^2$$
$$+ \frac{16}{\sigma_n^4} \sum_{k_1,k_2,k_3,k_4=1}^n (e_{k_1},e_{k_3})_\mathcal{H}(e_{k_2},e_{k_4})_\mathcal{H}(e_{k_1},e_{k_2})_\mathcal{H}(e_{k_3},e_{k_4})_\mathcal{H}$$
$$+ \frac{16}{\sigma_n^4} \sum_{k_1,k_2,k_3,k_4=1}^n (e_{k_1},e_{k_4})_\mathcal{H}(e_{k_2},e_{k_3})_\mathcal{H}(e_{k_1},e_{k_2})_\mathcal{H}(e_{k_3},e_{k_4})_\mathcal{H}. \quad (7.2.77)$$

The first term on the right-hand side of (7.2.77) equals 4 since

$$\sum_{k_1,k_2,k_3,k_4=1}^n (e_{k_1},e_{k_2})_\mathcal{H}^2(e_{k_3},e_{k_4})_\mathcal{H}^2 = \Big(\sum_{k_1,k_2=1}^n (e_{k_1},e_{k_2})_\mathcal{H}^2\Big)^2 = \Big(\sum_{k_1,k_2=1}^n (\gamma(k_1-k_2))^2\Big)^2 = \frac{\sigma_n^4}{4}$$

by using (7.2.76). The last two terms on the right-hand side of (7.2.77) are the same after interchanging the indices k_3 and k_4 in the last term. It now follows from (7.2.77) that

$$\mathbb{E}\|D\widetilde{F}_n\|_\mathcal{H}^4 = 4 + \frac{32}{\sigma_n^4} \sum_{k_1,k_2,k_3,k_4=1}^n (e_{k_1},e_{k_3})_\mathcal{H}(e_{k_2},e_{k_4})_\mathcal{H}(e_{k_1},e_{k_2})_\mathcal{H}(e_{k_3},e_{k_4})_\mathcal{H}$$

and hence

$$\text{Var}(\|D\widetilde{F}_n\|_\mathcal{H}^2) = \frac{32}{\sigma_n^4} \sum_{k_1,k_2,k_3,k_4=1}^n \gamma(k_1-k_3)\gamma(k_2-k_4)\gamma(k_1-k_2)\gamma(k_3-k_4) =: \frac{32}{\sigma_n^4} I_n.$$
$$(7.2.78)$$

Set

$$\gamma_n(h) = \begin{cases} \gamma(h), & \text{if } |h| < n, \\ 0, & \text{otherwise.} \end{cases}$$

Note that

$$I_n = \sum_{k_1,k_4=1}^{n} \left(\sum_{k_3 \in \mathbb{Z}} \gamma_n(k_1 - k_3)\gamma_n(k_3 - k_4)\right)\left(\sum_{k_2 \in \mathbb{Z}} \gamma_n(k_2 - k_4)\gamma_n(k_1 - k_2)\right)$$

$$= \sum_{j_1,j_2=1}^{n} \left((\gamma_n * \gamma_n)(j_1 - j_2)\right)^2 = \sum_{h=-(n-1)}^{n-1} (n - |h|)\left((\gamma_n * \gamma_n)(h)\right)^2$$

$$\leq n \sum_{h \in \mathbb{Z}} \left((\gamma_n * \gamma_n)(h)\right)^2 = n\|\gamma_n * \gamma_n\|_{\ell^2(\mathbb{Z})}^2. \qquad (7.2.79)$$

We shall use next Young's inequality: if $p, r, s \geq 1$ are such that $\frac{1}{r} + \frac{1}{s} = 1 + \frac{1}{p}$, then

$$\|u * v\|_{\ell^p(\mathbb{Z})} \leq \|u\|_{\ell^r(\mathbb{Z})}\|v\|_{\ell^s(\mathbb{Z})}. \qquad (7.2.80)$$

Applying the inequality with $p = 2$ and $r = s = 4/3$ to (7.2.79), we obtain that

$$I_n \leq n\left(\sum_{|h|<n} |\gamma(h)|^{4/3}\right)^3. \qquad (7.2.81)$$

By (2.8.2), $\gamma(h) \sim c|h|^{2\kappa-1}$, as $h \to \pm\infty$. For $\kappa \in (-1/2, 1/4)$, applying this to (7.2.81) yields

$$I_n \leq Cn \begin{cases} 1, & \text{if } \kappa \in (-1/2, 1/8), \\ (\log n)^3, & \text{if } \kappa = 1/8, \\ n^{8\kappa-1}, & \text{if } \kappa \in (1/8, 1/4). \end{cases} \qquad (7.2.82)$$

Finally, by using (7.2.78), (7.2.73) and (7.2.71), the bound (7.2.82) implies the desired inequality (7.2.74). □

7.2.5 Local Time of FBM

Let $B^\kappa = \{B^\kappa(t)\}_{t \in [0,a]}$ be FBM with parameter $\kappa \in (-1/2, 1/2)$. Consider the so-called occupation measure

$$\mu(t, A) = \int_0^t 1_A(B^\kappa(s))ds, \quad t \in [0, a], \ A \in \mathcal{B}(\mathbb{R}). \qquad (7.2.83)$$

It measures the total amount of time in $[0, t]$ that FBM B^κ spends in the set A. It is well known (e.g., Berman [131], Geman and Horowitz [384]) that the (random) measure $\mu(t, A)$, $A \in \mathcal{B}(\mathbb{R})$, has a density $\ell_t^\kappa(x)$ (almost surely), which has a continuous version in the time variable $t \in [0, a]$ and the space variable $x \in \mathbb{R}$. Moreover (Geman and Horowitz [384], Table 2, p. 62), $\ell_t^\kappa(x)$ has δ–Hölder continuous paths with $\delta < 1/2 - \kappa$ in time, and γ–Hölder continuous paths with $\gamma < (1/2 - \kappa)/(2\kappa + 1)$ in the space variable, provided $\kappa \geq -1/6$. Moreover, $\ell_t^\kappa(x)$ is absolutely continuous in x if $\kappa < -1/6$, it is continuously differentiable if $\kappa < -3/10$, and its smoothness increases when κ decreases. The local time $\ell_t^\kappa(x)$ is defined by the following property: for any continuous function g on $\mathbb{R}$ with compact support:

$$\int_0^t g(B^\kappa(s))ds = \int_{\mathbb{R}} g(x)\ell_t^\kappa(x)dx. \qquad (7.2.84)$$

A similar relation was given in (3.10.4). One can informally view $\ell_t^\kappa(x)$ as $\int_0^t \delta_{\{x\}}(B^\kappa(s))ds$, where $\delta_{\{x\}}(y)$ is the Dirac function with point mass at $y = x$.

The next result establishes the chaos expansion of the local time of FBM. Let

$$p_\epsilon(x) = \frac{1}{\sqrt{2\pi\epsilon}} e^{-\frac{x^2}{2\epsilon}}, \quad x \in \mathbb{R}, \tag{7.2.85}$$

by the density of the $\mathcal{N}(0, \epsilon)$ distribution. The multiple integrals I_n appearing below are with respect to the Brownian motion B (white noise) such that $B^\kappa(t) = \int_0^a f_t(u)dB(u) = I_1(f_t)$, where f_t is given by (6.2.3).

Proposition 7.2.17 *Let $\kappa \in (-1/2, 1/2)$. The local time $\ell_t^\kappa(x)$ of FBM has the following Wiener chaos expansion:*

$$\ell_t^\kappa(x) = \sum_{n=0}^\infty \frac{1}{n!} \int_0^t s^{-n(\kappa+1/2)} p_{s^{2\kappa+1}}(x) H_n(xs^{-\kappa-1/2}) I_n(f_s^{\otimes n}) ds, \tag{7.2.86}$$

where p is given by (7.2.85), H_n, $n \geq 1$, are the Hermite polynomials (4.1.1), and f_s is the kernel function defined in (6.2.3).

Sketch of Proof The basic idea is to compute the Wiener chaos expansion of

$$F := F_\epsilon(s) := p_\epsilon(B^\kappa(s) - x),$$

which can be thought as the Dirac's $\delta_{\{x\}}(B^\kappa(s))$ in the limit $\epsilon \to 0$. As discussed following (7.2.84), by integrating the Wiener chaos expansion of $F_\epsilon(s)$ from $s = 0$ to t and letting $\epsilon \to 0$, one can then expect to get in the limit the Wiener chaos expansion of $\int_0^t \delta_{\{x\}}(B^\kappa(s))ds$; that is, the local time of FBM. We thus need to compute $\int_0^t p_\epsilon(B^\kappa(s) - x)ds$.

To compute the Wiener chaos expansion of $F = F_\epsilon(s)$, we use the Stroock formula (C.2.15) in Appendix C, for which we need $D^n F$. By using (C.2.2) and since $B^\kappa(s) = I_1(f_s)$, we have

$$DF = D(p_\epsilon(B^\kappa(s) - x)) = D(p_\epsilon(I_1(f_s) - x)) = p_\epsilon^{(1)}(B^\kappa(s) - x)f_s$$

and thus

$$D^n F = D^n(p_\epsilon(B^\kappa(s) - x)) = p_\epsilon^{(n)}(B^\kappa(s) - x)f_s^{\otimes n}, \tag{7.2.87}$$

where $p_\epsilon^{(n)}$ denotes the nth derivative of p_ϵ. Observe from (4.1.1) that

$$p_\epsilon^{(n)}(y) = (-1)^n \epsilon^{-n/2} p_\epsilon(y) H_n\left(\frac{y}{\sqrt{\epsilon}}\right).$$

Note also that

$$\mathbb{E} p_\epsilon(B^\kappa(s) - x) = \int_\mathbb{R} \frac{1}{\sqrt{2\pi\epsilon}} e^{-\frac{(z-x)^2}{2\epsilon}} \frac{1}{\sqrt{2\pi s^{2\kappa+1}}} e^{-\frac{z^2}{2s^{2\kappa+1}}} dz = p_{s^{2\kappa+1}+\epsilon}(x),$$

since the integral is the convolution of the densities of $\mathcal{N}(0, \epsilon)$ and $\mathcal{N}(0, s^{2\kappa+1})$, and hence is the density $p_{s^{2\kappa+1}+\epsilon}$ of $\mathcal{N}(0, s^{2\kappa+1} + \epsilon)$. It follows that

$$\mathbb{E} p_\epsilon^{(n)}(B^\kappa(s) - x) = (-1)^n \frac{\partial^n}{\partial x^n} \mathbb{E} p_\epsilon(B^\kappa(s) - x) = (-1)^n p_{s^{2\kappa+1}+\epsilon}^{(n)}(x)$$
$$= (s^{2\kappa+1} + \epsilon)^{-n/2} p_{s^{2\kappa+1}+\epsilon}(x) H_n\left(\frac{x}{\sqrt{s^{2\kappa+1}+\epsilon}}\right).$$

By using the Stroock formula (C.2.15), we obtain the chaos expansion

$$p_\epsilon(B^\kappa(s) - x) = \sum_{n=0}^\infty \beta_{n,\epsilon}(s) I_n(f_s^{\otimes n}), \qquad (7.2.88)$$

where

$$\beta_{n,\epsilon}(s) = \mathbb{E} p_\epsilon^{(n)}(B^\kappa(s) - x) = (s^{2\kappa+1} + \epsilon)^{-n/2} p_{s^{2\kappa+1}+\epsilon}(x) H_n\left(\frac{x}{\sqrt{s^{2\kappa+1}+\epsilon}}\right). \qquad (7.2.89)$$

By integrating (7.2.88) from $s = 0$ to t, we obtain that

$$\int_0^t p_\epsilon(B^\kappa(s) - x) ds = \sum_{n=0}^\infty \int_0^t \beta_{n,\epsilon}(s) I_n(f_s^{\otimes n}) ds.$$

The argument proceeds by showing that, as $\epsilon \to 0$, the right-hand side converges to

$$\sum_{n=0}^\infty \int_0^t s^{-n(\kappa+1/2)} p_{s^{2\kappa+1}}(x) H_n\left(\frac{x}{\sqrt{s^{2\kappa+1}}}\right) I_n(f_s^{\otimes n}) ds \qquad (7.2.90)$$

and that the quantity (7.2.90) satisfies the relation (7.2.84) characterizing the local time $\ell_t^\kappa(x)$. See Coutin, Nualart, and Tudor [264]. □

When $x = 0$, the chaos expansion (7.2.86) can be simplified by using $p_{s^{2\kappa+1}}(0) = 1/\sqrt{2\pi s^{2\kappa+1}}$ and $H_n(0) = 0$ if n is odd, and $(-1)^{n/2}(n-1)!!$ if n is even. A different approach using the formula (7.2.96) below and leading to a more explicit chaos expansion of $\ell_t^\kappa(0)$ can be found in Hu and Øksendal [492]. The chaos expansion is used, for example, in Eddahbi, Lacayo, Solé, Vives, and Tudor [345] to prove more refined regularity properties of the local time of FBM.

As with the usual BM, the local time of FBM also appears in the Itô's formulas for convex (not necessarily twice continuously differentiable) functions. In fact, in the context of FBM, a more convenient quantity is the local time $L_t^\kappa(x)$ associated with the occupation measure

$$m(t, A) = (2\kappa + 1) \int_0^t 1_A(B^\kappa(s)) s^{2\kappa} ds, \qquad (7.2.91)$$

which satisfies

$$(2\kappa + 1) \int_0^t g(B^\kappa(s)) s^{2\kappa} ds = \int_\mathbb{R} g(x) L_t^\kappa(x) dx \qquad (7.2.92)$$

(cf. (7.2.84)). The relation between the occupation measures m in (7.2.91) and μ of FBM in (7.2.83) is

$$m(t, A) = (2\kappa + 1) \int_0^t s^{2\kappa} \mu(ds, A).$$

The local times $L_t^\kappa(x)$ and $\ell_t^\kappa(x)$ are related as

$$L_t^\kappa(x) = (2\kappa + 1)t^{2\kappa}\ell_t^\kappa(x) - 2\kappa(2\kappa + 1)\int_0^t s^{2\kappa-1}\ell_s^\kappa(x)ds, \qquad (7.2.93)$$

which can be seen by verifying that the right-hand side of (7.2.93) satisfies (7.2.92) as

$$\int_\mathbb{R} g(x)\Big((2\kappa + 1)t^{2\kappa}\ell_t^\kappa(x) - 2\kappa(2\kappa + 1)\int_0^t s^{2\kappa-1}\ell_s^\kappa(x)ds\Big)dx$$

$$= (2\kappa + 1)t^{2\kappa}\int_0^t g(B^\kappa(s))ds - 2\kappa(2\kappa + 1)\int_0^t s^{2\kappa-1}\int_0^s g(B^\kappa(u))du\,ds$$

$$= (2\kappa + 1)t^{2\kappa}\int_0^t g(B^\kappa(s))ds - (2\kappa + 1)\int_0^t g(B^\kappa(u))(t^{2\kappa} - u^{2\kappa})du$$

$$= (2\kappa + 1)\int_0^t g(B^\kappa(u))u^{2\kappa}du.$$

The following theorem gives an Itô's formula for FBM with $\kappa \in (-1/6, 1/2)$; that is, $H \in (1/3, 1)$.

Theorem 7.2.18 (*Itô's formula for convex functions of FBM.*) *Suppose that* $\kappa \in (-1/6, 1/2)$ *and* f *is a convex function such that the right derivative* f'_+ *is uniformly bounded. Then,*

$$f(B^\kappa(t)) = f(B^\kappa(0)) + \int_0^t f'_-(B^\kappa(s))dB^\kappa(s) + \frac{1}{2}\int_\mathbb{R} L_t^\kappa(x)f''(dx), \qquad (7.2.94)$$

where the measure $f''(dx)$ *denotes the second derivative of* f *in the sense of distributions and the integral with respect to FBM is the divergence integral.*

See Coutin et al. [264], Proposition 7, for the proof. An extension to the case $\kappa \in (-1/2, 0)$, with the divergence integral with respect to FBM in the larger domain $\text{Dom}^*\delta_a^\kappa$, can be found in Cheridito and Nualart [228]. A special case of (7.2.94) with $f(y) = |y - x|$ is Tanaka's formula:

$$|B^\kappa(t) - x| = |x| + \int_0^t \text{sign}(B^\kappa(s) - x)dB^\kappa(s) + L_t^\kappa(x). \qquad (7.2.95)$$

In fact, Tanaka's formula is proved first using the approach of Proposition 7.2.17. It then yields the Itô's formula (7.2.94) using the representation $f(y) = \alpha y + \beta + \frac{1}{2}\int_J |y-x|f''(dx)$ of a convex function f, where α and β are some real numbers, and J is the compact support of f''.

Further properties of the local time $\ell_t^\kappa(x)$ of FBM can be deduced by expressing it as the inverse Fourier transform of $\widehat{\mu}(t, u) := \int_\mathbb{R} e^{iux}\mu(t, dx) = \int_\mathbb{R} e^{iux}\ell_t^\kappa(x)dx = \int_0^t e^{iuB^\kappa(s)}ds$:

$$\ell_t^\kappa(x) = \frac{1}{2\pi}\int_\mathbb{R} e^{-ixu}\widehat{\mu}(t, u)du = \frac{1}{2\pi}\int_0^t \int_\mathbb{R} e^{-ixu+iuB^\kappa(s)}du\,ds. \qquad (7.2.96)$$

This formal relationship can be justified rigorously (Geman and Horowitz [384]). A consequence of (7.2.96) is the following formula for the moments of the local time: for $n \geq 1$,

$$\mathbb{E}(\ell_t^\kappa(x))^n = \frac{1}{(2\pi)^n} \int_{[0,t)^n} \int_{\mathbb{R}^n} e^{-ix \sum_{j=1}^n u_j} \mathbb{E} e^{i \sum_{j=1}^n u_j B^\kappa(s_j)} du_1 \ldots du_n ds_1 \ldots ds_n.$$
(7.2.97)

This yields, in particular,

$$\mathbb{E}\ell_t^\kappa(x) = \frac{1}{\sqrt{2\pi}} \int_0^t s^{-\kappa-1/2} e^{-\frac{s^{-2\kappa-1}x^2}{2}} ds,$$
(7.2.98)

$$\mathbb{E}(\ell_t^\kappa(x))^2 = \frac{1}{2\pi} \int_0^t \int_0^t \frac{s_1^{-\kappa-1/2} s_2^{-\kappa-1/2}}{\sqrt{1-\rho(s_1,s_2)^2}} e^{-\frac{x^2}{2} \frac{|s_1-s_2|^{-2\kappa-1}}{(1-\rho(s_1,s_2)^2)}} ds_1 ds_2,$$
(7.2.99)

where $\rho(s_1, s_2) = -\mathbb{E}B^\kappa(s_1)B^\kappa(s_2)/\sqrt{\mathbb{E}B^\kappa(s_1)^2 \mathbb{E}B^\kappa(s_2)^2}$ (Exercise 7.12). The formulas can be simplified further when $x = 0$.

7.3 Exercises

The symbols * and ** next to some exercises are explained in Section 2.12.

Exercise 7.1* Prove Lemma 7.1.4.

Exercise 7.2 Prove the following result used in Example 7.1.5. If D is viewed as an operation on $\mathcal{H}$–valued random variables, show that $(DB^\kappa)(u, v) = 1_{[0,u)}(v)$.

Exercise 7.3 Establish the following Itô's formula for $F(t, B^\kappa(t))$ under suitable assumptions:

$$F(t, B^\kappa(t)) = F(0, B^\kappa(0)) + \int_0^t \frac{\partial F}{\partial t}(u, B^\kappa(u)) du$$
$$+ \int_0^t \frac{\partial F}{\partial x}(u, B^\kappa(u)) dB^\kappa(u) + \frac{2\kappa+1}{2} \int_0^t \frac{\partial^2 F}{\partial x^2}(u, B^\kappa(u)) u^{2\kappa} du,$$

where $\frac{\partial F}{\partial t}, \frac{\partial F}{\partial x}$ and $\frac{\partial^2 F}{\partial x^2}$ refer to the partial derivatives of the function $F(t, x)$.

Exercise 7.4 Use (7.2.3) to show (7.2.6). Also, verify informally that the process $X(t)$ given by (7.2.2)–(7.2.5) satisfies the SDE (7.2.5).

Exercise 7.5 Prove that the solution to (7.2.21) is given by (7.2.30).

Exercise 7.6* Prove the relation (7.2.33). *Hint:* See Lemma 2 in Nourdin and Simon [765].

Exercise 7.7* Show that the distribution of the supremum of FBM $\sup_{t \in [0,a]} B^\kappa(a)$ has a bounded density with respect to the Lebesgue measure. *Hint:* See Lanjri Zadi and Nualart [593].

Exercise 7.8 In the proof of the convergence rate of numerical solutions of SDEs, show that the solution to (7.2.41) is given by (7.2.42) and satisfies (7.2.43).

Exercise 7.9 If $F = I_q(f)$ is a multiple integral of order q, show that

$$\mathbb{E} F^2 = \frac{1}{q} \mathbb{E} \|DF\|_{\mathcal{H}}^2.$$

Exercise 7.10 Prove the second inequality in (7.2.71). *Hint:* See Lemma 5.2.4 in Nourdin and Peccati [763].

Exercise 7.11** In Proposition 7.2.15, the rate of convergence to a standard normal law was found for the normalized and centered quadratic variation of FBM in (7.2.72). What about the rate of convergence of the normalized and centered log-variation of FBM defined by

$$\sum_{k=1}^{n} \log \left| B^{\kappa}\left(\frac{k}{n}\right) - B^{\kappa}\left(\frac{k-1}{n}\right) \right| ?$$

Exercise 7.12 Prove the moment formulae (7.2.98) and (7.2.99) for the local time of FBM.

Exercise 7.13** In this chapter, a stochastic analysis is developed for FBM. What would be a stochastic analysis for the Rosenblatt process or other Hermite processes? *Hint:* Some work in this direction can be found in Arras [45].

7.4 Bibliographical Notes

Section 7.1: The fact that FBM is not a semimartingale was proved, for example, in Rogers [859], with consequences of arbitrage opportunities in the FBM markets. The latter topic is explored further in Cheridito [227], Guasoni [435], Bender, Sottinen, and Valkeila [118], Czichowsky and Schachermayer [273]. Conditions for Hölder continuity of general Gaussian processes are studied in Azmoodeh, Sottinen, Viitasaari, and Yazigi [65].

The divergence integral and the Malliavin calculus for FBM and more general Gaussian processes were considered notably in Decreusefond and Üstünel [296], Alòs, Mazet, and Nualart [21, 22], Carmona, Coutin, and Montseny [203], Cheridito and Nualart [228], Decreusefond [294]. Other, essentially equivalent approaches are based Wick products (Duncan, Hu, and Pasik-Duncan [337]) and fractional white noise (Hu and Øksendal [493], Bender [116, 117], Elliott and van der Hoek [348], Biagini, Øksendal, Sulem, and Wallner [146]).

Among other approaches to stochastic integration for FBM are pathwise methods. Early attempts include Ciesielski, Kerkyacharian, and Roynette [243], Jumarie [538], Lin [634], Dai and Heyde [280], Zähle [1027], Ruzmaikina [874]. Pathwise integrals with respect to FBM can also be defined within the more general and recent approaches of rough paths (Lyons and Qian [655], Coutin and Qian [263]) and p–variations (Dudley and Norvaiša [330]). It should be noted that, unlike the divergence integral, the pathwise integrals with respect to FBM do not have zero mean in general.

A Stratonovich-type integral with respect to FBM is considered in Gradinaru, Russo, and Vallois [420] (see also references therein).

Stochastic integration for FBM is now also considered in several books, most notably in Nualart [769], Mishura [722], Biagini, Hu, Øksendal, and Zhang [147], Nourdin [760]. See also Hu [491], Mandrekar and Gawarecki [688].

Section 7.2: The regularity and other properties of the solutions of SDEs driven by FBM are also studied in Baudoin and Hairer [106], Baudoin and Ouyang [107], Baudoin, Ouyang, and Tindel [109], Baudoin, Nualart, Ouyang, and Tindel [108]. Similar issues are addressed for more general or other Gaussian driving processes in Cass and Friz [205], Cass, Hairer, Litterer, and Tindel [206], Hairer and Pillai [446]; for delay equations in Tindel, Tudor, and Viens [959], Neuenkirch, Nourdin, and Tindel [744], Ferrante and Rovira [363], León and Tindel [608]; for higher-dimensional systems in Driscoll [329]. Ergodicity of SDEs driven by FBM is considered in Hairer [444]. The regularity of the law of the supremum of FBM is studied in Lanjri Zadi and Nualart [593].

SDEs driven by FBM and Poisson point process are considered in Bai and Ma [73]. State space approach to fractional integral equations is studied in Buchmann and Klüppelberg [192].

There are a number of interesting works, not covered in Section 7.2, on statistical inference in nonlinear SDEs driven by FBM. See Gloter and Hoffmann [408], Tudor and Viens [964], Papavasiliou and Ladroue [791], Chronopoulou and Tindel [239], Lysy and Pillai [656], Beskos, Dureau, and Kalogeropoulos [139], Kubilius and Skorniakov [580].

Numerical approximations to SDEs driven by FBM are also studied in Neuenkirch [741, 742], Neuenkirch and Nourdin [743], Garrido-Atienza, Kloeden, and Neuenkirch [382], Gradinaru and Nourdin [419], Mishura and Shevchenko [723], Deya, Neuenkirch, and Tindel [304], Naganuma [738], Cohen, Panloup, and Tindel [256]. Most of these works also involve the Malliavin calculus.

The Malliavin calculus and Stein's method are the subject of the monographs by Nourdin and Peccati [763], Nourdin [760]. Several notable papers are Nualart and Peccati [771], Nourdin and Peccati [762]. The proof of Proposition 7.2.15 follows that of Theorem 6.3 in Nourdin [760].

The section on local time for FBM is based largely on the work of Coutin and Decreusefond [262], and the general treatment of local times in Geman and Horowitz [384]. See also Chapter 10 in Biagini et al. [147], Hu and Øksendal [492]. The so-called intersection local time for FBM is studied in Chen, Li, Rosiński, and Shao [225], Chen and Yan [216], Jung and Markowsky [540].

Among other applications not covered in Section 7.2 is the nonlinear filtering for FBM. See Coutin and Decreusefond [262], Amirdjanova [27], Xiong and Zhao [1018], Amirdjanova and Linn [28], Linn and Amirdjanova [636]. Decreusefond and Nualart [295] study the hitting times of FBM and other Gaussian processes. Ghosh, Roitershtein, and Weerasinghe [388] consider optimal control of systems driven by FBM. Nualart and Pérez-Abreu [772] study the eigenvalues of a matrix fractional Brownian motion. Finally, we also note the books by Rostek [864], Tudor [963], touching on stochastic calculus, FBM and applications.

8

Series Representations of Fractional Brownian Motion

In this chapter, we present several special series representations of FBM. They are interesting in their own right, shed further light on FBM, and can be used in simulating FBM, for the study of its sample paths and other (e.g., statistical) tasks. We recall

- The Karhunen–Loève decomposition of stochastic processes (Section 8.1)

Except the case of BM, an explicit form of the basis functions (in terms of known special functions or transformations) in the Karhunen–Loève decomposition of FBM is currently not known. Two other interesting and explicit series representations of FBM are also considered:

- A wavelet-based expansion of FBM (Section 8.2)
- The Paley–Wiener representation of FBM (Section 8.3)

Since we will sometimes make reference to Chapter 6, we will parameterize FBM B_H with the self-similarity parameter $H \in (0, 1)$ as

$$B^\kappa, \quad \kappa = H - \frac{1}{2} \in \left(-\frac{1}{2}, \frac{1}{2}\right).$$

BM B_H with $H = 1/2$ is now associated with $\kappa = 0$, and is denoted B^0. The material of Chapter 6 (more specifically, Sections 6.1–6.4) will be used only in Sections 8.2 and 8.3.

8.1 Karhunen–Loève Decomposition and FBM

We shall first recall the general Karhunen–Loève decomposition (Section 8.1.1), and then discuss it in the case of FBM (Section 8.1.3).

8.1.1 The Case of General Stochastic Processes

We recall here the Karhunen–Loève decomposition of general stochastic processes, following the formulation found in Ash and Gardner [52], Section 1.4. Let $\{X(t)\}_{t \in [0,a]}$ be a real-valued[1] stochastic process with $\mathbb{E}X(t) = 0$, $\mathbb{E}X(t)^2 < \infty$ and the covariance

$$K(t, s) = \mathbb{E}X(t)X(s), \quad s, t \in [0, a], \tag{8.1.1}$$

---

[1] Complex-valued processes can be considered as well.

which is assumed to be continuous on $[0, a] \times [0, a]$. Consider an integral operator

$$(Kf)(t) = \int_0^a K(t, s) f(s) ds, \quad t \in [0, a], \tag{8.1.2}$$

on $L^2[0, a]$. Since the operator is compact self-adjoint and by using the Hilbert–Schmidt theorem, there are eigenfunctions $\{\phi_n\}_{n \geq 1}$ and the corresponding eigenvalues $\{\lambda_n\}_{n \geq 1}$ satisfying

$$(K\phi_n)(t) = \int_0^a K(t, s)\phi_n(s) ds = \lambda_n \phi_n(t), \quad t \in [0, a] \tag{8.1.3}$$

(e.g., Gohberg, Goldberg, and Kaashoek [416], Section 5.1). The eigenfunctions are orthonormal in $L^2[0, a]$; that is,

$$\int_0^a \phi_m(s)\phi_n(s) ds = \begin{cases} 1, & \text{if } m = n, \\ 0, & \text{if } m \neq n. \end{cases} \tag{8.1.4}$$

Moreover, by Mercer's theorem,

$$K(t, s) = \sum_{n=1}^{\infty} \lambda_n \phi_n(t)\phi_n(s) \tag{8.1.5}$$

with the series converging absolutely and uniformly on $[0, a] \times [0, a]$ (e.g., Gohberg et al. [416], Section 5.2). The representation (8.1.5) leads to the following Karhunen–Loève decomposition (Ash and Gardner [52], Section 1.4.1; Riesz and Sz.-Nagy [850], Section 98).

Theorem 8.1.1 *In the setting described above,*

$$X(t) = \sum_{n=1}^{\infty} \sqrt{\lambda_n} \phi_n(t) Z_n, \quad t \in [0, a], \tag{8.1.6}$$

where

$$Z_n = \frac{1}{\sqrt{\lambda_n}} \int_0^a X(t)\phi_n(t) dt$$

are mean zero, variance one and uncorrelated random variables. The series (8.1.6) converges in $L^2(\Omega)$ uniformly in $t \in [0, a]$; that is,

$$\sup_{t \in [0,a]} \mathbb{E}\left| X(t) - \sum_{n=1}^{N} \sqrt{\lambda_n} \phi_n(t) Z_n \right|^2 \to 0, \tag{8.1.7}$$

as $N \to \infty$.

Note that changing a, changes the series expansion.

8.1.2 The Case of BM

The following example provides the well-known Karhunen–Loève decomposition of BM.

Example 8.1.2 (*Karhunen–Loève decomposition of BM*) For standard BM $B^0 = \{B^0(t)\}_{t \in [0,a]}$, the covariance is $K(t,s) = \min\{t,s\}$, $s,t \in [0,a]$. The eigenfunctions ϕ and eigenvalues λ satisfy the integral equation (8.1.3); that is,

$$\int_0^a \min\{t,s\}\phi(s)ds = \lambda \phi(t), \quad t \in [0,a], \tag{8.1.8}$$

that is,

$$\int_0^t s\phi(s)ds + t\int_t^a \phi(s)ds = \lambda \phi(t), \quad t \in [0,a]. \tag{8.1.9}$$

Differentiating (8.1.9) leads to

$$\int_t^a \phi(s)ds = \lambda \phi'(t) \tag{8.1.10}$$

and another differentiation to

$$-\phi(t) = \lambda \phi''(t). \tag{8.1.11}$$

This leads to the solution

$$\phi(t) = A \sin\left(\frac{t}{\sqrt{\lambda}}\right) + B \cos\left(\frac{t}{\sqrt{\lambda}}\right). \tag{8.1.12}$$

Setting $t=0$ in (8.1.9) gives $\phi(0)=0$ and hence $B=0$. Setting $t=a$ in (8.1.10) leads to $\phi'(a)=0$ and hence $\cos(a/\sqrt{\lambda})=0$ or $a/\sqrt{\lambda} = (2n-1)\pi/2$, $n \geq 1$. Thus, the eigenvalues and the normalized eigenfunctions are

$$\lambda_n = \frac{4a^2}{(2n-1)^2\pi^2}, \quad \phi_n(t) = \sqrt{\frac{2}{a}} \sin\frac{(2n-1)\pi t}{2a}, \tag{8.1.13}$$

for $n \geq 1$, where $A = \sqrt{2/a}$ ensures that $\int_0^a \phi_n(t)^2 dt = 1$. The Karhunen–Loève decomposition of BM is then

$$B^0(t) = \sqrt{\frac{2}{a}} \sum_{n=1}^{\infty} \frac{\sin((n-\frac{1}{2})\pi t/a)}{(n-\frac{1}{2})\pi/a} Z_n, \tag{8.1.14}$$

where Z_n, $n \geq 1$, are i.i.d. $\mathcal{N}(0,1)$ random variables.

8.1.3 The Case of FBM

The standard FBM $B^\kappa = \{B^\kappa(t)\}_{t \in [0,a]}$, $\kappa \in (-1/2, 1/2)$, has the covariance $K(t,s) = \frac{1}{2}(t^{2\kappa+1} + s^{2\kappa+1} - |t-s|^{2\kappa+1})$, $s,t \in [0,a]$. The eigenfunctions ϕ_n and eigenvalues λ_n satisfy the integral equation

$$\frac{1}{2}\int_0^a (t^{2\kappa+1} + s^{2\kappa+1} - |t-s|^{2\kappa+1})\phi_n(s)ds = \lambda_n \phi_n(t), \quad t \in [0,a]. \tag{8.1.15}$$

An explicit form for ϕ_n and λ_n satisfying (8.1.15) is currently not known. On the other hand, their asymptotic behavior is known as $n \to \infty$. As shown in Chigansky and Kleptsyna [236], Theorem 1.1, for $\kappa \neq 0$ and $a = 1$,

$$\lambda_n = \frac{\cos(\kappa\pi)\Gamma(2\kappa+2)}{(\nu_n)^{2\kappa+2}}, \quad n = 1, 2, \ldots, \tag{8.1.16}$$

where

$$v_n = \left(n - \frac{1}{2}\right) - \frac{\kappa\pi}{2} + O(n^{-1}), \quad \text{as } n \to \infty. \tag{8.1.17}$$

Moreover,

$$\phi_n(t) = \sqrt{2}\sin\left(v_n t + \frac{\kappa\pi}{4}\right) + \frac{\sqrt{2\kappa+2}}{\pi}\int_0^\infty \rho_0(u)\left((-1)^{n-1}e^{(t-1)v_n u} - ue^{-tv_n u}\right)du$$
$$+ r_n(t)n^{-1}, \quad n = 1, 2, \ldots, \tag{8.1.18}$$

where the residual $r_n(t)$ is bounded by a constant depending only on κ, and

$$\rho_0(u) = \frac{\sin\theta_0(u)}{\gamma_0(u)}\exp\left\{\frac{1}{\pi}\int_0^\infty \frac{\theta_0(v)}{v+u}dv\right\}, \quad u > 0,$$

with

$$\theta_0(u) = \arctan\frac{u^{-2\kappa-2}\sin(\kappa\pi)}{1 + u^{-2\kappa-1}\cos(\kappa\pi)},$$
$$\gamma_0^2(u) = (u + u^{-2\kappa-1}\cos(\kappa\pi))^2 + (u^{-2\kappa-1}\sin(\kappa\pi))^2. \tag{8.1.19}$$

See also Chigansky and Kleptsyna [235]. This can be restated for general a using Exercise 8.1.

8.2 Wavelet Expansion of FBM

Meyer, Sellan, and Taqqu [716] obtained another interesting expansion of FBM based on wavelets. We shall outline their construction in this section, by recalling first the idea of wavelet bases (Section 8.2.1), then by considering the so-called fractional wavelets (Section 8.2.2) and finally by giving the wavelet-based expansion of FBM. It should also be noted from the start that the derived wavelet expansion is special in the sense that the underlying wavelet basis is adapted to the covariance structure of FBM in order to obtain an expansion with uncorrelated wavelet coefficients.

Warning: Recall that detail (wavelet) coefficients were used in estimation of the LRD parameter in Section 2.10.4. The detail coefficients of a process X were defined in (2.10.20)–(2.10.21) as $d(j,k) = \int_\mathbb{R} X(t)2^{-j/2}\psi(2^{-j}t - k)dt$, where large j corresponds to large scales (low frequencies) and small (large negative) j corresponds to small scales (high frequencies). This choice of j is the convention used in the applied wavelet literature. In this section, we shall follow the mathematics convention where j is replaced by $-j$. In the mathematics convention, large j corresponds to small scales (high frequencies) and small (large negative) j corresponds to large scales (low frequencies). To avoid confusion, we shall write $d_{j,k}$ instead of $d(j,k)$.

8.2.1 Orthogonal Wavelet Bases

An orthonormal wavelet basis of $L^2(\mathbb{R})$ is a collection of functions

$$\phi(t-k), \quad k \in \mathbb{Z}; \quad 2^{j/2}\psi(2^j t - k), \quad j \geq 0, \ k \in \mathbb{Z}, \tag{8.2.1}$$

with special properties. In particular, the functions are orthogonal, have unit $L^2(\mathbb{R})$ norms and allow expanding any $f \in L^2(\mathbb{R})$ as

$$f(t) = \sum_{k=-\infty}^{\infty} a_{0,k} \phi(t-k) + \sum_{j=0}^{\infty} \sum_{k=-\infty}^{\infty} d_{j,k} 2^{j/2} \psi(2^j t - k), \qquad (8.2.2)$$

where

$$a_{0,k} = \int_{\mathbb{R}} f(u) \phi(u-k) du, \quad d_{j,k} = \int_{\mathbb{R}} f(u) 2^{j/2} \psi(2^j u - k) du. \qquad (8.2.3)$$

The function ψ is known as a *wavelet* (mother wavelet), and ϕ as a *scaling function* (father wavelet). Note that $\phi(t-k)$, $k \in \mathbb{Z}$, are the translated copies of ϕ, and that $2^{j/2} \psi(2^j t - k)$, $j \geq 0$, $k \in \mathbb{Z}$, are the translated and scaled copies of ψ. The index k refers to location, and the index j (or rather the quantity 2^j) to scale or resolution. Note that large positive j corresponds to fine (small) scales while large negative j to coarse (large) scales. The coefficients $a_{0,k}$ and $d_{j,k}$ are known, respectively, as the *approximation* and *detail (wavelet) coefficients*.

Moreover, the functions ϕ and ψ can be related across different scales. More precisely, there are sequences $\{h_n\}_{n \in \mathbb{Z}}$ and $\{g_n\}_{n \in \mathbb{Z}}$ such that

$$2^{-1/2} \phi(2^{-1} t) = \sum_{n=-\infty}^{\infty} h_n \phi(t - n), \qquad (8.2.4)$$

$$2^{-1/2} \psi(2^{-1} t) = \sum_{n=-\infty}^{\infty} g_n \phi(t - n). \qquad (8.2.5)$$

The sequences $h = \{h_n\}_{n \in \mathbb{Z}}$ and $g = \{g_n\}_{n \in \mathbb{Z}}$ are known as *conjugate mirror filters* (CMFs). The filter h is low-pass[2] and satisfies

$$|\widehat{h}(w)|^2 + |\widehat{h}(w+\pi)|^2 = 2, \quad \widehat{h}(0) = \sqrt{2}, \qquad (8.2.6)$$

where

$$\widehat{a}(w) = \sum_{n=-\infty}^{\infty} a_n e^{-inw}$$

denotes the Fourier transform of a sequence $a = \{a_n\}_{n \in \mathbb{Z}}$.[3] The filter g is high-pass and related to h as

$$\widehat{g}(w) = e^{-iw} \overline{\widehat{h}(w+\pi)}, \qquad (8.2.7)$$

where the bar indicates complex conjugation. See Exercise 8.2 for some other basic properties of CMFs h and g.

---

[2] A low-pass filter lets the low frequencies (large scales) through, and a high-pass filter lets the high frequencies (small scales) through.
[3] See Appendix A.1.1. When consulting the wavelet literature, it is important to be aware of the conventions used for the Fourier transforms.

As the name suggests, a wavelet ψ is oscillating and wave-like. This is also reflected in the number of zero moments: ψ has Q zero moments when

$$\int_{\mathbb{R}} \psi(u) u^q du = 0, \quad q = 0, \ldots, Q - 1 \qquad (8.2.8)$$

(cf. (2.10.19)). The larger the number of zero moments, the more oscillating ψ has to be. In contrast, a scaling function ϕ is usually such that

$$\widehat{\phi}(0) = \int_{\mathbb{R}} \phi(u) du = 1. \qquad (8.2.9)$$

The relations (8.2.4)–(8.2.5) can be used to write (8.2.2) as follows: for any $J \in \mathbb{Z}$,

$$f(t) = \sum_{k=-\infty}^{\infty} a_{J,k} 2^{J/2} \phi(2^J t - k) + \sum_{j=J}^{\infty} \sum_{k=-\infty}^{\infty} d_{j,k} 2^{j/2} \psi(2^j t - k),$$

$$= \sum_{j=-\infty}^{\infty} \sum_{k=-\infty}^{\infty} d_{j,k} 2^{j/2} \psi(2^j t - k), \qquad (8.2.10)$$

where

$$a_{J,k} = \int_{\mathbb{R}} f(u) 2^{J/2} \phi(2^J u - k) du. \qquad (8.2.11)$$

In the representation (8.2.10), the term $\sum_{k=-\infty}^{\infty} a_{J,k} 2^{J/2} \phi(2^J t - k)$ is thought as an approximation to f at scale J and the terms $\sum_{k=-\infty}^{\infty} d_{j,k} 2^{j/2} \psi(2^j t - k)$ as details at scales j. Moreover, if f is smooth, the approximation coefficients $\{a_{J,k}\}_{k \in \mathbb{Z}}$ approximate the function f as

$$2^{J/2} a_{J,k} \approx f\left(\frac{k}{2^J}\right). \qquad (8.2.12)$$

The convergence rate in (8.2.12) is exponential in J for smooth f (Exercise 8.3).

Fast Wavelet Transform (FWT)

The CMFs $h = \{h_n\}_{n \in \mathbb{Z}}$ and $g = \{g_n\}_{n \in \mathbb{Z}}$ play a central role in the so-called *Fast Wavelet Transform* (FWT), which relates the approximation coefficients $a_j = \{a_{j,k}\}_{k \in \mathbb{Z}}$ and the detail coefficients $d_j = \{d_{j,k}\}_{k \in \mathbb{Z}}$ across different scales. It is used *at decomposition* to expand a function in wavelets and *at reconstruction* to obtain it back using the wavelet coefficients.

At decomposition, the FWT uses

$$a_{j-1} = \downarrow_2 (\overline{h} * a_j), \quad d_{j-1} = \downarrow_2 (\overline{g} * a_j), \qquad (8.2.13)$$

where, for general sequences $x = \{x_n\}_{n \in \mathbb{Z}}$ and $y = \{y_n\}_{n \in \mathbb{Z}}$, $\overline{x}$ is the time reversion operation defined by $(\overline{x})_n = x_{-n}$, $\downarrow_2 x$ is the downsampling by 2 operation defined by $(\downarrow_2 x)_n = x_{2n}$ and $(x * y)_n = \sum_{k=-\infty}^{\infty} x_k y_{n-k}$ stands for the usual convolution. As illustration,

$$\downarrow_2 (\ldots x_{-2} \, x_{-1} \, x_0 \, x_1 \, x_2 \ldots) = (\ldots x_{-2} \, x_0 \, x_2 \ldots).$$

The FWT at decomposition thus allows obtaining the approximation and detail coefficients at coarser scale $j - 1$ from the approximation coefficients at finer scale j.

8.2 Wavelet Expansion of FBM

At reconstruction, the FWT uses

$$a_j = h * (\uparrow_2 a_{j-1}) + g * (\uparrow_2 d_{j-1}). \tag{8.2.14}$$

Here, $\uparrow_2 x$ is the upsampling by 2 operation defined by $(\uparrow_2 x)_n = x_k$ if $n = 2k$, and $= 0$ if $n = 2k + 1$, for $k \in \mathbb{Z}$. As illustration,

$$\uparrow_2 (\ldots x_{-2}\, x_{-1}\, x_0\, x_1\, x_2 \ldots) = (\ldots x_{-2}\, 0\, x_{-1}\, 0\, x_0\, 0\, x_1\, 0\, x_2 \ldots).$$

The FWT at reconstruction thus allows obtaining the approximation coefficients at finer scale j from the approximation and details coefficients at coarser scale $j - 1$.

In practice, one usually works only with the CMFs h and g, and the approximation and detail coefficients a_j and d_j. If the data f_k, $k = 1, \ldots, N$, are the function values $f(2^{-J}k)$ at scale J, the approximation coefficients would be taken as $2^{-J/2} f(2^{-J}k)$ in view of (8.2.12)[4] and the FWT at decomposition would be used to compute the available approximation and detail coefficients at coarser scales. The FWT operation is of complexity $O(N)$ (with the sample size N above) and hence is very fast. Likewise, in simulations, one would start with, e.g., the approximation coefficients a_0 at scale 0 and then use the FWT at reconstruction to obtain the approximation coefficients a_J at finer scale J, which are then taken for the values of $2^{-J/2} f(2^{-J}k)$ according to the approximation (8.2.12).

The pair of wavelet ψ and scaling function ϕ (and the associated CMFs h and g) satisfying the properties above are referred to as *Multiresolution Analysis* (MRA).[5]

Example 8.2.1 (*Haar MRA*) An example is the Haar MRA with

$$\phi(t) = 1_{[0,1)}(t), \quad \psi(t) = 1_{[1/2,1)}(t) - 1_{[0,1/2)}(t) = \begin{cases} -1, & t \in [0, 1/2), \\ 1, & t \in [1/2, 1), \\ 0, & \text{otherwise} \end{cases} \tag{8.2.15}$$

and the CMFs

$$h_n = \begin{cases} 2^{-1/2}, & n = 0, 1, \\ 0, & \text{otherwise}, \end{cases} \quad g_n = \begin{cases} -2^{-1/2}, & n = 0, \\ 2^{-1/2}, & n = 1, \\ 0, & \text{otherwise} \end{cases} \tag{8.2.16}$$

(Exercise 8.4). As in this example, the functions ϕ and ψ with bounded supports are associated with finite CMFs h and g (that is, having only a finite number of nonzero elements).

Note also that ϕ and ψ in the Haar MRA are non-smooth, and ψ has only one zero moment $Q = 1$ in the sense of (8.2.8).[6] Constructing smooth basis functions ϕ and ψ with compact supports and larger number of zero moments Q is, in fact, not an easy task. The celebrated Daubechies MRA provides examples of such bases for any desired number of zero moments $Q \geq 1$. When $Q = 1$, the Daubechies MRA is just the Haar MRA. When

---

[4] This approximation is sometimes referred to as a "wavelet crime" in the wavelet literature.
[5] Strictly speaking, the MRA refers to the sequence of approximation and detail subspaces of $L^2(\mathbb{R})$ induced by the wavelet and scaling functions.
[6] The degree of smoothness and the number of zero moments are related. See, e.g., Mallat [675], Section 7.2.

$Q \geq 2$, the corresponding CMFs can be computed only numerically. See Daubechies [287], Meyer [715], Mallat [675] for more details.

Example 8.2.2 (*Daubechies MRA with $Q = 2$ zero moments*) When the number of zero moments $Q = 2$, the CMF h in the Daubechies MRA is given (up to 12 decimal places) by $h_0 = 0.482962913145$, $h_1 = 0.836516303738$, $h_2 = 0.224143868042$ and $h_3 = -0.129409522551$ (and other h_n's being zero). The corresponding CMF g can be deduced from h by using (8.2.7). The functions ϕ and ψ can be approximated (computed numerically) by using the CMFs and the FWT at reconstruction as described above. More specifically, note that, for the function $f = \phi$,

$$a_{0,k} = \int_{\mathbb{R}} \phi(u)\phi(u-k)du = \begin{cases} 1, & k = 0, \\ 0, & k \neq 0 \end{cases} \quad d_{j,k} = \int_{\mathbb{R}} \phi(u)2^{-j/2}\psi(2^{-j}u - k)du = 0,$$

for $j \geq 0, k \in \mathbb{Z}$, by orthonormality. Similarly, for the function $f = \psi$,

$$a_{0,k} = \int_{\mathbb{R}} \psi(u)\phi(u-k)du = 0, \quad d_{0,k} = \int_{\mathbb{R}} \psi(u)\psi(u-k)du = \begin{cases} 1, & k = 0, \\ 0, & k \neq 0, \end{cases}$$

$$d_{j,k} = \int_{\mathbb{R}} \psi(u)2^{-j/2}\psi(2^{-j}u - k)du = 0,$$

for $j \geq 1, n \in \mathbb{Z}$. Having $a_0 = \{a_{0,k}\}_{k \in \mathbb{Z}}$ and $d_j = \{d_{j,k}\}_{k \in \mathbb{Z}}$, $j \geq 0$, the FWT at reconstruction given by (8.2.14) can be used to obtain the approximation coefficients $a_{J,k}, k \in \mathbb{Z}$, at finer and finer scales J. When normalized, $2^{J/2}a_{J,k}$ approximate $f(k/2^J)$ in view of (8.2.12). The plots of ϕ and ψ obtained through this procedure are given in Figure 8.1 for the Daubechies MRA with $Q = 2$ zero moments. A computationally more inclined and interested reader could try to implement the FWT procedure and to produce the plots in Figure 8.1.

Known examples of infinite MRAs are the Shannon MRA, the Battle–Lemarié MRA, the Meyer MRA and others. See, for example, Mallat [675].

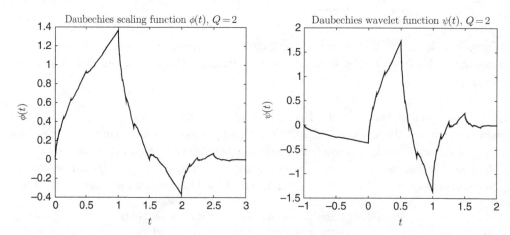

Figure 8.1 The Daubechies scaling and wavelet functions for $Q = 2$ zero moments.

The Meyer MRA

For technical convenience and following Meyer et al. [716], we shall work throughout with the Meyer MRA described as follows. Let $S(\mathbb{R})$ be the set of all infinitely differentiable functions $f(t)$ which satisfy

$$\lim_{|t|\to\infty} t^m \frac{d^n}{dt^n} f(t) = 0 \tag{8.2.17}$$

for any m and n; that is, the tails of f and its all derivatives decrease to 0 faster than any polynomial.

In the Meyer MRA, a scaling function ϕ is such that $\phi \in S(\mathbb{R})$ or, equivalently, $\widehat{\phi} \in S(\mathbb{R})$ (Exercise 8.5), and satisfies:

$$\begin{cases} \widehat{\phi}(x) \in [0, 1], \\ \widehat{\phi}(x) = 1 \quad \text{for } |x| \leq \frac{2\pi}{3}, \; \widehat{\phi}(x) = 0 \quad \text{for } |x| \geq \frac{4\pi}{3}, \\ \text{and } \widehat{\phi}(x) \text{ decreasing on } (\frac{2\pi}{3}, \frac{4\pi}{3}), \\ \widehat{\phi}(-x) = \widehat{\phi}(x), \\ \sum_{n=-\infty}^{\infty} (\widehat{\phi}(\xi + 2\pi n))^2 = 1. \end{cases} \tag{8.2.18}$$

The last condition in (8.2.18) ensures that $\phi(t-n)$, $n \in \mathbb{Z}$, are orthonormal. There are many choices for the function ϕ (Meyer [715], Section VIII). The wavelet function is defined through[7]

$$\widehat{\psi}(x) = e^{ix/2} \Big((\widehat{\phi}(x/2))^2 - (\widehat{\phi}(x))^2 \Big)^{1/2}. \tag{8.2.19}$$

The wavelet ψ is such that $\psi \in S(\mathbb{R})$ and

$$\widehat{\psi}(x) = 0 \quad \text{if } |x| \leq 2\pi/3 \text{ or } |x| \geq 8\pi/3. \tag{8.2.20}$$

In particular, all the moments of ψ are zero since $\int_{\mathbb{R}} t^n \psi(t) dt = \widehat{\psi}^{(n)}(0)/i^n = 0$ for all $n \geq 0$. In contrast, $\widehat{\phi}(x) = 1$ in the neighborhood of $x = 0$. In the following section, we use the Meyer MRA unless stated otherwise.

8.2.2 Fractional Wavelets

We outline first the idea behind a wavelet-based decomposition of FBM, and then examine the resulting wavelet basis, which we call "fractional wavelets." Recall from (6.2.29) that FBM can be represented as

$$B^{\kappa}(t) = \int_{\mathbb{R}} (I_-^{\kappa} 1_{[0,t)})(u) dB^0(u) =: \int_{\mathbb{R}} f_t(u) dB^0(u), \tag{8.2.21}$$

where

$$\mathbb{E} B^{\kappa}(t)^2 = \frac{c_{1,\kappa}^2}{\Gamma(\kappa+1)^2} \tag{8.2.22}$$

---

[7] Note that the convention for the Fourier transform used here is different from Meyer et al. [716].

and I_-^κ is a fractional integral or derivative of order κ. Since $f_t \in L^2(\mathbb{R})$, it can be expanded as in (8.2.2),

$$f_t(u) = \sum_{k=-\infty}^{\infty} a_{0,k}(t)\phi(u-k) + \sum_{j=0}^{\infty}\sum_{k=-\infty}^{\infty} d_{j,k}(t)2^{j/2}\psi(2^j u - k), \qquad (8.2.23)$$

where

$$a_{0,k}(t) = \int_{\mathbb{R}} f_t(u)\phi(u-k)du, \quad d_{j,k}(t) = \int_{\mathbb{R}} f_t(u)2^{j/2}\psi(2^j u - k)du. \qquad (8.2.24)$$

Substituting (8.2.23) into (8.2.21) yields the expansion

$$B^\kappa(t) = \sum_{k=-\infty}^{\infty} a_{0,k}(t)\eta_{0,k} + \sum_{j=0}^{\infty}\sum_{k=-\infty}^{\infty} d_{j,k}(t)\epsilon_{j,k}, \qquad (8.2.25)$$

where

$$\eta_{0,k} = \int_{\mathbb{R}} \phi(u-k)dB^0(u), \quad \epsilon_{j,k} = \int_{\mathbb{R}} 2^{j/2}\psi(2^j u - k)dB^0(u). \qquad (8.2.26)$$

By the orthonormality of the system $\phi(u-k)$, $2^{j/2}\psi(2^j u - k)$, $j \geq 0, k \in \mathbb{Z}$, the random variables $\eta_{0,k}, \epsilon_{j,k}, j \geq 0, k \in \mathbb{Z}$, are i.i.d. $\mathcal{N}(0,1)$.

Equivalent Expression of Detail Coefficients $d_{j,k}(t)$

We shall rewrite next the decomposition (8.2.25) so that it has the form of a wavelet expansion (8.2.2).[8] To do so, observe first that

$$d_{j,k}(t) = \int_{\mathbb{R}} f_t(u)2^{j/2}\psi(2^j u - k)du = \int_{\mathbb{R}} (I_-^\kappa 1_{[0,t)})(u)2^{j/2}\psi(2^j u - k)du$$
$$= \int_{\mathbb{R}} 1_{[0,t)}(s)(I_+^\kappa 2^{j/2}\psi(2^j \cdot -k))(s)ds = \int_{\mathbb{R}} 1_{[0,t)}(s)2^{-j(\kappa-1/2)}(I_+^\kappa \psi)(2^j s - k)ds, \qquad (8.2.27)$$

where we used the integration by parts formula (6.1.27) and the fact that, e.g., for $\kappa > 0$,

$$(I_+^\kappa \psi(2^j \cdot -k))(s) = \frac{1}{\Gamma(\kappa)}\int_{-\infty}^{s} \psi(2^j u - k)(s-u)^{\kappa-1}du$$
$$= \frac{2^{-j\kappa}}{\Gamma(\kappa)}\int_{-\infty}^{2^j s - k} \psi(v)(2^j s - k - v)^{\kappa-1}dv = 2^{-j\kappa}(I_+^\kappa \psi)(2^j s - k))$$

after making the change of variables $2^j u - k = v$. The relation (8.2.27) leads further to

$$d_{j,k}(t) = 2^{-j(\kappa-1/2)}\int_0^\infty (I_+^\kappa \psi)(2^j s - k)ds - 2^{-j(\kappa-1/2)}\int_t^\infty (I_+^\kappa \psi)(2^j s - k)ds$$
$$= 2^{-j(\kappa+1/2)}\int_{-k}^\infty (I_+^\kappa \psi)(u)du - 2^{-j(\kappa+1/2)}\int_{2^j t - k}^\infty (I_+^\kappa \psi)(u)du$$

---

[8] The reader not familiar with fractional integration, a topic introduced in Chapter 6, may jump directly to the definition (8.2.30) of the fractional wavelet $\widehat{\psi}^{(\alpha)}(x)$ and to the definition (8.2.34) of the fractional scaling function $\widehat{\phi}_\Delta^{(\alpha)}(x)$.

$$= 2^{-j(\kappa+1/2)}(I_+^1(I_+^\kappa\psi))(-k) - 2^{-j(\kappa+1/2)}(I_+^1(I_+^\kappa\psi))(2^j t - k)$$
$$= 2^{-j(\kappa+1/2)}(I_+^{\kappa+1}\psi)(-k) - 2^{-j(\kappa+1/2)}(I_+^{\kappa+1}\psi)(2^j t - k), \qquad (8.2.28)$$

where we made the change of variables $2^j s - k = u$ and also used the semigroup property (6.1.26). The presence of $I_+^{\kappa+1}\psi$ motivates the following definition.

Definition 8.2.3 A *fractional wavelet* of order $\alpha \in \mathbb{R}$ is defined as

$$\psi^{(\alpha)} = I_+^\alpha \psi \qquad (8.2.29)$$

or through the Fourier transform as

$$\widehat{\psi}^{(\alpha)}(x) = \widehat{\psi}(x)(-ix)^{-\alpha} \qquad (8.2.30)$$

(see Section 6.1.5).

Note that, for the Meyer MRA, $\widehat{\psi}^{(\alpha)} \in S(\mathbb{R})$ for all $\alpha \in \mathbb{R}$ since $\widehat{\psi} \in S(\mathbb{R})$ and $\widehat{\psi}(x) = 0$ in the neighborhood of $x = 0$ and $x = \infty$ (see (8.2.20)). Therefore, $\psi^{(\alpha)}$ is well-defined as the inverse of its Fourier transform and belongs to $S(\mathbb{R})$ (see Exercise 8.5).

In Section 8.2.4, we shall use (8.2.28) and hence the scaled and translated fractional wavelets to rewrite the detail coefficients $d_{j,k}(t)$ in (8.2.25). The approximation coefficients $a_{0,k}(t)$ in (8.2.25), on the other hand, cannot be dealt with in the same way as the detail coefficients $d_{j,k}(t)$ for two reasons. First, we are looking for a wavelet expansion where the coefficients $\eta_{0,k}$ approximate FBM $B^\kappa(k)$ (see (8.2.12)). This is certainly not possible in the expansion (8.2.25) since as k varies, $\eta_{0,k}$ are independent, whereas $B^\kappa(k)$ are dependent. Second, as indicated above, fractional wavelets are well-defined for the Meyer MRA for all $\alpha \in \mathbb{R}$ and hence can be used for $I_+^{\kappa+1}\psi$ in the representation (8.2.28). This is not the case for a scaling function ϕ. Since $\widehat{\phi}(x) = 1$ in the neighborhood of $x = 0$, $\widehat{\phi}(x)(-ix)^{-1-\kappa}$ is not in $L^2(\mathbb{R})$ and hence $I_+^{\kappa+1}\phi$ is not well-defined (at least, it cannot be defined through its Fourier transform). Both of these issues can be addressed simultaneously through the following modification.

Equivalent Expression for the Approximation Term

Informally, by arguing as for the detail coefficients (8.2.28) above, the approximation term in (8.2.25) is

$$\sum_{k=-\infty}^{\infty} a_{0,k}(t)\eta_{0,k} = \sum_{k=-\infty}^{\infty} (I_+^{\kappa+1}\phi)(-k)\eta_{0,k} - \sum_{k=-\infty}^{\infty} (I_+^{\kappa+1}\phi)(t-k)\eta_{0,k}.$$

Thus, up to a random shift and a change of sign, it is

$$Y_0(t) = \sum_{k=-\infty}^{\infty} (I_+^{\kappa+1}\phi)(t-k)\eta_{0,k}. \qquad (8.2.31)$$

Since the function $I_+^{\kappa+1}\phi$ is not well-defined as argued above, the term (8.2.31) will be modified as follows. We proceed heuristically to arrive at expressions which are well-defined.

By using the relation $\widehat{f(\cdot - k)}(x) = e^{ixk}\widehat{f}(x)$, the "Fourier transform" of $Y_0(t)$ is

$$\widehat{Y}_0(x) = \widehat{I_+^{\kappa+1}\phi}(x) \sum_{k=-\infty}^{\infty} e^{ixk}\eta_{0,k} = \frac{\widehat{\phi}(x)}{(-ix)^{\kappa+1}} \overline{\widehat{\eta_0}(x)}.$$

Write this further as

$$\widehat{Y}_0(x) = \widehat{\phi}_\Delta^{(\kappa+1)}(x)\overline{\widehat{S}_0(x)}, \tag{8.2.32}$$

where

$$\widehat{\phi}_\Delta^{(\kappa+1)}(x) = \widehat{\phi}(x)\left(\frac{1-e^{ix}}{-ix}\right)^{\kappa+1}, \quad \widehat{S}_0(x) = (1-e^{-ix})^{-(\kappa+1)}\widehat{\eta}(x).$$

Note the correction term $(1-e^{ix})^{\kappa+1}$ in $\widehat{\phi}_\Delta^{(\kappa+1)}(2^{-j}x)$ and its reciprocal in $\widehat{S}_0(x)$. Switching back to the time domain, Y_0 becomes

$$Y_0(t) = \sum_{k=-\infty}^{\infty} \phi_\Delta^{(\kappa+1)}(t-k)S_{0,k}. \tag{8.2.33}$$

This motivates the following definition.

Definition 8.2.4 A *fractional scaling function* of order $\alpha \in \mathbb{R}$ is defined through the Fourier transform as

$$\widehat{\phi}_\Delta^{(\alpha)}(x) = \widehat{\phi}(x)\left(\frac{1-e^{ix}}{-ix}\right)^\alpha. \tag{8.2.34}$$

Lemma 8.2.5 *For the Meyer MRA, the fractional scaling function $\phi_\Delta^{(\alpha)}$ is well-defined and belongs to $S(\mathbb{R})$ for all $\alpha \in \mathbb{R}$.*

Proof It is enough to show that $\widehat{\phi}_\Delta^{(\alpha)}(x)$, given by the right-hand side of (8.2.30), belongs to $S(\mathbb{R})$ (see Exercise 8.5). We implicitly assume that the function $u(x) := (1-e^{ix})/(-ix)$ is set to 1 at $x = 0$ so that it is continuous at this point. In particular, note that

$$u(x) = \frac{1-e^{ix}}{-ix} = \frac{1}{ix}\sum_{m=1}^{\infty}\frac{(ix)^m}{m!} = \sum_{m=1}^{\infty}\frac{(ix)^{m-1}}{m!}$$

and that $u(x)$ is infinitely differentiable. Observe that

$$\frac{d^n\widehat{\phi}_\Delta^{(\alpha)}}{dx^n}(x) = \frac{d^n}{dx^n}\left((u(x))^\alpha\widehat{\phi}(x)\right) = \sum_{k=0}^{n}\binom{n}{k}\frac{d^k}{dx^k}(u(x))^\alpha \frac{d^{n-k}\widehat{\phi}}{dx^{n-k}}(x).$$

Since $\widehat{\phi} \in S(\mathbb{R})$ and $\widehat{\phi}(x) = 0$ for $|x| > 4\pi/3$ by (8.2.18), it is enough to show that $(u(x))^\alpha$ is infinitely differentiable on $[-4\pi/3, 4\pi/3]$. This follows from the facts that $u(x) \neq 0$ for $x \in [-4\pi/3, 4\pi/3]$ and, as indicated above, $u(x)$ is infinitely differentiable. $\square$

Let $\{X_k\}_{k\in\mathbb{Z}}$ be a generic sequence and let B be the backward shift operator as in Section 2.4.1, so that $(I - B)X_k = X_k - X_{k-1}$. Then the inverse operator $(I - B)^{-1}$ involves summation. In the Fourier world, $(I - B)$ is expressed as $(1 - e^{-ix})$ and $(I - B)^{-1}$ is expressed as $(1 - e^{-ix})^{-1}$.

8.2 Wavelet Expansion of FBM

We shall apply this here to $X_k = S_{0,k}$. This will allow us to identify the coefficients $S_0 = \{S_{0,k}\}_{k\in\mathbb{Z}}$ in (8.2.33). Express $\widehat{S}_0(x)$ by using two successive operations as

$$\widehat{S}_0(x) = (1 - e^{-ix})^{-(\kappa+1)} \widehat{\eta}_0(x) = (1 - e^{-ix})^{-1}(1 - e^{-ix})^{-\kappa} \widehat{\eta}_0(x) =: (1 - e^{-ix})^{-1} \xi_0(x).$$

The first operation $(1-e^{-ix})^{-1}\xi_0(x)$ corresponds to taking a partial sum of the series ξ_0. The second operation $(1-e^{-ix})^{-\kappa}\widehat{\eta}_0(x)$, on the other hand, defines a Gaussian FARIMA$(0, \kappa, 0)$ series (see Proposition 2.4.3). Thus,

$$S_{0,k} = \begin{cases} \sum_{m=1}^{k} \xi_{0,m}, & k \geq 1, \\ 0, & k = 0, \\ -\sum_{m=k+1}^{0} \xi_{0,m}, & k \leq -1, \end{cases} \quad (8.2.35)$$

where

$$\xi_{0,k} = (I - B)^{-\kappa} \eta_{0,k} \quad (8.2.36)$$

is a Gaussian FARIMA$(0, \kappa, 0)$ series. Note also that, in contrast to $\eta_{0,k}$ as discussed above, $S_{0,k}$ in (8.2.35) can now be thought as approximation coefficients to FBM. The expansion of FBM follows by using fractional scaling and wavelet functions and will be given in Section 8.2.4 below.

Biorthogonal Systems

We provide here some further insight and information on fractional wavelet and scaling functions. According to the discussion above, the system of functions

$$2^{J/2}\phi_\Delta^{(\alpha)}(2^J t - k), \quad 2^{j/2}\psi^{(\alpha)}(2^j t - k), \quad j \geq J, k \in \mathbb{Z}, \quad (8.2.37)$$

with $\alpha = \kappa + 1$ will be used in the wavelet expansion of FBM B^κ. Note that, unless $\alpha = 0$, this is not an orthogonal system of functions. In fact, the functions (8.2.37) are biorthogonal to the functions

$$2^{J/2}\phi_\Delta^{(-\alpha)}(2^J t - k), \quad 2^{j/2}\psi^{(-\alpha)}(2^j t - k), \quad j \geq J, k \in \mathbb{Z}, \quad (8.2.38)$$

in the sense that

$$\int_\mathbb{R} \left[2^{J/2}\phi_\Delta^{(\alpha)}(2^J u - k_1)\right]\left[2^{J/2}\phi_\Delta^{(-\alpha)}(2^J u - k_2)\right] du = \begin{cases} 1, & k_1 = k_2, \\ 0, & \text{otherwise}, \end{cases} \quad (8.2.39)$$

$$\int_\mathbb{R} \left[2^{j_1/2}\psi^{(\alpha)}(2^{j_1} u - k_1)\right]\left[2^{j_2/2}\psi^{(-\alpha)}(2^{j_2} u - k_2)\right] du = \begin{cases} 1, & k_1 = k_2, j_1 = j_2, \\ 0, & \text{otherwise} \end{cases} \quad (8.2.40)$$

(Exercise 8.6).

Moreover, the functions (8.2.37) and (8.2.38) form the so-called biorthogonal wavelet bases. This means, in particular, that the biorthogonal systems (8.2.37) and (8.2.38) are the so-called Riesz bases[9] of $L^2(\mathbb{R})$, so that any $f \in L^2(\mathbb{R})$ can be expanded as

$$f(t) = \sum_{k=-\infty}^{\infty} c_{J,k} 2^{J/2} \phi_\Delta^{(\alpha)}(2^J t - k) + \sum_{j=J}^{\infty} \sum_{k=-\infty}^{\infty} e_{j,k} 2^{j/2} \psi^{(\alpha)}(2^j t - k), \quad (8.2.41)$$

---

[9] A family of functions $\{g_l\}_{l \in \mathbb{Z}}$ is a Riesz basis of $L^2(\mathbb{R})$ if (i) there are constants $C_2 \geq C_1 > 0$ such that $C_1 (\sum_{l \in \mathbb{Z}} |a_l|^2)^{1/2} \leq \|\sum_{l \in \mathbb{Z}} a_l g_l\|_{L^2(\mathbb{R})} \leq C_2 (\sum_{l \in \mathbb{Z}} |a_l|^2)^{1/2}$ for all sequences $\{a_l\}_{l \in \mathbb{Z}} \in \ell^2(\mathbb{Z})$, and (ii) the linear span of $\{e_l\}_{l \in \mathbb{Z}}$ is dense in $L^2(\mathbb{R})$.

where

$$c_{J,k} = \int_{\mathbb{R}} f(u) 2^{J/2} \phi_\Delta^{(-\alpha)}(2^J u - k) du, \quad e_{j,k} = \int_{\mathbb{R}} f(u) 2^{j/2} \psi^{(-\alpha)}(2^j u - k) du. \quad (8.2.42)$$

See Theorem 1 in Meyer et al. [716]. Note that the indices are α and $(-\alpha)$ in (8.2.41) and (8.2.42), respectively.

8.2.3 Fractional Conjugate Mirror Filters

As with orthogonal wavelet bases, there is also a FWT (also known as Fast Biorthogonal Wavelet Transform) relating the approximation coefficients $c_j = \{c_{j,k}\}_{k \in \mathbb{Z}}$ and the detail coefficients $e_j = \{e_{j,k}\}_{k \in \mathbb{Z}}$ in (8.2.41) across different scales. At decomposition, the FWT uses

$$c_{j-1} = \downarrow_2 (\overline{h^{(\alpha)}} * c_j), \quad e_{j-1} = \downarrow_2 (\overline{g^{(\alpha)}} * c_j) \quad (8.2.43)$$

(cf. (8.2.13)), and at reconstruction, it uses

$$c_j = h^{(-\alpha)} * (\uparrow_2 c_{j-1}) + g^{(-\alpha)} * (\uparrow_2 e_{j-1}) \quad (8.2.44)$$

(cf. (8.2.14)), where $h^{(\alpha)} = \{h_n^{(\alpha)}\}_{n \in \mathbb{Z}}$ and $g^{(\alpha)} = \{g_n^{(\alpha)}\}_{n \in \mathbb{Z}}$ appear in

$$2^{-1/2} \phi_\Delta^{(\alpha)}(2^{-1} t) = \sum_{n=-\infty}^{\infty} h_n^{(\alpha)} \phi_\Delta^{(\alpha)}(t - n), \quad (8.2.45)$$

$$2^{-1/2} \psi^{(\alpha)}(2^{-1} t) = \sum_{n=-\infty}^{\infty} g_n^{(\alpha)} \phi_\Delta^{(\alpha)}(t - n) \quad (8.2.46)$$

(cf. (8.2.4) and (8.2.5)). See Section 7.4.1 in Mallat [675] for more details. Note that the indices are α and $(-\alpha)$ in (8.2.43) and (8.2.44), respectively. It can be checked (Exercise 8.7) that $h^{(\alpha)}$ and $g^{(\alpha)}$ are given as in the following definition.

Definition 8.2.6 The *fractional conjugate mirror filters (CMFs)* $h^{(\alpha)}$ and $g^{(\alpha)}$ for $\alpha \in \mathbb{R}$ are defined through their Fourier transform as

$$\widehat{h^{(\alpha)}}(x) = 2^{-\alpha} \widehat{h}(x)(1 + e^{-ix})^\alpha, \quad (8.2.47)$$

$$\widehat{g^{(\alpha)}}(x) = 2^{-\alpha} \widehat{g}(x)(1 - e^{-ix})^{-\alpha}, \quad (8.2.48)$$

where h and g are CMFs associated with an orthogonal wavelet basis.

Set informally

$$\widehat{p}^{(\alpha)}(x) = (1 + e^{-ix})^\alpha, \quad \widehat{q}^{(\alpha)}(x) = (1 - e^{-ix})^{-\alpha}, \quad (8.2.49)$$

so that $\widehat{h^{(\alpha)}}(x) = 2^{-\alpha} \widehat{h}(x) \widehat{p}^{(\alpha)}(x)$ and $\widehat{g^{(\alpha)}}(x) = 2^{-\alpha} \widehat{g}(x) \widehat{q}^{(\alpha)}(x)$. Note that it is not immediately clear why the fractional CMFs are well-defined. For example, note that, with $\alpha = \kappa + 1$ in the case of FBM, $(1 - e^{-ix})^{-\alpha}$ does not even belong to $L^2[-\pi, \pi]$.

One way to use the Fourier transform in the definition is through the number of zero moments of the underlying MRA. It is known that, under mild assumptions, a wavelet ψ has Q zero moments in the sense of (8.2.8) if and only if

$$\widehat{h}(x) = (1 + e^{-ix})^Q \widehat{h_0}(x), \quad \widehat{g}(x) = (1 - e^{-ix})^Q \widehat{g_0}(x) \quad (8.2.50)$$

8.2 Wavelet Expansion of FBM

with $\sup_{x\in[-\pi,\pi]} |\widehat{h}_0(x)| < \infty$ and $\sup_{x\in[-\pi,\pi]} |\widehat{g}_0(x)| < \infty$ (e.g., Mallat [675], Theorem 7.4). For example, for the Daubechies MRA, the filters h_0 and g_0 have also finite length, and can be found in Daubechies [287], Table 6.2 on p. 196.

Substituting (8.2.50) into (8.2.47)–(8.2.48) leads to

$$\widehat{h}^{(\alpha)}(x) = 2^{-\alpha}\widehat{h}_0(x)(1+e^{-ix})^{\alpha+Q} = 2^{-\alpha}\widehat{h}_0(x)\,\widehat{p}^{(\alpha+Q)}(x), \tag{8.2.51}$$

$$\widehat{g}^{(\alpha)}(x) = 2^{-\alpha}\widehat{g}_0(x)(1-e^{-ix})^{-\alpha+Q} = 2^{-\alpha}\widehat{g}_0(x)\,\widehat{q}^{(\alpha-Q)}(x). \tag{8.2.52}$$

There are several consequences of writing $\widehat{h}^{(\alpha)}$ and $\widehat{g}^{(\alpha)}$ as in (8.2.51)–(8.2.52). First, note that $\widehat{p}^{(\alpha+Q)}$ and $\widehat{q}^{(\alpha-Q)}$ are in $L^2[-\pi,\pi]$ when $-(\alpha+Q) < 1/2$ and $\alpha-Q < 1/2$, respectively. This is because $|\widehat{p}^{(\alpha+Q)}(x)|^2$ has to be integrable around $x = \pm\pi$, which requires $2(\alpha+Q)+1 > 0$, and $|\widehat{q}^{(\alpha-Q)}(x)|^2$ has to be integrable around $x = 0$, which requires $2(-\alpha+Q)+1 > 0$. Note that combining the two conditions leads to

$$-Q - 1/2 < \alpha < Q + 1/2.$$

The conditions hold for any α for large enough Q, showing that in this case, $h^{(\alpha)}$ and $g^{(\alpha)}$ are well defined through their Fourier transforms.

Second, the filters $p^{(\alpha+Q)}$ and $q^{(-\alpha+Q)}$ decay much faster than the corresponding filters $p^{(\alpha)}$ and $q^{(\alpha)}$ when Q is taken large. Indeed, $p^{(-a)} = \{p_n^{(-a)}\}_{n\in\mathbb{Z}}$ and $q^{(a)} = \{q_n^{(a)}\}_{n\in\mathbb{Z}}$ are well-defined through their Fourier transforms when $a < -1/2$. Their elements are the coefficients in the Taylor expansions of $(1+z)^{-a}$ and $(1-z)^{-a}$, respectively. Hence, in view of (2.4.2) and (2.4.4), the elements satisfy

$$(-1)^n p_n^{(-a)} = q_n^{(a)} = \prod_{k=1}^{n} \frac{a+k-1}{k} = \frac{\Gamma(n+a)}{\Gamma(n+1)\Gamma(a)} \sim \frac{n^{a-1}}{\Gamma(a)}, \tag{8.2.53}$$

as $n \to \infty$ (with $a \in \mathbb{R}$ not an integer). This implies that

$$p_n^{(\alpha+Q)} \sim (-1)^n \frac{n^{-1-Q-\alpha}}{\Gamma(-Q-\alpha)}, \quad q_n^{(\alpha-Q)} \sim \frac{n^{-1-Q+\alpha}}{\Gamma(-Q+\alpha)}, \tag{8.2.54}$$

as $n \to \infty$. The decay of the filters $p^{(\alpha+Q)}$ and $q^{(-\alpha+Q)}$ is important when truncating them in practice at some a priori chosen cutoff level ϵ.

Table 8.1, borrowed from Pipiras [813], provides the lengths of the truncated (normalized) fractional CMFs $2^\alpha h^{(\alpha)}$ and $2^\alpha g^{(\alpha)}$ at several cutoff levels ϵ for $\alpha = 1.25$ (with $\alpha = \kappa + 1$ corresponding to $\kappa = 0.25$) and the underlying Daubechies MRA with $Q = 1, 3, 6$ and 10 number of zero moments. The length of a truncated filter $2^\alpha h^{(\alpha)}$ for example, is defined as

$$\min\{N \geq 0 : |2^\alpha h_n^{(\alpha)}| \leq \epsilon, \text{ for all } n \geq N\}.$$

Note that high values of Q decrease significantly the length of the truncated fractional CMFs, especially in the case of $2^\alpha g^{(\alpha)}$. Note, however, as seen from the table entries for $\epsilon = 10^{-4}$ for example, taking Q larger and larger will not necessarily make the truncated fractional CMFs shorter. This happens because, for example, for the Daubechies MRA, the elements of the filters h_0 and g_0 in (8.2.50) and (8.2.51)–(8.2.52) tend to become larger as Q increases (see Daubechies [287], Table 6.2 on p. 196).

The discussion above points to an interesting and important role played by the number of zero moments of a wavelet-based expansion in the context of FBM. Recall from Section

Table 8.1 *Lengths of truncated filters $2^\alpha h^{(\alpha)}$ and $2^\alpha g^{(\alpha)}$ at a cutoff ϵ with $\alpha = 1.25$ and the Daubechies MRA with Q vanishing moments.*

Filters	Cutoff ϵ	Length of a truncated filter			
		$Q=1$	$Q=3$	$Q=6$	$Q=10$
$2^\alpha h^{(\alpha)}$	10^{-4}	17	17	21	28
	10^{-7}	123	48	36	38
	10^{-10}	1009	164	70	56
	10^{-15}	34769	1425	257	121
$2^\alpha g^{(\alpha)}$	10^{-4}	$\approx 4 \cdot 10^4$	38	23	28
	10^{-7}	$\approx 4 \cdot 10^8$	422	57	42
	10^{-10}	$\approx 4 \cdot 10^{12}$	5162	171	72
	10^{-15}	$\approx 4 \cdot 10^{19}$	$\approx 3 \cdot 10^5$	1220	206

2.10.4 that the number of zero moments was also important when using orthogonal wavelet basis in estimation of the LRD parameter.

8.2.4 Wavelet-Based Expansion and Simulation of FBM

By gathering the results and following the discussion above, we deduce next a wavelet-based expansion of FBM.

Theorem 8.2.7 *Let $\kappa \in (-1/2, 1/2)$. FBM $B^\kappa = \{B^\kappa(t)\}_{t \in \mathbb{R}}$ in (8.2.21) admits the following wavelet expansion:*

$$B^\kappa(t) = \sum_{k=-\infty}^{\infty} S_{0,k} \phi_\Delta^{(\kappa+1)}(t-k) + \sum_{j=0}^{\infty} \sum_{k=-\infty}^{\infty} 2^{-j(\kappa+1/2)} \epsilon_{j,k} \psi^{(\kappa+1)}(2^j t - k) - b_0, \quad (8.2.55)$$

where the convergence is uniform on compact intervals of $\mathbb{R}$ almost surely, $\phi_\Delta^{(\kappa+1)}$ and $\psi^{(\kappa+1)}$ are fractional scaling and wavelet functions appearing in Definitions 8.2.3 and 8.2.4, respectively, $\{\epsilon_{j,k}\}_{j \geq J, k \in \mathbb{Z}}$ are i.i.d. $\mathcal{N}(0, 1)$ random variables, $\{S_{0,k}\}_{k \in \mathbb{Z}}$ is a partial sum process of a Gaussian FARIMA$(0, \kappa, 0)$ series defined in (8.2.35) (and independent of $\{\epsilon_{j,k}\}_{j \geq 0, k \in \mathbb{Z}}$), and b_0 is a random variable making the right-hand side of (8.2.55) zero at $t = 0$; that is,

$$b_0 = \sum_{k=-\infty}^{\infty} S_{0,k} \phi_\Delta^{(\kappa+1)}(-k) + \sum_{j=0}^{\infty} \sum_{k=-\infty}^{\infty} 2^{-j(\kappa+1/2)} \epsilon_{j,k} \psi^{(\kappa+1)}(-k). \quad (8.2.56)$$

Remark 8.2.8 In the case of BM with $\kappa = 0$ and when the underlying orthogonal wavelet basis is Haar, the expansion (8.2.61) is the celebrated midpoint displacement (Haar function) construction of BM due to Paul Lévy. See Section 2 in Meyer et al. [716] and also Section 2.3 in Karatzas and Shreve [549].

Sketch of Proof We shall use the fractional wavelets introduced in Section 8.2.2. In view of (8.2.25), (8.2.28) and (8.2.33), and by changing the signs of the random variables $\epsilon_{j,k}$ and $\eta_{0,k}$ in (8.2.25), we have

8.2 Wavelet Expansion of FBM

$$B^\kappa(t) = A_0(t) + D_0(t) - A_0(0) - D_0(0),$$

where

$$A_0(t) = \sum_{k=-\infty}^{\infty} S_{0,k}\phi_\Delta^{(\kappa+1)}(t-k), \quad D_0(t) = \sum_{j=0}^{\infty}\sum_{k=-\infty}^{\infty} 2^{-j(\kappa+1/2)}\epsilon_{j,k}\psi^{(\kappa+1)}(2^j t - k).$$

This holds rigorously and implies (8.2.55) as long as $D_0(t)$ is well-defined and converging uniformly on compact intervals as required, and

$$\sum_{k=-\infty}^{\infty} a_{0,k}(t)\eta_{0,k} = A_0(t) - A_0(0), \tag{8.2.57}$$

where $A_0(t)$ also converges uniformly on compact intervals. The coefficients $a_{0,k}(t)$ are defined in (8.2.24).

The equality (8.2.57) is proved in Lemma 9 and a discussion following it in Meyer et al. [716]. To show that $A_0(t)$ converges uniformly on compact intervals, note first that

$$|S_{0,k}| \leq C_0(1+|k|) \max_{m=-|k|+1,\ldots,|k|} |\xi_{0,m}|) \leq C(1+|k|)\sqrt{\log(2+|k|)} \tag{8.2.58}$$

for random constants C_0, C, where a FARIMA$(0, \kappa, 0)$ sequence $\xi_{0,m}$ in (8.2.35) is bounded above by using Lemma 8.2.9 below. For any $u \in S(\mathbb{R})$, real $\delta \geq 0$ and slowly varying function L at infinity, one can choose constants $C_1, C_2 > 0$ such that

$$\sum_{k=-\infty}^{\infty} u(\tau-k)(1+|k|)^\delta L(1+|k|) \leq C_1 \sum_{k=-\infty}^{\infty} (1+|\tau-k|)^{-q}(1+|k|)^\delta L(1+|k|)$$

$$\leq C_2(1+|\tau|)^\delta L(1+|\tau|), \tag{8.2.59}$$

by taking $q > 1 + \delta$ (Exercise 8.8). The bounds (8.2.58) and (8.2.59) with $u = \phi_\Delta^{(\kappa+1)}$, $\delta = 1$ and $L(1+|k|) = \sqrt{\log(2+|k|)}$ can be used to conclude that $|A_0(t)| \leq C(1+|t|)\sqrt{\log(2+|t|)}$ for a random constant C.

The uniform convergence of $D_0(t)$ on compact intervals can be proved similarly by using (8.2.59) (see the proof of Theorem 2 in Meyer et al. [716]). □

The next auxiliary and useful lemma was used in the proof above.

Lemma 8.2.9 *Let $\{\varepsilon_n\}_{n \in \mathbb{Z}}$ be $\mathcal{N}(0,1)$ random variables (which otherwise can be dependent). Then, there is a random constant $C(\omega)$ such that*

$$|\varepsilon_n(\omega)| \leq C(\omega)\sqrt{\log(2+|n|)}, \quad n \in \mathbb{Z}. \tag{8.2.60}$$

Proof Observe that, for $a > \sqrt{\log 2}$,

$$\sum_{n=3}^{\infty} \mathbb{P}(|\varepsilon_n| > a\sqrt{\log n}) \leq \sum_{n=3}^{\infty} \frac{2}{\sqrt{2\pi \log n} a} e^{-\frac{a^2 \log n}{2}} \leq \sum_{n=3}^{\infty} n^{-\frac{a^2}{2}} < \infty,$$

where we used the inequality $\mathbb{P}(\mathcal{N}(0,1) > x) \leq \frac{1}{\sqrt{2\pi}x}e^{-x^2/2}$ for $x > 0$. By the Borel–Cantelli Lemma, $|\varepsilon_n(\omega)| \leq a\sqrt{\log|n|} \leq a\sqrt{\log(2+|n|)}$ for $|n| > N(\omega)$ with a large enough $N(\omega)$. The relation (8.2.60) then holds with $C(\omega) = a + \max\{\epsilon_n(\omega) : n = -N(\omega), \ldots, N(\omega)\}/\sqrt{\log 2}$. □

Simulation of FBM

Note that the expansion (8.2.55) of FBM is (8.2.41) with $J = 0$, $\alpha = \kappa + 1$ and $f(t) = B^\kappa(t) + b_0$. It can also be written as: for any $J \geq 0$,

$$B^\kappa(t) = \sum_{k=-\infty}^{\infty} 2^{-J(\kappa+1/2)} S_{J,k} \phi_\Delta^{(\kappa+1)}(2^J t - k) + \sum_{j=J}^{\infty} \sum_{k=-\infty}^{\infty} 2^{-j(\kappa+1/2)} \epsilon_{j,k} \psi^{(\kappa+1)}(2^j t - k) - b_0, \quad (8.2.61)$$

where

$$S_{J,k} = 2^{J(\kappa+3/2)} \int_{\mathbb{R}} (B^\kappa(u) + b_0) \phi_\Delta^{(-\kappa-1)}(2^J u - k) du. \quad (8.2.62)$$

by using the first relation in (8.2.42). The expansion (8.2.61) of FBM is now exactly (8.2.41) with $\alpha = \kappa + 1$ and

$$f(t) = B^\kappa(t) + b_0, \quad c_{J,k} = 2^{-J(\kappa+1/2)} S_{J,k} 2^{-J/2}, \quad e_{j,k} = 2^{-j(\kappa+1/2)} \epsilon_{j,k} 2^{-j/2}. \quad (8.2.63)$$

The expansion (8.2.61) could be used to simulate FBM. The variables $S_{J,k}$ are defined for $J \in \mathbb{Z}$ and $k \in \mathbb{Z}$. We can interpolate to get a step function $S_{J,[2^J t]}$ defined say for $t \in [0, 1]$. The relation (8.2.62) can be used to show that, for any $\epsilon \in (0, \kappa + 1/2)$,

$$\sup_{t \in [0,1]} \left| 2^{-J(\kappa+1/2)} S_{J,[2^J t]} - (B^\kappa(t) + b_0) \right| \leq C 2^{-J(\kappa+1/2-\epsilon)} \quad (8.2.64)$$

almost surely, where the random variable C depends on κ, ϵ and the scaling function ϕ (Pipiras [813], Proposition 2.1). See also (8.2.12) and a related Exercise 8.3. The same result holds for the approximation of $B^\kappa(t)$ without b_0 if $S_{J,[2^J t]}$ is replaced by $S_{J,[2^J t]} - S_{J,0}$ (Pipiras [813], Corollary 2.1). Thus, suitably normalized and shifted coefficients $S_{J,[2^J t]}$ converge to FBM exponentially fast in J.

Substituting the coefficients c_j and e_j in (8.2.63) into (8.2.44) leads to the FWT at reconstruction given by

$$S_j = 2^{\kappa+1} h^{(-\kappa-1)} * (\uparrow_2 S_{j-1}) + 2^{\kappa+1} g^{(-\kappa-1)} * (\uparrow_2 \epsilon_{j-1}). \quad (8.2.65)$$

The FWT relation (8.2.65) could be used to generate the sequences $S_J = \{S_{J,k}\}_{k \in \mathbb{Z}}$ at finer and finer scales J from $S_0 = \{S_{0,k}\}_{k \in \mathbb{Z}}$ and $\epsilon_j = \{\epsilon_{j,k}\}_{k \in \mathbb{Z}}$, $j \geq 0$. Recall that $S_{0,k}$ is defined in (8.2.35) as a partial sum process of a FARIMA(0, κ, 0) sequence $\xi_{0,m}$. It can be shown that each S_J is a partial sum process of a Gaussian FARIMA(0, κ, 0) series as well (see Pipiras [813]). The relation (8.2.65) thus also provides a very fast (though approximate after truncating the fractional filters) method to generate partial sums of FARIMA(0, κ, 0) series. (See Pipiras [813] for further discussion.) The partial sums $S_J = \{S_{J,k}\}_{k \in \mathbb{Z}}$ of FARIMA(0, κ, 0) series, however, are special in the sense that they will converge to FBM almost surely as $J \to \infty$.

The left plot of Figure 8.2 presents the normalized and shifted approximations $S_{J,[2^J t]}$ to FBM $B^\kappa(t)$ on the interval $[0, 1]$ and for $\kappa = 0.25$. Linear interpolation instead of a step function is used. Thus, when $J = 0$, we get a straight line. In general, there are 2^J

8.3 Paley–Wiener Representation of FBM

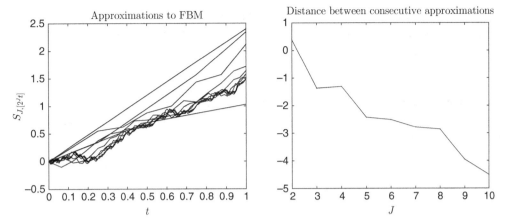

Figure 8.2 Left: Suitably normalized and shifted approximations $S_{J,[2^J t]}$ to FBM $B^\kappa(t)$ on the interval $[0, 1]$ and for $\kappa = 0.25$. Right: Uniform distances between the consecutive approximation on a log vertical scale.

connected line segments. The approximation improves as J increases. The right plot depicts the differences in uniform distance between the consecutive approximations on a log vertical scale. As in (8.2.64), the plot suggests that the convergence is exponentially fast in J.

8.3 Paley–Wiener Representation of FBM

An interesting expansion of FBM was obtained by Dzhaparidze and van Zanten [342]. The goal of this section is to sketch how this expansion is derived.

8.3.1 Complex-Valued FBM and its Representations

It is more convenient to work with complex-valued FBM defined as follows. Replace the Hermitian measure $\widehat{B}(dx)$ in the spectral representation (6.2.31) of FBM by the measure $\widetilde{B}(dx) = \widetilde{B}_1(dx) + i\widetilde{B}_2(dx)$, where $\widetilde{B}_1(dx)$ and $\widetilde{B}_2(dx)$ are independent Gaussian measures on $\mathbb{R}$ with control measures $dx/2$. A complex-valued FBM $\widetilde{B}^\kappa = \widetilde{B}_1^\kappa + i\widetilde{B}_2^\kappa$ is then defined as

$$\widetilde{B}^\kappa(t) = \frac{1}{c_{2,\kappa}} \int_\mathbb{R} \frac{e^{itx} - 1}{ix} |x|^{-\kappa} \widetilde{B}(dx), \qquad (8.3.1)$$

where $c_{2,\kappa} = (2\pi)^{1/2} c_{1,\kappa} / \Gamma(\kappa + 1)$ with the constant $c_{1,\kappa}$ appearing in (6.2.31). It can be checked that the real part process $\sqrt{2}\Re\widetilde{B}^\kappa = \sqrt{2}\widetilde{B}_1^\kappa$ and the imaginary part process $\sqrt{2}\Im\widetilde{B}^\kappa = \sqrt{2}\widetilde{B}_2^\kappa$ define two (dependent) standard FBMs (Exercise 8.9).

Let

$$\mu(dx) = c_{2,\kappa}^{-2} |x|^{-2\kappa} dx, \qquad (8.3.2)$$

$\widehat{1}_{[0,t)}(x) = \frac{e^{itx}-1}{ix}$ be the Fourier transform of $1_{[0,t)}$ and consider the space

$$\mathcal{L}_a = \overline{\operatorname{span}}\{\widehat{1}_{[0,t)} : t \in [0, a]\}, \qquad (8.3.3)$$

where the closure of the linear span is in $L^2(\mathbb{R}, \mu) = \{f : \mathbb{R} \to \mathbb{C} : \int_{\mathbb{R}} |f(x)|^2 \mu(dx) < \infty\}$, equipped with the inner product $(f, g)_{L^2(\mathbb{R}, \mu)} = \int_{\mathbb{R}} f(x)\overline{g(x)}\mu(dx)$ of $L^2(\mathbb{R}, \mu)$. The space $\mathcal{L}_a$ is isomorphic to the Gaussian space

$$\widetilde{\mathcal{H}}_a = \overline{\mathrm{span}}\{\widetilde{B}_H(t) : t \in [0, a]\}, \tag{8.3.4}$$

where the closure of the linear space is in $L^2(\Omega)$. Under the isomorphism, the function $\widehat{1}_{[0,t)}$ in $\mathcal{L}_a$ is associated with the random variable $\widetilde{B}^\kappa(t)$ in $\widetilde{\mathcal{H}}_a$. See Section 6.3.1 for the definitions of Gaussian spaces and isomorphisms.

If $\{\phi_n\}$ is any orthonormal basis in $\mathcal{L}_a$, then

$$\widehat{1}_{[0,t)}(x) = \sum_n c_n(t)\phi_n(x), \tag{8.3.5}$$

where

$$c_n(t) = \int_{\mathbb{R}} \widehat{1}_{[0,t)}(x)\overline{\phi_n(x)}\mu(dx). \tag{8.3.6}$$

Substituting (8.3.5) into (8.3.1) leads to the expansion

$$\widetilde{B}^\kappa(t) = \sum_n c_n(t) Z_n, \tag{8.3.7}$$

where $Z_n = \int_{\mathbb{R}} \phi_n(x) c_{2,\kappa}^{-1} |x|^{-\kappa} \widetilde{B}(dx)$ are uncorrelated complex-valued standard normal random variables (that is, satisfying $\mathbb{E}Z_n = 0$, $\mathbb{E}|Z_n|^2 = 1$ and $\mathbb{E}Z_n\overline{Z_m} = 0$). The convergence of the series (8.3.7) is in $L^2(\Omega)$ for each $t \in [0, a]$. Thus, the process defined by the right-hand side of (8.3.7) is a complex-valued FBM on $[0, a]$.

The expansion (8.3.7) holds for any orthonormal basis $\{\phi_n\}$ but the coefficients $c_n(t)$ in (8.3.6) do not necessarily have a closed form. Dzhaparidze and van Zanten [342] constructed a special orthonormal basis $\{\phi_n\}$ in $\mathcal{L}_a$ for which the coefficients $c_n(t)$ in (8.3.5)–(8.3.6) can be computed explicitly and which generalizes the Paley–Wiener expansion of BM (Paley and Wiener [788]). The latter is given by

$$\widetilde{B}^0(t) = \frac{1}{\sqrt{a}} \sum_{n=-\infty}^{\infty} \frac{e^{i2\pi nt/a} - 1}{i2\pi n/a} Z_n, \tag{8.3.8}$$

and defines a complex-valued BM. In the following section, we sketch the basis construction of Dzhaparidze and van Zanten [342].

8.3.2 Space $\mathcal{L}_a$ and its Orthonormal Basis

To work with the space $\mathcal{L}_a$ given in (8.3.3), it is convenient to introduce another space isomorphic to $\mathcal{L}_a$. Let $M^\kappa(t)$ be the fundamental martingale related to FBM $\widetilde{B}^\kappa$ through (see (6.3.49) and (6.3.50)):

$$M^\kappa(t) = \int_0^a m_t(u) d\widetilde{B}^\kappa(u), \tag{8.3.9}$$

where

$$m_t(u) = \frac{1}{\sigma_{2,\kappa}} u^{-\kappa} (I_{a-}^{-\kappa} 1_{[0,t)})(u) = \frac{u^{-\kappa}(t-u)^{-\kappa} 1_{[0,t)}(u)}{\sigma_{2,\kappa}\Gamma(1-\kappa)} \tag{8.3.10}$$

and $\sigma_{2,\kappa} = (2\kappa + 1)\Gamma(\kappa + 1)$. (The calculations below are based on second moments so that the complex-valued FBM in (8.3.9) will be viewed for simplicity as being real-valued.) Let

$$V(t) = \text{Var}(M^\kappa(t)) = \frac{\sigma_\kappa^2}{\sigma_{2,\kappa}^2(1-2\kappa)} t^{1-2\kappa} =: d_\kappa^2 t^{1-2\kappa} \tag{8.3.11}$$

be the quadratic variation appearing in (6.3.52). For later reference, note that

$$d_\kappa^2 = \frac{\Gamma(1-\kappa)}{(2\kappa+1)\Gamma(\kappa+1)\Gamma(2-2\kappa)} = \frac{\sqrt{\pi}}{(2\kappa+1)\Gamma(\kappa+1)\Gamma(3/2-\kappa)2^{1-2\kappa}}, \tag{8.3.12}$$

by using the identity $2^{1-2z}\sqrt{\pi}\Gamma(2z) = \Gamma(z)\Gamma(z+1/2)$. Note that $V(t) = t$ when $\kappa = 0$.

Inversion

The relation (8.3.9) can be inverted as done in (6.4.7)–(6.4.8) and other places in Sections 6.3 and 6.4. Informally, (8.3.9) yields

$$M^\kappa(t) = \int_0^a 1_{[0,t)}(u)\dot{M}^\kappa(u)du = \int_0^a 1_{[0,t)}(u)m_t(u)\widetilde{\dot{B}}^\kappa(u)du$$

$$= \int_0^a 1_{[0,t)}(u)\frac{1}{\sigma_{2,\kappa}}u^{-\kappa}(I_{a-}^{-\kappa}1_{[0,t)})(u)\widetilde{\dot{B}}^\kappa(u)du$$

$$= \int_0^a 1_{[0,t)}(s)\frac{1}{\sigma_{2,\kappa}}(I_{0+}^{-\kappa}u^{-\kappa}\widetilde{\dot{B}}^\kappa(u))(s)ds,$$

that is, $\dot{M}^\kappa(s) = (\sigma_{2,\kappa})^{-1}(I_{0+}^{-\kappa}u^{-\kappa}\widetilde{\dot{B}}^\kappa(u))(s)$ or $\sigma_{2,\kappa}u^\kappa(I_{0+}^\kappa \dot{M}^\kappa)(u) = \widetilde{\dot{B}}^\kappa(u)$ or

$$\widetilde{B}^\kappa(t) = \int_0^a x_t(s)dM^\kappa(s), \tag{8.3.13}$$

where

$$x_t(s) = \sigma_{2,\kappa}(I_{a-}^\kappa u^\kappa 1_{[0,t)}(u))(s) = \frac{\sigma_{2,\kappa}}{\Gamma(\kappa+1)}\left(t^\kappa(t-s)_+^\kappa - \kappa\int_s^t u^{\kappa-1}(u-s)_+^\kappa du\right), \tag{8.3.14}$$

by using (6.2.8).

Let now

$$\mathcal{K}_a = \overline{\text{span}}\{x_t : t \in [0,a]\}, \tag{8.3.15}$$

where the closure is in $L^2([0,a], dV)$. There is a natural isometry between $\mathcal{K}_a$ and $\widetilde{\mathcal{H}}_a$ which associates x_t to $\widetilde{B}^\kappa(t)$. Note that, under this isometry, $M^\kappa(t)$ is associated with $1_{[0,t)}$. So $1_{[0,t)} \in \mathcal{K}_a$ and hence $\mathcal{K}_a = L^2([0,a], dV)$.

The Three Isometric Hilbert Spaces

We now have three isometric Hilbert spaces associated with FBM:

- the linear Gaussian space $\widetilde{\mathcal{H}}_a = \overline{\text{span}}\{\widetilde{B}_H(t) : t \in [0,a]\}$,
- the frequency domain space $\mathcal{L}_a = \overline{\text{span}}\{\widehat{1}_{[0,t)} : t \in [0,a]\}$, and
- the space of integration kernels $\mathcal{K}_a = \overline{\text{span}}\{x_t : t \in [0,a]\}$.

The isometries between $\widetilde{\mathcal{H}}_a$ and $\mathcal{L}_a$, and between $\widetilde{\mathcal{H}}_a$ and $\mathcal{K}_a$ are determined by associating $\widehat{B^\kappa}(t)$ to $\widehat{1}_{[0,t)}$ and to x_t, respectively. These induce an isometry between the function spaces $\mathcal{K}_a$ and $\mathcal{L}_a$, which we denote by U:

$$U : \mathcal{K}_a \to \mathcal{L}_a. \tag{8.3.16}$$

Under the isometry $\mathcal{K}_a \to \widetilde{\mathcal{H}}_a$, the indicator function $1_{[0,t)} \in \mathcal{K}_a$ with $t \in [0,a]$ is mapped to the random variable $M_t \in \widetilde{\mathcal{H}}_a$. Under the spectral isometry $\widetilde{\mathcal{H}}_a \to \mathcal{L}_a$, $M_t \in \widetilde{\mathcal{H}}_a$ is mapped to the Fourier transform of the function m_t in (8.3.10); that is,

$$\widehat{m}_t(\lambda) = \int_{\mathbb{R}} e^{i\lambda u} m_t(u) du \tag{8.3.17}$$

in $\mathcal{L}_a$. Thus,

$$(U 1_{[0,t)})(\lambda) = \widehat{m}_t(\lambda). \tag{8.3.18}$$

The next goal is to rewrite the right-hand side of (8.3.18) and to characterize the isometry U acting on general function $f \in \mathcal{K}_a$.

Expression of Uf and $U^{-1}\psi$

It can be verified (Exercise 8.10) that

$$\widehat{m}_t(\lambda) = \begin{cases} \frac{\sqrt{\pi}}{(2\kappa+1)\Gamma(\kappa+1)} \left(\frac{t}{\lambda}\right)^{\frac{1}{2}-\kappa} e^{i\lambda t/2} J_{1/2-\kappa}\left(\frac{\lambda t}{2}\right), & \lambda \neq 0, \\ \frac{\sqrt{\pi}}{(2\kappa+1)\Gamma(\kappa+1)\Gamma(3/2-\kappa)2^{1-2\kappa}} t^{1-2\kappa}, & \lambda = 0, \end{cases} \tag{8.3.19}$$

where $J_{1/2-\kappa}$ is the Bessel function of the first kind of order $1/2 - \kappa$. Recall that for $\nu \neq -1, -2, \ldots$, the Bessel function J_ν of the first kind of order ν can be defined on the region $\{z \in \mathbb{C} : |\arg(z)| < \pi\}$ as the absolutely convergent sum

$$J_\nu(z) = \sum_{n=0}^{\infty} \frac{(-1)^n (z/2)^{\nu+2n}}{\Gamma(n+1)\Gamma(\nu+n+1)}. \tag{8.3.20}$$

To derive the relation (8.3.19), the following Poisson's integral formula for the Bessel function is used: when $\Re(\nu) > -1/2$,

$$J_\nu(z) = \frac{(z/2)^\nu}{\Gamma(\nu+1/2)\Gamma(1/2)} \int_{-1}^{1} (1-u^2)^{\nu-1/2} e^{izu} du \tag{8.3.21}$$

(Watson [999], Section 3, Eq. (3)). Note also that

$$\widehat{m}_t(0) = \frac{\sqrt{\pi}}{(2\kappa+1)\Gamma(\kappa+1)\Gamma(3/2-\kappa)2^{1-2\kappa}} t^{1-2\kappa} = d_\kappa^2 t^{1-2\kappa} = V(t), \tag{8.3.22}$$

by using (8.3.12) and (8.3.11).

In view of (8.3.18) and (8.3.22), write

$$(U 1_{[0,t)})(\lambda) = \int_0^t d\widehat{m}_u(\lambda) = \int_0^t \frac{d\widehat{m}_u(\lambda)}{d\widehat{m}_u(0)} d\widehat{m}_u(0) = \int_0^t \frac{d\widehat{m}_u(\lambda)}{d\widehat{m}_u(0)} dV(u). \tag{8.3.23}$$

By using the well-known formula for the Bessel functions

$$\frac{d}{dz}(z^\nu J_\nu(z)) = z^\nu J_{\nu-1}(z) \tag{8.3.24}$$

8.3 Paley–Wiener Representation of FBM

and the expression (8.3.19) of $\widehat{m}_t$, it can be checked that

$$\frac{d\widehat{m}_u(\lambda)}{d\widehat{m}_u(0)} = \frac{d\widehat{m}_u(\lambda)}{du}\left(\frac{d\widehat{m}_u(0)}{du}\right)^{-1} = \phi(\lambda u), \tag{8.3.25}$$

where

$$\phi(z) = \begin{cases} \Gamma(1/2-\kappa)\left(\frac{z}{4}\right)^{\kappa+1/2} e^{iz/2}\left(J_{-1/2-\kappa}(\frac{z}{2}) + iJ_{1/2-\kappa}(\frac{z}{2})\right), & z \neq 0, \\ 1, & z = 0 \end{cases} \tag{8.3.26}$$

(Exercise 8.11). Thus,

$$(U1_{[0,t)})(\lambda) = \int_0^a 1_{[0,t)}(u)\phi(\lambda u)dV(u) \tag{8.3.27}$$

for any $t \in [0, a]$. In fact, it can be shown (Dzhaparidze and van Zanten [342], Theorem 3.2) that, for $f \in \mathcal{K}_a$,

$$(Uf)(\lambda) = \int_0^a f(u)\phi(\lambda u)dV(u) \quad \text{a.e. } \mu(d\lambda) \tag{8.3.28}$$

and that

$$(U^{-1}\psi)(u) = \int_{\mathbb{R}} \psi(\lambda)\overline{\phi(u\lambda)}\mu(d\lambda) \quad \text{a.e. } dV(u), \tag{8.3.29}$$

for $\psi \in \mathcal{L}_a$ such that $\int_{\mathbb{R}} \|\phi(\cdot\lambda)\|_V |\psi(\lambda)|\mu(d\lambda) < \infty$.

Remark 8.3.1 Since

$$J_{1/2}(z) = \sqrt{\frac{2}{\pi z}}\sin(z), \quad J_{-1/2}(z) = \sqrt{\frac{2}{\pi z}}\cos(z), \quad z \neq 0 \ (z \in \mathbb{C}), \tag{8.3.30}$$

and $\Gamma(1/2) = \sqrt{\pi}$, it follows that $\phi(z) = e^{iz}$ in the standard BM case $\kappa = 0$. Thus, in this case, U is simply the Fourier transform. For general $\kappa \in (-1/2, 1/2)$, U can be viewed as a fractional version of the Fourier transform.

Remark 8.3.2 The following are some alternative representations of the quantities considered above in terms of fractional integrals and derivatives, provided the integrals are well defined:

$$\widehat{m}_t(\lambda) = \frac{1}{\Gamma(\kappa)} I_{0+}^{1-\kappa}(u^{-\kappa} e^{iu\lambda})(t),$$

$$(Uf)(\lambda) = \frac{1}{\Gamma(\kappa)} \mathcal{F}(u^{-\kappa}(I_{a-}^{-\kappa}f)(u))(\lambda),$$

$$(U^{-1}\psi)(t) = \Gamma(\kappa) I_{a-}^{\kappa}(u^{\kappa}(\mathcal{F}^{-1}\psi)(u))(t),$$

where $\mathcal{F}$ denotes the Fourier transform (Dzhaparidze and van Zanten [342], p. 623).

Reproducing Kernel on $\mathcal{L}_a$

An orthonormal basis of $\mathcal{L}_a = \overline{\text{span}}\{\widehat{1_{[0,t)}} : t \in [0,a]\}$ will be defined in terms of a reproducing kernel[10] on $\mathcal{L}_a$. The reproducing kernel is a function $S_a(w, \lambda)$ defined on $\mathbb{R} \times \mathbb{R}$ and satisfying the following reproducing kernel relation: for $\psi \in \mathcal{L}_a$,

$$\int_{\mathbb{R}} \psi(\lambda)\overline{S_a(w,\lambda)}\mu(d\lambda) = \psi(w). \tag{8.3.31}$$

The reproducing kernel $S_a(w, \lambda)$ is defined as

$$S_a(w, \lambda) = \int_0^a \overline{\phi(uw)}\phi(u\lambda)dV(u), \tag{8.3.32}$$

where ϕ is given by (8.3.26) and V by (8.3.11). It satisfies the reproducing kernel relation (8.3.31) since, by using Fubini's theorem, (8.3.29) and (8.3.28),

$$\int_{\mathbb{R}} \psi(\lambda)\overline{S_a(w,\lambda)}\mu(d\lambda) = \int_{\mathbb{R}} \psi(\lambda) \int_0^a \phi(uw)\overline{\phi(u\lambda)}dV(u)\mu(d\lambda)$$

$$= \int_0^a \phi(uw)\left(\int_{\mathbb{R}} \psi(\lambda)\overline{\phi(u\lambda)}\mu(d\lambda)\right)dV(u)$$

$$= \int_0^a \phi(uw)U^{-1}\psi(u)dV(u)$$

$$= U(U^{-1}\psi)(w) = \psi(w).$$

This is proved rigorously in Dzhaparidze and van Zanten [342], Theorem 4.1.

Note that the reproducing kernel can be written as

$$S_a(w, \lambda) = \int_0^a \overline{P(u,w)}P(u,\lambda)du, \tag{8.3.33}$$

where

$$P(u, \lambda) = \phi(u\lambda)\sqrt{\frac{dV(u)}{du}}. \tag{8.3.34}$$

By using the differential equations satisfied by $P(u, \lambda)$ and $P(u, \lambda)^* = e^{i\lambda u}\overline{P(u,\lambda)}$, Dzhaparidze and van Zanten [342] show that, for all $a > 0$ and $w, \lambda \in \mathbb{R}$,

$$i(\lambda - w)S_a(w, \lambda) = \overline{P(a,w)}P(a,\lambda) - \overline{P(a,w)^*}P(a,\lambda)^* \tag{8.3.35}$$

(Theorem 6.1 of Dzhaparidze and van Zanten [342]), and conclude that, for $w \neq \lambda$,

$$\frac{S_a(2w, 2\lambda)}{S_a(0,0)} = (1-2\kappa)\Gamma(1/2-\kappa)^2\left(\frac{a^2w\lambda}{4}\right)^{\kappa+1/2}e^{ia(\lambda-w)} \times$$

$$\times \frac{J_{-1/2-\kappa}(aw)J_{1/2-\kappa}(a\lambda) - J_{1/2-\kappa}(aw)J_{-1/2-\kappa}(a\lambda)}{a(\lambda-w)} \tag{8.3.36}$$

---

[10] Note that the notion here is that of a reproducing kernel and not of a reproducing kernel Hilbert space as in Section 6.3.7.

and, for $w \in \mathbb{R}$,

$$\frac{S_a(2w, 2w)}{S_a(0,0)} = (1 - 2\kappa)\Gamma(1/2 - \kappa)^2 \left(\frac{aw}{2}\right)^{2\kappa+1} \times$$

$$\times \left(J_{1/2-\kappa}(aw)^2 + \frac{2\kappa}{aw} J_{-1/2-\kappa}(aw) J_{1/2-\kappa}(aw) + J_{-1/2-\kappa}(aw)^2\right) \quad (8.3.37)$$

(Corollary 6.2 of Dzhaparidze and van Zanten [342]).

Orthonormal Basis in $\mathcal{L}_a$

The following result presents an orthonormal basis in $\mathcal{L}_a$ and the associated expansion formula (Dzhaparidze and van Zanten [342], Theorem 7.2). The zeros of the Bessel functions are well studied and understood, and can be computed in most mathematical software packages.

Let S_a be defined in (8.3.32), μ defined in (8.3.2) and V defined in (8.3.11).

Theorem 8.3.3 *Let* $\ldots < w_{-1} < w_0 = 0 < w_1 < \cdots$ *be the real zeros of* $J_{1/2-\kappa}$ *and, for* $n \in \mathbb{Z}$, *define the functions* ψ_n *on* $\mathbb{R}$ *by*

$$\psi_n(\lambda) = \frac{S_a(2w_n/a, \lambda)}{\|S_a(2w_n/a, \cdot)\|_\mu} \quad (8.3.38)$$

and set

$$\sigma(w_n)^{-2} = S_a\left(\frac{2w_n}{a}, \frac{2w_n}{a}\right)$$

$$= \begin{cases} (1 - 2\kappa)\Gamma(1/2 - \kappa)^2 (\frac{w_n}{2})^{2\kappa+1} J_{-1/2-\kappa}(w_n)^2 V(a), & w_n \neq 0, \\ V(a), & w_n = 0, \end{cases} \quad (8.3.39)$$

where V is defined in (8.3.11). The functions ψ_n form an orthonormal basis of $\mathcal{L}_a$, and every function $\psi \in \mathcal{L}_a$ can be expanded as

$$\psi(\lambda) = \sum_{n=-\infty}^{\infty} \sigma(w_n) \psi\left(\frac{2w_n}{a}\right) \psi_n(\lambda), \quad (8.3.40)$$

with the convergence taking place in $L^2(\mathbb{R}, \mu)$.

Observe that the functions ψ_n in (8.3.38) have unit norms by construction. They are also orthogonal. Indeed, for $n \neq m$,

$$\int_\mathbb{R} \psi_n(\lambda)\overline{\psi_m(\lambda)} \mu(d\lambda) = C_{n,m} \int_\mathbb{R} S_a\left(\frac{2w_n}{a}, \lambda\right) \overline{S_a\left(\frac{2w_m}{a}, \lambda\right)} \mu(d\lambda) = C_{n,m} S_a\left(\frac{2w_n}{a}, \frac{2w_m}{a}\right)$$

for a constant $C_{n,m} \neq 0$, by the reproducing kernel property (8.3.31). It remains to note that

$$S_a\left(\frac{2w_n}{a}, \frac{2w_m}{a}\right) = 0$$

by (8.3.36) since $J_{1/2-\kappa}(w_n) = 0$ and $J_{1/2-\kappa}(w_m) = 0$. The fact that $\{\psi_n\}$ is complete (and hence an orthonormal basis) is proved in Dzhaparidze and van Zanten [342], Theorem 7.2. Finally, note also that the expansion (8.3.40) follows from

$$\|S_a(2w_n/a, \cdot)\|_\mu = \sigma(w_n)^{-1}$$

and

$$\int_{\mathbb{R}} \psi(\lambda)\overline{\psi_n(\lambda)}\mu(d\lambda) = \sigma(w_n)\psi\left(\frac{2w_n}{a}\right)$$

by the the reproducing kernel property (8.3.31).

8.3.3 Expansion of FBM

The next result provides a Paley–Wiener expansion of FBM.

Theorem 8.3.4 *Let $\kappa \in (-1/2, 1/2)$ and $\ldots < w_{-1} < w_0 = 0 < w_1 < \cdots$ be the real zeros of $J_{1/2-\kappa}$. Let also Z_n, $n \in \mathbb{Z}$, be i.i.d. complex-valued Gaussian random variables with $\mathbb{E}Z_n = 0$ and $\mathbb{E}|Z_n|^2 = 1$, and let $\sigma(w_n)^2$ be given by (8.3.39). Then, the series*

$$\sum_{n \in \mathbb{Z}} \sigma(w_n) \frac{e^{i2w_n t/a} - 1}{i2w_n/a} Z_n \tag{8.3.41}$$

converges in $L^2(\Omega)$ for each $t \in [0, a]$ and defines a complex-valued FBM with index κ.

Proof We follow the general argument outlined in (8.3.5)–(8.3.7). The functions $\{\psi_n\}_{n \in \mathbb{Z}}$ given by (8.3.38) form an orthonormal basis of $\mathcal{L}_a$. As in (8.3.5), by using (8.3.40) with $\psi = \widehat{1}_{[0,t)}$, expand $\widehat{1}_{[0,t)} \in \mathcal{L}_a$ in the basis $\{\psi_n\}_{n \in \mathbb{Z}}$ as

$$\widehat{1}_{[0,t)}(x) = \sum_{n=-\infty}^{\infty} \sigma(w_n)\widehat{1}_{[0,t)}\left(\frac{2w_n}{a}\right)\psi_n(x) = \sum_{n=-\infty}^{\infty} \sigma(w_n)\frac{e^{i2w_n t/a} - 1}{i2w_n/a}\psi_n(x).$$

If $\widetilde{B}^\kappa$ is a complex-valued FBM, the general expansion (8.3.7) then yields

$$\widetilde{B}^\kappa(t) = \sum_{n=-\infty}^{\infty} \sigma(w_n)\frac{e^{i2w_n t/a} - 1}{i2w_n/a} Z_n. \qquad \square$$

In the case $\kappa = 0$, we have $w_n = \pi n$ (see (8.3.30)) and $\sigma(w_n)^{-2} = a$, where the latter follows from (8.3.39) by using $V(a) = a$ when $\kappa = 0$ and the expression (8.3.30) for the Bessel function $J_{-1/2}$. Hence, when $\kappa = 0$, the expansion (8.3.41) indeed becomes the classical Paley–Wiener expansion (8.3.8) of BM.

Taking the real part of the complex-valued FBM and the Paley–Wiener expansion (8.3.41) leads to real-valued FBM and the following result.

Corollary 8.3.5 *Let X, $\{Y_n^{(1)}\}_{n \geq 1}$ and $\{Y_n^{(2)}\}_{n \geq 1}$ be i.i.d. $\mathcal{N}(0, 1)$ random variables. Let V be defined in (8.3.11) and $\sigma(w_n)^2$ be given by (8.3.39). Then, the series*

$$\frac{t}{\sqrt{V(a)}} X + \sum_{n=1}^{\infty} \frac{\sigma(w_n)}{\sqrt{2}} \frac{\sin(2w_n t/a)}{w_n/a} Y_n^{(1)} + \sum_{n=1}^{\infty} \frac{\sigma(w_n)}{\sqrt{2}} \frac{\cos(2w_n t/a) - 1}{w_n/a} Y_n^{(2)} \quad (8.3.42)$$

converges for each t in $L^2(\Omega)$ and defines (real-valued) standard FBM with index κ.

Proof Denote the complex-valued variables in (8.3.41) as $Z_n = Z_n^{(1)} + i Z_n^{(2)}$, where $Z_n^{(1)}, Z_n^{(2)}$ are i.i.d. $\mathcal{N}(0, 1/2)$ variables. The real part of the process (8.3.41) can then be written as

$$\sum_{n=-\infty}^{\infty} \sigma(w_n) \frac{\sin(2w_n t/a)}{2w_n/a} Z_n^{(1)} + \sum_{n=-\infty}^{\infty} \sigma(w_n) \frac{\cos(2w_n t/a) - 1}{2w_n/a} Z_n^{(2)}. \quad (8.3.43)$$

By observing that $w_{-n} = -w_n$ and $w_0 = 0$, and by writing $Z_n^{(1)} + Z_{-n}^{(1)} = Y_n^{(1)}$ and $Z_n^{(2)} - Z_{-n}^{(2)} = Y_n^{(2)}$ for i.i.d. $\mathcal{N}(0, 1)$ variables $Y_n^{(1)}, Y_n^{(2)}$, and $\sigma(w_0) t Z_n^{(0)} = t Z_n^{(0)}/\sqrt{V(a)} = t X/\sqrt{2V(a)}$ for $\mathcal{N}(0, 1)$ random variable X by using (8.3.39), the process (8.3.43) can be written as

$$\frac{t}{\sqrt{2V(a)}} X + \sum_{n=1}^{\infty} \sigma(w_n) \frac{\sin(2w_n t/a)}{2w_n/a} Y_n^{(1)} + \sum_{n=1}^{\infty} \sigma(w_n) \frac{\cos(2w_n t/a) - 1}{2w_n/a} Y_n^{(2)}. \quad (8.3.44)$$

It remains to recall (see the discussion following (8.3.1)) that a standard FBM is obtained by multiplying the real part of the complex-valued FBM by $\sqrt{2}$, leading to the expansion (8.3.42). □

The convergence of the series (8.3.42) can be shown to be uniform in $t \in [0, a]$ with probability 1 by using the Itô–Nisio theorem (e.g., Kwapień and Woyczyński [586], Theorem 2.1.1). According to the theorem, since the summands of the series are independent across $n \geq 1$, for example the sequence

$$S_N(t) = \sum_{n=1}^{N} \frac{\sigma(w_n)}{\sqrt{2}} \frac{\sin(2w_n t/a)}{w_n/a} Y_n^{(1)}$$

converges uniformly on $[0, a]$ with probability 1 (that is, converges with probability 1 in $C[0, a]$ equipped with the uniform norm) if and only if S_N converges weakly in $C[0, a]$. For the weak convergence, since one already has the weak convergence of the finite-dimensional distributions (in fact, even the convergence in $L^2(\Omega)$), the weak convergence of the sequence S_N is equivalent to showing its tightness in $C[0, a]$. For a closely related sequence, this is proved in Theorem 4.5 of Dzhaparidze and van Zanten [341].

8.4 Exercises

Exercise 8.1 If $\phi_n = \phi_n^{(a)}$ and $\lambda_n = \lambda_n^{(a)}$ are, respectively, the eigenfunctions and eigenvalues for the integral equation (8.1.15) of FBM, show that

$$\phi_n^{(a)}(t) = \frac{1}{\sqrt{a}} \phi_n^{(1)}\left(\frac{t}{a}\right), \quad \lambda_n^{(a)} = a^{2\kappa+2} \lambda_n^{(1)}.$$

Hint: Use the $(2\kappa + 1)$-self-similarity of FBM.

Exercise 8.2 Show that the CMFs h and g satisfy the following relations:
$$|\widehat{g}(w)|^2 + |\widehat{g}(w+\pi)|^2 = 2, \quad \widehat{g}(w)\overline{\widehat{h}(w)} + \widehat{g}(w+\pi)\overline{\widehat{h}(w+\pi)} = 0.$$
Hint: Use (8.2.7) and (8.2.6).

Exercise 8.3 Suppose that $f \in L^2(\mathbb{R})$ is Hölder with exponent β; that is, $|f(t) - f(s)| \le c|t-s|^\beta$ for some constant c and all $s, t \in \mathbb{R}$. Suppose also that the scaling function ϕ satisfies (8.2.9) and is such that $\int_\mathbb{R} |u|^\beta |\phi(u)| du < \infty$. Show that
$$\sup_{k \in \mathbb{Z}} \left| 2^{J/2} a_{J,k} - f\left(\frac{k}{2^J}\right) \right| \le C 2^{-\beta J},$$
for some constant C. *Hint:* Write $2^{J/2} a_{J,k} - f(k/2^J) = 2^J \int_\mathbb{R} (f(u) - f(k/2^J)) \psi(u) du$.

Exercise 8.4 The Haar MRA is characterized by (8.2.15) and (8.2.16). Show that the Haar functions ϕ and ψ, and the CMFs h and g indeed satisfy the relations (8.2.4), (8.2.5) and (8.2.7).

Exercise 8.5 The space S consists of functions satisfying (8.2.17). Show that $f \in S$ if and only if $\widehat{f} \in S$.

Exercise 8.6 Prove the biorthogonality relations (8.2.39) and (8.2.40).

Exercise 8.7 By using (8.2.4) and (8.2.5), show that the fractional CMFs given in Definition 8.2.6 indeed satisfy the relations (8.2.45) and (8.2.45).

Exercise 8.8 Prove the second inequality in (8.2.59). *Hint:* For example, for $\tau > 0$ and integer for simplicity, split $\sum_{k=-\infty}^{\infty}$ into $\sum_{k=-\infty}^{0} + \sum_{k=0}^{\tau/2} + \sum_{k=\tau/2}^{\tau} + \sum_{k=\tau}^{2\tau} + \sum_{k=2\tau}^{\infty}$ and consider each of the sums separately.

Exercise 8.9 Consider the complex-valued FBM defined in (8.3.1). Show that the real part process $\sqrt{2}\Re \widetilde{B}^\kappa = \sqrt{2}\widetilde{B}_1^\kappa$ and the imaginary part process $\sqrt{2}\Im \widetilde{B}^\kappa = \sqrt{2}\widetilde{B}_2^\kappa$ define two (dependent) standard FBMs.

Exercise 8.10 Verify the relation (8.3.19) by using (8.3.20).

Exercise 8.11 Verify the relation (8.3.25).

8.5 Bibliographical Notes

Section 8.1: The asymptotics of the eigenvalues and eigenvectors for FBM in (8.1.16) and (8.1.18) also appear in Bronski [188, 189], with applications to small ball problems. A more statistical application can be found in Li, Hu, Chen, and Zhang [623], though with a mistake indicated in van Zanten [972] and a further response in Li [622]. A related problem of the covariance structure of a discrete time FBM is investigated in Gupta and Joshi [441]. Finally, we should also note the work of Maccone [659, 660], where the Karhunen–Loève expansion of a process related to FBM, and not that of FBM as claimed initially, was derived.

Section 8.2: As indicated a number of times in Section 8.2, the wavelet decomposition of FBM in Theorem 8.2.7 is due to Meyer et al. [716]. The idea of choosing a suitable (biorthogonal) wavelet basis so that detail coefficients of FBM and other processes become decorrelated can be traced at least to Benassi et al. [113], who used such wavelet decompositions to study sample properties of the processes (but who also did not specify the approximation term as in Theorem 8.2.7). Decompositions similar to that in Theorem 8.2.7 but based on splines appear in Blu and Unser [163], Unser and Blu [969]. Wavelet-based simulation of FBM was considered first in Abry and Sellan [7], and later revisited by Pipiras [813]. Rate optimality of the wavelet series approximations of FBM is studied in Ayache and Taqqu [62].

Similar wavelet-based decompositions and applications are considered for the Rosenblatt process in Pipiras [812], Abry and Pipiras [6], and for general stationary Gaussian processes in Didier and Pipiras [305, 306] (see also Unser and Blu [968]). Series representations of fractional Gaussian processes by trigonometric and Haar systems are studied in Ayache and Linde [61].

Section 8.3: This section is based on the work of Dzhaparidze and van Zanten [342]. A similar expansion of FBM, but involving the zeros of the Bessel functions $J_{-1/2-\kappa}$ and $J_{1/2-\kappa}$, was established in Dzhaparidze and van Zanten [341]. Dzhaparidze, van Zanten, and Zareba [344] provide a more general perspective on the expansion of FBM, based on a general theory valid for a large class of processes with stationary increments. Extensions to fractional Brownian sheet and optimality questions are studied in Dzhaparidze and van Zanten [343].

Other notes: A particular series expansion of FBM through the Lamperti transformation is considered in Baxevani and Podgórski [110]. Power series expansions for FBM are considered in Gilsing and Sottinen [389]. White and colored Gaussian noises as limits of sums of random dilations and translations of a single function appear in Gripenberg [432]. Functional quantization problems for FBM and other Gaussian processes are studied in Luschgy and Pagès [652, 653]. Construction of stationary self-similar generalized fields by random wavelet expansion can be found in Chi [234].

9

Multidimensional Models

The time series models and stochastic processes discussed in previous chapters consisted of real-valued random variables indexed by a subset of the real line $\mathbb{R}$. A number of applications require self-similar or long-range dependent *multidimensional* models indexed by a finite-dimensional space $\mathbb{R}^q$ or $\mathbb{Z}^q$, and taking values in a finite-dimensional space $\mathbb{R}^p$. That is, random processes X such that for fixed $\omega \in \Omega$,

$$X(\omega, \cdot) : \mathbb{R}^q \to \mathbb{R}^p \quad \text{or} \quad X(\omega, \cdot) : \mathbb{Z}^q \to \mathbb{R}^p$$
$$t \mapsto X(\omega, t) \quad\quad\quad\quad t \mapsto X(\omega, t).$$

We shall discuss such models in this chapter by considering extensions of the notions of self-similarity, fractional Brownian motion and long-range dependence. We shall focus on the cases

- $p \geq 1, q = 1$ (Sections 9.3 and 9.4), and
- $q \geq 1, p = 1$ (Sections 9.5 and 9.6).

We shall use the following terminology when referring to multidimensional models $X = \{X(t)\}$. When the index t belongs to a subset of a finite-dimensional space $\mathbb{R}^q$ or $\mathbb{Z}^q$, we shall refer to X as a *random field* (or a *spatial process*). When the variables $X(t)$ take values in a subset of a finite-dimensional space $\mathbb{R}^p$, we shall refer to X as a *vector process* (*motion*, when X has stationary increments). For example, in Section 9.3 below, we consider (vector) operator fractional Brownian motions which are extensions of fractional Brownian motion to processes taking values in $\mathbb{R}^p$.

We start in Section 9.1 with the fundamental notions behind multidimensional models. In Section 9.2, we define self-similarity in the multidimensional setting $q \geq 1, p \geq 1$, through the notion of operator self-similarity. Sections 9.3 and 9.4 focus on (vector) operator fractional Brownian motions and vector long-range dependence in the case $p \geq 1, q = 1$. We also discuss fractional cointegration. Sections 9.5 and 9.6 concern operator fractional Brownian fields and spatial long-range dependence in the case $q \geq 1, p = 1$. We thus consider the vector and spatial contexts separately. This is done in part for clarity, in part to highlight the differences between the two contexts, and in part because multidimensional models are not as well developed as one-dimensional models. Moreover, the two contexts are often considered separately in applications.

As noted above, we extend in this chapter the notion of self-similarity and fractional Brownian motion to both the vector and spatial settings. These extensions concern non-stationary processes. We also consider multivariate stationary time series with long-range dependence. As noted in Section 9.6, it is possible to provide a general notion of vector long-range dependence (see Section 9.4), but in the spatial context, there is no corresponding general definition, only particular cases. Of special importance are the isotropic case and some forms of anisotropic cases, described in Sections 9.6.1 and 9.6.2. One way to describe some of these stationary spatial time series is to use spherical harmonics.

For the convenience of the reader, we now list the various models considered.
In the case of *vector processes* ($p \geq 1, q = 1$):

- Vector operator FBM: vector *OFBM* (Section 9.3)
- Time reversible vector OFBM (Theorem 9.3.11)
- Vector FBM: *VFBM* (Section 9.3.3)
- Single-parameter vector OFBM (Example 9.3.23)
- Vector FARIMA(0, D, 0) (Section 9.4.2)
- Vector FGN (Section 9.4.3)

In the case of *random fields* ($q \geq 1, p = 1$):

- Operator fractional Brownian field: *OFBF* (Section 9.5)
- Isotropic fractional Brownian field: *isotropic FBF* (Example 9.5.18)
- Anisotropic fractional Brownian field: *anisotropic FBF* (Example 9.5.19)
- Fractional Brownian sheet (Example 9.5.20)
- Spatial LRD fileds with explicit autocovariances (Example 9.6.4)
- Spatial FARIMA (Example 9.6.5)

9.1 Fundamentals of Multidimensional Models

We recall here some fundamental notions behind multidimensional models, starting with basic facts for matrices (Section 9.1.1) and then distinguishing between models in the vector case (Section 9.1.2) and the spatial case (Section 9.1.3).

9.1.1 Basics of Matrix Analysis

We gather here some basic notions related to matrices as needed throughout this chapter. Our main reference is the excellent monograph on matrix analysis by Horn and Johnson [482]. The focus is on $p \times p$, $p \geq 1$, square matrices with complex-valued entries: that is, the elements of $\mathbb{C}^{p \times p}$. But the latter also include matrices with real-valued entries: that is, the elements of $\mathbb{R}^{p \times p}$. Matrices will usually be denoted in capitals such as A, B, etc.

For $A = (a_{jk})_{j,k=1,\ldots,p} \in \mathbb{C}^{p \times p}$, A^* stands for the *Hermitian transpose* of A; that is, the matrix $A^* = (\overline{a}_{kj})_{j,k=1,\ldots,p}$ with its (j, k) element being $\overline{a}_{kj}$, where the bar indicates complex conjugation. For example,

$$A = \begin{pmatrix} 1 & i \\ 2 & 3 \end{pmatrix}, \quad A^* = \begin{pmatrix} 1 & 2 \\ -i & 3 \end{pmatrix}.$$

This definition naturally extends to non-square matrices, including vectors. For $A \in \mathbb{R}^{p \times p}$, we have only transposition, namely, $A^* = A'$. Similarly, we define $\overline{A} = (\overline{a}_{jk})_{j,k=1,\ldots,p}$; that is, the matrix consisting of the complex conjugates of the elements of the matrix A. A matrix

$A \in \mathbb{C}^{p \times p}$ is *Hermitian symmetric* or simply *Hermitian* if $A^* = A$. For example, the 2×2 matrix

$$A = \begin{pmatrix} 1 & i \\ -i & 0 \end{pmatrix}$$

is Hermitian symmetric since $\bar{i} = -i$ and the diagonal entries 1 and 0 are real. A diagonal matrix will be denoted $\mathrm{diag}(a_1, \ldots, a_p)$ where $a_j \in \mathbb{C}$. In some instances, the a_js will be square matrices themselves and the matrix $\mathrm{diag}(a_1, \ldots, a_p)$ will be block diagonal.

The matrix $A \in \mathbb{C}^{p \times p}$ is *nonsingular* if its inverse A^{-1} exists, or equivalently if its rows (or columns) are linearly independent. It is *singular* otherwise.

An *eigenvalue* $\eta \in \mathbb{C}$ of a matrix $A \in \mathbb{C}^{p \times p}$ satisfies $Au = \eta u$ for a (nonzero) *eigenvector* $u \in \mathbb{C}^p$. A matrix A has at most p distinct eigenvalues. By saying that the eigenvalues of A are $\eta_1, \ldots, \eta_p$, we shall imply that these eigenvalues may not be all distinct. They appear in the so-called characteristic polynomial

$$\det(A - \eta I) = \prod_{k=1}^{p} (\eta_k - \eta),$$

and are obtained by setting $\det(A - \eta I) = 0$. We shall also use the so-called Jordan decomposition of $A \in \mathbb{C}^{p \times p}$ having the form

$$A = PJP^{-1}, \tag{9.1.1}$$

where P is a nonsingular $p \times p$ matrix, $J = \mathrm{diag}(J_{\lambda_1}, \ldots, J_{\lambda_N})$ is in Jordan canonical form with the Jordan blocks $J_{\lambda_1}, \ldots, J_{\lambda_N}$ on the diagonal and $\lambda_1, \ldots, \lambda_N$ are (possibly repeated) eigenvalues of A. A Jordan block J_λ is a matrix of the form

$$J_\lambda = \begin{pmatrix} \lambda & 0 & 0 & \ldots & 0 \\ 1 & \lambda & 0 & \ldots & 0 \\ 0 & 1 & \lambda & \ldots & 0 \\ \vdots & \vdots & \vdots & \ddots & \vdots \\ 0 & 0 & \ldots & 1 & \lambda \end{pmatrix}. \tag{9.1.2}$$

The matrix P is not unique but the Jordan block matrix J_λ is unique up to the ordering of the blocks.

A matrix $A \in \mathbb{R}^{p \times p}$ with real-valued entries can have both complex eigenvalues and complex eigenvectors. An Hermitian symmetric matrix A has real eigenvalues only. If the eigenvalues are nonnegative (positive, resp.), the Hermitian symmetric matrix A is called *positive semidefinite* (*positive definite*, resp.). In fact, positive semidefiniteness (positive definiteness, resp.) is well known to be equivalent to $u^*Au \geq 0$; that is, u^*Au is real and nonnegative for any $u \in \mathbb{C}$ ($u^*Au > 0$ for any nonzero $u \in \mathbb{C}$, resp.). For a positive semidefinite Hermitian symmetric matrix A, there is a unique positive semidefinite Hermitian symmetric matrix B such that $B^2 = A$ (Horn and Johnson [482], Theorem 7.2.6). One denotes $B = A^{1/2}$.

Note that $A = KK^*$ is always positive semidefinite since $u^*KK^*u = (u^*K)(u^*K)^* = |u^*K|^2 \geq 0$.

For $A \in \mathbb{C}^{p \times p}$, its matrix (operator) norm is defined as

$$\|A\| = \max_{\|u\|_2 = 1} \|Au\|_2, \tag{9.1.3}$$

where for a vector $x = (x_1, \ldots, x_p)' \in \mathbb{C}^p$, $\|x\|_2 = (\sum_{j=1}^{p} |x_j|^2)^{1/2}$ is the usual Euclidean norm.

9.1.2 Vector Setting

In the vector setting and in discrete time, we shall work with vector-valued ($\mathbb{R}^p$–valued),[1] second-order stationary time series $X = \{X_n\}_{n \in \mathbb{Z}}$. As in Section 2.1, such series have a constant mean vector $\mu_X = \mathbb{E}X_n \in \mathbb{R}^p$ and a $p \times p$ autocovariance matrix function

$$\gamma_X(m - n) = \mathbb{E}X_m X_n' - \mathbb{E}X_m \mathbb{E}X_n' = \mathbb{E}X_m X_n' - \mu_X \mu_X', \quad m, n \in \mathbb{Z}, \tag{9.1.4}$$

which depends only on the distance $m - n$. With $\gamma_X(\cdot) = (\gamma_{X,jk}(\cdot))_{j,k=1,\ldots,p}$, $\mu_X = (\mu_{X,1}, \ldots, \mu_{X,p})'$ and $X_n = (X_{1,n}, \ldots, X_{p,n})'$, the last expression can be written component-wise as

$$\gamma_{X,jk}(m - n) = \mathbb{E}X_{j,m} X_{k,n} - \mathbb{E}X_{j,m} \mathbb{E}X_{k,n} = \mathbb{E}X_{j,m} X_{k,n} - \mu_{X,j}\mu_{X,k}, \quad j, k = 1, \ldots, p.$$

Note that, by the second-order stationarity, the matrix $\gamma_X(n)$ satisfies

$$\gamma_X(n) = \mathbb{E}X_n X_0' - \mu_X \mu_X' = \mathbb{E}X_0 X_{-n}' - \mu_X \mu_X' = (\mathbb{E}X_{-n} X_0' - \mu_X \mu_X')' = \gamma_X(-n)', \tag{9.1.5}$$

and thus $\gamma_{X,jk}(n) = \gamma_{X,kj}(-n)$. The third equality in (9.1.5) follows by using transposition twice. The cross-covariances $\gamma_{X,jk}(n)$ need not be equal to $\gamma_{X,jk}(-n)$ when $j \neq k$. Therefore, in contrast to the univariate case, it is *not* true in general that

$$\gamma_X(n) = \gamma_X(-n), \quad n \in \mathbb{Z}, \tag{9.1.6}$$

or, equivalently, by using (9.1.5), that

$$\gamma_X(n) = \gamma_X(n)', \quad n \in \mathbb{Z}. \tag{9.1.7}$$

When (9.1.6)–(9.1.7) hold, the time series X is called *time reversible* (cf. (9.3.34)).

The spectral density $f_X(\lambda) = (f_{X,jk}(\lambda))_{j,k=1,\ldots,p}$, $\lambda \in (-\pi, \pi]$, is now matrix-valued, if it exists. It satisfies

$$\int_{-\pi}^{\pi} e^{in\lambda} f_X(\lambda) d\lambda = \gamma_X(n), \quad n \in \mathbb{Z}. \tag{9.1.8}$$

While in the univariate case, the spectral density $f_X(\lambda)$ is real and in fact nonnegative, this is not the case anymore in the vector case. Since (9.1.6) may not hold, $f_X(\lambda)$ is, in general, complex-valued. For fixed $\lambda \in (-\pi, \pi]$, the matrix $f_X(\lambda)$ is Hermitian symmetric and positive semidefinite. The function $f_X(\lambda)$, $\lambda \in (-\pi, \pi]$, is Hermitian; that is,

$$f_X(-\lambda) = \overline{f_X(\lambda)}, \quad \lambda \in (-\pi, \pi].$$

---

[1] Since a $r \times c$ matrix has r rows and c columns, each vector X_n is a $p \times 1$ matrix and its transpose X_n' is a $1 \times p$ matrix.

The spectral representation of the process X is now written as $X_n = \int_{(-\pi,\pi]} e^{in\lambda} Z(d\lambda)$ for a vector complex-valued random measure $Z(d\lambda)$ with $\mathbb{E}Z(d\lambda)Z(d\lambda)^* = f_X(\lambda)d\lambda$ and, more generally, $\mathbb{E}Z(d\lambda)Z(d\lambda)^* = F_X(d\lambda)$ where F_X is called the *spectral distribution function* (cf. Section 1.3.4). For example, if $p = 2$,

$$f_X(\lambda) = \begin{pmatrix} f_{X,11}(\lambda) & f_{X,12}(\lambda) \\ f_{X,21}(\lambda) & f_{X,22}(\lambda) \end{pmatrix},$$

where $f_{X,12}(\lambda)$ and $f_{X,21}(\lambda)$ are called the *cross spectrum*. The ratio

$$\mathcal{H}_{12}^2(\lambda) = \frac{|f_{12}(\lambda)|^2}{f_{11}(\lambda) f_{22}(\lambda)}$$

is interpreted as the *coherence* or *squared coherence function*. We have $0 \leq \mathcal{H}_{12}^2(\lambda) \leq 1$ by the Cauchy-Schwarz inequality, and thus a value close to one suggests a strong linear relationship between the comonents $Z_1(d\lambda)$ and $Z_2(d\lambda)$ of the vector $Z(d\lambda) = (Z_1(d\lambda), Z_2(d\lambda))'$. See Hannan [450], Brillinger [185], Brockwell and Davis [186], Lütkepohl [654] for more information on vector-valued time series X.

Warning: Two definitions of an autocovariance function are used in the literature. One definition is given in (9.1.4). The other definition is to set

$$\widetilde{\gamma}_X(m - n) = \mathbb{E}X_n X'_m - \mathbb{E}X_n \mathbb{E}X'_m = \mathbb{E}X_n X'_m - \mu_X \mu'_X, \quad m, n \in \mathbb{Z} \quad (9.1.9)$$

(e.g., Hannan [450]). Note that $\widetilde{\gamma}_X(m - n)$ is also a $p \times p$ matrix. The connection between (9.1.4) and (9.1.9) is $\widetilde{\gamma}_X(n) = \gamma_X(n)' = \gamma_X(-n)$, $n \in \mathbb{Z}$, in view of (9.1.5). Since the two definitions are not equivalent, it is important to be aware of which convention is used in a given source. For example, several works behind Section 9.4 (Kechagias and Pipiras [551], Helgason et al. [463]) use, in fact, the convention (9.1.9) and not (9.1.4). We also note that the choice of the convention (9.1.4) or (9.1.9) has implications on how spectral representations of time series are expressed. If (9.1.8) relates the autocovariance function $\widetilde{\gamma}_X$ and the spectral density $\widetilde{f}_X$, the spectral representation of a time series X would be written $X_n = \int_{(-\pi,\pi]} e^{-in\lambda} Z(d\lambda)$ for a vector complex-valued random measure $Z(d\lambda)$ with $\mathbb{E}Z(d\lambda)Z(d\lambda)^* = \widetilde{f}_X(\lambda)d\lambda$, since in this case $\widetilde{\gamma}_X(m - n) = \mathbb{E}X_n X'_m = \int_{(-\pi,\pi]} e^{i(m-n)\lambda} \widetilde{f}_X(\lambda)d\lambda$. On the other hand, under the convention (9.1.4) and as used in this book (see Section 1.3.4 and above), the spectral representation involves the deterministic kernel $e^{in\lambda}$, instead of $e^{-in\lambda}$.

Since $f_{X,jk}(\lambda)$, $j \neq k$, is complex-valued in general, it can be expressed in polar coordinates as

$$f_{X,jk}(\lambda) = \alpha_{X,jk}(\lambda) e^{i\phi_{X,jk}(\lambda)} \quad (9.1.10)$$

for real-valued functions $\alpha_{X,jk}(\lambda)$ and $\phi_{X,jk}(\lambda)$, known respectively as the amplitude spectrum and the phase spectrum of $\{X_{j,n}\}$ and $\{X_{k,n}\}$. In fact, there are two common but slightly different ways to represent the amplitude and phase spectra. On one hand, recall that a complex number $z = z_1 + iz_2 \in \mathbb{C}$, $z_1, z_2 \in \mathbb{R}$, can be represented as

$$z = |z|e^{i \operatorname{Arg}(z)}, \quad (9.1.11)$$

9.1 Fundamentals of Multidimensional Models

where the so-called principal argument function $\text{Arg}(z) \in (-\pi, \pi]$ by convention. Assume now that $z_1 \neq 0$. Note that

$$\text{Arg}(z) = \arctan\left(\frac{z_2}{z_1}\right) + \pi \,\text{sign}(z_2) 1_{\{z_1 < 0\}},$$

where by arctan we mean the principal value,[2] namely, the angle in $(-\pi/2, \pi/2)$. Indeed, if z_1 and z_2 have the same sign, then $\text{Arg}(z) = \arctan(z_2/z_1)$, but if $z_1 < 0$, then Arg and arctan differ by $+\pi$ if $z_2 > 0$ and by $-\pi$ if $z_2 < 0$. Since $e^0 = 1$ and $e^{\pm i\pi} = -1$, we can then also write

$$\begin{aligned}
z &= \sqrt{z_1^2 + z_2^2} \exp\left\{i(\arctan(\tfrac{z_2}{z_1}) + \pi\,\text{sign}(z_2) 1_{\{z_1<0\}})\right\} \\
&= \text{sign}(z_1)\sqrt{z_1^2 + z_2^2} \exp\left\{i \arctan(\tfrac{z_2}{z_1})\right\} \\
&= z_1 \sqrt{1 + \tfrac{z_2^2}{z_1^2}} \exp\left\{i \arctan(\tfrac{z_2}{z_1})\right\} = z_1 \sqrt{1 + \tan^2\left(\arctan \tfrac{z_2}{z_1}\right)} \exp\left\{i \arctan(\tfrac{z_2}{z_1})\right\} \\
&= z_1 \sqrt{1 + \tfrac{\sin^2\phi}{\cos^2\phi}} \exp\{i\phi\} = \tfrac{z_1}{\cos\phi} \exp\{i\phi\}, \quad \text{with } \phi = \arctan\left(\tfrac{z_2}{z_1}\right) \in \left(-\tfrac{\pi}{2}, \tfrac{\pi}{2}\right).
\end{aligned}$$
(9.1.12)

We thus get two specifications of the amplitude $\alpha_{X,jk}(\lambda)$ and the phase $\phi_{X,jk}(\lambda)$ of $f_{j,k}(\lambda)$ in (9.1.10). The first corresponds to (9.1.11) (e.g., Brockwell and Davis [186], p. 422), namely,

$$\alpha_{X,jk}(\lambda) = |f_{X,jk}(\lambda)|, \quad \phi_{X,jk}(\lambda) = \text{Arg}(f_{X,jk}(\lambda)). \tag{9.1.13}$$

The second specification is obtained by using (9.1.12) (e.g., Hannan [450], pp. 43–44):

$$\alpha_{X,jk}(\lambda) = \frac{\Re f_{X,jk}(\lambda)}{\cos \phi_{X,jk}(\lambda)}, \quad \phi_{X,jk}(\lambda) = \arctan \frac{\Im f_{X,jk}(\lambda)}{\Re f_{X,jk}(\lambda)}. \tag{9.1.14}$$

Note that, in the case (9.1.13), $\phi_{j,k}(\lambda) \in (-\pi, \pi]$ since this is the range of the Arg function, but in the case (9.1.14), $\phi_{X,jk}(\lambda) \in (-\pi/2, \pi/2)$ since this is the range of the arctan function.

9.1.3 Spatial Setting

In the spatial and discrete (lattice) setting, we focus on second-order stationary fields $X = \{X_n\}_{n \in \mathbb{Z}^q}$ taking values in $\mathbb{R}$. The autocovariance function of X is defined as

$$\gamma_X(m - n) = \mathbb{E} X_m X_n - \mathbb{E} X_m \mathbb{E} X_n = \mathbb{E} X_m X_n - \mu_X^2, \quad m, n \in \mathbb{Z}^q, \tag{9.1.15}$$

and depend only on $|m - n|$. The spectral density function $f_X(\lambda)$, $\lambda \in (-\pi, \pi]^q$, if it exists, satisfies

$$\int_{(-\pi, \pi]^q} e^{i \langle n, \lambda \rangle} f_X(\lambda) d\lambda = \gamma_X(n), \quad n \in \mathbb{Z}^q, \tag{9.1.16}$$

where $\langle \cdot, \cdot \rangle$ indicates the inner product in $\mathbb{R}^q$.

---

[2] The principal value is often denoted Arctan.

Remark 9.1.1 In the spatial setting, it is also common to consider second-order stationary random fields $X = \{X_u\}$ indexed by $u \in \mathbb{R}^q$. Such fields may be more appropriate (than those indexed by $\mathbb{Z}^q$) in modeling spatial phenomenon when data is collected at non-lattice locations. The results and discussion in the spatial setting found below also apply in this context with the following standard difference. The spectral density function $f_X(\lambda)$, if it exists, is now a function of $\lambda \in \mathbb{R}^q$ and satisfies

$$\int_{\mathbb{R}^q} e^{i\langle u, \lambda \rangle} f_X(\lambda) d\lambda = \gamma_X(u), \quad u \in \mathbb{R}^q.$$

See also Remark 1.3.8.

9.2 Operator Self-Similarity

Let us consider for a moment the general multidimensional setting

$$X(\omega, \cdot) : \mathbb{R}^q \to \mathbb{R}^p$$
$$t \mapsto X(\omega, t)$$

and ask what one means by self-similarity in the multidimensional setting. The following general definition will be useful.

Definition 9.2.1 A vector-valued ($\mathbb{R}^p$–valued) random field $\{X(t)\}_{t \in \mathbb{R}^q}$ is called *operator self-similar* with matrix exponents $E \in \mathbb{R}^{q \times q}$ and $H \in \mathbb{R}^{p \times p}$ if, for any $c > 0$,

$$\{X(c^E t)\}_{t \in \mathbb{R}^q} \stackrel{d}{=} \{c^H X(t)\}_{t \in \mathbb{R}^q}. \tag{9.2.1}$$

The relation (9.2.1) involves matrix exponentials $c^E = e^{(\log c)E}$ and $c^H = e^{(\log c)H}$ defined as follows. For $c > 0$ and general matrices $A, B \in \mathbb{C}^{m \times m}$, one sets

$$e^A := \sum_{k=0}^{\infty} \frac{A^k}{k!}, \tag{9.2.2}$$

$$c^B := e^{(\log c)B} := \sum_{k=0}^{\infty} \frac{(\log c)^k B^k}{k!}. \tag{9.2.3}$$

Note that e^A and c^B are matrices in $\mathbb{C}^{m \times m}$! Therefore in (9.2.1), c^H is a $p \times p$ matrix and hence $c^H X(t)$ is a $p \times 1$ matrix; that is, an $\mathbb{R}^p$–valued vector. Similarly, in (9.2.1), c^E is a $q \times q$ matrix, $c^E t$ is an $\mathbb{R}^q$–valued vector and $X(c^E t)$ is an $\mathbb{R}^p$–valued vector.

Figure 9.1 illustrates the mapping $S \ni x \mapsto c^A x$ in two dimensions for several choices of A and c, and for the ellipse $S = \{(x_1, x_2)' \in \mathbb{R}^2 : x_1^2 + 4x_2^2 = 1\}$ (the smallest ellipse in the three plots).

It is useful to discuss some basic but key properties of matrix exponentials. For a diagonal matrix $A = \text{diag}(a_1, \ldots, a_m) \in \mathbb{C}^{m \times m}$, we have

$$c^A = \text{diag}(c^{a_1}, \ldots, c^{a_m}).$$

In particular, if A is the identity matrix I, then c^A is the diagonal matrix cI. If

$$A = W A_0 W^{-1}$$

9.2 Operator Self-Similarity

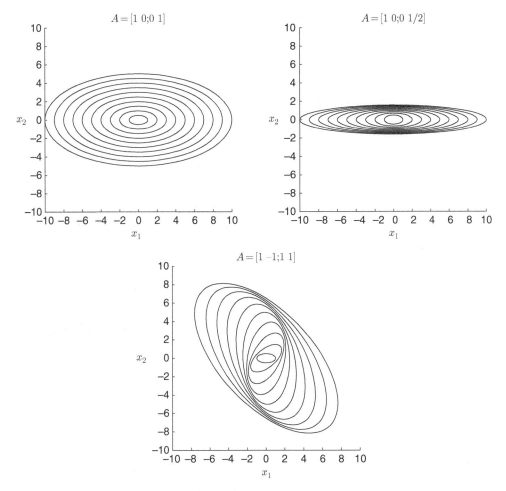

Figure 9.1 The mapping $S \ni x \mapsto c^A x$ in two dimensions for $A = [1\ 0;\ 0\ 1]$ (left), $A = [1\ 0;\ 0\ 1/2]$ (right) and $A = [1\ -1;\ 1\ 1]$ (bottom), $c = 1, 2, \ldots, 10$, and for the ellipse $S = \{(x_1, x_2)' \in \mathbb{R}^2 : x_1^2 + 4x_2^2 = 1\}$ (the smallest ellipse in the three plots).

for some invertible matrix W (with the inverse W^{-1}), then

$$c^A = \sum_{k=0}^{\infty} \frac{(\log c)^k A^k}{k!} = \sum_{k=0}^{\infty} \frac{(\log c)^k (W A_0 W^{-1})^k}{k!}$$
$$= \sum_{k=0}^{\infty} \frac{(\log c)^k (W A_0 W^{-1})(W A_0 W^{-1}) \ldots (W A_0 W^{-1})}{k!}$$
$$= \sum_{k=0}^{\infty} \frac{(\log c)^k W A_0^k W^{-1}}{k!} = W c^{A_0} W^{-1}. \tag{9.2.4}$$

In particular, if $A_0 = \mathrm{diag}(a_{0,1}, \ldots, a_{0,m}) \in \mathbb{C}^{m \times m}$ and thus A is diagonalizable over $\mathbb{C}$, then

$$c^A = W \operatorname{diag}(c^{a_{0,1}}, \ldots, c^{a_{0,m}}) W^{-1}.$$

It is *not* true in general that

$$c^{A+B} = c^A c^B = c^B c^A. \tag{9.2.5}$$

These relations hold, however, when A and B commute; that is, $AB = BA$.

To shed light on the operator self-similarity relation (9.2.1), suppose that E and H are diagonalizable over $\mathbb{R}$ with

$$E = W_E E_0 W_E^{-1}, \quad H = W_H H_0 W_H^{-1}, \tag{9.2.6}$$

where $W_E \in \mathbb{R}^{q \times q}$, $W_H \in \mathbb{R}^{p \times p}$, $E_0 = \operatorname{diag}(e_{0,1}, \ldots, e_{0,q}) \in \mathbb{R}^{q \times q}$ and $H_0 = \operatorname{diag}(h_{0,1}, \ldots, h_{0,p}) \in \mathbb{R}^{p \times p}$. One special case of (9.2.6) is when, for example, E is symmetric and W_E is orthogonal (that is, $W_E^{-1} = W_E'$). If (9.2.6) holds, by using (9.2.4), the relation (9.2.1) can be expressed as

$$\{X(W_E c^{E_0} W_E^{-1} t)\}_{t \in \mathbb{R}^q} \stackrel{d}{=} \{W_H c^{H_0} W_H^{-1} X(t)\}_{t \in \mathbb{R}^q}$$

or

$$\{W_H^{-1} X(W_E c^{E_0} W_E^{-1} t)\}_{t \in \mathbb{R}^q} \stackrel{d}{=} \{c^{H_0} W_H^{-1} X(t)\}_{t \in \mathbb{R}^q} \stackrel{d}{=} \{c^{H_0} W_H^{-1} X(W_E W_E^{-1} t)\}_{t \in \mathbb{R}^q}$$

or, after setting $s = W_E^{-1} t$,

$$\{W_H^{-1} X(W_E c^{E_0} s)\}_{s \in \mathbb{R}^q} \stackrel{d}{=} \{c^{H_0} W_H^{-1} X(W_E s)\}_{s \in \mathbb{R}^q}. \tag{9.2.7}$$

Thus, letting

$$Y(s) = W_H^{-1} X(W_E s), \quad s \in \mathbb{R}^q, \tag{9.2.8}$$

we get that

$$\{Y(c^{E_0} s)\}_{s \in \mathbb{R}^q} \stackrel{d}{=} \{c^{H_0} Y(s)\}_{s \in \mathbb{R}^q}. \tag{9.2.9}$$

In an expanded form, denoting $s = (s_1, \ldots, s_q)'$ and $Y(s) = (Y_1(s), \ldots, Y_p(s))'$, this can be expressed as

$$\left\{ \begin{pmatrix} Y_1 \begin{pmatrix} c^{e_{0,1}} s_1 \\ \vdots \\ c^{e_{0,q}} s_q \end{pmatrix} \\ \vdots \\ Y_p \begin{pmatrix} c^{e_{0,1}} s_1 \\ \vdots \\ c^{e_{0,q}} s_q \end{pmatrix} \end{pmatrix} \right\}_{s \in \mathbb{R}^q} \stackrel{d}{=} \left\{ \begin{pmatrix} c^{h_{0,1}} Y_1 \begin{pmatrix} s_1 \\ \vdots \\ s_q \end{pmatrix} \\ \vdots \\ c^{h_{0,p}} Y_p \begin{pmatrix} s_1 \\ \vdots \\ s_q \end{pmatrix} \end{pmatrix} \right\}_{s \in \mathbb{R}^q}. \tag{9.2.10}$$

- For vector processes (that is, $q = 1$), the relation (9.2.10) can be written as

$$\left\{ \begin{pmatrix} Y_1(c^{e_0} s) \\ \vdots \\ Y_p(c^{e_0} s) \end{pmatrix} \right\}_{s \in \mathbb{R}} \stackrel{d}{=} \left\{ \begin{pmatrix} c^{h_{0,1}} Y_1(s) \\ \vdots \\ c^{h_{0,p}} Y_p(s) \end{pmatrix} \right\}_{s \in \mathbb{R}}$$

or, supposing $e_0 \neq 0$,

$$\left\{ \begin{pmatrix} Y_1(cs) \\ \vdots \\ Y_p(cs) \end{pmatrix} \right\}_{s \in \mathbb{R}} \stackrel{d}{=} \left\{ \begin{pmatrix} c^{h_{0,1}/e_0} Y_1(s) \\ \vdots \\ c^{h_{0,p}/e_0} Y_p(s) \end{pmatrix} \right\}_{s \in \mathbb{R}}. \quad (9.2.11)$$

- For random fields (that is, $p = 1$), the relation (9.2.10) can be written as

$$\left\{ Y \begin{pmatrix} c^{e_{0,1}} s_1 \\ \vdots \\ c^{e_{0,q}} s_q \end{pmatrix} \right\}_{s \in \mathbb{R}^q} \stackrel{d}{=} \left\{ c^{h_0} Y \begin{pmatrix} s_1 \\ \vdots \\ s_q \end{pmatrix} \right\}_{s \in \mathbb{R}^q}$$

or, supposing $h_0 \neq 0$,

$$\left\{ Y \begin{pmatrix} c^{e_{0,1}/h_0} s_1 \\ \vdots \\ c^{e_{0,q}/h_0} s_q \end{pmatrix} \right\}_{s \in \mathbb{R}^q} \stackrel{d}{=} \left\{ c Y \begin{pmatrix} s_1 \\ \vdots \\ s_q \end{pmatrix} \right\}_{s \in \mathbb{R}^q}. \quad (9.2.12)$$

Relations (9.2.10)–(9.2.12) resemble the one-dimensional self-similarity relation (2.5.1). For example, (9.2.11) shows that each one-dimensional component process $Y_j(s)$ is self-similar in the sense of (2.5.1) with the scalar self-similarity parameter $H_j = h_{0,j}/e_0$. The self-similarity parameter may be different for different values of j. In (9.2.12), the self-similarity of the field Y holds even if it scales differently across the one-dimensional parameters s_j with self-similarity parameters $H_j = h_0/e_{0,j}$, $j = 1, \ldots, q$.

Finally, note that (9.2.8) corresponds to changing the coordinate systems in $\mathbb{R}^q$ and $\mathbb{R}^p$. Thus, after this change of coordinate systems, the process X becomes the process Y, and scales entry-wise.

9.3 Vector Operator Fractional Brownian Motions

We consider here extensions of FBM to the vector setting $p \geq 1, q = 1$. FBM was introduced in Definition 2.6.2 as a self-similar Gaussian process with stationary increments and $0 < H \leq 1$. We adopt an analogous definition in the vector setting, replacing self-similarity by operator self-similarity.

Definition 9.3.1 A vector-valued ($\mathbb{R}^p$–valued) process $\{X(t)\}_{t \in \mathbb{R}}$ is called *vector operator fractional Brownian motion* (vector OFBM) if it is Gaussian, has stationary increments and is operator self-similar.[3]

Stationarity of increments in Definition 9.3.1 means as in (2.5.2) that, for any $h \in \mathbb{R}$,

$$\{X(t+h) - X(h)\}_{t \in \mathbb{R}} \stackrel{d}{=} \{X(t) - X(0)\}_{t \in \mathbb{R}}.$$

For the definition of *operator self-similarity*, we shall take: for any $c > 0$,

$$\{X(ct)\}_{t \in \mathbb{R}} \stackrel{d}{=} \{c^H X(t)\}_{t \in \mathbb{R}}, \quad (9.3.1)$$

---

[3] Vector OFBM is often referred to simply as OFBM. We use the term *vector* OFBM to emphasize its difference from the operator fractional Brownian fields introduced below in Section 9.5.

that is, the relationship (9.2.1) with $E = 1$ and $H \in \mathbb{R}^{p \times p}$. We shall continue denoting a vector OFBM by $\{B_H(t)\}_{t \in \mathbb{R}}$. The properties of the matrix H will be described below.

Assumption We assume that a vector OFBM is *proper* in the sense that, for each $t \neq 0$, the distribution of $B_H(t)$ is not contained in a proper subspace of $\mathbb{R}^p$, and that it is continuous in distribution at all t.

In what follows, we will present integral representations, consider several special subclasses and discuss basic identifiability questions for vector OFBMs.

9.3.1 Integral Representations

The following result provides spectral domain representations of vector OFBMs. Compare it with the spectral representation when $p = 1$, given in Proposition 2.6.11.

Theorem 9.3.2 *Let $\{B_H(t)\}_{t \in \mathbb{R}}$ be a vector OFBM with exponent $H \in \mathbb{R}^{p \times p}$. Suppose that the eigenvalues h_k of H satisfy*

$$0 < \Re(h_k) < 1, \quad k = 1, \ldots, p. \tag{9.3.2}$$

Then, $\{B_H(t)\}_{t \in \mathbb{R}}$ admits the integral representation

$$\{B_H(t)\}_{t \in \mathbb{R}} \stackrel{d}{=} \left\{ \int_{\mathbb{R}} \frac{e^{itx} - 1}{ix} (x_+^{-D} A + x_-^{-D} \overline{A}) \widetilde{B}(dx) \right\}_{t \in \mathbb{R}} \tag{9.3.3}$$

for some $A \in \mathbb{C}^{p \times p}$. Here, the bar denotes component-wise complex conjugate,

$$D = H - \frac{1}{2} I \in \mathbb{R}^{p \times p} \tag{9.3.4}$$

and $\widetilde{B}(dx)$ is a vector random measure consisting of independent components $\widetilde{B}_k(dx)$, $k = 1, \ldots, p$, which are Hermitian Gaussian random measures with identical control measures $\mathbb{E}|\widetilde{B}_k(dx)|^2 = dx$.

Proof For notational simplicity, set $X = B_H$. Since X has stationary increments, we have

$$X(t) - X(s) = \int_{\mathbb{R}} \frac{e^{itx} - e^{isx}}{ix} \widetilde{Y}(dx), \tag{9.3.5}$$

$\widetilde{Y}(dx)$ is an orthogonal-increment[4] random measure with values in $\mathbb{C}^p$. The relation (9.3.5) can be proved following the approach for the one-dimensional case found in Doob [325], p. 550, under the assumption that $\mathbb{E}|X(t+h) - X(t)|^2 \to 0$ as $h \to 0$, i.e., X is $L^2(\Omega)$-continuous at every t (see also Yaglom [1021], p. 409, and Yaglom [1020], Theorem 7). The latter assumption is satisfied in our context because of the following. Property 2.1 in Maejima and Mason [667] states that, for an operator self-similar process Z with exponent H, if $\inf\{\Re(h_k) : k = 1, \ldots, p\} > 0$, then $Z(0) = 0$ a.s. Thus, in view of (9.3.2), $X(0) = 0$ a.s. So, by stationarity of the increments,

$$X(t+h) - X(t) \stackrel{d}{=} X(h) \stackrel{d}{\to} X(0) = 0, \quad h \to 0, \tag{9.3.6}$$

---
[4] A random measure $\widetilde{Y}$ has orthogonal increments if $\mathbb{E}\widetilde{Y}(E_1)\widetilde{Y}(E_2)^* = 0$ whenever $E_1 \cap E_2 = \emptyset$. Recall that $*$ denotes the Hermitian transpose.

9.3 Vector Operator Fractional Brownian Motions

and hence in $L^2(\Omega)$ since X is Gaussian. The last convergence in (9.3.6) either follows from the assumption of the continuity in distribution made following Definition 9.3.1 or can be proved directly under the assumption (9.3.2) as in Property 2.1 in Maejima and Mason [667]. This proves (9.3.5). In particular,

$$X(t) = \int_{\mathbb{R}} \frac{e^{itx} - 1}{ix} \widetilde{Y}(dx). \qquad (9.3.7)$$

Now let

$$F_X(dx) = \mathbb{E}\widetilde{Y}(dx)\widetilde{Y}(dx)^* \qquad (9.3.8)$$

be the vector spectral distribution of $\widetilde{Y}(dx)$ and note that it is a $p \times p$ matrix.[5]

The rest of the proof is in three steps:

1. show the existence of a spectral density function $f_X(x) = F_X(dx)/dx$,
2. decorrelate the measure $\widetilde{Y}(dx)$ component-wise by finding a filter based upon the spectral density function,
3. develop the form of the filter.

Step 1: Since X is operator self-similar with exponent H,

$$\{X(ct)\} \stackrel{d}{=} \left\{ c^H \int_{\mathbb{R}} \frac{e^{itx} - 1}{ix} \widetilde{Y}(dx) \right\}, \qquad (9.3.9)$$

for $c > 0$. On the other hand, through a change of variables $x = c^{-1}v$ in (9.3.7),

$$X(ct) = \int_{\mathbb{R}} \frac{e^{itv} - 1}{iv} c\widetilde{Y}(c^{-1}dv). \qquad (9.3.10)$$

The relations (9.3.9) and (9.3.10) provide two spectral representations for the process $\{X(ct)\}_{t \in \mathbb{R}}$. As a consequence of the uniqueness of the spectral distribution function of the stationary process $\{X(t) - X(t-1)\}_{t \in \mathbb{R}}$ and of the fact that $|(e^{ix} - 1)/(ix)|^2 > 0$, $x \in \mathbb{R} \setminus \{2\pi k, k \in \mathbb{Z}\}$, we obtain, in view of (9.3.8), that

$$c^H F_X(dx) c^{H^*} = c^2 F_X(c^{-1}dx), \quad c > 0, \qquad (9.3.11)$$

which can be written, by a change of variables, $c^H F_X(cdx) c^{H^*} = c F_X(dx) c$ or $F_X(cdx) = c^{I-H} F_X(dx) c^{(I-H)^*}$. Thus, for $c > 0$,

$$\int_{(0,1]} F_X(cdx) = F_X(0, c] = c^{I-H} F_X(0, 1] c^{(I-H)^*}, \qquad (9.3.12)$$

$$\int_{(-1,0]} F_X(cdx) = F_X(-c, 0] = c^{I-H} F_X(-1, 0] c^{(I-H)^*}. \qquad (9.3.13)$$

One can write $I - H$ in its Jordan decomposition form as $I - H = PJP^{-1}$ $P \in \mathbb{C}^{p \times p}$ is invertible and $J = \text{diag}(J_{1-h_{i_1}}, \ldots, J_{1-h_{i_s}})$ consists of Jordan blocks J_λ, defined in (9.1.2), and $i_1, \ldots, i_s$ is a set of indices. Then, $c^{I-H} = Pc^J P^{-1} = P\text{diag}(c^{J_{1-h_{i_1}}}, \ldots, c^{J_{1-h_{i_s}}})P^{-1}$. By the explicit formula for c^{J_λ} in Exercise 9.1, each individual entry $F_X(0, c]_{jk}$, $j, k = $

---

[5] Note also that we use the notation F_X and not F_Y.

$1, \ldots, p$, in the expression on the right-hand side of (9.3.12) is a linear combination (with complex weights) of terms of the form

$$\frac{(\log c)^m}{m!} c^{1-\overline{h}_r} \frac{(\log c)^n}{n!} c^{1-h_s}, \quad r, s = 1, \ldots, p, \quad m, n = 0, \ldots, p-1. \quad (9.3.14)$$

By (9.3.2), the eigenvalues h_k of H satisfy $0 < \Re(h_k) < 1$. Since (9.3.14) are differentiable in c over $(0, \infty)$, so are $F_X(0, c]_{jk}$. Hence, $F_X(c)$ is differentiable in c over $(0, \infty)$ as well, since $F_X(0, c]_{jk} = F_X(c)_{jk} - F_X(0)_{jk}$. The differentiability of F_X on $(-\infty, 0)$ follows from (9.3.13) and an analogous argument.

The above proves that F_X is differentiable and hence absolutely continuous on $\mathbb{R} \setminus \{0\}$. The function F_X will also be absolutely continuous on $\mathbb{R}$ if it is continuous at zero. Note that

$$F_X(-c, c] = c^{I-H} F_X(-1, 1] c^{(I-H)^*} \to 0,$$

as $c \to 0^+$. The limit holds because $\|c^{I-H}\| \to 0$, as $c \to 0^+$, $\|\cdot\|$ is the matrix norm defined in (9.1.3), which in turn follows from Proposition 2.1, (ii), in Maejima and Mason [667] under the assumption that $\Re(h_k) < 1, k = 1, \ldots, p$.

Step 2: Denote the spectral density of X by f_X. Since $|(1-e^{-ix})/(ix)|^2 f_X(x)$ is the spectral density of the stationary process $\{X(t) - X(t-1)\}_{t \in \mathbb{R}}$, $f_X(x)$ is a positive semidefinite Hermitian-symmetric matrix dx-a.e. (Hannan [450], Theorem 1, p. 34). Let

$$\widehat{a}(x) = f_X(x)^{1/2} \in \mathbb{C}^{p \times p} \quad (9.3.15)$$

be the (unique) positive semidefinite square root of $f_X(x)$ (see Section 9.1.1). Let also $\widetilde{B}(dx)$ be a complex-valued vector measure as in the statement of the theorem. Then, X can also be represented as

$$\{X(t)\} \stackrel{d}{=} \left\{ \int_{\mathbb{R}} \frac{e^{itx} - 1}{ix} \widehat{a}(x) \widetilde{B}(dx) \right\} \quad (9.3.16)$$

because

$$\mathbb{E}(\widehat{a}(x)\widetilde{B}(dx)\widetilde{B}(dx)^* \widehat{a}(x)^*) = \widehat{a}(x)\widehat{a}(x)^* dx = f_X(x) dx = F_X(dx),$$

and the processes on both sides of (9.3.16) are Gaussian and real-valued.

Step 3: It follows from the relation (9.3.11) in Step 1 that for every $c > 0$, $F_X(dx) = c^{I-H} F_X(c^{-1}dx) c^{I-H}$. By (9.3.15), $F_X(c^{-1}dx) = f_X(c^{-1}x) c^{-1}dx = \widehat{a}(c^{-1}x)\widehat{a}(c^{-1}x)^* c^{-1}dx$ and by (9.3.4), we have $(I - H) - (I/2) = (I/2) - H = -D$. Hence, for every $c > 0$,

$$\widehat{a}(x)\widehat{a}(x)^* = c^{-D} \widehat{a}\left(\frac{x}{c}\right) \widehat{a}\left(\frac{x}{c}\right)^* c^{-D^*} \quad dx\text{-a.e.} \quad (9.3.17)$$

The relation (9.3.17) also holds $dxdc$-a.e. This is a consequence of the following Fubini-based argument which is often used. If A denotes the set of x, c for which the two functions in (9.3.17) are equal, we have $\int_{\mathbb{R}} 1_A(x, c) dx = 0$ for every c by (9.3.17) and hence by Fubini's theorem, $\int_{\mathbb{R}} \int_{\mathbb{R}} 1_A(x, c) dx dc = 0$; that is, $A = \mathbb{R}^2$ $dxdc$-a.e. or the two functions are also equal $dxdc$-a.e.

Consider $x > 0$. A change of variables c to x/v leads to

$$\widehat{a}(x)\widehat{a}(x)^* = x^{-D} v^D \widehat{a}(v)\widehat{a}(v)^* v^{D^*} x^{-D^*} \quad dx\,dv\text{-a.e.}$$

Thus, one can choose $v_+ > 0$ such that

$$\widehat{a}(x)\widehat{a}(x)^* = x^{-D}v_+^D\widehat{a}(v_+)\widehat{a}(v_+)^*v_+^{D^*}x^{-D^*} \quad dx\text{-a.e.} \tag{9.3.18}$$

This means, in particular, that if we set, for dx-a.e. $x > 0$,

$$\widehat{\alpha}_+(x) = x^{-D}v_+^D\widehat{a}(v_+) = x_+^{-D}v_+^D\widehat{a}(v_+),$$

then $\widehat{\alpha}_+(x)\widehat{\alpha}_+(x)^* = \widehat{a}(x)\widehat{a}(x)^* = f_X(x)$ for $x > 0$.

We now suppose $x < 0$. Again by considering the stationary process $\{X(t)-X(t-1)\}_{t\in\mathbb{R}}$, and by applying Theorem 3 in Hannan [450], p. 41, which states that for continuity points $x_2 > x_1$ of F_X,

$$F_X(x_2) - F_X(x_1) = \lim_{T\to\infty} \frac{1}{2\pi} \int_{-T}^{T} (\mathbb{E}X(t)X(0)')\frac{e^{ix_2t} - e^{ix_1t}}{it}dt,$$

we conclude that $\overline{F_X(x_2)} - \overline{F_X(x_1)} = F_X(-x_1) - F_X(-x_2)$ and hence that $\int_{x_1}^{x_2} \overline{f_X(x)}dx = \int_{-x_2}^{-x_1} f_X(x)dx = \int_{x_1}^{x_2} f_X(-x)dx$. Since F_X has at most a countable number of discontinuity points (Exercise 9.3), the last equality holds for all $x_2 > x_1$ and hence $f_X(-x) = \overline{f_X(x)}$ a.e. dx. Thus,

$$\widehat{a}(-x)\widehat{a}(-x)^* = f_X(-x) = \overline{f_X(x)} = x^{-D}v_+^D\overline{\widehat{a}(v_+)\widehat{a}(v_+)^*}v_+^{D^*}x^{-D^*} \quad dx\text{-a.e.}$$

Hence, for $x < 0$, we can set

$$\widehat{\alpha}_-(x) = (-x)_+^{-D}v_+^D\overline{\widehat{a}(v_+)} = x_-^{-D}v_+^D\overline{\widehat{a}(v_+)}$$

and, for $x \in \mathbb{R}$, we have

$$\widehat{\alpha}(x) = x_+^{-D}v_+^D\widehat{a}(v_+) + x_-^{-D}v_+^D\overline{\widehat{a}(v_+)} \quad dx\text{-a.e.},$$

and

$$\widehat{\alpha}(x)\widehat{\alpha}(x)^* = f_X(x) \quad dx\text{-a.e.}$$

Therefore, we can use $\widehat{\alpha}$ in place of $\widehat{a}$ in the spectral representation (9.3.16) of X. In view (9.3.16), setting $A = v_+^D\widehat{a}(v_+)$, which is a $p \times p$ matrix, establishes the relation (9.3.3). □

Remark 9.3.3 As a consequence of Corollary 2.1 in Maejima and Mason [667], the eigenvalues h_k of the exponent H of a vector OFBM must satisfy $\Re(h_k) \leq 1$, $k = 1,\ldots,p$. However, the extension of the definition of vector OFBM to the case of H with at least one eigenvalue h_k satisfying $\Re(h_k) = 1$ can be delicate (see Didier and Pipiras [307], Remark 3.2).

Theorem 9.3.2 shows that vector OFBM is characterized by a (potentially non-unique; see Section 9.3.4) operator self-similarity exponent H and a matrix A. For the sake of simplicity, we will continue to use the notation B_H for vector OFBM instead of a more correct notation $B_{H,A}$.

We shall next provide integral representations of vector OFBMs in the time domain, which is done in Theorem 9.3.8 below. The key technical step in the proof is the calculation of the (entry-wise) Fourier transform of the kernels

$$(t-u)_{\pm}^D - (-u)_{\pm}^D = \exp(\log(t-u)_{\pm}D) - \exp(\log(-u)_{\pm}D), \tag{9.3.19}$$

which are the multivariate analogues of the corresponding univariate FBM time domain kernels.

It is natural and convenient to carry out this step in the framework of the so-called "primary matrix functions." The latter allows one to define naturally matrix analogues $f(D)$, $D \in \mathbb{R}^{p \times p}$, of univariate functions $f(d)$, $d \in \mathbb{R}$, and to say when two such matrix-valued functions are equal based on their univariate counterparts. We shall need to define, in particular, $\Gamma(D + I)$ and $\sin(D\pi/2)$ where D is a $p \times p$ matrix.

We shall now recall the definition of primary matrix functions (more details and properties can be found in Horn and Johnson [482], Sections 6.1 and 6.2; see also Higham [475]). Let $\Lambda \in \mathbb{C}^{p \times p}$ and consider its Jordan decomposition $\Lambda = PJP^{-1}$, $J = \mathrm{diag}(J_{\lambda_1}, \ldots, J_{\lambda_N})$ is in Jordan canonical form with the Jordan blocks $J_{\lambda_1}, \ldots, J_{\lambda_N}$, defined in (9.1.2) and of respective sizes $r_k \times r_k$, $k = 1, \ldots, N$, on the diagonal, and $\lambda_1, \ldots, \lambda_N$ are the eigenvalues of Λ.

Let $U \subseteq \mathbb{C}$ be an open set. Given a function $h : U \to \mathbb{C}$ and some $\Lambda \in \mathbb{C}^{p \times p}$ as above, consider the conditions:

(M1) $\lambda_k \in U, k = 1, \ldots, N$;
(M2) if $r_k > 1$ (with r_k referring to the size of J_{λ_k}), then $h(z)$ is analytic in a vicinity $U_k \ni \lambda_k$, $U_k \subseteq U$.

Let $\mathcal{M}_h = \{\Lambda \in \mathbb{C}^{p \times p} :$ conditions (M1) and (M2) hold at the eigenvalues $\lambda_1, \ldots, \lambda_N$ of $\Lambda\}$.

We now define the *primary matrix function* $h(\Lambda)$ associated with the scalar-valued function $h(z)$.

Definition 9.3.4 The *primary matrix function* $h : \mathcal{M}_h \to \mathbb{C}^{p \times p}$ is defined as

$$h(\Lambda) = Ph(J)P^{-1} = P \begin{pmatrix} h(J_{\lambda_1}) & \cdots & 0 \\ \vdots & \ddots & \vdots \\ 0 & \cdots & h(J_{\lambda_N}) \end{pmatrix} P^{-1},$$

where $h(J_{\lambda_k})$ is the following $r_k \times r_k$ matrix:

$$h(J_{\lambda_k}) = \begin{pmatrix} h(\lambda_k) & 0 & \cdots & 0 \\ h'(\lambda_k) & h(\lambda_k) & \cdots & 0 \\ \vdots & \ddots & \ddots & \vdots \\ \frac{h^{(r_k-1)}(\lambda_k)}{(r_k-1)!} & \cdots & h'(\lambda_k) & h(\lambda_k) \end{pmatrix}.$$

Thus one can express $h(\Lambda) = h(PJP^{-1})$ as $h(\Lambda) = Ph(J)P^{-1}$. Note also that the matrix $h(\Lambda)$ is invertible if $h(\lambda_k) \neq 0$, $k = 1, \ldots, N$. The following example illustrates the notion of a primary matrix function.

Example 9.3.5 Consider the Jordan block

$$J = \begin{pmatrix} 2 & 0 \\ 1 & 2 \end{pmatrix}$$

and let $h(x) = x^2$, so that $h'(x) = 2x$. Then

$$h(J) = \begin{pmatrix} h(2) & 0 \\ h'(2) & h(2) \end{pmatrix} = \begin{pmatrix} 4 & 0 \\ 4 & 4 \end{pmatrix},$$

which is also equal to J^2.

Example 9.3.6 (*Primary matrix functions*) (i) The function $h(z) = \sin(z)$ is analytic for $z \in \mathbb{C} =: U$. According to Definition 9.3.4, the primary matrix function $h(\Lambda) = \sin(\Lambda)$ can be defined for any matrix Λ as

$$\sin(\Lambda) = P \sin(J) P^{-1} = P \begin{pmatrix} \sin(J_{\lambda_1}) & \cdots & 0 \\ \vdots & \ddots & \vdots \\ 0 & \cdots & \sin(J_{\lambda_N}) \end{pmatrix} P^{-1},$$

where $\Lambda = PJP^{-1}$ is the Jordan decomposition of Λ, and

$$\sin(J_{\lambda_k}) = \begin{pmatrix} \sin(\lambda_k) & 0 & \cdots & 0 \\ \cos(\lambda_k) & \sin(\lambda_k) & \cdots & 0 \\ \vdots & \ddots & \ddots & \vdots \\ \frac{\sin^{(r_k-1)}(\lambda_k)}{(r_k-1)!} & \cdots & \cos(\lambda_k) & \sin(\lambda_k) \end{pmatrix}.$$

(ii) When $h(z) = e^z$, one can similarly define the primary matrix function $h(\Lambda) = e^{\Lambda}$ for any matrix Λ. It should be noted that the definition of the matrix exponential based on a series is equivalent to that based on primary matrix functions. See Exercise 9.2.

The functions $(t-u)_{\pm}^D$, $\Gamma(D+I)$, $|x|^{-D}$, $e^{\mp i \operatorname{sign}(x) D\pi/2}$ appearing in Proposition 9.3.7 below, and the functions $\sin(D\pi/2)$, $\cos(D\pi/2)$ appearing in Theorem 9.3.8 below are all defined as primary matrix functions.

The proof of the following technical result can be found in Didier and Pipiras [307], Proposition 3.1.

Proposition 9.3.7 *Under (9.3.2) and the condition (9.3.21) in Theorem 9.3.8 below,*

$$\int_{\mathbb{R}} e^{ixu} \Big((t-u)_{\pm}^D - (-u)_{\pm}^D \Big) du = \frac{e^{itx} - 1}{ix} |x|^{-D} \Gamma(D+I) e^{\mp i \operatorname{sign}(x) D\pi/2}. \quad (9.3.20)$$

Note that, when $p = 1$, the relation (9.3.20) agrees with (2.6.4) and (6.1.42) (together with (6.1.24) and (6.1.35)). But we emphasize again that the functions $(t-u)_{\pm}^D$, $\Gamma(D+I)$, $|x|^{-D}$, $e^{\mp i \operatorname{sign}(x) D\pi/2}$ appearing in (9.3.20) are all defined as primary matrix functions for general p.

Next, we construct time domain representations for vector OFBMs.

Theorem 9.3.8 *Let $\{B_H(t)\}_{t \in \mathbb{R}}$ be a vector OFBM with operator self-similarity exponent H having the spectral representation (9.3.3) with $A = A_1 + iA_2$, $A_1, A_2 \in \mathbb{R}^{p \times p}$.*

(i) Suppose that $H \in \mathbb{R}^{p \times p}$ has eigenvalues satisfying (9.3.2) and

$$\mathrm{Re}(h_k) \neq \frac{1}{2}, \quad k = 1, \ldots, p. \tag{9.3.21}$$

Then, there are matrices $M_+, M_- \in \mathbb{R}^{p \times p}$ such that

$$\{B_H(t)\}_{t \in \mathbb{R}} \stackrel{d}{=} \left\{ \int_\mathbb{R} \left(((t-u)_+^D - (-u)_+^D) M_+ + ((t-u)_-^D - (-u)_-^D) M_- \right) B(du) \right\}_{t \in \mathbb{R}}, \tag{9.3.22}$$

where $B(du)$ is a vector-valued random measure consisting of independent Gaussian random measures $B_k(du)$ with control measures $\mathbb{E} B_k(du) B_k(du)^* = du$. Moreover, the matrices M_+, M_- can be taken as

$$M_\pm = \sqrt{\frac{\pi}{2}} \left(\left(\sin \frac{D\pi}{2} \right)^{-1} \Gamma(D+I)^{-1} A_1 \pm \left(\cos \frac{D\pi}{2} \right)^{-1} \Gamma(D+I)^{-1} A_2 \right). \tag{9.3.23}$$

(ii) Suppose that $H = (1/2)I$. Then, there exist matrices $M, N \in \mathbb{R}^{p \times p}$ such that

$$\{B_H(t)\}_{t \in \mathbb{R}} \stackrel{d}{=} \left\{ \int_\mathbb{R} \left((\mathrm{sign}(t-u) - \mathrm{sign}(-u)) M + \log\left(\frac{|t-u|}{|u|} \right) N \right) B(du) \right\}_{t \in \mathbb{R}}, \tag{9.3.24}$$

where $B(du)$ is as in (9.3.22). Moreover, the matrices M, N can be taken as

$$M = \sqrt{\frac{\pi}{2}} A_1, \quad N = -\sqrt{\frac{2}{\pi}} A_2. \tag{9.3.25}$$

Proof (i) Denote the process on the right-hand side of (9.3.22) by X_H. By using the Jordan decomposition of D, one can show that X_H is well-defined (Exercise 9.5). To prove the theorem, it suffices to show that there are matrices $M_\pm$ such that the covariance structure of X_H matches that of the vector OFBM B_H given by its spectral representation (9.3.3) with $A = A_1 + i A_2$. By using Plancherel's identity (A.1.10) and (9.3.20), note first that

$$\mathbb{E} X_H(t) X_H(s)^* = \frac{1}{2\pi} \int_\mathbb{R} \frac{(e^{itx} - 1)(e^{-isx} - 1)}{|x|^2}$$
$$\times \left(|x|^{-D} \Gamma(D+I) [e^{-i\mathrm{sign}(x) D\pi/2} M_+ + e^{i\mathrm{sign}(x) D\pi/2} M_-] \right)$$
$$\cdot \left([M_+^* e^{i\mathrm{sign}(x) D^*\pi/2} + M_-^* e^{-i\mathrm{sign}(x) D^*\pi/2}] \Gamma(D+I)^* |x|^{-D^*} \right) dx,$$

where $*$ is just the transpose in the case of $X_H(t) \in \mathbb{R}^p$ and $M_+, M_-, D, \Gamma(D+I) \in \mathbb{R}^{p \times p}$. Meanwhile, for B_H, we have by (9.3.3),

$$\mathbb{E} B_H(t) B_H(s)^* = \int_\mathbb{R} \frac{(e^{itx} - 1)(e^{-isx} - 1)}{|x|^2} (x_+^{-D} A A^* x_+^{-D^*} + x_-^{-D} \overline{AA^*} x_-^{-D^*}) dx. \tag{9.3.26}$$

Thus, by using the relation $e^{i\Theta} = \cos(\Theta) + i \sin(\Theta)$, $\Theta \in \mathbb{R}^{p \times p}$, it is sufficient to find $M_\pm \in \mathbb{R}^{p \times p}$ such that

$$AA^* = \frac{1}{2\pi} \Gamma(D+I) (e^{-iD\pi/2} M_+ + e^{iD\pi/2} M_-)$$
$$\cdot (M_+^* e^{iD^*\pi/2} + M_-^* e^{-iD^*\pi/2}) \Gamma(D+I)^*$$

9.3 Vector Operator Fractional Brownian Motions

$$= \frac{1}{2\pi}\Gamma(D+I)\Big([\cos\frac{D\pi}{2} - i\sin\frac{D\pi}{2}]M_+ + [\cos\frac{D\pi}{2} + i\sin\frac{D\pi}{2}]M_-\Big)$$

$$\cdot \Big(M_+^*[\cos\frac{D^*\pi}{2} + i\sin\frac{D^*\pi}{2}] + M_-^*[\cos\frac{D^*\pi}{2} - i\sin\frac{D^*\pi}{2}]\Big)\Gamma(D+I)^*$$

$$= \frac{1}{2\pi}\Gamma(D+I)\Big(\sin\Big(\frac{D\pi}{2}\Big)(M_+ - M_-)(M_+^* - M_-^*)\sin\Big(\frac{D^*\pi}{2}\Big)$$

$$+ \cos\Big(\frac{D\pi}{2}\Big)(M_+ + M_-)(M_+^* + M_-^*)\cos\Big(\frac{D^*\pi}{2}\Big)$$

$$+ i\Big(\cos\Big(\frac{D\pi}{2}\Big)(M_+ + M_-)(M_+^* - M_-^*)\sin\Big(\frac{D^*\pi}{2}\Big)$$

$$- \sin\Big(\frac{D\pi}{2}\Big)(M_+ - M_-)(M_+^* + M_-^*)\cos\Big(\frac{D^*\pi}{2}\Big)\Big)\Big)\Gamma(D+I)^*. \quad (9.3.27)$$

On the other hand,

$$AA^* = (A_1 + iA_2)(A_1^* - iA_2^*) = (A_1A_1^* + A_2A_2^*) + i(A_2A_1^* - A_1A_2^*). \quad (9.3.28)$$

We now compare (9.3.28) and (9.3.27). A natural way to proceed is to view M_+ and M_- as solutions to the system

$$A_1 = \frac{1}{\sqrt{2\pi}}\Gamma(D+I)\sin\Big(\frac{D\pi}{2}\Big)(M_+ - M_-), \quad A_2 = \frac{1}{\sqrt{2\pi}}\Gamma(D+I)\cos\Big(\frac{D\pi}{2}\Big)(M_+ + M_-).$$
$$(9.3.29)$$

By assumption (9.3.21), $\sin(\frac{D\pi}{2})$, $\cos(\frac{D\pi}{2})$ and $\Gamma(D+I)$ are invertible (see the note following Definition 9.3.4). Then,

$$M_+ - M_- = \sqrt{2\pi}\Big(\sin\frac{D\pi}{2}\Big)^{-1}\Gamma(D+I)^{-1}A_1,$$

$$M_+ + M_- = \sqrt{2\pi}\Big(\cos\frac{D\pi}{2}\Big)^{-1}\Gamma(D+I)^{-1}A_2$$

and hence we obtain the solution given by (9.3.23).

(ii) In this case, $H = (1/2)I$ and one can readily compute the inverse Fourier transform of the integrand in (9.3.3), that is (up to $(2\pi)^{-1}$),

$$\int_{\mathbb{R}} e^{-iux}\frac{e^{itx}-1}{ix}\Big(1_{\{x>0\}}A + 1_{\{x<0\}}\overline{A}\Big)dx$$

$$= \int_{\mathbb{R}} \frac{\cos((t-u)x) - \cos(ux) + i(\sin((t-u)x) + \sin(ux))}{ix}\Big(1_{\{x>0\}}A + 1_{\{x<0\}}\overline{A}\Big)dx.$$
$$(9.3.30)$$

One can show (Exercise 9.6) that this becomes

$$-2\log\Big(\frac{|t-u|}{|u|}\Big)A_2 + (\text{sign}(t-u) - \text{sign}(-u))\pi A_1. \quad (9.3.31)$$

Then, by considering second moments and using Plancherel's identity, representation (9.3.24) holds with $M = (2\pi)^{-1/2}\pi A_1$ and $N = (2\pi)^{-1/2}(-2)A_2$. It is well-defined because the integrand comes from the inverse Fourier transform of a square-integrable function and hence is also square-integrable. □

Remark 9.3.9 When (9.3.21) does not hold and $H \neq (1/2)I$, the general form of time domain representations can be quite intricate. See Didier and Pipiras [307], Example 3.1.

Vector OFBMs provide a large class of processes indexed by matrices H and A. We examine next several interesting subclasses, the first subclass consisting of time reversible vector OFBMs, and the second class consisting of vector OFBMs with a diagonal self-similarity matrix H.

9.3.2 Time Reversible Vector OFBMs

Consider a vector OFBM B_H ($\mathbb{R}^p$-valued) whose exponent H has eigenvalues h_k with positive real parts. By using operator self-similarity and stationarity of increments, one can argue as in the one-dimensional case (see Proposition 2.5.6, (e)) that

$$\mathbb{E}B_H(t)B_H(s)^* + \mathbb{E}B_H(s)B_H(t)^*$$
$$= \mathbb{E}B_H(t)B_H(t)^* + \mathbb{E}B_H(s)B_H(s)^* - \mathbb{E}(B_H(t) - B_H(s))(B_H(t) - B_H(s))^*$$
$$= |t|^H \Gamma(1,1)|t|^{H^*} + |s|^H \Gamma(1,1)|s|^{H^*} - |t-s|^H \Gamma(1,1)|t-s|^{H^*}, \quad (9.3.32)$$
$$\Gamma(1,1) = \mathbb{E}B_H(1)B_H(1)^*$$

and we continue using $*$ instead of the prime for the transpose of real-valued matrices as in the previous section. In contrast to the one-dimensional case, it is *not* generally true that

$$\mathbb{E}B_H(t)B_H(s)^* = \mathbb{E}B_H(s)B_H(t)^*, \quad (9.3.33)$$

and hence the vector OFBM is not characterized by H and a matrix $\Gamma(1,1)$. As will be seen below, the condition (9.3.33) corresponds to time reversibility. We shall provide next conditions for vector OFBM to be time reversible.

Recall (cf. Section 9.1.2) that a process X is said to be *time reversible* if

$$\{X(t)\}_{t \in \mathbb{R}} \stackrel{d}{=} \{X(-t)\}_{t \in \mathbb{R}}. \quad (9.3.34)$$

When X is a zero-mean multivariate Gaussian stationary process, (9.3.34) is equivalent to

$$\mathbb{E}X(t)X(s)^* = \mathbb{E}X(-t)X(-s)^*, \quad s, t \in \mathbb{R},$$

which by adding $t + s$ to each argument on the right-hand side, is in turn equivalent to

$$\mathbb{E}X(t)X(s)^* = \mathbb{E}X(s)X(t)^*, \quad s, t \in \mathbb{R}.$$

The next proposition provides necessary and sufficient conditions for time reversibility in the case of Gaussian processes with stationary increments. It is stated without proof, since the latter is elementary.

Proposition 9.3.10 *Let X be an $\mathbb{R}^p$-valued Gaussian process with stationary increments and spectral representation*

$$\{X(t)\}_{t \in \mathbb{R}} \stackrel{d}{=} \left\{ \int_{\mathbb{R}} \frac{e^{itx} - 1}{ix} \widetilde{Y}(dx) \right\}_{t \in \mathbb{R}},$$

where $\widetilde{Y}(dx)$ is an orthogonal-increment random measure taking values in $\mathbb{C}^n$. The following statements are equivalent:

(i) X is time reversible;
(ii) $\mathbb{E}\widetilde{Y}(dx)\widetilde{Y}(dx)^* = \mathbb{E}\widetilde{Y}(-dx)\widetilde{Y}(-dx)^*$;
(iii) $\mathbb{E}X(t)X(s)^* = \mathbb{E}X(s)X(t)^*$, $s,t \in \mathbb{R}$.

The following result on time reversibility of vector OFBMs is a direct consequence of Proposition 9.3.10.

Theorem 9.3.11 *Let $\{B_H(t)\}_{t\in\mathbb{R}}$ be a vector OFBM with exponent H and the spectral representation (9.3.3). Let $A = A_1 + iA_2$, $A_1, A_2 \in \mathbb{R}^{p\times p}$. Then, B_H is time reversible if and only if*

$$AA^* = \overline{AA^*} \quad \text{or, equivalently,} \quad A_2 A_1^* = A_1 A_2^*. \tag{9.3.35}$$

Proof By (9.3.3) and Proposition 9.3.10, *(ii)*, time reversibility is equivalent to

$$\mathbb{E}\Big((x_+^{-D}A + x_-^{-D}\overline{A})\widetilde{B}(dx)\widetilde{B}(dx)^*(A^*x_+^{-D^*} + \overline{A^*}x_-^{-D^*})\Big)$$
$$= \mathbb{E}\Big((x_-^{-D}A + x_+^{-D}\overline{A})\widetilde{B}(-dx)\widetilde{B}(-dx)^*(A^*x_-^{-D^*} + \overline{A^*}x_+^{-D^*})\Big)$$

or

$$x_+^{-D}AA^*x_+^{-D^*} + x_-^{-D}\overline{AA^*}x_-^{-D^*} = x_-^{-D}AA^*x_-^{-D^*} + x_+^{-D}\overline{AA^*}x_+^{-D^*} \quad dx\text{-a.e.}$$

or, for $x > 0$,

$$x^{-D}AA^*x^{-D^*} = x^{-D}\overline{AA^*}x^{-D^*} \quad dx\text{-a.e.}$$

Since x^D is invertible for $x > 0$ (with the inverse x^{-D}), this is equivalent to the first relation in (9.3.35).

Let us now prove the statement in (9.3.35). $AA^* = \overline{AA^*}$ implies that AA^* is real-valued. But $AA^* = (A_1+iA_2)(A_1^*-iA_2^*) = A_1A_1^*+A_2A_2^*+i(A_2A_1^*-A_1A_2^*)$ so that $A_2A_1^* = A_1A_2^*$ as desired. □

Corollary 9.3.12 *The time reversible $\mathbb{R}^p$-valued $\{B_H(t)\}_{t\in\mathbb{R}}$ in Theorem 9.3.11 has covariance matrix-valued function*

$$\mathbb{E}B_H(t)B_H(s)^* = \int_\mathbb{R} \frac{(e^{itx}-1)(e^{-isx}-1)}{|x|^2}|x|^{-D}AA^*|x|^{-D^*}dx, \quad s,t \in \mathbb{R}. \tag{9.3.36}$$

Proof The result follows from (9.3.35) and (9.3.26). □

Corollary 9.3.13 *Let $\{B_H(t)\}_{t\in\mathbb{R}}$ be an $\mathbb{R}^p$-valued vector OFBM with the time domain representation given by (9.3.22), and exponent H satisfying (9.3.2) and (9.3.21). Then, B_H is time reversible if and only if*

$$\cos\left(\frac{D\pi}{2}\right)(M_+ + M_-)(M_+^* - M_-^*)\sin\left(\frac{D^*\pi}{2}\right)$$
$$= \sin\left(\frac{D\pi}{2}\right)(M_+ - M_-)(M_+^* + M_-^*)\cos\left(\frac{D^*\pi}{2}\right). \quad (9.3.37)$$

Proof As in the proof of Theorem 9.3.8, under (9.3.21) the matrices $\sin(D\pi/2)$, $\cos(D\pi/2)$ and $\Gamma(D+I)$ are invertible, and thus by using (9.3.29) to define $A = A_1 + iA_2$, one can equivalently reexpress the condition $A_2 A_1^* = A_1 A_2^*$ in (9.3.35) as (9.3.37). □

A consequence of Theorem 9.3.11 is that non-time reversible vector OFBMs can only emerge in the multivariate context, since in the univariate context condition (9.3.35) is always satisfied. Another elementary consequence of Proposition 9.3.10 is the following result, which partially justifies the interest in time reversibility in the case of vector OFBMs.

Proposition 9.3.14 *Let $\{B_H(t)\}_{t\in\mathbb{R}}$ be a vector OFBM with H satisfying (9.3.2). If $\{B_H(t)\}_{t\in\mathbb{R}}$ is time reversible, then its covariance structure is given by the function: for $s, t \in \mathbb{R}$,*

$$\mathbb{E} B_H(t) B_H(s)^* = \frac{1}{2}\Big(|t|^H \Gamma(1,1)|t|^{H^*} + |s|^H \Gamma(1,1)|s|^{H^*} - |t-s|^H \Gamma(1,1)|t-s|^{H^*}\Big), \quad (9.3.38)$$

where $\Gamma(1,1) = \mathbb{E} B_H(1) B_H(1)^*$. Conversely, a vector OFBM with covariance function (9.3.38) is time reversible.

Proof This follows from (9.3.32) and Proposition 9.3.10, *(iii)*. □

Remark 9.3.15 Relation (9.3.38) holds with $\Gamma(1,1)$ being equal to $\mathbb{E} B_H(1) B_H(1)^*$. There is a priori no reason to believe that, for a fixed exponent $H \in \mathbb{R}^{p\times p}$, every positive definite matrix $\Gamma(1,1) \in \mathbb{R}^{p\times p}$ leads to a valid covariance function (9.3.38) for time reversible vector OFBMs (see Exercise 9.7). This issue is a problem, for instance, in the context of simulation methods that require knowledge of the covariance function. For time reversible vector OFBMs with diagonalizable H, one natural way to parameterize $\Gamma(1,1)$ is through the formula (9.3.36). Moreover, as shown in Helgason, Pipiras, and Abry [465], Section 3.3.4, the expression (9.3.36) with $s = t = 1$ can be evaluated explicitly in terms of the parameters D and A.

9.3.3 Vector Fractional Brownian Motions

Another class of vector OFBMs of interest is when the operator self-similarity matrix H is diagonal.

Definition 9.3.16 An $\mathbb{R}^p$-valued vector OFBM with a diagonal operator self-similarity matrix $H = \mathrm{diag}(h_1, \ldots, h_p) \in \mathbb{R}^{p\times p}$ is called *vector fractional Brownian motion* (VFBM).

A VFBM $X = B_H$ with $H = \mathrm{diag}(h_1, \ldots, h_p)$ thus satisfies the relation: for $c > 0$,

$$\{X(ct)\}_{t\in\mathbb{R}} = \left\{ \begin{pmatrix} X_1(ct) \\ \vdots \\ X_p(ct) \end{pmatrix} \right\}_{t\in\mathbb{R}} \stackrel{d}{=} \left\{ \begin{pmatrix} c^{h_1} X_1(t) \\ \vdots \\ c^{h_p} X_p(t) \end{pmatrix} \right\}_{t\in\mathbb{R}} = \{\mathrm{diag}(c^{h_1}, \ldots, c^{h_p}) X(t)\}_{t\in\mathbb{R}}. \tag{9.3.39}$$

Processes X satisfying (9.3.39) will be called *vector self-similar* with vector self-similarity matrix $H = \mathrm{diag}(h_1, \ldots, h_p)$. Note that vector self-similar processes are naturally expected as limits of suitably normalized vectors. The following result is a simple extension of Lamperti's theorem (Theorem 2.8.5).

Theorem 9.3.17 *(i) If a vector-valued stochastic process Y satisfies*

$$f(\xi)^{-1}\bigl(Y(\xi t) + g(\xi)\bigr) \xrightarrow{fdd} X(t), \quad t \geq 0, \text{ as } \xi \to \infty, \tag{9.3.40}$$

for some proper vector-valued process $X(t)$, deterministic function $g(\xi) = (g_1(\xi), \ldots, g_p(\xi))' \in \mathbb{R}^p$ and $f(\xi) = \mathrm{diag}(f_1(\xi), \ldots, f_p(\xi))$ with some positive deterministic functions $f_k(\xi) \to \infty$, as $\xi \to \infty$, then there is $H = \mathrm{diag}(h_1, \ldots, h_p)$ with $h_k \geq 0$ such that

$$f(\xi) = \xi^H L(\xi), \quad g(\xi) = \xi^H L(\xi) w(\xi), \tag{9.3.41}$$

where $\xi^H = \mathrm{diag}(\xi^{h_1}, \ldots, \xi^{h_p})$, $L(\xi) = \mathrm{diag}(L_1(\xi), \ldots, L_p(\xi))$ with slowly varying functions $L_k(\xi)$ at infinity and $w(\xi) = (w_1(\xi), \ldots, w_p(\xi))'$ is a function which converges to a constant $w = (w_1, \ldots, w_p)'$ as $\xi \to \infty$. Moreover, the process $\{X(t) - w\}_{t\geq 0}$ is vector self-similar with the exponent H.

(ii) If $\{X(t) - w\}_{t\geq 0}$ is vector self-similar with exponent H, then there are functions $f(\xi)$, $g(\xi)$ and a process Y such that the above convergence (9.3.40) holds.

Proof (i) The relations in (9.3.41) follow from the one-dimensional Lamperti's theorem (Theorem 2.8.5) by considering the individual components of the relation (9.3.40). To see that $\{X(t) - w\}_{t\geq 0}$ is vector self-similar with the exponent H, suppose without loss of generality that $w = 0$. Observe that, for $c > 0$,

$$f(\xi)^{-1}\bigl(Y(\xi ct) + g(\xi)\bigr) \xrightarrow{fdd} X(ct).$$

On the other hand, by (9.3.41),

$$f(\xi)^{-1}\bigl(Y(\xi ct) + g(\xi)\bigr) = (f(\xi)^{-1} f(c\xi)) f(c\xi)^{-1}\bigl(Y(c\xi t) + g(c\xi)\bigr) + f(\xi)^{-1}\bigl(g(\xi) - g(c\xi)\bigr)$$

$$= c^H (L(\xi)^{-1} L(c\xi)) f(c\xi)^{-1}\bigl(Y(c\xi t) + g(c\xi)\bigr) + w(\xi) - c^H L(\xi)^{-1} L(c\xi) w(c\xi) \xrightarrow{fdd} c^H X(t),$$

as $\xi \to \infty$, since $L(\xi)^{-1} L(c\xi) \to 1$, $w(\xi) \to 0 (= w)$ and $f(c\xi)^{-1}(Y(c\xi t) + g(c\xi)) \xrightarrow{fdd} X(t)$. This implies that $X(ct)$ and $c^H X(t)$ have the same finite-dimensional distributions; that is, X is vector self-similar.

(ii) The proof is analogous to that of Theorem 2.8.5, (ii). $\square$

VFBMs underly many vector OFBMs through the change of coordinate systems as explained in (9.2.6)–(9.2.11). Being vector self-similar, VFBMs are also naturally expected as limits of component-wise normalized vector processes. Moreover, VFBMs can be readily used in applications as there is a way to estimate their parameters (Amblard and Coeurjolly [25]).

Note that each component $B_{H,k}$ of VFBM is a one-dimensional FBM, and its covariance function is

$$\mathbb{E} B_{H,k}(t) B_{H,k}(s) = \frac{\mathbb{E} B_{H,k}(1)^2}{2} \Big\{ |t|^{2h_k} + |s|^{2h_k} - |t-s|^{2h_k} \Big\}, \quad s, t \in \mathbb{R}, \qquad (9.3.42)$$

This is not the case for general vector OFBMs. Based on the following auxiliary result, we will provide an analogous relation for the cross-covariance function $\mathbb{E} B_{H,j}(t) B_{H,k}(s)$, $j \neq k$.

Lemma 9.3.18 *Let $H \in (0, 1)$ and $s, t \in \mathbb{R}$. Then,*

$$\int_{\mathbb{R}} \frac{(e^{itx} - 1)(e^{-isx} - 1)}{|x|^2} x_{\pm}^{-2H+1} dx = C_1(H) \Big(|t|^{2H} + |s|^{2H} - |t-s|^{2H} \Big)$$

$$\pm i \left\{ \begin{array}{ll} C_2(H)\Big(\text{sign}(t-s)|t-s|^{2H} - \text{sign}(t)|t|^{2H} + \text{sign}(s)|s|^{2H} \Big), & \text{if } H \neq \frac{1}{2} \\ -(t-s) \log|t-s| + t \log|t| - s \log|s|, & \text{if } H = \frac{1}{2} \end{array} \right\},$$
(9.3.43)

where

$$C_1(H) = \left\{ \begin{array}{ll} \frac{\Gamma(2-2H)\cos(H\pi)}{2H(1-2H)}, & \text{if } H \neq \frac{1}{2} \\ \frac{\pi}{2}, & \text{if } H = \frac{1}{2} \end{array} \right\}, \quad C_2(H) = \frac{\Gamma(2-2H)\sin(H\pi)}{2H(1-2H)}. \quad (9.3.44)$$

Proof Note that the relation (9.3.43) with x_- follows from (9.3.43) with x_+ after the change of variables $x \to (-x)$ and interchanging s and t. It is therefore enough to consider the integral in (9.3.43) with x_+; that is,

$$\int_0^\infty (e^{itx} - 1)(e^{-isx} - 1) x^{-2H-1} dx$$

$$= \int_0^\infty \Big(e^{i(t-s)x} - e^{itx} - e^{-isx} + 1 \Big) x^{-2H-1} dx = I_1 + i I_2,$$

where

$$I_1 = \int_0^\infty \Big(\cos((t-s)x) - \cos(tx) - \cos(sx) + 1 \Big) x^{-2H-1} dx,$$

$$I_2 = \int_0^\infty \Big(\sin((t-s)x) - \sin(tx) + \sin(sx) \Big) x^{-2H-1} dx.$$

To compute I_1, write

$$I_1 = \int_0^\infty \Big(\cos(|t-s|x) - \cos(|t|x) - \cos(|s|x) + 1 \Big) x^{-2H-1} dx$$

$$= \int_0^\infty (\cos(|t-s|x) - 1) x^{-2H-1} dx - \int_0^\infty (\cos(|t|x) - 1) x^{-2H-1} dx$$

9.3 Vector Operator Fractional Brownian Motions

$$-\int_0^\infty (\cos(|s|x) - 1)x^{-2H-1}dx$$
$$= \int_0^\infty (1 - \cos y)y^{-2H-1}dy\Big(|t|^{2H} + |s|^{2H} - |t-s|^{2H}\Big),$$

after the change of variables $|z|x = y$ in the integrals $\int_0^\infty(\cos(|z|x) - 1)x^{-2H-1}dx$, $z = t-s, t, s$. Observe now that, after integration by parts,

$$\int_0^\infty (1 - \cos y)y^{-2H-1}dy = \frac{1}{2H}\int_0^\infty \sin(y)\, y^{-2H}dy = C_1(H),$$

by using Formula 3.761.4 when $H \neq 1/2$, and Formula 3.782.2 when $H = 1/2$, in Gradshteyn and Ryzhik [421], pp. 436 and 447 (see also (2.6.10)).

To compute I_2, we consider the cases $H < 1/2$, $H > 1/2$ and $H = 1/2$ separately.

- When $H < 1/2$, I_2 can be written as

$$I_2 = \int_0^\infty \sin((t-s)x)x^{-2H-1}dx - \int_0^\infty \sin(tx)x^{-2H-1}dx + \int_0^\infty \sin(sx)x^{-2H-1}dx$$
$$= \text{sign}(t-s)\int_0^\infty \sin(|t-s|x)x^{-2H-1}dx - \text{sign}(t)\int_0^\infty \sin(|t|x)x^{-2H-1}dx$$
$$+ \text{sign}(s)\int_0^\infty \sin(|s|x)x^{-2H-1}dx$$
$$= \int_0^\infty \sin(y)\, y^{-2H-1}dy\Big(\text{sign}(t-s)|t-s|^{2H} - \text{sign}(t)|t|^{2H} + \text{sign}(s)|s|^{2H}\Big),$$

where $\int_0^\infty \sin(y)\, y^{-2H-1}dy = C_2(H)$ by Formula 3.761.4 in Gradshteyn and Ryzhik [421], p. 436.

- When $H > 1/2$, we have

$$I_2 = \int_0^\infty (\sin((t-s)x) - (t-s)x)x^{-2H-1}dx - \int_0^\infty (\sin(tx) - tx)x^{-2H-1}dx$$
$$+ \int_0^\infty (\sin(sx) - sx)x^{-2H-1}dx$$
$$= \int_0^\infty (\sin y - y)y^{-2H-1}dy\Big(\text{sign}(t-s)|t-s|^{2H} - \text{sign}(t)|t|^{2H} + \text{sign}(s)|s|^{2H}\Big)$$

and after integration by parts,

$$\int_0^\infty (\sin y - y)y^{-2H-1}dy = \frac{1}{2H}\int_0^\infty (\cos y - 1)y^{-2H}dy$$
$$= \frac{1}{2H(1-2H)}\int_0^\infty \sin(y)\, y^{1-2H}dy = C_2(H)$$

by Formula 3.761.4 in Gradshteyn and Ryzhik [421], p. 436.

- Finally, when $H = 1/2$, the above arguments do not apply since the integral $\int_0^\infty \sin(y)\, y^{-2}dy$ is not defined at $y = 0$, and the integral $\int_0^\infty (\sin y - y)y^{-2}dy$ is not defined at $y = +\infty$. In this case, we write I_2 as

$$I_2 = \int_0^\infty (\sin((t-s)x) - (t-s)x 1_{\{|t-s|x\le 1\}})x^{-2}dx - \int_0^\infty (\sin(tx) - tx 1_{\{|t|x\le 1\}})x^{-2}dx$$

$$+ \int_0^\infty (\sin(sx) - sx 1_{\{|s|x\le 1\}})x^{-2}dx + \int_0^\infty ((t-s)1_{\{|t-s|x\le 1\}} - t 1_{\{|t|x\le 1\}}$$

$$+ s 1_{\{|s|x\le 1\}})x^{-1}dx$$

$$= \int_0^\infty (\sin(y) - y 1_{\{y\le 1\}})y^{-2}dy \Big(\mathrm{sign}(t-s)|t-s| - \mathrm{sign}(t)|t| + \mathrm{sign}(s)|s|\Big)$$

$$+ \int_0^\infty ((t-s)1_{\{a\le x\le |t-s|^{-1}\}} - t 1_{\{a\le x\le |t|^{-1}\}}$$

$$+ s 1_{\{a\le x\le |s|^{-1}\}})x^{-1}dx + \int_0^a ((t-s) - t + s)x^{-1}dx =: I_{2,1} + I_{2,2} + I_{2,3},$$

where $a = \min\{|s|^{-1}, |t|^{-1}, |t-s|^{-1}\}$. Note that $I_{2,1} = 0$ since $\mathrm{sign}(t-s)|t-s| - \mathrm{sign}(t)|t| + \mathrm{sign}(s)|s| = (t-s) - t + s = 0$. Also $I_{2,3} = 0$. Hence,

$$I_2 = I_{2,2} = (t-s)\int_a^{|t-s|^{-1}} \frac{dx}{x} - t\int_a^{|t|^{-1}} \frac{dx}{x} + s\int_a^{|s|^{-1}} \frac{dx}{x}$$

$$= -(t-s)\log|t-s| + t\log|t| - s\log|s|.$$

This establishes (9.3.43) with x_+. □

The following proposition provides the cross-covariances $\mathbb{E}B_{H,j}(t)B_{H,k}(s)$ between the components of a VFBM.

Proposition 9.3.19 *Let* $\{B_H(t)\}_{t\in\mathbb{R}}$ *be a VFBM with exponent* $H = \mathrm{diag}(h_1,\ldots,h_p)$ *and the spectral representation (9.3.3) with matrix A. Then, for* $j\ne k$, $s,t\in\mathbb{R}$,

$$\mathbb{E}B_{H,j}(t)B_{H,k}(s) = \frac{\sigma_j\sigma_k}{2}(w_{jk}(t)|t|^{h_j+h_k} + w_{jk}(-s)|s|^{h_j+h_k} - w_{jk}(t-s)|t-s|^{h_j+h_k}), \tag{9.3.45}$$

where $\sigma_j^2 = \mathbb{E}B_{H,j}(1)^2$ *and*

$$w_{jk}(u) = \begin{cases} \rho_{jk} - \eta_{jk}\mathrm{sign}(u), & \text{if } h_j + h_k \ne 1, \\ \rho_{jk} - \eta_{jk}\mathrm{sign}(u)\log|u|, & \text{if } h_j + h_k = 1 \end{cases} \tag{9.3.46}$$

with $\rho_{jk} \in [-1,1]$ *and* $\eta_{jk} \in \mathbb{R}$. *Moreover, letting* $C_1(\cdot)$ *and* $C_2(\cdot)$ *be given in (9.3.44) and denoting* $AA^* = C = (c_{jk})_{j,k=1,\ldots,p}$, *we have*

$$\sigma_j\sigma_k\rho_{jk} = 4C_1\Big(\frac{h_j+h_k}{2}\Big)\Re(c_{jk}), \quad \sigma_j\sigma_k\eta_{jk} = 4C_2\Big(\frac{h_j+h_k}{2}\Big)\Im(c_{jk}) \tag{9.3.47}$$

if $h_j + h_k \ne 1$, *and*

$$\sigma_j\sigma_k\rho_{jk} = 4C_1\Big(\frac{1}{2}\Big)\Re(c_{jk}), \quad \sigma_j\sigma_k\eta_{jk} = 4\Im(c_{jk}) \tag{9.3.48}$$

if $h_j + h_k = 1$.

Proof As in (9.3.26),
$$\mathbb{E} B_H(t) B_H(s)^* = \int_{\mathbb{R}} \frac{(e^{itx}-1)(e^{-isx}-1)}{|x|^2}(x_+^{-D} A A^* x_+^{-D^*} + x_-^{-D} \overline{AA^*} x_-^{-D^*}) dx.$$

Since $d_j = h_j - 1/2$,
$$x_\pm^{-D} AA^* x_\pm^{-D^*} = (c_{jk} x_\pm^{-(h_j+h_k)+1})_{j,k=1,\ldots,p}. \quad (9.3.49)$$

It follows that
$$\mathbb{E} B_{H,j}(t) B_{H,k}(s) = \int_{\mathbb{R}} \frac{(e^{itx}-1)(e^{-isx}-1)}{|x|^2}(c_{jk} x_+^{-(h_j+h_k)+1} + \overline{c}_{jk} x_-^{-(h_j+h_k)+1}) dx.$$

If $h_j + h_k \neq 1$, Lemma 9.3.18 yields
$$\mathbb{E} B_{H,j}(t) B_{H,k}(s) = C_1\Big(\frac{h_j+h_k}{2}\Big)(c_{jk} + \overline{c}_{jk})\Big(|t|^{h_j+h_k} + |s|^{h_j+h_k} - |t-s|^{h_j+h_k}\Big)$$
$$+ i\, C_2\Big(\frac{h_j+h_k}{2}\Big)(c_{jk} - \overline{c}_{jk})\Big(\mathrm{sign}(t-s)|t-s|^{h_j+h_k}$$
$$- \mathrm{sign}(t)|t|^{h_j+h_k} + \mathrm{sign}(s)|s|^{h_j+h_k}\Big)$$
$$= (\alpha_{jk} - \beta_{jk}\mathrm{sign}(t))|t|^{h_j+h_k} + (\alpha_{jk} + \beta_{jk}\mathrm{sign}(s))|s|^{h_j+h_k}$$
$$- (\alpha_{jk} - \beta_{jk}\mathrm{sign}(t-s))|t-s|^{h_j+h_k},$$

where
$$\alpha_{jk} = 2C_1\Big(\frac{h_j+h_k}{2}\Big)\Re(c_{jk}), \quad \beta_{jk} = 2C_2\Big(\frac{h_j+h_k}{2}\Big)\Im(c_{jk}).$$

Note that $\mathbb{E} B_{H,j}(1) B_{H,k}(1) = 2\alpha_{jk}$. Hence, we can write
$$\alpha_{jk} = \frac{\sigma_j \sigma_k \rho_{jk}}{2}.$$

Since the covariance is $\mathbb{E} B_{H,j}(1) B_{H,k}(1) = 2\alpha_{jk}$, ρ_{jk} is the correlation between $B_{H,j}(1)$ and $B_{H,k}(1)$, and hence $\rho_{jk} \in [-1,1]$. Writing
$$\beta_{jk} = \frac{\sigma_j \sigma_k \eta_{jk}}{2}$$

with $\eta_{jk} \in \mathbb{R}$, we obtain (9.3.45) and (9.3.47) when $h_j + h_k \neq 1$. The proof in the case of $h_j + h_k = 1$ is similar using Lemma 9.3.18. □

Remark 9.3.20 Lavancier, Philippe, and Surgailis [597] used a different method and stronger assumptions on covariance structure to obtain the expression (9.3.45) directly from the definition of VFBMs.

The following result provides conditions for time reversibility of VFBMs in terms of the parameters in the cross-covariance function (9.3.45).

Proposition 9.3.21 *Let B_H be a VFBM with the cross-covariance structure (9.3.45). Then, B_H is time reversible if and only if $\eta_{jk} = 0$, $j,k = 1,\ldots,p$, $j \neq k$, in (9.3.46).*

Proof Proposition 9.3.14 implies that $w_{jk}(u)$ in (9.3.46) should not depend on u. □

The next result provides conditions for (9.3.42) and (9.3.45) to be a valid covariance structure: that is, the covariance structure of a VFBM.

Proposition 9.3.22 *The covariance structure given by (9.3.42) and (9.3.45) is that of a VFBM if and only if the matrix $Q = (q_{jk})_{j,k=1,\ldots,p}$ is positive semidefinite, where*

$$q_{jk} = \frac{\rho_{jk}}{C_1(\frac{h_j+h_k}{2})} + i\frac{\eta_{jk}}{C_2(\frac{h_j+h_k}{2})}, \quad \text{if } h_j + h_k \neq 1, \tag{9.3.50}$$

$$q_{jk} = \frac{\rho_{jk}}{C_1(\frac{1}{2})} + i\eta_{jk}, \quad \text{if } h_j + h_k = 1. \tag{9.3.51}$$

If $j = k$, it is assumed in (9.3.50) and (9.3.51) that $\rho_{jj} = 1$ and $\eta_{jj} = 0$.

Proof Suppose first that B_H is a VFBM and let $C = (c_{jk})_{j,k=1,\ldots,p} = AA^*$ as in (9.3.49). Hence the matrix C is positive semidefinite. It follows from (9.3.47) and (9.3.48) that

$$c_{jk} = \Re(c_{jk}) + i\Im(c_{jk}) = \frac{\sigma_j \sigma_k}{4} q_{jk}.$$

Since $C = (\frac{\sigma_j \sigma_k}{4} q_{jk})_{j,k=1,\ldots,p}$ is positive semidefinite, so is the matrix $Q = (q_{jk})_{j,k=1,\ldots,p}$.

Conversely, if Q is positive semidefinite, so is the matrix $C = (\frac{\sigma_j \sigma_k}{4} q_{jk})_{j,k=1,\ldots,p}$. Then, there is A such that $C = AA^*$ (see Section 9.1.1). The proof of Proposition 9.3.19 shows that the covariance structure (9.3.42) and (9.3.45) is that of a VFBM given by the spectral representation (9.3.3) with matrices H and A. □

When $p = 2$, the condition in Proposition 9.3.22 can be written in an alternative and more revealing form. See Exercise 9.8.

9.3.4 Identifiability Questions

One interesting fact about vector OFBMs and, more generally, vector operator self-similar processes is that an operator self-similarity matrix exponent H does not have to be unique. This fact has fundamental connections to algebra and matrix analysis. Our goal in this section is to give some idea of these connections and to present some fundamental results. We begin with a simple example illustrating the non-uniqueness of operator self-similarity exponents.

Example 9.3.23 (*Single-parameter vector OFBM*) Suppose $B_H = \{B_H(t)\}_{t\in\mathbb{R}}$ consists of p independent copies of a univariate FBM with parameter $h \in (0, 1)$. Note that B_H is a vector OFBM with an operator self-similarity exponent $H = hI$. It will be called a *single-parameter vector OFBM*. The matrix $H = hI$ is not the only operator self-similarity exponent of a single-parameter vector OFBM. Take $L \in so(p)$, where $so(p)$ stands for the space of skew-symmetric matrices (that is, matrices $M \in \mathbb{R}^{p\times p}$ such that $M^* = -M$). Then, for any $c > 0$, $c^L = e^{L \log(c)} \in O(p)$, where $O(p)$ is the space of orthogonal matrices (that is, matrices $M \in \mathbb{R}^{p\times p}$ such that $M^*M = I$). In particular, for any $c > 0$,

$$\{c^L B_H(t)\}_{t\in\mathbb{R}} \stackrel{d}{=} \{B_H(t)\}_{t\in\mathbb{R}},$$

since the processes on both sides are Gaussian and have the same covariance structure. To see that the covariance structure is the same, note that $\mathbb{E}(c^L B_H(t))(c^L B_H(s))^* = c^L \mathbb{E} B_H(t) B_H(s)^* (c^L)^* = \Gamma_h(t,s) c^L c^{L*} = \Gamma_h(t,s) c^{L-L} = \Gamma_h(t,s) c^0 = \Gamma_h(t,s) I$, $\Gamma_h(t,s) = \mathbb{E} B_h(t) B_h(s)$ is the covariance of a univariate FBM with parameter $h \in (0,1)$. Skew-symmetry was used in the step $c^L c^{L*} = c^{L-L}$. This shows that, for any $c > 0$,

$$\{B_H(ct)\}_{t \in \mathbb{R}} \stackrel{d}{=} \{c^H B_H(t)\}_{t \in \mathbb{R}} \stackrel{d}{=} \{c^H c^L B_H(t)\}_{t \in \mathbb{R}} = \{c^{H+L} B_H(t)\}_{t \in \mathbb{R}},$$

in the last step we used the fact that $H = hI$ and L commute. Thus, there is lack of uniqueness in the exponent since for any $L \in so(p)$, $H + L$ is another operator self-similarity exponent of a single-parameter vector OFBM.

It turns out (Hudson and Mason [499]) that, for an operator self-similar process, the class of its operator self-similarity exponents can be characterized in terms of the so-called symmetry group.

Definition 9.3.24 For a proper $\mathbb{R}^p$-valued process $\{X(t)\}_{t \in \mathbb{R}}$, its *symmetry group* is defined as

$$G_X = \left\{ C \in GL(p) : \{X(t)\}_{t \in \mathbb{R}} \stackrel{d}{=} \{CX(t)\}_{t \in \mathbb{R}} \right\}, \quad (9.3.52)$$

where $GL(p)$ is the multiplicative group of invertible matrices in $\mathbb{R}^{p \times p}$.

Note that, if X is a zero mean, Gaussian process, then

$$G_X = \left\{ C \in GL(p) : \mathbb{E} X(t) X(s)^* = C \mathbb{E} X(s) X(t)^* C^*, \text{ for all } s, t \in \mathbb{R} \right\}. \quad (9.3.53)$$

To state the main result, we need some notation. For an operator self-similar process, let $\mathcal{E}(X)$ denote the collection of all its operator self-similarity exponents. It will be characterized using

$$T(G_X) = \left\{ C : C = \lim_{n \to \infty} \frac{C_n - I}{d_n}, \text{ for some } \{C_n\} \subset G_X, 0 < d_n \to 0 \right\}, \quad (9.3.54)$$

the so-called *tangent space* of the symmetry group G_X.

Theorem 9.3.25 Let $X = \{X(t)\}_{t \in \mathbb{R}}$ be a proper operator self-similar process. Then, with the above notation,

$$\mathcal{E}(X) = H + T(G_X), \quad (9.3.55)$$

where H is any exponent of the process X.

For a proof of Theorem 9.3.25, see Hudson and Mason [499], Theorem 2.

Example 9.3.26 (*Single-parameter vector OFBM*) Let B_H be the single-parameter vector OFBM in Example 9.3.23. Its covariance function is $\mathbb{E} B_H(t) B_H(s) = \Gamma_h(t,s) I$, $\Gamma_h(t,s)$ is the covariance function of a univariate FBM with parameter h. In view of (9.3.53), the factors $\Gamma_h(t,s)$ cancel out, so that the symmetry group G_{B_H} turns out to be

$$G_{B_H} = \{C \in GL(p) : I = CC^*\}.$$

Consequently, $G_{B_H} = O(p)$, the space of orthogonal matrices. Moreover, it is known (e.g., Jurek and Mason [542], Example 1.5.7) that the tangent space of orthogonal matrices is the space of skew-symmetric matrices; that is, $T(O(p)) = so(p)$, and hence, by Theorem 9.3.25,

$$\mathcal{E}(B_H) = H + so(p),$$

which is in agreement with Example 9.3.23.

Symmetry groups, their structure and implications on the classes of exponents for vector OFBMs were studied in greater depth by Didier and Pipiras [308]. The results are formulated in terms of the matrix parameters H, $D = H - (1/2)I$ and A in the spectral domain representation (9.3.3) of vector OFBM. The following assumption was made: suppose $\Re(AA^*)$ has full rank, and hence can be factored as

$$\Re(AA^*) = S_R \Lambda_R^2 S_R^* =: W^2, \qquad (9.3.56)$$

with an orthogonal matrix $S_R \in \mathbb{R}^{p \times p}$, a diagonal $\Lambda_R \in \mathbb{R}^{p \times p}$ and a positive definite W (having inverse W^{-1}).

Denote the symmetry group of a vector OFBM B_H by G_H. It can be shown (Didier and Pipiras [308], Theorem 3.1) that

$$G_H = W \Big(\cap_{x>0} G(\Pi_x) \cap G(\Pi_I) \Big) W^{-1}, \qquad (9.3.57)$$

where W is defined in (9.3.56), $\Pi_I = W^{-1} \Im(AA^*) W^{-1}$,

$$\Pi_x = W^{-1} x^{-D} \Re(AA^*) x^{-D^*} W^{-1} = x^{-M} x^{-M^*}$$

with $M = W^{-1} D W$, and

$$G(\Pi) = \{O \in O(p) : O\Pi = \Pi O\} \qquad (9.3.58)$$

is the so-called *centralizer* of a matrix Π in the group $O(p)$ of orthogonal matrices.[6] A more refined analysis of the symmetry groups of vector OFBMs can thus be reduced to the study of centralizers and matrix commutativity. The latter is a well-known and well-studied problem in matrix analysis (e.g., Gantmacher [381]; Section 2.3 in Didier and Pipiras [308]).

Example 9.3.27 (*Single-parameter vector OFBM*) Let B_H be the single-parameter vector OFBM in Examples 9.3.23 and 9.3.26. It can be represented as (9.3.3) with $A = cI$ for a constant c. The relation (9.3.56) then holds with $W = cI$. The matrices Π_I and Π_x in (9.3.57) become $\Pi_I = 0$ (zero matrix) since $\Im(AA^*) = 0$ and $\Pi_x = x^{-2h+1} I$. Their centralizers are $G(\Pi_I) = G(\Pi_x) = O(p)$ (for the last equality since x^{-2h+1} is a scalar and I commutes with orthogonal matrices), leading to $G_H = O(p)$ as expected by using (9.3.57).

---

[6] Note that in (9.3.57), G_H, $G(\Pi_x)$ and $G(\Pi_I)$ are groups and thus collections of matrices, and that W is a matrix. For a collection G of matrices, the notation WGW^{-1} used on the right-hand side of (9.3.57) stands for the collections of matrices $\{WMW^{-1} : M \in G\}$.

By using the results on centralizers and matrix commutativity from matrix analysis, Didier and Pipiras [308] derived a number of results concerning the symmetry group G_H. Necessary and sufficient conditions were provided for G_H to be of maximal type: that is, to be conjugate to $O(p)$ in the sense that $G_H = WO(p)W^{-1}$ (Theorem 4.1 in Didier and Pipiras [308]). Sufficient conditions for G_H to be of minimal type (that is, $G_H = \{I, -I\}$) were given (Proposition 4.1 in Didier and Pipiras [308]). It was shown that the minimal type is the most "common" in a suitable sense.[7] A complete description of all possible symmetry groups in dimensions $p = 2$ and $p = 3$ was also provided, along with implications for the corresponding classes of exponents (Section 5 in Didier and Pipiras [308]).

Remark 9.3.28 Though operator self-similarity exponents are generally not unique as discussed above, several quantities related to them turn out to be unique. An important example is the real spectrum (that is, the real parts of the eigenvalues) of the exponents, whose uniqueness is established in e.g., Meerschaert and Scheffler [710], Corollary 11.1.7. The real spectrum determines the growth behavior of the operator self-similar process in any given radial direction; see Meerschaert and Scheffler [710], Section 11.2.

9.4 Vector Long-Range Dependence

We have focused so far on the vector operator FBM. We now turn to time series.

We are interested here in long-range dependence for vector-valued time series. The basics of stationary vector time series are recalled in Section 9.1.2. Definitions and basic properties of vector long-range dependence are given in Section 9.4.1. Several classes of models are considered in Sections 9.4.2 and 9.4.3. We consider vector FARIMA$(0, D, 0)$ in Section 9.4.2 and vector FGN in Section 9.4.3. Section 9.4.4 concerns the so-called *fractional cointegration*. As will be seen below, a number of new and interesting issues arise when one departs from the univariate case.

9.4.1 Definitions and Basic Properties

We begin with time and spectral domain conditions which extend conditions II and IV in Section 2.1 to the vector setting. In contrast to the conditions in Section 2.1, however, we do not consider general slowly varying functions, only those behaving as a constant asymptotically. One one hand, these slowly varying functions appear below in a (positive semidefinite) matrix form, and have, apparently, not been considered in the literature. On the other hand, the conditions we provide are most suitable for statistical inference. The suffix "-v" in the conditions refers to "vector", in order to distinguish them from the conditions in Section 2.1. There will also be a collection of parameters which we gather in a diagonal matrix

$$D = \text{diag}(d_1, \ldots, d_p)$$

and suppose that $d_j \in (0, 1/2)$, $j = 1, \ldots, p$.

---
[7] Minimal, or any finite symmetry groups, lead necessarily to unique exponents in view of the relation (9.3.55).

Condition II-v The autocovariance matrix function of the time series $X = \{X_n\}_{n\in\mathbb{Z}}$ satisfies: for $n \geq 1$,

$$\gamma_X(n) = n^{D-(1/2)I} L_2(n) n^{D-(1/2)I}, \qquad (9.4.1)$$

where L_2 is an $\mathbb{R}^{p\times p}$–valued function satisfying

$$L_2(u) \sim R, \quad \text{as } u \to +\infty, \qquad (9.4.2)$$

for some $R = (r_{jk})_{j,k=1,\ldots,p} \in \mathbb{R}^{p\times p}$. Equivalently, the relation (9.4.1) can be written component-wise as: for $j, k = 1, \ldots, p$,

$$\gamma_{X,jk}(n) = L_{2,jk}(n) n^{(d_j+d_k)-1} \sim r_{jk} n^{(d_j+d_k)-1}, \quad \text{as } n \to +\infty. \qquad (9.4.3)$$

Condition IV-v The time series $X = \{X_n\}_{n\in\mathbb{Z}}$ has a spectral density matrix satisfying: for $\lambda \in (0, \pi]$,

$$f_X(\lambda) = \lambda^{-D} L_4(\lambda) \lambda^{-D^*}, \qquad (9.4.4)$$

where $L_4(\lambda) \in \mathbb{C}^{p\times p}$ is Hermitian symmetric and positive semidefinite, satisfying

$$L_4(\lambda) \sim G, \quad \text{as } \lambda \to 0, \qquad (9.4.5)$$

where $G = (g_{jk})_{j,k=1,\ldots,p} \in \mathbb{C}^{p\times p}$ is also Hermitian symmetric and positive semidefinite. Equivalently, the relations (9.4.4) and (9.4.5) can be written component-wise as: for $j, k = 1, \ldots, p$,

$$f_{X,jk}(\lambda) = L_{4,jk}(\lambda) \lambda^{-(d_j+d_k)} \sim g_{jk} \lambda^{-(d_j+d_k)} =: \alpha_{jk} e^{i\phi_{jk}} \lambda^{-(d_j+d_k)}, \quad \text{as } \lambda \to 0, \qquad (9.4.6)$$

where $\alpha_{jk} \in \mathbb{R}$ and $\phi_{jk} \in (-\pi, \pi]$. See (9.4.7) and (9.4.8) below for two common specifications of α_{jk}, ϕ_{jk}.

Definition 9.4.1 A vector second-order stationary time series $X = \{X_n\}_{n\in\mathbb{Z}}$ is called *long-range dependent* (LRD, in short) if one of the non-equivalent conditions II-v and IV-v above holds with $d_j \in (0, 1/2)$, $j = 1, \ldots, p$.

The extension of condition I of Section 2.1 is discussed in Exercise 9.9. A number of comments regarding Definition 9.4.1 and conditions II-v and IV-v are in place.

First, note that the individual component series $\{X_{j,n}\}_{n\in\mathbb{Z}}$, $j = 1, \ldots, p$, of a vector LRD series are LRD with parameters d_j, $j = 1, \ldots, p$. (Note that $\phi_{jj} = 0$ in condition IV-v since the matrix G is Hermitian symmetric and hence has real-valued entries on the diagonal.) Another possibility would be to require that at least one of the individual series $\{X_{j,n}\}_{n\in\mathbb{Z}}$ be LRD. This could be achieved assuming in condition IV-v that $d_j \in [0, 1/2)$, $j = 1, \ldots, p$, and that at least one $d_j > 0$.

Second, note that the structure of (9.4.4) is such that $f_X(\lambda)$ is Hermitian symmetric and positive semidefinite. D^* appearing in (9.4.4) can be replaced by D since it is diagonal and consists of real-valued entries. We write D^* to make the semidefiniteness of $f_X(\lambda)$ more evident. The entries ϕ_{jk} in (9.4.6) are referred to as *phase parameters* of vector LRD series (around zero frequency) and α_{jk} as *amplitude parameters*. The role of phase parameters will become more apparent in Proposition 9.4.2 below where conditions II-v and IV-v are compared. Recall from (9.1.13)–(9.1.14) that there are two common but slightly different ways to represent the amplitude α_{jk} and the phase ϕ_{jk}, namely,

$$\alpha_{jk} = |g_{jk}|, \quad \phi_{jk} = \mathrm{Arg}(g_{jk}) \tag{9.4.7}$$

and

$$\alpha_{jk} = \frac{\Re g_{jk}}{\cos \phi_{jk}}, \quad \phi_{jk} = \arctan \frac{\Im g_{jk}}{\Re g_{jk}}. \tag{9.4.8}$$

In the case (9.4.8), $\phi_{jk} \in (-\pi/2, \pi/2)$.

It should also be noted that phase parameters are unique to the vector LRD case. Indeed, since they appear in the cross spectral density (spectrum) $f_{X,jk}(\lambda)$, $j \neq k$, the (nonzero) phases do not arise in the univariate case. Likewise, taking

$$\sum_{n=-\infty}^{\infty} \|\gamma_X(n)\|_2 < \infty \tag{9.4.9}$$

for the definition of vector *short-range dependent* (SRD) series, where $\|A\|_2 = (\sum_{j,k} |a_{jk}|^2)^{1/2}$ is the Frobenius norm of $A = (a_{jk})$, we have

$$f_X(\lambda) = \frac{1}{2\pi} \sum_{n=-\infty}^{\infty} e^{-in\lambda} \gamma_X(n) \tag{9.4.10}$$

(cf. Lemma 2.2.22). In particular, since $\gamma_X(n) \in \mathbb{R}^{p \times p}$, $f_X(0) = (2\pi)^{-1} \sum_{n=-\infty}^{\infty} \gamma_X(n)$ is in $\mathbb{R}^{p \times p}$ as well, and

$$f_X(\lambda) \sim G \; (= f_X(0)), \quad \text{as } \lambda \to 0, \tag{9.4.11}$$

where $G \in \mathbb{R}^{p \times p}$. The relation (9.4.11) corresponds to (9.4.6) with $d_j = 0$, $j = 1, \ldots, p$, and all phase parameters $\phi_{jk} = 0$.

Third, the squared coherence function $\mathcal{H}_{jk}^2(\lambda) = |f_{X,jk}(\lambda)|^2 / (f_{X,jj}(\lambda) f_{X,kk}(\lambda))$ satisfies $0 \leq \mathcal{H}_{jk}^2(\lambda) \leq 1$ by Cauchy–Schwarz (e.g., Brockwell and Davis [186], p. 436). As $\lambda \to 0$, this translates into

$$0 \leq \lim_{\lambda \to 0} \frac{|g_{jk}|^2 \lambda^{-2(d_j + d_k)}}{g_{jj} \lambda^{-2d_j} g_{kk} \lambda^{-2d_k}} = \frac{|g_{jk}|^2}{g_{jj} g_{kk}} \leq 1$$

and also explains why the choice of $\lambda^{-(d_j + d_k)}$ is natural for the cross-spectral density $f_{X,jk}(\lambda)$.

Fourth, note also that (9.4.6) is considered for $\lambda > 0$, $\lambda \to 0$. For real-valued λ, since $f_X(-\lambda) = \overline{f_X(\lambda)}$, (9.4.6) can be replaced by

$$f_{X,jk}(\lambda) = L_{4,jk}(\lambda) |\lambda|^{-(d_j + d_k)} \sim \alpha_{jk} e^{i \, \mathrm{sign}(\lambda) \phi_{jk}} |\lambda|^{-(d_j + d_k)}, \quad \text{as } \lambda \to 0. \tag{9.4.12}$$

It follows that $L_{4,jk}(-\lambda) = \overline{L_{4,jk}(\lambda)}$; that is,

$$\Re L_{4,jk}(-\lambda) = \Re L_{4,jk}(\lambda) \quad \text{and} \quad \Im L_{4,jk}(-\lambda) = -\Im L_{4,jk}(\lambda) \tag{9.4.13}$$

if both positive and negative λs are considered. (Note that since G is Hermitian symmetric, we have $\alpha_{jk} = \alpha_{kj} \in \mathbb{R}$.)

The following proposition relates conditions II-v and IV-v. Quasi-monotonic functions are defined in Definition 2.2.11.

Proposition 9.4.2 *(i) Suppose that the component functions $L_{2,jk}$ are quasi-monotone slowly varying. Then, condition II-v implies condition IV-v with*

$$g_{jk} = \frac{\Gamma(d_j + d_k)}{2\pi}\left\{(r_{jk}+r_{kj})\cos\left((d_j+d_k)\frac{\pi}{2}\right) - i(r_{jk}-r_{kj})\sin\left((d_j+d_k)\frac{\pi}{2}\right)\right\} \quad (9.4.14)$$

in the relation (9.4.6). In the specification (9.4.8) of $g_{jk} = \alpha_{jk}e^{i\phi_{jk}}$, one has

$$\phi_{jk} = -\arctan\left\{\frac{r_{jk} - r_{kj}}{r_{jk} + r_{kj}}\tan\left((d_j + d_k)\frac{\pi}{2}\right)\right\}, \quad (9.4.15)$$

$$\alpha_{jk} = \frac{\Gamma(d_j + d_k)(r_{jk} + r_{kj})\cos((d_j + d_k))\frac{\pi}{2}}{2\pi\cos(\phi_{jk})}. \quad (9.4.16)$$

(ii) Suppose that the component functions $\Re L_{4,jk}$, $\Im L_{4,jk}$ are quasi-monotone slowly varying. Then, condition IV-v implies condition II-v with

$$r_{jk} = 2\Gamma(1 - (d_j + d_k))\left\{\Re g_{jk}\sin\left((d_j + d_k)\frac{\pi}{2}\right) - \Im g_{jk}\cos\left((d_j + d_k)\frac{\pi}{2}\right)\right\} \quad (9.4.17)$$

in the relation (9.4.3).

Proof (i) As in Proposition 2.2.14, we have

$$f_{X,jk}(\lambda) = \frac{1}{2\pi}\sum_{n=-\infty}^{\infty} e^{-in\lambda}\gamma_{X,jk}(n).$$

(This is immediate by the proposition for the real part of $f_{X,jk}$, and the relation for the imaginary part can be deduced similarly.) Then, by using $\gamma_{X,jk}(-n) = \gamma_{X,kj}(n)$ (see (9.1.5)),

$$f_{X,jk}(\lambda) = \frac{1}{2\pi}\left\{\sum_{n=-\infty}^{\infty}\cos(n\lambda)\gamma_{X,jk}(n) - i\sum_{n=-\infty}^{\infty}\sin(n\lambda)\gamma_{X,jk}(n)\right\}$$

$$= \frac{1}{2\pi}\left\{\gamma_{X,jk}(0) + \sum_{n=1}^{\infty}\cos(n\lambda)(\gamma_{X,jk}(n) + \gamma_{X,kj}(n))\right\}$$

$$- \frac{i}{2\pi}\left\{\sum_{n=1}^{\infty}\sin(n\lambda)(\gamma_{X,jk}(n) - \gamma_{X,kj}(n))\right\}$$

$$= \frac{1}{2\pi}\left\{\gamma_{X,jk}(0) + \sum_{n=1}^{\infty}\cos(n\lambda)\frac{L_{2,jk}(n) + L_{2,kj}(n)}{n^{1-(d_j+d_k)}}\right\}$$

$$- \frac{i}{2\pi}\left\{\sum_{n=1}^{\infty}\sin(n\lambda)\frac{L_{2,jk}(n) - L_{2,kj}(n)}{n^{1-(d_j+d_k)}}\right\}.$$

It follows from Proposition A.2.1 in Appendix A.2 that

$$f_{X,jk}(\lambda) \sim \frac{\Gamma(d_j + d_k)}{2\pi}\lambda^{-(d_j+d_k)}\left\{(r_{jk} + r_{kj})\cos\left((d_j + d_k)\frac{\pi}{2}\right)\right.$$
$$\left. - i(r_{jk} - r_{kj})\sin\left((d_j + d_k)\frac{\pi}{2}\right)\right\},$$

showing (9.4.14). This also immediately yields (9.4.15)–(9.4.16) by using (9.4.8).

(ii) Note that

$$\gamma_{X,jk}(n) = \int_{-\pi}^{\pi} e^{in\lambda} f_{X,jk}(\lambda) d\lambda = \int_{-\pi}^{\pi} e^{in\lambda} L_{4,jk}(\lambda) |\lambda|^{-(d_j+d_k)} d\lambda$$

$$= \int_0^{\pi} e^{in\lambda} (\Re L_{4,jk}(\lambda) + i\Im L_{4,jk}(\lambda)) \lambda^{-(d_j+d_k)} d\lambda$$

$$+ \int_0^{\pi} e^{-in\lambda} (\Re L_{4,jk}(-\lambda) + i\Im L_{4,jk}(-\lambda)) \lambda^{-(d_j+d_k)} d\lambda$$

$$= 2 \int_0^{\pi} \cos(n\lambda) \Re L_{4,jk}(\lambda) \lambda^{-(d_j+d_k)} d\lambda - 2 \int_0^{\pi} \sin(n\lambda) \Im L_{4,jk}(\lambda) \lambda^{-(d_j+d_k)} d\lambda,$$

where we used (9.4.13). By Proposition A.2.2 in Appendix A.2 and an analogous result where cosine is replaced by sine, we get that

$$\gamma_{X,jk}(n) \sim 2\Gamma(1-(d_j+d_k))n^{(d_j+d_k)-1} \Big\{ \Re g_{jk} \cos\Big((1-(d_j+d_k))\frac{\pi}{2}\Big) - \Im g_{jk} \sin\Big((1-(d_j+d_k))\frac{\pi}{2}\Big) \Big\},$$

with g_{jk} defined in (9.4.6). This yields (9.4.3) with r_{jk} defined as in (9.4.17). □

The relation (9.4.15) sheds light on the phase parameters ϕ_{jk}. Note that $\phi_{jk} = 0$ if and only if $r_{jk} = r_{kj}$. In view of (9.1.5), the last property corresponds to $\gamma_{X,jk}(n)$ being symmetric at large positive and large negative lags, that is, as $n \to \infty$ and $n \to -\infty$.

We provide next examples of two vector LRD series: the vector FARIMA and the vector FGN series.

9.4.2 Vector FARIMA(0, D, 0) Series

The univariate FARIMA was defined in Section 2.4.1 and included the backshift operator B, defined as $Ba_n = a_{n-1}$. Here, we will also use the forward-shift operator B^{-1} as well, defined as $B^{-1}a_n = a_{n+1}$.

Let $D = \text{diag}(d_1, \ldots, d_p)$ with $d_j < 1/2$, $j = 1, \ldots, p$. Then,

$$(I-B)^{-D} = \begin{pmatrix} (I-B)^{-d_1} & 0 & \cdots & 0 \\ 0 & (I-B)^{-d_2} & \cdots & 0 \\ \vdots & \vdots & \ddots & \vdots \\ 0 & 0 & \cdots & (I-B)^{-d_p} \end{pmatrix}$$

and $(I-B^{-1})^{-D}$ is defined similarly, with B^{-1} replacing B. Let $Q_+ = (q_{jk}^+)$, $Q_- = (q_{jk}^-) \in \mathbb{R}^{p \times p}$, and $\{Z_n\}_{n \in \mathbb{Z}}$ be an $\mathbb{R}^p$-valued white noise series, satisfying $\mathbb{E} Z_n = 0$ and $\mathbb{E} Z_n Z_n' = I$.[8] Define a *vector FARIMA(0, D, 0) series* as

$$X_n = (I-B)^{-D} Q_+ Z_n + (I-B^{-1})^{-D} Q_- Z_n. \tag{9.4.18}$$

---

[8] Do not confuse the I in the operator $(I-B)^{-D}$ with the identity matrix I, appearing for example in $\mathbb{E} Z_n Z_n' = I$.

Note that Q_+ and Q_- introduce a dependence structure among the components of the noise vector Z. Since the second term in (9.4.18) includes B^{-1}, the series X_n is given by a two-sided linear representation (when Q_- is not identically zero). See also the discussion at the end of this section, and Exercise 9.9.

In the next result, we give the exact form of the autocovariance matrix function of the vector FARIMA$(0, D, 0)$ series in (9.4.18).

Proposition 9.4.3 *The (j, k) component $\gamma_{X,jk}(n)$ of the autocovariance matrix function $\gamma_X(n)$ of the vector FARIMA$(0, D, 0)$ series in (9.4.18) is given by*

$$\gamma_{X,jk}(n) = \frac{1}{2\pi}\Big(b^1_{jk}\gamma^1_{jk}(n) + b^2_{jk}\gamma^2_{jk}(n) + b^3_{jk}\gamma^3_{jk}(n) + b^4_{jk}\gamma^4_{jk}(n)\Big), \qquad (9.4.19)$$

where

$$b^1_{jk} = \sum_{t=1}^{p} q^-_{jt}q^-_{jt} = (Q^-(Q^-)^*)_{jk}, \quad b^3_{jk} = \sum_{t=1}^{p} q^+_{jt}q^+_{jt} = (Q^+(Q^+)^*)_{jk},$$
$$b^2_{jk} = \sum_{t=1}^{p} q^-_{jt}q^+_{jt} = (Q^-(Q^+)^*)_{jk}, \quad b^4_{jk} = \sum_{t=1}^{p} q^+_{jt}q^-_{jt} = (Q^+(Q^-)^*)_{jk}, \qquad (9.4.20)$$

and

$$\gamma^1_{jk}(n) = \gamma^3_{kj}(n) = 2\Gamma(1 - d_j - d_k)\sin(d_k\pi)\frac{\Gamma(n+d_k)}{\Gamma(n+1-d_j)},$$
$$\gamma^4_{jk}(n) = \gamma^2_{jk}(-n) = \begin{cases} 2\pi\frac{1}{\Gamma(d_j+d_k)}\frac{\Gamma(n+d_j+d_k)}{\Gamma(n+1)}, & n = 0, 1, 2, \ldots, \\ 0, & n = -1, -2, \ldots \end{cases} \qquad (9.4.21)$$

Proof By using Theorem 11.8.3 in Brockwell and Davis [186], the vector FARIMA$(0, D, 0)$ series in (9.4.18) has the spectral density matrix

$$f_X(\lambda) = \frac{1}{2\pi}G(\lambda)G(\lambda)^*, \qquad (9.4.22)$$

where $G(\lambda) = (1 - e^{-i\lambda})^{-D}Q_+ + (1 - e^{i\lambda})^{-D}Q_-$. This can be expressed componentwise as

$$f_{X,jk}(\lambda) = \frac{1}{2\pi}g_j(\lambda)g_k(\lambda)^*, \qquad (9.4.23)$$

where g_j is the jth row of G. Then, the (j, k) component of the autocovariance matrix is

$$\gamma_{X,jk}(n) = \int_0^{2\pi} e^{in\lambda}f_{X,jk}(\lambda)d\lambda = \frac{1}{2\pi}\int_0^{2\pi} e^{in\lambda}g_j(\lambda)g_k(\lambda)^*d\lambda$$
$$= \frac{1}{2\pi}\Big(b^1_{jk}\gamma^1_{jk}(n) + b^2_{jk}\gamma^2_{jk}(n) + b^3_{jk}\gamma^3_{jk}(n) + b^4_{jk}\gamma^4_{jk}(n)\Big), \qquad (9.4.24)$$

where $b^1_{jk}, b^2_{jk}, b^3_{jk}, b^4_{jk}$ are given in (9.4.20), and

$$\gamma^1_{jk}(n) = \gamma^3_{kj}(n) = \int_0^{2\pi} e^{in\lambda}(1 - e^{i\lambda})^{-d_j}(1 - e^{-i\lambda})^{-d_k}d\lambda,$$
$$\gamma^2_{jk}(n) = \int_0^{2\pi} e^{in\lambda}(1 - e^{i\lambda})^{-(d_j+d_k)}d\lambda, \quad \gamma^4_{jk}(n) = \int_0^{2\pi} e^{in\lambda}(1 - e^{-i\lambda})^{-(d_j+d_k)}d\lambda$$

9.4 Vector Long-Range Dependence

(note that γ^1 is expressed with the indices j, k but γ^3 is expressed with the reverse indices k, j). Note that $\gamma^4_{jk}(n) = \gamma^2_{jk}(-n)$ as stated in (9.4.21), so that it is enough to consider γ^2_{jk}, in addition to, for example, γ^1_{jk}.

We shall now use the relation

$$2\sin\left(\frac{\lambda}{2}\right)e^{\pm i(\lambda-\pi)/2} = 2\sin\left(\frac{\lambda}{2}\right)\left(\cos\left(\frac{\lambda}{2} - \frac{\pi}{2}\right) \pm i\sin\left(\frac{\lambda}{2} - \frac{\pi}{2}\right)\right)$$

$$= 2\sin\left(\frac{\lambda}{2}\right)\left(\sin\left(\frac{\lambda}{2}\right) \mp i\cos\left(\frac{\lambda}{2}\right)\right)$$

$$= 2\sin^2\left(\frac{\lambda}{2}\right) \mp i2\sin\left(\frac{\lambda}{2}\right)\cos\left(\frac{\lambda}{2}\right)$$

$$= (1 - \cos\lambda) \mp i\sin\lambda = 1 - e^{\pm i\lambda}. \tag{9.4.25}$$

We have

$$\gamma^1_{jk}(n) = \frac{e^{i(d_j - d_k)\pi/2}}{2^{d_j + d_k}} \int_0^{2\pi} e^{in\lambda} \sin^{-d_j - d_k}\left(\frac{\lambda}{2}\right) e^{i\lambda(d_k - d_j)/2} d\lambda$$

$$= \frac{2e^{i(d_j - d_k)\pi/2}}{2^{d_j + d_k}} \int_0^{\pi} e^{i\omega(2n + d_k - d_j)} \sin^{-d_j - d_k}(\omega) d\omega.$$

By using Formula 3.892.1 in Gradshteyn and Ryzhik [421], p. 485, we deduce that

$$\gamma^1_{jk}(n) = \frac{2e^{i(d_j - d_k)\pi/2}}{2^{d_j + d_k}} \frac{\pi e^{i\beta\pi/2}}{2^{\nu-1}\nu B(\frac{\nu+\beta+1}{2}, \frac{\nu-\beta+1}{2})},$$

where $\beta = 2n + d_k - d_j$ and $\nu = 1 - d_k - d_j$. Then,

$$\gamma^1_{jk}(n) = 2\pi(-1)^n \frac{\Gamma(1 - d_j - d_k)}{\Gamma(1 - d_j + n)\Gamma(1 - d_k - n)}. \tag{9.4.26}$$

Similar calculations yield

$$\gamma^2_{jk}(n) = 2\pi(-1)^n \frac{\Gamma(1 - d_j - d_k)}{\Gamma(1 - n)\Gamma(1 + n - d_j - d_k)}. \tag{9.4.27}$$

The relations (9.4.21) can now be deduced from (9.4.26) and (9.4.27) by using the identities $\Gamma(z)\Gamma(1-z) = \pi/\sin(\pi z)$ and $\Gamma(z)\Gamma(1-z) = (-1)^n\Gamma(n+z)\Gamma(1-n-z), 0 < z < 1$. □

By using the relation (2.4.3), observe that, as $n \to \infty$,

$$\gamma^1_{jk}(n) \sim 2\Gamma(1 - d_j - d_k)\sin(d_j\pi)\frac{\Gamma(n + d_j)}{\Gamma(n + 1 - d_k)} = \frac{2\pi \sin(d_j\pi)}{\Gamma(d_j + d_k)\sin((d_j + d_k)\pi)} n^{d_j + d_k - 1}, \tag{9.4.28}$$

$$\gamma^4_{jk}(n) \sim \frac{2\pi}{\Gamma(d_j + d_k)} n^{d_j + d_k - 1}. \tag{9.4.29}$$

These relations show that FARIMA$(0, D, 0)$ series satisfies condition II-v in Section 9.4.1. The relation (9.4.22) shows that the series also satisfies condition IV-v.

We now comment on the presence of the second term in (9.4.18). Note that $Q_- \equiv 0$ (that is, no second term in (9.4.18)) implies by (9.4.23) and (9.4.25) that

$$f_{X, jk}(\lambda) = \frac{(Q + Q^*_+)_{jk}}{2\pi}(1 - e^{-i\lambda})^{-d_j}(1 - e^{i\lambda})^{-d_k} \sim \frac{(Q + Q^*_+)_{jk}}{2\pi} e^{-i(d_j - d_k)\frac{\pi}{2}} \lambda^{-(d_j + d_k)},$$

as $\lambda \to 0$. Thus, such FARIMA$(0, D, 0)$ series necessarily have the *phase parameters*

$$\phi_{jk} = -(d_j - d_k)\frac{\pi}{2}. \tag{9.4.30}$$

The presence of Q_- in (9.4.18) allows one to obtain arbitrary phase parameters. To see this, note from (9.4.22) that

$$f_X(\lambda) \sim \lambda^{-D} W W^* \lambda^{-D^*}, \tag{9.4.31}$$

where

$$W = \frac{1}{2\pi}(e^{-iD\frac{\pi}{2}} Q_+ + e^{iD\frac{\pi}{2}} Q_-)$$

as $\lambda \to 0$, since $i = e^{i\pi/2}$. The right-hand side of (9.4.31) then captures the general behavior (9.4.4) in condition IV-v with a matrix $G = WW^*$ for some matrix W as above. But since G is positive semidefinite Hermitian symmetric, we have $G = ZZ^*$ for some $Z \in \mathbb{C}^{p \times p}$ (see Section 9.1.1). It remains to note that there are real-valued Q_+, Q_- such that Z equals W. (The latter observation follows from the fact that any $z \in \mathbb{C}$ can be expressed as $z = e^{-id\frac{\pi}{2}} q_+ + e^{id\frac{\pi}{2}} q_-$ for some $q_+, q_- \in \mathbb{R}$, since z can be viewed as a vector in the complex plane, and $e^{-id\frac{\pi}{2}}$ and $e^{id\frac{\pi}{2}}$ are linearly independent vectors.)

Remark 9.4.4 Though the presence of Q_- in (9.4.18) allows for general phase as indicated above, this can be achieved by more than one choice of the parameters Q_+, Q_- of vector FARIMA$(0, D, 0)$ series. For example, in the argument given above, Q_+, Q_- can be replaced by $Q_+ O$ and $Q_- O$ for an orthogonal matrix $O \in \mathbb{R}^{p \times p}$ satisfying $OO^* = I$. This identifiability issue is also apparent in the simplest bivariate case $p = 2$ where the specification (9.4.6) of bivariate LRD has six parameters (namely, $d_1, d_2, \alpha_{11}, \alpha_{12}, \alpha_{22}, \phi_{12}$), whereas the bivariate FARIMA model has eight parameters even after identification up to the products $Q_+ Q_+^*$ and $Q_- Q_-^*$ (namely, d_1, d_2, the three parameters of $Q_+ Q_+^*$ and the three parameters of $Q_- Q_-^*$). An identifiable bivariate FARIMA$(0, D, 0)$ model with six parameters and general phase is introduced in Kechagias and Pipiras [550], along with extensions incorporating AR and MA polynomials, estimation and application to a real data set.

Remark 9.4.5 When $Q_- \equiv 0$, FARIMA$(0, D, 0)$ series have a one-sided linear representation

$$X_n = \sum_{m=0}^{\infty} \Psi_m \epsilon_{n-m}, \tag{9.4.32}$$

where Ψ_m are coefficients in $\mathbb{R}^{q \times q}$. The entries $\psi_{jk,m}$ of these coefficients follow a power-law behavior in the sense that $\psi_{jk,m} \sim Cm^{d-1}$, as $m \to \infty$. In fact, as shown in Kechagias and Pipiras [551], Section 4.1, a one-sided time series with power-law coefficients necessarily has the phase parameters (9.4.30). One-sided linear representations can lead to general phase parameters when using a new class of coefficients, called *trigonometric power-law coefficients* (see Kechagias and Pipiras [551] for more information).

9.4.3 Vector FGN Series

The univariate FGN was defined in Section 2.8. We will focus here on the vector FGN. Let B_H be a vector FBM with a diagonal self-similarity matrix $H = \mathrm{diag}(h_1, \ldots, h_p) \in \mathbb{R}^{p \times p}$ (see Definition 9.3.16). In particular, we assume that B_H admits a spectral representation (9.3.3) with matrix A. Hence, its covariance is given in Proposition 9.3.19. In analogy to (2.8.3), we define a *vector fractional Gaussian noise* (FGN) series as

$$X_n = B_H(n) - B_H(n-1), \quad n \in \mathbb{Z}. \tag{9.4.33}$$

The series $\{X_n\}_{n \in \mathbb{Z}}$ is Gaussian, zero mean and stationary. For the autocovariance matrix function of $\{X_n\}_{n \in \mathbb{Z}}$, one can show by using the covariance of B_H in (9.3.45), that

$$\gamma_{X,jk}(n) = \frac{\sigma_j \sigma_k}{2} \left\{ w_{jk}(n+1)|n+1|^{h_j+h_k} + w_{jk}(n-1)|n-1|^{h_j+h_k} - 2w_{jk}(n)|n|^{h_j+h_k} \right\}, \tag{9.4.34}$$

where w_{jk} are defined in Proposition 9.3.19. By using (9.3.46), if $h_j + h_k \neq 1$ (and $n \neq 0$),

$$\gamma_{X,jk}(n) = \frac{\sigma_j \sigma_k}{2}(\rho_{jk} - \mathrm{sign}(n)\eta_{jk})\left\{|n+1|^{h_j+h_k} + |n-1|^{h_j+h_k} - 2|n|^{h_j+h_k}\right\} \tag{9.4.35}$$

and arguing as in the proof of Proposition 2.8.1,

$$\gamma_{X,jk}(n) \sim \frac{\sigma_j \sigma_k}{2}(\rho_{jk} - \mathrm{sign}(n)\eta_{jk})(h_j + h_k)(h_j + h_k - 1)|n|^{h_j+h_k-2}, \tag{9.4.36}$$

as $n \to \infty$. If $h_j + h_k = 1$ (and $|n| \geq 2$),

$$\gamma_{X,jk}(n) = -\frac{\sigma_j \sigma_k \eta_{jk}}{2}\mathrm{sign}(n)\left\{|n+1|\log|n+1| + |n-1|\log|n-1| - 2|n|\log|n|\right\}$$
$$\sim \frac{\sigma_j \sigma_k \eta_{jk}}{2}\mathrm{sign}(n)|n|^{-1}, \tag{9.4.37}$$

as $|n| \to \infty$. Deriving (9.4.34), (9.4.35)–(9.4.36) and (9.4.37) is left as an exercise (Exercise 9.4).

By using the spectral representation (9.3.3), one can show as in the proof of Proposition 2.8.1, that the spectral density matrix of $\{X_n\}_{n \in \mathbb{Z}}$ is given by

$$f_X(\lambda) = \frac{|1 - e^{-i\lambda}|^2}{|\lambda|^2} \sum_{n=-\infty}^{\infty} \left\{ (\lambda + 2\pi n)_+^{-D} C (\lambda + 2\pi n)_+^{-D} + (\lambda + 2\pi n)_-^{-D} \overline{C} (\lambda + 2\pi n)_-^{-D} \right\}, \tag{9.4.38}$$

where $C = (c_{jk}) = AA^*$ and $D = H - (1/2)I$. As $\lambda \to 0$, $f_X(\lambda) \sim \lambda^{-D} C \lambda^{-D}$ or, when written component-wise,

$$f_{X,jk}(\lambda) \sim c_{jk} \lambda^{-(d_j+d_k)} = c_{jk} \lambda^{-(h_j+h_k-1)}. \tag{9.4.39}$$

In view of (9.4.36) and (9.4.39), when $h_j \in (1/2, 1)$, $j = 1, \ldots, p$, the vector FGN series satisfies conditions II-v and IV-v with $d_j = h_j - 1/2$. As an exercise, the reader can also verify that the expression (9.4.36) for $\gamma_X(n)$ and the expression (9.4.39) for $f_X(\lambda)$ are consistent with (9.4.3), (9.4.6) and (9.4.17) relating g_{jk} to r_{jk}.

The class of vector FGN series captures the general asymptotic structure of the spectral density matrix given in condition IV-v. Indeed, if condition IV-v holds with a positive semidefinite Hermitian symmetric matrix G, then $G = AA^*$ for some $A \in \mathbb{C}^{p \times p}$ (see Section 9.1.1). But then the asymptotic relation (9.4.6) is (9.4.39) for vector FGN series when the underlying vector FBM has the matrix A in its spectral representation (9.3.3).

9.4.4 Fractional Cointegration

Fractional cointegration is another interesting phenomenon arising with vector LRD series (the other phenomenon being the existence of possibly nonzero phases around the origin, as discussed in Section 9.4.1). We introduce it here in the stationary context (see Section 9.8 for further notes). The following notation will be convenient. An LRD univariate series with parameter $d \in (0, 1/2)$ is denoted $I(d)$. An SRD univariate series is denoted $I(0)$. We adopt below condition IV stated in Section 2.1 and condition IV-v stated in Section 9.4 for the definitions of LRD in the univariate and multivariate contexts, respectively.

Example 9.4.6 (*Fractional cointegration for bivariate series*) Consider the $\mathbb{R}^2$-valued series

$$X_n = \begin{pmatrix} X_{1,n} \\ X_{2,n} \end{pmatrix} = \begin{pmatrix} u_n^{(0)} \\ a u_n^{(0)} + u_n^{(1)} \end{pmatrix}, \quad n \in \mathbb{Z}, \quad (9.4.40)$$

where $a \in \mathbb{R}$ ($a \neq 0$) is a constant, and $\{u_n^{(0)}\}$ and $\{u_n^{(1)}\}$ are uncorrelated $I(d_0)$ and $I(d_1)$ series, respectively, with $0 \leq d_1 < d_0 < 1/2$. Note that both component series $\{X_{1,n}\}$ and $\{X_{2,n}\}$ are $I(d_0)$. This is the case for $\{X_{1,n}\}$ by the assumption on $\{u_n^{(0)}\}$, and it holds for $\{X_{2,n}\}$ since $d_1 < d_0$. Though each component series $\{X_n\}$, $j = 1, 2$, is $I(d_0)$, the linear combination $X_{1,n} - (1/a)X_{2,n} = -(1/a)u_n^{(1)}$ is $I(d_1)$ with $d_1 < d_0$. This is precisely the idea of fractional cointegration: for a vector LRD series, each component series can be $I(d_0)$ but their linear combination can be $I(d_1)$ with $d_1 < d_0$; that is, can be less dependent (and possibly SRD when $d_1 = 0$).

Definition 9.4.7 Suppose $\{X_n\}_{n \in \mathbb{Z}}$ is an $\mathbb{R}^p$-valued LRD series such that each component series $\{X_{j,n}\}$, $j = 1, \ldots, p$, is $I(d_0)$ with $d_0 \in (0, 1/2)$. The series $\{X_n\}_{n \in \mathbb{Z}}$ is called *fractionally cointegrated* if there is $\beta \in \mathbb{R}^p$ such that $\beta' X_n$ is $I(d)$ with $0 \leq d < d_0$. A vector β is called a *cointegrating vector*.

If $\{X_n\}_{n \in \mathbb{Z}}$ is an $\mathbb{R}^p$-valued LRD series as in Definition 9.4.7, and $\beta \in \mathbb{R}^p$, note that $\beta' X_n$ cannot be expected $I(d)$ with $d > d_0$. Indeed, if $\{X_n\}_{n \in \mathbb{Z}}$ satisfies condition IV-v with the spectral density matrix $f_X(\lambda)$, then $\{\beta' X_n\}_{n \in \mathbb{Z}}$ has the spectral density

$$\beta' f_X(\lambda) \beta \sim (\beta' G \beta) \lambda^{-2d_0}, \quad (9.4.41)$$

as $\lambda \to 0$, where G is defined in (9.4.5). Since G is positive semidefinite, $\beta' G \beta \geq 0$. Then, either $\beta' G \beta > 0$ in which case $\{\beta' X_n\}_{n \in \mathbb{Z}}$ is $I(d_0)$, or $\beta' G \beta = 0$ in which case $\{\beta' X_n\}_{n \in \mathbb{Z}}$ cannot be $I(d)$ with $d > d_0$. Note also that this argument shows that, for a cointegrating vector β,

$$\beta' G \beta = 0. \quad (9.4.42)$$

Note that the converse does not have to be true; that is, if (9.4.42) holds, we can only deduce that $\{\beta' X_n\}_{n\in\mathbb{Z}}$ has the spectral density $\beta' f_X(\lambda)\beta$ satisfying $(\beta' f_X(\lambda)\beta)\lambda^{2d_0} \to 0$, whereas for a cointegrating vector β, we should have $\beta' f_X(\lambda)\beta \sim c\lambda^{-2d}$ with $d < d_0$.

There may be several, linearly independent cointegrating vectors. Their number has a special name.

Definition 9.4.8 The number of linearly independent cointegrating vectors, r, $1 \leq r \leq p-1$, is called the *cointegrating rank*.

Under additional assumptions, one can arrive at a more refined fractional cointegration model as follows. Suppose that the matrix G in condition IV-v is real-valued. Suppose also that every vector β satisfying (9.4.42) is a cointegrating vector. Then,

$$r = p - \text{rank}(G), \tag{9.4.43}$$

where $\text{rank}(G)$ denotes the rank of a matrix G. In particular, there are linearly independent vectors $\beta_{0,1}, \ldots, \beta_{0,p-r}, \eta_{0,1}, \ldots, \eta_{0,r} \in \mathbb{R}^p$ such that

$$\begin{pmatrix} \beta'_{0,1} \\ \vdots \\ \beta'_{0,p-r} \end{pmatrix} X_n = \begin{pmatrix} u^{(0)}_{1,n} \\ \vdots \\ u^{(0)}_{p-r,n} \end{pmatrix} =: u^{(0)}_n, \tag{9.4.44}$$

where each $\{u^{(0)}_{j,n}\}_{n\in\mathbb{Z}}$, $j = 1, \ldots, p-r$, is $I(d_0)$, and

$$\begin{pmatrix} \eta'_{0,1} \\ \vdots \\ \eta'_{0,r} \end{pmatrix} X_n = \begin{pmatrix} v^{(0)}_{1,n} \\ \vdots \\ v^{(0)}_{r,n} \end{pmatrix} =: v^{(0)}_n, \tag{9.4.45}$$

where for each $j = 1, \ldots, r$, $\{v^{(0)}_{j,n}\}$ is $I(d_{0,j})$ with $d_{0,j} < d_0$. Moreover, the $\mathbb{R}^{p-r}$-valued series $\{u^{(0)}_n\}$ is not fractionally cointegrated. We may also suppose without loss of generality that $d_{0,r} \leq \ldots \leq d_{0,1}$.

Let now $p_1 \geq 1$ be such that $d_{0,p_1+1} < d_{0,p_1} = \ldots = d_{0,1} =: d_1$; that is, the series $\{v^{(0)}_{1,n}\}, \ldots, \{v^{(0)}_{p_1,n}\}$ are all $I(d_1)$. Let r_1 be the cointegrating rank of $v^{(0)}_n$ and suppose as above that $r_1 = p_1 - \text{rank}(G_1)$, where $\widetilde{G}_1$ is the spectral density matrix of $v^{(0)}_n$. Then, as in (9.4.44) and (9.4.45), there are linearly independent vectors $\eta_{1,1}, \ldots, \eta_{1,p_1-r_1}, \eta_{1,p_1-r_1+1}, \ldots, \eta_{1,p_1}$ such that

$$\begin{pmatrix} \eta'_{1,1} \\ \vdots \\ \eta'_{1,p_1-r_1} \\ \eta'_{1,p_1-r_1+1} \\ \vdots \\ \eta'_{1,p_1} \end{pmatrix} v^{(0)}_n = \begin{pmatrix} u^{(1)}_{1,n} \\ \vdots \\ u^{(1)}_{p_1-r_1,n} \\ v^{(1)}_{1,n} \\ \vdots \\ v^{(1)}_{r_1,n} \end{pmatrix} =: \begin{pmatrix} u^{(1)}_n \\ v^{(1)}_n \end{pmatrix}, \tag{9.4.46}$$

where each component series $\{u_{j,n}^{(1)}\}$ of $\{u_n^{(1)}\}$ is $I(d_1)$, $\{u_n^{(1)}\}$ is not fractionally cointegrated, and a component series $\{v_{j,n}^{(1)}\}$ of $v_n^{(1)}$ is $I(d_{1,j})$ with $d_{1,j} < d_1$.

Let $a_0 = p - r$ and $a_1 = p_1 - r_1$. By combining (9.4.44), (9.4.45) and (9.4.46), there are $\beta_0 \in \mathbb{R}^{p \times a_0}$, $\beta_1 \in \mathbb{R}^{p \times a_1}$ and $\eta_2 \in \mathbb{R}^{p \times (p - a_0 - a_1)}$ with $(\beta_0, \beta_1, \eta_2)$ consisting of linearly independent vectors so that

$$\begin{pmatrix} \beta_0' \\ \beta_1' \\ \eta_2' \end{pmatrix} X_n = \begin{pmatrix} u_n^{(0)} \\ u_n^{(1)} \\ v_n^{(1)} \end{pmatrix}, \qquad (9.4.47)$$

where $\{u_n^{(0)}\}$, $\{u_n^{(1)}\}$ and $\{v_n^{(1)}\}$ are, respectively, $\mathbb{R}^{a_0}$-, $\mathbb{R}^{a_1}$- and $\mathbb{R}^{p-a_0-a_1}$-valued time series. Each component series of $\{u_n^{(0)}\}$ is $I(d_0)$. Each component series of $\{u_n^{(1)}\}$ is $I(d_1)$ with $d_1 < d_0$. Each component series of $\{v_n^{(1)}\}$ is $I(d)$ with $d < d_1$. Moreover, $\{u_n^{(0)}\}$ is not fractionally cointegrated, and neither is $\{u_n^{(1)}\}$.

By continuing the above procedure on $\{v_n^{(1)}\}$ and so on, and making suitable assumptions, we can conclude that there are $\beta_j \in \mathbb{R}^{p \times a_j}$, $j = 1, \ldots, s$, $a_0 + \cdots + a_s = p$, with $(\beta_j, j = 0, \ldots, s)$ consisting of linearly independent vectors such that

$$\begin{pmatrix} \beta_0' \\ \beta_1' \\ \beta_2' \\ \vdots \\ \beta_s' \end{pmatrix} X_n = \begin{pmatrix} u_n^{(0)} \\ u_n^{(1)} \\ \vdots \\ u_n^{(s)} \end{pmatrix} =: u_n, \qquad (9.4.48)$$

where $\{u_n^{(j)}\}$, $j = 1, \ldots, s$, are $\mathbb{R}^{a_j}$-valued series which are not fractionally cointegrated and whose components series are $I(d_j)$ with

$$0 \leq d_s < d_{s-1} < \cdots < d_1 < d_0. \qquad (9.4.49)$$

Since $B = (\beta_0, \ldots, \beta_s)$ consists of linearly independent vectors, the relation (9.4.48) can be inverted as $X_n = B^{-1} u_n =: A u_n$ and hence written as the following *fractional cointegration model*

$$X_n = A u_n = A_0 u_n^{(0)} + A_1 u_n^{(1)} + \cdots + A_s u_n^{(s)}, \qquad (9.4.50)$$

where A_j, $j = 0, \ldots, s$, are $\mathbb{R}^{p \times a_j}$ full rank matrices. Conversely, any series $\{X_n\}_{n \in \mathbb{Z}}$ given by (9.4.50) is fractionally cointegrated with the rank $r = p - a_0$.

There is a useful terminology of cointegrating subspaces related to the model (9.4.50). For any matrix A, let $\mathcal{M}(A)$ be the space spanned by the columns of A, and let $\mathcal{M}^\perp(A)$ denote the orthogonal complement of $\mathcal{M}(A)$. For the matrices $A_1, \ldots, A_s$ in (9.4.50), let $\mathcal{B}_0 = \mathcal{M}(A_0)$, and $\mathcal{B}_k$, $k = 1, \ldots, s$, be the subspaces such that

$$\mathcal{M}^\perp(A_0, \ldots, A_{k-1}) = \mathcal{M}^\perp(A_0, \ldots, A_k) \oplus \mathcal{B}_k \qquad (9.4.51)$$

and

$$\mathcal{B}_k \perp \mathcal{M}^\perp(A_0, \ldots, A_k). \qquad (9.4.52)$$

One has $\mathcal{B}_j \perp \mathcal{B}_k$, $j \neq k$, $j, k = 0, \ldots, s$, and

$$\mathbb{R}^p = \mathcal{B}_0 \oplus \mathcal{B}_1 \oplus \ldots \oplus \mathcal{B}_s. \tag{9.4.53}$$

Note also that a nonzero vector $\beta \in \mathcal{B}_k$, $k = 1, \ldots, s$, satisfies

$$\beta' A_l = 0, \ l = 0, \ldots, k-1, \quad \beta' A_k \neq 0. \tag{9.4.54}$$

The relation (9.4.54) implies that any nonzero vector $\beta \in \mathcal{B}_k$, $k = 1, \ldots, s$, produces a fractionally cointegrated error series $\{\beta' X_n\}_{n \in \mathbb{Z}}$ which is $I(d_k)$. The space $\mathcal{B}_0$, on the other hand, consists of non-cointegrating vectors. This motivates the following definition.

Definition 9.4.9 The spaces $\mathcal{B}_k$, $k = 1, \ldots, s$, defined in (9.4.51) and (9.4.52) above are called the *cointegrating spaces*.

The following simple example illustrates the terminology and notation introduced above.

Example 9.4.10 (*Fractional cointegration for bivariate series*) Let $\{X_n\}_{n \in \mathbb{Z}}$ be the series given in (9.4.40) of Example 9.4.6. It can be expressed as

$$X_n = \begin{pmatrix} u_n^{(0)} \\ a u_n^{(0)} + u_n^{(1)} \end{pmatrix} = \begin{pmatrix} 1 \\ a \end{pmatrix} u_n^{(0)} + \begin{pmatrix} 0 \\ 1 \end{pmatrix} u_n^{(1)} = A_0 u_n^{(0)} + A_1 u_n^{(1)},$$

where $\{u_n^{(0)}\}$ and $\{u_n^{(1)}\}$, respectively, $\mathbb{R}^{a_0}$- and $\mathbb{R}^{a_1}$-valued time series with $a_0 = 1$ and $a_1 = 1$, thus univariate time series. It has the representation (9.4.50) with $A_0 = (1 \ a)'$, $A_1 = (0 \ 1)'$. Since $p = 2$, X_n has cointegrating rank $r = p - a_1 = 2 - 1 = 1$. One has $X_n = A u_n$ and thus $B X_n = u_n$ where $B = A^{-1} = (B_0, B_1) = \begin{pmatrix} 1 & 0 \\ -a & 1 \end{pmatrix}$ and thus $s = 1$. Finally, the space $\mathcal{B}_0 = \mathcal{M}(A_0) = \{c(1 \ a)' : c \in \mathbb{R}\}$ and the single cointegrating space $\mathcal{B}_1 = \{c(-a \ 1)' : c \in \mathbb{R}\}$, consisting of vectors orthogonal to those in $\mathcal{B}_0$, are both one-dimensional.

Note also that the cointegration model (9.4.50) is naturally related to vector operator self-similarity in (9.3.1). The following example expands on this point.

Example 9.4.11 (*Fractional cointegration and vector OFBM*) Let B_H be a vector OFBM with the diagonal self-similarity matrix

$$H = \mathrm{diag}(\underbrace{h_0, \ldots, h_0}_{a_0}, \ldots, \underbrace{h_s, \ldots, h_s}_{a_s}), \tag{9.4.55}$$

where $a_0 + \cdots + a_{s-1} + a_s = p$ and

$$\frac{1}{2} \leq h_s < \cdots < h_0 < 1. \tag{9.4.56}$$

Let also B_H be given by its spectral representations (9.3.3) with a positive definite matrix AA^*. Consider the process

$$X(t) = Q B_H(t), \tag{9.4.57}$$

where $Q \in \mathbb{R}^{p \times p}$ is invertible. Note that the process X is zero mean, Gaussian and has stationary increments. Moreover, for $c > 0$,

$$\{X(ct)\}_{t \in \mathbb{R}} = \{QB_H(ct)\}_{t \in \mathbb{R}} \stackrel{d}{=} \{Qc^H B_H(t)\}_{t \in \mathbb{R}} = \{Qc^H Q^{-1} X(t)\}_{t \in \mathbb{R}}$$
$$= \{c^{QHQ^{-1}} X(t)\}_{t \in \mathbb{R}}, \qquad (9.4.58)$$

by using (9.2.4). Thus, X is vector OFBM with the (possibly non-diagonal) self-similarity matrix QHQ^{-1}. In view of (9.4.57), we can consider now the stationary vector FGN series

$$X_n = X(n) - X(n-1) = Q(B_H(n) - B_H(n-1)) =: Qu_n. \qquad (9.4.59)$$

The series $\{u_n\}$ in (9.4.59) has the properties of that in (9.4.48) with $d_j = h_j - 1/2$, $j = 1, \ldots, s$. The relation (9.4.59) corresponds to the fractional cointegration model (9.4.50) with A replaced by Q.

9.5 Operator Fractional Brownian Fields

We consider here extensions of FBM to the random fields setting $q \geq 1$, $p = 1$ (see Section 9.2). In this setting, recall that the operator self-similarity (9.2.1) reads, for any $c > 0$,

$$\{X(c^E t)\}_{t \in \mathbb{R}^q} \stackrel{d}{=} \{c^H X(t)\}_{t \in \mathbb{R}^q}, \qquad (9.5.1)$$

where X is real-valued, $E \in \mathbb{R}^{q \times q}$ and $H \in \mathbb{R}$. Note that replacing c by $c^{1/H}$ (supposing $H \neq 0$) and denoting $E_H = E/H$, the relation (9.5.1) is equivalent to

$$\{X(c^{E_H} t)\}_{t \in \mathbb{R}^q} \stackrel{d}{=} \{cX(t)\}_{t \in \mathbb{R}^q}, \qquad (9.5.2)$$

where $E_H \in \mathbb{R}^{q \times q}$. Though (9.5.2) is simpler and as general as (9.5.1), we shall take (9.5.1) for the definition of operator self-similarity in the random fields setting. One reason for this is that (9.5.1) is often considered in the literature with the constraint

$$\text{tr}(E) = q, \qquad (9.5.3)$$

where $\text{tr}(E)$ denotes the trace of the matrix E. The idea behind (9.5.3) is that when $E = I$ and hence $\text{tr}(E) = q$, the two sides of the relation (9.5.1) become $X(ct)$ and $c^H X(t)$, in which case H can be viewed as the usual self-similarity parameter along all the q axes of $\mathbb{R}^q$. Considering (9.5.1) also provides a more general formulation in the sense that results for (9.5.2) can be deduced readily from those for (9.5.1) by setting $H = 1$. It should also be noted that operator self-similarity (9.5.3) is often called *operator scaling* in the spatial setting and, for example, operator self-similar Gaussian random fields are known as operator scaling Gaussian random fields.

Definition 9.5.1 A (real-valued) random field $\{X(t)\}_{t \in \mathbb{R}^q}$ is called an *operator fractional Brownian field* (OFBF) if it is Gaussian, has stationary increments and is operator self-similar in the sense of (9.5.1) with exponents $E \in \mathbb{R}^{q \times q}$ and $H > 0$.

The stationarity of increments in Definition 9.5.1 means, as in (2.5.2), that, for any $h \in \mathbb{R}^q$,

$$\{X(t+h) - X(h)\}_{t \in \mathbb{R}^q} \stackrel{d}{=} \{X(t) - X(0)\}_{t \in \mathbb{R}^q}.$$

We shall denote an OFBF by $\{B_{E,H}(t)\}_{t\in\mathbb{R}^q}$. Integral representations and basic properties of OFBFs are given in Section 9.5.2, and several special classes and examples are considered in Section 9.5.3. The next section provides some useful facts on the so-called homogeneous functions which will play an important role.

Remark 9.5.2 As with operator self-similar processes (see Section 9.3.4 and, in particular, the relation (9.3.25)), the operator self-similarity exponents E are generally not unique for the same random field X and fixed H. More specifically, as shown under mild assumptions in, e.g., Didier, Meerschaert, and Pipiras [309],

$$\mathcal{E}_H(X) = E + T(G_X^{\text{dom}}),$$

where $\mathcal{E}_H(X)$ is the set of all operator self-similarity exponents E satisfying (9.5.1),

$$G_X^{\text{dom}} = \left\{ A \in GL(q) : \{X(t)\}_{t\in\mathbb{R}} \stackrel{d}{=} \{X(At)\}_{t\in\mathbb{R}} \right\}$$

is the domain symmetry group, and $T(G_X^{\text{dom}})$ denotes the tangent space (defined in (9.3.54)).

9.5.1 M–Homogeneous Functions

We will need the notion of homogeneous functions defined next.

Definition 9.5.3 Let M be an $\mathbb{R}^{q\times q}$ matrix. A function $\phi : \mathbb{R}^q \to \mathbb{C}$ is called M–*homogeneous* if, for all $c > 0$ and $x \in \mathbb{R}^q \setminus \{0\}$,

$$\phi(c^M x) = c\phi(x). \tag{9.5.4}$$

When $q = 1$ and $M \neq 0$ is a real number, an M–homogeneous function can be written as a power function

$$\phi(x) = c_1 x_+^{1/M} + c_2 x_-^{1/M}$$

for constants $c_1 = \phi(1)$ and $c_2 = \phi(-1)$, which follows by taking $c = x$, $x = 1$ for $x > 0$ and $c = -x$, $x = -1$ for $x < 0$ in (9.5.4). Thus $\phi(bx) = b^{1/M}\phi(x)$ if $b > 0$ and $x \neq 0$. If $M = 1$, one would get the usual homogeneous function.

Homogeneous functions can thus be viewed as generalizations of power functions to higher dimensions. In higher dimensions, however, they cannot be expressed in general through elementary power functions in a simple and explicit way. Power functions were used in the various representations of fractional Brownian motion (see Proposition 2.6.5 and Remark 2.6.8, and Proposition 2.6.11). Homogeneous functions will appear similarly in the representations of operator fractional Brownian fields (see Theorems 9.5.9 and 9.5.11 below).

For example, the rth norm $\phi(x) = \|x\|_r = (|x_1|^r + \cdots + |x_q^r|)^{1/r}$, $r \geq 1$, is M–homogeneous with M equal to the $q \times q$ identity matrix I since $\phi(c^I x) = \|c^I x\|_r = c\|x\|_r = c\phi(x)$.

Lemma 9.5.4 *Suppose that ϕ is M–homogeneous and let $a \in \mathbb{R}$, $a \neq 0$. Then $\psi(x) = (\phi(x))^a$ is (M/a)–homogeneous.*

Proof Indeed, note that

$$\psi(c^{M/a}x) = [\phi(c^{M/a}x)]^a = [\phi((c^{1/a})^M x)]^a = [c^{1/a}\phi(x)]^a = c\psi(x). \qquad \square$$

We will also need a "polar coordinate" representation of $x \in \mathbb{R}^q$. Let M be an $\mathbb{R}^{q \times q}$ matrix and suppose that its eigenvalues have positive real parts. By Lemma 6.1.5 in Meerschaert and Scheffler [710] and a change of variables $s = e^{-t}$,

$$\|x\|_0 = \int_0^\infty \|e^{-Mt}x\| dt = \int_0^1 \|s^M x\| s^{-1} ds \qquad (9.5.5)$$

defines a norm on $\mathbb{R}^q$, where $\|\cdot\|$ is any norm on $\mathbb{R}^q$. Moreover, with the norm $\|\cdot\|_0$ and the unit sphere $S_0 = \{x \in \mathbb{R}^q : \|x\|_0 = 1\}$, the mapping $\Psi : (0, \infty) \times S_0 \to \mathbb{R}^q \setminus \{0\}$, $\Psi(r, \theta) = r^M \theta$, is a homeomorphism.[9] Since Ψ is one-to-one, onto and has a continuous inverse, any $x \in \mathbb{R}^q \setminus \{0\}$ can be written uniquely as

$$x = (\tau_M(x))^M l_M(x), \qquad (9.5.6)$$

for some *radial part* $\tau_M(x) > 0$ and some *direction* $l_M(x) \in S_0$ such that $\tau_M(x)$ and $l_M(x)$ are continuous in x. Since $\tau_M(x) \to 0$, as $x \to 0$, one can extend $\tau_M(x)$ to 0 continuously by setting $\tau_M(0) = 0$.

Note from the uniqueness of the vector representation (9.5.6) that $\tau_M(-x) = \tau_M(x)$ and $l_M(-x) = -l_M(x)$. Similarly, for $c > 0$, since $c^M x = (c\tau_M(x))^M l_M(x)$, such uniqueness also yields

$$\tau_M(c^M x) = c\tau_M(x), \quad l_M(c^M x) = l_M(x). \qquad (9.5.7)$$

In particular, the function $\tau_M(x)$ is M–homogeneous. Note also that, for $x = r^M \theta$ with $r > 0$ and $\theta \in S_0$,

$$\tau_M(\theta) = 1, \quad \tau_M(r^M \theta) = r, \quad l_M(r^M \theta) = \theta, \qquad (9.5.8)$$

by the uniqueness of the polar coordinate representation.

Example 9.5.5 (*Polar coordinate representation associated with the identity matrix*) If $M = m_1 \cdot I$, where $m_1 > 0$ and I is the $q \times q$ identity matrix, then $\|x\|_0$ in (9.5.5) can be taken to be the Euclidean norm $\|x\|_2 = (|x_1|^2 + \cdots + |x_q|^2)^{1/2}$, and $S_0 = \{x \in \mathbb{R}^q : \|x\|_2 = 1\}$. Indeed, with $\|\cdot\| = m_1 \|\cdot\|_2$ and $M = m_1 \cdot I$ in (9.5.5), note that $\|x\|_0 = \|x\|_2 m_1 \int_0^\infty e^{-m_1 t} dt = \|x\|_2$. Moreover,

$$\tau_M(x) = \|x\|_2^{1/m_1}, \quad l_M(x) = \left(\frac{x_1}{\|x\|_2}, \ldots, \frac{x_q}{\|x\|_2}\right) = \frac{x}{\|x\|_2},$$

by the uniqueness of the polar coordinate representation.

Remark 9.5.6 The exponent M of a homogeneous function ϕ is not unique in general. By Theorem 5.2.13 in Meerschaert and Scheffler [710], the set of possible exponents has the form $M + T(S_\phi)$, where T denotes the tangent space as in (9.3.54) and $S_\phi = \{A : \phi(Ax) = \phi(x)\}$ is the so-called symmetry group of ϕ. For example, with $M = m_1 \cdot I$

---

[9] $\Psi : (0, \infty) \times S_0 \to \mathbb{R}^q \setminus \{0\}$, $\Psi(r, \theta) = r^M \theta$, is a homeomorphism if Ψ is continuous, one-to-one, onto and has a continuous inverse.

9.5 Operator Fractional Brownian Fields 511

and $\phi(x) = \|x\|_2^{1/m_1}$, $S_\phi = O(q)$, $T(S_\phi) = so(q)$ and the set of possible exponents is $M + so(q)$ (cf. Example 9.3.26). Thus, one can add to M any skewed symmetric matrix. For further discussion, see Remark 2.10 in Biermé, Meerschaert, and Scheffler [151].

Polar coordinate representations provide a convenient means of expressing M–homogeneous functions ϕ. Indeed, suppose $\phi(c^M x) = c\phi(x)$ and represent x through polar coordinates as $x = (\tau_M(x))^M l_M(x)$. Then, taking $c = \tau_M(x)^{-1}$ yields $\phi(\tau_M(x)^{-M} x) = \tau_M(x)^{-1} \phi(x)$ and hence

$$\phi(x) = \tau_M(x)\phi(\tau_M(x)^{-M} x) = \tau_M(x)\phi(l_M(x)), \qquad (9.5.9)$$

since $x = (\tau_M(x))^M l_M(x)$. In particular, ϕ is determined by its values on S_0 through $\phi(l_M(x))$. Conversely, by using (9.5.7), the function

$$\tau_M(x)\phi_0(l_M(x)) \quad \text{is } M\text{–homogeneous,} \qquad (9.5.10)$$

for any function ϕ_0 on S_0.

The proof of the next technical lemma illustrates the use of the polar coordinate representation (9.5.6). The lemma is used in the proof of Theorem 9.5.11 below.

Lemma 9.5.7 *Let M be an $\mathbb{R}^{q \times q}$ matrix with eigenvalues having positive real parts. If the function $h : \mathbb{R}^q \to \mathbb{C}$ is such that, for all $c > 0$,*

$$h(c^M x) = ch(x) \quad a.e. \ dx, \qquad (9.5.11)$$

then there is an M–homogeneous function ϕ such that $\phi(x) = h(x)$ a.e. dx.

Proof By the Fubini-based argument found following (9.3.17),

$$h(c^M x) = ch(x) \quad \text{a.e. } dcdx.$$

By using (9.5.6), we get

$$h(x) = c^{-1} h(c^M x) = c^{-1} h((c\tau_M(x))^M l_M(x)) \quad \text{a.e. } dcdx.$$

By making the change of variables (c, x) to $(c\tau_M(x)^{-1}, x)$, and using (9.5.6), we also have that

$$h(x) = c^{-1}\tau_M(x) h(c^M l_M(x)) \quad \text{a.e. } dcdx.$$

Now, there is c_0 such that the last relation holds a.e. dx; that is,

$$h(x) = \tau_M(x) c_0^{-1} h(c_0^M l_M(x)) \quad \text{a.e. } dx$$

or

$$h(x) = \phi(x) \quad \text{a.e. } dx,$$

where $\phi(x) = \tau_M(x)\phi_0(l_M(x))$ with $\phi_0(\theta) = c_0^{-1} h(c_0^M \theta)$ for $\theta \in S_0$. As noted in (9.5.10), the function ϕ is M–homogeneous. $\square$

Example 9.5.8 (*Large class of homogeneous functions*) Let $e_1, \ldots, e_q$ and $\theta_1, \ldots, \theta_q$ be, respectively, the eigenvalues and eigenvectors of a matrix M^*: that is, $M^*\theta_j = e_j\theta_j$. Suppose that $e_1, \ldots, e_q \in \mathbb{R} \setminus \{0\}$ and denote the inner product for $u, v \in \mathbb{R}^q$ by $\langle u, v \rangle = u^*v$. Then, for $C_1, \ldots, C_q \geq 0$ and $\rho > 0$, the function

$$\phi(x) = \Big(\sum_{j=1}^{q} C_j |\langle x, \theta_j \rangle|^{\rho/e_j}\Big)^{1/\rho}$$

is M–homogeneous. This follows from observing that

$$(c^M x)^* \theta_j = x^* c^{M^*} \theta_j = x^* \sum_{k=0}^{\infty} \frac{(\log c)^k (M^*)^k}{k!} \theta_j = x^* \sum_{k=0}^{\infty} \frac{(\log c)^k e_j^k}{k!} \theta_j = c^{e_j} x^* \theta_j.$$

Consequently,

$$\phi(c^M x) = \Big(\sum_{j=1}^{q} C_j |\langle c^M x, \theta_j \rangle|^{\rho/e_j}\Big)^{1/\rho} = c \Big(\sum_{j=1}^{q} C_j |\langle x, \theta_j \rangle|^{\rho/e_j}\Big)^{1/\rho} = c\phi(x).$$

We shall use below several other auxiliary results related to the notions introduced above. The first result is a useful integration in polar coordinates formula. As above, let $M \in \mathbb{R}^{q \times q}$ with eigenvalues having positive real parts $0 < a_1 < \cdots < a_q$. By Proposition 2.3 in Biermé et al. [151], for $f \in L^1(\mathbb{R}^q, dx)$,

$$\int_{\mathbb{R}^q} f(x) dx = \int_0^{\infty} \int_{S_0} f(r^M \theta) \sigma(d\theta) r^{\mathrm{tr}(M)-1} dr, \qquad (9.5.12)$$

for a unique finite Radon[10] measure σ on S_0, where $\mathrm{tr}(M)$ denotes the trace of the matrix M.

We shall also use several bounds on a norm in terms of the radial part. Let $\|\cdot\|$ be any norm on $\mathbb{R}^q$ and let x be as in (9.5.6), namely $x = (\tau_M(x))^M l_M(x)$. We will first bound $\|x\|$ when $\|x\| \leq 1$, then when $\|x\| \geq 1$.

Suppose first $\|x\| \leq 1$, so that $r = \tau_M(x) \leq 1$ and consider any $\theta \in S_0$. Then, as shown in Lemma 2.1 in Biermé et al. [151] (and by using the fact that all norms in $\mathbb{R}^q$ are equivalent),

$$C_1 \tau_M(x)^{a_q + \delta} \leq \|x\| \leq C_2 \tau_M(x)^{a_1 - \delta},$$

for arbitrarily small $\delta > 0$. This can be expressed in an expanded form as

$$C_1 r^{a_q + \delta} = C_1 \tau_M(x)^{a_q + \delta} \leq \|x\| = \|\tau_M(x)^M l(x)\| = \|r^M \theta\| \leq C_2 \tau_M(x)^{a_1 - \delta} = C_2 r^{a_1 - \delta}$$
$$(9.5.13)$$

for arbitrarily small $\delta > 0$. Suppose now $\|x\| \geq 1$, so that $r = \tau_M(x) \geq 1$ and again consider any $\theta \in S_0$. Then, as shown in Lemma 2.1 in Biermé et al. [151],

$$C_3 \tau_M(x)^{a_1 - \delta} \leq \|x\| \leq C_4 \tau_M(x)^{a_q + \delta},$$

---

[10] That is, a measure on the Borel sets of S_0 which is finite on compact sets of S_0, and is inner regular in the sense that, for any Borel set A of S_0 and $\epsilon > 0$, there is a compact set K_ϵ such that $\sigma(A \setminus K_\epsilon) < \epsilon$.

for arbitrarily small $\delta > 0$. In an expanded form, this can also be written as

$$C_3 r^{a_1-\delta} = C_3 \tau_M(x)^{a_1-\delta} \leq \|x\| = \|\tau_M(x)^M l(x)\| = \|r^M \theta\| \leq C_4 \tau_M(x)^{a_q+\delta} = C_4 r^{a_q+\delta} \tag{9.5.14}$$

for arbitrarily small $\delta > 0$.

9.5.2 Integral Representations

We shall provide below integral representations of OFBFs in the spectral and time domains. The following result provides a spectral domain representation of OFBF.

Theorem 9.5.9 *Let $B_{E,H} = \{B_{E,H}(t)\}_{t \in \mathbb{R}^q}$ be an OFBF with $E \in \mathbb{R}^{q \times q}$ whose eigenvalues have positive real parts and $H > 0$. Assume also that $E \neq H \cdot I$. Then, $B_{E,H}$ has a spectral representation*

$$\{B_{E,H}(t)\}_{t \in \mathbb{R}^q} \stackrel{d}{=} \left\{ \int_{\mathbb{R}^q} (e^{i \langle x,t \rangle} - 1) \widehat{B}_F(dx) \right\}_{t \in \mathbb{R}^q}, \tag{9.5.15}$$

where $\widehat{B}_F$ is an Hermitian Gaussian random measure on $\mathbb{R}^q$ with the control measure $\mathbb{E}|\widehat{B}_F(dx)|^2 = F(dx)$ such that

$$F(c^{-E^*} dx) = c^{2H} F(dx) \tag{9.5.16}$$

or, equivalently, the corresponding distribution function F is $(-E^*/(2H))$–homogeneous; that is,

$$F(c^{-E^*/(2H)} x) = c F(x). \tag{9.5.17}$$

If F is absolutely continuous with spectral density f, then $B_{E,H}$ can also be represented as

$$\{B_{E,H}(t)\}_{t \in \mathbb{R}^q} \stackrel{d}{=} \left\{ \int_{\mathbb{R}^q} (e^{i \langle x,t \rangle} - 1) g(x)^{-H-\text{tr}(E^*)/2} \widehat{B}(dx) \right\}_{t \in \mathbb{R}^q} \tag{9.5.18}$$

$$\stackrel{d}{=} \left\{ \int_{\mathbb{R}^q} (e^{i \langle x,t \rangle} - 1) \tau_{E^*}(x)^{-H-\text{tr}(E^*)/2} g_0(l_{E^*}(x)) \widehat{B}(dx) \right\}_{t \in \mathbb{R}^q}, \tag{9.5.19}$$

where $\widehat{B}$ is an Hermitian Gaussian random measure on $\mathbb{R}^q$ with the control measure dx, $\text{tr}(E^*)$ denotes the trace of the matrix E^*, g is a nonnegative, symmetric, E^*–homogeneous function and

$$g_0(y) = g(y)^{-H-\text{tr}(E^*)/2}. \tag{9.5.20}$$

Moreover, the function g_0 is $-E^*/(H + \text{tr}(E^*)/2)$–homogeneous, and

$$g_0(y) = f(y)^{1/2} \quad a.e. \ dy. \tag{9.5.21}$$

Conversely, suppose that

$$0 < H < \min\{\Re(e_j) : j = 1, \ldots, q\}, \tag{9.5.22}$$

where e_j are the eigenvalues of the matrix E. Suppose also that the function g_0 in (9.5.19) is bounded on S_0. Then, the right-hand side of (9.5.19) is a well-defined OFBF with exponents E and H.

Proof We first prove (9.5.15). An OFBF $B_{E,H}$ has stationary increments and hence can be represented as

$$\{B_{E,H}(t)\}_{t\in\mathbb{R}^q} \stackrel{d}{=} \left\{\int_{\mathbb{R}^q} (e^{i\langle x,t\rangle} - 1)\widehat{B}_F(dx) + \langle X, t\rangle + Y\right\}_{t\in\mathbb{R}^q}, \quad (9.5.23)$$

where $\widehat{B}_F(dx)$ is an Hermitian Gaussian random measure on $\mathbb{R}^q$ with a (symmetric) control measure $F(dx)$, X is a Gaussian $\mathbb{R}^q$–vector with mean vector $\mu = \mathbb{E}X$ and covariance matrix $A = \mathbb{E}XX'$, which is independent of $\widehat{B}_F(dx)$, and Y is a random variable (Yaglom [1021], p. 436, Yaglom [1020]). Since $B_{E,H}(0) = B_{E,H}(c^E \cdot 0) \stackrel{d}{=} cB_{E,H}(0) \to 0$, as $c \to 0$, we have $B_{E,H}(0) = 0$ a.s. Hence, $Y = 0$ a.s. in the representation (9.5.23).

The measure F, the vector μ and the matrix A are known to be unique (Yaglom [1020]). We will exploit this uniqueness to derive the homogeneity of F by using the operator self-similarity of $B_{E,H}$. On one hand, note that

$$\{B_{E,H}(c^E t)\}_{t\in\mathbb{R}^q} \stackrel{d}{=} \left\{\int_{\mathbb{R}^q} (e^{i\langle x, c^E t\rangle} - 1)\widehat{B}_F(dx) + \langle X, c^E t\rangle\right\}_{t\in\mathbb{R}^q}$$

$$= \left\{\int_{\mathbb{R}^q} (e^{i\langle c^{E^*} x, t\rangle} - 1)\widehat{B}_F(dx) + \langle c^{E^*} X, t\rangle\right\}_{t\in\mathbb{R}^q}$$

$$\stackrel{d}{=} \left\{\int_{\mathbb{R}^q} (e^{i\langle y, t\rangle} - 1)\widehat{B}_{F_E}(dy) + \langle c^{E^*} X, t\rangle\right\}_{t\in\mathbb{R}^q}, \quad (9.5.24)$$

by making the change of variables $c^{E^*}x = y$ ($x = c^{-E^*}y$), where the control measure F_E is defined as $F_E(dy) = F(c^{-E^*}dy)$. On the other hand, by the operator self-similarity, $\{B_{E,H}(c^E t)\}_{t\in\mathbb{R}^q}$ has the same distribution as $\{c^H B_{E,H}(t)\}_{t\in\mathbb{R}^q}$. The uniqueness then implies that

$$F(c^{-E^*}dy) = c^{2H} F(dy), \quad c^{E^*} X \stackrel{d}{=} c^H X. \quad (9.5.25)$$

Since $E \neq H \cdot I$ (or $E^* \neq H \cdot I$), the second relation in (9.5.25) holds only if $X = 0$ a.s. In terms of the spectral distribution function F, after setting $F(0) = 0$, the first relation in (9.5.25) becomes

$$F(c^{-E^*} x) = c^{2H} F(x) \quad (9.5.26)$$

and hence after changing c to $c^{1/(2H)}$ that $F(c^{-E^*/(2H)}x) = cF(x)$; that is, F is $(-E^*/(2H))$–homogeneous, as stated in (9.5.17).

We now turn to the representations (9.5.18) and (9.5.19) of $B_{E,H}(t)$. Suppose now that F is absolutely continuous with a spectral density f. The relation (9.5.26) can then be written as

$$\int_0^{c^{-E^*}x} f(y)dy = c^{2H} \int_0^x f(y)dy$$

or, by making the change of variables $y = c^{-E^*}z$ in the first integral and using the fact that $\det(c^E) = c^{\text{tr}(E^*)}$, as

$$\int_0^x f(c^{-E^*}z) c^{-\text{tr}(E^*)} dz = c^{2H} \int_0^x f(y)dy.$$

By differentiating the last relation with respect to x, we have that, for any $c > 0$,
$$f(c^{-E^*}x)c^{-\text{tr}(E^*)} = c^{2H}f(x) \quad \text{a.e. } dx$$

or by changing c to $1/c$,
$$f(c^{E^*}x) = c^{-2H-\text{tr}(E^*)}f(x) \quad \text{a.e. } dx.$$

Equivalently,
$$f(c^{E^*}x)^{1/(-2H-\text{tr}(E^*))} = cf(x)^{1/(-2H-\text{tr}(E^*))} \quad \text{a.e. } dx$$

or
$$h(c^{E^*}x) = ch(x) \quad \text{a.e. } dx$$

by setting $f(x) = h(x)^{-2H-\text{tr}(E^*)}$. By applying Lemma 9.5.7, there is an E^*–homogeneous function g such that $g(x) = h(x)$ a.e. dx. Hence,
$$g(x) = f(x)^{1/(-2H-\text{tr}(E^*))} \quad \text{a.e. } dx$$

or
$$f(x)^{1/2} = g(x)^{-H-\text{tr}(E^*)/2} \quad \text{a.e. } dx \qquad (9.5.27)$$

Combined with (9.5.23) and since $X = 0$ and $Y = 0$ as indicated above, this yields the representation (9.5.18). In turn, the representation (9.5.19) now follows from applying (9.5.9). The relation (9.5.21) is the same as (9.5.27), in view of (9.5.20).

Finally, we turn to the converse statement and show that the right-hand side of (9.5.19) is well-defined; that is,
$$\int_{\mathbb{R}^q} |e^{i\langle t,x\rangle} - 1|^2 \tau_{E^*}(x)^{-2H-\text{tr}(E^*)} |g_0(l_{E^*}(x))|^2 dx < \infty. \qquad (9.5.28)$$

By using the formula (9.5.12), setting $x = r^{E^*}\theta$ and using (9.5.8), the last integral becomes
$$\int_0^\infty \int_{S_0} |e^{i\langle t, r^{E^*}\theta\rangle} - 1|^2 r^{-2H-\text{tr}(E^*)} |g_0(\theta)|^2 \sigma(d\theta) r^{\text{tr}(E^*)-1} dr \qquad (9.5.29)$$

$$= \int_0^\infty \int_{S_0} |e^{i\langle t, r^{E^*}\theta\rangle} - 1|^2 r^{-2H-1} |g_0(\theta)|^2 \sigma(d\theta) dr, \qquad (9.5.30)$$

where $\sigma(d\theta)$ is a Radon measure. By using the inequality $|e^{iz} - 1| \leq \min\{|z|, 2\} \leq 2\min\{|z|, 1\}$ and (9.5.13), note that
$$|e^{i\langle t, r^{E^*}\theta\rangle} - 1|^2 \leq 4\min\{\langle t, r^{E^*}\theta\rangle^2, 1\} \leq 4\min\{\|t\|^2 \|r^{E^*}\theta\|^2, 1\}$$
$$\leq 4(1 + \|t\|^2)\min\{\|r^{E^*}\theta\|^2, 1\} \leq C(1 + \|t\|^2)\min\{r^{2(a_1-\delta)}, 1\},$$

for any $\delta \in (0, a_1 - H)$, where $a_1 = \min\{\Re(e_j) : j = 1, \ldots, q\}$. Therefore, the integral in (9.5.30) can be bounded by
$$C(1 + \|t\|^2)\Big(\sup_{\theta \in S_0} |g_0(\theta)|^2 \sigma(S_0)\Big)\Big(\int_0^\infty \min\{r^{2(a_1-\delta)}, 1\} r^{-2H-1} dr\Big).$$

This is finite because g_0 is assumed bounded on S_0, σ is finite, $H > 0$, so that $(-2H - 1) + 1 < 0$, and a_1 is assumed bigger than H, so that

$$2(a_1 - \delta) - 2H - 1 > -1. \qquad \square$$

Remark 9.5.10 In contrast to vector OFBM (cf. Theorem 9.3.2), the control measure F in the representation (9.5.15) of OFBF is not necessarily absolutely continuous. For example, for $q = 2$, the measure

$$F_1(dx) = F_1(dx_1, dx_2) = |x_1|^{-\frac{2}{e_1}-1} dx_1 \delta_{x_1}(dx_2), \quad e_1 > 1,$$

is not absolutely continuous, and satisfies (9.5.16) with

$$E^* = E = \mathrm{diag}(e_1, e_1) = e_1 \cdot I, \quad H = 1.$$

Take also any measure F_2 which is absolutely continuous and satisfies (9.5.16) with the same exponents E and H, and for which the corresponding OFBF is well-defined. Set $F = F_1 + F_2$, which is not absolutely continuous. The field given by (9.5.15) is then well-defined since

$$\int_{\mathbb{R}^2} |e^{i\langle x,t\rangle} - 1|^2 F_1(dx) = \int_{\mathbb{R}} |e^{ix_1(t_1+t_2)} - 1|^2 |x_1|^{-\frac{2}{e_1}-1} dx_1 < \infty,$$

supposing $e_1 > 1$. Hence, it is OFBF with the exponents E and H. (Note that the role of F_2 was just to ensure that the OFBF is proper. Indeed, without F_2, the resulting "OFBF" is zero for $t_1 = -t_2$.)

The next result provides a time domain representation of OFBF. In contrast to Theorem 9.3.8 concerning the time domain representations of vector OFBMs, there are technical difficulties in going from the spectral domain to the time domain representations for OFBFs. We make additional assumptions on the function in the spectral representation and the exponents E, H.

Theorem 9.5.11 *Let $B_{E,H} = \{B_{E,H}(t)\}_{t \in \mathbb{R}^q}$ be an OFBF with exponents E and H. Suppose that the assumptions of Theorem 9.5.9 hold and, in particular, that the control measure F is absolutely continuous. Suppose, in addition, that the function $g_0(y) = g(y)^{-H-\mathrm{tr}(E^*)/2}$ in Theorem 9.5.9 is differentiable a.e. in the sense that*

$$\frac{\partial^q}{\partial x_1 \ldots \partial x_q} g_0(x) = \widetilde{g}(x) \quad a.e. \ dx, \tag{9.5.31}$$

that $\widetilde{g}$ is bounded on S_0, that E is diagonalizable over $\mathbb{R}$ with real eigenvalues e_j, $j = 1, \ldots, q$, and that

$$H + \mathrm{tr}(E)/2 < q \min\{\Re(e_j) : j = 1, \ldots, q\}. \tag{9.5.32}$$

Then, there is a function $h : \mathbb{R}^q \to \mathbb{R}$ satisfying

(i) $h(t - \cdot) - h(-\cdot) \in L^2(\mathbb{R}^q)$, $t \in \mathbb{R}^q$,
(ii) h is $E/(H - \mathrm{tr}(E)/2)$-*homogeneous,*

and such that

$$\{B_{E,H}(t)\}_{t\in\mathbb{R}^q} \stackrel{d}{=} \Big\{\int_{\mathbb{R}^q}(h(t-u)-h(-u))B(du)\Big\}_{t\in\mathbb{R}^q}, \quad (9.5.33)$$

where $B(du)$ is a Gaussian random measure on $\mathbb{R}^q$ with the Lebesgue control measure $\mathbb{E}|B(du)|^2 = du$.

Remark 9.5.12 If the function h in (9.5.33) is nonnegative, then one can write $h = \phi^{H-\mathrm{tr}(E)/2}$ with a nonnegative E–homogeneous function ϕ (see Lemma 9.5.4), and the representation (9.5.33) as

$$\{B_{E,H}(t)\}_{t\in\mathbb{R}^q} \stackrel{d}{=} \Big\{\int_{\mathbb{R}^q}\Big(\phi(t-u)^{H-\mathrm{tr}(E)/2} - \phi(-u)^{H-\mathrm{tr}(E)/2}\Big)B(du)\Big\}_{t\in\mathbb{R}^q}. \quad (9.5.34)$$

Proof Step 1: We shall first show that it is enough to consider the case of a diagonal E. Indeed, for a general diagonalizable $E = WE_0 W^{-1}$ with $W \in \mathbb{R}^{q\times q}$ and diagonal $E_0 \in \mathbb{R}^{q\times q}$, the process $\widetilde{B}_{E_0,H}(t) := B_{E,H}(Wt)$ is an OFBF with the diagonal exponent E_0. Suppose that it can be represented as

$$\{\widetilde{B}_{E_0,H}(t)\}_{t\in\mathbb{R}^q} \stackrel{d}{=} \Big\{\int_{\mathbb{R}^q}(h_0(t-u)-h_0(-u))B(du)\Big\}_{t\in\mathbb{R}^q},$$

where h_0 is $E_0/(H-\mathrm{tr}(E_0)/2)$–homogeneous. Then,

$$\{B_{E,H}(t)\}_{t\in\mathbb{R}^q} \stackrel{d}{=} \Big\{\int_{\mathbb{R}^q}(h_0(W^{-1}t - u) - h_0(-u))B(du)\Big\}_{t\in\mathbb{R}^q}$$
$$\stackrel{d}{=} \Big\{\int_{\mathbb{R}^q}(h_0(W^{-1}(t-v)) - h_0(-W^{-1}v))|\det(W)|^{-1/2}B(dv)\Big\}_{t\in\mathbb{R}^q}$$
$$= \Big\{\int_{\mathbb{R}^q}(h(t-v) - h(-v))B(dv)\Big\}_{t\in\mathbb{R}^q}$$

after the change of variables $u = W^{-1}v$ (so that $du = |\det(W)|^{-1}dv$), where

$$h(v) = |\det(W)|^{-1/2}h_0(W^{-1}v).$$

By using the homogeneity of h_0 and (9.5.4), we get

$$h_0(c^{E_0}W^{-1}v) = h_0\Big((c^{H-\mathrm{tr}(E_0)/2})^{\frac{E_0}{H-\mathrm{tr}(E_0)/2}}W^{-1}v\Big) = c^{H-\mathrm{tr}(E_0)/2}h_0(W^{-1}v)$$

and since $\mathrm{tr}(E_0) = \mathrm{tr}(E)$, we have

$$h(c^E v) = |\det(W)|^{-1/2}h_0(W^{-1}c^E v) = |\det(W)|^{-1/2}h_0(W^{-1}Wc^{E_0}W^{-1}v)$$
$$= |\det(W)|^{-1/2}h_0(c^{E_0}W^{-1}v) = c^{H-\mathrm{tr}(E_0)/2}|\det(W)|^{-1/2}h_0(W^{-1}v) = c^{H-\mathrm{tr}(E)/2}h(v),$$

that is, the function h is $E/(H-\mathrm{tr}(E)/2)$–homogeneous. This shows that the conclusion of the theorem holds for E.

Step 2: Thus, suppose without loss of generality that E is diagonal. Let

$$\widehat{f_t}(x) = (e^{i\langle x,t\rangle} - 1)g_0(x) \quad (\in L^2(\mathbb{R}^q))$$

be the kernel function appearing in the spectral representation (9.5.19) of OFBF $B_{E,H}$ (with a nonnegative, symmetric function g_0). Then, by using Plancherel's identity (see (A.1.10)), $B_{E,H}$ can also be represented as

$$\{B_{E,H}(t)\}_{t\in\mathbb{R}^q} \stackrel{d}{=} \left\{ \int_{\mathbb{R}^q} f_t(u) B(du) \right\}_{t\in\mathbb{R}^q},$$

where

$$f_t(u) = \frac{1}{(2\pi)^q} \int_{\mathbb{R}^q} e^{-i\langle x,u\rangle} \widehat{f_t}(x) dx \quad (\in L^2(\mathbb{R}^q))$$

$$= \frac{1}{(2\pi)^q} \int_{\mathbb{R}^q} e^{-i\langle x,u\rangle} (e^{i\langle x,t\rangle} - 1) g_0(x) dx$$

$$= \frac{1}{(2\pi)^q} \int_{\mathbb{R}^q} (\cos(\langle x, t-u\rangle) - \cos(\langle x, -u\rangle)) g_0(x) dx. \qquad (9.5.35)$$

We want to express next $f_t(u)$ as $h(t-u) - h(-u)$, where h is $E/(H - \text{tr}(E)/2)$-homogeneous as stated in the theorem. Note that h cannot be defined directly as $h(v) = \int_{\mathbb{R}^q} \cos(\langle x, v\rangle) g_0(x) dx$ (see Remark 9.5.13 below).

Observe that, after the changes of variables for each orthant and the fact that g_0 is symmetric,

$$(2\pi)^q f_t(u) = \sum_\sigma \int_{[0,\infty)^q} (\cos(\langle x, (t-u)_\sigma\rangle) - \cos(\langle x, (-u)_\sigma\rangle)) g_0(x) dx =: \sum_\sigma f_{\sigma,t}(u),$$
(9.5.36)

where the sum is over all $\sigma = (\sigma_1, \ldots, \sigma_q)$ with $\sigma_j \in \{-1, 1\}$, and $(v)_\sigma = (\sigma_1 v_1, \ldots, \sigma_q v_q)$. It is thus enough to show that

$$f_{\sigma,t}(u) = h_\sigma(t-u) - h_\sigma(-u), \qquad (9.5.37)$$

for some $E/(H - \text{tr}(E)/2)$-homogeneous h_σ. For notational simplicity, set

$$b = \text{tr}(E) = \text{tr}(E^*).$$

Since in (9.5.20), g is E^*-homogeneous, we get that g_0 is $-E^*/(H+b/2)$-homogeneous by Lemma 9.5.4. Denote by $[x, \infty)$ the Cartesian product $[x_1, \infty) \times \cdots \times [x_q, \infty)$. Since g is assumed to be differentiable in the sense of (9.5.31), then by setting

$$g_0(x) = (-1)^q \int_{[x,\infty)} \widetilde{g}(y) dy, \qquad (9.5.38)$$

we get that $\widetilde{g}$ is $-E^*/(H + 3b/2)$-homogeneous (Exercise 9.11). Then,

$$f_{\sigma,t}(u) = (-1)^q \int_{[0,\infty)^q} (\cos(\langle x, (t-u)_\sigma\rangle) - \cos(\langle x, (-u)_\sigma\rangle)) \left(\int_{[x,\infty)} \widetilde{g}(y) dy \right) dx$$

$$= (-1)^q \int_{[0,\infty)^q} dy \widetilde{g}(y) \left(\int_{[0,y]} \cos(\langle x, (t-u)_\sigma\rangle) dx - \int_{[0,y]} \cos(\langle x, (-u)_\sigma\rangle) dx \right).$$
(9.5.39)

Observe now that

$$\int_{[0,y]} \cos(\langle x, v\rangle) dx = \Re \int_{[0,y]} e^{i\langle x,v\rangle} dx = \frac{1}{v_1 \ldots v_q} \Re \left(i^{-q} (e^{iy_1 v_1} - 1) \ldots (e^{iy_q v_q} - 1) \right).$$

Then, by (9.5.39), the relation (9.5.37) holds with

$$h_\sigma(v) = (-1)^q \int_{[0,\infty)^q} \frac{\widetilde{g}(y)}{v_1 \ldots v_q} \Re\Big(i^{-q}(e^{iy_1 v_1} - 1) \ldots (e^{iy_q v_q} - 1)\Big) dy$$

$$=: \int_{[0,\infty)^q} \frac{\widetilde{g}(y)}{v_1 \ldots v_q} k_{\sigma,v}(y) dy =: \frac{h_{0,\sigma}(v)}{v_1 \ldots v_q}. \quad (9.5.40)$$

Since E is diagonal, the function $h_\sigma(v)$ is $E^*/(H - b/2)$–homogeneous. Indeed, by letting $E^* = \mathrm{diag}(e_1^*, \ldots, e_q^*)$ and using

$$k_{\sigma, cE^*/(H-b/2)v}(y) = k_{\sigma,v}(c^{E^*/(H-b/2)}y),$$

note that

$$h_\sigma(c^{E^*/(H-b/2)}v) = h_\sigma(c^{e_1^*/(H-b/2)}v_1, \ldots, c^{e_q^*/(H-b/2)}v_q)$$

$$= \int_{[0,\infty)^q} \frac{\widetilde{g}(y)}{c^{b/(H-b/2)}v_1 \ldots v_q} k_{\sigma, cE^*/(H-b/2)v}(y) dy$$

$$= \int_{[0,\infty)^q} \frac{\widetilde{g}(y)}{c^{b/(H-b/2)}v_1 \ldots v_q} k_{\sigma,v}(c^{E^*/(H-b/2)}y) dy$$

$$= \int_{[0,\infty)^q} \frac{\widetilde{g}(c^{-E^*/(H-b/2)}x)}{c^{b/(H-b/2)}v_1 \ldots v_q} k_{\sigma,v}(x) c^{-b/(H-b/2)} dx$$

$$= \int_{[0,\infty)^q} c^{(H+3b/2)/(H-b/2)} \frac{\widetilde{g}(x)}{v_1 \ldots v_q} k_{\sigma,v}(x) c^{-2b/(H-b/2)} dx = c h_\sigma(v),$$

by using the $-E^*/(H + 3b/2)$–homogeneity of $\widetilde{g}$ and (9.5.40). It is then enough to show that the function $h_\sigma(v)$, or equivalently, the function $h_{0,\sigma}(v)$ in (9.5.40) is well-defined as a Lebesgue integral.

Note that, by the change of variables formula (9.5.12) with $f(y) = |\widetilde{g}(y)||k_{\sigma,v}(y)|$ and $M = -E^*/(H + 3b/2)$,

$$|h_{0,\sigma}(v)| \leq \int_{[0,\infty)^q} |\widetilde{g}(y)||k_{\sigma,v}(y)| dy = \int_0^\infty \int_{S_0} |\widetilde{g}(r^M\theta)||k_{\sigma,v}(r^M\theta)|\sigma(d\theta) r^{\mathrm{tr}(M)-1} dr$$

$$= \int_0^\infty dr\, r^{-\frac{b}{H+3b/2}-1} r\Big(\int_{S_0} |\widetilde{g}(\theta)||k_{\sigma,v}(r^{-\frac{E^*}{H+3b/2}}\theta)|\sigma(d\theta)\Big), \quad (9.5.41)$$

where we used the facts that $\widetilde{g}(r^{-\frac{E^*}{H+3b/2}}\theta) = r\widetilde{g}(\theta)$ by homogeneity and that

$$r^{\mathrm{tr}(M)} = r^{-\mathrm{tr}(E^*)/(H+3b/2)} = r^{-b/(H+3b/2)}.$$

Around $r = 0$, by bounding the integral over S_0 by a constant, the integrand behaves as $r^{-b/(H+3b/2)}$ and is integrable since

$$-\frac{b}{H+3b/2} + 1 = \frac{H+b/2}{H+3b/2} > 0$$

(by (9.5.22), $b = \mathrm{tr}(E^*) = \mathrm{tr}(E) > qH > 0$). For the behavior around $r = \infty$, observe that, for $y_1 \geq 0, \ldots, y_q \geq 0$, $|k_{\sigma,v}(y)| \leq C y_1 \ldots y_q \leq C(y_1 + \cdots + y_q)^q \leq C'\|y\|^q$. Then, the integrand in (9.5.41) is bounded by

$$r^{-\frac{b}{H+3b/2}}\|r^{-\frac{E}{H+3b/2}}\|^q \leq Cr^{-\frac{b}{H+3b/2}}r^{-\frac{q(a_1-\delta)}{H+3b/2}}$$

for arbitrarily small $\delta > 0$, where we used the bound (9.5.13) with $a_1 = \min\{\Re(e_j) : j = 1,\ldots,q\}$. The integrability around $r = \infty$ follows since

$$-\frac{b+qa_1}{H+3b/2}+1 = \frac{-qa_1+H+b/2}{H+3b/2} < 0,$$

in view of the assumption (9.5.32), which states that $H + b/2 < qa_1$. $\square$

Remark 9.5.13 It is illuminating to revisit the technical difficulties involved in the proof of Theorem 9.5.11 by considering the univariate context $q = 1$. In this case, one can pick the natural homogeneous specification $g_0(x) = c|x|^{-H-1/2}$ where $H = 1/E < 1$. Under such choice, the integral (9.5.35) can be recast as

$$c\int_{\mathbb{R}} (\cos(x(t-u)) - \cos(x(-u)))|x|^{-H-1/2}dx = \varphi(t-u) - \varphi(-u),$$

where

$$\varphi(v) = \begin{cases} c\int_{\mathbb{R}} \cos(xv)|x|^{-H-1/2}dx, & \text{if } H < 1/2, \\ c\int_{\mathbb{R}} (\cos(xv)-1)|x|^{-H-1/2}dx, & \text{if } H > 1/2. \end{cases} \quad (9.5.42)$$

The expression (9.5.42) is interpreted as involving an improper Riemann integral when $H < 1/2$ and a Lebesgue integral when $H > 1/2$. Neither interpretation carries over to the spatial case $q \geq 2$. In view of (9.5.20), the integral (9.5.35) is

$$\int_{\mathbb{R}^q} (\cos(\langle x, t-u\rangle) - \cos(\langle x, -u\rangle))g(x)^{-H-\text{tr}(E^*)/2}dx, \quad (9.5.43)$$

for an E^*-homogenous function g. On one hand, when $q \geq 2$, the homogeneous function g is no longer a simple power function as above, so that the integral

$$\int_{\mathbb{R}^q} \cos(\langle x, v\rangle)g(x)^{-H-\text{tr}(E^*)/2}dx$$

cannot be interpreted as an improper Riemann integral. On the other hand, one could subtract 1 from each cos in (9.5.43) and check when the integral $\int_{\mathbb{R}^q}(\cos(\langle x, v\rangle) - 1) g(x)^{-H-\text{tr}(E^*)/2}dx$ is well-defined in the Lebesgue sense; that is,

$$J = \int_{\mathbb{R}^q} |\cos(\langle x, v\rangle) - 1||g(x)^{-H-\text{tr}(E^*)/2}|dx < \infty. \quad (9.5.44)$$

To do so, we use (9.5.12) and the E^*-homogeneity of g to get

$$J = \int_0^\infty \int_{S_0} |\cos(\langle t, r^{E^*}\theta\rangle) - 1|r^{-H-\text{tr}(E^*)/2}|g(\theta)^{-H-\text{tr}(E^*)/2}|\sigma(d\theta)r^{\text{tr}(E^*)-1}dr.$$

We now adapt the argument following (9.5.30), with the difference that the integrand is raised to the power 1 and not 2. This leads to the integrability condition at $r = \infty$ as $-H-\text{tr}(E^*)/2+\text{tr}(E^*)-1 < -1$; that is, $\text{tr}(E^*) < 2H$. However, the condition (9.5.22) for the construction of a harmonizable representation requires $\min\{\Re(e_j) : j = 1, \ldots, q\} > H$. But this implies that

$$\mathrm{tr}(E^*) = \sum_j e_j = \sum_j \Re(e_j) \geq q \min\{\Re(e_j) : j = 1, \ldots, q\} > qH.$$

The conditions $\mathrm{tr}(E^*) < 2H$ and $\mathrm{tr}(E^*) > qH$ are not compatible for $q \geq 2$, and as a consequence, (9.5.44) can only play the role of the fractional filter $\varphi(v)$ in the univariate context, namely, when $q = 1$.

We also note that the condition (9.5.32) reduces to $H < 1/2$ when $q = 1$, and thus does not cover the case $H > 1/2$.

The next result provides conditions for (9.5.33) to be well-defined and hence be an OFBF with exponents E and H. The statement and proof follow Theorem 3.1 in Biermé et al. [151]. The following notion will be used.

Definition 9.5.14 Let $\beta > 0$ and $E \in \mathbb{R}^{q \times q}$. A continuous function $\phi : \mathbb{R}^q \to [0, \infty)$ is called (β, E)–*admissible* if $\phi(x) > 0$ for all $x \neq 0$ and, for any $0 < A < B$, there is a constant $C > 0$ such that, for any $A \leq \|y\| \leq B$,

$$\tau_E(x) \leq 1 \Rightarrow |\phi(x+y) - \phi(y)| \leq C\tau_E(x)^\beta, \qquad (9.5.45)$$

where $\tau_E(x)$ is the radial part of E, as in (9.5.6), where the matrix is denoted M.

Example 9.5.15 (*Admissible functions*) The E–homogeneous function ϕ in Example 9.5.8 can be shown to be (β, E)–admissible for $\beta < \min\{e_1, \rho\frac{e_1}{e_q}\}$ if $e_1 \leq \rho$, and $\beta = \rho$ if $e_1 > \rho$ (Exercise 9.13).

Theorem 9.5.16 *Let $\beta > 0$ and $E \in \mathbb{R}^{q \times q}$ with its eigenvalues having positive real parts. Let also $\phi : \mathbb{R}^q \to [0, \infty)$ be an E–homogeneous, (β, E)–admissible function with $\beta > H > 0$. Then, the random field*

$$\int_{\mathbb{R}^q} \left(\phi(t-u)^{H - \mathrm{tr}(E)/2} - \phi(-u)^{H - \mathrm{tr}(E)/2} \right) B(du), \quad t \in \mathbb{R}^q, \qquad (9.5.46)$$

is a well-defined OFBF with the exponents E and H, where $B(du)$ is a Gaussian random measure on $\mathbb{R}^q$ with the Lebesgue control measure $\mathbb{E}|B(du)|^2 = du$.

Proof It is enough to show that the random field (9.5.46) is well defined; that is, for fixed $t \in \mathbb{R}^q$,

$$\int_{\mathbb{R}^q} \left| \phi(t-u)^{H - \frac{b}{2}} - \phi(-u)^{H - \frac{b}{2}} \right|^2 du < \infty,$$

where $b = \mathrm{tr}(E)$. By the inequality $(a-b)^2 \leq 2a^2 + 2b^2$,

$$\left| \phi(t-u)^{H - \frac{b}{2}} - \phi(-u)^{H - \frac{b}{2}} \right|^2 \leq 2(\phi(t-u)^{2H-b} + \phi(-u)^{2H-b})$$
$$\leq C(\tau_E(t-u)^{2H-b} + \tau_E(u)^{2H-b}),$$

where we used the representation (9.5.9), the fact that τ is symmetric (see (9.5.7)) and the continuity of ϕ (in the definition of admissibility). For any $R > 0$, by using (9.5.12) and (9.5.8),

$$\int_{\tau_E(u)\leq R} \tau_E(u)^{2H-b} du = \int_0^\infty \int_{S_0} 1_{\{\tau_E(r^E\theta)\leq R\}} \tau_E(r^E\theta)^{2H-b} \sigma(d\theta) r^{b-1} dr$$

$$= \int_0^\infty \int_{S_0} 1_{\{r\tau_E(\theta)\leq R\}} r^{2H-b} \tau_E(\theta)^{2H-b} \sigma(d\theta) r^{b-1} dr = \sigma(S_0) \int_0^R r^{2H-1} dr < \infty.$$
(9.5.47)

One can show (Lemma 2.2 in Biermé et al. [151]) that $\{u : \tau_E(t-u) \leq R\} \subset \{u : \tau_E(u) \leq K(R+\tau_E(t))\}$ for some constant K. Then, by the change of variables and arguing as above,

$$\int_{\tau_E(u)\leq R} \tau_E(t-u)^{2H-b} du = \int_{\tau_E(t-u)\leq R} \tau_E(u)^{2H-b} du$$

$$\leq \int_{\tau_E(u)\leq K(R+\tau_E(t))} \tau_E(u)^{2H-b} du < \infty.$$

It remains to show that for some $R = R(t) > 0$, we have

$$\int_{\tau_E(u)>R} \left|\phi(t+u)^{H-\frac{b}{2}} - \phi(u)^{H-\frac{b}{2}}\right|^2 du < \infty.$$

Observe that for $\tau_E(u) > R$, we have by Definition 9.5.14 that $\phi(u) > 0$. Hence, we can write

$$\phi(t+u) = \phi\Big(\phi(u)^E(\phi(u)^{-E}t + \phi(u)^{-E}u)\Big) = \phi(u)\phi\big(\phi(u)^{-E}t + \phi(u)^{-E}u\big), \quad (9.5.48)$$

where in the last relation, we used the E-homogeneity of ϕ to write $\phi(c^E y) = c\phi(y)$, with $c = \phi(u)$ and $y = \phi(u)^{-E}(t+u)$. By using (9.5.9) and (9.5.6), observe also that $\|\phi(u)^{-E}u\| = \|\tau_E(u)^{-E}\phi(l_E(u))^{-E}\tau_E(u)^E l_E(u)\| = \|\phi(l_E(u))^{-E} l_E(u)\|$ and hence, since ϕ is continuous and positive by Definition 9.5.14, we have

$$0 < A \leq \|\phi(u)^{-E}u\| \leq B, \quad u \in \mathbb{R}^q. \quad (9.5.49)$$

Similarly, with $a_1 = \min\{\Re(e_j) : j = 1, \ldots, q\}$ and eigenvalues e_j of E, we have by (9.5.13) that $\|\phi(u)^{-E}t\| = \|(\phi(u)\tau_E(t))^{-E} l_E(t)\| \leq C(\phi(u)\tau_E(t))^{-(a_1-\delta)}$ for arbitrarily small $\delta > 0$, as long as $(\phi(u)\tau_E(t))^{-1} \leq 1$. This yields

$$\|\phi(u)^{-E}t\| \leq C'(\tau_E(u))^{-(a_1-\delta)} \leq 1, \quad (9.5.50)$$

for $\tau_E(u) > R$ with large enough R. Since ϕ is (β, E)–admissible and $\phi(\phi(u)^{-E}u) = \phi(u)^{-1}\phi(u) = 1$, the relations (9.5.49) and (9.5.50) imply that there is $C > 0$ such that

$$|\phi(\phi(u)^{-E}t + \phi(u)^{-E}u) - 1| = |\phi(\phi(u)^{-E}t + \phi(u)^{-E}u) - \phi(\phi(u)^{-E}u)|$$

$$\leq C\tau_E(\phi(u)^{-E}t)^\beta = C\phi(u)^{-\beta}\tau_E(t)^\beta, \quad (9.5.51)$$

where we also used the E-homogeneity of τ_E. Since ϕ is E-homogeneous, $\phi(u) = \phi(\tau_E(u)^E l_E(u)) = \tau_E(u)\phi(l_E(u))$, where $\phi(l_E(u))$ is bounded, and since $\beta > 0$, we can choose R large enough so that $C\phi(u)^{-\beta}\tau_E(t)^\beta < 1/2$ for all $\tau_E(u) > R$. By using (9.5.48),

$$|\phi(t+u)^{H-\frac{b}{2}} - \phi(u)^{H-\frac{b}{2}}| = \phi(u)^{H-\frac{b}{2}}\left|\phi(\phi(u)^{-E}t + \phi(u)^{-E}u)^{H-\frac{b}{2}} - 1^{H-\frac{b}{2}}\right|$$

$$= \phi(u)^{1-\frac{b}{2}}|H - \frac{b}{2}|v_*^{H-\frac{b}{2}-1}\left|\phi(\phi(u)^{-E}t + \phi(u)^{-E}u) - 1\right|$$

by applying the mean value theorem to the function $f(v) = v^{1-b/2}$, where v_* is a point between $v = 1$ and $v = \phi(\phi(u)^{-E}t + \phi(u)^{-E}u)$. Since $1/2 < v_* < 3/2$ by (9.5.51) and the observation following it for large enough R, we deduce that

$$|\phi(t+u)^{H-\frac{b}{2}} - \phi(u)^{H-\frac{b}{2}}| \leq C_0 \phi(u)^{H-\frac{b}{2}}\left|\phi(\phi(u)^{-E}t + \phi(u)^{-E}u) - 1\right|$$

$$\leq C_1 \phi(u)^{H-\frac{b}{2}-\beta}\tau_E(t)^\beta \leq C_2 \phi(u)^{H-\frac{b}{2}-\beta}$$

by (9.5.51), for all $\tau_E(u) > R$. Finally, by arguing in the same way as for (9.5.47) in the case $\tau_E(u) \leq R$ above, we conclude that

$$\int_{\tau_E(u)>R}\left|\phi(t+u)^{H-\frac{b}{2}} - \phi(u)^{H-\frac{b}{2}}\right|^2 du \leq C\int_{\tau_E(u)>R}\phi(u)^{2H-b-2\beta}du$$

$$\leq C'\int_{\tau_E(u)>R}\tau_E(u)^{2H-b-2\beta}du$$

$$= C''\int_R^\infty r^{2H-b-2\beta}r^{b-1}dr = C''\int_R^\infty r^{2H-2\beta-1}dr < \infty,$$

since $\beta > H$ by assumption. $\square$

9.5.3 Special Subclasses and Examples of OFBFs

It is instructive to describe first the covariance structure of an OFBF $B_{E,H}$. Setting $\sigma_\theta^2 = \mathbb{E}B_{E,H}(\theta)^2$, for $\theta \in S_0$, by the operator self-similarity (9.5.1) and the polar coordinate representation (9.5.6), we have that $\mathbb{E}B_{E,H}(t)^2 = \mathbb{E}B_{E,H}((\tau_E(t))^E l_E(t))^2 = \tau_E(t)^{2H}\mathbb{E}B_{E,H}(l_E(t))^2 = \tau_E(t)^{2H}\sigma_{l_E(t)}^2$. As in the univariate case, writing $2B_{E,H}(t)B_{E,H}(s) = B_{E,H}(t)^2 + B_{E,H}(s)^2 - (B_{E,H}(t) - B_{E,H}(s))^2$ and using the stationarity of increments, this yields the following result.

Proposition 9.5.17 *Let $B_{E,H} = \{B_{E,H}(t)\}_{t\in\mathbb{R}^q}$ be an OFBF. Then,*

$$\mathbb{E}B_{E,H}(t)B_{E,H}(s) = \frac{1}{2}\left\{\tau_E(t)^{2H}\sigma_{l_E(t)}^2 + \tau_E(s)^{2H}\sigma_{l_E(s)}^2 - \tau_E(t-s)^{2H}\sigma_{l_E(t-s)}^2\right\}, \quad s, t \in \mathbb{R}^q. \tag{9.5.52}$$

We next give several popular examples of OFBFs.

Example 9.5.18 (*Isotropic FBFs*) Take $E = E^* = e_1 \cdot I$, $H > 0$ with $e_1 > H$. By Example 9.5.5, one can take $\tau_{E^*}(x) = \|x\|_2^{1/e_1}$, $l_{E^*}(x) = x/\|x\|_2$. The spectral representation (9.5.19) of OFBF then becomes

$$\{B_{E,H}(t)\}_{t\in\mathbb{R}^q} \stackrel{d}{=} \left\{\int_{\mathbb{R}^q}(e^{i\langle x,t\rangle} - 1)\|x\|_2^{-\frac{H}{e_1}-\frac{q}{2}}g_0\left(\frac{x}{\|x\|_2}\right)\widehat{B}(dx)\right\}_{t\in\mathbb{R}^q}. \tag{9.5.53}$$

The OFBF in (9.5.53) is well-defined by Theorem 9.5.9 (supposing g_0 is bounded) and is called an *isotropic fractional Brownian field* (isotropic FBF). Note that it satisfies

$$\{B_{E,H}(c^{e_1}t)\}_{t\in\mathbb{R}^q} \stackrel{d}{=} \{c^H B_{E,H}(t)\}_{t\in\mathbb{R}^q}.$$

Equivalently, $B_{E,H}(ct)$ and $B_{E,H}(c^{H_E}t)$ have the same finite-dimensional distributions, where $H_E = H/e_1 < 1$.

When $g_0(\theta) = c$, an isotropic FBF is referred to as simply an FBF.[11] It has the spectral representation

$$\{B_{E,H}(t)\}_{t\in\mathbb{R}^q} \stackrel{d}{=} \left\{c_0 \int_{\mathbb{R}^q} (e^{i\langle x,t\rangle} - 1)\|x\|_2^{-\frac{H}{e_1}-\frac{q}{2}} \widehat{B}(dx)\right\}_{t\in\mathbb{R}^q} \quad (9.5.54)$$

and, by Proposition 9.5.17, the covariance structure

$$\mathbb{E} B_{E,H}(t) B_{E,H}(s) = \frac{\sigma^2}{2}\left\{\|t\|_2^{2H/e_1} + \|s\|_2^{2H/e_1} - \|t-s\|_2^{2H/e_1}\right\}, \quad s,t \in \mathbb{R}^q.$$

An approximate sample realization of the field (9.5.54) with $q = 2$, $e_1 = 4/3$ and $H = 1$ (equivalently, $e_1 = 1$ and $H = 3/4$) is given in Figure 9.2, left plot. It is obtained by discretizing the integral (9.5.54) and using the two-dimensional FFT to make the computations numerically more efficient.

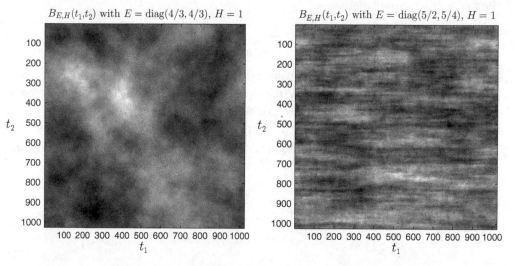

Figure 9.2 Sample realizations of OFBFs $B_{E,H}$. Left: The isotropic FBF (9.5.54) with $E = \mathrm{diag}(4/3, 4/3)$ and $H = 1$ (equivalently, $E = I$ and $H = 3/4$). Right: The anisotropic FBF (9.5.55) with $E = \mathrm{diag}(5/2, 5/4)$ and $H = 1$ (equivalently, $E = \mathrm{diag}(4/3, 2/3)$ with $\mathrm{tr}(E) = 2$ and $H = 8/15$). (The plots are in gray color map, where the white color represents the largest values and the black the smallest values.)

---

[11] Strictly speaking, only an FBF having the representation (9.5.54) could be called isotropic, since only for this field, the function g in its spectral representation depends solely on $\|x\|_2$. We call the fields having the representation (9.5.53) isotropic as well following Lavancier [595].

Example 9.5.19 (*Anisotropic FBFs*) OFBFs with $E = \text{diag}(e_1, \ldots, e_q) \neq e \cdot I$, $e_j > H > 0$, and hence not having the integral representation (9.5.53) are generally referred to as *anisotropic fractional Brownian fields* (anisotropic FBFs). For example, the random field

$$\int_{\mathbb{R}^q} (e^{i\langle x,t \rangle} - 1) \Big(\sum_{j=1}^{q} |x_j|^{\frac{1}{e_j}} \Big)^{-H - \frac{\text{tr}(E)}{2}} \widehat{B}(dx), \quad t \in \mathbb{R}^q, \tag{9.5.55}$$

is an anisotropic FBF. To see that it is a well-defined OFBF with the exponents $E = \text{diag}(e_1, \ldots, e_q)$ and H, use Theorem 9.5.9. The function g is E–homogeneous, since for $c > 0$, $g(c^E x) = \sum_{j=1}^{q} |c^{e_j} x_j|^{1/e_j} = c \sum_{j=1}^{q} |x_j|^{1/e_j} = cg(x)$. An approximate sample realization of the field (9.5.55) with $q = 2$ and $e_1 = 5/2$, $e_2 = 5/4$, $H = 1$ (equivalently, $e_1 = 4/3$, $e_2 = 2/3$, so that (9.5.3) holds, and $H = 8/15$), is given in Figure 9.2, right plot. Observe the anisotropic nature of the realization (cf. left plot).

The next example describes an interesting random field related to OFBFs.

Example 9.5.20 (*Fractional Brownian sheet*) Let $0 < H_1, \ldots, H_q < 1$. Consider a stochastic process X given by the spectral representation

$$\{X(t)\}_{t \in \mathbb{R}^q} \stackrel{d}{=} \Big\{ c \int_{\mathbb{R}^q} \prod_{j=1}^{q} \frac{e^{ix_j t_j} - 1}{|x_j|^{H_j + 1/2}} \widehat{B}(dx) \Big\}_{t \in \mathbb{R}^q}, \tag{9.5.56}$$

where $\widehat{B}(dx)$ is a Gaussian Hermitian random measure on $\mathbb{R}^q$ with the control measure dx. The function $(e^{ix_j t_j} - 1)/|x_j|^{H_j + 1/2}$ is the kernel function appearing in the spectral representation (2.6.9) of FBM with parameter H_j. As for the univariate FBM, the process (9.5.56) is well-defined, and has the covariance structure

$$\mathbb{E} X(t) X(s) = c^2 \Big(\prod_{j=1}^{q} c_2(H_j)^2 \Big) \prod_{j=1}^{q} \frac{1}{2} \Big\{ |t_j|^{2H_j} + |s_j|^{2H_j} - |t_j - s_j|^{2H_j} \Big\}, \tag{9.5.57}$$

where $c_2(H)$ appears in (2.6.9). It is called a *fractional Brownian sheet* (FBS).

FBS satisfies the operator self-similarity relation (9.5.1) with $E = \text{diag}(1/H_1, \ldots, 1/H_q)$ and $H = q$. Note, however, that FBS is not an OFBF as it does not have stationary increments. Even though FBS does not have stationary increments, it can still be used to construct discrete parameter stationary fields. For $j = 1, \ldots, q$, and a process $Y(t) = Y(t_1, \ldots, t_q)$, set

$$\Delta_j Y(t) = \Delta_j Y(t_1, \ldots, t_q) = Y(t_1, \ldots, t_j, \ldots, t_q) - Y(t_1, \ldots, t_j - 1, \ldots, t_q).$$

For FBS X, consider

$$U_n = \Delta_1 \ldots \Delta_q X(n), \quad n \in \mathbb{Z}^q.$$

One can check easily that $\{U_n\}_{n \in \mathbb{Z}^q}$ is stationary.

Finally, we also note that an OFBF analogue of FBS has the spectral representation

$$\Big\{ c \int_{\mathbb{R}^q} (e^{i\langle x,t \rangle} - 1) \prod_{j=1}^{q} |x_j|^{-H_j - 1/2} \widehat{B}(dx) \Big\}_{t \in \mathbb{R}^q}. \tag{9.5.58}$$

One can check that this OFBF is well-defined when $0 < H_j < 1$ (Exercise 9.10).

9.6 Spatial Long-Range Dependence

A general notion of vector LRD was given in Section 9.4. In contrast, a general definition of LRD is not yet available in the spatial context. One reason for this may be explained through the analogy to vector OFBMs. In the vector setting, a homogeneous function $\phi : \mathbb{R} \to \mathbb{C}^p$ satisfies $\phi(cu) = c^M \phi(u)$ and may be expressed explicitly as $\phi(v) = v_+^M \phi_1 + v_-^M \phi_2$ with $\phi_1, \phi_2 \in \mathbb{C}^p$, leading to the representations (9.3.3) and (9.3.22), and then to Definition 9.4.1 of vector LRD in Section 9.4. In the spatial setting, there are no such explicit representations for a homogeneous function $\phi : \mathbb{R}^q \to \mathbb{C}$ satisfying $\phi(c^E x) = c\phi(x)$. Since a general definition of spatial LRD is not available (at least for practical purposes), we shall focus on several special cases, namely, the isotropic case and examples of anisotropic spatial LRD. Even the isotropic case is already mathematically quite rich and interesting! The basics of stationary spatial time series are recalled in Section 9.1.3.

9.6.1 Definitions and Basic Properties

The following are conditions for spatial LRD in the special case of isotropy. The suffix "-s" in the conditions refers to "spatial", in order to distinguish them from the conditions in Sections 2.1 and 9.4. The suffix "-isot" in the conditions refers to "isotropic." Let $d \in (0, 1/2)$ be a parameter.

Condition II-s-isot The autocovariance function of the random field $X = \{X_n\}_{n \in \mathbb{Z}^q}$ satisfies

$$\gamma_X(n) = L_2(\|n\|_2) \|n\|_2^{(2d-1)q} b_2\left(\frac{n}{\|n\|_2}\right), \quad n \in \mathbb{Z}^q, \quad (9.6.1)$$

where L_2 is a slowly varying function at infinity, $b_2 : S^{q-1} \to (0, \infty)$ is a function on the unit sphere S^{q-1}, and $\|\cdot\|_2$ is the usual Euclidean norm.

Condition IV-s-isot The random field $X = \{X_n\}_{n \in \mathbb{Z}^q}$ has a spectral density satisfying

$$f_X(\lambda) = L_4(\|\lambda\|_2) \|\lambda\|_2^{-2dq} b_4\left(\frac{\lambda}{\|\lambda\|_2}\right), \quad \lambda \in (0, \pi]^q, \quad (9.6.2)$$

where L_4 is a slowly varying function at zero and $b_4 : S^{q-1} \to (0, \infty)$ is a function on the unit sphere S^{q-1}.

Definition 9.6.1 A second-order stationary random field $X = \{X_n\}_{n \in \mathbb{Z}^q}$ is called *isotropic long-range dependent* (isotropic LRD, in short) if one of the non-equivalent conditions II-s-isot and IV-s-isot above holds with $d \in (0, 1/2)$. (For the terminology used, see also the footnote in Example 9.5.18.)

The relationship between conditions II-s-isot and IV-s-isot was studied by Wainger [993] (for a more recent account, see Leonenko and Olenko [610]). It involves the so-called *spherical harmonics*, which are special functions on the unit sphere S^{q-1}. See, for example, Andrews et al. [35], Atkinson and Han [55]. The functions are labeled and denoted as $Y_{l,m}(u')$ with

$$u' = \frac{u}{\|u\|_2} \in S^{q-1},$$

where $l = 0, 1, 2, \ldots$ is the degree and $m = 1, \ldots, n_{q,l}$ is the index referring to spherical harmonics of degree l. They form an orthonormal basis with respect to the inner product

$$\int_{S^{q-1}} Y_{l,m}(u') Y_{l',m'}(u') \sigma(du') = \delta_{ll'} \delta_{mm'},$$

where $\sigma(du')$ denotes the measure over the sphere S^{q-1} which is invariant under orthogonal transformations and satisfies $\int_{S^{q-1}} \sigma(du') = 2\pi^{q/2}/\Gamma(q/2)$ (see, for example, (1.19) in Atkinson and Han [55]). The number $n_{q,l}$ is known (e.g., (2.10) in Atkinson and Han [55]) but will not be used here in a general form. For example, for $l \geq 1$,

$$n_{2,l} = 2 \ (q = 2), \quad n_{3,l} = 2l + 1 \ (q = 3),$$

whereas $n_{q,0} = 1$ for all $q \geq 1$ with $Y_{0,1}(u') \equiv (\Gamma(q/2)/2\pi^{q/2})^{1/2}$.
When $q = 2$, one can take $Y_{0,1}(u') = 1/(2\pi)^{1/2}$ and, for $l \geq 1$,

$$Y_{l,1}(u') = \frac{\cos(l\theta)}{\sqrt{\pi}}, \quad Y_{l,2}(u') = \frac{\sin(l\theta)}{\sqrt{\pi}}, \tag{9.6.3}$$

where $u' = (\cos(\theta), \sin(\theta))$, $\theta \in [0, 2\pi)$, is the polar representation of $u' \in S^1$ (e.g., Atkinson and Han [55], bottom of p. 22). When $q = 3$, one can take $Y_{0,1}(u') \equiv (\Gamma(3/2)/2\pi^{3/2})^{1/2}$ and, for $l \geq 1$, the $2l + 1$ spherical harmonics as (e.g., Andrews et al. [35], p. 457, after correcting several typos)

$$A_0 P_l^0(\cos\theta), \ A_m \cos(m\phi) P_l^m(\cos\theta), \ A_m \sin(m\phi) P_l^m(\cos\theta), \quad m = 1, \ldots, l, \tag{9.6.4}$$

where the normalizations are

$$A_0 = \sqrt{\frac{2l+1}{4\pi}}, \quad A_m = \sqrt{\frac{(l-m)!(2l+1)}{(l+m)!2\pi}},$$

$u' = (\cos(\theta)\sin(\phi), \sin(\theta)\sin(\phi), \cos(\phi))$, $\theta \in [0, 2\pi)$, $\phi \in [0, \pi)$, is the spherical representation of $u' \in S^2$, and $P_l^m(x)$ are the associated Legendre polynomials (e.g., (9.6.7) in Andrews et al. [35]). The functions in (9.6.4) are often also labeled as Y_l^0, Y_l^m, Y_l^{-m}, $m = 1, \ldots, l$.

Suppose that $b_2 \in C(S^{q-1})$ (with b_2 appearing in (9.6.1)) and consider its expansion in spherical harmonics

$$b_2(u') = \sum_{l,m} a(l,m) Y_{l,m}(u'), \quad u' \in S^{q-1}, \tag{9.6.5}$$

where the convergence is uniform in $C(S^{q-1})$ (e.g., Atkinson and Han [55]). If $b_2 \in C^\infty(S^{q-1})$, as assumed in Wainger [993], then the coefficients $a(l,m)$ have an exponential decay in l.

Suppose that $b_2 \in C^\infty(S^{q-1})$ and that L_2 is infinitely differentiable on $[0, \infty)$ such that $h_0(t) := L_2(t)$ and $h_j(t) := t h'_{j-1}(t)$, $j = 1, 2, \ldots$, belong to the Zygmund class.[12] Under

---

[12] A (positive) function f belongs to the Zygmund class if, for every $\delta > 0$, $t^\delta f(t)$ is ultimately increasing and $t^{-\delta} f(t)$ is ultimately decreasing. The Zygmund class is known to coincide with the so-called normalized slowly varying functions (Bingham et al. [157], p. 24).

these assumptions, Wainger [993], Theorems 6 and 7, showed that condition II-s-isot implies condition IV-s-isot with

$$L_4(t) = L_2\left(\frac{1}{t}\right) \tag{9.6.6}$$

and

$$b_4(u') = 2^{(1-2d)q}\pi^{-q}\sum_{l,m}(-i)^l a(l,m)\frac{\Gamma(\frac{1}{2}(l-(2d-2)q))}{\Gamma(\frac{1}{2}(l+(2d-1)q))}Y_{l,m}(u'), \tag{9.6.7}$$

where the coefficients $a(l,m)$ appear in (9.6.5).

Example 9.6.2 (*Relation between functions b_2 and b_4*) If $b_2(u') \equiv c_2$ (constant), then $a(0,1)$ is the only nonzero coefficient in (9.6.5) and the relation (9.6.7) reads

$$b_4(u') = 2^{(1-2d)q}\pi^{-q}\frac{\Gamma((1-d)q)}{\Gamma(\frac{q}{2}(2d-1))}c_2.$$

Example 9.6.3 (*Relation between functions b_2 and b_4*) Let $q = 3$,

$$u' = (\cos(\theta)\sin(\phi), \sin(\theta)\sin(\phi), \cos(\phi)), \quad \theta \in [0, 2\pi), \ \phi \in [0, \pi),$$

be the spherical representation of $u' \in S^2$ and $b_2(u') \equiv c_2(4 + 3\cos^2\phi)$. By using the notation following (9.6.4), one has $Y_0^0(u') \equiv \frac{1}{2\sqrt{5}}$ and $Y_2^0(u') = \frac{1}{4}\sqrt{\frac{5}{\pi}}(3\cos^2\phi - 1)$, so that

$$b_2(u') = a_0 Y_0^0(u') + a_2 Y_2^0(u')$$

with $a_0 = c_2 2\sqrt{5} \cdot 5$, $a_2 = c_2 4\sqrt{\frac{\pi}{5}}$, and hence the relation (9.6.7) reads

$$b_4(u') = 2^{(1-2d)3}\pi^{-3}\left(a_0\frac{\Gamma(3-3d)}{\Gamma(3d-3/2)}Y_0^0(u') + a_2\frac{\Gamma(4-3d)}{\Gamma(3d-1/2)}Y_2^0(u')\right)$$

$$= 2^{(1-2d)3}\pi^{-3}\frac{\Gamma(3-3d)}{\Gamma(3d-3/2)}\left(a_0 Y_0^0(u') + a_2\frac{1-d}{d-1/2}Y_2^0(u')\right),$$

by using the identity $\Gamma(1+x) = x\Gamma(x)$.

Conditions II-s-isot and IV-s-isot lead to the definition of isotropic LRD fields. *Anisotropic LRD* generally refers to random fields with slowly decaying autocovariance functions or spectral densities with a pole at 0 which do not have the form (9.6.1) or (9.6.2), respectively. We provide next several conditions in the spectral domain which have been taken as special cases of anisotropic LRD. The suffix "-anisot" in the condition refers to "anisotropic."

Condition IV-s-anisot The random field $X = \{X_n\}_{n\in\mathbb{Z}^q}$ has a spectral density satisfying one of the following conditions:

- For $d_1, \ldots, d_q \in (0, 1/2)$ and $C > 0$

$$f(\lambda) \sim C|\lambda_1|^{-2d_1}\ldots|\lambda_q|^{-2d_q}, \quad \text{as } \lambda \to 0. \tag{9.6.8}$$

- For $d \in (0, 1/2)$, $C > 0$ and $a \in \mathbb{R}^q$,

$$f(\lambda) \sim C|\langle a, \lambda\rangle|^{-2d}, \quad \text{as } \lambda \to 0. \tag{9.6.9}$$

- For an isotropic spectral density f_0 satisfying (9.6.2) and a nondegenerate matrix A,

$$f(\lambda) = f_0(\|Aw\|_2). \tag{9.6.10}$$

- For a function $d : S^{q-1} \to (0, 1/2)$ and $C > 0$,

$$f(\lambda) \sim C\|\lambda\|_2^{-2d(\lambda')q}, \quad \text{as } \lambda \to 0, \tag{9.6.11}$$

where $\lambda' = \lambda/\|\lambda\|_2 \in S^{q-1}$.
- For $q = 2$, $\beta_1, \beta_2 > 0$, $\beta_1\beta_2 < \beta_1 + \beta_2$, $c > 0$ and $C > 0$,

$$f(\lambda_1, \lambda_2) \sim \frac{C}{\||\lambda_1|^2 + |\lambda_2|^{2\beta_2/\beta_1}|^{\beta_1/2}}, \quad \text{as } (\lambda_1, \lambda_2) \to 0. \tag{9.6.12}$$

The conditions on the various parameters above are to ensure that the spectral density f is integrable around $\lambda = 0$ (Exercise 9.12).

Random fields satisfying (9.6.8) were considered in Lavancier [595], Beran, Ghosh, and Schell [125], Boissy, Bhattacharyya, Li, and Richardson [168], Guo, Lim, and Meerschaert [440]. The condition (9.6.8) states that there is potentially a different LRD parameter d_j in each direction λ_j. The condition (9.6.9) was considered in Lavancier [595, 596] and represents the divergence of the spectral density along the direction $\langle a, \lambda \rangle = 0$. The condition (9.6.10) can be found in Leonenko and Olenko [610], and the condition (9.6.11) in Bonami and Estrade [173], Bierme and Richard [150].

Random fields satisfying (9.6.12) were considered in Puplinskaitė and Surgailis [835]. The latter work also established interesting scaling properties of aggregate sums of random fields satisfying (9.6.12) or (9.6.8). More specifically, given a random field $X = \{X_n\}_{n \in \mathbb{Z}^2}$, assume that for any $\gamma > 0$, there is a nontrivial random field $V_\gamma = \{V_\gamma(x)\}_{x \in \mathbb{R}^2}$ and a normalization $A_N(\gamma) \to \infty$ such that

$$A_N(\gamma)^{-1} \sum_{n \in K_N^\gamma(x)} X_n \xrightarrow{fdd} V_\gamma(x), \quad x \in \mathbb{R}^2, \, N \to \infty, \tag{9.6.13}$$

where $K_N^\gamma(x) = K_N^\gamma(x_1, x_2) = \{n = (n_1, n_2) \in \mathbb{Z}^2 : 1 \leq n_1 \leq Nx_1, 1 \leq n_2 \leq N^\gamma x_2\}$. Puplinskaitė and Surgailis [835] say that X exhibits *scaling transition* if there is $\gamma_0 > 0$ such that

$$V_\gamma \stackrel{d}{=} V_+, \, \gamma > \gamma_0, \quad V_\gamma \stackrel{d}{=} V_-, \, \gamma < \gamma_0, \quad V_+ \stackrel{d}{\neq} aV_- \text{ (all } a > 0\text{)}, \tag{9.6.14}$$

that is, the distributions of the fields V_γ do not depend on $\gamma > \gamma_0$ and $\gamma < \gamma_0$, and are different up to a multiplicative constant. Under additional suitable assumptions, Puplinskaitė and Surgailis [835] showed that the random fields satisfying (9.6.12) or (9.6.8) exhibit scaling transition.

9.6.2 Examples

We discuss here several examples of classes of spatial LRD series.

Example 9.6.4 (*Spatial LRD fields with explicit autocovariances*) In several instances (e.g., modeling in geostatistics, or simulations), it is desirable to have an explicit form for the autocovariance function of a second-order stationary random field. For example, for what functions $\gamma : [0, \infty) \to \mathbb{R}$, does an isotropic function $\gamma(\|u\|_2)$ become an autocovariance function? This and related questions have been well-studied in the case of general stationary (not necessarily LRD) random fields $\{X_u\}$ index by $u \in \mathbb{R}^q$. (Stationary random fields $\{Y_n\}$ with explicit autocovariances and indexed by $n \in \mathbb{Z}^q$ can then be obtained by setting $Y_n = X_{n\Delta}$ for some $\Delta > 0$.) The following is one classical result in this direction.

It is known (e.g., Schoenberg [890], Yaglom [1020]) that $\gamma(\|u\|_2)$ is the autocovariance function of a stationary random field $\{X_u\}_{u \in \mathbb{R}^q}$ if and only if there is a finite measure $G(d\lambda)$ on $(0, \infty)$ such that

$$\gamma(r) = \int_0^\infty Y_q(\lambda r) G(d\lambda), \qquad (9.6.15)$$

where $Y_q(z)$ is the spherical Bessel function defined as

$$Y_1(z) = \cos z,$$
$$Y_q(z) = 2^{(q-2)/2} \Gamma\left(\frac{q}{2}\right) J_{(q-2)/2}(z) z^{(q-2)/2}, \quad z \geq 0, \ q \geq 2,$$

and for $\nu > -1/2$,

$$J_\nu(z) = \sum_{m=0}^\infty (-1)^m \left(\frac{z}{2}\right)^{2m+\nu} \frac{1}{m\Gamma(m+\nu+1)}, \quad z > 0,$$

is the Bessel function of the first kind of order ν. A function $\gamma(\|u\|_2)$ is the autocovariance function of some $\{X_u\}_{u \in \mathbb{R}^q}$ for *every* q if and only if

$$\gamma(r) = \int_0^\infty e^{-ar^2} F(da), \qquad (9.6.16)$$

for some measure $F(da)$ (that is, when $\gamma(r)$ is a normal scale mixture). A function $f(\|u\|_2^2)$ is the autocovariance function for *every* q if and only if f is completely monotone.[13]

The Cauchy family

$$\gamma(r) = \frac{1}{(1+r^\alpha)^{\beta/\alpha}}, \quad \alpha \in (0, 2], \ \beta > 0, \qquad (9.6.17)$$

is a well-known example of functions satisfying (9.6.16) (e.g., Schlather [888]). Setting $(2d-1)q = -\beta$, the corresponding random field is isotropic LRD when $\beta \in (0, q)$. When $\alpha = 2$, $\gamma(r) = f(r^2)$ with $f(r) = (1+r)^{-\beta/2}$, which can be checked to be completely monotone (Exercise 9.14).

Example 9.6.5 (*Anisotropic spatial FARIMA models*) Univariate FARIMA models considered in Section 2.4.2 can naturally be extended to the context of random fields in the following way. Consider the dimension $q = 2$ only (not to be confused with q in

---

[13] Recall that a function $f : [0, \infty) \to \mathbb{R}$ is completely monotone if it satisfies $(-1)^n f^{(n)}(t) \geq 0$ for any n and $t > 0$, where $f^{(n)}$ denotes the nth derivative of f.

9.6 Spatial Long-Range Dependence

FARIMA$(p, d, q))$ so that a random field is $X = \{X_{n_1,n_2}\}_{n_1,n_2 \in \mathbb{Z}}$. Consider the backshift operators B_1 and B_2 satisfying

$$B_1^k Y_{n_1,n_2} = Y_{n_1-k,n_2}, \quad B_2^k Y_{n_1,n_2} = Y_{n_1,n_2-k}, \quad k \in \mathbb{Z}, \tag{9.6.18}$$

that is, the backshift operators acting on the individual coordinates n_1 and n_2. Denote $I = B_1^0 = B_2^0$.

An *anisotropic spatial* FARIMA$((p_1, p_2), (d_1, d_2), (q_1, q_2))$ random field $X = \{X_{n_1,n_2}\}_{n_1,n_2 \in \mathbb{Z}}$ is defined as

$$X_{n_1,n_2} = (I - B_1)^{-d_1}(I - B_2)^{-d_2} \Psi_1(B_1) \Psi_2(B_2) Z_{n_1,n_2}, \tag{9.6.19}$$

where $\{Z_{n_1,n_2}\}_{n_1,n_2 \in \mathbb{Z}}$ is a spatial white noise satisfying $\mathbb{E} Z_{n_1,n_2} = 0$ and $\mathbb{E} Z_{n_1,n_2} Z_{m_1,m_2} = \sigma_Z^2 \delta_{n_1,m_1} \delta_{n_2,m_2}$,

$$\Psi_1(B_1) = \varphi_1^{-1}(B_1) \theta_1(B_1), \quad \Psi_2(B_2) = \varphi_2^{-1}(B_2) \theta_2(B_2),$$

and, for $j = 1, 2$,

$$\varphi_j(z) = 1 - \varphi_{j,1} z - \cdots - \varphi_{j,p_j} z^{p_j}, \quad \theta_j(z) = 1 + \theta_{j,1} z + \cdots + \theta_{j,q_j} z^{q_j}.$$

For the convergence of the field (9.6.19), it is assumed that $d_1, d_2 < 1/2$ and that the polynomials $\varphi_1(z), \varphi_2(z)$ do not have unit roots on the circle.

The spectral density of the field (9.6.19) is

$$f(\lambda_1, \lambda_2) = \frac{\sigma_Z^2}{4\pi^2} |1 - e^{-i\lambda_1}|^{-2d_1} |1 - e^{-i\lambda_2}|^{-2d_2} \left| \frac{\theta_1(e^{-i\lambda_1})}{\varphi_1(e^{-i\lambda_1})} \right|^2 \left| \frac{\theta_2(e^{-i\lambda_2})}{\varphi_2(e^{-i\lambda_2})} \right|^2. \tag{9.6.20}$$

When $d_1, d_2 \in (0, 1/2)$, it satisfies the condition (9.6.8) in the definition of anisotropic LRD. A different anisotropic FARIMA model is constructed in Example 9.6.7 below.

Example 9.6.6 (*Isotropic spatial FARIMA models*) The spatial FARIMA models considered in Example 9.6.5 are anisotropic as noted above. This is the result of the choice of the fractional integration operator $(I - B_1)^{-d_1}(I - B_2)^{-d_2}$ used in the definition (9.6.19). An isotropic FARIMA model can be obtained similarly by using a different fractional integration operator. More specifically, consider first the difference operator

$$\Delta(B_1, B_2) = \frac{1}{4}(I - B_1 + I - B_2 + I - B_1^{-1} + I - B_2^{-1}), \tag{9.6.21}$$

where $I = B_1^0 = B_2^0$, B_1, B_2 and their powers are defined in (9.6.18), so that

$$\Delta(B_1, B_2) Y_{n_1,n_2} = \frac{1}{4} \sum_{|k_1|+|k_2|=1} (Y_{n_1,n_2} - Y_{n_1+k_1,n_2+k_2})$$

$$= \frac{1}{4}(Y_{n_1,n_2} - Y_{n_1-1,n_2} + Y_{n_1,n_2} - Y_{n_1,n_2-1} + Y_{n_1,n_2} - Y_{n_1+1,n_2} + Y_{n_1,n_2} - Y_{n_1,n_2+1}).$$

An *isotropic spatial* FARIMA$((p_1, p_2), d, (q_1, q_2))$ random field $X = \{X_{n_1,n_2}\}_{n_1,n_2 \in \mathbb{Z}}$ can be defined as

$$X_{n_1,n_2} = \Delta(B_1, B_2)^{-d} \Psi_1(B_1) \Psi_2(B_2) Z_{n_1,n_2}, \tag{9.6.22}$$

where $d < 1/2$ and $\Psi_1(B_1)$, $\Psi_2(B_2)$, Z_{n_1,n_2} are as in (9.6.19). Since

$$|\Delta(e^{-i\lambda_1}, e^{-i\lambda_2})| = \frac{1}{2}|1 - \cos(\lambda_1) + 1 - \cos(\lambda_2)|,$$

an isotropic spatial FARIMA has spectral density

$$f(\lambda_1, \lambda_2) = \frac{\sigma_Z^2}{4\pi^2} 2^{2d} |1 - \cos(\lambda_1) + 1 - \cos(\lambda_2)|^{-2d} \left|\frac{\theta_1(e^{-i\lambda_1})}{\varphi_1(e^{-i\lambda_1})}\right|^2 \left|\frac{\theta_2(e^{-i\lambda_2})}{\varphi_2(e^{-i\lambda_2})}\right|^2. \quad (9.6.23)$$

When $d \in (0, 1/2)$, since $1 - \cos(\lambda) \sim \lambda^2/2$ as $\lambda \to 0$, it satisfies the condition (9.6.2) in the definition of isotropic LRD with the function b_4 being constant.

Example 9.6.7 (*Anisotropic spatial FARIMA models, cont'd*) Other anisotropic FARIMA models can be constructed as in Example 9.6.6 above by replacing the difference operator $\Delta(B_1, B_2)$ in (9.6.21) by its "anisotropic" counterpart. For example, one choice is to take

$$\Delta_\theta(B_1, B_2) = I - \theta B_1 - \frac{1-\theta}{2} B_1(B_2 + B_2^{-1}) = I - B_1 + (1-\theta)B_1\left(I - \frac{B_2 + B_2^{-1}}{2}\right), \quad (9.6.24)$$

where $\theta \in (0, 1)$. Note that

$$|\Delta_\theta(e^{-i\lambda_1}, e^{-i\lambda_2})| = |1 - e^{-i\lambda_1} + (1-\theta)e^{-i\lambda_1}(1 - \cos\lambda_2)|$$
$$= |e^{i\lambda_1} - 1 + (1-\theta)(1 - \cos\lambda_2)|$$
$$\sim |i\lambda_1 + (1-\theta)\lambda_2^2/2| = (\lambda_1^2 + (1-\theta)^2\lambda_2^4/4)^{1/2},$$

as $(\lambda_1, \lambda_2) \to 0$. Then, for example, the field

$$X_{n_1,n_2} = \Delta_\theta(B_1, B_2)^{-d} Z_{n_1,n_2}, \quad (9.6.25)$$

where $d < 3/2$ (see below) and Z_{n_1,n_2} is as in (9.6.19), has the spectral density

$$f(\lambda_1, \lambda_2) = \frac{\sigma_Z^2}{4\pi^2} |e^{i\lambda_1} - 1 + (1-\theta)(1 - \cos\lambda_2)|^{-d} \sim \frac{\sigma_Z^2}{4\pi^2} (\lambda_1^2 + (1-\theta)^2\lambda_2^4/4)^{-d}, \quad (9.6.26)$$

as $(\lambda_1, \lambda_2) \to 0$. When $d \in (0, 3/2)$, it satisfies the condition (9.6.12) in the definition of anisotropic LRD with $\beta_1 = 2d$, $\beta_2 = d$ (so that $\beta_1\beta_2 < \beta_1 + \beta_2$ when $d \in (0, 3/2)$), and can thus also be called an anisotropic FARIMA field.

9.7 Exercises

The symbols * and ** next to some exercises are explained in Section 2.12.

Exercise 9.1 Any matrix $A \in \mathbb{R}^{p \times p}$ has a Jordan decomposition $A = PJP^{-1}$, where $J = \mathrm{diag}(J_{\lambda_1}, \ldots, J_{\lambda_N})$ consists of Jordan blocks J_λ of size n_λ given by (9.1.2). Show that

$$z^{J_\lambda} = \begin{pmatrix} z^\lambda & 0 & 0 & \cdots & 0 \\ (\log z) z^\lambda & z^\lambda & 0 & \cdots & 0 \\ \frac{(\log z)^2}{2!} z^\lambda & (\log z) z^\lambda & z^\lambda & \cdots & 0 \\ \vdots & \vdots & \ddots & \ddots & 0 \\ \frac{(\log z)^{n_\lambda - 1}}{(n_\lambda - 1)!} z^\lambda & \frac{(\log z)^{n_\lambda - 2}}{(n_\lambda - 2)!} z^\lambda & \cdots & (\log z) z^\lambda & z^\lambda \end{pmatrix}. \qquad (9.7.1)$$

Hint: Use $z^{J_\lambda} = e^{(\log z) J_\lambda} = \sum_{k=0}^{\infty} (\log z)^k J_\lambda^k / k!$ and evaluate J_λ^k.

Exercise 9.2 Show that the definitions of e^A, $A \in \mathbb{R}^{p \times p}$, in (9.2.3) and as a primary matrix function through Definition 9.3.4 are equivalent.

Exercise 9.3 Let F_X be the spectral distribution matrix function of a stationary vector-valued process $X = \{X(t)\}_{t \in \mathbb{R}}$. Show that F_X has at most a countable number of discontinuity points. (This result was used in the proof of Theorem 9.3.2.)

Exercise 9.4 Verify the relations (9.4.34), (9.4.35)–(9.4.36) and (9.4.37).

Exercise 9.5* Show that the process on the right-hand side of (9.3.22) is well-defined under the condition of Theorem 9.3.8, (i); that is, for any $t \in \mathbb{R}$,

$$\int_{\mathbb{R}} \left| (t-u)_{\pm}^D - (-u)_{\pm}^D \right|^2 du < \infty.$$

Hint: It is enough to show this for a Jordan block $D = J_\lambda$ in (9.1.2), in which case $v_{\pm}^D$ can be expressed as in (9.7.1).

Exercise 9.6 Show that (9.3.30) can be written as (9.3.31). *Hint:* Use the following identities from Gradshteyn and Ryzhik [421]: $\int_{\mathbb{R}} 1_{\{x<0\}} \frac{\sin(ax)}{x} dx = \int_{\mathbb{R}} 1_{\{x>0\}} \frac{\sin(ax)}{x} dx = \frac{\pi}{2} \mathrm{sign}(a)$ (p. 423), $\int_{\mathbb{R}} 1_{\{x>0\}} \frac{\cos(ax) - \cos(bx)}{x} dx = \log \frac{|b|}{|a|}$ (p. 447), $\int_0^\infty \log(x) \sin(ax) \frac{dx}{x} = -\frac{\pi}{2}(C + \log(a))$, $a > 0$ (p. 594), and $\int_0^\infty \log(x)(\cos(ax) - \cos(bx)) \frac{dx}{x} = \log\left(\frac{a}{b}\right)\left(C + \frac{1}{2}\log(ab)\right)$, $a, b > 0$ (p. 594), where C is Euler's constant.

Exercise 9.7 For $p = 2$, $\Gamma = I$ and the exponent

$$H = \begin{pmatrix} 1/2 & 0 \\ -1 & 1/2 \end{pmatrix},$$

show that there does not exist a time reversible vector OFBM B_H such that $\mathbb{E} B_H(1) B_H(1)^* = \Gamma$. See Remark 9.3.15 for a related discussion. *Hint:* From Corollary 9.3.12,

$$\mathbb{E} B_H(1) B_H(1)^* = \int_{\mathbb{R}} \left| \frac{e^{ix}-1}{ix} \right|^2 |x|^{-D} A A^* |x|^{-D^*} dx,$$

where $D = H - (1/2)I$. Show that $\mathbb{E} B_H(1) B_H(1)^* = \Gamma = I$ is equivalent to

$$\begin{pmatrix} s_{11} r_0(d) & -s_{11} r_1(d) + s_{12} r_0(d) \\ -s_{11} r_1(d) + s_{12} r_0(d) & s_{11} r_2(d) - 2 s_{12} r_1(d) + s_{22} r_0(d) \end{pmatrix} = I,$$

where $AA^* =: (s_{jk})_{j,k=1,2}$,

$$r_k(d) = 4 \int_0^\infty \frac{1-\cos(x)}{x^2} (\log(x))^k x^{-2d} dx, \quad k=0,1,2.$$

Now, solve for s_{11}, s_{12} and s_{22} when $d=0$, and show that $s_{22} < 0$, contradicting the positive semidefiniteness of AA^*.

Exercise 9.8* When $p=2$, show that the condition in Proposition 9.3.22 can be reduced to $C_{12} \leq 1$, where

$$C_{12} = \frac{\Gamma(H_1+H_2+1)^2}{\Gamma(2H_1+1)\Gamma(2H_2+1)} \frac{\rho_{12}^2 \sin^2((H_1+H_2)\frac{\pi}{2}) + \eta_{12}^2 \cos^2((H_1+H_2)\frac{\pi}{2})}{\sin(H_1\pi)\sin(H_2\pi)}.$$

The constant C_{12} can be interpreted as the coherence function of the continuous-time stationary increment process $X_{H,\delta}(t) = B_H(t) - B_H(t-\delta)$, $t \in \mathbb{R}$; that is,

$$C_{12} = \frac{|f_{12}(\lambda)|^2}{f_{11}(\lambda) f_{22}(\lambda)},$$

where $(f_{jk}(\lambda))$ is the spectral density matrix of the stationary process $\{X_{H,\delta}(t)\}$. (The ratio above does not depend on λ or δ.) *Hint:* See Amblard, Coeurjolly, Lavancier, and Philippe [26], Proposition 9.

Exercise 9.9* Let $\{Z_n\}_{n \in \mathbb{Z}}$ be an $\mathbb{R}^p$-valued white noise, satisfying $\mathbb{E} Z_n = 0$ and $\mathbb{E} Z_n Z_n' = I$. Let also $\{\Psi_m = (\psi_{jk,m})_{j,k=1,\dots,p}\}_{m \in \mathbb{Z}} \subset \mathbb{R}^{p \times p}$ be such that $\psi_{jk,m} = L_{jk}(m) |m|^{d_j - 1}$, $m \in \mathbb{Z}$, where $d_j \in (0, 1/2)$ and $L(m) = (L_{jk}(m))_{j,k=1,\dots,p}$ is an $\mathbb{R}^{p \times p}$-valued function satisfying $L(m) \sim A^+$, as $m \to \infty$, and $L(m) \sim A^-$, as $m \to -\infty$, for some $A^+ = (\alpha_{jk}^+)_{j,k=1,\dots,p} \in \mathbb{R}^{q \times q}$, $A^- = (\alpha_{jk}^-)_{j,k=1,\dots,p} \in \mathbb{R}^{q \times q}$. Show that the time series $\{X_n\}$ given by a two-sided linear representation

$$X_n = \sum_{m=-\infty}^{\infty} \Psi_m Z_{n-m},$$

is LRD in the sense of (9.4.3) in condition II-v with

$$r_{jk} = \frac{\Gamma(d_j)\Gamma(d_k)}{\Gamma(d_j+d_k)} \left(c_{jk}^1 \frac{\sin(d_k \pi)}{\sin((d_j+d_k)\pi)} + c_{jk}^2 + c_{jk}^3 \frac{\sin(d_j \pi)}{\sin((d_j+d_k)\pi)} \right),$$

where

$$c_{jk}^1 = \sum_{t=1}^p \alpha_{jt}^- \alpha_{kt}^- = (A^-(A^-)^*)_{jk},$$

$$c_{jk}^2 = \sum_{t=1}^{p} \alpha_{jt}^+ \alpha_{kt}^- = (A^+(A^-)^*)_{jk},$$

$$c_{jk}^3 = \sum_{t=1}^{p} \alpha_{jt}^+ \alpha_{kt}^+ = (A^+(A^+)^*)_{jk}.$$

Hint: This is Proposition 3.1 in Kechagias and Pipiras [551], noting that a different convention (9.1.9) is used therein for the autocovariance function.

Exercise 9.10 Show that the OFBF is (9.5.58) is well-defined; that is,

$$\int_{\mathbb{R}^q} \left| (e^{i\langle x,t \rangle} - 1) \prod_{j=1}^{q} |x_j|^{-H_j - \frac{1}{2}} \right|^2 dx < \infty,$$

when $0 < H_j < 1$.

Exercise 9.11 Show that the function g_0 can indeed be written as in (9.5.38) where $\widetilde{g}$ is $-E^*/(H + 3\mathrm{tr}(E^*)/2)$–homogeneous.

Exercise 9.12 Verify that the conditions on the parameters found around (9.6.8)–(9.6.12) ensure that the corresponding spectral densities are integrable around $\lambda = 0$.

Exercise 9.13 Prove the statement in Example 9.5.15. *Hint:* See Corollary 2.12 in Biermé et al. [151].

Exercise 9.14 Check that the function $f(r) = (1+r)^{-\beta/2}$, $\beta > 0$, appearing in Example 9.6.4 is completely monotone.

Exercise 9.15** Formulate a definition of general spatial anisotropic LRD, even if it may not be necessarily practical as indicated in the beginning of Section 9.6.

9.8 Bibliographical Notes

Section 9.2: Operator self-similarity for stochastic processes finds its origins in a related notion, that of operator stability for probability distributions. A probability distribution μ on $\mathbb{R}^p$ is called *operator stable* if, for any $t > 0$,

$$\mu^t = t^B \mu * \delta(b(t)), \qquad (9.8.1)$$

where B is a matrix, $b : (0, \infty) \to \mathbb{R}^p$ is a function, and $*$ denotes the convolution.[14] Operator stable distributions arise as limit laws of operator normalized and centered sums of i.i.d. vectors. See, for example, the excellent monographs on the subject by Jurek and Mason [542], and Meerschaert and Scheffler [710].

---

[14] If ξ is a random vector with the probability distribution μ, then the right-hand side of (9.8.1) is the probability distribution of the random variable $t^B \xi + b(t)$. The left-hand side μ^t is the probability distribution whose characteristic function is $(\widehat{\mu})^t$, where $\widehat{\mu}$ is the characteristic function of μ.

On the one hand, the theories of operator stable distributions and operator self-similar processes bear lots of similarities (e.g., Mason [695]). On the other hand, first examples of operator self-similar processes were constructed as having independent increments whose distributions are operator stable (e.g., Hudson and Mason [499], Theorem 7). The latter is analogous to the connection between stable Lévy motions (which are self-similar) and stable distributions in one dimension.

In the vector setting $q = 1$, $p \geq 1$ (see (9.2.1)), the operator self-similarity was first considered in depth by Laha and Rohatgi [587], and Hudson and Mason [499]. In the multi-parameter setting $p = 1, q \geq 1$, the mathematical foundations of the operator self-similarity were laid in Biermé et al. [151].

For more recent fundamental work in the vector setting, see Meerschaert and Scheffler [709], Becker-Kern and Pap [111], Cohen, Meerschaert, and Rosiński [255]. This was preceded by influential work of Maejima and Mason [667], Maejima [666]. Operator self-similarity is also discussed in Meerschaert and Scheffler [710] (Section 11), Embrechts and Maejima [350] (Chapter 9).

Section 9.3: The material of this section is largely based on Didier and Pipiras [307, 308]. Operator fractional Brownian motions were also considered in Bahadoran, Benassi, and Dębicki [72], Pitt [824], with the latter early paper apparently little known in the area (while published before Laha and Rohatgi [587], and Hudson and Mason [499]). Sample path properties are studied in Mason and Xiao [696]. Vector fractional Brownian motions were also considered in Lavancier et al. [597], Amblard and Coeurjolly [25], Amblard et al. [26]. Convergence in law to operator fractional Brownian motion is studied in Dai [279]. Finally, for wavelet-based estimation of operator fractional Brownian motions, see Abry and Didier [5], Coeurjolly, Amblard, and Achard [252].

Section 9.4: Section 9.4.1 follows part of the work of Kechagias and Pipiras [551]. See also Kechagias and Pipiras [550]. One interesting problem raised in Remark 9.4.5 and addressed in Kechagias and Pipiras [551] is the construction of vector LRD series with one-sided linear representations (9.4.32) which have general phases. It is shown in Kechagias and Pipiras [551] that general phases could be obtained by taking the entries in the coefficient matrices ψ_m as linear combinations of what the authors called *trigonometric power-law coefficients*

$$m^{d_0-1} \cos(2\pi m^a), \quad m^{d_0-1} \sin(2\pi m^a),$$

for suitable parameter values d_0 and a.

In the bivariate context $p = 2$, vector LRD time series in the sense 1of Definition 9.4.1 with a general phase, and their estimation using an extension of the local Whittle method (see Section 10.6.1) were considered by Robinson [855]. A number of other authors considered vector LRD models and their estimation in the special case of zero phases $\phi_{jk} = 0$ (e.g., Lobato and Robinson [643], Lobato [642], Velasco [987], Robinson and Marinucci [857], Christensen and Nielsen [238], Nielsen [747, 748]). The case of the phases (9.4.30), as in one-sided FARIMA models, is another special case and was considered in, for example, Lobato [641], Robinson and Yajima [858], Shimotsu [905], Nielsen and Frederiksen [750], Nielsen [746]. The case of general ϕ is considered in Robinson [855] as indicated above, and without referring to ϕ explicitly in Robinson [852].

A more systematic study of fractional cointegration (Section 9.4.4) started with Marinucci and Robinson [691], Robinson and Yajima [858], though the notion itself has originated much earlier (e.g., Sowell [917], Dueker and Startz [331], Cheung and Lai [232], Baillie and Bollerslev [86]). This was followed by several fundamental papers by Chen and Hurvich [222, 223], and many other studies. To name but a few recent works, see Hualde and Robinson [497], Nielsen [749], Johansen and Nielsen [534, 535], Shimotsu [906],Łasak and Velasco [594]. Polynomial fractional cointegration is considered in Avarucci and Marinucci [57].

Multivariate LRD models also appear in Achard, Bassett, Meyer-Lindenberg, and Bullmore [14], Wendt, Scherrer, Abry, and Achard [1003] in the context of the so-called *fractal connectivity*. According to (9.4.6) in condition IV-v of the definition of vector LRD, the cross-spectral density $f_{jk}(\lambda)$ between the components j and k of the series satisfies

$$f_{jk}(\lambda) \sim G_{jk}\lambda^{-(d_j+d_k)},$$

as $\lambda \to 0$. Fractal connectivity postulates a more refined model where

$$f_{jk}(\lambda) \sim C_{jk}\lambda^{-d_{jk}}$$

with $d_{jk} \leq d_j + d_k$. The component j and k series are called *fractally connected* if $d_{jk} = d_j + d_k$ (and $C_{jk} \neq 0$). See also Sela and Hurvich [897], Kristoufek [579].

Section 9.5: The material here is largely based on Biermé et al. [151], and Baek, Didier, and Pipiras [70]. See also Clausel and Vedel [247].

Special cases of multiparameter OFBFs are found in the monographs by Leonenko [612], Cohen and Istas [253]. In particular, fractional Brownian sheet of Example 9.5.20 and related processes were studied in Ayache, Leger, and Pontier [63], Wang [995], Wang, Yan, and Yu [997], Wang [996], Makogin and Mishura [674]. For extensions, see Herbin and Merzbach [471], Clausel [246], Biermé, Lacaux, and Scheffler [152], Polisano, Clausel, Perrier, and Condat [827]. Interesting applications are considered in Benson, Meerschaert, Baeumer, and Scheffler [119], Roux, Clausel, Vedel, Jaffard, and Abry [871]. Estimation is considered by Lim, Meerschaert, and Scheffler [633]. Sample path properties were studied in, e.g., Kamont [548], Mason and Xiao [696], Xiao [1017], Li, Wang, and Xiao [626].

A number of other issues for fractional Gaussian fields are discussed in a survey paper by Lodhia, Sheffield, Sun, and Watson [644].

Section 9.6: The definition of isotropic multiparameter LRD follows Lavancier [595], Leonenko [612], Leonenko and Olenko [610]. Estimation in isotropic LRD models is considered in Anh and Lunney [36], Ludeña and Lavielle [650], Frías, Alonso, Ruiz-Medina, and Angulo [375].

Several references for anisotropic multiparameter LRD were given in Section 9.4.1. Estimation in anistropic LRD models is considered in Boissy et al. [168], Beran et al. [125], Guo et al. [440], Wang [994]. Invariance principles for anisotropic fields are considered in Lavancier [596], Puplinskaitė and Surgailis [835]. As noted around (9.6.13)–(9.6.14), the concept of scaling transition for random fields was considered by Puplinskaitė and Surgailis [835]. See also Pilipauskaitė and Surgailis [810], Puplinskaitė and Surgailis [836].

Construction of multiparameter stationary and other fields with explicit covariance functions (cf. Example 9.6.4) is a well-studied problem in spatial statistics (e.g., Chilès and

Delfiner [237], Sherman [904], Schlather [889]). With some focus on LRD, this problem was considered in, for example, Gneiting and Schlather [412].

Other notes: The vector and multiparameter settings were considered separately in Chapter 9. In the joint setting (both $p \geq 1$ and $q \geq 1$), the operator self-similar random fields and the analogues of fractional Brownian motion are studied in Li and Xiao [625], Baek et al. [70], Didier et al. [309], Didier, Meerschaert, and Pipiras [310]. See also Ma [658], Tafti and Unser [942].

Another interesting class of random fields are spatio-temporal models $\{X(t, u)\}$, where t is interpreted as a time variable and u as a spatial variable, e.g., in $\mathbb{R}^2$ (X itself takes values in $\mathbb{R}$). LRD spatio-temporal models are considered in, for example, Frías, Ruiz-Medina, Alonso, and Angulo [374, 376].

10

Maximum Likelihood Estimation Methods

Maximum likelihood estimation (MLE) methods are now among the most popular, best understood and most efficient estimation methods for long-range dependent time series. In this chapter, we provide a basic overview of the methods, as these are indispensable when working with long-range dependence, and improve on the heuristic methods discussed in Section 2.10. A more thorough treatment can be found in statistically oriented works (Palma [789], Giraitis et al.[406], Beran et al.[127]). We consider:

- The exact MLE in the time domain (Section 10.1)
- Whittle estimation in the spectral domain (Section 10.2.1)
- Autoregressive approximation (Section 10.2.2)
- The local Whittle estimation (Section 10.6.1)
- The broadband Whittle approach (Section 10.7)

10.1 Exact Gaussian MLE in the Time Domain

Suppose $\{X_n\}_{n\in\mathbb{Z}}$ is a Gaussian, zero mean, stationary series with autocovariance function $\gamma_\eta(n)$ and spectral density $f_\eta(\lambda)$. Here, η denotes a parameter vector, supposed to be unknown, which characterizes the series $\{X_n\}_{n\in\mathbb{Z}}$. (Unlike in the previous chapter, for notational simplicity, we shall suppress the subscript X in γ and f, by writing γ_η and f_η instead of $\gamma_{X,\eta}$ and $f_{X,\eta}$.) The vector η_0 will denote the *true value* of the parameter vector.

Example 10.1.1 (*Parametric FARIMA(p, d, q) family*) A Gaussian, zero mean, stationary FARIMA(p, d, q), $d \in (-1/2, 1/2)$, series is characterized by the spectral density

$$f_\eta(\lambda) = \frac{\sigma_Z^2}{2\pi}|1 - e^{-i\lambda}|^{-2d}\frac{|\theta(e^{-i\lambda})|^2}{|\varphi(e^{-i\lambda})|^2}, \quad \lambda \in (-\pi, \pi], \qquad (10.1.1)$$

where

$$\varphi(z) = 1 - \varphi_1 z - \cdots - \varphi_p z^p, \quad \theta(z) = 1 + \theta_1 z + \cdots + \theta_q z^q$$

and hence

$$\eta = (\sigma_Z^2, d, \varphi_1, \ldots, \varphi_p, \theta_1, \ldots, \theta_q)' \in \mathbb{R}^{p+q+2},$$

where prime indicates transpose (see Section 2.4.2). The autocovariance function $\gamma_\eta(n)$ does not have an explicit form in general but could be computed numerically in an efficient way (Sowell [918], Chung [242], Bertelli and Caporin [136], Doornik and Ooms [326]). See also Exercise 10.1.

Given N observations $X_1, \ldots, X_N$ of the process $\{X_n\}_{n \in \mathbb{Z}}$, form the vector $X = (X_1, \ldots, X_N)'$. The (Gaussian) *likelihood function* is defined as

$$L(\eta) = (2\pi)^{-\frac{N}{2}} \det(\Sigma_N(\eta))^{-\frac{1}{2}} \exp\left\{-\frac{1}{2} X' \Sigma_N(\eta)^{-1} X\right\},$$

where $\Sigma_N(\eta) = (\gamma_\eta(j-k))_{j,k=1,\ldots,N}$. The *log-likelihood* is

$$\ell(\eta) = -\frac{N}{2} \log 2\pi - \frac{1}{2} \log \det(\Sigma_N(\eta)) - \frac{1}{2} X' \Sigma_N(\eta)^{-1} X.$$

After multiplying by $-\frac{2}{N}$ and ignoring the constant term, the *maximum likelihood estimator* of η is defined as

$$\widehat{\eta}_{mle} = \underset{\eta}{\operatorname{argmin}}\, \ell_{exact}(\eta), \qquad (10.1.2)$$

where the (*scaled*) negative log-likelihood is

$$\ell_{exact}(\eta) = \frac{1}{N} \log \det(\Sigma_N(\eta)) + \frac{1}{N} X' \Sigma_N(\eta)^{-1} X. \qquad (10.1.3)$$

Remark 10.1.2 The series is assumed to have zero mean μ for simplicity. The unknown mean μ could be incorporated in the MLE as another parameter by replacing X by $X - \mu$. The resulting ML estimator of μ will be a linear combination of the components of the vector X. It is common, however, to estimate the mean μ by the sample mean $\overline{X}$ which is itself a linear combination of the components of X. When the components of X are dependent, it is likewise natural to consider the Best Linear Unbiased Estimator (BLUE) of the mean. It is known (Samarov and Taqqu [877]) that for LRD series, the BLUE provides only minimal gains in efficiency when compared to the sample mean (see also Exercise 10.2). This further justifies the simplification $\mu = 0$, we make for the MLE. It should also be noted that some asymptotic results for ML estimators have been derived when X is replaced by $X - \overline{X}$ or $X - \widehat{\mu}$ with a mean estimator $\widehat{\mu}$ satisfying suitable conditions (Dahlhaus [277, 278], Lieberman, Rosemarin, and Rousseau [631]). For further remarks, see also Remark 10.2.5. Finally, we also point out that in Section 10.2.1, the MLE is approximated in the frequency domain. In the latter domain, the approximate MLE is invariant to an additive shift of the data and hence $\mu = 0$ can be assumed.

The MLE $\widehat{\eta}_{mle}$ is computed through a numerical optimization. We are not going to discuss optimization methods here (see, for example, Lange [592]). But any optimization procedure involves computation of the (negative) log-likelihood $\ell_{exact}(\eta)$ for a given value η, which in turn requires computing $\Sigma_N(\eta)$, its inverse $\Sigma_N(\eta)^{-1}$ and the determinant $\det(\Sigma_N(\eta))$. Computing $\Sigma_N(\eta)$ may not be trivial itself (see Example 10.1.1). But it is the computation of $\Sigma_N(\eta)^{-1}$ and $\det(\Sigma_N(\eta))$ which presents real challenges, especially for larger sample sizes N.

Naïve ways to compute $\Sigma_N(\eta)^{-1}$, for example, through Cholesky decomposition, require $O(N^3)$ number of steps and are impractical already for moderate sample sizes N ($N \approx 500$). The computations could be sped up to the $O(N^2)$ number of steps by exploiting the Toeplitz structure of the covariance matrix $\Sigma_N(\eta)$, which allows one to deal with longer time series (up to $N \approx 10{,}000$). Two such popular methods, which are important for other considerations as well, are the Durbin–Levinson and innovations algorithms (e.g., Brockwell and Davis [186]).

- The *Durbin–Levinson algorithm* provides a recursive way (in n below), given an autocovariance function, to compute the coefficients $\varphi_{n1}, \ldots, \varphi_{nn}$ in

$$\widehat{X}_{n+1} = \varphi_{n1} X_n + \cdots + \varphi_{nn} X_1, \quad n = 1, 2, \ldots, \qquad (10.1.4)$$

where $\widehat{X}_{n+1}$ is the best linear one-step predictor of X_{n+1} in $L^2(\Omega)$ based on $X_1, \ldots, X_n$, and also obtain the mean squared error of the prediction

$$v_n = \mathbb{E}(X_{n+1} - \widehat{X}_{n+1})^2, \quad n = 1, 2, \ldots. \qquad (10.1.5)$$

One sets $\widehat{X}_1 = 0$ and $v_0 = \mathbb{E} X_1^2$ by convention.

- The *innovations algorithm* provides a recursive way (in n), given an autocovariance function, to compute the coefficients $\theta_{n1}, \ldots, \theta_{nn}$ in

$$\widehat{X}_{n+1} = \theta_{n1}(X_n - \widehat{X}_n) + \cdots + \theta_{nn}(X_1 - \widehat{X}_1), \quad n = 1, 2, \ldots, \qquad (10.1.6)$$

and the v_ns. Note that (10.1.6) can be written as

$$X_{n+1} = \theta_{nn}(X_1 - \widehat{X}_1) + \cdots + \theta_{n1}(X_n - \widehat{X}_n) + (X_{n+1} - \widehat{X}_{n+1}), \quad n = 1, 2, \ldots. \qquad (10.1.7)$$

The components $X_i - \widehat{X}_i$ and $X_j - \widehat{X}_j$ are uncorrelated for $i \neq j$, and hence

$$\mathbb{E}(X - \widehat{X})(X - \widehat{X})' = \mathrm{diag}(v_0, \ldots, v_{N-1}) =: D_N, \qquad (10.1.8)$$

where $\widehat{X} = (\widehat{X}_1, \ldots, \widehat{X}_N)'$.

For a detailed description of the Durbin–Levinson and innovations algorithms, see Brockwell and Davis [186], Chapter 5.

To see how these algorithms are used to get ℓ_{exact}, note that (10.1.7) can be expressed in a matrix form as

$$X = C_N(X - \widehat{X}), \qquad (10.1.9)$$

where

$$C_N = \begin{pmatrix} 1 & 0 & 0 & 0 & \cdots & 0 & 0 \\ \theta_{11} & 1 & 0 & 0 & \cdots & 0 & 0 \\ \theta_{22} & \theta_{21} & 1 & 0 & \cdots & 0 & 0 \\ \theta_{33} & \theta_{32} & \theta_{31} & 1 & \cdots & 0 & 0 \\ \vdots & \vdots & \vdots & \vdots & \ddots & \vdots & \vdots \\ \theta_{N-1,N-1} & \theta_{N-1,N-2} & \theta_{N-1,N-3} & \theta_{N-1,N-4} & \cdots & \theta_{N-1,1} & 1 \end{pmatrix}$$

and $\widehat{X} = (\widehat{X}_1, \ldots, \widehat{X}_N)'$ as above. By using (10.1.8), one has

$$\Sigma_N(\eta) = \mathbb{E} X X' = C_N \mathbb{E}(X - \widehat{X})(X - \widehat{X})' C_N' = C_N D_N C_N'.$$

Then,

$$X'\Sigma_N(\eta)^{-1}X = (X - \widehat{X})'C_N'(C_N D_N C_N')^{-1}C_N(X - \widehat{X})$$

$$= (X - \widehat{X})'D_N^{-1}(X - \widehat{X}) = \sum_{j=1}^{N} \frac{(X_j - \widehat{X}_j)^2}{v_{j-1}}$$

and

$$\det(\Sigma_N(\eta)) = \det(C_N)\det(D_N)\det(C_N') = \det(D_N) = \prod_{j=1}^{N} v_{j-1},$$

since the determinant of a triangular matrix equals the product of the elements in the diagonal. Hence, the log-likelihood (10.1.3) can be written as

$$\ell_{exact}(\eta) = \frac{1}{N}\sum_{j=1}^{N} \log v_{j-1} + \frac{1}{N}\sum_{j=1}^{N} \frac{(X_j - \widehat{X}_j)^2}{v_{j-1}}, \quad (10.1.10)$$

where the $X_j - \widehat{X}_j$s and v_js can be computed, for a fixed η and the resulting autocovariance function γ_η, by using either the Durbin-Levinson or the innovations algorithm (see also the discussion at the end of Example 10.1.1 for FARIMA models).

Another difficulty with the exact MLE often raised in the literature, even for small sample sizes, is that $\Sigma_N(\eta)$ can become ill-conditioned when the LRD parameter d is close to $1/2$, making the computation of the inverse numerically unstable. The degree of ill-conditioning can be measured by the ratio of the largest and smallest eigenvalues of the matrix, called the condition number (e.g., Gentle [385], Monahan [728]). A large condition number indicates that a matrix is ill-conditioned. For example, for FARIMA$(0, d, 0)$ series and $N = 100$, by using the explicit formula for the autocovariance function of FARIMA$(0, d, 0)$ series in Corollary 2.4.4, it can be checked numerically that the condition number is 8.35, 28.71, 133.96, 411.98 for $d = 0.2, 0.3, 0.4, 0.45$, respectively.

The difficulties with the exact MLE discussed above (computational issues for larger sample sizes and when LRD parameter approaches $1/2$) call for approximations of the likelihood. Two such approximations are discussed in the next section. Asymptotic results for the exact MLE will be discussed briefly in the context of those approximations.

10.2 Approximate MLE

Two approximations of the log-likelihood (10.1.3) are discussed below: the celebrated Whittle's approximation in the spectral domain (Section 10.2.1) and that based on the autoregressive representation (Section 10.2.2).

10.2.1 Whittle Estimation in the Spectral Domain

The Whittle's approximation deals with the two terms, $\frac{1}{N}\log\det(\Sigma_N(\eta))$ and $\frac{1}{N}X^T\Sigma_N(\eta)^{-1}X$, of the log-likelihood (10.1.3) separately.

10.2 Approximate MLE

Approximation of $\frac{1}{N} \log \det(\Sigma_N(\eta))$: This term is approximated by using Szegö's limit theorem; that is,

$$\lim_{N \to \infty} \frac{1}{N} \log \det(\Sigma_N(\eta)) = \frac{1}{2\pi} \int_{-\pi}^{\pi} \log(2\pi f_\eta(\lambda)) d\lambda \qquad (10.2.1)$$

(cf. the strong Szegö limit theorem in (3.11.101)). The presence of 2π next to $f_\eta(\lambda)$ is to account for the fact that the Fourier transform of $\{\gamma_\eta(n)\}$ (see Appendix A.1.1) is $2\pi f_\eta(\lambda)$, and not $f_\eta(\lambda)$.

Approximation of $\frac{1}{N} X^T \Sigma_N(\eta)^{-1} X$: Set $A_N(\eta) = (\alpha_\eta(j-k))_{j,k=1,\ldots,N}$, where

$$\alpha_\eta(j-k) = \frac{1}{(2\pi)^2} \int_{-\pi}^{\pi} e^{i(j-k)\lambda} \frac{1}{f_\eta(\lambda)} d\lambda,$$

that is, $\widehat{\alpha}_\eta(\lambda) = f_\eta(\lambda)^{-1}/(2\pi)$, where the hat indicates the Fourier tranform (see Appendix A.1.1). Then, for large N, $A_N(\eta)$ approximates $\Sigma_N(\eta)^{-1}$. Indeed, note that $\Sigma_N(\eta) = (\gamma_\eta(j-k))_{j,k=1,\ldots,N}$, where

$$\gamma_\eta(j-k) = \int_{-\pi}^{\pi} e^{i(j-k)\lambda} f_\eta(\lambda) d\lambda,$$

that is, $\widehat{\gamma}_\eta(\lambda) = 2\pi f_\eta(\lambda)$. Then, for large N, we expect that

$$(A_N(\eta) \Sigma_N(\eta))_{j,k} = \sum_{l=1}^{N} \alpha_\eta(j-l) \gamma_\eta(l-k) \approx \sum_{l=-\infty}^{\infty} \alpha_\eta(j-l) \gamma_\eta(l-k)$$

$$= \frac{1}{2\pi} \int_{-\pi}^{\pi} e^{-i(j-k)\lambda} \widehat{\alpha}_\eta(\lambda) \overline{\widehat{\gamma}_\eta(\lambda)} d\lambda = \frac{1}{2\pi} \int_{-\pi}^{\pi} e^{-i(j-k)\lambda} \frac{1}{f_\eta(\lambda)} f_\eta(\lambda) d\lambda = \begin{cases} 1, & \text{if } j = k, \\ 0, & \text{if } j \neq k, \end{cases}$$

by using Parseval's identity (A.1.3) and the relations between $\widehat{\alpha}_\eta(\lambda)$, $\widehat{\gamma}_\eta(\lambda)$ and $f_\eta(\lambda)$ noted above, where $\approx$ indicates an approximation for large N. This leads to the approximation

$$\frac{1}{N} X^T \Sigma_N(\eta)^{-1} X \approx \frac{1}{N} X^T A_N(\eta) X = \frac{1}{N} \sum_{j,k=1}^{N} \alpha_\eta(j-k) X_j X_k$$

$$= \frac{1}{(2\pi)^2} \int_{-\pi}^{\pi} \frac{\frac{1}{N} |\sum_{j=1}^{N} e^{ij\lambda} X_j|^2}{f_\eta(\lambda)} d\lambda = \frac{1}{2\pi} \int_{-\pi}^{\pi} \frac{I_X(\lambda)}{2\pi f_\eta(\lambda)} d\lambda, \qquad (10.2.2)$$

where $I_X(\lambda)$ is the periodogram (2.10.9) (see also the warning in Section 1.3.3).

Combining (10.2.1) and (10.2.2), consider the Whittle approximation of the log-likelihood (10.1.3)

$$\ell_{whittle}(\eta) = \frac{1}{2\pi} \int_{-\pi}^{\pi} \log(2\pi f_\eta(\lambda)) d\lambda + \frac{1}{2\pi} \int_{-\pi}^{\pi} \frac{I_X(\lambda)}{2\pi f_\eta(\lambda)} d\lambda \qquad (10.2.3)$$

and define

$$\widehat{\eta}_{whittle} = \underset{\eta}{\operatorname{argmin}}\, \ell_{whittle}(\eta). \qquad (10.2.4)$$

The discrete version of $\ell_{whittle}(\eta)$ is

$$\sum_{j=1}^{N}\log\left(2\pi f_\eta(\frac{2\pi j}{N})\right) + \sum_{j=1}^{N}\frac{I_X(\frac{2\pi j}{N})}{f_\eta(\frac{2\pi j}{N})}, \qquad (10.2.5)$$

which can be computed using the Fast Fourier Transform (FFT) in $O(N \log N)$ steps. (The 2π in the first integral in (10.2.3) and the first sum in (10.2.5) does not affect the optimization but we keep them for cancelation below.)

An important *special case* of a parametric family of spectral densities is

$$f_\eta(\lambda) = \frac{\sigma^2}{2\pi}|g_\upsilon(e^{-i\lambda})|^2 = \frac{\sigma^2}{2\pi}|\sum_{j=0}^{\infty}g_{\upsilon,j}e^{-ij\lambda}|^2 =: \frac{\sigma^2}{2\pi}h_\upsilon(\lambda), \qquad (10.2.6)$$

where $g_{\upsilon,0} = 1$, $\eta = (\sigma^2, \upsilon)'$ and $\upsilon \in \mathbb{R}^q$. Note that the parameter σ^2 appears as a multiplicative factor in (10.2.6). This special case (10.2.6) corresponds to a time series with a linear representation

$$X_n = \sum_{j=0}^{\infty}g_{\upsilon,j}Z_{n-j},$$

where $\{Z_n\}$ is a white noise series with variance $\sigma^2 = \sigma_Z^2$. For example, the parametric family of FARIMA(p,d,q) spectral densities (10.1.1) has the form (10.2.6) with $\upsilon = (d, \varphi_1, \ldots, \varphi_p, \theta_1, \ldots, \theta_q)'$. Note that g_υ is a function of $e^{-i\lambda}$ while h_υ is a function of λ.

In the case (10.2.6), one has

$$\int_{-\pi}^{\pi}\log h_\upsilon(\lambda)d\lambda = 0. \qquad (10.2.7)$$

This can be seen, on one hand, as a consequence of Kolmogorov's formula (e.g., Brockwell and Davis [186], Section 5.8), which asserts that $\int_{-\pi}^{\pi}\log f_\eta(\lambda)d\lambda = 2\pi\log(\sigma^2/2\pi)$. By (10.2.6), $\log f_\eta = \log(\sigma^2/2\pi) + \log h_\upsilon$ and therefore $\int_{-\pi}^{\pi}\log h_\upsilon(\lambda)d\lambda = \int_{-\pi}^{\pi}\log f_\eta(\lambda)d\lambda - \int_{-\pi}^{\pi}\log(\sigma^2/2\pi)d\lambda = 2\pi\log(\sigma^2/2\pi) - 2\pi\log(\sigma^2/2\pi) = 0$. On the other hand, this could informally be explained as follows. Note that $\log h_\upsilon(\lambda) = \log g_\upsilon(e^{-i\lambda}) + \log g_\upsilon(e^{i\lambda})$. Consider the expansion $g_\upsilon(z) = \sum_{j=0}^{\infty}g_{\upsilon,j}z^j$ in (10.2.6). Since $g_\upsilon(0) = g_{\upsilon,0} = 1$, one has $\log g_\upsilon(0) = 0$ and hence the expansion $\log g_\upsilon(z) = \sum_{k=1}^{\infty}a_{\upsilon,k}z^k$ starts at $k=1$. We now obtain that, for example, $\log g_\upsilon(e^{-i\lambda}) = \sum_{k=1}^{\infty}a_{\upsilon,k}e^{-ik\lambda}$ and hence that $\int_{-\pi}^{\pi}\log g_\upsilon(e^{-i\lambda})d\lambda = \sum_{k=1}^{\infty}a_{\upsilon,k}\int_{-\pi}^{\pi}e^{-ik\lambda}d\lambda = 0$.

By using (10.2.7), the log-likelihood (10.2.3) can be written as

$$\ell_{whittle}(\sigma^2, \upsilon) = \log\sigma^2 + \frac{1}{\sigma^2}\int_{-\pi}^{\pi}\frac{I_X(\lambda)}{2\pi h_\upsilon(\lambda)}d\lambda \qquad (10.2.8)$$

and the corresponding estimator of (σ^2, υ) as

$$(\widehat{\sigma}^2_{whittle}, \widehat{\upsilon}_{whittle}) = \underset{\sigma^2, \upsilon}{\operatorname{argmin}}\,\ell_{whittle}(\sigma^2, \upsilon). \qquad (10.2.9)$$

The minimization could also be carried over υ alone after solving for σ^2 (in terms of υ) and substituting it into $\ell_{whittle}(\sigma^2, \upsilon)$.

10.2 Approximate MLE

Another perspective on the log-likelihood in (10.2.8) is given in Exercise 10.3.

In the following result, we present the large sample asymptotics of the Whittle estimators (10.2.9) in the case (10.2.6). The following assumptions will be used on the functions h_υ defined in (10.2.6):

(A0) The parameter space Υ of υ is compact, the parameters $(\sigma, \upsilon) \in (0, \infty) \times \Upsilon$ determine the spectral density $\frac{\sigma^2}{2\pi} h_\upsilon$ uniquely, and for any $a > 0$, the function $1/(h_\upsilon(\lambda) + a)$ is continuous in $(\lambda, \upsilon) \in (-\pi, \pi) \times \Upsilon$. Moreover, the true value (σ_0, υ_0) lies in the interior of $(0, \infty) \times \Upsilon$.

There is, moreover, a compact ball Υ_0 centered at υ_0 such that:

(A1) $\int_{-\pi}^{\pi} \log h_\upsilon(\lambda) d\lambda \ (\equiv 0)$ is twice differentiable in $\upsilon \in \Upsilon_0$ under the integral sign.

(A2) The functions $h_\upsilon(\lambda)$, $\frac{\partial}{\partial \upsilon_j} h_\upsilon^{-1}(\lambda)$ and $\frac{\partial^2}{\partial \upsilon_r \partial \upsilon_s} h_\upsilon^{-1}(\lambda)$ are continuous at all $\lambda \in (-\pi, \pi) \setminus \{0\}$, $\upsilon \in \Upsilon_0$, for $r, s = 1, \ldots, q$.

(A3) There are $-1 < \alpha, \beta < 1$ such that $\alpha + \beta < 1/2$ and

$$h_{\upsilon_0}(\lambda) \leq C|\lambda|^{-\alpha}, \quad |\frac{\partial}{\partial \lambda} h_{\upsilon_0}(\lambda)| \leq C|\lambda|^{-\alpha-1}, \quad |\frac{\partial}{\partial \upsilon_j} h_{\upsilon_0}^{-1}(\lambda)| \leq C|\lambda|^{-\beta},$$

for $j = 1, \ldots, q, \lambda \in (-\pi, \pi) \setminus \{0\}$.

(A4) If $\alpha \leq 0$, then there is $G \in L^1(-\pi, \pi)$ such that $|\frac{\partial^2}{\partial \upsilon_r \partial \upsilon_s} h_\upsilon^{-1}(\lambda)| \leq G(\lambda)$, for $\lambda \in (-\pi, \pi) \setminus \{0\}$, $\upsilon \in \Upsilon_0$, $r, s = 1, \ldots, q$. If $\alpha > 0$, then $\frac{\partial^2}{\partial \upsilon_r \partial \upsilon_s} h_\upsilon^{-1}(\lambda)$ is continuous in $\lambda \in (-\pi, \pi) \setminus \{0\}$, $\upsilon \in \Upsilon_0$, for $r, s = 1, \ldots, q$.

Theorem 10.2.1 *Suppose that $\{X_n\}_{n \in \mathbb{Z}}$ has a linear representation*

$$X_n = \sum_{k=0}^{\infty} g_{\upsilon_0, k} \epsilon_{n-k},$$

where $\{\epsilon_n\}_{n \in \mathbb{Z}}$ consists of i.i.d. variables with zero mean, variance σ_0^2 and finite fourth moment. Suppose its spectral density belongs to the family given by (10.2.6) where the functions h_υ satisfy Assumptions (A0)–(A4) above. Then, as $N \to \infty$,

$$\sqrt{N}(\widehat{\upsilon}_{whittle} - \upsilon_0) \xrightarrow{d} \mathcal{N}(0, 4\pi C^{-1}(\upsilon_0)), \tag{10.2.10}$$

$$\sqrt{N}(\widehat{\sigma}_{whittle}^2 - \sigma_0^2) \xrightarrow{d} \mathcal{N}(0, \mathbb{E}\epsilon_0^4 - \sigma_0^4), \tag{10.2.11}$$

where $C(\upsilon) = (c_{rs}(\upsilon))_{r,s=1,\ldots,q}$ with

$$c_{rs}(\upsilon) = \int_{-\pi}^{\pi} \frac{\partial}{\partial \upsilon_r} \log h_\upsilon(\lambda) \frac{\partial}{\partial \upsilon_s} \log h_\upsilon(\lambda) d\lambda. \tag{10.2.12}$$

Remark 10.2.2 The assumptions (A0)–(A4) of Theorem 10.2.1 allow for LRD time series such as FARIMA(p, d, q) series in Example 10.1.1 (Proposition 8.3.1, Giraitis et al. [406]). The theorem shows that the rate $\sqrt{N}$ holds even for the LRD parameter estimator $\widehat{d}_{whittle}$. Note that the series $\{X_n\}$ is not necessarily Gaussian. It can also be shown that $(\widehat{\sigma}_{whittle}^2, \widehat{\upsilon}_{whittle})$ are jointly asymptotically normal with zero asymptotic covariances (Beran et al. [127], p. 425).

Remark 10.2.3 Under suitable assumptions, the asymptotic results (10.2.10)–(10.2.11) also hold for the exact Gaussian MLE estimators $(\widehat{\sigma}^2_{mle}, \widehat{\upsilon}_{mle}) = \widehat{\eta}_{mle}$ (Dahlhaus [277, 278], Lieberman et al. [631]), and for the Whittle estimators when using the discrete version of the log-likelihood (10.2.5) (see the discussion following Remark 2.2 in Dahlhaus [277], and Velasco and Robinson [988]).

Proof outline of Theorem 10.2.1 We provide here an outline of the proof of the theorem. A complete proof can be found in Giraitis et al. [406], Theorem 3.8.1. The proof is standard in the context of MLE though technical issues arise related to the convergence of functionals involving LRD series.

The MLE estimators $(\widehat{\sigma}^2_{whittle}, \widehat{\upsilon}_{whittle})$ arise from minimizing $\ell_{whittle}(\sigma^2, \upsilon)$ in (10.2.8). Note that, in view of (2.10.9), we have

$$\ell_{whittle}(\sigma^2, \upsilon) = \log \sigma^2 + \frac{1}{\sigma^2 2\pi N} \sum_{j,k=1}^N \beta_\upsilon(j-k) X_j X_k, \qquad (10.2.13)$$

where

$$\beta_\upsilon(n) = \int_{-\pi}^\pi e^{in\lambda} \frac{1}{h_\upsilon(\lambda)} d\lambda. \qquad (10.2.14)$$

We first consider the asymptotics of $\widehat{\upsilon}_{whittle}$. One can show that, under Assumption (A1), $\widehat{\upsilon}_{whittle}$ is consistent (Giraitis et al. [406], Theorem 8.2.1). Then, for large enough N and with probability approaching 1, $\widehat{\upsilon}_{whittle} \in \Upsilon_0$ and satisfies the first order condition

$$\dot{Q}(\widehat{\upsilon}_{whittle}) = 0,$$

where

$$\dot{Q}(\upsilon) = \frac{1}{N} \sum_{j,k=1}^N \dot{\beta}_\upsilon(j-k) X_j X_k \qquad (10.2.15)$$

and $\dot{\beta}_\upsilon = \partial \beta_\upsilon / \partial \upsilon$. (Note that, while $Q(\upsilon)$ is a scalar, $\dot{Q}(\upsilon)$ has the dimension of the vector υ.) By Taylor's expansion,

$$0 = \dot{Q}(\widehat{\upsilon}_{whittle}) = \dot{Q}(\upsilon_0) + \ddot{Q}(\upsilon^*)(\widehat{\upsilon}_{whittle} - \upsilon_0),$$

where $\ddot{Q} = \partial \dot{Q}/\partial \upsilon$ is a matrix and $|\upsilon^* - \upsilon_0| \leq |\widehat{\upsilon}_{whittle} - \upsilon_0|$. Hence,

$$\sqrt{N}(\widehat{\upsilon}_{whittle} - \upsilon_0) = -\sqrt{N}\big(\ddot{Q}(\upsilon^*)\big)^{-1} \dot{Q}(\upsilon_0). \qquad (10.2.16)$$

We next consider the convergence of $\ddot{Q}(\upsilon^*)$ and $\sqrt{N}\dot{Q}(\upsilon_0)$. The matrix $\ddot{Q}(\upsilon^*)$ consists of elements

$$\ddot{Q}(\upsilon^*)_{rs} = \frac{1}{N} \sum_{j,k=1}^N \frac{\partial^2}{\partial \upsilon_r \partial \upsilon_s} \beta_\upsilon(j-k) X_j X_k.$$

In view of (10.2.14) and (2.10.9), we have

$$\ddot{Q}(\upsilon^*)_{rs} = \int_{-\pi}^\pi I_X(\lambda) \frac{\partial^2}{\partial \upsilon_r \partial \upsilon_s}\left(\frac{1}{h_\upsilon(\lambda)}\right)\bigg|_{\upsilon=\upsilon^*} d\lambda.$$

Since v^* converges to v_0 in probability and $I_X(\lambda)$ is an estimator of $2\pi f_{(\sigma_0^2, v_0)}(\lambda)$, it is expected and can be shown that

$$\ddot{Q}(v^*)_{rs} \xrightarrow{p} 2\pi \int_{-\pi}^{\pi} f_{(\sigma_0^2, v_0)}(\lambda) \frac{\partial^2}{\partial v_r \partial v_s}\left(\frac{1}{h_v(\lambda)}\right)\bigg|_{v=v_0} d\lambda = \sigma_0^2 \int_{-\pi}^{\pi} h_v(\lambda) \frac{\partial^2}{\partial v_r \partial v_s}\left(\frac{1}{h_v(\lambda)}\right)\bigg|_{v=v_0} d\lambda$$

$$= \sigma_0^2 \int_{-\pi}^{\pi} \left(\frac{2}{h_v(\lambda)^2} \frac{\partial}{\partial v_r} h_v(\lambda) \frac{\partial}{\partial v_s} h_v(\lambda) - \frac{1}{h_v(\lambda)} \frac{\partial^2}{\partial v_r \partial v_s} h_v(\lambda)\right)\bigg|_{v=v_0} d\lambda. \quad (10.2.17)$$

Note that, by differentiating (10.2.7) twice inside the integral sign, we also have

$$0 = \int_{-\pi}^{\pi} \left(\frac{1}{h_v(\lambda)^2} \frac{\partial}{\partial v_r} h_v(\lambda) \frac{\partial}{\partial v_s} h_v(\lambda) - \frac{1}{h_v(\lambda)} \frac{\partial^2}{\partial v_r \partial v_s} h_v(\lambda)\right) d\lambda. \quad (10.2.18)$$

Combining (10.2.17) and (10.2.18) yields

$$\ddot{Q}(v^*)_{rs} \xrightarrow{p} \sigma_0^2 \int_{-\pi}^{\pi} \frac{1}{h_v(\lambda)^2} \frac{\partial}{\partial v_r} h_v(\lambda) \frac{\partial}{\partial v_s} h_v(\lambda)\bigg|_{v=v_0} d\lambda.$$

$$= \sigma_0^2 \int_{-\pi}^{\pi} \frac{\partial}{\partial v_r} \log h_v(\lambda) \frac{\partial}{\partial v_s} \log h_v(\lambda)\bigg|_{v=v_0} d\lambda = \sigma_0^2 c_{rs}(v_0), \quad (10.2.19)$$

where $c_{rs}(v)$ appears in (10.2.12).

The term $\sqrt{N} \dot{Q}(v_0)$, on the other hand, can be shown to be asymptotically normal. In view of (10.2.15), $\sqrt{N} \dot{Q}(v_0)$ is referred to as a quadratic form. The convergence of quadratic forms to a normal limit has attracted considerable attention in the literature on long-range dependent series. For $\sqrt{N} \dot{Q}(v_0)$, it can be shown that

$$\sqrt{N}(\dot{Q}(v_0) - \mathbb{E}\dot{Q}(v_0)) \xrightarrow{d} \mathcal{N}(0, 4\pi \sigma_0^4 C(v_0)) \quad (10.2.20)$$

and that

$$\sqrt{N} \mathbb{E}\dot{Q}(v_0) \to 0. \quad (10.2.21)$$

By using (10.2.20), (10.2.21) and (10.2.19), the relation (10.2.16) implies the desired convergence (10.2.10).

The asymptotic normality (10.2.20) is proved by showing first (Giraitis et al. [406], Lemma 6.3.1) that

$$N \operatorname{Var}(\dot{Q}(v_0) - R(v_0)) \to 0, \quad (10.2.22)$$

where

$$R(v) = \frac{1}{N} \sum_{k=1}^{N} Y_{1,k} Y_{2,k}, \quad Y_{j,k} = \sum_{m=-\infty}^{\infty} \ell_{j,v,m} \epsilon_{k-m}, \quad j = 1, 2, \quad (10.2.23)$$

and where the Fourier transforms of $\{\ell_{j,v_0,m}\}$ with m as variable, are

$$\widehat{\ell}_{1,v}(\lambda) = (2\pi)^{1/2} g_v(e^{-i\lambda}) \left|\frac{\partial}{\partial v} \frac{1}{h_v(\lambda)}\right|^{\frac{1}{2}},$$

$$\widehat{\ell}_{2,v}(\lambda) = (2\pi)^{1/2} g_v(e^{-i\lambda}) \left|\frac{\partial}{\partial v} \frac{1}{h_v(\lambda)}\right|^{\frac{1}{2}} \operatorname{sign}\left(\frac{\partial}{\partial v} \frac{1}{h_v(\lambda)}\right) \quad (10.2.24)$$

(the square root above, as well as the multiplications of $Y_{1,k} Y_{2,k}$ and by the sign function, are taken componentwise). Recall that g_v and h_v are related through (10.2.6). It is then enough to prove (10.2.20) with $\dot{Q}(v_0)$ replaced by $R(v_0)$. Note that, in contrast to the series X_k entering $\dot{Q}(v)$ which may be long-range dependent, the series $Y_{j,k}$ are expected to have shorter memory. For example, with $g_v(e^{-i\lambda}) = (1 - e^{-i\lambda})^{-d}$ corresponding to FARIMA$(0, d, 0)$ series and $v = d$, we have $|\widehat{\ell}_{1,v}(\lambda)|^2 = |\log|1 - e^{-i\lambda}||$ which is now bounded by $C|\lambda|^{-\delta}$ around $\lambda = 0$ for any arbitrarily small $\delta > 0$ (see Example 10.2.4 below).

We shall not prove the asymptotic normality (10.2.20) with $\dot{Q}(v_0)$ replaced by $R(v_0)$. We shall just argue for the asymptotic covariance matrix $4\pi\sigma_0^4 C(v_0)$. By using the superscript (r) to denote the rth element of a vector, note that

$$\text{Cov}\Big(\frac{1}{N^{1/2}}\sum_{k=1}^{N} Y_{1,k}^{(r)} Y_{2,k}^{(r)}, \frac{1}{N^{1/2}}\sum_{k=1}^{N} Y_{1,k}^{(s)} Y_{2,k}^{(s)}\Big)$$

$$= \frac{1}{N}\sum_{k=-(N-1)}^{N-1} (N - |k|)\text{Cov}\Big(Y_{1,0}^{(r)} Y_{2,0}^{(r)}, Y_{1,k}^{(s)} Y_{2,k}^{(s)}\Big) \to \sum_{k=-\infty}^{\infty} \text{Cov}\Big(Y_{1,0}^{(r)} Y_{2,0}^{(r)}, Y_{1,k}^{(s)} Y_{2,k}^{(s)}\Big) =: W_{rs},$$

where the first equality is obtained analogously as in the proof of Proposition 2.2.5. We show next that the asymptotic covariance W_{rs} is indeed $4\pi\sigma_0^4 c_{rs}(v_0)$.

Writing $Y_{1,k}^{(r)} Y_{2,k}^{(r)} = \sum_{m_1,m_2=-\infty}^{\infty} \ell_{1,v_0,k-m_1}^{(r)} \ell_{2,v_0,k-m_2}^{(r)} \epsilon_{m_1}\epsilon_{m_2}$, we have

$$W_{rs} = \sum_{k=-\infty}^{\infty} \sum_{m_1,m_2=-\infty}^{\infty} \sum_{n_1,n_2=-\infty}^{\infty} \ell_{1,v_0,-m_1}^{(r)} \ell_{2,v_0,-m_2}^{(r)} \ell_{1,v_0,k-n_1}^{(s)} \ell_{2,v_0,k-n_2}^{(s)} \text{Cov}(\epsilon_{m_1}\epsilon_{m_2}, \epsilon_{n_1}\epsilon_{n_2}).$$

Note that $\text{Cov}(\epsilon_{m_1}\epsilon_{m_2}, \epsilon_{n_1}\epsilon_{n_2}) \neq 0$ only in the following three cases:

(i) $m_1 = m_2 = n_1 = n_2$, (ii) $m_1 = n_1 \neq m_2 = n_2$, and (iii) $m_1 = n_2 \neq m_2 = n_1$.

Denote the contribution of these cases to W_{rs} by $W_{1,rs}$, $W_{2,rs}$ and $W_{3,rs}$, respectively.

Since $\text{Cov}(\epsilon_0\epsilon_0, \epsilon_0\epsilon_0) = \text{Var}(\epsilon_0^2)$, then setting $n = m_1 = m_2 = n_1 = n_2$, we have

$$W_{1,rs} = \text{Var}(\epsilon_0^2) \sum_{k=-\infty}^{\infty} \sum_{n=-\infty}^{\infty} \ell_{1,v_0,-n}^{(r)} \ell_{2,v_0,-n}^{(r)} \ell_{1,v_0,k-n}^{(s)} \ell_{2,v_0,k-n}^{(s)}$$

$$= \text{Var}(\epsilon_0^2) \sum_{n=-\infty}^{\infty} \ell_{1,v_0,n}^{(r)} \ell_{2,v_0,n}^{(r)} \sum_{k=-\infty}^{\infty} \ell_{1,v_0,k}^{(s)} \ell_{2,v_0,k}^{(s)}.$$

By using Parseval's identity (A.1.3), (10.2.24), (10.2.6) and (10.2.7), we conclude that

$$W_{1,rs} = \text{Var}(\epsilon_0^2) \frac{1}{2\pi}\int_{-\pi}^{\pi} \widehat{\ell}_{1,v_0}^{(r)}(\lambda)\overline{\widehat{\ell}_{2,v_0}^{(r)}(\lambda)} d\lambda \frac{1}{2\pi}\int_{-\pi}^{\pi} \widehat{\ell}_{1,v_0}^{(s)}(\lambda)\overline{\widehat{\ell}_{2,v_0}^{(s)}(\lambda)} d\lambda$$

$$= \text{Var}(\epsilon_0^2) \int_{-\pi}^{\pi} \Big(|g_v(e^{-i\lambda})|^2 \Big|\frac{\partial}{\partial v_r}\frac{1}{h_v(\lambda)}\Big| \text{sign}\Big(\frac{\partial}{\partial v_r}\frac{1}{h_v(\lambda)}\Big)\Big)\Big|_{v=v_0} d\lambda$$

$$\times \int_{-\pi}^{\pi} \Big(|g_v(e^{-i\lambda})|^2 \Big|\frac{\partial}{\partial v_s}\frac{1}{h_v(\lambda)}\Big| \text{sign}\Big(\frac{\partial}{\partial v_s}\frac{1}{h_v(\lambda)}\Big)\Big)\Big|_{v=v_0} d\lambda$$

$$= \text{Var}(\epsilon_0^2) \int_{-\pi}^{\pi} \Big(|g_v(e^{-i\lambda})|^2 \frac{1}{h_v(\lambda)^2}\Big|\frac{\partial}{\partial v_r}h_v(\lambda)\Big| \text{sign}\Big(\frac{\partial}{\partial v_r}h_v(\lambda)\Big)\Big)\Big|_{v=v_0} d\lambda$$

10.2 Approximate MLE

$$\times \int_{-\pi}^{\pi} \left(|g_\upsilon(e^{-i\lambda})|^2 \frac{1}{h_\upsilon(\lambda)^2} \left| \frac{\partial}{\partial \upsilon_s} h_\upsilon(\lambda) \right| \text{sign}\left(\frac{\partial}{\partial \upsilon_s} h_\upsilon(\lambda) \right) \right) \Big|_{\upsilon=\upsilon_0} d\lambda$$

$$= \text{Var}(\epsilon_0^2) \int_{-\pi}^{\pi} \frac{1}{h_\upsilon(\lambda)} \frac{\partial}{\partial \upsilon_r} h_\upsilon(\lambda) \Big|_{\upsilon=\upsilon_0} d\lambda \int_{-\pi}^{\pi} \frac{1}{h_\upsilon(\lambda)} \frac{\partial}{\partial \upsilon_s} h_\upsilon(\lambda) \Big|_{\upsilon=\upsilon_0} d\lambda$$

$$= \text{Var}(\epsilon_0^2) \frac{\partial}{\partial \upsilon_r} \int_{-\pi}^{\pi} \log h_\upsilon(\lambda) d\lambda \frac{\partial}{\partial \upsilon_s} \int_{-\pi}^{\pi} \log h_\upsilon(\lambda) d\lambda \Big|_{\upsilon=\upsilon_0} = 0.$$

This also means that cases (ii) and (iii) may include the terms with $m_1 = m_2 = n_1 = n_2$.

In case (ii) – that is, $m_1 = n_1 (= m) \neq m_2 = n_2 (= n)$ and including the case $m_1 = m_2 = n_1 = n_2$ as indicated above – we have

$$W_{2,rs} = (\mathbb{E}\epsilon_0^2)^2 \sum_{k=-\infty}^{\infty} \sum_{m=-\infty}^{\infty} \sum_{n=-\infty}^{\infty} \ell^{(r)}_{1,\upsilon_0,-m} \ell^{(r)}_{2,\upsilon_0,-n} \ell^{(s)}_{1,\upsilon_0,k-m} \ell^{(s)}_{2,\upsilon_0,k-n}$$

$$= (\mathbb{E}\epsilon_0^2)^2 \sum_{k=-\infty}^{\infty} \left(\sum_{m=-\infty}^{\infty} \ell^{(r)}_{1,\upsilon_0,-m} \ell^{(s)}_{1,\upsilon_0,k-m} \right) \left(\sum_{n=-\infty}^{\infty} \ell^{(r)}_{2,\upsilon_0,-n} \ell^{(s)}_{2,\upsilon_0,k-n} \right).$$

Note that the sum in the first parentheses is the convolution of $\{\ell^{(r)}_{1,\upsilon_0,-m}\}_{m\in\mathbb{Z}}$ and $\{\ell^{(s)}_{1,\upsilon_0,m}\}_{m\in\mathbb{Z}}$, and hence its Fourier transform is $\overline{\widehat{\ell}^{(r)}_{1,\upsilon_0}(\lambda)} \widehat{\ell}^{(s)}_{1,\upsilon_0}(\lambda)$. Similarly, the Fourier transform of the sequence in the second parentheses is $\overline{\widehat{\ell}^{(r)}_{2,\upsilon_0}(\lambda)} \widehat{\ell}^{(s)}_{2,\upsilon_0}(\lambda)$. By using Parseval's identity (A.1.3), we deduce that

$$W_{2,rs} = \frac{(\mathbb{E}\epsilon_0^2)^2}{2\pi} \int_{-\pi}^{\pi} \overline{\widehat{\ell}^{(r)}_{1,\upsilon_0}(\lambda)} \widehat{\ell}^{(s)}_{1,\upsilon_0}(\lambda) \widehat{\ell}^{(r)}_{2,\upsilon_0}(\lambda) \overline{\widehat{\ell}^{(s)}_{2,\upsilon_0}(\lambda)} d\lambda$$

$$= 2\pi (\mathbb{E}\epsilon_0^2)^2 \int_{-\pi}^{\pi} \frac{1}{h_\upsilon(\lambda)} \frac{\partial}{\partial \upsilon_r} h_\upsilon(\lambda) \frac{1}{h_\upsilon(\lambda)} \frac{\partial}{\partial \upsilon_s} h_\upsilon(\lambda) \Big|_{\upsilon=\upsilon_0} d\lambda$$

$$= 2\pi (\mathbb{E}\epsilon_0^2)^2 \int_{-\pi}^{\pi} \frac{\partial}{\partial \upsilon_r} \log h_\upsilon(\lambda) \frac{\partial}{\partial \upsilon_s} \log h_\upsilon(\lambda) \Big|_{\upsilon=\upsilon_0} d\lambda = 2\pi \sigma_0^4 c_{rs}(\upsilon_0).$$

Similarly, in case (iii) – that is, $m_1 = n_2 (= m) \neq m_2 = n_1 (= n)$ and including the case $m_1 = m_2 = n_1 = n_2$ as indicated above – we have

$$W_{3,rs} = (\mathbb{E}\epsilon_0^2)^2 \sum_{k=-\infty}^{\infty} \sum_{m=-\infty}^{\infty} \sum_{n=-\infty}^{\infty} \ell^{(r)}_{1,\upsilon_0,-m} \ell^{(r)}_{2,\upsilon_0,-n} \ell^{(s)}_{1,\upsilon_0,k-n} \ell^{(s)}_{2,\upsilon_0,k-m}$$

$$= (\mathbb{E}\epsilon_0^2)^2 \sum_{k=-\infty}^{\infty} \left(\sum_{m=-\infty}^{\infty} \ell^{(r)}_{1,\upsilon_0,-m} \ell^{(s)}_{2,\upsilon_0,k-m} \right) \left(\sum_{n=-\infty}^{\infty} \ell^{(r)}_{2,\upsilon_0,-n} \ell^{(s)}_{1,\upsilon_0,k-n} \right)$$

$$= \frac{(\mathbb{E}\epsilon_0^2)^2}{2\pi} \int_{-\pi}^{\pi} \overline{\widehat{\ell}^{(r)}_{1,\upsilon_0}(\lambda)} \widehat{\ell}^{(s)}_{2,\upsilon_0}(\lambda) \widehat{\ell}^{(r)}_{2,\upsilon_0}(\lambda) \overline{\widehat{\ell}^{(s)}_{1,\upsilon_0}(\lambda)} d\lambda$$

and hence $W_{3,rs} = W_{2,rs} = 2\pi \sigma_0^4 c_{rs}(\upsilon_0)$. It follows that $W_{rs} = W_{1,rs} + W_{2,rs} + W_{3,rs} = 4\pi \sigma_0^4 c_{rs}(\upsilon_0)$.

This concludes the outline of the proof of the asymptotic normality (10.2.20). Proving the convergence (10.2.21) is not immediate – see the proof of Theorem 8.3.1, Giraitis et al. [406], pp. 216–218. The proof of the asymptotic normality (10.2.11), on the other hand, is similar to that of (10.2.10) outlined above – see the proof of Theorem 8.3.1, Giraitis et al. [406], pp. 218–219. □

Example 10.2.4 The FARIMA$(0, d, 0)$ series have spectral density $f(\lambda) = \frac{\sigma^2}{2\pi}|1 - e^{-i\lambda}|^{-2d}$, corresponding to $\upsilon = d$, $g_\upsilon(e^{-i\lambda}) = (1 - e^{-i\lambda})^{-d}$ and $h_d(\lambda) = |1 - e^{-i\lambda}|^{-2d}$ in (10.2.6). Note that $\frac{\partial}{\partial d} \log h_d(\lambda) = -2\log|1 - e^{-i\lambda}|$ and hence that

$$c(d) = 4\int_{-\pi}^{\pi} (\log|1 - e^{-i\lambda}|)^2 d\lambda = 8\int_0^{\pi}\left(\log(2\sin\frac{\lambda}{2})\right)^2 d\lambda$$

$$= 16\int_0^{\pi}\left(\log(2\sin u)\right)^2 du = \frac{2\pi^3}{3},$$

where the last step follows by expanding the square $(\log(2\sin u))^2 = (\log 2 + \log\sin u)^2$ and using the formulas 4.224.3 and 4.224.7 in Gradshteyn and Ryzhik [421], p. 531. It follows that the variance in the asymptotic result (10.2.10) for $\widehat{d}_{whittle}$ is $4\pi(2\pi^3/3)^{-1} = 6/\pi^2$. See also Exercise 10.5.

Remark 10.2.5 The results presented in this section are asymptotic in nature, and apply to all considered estimation methods (exact MLE, Whittle approximation). How do the estimation methods compare for finite (small) samples? A nice and informative simulation study addressing this question can be found in Hauser [461]. See Section 10.9 for further discussion.

10.2.2 Autoregressive Approximation

Recall the expression (10.1.10) for the exact log-likelihood. The prediction $\widehat{X}_j$ is based on the observations $X_{j-1}, X_{j-2}, \ldots, X_1$. Consider first, however, the prediction $\widehat{X}_j^\infty$ based on the infinite past $X_{j-1}, X_{j-2}, \ldots$. The difference $X_j - \widehat{X}_j^\infty$ between the true value X_j and its prediction $\widehat{X}_j^\infty$ can be written as

$$X_j - \widehat{X}_j^\infty = X_j - \sum_{k=1}^{\infty} \pi_{\upsilon,k} X_{j-k}, \qquad (10.2.25)$$

where $\{\pi_{\upsilon,k}\}$ are defined formally through

$$1 - \sum_{k=1}^{\infty} \pi_{\upsilon,k} z^k = \frac{1}{\sum_{k=0}^{\infty} g_{\upsilon,k} z^k},$$

where $\{g_{\upsilon,k}\}$ is defined in (10.2.6). In other words, for the linear series $X_n = \sum_{j=0}^{\infty} g_{\upsilon,j} Z_{n-j}$, where $\{Z_n\}$ is a white noise series with variance $\sigma^2 = \sigma_Z^2$, the sequence $\{\pi_{\upsilon,k}\}$ appears in the inversion (*autoregressive representation*)

$$Z_n = X_n - \sum_{k=1}^{\infty} \pi_{\upsilon,k} X_{n-k}. \qquad (10.2.26)$$

If the observations are extended to the infinite past, the prediction error would be given by

$$\mathbb{E}(X_j - \widehat{X}_j^\infty)^2 = \sigma^2. \qquad (10.2.27)$$

The sequence $\{\pi_{v,k}\}$ could be computed efficiently and fast for many models of interest, for example, the FARIMA(p, d, q) family in Example 10.1.1.

Since the infinite past is not available in practice, we use instead $X_j - \widehat{X}_j$ and hence the approximate innovations $u_j(v) \approx X_j - \widehat{X}_j$, where

$$u_j(v) = X_j - \sum_{k=1}^{j-1} \pi_{v,k} X_{j-k}, \quad j = 1, \ldots, N. \qquad (10.2.28)$$

We also approximate the prediction errors v_{j-1} based on $\widehat{X}_j$ by σ^2. Then, an approximation of the log-likelihood (10.1.10) based on autoregressive representation is

$$\ell_{AR}(\sigma^2, v) = \log \sigma^2 + \frac{1}{\sigma^2 N} \sum_{j=1}^N u_j(v)^2. \qquad (10.2.29)$$

The corresponding ML estimators are defined as

$$(\widehat{\sigma}_{AR}^2, \widehat{v}_{AR}) = \operatorname*{argmin}_{\sigma^2, v} \ell_{AR}(\sigma^2, v). \qquad (10.2.30)$$

Note that $\widehat{v}_{AR}$ is obtained by minimizing the so-called conditional sum of squares $\sum_{j=1}^N u_j(v)^2$, and that

$$\widehat{\sigma}_{AR}^2 = \frac{1}{N} \sum_{j=1}^N u_j(\widehat{v}_{AR})^2. \qquad (10.2.31)$$

One can show that, under suitable assumptions, the estimators $\widehat{\sigma}_{AR}^2$, $\widehat{v}_{AR}$ satisfy the same asymptotic normality results as in Theorem 10.2.1 (Beran [122] and Section 5.5.5 in Beran et al. [127]). It should also be noted that these estimators can be applied to nonstationary series (Hualde and Robinson [498]).

10.3 Model Selection and Diagnostics

The estimation methods considered in Sections 10.1 and 10.2 assume that the model (order) is known (e.g., p and q in the FARIMA(p, d, q) class are known). In order to avoid overfitting, it is common to choose a model (order) using an information criterion wherein one minimizes

$$2\widehat{\ell} + \alpha s, \qquad (10.3.1)$$

where $\widehat{\ell}$ denotes the (non-scaled[1]) negative log-likelihood evaluated at the estimated parameter values, and s is the number of parameters used (e.g., $p+q+2$ for the FARIMA(p, d, q)

---

[1] By "non-scaled" we mean not divided by $N/2$ as in (10.1.3).

class, where 2 counts the parameters σ^2 and d). The value α in (10.3.1) depends on the criterion used:

$$
\begin{aligned}
&\text{AIC:} &&\alpha = 2 \\
&\text{BIC:} &&\alpha = 2\log N \\
&\text{HIC:} &&\alpha = c\log\log N
\end{aligned}
\quad (10.3.2)
$$

with $c > 1$ (AIC for Akaike Information Criterion, Akaike [17]; BIC for Bayesian Information Criterion, Schwarz [894]; HIC for Hannan Information Criterion, Hannan and Quinn [452]). The negative log-likelihood can be an approximate one as in (10.2.29), except that the latter should not be scaled by $N/2$ since (10.3.1) involves a non-scaled log-likelihood. Indeed, note that (10.2.29) is an approximation to (10.1.10) or, equivalently, (10.1.3). Thus, the approximate negative log-likelihood (10.2.29) needs to be multiplied by $N/2$; that is, the negative log-likelihood in (10.3.1) can be approximated by

$$\widehat{\ell} = \frac{N}{2}\ell_{AR}(\widehat{\sigma}_{AR}^2, \widehat{v}_{AR}) = \frac{N}{2}\log\widehat{\sigma}_{AR}^2 + \frac{N}{2}, \quad (10.3.3)$$

where we used (10.2.31). (Note that the last term $\frac{N}{2}$ can be ignored in minimizing (10.3.1).) The basic idea behind (10.3.1) is that the first term $2\widehat{\ell}$ decreases with the increasing number of parameters, whereas the second term αs acts as a penalty, which increases with the increasing number of parameters.

Some of the criteria above have been investigated for certain classes of LRD models. An empirical study of AIC in FARIMA(p, d, q) class was carried out in Crato and Ray [269]. Theoretical properties of AIC, BIC and HIC in FARIMA$(p, d, 0)$ class and the choice (10.3.3) were studied in Beran, Bhansali, and Ocker [124]. In particular, the order $\widehat{p}$ chosen by BIC and HIC was shown to be consistent, whereas the order chosen by AIC overestimates the true p with positive probability. A more recent work on the model selection can be found in Baillie, Kapetanios, and Papailias [90].

Once a time series model is chosen, its goodness of fit should be assessed through diagnostics checks. As with other statistical models, it is common to examine the normalized residuals. In the time series context, these are defined as

$$\widehat{\epsilon}_t = \frac{X_t - \widehat{X}_t(\widehat{\eta})}{v_{t-1}^{1/2}(\widehat{\eta})}, \quad (10.3.4)$$

where $\widehat{X}_t$ and v_{t-1} are defined as in (10.1.10) and evaluated at the estimated parameter η value. As in (10.2.28)–(10.2.31), the residuals could also be approximated through

$$\widehat{\epsilon}_t \approx \frac{u_t(\widehat{v}_{AR})}{\widehat{\sigma}_{AR}}. \quad (10.3.5)$$

If the model fits the data well, the residuals should be consistent with the white noise assumption. In the case of short-range dependent series, this is often assessed by examining the residual plot, and the portmanteau tests for residuals (Box–Pierce, Ljung–Box and others) based on the sample autocorrelation function coefficients of the residuals (see, for example, Brockwell and Davis [186]).

In the case of LRD series, these common diagnostics checks are problematic. The residuals (10.3.4) may be tedious to compute, especially for longer time series. The classical portmanteau tests involve only the sample autocorrelation function coefficients at a number of small lags, whereas LRD is intrinsically related to the autocorrelation function at large lags. A number of goodness-of-fit tests for LRD series were constructed to deal with these issues.

Chen and Deo [220] considered the test statistic

$$T_N(\widehat{\eta}) = \frac{\frac{2\pi}{N} \sum_{j=0}^{N-1} \widetilde{f}_e(\widehat{\eta}, \lambda_j)^2}{\left(\frac{2\pi}{N} \sum_{j=0}^{N-1} \widetilde{f}_e(\widehat{\eta}, \lambda_j)\right)^2}, \tag{10.3.6}$$

where

$$\widetilde{f}_e(\widehat{\eta}, \lambda) = \frac{1}{N} \sum_{k=1}^{N-1} \frac{W(\lambda - \lambda_k) I_X(\lambda_k)}{f(\widehat{\eta}, \lambda_k)}, \tag{10.3.7}$$

$I_X(\lambda)$ is the periodogram,[2] $\lambda_j = \frac{2\pi k}{N}$ are the Fourier frequencies, $\widehat{\eta}$ is an estimator of η, $\{f(\eta, \lambda)\}$ is a considered class of spectral densities (e.g., FARIMA(p, d, q) class) and W is the spectral window defined by

$$W(\lambda) = \frac{1}{2\pi} \sum_{|h|<N} k\left(\frac{h}{p_N}\right) e^{-ih\lambda},$$

where k is a kernel function (e.g., Bartlett $k(z) = (1-|z|)1_{[-1,1]}(z)$) and p_N is a bandwidth. Under suitable assumptions for both short-range and long-range dependent series, Chen and Deo [220] showed that, as $N \to \infty$,

$$\frac{N(T_N(\widehat{\eta}) - C_N(k))}{D_N(k)^{1/2}} \xrightarrow{d} \mathcal{N}(0, 1), \tag{10.3.8}$$

where

$$C_N(k) = \frac{1}{N\pi} \sum_{j=1}^{N-1} \left(1 - \frac{j}{N}\right) k^2\left(\frac{j}{p_N}\right) + \frac{1}{2\pi},$$

$$D_N(k) = \frac{2}{\pi^2} \sum_{j=1}^{N-2} \left(1 - \frac{j}{N}\right)\left(1 - \frac{j+1}{N}\right) k^4\left(\frac{j}{p_N}\right).$$

The asymptotic result (10.3.8) is used to devise a formal test where the null hypothesis is that the time series has the spectral density in the class of interest. One rejects the null hypothesis at a significance level α when the test statistic (the expression on the left-hand side of (10.3.8)) exceeds the critical value z_α, where z_α satisfies $\mathbb{P}(\mathcal{N}(0, 1) > z_\alpha) = \alpha$. Note that the test statistic (10.3.6) is not based on the model residuals and is easy to compute. It can nevertheless be viewed as a generalized portmanteau statistic involving *all* sample autocorrelation function coefficients of the residuals (see the discussion on pp. 384–385 in Chen and Deo [220]).

---

[2] Our definition of the periodogram differs by the factor 2π from that in Chen and Deo [220] – see also the warning in Section 1.3.3.

A statistic similar to (10.3.6) was also considered in Beran [120], with the correction in Deo and Chen [303]. Other testing procedures, based on the periodogram as well, can be found in Fay and Philippe [355], Delgado, Hidalgo, and Velasco [300].

10.4 Forecasting

A common objective of a time series analysis is (linear) forecasting (or prediction).[3] What about forecasting long-range dependent time series? These series are second-order stationary, and can be forecast in the general framework of second-order stationary series (e.g., Brockwell and Davis [186]). For example, a one-step forecast is defined as

$$\widehat{X}_{N+1} = \varphi_{N1} X_N + \varphi_{N2} X_{N-1} + \cdots + \varphi_{NN} X_1, \quad (10.4.1)$$

where the coefficients φ_{Ni} can be obtained from the Durbin–Levinson algorithm (cf. (10.1.4)) and the autocovariance function obtained from the fitted model. Similarly, a two-step forecast is defined by[4]

$$\widehat{X}_{N+2} = \varphi_{N+1,1} \widehat{X}_{N+1} + \varphi_{N+1,2} X_N + \cdots + \varphi_{N+1,N+1} X_1, \quad (10.4.2)$$

that is, as (10.4.1) but replacing the unobserved X_{N+1} by its forecast $\widehat{X}_{N+1}$, and so on. The mean-squared error of the one-step forecast is $v_N = \mathbb{E}(X_{N+1} - \widehat{X}_{N+1})^2$ (cf. (10.1.5)). It can be used to construct an approximate Gaussian confidence interval associated with the forecast $\widehat{X}_{N+1}$.

Though forecasting long-range dependent series fits in the general framework of second-order stationary series, there is also one natural feature that sets it apart. To explain this point, suppose N is large and consider a one-step forecast $\widehat{X}_{N+1}$ in (10.4.1). As in (10.2.26), we could write

$$X_{N+1} = \sum_{k=1}^{\infty} \pi_k X_{N+1-k} + Z_{N+1}. \quad (10.4.3)$$

If the forecast $\widehat{X}_{N+1}^{\infty}$ were based on the infinite past $X_N, X_{N-1}, \ldots$, then

$$\widehat{X}_{N+1}^{\infty} = \sum_{k=1}^{\infty} \pi_k X_{N+1-k}. \quad (10.4.4)$$

If N is assumed large, it is then expected that

$$\varphi_{Nk} \approx \pi_k. \quad (10.4.5)$$

On the other hand, for many LRD models with parameter $d \in (0, 1/2)$ (e.g., FARIMA$(0, d, 0)$ model for which $(I - B)^{-d} X_n = Z_n$), we have for large k,

$$\pi_k \approx C k^{-d-1}.$$

Hence, for large N,

$$\varphi_{Nk} \approx C k^{-d-1}. \quad (10.4.6)$$

---

[3] We do not distinguish between the terms "forecast" and "prediction." See also Section 6.4.2.
[4] We write $\varphi_{N+1,1}$ instead of $\varphi_{N+1 1}$ and so on to avoid possible confusion.

This shows that, for LRD series, the influence of the past observations decays as a power function if measured through the coefficients φ_{Nk}.

In fact, the view presented above draws a somewhat simplified picture. For example, as stated in Exercise 2.8, for FARIMA$(0, d, 0)$ series,

$$\varphi_{NN} = \frac{d}{N-d} \sim \frac{d}{N}, \qquad (10.4.7)$$

as $N \to \infty$. The same asymptotic behavior holds also for the entire class of long-range dependent FARIMA(p, d, q) models (Inoue [519]). Note that the decay in (10.4.7) is even slower than that in (10.4.6), with k replaced by N.

10.5 R Packages and Case Studies

We illustrate some of the methods presented above on real time series using R (R Core Team [838]). Several R packages are available to work with long-range dependent time series. We consider two of them below: `arfima` [976] and `fracdiff` [782], along with the use of the package `forecast` [1008] for the latter.

10.5.1 The ARFIMA Package

We consider first the `arfima` package. Figure 10.1 presents the time plot of the series `tmpyr` available in the package, and its sample autocorrelation function (ACF). The `tmpyr` series represents the central England mean yearly temperatures from 1659 to 1976. The number of observations is $N = 318$. The plots are obtained through the functions:

```
data(tmpyr)
plot(tmpyr)
acf(tmpyr,max.lag=100)
```

(The function `acf` is part of the general R stats package.)

Table 10.1 lists the AIC values for a number of fitted FARIMA(p, d, q) models. We focus on AIC (as opposed to other information criteria) since for the temperature time series, it

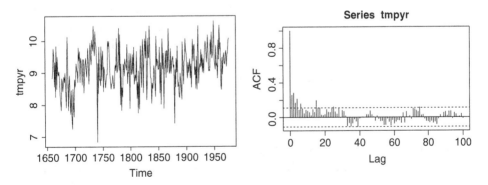

Figure 10.1 The time plot and the sample ACF of the central England mean yearly temperatures from 1659 to 1976.

Table 10.1 *The AIC values for the listed models.*

ARMA(p,q)			FARIMA(p,d,q)		
p	q	AIC	p	q	AIC
			0	0	−337.83
1	0	−319.56	1	0	−338.98
0	1	−312.18	0	1	−323.21
1	1	−337.99	1	1	**−339.75**
2	0	−335.37	2	0	−338.47
0	2	−326.26	0	2	−337.76

was considered in some early works on long-range dependence, e.g., Hosking [485]. For example, the AIC value for the FARIMA(1, d, 1) model is obtained through:

```
fit_1d1 <- arfima(tmpyr,order = c(1,0,1), numeach = c(2,1), dmean = FALSE)
fit_1d1
AIC(fit_1d1)
```

(the input `order = c(1,0,1)` sets the FARIMA(1, d, 1) model, the option `numeach = c(2,1)` sets the number of starting values for each parameter in the numerical optimization where the 2 refers to the ARMA parameters, the 1 to the long-range dependence parameter, and the option `dmean = FALSE` corresponds to the centered series where the maximization does not involve the mean). It is seen from the table (the entry in bold) that FARIMA(1, d, 1) is the model with the lowest AIC value. The model FARIMA(1, d, 0) has the second lowest AIC. The estimated parameter values (and the respective standard errors in parentheses) for the FARIMA(1, d, 1) fit are: $\widehat{d} = 0.278$ (0.05), $\widehat{\varphi}_1 = -0.748$ (0.147), $\widehat{\theta}_1 = 0.647$ (0.173),[5] $\widehat{\sigma}^2 = 0.335$ and the sample mean $\widehat{\mu} = 9.143$.

Figure 10.2 presents the prediction values and confidence intervals of the `tmpyr` series for five steps ahead based on the FARIMA(1, d, 1) model. The prediction and the plot is obtained through:

```
pred <- predict(fit_1d1, n.ahead = 5, bootpred = FALSE)
pred
plot(pred, numback = 25)
```

(the option `bootpred = FALSE` is to not produce bootstrap prediction and intervals, and the option `numback = 25` sets the number of the past observations of the original series to include in the plot). Two starting values were used for the likelihood when fitting the FARIMA(1, d, 1) model (see above). The top plot in Figure 10.2 corresponds to the starting value with the lowest AIC value and the bottom plot to the next lowest AIC value.

10.5.2 The FRACDIFF Package

We now turn to the `fracdiff` package. The function `fracdiff` in the package is one of the commonly used R functions to fit a FARIMA(p, d, q) model. Figure 10.3 presents the

---

[5] The convention used in the package has a different sign for the MA parameters θ. The numerical value given here follows our convention (see Example 2.3.3).

10.5 R Packages and Case Studies

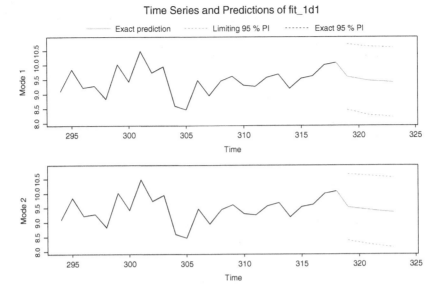

Figure 10.2 The five-step-ahead prediction of the central England mean yearly temperatures.

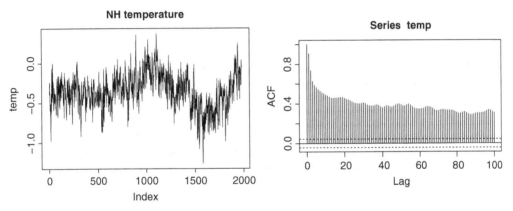

Figure 10.3 The time plot and the sample ACF of the Northern Hemisphere temperatures from 1 to 1979AD.

reconstructed annual NH (Northern Hemisphere) temperatures from 1 to 1979AD, analyzed by Moberg, Sonechkin, Holmgren, Datsenko, and Karlen [724], Mills [721].

Table 10.2 lists the AIC values for several FARIMA(p, d, q) models fitted to the Northern Hemisphere temperature data. For example, the entry for the FARIMA$(2, d, 1)$ model is obtained through:

```
fit_2d1 <- fracdiff(temp - mean(temp), nar = 2, nma = 1)
summary(fit_2d1)
```

The lowest AIC value is obtained for the FARIMA$(2, d, 2)$ model. The corresponding parameter estimates (and their standard errors in parentheses) are: $\widehat{d} = 0.44$ (0.02), $\widehat{\varphi}_1 = 0.32$ (0.03), $\widehat{\varphi}_2 = -0.25$ (0.025), $\widehat{\theta}_1 = 1.51$ (0.03), $\widehat{\theta}_2 = 0.603$ (0.025), together with the sample mean $\widehat{\mu} = -0.3537$ and the estimated standard deviation of the innovations $\widehat{\sigma} = 0.0484$.

Table 10.2 *The AIC values for the listed models.*

	FARIMA(p,d,q)			FARIMA(p,d,q)	
p	q	AIC	p	q	AIC
0	0	−3196.20	2	2	**−6353.06**
1	0	−4156.66	3	1	−6265.40
0	1	−5215.21	1	3	−6293.05
1	1	−5696.02	4	0	−5863.42
2	0	−5044.89	0	4	−6347.40
0	2	−6238.31	3	2	−6345.39
2	1	−6153.44	2	3	−6307.86
1	2	−5696.02	4	1	−6329.35
3	0	−6292.18	1	4	−6342.41
0	3	−5489.88			

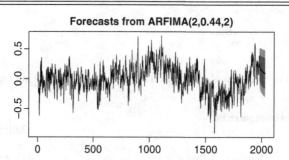

Figure 10.4 The five-step-ahead prediction of the Northern Hemisphere temperatures is depicted in the shaded region at the end of the graph.

Finally, Figure 10.4 depicts the forecasts of the demeaned time series for five steps ahead based on the fitted FARIMA$(2, d, 2)$ model. The plot is obtained through `plot(forecast(fit_2d2,h=50))`. The darker shaded band in the forecast region at the end of the graph corresponds to 80% confidence band. The wider band in lighter shade corresponds to 90% confidence band.

10.6 Local Whittle Estimation

Exact or approximate MLE is based on the possibly optimistic assumption that a specific model is correct. Moreover, for a parametric family of models with increasing number of parameters such as FARIMA(p, d, q) series in Example 10.1.1, MLE together with a model selection criterion will tend to favor models in the parametric family with fewer parameters. If one is specifically interested in the LRD parameter, one could use semi-parametric estimation, where the spectral density is specified around the frequency zero only. For example, the GPH estimator discussed in Section 2.10.3 is an example of a semi-parametric estimator. We discuss here another popular and more efficient semi-parametric estimator, called the *local Whittle estimator*.

10.6.1 Local Whittle Estimator

Recall the discrete Whittle approximation of the log-likelihood in (10.2.5). The idea of *local Whittle estimation* is to use the lowest m Fourier frequencies only, and consider

10.6 Local Whittle Estimation

$$\ell_{lw}(\eta) = \frac{1}{m} \sum_{j=1}^{m} \left(\log f_\eta(\frac{2\pi j}{N}) + \frac{\tilde{I}_X(\frac{2\pi j}{N})}{f_\eta(\frac{2\pi j}{N})} \right), \quad (10.6.1)$$

where we excluded the term with $\log(2\pi)$ resulting from the first sum of (10.2.5) and denote the periodogram normalized by 2π as

$$\tilde{I}_X(\lambda) = \frac{I_X(\lambda)}{2\pi} \quad (10.6.2)$$

for notational convenience throughout this section (see also the warning in Section 1.3.3). Assuming $f_\eta(\lambda) \sim c\lambda^{-2d}$, $0 < d < 1/2$, and $m/N \to 0$, the minimization of $\ell_{lw}(\eta)$ can be replaced by the minimization of

$$Q(c, d) = \frac{1}{m} \sum_{j=1}^{m} \left(\log(c\lambda_j^{-2d}) + \frac{\tilde{I}_X(\lambda_j)}{c\lambda_j^{-2d}} \right), \quad (10.6.3)$$

where $\lambda_j = \frac{2\pi j}{N}$. The parameter vector η of the spectral density $f_\eta(\lambda)$ now consists of the parameters c, d.

Differentiating with respect to c leads to

$$\frac{\partial Q}{\partial c}(c, d) = \frac{1}{m} \sum_{j=1}^{m} \left(c^{-1} - \frac{\tilde{I}_X(\lambda_j)}{c^2 \lambda_j^{-2d}} \right).$$

Setting this to zero gives

$$\hat{c} = \frac{1}{m} \sum_{j=1}^{m} \frac{\tilde{I}_X(\lambda_j)}{\lambda_j^{-2d}} =: G(d).$$

Then, the *local Whittle estimator* of d is defined as

$$\hat{d}_{lw} = \underset{d \in \Theta}{\operatorname{argmin}} \, Q(d), \quad (10.6.4)$$

where $\Theta \subset (-1/2, 1/2)$ and

$$Q(d) = \log G(d) - d\frac{2}{m} \sum_{j=1}^{m} \log \lambda_j = \log\left(\frac{1}{m} \sum_{j=1}^{m} \frac{\tilde{I}_X(\lambda_j)}{\lambda_j^{-2d}}\right) - d\frac{2}{m} \sum_{j=1}^{m} \log \lambda_j, \quad (10.6.5)$$

that is, by focusing on the first term in (10.6.3) and replacing c by $G(d)$.

It is useful to indicate first why the local Whittle estimator is expected to estimate the true parameter d_0. Assuming $\tilde{I}_X(\lambda_j) \approx f_{\eta_0}(\lambda_j) \sim c_0 \lambda_j^{-2d_0}$, the objective function in (10.6.5) is approximately

$$\log\left(\frac{1}{m} \sum_{j=1}^{m} c_0 \lambda_j^{-2(d_0-d)}\right) - d\frac{2}{m} \sum_{j=1}^{m} \log \lambda_j = R(\frac{m}{N}, c_0, d_0)$$

$$+ \log\left(\frac{1}{m} \sum_{j=1}^{m} (\frac{j}{m})^{-2(d_0-d)}\right) - d\frac{2}{m} \sum_{j=1}^{m} \log \frac{j}{m}$$

$$\approx R(\frac{m}{N}, c_0, d_0) + \log \int_0^1 x^{-2(d_0-d)} dx - 2d \int_0^1 \log x \, dx$$
$$= R(\frac{m}{N}, c_0, d_0) - \log(1 - 2(d_0 - d)) + 2d =: \overline{Q}(d),$$

where R is some function that does not depend on d. Note that the function $\overline{Q}(d)$ is minimized at $d = d_0$.

In the following result, we provide the asymptotic normality result for the local Whittle estimator $\widehat{d}_{lw}$, due to Robinson [853]. The following assumptions will be made:

(B1) For some $\rho \in (0, 2]$, the spectral density satisfies

$$f_{\eta_0}(\lambda) = c_0 \lambda^{-2d_0}(1 + O(\lambda^\rho)), \qquad (10.6.6)$$

as $\lambda \to 0$, where $c > 0$ and $d_0 \in [\Delta_1, \Delta_2]$ $(-1/2 < \Delta_1 < \Delta_2 < 1/2)$.

(B2) The series satisfies

$$X_n - \mathbb{E} X_0 = \sum_{j=0}^\infty \alpha_j \epsilon_{n-j}, \quad \sum_{j=0}^\infty \alpha_j^2 < \infty, \qquad (10.6.7)$$

where

$$\mathbb{E}(\epsilon_n | \mathcal{F}_{n-1}) = 0, \ \mathbb{E}(\epsilon_n^2 | \mathcal{F}_{n-1}) = 1, \ \mathbb{E}(\epsilon_n^3 | \mathcal{F}_{n-1}) = \mu_3 \text{ a.s.}, \ \mathbb{E}\epsilon_n^4 = \mu_4,$$

for finite constants μ_3, μ_4 and $\mathcal{F}_n = \sigma\{\epsilon_k, k \leq n\}$. Moreover, there exists a random variable ϵ such that $\mathbb{E}\epsilon^2 < \infty$ and $\mathbb{P}(|\epsilon_n| > x) \leq K\mathbb{P}(|\epsilon| > x)$ for some $K > 0$ and all $x \in \mathbb{R}$.

(B3) In a neighborhood of zero, $\alpha(\lambda) = \sum_{j=0}^\infty \alpha_j e^{-ij\lambda}$ is differentiable, and

$$\frac{d}{d\lambda}\alpha(\lambda) = O\Big(\frac{|\alpha(\lambda)|}{\lambda}\Big), \quad \text{as } \lambda \to 0.$$

(B4) As $N \to \infty$,[6]

$$\frac{1}{m} + \frac{m^{1+2\rho}}{N^{2\rho}} \to 0. \qquad (10.6.8)$$

Theorem 10.6.1 *Let $\widehat{d}_{lw}$ be the local Whittle estimator of d defined in (10.6.4). Under Assumptions (B1)–(B4) above,*

$$\sqrt{m}(\widehat{d}_{lw} - d_0) \xrightarrow{d} \mathcal{N}\Big(0, \frac{1}{4}\Big). \qquad (10.6.9)$$

Remark 10.6.2 Note the slower rate of convergence $\sqrt{m}$ when compared to $\sqrt{N}$ in (10.2.10). Note also the gain in efficiency when compared to the GPH estimator in (2.10.18). Here the asymptotic variance is $1/4$ whereas for the GPH it is $\pi^2/24 = 0.41$, which is greater. Thus, the gain in efficiency is $(1/4)/(\pi^2/24) = 1.64$.

---

[6] The corresponding Assumption (A4') in Robinson [853], p. 1641, carries a factor $(\log m)^2$ next to $m^{1+2\rho}$ in (10.6.8). This factor is not needed as shown by Andrews and Sun [34], p. 577.

Proof outline of Theorem 10.6.1 We provide here an outline of the proof of the theorem. A complete proof can be found in Robinson [853], Theorem 2. See also Theorem 2 and Corollary 1 in Andrews and Sun [34] for a slightly different approach in a more general context.

The first step in the proof is to show that $\widehat{d}_{lw}$ is consistent. We will not discuss the proof of consistency here (see Theorem 1 in Robinson [853], or Andrews and Sun [34] for a different approach). In view of (10.6.4)–(10.6.5), the first order condition and the Taylor's expansion yield

$$0 = \frac{\partial Q}{\partial d}(\widehat{d}_{lw}) = \frac{\partial Q}{\partial d}(d_0) + \frac{\partial^2 Q}{\partial^2 d}(d^*)(\widehat{d}_{lw} - d_0),$$

where $|d^* - d_0| \leq |\widehat{d}_{lw} - d_0|$. Hence,

$$\sqrt{m}(\widehat{d}_{lw} - d_0) = -\frac{1}{\frac{\partial^2 Q}{\partial^2 d}(d^*)} \sqrt{m} \frac{\partial Q}{\partial d}(d_0). \tag{10.6.10}$$

It is then enough to show that $\sqrt{m}\frac{\partial Q}{\partial d}(d_0) \xrightarrow{d} \mathcal{N}(0, 4)$ and $\frac{\partial^2 Q}{\partial d^2}(d^*) \xrightarrow{p} 4$.

Note that (10.6.5) implies

$$\frac{\partial Q}{\partial d}(d_0) = \Big(\frac{2}{m}\sum_{j=1}^{m} \log \lambda_j \frac{\widetilde{I}_X(\lambda_j)}{c_0 \lambda_j^{-2d_0}}\Big)\Big(\frac{1}{m}\sum_{j=1}^{m} \frac{\widetilde{I}_X(\lambda_j)}{c_0 \lambda_j^{-2d_0}}\Big)^{-1} - \frac{2}{m}\sum_{j=1}^{m} \log \lambda_j \tag{10.6.11}$$

$$= -\Big[\frac{1}{m}\sum_{j=1}^{m}(z_j - \overline{z})\Big(\frac{\widetilde{I}_X(\lambda_j)}{c_0 \lambda_j^{-2d_0}} - 1\Big)\Big]\Big(\frac{1}{m}\sum_{j=1}^{m} \frac{\widetilde{I}_X(\lambda_j)}{c_0 \lambda_j^{-2d_0}}\Big)^{-1}, \tag{10.6.12}$$

where $z_j = -2\log \lambda_j$ and $\overline{z} = \frac{1}{m}\sum_{j=1}^{m} z_j$, and we could add c_0 in (10.6.11) since it cancels out after multiplication with its reciprocal. In view of (2.10.15), one expects that

$$\frac{\widetilde{I}_X(\lambda_j)}{c_0 \lambda_j^{-2d_0}} \approx \xi_j, \tag{10.6.13}$$

where ξ_j are independent $\chi_2^2/2$ (or standard exponential) random variables. Since

$$\frac{1}{m}\sum_{j=1}^{m} \frac{\widetilde{I}_X(\lambda_j)}{c_0 \lambda_j^{-2d_0}} \approx \frac{1}{m}\sum_{j=1}^{m} \xi_j \approx \mathbb{E}\xi_j = 1,$$

we have

$$\sqrt{m}\frac{\partial Q}{\partial d}(d_0) \approx -\frac{1}{\sqrt{m}}\sum_{j=1}^{m}(z_j - \overline{z})(\xi_j - 1). \tag{10.6.14}$$

By the Lindeberg–Feller central limit theorem for independent, non-identically distributed random variables (Feller [360], Chapter VIII, Section 4), the right-hand side converges in distribution to $\mathcal{N}(0, \sigma^2)$, where

$$\sigma^2 = \lim_{m \to \infty} \frac{1}{m}\sum_{j=1}^{m}(z_j - \overline{z})^2 = 4,$$

by using Proposition 2.10.2.

Similarly, from (10.6.11),

$$\frac{\partial^2 Q}{\partial d^2}(d) = 4\frac{\left(\frac{1}{m}\sum_{j=1}^m (\log \lambda_j)^2 \frac{\widetilde{I}_X(\lambda_j)}{c_0 \lambda_j^{-2d}}\right)\left(\frac{1}{m}\sum_{j=1}^m \frac{\widetilde{I}_X(\lambda_j)}{c_0 \lambda_j^{-2d}}\right) - \left(\frac{1}{m}\sum_{j=1}^m \log \lambda_j \frac{\widetilde{I}_X(\lambda_j)}{c_0 \lambda_j^{-2d}}\right)^2}{\left(\frac{1}{m}\sum_{j=1}^m \frac{\widetilde{I}_X(\lambda_j)}{c_0 \lambda_j^{-2d}}\right)^2}$$

$$= 4\frac{\left(\frac{1}{m}\sum_{j=1}^m (\log j)^2 \frac{\widetilde{I}_X(\lambda_j)}{c_0 \lambda_j^{-2d}}\right)\left(\frac{1}{m}\sum_{j=1}^m \frac{\widetilde{I}_X(\lambda_j)}{c_0 \lambda_j^{-2d}}\right) - \left(\frac{1}{m}\sum_{j=1}^m \log j \frac{\widetilde{I}_X(\lambda_j)}{c_0 \lambda_j^{-2d}}\right)^2}{\left(\frac{1}{m}\sum_{j=1}^m \frac{\widetilde{I}_X(\lambda_j)}{c_0 \lambda_j^{-2d}}\right)^2},$$

where the last equality follows by simple calculation using $\lambda_j = 2\pi j/N$. Since d^* approaches d_0 in probability and using the approximation (10.6.13), it is expected that

$$\frac{\partial^2 Q}{\partial d^2}(d^*) \approx 4\frac{\left(\frac{1}{m}\sum_{j=1}^m (\log j)^2 \xi_j\right)\left(\frac{1}{m}\sum_{j=1}^m \xi_j\right) - \left(\frac{1}{m}\sum_{j=1}^m (\log j)\xi_j\right)^2}{\left(\frac{1}{m}\sum_{j=1}^m \xi_j\right)^2}. \quad (10.6.15)$$

Since $\mathrm{Var}(m^{-1}\sum_{j=1}^m (\log j)^k \xi_j) = Cm^{-2}\sum_{j=1}^m (\log j)^{2k} \sim Cm^{-2}\int_1^m (\log x)^{2k} dx \sim Cm^{-1}(\log m)^{2k}$, for $k = 0, 1, 2$, we have

$$\frac{1}{m}\sum_{j=1}^m (\log j)^k \xi_j - \frac{1}{m}\sum_{j=1}^m (\log j)^k = O_p\left(\frac{(\log m)^k}{m^{1/2}}\right) \to 0,$$

as $m \to \infty$, so we can approximate the first term on the left-hand side by the second term. We also have $m^{-1}\sum_{j=1}^m \xi_j \to \mathbb{E}\xi_j = 1$ as $m \to \infty$. Hence, the right-hand side of (10.6.15) behaves in probability as

$$4\left(\frac{1}{m}\sum_{j=1}^m (\log j)^2 - \left(\frac{1}{m}\sum_{j=1}^m \log j\right)^2\right) = 4\frac{1}{m}\sum_{j=1}^m \left(\log j - \frac{1}{m}\sum_{j=1}^m \log j\right)^2 = \frac{1}{m}\sum_{j=1}^m (z_j - \bar{z})^2 \to 4,$$

where $z_j = -2\log\frac{2\pi j}{N}$ and using Proposition 2.10.2 for the last convergence. $\square$

In practice, the local Whittle estimation is accompanied by a plot of the estimated values $\widehat{d}_{lw}$ against the number of lower frequencies m used. Such plot is referred to as the (local) *Whittle plot*. Since the numerical optimization procedure can find minimizers d outside the primary range $(-1/2, 1/2)$, especially when m is small, the vertical axis in the Whittle plot is often larger than this range to indicate such values of d. For example, the left plot of Figure 10.5 presents the Whittle plot for a Gaussian FARIMA$(0, 0.4, 0)$ series of length $N = 1024$. The 95% asymptotic confidence bands, based on Theorem 10.6.1, are also plotted in dash-dot lines. The right plot of the figure presents the Whittle plots for 20 realizations of the series.

10.6.2 Bandwidth Selection

The estimated values $\widehat{d}_{lw}$ in the Whittle plots of Figure 10.5 are close to the true value $d = 0.4$ for all numbers of lower frequencies m used. This is not necessarily a typical behavior.

10.6 Local Whittle Estimation

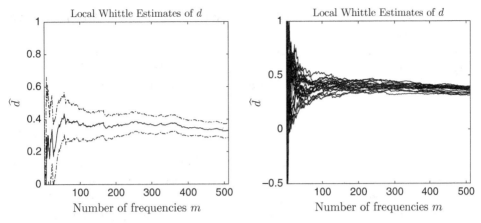

Figure 10.5 Local Whittle plots for Gaussian FARIMA(0, 0.4, 0) series of length $N = 1024$. Left: one realization with confidence bands. Right: 20 realizations.

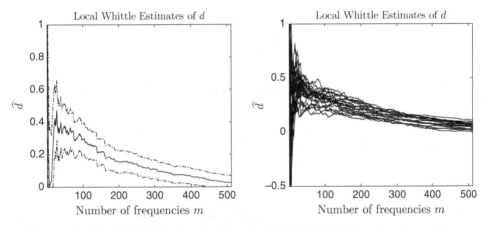

Figure 10.6 Local Whittle plots for Gaussian FARIMA(0, d, 1) series of length $N = 1024$ with $d = 0.4$ and $\theta_1 = -2$. Left: one realization with confidence bands. Right: 20 realizations.

For example, Figure 10.6 presents similar Whittle plots for a Gaussian FARIMA(0, d, 1) series with $d = 0.4$ and the MA parameter $\theta_1 = -2$. The estimated value $\widehat{d}_{lw}$ is now much more sensitive to the choice of the number of frequencies m. The value m is referred to as the *bandwidth*. How does one choose the bandwidth in such situations?

The reason that the choice of the bandwidth is not immediate, for example, in Figure 10.6, is two-fold. On one hand, for larger bandwidth m, the spectral density may deviate from the power law behavior stipulated by the model. This translates into a bias of the local Whittle estimator, as in Figure 10.6. On the other hand, for smaller m, no bias is expected but the variance of the local Whittle estimator is now larger. This phenomenon is quite typical in many estimation problems and is known as the *bias-variance tradeoff*.

A bandwidth m should be chosen to balance the bias-variance tradeoff. If the decision on m needs to be made from a single time series realization and hence from one single Whittle plot, a basic idea is to choose a region of m where the plot "flattens out" for smaller values of m and the variance is not "too large" yet. In the left plot of Figure 10.6, this would be a rough range of m from 25 to 75.

The bias-variance tradeoff could also be examined in more mathematical terms, leading to plug-in (automatic) choices of the bandwidth if such are desired (e.g., when dealing with many realizations of time series). More specifically, suppose that the spectral density of the time series satisfies

$$f_{\eta_0}(\lambda) = c_0 \lambda^{-2d_0}(1 + c_\rho \lambda^\rho + o(\lambda^\rho)), \quad \rho > 0, \tag{10.6.16}$$

as $\lambda \to 0$. The value $\rho = 2$ refers to the so-called *smooth* models such as FARIMA series. *Non-smooth* models are also of interest. For example, consider the time series model $X_n + Y_n$, where X_n has the spectral density $c_0 \lambda^{-2d_0}$ and Y_n is a white noise. (Similar models are considered in the context of stochastic volatility.) Then, the spectral density of $X_n + Y_n$ is $c_0 \lambda^{-2d_0} + \sigma_Y^2/(2\pi)$ and hence satisfies (10.6.16) with $\rho = 2d_0$ (depending on the parameter d_0!).

Assuming from (10.6.10) and (10.6.12) that

$$\widehat{d}_{lw} - d_0 \approx -\frac{1}{4m}\sum_{j=1}^m (z_j - \overline{z})\Big(\frac{\widetilde{I}_X(\lambda_j)}{c_0 \lambda_j^{-2d_0}} - 1\Big),$$

where $z_j = -2\log \lambda_j = -2\log\frac{2\pi j}{N}$, and supposing that the difference between $f_{\eta_0}(\lambda_j)$ and $\mathbb{E}\widetilde{I}_X(\lambda_j)$ is negligible, we have

$$\widehat{d}_{lw} - d_0 \approx -\frac{1}{4m}\sum_{j=1}^m (z_j - \overline{z})\Big(\frac{\widetilde{I}_X(\lambda_j)}{c_0 \lambda_j^{-2d_0}} - \frac{\mathbb{E}\widetilde{I}_X(\lambda_j)}{c_0 \lambda_j^{-2d_0}}\Big) - \frac{1}{4m}\sum_{j=1}^m (z_j - \overline{z})\Big(\frac{f_{\eta_0}(\lambda_j)}{c_0 \lambda_j^{-2d_0}} - 1\Big). \tag{10.6.17}$$

The first term in (10.6.17) has zero mean, and its variance behaves as $1/(4m)$ by (10.6.9). The second term in (10.6.17), on the other hand, is deterministic and is responsible for the bias of $\widehat{d}_{lw}$. According to (10.6.16), it behaves as

$$\frac{1}{4m}\sum_{j=1}^m (z_j - \overline{z})\Big(\frac{f_{\eta_0}(\lambda_j)}{c_0 \lambda_j^{-2d_0}} - 1\Big) \approx \frac{1}{4m}\sum_{j=1}^m (z_j - \overline{z}) c_\rho \lambda_j^\rho$$

$$= \Big(\frac{m}{N}\Big)^\rho \frac{(2\pi)^\rho c_\rho}{4} \frac{1}{m}\sum_{j=1}^m (z_j - \overline{z})\Big(\frac{j}{m}\Big)^\rho \approx -\Big(\frac{m}{N}\Big)^\rho \frac{(2\pi)^\rho c_\rho}{4} \frac{2}{m}\sum_{j=1}^m \Big(\log(\frac{j}{m}) + 1\Big)\Big(\frac{j}{m}\Big)^\rho$$

$$\approx -\Big(\frac{m}{N}\Big)^\rho \frac{(2\pi)^\rho c_\rho}{2} \int_0^1 (\log x + 1) x^\rho dx = -\Big(\frac{m}{N}\Big)^\rho \frac{(2\pi)^\rho c_\rho}{2} \frac{\rho}{(\rho+1)^2}, \tag{10.6.18}$$

where we used Proposition 2.10.2 to replace $z_j - \overline{z}$ and integration by parts in order to compute the integral above. The mean squared error (MSE) of $\widehat{d}_{lw}$ is then approximately equal to

$$\mathbb{E}(\widehat{d}_{lw} - d_0)^2 \approx \frac{1}{4m} + \frac{\rho^2 c_\rho^2 (2\pi)^{2\rho}}{4(\rho+1)^4}\Big(\frac{m}{N}\Big)^{2\rho}, \tag{10.6.19}$$

where the first term accounts for the asymptotic variance and the second term for (the square of) the asymptotic bias. The optimal bandwidth m that balances the variance and the bias is (by differentiating and setting to 0 to solve for m)

$$m_{opt} = \left(\frac{(\rho+1)^4}{2\rho^3 c_\rho^2 (2\pi)^{2\rho}}\right)^{\frac{1}{2\rho+1}} N^{\frac{2\rho}{2\rho+1}}. \tag{10.6.20}$$

Note also that $m = m_{opt}$ is such that, up to a multiplicative constant, the asymptotic variance $1/(4m)$ is equal to the asymptotic bias squared, expressed through the last term in (10.6.19). Indeed, this fact follows since by differentiating (10.6.19), the powers of m in both terms are reduced by 1.

Note that m_{opt} is of little use when ρ is unknown. If one is willing to assume a smooth model corresponding to $\rho = 2$, then

$$m_{opt} = KN^{\frac{4}{5}}, \tag{10.6.21}$$

where K is a constant depending on an unknown c_ρ. The constant K or c_ρ can be estimated through an iterative procedure (Henry and Robinson [470], Henry [468]) or a least squares method (Delgado and Robinson [299], Henry [468]).

Note also that with (10.6.21) when $\rho = 2$ or with (10.6.20) for general ρ, the bias is of the order $1/\sqrt{m}$ (see the observation following (10.6.20)) and thus is not negligible in the limit when multiplied by $\sqrt{m}$ (cf. Theorem 10.6.1). When $\rho = 2$ and m is m_{opt} in (10.6.21), the limit result (10.6.9) becomes

$$\sqrt{m_{opt}}(\widehat{d}_{lw} - d_0) - \frac{4\pi^2 c_\rho K^2}{9} \overset{d}{\to} \mathcal{N}\left(0, \frac{1}{4}\right), \tag{10.6.22}$$

where the bias $\frac{4\pi^2 c_\rho K^2}{9}$ is that in (10.6.18) with $\rho = 2$, $m = KN^{\frac{4}{5}}$ and the change of sign in view of (10.6.17) (see also Corollary 1 and Comment 1 in Andrews and Sun [34]). In practice, after estimating c_ρ, the bias can be removed by considering $\widehat{d}_{lw} - 4\pi^2 c_\rho K^2/(9\sqrt{m_{opt}})$.

10.6.3 Bias Reduction and Rate Optimality

The bias $\frac{4\pi^2 c_\rho K^2}{9}$ in (10.6.22) is not only characteristic of the bandwidth (10.6.21). In fact, it creates a well-known challenge to estimating d through the local Whittle method. According to (10.6.19), the bias is approximately

$$\frac{\rho c_\rho (2\pi)^\rho}{2(\rho+1)^2} \left(\frac{m}{N}\right)^\rho$$

and can be nonnegligible for a range of bandwidths m.

If one is willing to assume $\rho = 2$, then the bias can be removed after estimating c_ρ as discussed above. Another possibility is to use tapering where the periodogram is replaced by

$$\frac{1}{2\pi \sum_n h_n^2} \left| \sum_{n=1}^N X_n h_n e^{-in\lambda} \right|^2, \tag{10.6.23}$$

where h_n is the taper, e.g., the cosine bell taper

$$h_n = 0.5\left(1 - \cos\frac{2\pi(n - 1/2)}{N}\right), \quad n = 1, 2, \ldots, N.$$

Tapering is well-known to reduce bias in estimating the spectral density, and leads to smaller bias in the local Whittle estimation of d (Hurvich and Chen [506]). One downside is that the asymptotic variance gets slightly inflated.

Another interesting approach to bias reduction was suggested by Andrews and Sun [34]. Assume that the spectral density satisfies

$$f_{\eta_0}(\lambda) = c_0 \lambda^{-2d_0}\left(1 + \sum_{k=1}^{q}\frac{b_k \lambda^{2k}}{(2k)!} + o(\lambda^{2q})\right), \qquad (10.6.24)$$

as $\lambda \to 0$. This coincides with the specification (10.6.16) with $\rho = 2q$. When $q = 1$, we get $\rho = 2$. The choice $q \geq 2$ in (10.6.24) assumes that $f_{\eta_0}(\lambda)\lambda^{2d_0}$ is smoother. The bias-reduced local Whittle estimator is obtained by minimizing

$$Q(c, d, \theta_1, \ldots, \theta_{q-1}) = \frac{1}{m}\sum_{j=1}^{m}\left(\log(c\lambda_j^{-2d} e^{\sum_{k=1}^{q-1}\theta_k \lambda_j^{2k}}) + \frac{\widetilde{I}_X(\lambda_j)}{c\lambda_j^{-2d} e^{\sum_{k=1}^{q-1}\theta_k \lambda_j^{2k}}}\right)$$

(cf. (10.6.3)). The estimator is also called the *local polynomial Whittle estimator*, in analogy to local polynomial kernel smoothing (Fan and Gijbels [353]). As with local polynomial kernel smoothing, the basic idea is that the log of the spectral density is now modeled around 0 as

$$\log(c\lambda_j^{-2d} e^{\sum_{k=1}^{q-1}\theta_k \lambda_j^{2k}}) = \log c - 2d\log\lambda_j + \sum_{k=1}^{q-1}\theta_k \lambda_j^{2k}.$$

The bias in estimating d is reduced by capturing deviations from a straight line in d in the log-log plot through a higher-order polynomial.

Under (10.6.24), the bias of the local polynomial Whittle estimator is of the order $(m/N)^{2q}$, which decreases as a power of (m/N) as q increases. On the downside, however, the asymptotic variance of the local polynomial Whittle estimator increases as $q \to \infty$ Andrews and Sun [34], Comment 1 on p. 582), making the choice between a reduced bias and a larger asymptotic variance quite delicate. Usually a large sample size is needed to benefit from the bias-reduced estimation.

Finally, we also note that the issue of bias is naturally related to the rate of convergence of the local Whittle estimators. With the optimal bandwidth (10.6.20), the optimal rate of convergence is

$$m_{opt}^{-1/2} \quad \text{or} \quad N^{-\frac{\rho}{1+2\rho}}. \qquad (10.6.25)$$

For example, when $\rho = 2$, the rate is $N^{-2/5}$. When the bias is reduced, one can expect a better convergence rate. Indeed, by balancing the orders of the asymptotic variance $1/m$

and the squared bias $(m/N)^{4q}$ as indicated above,[7] the optimal bandwidth is $m_{opt} = CN^{4q/(1+4q)}$. The rate of convergence $m_{opt}^{-1/2}$ is

$$N^{-\frac{2q}{1+4q}}.$$

Note that this rate approaches the parametric rate $N^{-1/2}$ as $q \to \infty$.

10.7 Broadband Whittle Approach

In modeling LRD time series, the spectral density can be viewed as

$$f_\eta(\lambda) = \lambda^{-2d} f^*(\lambda), \tag{10.7.1}$$

where $f^*(\lambda)$ approaches a constant c as $\lambda \to 0$ (and λ^{-2d} can be replaced by $|1-e^{-i\lambda}|^{-2d}$). Here, λ^{-2d} accounts for long-range dependence and $f^*(\lambda)$ captures short-range dependence effects. As in Sections 10.1 and 10.2, we can assume a parametric form for $f^*(\lambda)$, e.g., that of ARMA(p,q) family. The MLE of d then converges to the true value at the parametric rate $N^{-1/2}$. This is a *parametric* and *global* approach. Alternatively, as in Section 10.6, the behavior of $f^*(\lambda)$ can be specified around $\lambda = 0$ only. Assuming $f^*(\lambda)$ is smooth at $\lambda = 0$, d can then be estimated at the rate $N^{-2/5}$ as in (10.6.25) with $\rho = 2$. This is a *semi-parametric* or *local* (*narrow-band*) approach.

There is one other interesting approach, called *broadband*, where $f^*(\lambda)$ is estimated by using a truncated series expansion. More specifically, suppose that $\log f^*(\lambda)$ has a Fourier cosine expansion

$$f^*(\lambda) = \exp\left\{\sum_{l=0}^\infty a_l \cos(l\lambda)\right\}, \tag{10.7.2}$$

which can also be written as[8]

$$f^*(\lambda) = \frac{\sigma^2}{2\pi} \exp\left\{\sum_{l=1}^\infty a_l \cos(l\lambda)\right\} \tag{10.7.3}$$

(after setting $\exp\{a_0\} = \sigma^2/(2\pi)$). By truncating the series in (10.7.2)–(10.7.3) to a finite order p, replace the Whittle approximation of the log-likelihood in (10.2.5) by

$$\ell_{bw,p}(\eta) = \sum_{j=1}^N \log\left(2\pi f_{p,\eta}(\frac{2\pi j}{N})\right) + \sum_{j=1}^N \frac{I_X(\frac{2\pi j}{N})}{2\pi f_{p,\eta}(\frac{2\pi j}{N})}, \tag{10.7.4}$$

where $\eta = (d, a_0, a_1, \ldots, a_p)$ and

$$f_{p,\eta}(\lambda) = \lambda^{-2d} \exp\left\{\sum_{l=0}^p a_l \cos(l\lambda)\right\}.$$

---

[7] Again, minimize $C_1/m + C_2(m/N)^{4q}$ with respect to m by differentiating and setting to 0.
[8] σ^2 in (10.7.3) plays the same role as in the relation (10.2.6) and the discussion following it.

The objective function $\ell_{bw,p}(\eta)$ approaches the discrete Whittle approximation of the log-likelihood in (10.2.5) as $p \to \infty$ since $f_{p,\eta}(\lambda)$ becomes $\lambda^{-2d} f^*(\lambda)$. The function $\ell_{bw,p}(\eta)$ is convex in its parameters (Exercise 10.8), and is amenable to numerical minimization. The zero vector is often suggested for the initial estimators of the parameters for the nonlinear optimization. Set

$$(\widehat{d}_{bw}, \widehat{a}_0, \ldots, \widehat{a}_p) = \widehat{\eta}_{bw} = \operatorname*{argmin}_{\eta} \ell_{bw,p}(\eta). \quad (10.7.5)$$

In the next result, due to Narukawa and Matsuda [740], we provide the asymptotics of the broadband Whittle estimator $\widehat{d}_{bw}$ defined above. The following assumptions will be made:[9]

(C1) The Fourier series (10.7.2) of $\log f^*(\lambda)$ is such that $\sum_{l=1}^{\infty} l^{\varrho} |a_l| < \infty$, with $\varrho > 9/2$.

(C2) The series $\{X_n\}_{n \in \mathbb{Z}}$ can be expressed through a linear representation $X_n = \mu + \sum_{j=0}^{\infty} \alpha_j \epsilon_{n-j}$, $\sum_{j=0}^{\infty} \alpha_j^2 < \infty$, where $\{\epsilon_n\}$ is a sequence of i.i.d. random variables with $\mathbb{E}\epsilon_n = 0$, $\mathbb{E}\epsilon_n^4 < \infty$.

(C3) The true parameter value $\eta_0 = (d_0, a_{0,0}, a_{0,1}, \ldots)$ lies in the interior of the parameter space $\{d \in [0, 1/2), a_0, a_1, \ldots \in \mathbb{R}, |a_j| \le C j^{-\delta}\}$ where $0 < C < \infty$ and $\delta > 1$. Moreover, for any parameter values $(d_1, a_0^1, a_1^1, \ldots)$ and $(d_2, a_0^2, a_1^2, \ldots)$ from the parameter space with $d_1 \ne d_2$, the functions $\lambda^{-2d_1} \exp\{\sum_{l=1}^{\infty} a_l^1 \cos(l\lambda)\}$ and $\lambda^{-2d_2} \exp\{\sum_{l=1}^{\infty} a_l^2 \cos(l\lambda)\}$ are different on a set of λs with positive Lebesgue measure.

(C4) $(p =) p_N \to \infty$, $p_N/N \to 0$ is such that

$$\frac{p_N^{2\varrho}}{N} \to \infty, \quad \frac{p_N^9 (\log N)^4}{N} \to 0$$

with $\varrho > 9/2$.

Theorem 10.7.1 *Under Assumptions (C1)–(C4) above,*

$$\sqrt{\frac{N}{p_N}} (\widehat{d}_{lw} - d_0) \xrightarrow{d} \mathcal{N}(0, 1). \quad (10.7.6)$$

See Narukawa and Matsuda [740] for a proof. If $p_N = N^r$, the condition (C4) above is satisfied when $1/(2\varrho) < r < 1/9$. The convergence rate then is

$$N^{\frac{1}{2} - \frac{r}{2}},$$

which approaches the parametric rate $N^{1/2}$ as $r \to 0$, possible when $\varrho \to \infty$. Note also that requiring $\varrho \to \infty$ here is very different from supposing $q \to \infty$ in the local polynomial Whittle estimation in Section 10.6.3 (see (10.6.24) and the discussion thereafter). Indeed, under smoothness assumptions on f^* (Exercise 10.9), the coefficients a_j in (10.7.2) are expected to decay exponentially fast and hence satisfy the condition (C1) above for arbitrarily large ϱ.

---

[9] Strictly speaking, in Narukawa and Matsuda [740], λ^{-2d} is replaced by $|1 - e^{-i\lambda}|^{-2d}$ in (10.7.1).

10.8 Exercises

The symbols * and ** next to some exercises are explained in Section 2.12.

Exercise 10.1 Corollary 2.4.4 provides an explicit formula for the autocovariance function (ACVF) of a FARIMA$(0, d, 0)$ series. Provide an explicit formula for the ACVF of a FARIMA$(0, d, 1)$ series. Also, suggest a numerical procedure for evaluating the ACVF of a FARIMA$(1, d, 0)$ series.

Exercise 10.2 Suppose that $\{X_n\}_{n \in \mathbb{Z}}$ is a stationary time series with autocovariance function $\gamma_X(n)$ and constant mean μ_X. The Best Linear Unbiased Estimator (BLUE) of μ_X is defined as $\widehat{\mu}_{blue} = \sum_{n=1}^{N} w_n Y_n$, $w_n \in \mathbb{R}$, such that Var$(\widehat{\mu}_{blue})$ is minimal subject to the condition $\sum_{n=1}^{N} w_n = 1$ (ensuring $\mathbb{E}\widehat{\mu}_{blue} = \mu_X$). Show that the weight vector $w = (w_1, \ldots, w_N)'$ is given by

$$w = \frac{\Sigma_N \mathbf{1}}{\mathbf{1}' \Sigma_N \mathbf{1}},$$

where $\mathbf{1} = (1, \ldots, 1)'$ and $\Sigma_N = (\gamma_X(j-k))_{j,k=1,\ldots,N}$. Show also that Var$(\widehat{\mu}_{blue}) = (\mathbf{1}' \Sigma_N \mathbf{1})^{-1}$. For moderate N and FARIMA$(0, d, 0)$ series, compare numerically the ratio of Var$(\widehat{\mu}_{blue})$ to Var$(\overline{X})$, where $\overline{X}$ is the sample mean.

Exercise 10.3 Consider the log-likelihood in (10.2.8) and suppose for simplicity that σ^2 is known. The objective is then to minimize

$$\int_{-\pi}^{\pi} \frac{I_X(\lambda)}{h_\upsilon(\lambda)} d\lambda$$

with respect to υ. At the population level, that is, replacing $I_X(\lambda)$ by the spectral density, this is equivalent to minimizing

$$F(\upsilon) = \int_{-\pi}^{\pi} \frac{h_{\upsilon_0}(\lambda)}{h_\upsilon(\lambda)} d\lambda,$$

where υ_0 refers to the true parameter value. Explain why the function $F(\upsilon)$ is expected to reach its minimum at $\upsilon = \upsilon_0$ under mild assumptions? *Hint:* Use Parseval's identity $(1/2\pi) \int_{-\pi}^{\pi} |\widehat{r}(\lambda)|^2 d\lambda = \sum_{n=-\infty}^{\infty} r_n^2$, where $r_n = 1/2\pi \int_{-\pi}^{\pi} e^{in\lambda} \widehat{r}(\lambda) d\lambda$ are the Fourier coefficients of $\widehat{r}(\lambda)$.

Exercise 10.4 The entries $c_{rs}(\upsilon)$ of the limiting covariance matrix in (10.2.12) are sometimes written as

$$c_{rs}(\upsilon) = \int_{-\pi}^{\pi} h_\upsilon(\lambda) \frac{\partial^2 h_\upsilon^{-1}}{\partial \upsilon_r \upsilon_s}(\lambda) d\lambda.$$

Show that the two expressions are equivalent.

Exercise 10.5* In Example 10.2.4, the asymptotic variance was computed for the Whittle estimator of d in FARIMA$(0, d, 0)$ model. What is the asymptotic variance of the Whittle estimator of d in FARIMA(p, d, q) models?

Exercise 10.6 The functions arfima and fracdiff in the respective R packages arfima and fracdiff (cf. Section 10.5) use different methods in implementing MLE. Fit a FARIMA(1, d, 1) model to tmpyr data from the arfima package using the function fracdiff, and compare the fitted model to that obtained by using the function arfima in Section 10.5.

Exercise 10.7 The R package longmemo contains the celebrated time series NileMin of yearly minimal water levels of the Nile river for the years 622 to 1281, measured at the Roda gauge near Cairo. Fit a FARIMA(p, d, q) model to this time series using the function fracdiff (cf. Section 10.5) and provide its forecasts for 20 steps ahead.

Exercise 10.8 Show that the function $\ell_{bw,p}(\eta)$ in (10.7.4) is convex in η.

Exercise 10.9 Suppose $f^*(\lambda)$ in (10.7.1) is the spectral density of ARMA(p, q) time series; that is,

$$f^*(\lambda) = \frac{\sigma^2}{2\pi} \frac{|(1 - s_1^{-1}e^{-i\lambda})\ldots(1 - s_q^{-1}e^{-i\lambda})|^2}{|(1 - r_1^{-1}e^{-i\lambda})\ldots(1 - r_p^{-1}e^{-i\lambda})|^2},$$

where the roots satisfy $|s_n|^{-1} < 1, n = 1, \ldots, q$, and $|r_m|^{-1} < 1, m = 1, \ldots, p$.[10] Show that the Fourier coefficients of the function $\log f^*(\lambda)$ decay exponentially. *Hint:* Use the fact that $\log(1 - z) = -\sum_{l=1}^{\infty} z^l/l$ for complex $|z| < 1$.

10.9 Bibliographical Notes

Section 10.1: To implement MLE in practice, it is critical to be able to compute numerically the autocovariance function (ACVF) of the fitted model. As indicated in Example 10.1.1, for FARIMA(p, d, q) models, the calculation of ACVF was considered in Sowell [918], Chung [242], Bertelli and Caporin [136], Doornik and Ooms [326]. See also Luceño [649] for an earlier work, Sela and Hurvich [896], Pai and Ravishanker [786, 785], Tsay [961] for extensions to the multivariate case, and Chen, Hurvich, and Lu [224] for related work. Several of these references (Sowell [918], Doornik and Ooms [326]) also touch upon other practical implementation issues behind MLE. An issue especially relevant in the multivariate case is being able to parameterize the fitted models in such a way that they are stationary (and causal and invertible, if desired) for the considered and hence estimated parameter values; see Ansley and Kohn [38], Roy, McElroy, and Linton [872].

Section 10.2: The Whittle approximation of the likelihood in the spectral domain goes back to Whittle [1005, 1006], including the multivariate case. See also Calder and Davis [197] for a historical perspective. For univariate short-range dependent series, the asymptotics of exact ML and Whittle estimators are established in, for example, Hannan [451].

For long-range dependent series, Yajima [1022] was the first to study the asymptotics of exact ML and Whittle estimators for Gaussian FARIMA(0, d, 0) model (though the Gaussianity assumption was made for simplicity). These estimators were studied under quite

---

[10] These conditions correspond to invertible and causal ARMA time series.

general assumptions for Gaussian series in Fox and Taqqu [368] and Dahlhaus [277, 278]. The case of linear time series was considered in Giraitis and Surgailis [399]. The latter work provides the basis for the proof of Theorem 10.2.1, as well as Theorem 3.8.1 in Giraitis et al. [406]. For related developments, see also Heyde and Gay [473] concerning the multiparameter case, Hosoya [487] concerning multivariate time series, Velasco and Robinson [988] for extensions to the non-stationary case, and Andrews and Lieberman [33], Lieberman and Phillips [627], Lieberman, Rousseau, and Zucker [630], Lieberman et al. [631] for finer studies of MLE. The Whittle estimation for Gaussian subordinated times series is considered in Giraitis and Taqqu [402].

The convergence of quadratic forms of long-range dependent series to a normal limit plays a central role in the proof of Theorem 10.2.1. The approach taken in the proof is due to Giraitis and Surgailis [399]. For other work related to quadratic forms of long-range dependent series, see Avram [58], Fox and Taqqu [369], Ginovian [390], Giraitis and Taqqu [401], Horváth and Shao [483], Ginovyan and Sahakyan [391], Wu and Shao [1014], Avram, Leonenko, and Sakhno [60], Bai, Ginovyan, and Taqqu [83, 84]. A different approach, based on Stein's method and the Malliavin calculus, is taken in Nourdin and Peccati [761].

The question about the finite-sample performance of the ML estimators is raised in Remark 10.2.5. Hauser [461] provides a simulation study comparing the performance of the exact maximum likelihood (EML) for the demeaned data, the associated modified profile likelihood (MPL) and the Whittle estimator with (WLT) and without (WL) tapered data. Quoting from the abstract of Hauser [461]: "The tapered version of the Whittle likelihood turns out to be a reliable estimator for ARMA and ARFIMA models. [...] The modified profile likelihood is [...] either equivalent to the EML or more favorable than the EML. For fractionally integrated models, particularly, it dominates clearly the EML. The WL has serious deficiencies for large ranges of parameters, and so cannot be recommended in general. The EML, on the other hand, should only be used with care for fractionally integrated models due to its potential large negative bias of the fractional integration parameter." For other studies of finite sample properties, see also Sowell [918], Cheung and Diebold [231], Taqqu and Teverovsky [949], Bisaglia and Guégan [160], Reisen, Abraham, and Lopes [843], Nielsen and Frederiksen [751], Boutahar, Marimoutou, and Nouira [179]. Some of these studies also include the comparison to "local" estimation methods, the local Whittle estimator in Section 10.6 being an example.

Section 10.3: For other studies on model selection for long-range dependent series, see Bhansali and Kokoszka [141], Bisaglia [158], Carbonez [202]. For related work on estimation in misspecified models, see Chen and Deo [221].

Section 10.4: For theoretical studies concerning prediction of long-range dependent series, see Bhansali and Kokoszka [142, 143], Godet [414, 415]. Empirical considerations can be found in Brodsky and Hurvich [187], Hurvich [503], Ellis and Wilson [349], Bhardwaj and Swanson [145]. Brodsky and Hurvich [187], Basak, Chan, and Palma [105] compare the prediction performance of competing LRD and ARMA(1,1) models in both theory and applications. A critical look at the forecasting of FARIMA models can be found in Ellis and Wilson [349]; see also references therein.

For the behavior of the prediction coefficients, see also Inoue and Kasahara [521], Dębowski [292], and related interesting review papers by Bingham [155, 156].

Section 10.5: Several other R packages are available to analyze long-range dependent series. The package longmemo [895] implements some methods discussed by Beran [121]. The package afmtools [259] is another package for working with FARIMA models. The package FGN [975] provides synthesis and analysis tools around the fractional Gaussian noise model. See also the package fArma [1016] which includes a number of simulation and "local" estimation methods.

Time series modeling based on FARIMA models is also available through the commercial SAS/IML software [516].

Two temperature datasets were considered in Section 10.5 for illustration purposes. There are many other applications where long-range dependent time series are used. We shall not provide here a comprehensive review of all the many applications. We shall indicate next a few works of historical and other interests, where the focus is on the application at hand and model fitting in particular, rather than on methodological issues.

For applications and model fitting in hydrology, see, for example, Hosking [485], Noakes, Hipel, McLeod, Jimenez, and Yakowitz [754], Montanari, Rosso, and Taqqu [730]. For applications in internet traffic modeling, see, for example, Scherrer et al. [887] where the reference is chosen for a more explicit modeling of the traffic, as opposed to the many empirical studies providing evidence of long-range dependence. For applications in economics and finance, see, for example, Diebold and Rudebusch [311], Cheung [230], Harvey [457]. Another often cited and early application to wind power can be found in Haslett and Raftery [458].

Section 10.6: The local Whittle estimator was suggested in Künsch [582], and justified theoretically in the seminal work by Robinson [853]. (As indicated in Section 10.6.1, the estimator was considered under slightly weaker assumptions in Andrews and Sun [34].) Extensions to nonstationary long-range dependent series (associated with the value $d \geq 1/2$) are considered in Velasco [984], Shimotsu and Phillips [907, 908], Abadir, Distaso, and Giraitis [2]. When the Whittle estimator is viewed in the M–estimation framework, see Robinson and Henry [856].

In addition to Delgado and Robinson [299], Henry and Robinson [470], Henry [468] mentioned in Section 10.6.2, see also Hassler and Olivares [459] for a recent empirical study of bandwidth selection.

Bias reduction and rate optimality is discussed in a nice review paper by Henry [469]. Giraitis, Robinson, and Samarov [403, 404] is another influential but somewhat technical study on rate optimality (see also Hidalgo and Yajima [474]). For another work on tapering, see Chen [219]. The work of Andrews and Sun [34] is continued in Guggenberger and Sun [439]. See also Giraitis and Robinson [395], Lieberman and Phillips [627, 628] for Edgeworth expansions, Soulier [916] for best attainable rates in estimation, and Baillie et al. [90] for further discussion on bias.

Empirical studies and comparison of several available semi-parametric estimators, including the local Whittle estimator, can be found in the references given in the notes to Section 10.2.

Section 10.7: A broadband approach similar to that described in Section 10.7, where the exponential in (10.7.2) is replaced by the square of a Fourier transform itself, was taken by Bhansali, Giraitis, and Kokoszka [144]. An application of the broadband local Whittle estimation to testing for long-range dependence can be found in Narukawa [739].

Other notes: Another local semi-parametric estimator, the GPH estimator, was given in Section 2.10.3. It should be noted that the two estimators, the GPH and the local Whittle, have been developed very much in parallel. For example, following the presentation for the Whittle estimator, the GPH estimator (possibly involving tapering) was studied by Hurvich and Ray [508], Velasco [985], Nouira, Boutahar, and Marimoutou [759]. The bias of the GPH estimator and the choice of the bandwidth were considered in Agiakloglou et al.[15], Hurvich and Beltrao [504], Hurvich and Deo [507], Hurvich et al.[510]. The local polynomial modification of the GPH estimator can be found in Guggenberger and Sun [439]. The broadband estimator associated with the GPH and based on fitting a fractional exponential (FEXP) model was studied in Moulines and Soulier [734], Hurvich [502], Hurvich and Brodsky [505], Hurvich, Moulines, and Soulier [511], Iouditsky, Moulines, and Soulier [523].

A number of other topics on estimation of long-range dependent time series were not included in Chapter 10. Some of these are discussed in connection to Chapters 2 and 9. Several others are listed below for interested readers and a few references are provided without aiming at being exhaustive.

Estimation of the so-called impulse response coefficients in LRD series is considered in Baillie and Kapetanios [88], Baillie, Kapetanios, and Papailias [91].

FARIMA modeling with stable innovations is studied in Kokoszka and Taqqu [571, 573], Burnecki and Sikora [194]. Nonlinear long-range dependent time series and their estimation, of particular interest in modeling stochastic volatility, are considered in Deo, Hsieh, Hurvich, and Soulier [302], Shao and Wu [902]. For seasonal long-range dependent models, see Arteche and Robinson [51], Oppenheim, Ould Haye, and Viano [781], Palma and Chan [790], Reisen, Rodrigues, and Palma [844], Philippe, Surgailis, and Viano [808]. State space modeling of long-range dependent series can be found in Chan and Palma [215].

It is known that long-range dependence is confused easily with non-stationary-like behavior, such as shifting means. Distinguishing the two phenomena has attracted a lot of attention. See, for example, Berkes et al. [130], Bisaglia and Gerolimetto [159], Iacone [513], Qu [837], Baek and Pipiras [69], McCloskey and Perron [701], Dehling, Rooch, and Taqqu [297]. (See also the notes to Section 3.6 above.) Prediction performance of competing models, LRD and series with breaks, is investigated in Gabriel and Martins [378]. An approach to modeling trends, long-range dependence and nonstationarity, through SEMIFAR models, is considered in Beran and Feng [123]. See also Roueff and von Sachs [868], Sen, Preuss, and Dette [898].

Bayesian methods for long-range dependent time series are considered in Pai and Ravishanker [787], Liseo, Marinucci, and Petrella [638], Holan, McElroy, and Chakraborty [480], Graves, Gramacy, Franzke, and Watkins [425]. Robust estimation is discussed in Lévy-Leduc, Boistard, Moulines, Taqqu, and Reisen [619]. Parametric and nonparametric regressions with long-range dependent errors are considered in Sections 11 and 12 of Giraitis et al. [406], Section 7 of Beran et al. [127].

Bootstrap and other resampling schemes are used in Hall, Jing, and Lahiri [447], Bisaglia, Bordignon, and Cecchinato [161], Conti, De Giovanni, Stoev, and Taqqu [258], Kim and Nordman [556, 557], Zhang, Ho, Wendler, and Wu [1028], Grose, Martin, and Poskitt [434], Poskitt [829], Poskitt, Grose, and Martin [830], Shao [901]. See also Section 10 in Beran et al. [127]. Subsampling, also called "block sampling," is a form of bootstrap. When used together with self-normalization, it provides an effective method for approximating an asymptotic distribution function without having to know potential nuisance parameters, for example an unspecified Hurst index H when dealing with long-range dependent processes. For details, see Betken and Wendler [140], Bai, Taqqu, and Zhang [85], Bai and Taqqu [81].

Testing for long-range dependence in the time domain is considered in Hassler, Rodrigues, and Rubia [460]. Studies of the asymptotic distributions of the sample mean, autocovariances and autocorrelations of LRD series go back to Hosking [486]. See also Lévy-Leduc, Boistard, Moulines, Taqqu, and Reisen [620], Wu, Huang, and Zheng [1015]. Estimation of inverse autocovariances matrices is considered in Ing, Chiou, and Guo [517].

Finally, empirical graphical methods for estimating the long-range dependence parameter can be found in Taqqu, Teverovsky, and Willinger [951] and Taqqu and Teverovsky [948]. A wavelet Whittle estimator of the long-range dependence parameter is described in Moulines et al. [737], with the extension to the multivariate setting in Achard and Gannaz [13]. A comparison between Fourier and wavelet techniques can be found in Faÿ, Moulines, Roueff, and Taqqu [358].

Appendix A

Auxiliary Notions and Results

We gather here a number of auxiliary notions and results concerning Fourier transforms (Appendices A.1 and A.2), convergence of measures (Appendix A.3) and heavy-tailed random variables (Appendix A.4).

A.1 Fourier Series and Fourier Transforms

We present here basic facts about the Fourier series (Section A.1.1) and the Fourier transforms (Section A.1.2), as used throughout the book.

A.1.1 Fourier Series and Fourier Transform for Sequences

If $F \in L^2(-\pi, \pi]$ is a complex- or real-valued function, let

$$f_k = \frac{1}{2\pi} \int_{-\pi}^{\pi} e^{ik\lambda} F(\lambda) d\lambda, \quad k \in \mathbb{Z}. \tag{A.1.1}$$

It is well known (e.g., Dym and McKean [340], Zygmund [1030]) that

$$F(\lambda) = \sum_{k=-\infty}^{\infty} f_k e^{-ik\lambda} \tag{A.1.2}$$

with the convergence of the series in $L^2(-\pi, \pi]$. The right-hand side of (A.1.2) is known as the *Fourier series* of the function F, and the coefficients $f_k, k \in \mathbb{Z}$, in (A.1.1) are known as the Fourier coefficients of F. The equality in (A.1.2) is understood in the a.e. sense, up to which the functions are identified in the space $L^2(-\pi, \pi]$. But, for example, if F is continuously differentiable, then the Fourier series converges to F uniformly (Dym and McKean [340], Theorem 2, p. 31).

Warning: We should also note that the convention used for the Fourier series and the Fourier coefficients varies in the literature. For example, $e^{-ik\lambda}$ and $e^{ik\lambda}$ may be interchanged above, or the Fourier series be defined as $\sum_{k=-\infty}^{\infty} f_k \frac{e^{-ik\lambda}}{\sqrt{2\pi}}$ and the Fourier coefficients as $f_k = \int_{-\pi}^{\pi} \frac{e^{ik\lambda}}{\sqrt{2\pi}} F(\lambda) d\lambda$, etc. In fact, in a few places of this book, e.g., in the discussion following (3.11.100), in order to keep in line with the traditional approach used in that setting, we have adopted another convention for the Fourier series and coefficients than that in (A.1.1)–(A.1.2). In these cases, we continue using the terms Fourier series and Fourier coefficients. The same applies to the Fourier transform discussed below.

Example A.1.1 Recall that in Chapter 1, in relations (1.3.2) and (1.3.1), we let

$$\gamma_X(h) = \int_{-\pi}^{\pi} e^{ih\lambda} f_X(\lambda) d\lambda, \ k \in \mathbb{Z}, \quad \text{and} \quad f_X(\lambda) = \frac{1}{2\pi} \sum_{h=-\infty}^{\infty} e^{-ih\lambda} \gamma_X(h), \ \lambda \in (-\pi, \pi],$$

where $f_X(\lambda)$ is the spectral density and $\gamma_X(k)$ is the autocovariance function. Hence, the autocovariances $\gamma_X(k)$ are the Fourier coefficients of the function $2\pi f_X(\lambda)$ and the spectral density is the function whose Fourier coefficients are $\gamma_X(k)/(2\pi)$.

If

$$G(\lambda) = \sum_{k=-\infty}^{\infty} g_k e^{-ik\lambda}$$

is the Fourier series of another function $G \in L^2(-\pi, \pi]$, then the following useful *Parseval's identity* (theorem) holds:

$$\frac{1}{2\pi} \int_{-\pi}^{\pi} F(\lambda) \overline{G(\lambda)} d\lambda = \sum_{k=-\infty}^{\infty} f_k \overline{g}_k, \qquad (A.1.3)$$

where the bar indicates the complex conjugation. For $F = G$, in particular,

$$\frac{1}{2\pi} \int_{-\pi}^{\pi} |F(\lambda)|^2 d\lambda = \sum_{k=-\infty}^{\infty} |f_k|^2 \qquad (A.1.4)$$

and the right-hand side is finite; that is, $\{f_k\}_{k \in \mathbb{Z}} \in \ell^2(\mathbb{Z})$, where $\ell^2(\mathbb{Z})$ consists of square summable sequences of complex or real coefficients.

Instead of starting with square-integrable functions as above, one could be given a square summable sequence $f = \{f_k\}_{k \in \mathbb{Z}} \in \ell^2(\mathbb{Z})$. Its *Fourier transform* $\widehat{f}$ or $f(e^{-i\cdot})$ is defined as the function

$$\widehat{f}(\lambda) := f(e^{-i\lambda}) := \sum_{k=-\infty}^{\infty} f_k e^{-ik\lambda}, \quad \lambda \in (-\pi, \pi], \qquad (A.1.5)$$

and belongs to $L^2(-\pi, \pi]$. Parseval's identity (A.1.3) (and (A.1.4)) can then be written in terms of the Fourier transforms $\widehat{f}(\lambda)$ and $\widehat{g}(\lambda)$.

Given two sequences $f = \{f_k\}_{k \in \mathbb{Z}} \in \ell^2(\mathbb{Z})$ and $g = \{g_k\}_{k \in \mathbb{Z}} \in \ell^2(\mathbb{Z})$, their *convolution* is the sequence $f * g = \{(f * g)_k\}_{k \in \mathbb{Z}}$ defined as

$$(f * g)_k = \sum_{n=-\infty}^{\infty} f_n g_{k-n}. \qquad (A.1.6)$$

The convolution is well-defined for $f, g \in \ell^2(\mathbb{Z})$ by using the Cauchy–Schwarz inequality. Moreover, since $\sum_k \sum_n |f_n| |g_{k-n}| = \sum_k |f_k| \sum_m |g_m|$, we have $f * g \in \ell^1(\mathbb{Z})$ for $f, g \in \ell^1(\mathbb{Z})$. By the generalized Minkowski's inequality (6.1.21), we also have

$$\left(\sum_k \left(\sum_n |f_n| |g_{k-n}| \right)^2 \right)^{1/2} \leq \sum_n |f_n| \left(\sum_m |g_m|^2 \right)^{1/2},$$

so that $f * g \in \ell^2(\mathbb{Z})$ for $f \in \ell^1(\mathbb{Z})$, $g \in \ell^2(\mathbb{Z})$. If $f, g \in \ell^1(\mathbb{Z})$, then

$$\widehat{(f * g)}(\lambda) = \widehat{f}(\lambda)\widehat{g}(\lambda), \tag{A.1.7}$$

since $\sum_k \sum_n f_n g_{k-n} e^{-ik\lambda} = \sum_n f_n e^{-in\lambda} \sum_m g_m e^{-im\lambda}$ after changing the order of summation and setting $k - n = m$. The relation (A.1.7) also holds for $f \in \ell^1(\mathbb{Z}), g \in \ell^2(\mathbb{Z})$. This follows by taking $g^{(n)} \in \ell^1(\mathbb{Z})$ satisfying $g^{(n)} \to g$ in $\ell^2(\mathbb{Z})$ (e.g., $g_k^{(n)} = g_k$ for $|k| \leq n$ and $= 0$ for $|k| > n$), by noting that $\widehat{f * g^{(n)}} = \widehat{f}\widehat{g^{(n)}}$ by (A.1.7) and finally by passing n to the limit in the last relation. Indeed, in the limit of $n \to \infty$, $g^{(n)} \to g$ in $\ell^2(\mathbb{Z})$ so that $\widehat{g^{(n)}} \to \widehat{g}$ in $L^2(-\pi, \pi]$ (in view of (A.1.4)) and hence $\widehat{f}\widehat{g^{(n)}} \to \widehat{f}\widehat{g}$ in $L^2(-\pi, \pi]$ (since $|\widehat{f}| \leq \sum_n |f_n| < \infty$), and similarly $f * g^{(n)} \to f * g$ in $\ell^2(\mathbb{Z})$ (by using the generalized Minkowski's inequality as above) so that $\widehat{f * g^{(n)}} \to \widehat{f * g}$ in $L^2(-\pi, \pi]$.

A.1.2 Fourier Transform for Functions

Throughout the book, unless specified otherwise, we work with the $L^2(\mathbb{R})$–Fourier transform. For a complex- or real-valued function $f \in L^2(\mathbb{R})$, its *Fourier transform* is formally defined as

$$\widehat{f}(x) = \int_{\mathbb{R}} e^{ixu} f(u) du \tag{A.1.8}$$

and, by construction, belongs to $L^2(\mathbb{R})$ (e.g., Dym and McKean [340]). Moreover, if $f \in L^1(\mathbb{R}) \cap L^2(\mathbb{R})$, then the Fourier integral in (A.1.8) can be interpreted in the Lebesgue sense. As in the "Warning" for the Fourier series in Section A.1.1, the convention used to define the Fourier transform varies in the literature.

The Fourier transform can be inverted back to the original function through the *inversion formula*

$$f(u) = \frac{1}{2\pi} \int_{\mathbb{R}} e^{-iux} \widehat{f}(x) dx. \tag{A.1.9}$$

Plancherel's identity (theorem) for the Fourier transform states that, for $f, g \in L^2(\mathbb{R})$,

$$\int_{\mathbb{R}} f(u)\overline{g(u)} du = \frac{1}{2\pi} \int_{\mathbb{R}} \widehat{f}(x)\overline{\widehat{g}(x)} dx, \tag{A.1.10}$$

where the bar indicates the complex conjugation, and

$$\|f\|^2_{L^2(\mathbb{R})} = \int_{\mathbb{R}} |f(u)|^2 du = \frac{1}{2\pi} \int_{\mathbb{R}} |\widehat{f}(x)|^2 dx = \frac{1}{2\pi} \|\widehat{f}\|^2_{L^2(\mathbb{R})} \tag{A.1.11}$$

in the special case $f = g$ (cf. (A.1.3) and (A.1.4)), when it is also referred to as *Parseval's identity*.

The *convolution* of two functions $f, g \in L^2(\mathbb{R})$ is defined as the function

$$(f * g)(u) = \int_{\mathbb{R}} f(v)g(u - v) dv. \tag{A.1.12}$$

It is well-defined pointwise for $f, g \in L^2(\mathbb{R})$ by using the Cauchy-Schwarz inequality. It can also be defined a.e. du for $f, g \in L^1(\mathbb{R})$ since $\int_{\mathbb{R}} \int_{\mathbb{R}} |f(v)| |g(u - v)| dv du =$

Table A.1 *The conditions on f, g for the convolution $f * g$ to be well-defined, and the properties that the resulting convolution satisfies.*

Conditions	$f * g$ defined	$f * g$ belongs to	$\widehat{f * g} = \widehat{f}\widehat{g}$ satisfied
$f, g \in L^2(\mathbb{R})$	pointwise	–	–
$f, g \in L^1(\mathbb{R})$	a.e.	$L^1(\mathbb{R})$	Yes
$f \in L^2(\mathbb{R}), g \in L^1(\mathbb{R})$	a.e.	$L^2(\mathbb{R})$	Yes

$\int_{\mathbb{R}} |f(v)|dv \int_{\mathbb{R}} |g(w)|dw < \infty$. The last relation shows that $f * g \in L^1(\mathbb{R})$ for $f, g \in L^1(\mathbb{R})$. By the generalized Minkowski's inequality (6.1.21), we also have

$$\left(\int_{\mathbb{R}} \left(\int_{\mathbb{R}} |f(v)| |g(u-v)|dv\right)^2\right)^{1/2} \leq \int_{\mathbb{R}} |f(v)|dv \left(\int_{\mathbb{R}} |g(w)|^2 dw\right)^{1/2}$$

so that $f * g \in L^2(\mathbb{R})$ for $f \in L^1(\mathbb{R}), g \in L^2(\mathbb{R})$. For $f, g \in L^1(\mathbb{R})$,

$$\widehat{(f * g)}(x) = \widehat{f}(x)\widehat{g}(x) \tag{A.1.13}$$

since $\int_{\mathbb{R}} e^{ixu} \int_{\mathbb{R}} f(v) g(u-v) dv du = \int_{\mathbb{R}} e^{ixv} f(v) dv \int_{\mathbb{R}} e^{ixw} g(w) dw$ after changing the order of integration and setting $u - v = w$ (cf. (A.1.7)). The relation (A.1.13) also holds for $f \in L^1(\mathbb{R}), g \in L^2(\mathbb{R})$. To see this, take $g_n(u) = g(u) 1_{\{|u| \leq n\}}$ so that $g_n \in L^1(\mathbb{R})$ and $g_n \to g$ in $L^2(\mathbb{R})$. Then, $\widehat{(f * g_n)} = \widehat{f}\widehat{g_n}$ in view of (A.1.13) and it remains to let $n \to \infty$ in the latter relation. Since $g_n \to g$ in $L^2(\mathbb{R})$, we have $\widehat{g_n} \to \widehat{g}$ in $L^2(\mathbb{R})$ and hence $\widehat{f}\widehat{g_n} \to \widehat{f}\widehat{g}$ in $L^2(\mathbb{R})$ (since $|\widehat{f}| \leq \int_{\mathbb{R}} |f(u)|du < \infty$). Similarly, by using the generalized Minkowski's inequality as above, $f * g_n \to f * g$ in $L^2(\mathbb{R})$, so that $\widehat{f * g_n} \to \widehat{f * g}$ in $L^2(\mathbb{R})$.

The various conditions on the functions f and g for the convolution $f * g$ to be defined and the properties that the resulting convolution satisfies are summarized in Table A.1.

The notions and the results above extend to functions defined on $\mathbb{R}^q$. For a complex- or real-valued function $f \in L^2(\mathbb{R}^q), q \geq 1$, its *Fourier transform* is formally defined as

$$\widehat{f}(x_1, \ldots, x_q) = \int_{\mathbb{R}^q} e^{i(x_1 u_1 + \cdots + x_q u_q)} f(u_1, \ldots, u_q) du_1 \ldots du_q \tag{A.1.14}$$

and, by construction, belongs to $L^2(\mathbb{R}^q)$. The inversion formula reads

$$f(u_1, \ldots, u_q) = \frac{1}{(2\pi)^q} \int_{\mathbb{R}^q} e^{-i(u_1 x_1 + \cdots + u_q x_q)} \widehat{f}(x_1, \ldots, x_q) dx_1 \ldots dx_q \tag{A.1.15}$$

and Plancherel's identity is

$$\int_{\mathbb{R}^q} f(u_1, \ldots, u_q) \overline{g(u_1, \ldots, u_q)} du_1 \ldots du_q$$

$$= \frac{1}{(2\pi)^q} \int_{\mathbb{R}^q} \widehat{f}(x_1, \ldots, x_q) \overline{\widehat{g}(x_1, \ldots, x_q)} dx_1 \ldots dx_q, \tag{A.1.16}$$

where the bar indicates the complex conjugation, and Parseval's identity is

$$\|f\|^2_{L^2(\mathbb{R}^q)} = \frac{1}{(2\pi)^q} \|\widehat{f}\|^2_{L^2(\mathbb{R}^q)}. \tag{A.1.17}$$

A.2 Fourier Series of Regularly Varying Sequences

As noted in Section 2.2.5, taking the Fourier transform of a regularly varying function can be a delicate operation, even when the corresponding slowly varying function is asymptotically constant. The assumption of quasi-monotonicity (see Definition 2.2.11) allows the Fourier transform to have the expected asymptotic behavior. The following proposition relates the asymptotic behavior of a regularly varying function to that of its Fourier transform.

Proposition A.2.1 *Suppose L is a quasi-monotone slowly varying function. Let $g(u)$ stand for $\cos(u)$, $\sin(u)$, or e^{iu}, and $0 < p < 1$. Then, the following series converges conditionally for all $\lambda \in (0, \pi]$, and*

$$\sum_{k=1}^{\infty} g(k\lambda)\frac{L(k)}{k^p} \sim \lambda^{p-1} L\left(\frac{1}{\lambda}\right)\Gamma(1-p)g\left((1-p)\frac{\pi}{2}\right), \quad \text{as } \lambda \to 0. \tag{A.2.1}$$

Proof We only consider the case $g(u) = \cos(u)$. Let $S_n(\lambda) = \sum_{k=1}^{n} \cos(k\lambda)$ and observe that

$$|S_n(\lambda)| = \left|\Re\left(\sum_{k=1}^{n} e^{ik\lambda}\right)\right| \leq \left|\sum_{k=1}^{n} e^{ik\lambda}\right| = \left|\frac{e^{i\lambda} - e^{i\lambda(n+1)}}{1 - e^{i\lambda}}\right| \leq \frac{2}{|1-e^{i\lambda}|} = \frac{1}{\sin\frac{\lambda}{2}}. \tag{A.2.2}$$

Let also $U_n(\lambda) = \sum_{k=n}^{\infty} \cos(k\lambda)/k^p$. By the summation by parts formula

$$\sum_{k=n}^{m} a_k b_k = A_m b_m - A_{n-1} b_n + \sum_{k=n}^{m-1} A_k (b_k - b_{k+1}), \tag{A.2.3}$$

where $a_k, b_k \in \mathbb{R}$ and $A_n = \sum_{k=1}^{n} a_k$ (Exercise 2.2), the fact that $\lim_{m\to\infty} \frac{|S_m(\lambda)|}{m^p} = 0$, for $\lambda \in (0, \pi]$, and the inequality (A.2.2),

$$|U_n(\lambda)| \leq \frac{|S_{n-1}(\lambda)|}{n^p} + \sum_{k=n}^{\infty}\left(\frac{1}{k^p} - \frac{1}{(k+1)^p}\right)|S_k(\lambda)| \leq \frac{2}{n^p \sin\frac{\lambda}{2}}. \tag{A.2.4}$$

By using the summation by parts formula

$$\sum_{k=n}^{m} a_k b_k = \widetilde{A}_n b_n - \widetilde{A}_{m+1} b_m + \sum_{k=n+1}^{m} \widetilde{A}_k (b_k - b_{k-1}), \tag{A.2.5}$$

where $a_k, b_k \in \mathbb{R}$ and $\widehat{A}_n = \sum_{k=n}^{\infty} a_k$ (Exercise 2.2), and (2.2.18), we obtain that

$$\left|\sum_{k=n}^{m} \cos(k\lambda)\frac{L(k)}{k^p}\right| \leq |L(n)U_n(\lambda)| + |L(m)U_{m+1}(\lambda)| + \sum_{k=n+1}^{m} |U_k(\lambda)||L(k) - L(k-1)|$$

$$\leq \frac{2}{\sin\frac{\lambda}{2}}\left(\frac{|L(n)|}{n^p} + \frac{|L(m)|}{m^p} + \int_n^{\infty} u^{-p}|dL(u)|\right) = \frac{2}{\sin\frac{\lambda}{2}}\left(\frac{|L(n)|}{n^p}(1 + O(1)) + \frac{|L(m)|}{m^p}\right). \tag{A.2.6}$$

The conditional convergence now follows letting $n, m \to \infty$.

It is now sufficient to show that, as $\lambda \to 0$,

$$\sum_{k=1}^{\infty} \frac{\cos(k\lambda)}{k^p} \sim \lambda^{p-1}\Gamma(1-p)\cos\left((1-p)\frac{\pi}{2}\right) \tag{A.2.7}$$

and

$$T(\lambda) = \sum_{k=1}^{\infty} \cos(k\lambda)\frac{L(k) - L(\frac{1}{\lambda})}{k^p} \sim o(\lambda^{p-1}L(1/\lambda)). \tag{A.2.8}$$

The relation (A.2.7) is stated as Exercise 2.4. For (A.2.8), split the sum $T(\lambda)$ into three sums $T_1(\lambda)$, $T_2(\lambda)$ and $T_3(\lambda)$ over $k \leq [a/\lambda]$, $[a/\lambda] + 1 \leq k \leq [b/\lambda]$ and $[b/\lambda] + 1 \leq k$, respectively, where $0 < a \leq 1 \leq b < \infty$. The relation $T_2(\lambda) = o(\lambda^{p-1}L(1/\lambda))$ follows by using the uniform convergence theorem in (2.2.4). Indeed, note that

$$\left|\frac{T_2(\lambda)}{\lambda^{p-1}L(1/\lambda)}\right| \leq \sum_{k=[a/\lambda]+1}^{[b/\lambda]} \left|\frac{L(k\lambda(1/\lambda))}{L(1/\lambda)} - 1\right|(k\lambda)^{-p}\lambda$$

$$\leq \sup_{a\leq u\leq b}\left|\frac{L(u(1/\lambda))}{L(1/\lambda)} - 1\right| \sum_{k=[a/\lambda]+1}^{[b/\lambda]} (k\lambda)^{-p}\lambda.$$

As $\lambda \to 0$, the supremum above converges to 0 by the uniform convergence theorem in (2.2.4) and the sum converges to $\int_a^b u^{-p}du$. For the sum $T_3(\lambda)$, by using summation by parts as in (A.2.6),

$$|T_3(\lambda)| \leq |U_{[b/\lambda]+1}(\lambda)||L([b/\lambda]+1) - L(1/\lambda)| + \sum_{k=[b/\lambda]+2}^{\infty} |U_k(\lambda)||L(k) - L(k-1)|$$

$$\leq \frac{2}{\sin\frac{\lambda}{2}}\left\{\frac{1}{[b/\lambda]^p}|L([b/\lambda]+1) - L(1/\lambda)| + \int_{b/\lambda}^{\infty} u^{-p}|dL(u)|\right\}$$

$$= \frac{2}{\sin\frac{\lambda}{2}}\left\{\frac{1}{[b/\lambda]^p}||L([b/\lambda]+1) - L(1/\lambda)| + \frac{L(b/\lambda)}{(b/\lambda)^p}(1 + O(1))\right\}.$$

Taking b large enough and fixed, this shows that $T_3(\lambda)/(\lambda^{p-1}L(1/\lambda))$ is arbitrarily small as $\lambda \to 0$. The term $T_1(\lambda)$ can be dealt with similarly. □

Proposition A.2.1 is used in the proof of Proposition 2.2.14.

Proof of Proposition 2.2.14 For the function $f_X(\lambda)$ in (2.2.19) to be a spectral density of the series X, it is enough to show that $f_X(\lambda) \geq 0$, for $\lambda \in (-\pi, \pi] \setminus \{0\}$, and that

$$\int_{-\pi}^{\pi} e^{in\lambda} f_X(\lambda)d\lambda = \gamma_X(n). \tag{A.2.9}$$

For nonnegativeness of $f_X(\lambda)$, observe that

A.2 Fourier Series of Regularly Varying Sequences

$$0 \leq \frac{1}{2\pi K}\Big|\sum_{k=1}^{K} e^{-ik\lambda}\gamma_X(k)\Big|^2 = \frac{1}{2\pi K}\sum_{k,l=1}^{K} e^{-ik\lambda}\gamma_X(k-l)e^{il\lambda}$$

$$= \frac{1}{2\pi}\Big\{\gamma_X(0) + 2\sum_{k=1}^{K}\cos(k\lambda)\gamma_X(k)\Big(1 - \frac{k}{K}\Big)\Big\}.$$

It is enough to show that the last expression converges (conditionally) to $f_X(\lambda)$, for each $\lambda \in (-\pi, \pi] \setminus \{0\}$. In view of Proposition A.2.1, $\frac{1}{2\pi}(\gamma_X(0) + 2\sum_{k=1}^{K}\cos(k\lambda)\gamma_X(k))$ converges conditionally to $\frac{1}{2\pi}(\gamma_X(0) + 2\sum_{k=1}^{\infty}\cos(k\lambda)\gamma_X(k))$, which is the function $f_X(\lambda)$. Hence, it is enough to prove that

$$V_K(\lambda) = \sum_{k=1}^{K}\cos(k\lambda)\gamma_X(k)\frac{k}{K} = \frac{1}{K}\sum_{k=1}^{K}\cos(k\lambda)k^{2d}L_2(k) \to 0,$$

as $K \to \infty$. To prove the latter convergence, proceed as in the proof of Proposition A.2.1. Let $Z_n(\lambda) = \sum_{k=1}^{n}\cos(k\lambda)k^{2d}$ and observe that the summation here is $\sum_{k=1}^{n}$ instead of $\sum_{k=n+1}^{\infty}$ as in (A.2.4). By the summation by parts, we have

$$|Z_n(\lambda)| \leq \frac{2n^{2d}}{|\sin\frac{\lambda}{2}|}.$$

Then, proceeding as for (A.2.6),

$$|V_K(\lambda)| \leq \frac{1}{K}\Big\{|Z_K(\lambda)L_2(K)| + \sum_{k=1}^{K-1}|Z_k(\lambda)||L_2(k) - L_2(k-1)|\Big\}$$

$$\leq \frac{2}{|\sin\frac{\lambda}{2}|}\Big\{K^{2d-1}|L_2(K)| + \frac{1}{K}\sum_{k=1}^{K-1}k^{2d}|L_2(k) - L_2(k-1)|\Big\}$$

$$\leq \frac{2}{|\sin\frac{\lambda}{2}|}K^{2d-1}|L_2(K)|(1 + O(1)) \to 0,$$

as $K \to \infty$. Thus, $f_X(\lambda)$ is nonnegative.

We now prove (A.2.9). By inserting the expression (2.2.19) of $f_X(\lambda)$ in the relation (A.2.9), we get

$$\int_{-\pi}^{\pi} e^{in\lambda}f_X(\lambda)d\lambda = \int_{-\pi}^{\pi} e^{in\lambda}\frac{1}{2\pi}\sum_{j=-\infty}^{\infty}e^{-ij\lambda}\gamma_X(j)d\lambda$$

$$= \gamma_X(n) + \frac{1}{2\pi}\int_{-\pi}^{\pi}\sum_{k=1}^{\infty}e^{-ik\lambda}\gamma_X(k+n)d\lambda + \frac{1}{2\pi}\int_{-\pi}^{\pi}\sum_{k=1}^{\infty}e^{ik\lambda}\gamma_X(k-n)d\lambda.$$

Thus (A.2.9) is equivalent to showing that the last two integrals vanish. Since the argument for both is similar, we shall show that

$$\int_{-\pi}^{\pi}\sum_{k=1}^{\infty}e^{-ik\lambda}\gamma_{X,n}(k)d\lambda = 0, \tag{A.2.10}$$

where $\gamma_{X,n}(k) = \gamma_X(k+n) = L_2(k+n)(k+n)^{2d-1} = L_2(k+n)(1+n/k)^{2d-1}k^{2d-1} =: L_{2,n}(k)k^{2d-1}$.

Consider the slowly varying function $L_{2,n}(u) = L_2(u+n)(1+n/u)^{2d-1}$. The function $L_2(u+n)$ is quasi-monotone because the function $L_2(u)$ is quasi-monotone by assumption. The function $(1+n/u)^{2d-1}$, moreover, is quasi-monotone by Example 2.2.13. Since the product of two quasi-monotone slowly varying functions is also quasi-monotone (e.g., Corollary 2.7.4 in Bingham et al.[157]), we conclude that the function $L_{2,n}(u)$ is quasi-monotone.

Now decompose $\int_{-\pi}^{\pi} = \int_{-\pi}^{-\epsilon} + \int_{-\epsilon}^{\epsilon} + \int_{\epsilon}^{\pi}$ in (A.2.10) and use Proposition A.2.1 to evaluate

$$\int_{-\epsilon}^{\epsilon} \Big| \sum_{k=1}^{\infty} e^{-ik\lambda} \gamma_{X,n}(k) \Big| d\lambda = \int_{-\epsilon}^{\epsilon} \Big| \sum_{k=1}^{\infty} e^{-ik\lambda} L_{2,n}(k) k^{2d-1} \Big| d\lambda \to 0,$$

as $\epsilon \to 0$, since, in particular, $\lambda^{-2d} L_{2,n}(1/\lambda)$, which appears in (A.2.1), is integrable around $\lambda = 0$. On the other hand, observe that

$$\Big(\int_{-\pi}^{-\epsilon} + \int_{\epsilon}^{\pi} \Big) \sum_{k=1}^{K} e^{-ik\lambda} \gamma_{X,n}(k) d\lambda = \sum_{k=1}^{K} \Big(\Big(\int_{-\pi}^{-\epsilon} + \int_{\epsilon}^{\pi} \Big) e^{-ik\lambda} d\lambda \Big) \gamma_{X,n}(k)$$

$$= 2 \sum_{k=1}^{K} \Big(\int_{\epsilon}^{\pi} \cos(k\lambda) d\lambda \Big) \gamma_{X,n}(k) = -2 \sum_{k=1}^{K} \sin(k\epsilon) k^{2d-2} L_{2,n}(k).$$

The right-hand side of this relation converges absolutely to $-2 \sum_{k=1}^{\infty} \sin(k\epsilon) k^{2d-2} L_{2,n}(k)$, since $2d - 1 < 0$. For the left-hand side, as in (A.2.6), we have $|\sum_{k=1}^{K} e^{-ik\lambda} \gamma_{X,n}(k)| \leq C/|\sin \frac{\lambda}{2}|$. Since the bound is an integrable function over $(-\pi, -\epsilon) \cup (\epsilon, \pi)$, the left-hand side converges to $(\int_{-\pi}^{-\epsilon} + \int_{\epsilon}^{\pi}) \sum_{k=1}^{\infty} e^{-ik\lambda} \gamma_{X,n}(k) d\lambda$. We thus have

$$\Big| \Big(\int_{-\pi}^{-\epsilon} + \int_{\epsilon}^{\pi} \Big) \sum_{k=1}^{\infty} e^{-ik\lambda} \gamma_{X,n}(k) d\lambda \Big| \leq 2 \sum_{k=1}^{\infty} |\sin(k\epsilon)| k^{2d-2} |L_{2,n}(k)| \to 0,$$

as $\epsilon \to 0$, where the convergence follows by the dominated convergence theorem. This establishes (A.2.10). □

Proposition A.2.1 has the following converse.

Proposition A.2.2 *Suppose that $L(1/u)$ is a quasi-monotone slowly varying function on $(1/\pi, \infty)$. Let $0 < p < 1$. Then,*

$$\int_0^{\pi} \cos(n\lambda) \lambda^{p-1} L(\lambda) d\lambda \sim n^{-p} L\Big(\frac{1}{n}\Big) \Gamma(p) \cos\Big(\frac{p\pi}{2}\Big), \quad as\ n \to \infty.$$

Proof The proof is similar to that of Proposition A.2.1 and is left as an exercise. □

The next result is a partial extension of Proposition A.2.1 to the case $p \in (1, 3)$ ($p \neq 2$). We formulate the result in terms $\alpha = p - 1$.

Proposition A.2.3 *Let p_k, $k \geq 1$, be a sequence of probabilities ($p_k \geq 0$, $\sum_{k=1}^{\infty} p_k = 1$). Suppose L is a slowly varying function at infinity and $\alpha \in (0, 2)$ ($\alpha \neq 1$). Then,*

$$\sum_{k=n}^{\infty} p_k = n^{-\alpha} L(n) \tag{A.2.11}$$

if and only if

$$\sum_{k=1}^{\infty} p_k e^{ik\lambda} - 1 = i\gamma\lambda - C_\alpha |\lambda|^\alpha L(|\lambda|^{-1})\left(1 - i \operatorname{sign}(\lambda) \tan\left(\frac{\alpha\pi}{2}\right)\right) + o(|\lambda|^\alpha L(|\lambda|^{-1})), \quad \lambda \to 0, \tag{A.2.12}$$

where

$$C_\alpha = \Gamma(1-\alpha)\cos\left(\frac{\alpha\pi}{2}\right), \quad \gamma = \begin{cases} 0, & \text{if } \alpha \in (0,1), \\ \sum_{k=1}^{\infty} kp_k, & \text{if } \alpha \in (1,2). \end{cases} \tag{A.2.13}$$

Proof Let X be a random variable such that $\mathbb{P}(X = k) = p_k$, $k \geq 1$, and $\psi(\lambda) = \mathbb{E}e^{i\lambda X} = \sum_{k=1}^{\infty} p_k e^{ik\lambda}$ be the characteristic function of X. The condition (A.2.11) is necessary and sufficient for the distribution of X to be attracted to a totally right-skewed α-stable distribution. This is known to be equivalent to the logarithm $\log \psi(\lambda)$ of the characteristic function to have the behavior on the right-hand side of (A.2.12) (see Ibragimov and Linnik [515], Theorem 2.6.5 and the remarks on pp. 87–91). See also Appendix A.4. It remains to note that $\psi(\lambda) \to 1$, as $\lambda \to 0$, and hence $\log \psi(\lambda) \sim \psi(\lambda) - 1$, as $\lambda \to 0$. □

The case $p = 2$ (i.e., $\alpha = 1$) is special and is treated in Aaronson and Denker [1].

A.3 Weak and Vague Convergence of Measures

We gather here some basic facts on the convergence of measures that are used in this book.

A.3.1 The Case of Probability Measures

Let S be a Polish space, that is, a separable complete metric space, and let also $\mathcal{S}$ be the σ–field generated by open sets of S. For example, $S = \mathbb{R}$ equipped with the usual Euclidean metric is a Polish space, in which case $\mathcal{S} = \mathcal{B}(\mathbb{R})$ is the σ–field consisting of Borel sets. Similarly, $S = \mathbb{R}^\ell$, $\ell \geq 1$, with $\mathcal{S} = \mathcal{B}(\mathbb{R}^\ell)$ is a Polish space, and so is $S = C[0, 1]$, the space of continuous functions on $[0, 1]$, equipped with the metric induced by the supremum norm.

If P, P_n, $n \geq 1$, are probability measures on $(S, \mathcal{S})$, then P_n, $n \geq 1$, is said to converge *weakly* to P, denoted $P_n \xrightarrow{d} P$, if

$$\int_S f(x) P_n(dx) \to \int_S f(x) P(dx), \tag{A.3.1}$$

as $n \to \infty$, for every $f \in C_b(S)$; that is, every real-valued, continuous, bounded function on S. The weak convergence condition (A.3.1) is well known to be equivalent to a number of other conditions. See, for example, Billingsley [154], Theorem 2.1, p. 16.

If $\xi, \xi_n, n \geq 1$, are random variables (vectors or processes, resp.) on a probability space $(\Omega, \mathcal{F}, \mathbb{P})$, they induce the probability measures $P = \mathbb{P} \circ \xi^{-1}$, $P_n = \mathbb{P} \circ \xi_n^{-1}, n \geq 1$, on $(\mathbb{R}, \mathcal{B}(\mathbb{R}))$ $((\mathbb{R}^\ell, \mathcal{B}(\mathbb{R}^\ell)))$ or some space of functions such as $(C[0, 1], \mathcal{C}[0, 1])$, resp.). In this case, $\xi_n, n \geq 1$, is said to converge in *distribution* to ξ, denoted $\xi_n \xrightarrow{d} \xi$, if the sequence of the corresponding probability measures $P_n = \mathbb{P} \circ \xi_n^{-1}, n \geq 1$, converges weakly to $P = \mathbb{P} \circ \xi^{-1}$.

It is well known (e.g., Billingsley [154]) that in the case of random variables, the convergence in distribution is equivalent to the convergence of characteristic functions, that is,

$$\mathbb{E}e^{ia\xi_n} \to \mathbb{E}e^{ia\xi}, \quad \text{for } a \in \mathbb{R}, \tag{A.3.2}$$

or the convergence of the corresponding cumulative distribution functions at the points of the continuity of the limit; that is, $\mathbb{P}(\xi_n \leq a) \to \mathbb{P}(\xi \leq a)$ for every continuity point a of $\mathbb{P}(\xi \leq a)$.

For random vectors $\xi = (\xi^{(1)}, \ldots, \xi^{(\ell)})'$, $\xi_n = (\xi_n^{(1)}, \ldots, \xi_n^{(\ell)})'$, the convergence in distribution is known to be equivalent to the convergence in distribution

$$\sum_{j=1}^{\ell} a_j \xi_n^{(j)} \xrightarrow{d} \sum_{j=1}^{\ell} a_j \xi^{(j)}, \quad a_j \in \mathbb{R}, \tag{A.3.3}$$

for all possible linear combinations of the corresponding vectors. This is known as the *Cramér–Wold device*.

A.3.2 The Case of Locally Finite Measures

We discuss here the convergence of locally finite measures on a finite-dimensional Euclidean space as it is used in Chapter 5. Thus, let $\mu, \mu_n, n \geq 1$, be locally finite measures on $(\mathbb{R}^\ell, \mathcal{B}(\mathbb{R}^\ell))$: that is, measures such that $\mu(B) < \infty$ and $\mu_n(B) < \infty$ for every *bounded* $B \in \mathcal{B}(\mathbb{R}^\ell)$. The sequence $\mu_n, n \geq 1$, is said to converge *vaguely* (also *locally weakly*) to μ, denoted $\mu_n \xrightarrow{v} \mu$, if

$$\int_{\mathbb{R}^\ell} f(x) \mu_n(dx) \to \int_{\mathbb{R}^\ell} f(x) \mu(dx), \tag{A.3.4}$$

for every continuous f with *bounded support*. The sequence $\mu_n, n \geq 1$, is said to converge *weakly* to μ, denoted $\mu_n \xrightarrow{d} \mu$, if it converges vaguely to μ and also if

$$\lim_{A \to \infty} \sup_n \mu_n(|x| > A) = 0. \tag{A.3.5}$$

If the μ_ns are probability measures P_ns as in (A.3.1), then we recover the notion of weak convergence of probability measures defined in Section A.3.1.

The following auxiliary lemma was used in Section 5.3. It can be proved as the well-known result on the equivalence of the weak convergence of measures and the convergence of their Fourier transforms. The proof is omitted but can be found in Dobrushin and Major [321], Lemma 2, and in Major [673], Lemma 8.4.

Lemma A.3.1 *Let μ_N, $N \geq 1$, be a sequence of finite measures on $(\mathbb{R}^\ell, \mathcal{B}(\mathbb{R}^\ell))$ such that $\mu_N(\mathbb{R}^\ell \setminus [-C_N\pi, C_N\pi]^\ell) = 0$, with some sequence $C_N \to \infty$. Define*

$$\phi_N(t) = \phi_N(t_1, \ldots, t_\ell) = \int_{\mathbb{R}^\ell} e^{i(\frac{\lfloor C_N t_1 \rfloor}{C_N}x_1 + \cdots + \frac{\lfloor C_N t_\ell \rfloor}{C_N}x_\ell)} \mu_N(dx_1, \ldots, dx_\ell). \quad (A.3.6)$$

If, for every $t \in \mathbb{R}^\ell$, the sequence $\phi_N(t)$ tends to a function $\phi(t)$ continuous at the origin, then μ_N tends weakly to a finite measure μ_0. Moreover, $\phi(t)$ is the Fourier transform of μ_0.

A.4 Stable and Heavy-Tailed Random Variables and Series

We first recall the definition of a (non-Gaussian) stable random variable.

Definition A.4.1 A random variable X is said to be α-*stable* with the index of stability $\alpha \in (0, 2)$, scale parameter $\sigma > 0$, skewness parameter $\beta \in [-1, 1]$ and shift parameter $\mu \in \mathbb{R}$ if its characteristic function has the following form: for $\theta \in \mathbb{R}$,

$$\mathbb{E}e^{i\theta X} = \begin{cases} \exp\left\{-\sigma^\alpha |\theta|^\alpha (1 - i\beta \operatorname{sign}(\theta) \tan \frac{\pi\alpha}{2}) + i\mu\theta\right\}, & \text{if } \alpha \neq 1, \\ \exp\left\{-\sigma |\theta|(1 + i\beta \frac{2}{\pi} \operatorname{sign}(\theta) \log |\theta|) + i\mu\theta\right\}, & \text{if } \alpha = 1, \end{cases} \quad (A.4.1)$$

where $\operatorname{sign}(\theta) = 1$ if $\theta > 0$, $= 0$ if $\theta = 0$, and $= -1$ if $\theta < 0$. If $\beta = 1$ ($\beta = -1$, resp.), an α-stable random variable is called *totally skewed to the right* (*left*, resp.)

To denote an α-stable random variable X in Definition A.4.1, we write

$$X \sim S_\alpha(\sigma, \beta, \mu).$$

When $\mu = 0$, $\beta = 0$, X is said to be a *symmetric α-stable* ($S\alpha S$, in short) random variable. Thus, a $S\alpha S$ random variable has a characteristic function of the form

$$\mathbb{E}e^{i\theta X} = e^{-\sigma^\alpha |\theta|^\alpha}, \quad \theta \in \mathbb{R}. \quad (A.4.2)$$

When $\alpha = 2$, which is the case excluded here, a stable random variable is Gaussian. For more information on stable distributions, see Zolotarev [1029], Samorodnitsky and Taqqu [883]. A direct approach to stable distributions can be found in Pitman and Pitman [823].

Stable random variables arise as the only possible weak limits of normalized and centered partial sums

$$\frac{\sum_{k=1}^n X_k - A_n}{B_n}, \quad (A.4.3)$$

where X_k, $k \geq 1$, are i.i.d. random variables, $0 < B_n \to \infty$ and $A_n \in \mathbb{R}$. The distribution of X_1 is then said to be in the domain of attraction of the stable random variable. The sufficient and necessary conditions for (A.4.3) to converge in distribution to an α-stable distribution (for some A_n, B_n) are

$$\mathbb{P}(X_k > x) = (c_1 + o(1))h(x)x^{-\alpha}, \quad \mathbb{P}(X_k < -x) = (c_2 + o(1))h(x)x^{-\alpha}, \quad (A.4.4)$$

as $x \to \infty$, where $h(x)$ is a slowly varying functions at infinity and $c_1, c_2 \geq 0$ are two constants such that $c_1 + c_2 > 0$ (Gnedenko and Kolmogorov [409]). When $\alpha \neq 1$, the conditions in (A.4.4) are equivalent to

$$\mathbb{E}e^{i\theta X_k} = \exp\left\{-\sigma^\alpha |\theta|^\alpha h(|\theta|^{-1})(1 - i\beta\,\text{sign}(\theta)\tan\frac{\pi\alpha}{2}) + i\mu\theta + o(|\theta|^\alpha h(|\theta|^{-1}))\right\}, \quad (A.4.5)$$

as $\theta \to 0$, where

$$\beta = \frac{c_1 - c_2}{c_1 + c_2}, \quad \sigma^\alpha = \Gamma(1-\alpha)(c_1 + c_2)\cos\left(\frac{\alpha\pi}{2}\right), \quad \mu = \begin{cases} 0, & 0 < \alpha < 1, \\ \mathbb{E}X_k, & 1 < \alpha < 2 \end{cases} \quad (A.4.6)$$

(Ibragimov and Linnik [515], Theorem 2.6.5). There is an analogous but more delicate characterization in the case $\alpha = 1$ as well (Aaronson and Denker [1]). It coincides with (A.4.5) when the distribution of X_k is symmetric. Moreover, assuming (A.4.4) or (A.4.5), the partial sums (A.4.3) converge to an α-stable distribution with A_n and B_n satisfying

$$nh(B_n) = B_n^\alpha, \quad A_n = \begin{cases} 0, & 0 < \alpha < 1, \\ (\mathbb{E}X_k)n, & 1 < \alpha < 2. \end{cases} \quad (A.4.7)$$

The limiting stable distribution is $S_\alpha(\sigma, \beta, 0)$, where σ and β are related to c_1, c_2 and α through (A.4.6). When $\alpha = 1$, the sequence B_n can be chosen in the same way but the situation with A_n is more involved in general (Aaronson and Denker [1]). But A_n can be taken as μn at least for symmetric X_k. Otherwise, in the general case, A_n is a *nonlinear* function of n.

We also note that B_n satisfying (A.4.7) can be taken as

$$B_n \sim n^{1/\alpha} h_\alpha(n)^{1/\alpha}, \quad (A.4.8)$$

where h_α is a slowly varying function. Such function h_α can be constructed as follows. Note that it satisfies, equivalently,

$$h(n^{1/\alpha} h_\alpha(n)^{1/\alpha}) \sim h_\alpha(n), \quad (A.4.9)$$

as $n \to \infty$. Now let $f(x) = x^\alpha (1/h(x))$ which is a regularly varying function. By Theorem 1.5.12 in Bingham et al. [157], there is a regularly varying function $g(x) = x^{1/\alpha} l(x)$ with a slowly varying function $l(x)$ such that $f(g(x)) \sim x$, as $x \to \infty$. Hence,

$$f(g(x)) = \frac{g(x)^\alpha}{h(g(x))} = \frac{(x^{1/\alpha} l(x))^\alpha}{h(x^{1/\alpha} l(x))} \sim x.$$

This shows that

$$(l(x))^{-\alpha} h(x^{1/\alpha} l(x)) \sim 1, \quad \text{as } x \to \infty.$$

Then, $h_\alpha(x)$ can be taken as $l(x)^\alpha$. Theorem 1.5.12 in Bingham et al. [157] also yields the uniqueness of the function g, or the slowly varying function l, within asymptotic equivalence.

We shall take the conditions in (A.4.4) for a definition of α–heavy-tailed random variables and series.

Definition A.4.2 Let $\alpha \in (0, 2)$. A random variable X_k is called *heavy-tailed* or *α–heavy-tailed* if it satisfies the conditions in (A.4.4). A time series $\{X_k\}$ is called *α–heavy-tailed* if each random variable X_k is α–heavy-tailed.

An α–heavy-tailed random variable has infinite second moment for all $\alpha < 2$, and infinite mean for all $\alpha < 1$.

By the discussion proceeding Definition A.4.2, if X_k, $k \geq 1$, are i.i.d. α–heavy-tailed random variables, then

$$\frac{\sum_{k=1}^{n} X_k - A_n}{B_n} \xrightarrow{d} S_\alpha(\sigma, \beta, \mu), \tag{A.4.10}$$

for suitable A_n, B_n and σ, β, μ. This can also be stated as

$$\frac{\sum_{k=1}^{[Nt]} X_k - A(Nt)}{B(N)} \xrightarrow{fdd} L_\alpha(t), \quad t \geq 0, \tag{A.4.11}$$

where $\xrightarrow{fdd}$ stands for the convergence of finite-dimensional distributions, and L_α is an α-stable Lévy motion such that $L_\alpha(1)$ has the same distribution as $S_\alpha(\sigma, \beta, \mu)$. That is, L_α is the process with independent stationary increments such that $L_\alpha(t)$ has the same distribution as $t^{1/\alpha} S_\alpha(\sigma, \beta, \mu)$.

As introduced in Definition A.4.2, there is a wide variety of α–heavy-tailed series $\{X_k\}$. In fact, the definition is quite broad as it does not specify the dependence structure in the series $\{X_k\}$. The examples of α–heavy-tailed series considered in this volume include $S\alpha S$ time series, usually defined through the integrals with respect to $S\alpha S$ random measures (Section B.1.3), or linear time series with i.i.d. heavy-tailed innovations as in Example 2.9.2.

Appendix B

Integrals with Respect to Random Measures

Single and multiple integrals with respect to random measures are used throughout the book. We discuss here their constructions and basic properties. Single integrals are considered in Section B.1, and multiple integrals in Section B.2.

B.1 Single Integrals with Respect to Random Measures

Let $(E, \mathcal{E})$ be a measurable space: that is, E is a set and $\mathcal{E}$ is a σ–field of subsets of E. Single integrals with respect to random measures are written as

$$\int_E f(x) M(dx) =: I(f). \tag{B.1.1}$$

Here, f is a *deterministic*, possibly complex-valued measurable function on E. M is a suitable, possibly complex-valued *random measure* on $(E, \mathcal{E})$.

Four kinds of random measures M and the resulting integrals (B.1.1) are considered in Sections B.1.1–B.1.5:

- random measures with orthogonal increments,
- Gaussian measures,
- stable measures,
- Poisson measures, and
- Lévy measures.

In all these cases, a random measure M is defined with respect to a deterministic measure m on $(E, \mathcal{E})$, called a *control measure* of M. Let

$$\mathcal{E}_0 = \{A \in \mathcal{E} : m(A) < \infty\}. \tag{B.1.2}$$

A random measure M is defined as a set function $M : \mathcal{E}_0 \mapsto L^0(\Omega)$ having certain properties. In particular, a random measure M is σ–*additive* in the sense that

$$M\Big(\bigcup_{n=1}^\infty A_n\Big) = \sum_{n=1}^\infty M(A_n) \quad \text{a.s.} \tag{B.1.3}$$

for any pairwise disjoint $A_n \in \mathcal{E}_0$, $n \geq 1$, such that $\cup_{n=1}^\infty A_n \in \mathcal{E}_0$. In some cases, M is assumed to be *independently scattered* in the sense that if $A_1, \ldots, A_n \in \mathcal{E}_0$ are pairwise disjoint, then $M(A_1), \ldots, M(A_n)$ are independent. Additional assumptions are made depending on the kinds of measures M considered.

Integrals (B.1.1) are defined following the same idea. A function f is called *simple* (or *elementary*) if it can be written as

$$f(x) = \sum_{k=1}^{n} a_k 1_{A_k}(x), \qquad (B.1.4)$$

where $a_k \in \mathbb{C}$ and $A_k \in \mathcal{E}_0$, $k = 1, \ldots, n$, are pairwise disjoint. Denote the set of simple functions by $\mathcal{S}$.[1] For simple function $f \in \mathcal{S}$, the integral $I(f)$ in (B.1.1) is defined as

$$I(f) = \sum_{k=1}^{n} a_k M(A_k). \qquad (B.1.5)$$

This definition does not depend (in the a.s. sense) on the representation (B.1.4) of simple function f if M is σ–additive. Moreover,

$$I(af + bg) = aI(f) + bI(g) \quad \text{a.s.} \qquad (B.1.6)$$

for any $a, b \in \mathbb{C}$, $f, g \in \mathcal{S}$.

B.1.1 Integrals with Respect to Random Measures with Orthogonal Increments

The focus here is on the following random measures.

Definition B.1.1 A set function $Z : \mathcal{E}_0 \mapsto L^0(\Omega)$ with control measure m is called a *random measure with orthogonal increments* on $(E, \mathcal{E})$ if Z is σ–additive (see (B.1.3)) and

$$\mathbb{E} Z(A) = 0, \qquad (B.1.7)$$

$$\mathbb{E} Z(A_1)\overline{Z(A_2)} = m(A_1 \cap A_2), \qquad (B.1.8)$$

for any $A, A_1, A_2 \in \mathcal{E}_0$.

For an elementary function $f \in \mathcal{S}$, set $I(f)$ as in (B.1.5). For $f, g \in \mathcal{S}$, by using (B.1.8), one has

$$\mathbb{E} I(f)\overline{I(g)} = \int_E f(x)\overline{g(x)} m(dx) = (f, g)_{L^2(E,m)}. \qquad (B.1.9)$$

The definition of $I(f)$ can now be extended to functions $f \in L^2(E, m)$ as follows. For $f \in L^2(E, m)$, there is a sequence of simple functions f_n, $n \geq 1$, such that $\|f - f_n\|_{L^2(E,m)} \to 0$, as $n \to \infty$. Then, f_n, $n \geq 1$, is also a Cauchy sequence in $L^2(E, m)$. Hence, by using (B.1.6) and (B.1.9),

$$\mathbb{E}|I(f_n) - I(f_m)|^2 = \mathbb{E}|I(f_n - f_m)|^2 = \|f_n - f_m\|^2_{L^2(E,m)} \to 0,$$

as $n, m \to \infty$. This shows that $I(f_n)$, $n \geq 1$, is a Cauchy sequence in $L^2(\Omega)$. Hence, it has a limit in $L^2(\Omega)$ which is defined as the integral $I(f)$:

$$I(f) = \lim_{n \to \infty} (L^2(\Omega)) I(f_n). \qquad (B.1.10)$$

---

[1] Strictly speaking, $\mathcal{S}$ also depends on m.

The limit (B.1.10) does not depend on the approximating sequence f_n. Indeed, if $f_{1,n}$ and $f_{2,n}$ are two such sequences, then $\|f_{1,n} - f_{2,n}\|_{L^2(E,m)} \leq \|f - f_{1,n}\|_{L^2(E,m)} + \|f - f_{2,n}\|_{L^2(E,m)} \to 0$ and hence $\mathbb{E}|I(f_{1,n}) - I(f_{2,n})|^2 \to 0$. The relations (B.1.6) and (B.1.9) continue to hold for all $f, g \in L^2(E, m)$. Indeed, the relations hold for approximating sequences of simple functions f_n, g_n and follow for $f, g \in L^2(E, m)$ by passing to the limit $n \to \infty$.

B.1.2 Integrals with Respect to Gaussian Measures

A Gaussian random measure is a particular case of random measures with orthogonal increments.

Definition B.1.2 A random measure W on $(E, \mathcal{E})$ with orthogonal increments is called *Gaussian* if any vector $(W(A_1), \ldots, W(A_n))$, $A_k \in \mathcal{E}_0$, $k = 1, \ldots, n$, is multivariate Gaussian.

The integral $I(f)$ with respect to Gaussian measure W is defined as in Section B.1.1 for all $f \in L^2(E, m)$. An additional property of these integrals is that any vector $(I(f_1), \ldots, I(f_n))$, $f_k \in L^2(E, m)$, $k = 1, \ldots, n$, is also multivariate Gaussian.

Among complex-valued measures, the following Gaussian measures on $(\mathbb{R}, \mathcal{B}(\mathbb{R}))$ are of particular interest. Let m be a *symmetric* measure on $(\mathbb{R}, \mathcal{B}(\mathbb{R}))$ in the sense that

$$m(A) = m(-A), \quad \text{for } A \in \mathcal{B}(\mathbb{R}), \tag{B.1.11}$$

where

$$-A = \{x \in \mathbb{R} : (-x) \in A\}.$$

Definition B.1.3 An *Hermitian Gaussian random measure* Z on $(\mathbb{R}, \mathcal{B}(\mathbb{R}))$ is a complex-valued Gaussian measure on $(\mathbb{R}, \mathcal{B}(\mathbb{R}))$ with a symmetric control measure m such that

$$\overline{Z(A)} = Z(-A), \quad A \in \mathcal{B}(\mathbb{R})_0. \tag{B.1.12}$$

The relations (B.1.11) and (B.1.12) are often written as $m(dx) = m(-dx)$ and $\overline{Z(dx)} = Z(-dx)$, respectively. Here are some properties of Z. Suppose in the following that the sets belong to $\mathcal{B}(\mathbb{R})_0$.

- $\mathbb{E}Z(A) = 0$, $\mathbb{E}Z(A_1)\overline{Z(A_2)} = m(A_1 \cap A_2)$.
  This follows from the fact that Z is a complex-valued random measure.
- If $A \cap (-A) = \emptyset$, then $\mathbb{E}Z(A)^2 = 0$.
  Indeed, $\mathbb{E}Z(A)^2 = \mathbb{E}Z(A)\overline{Z(-A)} = m(A \cap (-A)) = 0$.
- If $A \cap (-A) = \emptyset$, then $Z(A)$ and $Z(-A)$ are uncorrelated each with variance $m(A)/2$.
  Indeed, $\mathbb{E}Z(A)\overline{Z(-A)} = m(A \cap (-A)) = 0$, hence uncorrelated. Moreover, $\mathbb{E}|Z(A)|^2 = \mathbb{E}|Z(-A)|^2$ and $\mathbb{E}|Z(A)|^2 + \mathbb{E}|Z(-A)|^2 = m(A) + m(-A) = 2m(A)$, so that $Z(A)$ and $Z(-A)$ have each variance $m(A)/2$.

- $\Re Z(A)$ and $\Im Z(B)$ are independent.
  This follows from

$$\begin{aligned}
\mathbb{E}\Re Z(A)\Im Z(B) &= \frac{1}{4i}\mathbb{E}\Big(Z(A) + \overline{Z(A)}\Big)\Big(Z(B) - \overline{Z(B)}\Big) \\
&= \frac{1}{4i}\mathbb{E}\Big(Z(A) + Z(-A)\Big)\Big(\overline{Z(-B)} - \overline{Z(B)}\Big) \\
&= \frac{1}{4i}\Big(m(A \cap (-B)) - m(A \cap B) + m((-A) \cap (-B)) - m((-A) \cap B)\Big) \\
&= \frac{1}{4i}\Big(m(A \cap (-B)) - m(A \cap B) + m(A \cap B) - m(A \cap (-B))\Big) = 0,
\end{aligned}$$

(B.1.13)

by using (B.1.11).

- If $A_1 \cup (-A_1), \ldots, A_n \cup (-A_n)$ are disjoint, then $Z(A_1), \ldots, Z(A_n)$ are independent.
  This follows from

$$\begin{aligned}
\mathbb{E}\Re Z(A_i)\Re Z(A_j) &= \mathbb{E}\Im Z(A_i)\Im Z(A_j) = \mathbb{E}\Re Z(A_i)\Im Z(A_j) \\
&= \mathbb{E}\Im Z(A_i)\Re Z(A_j) = 0, \quad i \neq j,
\end{aligned}$$

which can be proved by arguing as in (B.1.13).

- If $A \cap (-A) = \emptyset$, then $\Re Z(A)$ and $\Im Z(A)$ are independent with mean zero and variance $m(A)/2$.
  To see this, observe that on one hand,

$$0 = \mathbb{E}Z(A)\overline{Z(-A)} = \mathbb{E}Z(A)^2 = \mathbb{E}(\Re Z(A) + i\Im Z(A))^2 = \mathbb{E}(\Re Z(A))^2 - \mathbb{E}(\Im Z(A))^2$$

and on the other hand,

$$m(A) = \mathbb{E}Z(A)\overline{Z(A)} = \mathbb{E}(\Re Z(A))^2 + \mathbb{E}(\Im Z(A))^2.$$

If $g \in L^2(\mathbb{R}, m)$ is a deterministic function satisfying $\overline{g(x)} = g(-x)$, $x \in \mathbb{R}$ (that is, an Hermitian function), then

$$\int_{\mathbb{R}} g(x) Z(dx)$$

is always real-valued. This follows from

$$\overline{\int_{\mathbb{R}} g(x) Z(dx)} = \int_{\mathbb{R}} \overline{g(x) Z(dx)} = \int_{\mathbb{R}} g(-x) Z(-dx) = \int_{\mathbb{R}} g(x) Z(dx).$$

Hermitian random measures are also of interest because of the following well-known fact. Let B be a Gaussian random measure on $(\mathbb{R}, \mathcal{B}(\mathbb{R}))$ with the Lebesgue control measure du, and $\widehat{B}$ be an Hermitian Gaussian measure on $(\mathbb{R}, \mathcal{B}(\mathbb{R}))$ with the Lebesgue control measure dx. Let also $f \in L^2(\mathbb{R}, du)$ and

$$\widehat{f}(x) = \int_{\mathbb{R}} e^{ixu} f(u) du, \quad f(u) = \frac{1}{2\pi} \int_{\mathbb{R}} e^{-ixu} \widehat{f}(x) dx$$

be the corresponding $L^2(\mathbb{R})$–Fourier transforms (see Appendix A.1.2). Then,

$$\int_{\mathbb{R}} f(u) B(du) \stackrel{d}{=} \frac{1}{\sqrt{2\pi}} \int_{\mathbb{R}} \widehat{f}(x) \widehat{B}(dx). \tag{B.1.14}$$

The relation follows by observing that both sides of (B.1.14) are real-valued Gaussian variables with zero mean and the same variance. Indeed, to see that the variance is the same, observe by using Parseval's identity (A.1.11) that

$$\mathbb{E}\left(\int_{\mathbb{R}} f(u)B(du)\right)^2 = \int_{\mathbb{R}} |f(u)|^2 du = \frac{1}{2\pi}\int_{\mathbb{R}} |\widehat{f}(x)|^2 dx = \mathbb{E}\left(\frac{1}{\sqrt{2\pi}}\int_{\mathbb{R}} \widehat{f}(x)\widehat{B}(dx)\right)^2.$$

The integral on the left-hand side of (B.1.14) is said to be in the *time domain* and that on the right-hand side of (B.1.14) in the *spectral domain*.

B.1.3 Integrals with Respect to Stable Measures

Recall Definition A.4.1 of α-stable random variables and the notation $S_\alpha(\sigma, \beta, \mu)$. Stable random measures are defined as follows.

Definition B.1.4 A set function $M : \mathcal{E}_0 \mapsto L^0(\Omega)$ is called an *α-stable random measure*, $\alpha \in (0, 2)$, on $(E, \mathcal{E})$ with control measure m and skewness intensity (function) $\beta : E \mapsto [-1, 1]$ if M is σ–additive (see (B.1.3)), independently scattered and such that

$$M(A) \sim S_\alpha\left((m(A))^{1/\alpha}, \frac{\int_A \beta(x) m(dx)}{m(A)}, 0\right) \tag{B.1.15}$$

The stable measure M is called *$S\alpha S$ random measure* with control measure m if $M(A)$ is $S\alpha S$ with scale parameter $(m(A))^{1/\alpha}$. If $\beta \equiv 1$ ($\beta \equiv -1$, resp.), the stable measure is called *totally skewed to the right (left, resp.)*

For a simple function $f \in \mathcal{S}$, set $I(f)$ as in (B.1.5). For $f \in \mathcal{S}$, one can show (Samorodnitsky and Taqqu [883], p. 119) that

$$I(f) \sim S_\alpha(\sigma, \beta, \mu), \tag{B.1.16}$$

where

$$\sigma^\alpha = \int_E |f(x)|^\alpha m(dx), \tag{B.1.17}$$

$$\beta = \frac{\int_E \beta(x)(f(x))^{\langle\alpha\rangle} m(dx)}{\int_E |f(x)|^\alpha m(dx)}, \tag{B.1.18}$$

with $a^{\langle\alpha\rangle} = |a|^\alpha \text{sign}(a)$, and

$$\mu = \begin{cases} 0, & \text{if } \alpha \neq 1, \\ -\frac{2}{\pi}\int_E f(x)\beta(x)\log|f(x)|m(dx), & \text{if } \alpha = 1. \end{cases} \tag{B.1.19}$$

For a $S\alpha S$ random measure and $f \in \mathcal{S}$,

$$I(f) \sim S_\alpha(\sigma, 0, 0), \tag{B.1.20}$$

where σ is defined in (B.1.17).

We have defined $I(f)$ for simple functions. The definition of $I(f)$ can be extended to functions $f \in F$, where $f : E \to \mathbb{R}$ and

$$F = \begin{cases} L^\alpha(E, m), & \text{if } \alpha \neq 1, \\ L^1(E, m) \cap \{f : \int_E |f(x)\beta(x) \log|f(x)||m(dx) < \infty\}, & \text{if } \alpha = 1. \end{cases} \quad (B.1.21)$$

For $f \in F$, the basic idea is to consider simple functions $f_n \in \mathcal{S}$ such that $f_n \to f$ a.e. and $|f_n(x)| \leq g(x)$ for all x, n and some $g \in F$. One can show (Samorodnitsky and Taqqu [883], pp. 121–125) that $I(f_n)$, $n \geq 1$, is a Cauchy sequence in probability and hence has a limit in probability. One then sets

$$I(f) = \lim_{n \to \infty} (P) I(f_n). \quad (B.1.22)$$

One can show (Samorodnitsky and Taqqu [883], pp. 121–125) that the limit does not depend on the approximating sequence f_n, and that the relations (B.1.6) and (B.1.16)–(B.1.19) continue to hold for $f \in F$.

The definition of F in the case $\alpha = 1$ is slightly more complicated than in the case $\alpha \neq 1$ because of (B.1.19). If the random measure M is $S\alpha S$, however, then

$$F = L^\alpha(E, m), \quad \text{for } 0 < \alpha < 2.$$

B.1.4 Integrals with Respect to Poisson Measures

A Poisson random measure is defined as follows.[2]

Definition B.1.5 A set function $N : \mathcal{E}_0 \mapsto L^0(\Omega)$ is called a *Poisson random measure* on $(E, \mathcal{E})$ with control (intensity) measure n if N is σ–additive (see (B.1.3)), independently scattered and $N(A)$, $A \in \mathcal{E}_0$, is a Poisson random variable with mean $n(A)$; that is,

$$\mathbb{P}(N(A) = k) = e^{-n(A)} \frac{(n(A))^k}{k!}, \quad k = 1, 2, \ldots. \quad (B.1.23)$$

One can think of a Poisson random measure N as a point process on E: for $A \in \mathcal{E}_0$, $N(A)$ can be regarded as the (random) number of points falling in A. Existence of Poisson random measures is discussed in, e.g., Kingman [558], p. 23, and shown under the assumption that n is non-atomic (that is, $n(\{x\}) = 0$ for every $x \in E$) and σ–finite.

For a simple function f, the integral

$$I(f) = \int_E f(x) N(dx) \quad (B.1.24)$$

is defined as in (B.1.5). One can check (e.g., Kingman [558], p. 26) that $I(f)$ has the characteristic function

$$\mathbb{E} e^{i\theta I(f)} = \exp\left\{\int_E (e^{i\theta f(x)} - 1) n(dx)\right\}, \quad \theta \in \mathbb{R}. \quad (B.1.25)$$

---

[2] For Poisson random measures and their control measures, we write N instead of M and n instead of m.

We also note that by (B.1.5), the integral (B.1.24) for a simple function f can be written as $\sum_{X \in \Pi} f(X)$, where Π denotes the random points of the Poisson random measure on $(E, \mathcal{E})$. The definition of $I(f)$ can be extended to more general functions $f : E \to \mathbb{R}$ by using Campbell's theorem (Kingman [558], p. 28), so that the relations (B.1.6) and (B.1.25) continue to hold. The theorem states that the sum $\sum_{X \in \Pi} f(X)$ $(= \int_E f(x) N(dx))$ converges absolutely with probability one if and only if

$$\int_E (|f(x)| \wedge 1) n(dx) < \infty, \tag{B.1.26}$$

in which case we set $I(f) = \int_E f(x) N(dx) := \sum_{X \in \Pi} f(X)$. Moreover, by the same Campbell's theorem,

$$\mathbb{E} I(f) = \int_E f(x) n(dx) \tag{B.1.27}$$

when $\int_E |f(x)| n(dx) < \infty$, and

$$\mathrm{Var}(I(f)) = \int_E (f(x))^2 n(dx) \tag{B.1.28}$$

when $\int_E |f(x)|^2 n(dx) < \infty$.

Another interesting way to define the integral $\int_E f(x) N(dx)$ can be found in Rajput and Rosiński [840]. Suppose that there is a sequence of sets $E_n \in \mathcal{E}_0$ such that $\cup_{n=1}^\infty E_n = E$. An $\sigma(\mathcal{E}_0)$–measurable function f is called N–integrable if the limit in probability of $\int_A f_m(x) N(dx)$ exists for every $A \in \sigma(\mathcal{E}_0)$ and simple functions f_m such that $f_m \to f$ a.e. dn. For N–integrable function, one sets $\int_A f(x) N(dx) = \lim(\mathbb{P}) \int_A f_m(x) N(dx)$. By applying Theorem 2.7 in Rajput and Rosiński [840] (with $a(s) \equiv 1$, the point mass $\rho(s, dx) = \delta_{\{1\}}(dx)$ at $x = 1$, $\sigma^2(s) \equiv 0$ and $\lambda(ds) = n(ds)$ associated with the Poisson random measure N; see their Proposition 2.7 and compare with the relation (B.1.25) above), a function f is N–integrable if and only if

$$\int_E |\tau(f(x))| n(dx) = \int_E (|f(x)| \wedge 1) n(dx) < \infty, \quad \int_E (|f(x)|^2 \wedge 1) n(dx) < \infty, \tag{B.1.29}$$

where

$$\tau(u) = \begin{cases} u, & \text{if } |u| \leq 1, \\ \mathrm{sign}(u), & \text{if } |u| > 1. \end{cases} \tag{B.1.30}$$

Note that (B.1.29) is stronger than (B.1.26). But the condition (B.1.26) is more natural in defining the integral in view of (B.1.25), where the exponent is well-defined under the condition (B.1.26) by using the inequality $|e^{iy} - 1| \leq 2(|y| \wedge 1)$, $y \in \mathbb{R}$.

The compensated version of the Poisson random measure N is the measure $\widetilde{N}$ defined as $\widetilde{N}(A) = N(A) - n(A)$, $A \in \mathcal{E}_0$, for which $\mathbb{E} \widetilde{N}(A) = 0$. There is a useful extension of integrals (B.1.24) with respect to a Poisson measure to integrals with respect to a compensated Poisson measure $\widetilde{N}$. Such integrals are written as

$$\widetilde{I}(f) = \int_E f(x)(N(dx) - n(dx)) = \int_E f(x) \widetilde{N}(dx), \tag{B.1.31}$$

where N is a Poisson random measure with control measure n, and $\widetilde{N}$ is the compensated Poisson random measure. For simple functions f, $\widetilde{I}(f)$ is defined as

$$\widetilde{I}(f) = \int_E f(x)N(dx) - \int_E f(x)n(dx) = I(f) - \int_E f(x)n(dx) \tag{B.1.32}$$

and has the characteristic function

$$\mathbb{E}e^{i\theta \widetilde{I}(f)} = \exp\left\{\int_E (e^{i\theta f(x)} - 1 - i\theta f(x))n(dx)\right\}, \quad \theta \in \mathbb{R}. \tag{B.1.33}$$

As above, we will refer to Rajput and Rosiński [840] to extend the integral $\widetilde{I}(f) = \int_E f(x)\widetilde{N}(dx)$ to more general $\sigma(\mathcal{E}_0)$–measurable functions f, so that (B.1.6) and (B.1.33) continue to hold. By applying Theorem 2.7 in Rajput and Rosiński [840] (with $a(s) \equiv 0$, the point mass $\rho(s, dx) = \delta_{\{1\}}(dx)$ at $x = 1$, $\sigma^2(s) \equiv 0$ and $\lambda(ds) = n(ds)$ associated with the compensated Poisson random measure $\widetilde{N}$; see their Proposition 2.7 and compare with the relation (B.1.33) above), a function f is $\widetilde{N}$–integrable if and only if

$$\int_E |\tau(f(x)) - f(x)|n(dx) < \infty, \quad \int_E (|f(x)|^2 \wedge 1)n(dx) < \infty, \tag{B.1.34}$$

where τ is given in (B.1.30). By the following lemma, the conditions in (B.1.34) are equivalent to the condition

$$\int_E (|f(x)|^2 \wedge |f(x)|)n(dx) < \infty. \tag{B.1.35}$$

Lemma B.1.6 *The two conditions in (B.1.34) are equivalent to the condition (B.1.35).*

Proof We first show that (B.1.35) implies the two conditions in (B.1.34). For the second condition in (B.1.34), this follows from the inequality $|u|^2 \wedge 1 \leq |u|^2 \wedge |u|$, $u \in \mathbb{R}$, with $u = f(x)$. For the first condition in (B.1.34), this follows from the inequality $|\tau(u) - u| \leq |u|^2 \wedge |u|$, $u \in \mathbb{R}$, with $u = f(x)$, after observing that $|\tau(u) - u| = 0$ if $|u| \leq 1$, and $= |\text{sign}(u) - u| = |1 - |u|| \leq |u|$ if $|u| > 1$. The fact that the conditions in (B.1.34) imply (B.1.35) follows similarly by observing that

$$|u|^2 \wedge |u| = |u|^2 1_{\{|u| \leq 1\}} + |u| 1_{\{|u| > 1\}} = (|u|^2 \wedge 1) 1_{\{|u| \leq 1\}} + (|u| - 1 + (|u|^2 \wedge 1)) 1_{\{|u| > 1\}}$$

$$= (|u|^2 \wedge 1) + (|u| - 1) 1_{\{|u| > 1\}} = (|u|^2 \wedge |u|) + |\tau(u) - u|,$$

since for the last equality, $|\tau(u) - u| = 0$ if $|u| \leq 1$, and $= |\text{sign}(u) - u| = |u| - 1$ if $|u| > 1$, as noted above. $\square$

The integral $\widetilde{I}(f) = \int_E f(x)\widetilde{N}(dx)$ can be extended to other classes of integrands (e.g., those f satisfying $\int_E |f(x)|^2 n(dx) < \infty$; see the footnotes on p. 158 and Exercises 3.22 and 3.23 in Samorodnitsky and Taqqu [883]) but the condition (B.1.35) is more natural in defining the integral in view of (B.1.33), where the exponent is well-defined under the condition (B.1.35) by using the inequality $|e^{iy} - 1 - iy| \leq 2(|y|^2 \wedge |y|)$, $y \in \mathbb{R}$.

By applying Theorem 3.3 in Rajput and Rosiński [840] (with the choices of $a(s)$, $\rho(s, dx)$, $\sigma^2(s)$ and $\lambda(ds)$ indicated above), the integral $\widetilde{I}(f)$ has finite mean when

$\int_E |f(x)| n(dx) < \infty$. Then, under this condition, differentiating (B.1.33) with respect to θ and setting $\theta = 0$ yields $\mathbb{E}\tilde{I}(f) = 0$.

Integrals with respect to stable measures can be related to integrals with respect to Poisson measures. Let M be an α-stable random measure on $(E, \mathcal{E})$ with control measure m, index of stability $\alpha \in (0, 2)$ and skewness intensity β. Consider a Poisson random measure $N(dx, du)$ on $E \times \mathbb{R}_0$, $\mathbb{R}_0 = \mathbb{R} \setminus \{0\}$, with control measure

$$n(dx, du) = \begin{cases} (1 + \beta(x)) m(dx) \frac{du}{|u|^{\alpha+1}}, & \text{if } u > 0, \\ (1 - \beta(x)) m(dx) \frac{du}{|u|^{\alpha+1}}, & \text{if } u < 0. \end{cases} \quad (B.1.36)$$

Observe that the space E in Definition B.1.5 is now $E \times \mathbb{R}_0$. Let also

$$c_\alpha = \Big(\frac{2\Gamma(2-\alpha)}{\alpha(1-\alpha)} \cos \frac{\alpha\pi}{2} \Big)^{1/\alpha}. \quad (B.1.37)$$

One can then show (Samorodnitsky and Taqqu [883], Theorem 3.12.2, p. 156) that

$$c_\alpha \int_E f(x) M(dx) \stackrel{d}{=} \int_E \int_{\mathbb{R}_0} f(x) u N(dx, du), \quad (B.1.38)$$

if $\alpha \in (0, 1)$, and

$$c_\alpha \int_E f(x) M(dx) \stackrel{d}{=} \int_E \int_{\mathbb{R}_0} f(x) u (N(dx, du) - n(dx, du)), \quad (B.1.39)$$

if $\alpha \in (1, 2)$. There is a similar but more involved formula when $\alpha = 1$.

B.1.5 Integrals with Respect to Lévy Measures

A Lévy random measure is defined as follows. For simplicity, we consider only the symmetric case.

Definition B.1.7 A set function $Z : \mathcal{E}_0 \mapsto L^0(\Omega)$ is called a *Lévy random measure* on $(E, \mathcal{E})$ with control measure m on E and symmetric Lévy measure ρ on $\mathbb{R}$ if Z is σ-additive (see (B.1.3)), independently scattered and $Z(A)$, $A \in \mathcal{E}_0$, is a symmetric infinitely divisible random variable with symmetric Lévy measure $m(A)\rho$ (and without Gaussian part), that is, having the characteristic function

$$\mathbb{E} e^{i\theta Z(A)} = \exp\Big\{m(A) \int_{\mathbb{R}} (e^{i\theta u} - 1) \rho(du)\Big\} = \exp\Big\{-m(A) \int_{\mathbb{R}} (1 - \cos(\theta u)) \rho(du)\Big\}, \ \theta \in \mathbb{R}, \quad (B.1.40)$$

where the Lévy measure ρ satisfies $\int_{\mathbb{R}} (|u| \wedge 1) \rho(du) < \infty$.

When $\rho(du) = c|u|^{-\alpha-1} du$, $\alpha \in (0, 2)$, a symmetric Lévy random measure is a symmetric α-stable random measure. In addition to symmetric stable distributions, two-sided exponential, Student's t, Cauchy, symmetrized gamma and other distributions are symmetric infinitely divisible (e.g., Sato [886], Steutel and van Harn [923]).

If $f : E \to \mathbb{R}$ is a measurable function such that

$$\int_E \int_{\mathbb{R}} (u^2 (f(x))^2 \wedge 1) \rho(du) m(dx) < \infty, \quad (B.1.41)$$

then the integral

$$I(f) = \int_E f(x) Z(dx) \tag{B.1.42}$$

can be defined in a natural way extending (B.1.5) so that it is a symmetric infinitely divisible random variable with the characteristic function

$$\mathbb{E} e^{i\theta I(f)} = \exp\left\{ -\int_E \int_{\mathbb{R}} \Big(1 - \cos(\theta u f(x))\Big) \rho(du) m(dx) \right\}, \quad \theta \in \mathbb{R}. \tag{B.1.43}$$

See Rajput and Rosiński [840], who also consider the general, not necessarily symmetric case.

In view of (B.1.25), the integral (B.1.42) can also be represented as

$$I(f) = \int_E f(x) Z(dx) \stackrel{d}{=} \int_E \int_{\mathbb{R}} f(x) u N(dx, du), \tag{B.1.44}$$

where N is a Poisson random measure on $E \times \mathbb{R}$ with the intensity measure $m(dx)\rho(du)$. When $\int_E \int_{\mathbb{R}} (f(x))^2 u^2 m(dx)\rho(du) = \int_{\mathbb{R}} u^2 \rho(du) \int_E (f(x))^2 m(dx) < \infty$,

$$\mathrm{Var}(I(f)) = \int_E \int_{\mathbb{R}} (f(x))^2 u^2 m(dx)\rho(du) = \int_{\mathbb{R}} u^2 \rho(du) \int_E (f(x))^2 m(dx) < \infty. \tag{B.1.45}$$

Moreover, by symmetry, $\mathbb{E} I(f) = 0$.

B.2 Multiple Integrals with Respect to Gaussian Measures

We recall here the definition and basic properties of of multiple integrals with respect to Gaussian measures, also known as multiple Wiener-Itô integrals. If W is a *real-valued* Gaussian measure on $(E, \mathcal{E})$ with control measure m (see Section B.1.2), these integrals are written as

$$\int_{E^k}' f(u_1, \ldots, u_k) W(du_1) \ldots W(du_k) =: I_k(f), \tag{B.2.1}$$

where $f : E^k \to \mathbb{R}$ is a deterministic, real-valued function. The prime in $\int_{E^k}'$ refers to the fact that integration excludes diagonals. When $E = \mathbb{R}$, we shall also consider multiple integrals of complex-valued functions with respect to complex-valued Hermitian Gaussian measures. The measure m is assumed to be non-atomic; that is, $m(\{u\}) = 0$ for every $u \in E$.

The integral (B.2.1) is first defined on the set $\mathcal{S}^k$ of simple functions. A function $f : E^k \mapsto \mathbb{R}$ is *simple* if it has the form

$$f(u_1, \ldots, u_k) = \sum_{i_1, \ldots, i_k = 1}^{n} a_{i_1, \ldots, i_k} 1_{A_{i_1} \times \ldots \times A_{i_k}}(u_1, \ldots, u_k), \tag{B.2.2}$$

where $A_1, \ldots, A_k \in \mathcal{E}_0$ are pairwise disjoint, and real-valued coefficients $a_{i_1, \ldots, i_k}$ are zero if any two of the indices $i_1, \ldots, i_k$ are equal. The last property ensures that the function f vanishes on the diagonals $\{u_i = u_j : i \neq j\}$. For $f \in \mathcal{S}^k$, one sets

$$I_k(f) = \sum_{i_1, \ldots, i_k = 1}^{n} a_{i_1, \ldots, i_k} W(A_{i_1}) \ldots W(A_{i_k}). \tag{B.2.3}$$

One can show (Nualart [769], p. 9) that the definition (B.2.3) does not depend on the particular representation of f and that the following properties hold: for $f, g \in \mathcal{S}^k, a, b \in \mathbb{R}$,

$$I_k(af + bg) = aI_k(f) + bI_k(g) \tag{B.2.4}$$

and

$$I_k(f) = I_k(\widetilde{f}), \tag{B.2.5}$$

where $\widetilde{f}$ denotes the symmetrization of f defined by

$$\widetilde{f}(u_1, \ldots, u_k) = \frac{1}{k!} \sum_\sigma f(u_{\sigma(1)}, \ldots, u_{\sigma(k)}), \tag{B.2.6}$$

with the sum over all permutations σ of $\{1, \ldots, k\}$. This is because $W(A_{i_1}) \ldots W(A_{i_k})$ is invariant under a permutation of $(1, \ldots, k)$. Moreover,

$$\mathbb{E} I_k(f) = 0, \tag{B.2.7}$$

$$\mathbb{E} I_k(f) I_q(g) = \begin{cases} 0, & \text{if } k \neq q, \\ k!(\widetilde{f}, \widetilde{g})_{L^2(E^k, m^k)}, & \text{if } k = q. \end{cases} \tag{B.2.8}$$

Relation (B.2.7) results from the fact that $a_{i_1, \ldots, i_k} \mathbb{E} W(A_{i_1}) \ldots W(A_{i_k}) = 0$ in (B.2.3) since the coefficients $a_{i_1, \ldots, i_k}$ are zero if any of two indices are equal. To verify (B.2.8), denote by $i_1, \ldots, i_k$ and $j_1, \ldots, j_q$ the indices in the summation of $I_k(f)$ and $I_q(g)$ respectively. If $k > q$, then there is always at least one index i' among $i_1, \ldots, i_k$ which is different from $j_1, \ldots, j_q$. Since the corresponding $W(A_{i'})$ is independent of the other $W(A)$s and since $\mathbb{E} W(A_{i'}) = 0$, one gets $\mathbb{E} I_k(f) I_q(g) = 0$. If $k = q$, then $\mathbb{E} W(A_{i_1}) \ldots W(A_{i_k}) W(A_{j_1}) \ldots W(A_{j_k}) = m(A_{i_1}) \ldots m(A_{i_k})$ for any of the $k!$ permutations $j_1, \ldots, j_k$ of $i_1, \ldots, i_k$, resulting in $\mathbb{E} I_k(f) I_q(g) = k!(\widetilde{f}, \widetilde{g})_{L^2(E^k, m^k)}$.

Note that the symmetrization $\widetilde{f}$ is a *symmetric* function in the sense that it is the same after interchanging any two of its arguments. A useful consequence of (B.2.8) is that

$$\mathbb{E} I_k(f)^2 = k! \|\widetilde{f}\|^2_{L^2(E^k, m^k)} \leq k! \|f\|^2_{L^2(E^k, m^k)}. \tag{B.2.9}$$

The second inequality follows by applying the triangle inequality to (B.2.6).

The definition (B.2.3) of $I_k(f)$ is extended to functions $f \in L^2(E^k, m^k)$ (Nualart [769], p. 7). For $f \in L^2(E^k, m^k)$, there is a sequence of simple functions $f_n \in \mathcal{S}^k$ such that $\|f - f_n\|_{L^2(E^k, m^k)} \to 0$, as $n \to \infty$. Then, one can show that $I_k(f_n), n \geq 1$, converges in $L^2(\Omega)$. The limit is taken as the multiple integral of f:

$$I_k(f) = \lim_{n \to \infty} (L^2(\Omega)) I_k(f_n). \tag{B.2.10}$$

It satisfies the properties (B.2.4)–(B.2.9).

• **Change of variables formula:** Another useful property of multiple integrals is the following formula for the change of variables. Let W_1 and W_2 be two real-valued Gaussian measures on $(E, \mathcal{E})$ with control measures m_1 and m_2, respectively. Denote the corresponding multiple integrals by $I_{1,k}(f_1)$ and $I_{2,k}(f_2)$. Suppose m_1 is absolutely continuous with respect to m_2. Let $h : E \mapsto \mathbb{R}$ be such that

$$|h(u)|^2 = \frac{dm_1}{dm_2}(u), \quad u \in E.$$

Let also $f_{1,n} \in L^2(E^n, m_1^n)$, $n = 1, \ldots, k$, and set

$$f_{2,n}(u_1, \ldots, u_n) = f_{1,n}(u_1, \ldots, u_n) h(u_1) \ldots h(u_n), \quad n = 1, \ldots, k.$$

The change of variables formula states that

$$\left(I_{1,1}(f_{1,1}), I_{1,2}(f_{1,2}), \ldots, I_{1,k}(f_{1,k}) \right) \stackrel{d}{=} \left(I_{2,1}(f_{2,1}), I_{2,2}(f_{2,2}), \ldots, I_{2,k}(f_{2,k}) \right).$$
(B.2.11)

Multiple integrals are also defined with respect to Hermitian Gaussian measures. Let Z be an Hermitian Gaussian measure on $(\mathbb{R}, \mathcal{B}(\mathbb{R}))$ with symmetric control measure m (see Definition B.1.3), and $g \in L^2(\mathbb{R}^k, m^k)$ be a deterministic *Hermitian function*, that is, satisfying

$$\overline{g(x_1, \ldots, x_k)} = g(-x_1, \ldots, -x_k).$$
(B.2.12)

Proceeding as for the multiple integrals (B.2.1), one can similarly define a multiple integral of g with respect to an Hermitian Gaussian measure Z by starting with an Hermitian function

$$g(x_1, \ldots, x_k) = \sum_{i_1=-n}^{n} \cdots \sum_{i_k=-n}^{n} a_{i_1, \ldots, i_k} 1_{A_{i_1} \times \cdots \times A_{i_k}}(x_1, \ldots, x_k),$$
(B.2.13)

where $a_{i_1, \ldots, i_k}$ are assumed to be zero if any of the indices $i_1, \ldots, i_k$ are either equal or the negative of the other, $A_{i_j} = -A_{-i_j}$ and $A_{i_j} \cap A_{i_k} = \emptyset$ if $j \neq k$ (Peccati and Taqqu [797], pp. 161–164[3]). The integral is denoted

$$\int_{\mathbb{R}^k}'' g(x_1, \ldots, x_k) Z(dx_1) \ldots Z(dx_k) =: \widehat{I}_k(g),$$
(B.2.14)

where the double prime on $\int_{\mathbb{R}^k}''$ indicates that the diagonals $\{x_i = \pm x_j, i \neq j\}$ are excluded in integration. The multiple integral (B.2.14) is real-valued and has properties analogous to (B.2.4)–(B.2.9).

• **Change of variables formula in the Hermitian case:** There is a formula analogous to (B.2.11) for the change of variables in multiple integrals (B.2.14). Let Z_1 and Z_2 be two Hermitian Gaussian measures on $(\mathbb{R}, \mathcal{B}(\mathbb{R}))$ with symmetric control measures m_1 and m_2, respectively. Denote the corresponding multiple integrals by $\widehat{I}_{1,k}(g_1)$ and $\widehat{I}_{2,k}(g_2)$. Suppose m_1 is absolutely continuous with respect to m_2. Let $h : \mathbb{R} \mapsto \mathbb{C}$ be an Hermitian function such that

$$|h(x)|^2 = \frac{dm_1}{dm_2}(x), \quad x \in \mathbb{R}.$$

Let also $g_{1,n} \in L^2(\mathbb{R}^n, m_1^n)$, $k = 1, \ldots, n$, be Hermitian functions and set

$$g_{2,n}(x_1, \ldots, x_n) = g_{1,n}(x_1, \ldots, x_n) h(x_1) \ldots h(x_n), \quad n = 1, \ldots, k.$$

Then,

$$\left(\widehat{I}_{1,1}(g_{1,1}), \widehat{I}_{1,2}(g_{1,2}), \ldots, \widehat{I}_{1,k}(g_{1,k}) \right) \stackrel{d}{=} \left(\widehat{I}_{2,1}(g_{2,1}), \widehat{I}_{2,2}(g_{2,2}), \ldots, \widehat{I}_{2,k}(g_{2,k}) \right).$$
(B.2.15)

---

[3] Note the following typos in that reference on p. 162: $c_{j_1}, \ldots, c_{j_q}$ should be $c_{j_1, \ldots, j_q}$ and these coefficients should be assumed to be 0 if any of the indices are either equal or the negative of the other.

See Major [672], Theorem 4.4 for a proof.

• **Relationship between time and spectral domains:** Another useful property of integrals (B.2.14) is the following. Let B be a Gaussian random measure on $(\mathbb{R}, \mathcal{B}(\mathbb{R}))$ with the Lebesgue control measure du, and $\widehat{B}$ be an Hermitian Gaussian measure on $(\mathbb{R}, \mathcal{B}(\mathbb{R}))$ with the Lebesgue control measure dx. Denote the corresponding multiple integrals by I_k and $\widehat{I}_k$. Let also $f_n \in L^2(\mathbb{R}^n, du_1 \ldots du_n)$, $n = 1, \ldots, k$, and $\widehat{f}_n(x_1, \ldots, x_n) = \int_{\mathbb{R}} e^{i(x_1 u_1 + \cdots + x_n u_n)} f_n(u_1, \ldots, u_n) du_1 \ldots du_n$ be their $L^2(\mathbb{R})$–Fourier transforms. Then (Proposition 9.3.1 in Peccati and Taqqu [797]),

$$\left(I_1(f_1), \ldots, I_k(f_k)\right) \stackrel{d}{=} \left(\frac{1}{(2\pi)^{1/2}} \widehat{I}_1(\widehat{f}_1), \ldots, \frac{1}{(2\pi)^{k/2}} \widehat{I}_k(\widehat{f}_k)\right). \quad (B.2.16)$$

This relation generalizes (B.1.14). As for that relation, the left-hand side of (B.2.16) is said to be in the *time domain* and that on the right-hand side of (B.2.16) in the *spectral domain*.

• **Multiplication of multiple integrals:** Finally, we include few other useful facts about multiple integrals, which are used in the book. The following formula concerns the multiplication of multiple integrals. For functions $f \in L^2(E^p, m^p)$ and $g \in L^2(E^q, m^q)$, the *contraction* $f \otimes_r g \in L^2(E^{p+q-2r}, m^{p+q-2r})$ of r indices, $1 \le r \le p \wedge q$, is defined as

$$(f \otimes_r g)(u_1, \ldots, u_{p+q-2r})$$
$$= \int_{E^r} f(u_1, \ldots, u_{p-r}, s_1, \ldots, s_r) g(u_{p-r+1}, \ldots, u_{p+q-2r}, s_1, \ldots, s_r) m(ds_1) \ldots m(ds_r). \quad (B.2.17)$$

For $r = 0$, $f \otimes_0 g$ is the *tensor product* $f \otimes g \in L^2(E^{p+q}, m^{p+q})$ defined as

$$(f \otimes g)(u_1, \ldots, u_p, u_{p+1}, \ldots, u_{p+q}) = f(u_1, \ldots, u_p) g(u_{p+1}, \ldots, u_{p+q}). \quad (B.2.18)$$

We write $f^{\otimes n} = f \otimes \ldots \otimes f$ (n times). For symmetric $f \in L^2(E^p, m^p)$ and $g \in L^2(E^q, m^q)$, one then has

$$I_p(f) I_q(g) = \sum_{r=0}^{p \wedge q} r! \binom{p}{r} \binom{q}{r} I_{p+q-2r}(f \otimes_r g) \quad (B.2.19)$$

(Nualart [769], Proposition 1.1.3). In (B.2.19), $p + q - 2r = 0$ for $p = q = r$, $f \otimes_r g$ is a constant and $I_0(c)$ is interpreted as c for a constant c. A special case of (B.2.19) is $q = 1$ for which

$$I_p(f) I_1(g) = I_{p+1}(f \otimes g) + p I_{p-1}(f \otimes_1 g). \quad (B.2.20)$$

There is an analogous formula for multiple integrals with respect to Hermitian Gaussian measures. For Hermitian functions $f \in L^2(\mathbb{R}^p, m^p)$ and $g \in L^2(\mathbb{R}^q, m^q)$, the *contraction* $f \overline{\otimes}_r g \in L^2(E^{p+q-2r}, m^{p+q-2r})$, $1 \le r \le p \wedge q$, is defined as

$$(f \overline{\otimes}_r g)(x_1, \ldots, x_{p+q-2r})$$
$$= \int_{\mathbb{R}^r} f(x_1, \ldots, x_{p-r}, z_1, \ldots, z_r) g(x_{p-r+1}, \ldots, x_{p+q-2r}, z_1, \ldots, z_r) m(dz_1) \ldots m(dz_r). \quad (B.2.21)$$

B.2 Multiple Integrals with Respect to Gaussian Measures

For symmetric, Hermitian $f \in L^2(\mathbb{R}^p, m^p)$ and $g \in L^2(\mathbb{R}^q, m^q)$, one has

$$\widehat{I}_p(f)\widehat{I}_q(g) = \sum_{r=0}^{p \wedge q} r! \binom{p}{r}\binom{q}{r} \widehat{I}_{p+q-2r}(f \overline{\otimes}_r g) \qquad (B.2.22)$$

(Peccati and Taqqu [797], Eq. (9.1.8)) and, in the special case $q = 1$,

$$\widehat{I}_p(f)\widehat{I}_1(g) = \widehat{I}_{p+1}(f \otimes g) + p\widehat{I}_{p-1}(f \overline{\otimes}_1 g). \qquad (B.2.23)$$

- **Connections to Itô stochastic integrals:** We have also used connections of multiple integrals to Itô stochastic integrals. More specifically, for $f \in L^2([0,\infty)^n, (du)^n)$, let $I_n(f)$ be a multiple integral with respect to a Gaussian measure on $[0,\infty)$ with control measure du. Let also $\{B(u)\}_{u \geq 0}$ be the usual Brownian motion. Then, for symmetric $f \in L^2([0,\infty)^n, (du)^n)$, one has

$$I_n(f) = n! \int_0^\infty \left\{ \int_0^{t_n} \cdots \left\{ \int_0^{t_2} f_n(t_1, \ldots, t_n) dB(t_1) \right\} \ldots dB(t_{n-1}) \right\} dB(t_n), \qquad (B.2.24)$$

where the last integral is an iterated Itô integral (Nualart [769], Eq. (1.27)).

- **Stochastic Fubini theorem:** Finally, we also mention the following stochastic Fubini theorem used in Chapter 4. Let (S, μ) be a measure space and $f(s, u_1, \ldots, u_k)$ be a deterministic function on $S \times \mathbb{R}^k$ such that

$$\int_S \left(\int_{\mathbb{R}^k} |f(s, u_1, \ldots, u_k)|^2 m(du_1) \ldots m(du_k) \right)^{1/2} \mu(ds) < \infty.$$

Then,

$$\int_{\mathbb{R}^k}' \left\{ \int_S f(s, u_1, \ldots, u_k) \mu(ds) \right\} W(du_1) \ldots W(du_k)$$

$$= \int_S \left\{ \int_{\mathbb{R}^k}' f(s, u_1, \ldots, u_k) W(du_1) \ldots W(du_k) \right\} \mu(ds) \qquad (B.2.25)$$

a.s. (Theorem 2.1 in Pipiras and Taqqu [819], or Proposition 5.13.1 in Peccati and Taqqu [797]).

Appendix C

Basics of Malliavin Calculus

We recall here the basics of the Malliavin calculus which we illustrate with a number of examples. The Malliavin calculus is used in connection to fractional Brownian motion in Chapter 7. For a much more thorough treatment, see for example Nualart [769].

C.1 Isonormal Gaussian Processes

Let $\mathcal{H}$ be a real separable Hilbert space with scalar product denoted by $(\cdot,\cdot)_{\mathcal{H}}$. A fundamental notion related to the Hilbert space $\mathcal{H}$ is that of an isonormal Gaussian process.

Definition C.1.1 A stochastic process W defined by $W = \{W(h), h \in \mathcal{H}\}$ is called an *isonormal Gaussian process* if W is a centered Gaussian family of random variables such that $\mathbb{E}(W(h)W(g)) = (f,g)_{\mathcal{H}}$ for all $h, g \in \mathcal{H}$.

The mapping $h \mapsto W(h)$ is linear since for any $a_1, a_2 \in \mathbb{R}, h_1, h_2 \in \mathcal{H}$,

$$\mathbb{E}\Big(W(a_1 h_1 + a_2 h_2) - (a_1 W(h_1) + a_2 W(h_2))\Big)^2 = 0.$$

Example C.1.2 A typical example is $\mathcal{H} = L^2(E, \mathcal{E}, m)$ equipped with the inner product

$$(f,g)_{\mathcal{H}} = \int_E f(u)g(u)m(du),$$

where $(E, \mathcal{E})$ is a measurable space and m is a σ-finite measure without atoms. This is known as the *white noise* case. For example, one could take $E = [0, T]$ and the Lebesgue measure m, in which case $W(1_{[0,t)}) =: W(t)$ is a standard Brownian motion. Note also that while the standard Brownian motion $\{W(t), t \in [0, T]\}$ is not isonormal because $\mathbb{E}W(s)W(t) = \min\{s, t\}$ is not an inner product on the space $[0, T]$ (the function $\min\{s, t\}$ is not linear in its arguments), the stochastic process $W = \{W(h) = \int_0^T h(t)dW(t)\}$ is an isonormal Gaussian process with $\mathcal{H} = L^2([0, T], m)$, where m is the Lebesgue measure. Thus $W(h)$ is a Gaussian random variable with mean zero and variance $\int_0^T h(t)^2 dt$. Another similar example, but not of a white noise, can be constructed in terms of fractional Brownian motion (Section 7.1).

There are three fundamental notions: the Malliavin derivative D, its adjoint or Malliavin (divergence) integral δ and the Ornstein–Uhlenbeck generator L. The operators D, δ, L will be defined below.

C.2 Derivative Operator

The first notion is that of the derivative operator (the *Malliavin derivative*). Let $C_p^\infty(\mathbb{R}^n)$ be the set of all infinitely continuously differentiable functions $f : \mathbb{R}^n \to \mathbb{R}$ such that f and all its partial derivatives have polynomial growth. Let also $\mathcal{S}$ denote the class of "smooth" random variables having the form

$$F = f(W(h_1), \ldots, W(h_n)), \tag{C.2.1}$$

where $f \in C_p^\infty(\mathbb{R}^n)$, $h_1, \ldots, h_n \in \mathcal{H}$ and $n \geq 1$. The random variable F is said to be "smooth" because the function f is smooth.

Definition C.2.1 The (Malliavin) derivative of a smooth random variable F of the form (C.2.1) is the $\mathcal{H}$–valued random variable given by

$$DF = \sum_{k=1}^n \left(\partial_k f(W(h_1), \ldots, W(h_n))\right) h_k, \tag{C.2.2}$$

where $\partial_k f = \frac{\partial f}{\partial x_k}$.

Remark C.2.2 The definition of DF can be shown not to depend on the representation (C.2.1) of F. See Kunze [583].

Remark C.2.3 Suppose that $\mathcal{H}$ is a space of functions. Then, in (C.2.1), F is a scalar random variable but its derivative DF in (C.2.2) is a random function, since it is a weighted average of the functions $h_1, \ldots, h_n$.

Example C.2.4 If $n = 1$ and $f(x) = x^r$, $r \geq 1$, then $D(W(h))^r = r(W(h))^{r-1} h$, for any $h \in \mathcal{H}$, and in particular, $DW(h) = h$.

Example C.2.5 In the white noise case with $\mathcal{H} = L^2([0, T], m)$ and the Lebesgue measure m, $W(1_{[0,t)}) =: W(t)$ is a standard Brownian motion (see Example C.1.2). Then, with $h_k(u) = 1_{[0,t_k)}(u)$, F in (C.2.1) is the random variable $F = f(W(t_1), \ldots, W(t_n))$ and its derivative is the random function

$$(DF)(u) = D_u F = \sum_{k=1}^n \left(\partial_k f(W(t_1), \ldots, W(t_n))\right) 1_{[0,t_k)}(u).$$

The shorter notation $D_u F$ is usually preferred. For example,

$$(Df(W(t)))(u) = D_u f(W(t)) = f'(W(t)) 1_{[0,t)}(u). \tag{C.2.3}$$

A fundamental relation is

$$\mathbb{E}(F W(h)) = \mathbb{E}(DF, h)_\mathcal{H} \tag{C.2.4}$$

for $f \in \mathcal{S}$ and $h \in \mathcal{H}$ (Lemma 1.2.1 in Nualart [769]). It is easy to see that the relation (C.2.4) holds for example if $F = f(W(g))$ and $\mathbb{E} W(g)^2 = 1$, since

$$\mathbb{E}(DF, h)_{\mathcal{H}} = \mathbb{E}[f'(W(g))](g, h)_{\mathcal{H}} = \mathbb{E}[f(W(g))W(g)](g, h)_{\mathcal{H}}$$
$$= \mathbb{E}[f(W(g))W(h)] = \mathbb{E}[FW(h)],$$

where the second equality follows by using the following relation

$$\int_{\mathbb{R}} f'(x) \frac{1}{\sqrt{2\pi}} e^{-x^2/2} dx = \int_{\mathbb{R}} x f(x) \frac{1}{\sqrt{2\pi}} e^{-x^2/2} dx, \tag{C.2.5}$$

which can be proved by integration by parts (see Lemma 1.1.1 of Nourdin and Peccati [763]), and the third equality follows from the fact that $(W(g), W(h))$ has the same distribution as $(W(g), (g, h)_{\mathcal{H}} W(g) + \epsilon)$ for a zero mean, Gaussian random variable ϵ independent of $W(g)$, as can be verified by computing the covariances.

Remark C.2.6 Formula (C.2.4) is viewed as an *integration by parts formula* since it is based on (C.2.5). It is interesting to note the difference between the integration by parts formula $\int fg' = fg - \int f'g$ valid for Lebesgue measure and the corresponding formula (C.2.5), which involves the Gaussian measure $\gamma(dx) = e^{-x^2/2} dx / \sqrt{2\pi}$, and which can be rewritten conveniently as $\int_{\mathbb{R}} 1 f'(x) \gamma(dx) = \int_{\mathbb{R}} x f(x) \gamma(dx)$. Thus, in the Gaussian integration by parts setting, both 1 and $f'(x)$ are integrated to become x and $f(x)$ respectively. This is also what one observes in (C.2.4) where h and DF are both integrated to become $W(h)$ and F, respectively.

For any $p \geq 1$, denote the domain of the derivative operator D in $L^p(\Omega)$ by $\mathbb{D}^{1,p}$, where $\mathbb{D}^{1,p}$ is the closure of the class of smooth random variables $\mathcal{S}$ with respect to the norm

$$\|F\|_{1,p} = \left(\mathbb{E}|F|^p + \mathbb{E}\|DF\|_{\mathcal{H}}^p \right)^{\frac{1}{p}}, \tag{C.2.6}$$

which involves the pth moment of the random variable F and also the pth moment of the random variable $\|DF\|_{\mathcal{H}}$. The space $\mathbb{D}^{1,p}_{\text{loc}}$ consists of the random variables F, for which there exists a sequence $\{(\Omega_n, F_n), n \geq 1\} \subset \mathcal{F} \times \mathbb{D}^{1,p}$ such that (i) $\Omega_n \uparrow \Omega$ a.s., and (ii) $F = F_n$ a.s. on Ω_n. One says that (Ω_n, F_n) "localizes" F in $\mathbb{D}^{1,p}$. Then, DF is defined by $DF = DF_n$ on $\Omega_n, n \geq 1$.

The space $\mathbb{D}^{1,2}$ is the one used most frequently. The associated norm (C.2.6) becomes

$$\|F\|_{1,2} = \left(\mathbb{E}|F|^2 + \mathbb{E}\|DF\|_{\mathcal{H}}^2 \right)^{\frac{1}{2}}.$$

There are a number of ways to gain insight of the elements of $\mathbb{D}^{1,2}$. If $\phi : \mathbb{R}^m \to \mathbb{R}$ is a Lipschitz function and $F_i \in \mathbb{D}^{1,2}$ (in particular, $F_i \in \mathcal{S}$), $i = 1, \ldots, m$, then $\phi(F_1, \ldots, F_m) \in \mathbb{D}^{1,2}$. For example, since the function $\phi(x) = |x|$ is Lipschitz, then $|W(h)| \in \mathbb{D}^{1,2}$.

In the case when ϕ is continuously differentiable, one has the *chain rule*

$$D\phi(F_1, \ldots, F_m) = \sum_{k=1}^{n} \frac{\partial}{\partial x_k} \phi(F_1, \ldots, F_m) DF_k \tag{C.2.7}$$

(Propositions 1.2.3 and 1.2.4 in Nualart [769]). If $F_n \in \mathbb{D}^{1,2}$ converge to F in $L^2(\Omega)$ and are such that $\sup_n \mathbb{E}\|DF_n\|_{\mathcal{H}}^2 < \infty$, then $F \in \mathbb{D}^{1,2}$ (Lemma 1.2.3 in Nualart [769]).

The next example gathers some results on the derivative and the space $\mathbb{D}^{1,2}$ in the white noise case.

Example C.2.7 Consider the case where the isonormal process is a white noise (see Example C.1.2). Then, the derivative is a stochastic process denoted $(DF)(u)$ or $D_u F$. Suppose F is a square integrable random variable expressed through its Wiener chaos expansion

$$F = \sum_{n=0}^{\infty} I_n(f_n), \qquad (C.2.8)$$

where I_n are multiple integrals as in Appendix B.2 and $f_n \in L^2(E^n, m^n)$ are symmetric functions. Then, $F \in \mathbb{D}^{1,2}$ if and only if $\sum_{n=1}^{\infty} n\mathbb{E}|I_n(f)|^2 < \infty$ (Proposition 1.2.2 in Nualart [769]). Moreover, if $F \in \mathbb{D}^{1,2}$, then

$$D_u F = \sum_{n=1}^{\infty} n I_{n-1}(f_n(\cdot, u)) \qquad (C.2.9)$$

(Proposition 1.2.7 in Nualart [769]). For example,

$$D_u \left(\int_{\mathbb{R}^2}' f(v_1, v_2) W(dv_1) W(dv_2) \right) = 2 \int_{\mathbb{R}} f(v, u) W(dv),$$

which is a random function of u, or

$$D_u \left(\int_{\mathbb{R}^3}' f(v_1, v_2, v_3) W(dv_1) W(dv_2) W(dv_3) \right) = 3 \int_{\mathbb{R}^2}' f(v_1, v_2, u) W(dv_1) W(dv_2).$$

In (C.2.9), one integrates over $n - 1$ variables, freeing the nth variable, so that one obtains a random function. Since f is symmetric, it does not matter which variable becomes free.

One can define the iteration of the derivative operator D in such a way that for a smooth random variable F, the iterated derivative $D^k F$ is a random variable with values in $\mathcal{H}^{\otimes k}$. Then, for every $p \geq 1$ and $k \geq 1$, consider the norm on $\mathcal{S}$ defined by

$$\|F\|_{k,p} = \left(\mathbb{E}|F|^p + \sum_{j=1}^{k} \mathbb{E}\|D^j F\|_{\mathcal{H}^{\otimes j}}^p \right)^{\frac{1}{p}}. \qquad (C.2.10)$$

Denote by $\mathbb{D}^{k,p}$ the completion of $\mathcal{S}$ with respect to the norm $\|\cdot\|_{k,p}$.

Example C.2.8 Consider the white noise case with $\mathcal{H} = L^2([0, T], m)$ and the Lebesgue measure m as in Examples C.1.2 and C.2.5. The Malliavin derivative of $f(W(t))$ was computed at the end of Example C.2.5 as $f'(W(t)) 1_{[0,t)}(v)$. The second Malliavin derivative is then

$$(D^2 f(W(t)))(u, v) = (Df'(W(t)))(u) 1_{[0,t)}(v) = f''(W(t)) 1_{[0,t)}(u) 1_{[0,t)}(v),$$

which is now a random variable with values in $\mathcal{H}^2$.

The above definitions applied to real-valued random variables F can be extended to

$$\mathbb{D}^{k,p}(\mathcal{G}) \subset L^0(\Omega; \mathcal{G}) = \{\mathcal{G}\text{-valued random variables } F : \Omega \to \mathcal{G}\}, \qquad (C.2.11)$$

where $\mathcal{G}$ is a real separable Hilbert space, for example, $\mathcal{G} = \mathbb{R}^d$ or some function space. The corresponding family $\mathcal{S}_\mathcal{G}$ now consists of smooth random variables given by

$$F = \sum_{j=1}^n F_j g_j, \quad F_j \in \mathcal{S}, \ g_j \in \mathcal{G}. \qquad (C.2.12)$$

The derivative operator is defined as

$$D^k F = \sum_{j=1}^n (D^k F_j) \otimes g_j, \quad k \geq 1, \qquad (C.2.13)$$

and the norm on $\mathcal{S}_V$ is defined as

$$\|F\|_{k,p,V} = \left(\mathbb{E} \|F\|_\mathcal{G}^p + \sum_{j=1}^k \mathbb{E} \|D^j F\|_{\mathcal{H}^{\otimes j} \otimes \mathcal{G}}^p \right)^{\frac{1}{p}}. \qquad (C.2.14)$$

The space $\mathbb{D}^{k,p}(\mathcal{G})$ is then the completion of $\mathcal{S}_\mathcal{G}$ with respect to $\|\cdot\|_{k,p,\mathcal{G}}$. Note that, if $\mathcal{H}$ and $\mathcal{G}$ can be thought of as function spaces, then F in (C.2.12) can be written as $F(v) = \sum_{j=1}^n F_j g_j(v)$ and $D^k F$ in (C.2.13) becomes

$$(D^k F)(u_1, \ldots, u_k, v) = \sum_{j=1}^n (D^k F_j)(u_1, \ldots, u_k) g_j(v).$$

In other words, to compute $D^k F$, one can fix v and apply D^k to F_j alone.

Example C.2.9 Consider the white noise case with $\mathcal{H} = L^2([0, T], m)$ and the Lebesgue measure m as in Examples C.1.2, C.2.5 and C.2.8. The Malliavin derivative $(Df(W(t)))(u) = f'(W(t))1_{[0,t)}(u)$ is an $\mathcal{H}$-valued random variable. Taking $\mathcal{G} = \mathcal{H}$ above, we have

$$\Big(D(Df(W(t)))(\cdot)\Big)(u, v) = \Big(Df'(W(t))1_{[0,t)}(\cdot)\Big)(u, v)$$
$$= (Df'(W(t)))(u)1_{[0,t)}(u) = f''(W(t))1_{[0,t)}(u)1_{[0,t)}(v),$$

by the definition (C.2.12)–(C.2.13) (strictly speaking, supposing $f' \in C_p^\infty(\mathbb{R})$, so that $f'(W(t)) \in \mathcal{S}$). As noted above, the expression is the same as for $D^2 f(W(t))$ in Example C.2.8, which is obtained by taking the Malliavin derivative of $(Df(W(t)))(v) = f'(W(t))1_{[0,t)}(v)$ for fixed v.

Remark C.2.10 In the white noise case, there is a useful Stroock formula for the functions f_n appearing in the Wiener chaos expansion (C.2.8) of a random variable F. More specifically, if $F = \sum_{n=0}^\infty I_n(f_n)$ belongs to $\cap_k \mathbb{D}^{k,2}$, then

$$f_n = \frac{1}{n!} \mathbb{E}(D^n F) \qquad (C.2.15)$$

(e.g., Nualart [769], Exercise 1.2.6).

C.3 Divergence Integral

We now turn to the adjoint operator of D. It is called the *Skorokhod* or *divergence operator* (also called the *divergence* or *Skorokhod integral*), and is denoted δ. It plays a central role in the Malliavin calculus. As mentioned below, it satisfies the duality relation $\mathbb{E}(F\delta(X)) = \mathbb{E}(DF, X)_{\mathcal{H}}$ for any random variable $F \in \mathbb{D}^{1,2}$. Here $X \in L^2(\Omega; \mathcal{H})$, where $\mathcal{H}$ is a Hilbert space; that is, X is random, $\mathcal{H}$-valued and such that $\mathbb{E}\|X\|_{\mathcal{H}}^2 < \infty$. The space $L^2(\Omega; \mathcal{H})$ is itself a Hilbert space with inner product $(X, Y)_{L^2(\Omega;\mathcal{H})} = \mathbb{E}(X, Y)_{\mathcal{H}}$. The derivative operator D is a closed and unbounded operator on the dense subset $\mathbb{D}^{1,2}$ of $L^2(\Omega)$ with values in $L^2(\Omega; \mathcal{H})$. On the other hand, the divergence operator δ acts on random elements $X \in L^2(\Omega; \mathcal{H})$ and takes values in $L^2(\Omega)$. Thus,

$$D : \mathbb{D}^{1,2}(\subset L^2(\Omega)) \to L^2(\Omega; \mathcal{H}),$$
$$\delta : \text{Dom}\,\delta(\subset L^2(\Omega; \mathcal{H})) \to L^2(\Omega).$$

Definition C.3.1 Let δ be the adjoint operator of D; that is, δ is an unbounded operator on $L^2(\Omega; \mathcal{H})$ with values in $L^2(\Omega)$ such that

(i) The domain of δ, denoted by $\text{Dom}\,\delta$, is the set of $\mathcal{H}$-valued square integrable random variables $X \in L^2(\Omega; \mathcal{H})$ such that

$$|\mathbb{E}(DF, X)_{\mathcal{H}}| \leq c(\mathbb{E}|F|^2)^{1/2}, \tag{C.3.1}$$

for all random variables $F \in \mathbb{D}^{1,2}$, where c is some constant depending on X.

(ii) If X belongs to $\text{Dom}\,\delta$, then $\delta(X)$ is the element of $L^2(\Omega)$ characterized by

$$\mathbb{E}(F\delta(X)) = \mathbb{E}(DF, X)_{\mathcal{H}}, \tag{C.3.2}$$

for any random variable $F \in \mathbb{D}^{1,2}$.

Observe that (C.3.2) is consistent with the integration by parts formula (C.2.4), setting $\delta(h) = W(h)$, $h \in \mathcal{H}$. In this case, $X \in L^2(\Omega; \mathcal{H})$ is denoted h and is not random.

Some basic properties of the divergence operator are as follows:

- For any $X \in \text{Dom}\,\delta$,

$$\mathbb{E}\delta(X) = 0. \tag{C.3.3}$$

This follows by taking $F \equiv 1$ in (C.3.2) since $DF = 0$.
- If $X = Fh$, where F is a random variable in $\mathcal{S}$ and h is a function in $\mathcal{H}$, then $X \in \text{Dom}\,\delta$ and

$$\delta(X) = FW(h) - (DF, h)_{\mathcal{H}}. \tag{C.3.4}$$

To show (C.3.4), we prove only the condition (ii) of Definition C.3.1. To avoid confusion, replace F by G in the relation (C.3.4), so that it becomes

$$\delta(Gh) = GW(h) - (DG, h)_{\mathcal{H}}. \tag{C.3.5}$$

Note that, for $G \in \mathcal{S}$,

$$\mathbb{E}(DG, Fh)_{\mathcal{H}} = \mathbb{E}(F(DG, h)_{\mathcal{H}}) = \mathbb{E}(FDG, h)_{\mathcal{H}}$$

$$= \mathbb{E}(D(FG), h)_{\mathcal{H}} - \mathbb{E}(GDF, h)_{\mathcal{H}} = \mathbb{E}(D(FG), h)_{\mathcal{H}} - \mathbb{E}(G(DF, h)_{\mathcal{H}}) \quad \text{(C.3.6)}$$

by using the fact that

$$D(FG) = F(DG) + G(DF)$$

for $F, G \in \mathcal{S}$. The relation (C.2.4) applied to $\mathbb{E}(D(FG), h)_{\mathcal{H}}$ on the right-hand side of (C.3.6) yields

$$\mathbb{E}(DG, Fh)_{\mathcal{H}} = \mathbb{E}FGW(h) - \mathbb{E}G(DF, h)_{\mathcal{H}} = \mathbb{E}\Big(G(FW(h) - (DF, h)_{\mathcal{H}})\Big).$$

Now replace Fh on the left-hand side by X and replace $FW(h) - (DF, h)_{\mathcal{H}}$ on the right-hand side by $\delta(X)$ as well, and observe that one obtains $\mathbb{E}(DG, X)_{\mathcal{H}} = \mathbb{E}(G\delta(X))$ which is (C.3.2) for any $G \in \mathcal{S}$ replacing F. A limit argument allows to deduce the same relation for any $G \in \mathbb{D}^{1,2}$.

- If $F \in \mathbb{D}^{1,2}$, $X \in \mathrm{Dom}\,\delta$ and $FX \in L^2(\Omega; \mathcal{H})$, then $FX \in \mathrm{Dom}\,\delta$ and the following equality is true:

$$\delta(FX) = F\delta(X) - (DF, X)_{\mathcal{H}}, \quad \text{(C.3.7)}$$

provided the right-hand side of the above equation is square integrable (Proposition 1.3.3 in Nualart [769]).

- The space $\mathbb{D}^{1,2}(\mathcal{H})$, introduced in (C.2.11), is included in $\mathrm{Dom}\,\delta$. If $X, Y \in \mathbb{D}^{1,2}(\mathcal{H})$ and hence belong to $\mathrm{Dom}\,\delta$, it can be shown that the covariance of $\delta(X)$ and $\delta(Y)$ is given by

$$\mathbb{E}\delta(X)\delta(Y) = \mathbb{E}(X, Y)_{\mathcal{H}} + \mathbb{E}(DX, (DY)^*)_{\mathcal{H} \otimes \mathcal{H}}, \quad \text{(C.3.8)}$$

where $(DY)^*$ indicates the adjoint[1] of DY in the Hilbert space $\mathcal{H} \otimes \mathcal{H}$ (Proposition 1.3.1 in Nualart [769]). In particular,

$$\mathbb{E}\delta(X)^2 = \mathbb{E}\|X\|_{\mathcal{H}}^2 + \mathbb{E}(DX, (DX)^*)_{\mathcal{H} \otimes \mathcal{H}} \quad \text{(C.3.9)}$$

and hence

$$\mathbb{E}\delta(X)^2 \leq \mathbb{E}\|X\|_{\mathcal{H}}^2 + \mathbb{E}\|DX\|_{\mathcal{H} \otimes \mathcal{H}}^2. \quad \text{(C.3.10)}$$

The relation (C.3.10) implies that δ is a continuous mapping from $\mathbb{D}^{1,2}(\mathcal{H})$ into $L^2(\Omega)$.

C.4 Generator of the Ornstein–Uhlenbeck Semigroup

Finally, we also use the so-called *generator of the Ornstein–Uhlenbeck semigroup*, defined as follows: for $F \in L^2(\Omega)$,

$$LF = -\sum_{n=0}^{\infty} n J_n F, \quad \text{(C.4.1)}$$

provided the series converges in $L^2(\Omega)$, where J_n denotes the orthogonal projection on the so-called nth Wiener chaos, which is the closed linear subspace of $L^2(\Omega)$ generated by the

---

[1] Recall once again that in a Hilbert space $\mathcal{G}$, the adjoint of $g \in \mathcal{G}$ is the element $g^* \in \mathcal{G}$ satisfying $(g, f)_{\mathcal{G}} = (f, g^*)_{\mathcal{G}}$ for any $f \in \mathcal{G}$. In particular, $(f, f^*)_{\mathcal{G}} = \|f\|_{\mathcal{G}}^2$.

C.4 Generator of the Ornstein–Uhlenbeck Semigroup

random variables $\{H_n(W(h)), h \in \mathcal{H}, \|h\|_{\mathcal{H}} = 1\}$ where H_n is the Hermite polynomial of order n. The operator L^{-1}, called the pseudo-inverse of L, is defined as follows: for $F \in L^2(\Omega)$,

$$L^{-1}F = -\sum_{n=1}^{\infty} \frac{1}{n} J_n F. \tag{C.4.2}$$

The connecting relation between the three operators D, δ and L is

$$LF = -\delta(DF) \tag{C.4.3}$$

(Nualart [769], Proposition 1.4.3) and indeed, $F \in L^2(\Omega)$ belongs to the domain of L if and only if $F \in \mathbb{D}^{1,2}$ and DF belongs to the domain of δ.

Appendix D

Other Notes and Topics

A number of new books on long-range dependence and self-similarity have been published in the last ten years or so. On the statistics and modeling side, these include (in alphabetical order):

- Beran et al. [127], Cohen and Istas [253], Giraitis et al. [406], Leonenko [612], Palma [789].

On the probability side, these include:

- Berzin, Latour, and León [138], Biagni et al. [146], Embrechts and Maejima [350], Major [673], Mishura [722], Nourdin [760], Nourdin and Peccati [763], Nualart [769], Peccati and Taqqu [797], Rao [842], Samorodnitsky [879, 880], Tudor [963].

On the application side, these include:

- Dmowska and Saltzman [319] regarding applications in geophysics, Park and Willinger [794] and Sheluhin et al. [903] in connection to telecommunications, Robinson [854] and Teyssière and Kirman [958] in connection to economics and finance.

Collections of articles include:

- Doukhan, Oppenheim, and Taqqu [328], Rangarajan and Ding [841].

See also the following books of related interest:

- Houdré and Pérez-Abreu [488], Janson [529], Marinucci and Peccati [689], Meerschaert and Sikorskii [712], Terdik [954].

We have not covered in this monograph a number of other interesting topics related to long-range dependence and/or self-similarity, including:

- *Long-range dependence for point processes:* In a number of applications, data represent events recorded in time (e.g., spike trains in brain science, packet arrivals in computer networks, and so on). Such data can be modeled through point processes (e.g., Daley and Vere-Jones [282, 283]). Long-range dependence for point processes was defined and studied in Daley [281], Daley and Vesilo [284], Daley, Rolski, and Vesilo [285]. See also the monograph of Lowen and Teich [648].
- *Other nonlinear time series with long-range dependence:* A large portion of this monograph (in particular, Chapter 2) focuses on linear time series exhibiting long-range dependence, with the exception of the non-linear time series defined as functions of linear

series in Chapter 5. A number of other nonlinear time series with long-range dependence features have been considered in the literature, especially in the context of financial time series, e.g., FIGARCH and related models (Ding, Granger, and Engle [316], Baillie, Bollerslev, and Mikkelsen [89], Andersen and Bollerslev [29], Andersen, Bollerslev, Diebold, and Ebens [30], Comte and Renault [257], Tayefi and Ramanathan [953], Deo et al. [302]). See also Shao and Wu [902], Wu and Shao [1013], Baillie and Kapetanios [87].

In the context of financial time series, long-range dependence has been associated to their volatility process. Estimation issues in related models are also discussed in Arteche [47, 48], Frederiksen and Nielsen [371], Frederiksen, Nielsen, and Nielsen [373], Frederiksen and Nielsen [372], Hurvich and Ray [509], Hurvich, Moulines, and Soulier [512], Lieberman and Phillips [629].

A review of long-range dependent count (discrete-valued) time series can be found in Lund, Holan, and Livsey [651].

- *Structure of stable self-similar processes with stationary increments:* We have already mentioned in Section 2.6 that, in contrast to the Gaussian case of FBM, there are infinitely many different $S\alpha S$ H–self-similar processes with stationary increments (for foxed α and H). But we emphasize here that the structure of these processes, at least of their large class known as mixed moving averages, can be described by relating these processes to deterministic non-singular flows. Various classes of non-singular flows (dissipative and conservative, identity, cyclic, positive and null, and so on) can then give rise to different classes of the corresponding processes. A review can be found in Pipiras and Taqqu [820].

- *Large deviations and queueing applications of FBM:* With the emerging relevance of long-range dependence and self-similarity for modern communication networks such as the Internet (see Section 3.3 and historical notes in Section 3.15), several authors considered implications of the related models on the behavior of queueing systems. For example, the tail probabilities of queue sizes were studied for LRD input processes such as FBM. These large deviation problems for FBM were first considered in Norros [756], Duffield and O'Connell [332], Massoulie and Simonian [697]. For a more recent account, see the monograph by Mandjes [687].

- *Multifractional motions:* Multifractional Brownian motions (MBMs) extend the usual FBM $B_H = \{B_H(t)\}_{t \in \mathbb{R}}$ by allowing their self-similarity parameter $H \in (0, 1)$ to depend on time; that is, $H = H(t)$. These processes are defined by replacing H in the time or spectral domain representation of FBM (see (2.6.1) and (2.6.9)) by a deterministic function $H(t)$. MBMs were defined through the time domain representations by Peltier and Véhel [804], and through the spectral domain representations by Benassi et al. [113]. See Stoev and Taqqu [930] for comparison of the two definitions. The main interest in MBM is that $H(t)$ is its Hölder exponent, which can thus vary from point to point. For other work related to MBMs, see Benassi, Cohen, and Istas [114], Coeurjolly [251], Lebovits and Véhel [602], Balança [93], to name but a few. Analogous extensions of linear fractional stable motion, known as linear multifractional stable motions, were considered in Stoev and Taqqu [925, 928].

- *Multifractal processes:* Multifractal processes present another interesting extension of FBMs and, more generally, self-similar processes. Note that FBM $B_H = \{B_H(t)\}_{t \in \mathbb{R}}$

satisfies $\mathbb{E}|B_H(t)|^q = C_q|t|^{\zeta(q)}$ for $q > 0$ with a *linear* function $\zeta(q) = Hq$. Multifractal processes $X = \{X(t)\}_{t \in \mathbb{R}}$, on the other hand, satisfy the scaling relation $\mathbb{E}|X(t)|^q \sim C_q|t|^{\zeta(q)}$ as $t \to 0$ for a range of q's with a *non-linear* exponent function $\zeta(q)$.[1] For example, the so-called infinitely divisible cascades of Bacry and Muzy [67] satisfy the scaling relation exactly as $\mathbb{E}|X(t)|^q = C_q|t|^{\zeta(q)}$ for $0 < t < t_c < \infty$. But many other models of multifractal processes are available as well (e.g., multiplicative cascades in Mandelbrot [677], Kahane and Peyrière [544], Ossiander and Waymire [783], Barral [102], compound Poisson cascades in Barral and Mandelbrot [103], self-similar processes in multifractal time in Mandelbrot [680], multifractal random walks with respect to self-similar processes in Bacry, Delour, and Muzy [68], Abry, Chainais, Coutin, and Pipiras [11]). See also a review article by Riedi [849]. Recent work on estimation of the exponent function through the so-called wavelet leaders can be found in Jaffard, Lashermes, and Abry [527], Jaffard, Melot, Leonarduzzi, Wendt, Abry, Roux, and Torres [528], Leonarduzzi, Wendt, Abry, abd C. Melot, Roux, and Torres [609].

- *Mixing and long-range dependence:* Long-range dependent series provide interesting (counter) examples in the study of mixing properties of stationary time series. See Bradley [181] for various definitions of mixing properties. In fact, the noncentral limit theorems have originated with Rosenblatt [860] who noted that certain Gaussian long-range dependent series are mixing but not strongly mixing. (If a Gaussian long-range dependent series $\{X_n\}$ were strongly mixing, then the series $\{X_n^2 - \mathbb{E}X_n^2\}$ would be strongly mixing as well, and hence satisfy the central limit theorem, which is not necessarily the case in view of the results of Chapter 5.) For recent work in this direction, see Bai and Taqqu [80]. They show that short-range dependent processes subordinated to Gaussian long-range dependent processes may not be strong mixing.

- *Self-similar Markov processes:* Starting with a seminal work of Lamperti [590], considerable progress has been made in characterizing self-similar Markov processes, and in understanding their properties. A nice review is provided by Pardo and Rivero [792]. The techniques used in the area rely on the Markovian property, often involve Lévy processes and stand in contrast to those used in this book. Note, in particular, that most self-similar processes considered in this book are non-Markovian. Likewise, most self-similar Markov processes considered in the literature do not have stationary increments, the property which is often assumed for self-similar processes in this book. Self-similar scaling limits of Markov chains are considered in Haas and Miermont [443], Bertoin and Kortchemski [137].

---

[1] There is a complementary perspective on multifractals, where $\zeta(q)$ is related to the Legendre transform of the so-called multifractal spectrum $D(h)$ of the process: that is, the function

$$D(h) = \dim_H\{t : h_X(t) = h\},$$

where $\dim_H$ stands for the Hausdorff dimension (e.g., Falconer [352]) and $h_X(t)$ denotes the Hölder (or some other) exponent measuring smoothness of the process X at time t. (This is also summing that the equality above holds almost surely, since X is a random process.) For more information on this perspective, see Riedi [849].

Bibliography

[1] J. Aaronson and M. Denker. Characteristic functions of random variables attracted to 1-stable laws. *The Annals of Probability*, 26(1):399–415, 1998. (Cited on pages 583 and 586.)

[2] K. M. Abadir, W. Distaso, and L. Giraitis. Nonstationarity-extended local Whittle estimation. *Journal of Econometrics*, 141(2):1353–1384, 2007. (Cited on page 572.)

[3] K. M. Abadir, W. Distaso, and L. Giraitis. Two estimators of the long-run variance: beyond short memory. *Journal of Econometrics*, 150(1):56–70, 2009. (Cited on page 28.)

[4] J. Abate and W. Whitt. The Fourier-series method for inverting transforms of probability distributions. *Queueing Systems. Theory and Applications*, 10(1-2):5–87, 1992. (Cited on page 271.)

[5] P. Abry and G. Didier. Wavelet estimation of operator fractional Brownian motions. To appear in *Bernoulli*. Preprint, 2015. (Cited on page 536.)

[6] P. Abry and V. Pipiras. Wavelet-based synthesis of the Rosenblatt process. *Signal Processing*, 86(9):2326–2339, 2006. (Cited on page 465.)

[7] P. Abry and F. Sellan. The wavelet-based synthesis for the fractional Brownian motion proposed by F. Sellan and Y. Meyer: Remarks and fast implementation. *Applied and Computational Harmonic Analysis*, 3(4):377–383, 1996. (Cited on page 465.)

[8] P. Abry and D. Veitch. Wavelet analysis of long range dependent traffic. *IEEE Transactions on Information Theory*, 44(1):2–15, 1998. (Cited on page 111.)

[9] P. Abry, D. Veitch, and P. Flandrin. Long-range dependence: revisiting aggregation with wavelets. *Journal of Time Series Analysis*, 19(3):253–266, 1998. (Cited on page 112.)

[10] P. Abry, P. Flandrin, M. S. Taqqu, and D. Veitch. Self-similarity and long-range dependence through the wavelet lens. In P. Doukhan, G. Oppenheim, and M. S. Taqqu, editors, *Theory and Applications of Long-Range Dependence*, pages 527–556. Birkhäuser, 2003. (Cited on page 112.)

[11] P. Abry, P. Chainais, L. Coutin, and V. Pipiras. Multifractal random walks as fractional Wiener integrals. *IEEE Transactions on Information Theory*, 55(8):3825–3846, 2009. (Cited on page 612.)

[12] P. Abry, P. Borgnat, F. Ricciato, A. Scherrer, and D. Veitch. Revisiting an old friend: on the observability of the relation between long range dependence and heavy tail. *Telecommunication Systems*, 43(3-4):147–165, 2010. (Cited on page 224.)

[13] S. Achard and I. Gannaz. Multivariate wavelet Whittle estimation in long-range dependence. *Journal of Time Series Analysis*, 2015. (Cited on page 574.)

[14] S. Achard, D. S. Bassett, A. Meyer-Lindenberg, and E. Bullmore. Fractal connectivity of long-memory networks. *Physical Review E*, 77:036104, Mar 2008. (Cited on page 537.)

[15] C. Agiakloglou, P. Newbold, and M. Wohar. Bias in an estimator of the fractional difference parameter. *Journal of Time Series Analysis*, 14(3):235–246, 1993. (Cited on pages 111 and 573.)

[16] N. U. Ahmed and C. D. Charalambous. Filtering for linear systems driven by fractional Brownian motion. *SIAM Journal on Control and Optimization*, 41(1):313–330, 2002. (Cited on page 395.)

[17] H. Akaike. Information theory and an extension of the maximum likelihood principle. In *Second International Symposium on Information Theory (Tsahkadsor, 1971)*, pages 267–281. Budapest: Akadémiai Kiadó, 1973. (Cited on page 552.)

[18] J. M. P. Albin. A note on Rosenblatt distributions. *Statistics & Probability Letters*, 40(1):83–91, 1998. (Cited on page 281.)

[19] D. W. Allan. Statistics of atomic frequency standards. *Proceedings of the IEEE*, 54(2):221–230,1966. (Cited on page 112.)

[20] E. Alòs and D. Nualart. Stochastic integration with respect to the fractional Brownian motion. *Stochastics and Stochastics Reports*, 75(3):129–152, 2003. (Cited on page 412.)

[21] E. Alòs, O. Mazet, and D. Nualart. Stochastic calculus with respect to fractional Brownian motion with Hurst parameter lesser than $\frac{1}{2}$. *Stochastic Processes and their Applications*, 86(1):121–139, 2000. (Cited on page 435.)

[22] E. Alòs, O. Mazet, and D. Nualart. Stochastic calculus with respect to Gaussian processes. *The Annals of Probability*, 29(2):766–801, 2001. (Cited on page 435.)

[23] F. Altissimo, B. Mojon, and P. Zaffaroni. Can aggregation explain the persistence of inflation? *Journal of Monetary Economics*, 56(2):231–241, 2009. (Cited on page 224.)

[24] E. Alvarez-Lacalle, B. Dorow, J.-P. Eckmann, and E. Moses. Hierarchical structures induce long-range dynamical correlations in written texts. *Proceedings of the National Academy of Sciences*, 103(21):7956–7961, 2006. (Cited on page 228.)

[25] P.-O. Amblard and J.-F. Coeurjolly. Identification of the multivariate fractional Brownian motion. *Signal Processing, IEEE Transactions on*, 59(11):5152–5168, nov. 2011. (Cited on pages 488 and 536.)

[26] P.-O. Amblard, J.-F. Coeurjolly, F. Lavancier, and A. Philippe. Basic properties of the multivariate fractional Brownian motion. *Séminaires & Congrès*, 28:65–87, 2012. (Cited on pages 534 and 536.)

[27] A. Amirdjanova. Nonlinear filtering with fractional Brownian motion. *Applied Mathematics and Optimization*, 46(2-3):81–88, 2002. Special issue dedicated to the memory of Jacques-Louis Lions. (Cited on page 436.)

[28] A. Amirdjanova and M. Linn. Stochastic evolution equations for nonlinear filtering of random fields in the presence of fractional Brownian sheet observation noise. *Computers & Mathematics with Applications. An International Journal*, 55(8):1766–1784, 2008. (Cited on page 436.)

[29] T. G. Andersen and T. Bollerslev. Intraday periodicity and volatility persistence in financial markets. *Journal of Empirical Finance*, 4(2):115–158, 1997. (Cited on page 611.)

[30] T. G. Andersen, T. Bollerslev, F. X. Diebold, and H. Ebens. The distribution of realized stock return volatility. *Journal of Financial Economics*, 61(1):43–76, 2001. (Cited on page 611.)

[31] J. Andersson. An improvement of the GPH estimator. *Economics Letters*, 77(1):137–146, 2002. (Cited on page 111.)

[32] D. W. K. Andrews and P. Guggenberger. A bias-reduced log-periodogram regression estimator for the long-memory parameter. *Econometrica*, 71(2):675–712, 2003. (Cited on page 111.)

[33] D. W. K. Andrews and O. Lieberman. Valid Edgeworth expansions for the Whittle maximum likelihood estimator for stationary long-memory Gaussian time series. *Econometric Theory*, 21(4):710–734, 2005. (Cited on page 571.)

[34] D. W. K. Andrews and Y. Sun. Adaptive local polynomial Whittle estimation of long-range dependence. *Econometrica*, 72(2):569–614, 2004. (Cited on pages 560, 561, 565, 566, and 572.)

[35] G. E. Andrews, R. Askey, and R. Roy. *Special Functions*, volume 71 of *Encyclopedia of Mathematics and its Applications*. Cambridge: Cambridge University Press, 1999. (Cited on pages 280, 526, and 527.)

[36] V. V. Anh and K. E. Lunney. Parameter estimation of random fields with long-range dependence. *Mathematical and Computer Modelling*, 21(9):67–77, 1995. (Cited on page 537.)

[37] V. V. Anh, N. Leonenko, and A. Olenko. On the rate of convergence to Rosenblatt-type distribution. *Journal of Mathematical Analysis and Applications*,425(1):111–132, 2015. (Cited on page 343.)

[38] C. F. Ansley and R. Kohn. A note on reparameterizing a vector autoregressive moving average model to enforce stationarity. *Journal of Statistical Computation and Simulation*, 24(2):99–106, 1986. (Cited on page 570.)

[39] N. Antunes, V. Pipiras, P. Abry, and D. Veitch. Small and large scale behavior of moments of Poisson cluster processes. Preprint, 2016. (Cited on page 224.)

[40] D. Applebaum. *Lévy Processes and Stochastic Calculus*, 2nd edition volume 116 of *Cambridge Studies in Advanced Mathematics*. Cambridge: Cambridge University Press, 2009. (Cited on page 268.)

[41] V. F. Araman and P. W. Glynn. Fractional Brownian motion with $H < \frac{1}{2}$ as a limit of scheduled traffic. *Journal of Applied Probability*, 49(3):710–718, 2012. (Cited on page 225.)

[42] M. A. Arcones. Limit theorems for non-linear functionals of a stationary Gaussian sequence of vectors. *The Annals of Probability*, 22:2242–2274, 1994. (Cited on page 344.)

[43] M. A. Arcones. Distributional limit theorems over a stationary Gaussian sequence of random vectors. *Stochastic Processes and their Applications*, 88(1):135–159, 2000. (Cited on page 344.)

[44] B. Arras. On a class of self-similar processes with stationary increments in higher order Wiener chaoses. *Stochastic Processes and their Applications*, 124(7):2415–2441, 2014. (Cited on page 281.)

[45] B. Arras. A white noise approach to stochastic integration with respect to the Rosenblatt process. *Potential Analysis*, 43(4):547–591, 2015. (Cited on page 435.)

[46] R. Arratia. The motion of a tagged particle in the simple symmetric exclusion system on **Z**. *The Annals of Probability*, 11(2):362–373, 1983. (Cited on pages 168, 169, and 172.)

[47] J. Arteche. Gaussian semiparametric estimation in long memory in stochastic volatility and signal plus noise models. *Journal of Econometrics*, 119(1):131–154, 2004. (Cited on page 611.)

[48] J. Arteche. Semiparametric inference in correlated long memory signal plus noise models. *Econometric Reviews*, 31(4):440–474, 2012. (Cited on page 611.)

[49] J. Arteche and J. Orbe. Bootstrap-based bandwidth choice for log-periodogram regression. *Journal of Time Series Analysis*, 30(6):591–617, 2009. (Cited on page 111.)

[50] J. Arteche and J. Orbe. Using the bootstrap for finite sample confidence intervals of the log periodogram regression. *Computational Statistics & Data Analysis*, 53(6):1940–1953, 2009. (Cited on page 111.)

[51] J. Arteche and P. M. Robinson. Semiparametric inference in seasonal and cyclical long memory processes. *Journal of Time Series Analysis*, 21(1):1–25, 2000. (Cited on page 573.)

[52] R. B. Ash and M. F. Gardner. *Topics in Stochastic Processes*. Probability and Mathematical Statistics, Vol. 27. New York, London: Academic Press [Harcourt Brace Jovanovich, Publishers], 1975. (Cited on pages 437 and 438.)

[53] A. Astrauskas. Limit theorems for sums of linearly generated random variables. *Lithuanian Mathematical Journal*, 23(2):127–134, 1983. (Cited on pages 82, 110, and 111.)

[54] A. Astrauskas, J. B. Levy, and M. S. Taqqu. The asymptotic dependence structure of the linear fractional Lévy motion. *Lithuanian Mathematical Journal*, 31(1):1–19, 1991. (Cited on pages 83 and 110.)

[55] K. Atkinson and W. Han. *Spherical Harmonics and Approximations on the Unit Sphere: an Introduction*, volume 2044 of *Lecture Notes in Mathematics*. Heidelberg: Springer, 2012. (Cited on pages 526 and 527.)

[56] M. Ausloos and D. H. Berman. A multivariate Weierstrass-Mandelbrot function. *Proceedings of the Royal Society. London. Series A. Mathematical, Physical and Engineering Sciences*, 400(1819):331–350, 1985. (Cited on page 227.)

[57] M. Avarucci and D. Marinucci. Polynomial cointegration between stationary processes with long memory. *Journal of Time Series Analysis*, 28(6):923–942, 2007. (Cited on page 537.)

[58] F. Avram. On bilinear forms in Gaussian random variables and Toeplitz matrices. *Probability Theory and Related Fields*, 79(1):37–45, 1988. (Cited on page 571.)

[59] F. Avram and M. S. Taqqu. Noncentral limit theorems and Appell polynomials. *The Annals of Probability*, 15:767–775, 1987. (Cited on page 343.)

[60] F. Avram, N. Leonenko, and L. Sakhno. On a Szegő type limit theorem, the Hölder-Young-Brascamp-Lieb inequality, and the asymptotic theory of integrals and quadratic forms of stationary fields. *ESAIM. Probability and Statistics*, 14:210–255, 2010. (Cited on page 571.)

[61] A. Ayache and W. Linde. Series representations of fractional Gaussian processes by trigonometric and Haar systems. *Electronic Journal of Probability*, 14:no. 94, 2691–2719, 2009. (Cited on page 465.)

[62] A. Ayache and M. S. Taqqu. Rate optimality of wavelet series approximations of fractional Brownian motion. *The Journal of Fourier Analysis and Applications*, 9(5):451–471, 2003. (Cited on page 465.)

[63] A. Ayache, S. Leger, and M. Pontier. Drap brownien fractionnaire. *Potential Analysis*, 17(1):31–43, 2002. (Cited on page 537.)

[64] M. Azimmohseni, A. R. Soltani, and M. Khalafi. Simulation of real discrete time Gaussian multivariate stationary processes with given spectral densities. *Journal of Time Series Analysis*, 36(6):783–796, 2015. (Cited on page 112.)

[65] E. Azmoodeh, T. Sottinen, L. Viitasaari, and A. Yazigi. Necessary and sufficient conditions for Hölder continuity of Gaussian processes. *Statistics & Probability Letters*, 94:230–235, 2014. (Cited on page 435.)

[66] F. Baccelli and A. Biswas. On scaling limits of power law shot-noise fields. *Stochastic Models*, 31(2):187–207, 2015. (Cited on page 225.)

[67] E. Bacry and J.-F. Muzy. Log-infinitely divisible multifractal processes. *Communications in Mathematical Physics*, 236(3):449–475, 2003. (Cited on page 612.)

[68] E. Bacry, J. Delour, and J.-F. Muzy. Multifractal random walk. *Physical Review E*, 64(2):026103, 2001. (Cited on page 612.)

[69] C. Baek and V. Pipiras. On distinguishing multiple changes in mean and long-range dependence using local Whittle estimation. *Electronic Journal of Statistics*, 8(1):931–964, 2014. (Cited on pages 225 and 573.)

[70] C. Baek, G. Didier, and V. Pipiras. On integral representations of operator fractional Brownian fields. *Statistics & Probability Letters*, 92:190–198, 2014. (Cited on pages 537 and 538.)

[71] C. Baek, N. Fortuna, and V. Pipiras. Can Markov switching model generate long memory? *Economics Letters*, 124(1):117–121, 2014. (Cited on pages 162 and 225.)

[72] C. Bahadoran, A. Benassi, and K. Dębicki. Operator-self-similar Gaussian processes with stationary increments. Preprint, 2003. (Cited on page 536.)

[73] L. Bai and J. Ma. Stochastic differential equations driven by fractional Brownian motion and Poisson point process. *Bernoulli*, 21(1):303–334, 2015. (Cited on page 436.)

[74] S. Bai and M. S. Taqqu. Multivariate limit theorems in the context of long-range dependence. *Journal of Time Series Analysis*, 34 (6): 717–743, 2013. (Cited on pages 320 and 344.)

[75] S. Bai and M. S. Taqqu. Multivariate limits of multilinear polynomial-form processes with long memory. *Statistics & Probability Letters*, 83(11):2473–2485, 2013. (Cited on pages 327 and 344.)

[76] S. Bai and M. S. Taqqu. Structure of the third moment of the generalized Rosenblatt distribution. *Statistics & Probability Letters*, 94:144–152, 2014. (Cited on page 344.)

[77] S. Bai and M. S. Taqqu. Generalized Hermite processes, discrete chaos and limit theorems. *Stochastic Processes and their Applications*, 124:1710–1739, 2014. (Cited on pages 276, 277, 281, and 344.)

[78] S. Bai and M. S. Taqqu. Convergence of long-memory discrete k-th order Volterra processes. *Stochastic Processes and Their Applications*, 125(5):2026–2053, 2015. (Cited on page 344.)

[79] S. Bai and M. S. Taqqu. The universality of homogeneous polynomial forms and critical limits. *Journal of Theoretical Probability*, 29(4):1710–1727, 2016. (Cited on page 344.)

[80] S. Bai and M. S. Taqqu. Short-range dependent processes subordinated to the Gaussian may not be strong mixing. *Statistics & Probability Letters*, 110:198–200, 2016. (Cited on page 612.)

[81] S. Bai and M. S. Taqqu. On the validity of resampling methods under long memory. To appear in *The Annals of Statistics*, 2017. (Cited on page 574.)

[82] S. Bai and M. S. Taqqu. The impact of diagonals of polynomial forms on limit theorems with long memory. *Bernoulli*, 23(1):710–742, 2017. (Cited on page 344.)

[83] S. Bai, M. S. Ginovyan, and M. S. Taqqu. Functional limit theorems for Toeplitz quadratic functionals of continuous time Gaussian stationary processes. *Statistics & Probability Letters*, 104:58–67, 2015. (Cited on page 571.)

[84] S. Bai, M. S. Ginovyan, and M. S. Taqqu. Limit theorems for quadratic forms of Lévy-driven continuous-time linear processes. *Stochastic Processes and their Applications*, 126(4):1036–1065, 2016. (Cited on page 571.)

[85] S. Bai, M. S. Taqqu, and T. Zhang. A unified approach to self-normalized block sampling. *Stochastic Processes and their Applications*, 126(8):2465–2493, 2016. (Cited on page 574.)

[86] R. T. Baillie and T. Bollerslev. Contintegration, fractional cointegration, and exchange rate dynamics. *The Journal of Finance*, 49:737–745, 1994. (Cited on page 537.)

[87] R. T. Baillie and G. Kapetanios. Testing for neglected nonlinearity in long-memory models. *Journal of Business & Economic Statistics*, 25(4):447–461, 2007. (Cited on page 611.)

[88] R. T. Baillie and G. Kapetanios. Estimation and inference for impulse response functions from univariate strongly persistent processes. *The Econometrics Journal*, 16(3):373–399, 2013. (Cited on page 573.)

[89] R. T. Baillie, T. Bollerslev, and H. O. Mikkelsen. Fractionally integrated generalized autoregressive conditional heteroskedasticity. *Journal of Econometrics*, 74(1):3–30, 1996. (Cited on page 611.)

[90] R. T. Baillie, G. Kapetanios, and F. Papailias. Modified information criteria and selection of long memory time series models. *Computational Statistics and Data Analysis*, 76:116–131, 2014. (Cited on pages 552 and 572.)

[91] R. T. Baillie, G. Kapetanios, and F. Papailias. Inference for impulse response coefficients from multivariate fractionally integrated processes. *Econometric Reviews*, 2016. (Cited on page 573.)

[92] R. M. Balan. Recent advances related to SPDEs with fractional noise. In R. C. Dalang, M. Dozzi, and F. Russo, editors, *Seminar on Stochastic Analysis, Random Fields and Applications VII*, volume 67 of *Progress in Probability*, pages 3–22. Basel: Springer, 2013. (Cited on page 227.)

[93] P. Balança. Some sample path properties of multifractional Brownian motion. *Stochastic Processes and their Applications*, 125(10):3823–3850, 2015. (Cited on page 611.)

[94] C. Barakat, P. Thiran, G. Iannaccone, C. Diot, and P. Owezarski. Modeling Internet backbone traffic at the flow level. *IEEE Transactions on Signal Processing*, 51(8):2111–2124, 2003. (Cited on page 224.)

[95] Ph. Barbe and W. P. McCormick. Invariance principles for some FARIMA and nonstationary linear processes in the domain of a stable distribution. *Probab. Theory Related Fields*, 154(3-4):429–476, 2012. (Cited on page 111.)

[96] J.-M. Bardet. Testing for the presence of self-similarity of Gaussian time series having stationary increments. *Journal of Time Series Analysis*, 21:497–515, 2000. (Cited on page 112.)

[97] J.-M. Bardet. Statistical study of the wavelet analysis of fractional Brownian motion. *IEEE Transactions on Information Theory*, 48:991–999, 2002. (Cited on page 112.)

[98] J.-M. Bardet and H. Bibi. Adaptive semiparametric wavelet estimator and goodness-of-fit test for long-memory linear processes. *Electronic Journal of Statistics*, 6:2383–2419, 2012. (Cited on page 112.)

[99] J.-M. Bardet, H. Bibi, and A. Jouini. Adaptive wavelet-based estimator of the memory parameter for stationary Gaussian processes. *Bernoulli*, 14(3):691–724, 2008. (Cited on page 112.)

[100] X. Bardina and K. Es-Sebaiy. An extension of bifractional Brownian motion. *Communications on Stochastic Analysis*, 5(2):333–340, 2011. (Cited on page 54.)

[101] O. E. Barndorff-Nielsen and N. N. Leonenko. Spectral properties of superpositions of Ornstein-Uhlenbeck type processes. *Methodology and Computing in Applied Probability*, 7(3):335–352, 2005. (Cited on page 223.)

[102] J. Barral. Mandelbrot cascades and related topics. In *Geometry and Analysis of Fractals*, volume 88 of *Springer Proc. Math. Stat.*, pages 1–45. Heidelberg: Springer, 2014. (Cited on page 612.)

[103] J. Barral and B. Mandelbrot. Multifractal products of cylindrical pulses. *Probability Theory and Related Fields*, 124(3):409–430, 2002. (Cited on page 612.)

[104] R. J. Barton and H. V. Poor. Signal detection in fractional Gaussian noise. *IEEE Transactions on Information Theory*, 34(5):943–959, 1988. (Cited on page 395.)

[105] G. K. Basak, N. H. Chan, and W. Palma. The approximation of long-memory processes by an arma model. *Journal of Forecasting*, 20(6):367–389, 2001. (Cited on page 571.)

[106] F. Baudoin and M. Hairer. A version of Hörmander's theorem for the fractional Brownian motion. *Probability Theory and Related Fields*, 139(3-4):373–395, 2007. (Cited on page 436.)

[107] F. Baudoin and C. Ouyang. Small-time kernel expansion for solutions of stochastic differential equations driven by fractional Brownian motions. *Stochastic Processes and their Applications*, 121(4):759–792, 2011. (Cited on page 436.)

[108] F. Baudoin, E. Nualart, C. Ouyang, and S. Tindel. On probability laws of solutions to differential systems driven by a fractional Brownian motion. *The Annals of Probability*, 44(4):2554–2590, 2016. (Cited on page 436.)

[109] F. Baudoin, C. Ouyang, and S. Tindel. Upper bounds for the density of solutions to stochastic differential equations driven by fractional Brownian motions. *Annales de l'Institut Henri Poincaré Probabilités et Statistiques*, 50(1):111–135, 2014. (Cited on page 436.)

[110] A. Baxevani and K. Podgórski. Lamperti transform and a series decomposition of fractional Brownian motion. Preprint, 2007. (Cited on page 465.)

[111] P. Becker-Kern and G. Pap. Parameter estimation of selfsimilarity exponents. *Journal of Multivariate Analysis*, 99(1):117–140, 2008. (Cited on page 536.)

[112] P. Becker-Kern, M. M. Meerschaert, and H.-P. Scheffler. Limit theorems for coupled continuous time random walks. *The Annals of Probability*, 32(1B):730–756, 2004. (Cited on page 227.)

[113] A. Benassi, S. Jaffard, and D. Roux. Elliptic Gaussian random processes. *Revista Matemática Iberoamericana*, 13(1):19–90, 1997. (Cited on pages 145, 465, and 611.)

[114] A. Benassi, S. Cohen, and J. Istas. Identifying the multifractional function of a Gaussian process. *Statistics & Probability Letters*, 39(4):337–345, 1998. (Cited on page 611.)

[115] A. Benassi, S. Cohen, and J. Istas. Identification and properties of real harmonizable fractional Lévy motions. *Bernoulli*, 8(1):97–115, 2002. (Cited on page 145.)

[116] C. Bender. An S-transform approach to integration with respect to a fractional Brownian motion. *Bernoulli*, 9(6):955–983, 2003. (Cited on page 435.)

[117] C. Bender. An Itô formula for generalized functionals of a fractional Brownian motion with arbitrary Hurst parameter. *Stochastic Processes and their Applications*, 104(1):81–106, 2003. (Cited on page 435.)

[118] C. Bender, T. Sottinen, and E. Valkeila. Arbitrage with fractional Brownian motion? *Theory of Stochastic Processes*, 13(1-2):23–34, 2007. (Cited on page 435.)

[119] D. A. Benson, M. M. Meerschaert, B. Baeumer, and H.-P. Scheffler. Aquifer operator scaling and the effect on solute mixing and dispersion. *Water Resources Research*, 42(1):W01415, 2006. (Cited on page 537.)

[120] J. Beran. A goodness of fit test for time series with long-range dependence. *Journal of the Royal Statistical Society, Series B*, 54:749–760, 1992. (Cited on page 554.)

[121] J. Beran. *Statistics for Long-Memory Processes*. New York: Chapman & Hall, 1994. (Cited on pages 109 and 572.)

[122] J. Beran. Maximum likelihood estimation of the differencing parameter for invertible short and long memory autoregressive integrated moving average models. *Journal of the Royal Statistical Society. Series B. Methodological*, 57(4):659–672, 1995. (Cited on page 551.)

[123] J. Beran and Y. Feng. SEMIFAR models—a semiparametric approach to modelling trends, long-range dependence and nonstationarity. *Computational Statistics & Data Analysis*, 40(2):393–419, 2002. (Cited on page 573.)

[124] J. Beran, R. J. Bhansali, and D. Ocker. On unified model selection for stationary and nonstationary short- and long-memory autoregressive processes. *Biometrika*, 85(4):921–934, 1998. (Cited on page 552.)

[125] J. Beran, S. Ghosh, and D. Schell. On least squares estimation for long-memory lattice processes. *Journal of Multivariate Analysis*, 100(10):2178–2194, 2009. (Cited on pages 529 and 537.)

[126] J. Beran, M. Schützner, and S. Ghosh. From short to long memory: aggregation and estimation. *Computational Statistics and Data Analysis*, 54(11):2432–2442, 2010. (Cited on page 223.)

[127] J. Beran, Y. Feng, S. Ghosh, and R. Kulik. *Long-Memory Processes: Probabilistic Properties and Statistical Methods*. Springer, 2013. (Cited on pages 109, 539, 545, 551, 573, 574, and 610.)

[128] J. Beran, S. Möhrle, and S. Ghosh. Testing for Hermite rank in Gaussian subordination processes. *Journal of Computational and Graphical Statistics*, 2016. (Cited on page 343.)

[129] C. Berg. The cube of a normal distribution is indeterminate. *The Annals of Probability*, 16(2):910–913, 1988. (Cited on page 280.)

[130] I. Berkes, L. Horváth, P. Kokoszka, and Q.-M. Shao. On discriminating between long-range dependence and changes in mean. *The Annals of Statistics*, 34(3):1140–1165, 2006. (Cited on pages 225 and 573.)

[131] S. M. Berman. Local nondeterminism and local times of Gaussian processes. *Indiana University Mathematics Journal*, 23:69–94, 1973/74. (Cited on page 430.)

[132] S. M. Berman. High level sojourns for strongly dependent Gaussian processes. *Zeitschrift für Wahrscheinlichkeitstheorie und verwandte Gebiete*, 50:223–236, 1979. (Cited on page 343.)

[133] S. M. Berman. Sojourns of vector Gaussian processes inside and outside spheres. *Zeitschrift für Wahrscheinlichkeitstheorie und verwandte Gebiete*, 66:529–542, 1984. (Cited on page 344.)

[134] S. M. Berman. *Sojourns and Extremes of Stochastic Processes*. The Wadsworth & Brooks/Cole Statistics/Probability Series. Pacific Grove, CA: Wadsworth & Brooks/Cole Advanced Books & Software, 1992. (Cited on page 178.)

[135] M. V. Berry and Z. V. Lewis. On the Weierstrass-Mandelbrot fractal function. *Proceedings of the Royal Society of London*, A370:459–484, 1980. (Cited on page 227.)

[136] S. Bertelli and M. Caporin. A note on calculating autocovariances of long-memory processes. *Journal of Time Series Analysis*, 23(5):503–508, 2002. (Cited on pages 539 and 570.)

[137] J. Bertoin and I. Kortchemski. Self-similar scaling limits of Markov chains on the positive integers. *The Annals of Applied Probability*, 26(4):2556–2595, 2016. (Cited on page 612.)

[138] C. Berzin, A. Latour, and J. R. León. *Inference on the Hurst Parameter and the Variance of Diffusions Driven by Fractional Brownian Motion*. Lecture Notes in Statistics. Springer International Publishing, 2014. (Cited on page 610.)

[139] A. Beskos, J. Dureau, and K. Kalogeropoulos. Bayesian inference for partially observed stochastic differential equations driven by fractional Brownian motion. *Biometrika*, 102(4):809–827, 2015. (Cited on page 436.)

[140] A. Betken and M. Wendler. Subsampling for general statistics under long range dependence. Preprint, 2015. (Cited on page 574.)

[141] R. J. Bhansali and P. S. Kokoszka. Estimation of the long-memory parameter: a review of recent developments and an extension. In *Selected Proceedings of the Symposium on Inference for Stochastic Processes (Athens, GA, 2000)*, volume 37 of *IMS Lecture Notes Monogr. Ser.*, pages 125–150. Beachwood, OH: Inst. Math. Statist., 2001. (Cited on page 571.)

[142] R. J. Bhansali and P. S. Kokoszka. Computation of the forecast coefficients for multistep prediction of long-range dependent time series. *International Journal of Forecasting*, 18(2):181–206, 2002. (Cited on page 571.)

[143] R. J. Bhansali and P. S. Kokoszka. Prediction of long-memory time series. In *Theory and Applications of Long-Range Dependence*, pages 355–367. Boston, MA: Birkhäuser Boston, 2003. (Cited on page 571.)

[144] R. J. Bhansali, L. Giraitis, and P. S. Kokoszka. Estimation of the memory parameter by fitting fractionally differenced autoregressive models. *Journal of Multivariate Analysis*, 97(10):2101–2130, 2006. (Cited on page 573.)

[145] G. Bhardwaj and N. R. Swanson. An empirical investigation of the usefulness of ARFIMA models for predicting macroeconomic and financial time series. *Journal of Econometrics*, 131(1-2):539–578, 2006. (Cited on page 571.)

[146] F. Biagini, B. Øksendal, A. Sulem, and N. Wallner. An introduction to white-noise theory and Malliavin calculus for fractional Brownian motion. *Proceedings of The Royal Society of London. Series A. Mathematical, Physical and Engineering Sciences*, 460(2041):347–372, 2004. Stochastic analysis with applications to mathematical finance. (Cited on pages 435 and 610.)

[147] F. Biagini, Y. Hu, B. Øksendal, and T. Zhang. *Stochastic Calculus for Fractional Brownian Motion and Applications*. Probability and its Applications (New York). London: Springer-Verlag London, Ltd., 2008. (Cited on page 436.)

[148] H. Biermé and A. Estrade. Poisson random balls: self-similarity and X-ray images. *Advances in Applied Probability*, 38(4):853–872, 2006. (Cited on page 224.)

[149] H. Biermé and F. Moisan, L.and Richard. A turning-band method for the simulation of anisotropic fractional Brownian fields. *Journal of Computational and Graphical Statistics*, 24(3):885–904, 2015. (Cited on page 112.)

[150] H. Biermé and F. Richard. Estimation of anisotropic Gaussian fields through Radon transform. *ESAIM. Probability and Statistics*, 12:30–50 (electronic), 2008. (Cited on page 529.)

[151] H. Biermé, M. M. Meerschaert, and H.-P. Scheffler. Operator scaling stable random fields. *Stochastic Processes and their Applications*, 117(3):312–332, 2007. (Cited on pages 511, 512, 521, 522, 535, 536, and 537.)

[152] H. Biermé, C. Lacaux, and H.-P. Scheffler. Multi-operator scaling random fields. *Stochastic Processes and their Applications*, 121(11):2642–2677, 2011. (Cited on page 537.)

[153] H. Biermé, O. Durieu, and Y. Wang. Invariance principles for operator-scaling Gaussian random fields. To appear in *The Annals of Applied Probability*. Preprint, 2015. (Cited on page 226.)

[154] P. Billingsley. *Convergence of Probability Measures*, 2nd edition. Wiley Series in Probability and Statistics: Probability and Statistics. New York: John Wiley & Sons Inc., 1999. A Wiley-Interscience Publication. (Cited on pages 219, 292, 296, 583, and 584.)

[155] N. H. Bingham. Szegö's theorem and its probabilistic descendants. *Probability Surveys*, 9:287–324, 2012. (Cited on page 571.)

[156] N. H. Bingham. Multivariate prediction and matrix Szegö theory. *Probability Surveys*, 9:325–339, 2012. (Cited on page 571.)

[157] N. H. Bingham, C. M. Goldie, and J. L. Teugels. *Regular Variation*, volume 27 of *Encyclopedia of Mathematics and its Applications*. Cambridge: Cambridge University Press, 1987. (Cited on pages 17, 19, 20, 25, 70, 109, 130, 176, 527, 582, and 586.)

[158] L. Bisaglia. Model selection for long-memory models. *Quaderni di Statistica*, 4:33–49, 2002. (Cited on page 571.)

[159] L. Bisaglia and M. Gerolimetto. An empirical strategy to detect spurious effects in long memory and occasional-break processes. *Communications in Statistics - Simulation and Computation*, 38:172–189, 2009. (Cited on page 573.)

[160] L. Bisaglia and D. Guégan. A comparison of techniques of estimation in long-memory processes. *Computational Statistics and Data Analysis*, 27:61–81, 1998. (Cited on page 571.)

[161] L. Bisaglia, S. Bordignon, and N. Cecchinato. Bootstrap approaches for estimation and confidence intervals of long memory processes. *Journal of Statistical Computation and Simulation*, 80(9-10):959–978, 2010. (Cited on page 573.)

[162] J. Blath, A. González Casanova, N. Kurt, and D. Spanò. The ancestral process of long-range seed bank models. *Journal of Applied Probability*, 50(3):741–759, 2013. (Cited on page 226.)

[163] T. Blu and M. Unser. Self-similarity. II. Optimal estimation of fractal processes. *IEEE Transactions on Signal Processing*, 55(4):1364–1378, 2007. (Cited on page 465.)

[164] P. Bocchinia and G. Deodatis. Critical review and latest developments of a class of simulation algorithms for strongly non-Gaussian random fields. *Probabilistic Engineering Mechanics*, 23:393–407, 2008. (Cited on page 344.)

[165] D. C. Boes and J. D. Salas. Nonstationarity of the mean and the Hurst phenomenon. *Water Resources Research*, 14(1):135–143, 1978. (Cited on page 225.)

[166] L. V. Bogachev. Random walks in random environments. In J.-P. Francoise, G. Naber, and S. T. Tsou, editors, *Encyclopedia of Mathematical Physics*, volume 4, pages 353–371. Oxford: Elsevier, 2006. (Cited on page 178.)

[167] V. I. Bogachev. *Gaussian Measures*, volume 62 of *Mathematical Surveys and Monographs*. Providence, RI: American Mathematical Society, 1998. (Cited on pages 409 and 422.)

[168] Y. Boissy, B. B. Bhattacharyya, X. Li, and G. D. Richardson. Parameter estimates for fractional autoregressive spatial processes. *The Annals of Statistics*, 33(6):2553–2567, 2005. (Cited on pages 529 and 537.)

[169] T. Bojdecki and A. Talarczyk. Particle picture interpretation of some Gaussian processes related to fractional Brownian motion. *Stochastic Processes and their Applications*, 122(5):2134–2154, 2012. (Cited on page 228.)

[170] T. Bojdecki, L. G. Gorostiza, and A. Talarczyk. Limit theorems for occupation time fluctuations of branching systems. I. Long-range dependence. *Stochastic Processes and their Applications*, 116(1):1–18, 2006. (Cited on page 228.)

[171] T. Bojdecki, L. G. Gorostiza, and A. Talarczyk. Limit theorems for occupation time fluctuations of branching systems. II. Critical and large dimensions. *Stochastic Processes and their Applications*, 116(1):19–35, 2006. (Cited on page 228.)

[172] T. Bojdecki, L. G. Gorostiza, and A. Talarczyk. From intersection local time to the Rosenblatt process. *Journal of Theoretical Probability*, 28(3):1227–1249, 2015. (Cited on page 228.)

[173] A. Bonami and A. Estrade. Anisotropic analysis of some Gaussian models. *The Journal of Fourier Analysis and Applications*, 9(3):215–236, 2003. (Cited on page 529.)

[174] P. Bondon and W. Palma. A class of antipersistent processes. *Journal of Time Series Analysis*, 28(2):261–273, 2007. (Cited on page 109.)

[175] A. Borodin and I. Corwin. Macdonald processes. *Probability Theory and Related Fields*, 158(1-2):225–400, 2014. (Cited on page 227.)

[176] A. Böttcher and B. Silbermann. Toeplitz matrices and determinants with Fisher-Hartwig symbols. *Journal of Functional Analysis*, 63(2):178–214, 1985. (Cited on page 211.)

[177] J.-P. Bouchaud and A. Georges. Anomalous diffusion in disordered media: statistical mechanisms, models and physical applications. *Physics Reports*, 195(4-5):127–293, 1990. (Cited on page 227.)

[178] J.-P. Bouchaud, A. Georges, J. Koplik, A. Provata, and S. Redner. Superdiffusion in random velocity fields. *Physical Review Letters*, 64(21):2503, 1990. (Cited on page 226.)

[179] M. Boutahar, V. Marimoutou, and L. Nouira. Estimation methods of the long memory parameter: Monte Carlo analysis and application. *Journal of Applied Statistics*, 34(3-4):261–301, 2007. (Cited on page 571.)

[180] C. Boutillier and B. de Tilière. The critical $\mathbf{Z}$-invariant Ising model via dimers: the periodic case. *Probability Theory and Related Fields*, 147(3-4):379–413, 2010. (Cited on page 227.)

[181] R. C. Bradley. Basic properties of strong mixing conditions. A survey and some open questions. *Probability Surveys*, 2:107–144, 2005. ISSN 1549-5787. Update of, and a supplement to, the 1986 original. (Cited on page 612.)

[182] M. Bramson and D. Griffeath. Renormalizing the 3-dimensional voter model. *The Annals of Probability*, 7(3):418–432, 1979. (Cited on page 226.)

[183] J.-C. Breton and I. Nourdin. Error bounds on the non-normal approximation of Hermite power variations of fractional Brownian motion. *Electronic Communications in Probability*, 13:482–493, 2008. (Cited on page 281.)

[184] P. Breuer and P. Major. Central limit theorems for non-linear functionals of Gaussian fields. *Journal of Multivariate Analysis*, 13:425–441, 1983. (Cited on pages 282, 343, and 344.)

[185] D. R. Brillinger. *Time Series*, volume 36 of *Classics in Applied Mathematics*. Philadelphia, PA: Society for Industrial and Applied Mathematics (SIAM), 2001. Data Analysis and Theory, Reprint of the 1981 edition. (Cited on page 470.)

[186] P. J. Brockwell and R. A. Davis. *Time Series: Theory and Methods*, 2nd edition. New York: Springer Series in Statistics. Springer-Verlag, 1991. (Cited on pages xxi, 9, 14, 28, 32, 33, 37, 39, 109, 337, 470, 471, 497, 500, 541, 544, 552, and 554.)

[187] J. Brodsky and C. M. Hurvich. Multi-step forecasting for long-memory processes. *Journal of Forecasting*, 18:59–75, 1999. (Cited on page 571.)

[188] J. C. Bronski. Asymptotics of Karhunen-Loeve eigenvalues and tight constants for probability distributions of passive scalar transport. *Communications in Mathematical Physics*, 238(3):563–582, 2003. (Cited on page 464.)

[189] J. C. Bronski. Small ball constants and tight eigenvalue asymptotics for fractional Brownian motions. *Journal of Theoretical Probability*, 16(1):87–100, 2003. (Cited on page 464.)

[190] A. Brouste and M. Kleptsyna. Asymptotic properties of MLE for partially observed fractional diffusion system. *Statistical Inference for Stochastic Processes*, 13(1):1–13, 2010. (Cited on page 395.)

[191] A. Brouste, J. Istas, and S. Lambert-Lacroix. On fractional Gaussian random fields simulations. *Journal of Statistical Software*, 23(1):1–23, 2007. (Cited on page 112.)

[192] B. Buchmann and C. Klüppelberg. Fractional integral equations and state space transforms. *Bernoulli*, 12(3):431–456, 2006. (Cited on page 436.)

[193] A. Budhiraja, V. Pipiras, and X. Song. Admission control for multidimensional workload input with heavy tails and fractional Ornstein-Uhlenbeck process. *Advances in Applied Probability*, 47:1–30, 2015. (Cited on page 224.)

[194] K. Burnecki and G. Sikora. Estimation of FARIMA parameters in the case of negative memory and stable noise. *IEEE Transactions on Signal Processing*, 61(11):2825–2835, 2013. (Cited on pages 111 and 573.)

[195] K. Burnecki, M. Maejima, and A. Weron. The Lamperti transformation for self-similar processes. *Yokohama Mathematical Journal*, 44(1):25–42, 1997. (Cited on page 111.)

[196] M. Çağlar. A long-range dependent workload model for packet data traffic. *Mathematics of Operations Research*, 29(1):92–105, 2004. (Cited on page 224.)

[197] M. Calder and R. A. Davis. Introduction to Whittle (1953) "The Analysis of Multiple Stationary Time Series". In S. Kotz and N. L. Johnson, editors, *Breakthroughs in Statistics*, volume 3, pages 141–148. Springer-Verlag, 1997. (Cited on page 570.)

[198] F. Camia, C. Garban, and C. M. Newman. Planar Ising magnetization field I. Uniqueness of the critical scaling limit. *The Annals of Probability*, 43(2):528–571, 2015. (Cited on page 227.)

[199] F. Camia, C. Garban, and C. M. Newman. Planar Ising magnetization field II. Properties of the critical and near-critical scaling limits. *Annales de l'Institut Henri Poincaré Probabilités et Statistiques*, 52(1):146–161, 2016. (Cited on page 227.)

[200] B. Candelpergher, M. Miniconi, and F. Pelgrin. Long-memory process and aggregation of AR(1) stochastic processes: A new characterization. Preprint, 2015. (Cited on page 224.)

[201] G. M. Caporale, J. Cuñado, and L. A. Gil-Alana. Modelling long-run trends and cycles in financial time series data. *Journal of Time Series Analysis*, 34(3):405–421, 2013. (Cited on page 110.)

[202] K. A. E. Carbonez. Model selection and estimation of long-memory time-series models. *Review of Business and Economics*, XLIV(4):512–554, 2009. (Cited on page 571.)

[203] P. Carmona, L. Coutin, and G. Montseny. Stochastic integration with respect to fractional Brownian motion. *Annales de l'Institut Henri Poincaré. Probabilités et Statistiques*, 39(1):27–68, 2003. (Cited on page 435.)

[204] K. J. E. Carpio and D. J. Daley. Long-range dependence of Markov chains in discrete time on countable state space. *Journal of Applied Probability*, 44(4):1047–1055, 2007. (Cited on page 109.)

[205] T. Cass and P. Friz. Densities for rough differential equations under Hörmander's condition. *Annals of Mathematics. Second Series*, 171(3):2115–2141, 2010. (Cited on page 436.)

[206] T. Cass, M. Hairer, C. Litterer, and S. Tindel. Smoothness of the density for solutions to Gaussian rough differential equations. *The Annals of Probability*, 43(1):188–239, 2015. (Cited on page 436.)

[207] A. Castaño-Martínez and F. López-Blázquez. Distribution of a sum of weighted noncentral chi-square variables. *Test*, 14(2):397–415, 2005. (Cited on page 271.)

[208] D. Celov, R. Leipus, and A. Philippe. Time series aggregation, disaggregation, and long memory. *Lietuvos Matematikos Rinkinys*, 47(4):466–481, 2007. (Cited on pages 117, 221, and 223.)

[209] D. Chambers and E. Slud. Central limit theorems for nonlinear functionals of stationary Gaussian processes. *Probability Theory and Related Fields*, 80:323–346, 1989. (Cited on page 343.)

[210] D. Chambers and E. Slud. Necessary conditions for nonlinear functionals of Gaussian processes to satisfy central limit theorems. *Stochastic Processes and their Applications*, 32(1):93–107, 1989. (Cited on page 343.)

[211] M. J. Chambers. The simulation of random vector time series with given spectrum. *Mathematical and Computer Modelling*, 22(2):1–6, 1995. (Cited on page 112.)

[212] M. J. Chambers. Long memory and aggregation in macroeconomic time series. *International Economic Review*, 39(4):1053–1072, 1998. Symposium on Forecasting and Empirical Methods in Macroeconomics and Finance. (Cited on page 223.)

[213] G. Chan and A. T. A. Wood. An algorithm for simulating stationary Gaussian random fields. *Applied Statistics, Algorithm Section*, 46:171–181, 1997. (Cited on page 112.)

[214] G. Chan and A. T. A. Wood. Simulation of stationary Gaussian vector fields. *Statistics and Computing*, 9(4):265–268, 1999. (Cited on page 112.)

[215] N. H. Chan and W. Palma. State space modeling of long-memory processes. *The Annals of Statistics*, 26(2):719–740, 1998. (Cited on page 573.)

[216] C. Chen and L. Yan. Remarks on the intersection local time of fractional Brownian motions. *Statistics & Probability Letters*, 81(8):1003–1012, 2011. (Cited on page 436.)

[217] L.-C. Chen and A. Sakai. Critical two-point functions for long-range statistical-mechanical models in high dimensions. *The Annals of Probability*, 43(2):639–681, 2015. (Cited on page 227.)

[218] L. H. Y. Chen, L. Goldstein, and Q.-M. Shao. *Normal Approximation by Stein's Method*. Probability and its Applications (New York). Heidelberg: Springer, 2011. (Cited on page 426.)

[219] W. W. Chen. Efficiency in estimation of memory. *Journal of Statistical Planning and Inference*, 140(12):3820–3840, 2010. (Cited on page 572.)

[220] W. W. Chen and R. S. Deo. A generalized Portmanteau goodness-of-fit test for time series models. *Econometric Theory*, 20(2):382–416, 2004. (Cited on page 553.)

[221] W. W. Chen and R. S. Deo. Estimation of mis-specified long memory models. *Journal of Econometrics*, 134(1):257–281, 2006. ISSN 0304-4076. (Cited on page 571.)

[222] W. W. Chen and C. M. Hurvich. Semiparametric estimation of multivariate fractional cointegration. *Journal of the American Statistical Association*, 98(463):629–642, 2003. (Cited on page 537.)

[223] W. W. Chen and C. M. Hurvich. Semiparametric estimation of fractional cointegrating subspaces. *The Annals of Statistics*, 34(6):2939–2979, 2006. (Cited on page 537.)

[224] W. W. Chen, C. M. Hurvich, and Y. Lu. On the correlation matrix of the discrete Fourier transform and the fast solution of large Toeplitz systems for long-memory time series. *Journal of the American Statistical Association*, 101(474):812–822, 2006. (Cited on page 570.)

[225] X. Chen, W. V. Li, J. Rosiński, and Q.-M. Shao. Large deviations for local times and intersection local times of fractional Brownian motions and Riemann-Liouville processes. *The Annals of Probability*, 39(2):729–778, 2011. (Cited on page 436.)

[226] Z. Chen, L. Xu, and D. Zhu. Generalized continuous time random walks and Hermite processes. *Statistics & Probability Letters*, 99:44–53, 2015. (Cited on page 228.)

[227] P. Cheridito. Arbitrage in fractional Brownian motion models. *Finance and Stochastics*, 7(4):533–553, 2003. (Cited on page 435.)

[228] P. Cheridito and D. Nualart. Stochastic integral of divergence type with respect to fractional Brownian motion with Hurst parameter $H \in (0, \frac{1}{2})$. *Annales de l'Institut Henri Poincaré. Probabilités et Statistiques*, 41(6):1049–1081, 2005. (Cited on pages 406, 433, and 435.)

[229] P. Cheridito, H. Kawaguchi, and M. Maejima. Fractional Ornstein-Uhlenbeck processes. *Electronic Journal of Probability*, 8:no. 3, 14 pp. (electronic), 2003. (Cited on pages 384 and 394.)

[230] Y.-W. Cheung. Long memory in foreign exchange rates. *Journal of Business and Economic Statistics*, 11:93–101, 1993. (Cited on page 572.)

[231] Y.-W. Cheung and F. X. Diebold. On maximum likelihood estimation of the differencing parameter of fractionally-integrated noise with unknown mean. *Journal of Econometrics*, 62:301–316, 1994. (Cited on page 571.)

[232] Y.-W. Cheung and K. S. Lai. A fractional cointegration analysis of purchasing power parity. *Journal of Business & Economic Statistics*, 11(1):103–112, 1993. (Cited on page 537.)

[233] G. Chevillon and S. Mavroeidis. Learning can generate long memory. To appear in *Journal of Econometrics*. Preprint, 2013. (Cited on page 228.)

[234] Z. Chi. Construction of stationary self-similar generalized fields by random wavelet expansion. *Probability Theory and Related Fields*, 121(2):269–300, 2001. (Cited on page 465.)

[235] P. Chigansky and M. Kleptsyna. Spectral asymptotics of the fractional Brownian covariance operator. Preprint, 2015. (Cited on page 440.)

[236] P. Chigansky and M. Kleptsyna. Asymptotics of the Karhunen-Loève expansion for the fractional Brownian motion. Preprint, 2016. (Cited on page 439.)

[237] J.-P. Chilès and P. Delfiner. *Geostatistics*, 2nd edition. Wiley Series in Probability and Statistics. Hoboken, NJ: John Wiley & Sons Inc., 2012. Modeling spatial uncertainty. (Cited on page 538.)

[238] B. J. Christensen and M. Ø Nielsen. Asymptotic normality of narrow-band least squares in the stationary fractional cointegration model and volatility forecasting. *Journal of Econometrics*, 133(1):343–371, 2006. (Cited on page 536.)

[239] A. Chronopoulou and S. Tindel. On inference for fractional differential equations. *Statistical Inference for Stochastic Processes*, 16(1):29–61, 2013. (Cited on page 436.)

[240] A. Chronopoulou, C. A. Tudor, and F. G. Viens. Application of Malliavin calculus to long-memory parameter estimation for non-Gaussian processes. *Comptes Rendus Mathématique. Académie des Sciences. Paris*, 347(11-12):663–666, 2009. (Cited on page 281.)

[241] A. Chronopoulou, C. A. Tudor, and F. G. Viens. Self-similarity parameter estimation and reproduction property for non-Gaussian Hermite processes. *Communications on Stochastic Analysis*, 5(1):161–185, 2011. (Cited on page 281.)

[242] C.-F. Chung. A note on calculating the autocovariances of the fractionally integrated ARMA models. *Economics Letters*, 45(3):293–297, 1994. (Cited on pages 539 and 570.)

[243] Z. Ciesielski, G. Kerkyacharian, and B. Roynette. Quelques espaces fonctionnels associés à des processus gaussiens. *Studia Mathematica*, 107(2):171–204, 1993. (Cited on page 435.)

[244] R. Cioczek-Georges and B. B. Mandelbrot. Alternative micropulses and fractional Brownian motion. *Stochastic Processes and their Applications*, 64:143–152, 1996. (Cited on page 224.)

[245] B. Cipra. Statistical physicists phase out a dream. *Science*, 288(5471):1561–1562, 2000. (Cited on page 181.)

[246] M. Clausel. Gaussian fields satisfying simultaneous operator scaling relations. In *Recent Developments in Fractals and Related Fields*, Appl. Numer. Harmon. Anal., pages 327–341. Boston, MA: Birkhäuser Boston Inc., 2010. (Cited on page 537.)

[247] M. Clausel and B. Vedel. Explicit construction of operator scaling Gaussian random fields. *Fractals*, 19(1):101–111, 2011. (Cited on page 537.)

[248] M. Clausel, F. Roueff, M. S. Taqqu, and C. Tudor. Large scale behavior of wavelet coefficients of non-linear subordinated processes with long memory. *Applied and Computational Harmonic Analysis*, 32(2):223–241, 2012. (Cited on page 112.)

[249] M. Clausel, F. Roueff, M. S. Taqqu, and C. Tudor. High order chaotic limits of wavelet scalograms under long-range dependence. *ALEA-Latin American Journal of Probability and Mathematical Statistics*, 10(2):979–1011, 2013. (Cited on page 112.)

[250] M. Clausel, F. Roueff, M. S. Taqqu, and C. Tudor. Wavelet estimation of the long memory parameter for Hermite polynomial of Gaussian processes. *ESAIM: Probability and Statistics*, 18:42–76, 2014. (Cited on page 112.)

[251] J.-F. Coeurjolly. Identification of multifractional Brownian motion. *Bernoulli*, 11(6):987–1008, 2005. (Cited on page 611.)

[252] J.-F. Coeurjolly, P.-O. Amblard, and S. Achard. Wavelet analysis of the multivariate fractional Brownian motion. *ESAIM. Probability and Statistics*, 17:592–604, 2013. (Cited on page 536.)

[253] S. Cohen and J. Istas. *Fractional Fields and Applications*, volume 73 of *Mathématiques & Applications (Berlin) [Mathematics & Applications]*. Heidelberg: Springer, 2013. With a foreword by S. Jaffard. (Cited on pages 65, 110, 537, and 610.)

[254] S. Cohen and G. Samorodnitsky. Random rewards, fractional Brownian local times and stable self-similar processes. *The Annals of Applied Probability*, 16(3):1432–1461, 2006. (Cited on page 226.)

[255] S. Cohen, M. M. Meerschaert, and J. Rosiński. Modeling and simulation with operator scaling. *Stochastic Processes and their Applications*, 120(12):2390–2411, 2010. (Cited on page 536.)

[256] S. Cohen, F. Panloup, and S. Tindel. Approximation of stationary solutions to SDEs driven by multiplicative fractional noise. *Stochastic Processes and their Applications*, 124(3):1197–1225, 2014. (Cited on page 436.)

[257] F. Comte and E. Renault. Long memory in continuous-time stochastic volatility models. *Mathematical Finance*, 8(4):291–323, 1998. (Cited on page 611.)

[258] P. L. Conti, L. De Giovanni, S. A. Stoev, and M. S. Taqqu. Confidence intervals for the long memory parameter based on wavelets and resampling. *Statistica Sinica*, 18(2):559–579, 2008. (Cited on page 573.)

[259] J. Contreras-Reyes, G. M. Goerg, and W. Palma. *R package afmtools: Estimation, Diagnostic and Forecasting Functions for ARFIMA models (version 0.1.7)*. Chile: Universidad de Chile, 2011. URL http://cran.r-project.org/web/packages/afmtools. (Cited on page 572.)

[260] D. Conus, M. Joseph, D. Khoshnevisan, and S.-Y. Shiu. Initial measures for the stochastic heat equation. *Annales de l'Institut Henri Poincaré Probabilités et Statistiques*, 50(1):136–153, 2014. (Cited on page 227.)

[261] J. B. Conway. *A Course in Functional Analysis*, 2nd edition, volume 96 of *Graduate Texts in Mathematics*. New York: Springer-Verlag, 1990. (Cited on page 260.)

[262] L. Coutin and L. Decreusefond. Abstract nonlinear filtering theory in the presence of fractional Brownian motion. *The Annals of Applied Probability*, 9(4): 1058–1090, 1999. (Cited on page 436.)

[263] L. Coutin and Z. Qian. Stochastic analysis, rough path analysis and fractional Brownian motions. *Probability Theory and Related Fields*, 122(1):108–140, 2002. (Cited on page 435.)

[264] L. Coutin, D. Nualart, and C. A. Tudor. Tanaka formula for the fractional Brownian motion. *Stochastic Processes and their Applications*, 94(2):301–315, 2001. (Cited on pages 432 and 433.)

[265] D. R. Cox. Long-range dependence: a review. In H.A. David and H.T. David, editors, *Statistics: An Appraisal*, pages 55–74. Iowa State University Press, 1984. (Cited on page 225.)

[266] D.R. Cox and M.W. Townsend. The use of correlograms for measuring yarn irregularity. *Journal of the Textile Institute Proceedings*, 42(4):P145–P151, 1951. (Cited on pages 109 and 225.)

[267] P. F. Craigmile. Simulating a class of stationary Gaussian processes using the Davies-Harte algorithm, with application to long memory processes. *Journal of Time Series Analysis*, 24(5):505–511, 2003. (Cited on page 112.)

[268] P. F. Craigmile, P. Guttorp, and D. B. Percival. Wavelet-based parameter estimation for polynomial contaminated fractionally differenced processes. *IEEE Transactions on Signal Processing*, 53(8, part 2):3151–3161, 2005. (Cited on page 112.)

[269] N. Crato and B. K. Ray. Model selection and forecasting for long-range dependent processes. *Journal of Forecasting*, 15:107–125, 1996. (Cited on page 552.)

[270] M. E. Crovella and A. Bestavros. Self-similarity in World Wide Web traffic: evidence and possible causes. *IEEE/ACM Transactions on Networking*, 5(6):835–846, 1997. (Cited on page 121.)

[271] J. D. Cryer and K. S. Chan. *Time Series Analysis: With Applications in R*. Springer Texts in Statistics. Springer, 2008. (Cited on page 14.)

[272] J. Cuzick. A central limit theorem for the number of zeros of a stationary Gaussian process. *The Annals of Probability*, 4(4):547–556, 1976. (Cited on page 343.)

[273] C. Czichowsky and W. Schachermayer. Portfolio optimisation beyond semimartingales: shadow prices and fractional Brownian motion. To appear in *The Annals of Applied Probability*. Preprint, 2015. (Cited on page 435.)

[274] D. Dacunha-Castelle and L. Fermín. Disaggregation of long memory processes on C^∞ class. *Electronic Communications in Probability*, 11:35–44 (electronic), 2006. (Cited on page 223.)

[275] D. Dacunha-Castelle and L. Fermín. Aggregation of autoregressive processes and long memory. Preprint, 2008. (Cited on page 223.)

[276] D. Dacunha-Castelle and G. Oppenheim. Mixtures, aggregation and long memory. Preprint, 2001. (Cited on page 223.)

[277] R. Dahlhaus. Efficient parameter estimation for self similar processes. *The Annals of Statistics*, 17(4):1749–1766, 1989. (Cited on pages 540, 546, and 571.)

[278] R. Dahlhaus. Correction: Efficient parameter estimation for self-similar processes. *The Annals of Statistics*, 34(2):1045–1047, 2006. (Cited on pages 540, 546, and 571.)

[279] H. Dai. Convergence in law to operator fractional Brownian motions. *Journal of Theoretical Probability*, 26(3):676–696, 2013. (Cited on page 536.)

[280] W. Dai and C. C. Heyde. Itô's formula with respect to fractional Brownian motion and its application. *Journal of Applied Mathematics and Stochastic Analysis*, 9(4):439–448, 1996. (Cited on page 435.)

[281] D. J. Daley. The Hurst index of long-range dependent renewal processes. *The Annals of Probability*, 27(4):2035–2041, 1999. (Cited on page 610.)

[282] D. J. Daley and D. Vere-Jones. *An Introduction to the Theory of Point Processes. Vol. I*, 2nd edition. Probability and its Applications (New York). New York: Springer-Verlag, 2003. Elementary Theory and Methods. (Cited on page 610.)

[283] D. J. Daley and D. Vere-Jones. *An Introduction to the Theory of Point Processes. Vol. II*, 2nd edition. Probability and its Applications (New York). New York: Springer, 2008. General theory and structure. (Cited on page 610.)

[284] D. J. Daley and R. Vesilo. Long range dependence of point processes, with queueing examples. *Stochastic Processes and their Applications*, 70(2):265–282, 1997. (Cited on page 610.)

[285] D. J. Daley, T. Rolski, and R. Vesilo. Long-range dependent point processes and their Palm-Khinchin distributions. *Advances in Applied Probability*, 32(4):1051–1063, 2000. (Cited on page 610.)

[286] J. Damarackas and V. Paulauskas. Properties of spectral covariance for linear processes with infinite variance. *Lithuanian Mathematical Journal*, 54(3):252–276, 2014. (Cited on page 111.)

[287] I. Daubechies. *Ten Lectures on Wavelets*. SIAM Philadelphia, 1992. CBMS-NSF series, Volume 61. (Cited on pages 444 and 451.)

[288] B. D'Auria and S. I. Resnick. The influence of dependence on data network models. *Advances in Applied Probability*, 40(1):60–94, 2008. (Cited on page 224.)

[289] J. Davidson and P. Sibbertsen. Generating schemes for long memory processes: regimes, aggregation and linearity. *Journal of Econometrics*, 128(2):253–282, 2005. (Cited on page 224.)

[290] R. B. Davies and D. S. Harte. Tests for Hurst effect. *Biometrika*, 74(4):95–101, 1987. (Cited on page 112.)

[291] Yu. A. Davydov. The invariance principle for stationary processes. *Theory of Probability and its Applications*, 15:487–498, 1970. (Cited on page 111.)

[292] Ł. Dębowski. On processes with hyperbolically decaying autocorrelations. *Journal of Time Series Analysis*, 32:580–584, 2011. (Cited on page 571.)

[293] A. De Masi and P. A. Ferrari. Flux fluctuations in the one dimensional nearest neighbors symmetric simple exclusion process. *Journal of Statistical Physics*, 107(3-4):677–683, 2002. (Cited on page 168.)

[294] L. Decreusefond. Stochastic integration with respect to Volterra processes. *Annales de l'Institut Henri Poincaré. Probabilités et Statistiques*, 41(2):123–149, 2005. (Cited on page 435.)

[295] L. Decreusefond and D. Nualart. Hitting times for Gaussian processes. *The Annals of Probability*, 36(1):319–330, 2008. (Cited on page 436.)

[296] L. Decreusefond and A. S. Üstünel. Stochastic analysis of the fractional Brownian motion. *Potential Analysis*, 10:177–214, 1999. (Cited on pages 395 and 435.)

[297] H. Dehling, A. Rooch, and M. S. Taqqu. Non-parametric change-point tests for long-range dependent data. *Scandinavian Journal of Statistics. Theory and Applications*, 40(1):153–173, 2013. (Cited on page 573.)

[298] P. Deift, A. Its, and I. Krasovsky. Asymptotics of Toeplitz, Hankel, and Toeplitz+Hankel determinants with Fisher-Hartwig singularities. *Annals of Mathematics. Second Series*, 174(2):1243–1299, 2011. (Cited on page 211.)

[299] M. A. Delgado and P. M. Robinson. Optimal spectral bandwidth for long memory. *Statistica Sinica*, 6(1):97–112, 1996. (Cited on pages 111, 565, and 572.)

[300] M. A. Delgado, J. Hidalgo, and C. Velasco. Bootstrap assisted specification tests for the ARFIMA model. *Econometric Theory*, 27(5):1083–1116, 2011. (Cited on page 554.)

[301] A. Dembo, C. L. Mallows, and L. A. Shepp. Embedding nonnegative definite Toeplitz matrices in nonnegative definite circulant matrices, with applications to covariance estimation. *IEEE Transactions on Information Theory*, 35:1206–1212, 1989. (Cited on page 112.)

[302] R. Deo, M. Hsieh, C. M. Hurvich, and P. Soulier. Long memory in nonlinear processes. In *Dependence in Probability and Statistics*, volume 187 of *Lecture Notes in Statist.*, pages 221–244. New York: Springer, 2006. (Cited on pages 573 and 611.)

[303] R. S. Deo and W. W. Chen. On the integral of the squared periodogram. *Stochastic Processes and their Applications*, 85(1):159–176, 2000. (Cited on page 554.)

[304] A. Deya, A. Neuenkirch, and S. Tindel. A Milstein-type scheme without Lévy area terms for SDEs driven by fractional Brownian motion. *Annales de l'Institut Henri Poincaré Probabilités et Statistiques*, 48(2):518–550, 2012. (Cited on page 436.)

[305] G. Didier and V. Pipiras. Gaussian stationary processes: adaptive wavelet decompositions, discrete approximations, and their convergence. *The Journal of Fourier Analysis and Applications*, 14(2):203–234, 2008. (Cited on page 465.)

[306] G. Didier and V. Pipiras. Adaptive wavelet decompositions of stationary time series. *Journal of Time Series Analysis*, 31(3):182–209, 2010. (Cited on page 465.)

[307] G. Didier and V. Pipiras. Integral representations and properties of operator fractional Brownian motions. *Bernoulli*, 17(1):1–33, 2011. (Cited on pages 479, 481, 484, and 536.)

[308] G. Didier and V. Pipiras. Exponents, symmetry groups and classification of operator fractional Brownian motions. *Journal of Theoretical Probability*, 25(2):353–395, 2012. (Cited on pages 494, 495, and 536.)

[309] G. Didier, M. M. Meerschaert, and V. Pipiras. The exponents of operator self-similar random fields. *Journal of Mathematical Analysis and Applications*, 448(2):1450–1466, 2017. (Cited on pages 509 and 538.)

[310] G. Didier, M. M. Meerschaert, and V. Pipiras. Domain and range symmetries of operator fractional Brownian fields. Preprint, 2016. (Cited on page 538.)

[311] F. Diebold and G. Rudebusch. Long memory and persistence in aggregate output. *Journal of Monetary Economics*, 24:189–209, 1989. (Cited on page 572.)

[312] F. X. Diebold and A. Inoue. Long memory and regime switching. *Journal of Econometrics*, 105(1):131–159, 2001. (Cited on pages 162, 222, and 225.)

[313] C. R. Dietrich and G. N. Newsam. A fast and exact simulation for multidimensional Gaussian stochastic simulations. *Water Resources Research*, 29(8):2861–2869, 1993. (Cited on page 112.)

[314] C. R. Dietrich and G. N. Newsam. Fast and exact simulation of stationary Gaussian processes through circulant embedding of the covariance matrix. *SIAM Journal on Scientific Computing*, 18(4):1088–1107, 1997. (Cited on page 112.)

[315] R. W. Dijkerman and R. R. Mazumdar. On the correlation structure of the wavelet coefficients of fractional Brownian motion. *IEEE Trans. Inform. Theory*, 40(5):1609–1612, 1994. (Cited on page 112.)

[316] Z. Ding, C. W. J. Granger, and R. F. Engle. A long memory property of stock market returns and a new model. *Journal of Empirical Finance*, 1(1):83–106, 1993. (Cited on page 611.)

[317] G. S. Dissanayake, M. S. Peiris, and T. Proietti. State space modeling of Gegenbauer processes with long memory. *Computational Statistics and Data Analysis*, 2016. (Cited on page 110.)

[318] I. Dittmann and C. W. J. Granger. Properties of nonlinear transformations of fractionally integrated processes. *Journal of Econometrics*, 110(2):113–133, 2002. Long memory and nonlinear time series, Cardiff, 2000. (Cited on page 344.)

[319] R. Dmowska and B. Saltzman. *Long-Range Persistence in Geophysical Time Series*. Advances in Geophysics. Elsevier Science, 1999. (Cited on pages 111 and 610.)

[320] R. L. Dobrushin. Gaussian and their subordinated self-similar random generalized fields. *The Annals of Probability*, 7:1–28, 1979. (Cited on page 281.)

[321] R. L. Dobrushin and P. Major. Non-central limit theorems for nonlinear functionals of Gaussian fields. *Zeitschrift für Wahrscheinlichkeitstheorie und Verwandte Gebiete*, 50(1):27–52, 1979. (Cited on pages 280, 343, 344, and 584.)

[322] C. Dombry and N. Guillotin-Plantard. A functional approach for random walks in random sceneries. *Electronic Journal of Probability*, 14:1495–1512, 2009. (Cited on pages 179 and 226.)

[323] C. Dombry and I. Kaj. The on-off network traffic model under intermediate scaling. *Queueing Systems*, 69(1):29–44, 2011. (Cited on page 224.)

[324] C. Dombry and I. Kaj. Moment measures of heavy-tailed renewal point processes: asymptotics and applications. *ESAIM. Probability and Statistics*, 17:567–591, 2013. (Cited on page 224.)

[325] J. L. Doob. *Stochastic Processes*. Wiley Classics Library. John Wiley & Sons Inc., New York, 1953. Reprint of the 1953 original, a Wiley-Interscience Publication. (Cited on page 476.)

[326] J. A. Doornik and M. Ooms. Computational aspects of maximum likelihood estimation of autoregressive fractionally integrated moving average models. *Computational Statistics & Data Analysis*, 42(3):333–348, 2003. (Cited on pages 539 and 570.)

[327] R. Douc, E. Moulines, and D. S. Stoffer. *Nonlinear Time Series*. Chapman & Hall/CRC Texts in Statistical Science Series. Chapman & Hall/CRC, Boca Raton, FL, 2014. Theory, methods, and applications with R examples. (Cited on page 14.)

[328] P. Doukhan, G. Oppenheim, and M. S. Taqqu, editors. *Theory and Applications of Long-Range Dependence*. Boston: Birkhäuser, 2003. (Cited on page 610.)

[329] P. Driscoll. Smoothness of densities for area-like processes of fractional Brownian motion. *Probability Theory and Related Fields*, 155(1-2):1–34, 2013. (Cited on page 436.)

[330] R. M. Dudley and R. Norvaiša. *Concrete Functional Calculus*. Springer Monographs in Mathematics. New York: Springer, 2011. (Cited on page 435.)

[331] M. Dueker and R. Startz. Maximum-likelihood estimation of fractional cointegration with an application to US and Canadian bond rates. *Review of Economics and Statistics*, 80(3):420–426, 1998. (Cited on page 537.)

[332] N. G. Duffield and N. O'Connell. Large deviations and overflow probabilities for the general single-server queue, with applications. *Mathematical Proceedings of the Cambridge Philosophical Society*, 118(2):363–374, 1995. (Cited on page 611.)

[333] J. A. Duffy. A uniform law for convergence to the local times of linear fractional stable motions. *The Annals of Applied Probability*, 26(1):45–72, 2016. (Cited on page 111.)

[334] T. E. Duncan. Prediction for some processes related to a fractional Brownian motion. *Statistics & Probability Letters*, 76(2):128–134, 2006. (Cited on page 395.)

[335] T. E. Duncan. Mutual information for stochastic signals and fractional Brownian motion. *IEEE Transactions on Information Theory*, 54(10):4432–4438, 2008. (Cited on page 395.)

[336] T. E. Duncan and H. Fink. Corrigendum to "Prediction for some processes related to a fractional Brownian motion". *Statistics & Probability Letters*, 81(8):1336–1337, 2011. (Cited on page 395.)

[337] T. E. Duncan, Y. Hu, and B. Pasik-Duncan. Stochastic calculus for fractional Brownian motion. I. Theory. *SIAM Journal on Control and Optimization*, 38(2):582–612, 2000. (Cited on page 435.)

[338] O. Durieu and Y. Wang. From infinite urn schemes to decompositions of self-similar Gaussian processes. *Electronic Journal of Probability*, 21, paper no. 43. (Cited on page 226.)

[339] D. Dürr, S. Goldstein, and J. L. Lebowitz. Asymptotics of particle trajectories in infinite one-dimensional systems with collisions. *Communications on Pure and Applied Mathematics*, 38(5):573–597, 1985. (Cited on pages 163, 165, and 225.)

[340] H. Dym and H. P. McKean. *Fourier Series and Integrals*. New York: Academic Press, 1972. (Cited on pages 575 and 577.)

[341] K. Dzhaparidze and H. van Zanten. A series expansion of fractional Brownian motion. *Probability Theory and Related Fields*, 130(1):39–55, 2004. (Cited on pages 463 and 465.)

[342] K. Dzhaparidze and H. van Zanten. Krein's spectral theory and the Paley-Wiener expansion for fractional Brownian motion. *The Annals of Probability*, 33(2)620–644, 2005. (Cited on pages 455, 456, 459, 460, 461, 462, and 465.)

[343] K. Dzhaparidze and H. van Zanten. Optimality of an explicit series expansion of the fractional Brownian sheet. *Statistics & Probability Letters*, 71(4):295–301, 2005. (Cited on page 465.)

[344] K. Dzhaparidze, H. van Zanten, and P. Zareba. Representations of fractional Brownian motion using vibrating strings. *Stochastic Processes and their Applications*, 115(12):1928–1953, 2005. (Cited on page 465.)

[345] M. Eddahbi, R. Lacayo, J. L. Solé, J. Vives, and C. A. Tudor. Regularity of the local time for the d-dimensional fractional Brownian motion with N-parameters. *Stochastic Analysis and Applications*, 23(2):383–400, 2005. (Cited on page 432.)

[346] E. Eglói. *Dilative Stability*. PhD thesis, University of Debrecen, 2008. (Cited on page 110.)

[347] T. Ehrhardt and B. Silbermann. Toeplitz determinants with one Fisher-Hartwig singularity. *Journal of Functional Analysis*, 148(1):229–256, 1997. (Cited on page 211.)

[348] R. J. Elliott and J. van der Hoek. A general fractional white noise theory and applications to finance. *Mathematical Finance. An International Journal of Mathematics, Statistics and Financial Economics*, 13(2):301–330, 2003. (Cited on page 435.)

[349] C. Ellis and P. Wilson. Another look at the forecast performance of ARFIMA models. *International Review of Financial Analysis*, 13(1):63–81, 2004. (Cited on page 571.)

[350] P. Embrechts and M. Maejima. *Selfsimilar Processes*. Princeton Series in Applied Mathematics. Princeton University Press, 2002. (Cited on pages 44, 110, 536, and 610.)

[351] N. Enriquez. A simple construction of the fractional Brownian motion. *Stochastic Processes and Their Applications*, 109(2):203–223, 2004. (Cited on pages 222 and 224.)

[352] K. Falconer. *Fractal Geometry*. John Wiley & Sons, Ltd., Chichester, third edition, 2014. Mathematical Foundations and Applications. (Cited on page 612.)

[353] J. Fan and I. Gijbels. *Local Polynomial Modelling and its Applications*, volume 66 of *Monographs on Statistics and Applied Probability*. London: Chapman & Hall, 1996. (Cited on page 566.)

[354] V. Fasen. Modeling network traffic by a cluster Poisson input process with heavy and light-tailed file sizes. *Queueing Systems*, 66(4):313–350, 2010. (Cited on page 224.)

[355] G. Fay and A. Philippe. Goodness-of-fit test for long range dependent processes. *European Series in Applied and Industrial Mathematics (ESAIM). Probability and Statistics*, 6:239–258 (electronic), 2002. ISSN 1292-8100. New directions in time series analysis (Luminy, 2001). (Cited on page 554.)

[356] G. Faÿ, B. González-Arévalo, T. Mikosch, and G. Samorodnitsky. Modeling teletraffic arrivals by a Poisson cluster process. *Queueing Systems*, 54(2):121–140, 2006. (Cited on page 224.)

[357] G. Faÿ, E. Moulines, F. Roueff, and M. S. Taqqu. Estimators of long-memory: Fourier versus wavelets. *Journal of Econometrics*, 151(2):159–177, 2009. (Cited on page 112.)

[358] G. Faÿ, E. Moulines, F. Roueff, and M. S. Taqqu. Estimators of long-memory: Fourier versus wavelets. *The Journal of Econometrics*, 151(2):159–177, 2009. (Cited on page 574.)

[359] W. Feller. The asymptotic distribution of the range of sums of independent random variables. *Annals of Mathematical Statistics*, 22:427–432, 1951. (Cited on pages 87 and 111.)

[360] W. Feller. *An Introduction to Probability Theory and its Applications. Vol. II.* New York: John Wiley & Sons Inc., New York, 1966. (Cited on pages 92 and 561.)

[361] W. Feller. *An Introduction to Probability Theory and its Applications. Vol. I*, 3rd edition. New York: John Wiley & Sons Inc., New York, 1968. (Cited on pages 35 and 109.)

[362] Y. Feng and J. Beran. Filtered log-periodogram regression of long memory processes. *Journal of Statistical Theory and Practice*, 3(4):777–793, 2009. (Cited on page 111.)

[363] M. Ferrante and C. Rovira. Stochastic delay differential equations driven by fractional Brownian motion with Hurst parameter $H > \frac{1}{2}$. *Bernoulli*, 12(1):85–100, 2006. (Cited on page 436.)

[364] H. Fink, C. Klüppelberg, and M. Zähle. Conditional distributions of processes related to fractional Brownian motion. *Journal of Applied Probability*, 50(1):166–183, 2013. (Cited on page 395.)

[365] M. E. Fisher and R. E. Hartwig. Toeplitz determinants: some applications, theorems, and conjectures. In K. E. Shuler, editor, *Advances in Chemical Physics: Stochastic Processes in Chemical Physics*, volume 15, pages 333–353. Hoboken, NJ: John Wiley & Sons, Inc., 1968. (Cited on page 211.)

[366] P. Flandrin. Wavelet analysis and synthesis of fractional Brownian motion. *IEEE Transactions on Information Theory*, 38:910–917, 1992. (Cited on page 112.)

[367] P. Flandrin, P. Borgnat, and P.-O. Amblard. From stationarity to self-similarity, and back: Variations on the Lamperti transformation. In G. Rangarajan and M. Ding, editors, *Processes with Long-Range Correlations*, pages 88–117. Springer, 2003. (Cited on page 111.)

[368] R. Fox and M. S. Taqqu. Large-sample properties of parameter estimates for strongly dependent stationary Gaussian time series. *The Annals of Statistics*, 14:517–532, 1986. (Cited on page 571.)

[369] R. Fox and M. S. Taqqu. Central limit theorems for quadratic forms in random variables having long-range dependence. *Probability Theory and Related Fields*, 74:213–240, 1987. (Cited on page 571.)

[370] B. Franke and T. Saigo. A self-similar process arising from a random walk with random environment in random scenery. *Bernoulli*, 16(3):825–857, 2010. (Cited on page 226.)

[371] P. Frederiksen and F. S. Nielsen. Testing for long memory in potentially nonstationary perturbed fractional processes. *Journal of Financial Econometrics*, page nbs027, 2013. (Cited on page 611.)

[372] P. Frederiksen and M. Ø. Nielsen. Bias-reduced estimation of long-memory stochastic volatility. *Journal of Financial Econometrics*, 6(4):496–512, 2008. (Cited on page 611.)

[373] P. Frederiksen, F. S. Nielsen, and M. Ø. Nielsen. Local polynomial Whittle estimation of perturbed fractional processes. *Journal of Econometrics*, 167(2):426–447, 2012. (Cited on page 611.)

[374] M. P. Frías, M. D. Ruiz-Medina, F. J. Alonso, and J. M. Angulo. Spatiotemporal generation of long-range dependence models and estimation. *Environmetrics*, 17(2):139–146, 2006. (Cited on page 538.)

[375] M. P. Frías, F. J. Alonso, M. D. Ruiz-Medina, and J. M. Angulo. Semiparametric estimation of spatial long-range dependence. *Journal of Statistical Planning and Inference*, 138(5):1479–1495, 2008. (Cited on page 537.)

[376] M. P. Frías, M. D. Ruiz-Medina, F. J. Alonso, and J. M. Angulo. Spectral-marginal-based estimation of spatiotemporal long-range dependence. *Communications in Statistics. Theory and Methods*, 38(1-2):103–114, 2009. (Cited on page 538.)

[377] Y. Fyodorov, B. Khoruzhenko, and N. Simm. Fractional Brownian motion with Hurst index $H = 0$ and the Gaussian Unitary Ensemble. *The Annals of Probability*, 44(4):2980–3031, 2016. (Cited on page 110.)

[378] V. J. Gabriel and L. F. Martins. On the forecasting ability of ARFIMA models when infrequent breaks occur. *The Econometrics Journal*, 7(2):455–475, 2004. (Cited on page 573.)

[379] R. Gaigalas. A Poisson bridge between fractional Brownian motion and stable Lévy motion. *Stochastic Processes and their Applications*, 116(3):447–462, 2006. (Cited on page 145.)

[380] R. Gaigalas and I. Kaj. Convergence of scaled renewal processes and a packet arrival model. *Bernoulli*, 9(4):671–703, 2003. (Cited on pages 145 and 224.)

[381] F. R. Gantmacher. *The Theory of Matrices. Vols. 1, 2*. Translated by K. A. Hirsch. New York: Chelsea Publishing Co., 1959. (Cited on page 494.)

[382] M. J. Garrido-Atienza, P. E. Kloeden, and A. Neuenkirch. Discretization of stationary solutions of stochastic systems driven by fractional Brownian motion. *Applied Mathematics and Optimization. An International Journal with Applications to Stochastics*, 60(2):151–172, 2009. (Cited on page 436.)

[383] D. Gasbarra, T. Sottinen, and E. Valkeila. Gaussian bridges. In *Stochastic analysis and applications*, volume 2 of *Abel Symp.*, pages 361–382. Berlin: Springer, 2007. (Cited on page 395.)

[384] D. Geman and J. Horowitz. Occupation densities. *The Annals of Probability*, 8(1):1–67, 1980. (Cited on pages 430, 433, and 436.)

[385] J. E. Gentle. *Numerical Linear Algebra for Applications in Statistics*. Statistics and Computing. New York: Springer-Verlag, 1998. (Cited on page 542.)

[386] M. G. Genton, O. Perrin, and M. S. Taqqu. Self-similarity and Lamperti transformation for random fields. *Stochastic Models*, 23(3):397–411, 2007. (Cited on page 111.)

[387] J. Geweke and S. Porter-Hudak. The estimation and application of long memory time series models. *Journal of Time Series Analysis*, 4(4):221–238, 1983. (Cited on pages 89 and 111.)

[388] A. P. Ghosh, A. Roitershtein, and A. Weerasinghe. Optimal control of a stochastic processing system driven by a fractional Brownian motion input. *Advances in Applied Probability*, 42(1):183–209, 2010. (Cited on page 436.)

[389] H. Gilsing and T. Sottinen. Power series expansions for fractional Brownian motions. *Theory of Stochastic Processes*, 9(3-4):38–49, 2003. (Cited on page 465.)

[390] M. S. Ginovian. On Toeplitz type quadratic functionals of stationary Gaussian processes. *Probability Theory and Related Fields*, 100(3):395–406, 1994. (Cited on page 571.)

[391] M. S. Ginovyan and A. A. Sahakyan. Limit theorems for Toeplitz quadratic functionals of continuous-time stationary processes. *Probability Theory and Related Fields*, 138(3-4):551–579, 2007. (Cited on page 571.)

[392] L. Giraitis. Central limit theorem for functionals of a linear process. *Lithuanian Mathematical Journal*, 25:25–35, 1985. (Cited on page 343.)

[393] L. Giraitis. Central limit theorem for polynomial forms I. *Lithuanian Mathematical Journal*, 29:109–128, 1989. (Cited on page 343.)

[394] L. Giraitis. Central limit theorem for polynomial forms II. *Lithuanian Mathematical Journal*, 29:338–350, 1989. (Cited on page 343.)

[395] L. Giraitis and P. M. Robinson. Edgeworth expansions for semiparametric Whittle estimation of long memory. *The Annals of Statistics*, 31(4):1325–1375, 2003. (Cited on page 572.)

[396] L. Giraitis and D. Surgailis. CLT and other limit theorems for functionals of Gaussian processes. *Probability Theory and Related Fields*, 70:191–212, 1985. (Cited on page 343.)

[397] L. Giraitis and D. Surgailis. Multivariate Appell polynomials and the central limit theorem. In E. Eberlein and M. S. Taqqu, editors, *Dependence in Probability and Statistics*, pages 21–71. New York: Birkhäuser, 1986. (Cited on page 343.)

[398] L. Giraitis and D. Surgailis. Limit theorem for polynomials of linear processes with long range dependence. *Lietuvos Matematikos Rinkinys*, 29:128–145, 1989. (Cited on page 343.)

[399] L. Giraitis and D. Surgailis. A central limit theorem for quadratic forms in strongly dependent linear variables and application to asymptotical normality of Whittle's estimate. *Probability Theory and Related Fields*, 86:87–104, 1990. (Cited on page 571.)

[400] L. Giraitis and D. Surgailis. Central limit theorem for the empirical process of a linear sequence with long memory. *Journal of Statistical Planning and Inference*, 80(1-2):81–93, 1999. (Cited on page 344.)

[401] L. Giraitis and M. S. Taqqu. Central limit theorems for quadratic forms with time-domain conditions. *The Annals of Probability*, 26:377–398, 1998. (Cited on page 571.)

[402] L. Giraitis and M. S. Taqqu. Whittle estimator for finite-variance non-Gaussian time series with long memory. *The Annals of Statistics*, 27(1):178–203, 1999. (Cited on page 571.)

[403] L. Giraitis, P. M. Robinson, and A. Samarov. Rate optimal semiparametric estimation of the memory parameter of the Gaussian time series with long range dependence. *Journal of Time Series Analysis*, 18(1):49–60, 1997. (Cited on page 572.)

[404] L. Giraitis, P. M. Robinson, and A. Samarov. Adaptive semiparametric estimation of the long memory parameter. *Journal of Multivariate Analysis*, 72:183–207, 2000. (Cited on page 572.)

[405] L. Giraitis, P. Kokoszka, R. Leipus, and G. Teyssière. Rescaled variance and related tests for long memory in volatility and levels. *Journal of Econometrics*, 112(2):265–294, 2003. (Cited on page 111.)

[406] L. Giraitis, H. L. Koul, and D. Surgailis. *Large Sample Inference for Long Memory Processes*. London: Imperial College Press, 2012. (Cited on pages 310, 311, 325, 539, 545, 546, 547, 549, 550, 571, 573, and 610.)

[407] R. Gisselquist. A continuum of collision process limit theorems. *The Annals of Probability*, 1:231–239, 1973. (Cited on pages 163 and 225.)

[408] A. Gloter and M. Hoffmann. Stochastic volatility and fractional Brownian motion. *Stochastic Processes and their Applications*, 113(1):143–172, 2004. (Cited on page 436.)

[409] B. V. Gnedenko and A. N. Kolmogorov. *Limit distributions for sums of independent random variables*. Readiing, MA: Addison-Wesley, 1954. (Cited on page 586.)

[410] A. Gnedin, B. Hansen, and J. Pitman. Notes on the occupancy problem with infinitely many boxes: general asymptotics and power laws. *Probability Surveys*, 4:146–171, 2007. (Cited on page 228.)

[411] T. Gneiting. Power-law correlations, related models for long-range dependence and their simulation. *Journal of Applied Probability*, 37(4):1104–1109, 2000. (Cited on page 112.)

[412] T. Gneiting and M. Schlather. Stochastic models that separate fractal dimension and the Hurst effect. *SIAM Review*, 46(2):269–282 (electronic), 2004. (Cited on page 538.)

[413] T. Gneiting, H. Ševčíková, D. B. Percival, M. Schlather, and Y. Jiang. Fast and exact simulation of large Gaussian lattice systems in R^2: Exploring the limits. *Journal of Computational and Graphical Statistics*, 15(3):483–501, 2006. (Cited on page 112.)

[414] F. Godet. Linear prediction of long-range dependent time series. *ESAIM. Probability and Statistics*, 13:115–134, 2009. ISSN 1292-8100. (Cited on page 571.)

[415] F. Godet. Prediction of long memory processes on same-realisation. *Journal of Statistical Planning and Inference*, 140(4):907–926, 2010. (Cited on page 571.)

[416] I. Gohberg, S. Goldberg, and M. A. Kaashoek. *Basic Classes of Linear Operators*. Basel: Birkhäuser Verlag, 2003. (Cited on page 438.)

[417] E. Gonçalves and C. Gouriéroux. Agrégation de processus autorégressifs d'ordre 1. *Annales d'Économie et de Statistique*, 12:127–149, 1988. (Cited on page 223.)

[418] V. V. Gorodetskii. On convergence to semi-stable Gaussian processes. *Theory of Probability and its Applications*, 22:498–508, 1977. (Cited on page 111.)

[419] M. Gradinaru and I. Nourdin. Milstein's type schemes for fractional SDEs. *Annales de l'Institut Henri Poincaré Probabilités et Statistiques*, 45(4):1085–1098, 2009. (Cited on page 436.)

[420] M. Gradinaru, F. Russo, and P. Vallois. Generalized covariations, local time and Stratonovich Itô's formula for fractional Brownian motion with Hurst index $H \geq \frac{1}{4}$. *The Annals of Probability*, 31(4):1772–1820, 2003. (Cited on page 435.)

[421] I. S. Gradshteyn and I. M. Ryzhik. *Table of Integrals, Series, and Products*, 7th edition. Amsterdam: Elsevier/Academic Press, 2007. Translated from the Russian, Translation edited and with a preface

by Alan Jeffrey and Daniel Zwillinger, With one CD-ROM (Windows, Macintosh and UNIX). (Cited on pages 27, 30, 38, 50, 93, 98, 214, 295, 389, 489, 501, 533, and 550.)

[422] C. W. J. Granger. Long memory relationships and the aggregation of dynamic models. *Journal of Econometrics*, 14(2):227–238, 1980. (Cited on page 223.)

[423] C. W. J. Granger and N. Hyung. Occasional structural breaks and long memory with an application to the S&P 500 absolute stock returns. *Journal of Empirical Finance*, 11(3):399–421, 2004. (Cited on page 225.)

[424] C. W. J. Granger and R. Joyeux. An introduction to long-memory time series and fractional differencing. *Journal of Time Series Analysis*, 1:15–30, 1980. (Cited on page 109.)

[425] T. Graves, R. B. Gramacy, C. L. E. Franzke, and N. W. Watkins. Efficient Bayesian inference for ARFIMA processes. *Nonlinear Processes in Geophysics*, 22:679–700, 2015. (Cited on page 573.)

[426] T. Graves, R. B. Gramacy, N. W. Watkins, and C. L. E. Franzke. A brief history of long memory: Hurst, Mandelbrot and the road to ARFIMA. Preprint, 2016. (Cited on page 109.)

[427] H. L. Gray, N.-F. Zhang, and W. A. Woodward. On generalized fractional processes. *Journal of Time Series Analysis*, 10(3):233–257, 1989. (Cited on page 110.)

[428] H. L. Gray, N.-F. Zhang, and W. A. Woodward. A correction: "On generalized fractional processes" [J. Time Ser. Anal. **10** (1989), no. 3, 233–257. *Journal of Time Series Analysis*, 15(5):561–562, 1994. (Cited on page 110.)

[429] U. Grenander. *Abstract Inference*. New York: John Wiley & Sons, 1981. (Cited on pages 384 and 385.)

[430] U. Grenander and G. Szegö. *Toeplitz Forms and their Applications*. California Monographs in Mathematical Sciences. Berkeley: University of California Press, 1958. (Cited on page 337.)

[431] M. Grigoriu. Simulation of stationary non-Gaussian translation processes. *Journal of Engineering Mechanics*, 124(2):121–126, 1998. (Cited on page 344.)

[432] G. Gripenberg. White and colored Gaussian noises as limits of sums of random dilations and translations of a single function. *Electronic Communications in Probability*, 16:507–516, 2011. (Cited on page 465.)

[433] G. Gripenberg and I. Norros. On the prediction of fractional Brownian motion. *Journal of Applied Probability*, 33:400–410, 1996. (Cited on page 395.)

[434] S. D. Grose, G. M. Martin, and D. S. Poskitt. Bias correction of persistence measures in fractionally integrated models. *Journal of Time Series Analysis*, 36(5):721–740, 2015. (Cited on page 574.)

[435] P. Guasoni. No arbitrage under transaction costs, with fractional Brownian motion and beyond. *Mathematical Finance. An International Journal of Mathematics, Statistics and Financial Economics*, 16(3):569–582, 2006. (Cited on page 435.)

[436] J. A. Gubner. Theorems and fallacies in the theory of long-range-dependent processes. *IEEE Transactions on Information Theory*, 51(3):1234–1239, 2005. (Cited on pages 27 and 106.)

[437] D. Guégan. How can we define the concept of long memory? An econometric survey. *Econometric Reviews*, 24(2):113–149, 2005. (Cited on page 109.)

[438] C. A. Guerin, H. Nyberg, O. Perrin, S. Resnick, H. Rootzén, and C. Stărică. Empirical testing of the infinite source Poisson data traffic model. *Stochastic Models*, 19:56–199, 2003. (Cited on pages 121 and 224.)

[439] P. Guggenberger and Y. Sun. Bias-reduced log-periodogram and Whittle estimation of the long-memory parameter without variance inflation. *Econometric Theory*, 22(5):863–912, 2006. ISSN 0266-4666. (Cited on pages 111, 572, and 573.)

[440] H. Guo, C. Y. Lim, and M. M. Meerschaert. Local Whittle estimator for anisotropic random fields. *Journal of Multivariate Analysis*, 100(5):993–1028, 2009. (Cited on pages 529 and 537.)

[441] A. Gupta and S. Joshi. Some studies on the structure of covariance matrix of discrete-time fBm. *IEEE Transactions on Signal Processing*, 56(10, part 1):4635–4650, 2008. (Cited on page 464.)

[442] A. Gut. *Probability: A Graduate Course*. Springer Texts in Statistics. New York: Springer, 2005. (Cited on pages 300, 314, and 317.)

[443] B. Haas and G. Miermont. Self-similar scaling limits of non-increasing Markov chains. *Bernoulli*, 17(4):1217–1247, 2011. (Cited on page 612.)

[444] M. Hairer. Ergodicity of stochastic differential equations driven by fractional Brownian motion. *The Annals of Probability*, 33(2):03–758, 2005. (Cited on page 436.)

[445] M. Hairer. Solving the KPZ equation. *Annals of Mathematics. Second Series*, 178(2):559–664, 2013. (Cited on page 227.)

[446] M. Hairer and N. S. Pillai. Regularity of laws and ergodicity of hypoelliptic SDEs driven by rough paths. *The Annals of Probability*, 41(4):2544–2598, 2013. (Cited on page 436.)

[447] P. Hall, B.-Y. Jing, and S. N. Lahiri. On the sampling window method for long-range dependent data. *Statistica Sinica*, 8(4):1189–1204, 1998. (Cited on page 573.)

[448] K. H. Hamed. Improved finite-sample hurst exponent estimates using rescaled range analysis. *Water Resources Research*, 43(4), 2007. (Cited on page 111.)

[449] A. Hammond and S. Sheffield. Power law Pólya's urn and fractional Brownian motion. *Probability Theory and Related Fields*, 157(3-4):691–719, 2013. (Cited on pages 172, 173, 177, and 226.)

[450] E. J. Hannan. *Multiple Time Series*. John Wiley and Sons, Inc., 1970. (Cited on pages 470, 471, 478, and 479.)

[451] E. J. Hannan. The asymptotic theory of linear time series models. *Journal of Applied Probability*, 10:130–145, 1973. (Cited on page 570.)

[452] E. J. Hannan and B. G. Quinn. The determination of the order of an autoregression. *Journal of the Royal Statistical Society. Series B. Methodological*, 41(2):190–195, 1979. (Cited on page 552.)

[453] T. Hara. Decay of correlations in nearest-neighbor self-avoiding walk, percolation, lattice trees and animals. *The Annals of Probability*, 36(2):530–593, 2008. (Cited on page 227.)

[454] G. H. Hardy. Weierstrass's non-differentiable function. *Transactions of the American Mathematical Society*, 17:322–323, 1916. (Cited on page 216.)

[455] T. E. Harris. Diffusion with "collisions" between particles. *Journal of Applied Probability*, 2:323–338, 1965. (Cited on pages 163 and 225.)

[456] T. E. Harris. A correlation inequality for Markov processes in partially ordered state spaces. *The Annals of Probability*, 5(3):451–454, 1977. (Cited on page 172.)

[457] A. C. Harvey. 16 - long memory in stochastic volatility. In J. Knight and S. Satchell, editors, *Forecasting Volatility in the Financial Markets (Third Edition)*, Quantitative Finance, pages 351–363. Oxford: Butterworth-Heinemann, 2007. (Cited on page 572.)

[458] J. Haslett and A. E. Raftery. Space-time modelling with long-memory dependence: assessing Ireland's wind power resource. *Applied Statistics*, 38:1–50, 1989. Includes discussion. (Cited on page 572.)

[459] U. Hassler and M. Olivares. Semiparametric inference and bandwidth choice under long memory: experimental evidence. *İstatistik. Journal of the Turkish Statistical Association*, 6(1):27–41, 2013. (Cited on page 572.)

[460] U. Hassler, P. M. M. Rodrigues, and A. Rubia. Testing for general fractional integration in the time domain. *Econometric Theory*, 25(6):1793–1828, 2009. (Cited on page 574.)

[461] M. A. Hauser. Maximum likelihood estimators for ARMA and ARFIMA models: a Monte Carlo study. *Journal of Statistical Planning and Inference*, 80(1-2):229–255, 1999. (Cited on pages 550 and 571.)

[462] D. Heath, S. Resnick, and G. Samorodnitsky. Heavy tails and long range dependence in on/off processes and associated fluid models. *Mathematics of Operations Research*, 23:146–165, 1998. (Cited on page 224.)

[463] H. Helgason, V. Pipiras, and P. Abry. Fast and exact synthesis of stationary multivariate Gaussian time series using circulant embedding. *Signal Processing*, 91(5):1123–1133, 2011. (Cited on pages 112 and 470.)

[464] H. Helgason, V. Pipiras, and P. Abry. Synthesis of multivariate stationary series with prescribed marginal distributions and covariance using circulant matrix embedding. *Signal Processing*, 91:1741–1758, 2011. (Cited on pages 339 and 344.)

[465] H. Helgason, V. Pipiras, and P. Abry. Fast and exact synthesis of stationary multivariate Gaussian time series using circulant embedding. *Signal Processing*, 91(5):1123–1133, 2011. (Cited on page 486.)

[466] H. Helgason, V. Pipiras, and P. Abry. Smoothing windows for the synthesis of Gaussian stationary random fields using circulant matrix embedding. *Journal of Computational and Graphical Statistics*, 23(3):616–635, 2014. (Cited on page 112.)

[467] P. Henrici. *Applied and Computational Complex Analysis*. New York: Wiley-Interscience [John Wiley & Sons], 1974. Volume 1: Power series—integration—conformal mapping—location of zeros, Pure and Applied Mathematics. (Cited on page 340.)

[468] M. Henry. Robust automatic bandwidth for long memory. *Journal of Time Series Analysis*, 22(3):293–316, 2001. (Cited on pages 565 and 572.)

[469] M. Henry. Bandwidth choice, optimal rates and adaptivity in semiparametric estimation of long memory. In *Long Memory in Economics*, pages 157–172. Berlin: Springer, 2007. (Cited on page 572.)

[470] M. Henry and P. M. Robinson. Bandwidth choice in Gaussian semiparametric estimation of long range dependence. In P. M. Robinson and M. Rosenblatt, editors, *Athens Conference on Applied Probability and Time Series Analysis. Volume II: Time Series Analysis in Memory of E. J. Hannan*, pages 220–232, New York: Springer-Verlag, 1996. Lecture Notes in Statistics, **115**. (Cited on pages 565 and 572.)

[471] E. Herbin and E. Merzbach. Stationarity and self-similarity characterization of the set-indexed fractional Brownian motion. *Journal of Theoretical Probability*, 22(4):1010–1029, 2009. (Cited on page 537.)

[472] C. C. Heyde. On modes of long-range dependence. *Journal of Applied Probability*, 39(4): 882–888, 2002. (Cited on page 109.)

[473] C. C. Heyde and R. Gay. Smoothed periodogram asymptotics and estimation for processes and fields with possible long-range dependence. *Stochastic Processes and their Applications*, 45:169–182, 1993. (Cited on page 571.)

[474] J. Hidalgo and Y. Yajima. Semiparametric estimation of the long-range parameter. *Annals of the Institute of Statistical Mathematics*, 55(4):705–736, 2003. (Cited on page 572.)

[475] N. J. Higham. *Functions of Matrices*. Philadelphia, PA: Society for Industrial and Applied Mathematics (SIAM), 2008. Theory and computation. (Cited on page 480.)

[476] H. C. Ho and T. Hsing. On the asymptotic expansion of the empirical process of long memory moving averages. *The Annals of Statistics*, 24:992–1024, 1996. (Cited on page 344.)

[477] H. C. Ho and T. Hsing. Limit theorems for functionals of moving averages. *The Annals of Probability*, 25:1636–1669, 1997. (Cited on pages 313, 315, and 344.)

[478] H. C. Ho and T. C. Sun. A central limit theorem for non-instantaneous filters of a stationary Gaussian process. *Journal of Multivariate Analysis*, 22:144–155, 1987. (Cited on pages 343 and 344.)

[479] N. Hohn, D. Veitch, and P. Abry. Cluster processes: a natural language for network traffic. *IEEE Transactions on Signal Processing*, 51 (8):2229–2244, 2003. (Cited on page 224.)

[480] S. Holan, T. McElroy, and So. Chakraborty. A Bayesian approach to estimating the long memory parameter. *Bayesian Analysis*, 4(1):159–190, 2009. (Cited on page 573.)

[481] R. Holley and D. W. Stroock. Central limit phenomena of various interacting systems. *Annals of Mathematics. Second Series*, 110(2):333–393, 1979. (Cited on page 226.)

[482] R. Horn and C. Johnson. *Topics in Matrix Analysis*. New York, NY: Cambridge University Press, 1991. (Cited on pages 467, 468, and 480.)

[483] L. Horváth and Q.-M. Shao. Limit theorems for quadratic forms with applications to Whittle's estimate. *The Annals of Applied Probability*, 9(1):146–187, 1999. (Cited on page 571.)

[484] J. R. M. Hosking. Fractional differencing. *Biometrika*, 68(1):165–176, 1981. (Cited on pages 43 and 109.)

[485] J. R. M. Hosking. Modeling persistence in hydrological time series using fractional differencing. *Water Resources Research*, 20:1898–1908, 1984. (Cited on pages 556 and 572.)

[486] J. R. M. Hosking. Asymptotic distributions of the sample mean, autocovariances, and autocorrelations of long-memory time series. *Journal of Econometrics*, 73(1):261–284, 1996. (Cited on page 574.)

[487] Y. Hosoya. The quasi-likelihood approach to statistical inference on multiple time-series with long-range dependence. *Journal of Econometrics*, 73:217–236, 1996. (Cited on page 571.)

[488] C. Houdré and V. Pérez-Abreu, editors. *Chaos expansions, multiple Wiener-Itô integrals and their applications*. Probability and Stochastics Series. Boca Raton, FL: CRC Press, 1994. Papers from the workshop held in Guanajuato, July 27–31, 1992. (Cited on page 610.)

[489] C. Houdré and J. Villa. An example of infinite dimensional quasi-helix. In *Stochastic models (Mexico City, 2002)*, volume 336 of *Contemp. Math.*, pages 195–201. Providence, RI: Amer. Math. Soc., 2003. (Cited on pages 54, 55, and 110.)

[490] T. Hsing. Linear processes, long-range dependence and asymptotic expansions. *Statistical Inference for Stochastic Processes*, 3 (1-2): 19–29, 2000. 19th "Rencontres Franco-Belges de Statisticiens" (Marseille, 1998). (Cited on page 344.)

[491] Y. Hu. Integral transformations and anticipative calculus for fractional Brownian motions. *Memoirs of the American Mathematical Society*, 175(825), 2005. (Cited on page 436.)

[492] Y. Hu and B. Øksendal. Chaos expansion of local time of fractional Brownian motions. *Stochastic Analysis and Applications*, 20(4):815–837, 2002. (Cited on pages 432 and 436.)

[493] Y. Hu and B. Øksendal. Fractional white noise calculus and applications to finance. *Infinite Dimensional Analysis, Quantum Probability and Related Topics*, 6(1):1–32, 2003. (Cited on page 435.)

[494] Y. Hu, J. Huang, D. Nualart, and S. Tindel. Stochastic heat equations with general multiplicative Gaussian noises: Hölder continuity and intermittency. *Electronic Journal of Probability*, 20:no. 55, 50 pp. (electronic), 2003. (Cited on page 227.)

[495] Y. Hu, Y. Liu, and D. Nualart. Rate of convergence and asymptotic error distribution of Euler approximation schemes for fractional diffusions. *The Annals of Applied Probability*, 26(2):1147–1207, 2016. (Cited on pages 418, 419, and 424.)

[496] Y. Hu, D. Nualart, S. Tindel, and F. Xu. Density convergence in the Breuer-Major theorem for Gaussian stationary sequences. *Bernoulli*, 21(4):2336–2350, 2015. (Cited on page 343.)

[497] J. Hualde and P. M. Robinson. Semiparametric inference in multivariate fractionally cointegrated systems. *Journal of Econometrics*, 157(2):492–511, 2010. (Cited on page 537.)

[498] J. Hualde and P. M. Robinson. Gaussian pseudo-maximum likelihood estimation of fractional time series models. *The Annals of Statistics*, 39(6):3152–3181, 2011. (Cited on page 551.)

[499] W. N. Hudson and J. D. Mason. Operator-self-similar processes in a finite-dimensional space. *Transactions of the American Mathematical Society*, 273(1):281–297, 1982. (Cited on pages 493 and 536.)

[500] Y. Huh, J.S. Kim, S. H. Kim, and M.W. Suh. Characterizing yarn thickness variation by correlograms. *Fibers and Polymers*, 6(1):66–71, 2005. (Cited on page 225.)

[501] H. E. Hurst. Long-term storage capacity of reservoirs. *Transactions of the American Society of Civil Engineers*, 116:770–779, 1951. (Cited on pages 84, 108, and 111.)

[502] C. M. Hurvich. Model selection for broadband semiparametric estimation of long memory in time series. *Journal of Time Series Analysis*, 22(6):679–709, 2001. (Cited on page 573.)

[503] C. M. Hurvich. Multistep forecasting of long memory series using fractional exponential models. *International Journal of Forecasting*, 18(2):167–179, 2002. (Cited on page 571.)

[504] C. M. Hurvich and K. I. Beltrao. Automatic semiparametric estimation of the memory parameter of a long-memory time series. *Journal of Time Series Analysis*, 15:285–302, 1994. (Cited on page 573.)

[505] C. M. Hurvich and J. Brodsky. Broadband semiparametric estimation of the memory parameter of a long-memory time series using fractional exponential models. *Journal of Time Series Analysis*, 22(2):221–249, 2001. (Cited on page 573.)

[506] C. M. Hurvich and W. W. Chen. An efficient taper for potentially overdifferenced long-memory time series. *Journal of Time Series Analysis*, 21(2):155–180, 2000. (Cited on page 566.)

[507] C. M. Hurvich and R. S. Deo. Plug-in selection of the number of frequencies in regression estimates of the memory parameter of a long-memory time series. *Journal of Time Series Analysis*, 20:331–341, 1999. (Cited on pages 111 and 573.)

[508] C. M. Hurvich and B. K. Ray. Estimation of the memory parameter for nonstationary or noninvertible fractionally integrated processes. *Journal of Time Series Analysis*, 16(1):17–41, 1995. (Cited on page 573.)

[509] C. M. Hurvich and B. K. Ray. The local Whittle estimator of long-memory stochastic volatility. *Journal of Financial Econometrics*, 1(3):445–470, 2003. (Cited on page 611.)

[510] C. M. Hurvich, R. Deo, and J. Brodsky. The mean squared error of Geweke and Porter-Hudak's estimator of the memory parameter of a long-memory time series. *Journal of Time Series Analysis*, 19:19–46, 1998. (Cited on pages 111 and 573.)

[511] C. M. Hurvich, E. Moulines, and P. Soulier. The FEXP estimator for potentially non-stationary linear time series. *Stochastic Processes and their Applications*, 97(2):307–340, 2002. (Cited on page 573.)

[512] C. M. Hurvich, E. Moulines, and P. Soulier. Estimating long memory in volatility. *Econometrica*, 73(4):1283–1328, 2005. (Cited on page 611.)

[513] F. Iacone. Local Whittle estimation of the memory parameter in presence of deterministic components. *Journal of Time Series Analysis*, 31(1):37–49, 2010. (Cited on pages 225 and 573.)

[514] I. A. Ibragimov. An estimate for the spectral function of a stationary Gaussian process. *Teor. Verojatnost. i Primenen*, 8:391–430, 1963. (Cited on page 343.)

[515] I. A. Ibragimov and Yu. V. Linnik. *Independent and Stationary Sequences of Random Variables*. The Netherlands: Wolters-Nordhoff, 1971. (Cited on pages 583 and 586.)

[516] SAS Institute Inc. *SAS/IML 9.2 Users Guide*. Cary, NC: SAS Institute Inc., 2008. (Cited on page 572.)

[517] C.-K. Ing, H.-T. Chiou, and M. Guo. Estimation of inverse autocovariance matrices for long memory processes. *Bernoulli*, 22(3):1301–1330, 2016. (Cited on page 574.)

[518] A. Inoue. Abel-Tauber theorems for Fourier-Stieltjes coefficients. *Journal of Mathematical Analysis and Applications*, 211(2):460–480, 1997. (Cited on page 109.)

[519] A. Inoue. Asymptotic behavior for partial autocorrelation functions of fractional ARIMA processes. *The Annals of Applied Probability*, 12(4):1471–1491, 2002. (Cited on page 555.)

[520] A. Inoue. AR and MA representation of partial autocorrelation functions, with applications. *Probability Theory and Related Fields*, 140(3-4):523–551, 2008. (Cited on page 109.)

[521] A. Inoue and Y. Kasahara. Explicit representation of finite predictor coefficients and its applications. *The Annals of Statistics*, 34(2):973–993, 2006. (Cited on page 571.)

[522] A. Inoue, Y. Nakano, and V. Anh. Linear filtering of systems with memory and application to finance. *Journal of Applied Mathematics and Stochastic Analysis. JAMSA*, pages Art. ID 53104, 26, 2006. (Cited on page 395.)

[523] A. Iouditsky, E. Moulines, and P. Soulier. Adaptive estimation of the fractional differencing coefficient. *Bernoulli*, 7(5):699–731, 2001. (Cited on page 573.)

[524] K. Itô. Multiple Wiener integral. *Journal of the Mathematical Society of Japan*, 3:157–169, 1951. (Cited on page 280.)

[525] A. V. Ivanov and N. N. Leonenko. *Statistical Analysis of Random Fields*. Dordrecht/Boston/London: Kluwer Academic Publishers, 1989. Translated from the Russian, 1986 edition. (Cited on page 344.)

[526] A. Jach and P. Kokoszka. Wavelet-domain test for long-range dependence in the presence of a trend. *Statistics. A Journal of Theoretical and Applied Statistics*, 42(2):101–113, 2008. (Cited on page 225.)

[527] S. Jaffard, B. Lashermes, and P. Abry. Wavelet leaders in multifractal analysis. In *Wavelet analysis and applications*, Appl. Numer. Harmon. Anal., pages 201–246. Basel: Birkhäuser, 2007. (Cited on page 612.)

[528] S. Jaffard, C. Melot, R. Leonarduzzi, H. Wendt, P. Abry, S. G. Roux, and M. E. Torres. p-exponent and p-leaders, part i: Negative pointwise regularity. *Physica A: Statistical Mechanics and its Applications*, 448:300–318, 2016. (Cited on page 612.)

[529] S. Janson. *Gaussian Hilbert Spaces*, volume 129 of *Cambridge Tracts in Mathematics*. Cambridge: Cambridge University Press, 1997. (Cited on pages 276, 281, and 610.)

[530] M. D. Jara. Nonequilibrium scaling limit for a tagged particle in the simple exclusion process with long jumps. *Communications on Pure and Applied Mathematics*, 62(2):198–214, 2009. (Cited on page 226.)

[531] M. D. Jara and T. Komorowski. Limit theorems for some continuous-time random walks. *Advances in Applied Probability*, 43(3):782–813, 2011. (Cited on page 228.)

[532] M. D. Jara and C. Landim. Nonequilibrium central limit theorem for a tagged particle in symmetric simple exclusion. *Annales de l'Institut Henri Poincaré. Probabilités et Statistiques*, 42(5):567–577, 2006. (Cited on page 226.)

[533] M. D. Jara, C. Landim, and S. Sethuraman. Nonequilibrium fluctuations for a tagged particle in mean-zero one-dimensional zero-range processes. *Probability Theory and Related Fields*, 145(3-4):565–590, 2009. (Cited on page 226.)

[534] S. Johansen and M. Ø. Nielsen. Likelihood inference for a nonstationary fractional autoregressive model. *Journal of Econometrics*, 158(1):51–66, 2010. (Cited on page 537.)

[535] S. Johansen and M. Ø. Nielsen. Likelihood inference for a fractionally cointegrated vector autoregressive model. *Econometrica*, 80(6):2667–2732, 2012. (Cited on page 537.)

[536] M. Jolis and N. Viles. Continuity in the Hurst parameter of the law of the Wiener integral with respect to the fractional Brownian motion. *Statistics & Probability Letters*, 80(7-8):566–572, 2010. (Cited on page 395.)

[537] C. Jost. On the connection between Molchan-Golosov and Mandelbrot-Van Ness representations of fractional Brownian motion. *Journal of Integral Equations and Applications*, 20(1):93–119, 2008. (Cited on page 395.)

[538] G. Jumarie. Stochastic differential equations with fractional Brownian motion input. *International Journal of Systems Science*, 24(6):1113–1131, 1993. (Cited on page 435.)

[539] P. Jung and G. Markowsky. Random walks at random times: convergence to iterated Lévy motion, fractional stable motions, and other self-similar processes. *The Annals of Probability*, 41(4):2682–2708, 2013. (Cited on page 226.)

[540] P. Jung and G. Markowsky. Hölder continuity and occupation-time formulas for fBm self-intersection local time and its derivative. *Journal of Theoretical Probability*, 28(1):299–312, 2015. (Cited on page 436.)

[541] P. Jung, T. Owada, and G. Samorodnitsky. Functional central limit theorem for negatively dependent heavy-tailed stationary infinitely divisible processes generated by conservative flows. Preprint, 2015. (Cited on page 228.)

[542] Z. J. Jurek and J. D. Mason. *Operator-Limit Distributions in Probability Theory*. Wiley Series in Probability and Mathematical Statistics: Probability and Mathematical Statistics. New York: John Wiley & Sons Inc., 1993. A Wiley-Interscience Publication. (Cited on pages 494 and 535.)

[543] T. Kaarakka and P. Salminen. On fractional Ornstein-Uhlenbeck processes. *Communications on Stochastic Analysis*, 5(1):121–133, 2011. (Cited on page 395.)

[544] J.-P. Kahane and J. Peyrière. Sur certaines martingales de Benoit Mandelbrot. *Advances in Mathematics*, 22(2):131–145, 1976. (Cited on page 612.)

[545] I. Kaj. *Stochastic Modeling in Broadband Communications Systems*. SIAM Monographs on Mathematical Modeling and Computation. Philadelphia, PA: Society for Industrial and Applied Mathematics (SIAM), 2002. (Cited on page 224.)

[546] I. Kaj and M. S. Taqqu. Convergence to fractional Brownian motion and to the Telecom process: the integral representation approach. In *In and out of equilibrium. 2*, volume 60 of *Progr. Probab.*, pages 383–427. Basel: Birkhäuser, 2008. (Cited on pages 149 and 224.)

[547] I. Kaj, L. Leskelä, I. Norros, and V. Schmidt. Scaling limits for random fields with long-range dependence. *The Annals of Probability*, 35(2):528–550, 2007. (Cited on page 224.)

[548] A. Kamont. On the fractional anisotropic Wiener field. *Probability and Mathematical Statistics*, 16(1):85–98, 1996. (Cited on page 537.)

[549] I. Karatzas and S. E. Shreve. *Brownian Motion and Stochastic Calculus*, 2nd edition, volume 113 of *Graduate Texts in Mathematics*. New York: Springer-Verlag, 1991. (Cited on pages 14 and 452.)

[550] S. Kechagias and V. Pipiras. Identification, estimation and applications of a bivariate long-range dependent time series model with general phase. Preprint, 2015. (Cited on pages 502 and 536.)

[551] S. Kechagias and V. Pipiras. Definitions and representations of multivariate long-range dependent time series. *Journal of Time Series Analysis*, 36(1):1–25, 2015. (Cited on pages 470, 502, 535, and 536.)

[552] M. Kendall and A. Stuart. *The Advanced Theory of Statistics. Vol. 1*, 4th edition. New York: Macmillan Publishing Co., Inc., 1977. Distribution Theory. (Cited on page 332.)

[553] H. Kesten and F. Spitzer. A limit theorem related to a new class of self-similar processes. *Zeitschrift für Wahrscheinlichkeitstheorie und verwandte Gebiete*, 50:5–25, 1979. (Cited on pages 177 and 226.)

[554] H. Kesten, M. V. Kozlov, and F. Spitzer. A limit law for random walk in a random environment. *Compositio Mathematica*, 30:145–168, 1975. (Cited on page 178.)

[555] C. S. Kim and P. C. Phillips. Log periodogram regression: the nonstationary case, 2006. Cowles Foundation Discussion Paper. (Cited on page 111.)

[556] Y. M. Kim and D. J. Nordman. Properties of a block bootstrap under long-range dependence. *Sankhya A. Mathematical Statistics and Probability*, 73(1):79–109, 2011. (Cited on page 574.)

[557] Y. M. Kim and D. J. Nordman. A frequency domain bootstrap for Whittle estimation under long-range dependence. *J. Multivariate Anal.*, 115:405–420, 2013. (Cited on page 574.)

[558] J. F. C. Kingman. *Poisson Processes*, volume 3 of *Oxford Studies in Probability*. New York: The Clarendon Press Oxford University Press, 1993. Oxford Science Publications. (Cited on pages 125, 127, 593, and 594.)

[559] C. Kipnis. Central limit theorems for infinite series of queues and applications to simple exclusion. *The Annals of Probability*, 14(2):397–408, 1986. (Cited on page 226.)

[560] C. Kipnis and S. R. S. Varadhan. Central limit theorem for additive functionals of reversible Markov processes and applications to simple exclusions. *Communications in Mathematical Physics*, 104(1):1–19, 1986. (Cited on page 226.)

[561] J. Klafler and I. M. Sokolov. Anomalous diffusion spreads its wings. *Physics World*, 18 (8): 29–32, 2005. (Cited on page 227.)

[562] F. C. Klebaner. *Introduction to Stochastic Calculus with Applications*, 2nd edition. London: Imperial College Press, 2005. (Cited on page 382.)

[563] V Klemeš. The Hurst pheomenon: a puzzle? *Water Resources Research*, 10(4):675–688, 1974. (Cited on pages 111 and 225.)

[564] M. L. Kleptsyna and A. Le Breton. Extension of the Kalman-Bucy filter to elementary linear systems with fractional Brownian noises. *Statistical Inference for Stochastic Processes*, 5(3):249–271, 2002. (Cited on page 395.)

[565] M. L. Kleptsyna, P. E. Kloeden, and V. V. Anh. Linear filtering with fractional Brownian motion. *Stochastic Analysis and Applications*, 16(5):907–914, 1998. (Cited on page 395.)

[566] M. L. Kleptsyna, A. Le Breton, and M.-C. Roubaud. Parameter estimation and optimal filtering for fractional type stochastic systems. *Statistical Inference for Stochastic Processes*, 3(1-2):173–182, 2000. 19th "Rencontres Franco-Belges de Statisticiens" (Marseille, 1998). (Cited on page 395.)

[567] F. Klingenhöfer and M. Zähle. Ordinary differential equations with fractal noise. *Proceedings of the American Mathematical Society*, 127(4):1021–1028, 1999. (Cited on page 414.)

[568] C. Klüppelberg and C. Kühn. Fractional Brownian motion as a weak limit of Poisson shot noise processes—with applications to finance. *Stochastic Processes and their Applications*, 113(2):333–351, 2004. (Cited on page 225.)

[569] C. Klüppelberg and T. Mikosch. Explosive Poisson shot noise processes with applications to risk reserves. *Bernoulli*, 1(1-2):125–147, 1995. (Cited on page 225.)

[570] C. Klüppelberg, T. Mikosch, and A. Schärf. Regular variation in the mean and stable limits for poisson shot noise. *Bernoulli*, 9(3):467–496, 2003. (Cited on page 225.)

[571] P. S. Kokoszka and M. S. Taqqu. Fractional ARIMA with stable innovations. *Stochastic Processes and their Applications*, 60:19–47, 1995. (Cited on pages 82, 83, 107, 111, and 573.)

[572] P. S. Kokoszka and M. S. Taqqu. Infinite variance stable moving averages with long memory. *Journal of Econometrics*, 73:79–99, 1996. (Cited on pages 82 and 111.)

[573] P. S. Kokoszka and M. S. Taqqu. Parameter estimation for infinite variance fractional ARIMA. *The Annals of Statistics*, 24:1880–1913, 1996. (Cited on pages 82, 111, and 573.)

[574] P. S. Kokoszka and M. S. Taqqu. Can one use the Durbin-Levinson algorithm to generate infinite variance fractional ARIMA time series. *Journal of Time Series Analysis*, 22(3):317–337, 2001. (Cited on page 107.)

[575] A. N. Kolmogorov. Wienersche Spiralen und einige andere interessante Kurven im Hilbertschen Raum. *Comptes Rendus (Doklady) de l'Académie des Sciences de l'URSS (N.S.)*, 26:115–118, 1940. (Cited on page 110.)

[576] T. Konstantopoulos and S.-J. Lin. Macroscopic models for long-range dependent network traffic. *Queueing Systems*, 28:215–243, 1998. (Cited on page 224.)

[577] S. Kotz, N. Balakrishnan, and N. L. Johnson. *Continuous Multivariate Distributions. Vol. 1*, 2nd edition. New York: Wiley Series in Probability and Statistics: Applied Probability and Statistics. Wiley-Interscience, 2000. Models and applications. (Cited on page 332.)

[578] D. Koutsoyiannis. Uncertainty, entropy, scaling and hydrological statistics. 1. Marginal distributional properties of hydrological processes and state scaling/Incertitude, entropie, effet d'échelle et propriétés stochastiques hydrologiques. 1. Propriétés distributionnelles marginales des processus hydrologiques et échelle d'état. *Hydrological Sciences Journal*, 50(3):381–404, 2005. (Cited on page 228.)

[579] L. Kristoufek. Testing power-law cross-correlations: rescaled covariance test. *The European Physical Journal B. Condensed Matter and Complex Systems*, 86(10):86:418, 11, 2013. (Cited on page 537.)

[580] K. Kubilius and V. Skorniakov. On some estimators of the Hurst index of the solution of SDE driven by a fractional Brownian motion. *Statistics & Probability Letters*, 109:159–167, 2016. (Cited on page 436.)

[581] R. Kulik and P. Soulier. Limit theorems for long-memory stochastic volatility models with infinite variance: partial sums and sample covariances. *Advances in Applied Probability*, 44(4):1113–1141, 2012. (Cited on page 111.)

[582] H. Künsch. Statistical aspects of self-similar processes. *Proceedings of the First World Congress of the Bernoulli Society*, 1:67–74, 1987. (Cited on page 572.)

[583] M. Kunze. An Introduction to Malliavin Calculus. Lecture Notes, 2013. (Cited on page 603.)

[584] M. Kuronen and L. Leskelä. Hard-core thinnings of germ-grain models with power-law grain sizes. *Advances in Applied Probability*, 45(3):595–625, 2013. (Cited on page 224.)

[585] T. G. Kurtz. Limit theorems for workload input models. In F. P. Kelly, S. Zachary, and I. Ziedins, editors, *Stochastic Networks: Theory and Applications*, pages 339–366. Oxford: Clarendon Press, 1996. (Cited on page 224.)

[586] S. Kwapień and N. A. Woyczyński. *Random Series and Stochastic Integrals: Single and Multiple*. Boston: Birkhäuser, 1992. (Cited on page 463.)

[587] R. G. Laha and V. K. Rohatgi. Operator self-similar stochastic processes in $\mathbf{R}_d$. *Stochastic Processes and their Applications*, 12(1):73–84, 1982. (Cited on page 536.)

[588] S. N. Lahiri and P. M. Robinson. Central limit theorems for long range dependent spatial linear processes. *Bernoulli*, 22(1):345–375, 2016. (Cited on page 344.)

[589] J. Lamperti. Semi-stable stochastic processes. *Transactions of the American Mathematical Society*, 104:62–78, 1962. (Cited on pages 66, 68, 69, 110, and 111.)

[590] J. Lamperti. Semi-stable Markov processes. I. *Zeitschrift für Wahrscheinlichkeitstheorie und verwandte Gebiete*, 22:205–225, 1972. (Cited on page 612.)

[591] J. A. Lane. The central limit theorem for the Poisson shot-noise process. *Journal of Applied Probability*, 21(2):287–301, 1984. (Cited on page 225.)

[592] K. Lange. *Optimization*. Springer Texts in Statistics. New York: Springer-Verlag, 2004. (Cited on page 540.)

[593] N. Lanjri Zadi and D. Nualart. Smoothness of the law of the supremum of the fractional Brownian motion. *Electronic Communications in Probability*, 8:102–111 (electronic), 2003. (Cited on pages 434 and 436.)

[594] K. Łasak and C. Velasco. Fractional cointegration rank estimation. *Journal of Business & Economic Statistics*, 33(2):241–254, 2015. (Cited on page 537.)

[595] F. Lavancier. Long memory random fields. In *Dependence in Probability and Statistics*, volume 187 of *Lecture Notes in Statist.*, pages 195–220. New York: Springer, 2006. (Cited on pages 524, 529, and 537.)

[596] F. Lavancier. Invariance principles for non-isotropic long memory random fields. *Statistical Inference for Stochastic Processes*, 10(3):255–282, 2007. (Cited on pages 529 and 537.)

[597] F. Lavancier, A. Philippe, and D. Surgailis. Covariance function of vector self-similar processes. *Statistics & Probability Letters*, 79(23):2415–2421, 2009. (Cited on pages 491 and 536.)

[598] G. F. Lawler. *Conformally Invariant Processes in the Plane*, volume 114 of *Mathematical Surveys and Monographs*. Providence, RI: American Mathematical Society, 2005. (Cited on page 227.)

[599] A. J. Lawrence and N. T. Kottegoda. Stochastic modelling of riverflow time series. *Journal of the Royal Statistical Society*, A 140(1):1–47, 1977. (Cited on page 111.)

[600] A. Le Breton. Filtering and parameter estimation in a simple linear system driven by a fractional Brownian motion. *Statistics & Probability Letters*, 38(3):263–274, 1998. (Cited on page 395.)

[601] N. N. Lebedev. *Special Functions and their Applications*. Dover Publications Inc., New York, 1972. Revised edition, translated from the Russian and edited by Richard A. Silverman, Unabridged and corrected republication. (Cited on pages 266, 280, and 363.)

[602] J. Lebovits and J. L. Véhel. White noise-based stochastic calculus with respect to multifractional Brownian motion. *Stochastics*, 86(1):87–124, 2014. (Cited on page 611.)

[603] P. Lei and D. Nualart. A decomposition of the bifractional Brownian motion and some applications. *Statistics & Probability Letters*, 79(5):619–624, 2009. (Cited on pages 55, 56, and 110.)

[604] R. Leipus and D. Surgailis. Random coefficient autoregression, regime switching and long memory. *Advances in Applied Probability*, 35(3):737–754, 2003. (Cited on page 223.)

[605] R. Leipus, G. Oppenheim, A. Philippe, and M.-C. Viano. Orthogonal series density estimation in a disaggregation scheme. *Journal of Statistical Planning and Inference*, 136(8):2547–2571, 2006. (Cited on page 223.)

[606] R. Leipus, A. Philippe, D. Puplinskaitė, and D. Surgailis. Aggregation and long memory: recent developments. *Journal of the Indian Statistical Association*, 52(1):81–111, 2014. (Cited on page 224.)

[607] W. E. Leland, M. S. Taqqu, W. Willinger, and D. V. Wilson. On the self-similar nature of Ethernet traffic (Extended version). *IEEE/ACM Transactions on Networking*, 2:1–15, 1994. (Cited on page 224.)

[608] J. A. León and S. Tindel. Malliavin calculus for fractional delay equations. *Journal of Theoretical Probability*, 25(3):854–889, 2012. (Cited on page 436.)

[609] R. Leonarduzzi, H. Wendt, P. Abry, S. Jaffard, C. Melot, S. G. Roux, and M. E. Torres. p-exponent and p-leaders, part ii: Multifractal analysis. Relations to detrended fluctuation analysis. *Physica A: Statistical Mechanics and its Applications*, 448:319–339, 2016. (Cited on page 612.)

[610] N. Leonenko and A. Olenko. Tauberian and Abelian theorems for long-range dependent random fields. *Methodology and Computing in Applied Probability*, 15(4):715–742, 2013. (Cited on pages 526, 529, and 537.)

[611] N. Leonenko and E. Taufer. Convergence of integrated superpositions of Ornstein-Uhlenbeck processes to fractional Brownian motion. *Stochastics. An International Journal of Probability and Stochastic Processes*, 77(6):477–499, 2005. (Cited on page 224.)

[612] N. N. Leonenko. *Limit Theorems for Random Fields with Singular Spectrum*. Kluwer, 1999. (Cited on pages 344, 537, and 610.)

[613] N. N. Leonenko and A. Ya. Olenko. Tauberian theorems for correlation functions and limit theorems for spherical averages of random fields. *Random Operators and Stochastic Equations*, 1(1):58–68, 1992. (Cited on page 344.)

[614] N. N. Leonenko and V. N. Parkhomenko. Central limit theorem for non-linear transforms of vector Gaussian fields. *Ukrainian Mathematical Journal*, 42(8):1057–1063, 1990. (Cited on page 344.)

[615] N. N. Leonenko, M. D. Ruiz-Medina, and M. S. Taqqu. Non-central limit theorems for random fields subordinated to gamma-correlated random fields. To appear in *Bernoulli*, 2017. (Cited on page 343.)

[616] N. N. Leonenko, M. D. Ruiz-Medina, and M. S. Taqqu. Rosenblatt distribution subordinated to Gaussian random fields with long-range dependence. *Stochastic Analysis and Applications*, 35(1):144–177, 2017. (Cited on page 343.)

[617] J. B. Levy and M. S. Taqqu. Renewal reward processes with heavy-tailed interrenewal times and heavy-tailed rewards. *Bernoulli*, 6(1):23–44, 2000. (Cited on page 224.)

[618] C. Lévy-Leduc and M. S. Taqqu. Long-range dependence and the rank of decompositions. In D. Carfi, M. L. Lapidus, E. P. J. Pearse, and M. van Frankenhuijsen, editors, *Fractal geometry and dynamical systems in pure and applied mathematics. II. Fractals in applied mathematics*, volume 601 of *Contemp. Math.*, pages 289–305. American Mathematical Society, 2013. (Cited on pages 343 and 344.)

[619] C. Lévy-Leduc, H. Boistard, E. Moulines, M. S. Taqqu, and V. A. Reisen. Large sample behaviour of some well-known robust estimators under long-range dependence. *Statistics. A Journal of Theoretical and Applied Statistics*, 45(1):59–71, 2011. (Cited on page 573.)

[620] C. Lévy-Leduc, H. Boistard, E. Moulines, M. S. Taqqu, and V. A. Reisen. Robust estimation of the scale and of the autocovariance function of Gaussian short- and long-range dependent processes. *Journal of Time Series Analysis*, 32(2):135–156, 2011. (Cited on page 574.)

[621] A. Lewbel. Aggregation and simple dynamics. *The American Economic Review*, 84(4):905–918, 1994. (Cited on page 224.)

[622] L. Li. Response to comments on "PCA based Hurst exponent estimator for fBm signals under disturbances". ArXiv preprint arXiv:0805.3002, 2010. (Cited on page 464.)

[623] L. Li, J. Hu, Y. Chen, and Y. Zhang. PCA based Hurst exponent estimator for fBm signals under disturbances. *IEEE Transactions on Signal Processing*, 57(7):2840–2846, 2009. (Cited on page 464.)

[624] W. Li, C. Yu, A. Carriquiry, and W. Kliemann. The asymptotic behavior of the R/S statistic for fractional Brownian motion. *Statistics & Probability Letters*, 81(1):83–91, 2011. (Cited on page 111.)

[625] Y. Li and Y. Xiao. Multivariate operator-self-similar random fields. *Stochastic Processes and their Applications*, 121(6):1178–1200, 2011. (Cited on page 538.)

[626] Y. Li, W. Wang, and Y. Xiao. Exact moduli of continuity for operator-scaling Gaussian random fields. *Bernoulli*, 21(2):930–956, 2015. (Cited on page 537.)

[627] O. Lieberman and P. C. B. Phillips. Expansions for the distribution of the maximum likelihood estimator of the fractional difference parameter. *Econometric Theory*, 20(3):464–484, 2004. (Cited on pages 571 and 572.)

[628] O. Lieberman and P. C. B. Phillips. Expansions for approximate maximum likelihood estimators of the fractional difference parameter. *The Econometrics Journal*, 8(3):367–379, 2005. (Cited on page 572.)

[629] O. Lieberman and P. C. B. Phillips. Refined inference on long memory in realized volatility. *Econometric Reviews*, 27(1-3) 254–267, 2008. (Cited on page 611.)

[630] O. Lieberman, J. Rousseau, and D. M. Zucker. Small-sample likelihood-based inference in the ARFIMA model. *Econometric Theory*, 16(2):231–248, 2000. (Cited on page 571.)

[631] O. Lieberman, R. Rosemarin, and J. Rousseau. Asymptotic theory for maximum likelihood estimation of the memory parameter in stationary Gaussian processes. *Econometric Theory*, 28(2):457–470, 2012. (Cited on pages 540, 546, and 571.)

[632] T. M. Liggett. *Interacting Particle Systems*. Classics in Mathematics. Berlin: Springer-Verlag, 2005. Reprint of the 1985 original. (Cited on pages 168 and 226.)

[633] C. Y. Lim, M. M. Meerschaert, and H.-P. Scheffler. Parameter estimation for operator scaling random fields. *Journal of Multivariate Analysis*, 123:172–183, 2014. (Cited on page 537.)

[634] S. J. Lin. Stochastic analysis of fractional Brownian motions. *Stochastics and Stochastics Reports*, 55(1-2):121–140, 1995. (Cited on page 435.)

[635] G. Lindgren. *Stationary Stochastic Processes*. Chapman & Hall/CRC Texts in Statistical Science Series. Boca Raton, FL: CRC Press, 2013. Theory and applications. (Cited on pages 14 and 398.)

[636] M. Linn and A. Amirdjanova. Representations of the optimal filter in the context of nonlinear filtering of random fields with fractional noise. *Stochastic Processes and their Applications*, 119(8):2481–2500, 2009. (Cited on page 436.)

[637] M. Lippi and P. Zaffaroni. Aggregation of simple linear dynamics: exact asymptotics results. Econometrics Discussion Paper 350, STICERD-LSE, 1998. (Cited on page 223.)

[638] Br. Liseo, D. Marinucci, and L. Petrella. Bayesian semiparametric inference on long-range dependence. *Biometrika*, 88(4):1089–1104, 2001. (Cited on page 573.)

[639] B. Liu and D. Munson, Jr. Generation of a random sequence having a jointly specified marginal distribution and autocovariance. *Acoustics, Speech and Signal Processing, IEEE Transactions on*, 30(6):973–983, Dec 1982. (Cited on page 344.)

[640] J. Livsey, R. Lund, S. Kechagias, and V. Pipiras. Multivariate count time series with flexible autocovariances. Preprint, 2016. (Cited on page 340.)

[641] I. N. Lobato. Consistency of the averaged cross-periodogram in long memory series. *Journal of Time Series Analysis*, 18(2):137–155, 1997. (Cited on page 536.)

[642] I. N. Lobato. A semiparametric two-step estimator in a multivariate long memory model. *Journal of Econometrics*, 90(1):129–153, 1999. (Cited on page 536.)

[643] I. N. Lobato and P. M. Robinson. A nonparametric test for I(0). *Review of Economic Studies*, 65(3):475–495, 1998. (Cited on page 536.)

[644] A. Lodhia, S. Sheffield, X. Sun, and S. S. Watson. Fractional Gaussian fields: A survey. *Probability Surveys*, 13:1–56, 2016. (Cited on page 537.)

[645] L. López-Oliveros and S. I. Resnick. Extremal dependence analysis of network sessions. *Extremes*, 14(1):1–28, 2011. (Cited on page 121.)

[646] S. B. Lowen. Efficient generation of fractional Brownian motion for simulation of infrared focal-plane array calibration drift. *Methodology and Computing in Applied Probability*, 1 (4): 445–456, 1999. (Cited on page 112.)

[647] S. B. Lowen and M. C. Teich. Power-law shot noise. *IEEE Transactions on Information Theory*, IT-36(6):1302–1318, 1990. (Cited on page 225.)

[648] S. B. Lowen and M. C. Teich. *Fractal-Based Point Processes*. Wiley Series in Probability and Statistics. Hoboken, NJ: Wiley-Interscience [John Wiley & Sons], 2005. (Cited on pages 225 and 610.)

[649] A. Luceño. A fast likelihood approximation for vector general linear processes with long series: application to fractional differencing. *Biometrika*, 83(3):603–614, 1996. (Cited on page 570.)

[650] C. Ludeña and M. Lavielle. The Whittle estimator for strongly dependent stationary Gaussian fields. *Scandinavian Journal of Statistics. Theory and Applications*, 26(3):433–450, 1999. (Cited on page 537.)

[651] R. B. Lund, S. H. Holan, and J. Livsey. Long memory discrete-valued time series. In R. A. Davis, S. H. Holan, R. Lund, and N. Ravishanker, editors, *Handbook of Discrete-Valued Time Series*, pages 447–458. CRC Press, 2015. (Cited on page 611.)

[652] H. Luschgy and G. Pagès. Functional quantization of Gaussian processes. *Journal of Functional Analysis*, 196(2):486–531, 2002. (Cited on page 465.)

[653] H. Luschgy and G. Pagès. High-resolution product quantization for Gaussian processes under sup-norm distortion. *Bernoulli*, 13(3):653–671, 2007. (Cited on page 465.)

[654] H. Lütkepohl. *New Introduction to Multiple Time Series Analysis*. Berlin: Springer-Verlag, 2005. (Cited on page 470.)

[655] T. Lyons and Z. Qian. *System Control and Rough Paths*. Oxford Mathematical Monographs. Oxford: Oxford University Press, 2002. Oxford Science Publications. (Cited on page 435.)

[656] M. Lysy and N. S. Pillai. Statistical inference for stochastic differential equations with memory. Preprint, 2013. (Cited on page 436.)

[657] C. Ma. Correlation models with long-range dependence. *Journal of Applied Probability*, 39(2):370–382, 2002. (Cited on page 109.)

[658] C. Ma. Vector random fields with long-range dependence. *Fractals*, 19(2):249–258, 2011. (Cited on page 538.)

[659] C. Maccone. Eigenfunction expansion for fractional Brownian motions. *Il Nuovo Cimento. B. Serie 11*, 61(2):229–248, 1981. ISSN 0369-4100. (Cited on page 465.)

[660] C. Maccone. On the fractional Brownian motions $B_{LH}(t)$ and on the process $B(t^{2H})$. *Lettere al Nuovo Cimento. Rivista Internazionale della Società Italiana di Fisica. Serie 2*, 36(2):33–34, 1983. (Cited on page 465.)

[661] M. Maejima. Some sojourn time problems for strongly dependent Gaussian processes. *Zeitschrift für Wahrscheinlichkeitstheorie und verwandte Gebiete*, 57:1–14, 1981. (Cited on page 343.)

[662] M. Maejima. Some limit theorems for sojourn times of strongly dependent Gaussian processes. *Zeitschrift für Wahrscheinlichkeitstheorie und verwandte Gebiete*, 60:359–380, 1982. (Cited on page 343.)

[663] M. Maejima. Sojourns of multidimensional Gaussian processes with dependent components. *Yokohama Math. J.*, 33:121–130, 1985. (Cited on page 344.)

[664] M. Maejima. A remark on self-similar processes with stationary increments. *The Canadian Journal of Statistics*, 14(1):81–82, 1986. (Cited on pages 108 and 110.)

[665] M. Maejima. Some sojourn time problems for 2-dimensional Gaussian processes. *Journal of Multivariate Analysis*, 18:52–69, 1986. (Cited on page 344.)

[666] M. Maejima. Norming operators for operator-self-similar processes. In *Stochastic Processes and Related Topics*, Trends Math., pages 287–295. Boston, MA: Birkhäuser Boston, 1998. (Cited on page 536.)

[667] M. Maejima and J. Mason. Operator-self-similar stable processes. *Stochastic Processes and their Applications*, 54:139–163, 1994. (Cited on pages 476, 477, 478, 479, and 536.)

[668] M. Maejima and C. A. Tudor. Selfsimilar processes with stationary increments in the second Wiener chaos. *Probability and Mathematical Statistics*, 32(1):167–186, 2012. (Cited on page 281.)

[669] M. Maejima and C. A. Tudor. On the distribution of the Rosenblatt process. *Statistics & Probability Letters*, 83(6):1490–1495, 2013. (Cited on page 281.)

[670] M. Magdziarz. Correlation cascades, ergodic properties and long memory of infinitely divisible processes. *Stochastic Processes and their Applications*, 119(10):3416–3434, 2009. (Cited on page 111.)

[671] M. Magdziarz, A. Weron, K. Burnecki, and J. Klafter. Fractional Brownian motion versus the continuous-time random walk: A simple test for subdiffusive dynamics. *Physical Review Letters*, 103(18):180602, 2009. (Cited on page 228.)

[672] P. Major. *Multiple Wiener-Itô Integrals*, volume 849 of *Lecture Notes in Mathematics*. Berlin: Springer, 1981. With applications to limit theorems. (Cited on pages 280 and 600.)

[673] P. Major. *Multiple Wiener-Itô Integrals*, volume 849 of *Lecture Notes in Mathematics*. Springer, Cham, second edition, 2014. With applications to limit theorems. (Cited on pages 280, 281, 584, and 610.)

[674] V. Makogin and Y. Mishura. Example of a Gaussian self-similar field with stationary rectangular increments that is not a fractional Brownian sheet. *Stochastic Analysis and Applications*, 33(3):413–428, 2015. (Cited on page 537.)

[675] S. Mallat. *A Wavelet Tour of Signal Processing*. Boston: Academic Press, 1998. (Cited on pages 443, 444, 450, and 451.)

[676] B. B. Mandelbrot. Long-run linearity, locally Gaussian processes, H-spectra and infinite variances. *International Economic Review*, 10:82–113, 1969. (Cited on page 224.)

[677] B. B. Mandelbrot. Intermittent turbulence in self-similar cascades; divergence of high moments and dimension of the carrier. *Journal of Fluid Mechanics*, 62:331–358, 1974. (Cited on page 612.)

[678] B. B. Mandelbrot. Limit theorems on the self-normalized range for weakly and strongly dependent processes. *Zeitschrift für Wahrscheinlichkeitstheorie und verwandte Gebiete*, 31:271–285, 1975. (Cited on page 111.)

[679] B. B. Mandelbrot. *The Fractal Geometry of Nature*. New York: W. H. Freeman and Co., 1982. (Cited on pages 217 and 227.)

[680] B. B. Mandelbrot. A multifractal walk down Wall Street. *Scientific American*, pages 70–73, February 1999. (Cited on page 612.)

[681] B. B. Mandelbrot. *The Fractalist*. New York: Pantheon Books, 2012. Memoir of a scientific maverick, With an afterword by Michael Frame. (Cited on page xxii.)

[682] B. B. Mandelbrot and J. W. Van Ness. Fractional Brownian motions, fractional noises and applications. *SIAM Review*, 10:422–437, 1968. (Cited on pages 109, 110, 111, and 395.)

[683] B. B. Mandelbrot and J. R. Wallis. Noah, Joseph and operational hydrology. *Water Resources Research*, 4:909–918, 1968. (Cited on page 111.)

[684] B .B. Mandelbrot and J. R. Wallis. Computer experiments with fractional Gaussian noises, Parts 1,2,3. *Water Resources Research*, 5:228–267, 1969. (Cited on pages 109 and 111.)

[685] B. B. Mandelbrot and J. R. Wallis. Some long-run properties of geophysical records. *Water Resources Research*, 5:321–340, 1969. (Cited on page 111.)

[686] B. B. Mandelbrot and J. R. Wallis. Robustness of the rescaled range R/S in the measurement of noncyclic long-run statistical dependence. *Water Resources Research*, 5:967–988, 1969. (Cited on page 111.)

[687] M. Mandjes. *Large Deviations for Gaussian Queues*. John Wiley & Sons, Ltd., Chichester, 2007. Modelling Communication Networks. (Cited on page 611.)

[688] V. S. Mandrekar and L. Gawarecki. *Stochastic analysis for Gaussian random processes and fields*, volume 145 of *Monographs on Statistics and Applied Probability*. Boca Raton, FL: CRC Press, 2016. With applications. (Cited on page 436.)

[689] D. Marinucci and G. Peccati. *Random Fields on the Sphere: Representation, Limit Theorems and Cosmological Applications*, volume 389. Cambridge University Press, 2011. (Cited on page 610.)

[690] D. Marinucci and P. M. Robinson. Alternative forms of fractional Brownian motion. *Journal of Statistical Planning and Inference*, 80(1-2):111–122, 1999. (Cited on page 110.)

[691] D. Marinucci and P. M. Robinson. Semiparametric fractional cointegration analysis. *Journal of Econometrics*, 105(1):225–247, 2001. (Cited on page 537.)

[692] T. Marquardt. Fractional Lévy processes with an application to long memory moving average processes. *Bernoulli*, 12(6):1099–1126, 2006. (Cited on page 110.)

[693] G. Maruyama. Nonlinear functionals of Gaussian stationary processes and their applications. In G. Maruyama and J.V. Prohorov, editors, *Proceedings of the Third Japan-URSS symposium on probability theory*, volume 550 of *Lecture notes in Mathematics*, pages 375–378, New York: Springer Verlag, 1976. (Cited on page 343.)

[694] G. Maruyama. Wiener functionals and probability limit theorems. I. The central limit theorems. *Osaka Journal of Mathematics*, 22(4):697–732, 1985. (Cited on page 343.)

[695] J. Mason. A comparison of the properties of operator-stable distributions and operator-self-similar processes. *Colloquia Mathematica Societatis János Bolyai*, 36:751–760, 1984. (Cited on page 536.)

[696] J. Mason and M. Xiao. Sample path properties of operator-self-similiar Gaussian random fields. *Theory of Probability and its Applications*, 46(1):58–78, 2002. (Cited on pages 536 and 537.)

[697] L. Massoulie and A. Simonian. Large buffer asymptotics for the queue with fractional Brownian input. *Journal of Applied Probability*, 36(3):894–906, 1999. (Cited on page 611.)

[698] M. Matsui and N.-R. Shieh. The Lamperti transforms of self-similar Gaussian processes and their exponentials. *Stochastic Models*, 30(1):68–98, 2014. (Cited on page 111.)

[699] K. Maulik, S. Resnick, and H. Rootzén. Asymptotic independence and a network traffic model. *Journal of Applied Probability*, 39(4):671–699, 2002. (Cited on page 224.)

[700] L. Mayoral. Heterogeneous dynamics, aggregation and the persistence of economic shocks. *International Economic Review*, 54(5):1295–1307, 2013. (Cited on page 224.)

[701] A. McCloskey and P. Perron. Memory parameter estimation in the presence of level shifts and deterministic trends. *Econometric Theory*, FirstView: 1–42, 10 2013. (Cited on pages 225 and 573.)

[702] B. M. McCoy. *Advanced Statistical Mechanics*, volume 146 of *International Series of Monographs on Physics*. Oxford: Oxford University Press, 2010. (Cited on pages 183 and 204.)

[703] B. M. McCoy and T. T. Wu. *The Two-Dimensional Ising Model*. Cambridge, MA: Harvard University Press, 1973. (Cited on pages 184, 187, 193, 194, 206, 209, 213, and 223.)

[704] T. McElroy and A. Jach. Tail index estimation in the presence of long-memory dynamics. *Computational Statistics & Data Analysis*, 56(2):266–282, 2012. (Cited on page 111.)

[705] T. S. McElroy and S. H. Holan. On the computation of autocovariances for generalized Gegenbauer processes. *Statistica Sinica*, 22(4):1661–1687, 2012. (Cited on page 110.)

[706] T. S. McElroy and S. H. Holan. Computation of the autocovariances for time series with multiple long-range persistencies. *Computational Statistics and Data Analysis*, 101:44–56, 2016. (Cited on page 110.)

[707] M. M. Meerschaert. Fractional calculus, anomalous diffusion, and probability. In *Fractional dynamics*, pages 265–284. Hackensack, NJ: World Sci. Publ., 2012. (Cited on page 396.)

[708] M. M. Meerschaert and F. Sabzikar. Stochastic integration for tempered fractional Brownian motion. *Stochastic Processes and their Applications*, 124(7):2363–2387, 2014. (Cited on page 395.)

[709] M. M. Meerschaert and H.-P. Scheffler. Spectral decomposition for operator self-similar processes and their generalized domains of attraction. *Stochastic Processes and their Applications*, 84(1):71–80, 1999. (Cited on page 536.)

[710] M. M. Meerschaert and H.-P. Scheffler. *Limit Distributions for Sums of Independent Random Vectors: Heavy Tails in Theory and Practice*. Wiley Series in Probability and Statistics. New York: John Wiley & Sons Inc., 2001. (Cited on pages 495, 510, 535, and 536.)

[711] M. M. Meerschaert and H.-P. Scheffler. Triangular array limits for continuous time random walks. *Stochastic Processes and their Applications*, 118(9):1606–1633, 2008. (Cited on page 227.)

[712] M. M. Meerschaert and A. Sikorskii. *Stochastic Models for Fractional Calculus*, volume 43 of *de Gruyter Studies in Mathematics*. Berlin: Walter de Gruyter & Co., 2012. (Cited on pages 352, 396, and 610.)

[713] M. M. Meerschaert, E. Nane, and Y. Xiao. Correlated continuous time random walks. *Statistics & Probability Letters*, 79(9):1194–1202, 2009. (Cited on page 228.)

[714] F. G. Mehler. Ueber die Entwicklung einer Function von beliebig vielen Variablen nach Laplaceschen Functionen höherer Ordnung. *Journal für die Reine und Angewandte Mathematik. [Crelle's Journal]*, 66:161–176, 1866. (Cited on page 342.)

[715] Y. Meyer. *Wavelets and Operators*, volume 37 of *Cambridge Studies in Advanced Mathematics*. Cambridge: Cambridge University Press, 1992. (Cited on pages 444 and 445.)

[716] Y. Meyer, F. Sellan, and M. S. Taqqu. Wavelets, generalized white noise and fractional integration: the synthesis of fractional Brownian motion. *The Journal of Fourier Analysis and Applications*, 5(5):465–494, 1999. (Cited on pages 440, 445, 450, 452, 453, and 465.)

[717] T. Mikosch and G. Samorodnitsky. Scaling limits for cumulative input processes. *Mathematics of Operations Research*, 32(4):890–918, 2007. (Cited on page 224.)

[718] T. Mikosch and C. Stărică. Changes of structure in financial time series and the GARCH model. *REVSTAT Statistical Journal*, 2(1):41–73, 2004. (Cited on page 225.)

[719] T. Mikosch, S. Resnick, H. Rootzén, and A. Stegeman. Is network traffic approximated by stable Lévy motion or fractional Brownian motion? *The Annals of Applied Probability*, 12(1):23–68, 2002. (Cited on page 224.)

[720] J. Militký and S. Ibrahim. Complex characterization of yarn unevenness. In X. Zeng, Y. Li, D. Ruan, and L. Koehl, editors, *Computational Textile*, volume 55 of *Studies in Computational Intelligence*, pages 57–73. Berlin/Heidelberg: Springer, 2007. (Cited on page 225.)

[721] T. C. Mills. Time series modelling of two millennia of northern hemisphere temperatures: long memory or shifting trends? *Journal of the Royal Statistical Society. Series A. Statistics in Society*, 170(1):83–94, 2007. (Cited on pages 225 and 557.)

[722] Y. S. Mishura. *Stochastic Calculus for Fractional Brownian Motion and Related Processes*, volume 1929 of *Lecture Notes in Mathematics*. Berlin: Springer-Verlag, 2008. (Cited on pages 436 and 610.)

[723] Y. S. Mishura and G. M. Shevchenko. Rate of convergence of Euler approximations of solution to mixed stochastic differential equation involving Brownian motion and fractional Brownian motion. *Random Operators and Stochastic Equations*, 19(4):387–406, 2011. (Cited on page 436.)

[724] A. Moberg, D. M. Sonechkin, K. Holmgren, N. M. Datsenko, and W. Karlen. Highly variable northern hemisphere temperatures reconstructed from low- and high-resolution proxy data. *Nature*, 433:613–617, 2005. (Cited on page 557.)

[725] G. M. Molčan and Ju. I. Golosov. Gaussian stationary processes with asymptotically a power spectrum. *Doklady Akademii Nauk SSSR*, 184:546–549, 1969. (Cited on page 395.)

[726] G. M. Molchan. Gaussian processes with spectra which are asymptotically equivalent to a power of λ. *Theory of Probability and Its Applications*, 14:530–532, 1969. (Cited on page 395.)

[727] G. M. Molchan. Linear problems for fractional Brownian motion: a group approach. *Theory of Probability and Its Applications*, 47(1):69–78, 2003. (Cited on page 395.)

[728] J. F. Monahan. *Numerical Methods of Statistics*, 2nd edition. Cambridge Series in Statistical and Probabilistic Mathematics. Cambridge: Cambridge University Press, 2011. (Cited on page 542.)

[729] A. Montanari. Longe range dependence in Hydrology. In P. Doukhan, G. Oppenheim, and M. S. Taqqu, editors, Theory and Applications of Long-range Dependence, pages 461–472. Birkhäuser, 2003. (Cited on page 111.)

[730] A. Montanari, R. Rosso, and M. S. Taqqu. Some long-run properties of rainfall records in Italy. *Journal of Geophysical Research – Atmospheres*, 101(D23):431–438, 1996. (Cited on page 572.)

[731] A. Montanari, R. Rosso, and M. S. Taqqu. Fractionally differenced ARIMA models applied to hydrologic time series: identification, estimation and simulation. *Water Resources Research*, 33:1035–1044, 1997. (Cited on page 111.)

[732] T. Mori and H. Oodaira. The law of the iterated logarithm for self-similar processes represented by multiple Wiener integrals. *Probability Theory and Related Fields*, 71(3):367–391, 1986. (Cited on page 281.)

[733] P. Mörters and Y. Peres. *Brownian Motion*. Cambridge Series in Statistical and Probabilistic Mathematics. Cambridge: Cambridge University Press, 2010. With an appendix by Oded Schramm and Wendelin Werner. (Cited on page 14.)

[734] E. Moulines and P. Soulier. Broadband log-periodogram regression of time series with long-range dependence. *The Annals of Statistics*, 27(4):1415–1439, 1999. (Cited on page 573.)

[735] E. Moulines, F. Roueff, and M. S. Taqqu. Central limit theorem for the log-regression wavelet estimation of the memory parameter in the Gaussian semi-parametric context. *Fractals*, 15(4):301–313, 2007. (Cited on page 112.)

[736] E. Moulines, F. Roueff, and M. S. Taqqu. On the spectral density of the wavelet coefficients of long-memory time series with application to the log-regression estimation of the memory parameter. *Journal of Time Series Analysis*, 28(2):155–187, 2007. (Cited on page 112.)

[737] E. Moulines, F. Roueff, and M. S. Taqqu. A wavelet Whittle estimator of the memory parameter of a non-stationary Gaussian time series. *The Annals of Statistics*, 36(4):1925–1956, 2008. (Cited on pages 112 and 574.)

[738] N. Naganuma. Asymptotic error distributions of the cranknicholson scheme for SDEs driven by fractional Brownian motion. *Journal of Theoretical Probability*, 28(3):1082–1124, 2015. (Cited on page 436.)

[739] M. Narukawa. On semiparametric testing of I(d) by FEXP models. *Communications in Statistics. Theory and Methods*, 42(9):1637–1653, 2013. (Cited on page 573.)

[740] M. Narukawa and Y. Matsuda. Broadband semi-parametric estimation of long-memory time series by fractional exponential models. *Journal of Time Series Analysis*, 32(2):175–193, 2011. (Cited on page 568.)

[741] A. Neuenkirch. Optimal approximation of SDE's with additive fractional noise. *Journal of Complexity*, 22(4):459–474, 2006. (Cited on page 436.)

[742] A. Neuenkirch. Optimal pointwise approximation of stochastic differential equations driven by fractional Brownian motion. *Stochastic Processes and their Applications*, 118(12):2294–2333, 2008. (Cited on page 436.)

[743] A. Neuenkirch and I. Nourdin. Exact rate of convergence of some approximation schemes associated to SDEs driven by a fractional Brownian motion. *Journal of Theoretical Probability*, 20(4):871–899, 2007. (Cited on page 436.)

[744] A. Neuenkirch, I. Nourdin, and S. Tindel. Delay equations driven by rough paths. *Electronic Journal of Probability*, 13: no. 67, 2031–2068, 2008. (Cited on page 436.)

[745] J. M. Nichols, C. C. Olson, J. V. Michalowicz, and F. Bucholtz. A simple algorithm for generating spectrally colored, non-Gaussian signals. *Probabilistic Engineering Mechanics*, 25(3):315–322, 2010. (Cited on page 344.)

[746] F. S. Nielsen. Local Whittle estimation of multi-variate fractionally integrated processes. *Journal of Time Series Analysis*, 32(3):317–335, 2011. (Cited on page 536.)

[747] M. Ø. Nielsen. Local empirical spectral measure of multivariate processes with long range dependence. *Stochastic Processes and their Applications*, 109(1):145–166, 2004. (Cited on page 536.)

[748] M. Ø. Nielsen. Local Whittle analysis of stationary fractional cointegration and the implied-realized volatility relation. *Journal of Business & Economic Statistics*, 25(4):427–446, 2007. (Cited on page 536.)

[749] M. Ø. Nielsen. Nonparametric cointegration analysis of fractional systems with unknown integration orders. *Journal of Econometrics*, 155(2):170–187, 2010. (Cited on page 537.)

[750] M. Ø. Nielsen and P. Frederiksen. Fully modified narrow-band least squares estimation of weak fractional cointegration. *The Econometrics Journal*, 14(1):77–120, 2011. (Cited on page 536.)

[751] M. Ø. Nielsen and P. H. Frederiksen. Finite sample comparison of parametric, semiparametric, and wavelet estimators of fractional integration. *Econometric Reviews*, 24(4):405–443, 2005. (Cited on page 571.)

[752] M. Niss. History of the Lenz-Ising model 1920–1950: from ferromagnetic to cooperative phenomena. *Archive for History of Exact Sciences*, 59(3):267–318, 2005. (Cited on page 227.)

[753] M. Niss. History of the Lenz-Ising model 1950–1965: from irrelevance to relevance. *Archive for History of Exact Sciences*, 63(3):243–287, 2009. (Cited on page 227.)

[754] D. J. Noakes, K. W. Hipel, A. I. McLeod, C. Jimenez, and S. Yakowitz. Forecasting annual geophysical time series. *International Journal of Forecasting*, 4(1):103–115, 1998. (Cited on page 572.)

[755] J. P. Nolan. *Stable Distributions - Models for Heavy Tailed Data*. Boston: Birkhauser, 2014. (Cited on page 111.)

[756] I. Norros. A storage model with self-similar input. *Queueing Systems And Their Applications*, 16:387–396, 1994. (Cited on page 611.)

[757] I. Norros and E. Saksman. Local independence of fractional Brownian motion. *Stochastic Processes and their Applications*, 119(10):3155–3172, 2009. (Cited on page 395.)

[758] I. Norros, E. Valkeila, and J. Virtamo. An elementary approach to a Girsanov formula and other analytical results on fractional Brownian motions. *Bernoulli*, 15:571–587, 1999. (Cited on page 395.)

[759] L. Nouira, M. Boutahar, and V. Marimoutou. The effect of tapering on the semiparametric estimators for nonstationary long memory processes. *Statistical Papers*, 50(2):225–248, 2009. (Cited on page 573.)

[760] I. Nourdin. *Selected Aspects of Fractional Brownian Motion*, volume 4 of *Bocconi & Springer Series*. Springer, Milan; Bocconi University Press, Milan, 2012. (Cited on pages 305, 343, 344, 436, and 610.)

[761] I. Nourdin and G. Peccati. Stein's method and exact Berry-Esseen asymptotics for functionals of Gaussian fields. *The Annals of Probability*, 37(6):2231–2261, 2009. (Cited on page 571.)

[762] I. Nourdin and G. Peccati. Stein's method on Wiener chaos. *Probability Theory and Related Fields*, 145(1-2):75–118, 2009. (Cited on pages 344 and 436.)

[763] I. Nourdin and G. Peccati. *Normal Approximations with Malliavin Calculus*, volume 192 of *Cambridge Tracts in Mathematics*. Cambridge: Cambridge University Press, 2012. From Stein's Method to Universality. (Cited on pages 305, 343, 344, 426, 427, 435, 436, 604, and 610.)

[764] I. Nourdin and J. Rosiński. Asymptotic independence of multiple Wiener-Itô integrals and the resulting limit laws. *The Annals of Probability*, 42(2):497–526, 2014. (Cited on pages 324 and 344.)

[765] I. Nourdin and T. Simon. On the absolute continuity of one-dimensional SDEs driven by a fractional Brownian motion. *Statistics & Probability Letters*, 76(9):907–912, 2006. (Cited on pages 413, 414, 418, and 434.)

[766] I. Nourdin, D. Nualart, and C. A. Tudor. Central and non-central limit theorems for weighted power variations of fractional Brownian motion. *Annales de l'Institut Henri Poincaré Probabilités et Statistiques*, 46(4):1055–1079, 2010. (Cited on page 280.)

[767] I. Nourdin, G. Peccati, and G. Reinert. Invariance principles for homogeneous sums: universality of Gaussian Wiener chaos. *The Annals of Probability*, 38(5):1947–1985, 2010. (Cited on page 343.)

[768] I. Nourdin, D. Nualart, and G. Peccati. Strong asymptotic independence on Wiener chaos. *Proceedings of the American Mathematical Society*, 144(2):875–886, 2016. (Cited on pages 323 and 344.)

[769] D. Nualart. *The Malliavin Calculus and Related Topics*, 2nd edition. Probability and its Applications (New York). Berlin: Springer-Verlag, 2006. (Cited on pages 280, 408, 409, 412, 413, 414, 415, 436, 598, 600, 601, 602, 603, 604, 605, 606, 608, 609, and 610.)

[770] D. Nualart. *Normal approximations with Malliavin calculus* [book review of mr2962301]. American Mathematical Society. Bulletin. New Series, 51(3):491–497, 2014. (Cited on page 426.)

[771] D. Nualart and G. Peccati. Central limit theorems for sequences of multiple stochastic integrals. *The Annals of Probability*, 33(1):177–193, 2005. (Cited on pages 282, 305, 324, 343, 344, and 436.)

[772] D. Nualart and V. Pérez-Abreu. On the eigenvalue process of a matrix fractional Brownian motion. *Stochastic Processes and their Applications*, 124(12):4266–4282, 2014. (Cited on page 436.)

[773] D. Nualart and A. Răşcanu. Differential equations driven by fractional Brownian motion. *Universitat de Barcelona. Collectanea Mathematica*, 53(1):55–81, 2002. (Cited on page 413.)

[774] D. Nualart and B. Saussereau. Malliavin calculus for stochastic differential equations driven by a fractional Brownian motion. *Stochastic Processes and their Applications*, 119(2):391–409, 2009. (Cited on pages 414 and 415.)

[775] C. J. Nuzman and H. V. Poor. Reproducing kernel Hilbert space methods for wide-sense self-similar processes. *The Annals of Applied Probability*, 11(4):1199–1219, 2001. (Cited on page 395.)

[776] G. L. O'Brien and W. Vervaat. Self-similar processes with stationary increments generated by point processes. *The Annals of Probability*, 13(1):28–52, 1985. (Cited on page 111.)

[777] B. Oğuz and V. Anantharam. Hurst index of functions of long-range-dependent Markov chains. *Journal of Applied Probability*, 49(2):451–471, 2012. (Cited on page 109.)

[778] A. Ohanissian, J. R. Russell, and R. S. Tsay. True or spurious long memory? A new test. *Journal of Business & Economic Statistics*, 26:161–175, 2008. (Cited on page 225.)

[779] L. Onsager. Crystal statistics. I. A two-dimensional model with an order-disorder transition. *Phys. Rev. (2)*, 65:117–149, 1944. (Cited on page 226.)

[780] G. Oppenheim and M.-C. Viano. Aggregation of random parameters Ornstein-Uhlenbeck or AR processes: some convergence results. *Journal of Time Series Analysis*, 25(3):335–350, 2004. (Cited on page 223.)

[781] G. Oppenheim, M. Ould Haye, and M.-C. Viano. Long memory with seasonal effects. *Statistical Inference for Stochastic Processes*, 3(1-2):53–68, 2000. 19th "Rencontres Franco-Belges de Statisticiens" (Marseille, 1998). (Cited on page 573.)

[782] S original by C. Fraley, U.Washington, Seattle. R port by F. Leisch at TU Wien; since 2003-12: M. Maechler; fdGPH, fdSperio, etc by V. Reisen, and A. Lemonte. *fracdiff: Fractionally differenced ARIMA aka ARFIMA(p,d,q) models*, 2012. URL http://CRAN.R-project.org/package=fracdiff. R package version 1.4-2. (Cited on page 555.)

[783] M. Ossiander and E. C. Waymire. Statistical estimation for multiplicative cascades. *The Annals of Statistics*, 28(6):1533–1560, 2000. (Cited on page 612.)

[784] T. Owada and G. Samorodnitsky. Functional central limit theorem for heavy tailed stationary infinitely divisible processes generated by conservative flows. *The Annals of Probability*, 43(1):240–285, 2015. (Cited on page 228.)

[785] J. Pai and N. Ravishanker. A multivariate preconditioned conjugate gradient approach for maximum likelihood estimation in vector long memory processes. *Statistics & Probability Letters*, 79(9):1282–1289, 2009. (Cited on page 570.)

[786] J. Pai and N. Ravishanker. Maximum likelihood estimation in vector long memory processes via EM algorithm. *Computational Statistics & Data Analysis*, 53(12):4133–4142, 2009. (Cited on page 570.)

[787] J. S. Pai and N. Ravishanker. Bayesian analysis of autoregressive fractionally integrated moving-average processes. *Journal of Time Series Analysis*, 19(1):99–112, 1998. (Cited on page 573.)

[788] R. E. A. C. Paley and N. Wiener. *Fourier Transforms in the Complex Domain*, volume 19. American Mathematical Soc., 1934. (Cited on page 456.)

[789] W. Palma. *Long-Memory Time Series*. NJ: Wiley Series in Probability and Statistics. Hoboken, Wiley-Interscience [John Wiley & Sons], 2007. Theory and Methods. (Cited on pages 43, 539, and 610.)

[790] W. Palma and N. H. Chan. Efficient estimation of seasonal long-range-dependent processes. *Journal of Time Series Analysis*, 26(6):863–892, 2005. (Cited on page 573.)

[791] A. Papavasiliou and C. Ladroue. Parameter estimation for rough differential equations. *The Annals of Statistics*, 39(4):2047–2073, 2011. (Cited on page 436.)

[792] J. C. Pardo and V. Rivero. Self-similar Markov processes. *Sociedad Matemática Mexicana. Boletí n. Tercera Serie*, 19(2):201–235, 2013. (Cited on page 612.)

[793] C. Park, F. Godtliebsen, M. S. Taqqu, S. Stoev, and J. S. Marron. Visualization and inference based on wavelet coefficients, SiZer and SiNos. *Computational Statistics & Data Analysis*, 51(12):5994–6012, 2007. (Cited on page 112.)

[794] K. Park and W. Willinger, editors. *Self-Similar Network Traffic and Performance Evaluation*. J. Wiley & Sons, Inc., New York, 2000. (Cited on pages 224 and 610.)

[795] V. Paulauskas. On α-covariance, long, short and negative memories for sequences of random variables with infinite variance. Preprint, 2013. (Cited on page 111.)

[796] V. Paulauskas. Some remarks on definitions of memory for stationary random processes and fields. Preprint, 2016. (Cited on page 111.)

[797] G. Peccati and M. S. Taqqu. *Wiener Chaos: Moments, Cumulants and Diagrams*, volume 1 of *Bocconi & Springer Series*. Springer, 2011. (Cited on pages 242, 243, 245, 259, 265, 281, 305, 324, 343, 599, 600, 601, and 610.)

[798] G. Peccati and C. Tudor. Gaussian limits for vector-valued multiple stochastic integrals. *Séminaire de Probabilités XXXVIII*, pages 219–245, 2005. (Cited on page 344.)

[799] G. Peccati and C. Zheng. Multi-dimensional Gaussian fluctuations on the Poisson space. *Electronic Journal of Probability*, 15(48):1487–1527, 2010. (Cited on page 343.)

[800] M. Peligrad. Central limit theorem for triangular arrays of non-homogeneous Markov chains. *Probability Theory and Related Fields*, 154(3-4):409–428, 2012. (Cited on page 161.)

[801] M. Peligrad and H. Sang. Asymptotic properties of self-normalized linear processes with long memory. *Econometric Theory*, 28(3):548–569, 2012. (Cited on page 111.)

[802] M. Peligrad and H. Sang. Central limit theorem for linear processes with infinite variance. *Journal of Theoretical Probability*, 26(1):222–239, 2013. (Cited on page 111.)

[803] M. Peligrad and S. Sethuraman. On fractional Brownian motion limits in one dimensional nearest-neighbor symmetric simple exclusion. *ALEA. Latin American Journal of Probability and Mathematical Statistics*, 4:245–255, 2008. (Cited on page 172.)

[804] R. F. Peltier and J. L. Véhel. Multifractional Brownian motion: definition and preliminary results. Technical Report 2645, Institut National de Recherche en Informatique et an Automatique, INRIA, Le Chesnay, France, 1995. (Cited on page 611.)

[805] L. Peng. Semi-parametric estimation of long-range dependence index in infinite variance time series. *Statistics & Probability Letters*, 51(2):101–109, 2001. (Cited on page 111.)

[806] D. B. Percival. Exact simulation of complex-valued Gaussian stationary processes via circulant embedding. *Signal Processing*, 86(7):1470–1476, 2006. (Cited on page 112.)

[807] E. Perrin, R. Harba, C. Berzin-Joseph, I. Iribarren, and A. Bonami. nth-Order fractional Brownian motion and fractional Gaussian noises. *IEEE Transactions on Signal Processing*, 49(5):1049–1059, 2001. (Cited on page 110.)

[808] A. Philippe, D. Surgailis, and M.-C. Viano. Almost periodically correlated processes with long memory. In *Dependence in probability and statistics*, volume 187 of *Lecture Notes in Statist.*, pages 159–194. New York: Springer, 2006. (Cited on page 573.)

[809] J. Picard. Representation formulae for the fractional Brownian motion. In *Séminaire de Probabilités XLIII*, volume 2006 of *Lecture Notes in Math.*, pages 3–70. Berlin: Springer, 2011. (Cited on page 395.)

[810] D. Pilipauskaitė and D. Surgailis. Scaling transition for nonlinear random fields with long-range dependence. To appear in *Stochastic Processes and their Applications*. Preprint, 2016. (Cited on page 537.)

[811] M. S. Pinsker and A. M. Yaglom. On linear extrapolation of random processes with stationary nth increments. *Doklady Akad. Nauk SSSR (N.S.)*, 94:385–388, 1954. (Cited on page 110.)

[812] V. Pipiras. Wavelet-type expansion of the Rosenblatt process. *The Journal of Fourier Analysis and Applications*, 10(6):599–634, 2004. (Cited on page 465.)

[813] V. Pipiras. Wavelet-based simulation of fractional Brownian motion revisited. *Applied and Computational Harmonic Analysis*, 19(1):49–60, 2005. (Cited on pages 451, 454, and 465.)

[814] V. Pipiras and M. S. Taqqu. Convergence of the Weierstrass-Mandelbrot process to fractional Brownian motion. *Fractals*, 8:369–384, 2000. (Cited on page 227.)

[815] V. Pipiras and M. S. Taqqu. Integration questions related to fractional Brownian motion. *Probability Theory and Related Fields*, 118(2):251–291, 2000. (Cited on pages 358, 372, 374, 384, and 395.)

[816] V. Pipiras and M. S. Taqqu. The Weierstass-Mandelbrot process provides a series approximation to the harmonizable fractional stable motion. In C. Bandt, S. Graf, and M. Zähle, editors, *Fractal Geometry and Stochastics II*, pages 161–179. Birkhäuser, 2000. (Cited on page 227.)

[817] V. Pipiras and M. S. Taqqu. Are classes of deterministic integrands for fractional Brownian motion on an interval complete? *Bernoulli*, 7(6):873–897, 2001. (Cited on pages 364, 375, 377, and 395.)

[818] V. Pipiras and M. S. Taqqu. Deconvolution of fractional Brownian motion. *Journal of Time Series Analysis*, 23(4):487–501, 2002. (Cited on page 395.)

[819] V. Pipiras and M. S. Taqqu. Regularization and integral representations of Hermite processes. *Statistics & Probability Letters*, 80(23-24):2014–2023, 2010. (Cited on pages 241, 280, and 601.)

[820] V. Pipiras and M. S. Taqqu. Stable self-similar processes with stationary increments. Preprint, 2015. (Cited on pages 60, 63, 110, and 611.)

[821] V. Pipiras and M. S. Taqqu. Long-range dependence of the two-dimensional Ising model at critical temperature. In M. Frame and N. Cohen, editors, *Benoit Mandelbrot: A Life in Many Dimensions*, pages 399–440. World Scientific, 2015. (Cited on page 227.)

[822] V. Pipiras, M. S. Taqqu, and J. B. Levy. Slow, fast and arbitrary growth conditions for renewal reward processes when both the renewals and the rewards are heavy-tailed. *Bernoulli*, 10:121–163, 2004. (Cited on page 224.)

[823] E. J. G. Pitman and J. Pitman. A direct approach to the stable distributions. *Advances in Applied Probability*, 48(A):261–282, 2016. (Cited on page 585.)

[824] L. D. Pitt. Scaling limits of Gaussian vectors fields. *Journal of Multivariate Analysis*, 8(1):45–54, 1978. (Cited on page 536.)

[825] I. Podlubny. *Fractional Differential Equations*, volume 198 of *Mathematics in Science and Engineering*. San Diego, CA: Academic Press, Inc., 1999. An introduction to fractional derivatives, fractional differential equations, to methods of their solution and some of their applications. (Cited on page 352.)

[826] I. Podlubny. Geometric and physical interpretation of fractional integration and fractional differentiation. *Fractional Calculus & Applied Analysis*, 5(4):367–386, 2002. Dedicated to the 60th anniversary of Prof. Francesco Mainardi. (Cited on page 395.)

[827] K. Polisano, M. Clausel, V. Perrier, and L. Condat. Texture modeling by Gaussian fields with prescribed local orientation. In *Image Processing (ICIP), 2014 IEEE International Conference on*, pages 6091–6095, 2014. (Cited on page 537.)

[828] M. Polyak. Feynman diagrams for pedestrians and mathematicians. In *Graphs and Patterns in Mathematics and Theoretical Physics*, volume 73 of *Proc. Sympos. Pure Math.*, pages 15–42. Providence, RI: Amer. Math. Soc., 2005. (Cited on page 281.)

[829] D. S. Poskitt. Properties of the sieve bootstrap for fractionally integrated and non-invertible processes. *Journal of Time Series Analysis*, 29(2):224–250, 2008. (Cited on page 574.)

[830] D. S. Poskitt, S. D. Grose, and G. M. Martin. Higher-order improvements of the sieve bootstrap for fractionally integrated processes. *Journal of Econometrics*, 188(1):94–110, 2015. (Cited on page 574.)

[831] B. L. S. Prakasa Rao. *Statistical Inference for Fractional Diffusion Processes*. Wiley Series in Probability and Statistics. Chichester: John Wiley & Sons Ltd., 2010. (Cited on page 395.)

[832] R. Price. A useful theorem for nonlinear devices having Gaussian inputs. *IRE Trans.*, IT-4:69–72, 1958. (Cited on page 333.)

[833] M. B. Priestley. *Spectral Analysis and Time Series. Vol. 1*. London-New York: Academic Press, Inc. [Harcourt Brace Jovanovich, Publishers], 1981. Univariate Series, Probability and Mathematical Statistics. (Cited on page 14.)

[834] P. E. Protter. *Stochastic Integration and Differential Equations*, 2nd edition, volume 21 of *Stochastic Modelling and Applied Probability*. Berlin: Springer-Verlag, 2005. Version 2.1, Corrected third printing. (Cited on pages 397 and 398.)

[835] D. Puplinskaitė and D. Surgailis. Scaling transition for long-range dependent Gaussian random fields. *Stochastic Processes and their Applications*, 125(6):2256–2271, 2015. (Cited on pages 529 and 537.)

[836] D. Puplinskaitė and D. Surgailis. Aggregation of autoregressive random fields and anisotropic long-range dependence. *Bernoulli*, 22(4):2401–2441, 2016. (Cited on page 537.)

[837] Z. Qu. A test against spurious long memory. *Journal of Business and Economic Statistics*, 29(9):423–438, 2011. (Cited on pages 225 and 573.)

[838] R Core Team. *R: A Language and Environment for Statistical Computing*. Vienna, Austria: R Foundation for Statistical Computing, 2013. URL www.R-project.org/. (Cited on page 555.)

[839] S. T. Rachev and S. Mittnik. *Stable Paretian Models in Finance*. New York: Wiley, 2000. (Cited on page 111.)

[840] B. S. Rajput and J. Rosiński. Spectral representation of infinitely divisible processes. *Probability Theory and Related Fields*, 82: 451–487, 1989. (Cited on pages 594, 595, and 597.)

[841] G. Rangarajan and M. Ding. *Processes with Long-Range Correlations: Theory and Applications*, volume 621. Springer Science & Business Media, 2003. (Cited on pages 228 and 610.)

[842] B. L. S. P. Rao. *Statistical Inference for Fractional Diffusion Processes*. Wiley Series in Probability and Statistics. Wiley, 2011. (Cited on page 610.)

[843] V. Reisen, B. Abraham, and S. Lopes. Estimation of parameters in ARFIMA processes: a simulation study. *Communications in Statistics. Simulation and Computation*, 30(4):787–803, 2001. (Cited on page 571.)

[844] V. A. Reisen, A. L. Rodrigues, and W. Palma. Estimating seasonal long-memory processes: a Monte Carlo study. *Journal of Statistical Computation and Simulation*, 76 (4):305–316, 2006. (Cited on page 573.)

[845] S. Resnick. *Adventures in Stochastic Processes*. Boston, MA: Birkhäuser Boston Inc., 1992. (Cited on page 87.)

[846] S. I. Resnick. *A Probability Path*. Boston, MA: Birkhäuser Boston Inc., 1999. (Cited on page 68.)

[847] S. I. Resnick. *Heavy-Tail Phenomena*. Springer Series in Operations Research and Financial Engineering. New York: Springer, 2007. Probabilistic and statistical modeling. (Cited on page 129.)

[848] D. Revuz and M. Yor. *Continuous Martingales and Brownian Motion*. Berlin: Springer-Verlag, 1999. (Cited on page 398.)

[849] R. H. Riedi. Multifractal processes. In *Theory and applications of long-range dependence*, pages 625–716. Boston, MA: Birkhäuser Boston, 2003. (Cited on page 612.)

[850] F. Riesz and B. Sz.-Nagy. *Functional Analysis*. New York: Frederick Ungar Publishing Co., 1955. Translated by Leo F. Boron. (Cited on page 438.)

[851] P. M. Robinson. Statistical inference for a random coefficient autoregressive model. *Scandinavian Journal of Statistics. Theory and Applications*, 5(3):163–168, 1978. (Cited on page 223.)

[852] P. M. Robinson. Log-periodogram regression of time series with long range dependence. *The Annals of Statistics*, 23(3):1048–1072, 1995. (Cited on pages 93, 111, and 536.)

[853] P. M. Robinson. Gaussian semiparametric estimation of long range dependence. *The Annals of Statistics*, 23:1630–1661, 1995. (Cited on pages 560, 561, and 572.)

[854] P. M. Robinson. *Time Series with Long Memory*. Advanced texts in econometrics. Oxford University Press, 2003. (Cited on page 610.)

[855] P. M. Robinson. Multiple local Whittle estimation in stationary systems. *The Annals of Statistics*, 36(5):2508–2530, 2008. (Cited on page 536.)

[856] P. M. Robinson and M. Henry. Higher-order kernel semiparametric M-estimation of long memory. *Journal of Econometrics*, 114(1):1–27, 2003. (Cited on page 572.)

[857] P. M. Robinson and D. Marinucci. Semiparametric frequency domain analysis of fractional cointegration. In P. M. Robinson, editor, *Time Series with Long Memory*. Oxford: Oxford University Press, 2003. (Cited on page 536.)

[858] P. M. Robinson and Y. Yajima. Determination of cointegrating rank in fractional systems. *Journal of Econometrics*, 106(2):217–241, 2002. (Cited on pages 536 and 537.)

[859] L. C. G. Rogers. Arbitrage with fractional Brownian motion. *Mathematical Finance*, 7:95–105, 1997. (Cited on page 435.)

[860] M. Rosenblatt. Independence and dependence. In *Proceedings of the 4th Berkeley Symposium Mathematical Statistics and Probability*, pages 431–443. University of California Press, 1961. (Cited on pages 110, 281, 343, and 612.)

[861] M. Rosenblatt. Some limit theorems for partial sums of quadratic forms in stationary Gaussian variables. *Zeitschrift für Wahrscheinlichkeitstheorie und verwandte Gebiete*, 49:125–132, 1979. (Cited on page 343.)

[862] W. A. Rosenkrantz and J. Horowitz. The infinite source model for internet traffic: statistical analysis and limit theorems. *Methods and Applications of Analysis*, 9(3):445–461, 2002. Special issue dedicated to Daniel W. Stroock and Srinivasa S. R. Varadhan on the occasion of their 60th birthday. (Cited on page 224.)

[863] H. Rost and M. E. Vares. Hydrodynamics of a one-dimensional nearest neighbor model. In *Particle systems, random media and large deviations (Brunswick, Maine, 1984)*, volume 41 of *Contemp. Math.*, pages 329–342. Providence, RI: Amer. Math. Soc., 1985. (Cited on page 168.)

[864] S. Rostek. *Option Pricing in Fractional Brownian Markets*, volume 622 of *Lecture Notes in Economics and Mathematical Systems*. Berlin: Springer-Verlag, Berlin, 2009. With a foreword by Rainer Schöbel. (Cited on page 436.)

[865] G.-C. Rota and T. C. Wallstrom. Stochastic integrals: a combinatorial approach. *The Annals of Probability*, 25(3):1257–1283, 1997. (Cited on page 281.)

[866] F. Roueff and M. S. Taqqu. Asymptotic normality of wavelet estimators of the memory parameter for linear processes. *Journal of Time Series Analysis*, 30(5):534–558, 2009. (Cited on page 112.)

[867] F. Roueff and M. S. Taqqu. Central limit theorems for arrays of decimated linear processes. *Stochastic Processes and their Applications*, 119(9):3006–3041, 2009. (Cited on page 112.)

[868] F. Roueff and R. von Sachs. Locally stationary long memory estimation. *Stochastic Processes and their Applications*, 121(4):813–844, 2011. (Cited on page 573.)

[869] M. Roughan and D. Veitch. Measuring long-range dependence under changing traffic conditions. In *Proceedings of IEEE INFOCOM*, volume 3, pages 1513–1521, Mar 1999. (Cited on page 225.)

[870] M. Roughan, D. Veitch, and P. Abry. Real-time estimation of the parameters of long-range dependence. *IEEE/ACM Transactions on Networking (TON)*, 8(4):467–478, 2000. (Cited on page 112.)

[871] S. G. Roux, M. Clausel, B. Vedel, S. Jaffard, and P. Abry. Self-similar anisotropic texture analysis: the hyperbolic wavelet transform contribution. *IEEE Transactions on Image Processing*, 22(11):4353–4363, 2013. (Cited on page 537.)

[872] A. Roy, T. S. McElroy, and P. Linton. Estimation of causal invertible VARMA models. Preprint, 2014. (Cited on page 570.)

[873] F. Russo and C. A. Tudor. On bifractional Brownian motion. *Stochastic Processes and their Applications*, 116(5):830–856, 2006. (Cited on page 110.)

[874] A. A. Ruzmaikina. Stieltjes integrals of Hölder continuous functions with applications to fractional Brownian motion. *Journal of Statistical Physics*, 100(5-6):1049–1069, 2000. (Cited on page 435.)

[875] F. Sabzikar. Tempered Hermite process. *Modern Stochastics. Theory and Applications*, 2:327–341, 2015. (Cited on page 344.)

[876] A. Sakai. Lace expansion for the Ising model. *Communications in Mathematical Physics*, 272(2):283–344, 2007. (Cited on page 227.)

[877] A. Samarov and M. S. Taqqu. On the efficiency of the sample mean in long memory noise. *Journal of Time Series Analysis*, 9:191–200, 1988. (Cited on page 540.)

[878] S. G. Samko, A. A. Kilbas, and O. I. Marichev. *Fractional Integrals and Derivatives*. Yverdon: Gordon and Breach Science Publishers, 1993. Theory and Applications. (Cited on pages 348, 354, 356, 357, 359, 376, 377, 379, 381, 395, 401, and 406.)

[879] G. Samorodnitsky. Long range dependence. *Foundations and Trends® in Stochastic Systems*, 1(3):163–257, 2006. (Cited on pages 27, 77, 107, and 610.)

[880] G. Samorodnitsky. *Stochastic Processes and Long Range Dependence*. Springer, 2016. (Cited on page 610.)

[881] G. Samorodnitsky and M. S. Taqqu. The various linear fractional Lévy motions. In T. W. Anderson, K. B. Athreya, and D. L. Iglehart, editors, *Probability, Statistics and Mathematics: Papers in Honor of Samuel Karlin*, pages 261–270, Boston: Academic Press, 1989. (Cited on page 110.)

[882] G. Samorodnitsky and M. S. Taqqu. Linear models with long-range dependence and finite or infinite variance. In D. Brillinger, P. Caines, J. Geweke, E. Parzen, M. Rosenblatt, and M. S. Taqqu, editors, *New Directions in Time Series Analysis, Part II*, pages 325–340. IMA Volumes in Mathematics and its Applications, Volume 46, New York: Springer-Verlag, 1992. (Cited on page 110.)

[883] G. Samorodnitsky and M. S. Taqqu. *Stable Non-Gaussian Random Processes*. Stochastic Modeling. New York: Chapman & Hall, 1994. Stochastic models with infinite variance. (Cited on pages 61, 62, 83, 84, 585, 592, 593, 595, and 596.)

[884] M. V. Sánchez de Naranjo. Non-central limit theorems for nonlinear functionals of k Gaussian fields. *Journal of Multivariate Analysis*, 44(2):227–255, 1993. (Cited on page 344.)

[885] M. V. Sánchez de Naranjo. A central limit theorem for non-linear functionals of stationary Gaussian vector processes. *Statistics & Probability Letters*, 22(3):223–230, 1995. (Cited on page 344.)

[886] K.-i. Sato. *Lévy Processes and Infinitely Divisible Distributions*, volume 68 of *Cambridge Studies in Advanced Mathematics*. Cambridge: Cambridge University Press, 2013. Translated from the 1990 Japanese original, Revised edition of the 1999 English translation. (Cited on page 596.)

[887] A. Scherrer, N. Larrieu, P. Owezarski, P. Borgnat, and P. Abry. Non-Gaussian and long memory statistical characterizations for internet traffic with anomalies. *Dependable and Secure Computing, IEEE Transactions on*, 4(1):56–70, 2007. (Cited on pages 344 and 572.)

[888] M. Schlather. Some covariance models based on normal scale mixtures. *Bernoulli*, 16(3):780–797, 2010. (Cited on page 530.)

[889] M. Schlather. Construction of covariance functions and unconditional simulation of random fields. In E. Porcu, J. M. Montero, and M. Schlather, editors, *Advances and Challenges in Space-time Modelling of Natural Events*, Lecture Notes in Statistics, pages 25–54. Berlin, Heidelberg: Springer, 2012. (Cited on page 538.)

[890] I. J. Schoenberg. Metric spaces and completely monotone functions. *Annals of Mathematics. Second Series*, 39(4):811–841, 1938. (Cited on page 530.)

[891] H. Schönfeld. Beitrag zum 1/f-gesetz beim rauschen von halbleitern. *Zeitschrift für Naturforschung A*, 10 (4): 291–300, 1955. (Cited on page 225.)

[892] O. Schramm. Scaling limits of loop-erased random walks and uniform spanning trees. *Israel Journal of Mathematics*, 118:221–288, 2000. ISSN 0021-2172. (Cited on page 227.)

[893] K. J. Schrenk, N. Posé, J. J. Kranz, L. V. M. van Kessenich, N. A. M. Araújo, and H. J. Herrmann. Percolation with long-range correlated disorder. *Physical Review E*, 88(5):052102, 2013. (Cited on page 227.)

[894] G. Schwarz. Estimating the dimension of a model. *The Annals of Statistics*, 6(2):461–464, 1978. (Cited on page 552.)

[895] S scripts originally by J. Beran; Datasets via B. Whitcher Toplevel R functions and much more by M. Maechler. *longmemo: Statistics for Long-Memory Processes (Jan Beran) – Data and Functions*, 2011. URL http://CRAN.R-project.org/package=longmemo. R package version 1.0-0. (Cited on page 572.)

[896] R. J. Sela and C. M. Hurvich. Computationally efficient methods for two multivariate fractionally integrated models. *Journal of Time Series Analysis*, 30(6):631–651, 2009. (Cited on page 570.)

[897] R. J. Sela and C. M. Hurvich. The averaged periodogram estimator for a power law in coherency. *Journal of Time Series Analysis*, 33(2):340–363, 2012. (Cited on page 537.)

[898] K. Sen, P. Preuss, and H. Dette. Measuring stationarity in long-memory processes. To appear in *Statistica Sinica*. Preprint, 2013. (Cited on page 573.)

[899] S. Sethuraman. Diffusive variance for a tagged particle in $d \leq 2$ asymmetric simple exclusion. *ALEA. Latin American Journal of Probability and Mathematical Statistics*, 1:305–332, 2006. (Cited on page 226.)

[900] S. Sethuraman, S. R. S. Varadhan, and H.-T. Yau. Diffusive limit of a tagged particle in asymmetric simple exclusion processes. *Communications on Pure and Applied Mathematics*, 53(8):972–1006, 2000. (Cited on page 226.)

[901] X. Shao. Self-normalization for time series: a review of recent developments. *Journal of the American Statistical Association*, 110(512):1797–1817, 2015. (Cited on page 574.)

[902] X. Shao and W. B. Wu. Local Whittle estimation of fractional integration for nonlinear processes. *Econometric Theory*, 23(5):899–929, 2007. (Cited on pages 573 and 611.)

[903] O. Sheluhin, S. Smolskiy, and A. Osin. *Self-Similar Processes in Telecommunications*. Wiley, 2007. (Cited on pages 224 and 610.)

[904] M. Sherman. *Spatial Statistics and Spatio-Temporal Data*. Wiley Series in Probability and Statistics. Chichester: John Wiley & Sons Ltd., 2011. Covariance functions and directional properties. (Cited on page 538.)

[905] K. Shimotsu. Gaussian semiparametric estimation of multivariate fractionally integrated processes. *Journal of Econometrics*, 137(2):277–310, 2007. (Cited on page 536.)

[906] K. Shimotsu. Exact local Whittle estimation of fractionally cointegrated systems. *Journal of Econometrics*, 169(2):266–278, 2012. (Cited on page 537.)

[907] K. Shimotsu and P. C. B. Phillips. Exact local Whittle estimation of fractional integration. *The Annals of Statistics*, 33 (4): 1890–1933, 2005. (Cited on page 572.)

[908] K. Shimotsu and P. C. B. Phillips. Local Whittle estimation of fractional integration and some of its variants. *Journal of Econometrics*, 130(2):209–233, 2006. (Cited on page 572.)

[909] R. H. Shumway and D. S. Stoffer. *Time Series Analysis and Its Applications*, 3rd edition. Springer Texts in Statistics. New York: Springer, 2011. With R examples. (Cited on page 14.)

[910] E. V. Slud. The moment problem for polynomial forms in normal random variables. *The Annals of Probability*, 21(4):2200–2214, 1993. (Cited on pages 280 and 281.)

[911] A. Sly and C. Heyde. Nonstandard limit theorem for infinite variance functionals. *The Annals of Probability*, 36(2):796–805, 2008. (Cited on page 343.)

[912] S. Smirnov. Conformal invariance in random cluster models. II. Scaling limit of the interface. *Annals of Mathematics*, 172(2):1453–1467, 2010. (Cited on page 227.)

[913] A. Smith. Level shifts and the illusion of long memory in economic time series. *Journal of Business & Economic Statistics*, 23(3):321–335, 2005. (Cited on page 225.)

[914] P. J. Smith. A recursive formulation of the old problem of obtaining moments from cumulants and vice versa. *The American Statistician*, 49(2):217–218, 1995. (Cited on page 265.)

[915] T. Sottinen. On Gaussian processes equivalent in law to fractional Brownian motion. *Journal of Theoretical Probability*, 17(2):309–325, 2004. (Cited on page 395.)

[916] P. Soulier. Best attainable rates of convergence for the estimation of the memory parameter. In *Dependence in probability and statistics*, volume 200 of *Lecture Notes in Statist.*, pages 43–57. Berlin: Springer, 2010. (Cited on page 572.)

[917] F. Sowell. *Fractionally integrated vector time series*. Ph.D. dissertation, Duke University, 1986. (Cited on page 537.)

[918] F. B. Sowell. Maximum likelihood estimation of stationary univariate fractionally integrated time series models. *Journal of Econometrics*, 53:165–188, 1992. (Cited on pages 539, 570, and 571.)

[919] F. Spitzer. Uniform motion with elastic collision of an infinite particle system. *Journal of Mathematics and Mechanics*, 18:973–989, 1968/1969. (Cited on pages 163 and 225.)

[920] M. L. Stein. Local stationarity and simulation of self-affine intrinsic random functions. *IEEE Transactions on Information Theory*, 47 (4): 1385–1390, 2001. (Cited on page 112.)

[921] M. L. Stein. Fast and exact simulation of fractional Brownian surfaces. *Journal of Computational and Graphical Statistics*, 11(3):587–599, 2002. (Cited on page 112.)

[922] M. L. Stein. Simulation of Gaussian random fields with one derivative. *Journal of Computational and Graphical Statistics*, 21(1):155–173, 2012. (Cited on page 112.)

[923] F. W. Steutel and K. van Harn. *Infinite Divisibility of Probability Distributions on the Real Line*, volume 259 of *Monographs and Textbooks in Pure and Applied Mathematics*. New York: Marcel Dekker, Inc., 2004. (Cited on page 596.)

[924] S. Stoev and M. S. Taqqu. Wavelet estimation for the Hurst parameter in stable processes. In G. Rangarajan and M. Ding, editors, *Processes with Long-Range Correlations: Theory and Applications*, pages 61–87, Berlin: Springer-Verlag, 2003. Lecture Notes in Physics 621. (Cited on page 112.)

[925] S. Stoev and M. S. Taqqu. Stochastic properties of the linear multifractional stable motion. *Advances in Applied Probability*, 36(4):1085–1115, 2004. (Cited on pages 110 and 611.)

[926] S. Stoev and M. S. Taqqu. Simulation methods for linear fractional stable motion and FARIMA using the fast Fourier transform. *Fractals*, 12(1):95–121, 2004. (Cited on pages 61 and 82.)

[927] S. Stoev and M. S. Taqqu. Asymptotic self-similarity and wavelet estimation for long-range dependent FARIMA time series with stable innovations. *Journal of Time Series Analysis*, 26(2):211–249, 2005. (Cited on page 112.)

[928] S. Stoev and M. S. Taqqu. Path properties of the linear multifractional stable motion. *Fractals*, 13(2):157–178, 2005. (Cited on pages 110 and 611.)

[929] S. Stoev, M. S. Taqqu, C. Park, G. Michailidis, and J. S. Marron. LASS: a tool for the local analysis of self-similarity. *Computational Statistics and Data Analysis*, 50(9):2447–2471, 2006. (Cited on page 112.)

[930] S. A. Stoev and M. S. Taqqu. How rich is the class of multifractional Brownian motions? *Stochastic Processes and their Applications*, 116(2):200–221, 2006. (Cited on page 611.)

[931] T. C. Sun. A central limit theorem for non-linear functions of a normal stationary process. *Journal of Mathematics and Mechanics*, 12:945–978, 1963. (Cited on page 343.)

[932] T. C. Sun. Some further results on central limit theorems for non-linear functions of a normal stationary process. *Journal of Mathematics and Mechanics*, 14:71–85, 1965. (Cited on page 343.)

[933] X. Sun and F. Guo. On integration by parts formula and characterization of fractional Ornstein-Uhlenbeck process. *Statistics & Probability Letters*, 107:170–177, 2015. (Cited on page 395.)

[934] D. Surgailis. Convergence of sums of nonlinear functions of moving averages to self-similar processes. *Doklady Akademii Nauk SSSR*, 257(1):51–54, 1981. (Cited on page 343.)

[935] D. Surgailis. Zones of attraction of self-similar multiple integrals. *Litovsk. Mat. Sb.*, 22(3):185–201, 1982. (Cited on pages 307, 309, 311, and 343.)

[936] D. Surgailis. Long-range dependence and Appell rank. *The Annals of Probability*, 28(1):478–497, 2000. (Cited on page 343.)

[937] D. Surgailis. CLTs for polynomials of linear sequences: diagram formula with illustrations. In *Theory and Applications of Long-Range Dependence*, pages 111–127. Boston, MA: Birkhäuser Boston, 2003. (Cited on pages 244 and 343.)

[938] D. Surgailis and M. Vaičiulis. Convergence of Appell polynomials of long range dependent moving averages in martingale differences. *Acta Applicandae Mathematicae*, 58(1-3):343–357, 1999. Limit theorems of probability theory (Vilnius, 1999). (Cited on page 343.)

[939] G. Szegö. On certain Hermitian forms associated with the Fourier series of a positive function. *Comm. Sém. Math. Univ. Lund [Medd. Lunds Univ. Mat. Sem.]*, 1952 (Tome Supplementaire): 228–238, 1952. (Cited on page 211.)

[940] A.-S. Sznitman. Topics in random walks in random environment. In *School and Conference on Probability Theory*, ICTP Lect. Notes, XVII, pages 203–266 (electronic). Abdus Salam Int. Cent. Theoret. Phys., Trieste, 2004. (Cited on page 178.)

[941] J. Szulga and F. Molz. The Weierstrass-Mandelbrot process revisited. *Journal of Statistical Physics*, 104(5-6):1317–1348, 2001. (Cited on page 227.)

[942] P. D. Tafti and M. Unser. Fractional Brownian vector fields. *Multiscale Modeling & Simulation. A SIAM Interdisciplinary Journal*, 8(5):1645–1670, 2010. (Cited on page 538.)

[943] M. S. Taqqu. Weak convergence to fractional Brownian motion and to the Rosenblatt process. *Zeitschrift für Wahrscheinlichkeitstheorie und verwandte Gebiete*, 31:287–302, 1975. (Cited on pages 110, 281, and 343.)

[944] M. S. Taqqu. Law of the iterated logarithm for sums of non-linear functions of the Gaussian variables that exhibit a long range dependence. *Zeitschrift für Wahrscheinlichkeitstheorie und verwandte Gebiete*, 40:203–238, 1977. (Cited on pages 279 and 281.)

[945] M. S. Taqqu. Convergence of integrated processes of arbitrary Hermite rank. *Zeitschrift für Wahrscheinlichkeitstheorie und verwandte Gebiete*, 50:53–83, 1979. (Cited on pages 280 and 343.)

[946] M. S. Taqqu. The Rosenblatt process. In R. A. Davis, K.-. Lii, and D. N. Politis, editors, *Selected Works of Murray Rosenblatt*, Selected Works in Probability and Statistics, pages 29–45. New York: Springer, 2011. (Cited on page 281.)

[947] M. S. Taqqu and J. Levy. Using renewal processes to generate long-range dependence and high variability. In E. Eberlein and M. S. Taqqu, editors, *Dependence in Probability and Statistics*, pages 73–89, Boston: Birkhäuser, 1986. (Cited on page 224.)

[948] M. S. Taqqu and V. Teverovsky. Semi-parametric graphical estimation techniques for long-memory data. In P. M. Robinson and M. Rosenblatt, editors, *Athens Conference on Applied Probability and Time Series Analysis. Volume II: Time Series Analysis in Memory of E. J. Hannan*, volume 115 of *Lecture Notes in Statistics*, pages 420–432, New York: Springer-Verlag, 1996. (Cited on page 574.)

[949] M. S. Taqqu and V. Teverovsky. Robustness of Whittle-type estimates for time series with long-range dependence. *Stochastic Models*, 13:723–757, 1997. (Cited on page 571.)

[950] M. S. Taqqu and R. Wolpert. Infinite variance self-similar processes subordinate to a Poisson measure. *Zeitschrift für Wahrscheinlichkeitstheorie und verwandte Gebiete*, 62:53–72, 1983. (Cited on page 111.)

[951] M. S. Taqqu, V. Teverovsky, and W. Willinger. Estimators for long-range dependence: an empirical study. *Fractals*, 3 (4): 785–798, 1995. Reprinted in *Fractal Geometry and Analysis*, C. J. G. Evertsz, H.-O. Peitgen and R. F. Voss, editors. Singapore: World Scientific Publishing Co., 1996. (Cited on page 574.)

[952] M. S. Taqqu, W. Willinger, and R. Sherman. Proof of a fundamental result in self-similar traffic modeling. *Computer Communications Review*, 27(2):5–23, 1997. (Cited on page 224.)

[953] M. Tayefi and T. V. Ramanathan. An overview of FIGARCH and related time series models. *Austrian Journal of Statistics*, 41(3):175–196, 2012. (Cited on page 611.)

[954] G. Terdik. *Bilinear Stochastic Models and Related Problems of Nonlinear Time Series Analysis: A Frequency Domain Approach*, volume 142. Springer Science & Business Media, 1999. (Cited on page 610.)

[955] G. Terdik. Long range dependence in third order for non-Gaussian time series. In *Advances in directional and linear statistics*, pages 281–304. Heidelberg: Physica-Verlag/Springer, 2011. (Cited on page 109.)

[956] A. H. Tewfik and M. Kim. Correlation structure of the discrete wavelet coefficients of fractional Brownian motions. *IEEE Transactions on Information Theory*, IT-38(2):904–909, 1992. (Cited on page 112.)

[957] G. Teyssière and P. Abry. Wavelet analysis of nonlinear long-range dependent processes. Applications to financial time series. In *Long Memory in Economics*, pages 173–238. Berlin: Springer, 2007. (Cited on page 112.)

[958] G. Teyssière and A. P. Kirman. *Long Memory in Economics*. Berlin, Heidelberg: Springer 2006. (Cited on page 610.)

[959] S. Tindel, C. A. Tudor, and F. Viens. Stochastic evolution equations with fractional Brownian motion. *Probability Theory and Related Fields*, 127(2):186–204, 2003. (Cited on page 436.)

[960] S. Torres, C. A. Tudor, and F. G. Viens. Quadratic variations for the fractional-colored stochastic heat equation. *Electronic Journal of Probability*, 19: no. 76, 51, 2014. (Cited on page 227.)

[961] W.-J. Tsay. Maximum likelihood estimation of stationary multivariate ARFIMA processes. *Journal of Statistical Computation and Simulation*, 80(7-8):729–745, 2010. (Cited on page 570.)

[962] C. A. Tudor. Analysis of the Rosenblatt process. *ESAIM. Probability and Statistics*, 12:230–257, 2008. (Cited on page 280.)

[963] C. A. Tudor. *Analysis of Variations for Self-Similar Processes*. Probability and its Applications (New York). Springer, Cham, 2013. A Stochastic Calculus Approach. (Cited on pages 343, 436, and 610.)

[964] C. A. Tudor and F. G. Viens. Statistical aspects of the fractional stochastic calculus. *The Annals of Statistics*, 35(3):1183–1212, 2007. (Cited on page 436.)

[965] C. A. Tudor and Y. Xiao. Sample path properties of bifractional Brownian motion. *Bernoulli*, 13(4):1023–1052, 2007. (Cited on page 110.)

[966] C. A. Tudor and Y. Xiao. Sample paths of the solution to the fractional-colored stochastic heat equation. *Stochastics and Dynamics*, 17(1):1750004, 2017. (Cited on page 227.)

[967] V. V. Uchaikin and V. M. Zolotarev. *Chance and Stability*. Modern Probability and Statistics. Utrecht: VSP, 1999. Stable distributions and their applications, With a foreword by V. Yu. Korolev and Zolotarev. (Cited on pages 111 and 227.)

[968] M. Unser and T. Blu. Cardinal exponential splines. I. Theory and filtering algorithms. *IEEE Transactions on Signal Processing*, 53(4):1425–1438, 2005. (Cited on page 465.)

[969] M. Unser and T. Blu. Self-similarity. I. Splines and operators. *IEEE Transactions on Signal Processing*, 55(4):1352–1363, 2007. (Cited on page 465.)

[970] A. Van der Ziel. Flicker noise in electronic devices. *Advances in Electronics and Electron Physics*, 49:225–297, 1979. (Cited on page 225.)

[971] K. van Harn and F. W. Steutel. Integer-valued self-similar processes. *Communications in Statistics. Stochastic Models*, 1(2):191–208, 1985. (Cited on page 110.)

[972] H. van Zanten. Comments on "PCA based Hurst exponent estimator for fBm signals under disturbances". *IEEE Transactions on Signal Processing*, 58(8):4466–4467, 2010. (Cited on page 464.)

[973] S. R. S. Varadhan. Self-diffusion of a tagged particle in equilibrium for asymmetric mean zero random walk with simple exclusion. *Annales de l'Institut Henri Poincaré. Probabilités et Statistiques*, 31(1):273–285, 1995. (Cited on page 226.)

[974] D. E. Varberg. Convergence of quadratic forms in independent random variables. *Annals of Mathematical Statistics*, 37:567–576, 1966. (Cited on page 281.)

[975] J. Veenstra and A. I. McLeod. Hyperbolic decay time series models. *In press*, 2012. (Cited on page 572.)

[976] J. Q. Veenstra. *Persistence and Anti-persistence: Theory and Software*. Ph.D. thesis, Western University, 2012. (Cited on page 555.)

[977] M. S. Veillette and M. S. Taqqu. A technique for computing the PDFs and CDFs of nonnegative infinitely divisible random variables. *Journal of Applied Probability*, 48(1):217–237, 2011. (Cited on page 271.)

[978] M. S. Veillette and M. S. Taqqu. Berry-Esseen and Edgeworth approximations for the normalized tail of an infinite sum of independent weighted gamma random variables. *Stochastic Processes and their Applications*, 122(3):885–909, 2012. (Cited on page 271.)

[979] M. S. Veillette and M. S. Taqqu. Properties and numerical evaluation of the Rosenblatt distribution. *Bernoulli*, 19(3):982–1005, 2013. (Cited on pages 264, 265, 266, 267, 270, 271, 272, 280, and 281.)

[980] M. S. Veillette and M. S. Taqqu. Supplement to "Properties and numerical evaluation of the Rosenblatt distribution". 2013. (Cited on pages 266, 271, 272, and 281.)

[981] D. Veitch and P. Abry. A wavelet-based joint estimator of the parameters of long-range dependence. *IEEE Transactions on Information Theory*, 45(3):878–897, 1999. (Cited on page 111.)

[982] D. Veitch, M. S. Taqqu, and P. Abry. On the automatic selection of the onset of scaling. *Fractals*, 11(4):377–390, 2003. (Cited on page 112.)

[983] D. Veitch, A. Gorst-Rasmussen, and A. Gefferth. Why FARIMA models are brittle. *Fractals*, 21(2):1350012, 12, 2013. (Cited on page 109.)

[984] C. Velasco. Gaussian semiparametric estimation of non-stationary time series. *Journal of Time Series Analysis*, 20(1):87–127, 1999. (Cited on page 572.)

[985] C. Velasco. Non-stationary log-periodogram regression. *Journal of Econometrics*, 91(2):325–371, 1999. (Cited on pages 111 and 573.)

[986] C. Velasco. Non-Gaussian log-periodogram regression. *Econometric Theory*, 16:44–79, 2000. (Cited on page 111.)

[987] C. Velasco. Gaussian semi-parametric estimation of fractional cointegration. *Journal of Time Series Analysis*, 24(3):345–378, 2003. (Cited on page 536.)

[988] C. Velasco and P. M. Robinson. Whittle pseudo-maximum likelihood estimation for nonstationary time series. *Journal of the American Statistical Association*, 95(452):1229–1243, 2000. (Cited on pages 546 and 571.)

[989] S. Veres and M. Boda. The chaotic nature of TCP congestion control. In *Proceedings of IEEE INFOCOM*, volume 3, pages 1715–1723, 2000. (Cited on page 225.)

[990] W. Vervaat. Sample paths of self-similar processes with stationary increments. *The Annals of Probability*, 13:1–27, 1985. (Cited on pages 108 and 110.)

[991] W. Vervaat. Properties of general self-similar processes. *Bulletin of the International Statistical Institute*, 52(Book 4):199–216, 1987. (Cited on page 110.)

[992] B. Vollenbröker. Strictly stationary solutions of ARMA equations with fractional noise. *Journal of Time Series Analysis*, 33(4):570–582, 2012. (Cited on page 109.)

[993] S. Wainger. Special trigonometric series in k-dimensions. *Memoirs of the American Mathematical Society*, 59:102, 1965.(Cited on pages 526, 527, and 528.)

[994] L. Wang. Memory parameter estimation for long range dependent random fields. *Statistics & Probability Letters*, 79(21):2297–2306, 2009. (Cited on page 537.)

[995] W. Wang. Almost-sure path properties of fractional Brownian sheet. *Annales de l'Institut Henri Poincaré. Probabilités et Statistiques*, 43(5):619–631, 2007. (Cited on page 537.)

[996] Y. Wang. An invariance principle for fractional Brownian sheets. *Journal of Theoretical Probability*, 27(4):1124–1139, 2014. (Cited on page 537.)

[997] Z. Wang, L. Yan, and X. Yu. Weak approximation of the fractional Brownian sheet from random walks. *Electronic Communications in Probability*, 18:no. 90, 13, 2013. (Cited on page 537.)

[998] L. M. Ward and P. E Greenwood. 1/f noise. *Scholarpedia*, 2(12):1537, 2007. revision #90924. (Cited on page 227.)

[999] G. N. Watson. *A Treatise on the Theory of Bessel Functions*. Cambridge Mathematical Library. Cambridge: Cambridge University Press, 1995. Reprint of the second (1944) edition. (Cited on page 458.)

[1000] K. Weierstrass. Über continuirliche Functionen eines reellen Arguments, die für keinen Werth des letzteren einen bestimmten Differentialquotienten besitzen. Presented at the Königlische Akademie der Wissenschaften in Berlin on 18 July 1872. Published in Volume 2 of his complete works, pages 71-74, see Weierstrass (1894-1927), 1872. (Cited on page 216.)

[1001] K. Weierstrass. *Matematische Werke*. Berlin and Leipzig: Mayer & Muller, 1894–1927. 7 volumes. (Cited on page 216.)

[1002] H. L. Weinert. *Reproducing Kernel Hilbert Spaces: Applications in Statistical Signal Processing*. Stroudsburg, PA: Hutchinson Ross, 1982. (Cited on page 384.)

[1003] H. Wendt, A. Scherrer, P. Abry, and S. Achard. Testing fractal connectivity in multivariate long memory processes. In *Acoustics, Speech and Signal Processing, 2009. ICASSP 2009. IEEE International Conference on*, pages 2913–2916. IEEE, 2009. (Cited on page 537.)

[1004] W. Werner. Random planar curves and Schramm-Loewner evolutions. In *Lectures on probability theory and statistics*, volume 1840 of *Lecture Notes in Math.*, pages 107–195. Berlin: Springer, 2004. (Cited on page 227.)

[1005] P. Whittle. *Hypothesis Testing in Time Series Analysis*. New York: Hafner, 1951. (Cited on page 570.)

[1006] P. Whittle. The analysis of multiple stationary time series. *Journal of the Royal Statistical Society. Series B. Methodological*, 15:125–139, 1953. (Cited on page 570.)

[1007] P. Whittle. On the variation of yield variance with plot size. *Biometrika*, 43:337–343, 1956. (Cited on page 109.)

[1008] R. J. Hyndman with contributions from G. Athanasopoulos, S. Razbash, D. Schmidt, Z. Zhou, Y. Khan, and C. Bergmeir. *forecast: Forecasting functions for time series and linear models*, 2013. URL http://CRAN.R-project.org/package=forecast. R package version 4.8. (Cited on page 555.)

[1009] R. L. Wolpert and M. S. Taqqu. Fractional Ornstein–Uhlenbeck Lévy processes and the Telecom process: Upstairs and downstairs. *Signal Processing*, 85(8):1523–1545, 2005. (Cited on page 111.)

[1010] A. T. A. Wood and G. Chan. Simulation of stationary Gaussian processes in $[0, 1]^d$. *Journal of Computational and Graphical Statistics*, 3(4):409–432, 1994. (Cited on page 112.)

[1011] G. Wornell. *Signal Processing with Fractals: A Wavelet-Based Approach*. Upper Saddle River, NJ: Prentice Hall PTR, 1996. (Cited on page 112.)

[1012] W. B. Wu. Empirical processes of long-memory sequences. *Bernoulli*, 9(5):809–831, 2003. (Cited on page 344.)

[1013] W. B. Wu and X. Shao. Invariance principles for fractionally integrated nonlinear processes. In *Recent Developments in Nonparametric Inference and Probability*, volume 50 of *IMS Lecture Notes Monogr. Ser.*, pages 20–30. Beachwood, OH: Inst. Math. Statist., 2006. (Cited on page 611.)

[1014] W. B. Wu and X. Shao. A limit theorem for quadratic forms and its applications. *Econometric Theory*, 23(5):930–951, 2007. (Cited on page 571.)

[1015] W. B. Wu, Y. Huang, and W. Zheng. Covariances estimation for long-memory processes. *Advances in Applied Probability*, 42(1): 137–157, 2010. (Cited on page 574.)

[1016] D. Wuertz, many others, and see the SOURCE file. *fArma: ARMA Time Series Modelling*, 2013. URL http://CRAN.R-project.org/package=fArma. R package version 3010.79. (Cited on page 572.)

[1017] Y. Xiao. Sample path properties of anisotropic Gaussian random fields. In *A minicourse on stochastic partial differential equations*, volume 1962 of *Lecture Notes in Math.*, pages 145–212. Berlin: Springer, 2009. (Cited on page 537.)

[1018] J. Xiong and X. Zhao. Nonlinear filtering with fractional Brownian motion noise. *Stochastic Analysis and Applications*, 23(1):55–67, 2005. (Cited on page 436.)

[1019] A. M. Yaglom. Correlation theory of processes with stationary random increments of order n. *Matematicheski Sbornik*, 37:141–196, 1955. English translation in *American Mathematical Society Translations Series 2* **8**(1958), 87-141. (Cited on page 110.)

[1020] A. M. Yaglom. Some classes of random fields in n-dimensional space, related to stationary random processes. *Theory of Probability and its Applications*, II (3):273–320, 1957. (Cited on pages 476, 514, and 530.)

[1021] A. M. Yaglom. *Correlation Theory of Stationary and Related Random Functions. Volume I: Basic Results*. Springer Series in Statistics. Springer, 1987. (Cited on pages 476 and 514.)

[1022] Y. Yajima. On estimation of long-memory time series models. *Australian Journal of Statistics*, 27 (3): 303–320, 1985. (Cited on page 570.)

[1023] C. Y. Yau and R. A. Davis. Likelihood inference for discriminating between long-memory and change-point models. *Journal of Time Series Analysis*, 33(4):649–664, 2012. (Cited on page 225.)

[1024] L. C. Young. An inequality of the Hölder type, connected with Stieltjes integration. *Acta Mathematica*, 67(1):251–282, 1936. (Cited on page 408.)

[1025] P. Zaffaroni. Contemporaneous aggregation of linear dynamic models in large economies. *Journal of Econometrics*, 120(1):75–102, 2004. (Cited on page 223.)

[1026] P. Zaffaroni. Aggregation and memory of models of changing volatility. *Journal of Econometrics*, 136(1):237–249, 2007. ISSN 0304-4076. (Cited on page 223.)

[1027] M. Zähle. Integration with respect to fractal functions and stochastic calculus. I. *Probability Theory and Related Fields*, 111(3):333–374, 1998. (Cited on page 435.)

[1028] T. Zhang, H.-C. Ho, M. Wendler, and W. B. Wu. Block sampling under strong dependence. *Stochastic Processes and their Applications*, 123(6):2323–2339, 2013. (Cited on page 574.)

[1029] V. M. Zolotarev. *One-dimensional Stable Distributions*, volume 65 of *"Translations of mathematical monographs"*. American Mathematical Society, 1986. Translation from the original 1983 Russian edition. (Cited on pages 111 and 585.)

[1030] A. Zygmund. *Trigonometric Series. 2nd ed. Vols. I, II*. New York: Cambridge University Press, 1959. (Cited on pages 291, 353, and 575.)

Index

$1/f$ noise, 227
$S\alpha S$ Lévy motion, 59, 60
   convergence to, 166

Abelian theorem, 109
aggregated variance method, 88
aggregation of SRD series, 113
   beta-like mixture distribution, 116
   disaggregation problem, 117
   long-range dependence, 115
   mixture distribution, 117
anomalous diffusion, 227, 396
antipersistent series, 31, 38, 40, 72, 109
Appell's hypergeometric function, 390
applications of LRD, 572, 610
autocorrelation function (ACF), 4
autocovariance function (ACVF), 4
   AR series, 6
   FARIMA$(0, d, 0)$ series, 37
   FARIMA(p, d, q) series, 43, 569, 570
   FGN, 23, 67
   function of Gaussian series, 285
   MA series, 5
   spatial setting, 471
   vector FARIMA$(0, D, 0)$ series, 500
   vector setting, 469
   white noise, 5
autoregressive (AR) series, 5, 8, 9, 339
   aggregation, 114
autoregressive moving average (ARMA) series, 6, 31, 570
   characteristic polynomial, 32
   short-range dependence, 31

backshift operator, 31, 357, 531
   forward-shift operator, 499
Barnes G-function, 212
Bessel function of the first kind, 458
beta function, 24
bifractional Brownian motion (biFBM), 53, 110, 216
Boltzmann (Gibbs) distribution, *see* Ising model
bridge process, 86

Brownian bridge, 87
Brownian motion
   convergence to, 146
Brownian motion (BM), 2, 49, 54, 59, 62, 166
   convergence to, 118, 158, 221, 298, 316
   integral representations, 13
   integrated, 166
   Karhunen–Loève decomposition, 438
   Lamperti transformation, 66
   Paley–Wiener expansion, 456, 462
   two-parameter, 215
   wavelet-based expansion (midpoint displacement), 452
   Wiener integral, 371

Campbell's theorem, 125
Caputo, *see* fractional derivative
Cauchy determinant, 213
Cauchy's integral formula, 206
central limit theorem, 298
   fourth moment condition, 305
   functional of Markov chains, 161
   Lindeberg–Feller, 92
   martingale, 177
   multilinear processes, 327
   multivariate extension, 316
   multivariate extension (mixed case), 322
characteristic polynomial
   ARMA series, 32
   matrix, 468
Cholesky, *see* generation of Gaussian series
circulant matrix, 100, 201
   eigenvectors and eigenvalues, 100
codifference, 83
completely monotone function, 530, 535
contraction, 305, 323, 600
convergence
   conditional, 25
   Cramér–Wold device, 584
   in $L^2(\Omega)$–sense, 5
   in distribution, of random variables, vectors and processes, 584

vague (locally weak), of locally finite measures, 584
weak, of locally finite measures, 584
weak, of probability measures, 583
convergence to types theorem, 68
convolution
   Fourier transform, 576, 577
   functions, 577
   sequences, 7, 576
correlated random walk (CRW), 117, 162, 173
covariance function, 2
   biFBM, 53
   covariance function, 216
   FBM, 47
   OU process, 13
   VFBM, 490
cumulant, 244
   joint, 244

diagram
   association with multigraphs, 247
   connected, 244
   definition, 241
   formula for cumulants of Hermite polynomials of Gaussian variables, 253
   formula for cumulants of multiple integrals, 245
   formula for moments of Hermite polynomials of Gaussian variables, 251
   formula for moments of multiple integrals (spectral domain), 244
   formula for moments of multiple integrals (time domain), 242
   Gaussian, 242
   non-flat, 242
difference (differencing) operator, 3, 36
digamma function, 98
discrete-chaos process, *see* multilinear process
Durbin–Levinson algorithm, 107, 541

elastic collision of particles, 162, 168
entire function, 306
ergodic theorem, 398, 405
estimation in LRD series
   aggregated variance method, 88
   Bayesian, 573
   bootstrap, 574
   impulse response coefficients, 573
   KPSS method, 111
   maximum likelihood, *see* maximum likelihood estimation (MLE)
   R/S method, 84
   regression in spectral domain (GPH estimation), 88
   regression with LRD errors, 573
   resampling (bootstrap), 574
   seasonal models, 573
   self-normalization, 574
   state space modeling, 573
   subsampling, 574
   V/S method, 111
   wavelet-based, 93
estimation of ACVF and ACF, 574
estimation of mean, 540
   best linear unbiased estimator (BLUE), 540, 569
Euler's constant, 214
exponent rank, 306

FARIMA models, 35
   data applications, 555
   disaggregation, 117
   equivalence of LRD conditions, 43
   FARIMA$(0, d, 0)$ series, 35, 542, 550
   FARIMA(p, d, q) series, 42, 539
   generation, 105
   heavy-tailed, 82, 573
   invertibility, 38
   principal range, 41
   spatial FARIMA, 530–532
   time plots, 40, 82
   vector FARIMA series, 499
   wavelet-based expansion of FBM, 449
fast Fourier transform (FFT), 9, 61, 101, 524, 544
fast wavelet transform (FWT), 97, 442, 454
   complexity, 443
   fast biorthogoal wavelet transform, 450
Fernique's inequality, 409, 422
filter
   high-pass, 441
   linear, 8, 33, 39
   low-pass, 441
finite-dimensional distributions, 1
Fourier frequencies, 9, 558
Fourier series
   ACVF and spectral density, 576
   convolution, 577
   definition, 575
   Fourier coefficients, 575
   Fourier transform, 576
   Parseval's identity, 576
   regularly varying sequences, 579, 583
Fourier transform (functions)
   convolution, 578
   definition, 577
   extensions, 578
   inversion formula, 577
   Parseval's identity, 577
   Plancherel's identity, 577
fourth moment condition, 305
fractional Brownian motion (FBM)
   pth variation, 398
   arbitrage, 435
   complex-valued, 455
   convergence to, 64, 71, 72, 119, 132, 152, 164, 166, 172, 177, 220
   deconvolution formula, 383
   definition, 47

Fernique's inequality, 409, 422
Gaussian space, 369
geometric, 412, 425
Girsanov's formula, 387
Hölder continuity, 398
intersection local time, 436
Karhunen–Loève decomposition, 439
kernel function on interval, 359, 361, 363
large deviations, 611
linear filtering, 392
linear span and its closure, 369
local time, 430, 433
not semimartingale, 398
operator field, *see* operator fractional Brownian field (OFBF)
Paley–Wiener expansion, 462
parametrization, 359
prediction formula, 389
quadratic variation, 427
realizations, 53
representation on interval, 364
RKHS, 385
spectral domain representation, 51, 368
standard, 47
tempered, 395
time domain representation, 48, 49, 63, 368
types I and II, 110
vector operator, *see* vector operator fractional Brownian motion (vector OFBM)
wavelet-based expansion, 452
wavelet-based simulation, 454
fractional Brownian sheet (FBS), 465, 525
fractional cointegration
cointegrating rank, 505
cointegrating spaces, 507
cointegrating vector, 504
definition, 504
fractional cointegration model, 506
fractionally cointegrated, 504
vector OFBM, 507
fractional derivative
Caputo, 352
composition with fractional integral, 350, 356
exponential function, 353
Fourier transform, 358
function of two variables, 400
Grünwald–Letnikov, 359
in representation of FBM, 364, 368
indicator functions, 353, 356, 358
Liouville (on real line), 352, 356
Marchaud (on real line), 355, 356
partie finie (finite part), 356
power functions, 350
renormalization, 355, 357
Riemann–Liouville (on interval), 349
weighted, 375
fractional diffusion equation, 396
fractional Gaussian noise (FGN), 23, 68, 108

ergodic, 398
vector, 503, 508
fractional integral
composition with fractional derivative, 350
exponential function, 353
Fourier transform, 358
fractional integration by parts, 347, 354
function of two variables, 400
in representation of FBM, 364, 368
indicator functions, 353, 358
Liouville (on real line), 352
power functions, 347
reflection property, 347, 354
Riemann–Liouville (on interval), 346
semigroup property, 347, 354
fractional Ornstein–Uhlenbeck (OU) process, 384, 394
fractional Wiener integral
class of integrands, 370
classes of integrands for $\kappa \in (-1/2, 0)$, 375, 376, 379
classes of integrands for $\kappa \in (0, 1/2)$, 372, 377, 378
classes of integrands on real line, 383
completeness, 370–372, 375, 377, 378
connection to RKHS, 384
covariance, 374, 377
deconvolution formula, 383
definition, 370
definition for $\kappa \in (-1/2, 0)$, 375
definition for $\kappa \in (0, 1/2)$, 374
elementary (step) function, 370
fractional OU process, 384, 394
fundamental martingale, 381, 386, 456
Gaussian space, 369
Girsanov's formula, 387
linear filtering, 392
prediction formula, 389
Fubini-based argument, 478
functions of Gaussian variables, 282
covariance, 284, 331
exponent rank, 306
Hermite rank, 285
long-range dependence, 286
matching marginal distribution, 329
power rank, 313

gamma function, 24
Gauss hypergeometric function, 266, 363
Gaussian subordination, *see* functions of Gaussian variables
Gegenbauer ARMA (GARMA) models, 109
G frequency (Gegenbauer frequency), 110
generalized dominated convergence theorem, 80
generalized Hermite process, 276
kernel, 274
generalized inverse, 129, 329
regularly varying function, 132, 586

generalized Riemann zeta function, 98
generation of Gaussian series
   Cholesky method, 100
   circulant matrix embedding method, 100
   extensions, 112
generation of non-Gaussian series, 328
   circulant matrix embedding, 337
   matching autocovariance, 336
   matching marginal distribution, 329
   Price theorem, 333
geometric FBM, 412, 425
Girsanov's formula, 387
GPH estimation
   estimator, 89, 560, 573

Hölder continuity, 379
Hölder norm, 424
heavy-tailed random variable
   definition, 586
   infinite source Poisson model, 121
heavy-tailed time series
   definition, 586
   FARIMA series, 82
   linear, 77
   long-range dependence, 76, 82
   short-range dependence, 77, 82
Hermite expansion, 284
   coefficients, 284, 341
   exponential function, 331
   Hermite rank, 285
   indicator function, 332
Hermite polynomial
   connection to multiple integrals, 230
   definition, 229
   recursion property, 231, 279
Hermite process
   convergence to, 288, 307, 313
   cumulants, 258
   generalized, 276
   moments, 254
   representation on interval, 234, 240
   representation on positive half-axis, 237, 239, 240
   spectral domain representation, 234, 239, 240
   time domain representation, 232, 239, 240
Hermite rank, 285
Hermitian, *see* matrix analysis
Hermitian function, 599
hierarchical model, 154
Hilbert space, 370
Hilbert–Schmidt theorem, 438
homogeneous function, 509
Hurst index (parameter), 44

inequality
   generalized Minkowski's, 353
   geometric and arithmetic means, 303
   Young's, 430

infinite source Poisson model, 120
   convergence to BM, 146
   convergence to FBM, 132
   convergence to intermediate Telecom process, 144
   convergence to stable Lévy motion, 140, 147
   convergence to Telecom process, 133
   input rate regimes (fast, slow and intermediate), 129
   long-range dependence, 128
   related models, 224
   workload process, 123
infrared correction, 49, 217
inner product space, 370
   complete, 370
innovations algorithm, 107, 541
integral representation, 11
integrated series, 6
Ising model, 180
   critical temperature, 182
   Boltzmann (Gibbs) distribution, 182
   boundary conditions, 183
   configurations, 182
   counting lattice, 187
   Curie temperature, 183
   dimer, 189
   energy (Hamiltonian), 182
   external magnetic field, 182
   ferromagnetism, 183
   inverse temperature, 182
   Ising lattice, 181
   long-range dependence, 183
   one-dimensional, 223
   Pfaffian, 189, 192
   phase transition, 182
   spins, 183
   strong Szegö limit theorem, 209
   temperature, 182
   thermodynamic limit, 181
Itô's formulas, 407, 433
Itô–Nisio theorem, 463

Karamata's theorem, 20
Karhunen–Loève decomposition, 438
   BM, 438
   FBM, 439
Kolmogorov's continuity criterion, 398
Kolmogorov's formula, 544

Lévy, *see* random measure
Lamperti transformation, 66
   BM, 66
Lamperti's limit theorem, 69
   vector setting, 487
Langevin stochastic differential equation, 3, 384
large deviations of FBM, 611
Leibniz's rule for differentiation under integral sign, 349

linear filter, 8, 33, 39
linear fractional Lévy motion, 63
linear fractional stable motion (LFSM)
   convergence to, 65, 78
   definition, 60
   integral representation, 59
linear fractional stable noise (LFSN), 77, 83
linear time series, 6
   heavy-tailed, 77
Liouville, *see* fractional derivative, *see* fractional integral
local time, 178, 430
local Whittle estimation
   bandwidth selection, 562
   bias reduction, 565
   bias-variance tradeoff, 563
   estimator, 559
   GPH estimator, 560
   large sample asymptotics, 560
   local polynomial estimator, 566
   mean squared error (MSE), 564
   non-smooth models, 564
   optimal bandwidth, 565
   optimal rate, 566
   rate optimality, 565
   smooth models, 564
   tapering, 565
   Whittle plot, 562
log-fractional stable motion (log-FSM), 62, 108
long memory, *see* long-range dependence
long-range dependence (LRD)
   conditions I-V, 17
   connections to self-similar processes, 68
   definition, 18
   infinite variance, 76
   long-run variance, 28
   mixing, 612
   non-linear time series, 610
   nonstationarity, 573
   origins, 108
   parameter, 18
   point processes, 610
   spatial, *see* spatial long-range dependence (LRD)
   vector, *see* vector long-range dependence (LRD)
long-run variance
   LRD series, 23, 28
   SRD series, 31

Malliavin calculus
   chain rule, 415, 604
   chaos expansion, 431, 605
   derivative operator, 400, 603, 606
   divergence (Skorokhod) integral, 399, 607
   generator of OU semigroup, 426, 608
   integration by parts formula, 604
   isonormal Gaussian process, 399, 602
   smooth random variables, 603
   space $\mathbb{D}^{1,p}$, 604
   space $\mathbb{D}^{1,p}_{\text{loc}}$, 604
   space $\mathbb{D}^{k,p}$, 605
   Stroock formula, 431, 606
   white noise case, 602, 603, 605, 606
Marchaud, *see* fractional derivative
Markov chain, 33
   β–regular, 228
   central limit theorem, 161
   short-range dependence, 33
Markov switching model, 156
   convergence to BM, 158
matrix analysis
   centralizer, 494
   characteristic polynomial, 468
   diagonal matrix, 468
   eigenvalue, 468
   eigenvector, 468
   Hermitian symmetric matrix, 468
   Hermitian transpose, 467
   Jordan block, 468
   Jordan decomposition, 468, 477, 533
   matrix exponential, 472, 533
   matrix norm, 469
   positive (semi)definite matrix, 468
   primary matrix function, 480
   singular, 468
   square root of positive semidefinite matrix, 468, 478
maximum likelihood estimation (MLE)
   applications, 572
   autoregressive approximation, 550
   best linear predictor, 541
   broadband Whittle, 567
   Durbin–Levinson algorithm, 541
   exact Gaussian MLE, 539
   FARIMA$(0, d, 0)$ series, 550
   finite (small) samples, 550, 571
   forecasting, 554
   goodness of fit, 552
   ill conditioning, 542
   implementation, 570
   information criteria (AIC, BIC, HIC), 552
   innovations algorithm, 541
   large sample asymptotics, 545, 551, 570
   likelihood function, 540
   local Whittle, *see* local Whittle estimation
   maximum likelihood estimator, 540
   model selection, 551
   portmanteau tests, 552
   quadratic form, 547
   R packages, 555, 571
   spectral window, 553
   temperature data, 555, 557
   Whittle's approximation and estimation, 542, 569
Mehler's formula, 342
method of moments, 300, 317

mixing, 612
mixture of correlated random walks, 117
   convergence to FBM, 119
moving average (MA) series, 5
multifractal processes, 611
multifractional motions, 611
multigraph
   k-regular, 246
   (non-)standard, 302
   association with diagrams, 247
   connected, 248
   cumulants of Hermite process, 258
   cumulants of multiple integral of order two, 259
   definition, 246
   degree, 246
   formula for cumulants of Hermite polynomials of Gaussian variables, 253
   formula for cumulants of multiple integrals, 250
   formula for moments of Hermite polynomials of Gaussian variables, 253
   formula for moments of multiple integrals, 250
   moments of Hermite process, 254
   multiplicity number, 246
   pair sequence (set of lines), 246
multilinear process, 324
   limit theorem, 327
multinomial occupancy scheme, 228
multiple integral of order two, 259
   characteristic function, 262
   cumulants, 259
   series representation, 261
multiple integrals with respect to Gaussian measures
   change of variables formula, 599
   connection between time and spectral domains, 600
   connection to Itô stochastic integrals, 601
   contraction, 600
   covariance, 598
   definition, 598
   Hermitian case, 599
   Hermitian function, 599
   product, 600, 601
   simple function, 597
   stochastic Fubini theorem, 601
   symmetrization, 598
   tensor product, 600
multiresolution analysis (MRA), *see* wavelet analysis

Nile river time series, 87, 93, 99, 108, 570
non-central limit theorem, 288
   linear time series (entire functions), 307
   linear time series (martingale differences), 313
   multilinear processes, 327
   multivariate extension, 320
   multivariate extension (mixed case), 322

ON/OFF model, 224
operator fractional Brownian field (OFBF)
   admissible functions, 521
   anisotropic fractional Brownian field (anisotropic FBF), 525
   definition, 508
   domain symmetry group, 509
   fractional Brownian sheet (FBS), 525
   homogeneous function, 509, 511
   identifiability, 509
   isotropic fractional Brownian field (isotropic FBF), 524
   operator scaling, 508
   polar coordinate representation, 510
   sample path properties, 537
   spectral domain representation, 513
   symmetry group, 510
   tangent space of symmetry group, 509
   time domain representation, 516, 521
operator scaling Gaussian random field, *see* operator fractional Brownian field (OFBF)
operator self-similarity, 472
   operator stable distribution, 535
operator stable distribution, 535
order statistics, *see* simple symmetric exclusion
Ornstein–Uhlenbeck (OU) process, 2, 3
   integral representations, 13
   Lamperti transformation, 66
orthant probability for normal vector, 332

Pólya's urn, 172
   ancestral line, 173
   convergence to FBM, 177
   correlated random walk, 173
Paley–Wiener expansion of FBM, 462
   Bessel function of the first kind, 458
   BM, 456, 462
   fundamental martingale, 456
   isometric Hilbert spaces, 457
   uniform convergence, 463
   zeros of Bessel functions, 461
Parseval's identity, 576, 578
partial autocorrelation function, 107
periodogram, 9
Pfaffian, *see* Ising model
Plancherel's identity, 578
Pochhammer symbol, 363
Poisson, *see* random measure
polar coordinate representation, *see* operator fractional Brownian field (OFBF)
Potter's bound, 19
Potters bound, 21
power rank, 313
Price theorem, 333

quadratic variation, 382, 398
quasi-monotonicity, 25, 497, 579

R/S method, 84
    R/S plot (poxplot), 87
    R/S statistic, 85
random field, 466
    conditions for ACVF, 530
    stationary, 471
random measure
    σ-additive, 588
    compensated Poisson, 594
    control measure, 588
    Gaussian, 590
    Hermitian Gaussian, 590
    independently scattered, 588
    Lévy, 596
    Poisson, 593
    stable, 592
    with orthogonal increments, 589
random walk, 6
    in random environment, 178
    in random scenery, 177
reduction theorem, 292
regime switching, 156
regression in spectral domain, *see* GPH estimation
regularly varying function, 16
    uniform convergence theorem, 19
renewal process, 224
renewal-reward process, 224
reproducing kernel Hilbert space (RKHS), 384
    FBM, 385
reversion operation, 340
Riemann zeta function, 267
    generalized, 267
Riemann–Liouville, *see* fractional integral, *see* fractional derivative
Riemann–Stieltjes integral, 408
Rosenblatt distribution
    CDF, 270, 272
    CDF plots, 272
    CDF table, 273
    characteristic function, 265
    cumulants, 265
    definition, 264
    Lévy–Khintchine representation, 268
    PDF, 270
    PDF plots, 272
    series representation, 266
Rosenblatt process
    characteristic function, 263
    convergence to, 74
    cumulants, 263
    definition, 58, 263
    integral representations, 57

Schramm–Loewner evolution (SLE), 227
self-similar Markov processes, 612
self-similar process
    asymptotically at infinity, 64, 145
    basic properties, 45

    definition, 43
    locally asymptotically, 65, 145
    operator, *see* operator self-similarity
    second-order, 144
    self-similarity parameter, 44
semimartingale, 398
short-range dependence (SRD)
    ARMA series, 31
    definition, 30
    infinite variance, 76
    Markov chain, 33
    vector, 497
shot noise model, 149
    convergence to FBM, 152
    long-range dependence, 151
    shots, 150
simple symmetric exclusion, 167
    convergence to FBM, 172
    extensions, 225
    order statistics, 168
    stirring motion, 168
    tagged particle, 167
single integral with respect to random measure
    compensated Poisson measure, 595
    connection for Lévy and Poisson measures, 597
    connection for stable and Poisson measures, 596
    Gaussian measure, 590
    Lévy measure, 597
    Poisson measure, 594
    random measure with orthogonal increments, 590
    simple (elementary) function, 589
    spectral domain, 592
    stable measure, 593
    time domain, 592
slowly varying function, 16
    Karamata's theorem, 20
    Potter's bound, 19
    quasi-monotone, 25
    Zygmund class, 25
spatial long-range dependence (LRD)
    anisotropic, 528
    Cauchy family, 530
    condition IV-s-anisot, 528
    conditions II-s-isot and IV-s-isot, 526
    isotropic, 526
    scaling transition, 529
    spherical harmonics, 526
spatial process, *see* random field
spectral density, 7
    AR series, 8, 9
    FARIMA$(0, d, 0)$ series, 37
    FARIMA(p, d, q) series, 43
    FGN, 67
    spatial setting, 471
    vector setting, 469
    white noise, 7
spectral domain, 7

spectral measure, 10
spectral representation, 9
stable Lévy motion, 123
   transition probability, 396
stable random measure, 592
stable random variable
   definition, 585
   domain of attraction, 585
   symmetric α-stable ($S\alpha S$), 585
stable self-similar processes with stationary
     increments, 611
stationarity
   fields, 471
   strict, 3
   vector, 469
   weak or second order, 4
stationarity of increments, 3, 44
Stein's method, 426
   quadratic variation of FBM, 427
   total variation distance, 426
Stirling's formula, 36
stirring motion, *see* simple symmetric exclusion
stochastic differential equation (SDE) for FBM, 413
   geometric FBM, 425
   derivative of solution, 414
   Doss–Sussmann transformation, 413
   ergodicity, 436
   Euler scheme, 419
   existence and uniqueness of solution, 413
   modified Euler scheme, 419
   numerical solutions, 418
   rate of convergence of numerical solutions, 419
   regularity of law of solution, 418
   statistical inference, 436
stochastic Fubini theorem, *see* multiple integrals
     with respect to Gaussian measures
stochastic heat equation, 215
   biFBM, 216
stochastic integration for FBM, 397
   chaos expansion of local time, 431
   derivative of solution of SDE, 414
   divergence integral, 399
   domain of divergence integral, 400
   Doss–Sussmann transformation, 413
   extended domain of divergence integral, 406
   fractional white noise, 435
   isonormal Gaussian process, 399
   Itô's formula, 408, 409, 412, 434
   Itô's formula for convex functions, 433
   Itô's formula for geometric FBM, 412
   moments of local time, 434
   nonlinear filtering, 436
   numerical solutions of SDE, 418
   optimal control, 436
   quadratic variation, 427
   regularity of law of solution of SDE, 418
   regularity of laws, 414
   Riemann–Stieltjes integral, 408
   rough paths, 435
   self-integration, 402, 406
   Stein's method, 426
   stochastic differential equation (SDE), 413
   Stratonovich integral, 408
   supremum of FBM, 434
   Tanaka's formula, 433
   tensor product Hilbert space, 400, 401
   Wick product, 435
stochastic partial differential equation, 215, 227
   parabolic Anderson model, 227
stochastic process, 1
   Gaussian, 2
   linear, 7
   stationary (strictly), 3
   stationary (weakly or second order), 4
   stationary increment (strictly), 3, 44
strong Szegö limit theorem
   classical, 210
   with singularity, 211
summation by parts formula, 579
supremum norm, 424
Szegö's limit theorem, 543

tagged particle, *see* simple symmetric exclusion
Tanaka's formula, 433
tangent space, *see* vector operator fractional
     Brownian motion (vector OFBM)
Tauberian theorem, 109
Telecom process
   convergence to, 133
   definition, 62, 122
   intermediate, 123
tensor product, 600
time domain, 4
time lag, 4
time reversibility, *see* vector operator fractional
     Brownian motion (vector OFBM)
time series, 2
   linear, 6
   nonlinear, 610
Toeplitz matrix, 210
total variation distance, 426

uniform convergence theorem, 19

vague convergence, 292, 584
vector long-range dependence (LRD)
   amplitude parameters, 496
   conditions II-v and IV-v, 495
   definition, 496
   fractal connectivity, 537
   fractional cointegration, 504
   linear representations, 502, 534
   phase parameters, 496
   trigonometric power-law coefficients, 502, 536

vector operator fractional Brownian motion (vector OFBM)
   continuity in distribution, 476
   convergence to, 536
   definition, 475
   fractional cointegration, 507
   identifiability, 492
   Lamperti's theorem, 487
   primary matrix function, 480
   proper, 476
   real spectrum, 495
   sample path properties, 536
   spectral domain representation, 476
   symmetry group, 493
   tangent space of symmetry group, 493
   time domain representation, 481
   time reversibility, 485
   vector fractional Brownian motion (VFBM), 486
vector process, 466
vector-valued stationary series, 469
   amplitude spectrum, 470
   coherence, 470, 497
   definitions, 470
   phase spectrum, 470
   time reversibility, 469

wavelet analysis
   approximation coefficients, 441
   biorthogonal wavelet basis, 449
   conjugate mirror filters (CMFs), 441
   Daubechies MRA, 444
   downsampling by 2 operation, 442
   expansion of BM, 452
   expansion of FBM, 452
   fast wavelet transform (FWT), 97, 442
   fractional CMFs, 450
   fractional scaling function, 448
   fractional wavelet, 447
   Haar MRA, 443
   Meyer MRA, 445
   multiresolution analysis (MRA), 443
   orthonormal wavelet basis, 440
   scaling function, 441
   simulation of FBM, 454
   upsampling by 2 operation, 443
   wavelet, 94, 441
   wavelet (detail) coefficients, 94, 441
   wavelet crime, 443
   zero moments, 94, 442, 450
wavelet Whittle estimation, 574
wavelet-based estimation, 93
   decorrelation property, 96
   log-scale diagram, 99
Weierstrass function, 216
   convergence to FBM, 220
   Weirstrass–Mandelbrot process, 217
white noise (WN), 4, 7
Wiener integral, 371
Wiener process, *see* Brownian motion
Wiener–Hopf factorization, 211

Young's inequality, 430

Zygmund class, 25, 527